Integrated Circuit Design

Fourth Edition

MOSIS SUBM design rules (3 metal, 1 poly with stacked vias & alternate contact rules)			
Layer	**Rule**	**Description**	**Rule (λ)**
N-well	1.1	Width	12
	1.2	Spacing to well at different potential	18
	1.3	Spacing to well at same potential	6
Active (diffusion)	2.1	Width	3
	2.2	Spacing to active	3
	2.3	Source/drain surround by well	6
	2.4	Substrate/well contact surround by well	3
	2.5	Spacing to active of opposite type	4
Poly	3.1	Width	2
	3.2	Spacing to poly over field oxide	3
	3.2a	Spacing to poly over active	3
	3.3	Gate extension beyond active	2
	3.4	Active extension beyond poly	3
	3.5	Spacing of poly to active	1
Select (n or p)	4.1	Spacing from substrate/well contact to gate	3
	4.2	Overlap of active	2
	4.3	Overlap of substrate/well contact	1
	4.4	Spacing to select	2
Contact (to poly or active)	5.1, 6.1	Width (exact)	2×2
	5.2b, 6.2b	Overlap by poly or active	1
	5.3, 6.3	Spacing to contact	3
	5.4, 6.4	Spacing to gate	2
	5.5b	Spacing of poly contact to other poly	5
	5.7b, 6.7b	Spacing to active/poly for multiple poly/active contacts	3
	6.8b	Spacing of active contact to poly contact	4
Metal1, Metal2	7.1, 9.1	Width	3
	7.2, 9.2	Spacing to same layer of metal	3
	7.3, 8.3, 9.3	Overlap of contact or via	1
	7.4, 9.4	Spacing to metal for lines wider than 10 λ	6
Via1, Via2	8.1, 14.1	Width (exact)	2×2
	8.2, 14.2	Spacing to via on same layer	3
Metal3	15.1	Width	5
	15.2	Spacing to metal3	3
	15.3	Overlap of via2	2
	15.4	Spacing to metal for lines wider than 10 λ	6
Overglass Cut	10.1	Width of bond pad opening	60 μm
	10.2	Width of probe pad opening	20 μm
	10.3	Metal3 overlap of overglass cut	6 μm
	10.4	Spacing of pad metal to unrelated metal	30 μm
	10.5	Spacing of pad metal to active or poly	15 μm

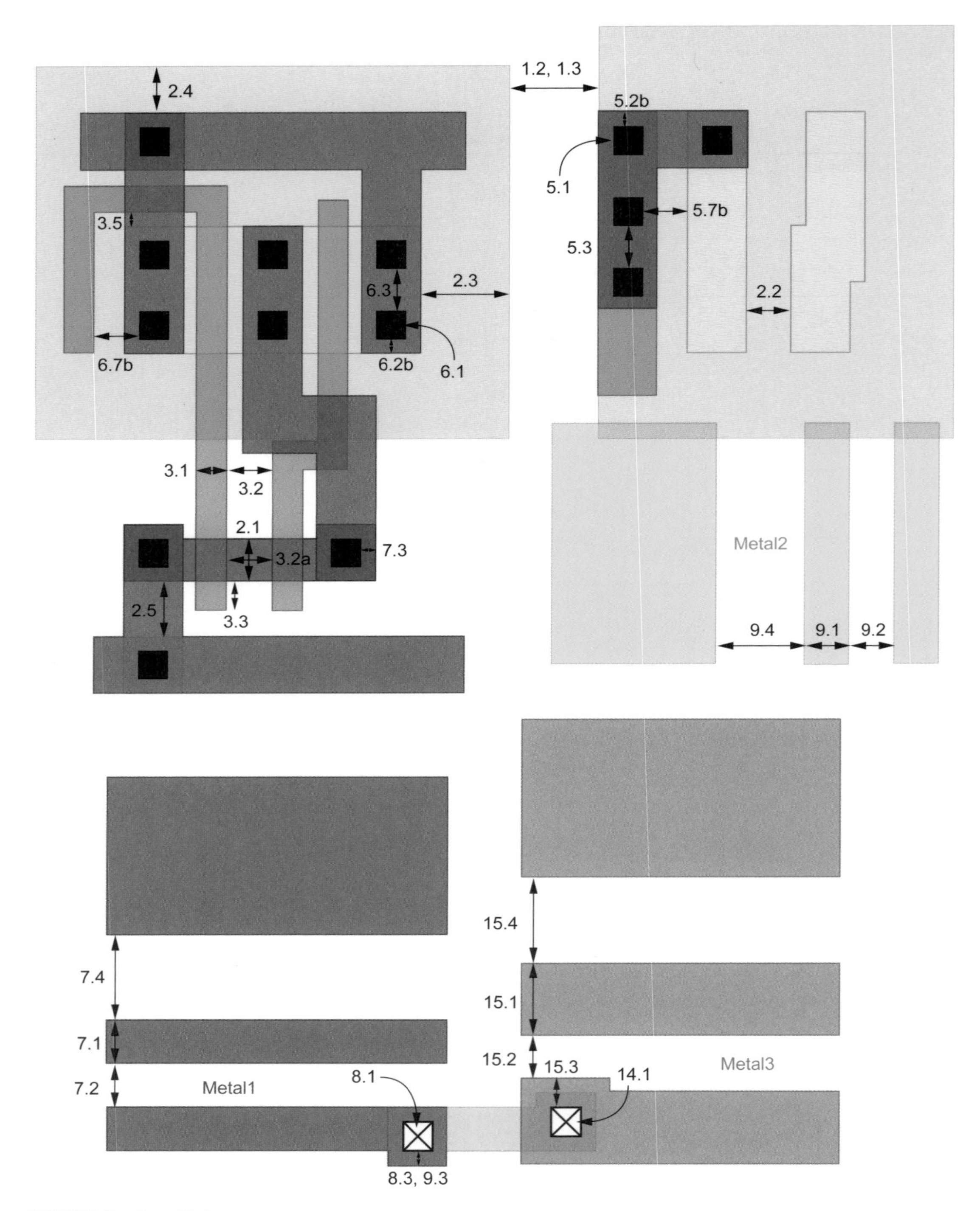

MOSIS Design Rules

Integrated Circuit Design

Fourth Edition

Neil H. E. Weste
Macquarie University and
The University of Adelaide

David Money Harris
Harvey Mudd College

Boston Columbus Indianapolis New York San Francisco Upper Saddle River
Amsterdam Cape Town Dubai London Madrid Milan Munich Paris Montreal Toronto
Delhi Mexico City Sao Paulo Sydney Hong Kong Seoul Singapore Taipei Tokyo

Editor in Chief: Michael Hirsch
Acquisitions Editor: Matt Goldstein
Editorial Assistant: Chelsea Bell
Managing Editor: Jeffrey Holcomb
Senior Production Project Manager: Marilyn Lloyd
Media Producer: Katelyn Boller
Director of Marketing: Margaret Waples
Marketing Coordinator: Kathryn Ferranti
Senior Manufacturing Buyer: Carol Melville
Senior Media Buyer: Ginny Michaud
Text Designer: Susan Raymond
Cover Designer: Jodi Notowitz
Full Service Vendor: Gillian Hall/The Aardvark Group Publishing Service
Copyeditor: Kathleen Cantwell, C4 Technologies
Proofreader: Holly McLean-Aldis
Indexer: Jack Lewis
Printer/Binder: Edwards Brothers Malloy
Cover Printer: Lehigh-Phoenix Color/Hagerstown

Credits and acknowledgments borrowed from other sources and reproduced with permission in this textbook appear on appropriate page within text or on page 751.

The interior of this book was set in Adobe Caslon and Trade Gothic.

ISBN 10: 0-321-69694-8
ISBN 13: 978-0-321-69694-6
10 9 8 7 6 5 4 3—EBM—14 13

To Avril, Melissa, Tamara, Nicky, Jocelyn,
Makayla, Emily, Danika, Dan and Simon
N. W.

To Jennifer, Samuel, and Abraham
D. M. H.

Contents

Chapter 6 **Scaling, Reliability, and Variability**

Chapter 7 **SPICE**

Chapter 8 **Gates**

Chapter 10 **Datapaths**

Chapter 11 **Memories**

Chapter 13 **Methodology**

Chapter 14 **Test**

Chapter 15 **Fabrication**

Preface

In the two-and-a-half decades since the first edition of this book was published, CMOS technology has claimed the preeminent position in modern electrical system design. It has enabled the widespread use of wireless communication, the Internet, and personal computers. No other human invention has seen such rapid growth for such a sustained period. The transistor counts and clock frequencies of state-of-the-art chips have grown by orders of magnitude.

	1st Edition	2nd Edition	3rd Edition	4th Edition
Year	1985	1993	2004	2010
Transistor Counts	10^5–10^6	10^6–10^7	10^8–10^9	10^9–10^{10}
Clock Frequencies	10^7	10^8	10^9	10^9
Worldwide Market	$25B	$60B	$170B	$250B

This edition has been heavily revised to reflect the rapid changes in integrated circuit design over the past six years. While the basic principles are largely the same, power consumption and variability have become primary factors for chip design. The book has been reorganized to emphasize the key factors: delay, power, interconnect, and robustness. Other chapters have been reordered to reflect the order in which we teach the material.

How to Use This Book

This book intentionally covers more breadth and depth than any course would cover in a semester. It is accessible for a first undergraduate course in VLSI, yet detailed enough for advanced graduate courses and is useful as a reference to the practicing engineer. You are encouraged to pick and choose topics according to your interest. Chapter 1 previews the entire field, while subsequent chapters elaborate on specific topics. Sections are marked with the "Optional" icon (shown here in the margin) if they are not needed to understand subsequent sections. You may skip them on a first reading and return when they are relevant to you.

We have endeavored to include figures whenever possible ("a picture is worth a thousand words") to trigger your thinking. As you encounter examples throughout the text, we urge you to think about them before reading the solutions. We have also provided extensive references for those who need to delve deeper into topics introduced in this text. We have emphasized the best practices that are used in industry and warned of pitfalls and fallacies. Our judgments about the merits of circuits may become incorrect as technology and applications change, but we believe it is the responsibility of a writer to attempt to call out the most relevant information.

Supplements

Numerous supplements are available on the Companion Web site for the book, `www.cmosvlsi.com`. Supplements to help students with the course include:

- A lab manual with laboratory exercises involving the design of an 8-bit microprocessor covered in Chapter 1.
- A collection of links to VLSI resources including open-source CAD tools and process parameters.
- A student solutions manual that includes answers to odd-numbered problems.

Supplements to help instructors with the course include:

- A sample syllabus.
- Lecture slides for an introductory VLSI course.
- An instructor's manual with solutions.

These materials have been prepared exclusively for professors using the book in a course. Please send email to `computing@aw.com` for information on how to access them.

Acknowledgments

We are indebted to many people for their reviews, suggestions, and technical discussions. These people include: Bharadwaj "Birdy" Amrutur, Mark Anders, Adnan Aziz, Jacob Baker, Kaustav Banerjee, Steve Bibyk, David Blaauw, Erik Brunvand, Neil Burgess, Wayne Burleson, Robert Drost, Jo Ebergen, Sarah Harris, Jacob Herbold, Ron Ho, David Hopkins, Mark Horowitz, Steven Hsu, Tanay Karnik, Omid Kaveh, Matthew Keeter, Ben Keller, Ali Keshavarzi, Brucek Khailany, Jaeha Kim, Volkan Kursun, Simon Knowles, Ram Krishnamurthy, Austin Lee, Ana Sonia Leon, Shih-Lien Lu, Sanu Mathew, Aleksandar Milenkovic, Sam Naffziger, Braden Phillips, Stefan Rusu, Justin Schauer, James Stine, Jason Stinson, Aaron Stratton, Ivan Sutherland, Jim Tschanz, Alice Wang, Gu-Yeon Wei, and Peiyi Zhao. We apologize in advance to anyone we overlooked.

MOSIS and IBM kindly provided permission to use nanometer SPICE models for many examples. Nathaniel Pinckney spent a summer revising the laboratory exercises and updating simulations. Jaeha Kim contributed new sections on phase-locked loops and high-speed I/O for Chapter 13. David would like to thank Bharadwaj Amrutur of the Indian Institute of Science and Braden Phillips of the University of Adelaide for hosting him during two productive summers of writing.

Addison-Wesley has done an admirable job with the grueling editorial and production process. We would particularly like to thank our editor, Matt Goldstein, and our compositor, Gillian Hall.

Sally Harris has been editing family books since David was an infant on her lap. She read the page proofs with amazing attention to detail and unearthed hundreds of errors.

This book would not have existed without the support of our families. David would particularly like to thank his wife Jennifer and sons Abraham and Samuel for enduring two summers of absence while writing, and to our extended family for their tremendous assistance.

We have become painfully aware of the ease with which mistakes creep into a book. Scores of 3rd edition readers have reported bugs that are now corrected. Despite our best efforts at validation, we are confident that we have introduced a similar number of new errors. Please check the errata sheet at `www.cmosvlsi.com/errata.pdf` to see if the bug has already been reported. Send your reports to `bugs@cmosvlsi.com`.

N. W.
D. M. H.
January 2010

Welcome to VLSI

1

1.1 A Brief History

In 1958, Jack Kilby built the first integrated circuit flip-flop with two transistors at Texas Instruments. In 2008, Intel's Itanium microprocessor contained more than 2 billion transistors and a 16 Gb Flash memory contained more than 4 billion transistors. This corresponds to a compound annual growth rate of 53% over 50 years. No other technology in history has sustained such a high growth rate lasting for so long.

This incredible growth has come from steady miniaturization of transistors and improvements in manufacturing processes. Most other fields of engineering involve tradeoffs between performance, power, and price. However, as transistors become smaller, they also become faster, dissipate less power, and are cheaper to manufacture. This synergy has not only revolutionized electronics, but also society at large.

The processing performance once dedicated to secret government supercomputers is now available in disposable cellular telephones. The memory once needed for an entire company's accounting system is now carried by a teenager in her iPod. Improvements in integrated circuits have enabled space exploration, made automobiles safer and more fuel-efficient, revolutionized the nature of warfare, brought much of mankind's knowledge to our Web browsers, and made the world a flatter place.

Figure 1.1 shows annual sales in the worldwide semiconductor market. Integrated circuits became a \$100 billion/year business in 1994. In 2007, the industry manufactured approximately 6 quintillion (6×10^{18}) transistors, or nearly a billion for every human being on the planet. Thousands of engineers have made their fortunes in the field. New fortunes lie ahead for those with innovative ideas and the talent to bring those ideas to reality.

During the first half of the twentieth century, electronic circuits used large, expensive, power-hungry, and unreliable vacuum tubes. In 1947, John Bardeen and Walter Brattain built the first functioning point contact transistor at Bell Laboratories, shown in Figure 1.2(a) [Riordan97]. It was nearly classified as a military secret, but Bell Labs publicly introduced the device the following year.

> *We have called it the Transistor, T-R-A-N-S-I-S-T-O-R, because it is a resistor or semiconductor device which can amplify electrical signals as they are transferred through it from input to output terminals. It is, if you will, the electrical equivalent of a vacuum tube amplifier. But there the similarity ceases. It has no vacuum, no filament, no glass tube. It is composed entirely of cold, solid substances.*

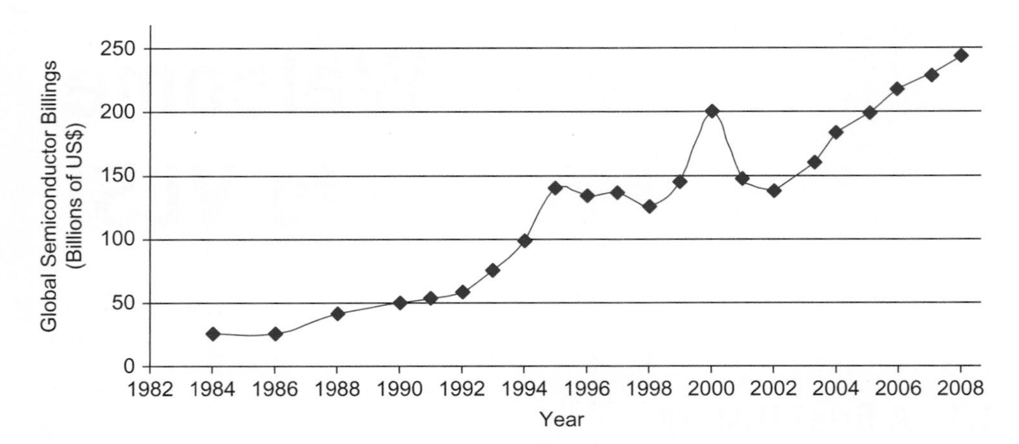

FIGURE 1.1 Size of worldwide semiconductor market (Courtesy of Semiconductor Industry Association.)

Ten years later, Jack Kilby at Texas Instruments realized the potential for miniaturization if multiple transistors could be built on one piece of silicon. Figure 1.2(b) shows his first prototype of an integrated circuit, constructed from a germanium slice and gold wires.

The invention of the transistor earned the Nobel Prize in Physics in 1956 for Bardeen, Brattain, and their supervisor William Shockley. Kilby received the Nobel Prize in Physics in 2000 for the invention of the integrated circuit.

Transistors can be viewed as electrically controlled switches with a control terminal and two other terminals that are connected or disconnected depending on the voltage or current applied to the control. Soon after inventing the point contact transistor, Bell Labs developed the bipolar junction transistor. Bipolar transistors were more reliable, less noisy, and more power-efficient. Early integrated circuits primarily used bipolar transistors. Bipolar transistors require a small current into the control (base) terminal to switch much larger currents between the other two (emitter and collector) terminals. The quiescent power dissipated by these base currents, drawn even when the circuit is not switching,

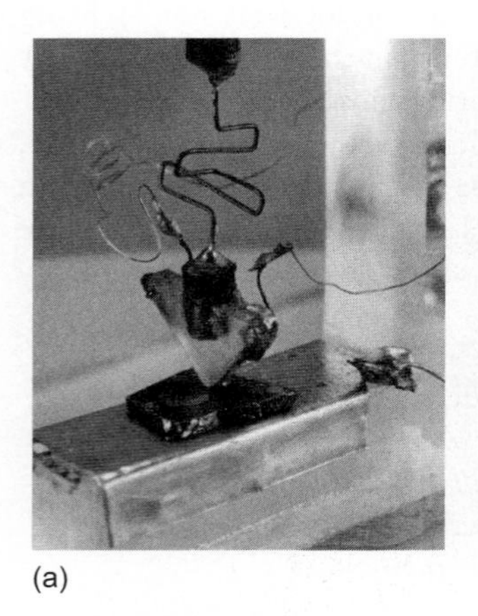
(a)

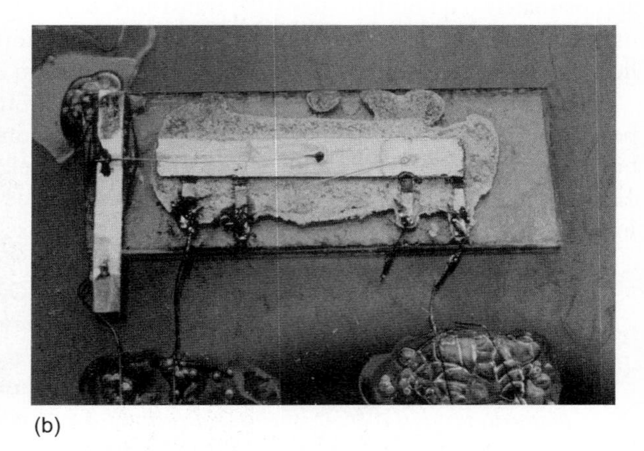
(b)

FIGURE 1.2 (a) First transistor (Property of AT&T Archives. Reprinted with permission of AT&T.) and (b) first integrated circuit (Courtesy of Texas Instruments.)

limits the maximum number of transistors that can be integrated onto a single die. By the 1960s, Metal Oxide Semiconductor Field Effect Transistors (MOSFETs) began to enter production. MOSFETs offer the compelling advantage that they draw almost zero control current while idle. They come in two flavors: nMOS and pMOS, using n-type and p-type silicon, respectively. The original idea of field effect transistors dated back to the German scientist Julius Lilienfield in 1925 [US patent 1,745,175] and a structure closely resembling the MOSFET was proposed in 1935 by Oskar Heil [British patent 439,457], but materials problems foiled early attempts to make functioning devices.

In 1963, Frank Wanlass at Fairchild described the first logic gates using MOSFETs [Wanlass63]. Fairchild's gates used both nMOS and pMOS transistors, earning the name Complementary Metal Oxide Semiconductor, or CMOS. The circuits used discrete transistors but consumed only nanowatts of power, six orders of magnitude less than their bipolar counterparts. With the development of the silicon planar process, MOS integrated circuits became attractive for their low cost because each transistor occupied less area and the fabrication process was simpler [Vadasz69]. Early commercial processes used only pMOS transistors and suffered from poor performance, yield, and reliability. Processes using nMOS transistors became common in the 1970s [Mead80]. Intel pioneered nMOS technology with its 1101 256-bit static random access memory and 4004 4-bit microprocessor, as shown in Figure 1.3. While the nMOS process was less expensive than CMOS, nMOS logic gates still consumed power while idle. Power consumption became a major issue in the 1980s as hundreds of thousands of transistors were integrated onto a single die. CMOS processes were widely adopted and have essentially replaced nMOS and bipolar processes for nearly all digital logic applications.

In 1965, Gordon Moore observed that plotting the number of transistors that can be most economically manufactured on a chip gives a straight line on a semilogarithmic scale [Moore65]. At the time, he found transistor count doubling every 18 months. This observation has been called *Moore's Law* and has become a self-fulfilling prophecy. Figure 1.4 shows that the number of transistors in Intel microprocessors has doubled every 26 months since the invention of the 4004. Moore's Law is driven primarily by *scaling* down the size of transistors and, to a minor extent, by building larger chips. The level of integration of chips has been classified as small-scale, medium-scale, large-scale, and very large-scale. *Small-scale integration* (SSI) circuits, such as the 7404 inverter, have fewer than 10

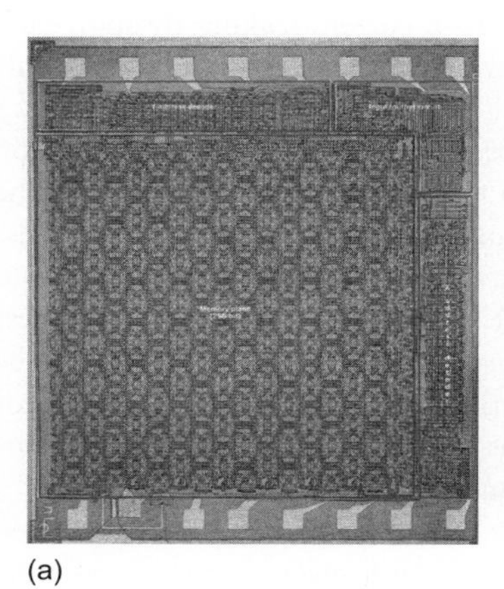

(a)

(b)

FIGURE 1.3 (a) Intel 1101 SRAM (© IEEE 1969 [Vadasz69]) and (b) 4004 microprocessor (Courtesy of Intel Corporation.)

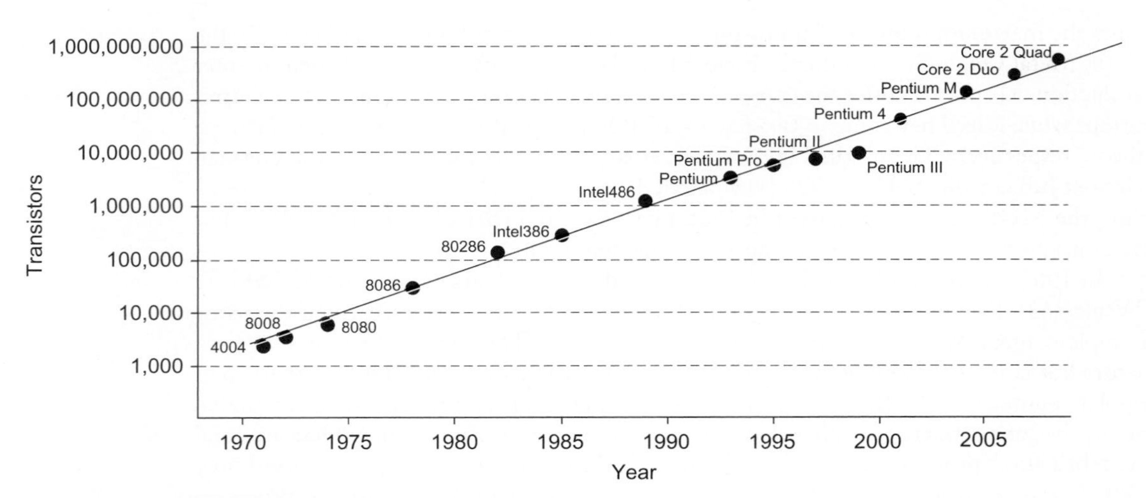

FIGURE 1.4 Transistors in Intel microprocessors [Intel10]

gates, with roughly half a dozen transistors per gate. *Medium-scale integration* (MSI) circuits, such as the 74161 counter, have up to 1000 gates. *Large-scale integration* (LSI) circuits, such as simple 8-bit microprocessors, have up to 10,000 gates. It soon became apparent that new names would have to be created every five years if this naming trend continued and thus the term *very large-scale integration* (VLSI) is used to describe most integrated circuits from the 1980s onward. A corollary of Moore's law is *Dennard's Scaling Law* [Dennard74]: as transistors shrink, they become faster, consume less power, and are cheaper to manufacture. Figure 1.5 shows that Intel microprocessor clock frequencies have doubled roughly every 34 months.This frequency scaling hit the power wall around 2004, and clock frequencies have leveled off around 3 GHz. Computer performance, measured in time to run an application, has advanced even more than raw clock speed. Presently, the performance is driven by the number of cores on a chip rather than by the clock. Even though an individual CMOS transistor uses very little energy each time it switches, the enormous number of transistors switching at very high rates of speed have made power consumption a major design consideration again. Moreover, as transistors have become so small, they cease to turn completely OFF. Small amounts of current leaking through each transistor now lead to significant power consumption when multiplied by millions or billions of transistors on a chip.

The feature size of a CMOS manufacturing process refers to the minimum dimension of a transistor that can be reliably built. The 4004 had a feature size of 10 μm in 1971. The Core 2 Duo had a feature size of 45 nm in 2008. Manufacturers introduce a new process generation (also called a technology node) every 2–3 years with a 30% smaller feature size to pack twice as many transistors in the same area. Figure 1.6 shows the progression of process generations. Feature sizes down to 0.25 μm are generally specified in microns (10^{-6} m), while smaller feature sizes are expressed in nanometers (10^{-9} m). Effects that were relatively minor in micron processes, such as transistor leakage, variations in characteristics of adjacent transistors, and wire resistance, are of great significance in nanometer processes.

Moore's Law has become a self-fulfilling prophecy because each company must keep up with its competitors. Obviously, this scaling cannot go on forever because transistors cannot be smaller than atoms. Dennard scaling has already begun to slow. By the 45 nm

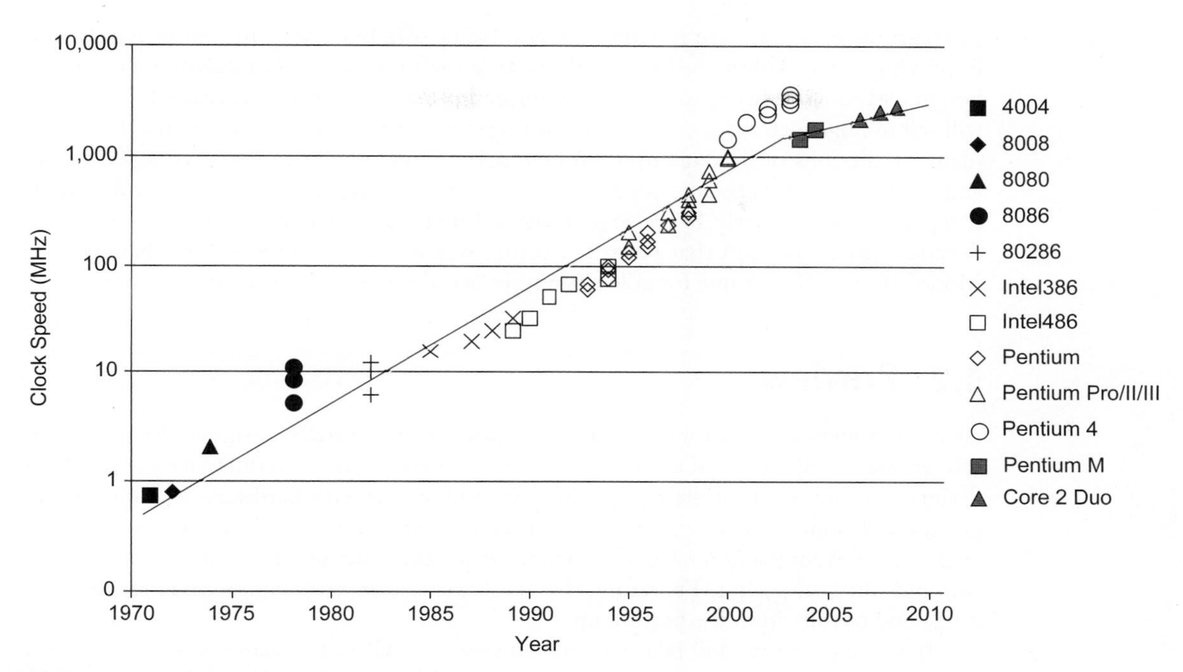

FIGURE 1.5 Clock frequencies of Intel microprocessors

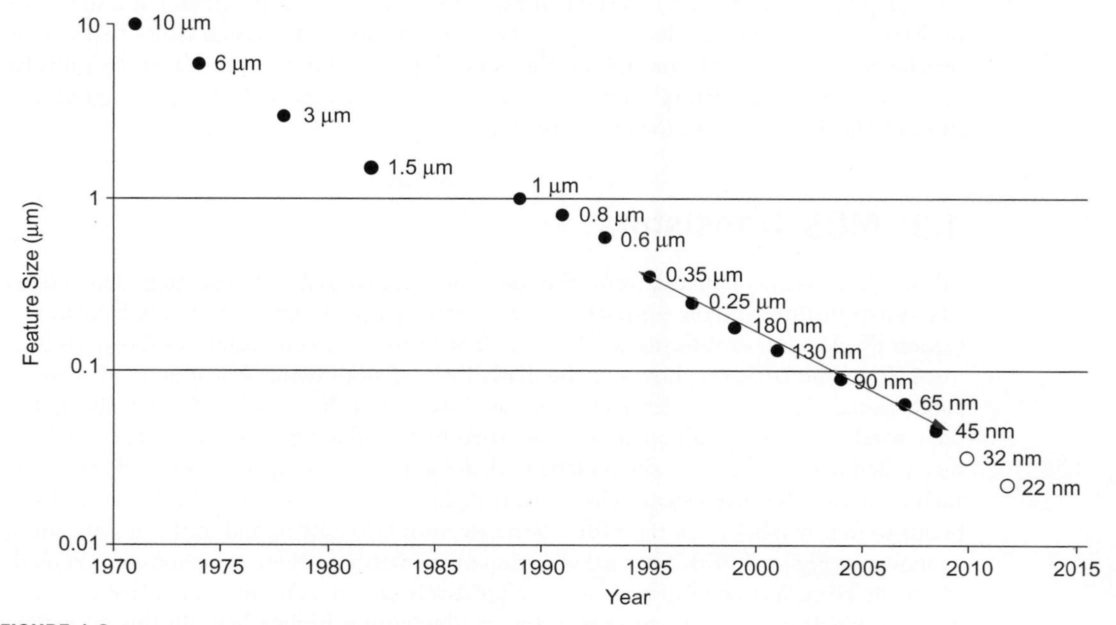

FIGURE 1.6 Process generations. Future predictions from [SIA2007].

generation, designers are having to make trade-offs between improving power and improving delay. Although the cost of printing each transistor goes down, the one-time design costs are increasing exponentially, relegating state-of-the-art processes to chips that will sell in huge quantities or that have cutting-edge performance requirements. However, many predictions of fundamental limits to scaling have already proven wrong. Creative engineers and material scientists have billions of dollars to gain by getting ahead of their competitors. In the early 1990s, experts agreed that scaling would continue for at least a decade but that beyond that point the future was murky. In 2009, we still believe that Moore's Law will continue for at least another decade. The future is yours to invent.

1.2 Preview

As the number of transistors on a chip has grown exponentially, designers have come to rely on increasing levels of automation to seek corresponding productivity gains. Many designers spend much of their effort specifying functions with hardware description languages and seldom look at actual transistors. Nevertheless, chip design is not software engineering. Addressing the harder problems requires a fundamental understanding of circuit and physical design. Therefore, this book focuses on building an understanding of integrated circuits from the bottom up.

In this chapter, we will take a simplified view of CMOS transistors as switches. With this model we will develop CMOS logic gates and latches. CMOS transistors are mass-produced on silicon wafers using lithographic steps much like a printing press process. We will explore how to lay out transistors by specifying rectangles indicating where dopants should be diffused, polysilicon should be grown, metal wires should be deposited, and contacts should be etched to connect all the layers. By the middle of this chapter, you will understand all the principles required to design and lay out your own simple CMOS chip. The chapter concludes with an extended example demonstrating the design of a simple 8-bit MIPS microprocessor chip. The processor raises many of the design issues that will be developed in more depth throughout the book. The best way to learn VLSI design is by doing it. A set of laboratory exercises are available at `www.cmosvlsi.com` to guide you through the design of your own microprocessor chip.

1.3 MOS Transistors

Silicon (Si), a semiconductor, forms the basic starting material for most integrated circuits [Tsividis99]. Pure silicon consists of a three-dimensional *lattice* of atoms. Silicon is a Group IV element, so it forms covalent bonds with four adjacent atoms, as shown in Figure 1.7(a). The lattice is shown in the plane for ease of drawing, but it actually forms a cubic crystal. As all of its valence electrons are involved in chemical bonds, pure silicon is a poor conductor. The conductivity can be raised by introducing small amounts of impurities, called *dopants*, into the silicon lattice. A dopant from Group V of the periodic table, such as arsenic, has five valence electrons. It replaces a silicon atom in the lattice and still bonds to four neighbors, so the fifth valence electron is loosely bound to the arsenic atom, as shown in Figure 1.7(b). Thermal vibration of the lattice at room temperature is enough to set the electron free to move, leaving a positively charged As^+ ion and a free electron. The free electron can carry current so the conductivity is higher. We call this an *n*-type

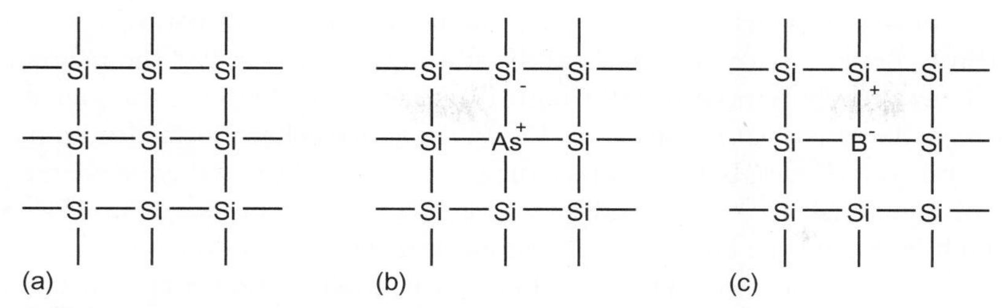

FIGURE 1.7 Silicon lattice and dopant atoms

semiconductor because the free carriers are negatively charged electrons. Similarly, a Group III dopant, such as boron, has three valence electrons, as shown in Figure 1.7(c). The dopant atom can borrow an electron from a neighboring silicon atom, which in turn becomes short by one electron. That atom in turn can borrow an electron, and so forth, so the missing electron, or *hole*, can propagate about the lattice. The hole acts as a positive carrier so we call this a *p*-type semiconductor.

A junction between p-type and n-type silicon is called a *diode*, as shown in Figure 1.8. When the voltage on the p-type semiconductor, called the *anode*, is raised above the n-type *cathode*, the diode is *forward biased* and current flows. When the anode voltage is less than or equal to the cathode voltage, the diode is *reverse biased* and very little current flows.

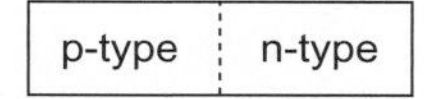

Anode Cathode

FIGURE 1.8 p-n junction diode structure and symbol

A Metal-Oxide-Semiconductor (*MOS*) structure is created by superimposing several layers of conducting and insulating materials to form a sandwich-like structure. These structures are manufactured using a series of chemical processing steps involving oxidation of the silicon, selective introduction of dopants, and deposition and etching of metal wires and contacts. Transistors are built on nearly flawless single crystals of silicon, which are available as thin flat circular wafers of 15–30 cm in diameter. CMOS technology provides two types of transistors (also called *devices*): an n-type transistor (*nMOS*) and a p-type transistor (*pMOS*). Transistor operation is controlled by electric fields so the devices are also called Metal Oxide Semiconductor Field Effect Transistors (*MOSFETs*) or simply *FETs*. Cross-sections and symbols of these transistors are shown in Figure 1.9. The n+ and p+ regions indicate heavily doped n- or p-type silicon.

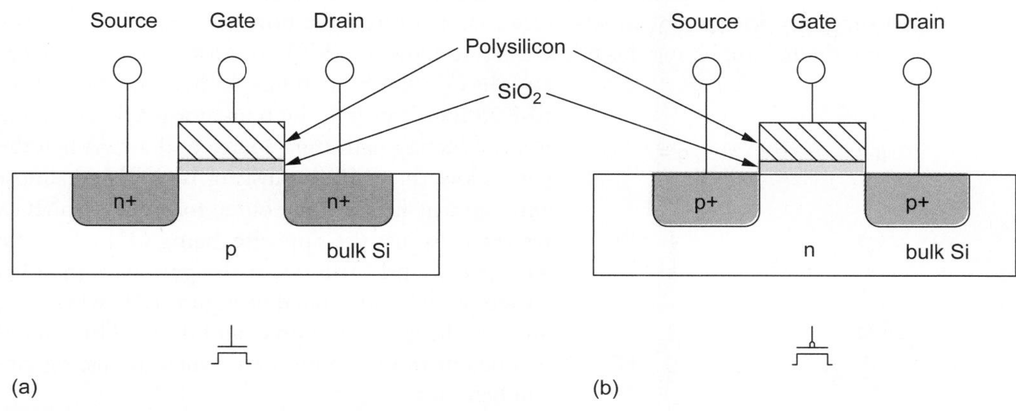

FIGURE 1.9 nMOS transistor (a) and pMOS transistor (b)

Each transistor consists of a stack of the conducting *gate*, an insulating layer of silicon dioxide (SiO_2, better known as glass), and the silicon wafer, also called the *substrate*, *body*, or *bulk*. Gates of early transistors were built from metal, so the stack was called metal-oxide-semiconductor, or MOS. Since the 1970s, the gate has been formed from polycrystalline silicon (*polysilicon*), but the name stuck. (Interestingly, metal gates reemerged in 2007 to solve materials problems in advanced manufacturing processes.) An nMOS transistor is built with a p-type body and has regions of n-type semiconductor adjacent to the gate called the *source* and *drain*. They are physically equivalent and for now we will regard them as interchangeable. The body is typically grounded. A pMOS transistor is just the opposite, consisting of p-type source and drain regions with an n-type body. In a CMOS technology with both flavors of transistors, the substrate is either n-type or p-type. The other flavor of transistor must be built in a special *well* in which dopant atoms have been added to form the body of the opposite type.

The gate is a control input: It affects the flow of electrical current between the source and drain. Consider an nMOS transistor. The body is generally grounded so the p–n junctions of the source and drain to body are reverse-biased. If the gate is also grounded, no current flows through the reverse-biased junctions. Hence, we say the transistor is OFF. If the gate voltage is raised, it creates an electric field that starts to attract free electrons to the underside of the Si–SiO_2 interface. If the voltage is raised enough, the electrons outnumber the holes and a thin region under the gate called the *channel* is inverted to act as an n-type semiconductor. Hence, a conducting path of electron carriers is formed from source to drain and current can flow. We say the transistor is ON.

For a pMOS transistor, the situation is again reversed. The body is held at a positive voltage. When the gate is also at a positive voltage, the source and drain junctions are reverse-biased and no current flows, so the transistor is OFF. When the gate voltage is lowered, positive charges are attracted to the underside of the Si–SiO_2 interface. A sufficiently low gate voltage inverts the channel and a conducting path of positive carriers is formed from source to drain, so the transistor is ON. Notice that the symbol for the pMOS transistor has a bubble on the gate, indicating that the transistor behavior is the opposite of the nMOS.

The positive voltage is usually called V_{DD} or POWER and represents a logic 1 value in digital circuits. In popular logic families of the 1970s and 1980s, V_{DD} was set to 5 volts. Smaller, more recent transistors are unable to withstand such high voltages and have used supplies of 3.3 V, 2.5 V, 1.8 V, 1.5 V, 1.2 V, 1.0 V, and so forth. The low voltage is called GROUND (GND) or V_{SS} and represents a logic 0. It is normally 0 volts.

In summary, the gate of an MOS transistor controls the flow of current between the source and drain. Simplifying this to the extreme allows the MOS transistors to be viewed as simple ON/OFF switches. When the gate of an nMOS transistor is 1, the transistor is ON and there is a conducting path from source to drain. When the gate is low, the nMOS transistor is OFF and almost zero current flows from source to drain. A pMOS transistor is just the opposite, being ON when the gate is low and OFF when the gate is high. This switch model is illustrated in Figure 1.10, where *g*, *s*, and *d* indicate gate, source, and drain. This model will be our most common one when discussing circuit behavior.

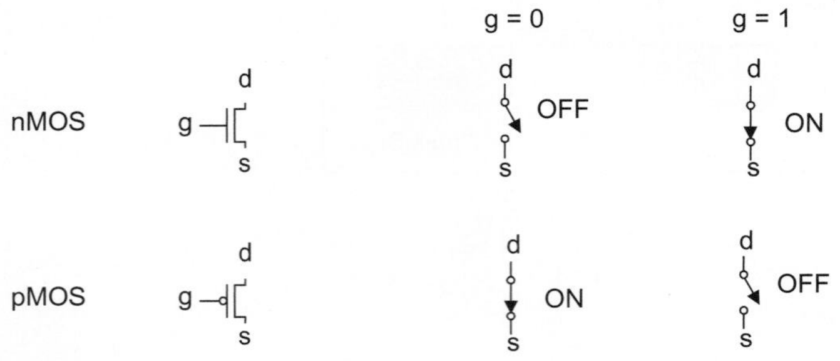

FIGURE 1.10 Transistor symbols and switch-level models

1.4 CMOS Logic

1.4.1 The Inverter

Figure 1.11 shows the schematic and symbol for a CMOS inverter or NOT gate using one nMOS transistor and one pMOS transistor. The bar at the top indicates V_{DD} and the triangle at the bottom indicates GND. When the input A is 0, the nMOS transistor is OFF and the pMOS transistor is ON. Thus, the output Y is pulled up to 1 because it is connected to V_{DD} but not to GND. Conversely, when A is 1, the nMOS is ON, the pMOS is OFF, and Y is pulled down to '0.' This is summarized in Table 1.1.

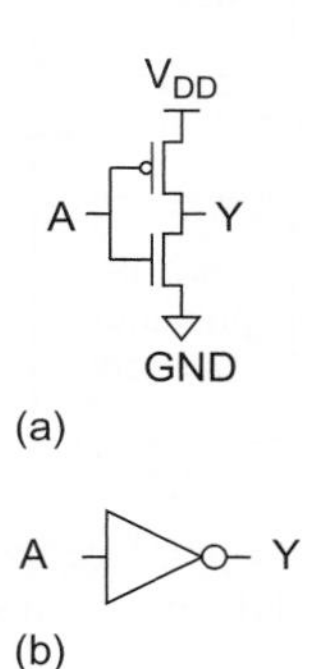

FIGURE 1.11 Inverter schematic (a) and symbol (b) $Y = \overline{A}$

TABLE 1.1 Inverter truth table

A	Y
0	1
1	0

1.4.2 The NAND Gate

Figure 1.12(a) shows a 2-input CMOS NAND gate. It consists of two series nMOS transistors between Y and GND and two parallel pMOS transistors between Y and V_{DD}. If either input A or B is 0, at least one of the nMOS transistors will be OFF, breaking the path from Y to GND. But at least one of the pMOS transistors will be ON, creating a path from Y to V_{DD}. Hence, the output Y will be 1. If both inputs are 1, both of the nMOS transistors will be ON and both of the pMOS transistors will be OFF. Hence, the output will be 0. The truth table is given in Table 1.2 and the symbol is shown in Figure 1.12(b). Note that by DeMorgan's Law, the inversion bubble may be placed on either side of the gate. In the figures in this book, two lines intersecting at a T-junction are connected. Two lines crossing are connected if and only if a dot is shown.

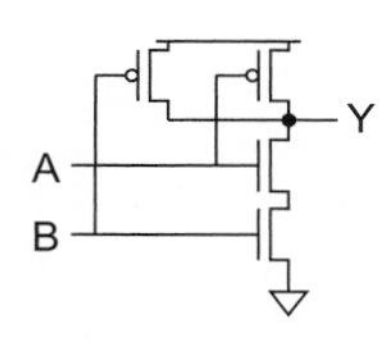

(a)

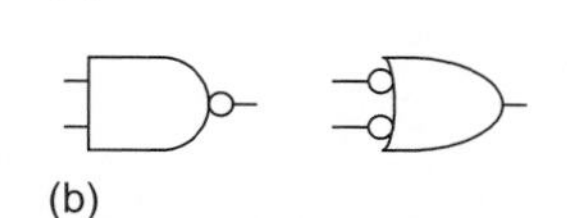

(b)

FIGURE 1.12 2-input NAND gate schematic (a) and symbol (b) $Y = \overline{A \cdot B}$

TABLE 1.2 NAND gate truth table

A	B	Pull-Down Network	Pull-Up Network	Y
0	0	OFF	ON	1
0	1	OFF	ON	1
1	0	OFF	ON	1
1	1	ON	OFF	0

k-input NAND gates are constructed using k series nMOS transistors and k parallel pMOS transistors. For example, a 3-input NAND gate is shown in Figure 1.13. When any of the inputs are 0, the output is pulled high through the parallel pMOS transistors. When all of the inputs are 1, the output is pulled low through the series nMOS transistors.

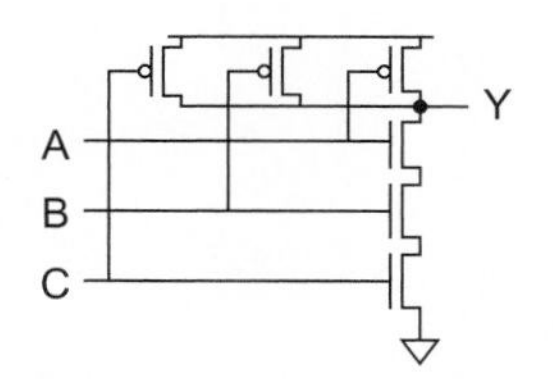

FIGURE 1.13 3-input NAND gate schematic $Y = \overline{A \cdot B \cdot C}$

1.4.3 CMOS Logic Gates

The inverter and NAND gates are examples of *static CMOS logic gates*, also called *complementary CMOS gates*. In general, a static CMOS gate has an nMOS *pull-down network* to connect the output to 0 (GND) and pMOS *pull-up network* to connect the output to 1 (V_{DD}), as shown in Figure 1.14. The networks are arranged such that one is ON and the other OFF for any input pattern.

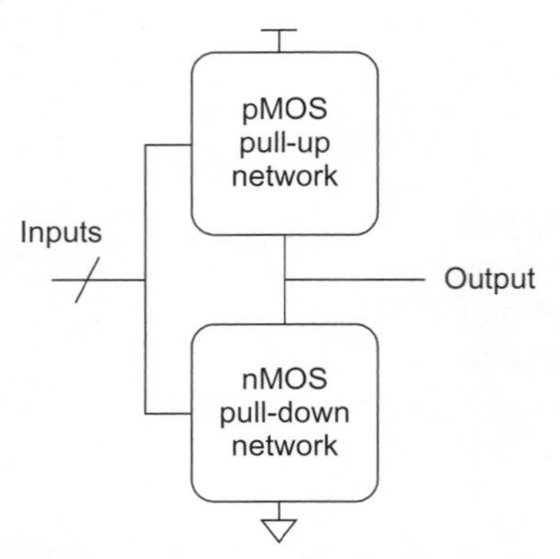

FIGURE 1.14 General logic gate using pull-up and pull-down networks

The pull-up and pull-down networks in the inverter each consist of a single transistor. The NAND gate uses a series pull-down network and a parallel pull-up network. More elaborate networks are used for more complex gates. Two or more transistors in series are ON only if all of the series transistors are ON. Two or more transistors in parallel are ON if any of the parallel transistors are ON. This is illustrated in Figure 1.15 for nMOS and pMOS transistor pairs. By using combinations of these constructions, CMOS combinational gates can be constructed. Although such static CMOS gates are most widely used, Chapter 8 explores alternate ways of building gates with transistors.

In general, when we join a pull-up network to a pull-down network to form a logic gate as shown in Figure 1.14, they both will attempt to exert a logic level at the output. The possible levels at the output are shown in Table 1.3. From this table it can be seen that the output of a CMOS logic gate can be in four states. The 1 and 0 levels have been encountered with the inverter and NAND gates, where either the pull-up or pull-down is OFF and the other structure is ON. When both pull-up and pull-down are OFF, the *high-impedance* or *floating* Z output state results. This is of importance in multiplexers, memory elements, and tristate bus drivers. The *crowbarred* (or *contention*) X level exists when both pull-up and pull-down are simultaneously turned ON. Contention between the two networks results in an indeterminate output level and dissipates static power. It is usually an unwanted condition.

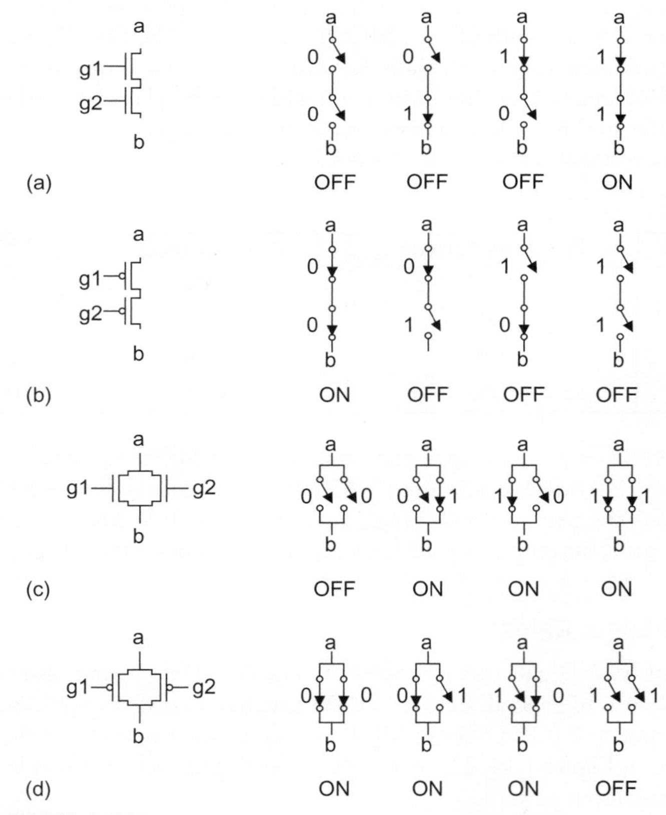

FIGURE 1.15 Connection and behavior of series and parallel transistors

TABLE 1.3 Output states of CMOS logic gates

	pull-up OFF	pull-up ON
pull-down OFF	Z	1
pull-down ON	0	crowbarred (X)

1.4.4 The NOR Gate

A 2-input NOR gate is shown in Figure 1.16. The nMOS transistors are in parallel to pull the output low when either input is high. The pMOS transistors are in series to pull the output high when both inputs are low, as indicated in Table 1.4. The output is never crowbarred or left floating.

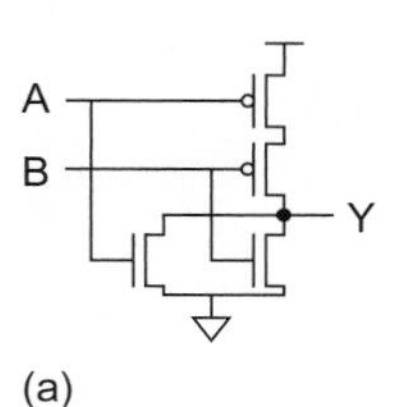

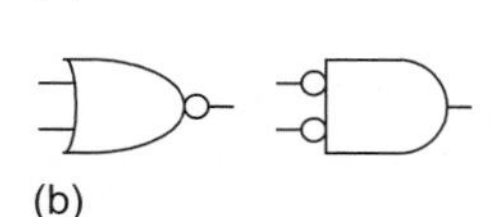

FIGURE 1.16 2-input NOR gate schematic (a) and symbol (b) $Y = \overline{A + B}$

TABLE 1.4 NOR gate truth table

A	*B*	*Y*
0	0	1
0	1	0
1	0	0
1	1	0

Example 1.1

Sketch a 3-input CMOS NOR gate.

SOLUTION: Figure 1.17 shows such a gate. If any input is high, the output is pulled low through the parallel nMOS transistors. If all inputs are low, the output is pulled high through the series pMOS transistors.

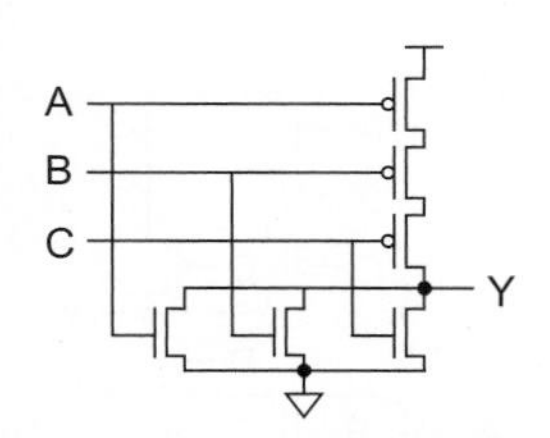

FIGURE 1.17 3-input NOR gate schematic $Y = \overline{A + B + C}$

1.4.5 Compound Gates

A *compound gate* performing a more complex logic function in a single stage of logic is formed by using a combination of series and parallel switch structures. For example, the derivation of the circuit for the function $Y = \overline{(A \cdot B) + (C \cdot D)}$ is shown in Figure 1.18. This function is sometimes called AND-OR-INVERT-22, or AOI22 because it performs the NOR of a pair of 2-input ANDs. For the nMOS pull-down network, take the uninverted expression $((A \cdot B) + (C \cdot D))$ indicating when the output should be pulled to '0.' The AND expressions $(A \cdot B)$ and $(C \cdot D)$ may be implemented by series connections of switches, as shown in Figure 1.18(a). Now ORing the result requires the parallel connection of these two structures, which is shown in Figure 1.18(b). For the pMOS pull-up network, we must compute the complementary expression using switches that turn on with inverted polarity. By DeMorgan's Law, this is equivalent to interchanging AND and OR operations. Hence, transistors that appear in series in the pull-down network must appear in parallel in the pull-up network. Transistors that appear in parallel in the pull-down network must appear in series in the pull-up network. This principle is called *conduction complements* and has already been used in the design of the NAND and NOR gates. In the pull-up network, the parallel combination of A and B is placed in series with the parallel combination of C and D. This progression is evident in Figure 1.18(c) and Figure 1.18(d). Putting the networks together yields the full schematic (Figure 1.18(e)). The symbol is shown in Figure 1.18(f).

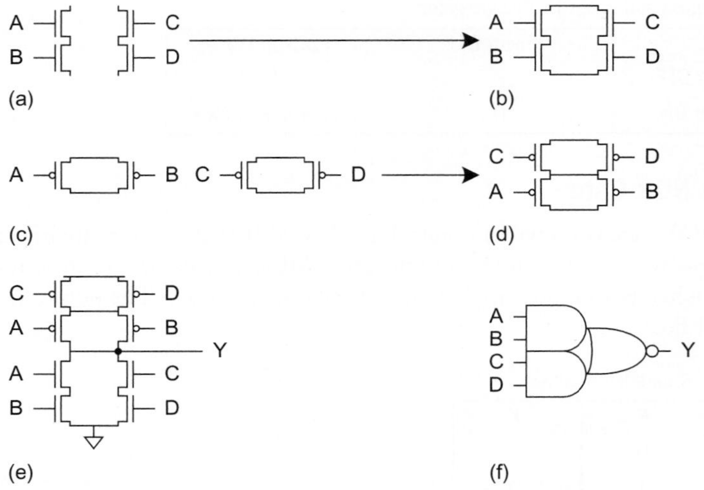

FIGURE 1.18 CMOS compound gate for function $Y = \overline{(A \cdot B) + (C \cdot D)}$

This AOI22 gate can be used as a 2-input inverting multiplexer by connecting $C = \overline{A}$ as a select signal. Then, $Y = \overline{B}$ if C is 0, while $Y = \overline{D}$ if C is 1. Section 1.4.8 shows a way to improve this multiplexer design.

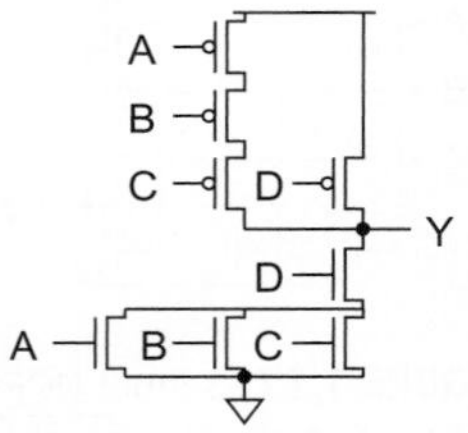

FIGURE 1.19 CMOS compound gate for function $Y = \overline{(A + B + C) \cdot D}$

Example 1.2

Sketch a static CMOS gate computing $Y = \overline{(A + B + C) \cdot D}$.

SOLUTION: Figure 1.19 shows such an OR-AND-INVERT-3-1 (OAI31) gate. The nMOS pull-down network pulls the output low if D is 1 and either A or B or C are 1, so D is in series with the parallel combination of A, B, and C. The pMOS pull-up network is the conduction complement, so D must be in parallel with the series combination of A, B, and C.

1.4.6 Pass Transistors and Transmission Gates

The *strength* of a signal is measured by how closely it approximates an ideal voltage source. In general, the stronger a signal, the more current it can source or sink. The power supplies, or *rails*, (V_{DD} and GND) are the source of the strongest 1s and 0s.

An nMOS transistor is an almost perfect switch when passing a 0 and thus we say it passes a *strong* 0. However, the nMOS transistor is imperfect at passing a 1. The high voltage level is somewhat less than V_{DD}, as will be explained in Section 2.5.4. We say it passes a *degraded* or *weak* 1. A pMOS transistor again has the opposite behavior, passing strong 1s but degraded 0s. The transistor symbols and behaviors are summarized in Figure 1.20 with g, s, and d indicating gate, source, and drain.

When an nMOS or pMOS is used alone as an imperfect switch, we sometimes call it a *pass transistor*. By combining an nMOS and a pMOS transistor in parallel (Figure 1.21(a)), we obtain a switch that turns on when a 1 is applied to g (Figure 1.21(b)) in which 0s and 1s are both passed in an acceptable fashion (Figure 1.21(c)). We term this a *transmission gate* or *pass gate*. In a circuit where only a 0 or a 1 has to be passed, the appropriate transistor (n or p) can be deleted, reverting to a single nMOS or pMOS device.

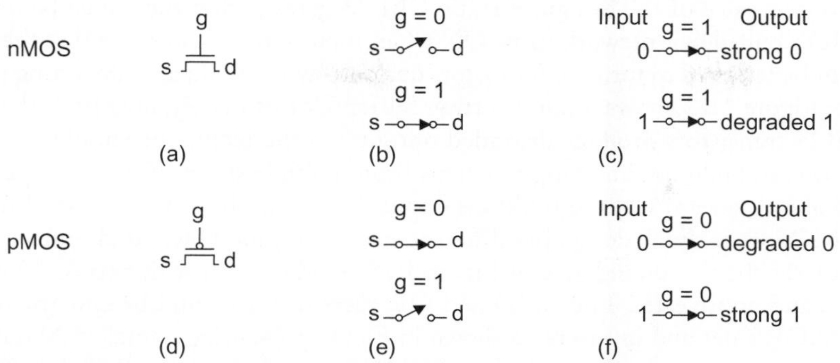

FIGURE 1.20 Pass transistor strong and degraded outputs

Note that both the control input and its complement are required by the transmission gate. This is called *double rail* logic. Some circuit symbols for the transmission gate are shown in Figure 1.21(d).[1] None are easier to draw than the simple schematic, so we will use the schematic version to represent a transmission gate in this book.

In all of our examples so far, the inputs drive the gate terminals of nMOS transistors in the pull-down network and pMOS transistors in the complementary pull-up network, as was shown in Figure 1.14. Thus, the nMOS transistors only need to pass 0s and the pMOS only pass 1s, so the output is always strongly driven and the levels are never degraded. This is called a *fully restored* logic gate and simplifies circuit design considerably. In contrast to other forms of logic, where the pull-up and pull-down switch networks have to be ratioed in some manner, static CMOS gates operate correctly independently of the physical sizes of the transistors. Moreover, there is never a path through 'ON' transistors from the 1 to the 0 supplies for any combination of inputs (in contrast to single-channel MOS, GaAs technologies, or bipolar). As we will find in subsequent chapters, this is the basis for the low static power dissipation in CMOS.

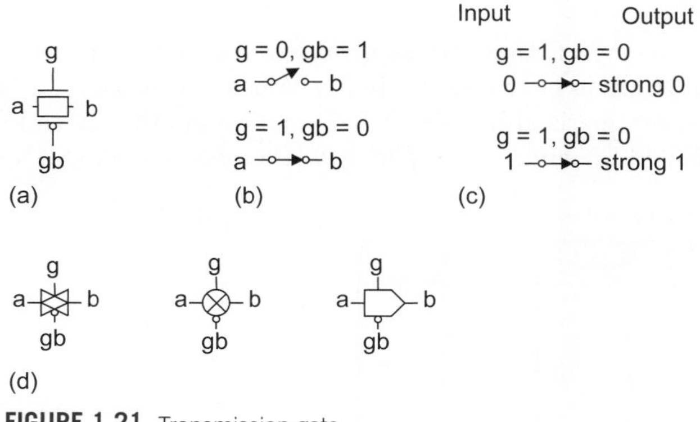

FIGURE 1.21 Transmission gate

[1]We call the left and right terminals *a* and *b* because each is technically the source of one of the transistors and the drain of the other.

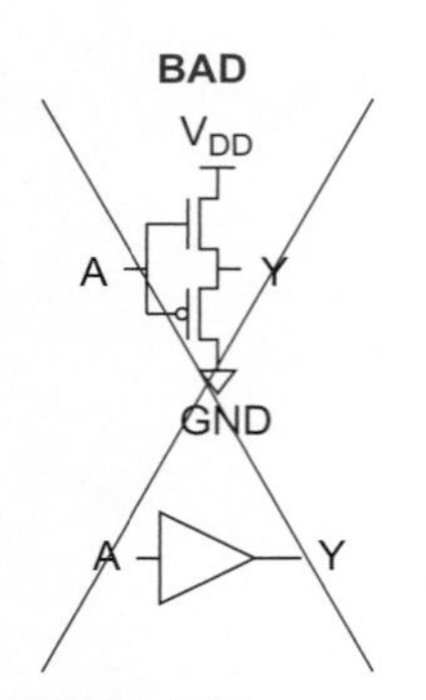

FIGURE 1.22
Bad noninverting buffer

A consequence of the design of static CMOS gates is that they must be inverting. The nMOS pull-down network turns ON when inputs are 1, leading to 0 at the output. We might be tempted to turn the transistors upside down to build a noninverting gate. For example, Figure 1.22 shows a noninverting buffer. Unfortunately, now both the nMOS and pMOS transistors produce degraded outputs, so the technique should be avoided. Instead, we can build noninverting functions from multiple stages of inverting gates. Figure 1.23 shows several ways to build a 4-input AND gate from two levels of inverting static CMOS gates. Each design has different speed, size, and power trade-offs.

Similarly, the compound gate of Figure 1.18 could be built with two AND gates, an OR gate, and an inverter. The AND and OR gates in turn could be constructed from NAND/NOR gates and inverters, as shown in Figure 1.24, using a total of 20 transistors, as compared to eight in Figure 1.18. Good CMOS logic designers exploit the efficiencies of compound gates rather than using large numbers of AND/OR gates.

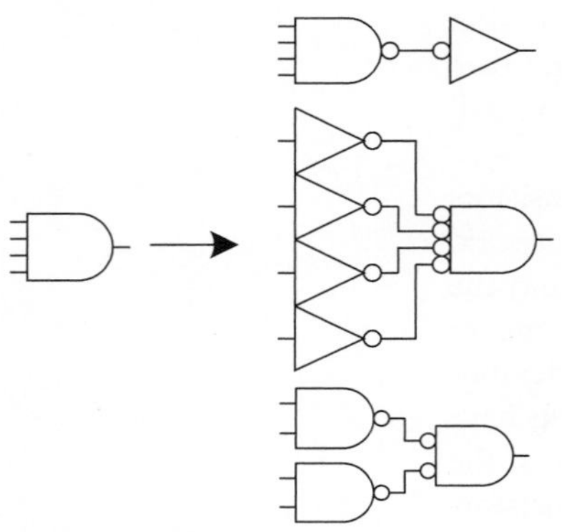

FIGURE 1.23 Various implementations of a CMOS 4-input AND gate

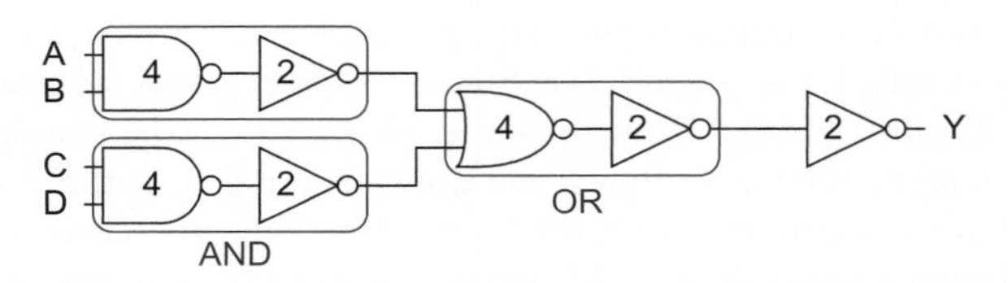

FIGURE 1.24 Inefficient discrete gate implementation of AOI22 with transistor counts indicated

FIGURE 1.25
Tristate buffer symbol

1.4.7 Tristates

Figure 1.25 shows symbols for a *tristate buffer*. When the enable input EN is 1, the output *Y* equals the input *A*, just as in an ordinary buffer. When the enable is 0, *Y* is left floating (a 'Z' value). This is summarized in Table 1.5. Sometimes both true and complementary enable signals EN and $\overline{\text{EN}}$ are drawn explicitly, while sometimes only EN is shown.

TABLE 1.5 Truth table for tristate

EN / $\overline{\text{EN}}$	*A*	*Y*
0 / 1	0	Z
0 / 1	1	Z
1 / 0	0	0
1 / 0	1	1

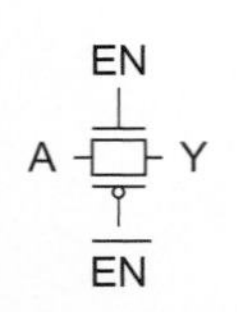

FIGURE 1.26
Transmission gate

The transmission gate in Figure 1.26 has the same truth table as a tristate buffer. It only requires two transistors but it is a *nonrestoring* circuit. If the input is noisy or otherwise degraded, the output will receive the same noise. We will see in Section 3.4.2 that the delay of a series of nonrestoring gates increases rapidly with the number of gates.

Figure 1.27(a) shows a *tristate inverter*. The output is actively driven from V_{DD} or GND, so it is a restoring logic gate. Unlike any of the gates considered so far, the tristate inverter does not obey the conduction complements rule because it allows the output to float under certain input combinations. When EN is 0 (Figure 1.27(b)), both enable transistors are OFF, leaving the output floating. When EN is 1 (Figure 1.27(c)), both enable transistors are ON. They are conceptually removed from the circuit, leaving a simple inverter. Figure 1.27(d) shows symbols for the tristate inverter. The complementary enable signal can be generated internally or can be routed to the cell explicitly. A tristate buffer can be built as an ordinary inverter followed by a tristate inverter.

FIGURE 1.27 Tristate Inverter

Tristates were once commonly used to allow multiple units to drive a common bus, as long as exactly one unit is enabled at a time. If multiple units drive the bus, contention occurs and power is wasted. If no units drive the bus, it can float to an invalid logic level that causes the receivers to waste power. Moreover, it can be difficult to switch enable signals at exactly the same time when they are distributed across a large chip. Delay between different enables switching can cause contention. Given these problems, multiplexers are now preferred over tristate busses.

1.4.8 Multiplexers

Multiplexers are key components in CMOS memory elements and data manipulation structures. A *multiplexer* chooses the output from among several inputs based on a select signal. A 2-input, or 2:1 multiplexer, chooses input $D0$ when the select is 0 and input $D1$ when the select is 1. The truth table is given in Table 1.6; the logic function is $Y = \overline{S} \cdot D0 + S \cdot D1$.

TABLE 1.6 Multiplexer truth table

$S/\overline{S}$	$D1$	$D0$	Y
0 / 1	X	0	0
0 / 1	X	1	1
1 / 0	0	X	0
1 / 0	1	X	1

Two transmission gates can be tied together to form a compact 2-input multiplexer, as shown in Figure 1.28(a). The select and its complement enable exactly one of the two transmission gates at any given time. The complementary select $\overline{S}$ is often not drawn in the symbol, as shown in Figure 1.28(b).

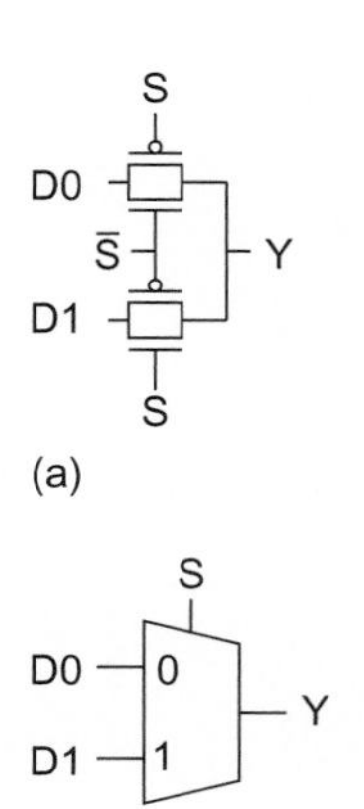

FIGURE 1.28 Transmission gate multiplexer

Again, the transmission gates produce a nonrestoring multiplexer. We could build a restoring, inverting multiplexer out of gates in several ways. One is the compound gate of Figure 1.18(e), connected as shown in Figure 1.29(a). Another is to gang together two tristate inverters, as shown in Figure 1.29(b). Notice that the schematics of these two approaches are nearly identical, save that the pull-up network has been slightly simplified and permuted in Figure 1.29(b). This is possible because the select and its complement are mutually exclusive. The tristate approach is slightly more compact and faster because it

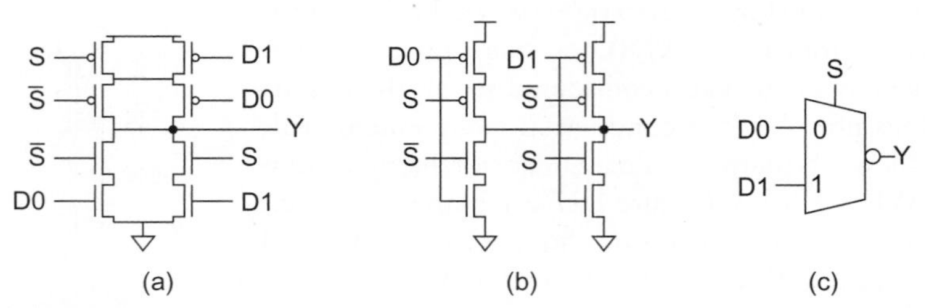

FIGURE 1.29 Inverting multiplexer

requires less internal wire. Again, if the complementary select is generated within the cell, it is omitted from the symbol (Figure 1.29(c)).

Larger multiplexers can be built from multiple 2-input multiplexers or by directly ganging together several tristates. The latter approach requires decoded enable signals for each tristate; the enables should switch simultaneously to prevent contention. 4-input (4:1) multiplexers using each of these approaches are shown in Figure 1.30. In practice, both inverting and noninverting multiplexers are simply called multiplexers or muxes.

1.4.9 Sequential Circuits

So far, we have considered *combinational circuits*, whose outputs depend only on the current inputs. *Sequential circuits* have memory: their outputs depend on both current and previous inputs. Using the combinational circuits developed so far, we can now build sequential circuits such as latches and flip-flops. These elements receive a clock, *CLK*, and a data input, *D*, and produce an output, *Q*. A *D latch* is *transparent* when *CLK* = 1, meaning that *Q* follows *D*. It becomes *opaque* when *CLK* = 0, meaning *Q* retains its previous value and ignores changes in *D*. An *edge-triggered flip-flop* copies *D* to *Q* on the rising edge of *CLK* and remembers its old value at other times.

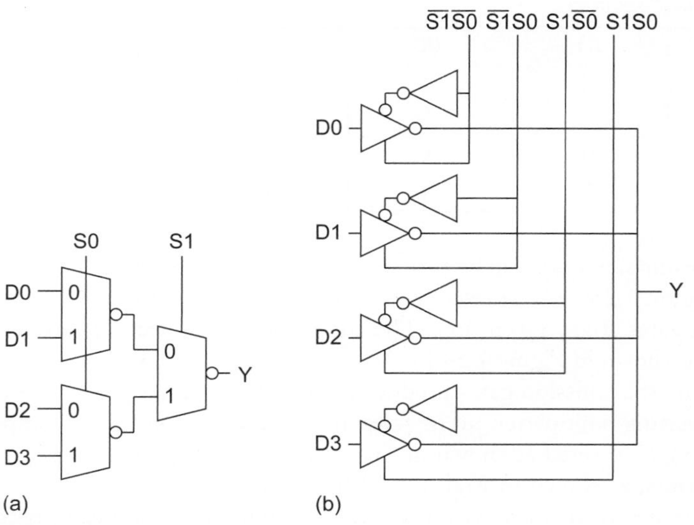

FIGURE 1.30 4:1 multiplexer

1.4.9.1 Latches A *D* latch built from a 2-input multiplexer and two inverters is shown in Figure 1.31(a). The multiplexer can be built from a pair of transmission gates, shown in Figure 1.31(b), because the inverters are restoring. This latch also produces a complementary output, $\overline{Q}$. When *CLK* = 1, the latch is transparent and *D* flows through to *Q* (Figure 1.31(c)). When *CLK* falls to 0, the latch becomes opaque. A feedback path around the inverter pair is established (Figure 1.31(d)) to hold the current state of *Q* indefinitely.

The *D* latch is also known as a *level-sensitive latch* because the state of the output is dependent on the level of the clock signal, as shown in Figure 1.31(e). The latch shown is a positive-level-sensitive latch, represented by the symbol in Figure 1.31(f). By inverting the control connections to the multiplexer, the latch becomes negative-level-sensitive.

1.4.9.2 Flip-Flops By combining two level-sensitive latches, one negative-sensitive and one positive-sensitive, we construct the edge-triggered flip-flop shown in Figure 1.32(a–b). The first latch stage is called the *master* and the second is called the *slave*.

While *CLK* is low, the master negative-level-sensitive latch output ($\overline{QM}$) follows the *D* input while the slave positive-level-sensitive latch holds the previous value (Figure 1.32(c)). When the clock transitions from 0 to 1, the master latch becomes opaque and holds the *D* value at the time of the clock transition. The slave latch becomes transparent, passing the stored master value ($\overline{QM}$) to the output of the slave latch (*Q*). The *D* input is blocked from affecting the output because the master is disconnected from the *D* input (Figure 1.32(d)). When the clock transitions from 1 to 0, the slave latch holds its value and the master starts sampling the input again.

While we have shown a transmission gate multiplexer as the input stage, good design practice would buffer the input and output with inverters, as shown in Figure 1.32(e), to

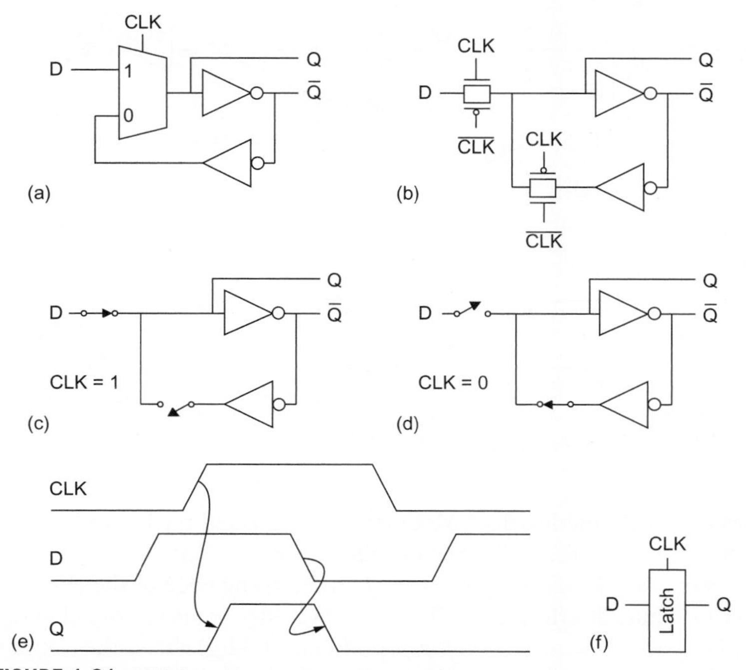

FIGURE 1.31 CMOS positive-level-sensitive *D* latch

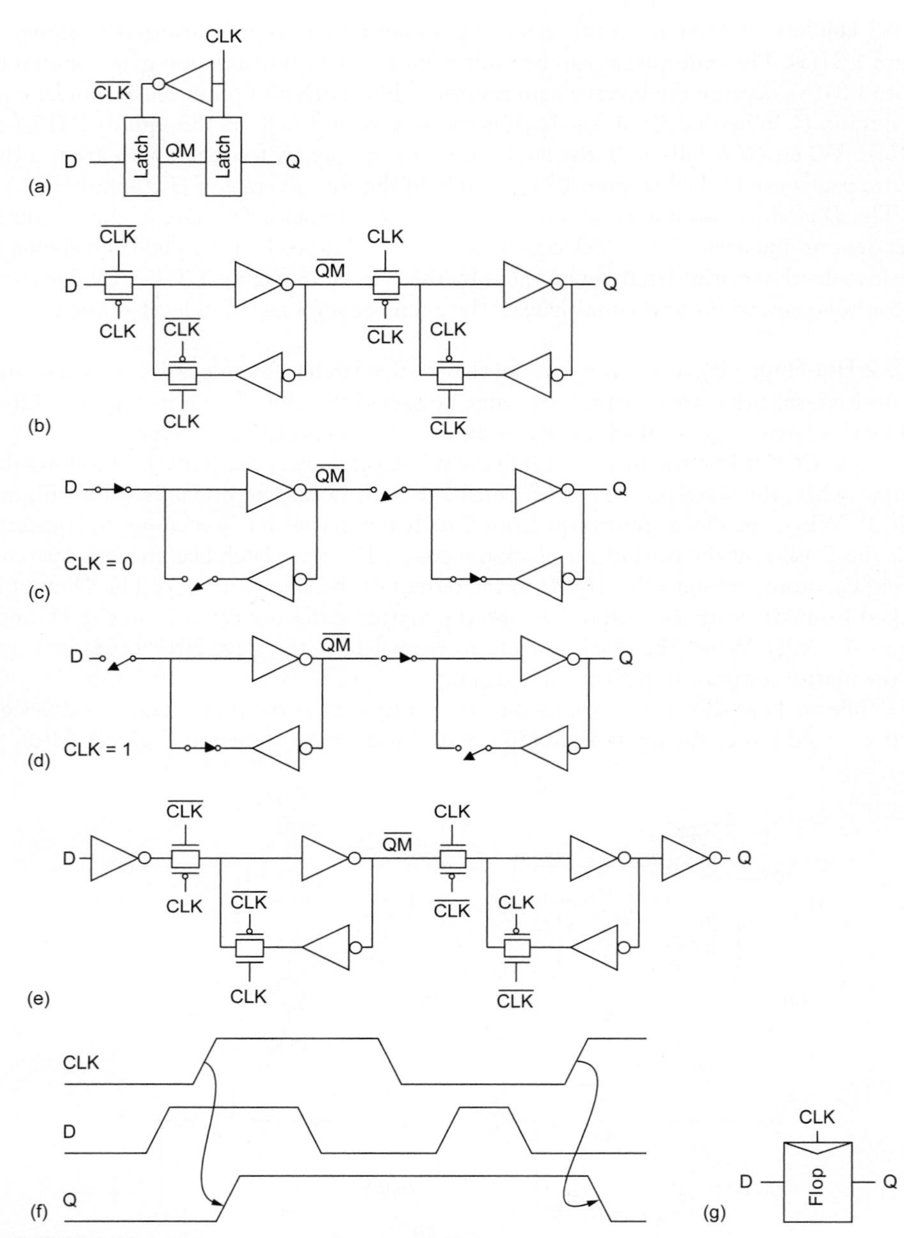

FIGURE 1.32 CMOS positive-edge-triggered *D* flip-flop

preserve what we call "modularity." Modularity is explained further in Section 1.6.2 and robust latches and registers are discussed further in Section 9.3.

In summary, this flip-flop copies *D* to *Q* on the rising edge of the clock, as shown in Figure 1.32(f). Thus, this device is called a positive-edge triggered flip-flop (also called a *D flip-flop*, *D register*, or *master–slave flip-flop*). Figure 1.32(g) shows the circuit symbol for the flip-flop. By reversing the latch polarities, a negative-edge triggered flip-flop may be constructed. A collection of D flip-flops sharing a common clock input is called a *register*. A register is often drawn as a flip-flop with multi-bit D and Q busses.

In Section 9.2.5 we will see that flip-flops may experience hold-time failures if the system has too much *clock skew*, i.e., if one flip-flop triggers early and another triggers late because of variations in clock arrival times. In industrial designs, a great deal of effort is devoted to timing simulations to catch hold-time problems. When design time is more important (e.g., in class projects), hold-time problems can be avoided altogether by distributing a two-phase nonoverlapping clock. Figure 1.33 shows the flip-flop clocked with two nonoverlapping phases. As long as the phases never overlap, at least one latch will be opaque at any given time and hold-time problems cannot occur.

1.5 CMOS Fabrication and Layout

Now that we can design logic gates and registers from transistors, let us consider how the transistors are built. Designers need to understand the physical implementation of circuits because it has a major impact on performance, power, and cost.

Transistors are fabricated on thin silicon wafers that serve as both a mechanical support and an electrical common point called the *substrate*. We can examine the physical layout of transistors from two perspectives. One is the top view, obtained by looking down on a wafer. The other is the cross-section, obtained by slicing the wafer through the middle of a transistor and looking at it edgewise. We begin by looking at the cross-section of a complete CMOS inverter. We then look at the top view of the same inverter and define a set of masks used to manufacture the different parts of the inverter. The size of the transistors and wires is set by the mask dimensions and is limited by the resolution of the manufacturing process. Continual advancements in this resolution have fueled the exponential growth of the semiconductor industry.

1.5.1 Inverter Cross-Section

Figure 1.34 shows a cross-section and corresponding schematic of an inverter. (See the inside front cover for a color cross-section.) In this diagram, the inverter is built on a p-type substrate. The pMOS transistor requires an n-type body region, so an n-well is diffused into the substrate in its vicinity. As described in Section 1.3, the nMOS transistor

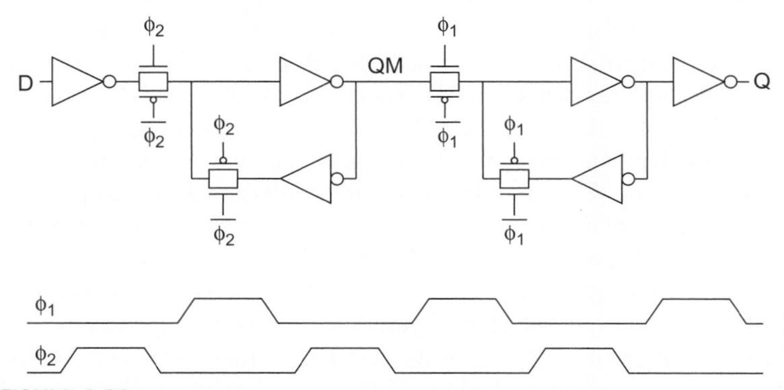

FIGURE 1.33 CMOS flip-flop with two-phase nonoverlapping clocks

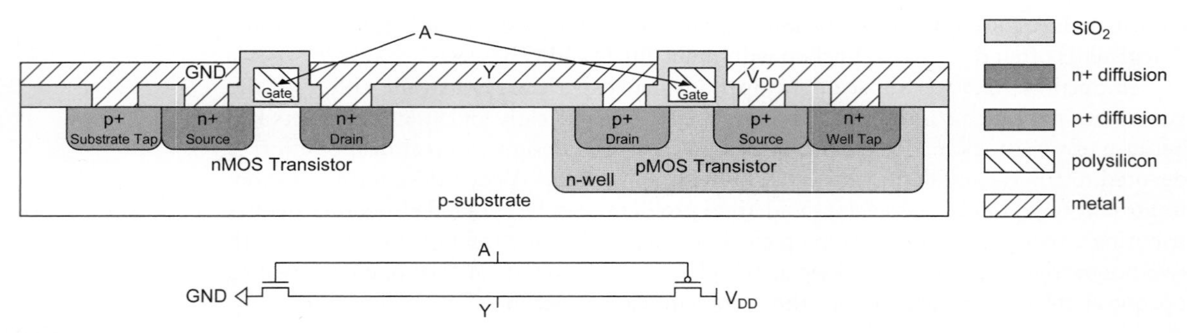

FIGURE 1.34 Inverter cross-section with well and substrate contacts. Color version on inside front cover.

has heavily doped n-type source and drain regions and a polysilicon gate over a thin layer of silicon dioxide (SiO_2, also called *gate oxide*). n+ and p+ diffusion regions indicate heavily doped n-type and p-type silicon. The pMOS transistor is a similar structure with p-type source and drain regions. The polysilicon gates of the two transistors are tied together somewhere off the page and form the input A. The source of the nMOS transistor is connected to a metal ground line and the source of the pMOS transistor is connected to a metal V_{DD} line. The drains of the two transistors are connected with metal to form the output Y. A thick layer of SiO_2 called *field oxide* prevents metal from shorting to other layers except where contacts are explicitly etched.

A junction between metal and a lightly doped semiconductor forms a *Schottky diode* that only carries current in one direction. When the semiconductor is doped more heavily, it forms a good ohmic contact with metal that provides low resistance for bidirectional current flow. The substrate must be tied to a low potential to avoid forward-biasing the p-n junction between the p-type substrate and the n+ nMOS source or drain. Likewise, the n-well must be tied to a high potential. This is done by adding heavily doped substrate and well contacts, or *taps*, to connect GND and V_{DD} to the substrate and n-well, respectively.

1.5.2 Fabrication Process

For all their complexity, chips are amazingly inexpensive because all the transistors and wires can be printed in much the same way as books. The fabrication sequence consists of a series of steps in which layers of the chip are defined through a process called *photolithography*. Because a whole wafer full of chips is processed in each step, the cost of the chip is proportional to the chip area, rather than the number of transistors. As manufacturing advances allow engineers to build smaller transistors and thus fit more in the same area, each transistor gets cheaper. Smaller transistors are also faster because electrons don't have to travel as far to get from the source to the drain, and they consume less energy because fewer electrons are needed to charge up the gates! This explains the remarkable trend for computers and electronics to become cheaper and more capable with each generation.

The inverter could be defined by a hypothetical set of six masks: n-well, polysilicon, n+ diffusion, p+ diffusion, contacts, and metal (for fabrication reasons discussed in Chapter 15, the actual mask set tends to be more elaborate). Masks specify where the components will be manufactured on the chip. Figure 1.35(a) shows a top view of the six masks. (See also the inside front cover for a color picture.) The cross-section of the inverter from Figure 1.34 was taken along the dashed line. Take some time to convince yourself how the top view and cross-section relate; this is critical to understanding chip layout.

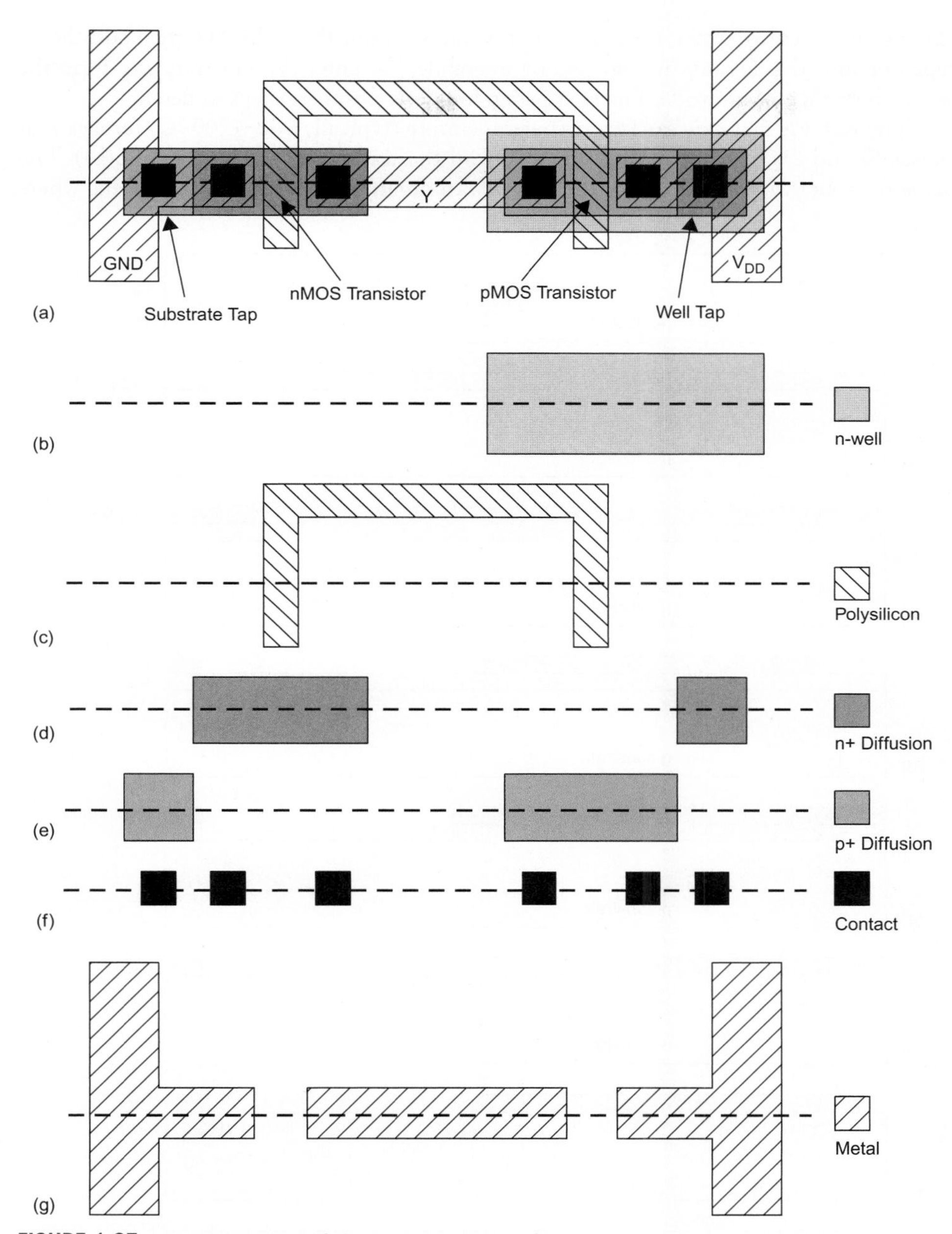

FIGURE 1.35 Inverter mask set. Color version on inside front cover.

Consider a simple fabrication process to illustrate the concept. The process begins with the creation of an n-well on a bare p-type silicon wafer. Figure 1.36 shows cross-sections of the wafer after each processing step involved in forming the n-well; Figure 1.36(a) illustrates the bare substrate before processing. Forming the n-well requires adding enough Group V dopants into the silicon substrate to change the substrate from p-type to n-type in the region of the well. To define what regions receive n-wells, we grow a protective layer of oxide over the entire wafer, then remove it where we want the wells. We then add the n-type dopants; the dopants are blocked by the oxide, but enter the substrate and form the wells where there is no oxide. The next paragraph describes these steps in detail.

The wafer is first *oxidized* in a high-temperature (typically 900–1200 °C) furnace that causes Si and O_2 to react and become SiO_2 on the wafer surface (Figure 1.36(b)). The oxide must be *patterned* to define the n-well. An organic photoresist[2] that softens where

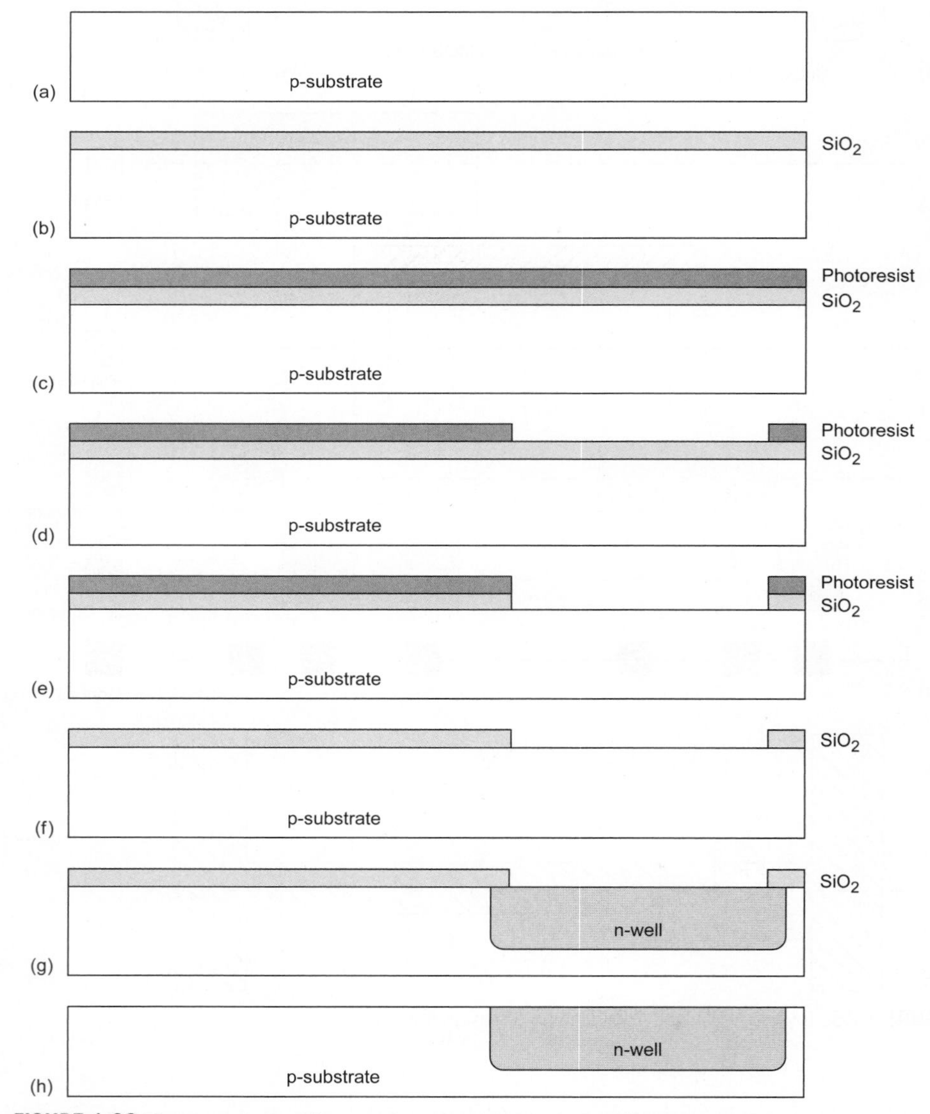

FIGURE 1.36 Cross-sections while manufacturing the n-well

[2]Engineers have experimented with many organic polymers for photoresists. In 1958, Brumford and Walker reported that Jello™ could be used for masking. They did extensive testing, observing that "various Jellos™ were evaluated with lemon giving the best result."

exposed to light is spun onto the wafer (Figure 1.36(c)). The photoresist is exposed through the n-well mask (Figure 1.35(b)) that allows light to pass through only where the well should be. The softened photoresist is removed to expose the oxide (Figure 1.36(d)). The oxide is etched with hydrofluoric acid (HF) where it is not protected by the photoresist (Figure 1.36(e)), then the remaining photoresist is stripped away using a mixture of acids called *piranha etch* (Figure 1.36(f)). The well is formed where the substrate is not covered with oxide. Two ways to add dopants are diffusion and ion implantation. In the *diffusion* process, the wafer is placed in a furnace with a gas containing the dopants. When heated, dopant atoms diffuse into the substrate. Notice how the well is wider than the hole in the oxide on account of *lateral* diffusion (Figure 1.36(g)). With *ion implantation*, dopant ions are accelerated through an electric field and blasted into the substrate. In either method, the oxide layer prevents dopant atoms from entering the substrate where no well is intended. Finally, the remaining oxide is stripped with HF to leave the bare wafer with wells in the appropriate places.

The transistor gates are formed next. These consist of polycrystalline silicon, generally called *polysilicon*, over a thin layer of oxide. The thin oxide is grown in a furnace. Then the wafer is placed in a reactor with silane gas (SiH_4) and heated again to grow the polysilicon layer through a process called *chemical vapor deposition*. The polysilicon is heavily doped to form a reasonably good conductor. The resulting cross-section is shown in Figure 1.37(a). As before, the wafer is patterned with photoresist and the polysilicon mask (Figure 1.35(c)), leaving the polysilicon gates atop the thin gate oxide (Figure 1.37(b)).

The n+ regions are introduced for the transistor active area and the well contact. As with the well, a protective layer of oxide is formed (Figure 1.37(c)) and patterned with the n-diffusion mask (Figure 1.35(d)) to expose the areas where the dopants are needed (Figure 1.37(d)). Although the n+ regions in Figure 1.37(e) are typically formed with ion implantation, they were historically diffused and thus still are often called *n-diffusion*. Notice that the polysilicon gate over the nMOS transistor blocks the diffusion so the source and drain are separated by a channel under the gate. This is called a *self-aligned* process because the source and drain of the transistor are automatically formed adjacent to the gate without the need to precisely align the masks. Finally, the protective oxide is stripped (Figure 1.37(f)).

The process is repeated for the p-diffusion mask (Figure 1.35(e)) to give the structure of Figure 1.38(a). Oxide is used for masking in the same way, and thus is not shown. The field oxide is grown to insulate the wafer from metal and patterned with the contact mask (Figure 1.35(f)) to leave contact cuts where metal should attach to diffusion or polysilicon (Figure 1.38(b)). Finally, aluminum is sputtered over the entire wafer, filling the contact cuts as well. Sputtering involves blasting aluminum into a vapor that evenly coats the wafer. The metal is patterned with the metal mask (Figure 1.35(g)) and plasma etched to remove metal everywhere except where wires should remain (Figure 1.38(c)). This completes the simple fabrication process.

Modern fabrication sequences are more elaborate because they must create complex doping profiles around the channel of the transistor and print features that are smaller than the wavelength of the light being used in lithography. However, masks for these elaborations can be automatically generated from the simple set of masks we have just examined. Modern processes also have 5–10+ layers of metal, so the metal and contact steps must be repeated for each layer. Chip manufacturing has become a commodity, and many different foundries will build designs from a basic set of masks.

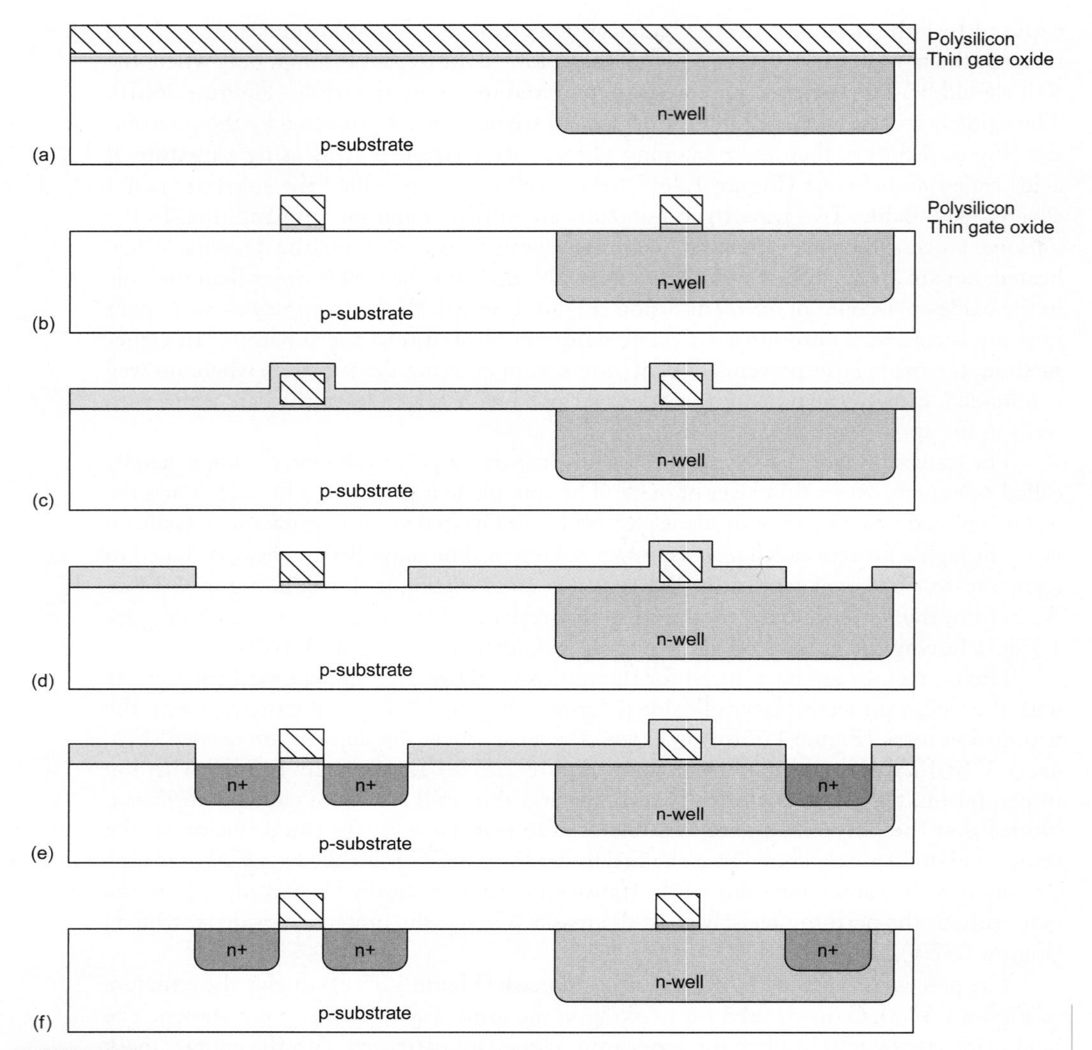

FIGURE 1.37 Cross-sections while manufacturing polysilicon and n-diffusion

1.5.3 Layout Design Rules

Layout design rules describe how small features can be and how closely they can be reliably packed in a particular manufacturing process. Industrial design rules are usually specified in microns. This makes migrating from one process to a more advanced process or a different foundry's process difficult because not all rules scale in the same way.

Universities sometimes simplify design by using scalable design rules that are conservative enough to apply to many manufacturing processes. Mead and Conway [Mead80] popularized scalable design rules based on a single parameter, λ, that characterizes the resolution of the process. λ is generally half of the minimum drawn transistor channel length. This length is the distance between the source and drain of a transistor and is set by the minimum width of a polysilicon wire. For example, a 180 nm process has a minimum polysilicon width (and hence transistor length) of 0.18 μm and uses design rules with

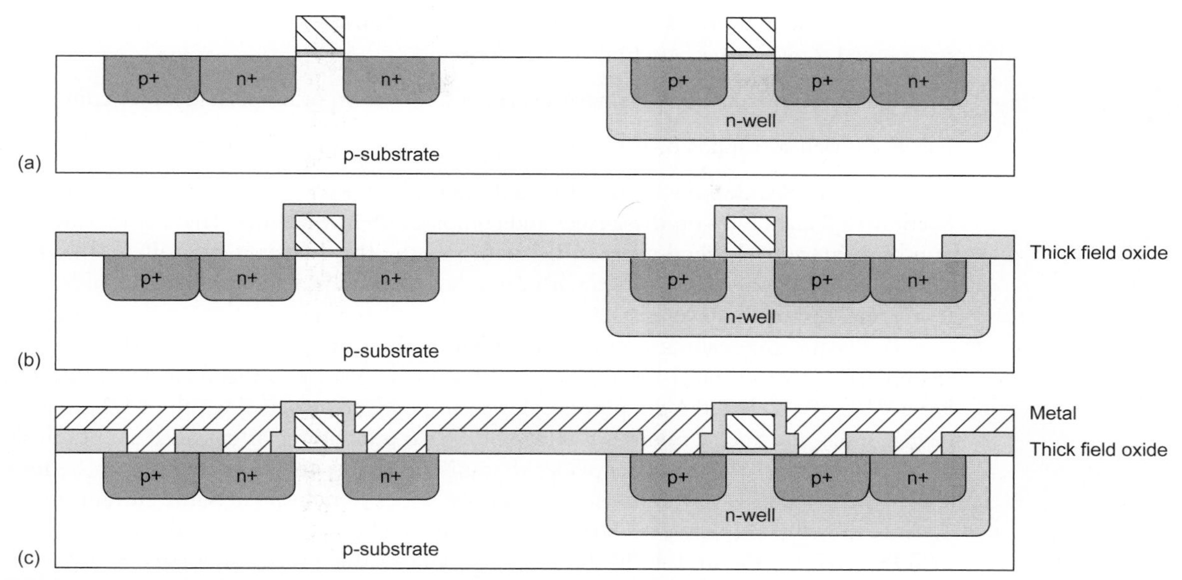

FIGURE 1.38 Cross-sections while manufacturing p-diffusion, contacts, and metal

$\lambda = 0.09\ \mu$m.[3] Lambda-based rules are necessarily conservative because they round up dimensions to an integer multiple of λ. However, they make scaling layout trivial; the same layout can be moved to a new process simply by specifying a new value of λ. This chapter will present design rules in terms of λ. The potential density advantage of micron rules is sacrificed for simplicity and easy scalability of lambda rules. Designers often describe a process by its *feature size*. Feature size refers to minimum transistor length, so λ is half the feature size.

Unfortunately, below 180 nm, design rules have become so complex and process-specific that scalable design rules are difficult to apply. However, the intuition gained from a simple set of scalable rules is still a valuable foundation for understanding the more complex rules. Chapter 15 will examine some of these process-specific rules in more detail.

The MOSIS service [Piña02] is a low-cost prototyping service that collects designs from academic, commercial, and government customers and aggregates them onto one mask set to share overhead costs and generate production volumes sufficient to interest fabrication companies. MOSIS has developed a set of scalable lambda-based design rules that covers a wide range of manufacturing processes. The rules describe the minimum width to avoid breaks in a line, minimum spacing to avoid shorts between lines, and minimum overlap to ensure that two layers completely overlap.

A conservative but easy-to-use set of design rules for layouts with two metal layers in an n-well process is as follows:

- Metal and diffusion have minimum width and spacing of 4 λ.
- Contacts are 2 $\lambda \times 2\ \lambda$ and must be surrounded by 1 λ on the layers above and below.
- Polysilicon uses a width of 2 λ.

[3]Some 180 nm lambda-based rules actually set $\lambda = 0.10\ \mu$m, then shrink the gate by 20 nm while generating masks. This keeps 180 nm gate lengths but makes all other features slightly larger.

- Polysilicon overlaps diffusion by 2 λ where a transistor is desired and has a spacing of 1 λ away where no transistor is desired.
- Polysilicon and contacts have a spacing of 3 λ from other polysilicon or contacts.
- N-well surrounds pMOS transistors by 6 λ and avoids nMOS transistors by 6 λ.

Figure 1.39 shows the basic MOSIS design rules for a process with two metal layers. Section 15.3 elaborates on these rules and compares them with industrial design rules.

In a three-level metal process, the width of the third layer is typically 6 λ and the spacing 4 λ. In general, processes with more layers often provide thicker and wider top-level metal that has a lower resistance.

Transistor dimensions are often specified by their Width/Length (W/L) ratio. For example, the nMOS transistor in Figure 1.39 formed where polysilicon crosses n-diffusion has a W/L of 4/2. In a 0.6 μm process, this corresponds to an actual width of 1.2 μm and a length of 0.6 μm. Such a minimum-width contacted transistor is often called a unit transistor.[4] pMOS transistors are often wider than nMOS transistors because holes move more slowly than electrons so the transistor has to be wider to deliver the same current. Figure 1.40(a) shows a unit inverter layout with a unit nMOS transistor and a double-sized pMOS transistor. Figure 1.40(b) shows a schematic for the inverter annotated with Width/Length for each transistor. In digital systems, transistors are typically chosen to have the minimum possible length because short-channel transistors are faster, smaller, and consume less power. Figure 1.40(c) shows a shorthand we will often use, specifying multiples of unit width and assuming minimum length.

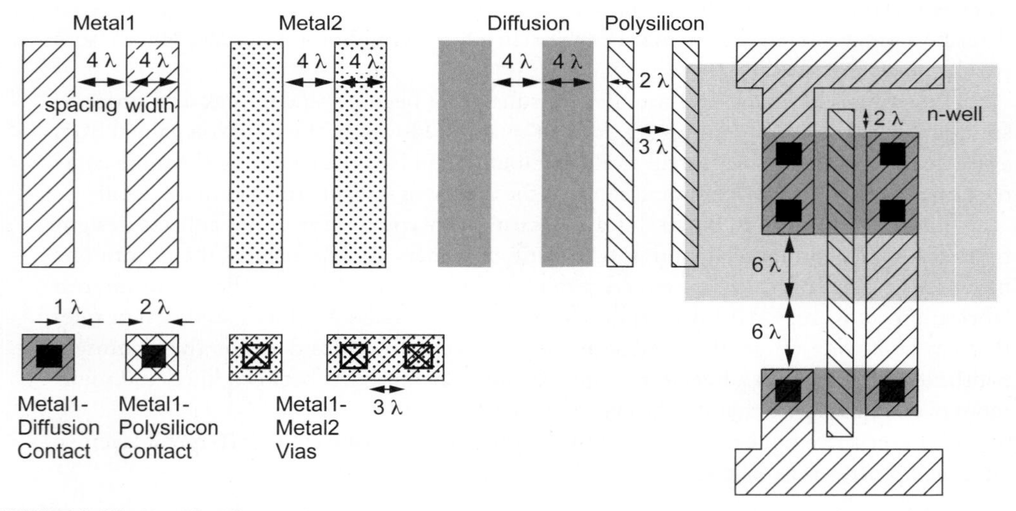

FIGURE 1.39 Simplified λ-based design rules

[4]Such small transistors in modern processes often behave slightly differently than their wider counterparts. Moreover, the transistor will not operate if either contact is damaged. Industrial designers often use a transistor wide enough for two contacts (9 λ) as the unit transistor to avoid these problems.

1.5.4 Gate Layouts

A good deal of ingenuity can be exercised and a vast amount of time wasted exploring layout topologies to minimize the size of a gate or other *cell* such as an adder or memory element. For many applications, a straightforward layout is good enough and can be automatically generated or rapidly built by hand. This section presents a simple layout style based on a "line of diffusion" rule that is commonly used for standard cells in automated layout systems. This style consists of four horizontal strips: metal ground at the bottom of the cell, n-diffusion, p-diffusion, and metal power at the top. The power and ground lines are often called *supply rails*. Polysilicon lines run vertically to form transistor gates. Metal wires within the cell connect the transistors appropriately.

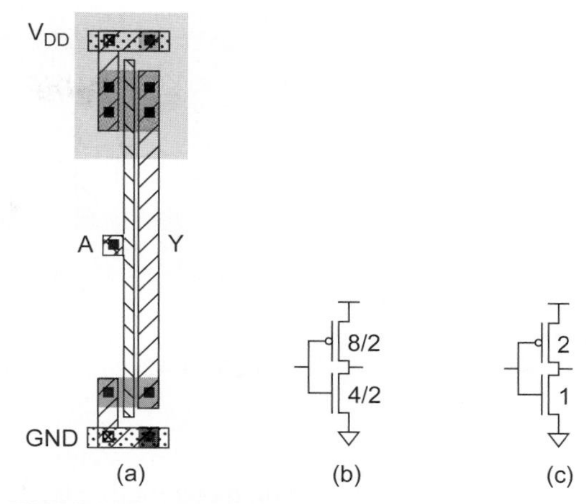

FIGURE 1.40 Inverter with dimensions labeled

Figure 1.41(a) shows such a layout for an inverter. The input *A* can be connected from the top, bottom, or left in polysilicon. The output *Y* is available at the right side of the cell in metal. Recall that the p-substrate and n-well must be tied to ground and power, respectively. Figure 1.41(b) shows the same inverter with well and substrate taps placed under the power and ground rails, respectively. Figure 1.42 shows a 3-input NAND gate. Notice how the nMOS transistors are connected in series while the pMOS transistors are connected in parallel. Power and ground extend 2 λ on each side so if two gates were abutted the contents would be separated by 4 λ, satisfying design rules. The height of the cell is 36 λ, or 40 λ if the 4 λ space between the cell and another wire above it is counted. All these examples use transistors of width 4 λ. Choice of transistor width is addressed further in Chapters 3–4.

These cells were designed such that the gate connections are made from the top or bottom in polysilicon. In contemporary standard cells, polysilicon is generally not used as a routing layer so the cell must allow metal2 to metal1 and metal1 to polysilicon contacts

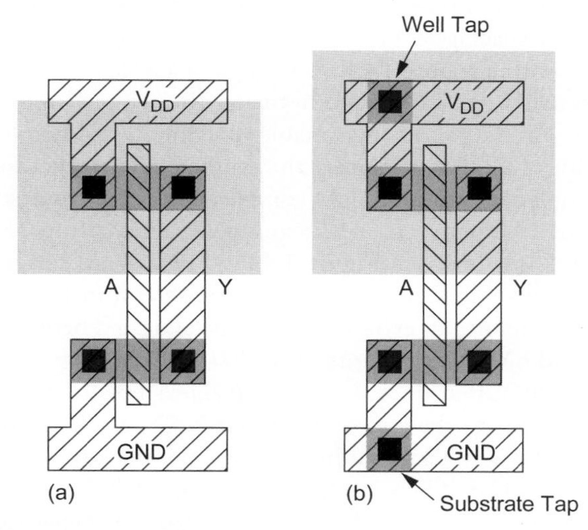

FIGURE 1.41 Inverter cell layout

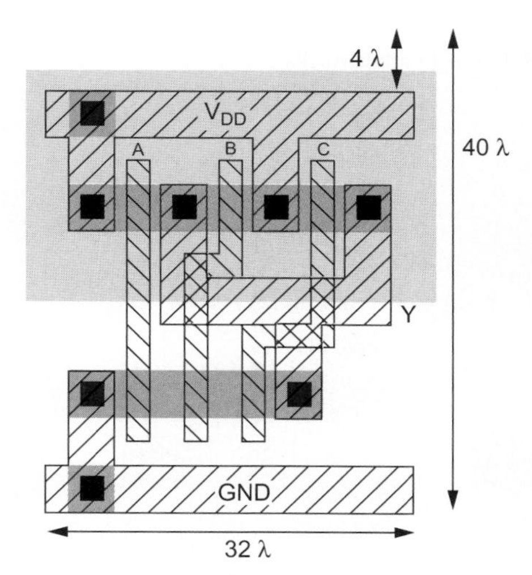

FIGURE 1.42 3-input NAND standard cell gate layouts

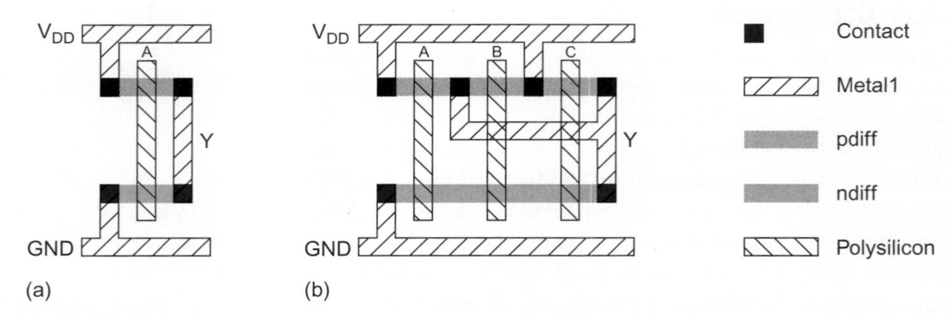

FIGURE 1.43 Stick diagrams of inverter and 3-input NAND gate. Color version on inside front cover.

to each gate. While this increases the size of the cell, it allows free access to all terminals on metal routing layers.

1.5.5 Stick Diagrams

Because layout is time-consuming, designers need fast ways to plan cells and estimate area before committing to a full layout. *Stick diagrams* are easy to draw because they do not need to be drawn to scale. Figure 1.43 and the inside front cover show stick diagrams for an inverter and a 3-input NAND gate. While this book uses stipple patterns, layout designers use dry-erase markers or colored pencils.

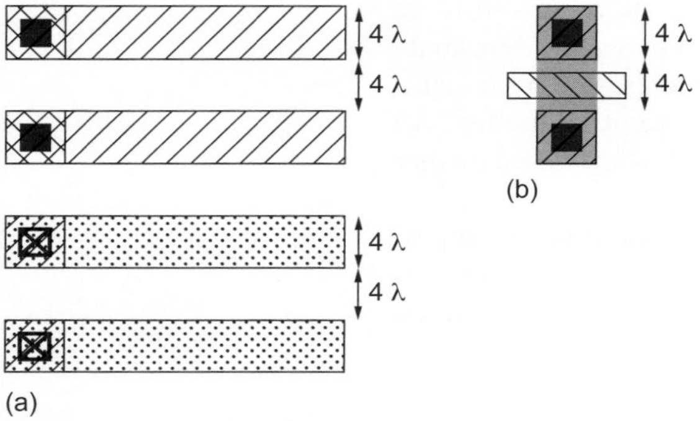

With practice, it is easy to estimate the area of a layout from the corresponding stick diagram even though the diagram is not to scale. Although schematics focus on transistors, layout area is usually determined by the metal wires. Transistors are merely widgets that fit under the wires. We define a *routing track* as enough space to place a wire and the required spacing to the next wire. If our wires have a width of 4 λ and a spacing of 4 λ to the next wire, the track *pitch* is 8 λ, as shown in Figure 1.44(a). This pitch also leaves room for a transistor to be placed between the wires (Figure 1.44(b)). Therefore, it is reasonable to estimate the height and width of a cell by counting the number of metal tracks and multiplying by 8 λ. A slight complication is the required spacing of 12 λ between nMOS and pMOS transistors set by the well, as shown in Figure 1.45(a). This space can be occupied by an additional track of wire, shown in Figure 1.45(b). Therefore, an extra track must be allocated between nMOS and pMOS transistors regardless of whether wire is actually used in that track. Figure 1.46 shows how to count tracks to estimate the size of a 3-input NAND. There are four vertical wire tracks, multiplied by 8 λ per track to give a cell width of 32 λ. There are five horizontal tracks, giving a cell height of 40 λ. Even though the horizontal tracks are not drawn to scale, they are still easy to count. Figure 1.42

FIGURE 1.44 Pitch of routing tracks

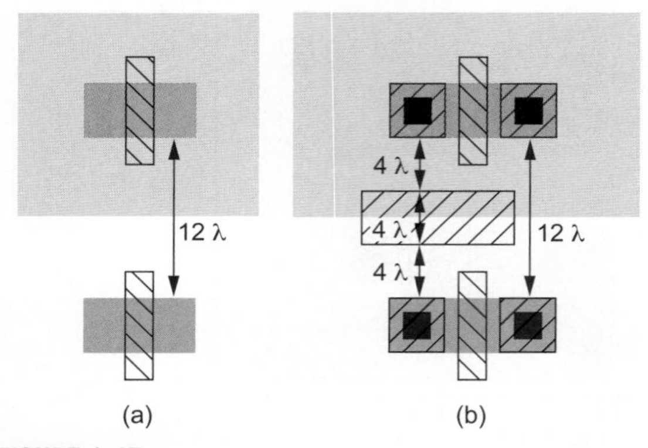

FIGURE 1.45 Spacing between nMOS and pMOS transistors

shows that the actual NAND gate layout matches the dimensions predicted by the stick diagram. If transistors are wider than 4 λ, the extra width must be factored into the area estimate. Of course, these estimates are oversimplifications of the complete design rules and a trial layout should be performed for truly critical cells.

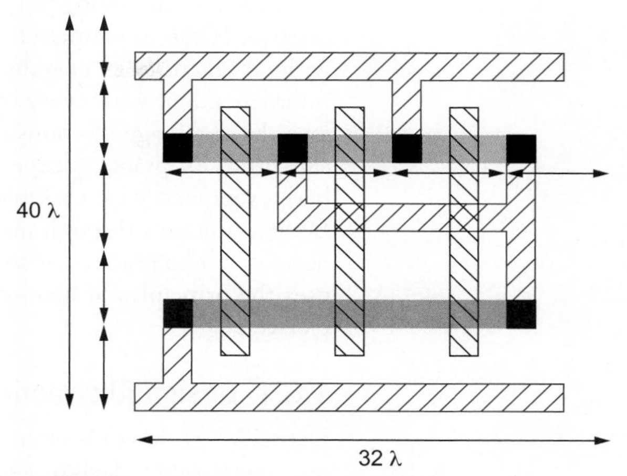

FIGURE 1.46 3-input NAND gate area estimation

Example 1.3

Sketch a stick diagram for a CMOS gate computing $Y = \overline{(A + B + C) \cdot D}$ (see Figure 1.18) and estimate the cell width and height.

SOLUTION: Figure 1.47 shows a stick diagram. Counting horizontal and vertical pitches gives an estimated cell size of 40 by 48 λ.

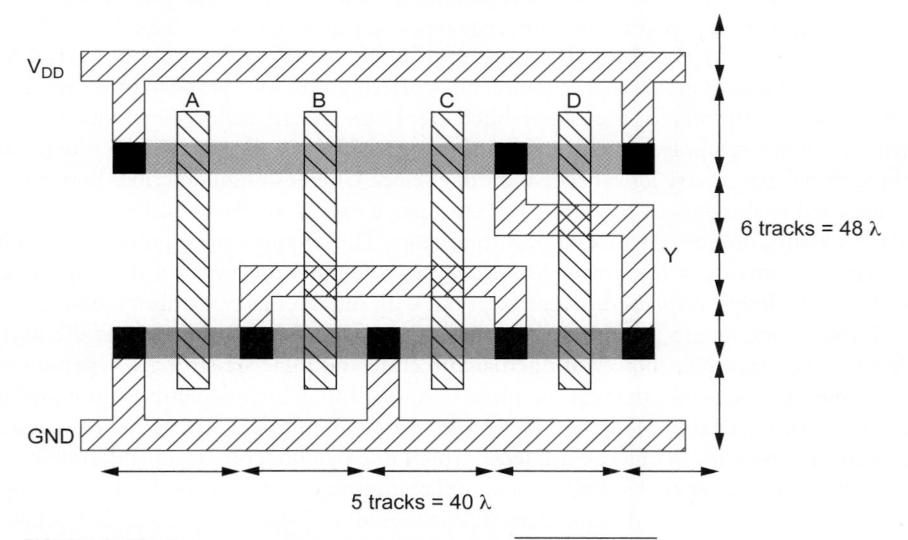

FIGURE 1.47 CMOS compound gate for function $Y = \overline{(A + B + C) \cdot D}$

1.6 Design Partitioning

By this point, you know that MOS transistors behave as voltage-controlled switches. You know how to build logic gates out of transistors. And you know how transistors are fabricated and how to draw a layout that specifies how transistors should be placed and connected together. You know enough to start building your own simple chips.

The greatest challenge in modern VLSI design is not in designing the individual transistors but rather in managing system complexity. Modern *System-On-Chip* (SOC) designs combine memories, processors, high-speed I/O interfaces, and dedicated application-specific logic on a single chip. They use hundreds of millions or billions of transistors and cost tens of millions of dollars (or more) to design. The implementation must be divided among large teams of engineers and each engineer must be highly productive. If the implementation is too rigidly partitioned, each block can be optimized without regard to its neighbors, leading to poor system results. Conversely, if every task is interdependent with every other task, design will progress too slowly. Design managers face the challenge of choosing a suitable trade-off between these extremes. There is no substitute for practical experience in making these choices, and talented engineers who have experience with multiple designs are very important to the success of a large project. Design proceeds through multiple levels of abstraction, hiding details until they become necessary. The practice of *structured design*, which is also used in large software projects, uses the principles of hierarchy, regularity, modularity, and locality to manage the complexity.

1.6.1 Design Abstractions

Digital VLSI design is often partitioned into five levels of abstractions: *architecture* design, *microarchitecture* design, *logic* design, *circuit* design, and *physical* design. Architecture describes the functions of the system. For example, the x86 microprocessor architecture specifies the instruction set, register set, and memory model. Microarchitecture describes how the architecture is partitioned into registers and functional units. The 80386, 80486, Pentium, Pentium II, Pentium III, Pentium 4, Core, Core 2, Atom, Cyrix MII, AMD Athlon, and Phenom are all microarchitectures offering different performance / transistor count / power trade-offs for the x86 architecture. Logic describes how functional units are constructed. For example, various logic designs for a 32-bit adder in the x86 integer unit include ripple carry, carry lookahead, and carry select. Circuit design describes how transistors are used to implement the logic. For example, a carry lookahead adder can use static CMOS circuits, domino circuits, or pass transistors. The circuits can be tailored to emphasize high performance or low power. Physical design describes the layout of the chip. Analog and RF VLSI design involves the same steps but with different layers of abstraction.

These elements are inherently interdependent and all influence each of the design objectives. For example, choices of microarchitecture and logic are strongly dependent on the number of transistors that can be placed on the chip, which depends on the physical design and process technology. Similarly, innovative circuit design that reduces a cache access from two cycles to one can influence which microarchitecture is most desirable. The choice of clock frequency depends on a complex interplay of microarchitecture and logic, circuit design, and physical design. Deeper pipelines allow higher frequencies but consume more power and lead to greater performance penalties when operations early in the pipeline are dependent on those late in the pipeline. Many functions have various logic and circuit designs trading speed for area, power, and design effort. Custom physical design allows more compact, faster circuits and lower manufacturing costs, but involves an enormous labor cost. Automatic layout with CAD systems reduces the labor and achieves faster times to market.

To deal with these interdependencies, microarchitecture, logic, circuit, and physical design must occur, at least in part, in parallel. Microarchitects depend on circuit and physical design studies to understand the cost of proposed microarchitectural features. Engineers are sometimes categorized as "short and fat" or "tall and skinny" (nothing personal, we assure you!). Tall, skinny engineers understand something about a broad range of topics. Short, fat engineers understand a large amount about a narrow field. Digital VLSI design favors the tall, skinny engineer who can evaluate how choices in one part of the system impact other parts of the system.

1.6.2 Structured Design

Hierarchy is a critical tool for managing complex designs. A large system can be partitioned hierarchically into multiple *cores*. Each core is built from various *units*. Each unit in turn is composed of multiple *functional blocks*.[5] These blocks in turn are built from *cells*, which ultimately are constructed from transistors. The system can be more easily understood at the top level by viewing components as black boxes with well-defined interfaces and functions rather than looking at each individual transistor. Logic, circuit, and physical views of the design should share the same hierarchy for ease of verification. A design hierarchy can be viewed as a tree structure with the overall chip as the *root* and the primitive cells as *leafs*.

Regularity aids the management of design complexity by designing the minimum number of different blocks. Once a block is designed and verified, it can be reused in many places. *Modularity* requires that the blocks have well-defined interfaces to avoid unanticipated interactions. *Locality* involves keeping information where it is used, physically and temporally. Structured design is discussed further in Section 13.2.

1.6.3 Behavioral, Structural, and Physical Domains

An alternative way of viewing design partitioning is shown with the Y-chart shown in Figure 1.48 [Gajski83, Kang03]. The radial lines on the Y-chart represent three distinct design domains: behavioral, structural, and physical. These domains can be used to describe the design of almost any artifact and thus form a general taxonomy for describing

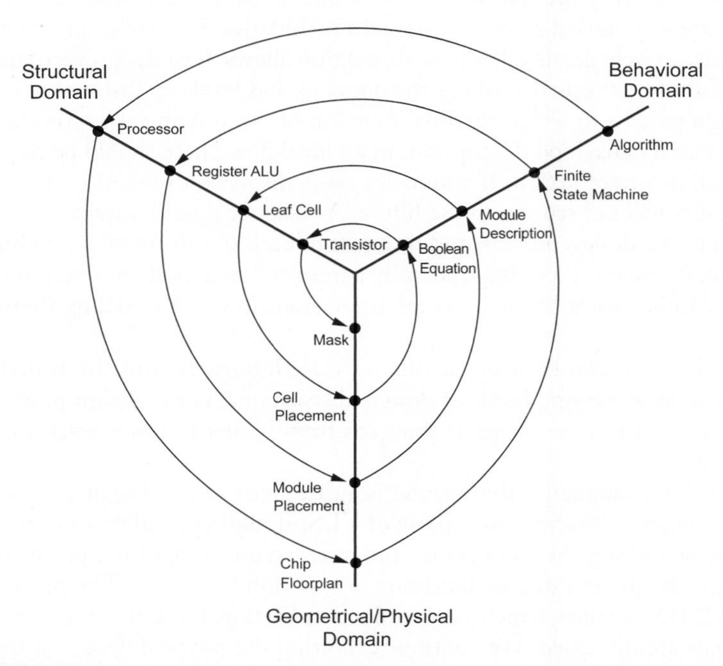

FIGURE 1.48 Y Diagram (Reproduced from [Kang03] with permission of The McGraw-Hill Companies.)

[5]Some designers refer to both units and functional blocks as *modules*.

the design process. Within each domain there are a number of levels of design abstraction that start at a very high level and descend eventually to the individual elements that need to be aggregated to yield the top level function (i.e., transistors in the case of chip design).

The behavioral domain describes what a particular system does. For instance, at the highest level we might specify a telephone touch-tone generator. This behavior can be successively refined to more precisely describe what needs to be done in order to build the tone generator (i.e., the frequencies desired, output levels, distortion allowed, etc.).

At each abstraction level, a corresponding structural description can be developed. The structural domain describes the interconnection of modules necessary to achieve a particular behavior. For instance, at the highest level, the touch-tone generator might consist of a keypad, a tone generator chip, an audio amplifier, a battery, and a speaker. Eventually at lower levels of abstraction, the individual gate and then transistor connections required to build the tone generator are described.

For each level of abstraction, the physical domain description explains how to physically construct that level of abstraction. At high levels, this might consist of an engineering drawing showing how to put together the keypad, tone generator chip, battery, and speaker in the associated housing. At the top chip level, this might consist of a floorplan, and at lower levels, the actual geometry of individual transistors.

The design process can be viewed as making transformations from one domain to another while maintaining the equivalency of the domains. Behavioral descriptions are transformed to structural descriptions, which in turn are transformed to physical descriptions. These transformations can be manual or automatic. In either case, it is normal design practice to verify the transformation of one domain to the other. This ensures that the design intent is carried across the domain boundaries. Hierarchically specifying each domain at successively detailed levels of abstraction allows us to design very large systems.

The reason for strictly describing the domains and levels of abstraction is to define a precise design process in which the final function of the system can be traced all the way back to the initial behavioral description. In an ideal flow, there should be no opportunity to produce an incorrect design. If anomalies arise, the design process is corrected so that those anomalies will not reoccur in the future. A designer should acquire a rigid discipline with respect to the design process, and be aware of each transformation and how and why it is failproof. Normally, these steps are fully automated in a modern design process, but it is important to be aware of the basis for these steps in order to debug them if they go astray.

The Y diagram can be used to illustrate each domain and the transformations between domains at varying levels of design abstraction. As the design process winds its way from the outer to inner rings, it proceeds from higher to lower levels of abstraction and hierarchy.

Most of the remainder of this chapter is a case study in the design of a simple microprocessor to illustrate the various aspects of VLSI design applied to a nontrivial system. We begin by describing the architecture and microarchitecture of the processor. We then consider logic design and discuss hardware description languages. The processor is built with static CMOS circuits, which we examined in Section 1.4; transistor-level design and netlist formats are discussed. We continue exploring the physical design of the processor including floorplanning and area estimation. Design verification is critically important and happens at each level of the hierarchy for each element of the design. Finally, the layout is converted into masks so the chip can be manufactured, packaged, and tested.

1.7 Example: A Simple MIPS Microprocessor

We consider an 8-bit subset of the MIPS microprocessor architecture [Patterson04, Harris07] because it is widely studied and is relatively simple, yet still large enough to illustrate hierarchical design. This section describes the architecture and the multicycle microarchitecture we will be implementing. If you are not familiar with computer architecture, you can regard the MIPS processor as a black box and skip to Section 1.8.

A set of laboratory exercises is available at `www.cmosvlsi.com` in which you can learn VLSI design by building the microprocessor yourself using a free open-source CAD tool called *Electric* or with commercial design tools from Cadence and Synopsys.

1.7.1 MIPS Architecture

The MIPS32 architecture is a simple 32-bit RISC architecture with relatively few idiosyncrasies. Our subset of the architecture uses 32-bit instruction encodings but only eight 8-bit general-purpose registers named `$0`–`$7`. We also use an 8-bit program counter (PC). Register `$0` is hardwired to contain the number 0. The instructions are `ADD`, `SUB`, `AND`, `OR`, `SLT`, `ADDI`, `BEQ`, `J`, `LB`, and `SB`.

The function and encoding of each instruction is given in Table 1.7. Each instruction is encoded using one of three templates: R, I, and J. R-type instructions (*register*-based) are used for arithmetic and specify two source registers and a destination register. I-type instructions are used when a 16-bit constant (also known as an *immediate*) and two registers must be specified. J-type instructions (*jumps*) dedicate most of the instruction word to a 26-bit jump destination. The format of each encoding is defined in Figure 1.49. The six most significant bits of all formats are the operation code (`op`). R-type instructions all share `op` = 000000 and use six more `funct` bits to differentiate the functions.

TABLE 1.7 MIPS instruction set (subset supported)

Instruction	Function		Encoding	op	funct
`add $1, $2, $3`	addition:	`$1 <- $2 + $3`	R	000000	100000
`sub $1, $2, $3`	subtraction:	`$1 <- $2 – $3`	R	000000	100010
`and $1, $2, $3`	bitwise and:	`$1 <- $2 and $3`	R	000000	100100
`or $1, $2, $3`	bitwise or:	`$1 <- $2 or $3`	R	000000	100101
`slt $1, $2, $3`	set less than:	`$1 <- 1 if $2 < $3` `$1 <- 0 otherwise`	R	000000	101010
`addi $1, $2, imm`	add immediate:	`$1 <- $2 + imm`	I	001000	n/a
`beq $1, $2, imm`	branch if equal:	`PC <- PC + imm × 4`[a]	I	000100	n/a
`j destination`	jump:	`PC <- destination`[a]	J	000010	n/a
`lb $1, imm($2)`	load byte:	`$1 <- mem[$2 + imm]`	I	100000	n/a
`sb $1, imm($2)`	store byte:	`mem[$2 + imm] <- $1`	I	101000	n/a

a. Technically, MIPS addresses specify bytes. Instructions require a 4-byte word and must begin at addresses that are a multiple of four. To most effectively use instruction bits in the full 32-bit MIPS architecture, branch and jump constants are specified in words and must be multiplied by four (shifted left 2 bits) to be converted to byte addresses.

Format	Example	Encoding					
		6	5	5	5	5	6
R	add $rd, $ra, $rb	0	ra	rb	rd	0	funct
		6	5	5	16		
I	beq $ra, $rb, imm	op	ra	rb	imm		
		6	26				
J	j dest	op	dest				

FIGURE 1.49 Instruction encoding formats

We can write programs for the MIPS processor in *assembly language*, where each line of the program contains one instruction such as ADD or BEQ. However, the MIPS hardware ultimately must read the program as a series of 32-bit numbers called *machine language*. An *assembler* automates the tedious process of translating from assembly language to machine language using the encodings defined in Table 1.7 and Figure 1.49. Writing nontrivial programs in assembly language is also tedious, so programmers usually work in a *high-level language* such as C or Java. A *compiler* translates a program from high-level language *source code* into the appropriate machine language *object code*.

Example 1.4

Figure 1.50 shows a simple C program that computes the nth Fibonacci number f_n defined recursively for $n > 0$ as $f_n = f_{n-1} + f_{n-2}$, $f_{-1} = -1$, $f_0 = 1$. Translate the program into MIPS assembly language and machine language.

SOLUTION: Figure 1.51 gives a commented assembly language program. Figure 1.52 translates the assembly language to machine language.

```
int fib(void)
{
   int n = 8;                  /* compute nth Fibonacci number */
   int f1 = 1, f2 = -1;        /* last two Fibonacci numbers */

   while (n != 0) {            /* count down to n = 0 */
     f1 = f1 + f2;
     f2 = f1 - f2;
     n = n - 1;
   }
   return f1;
```

FIGURE 1.50 C Code for Fibonacci program

1.7.2 Multicycle MIPS Microarchitecture

We will implement the multicycle MIPS microarchitecture given in Chapter 5 of [Patterson04] and Chapter 7 of [Harris07] modified to process 8-bit data. The microarchitecture is illustrated in Figure 1.53. Light lines indicate individual signals while heavy

```
# fib.asm
# Register usage: $3: n $4: f1 $5: f2
# return value written to address 255
fib:  addi $3, $0, 8      # initialize n=8
      addi $4, $0, 1      # initialize f1 = 1
      addi $5, $0, -1     # initialize f2 = -1
loop: beq $3, $0, end     # Done with loop if n = 0
      add $4, $4, $5      # f1 = f1 + f2
      sub $5, $4, $5      # f2 = f1 - f2
      addi $3, $3, -1     # n = n - 1
      j loop              # repeat until done
end:  sb $4, 255($0)      # store result in address 255
```

FIGURE 1.51 Assembly language code for Fibonacci program

Instruction	Binary Encoding	Hexadecimal Encoding
addi $3, $0, 8	001000 00000 00011 0000000000001000	20030008
addi $4, $0, 1	001000 00000 00100 0000000000000001	20040001
addi $5, $0, -1	001000 00000 00101 1111111111111111	2005ffff
beq $3, $0, end	000100 00011 00000 0000000000000100	10600004
add $4, $4, $5	000000 00100 00101 00100 00000 100000	00852020
sub $5, $4, $5	000000 00100 00101 00101 00000 100010	00852822
addi $3, $3, -1	001000 00011 00011 1111111111111111	2063ffff
j loop	000010 00000000000000000000000011	08000003
sb $4, 255($0)	101000 00000 00100 0000000011111111	a00400ff

FIGURE 1.52 Machine language code for Fibonacci program

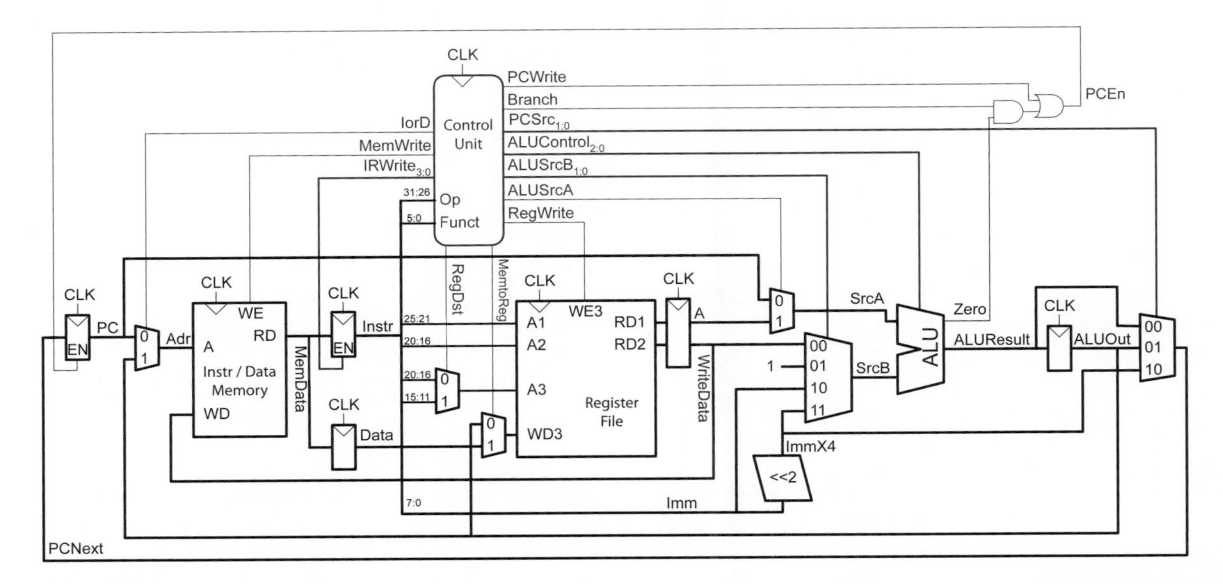

FIGURE 1.53 Multicycle MIPS microarchitecture. Adapted from [Patterson04] and [Harris07] with permission from Elsevier.

lines indicate busses. The control logic and signals are highlighted in blue while the datapath is shown in black. Control signals generally drive multiplexer select signals and register enables to tell the datapath how to execute an instruction.

Instruction execution generally flows from left to right. The program counter (PC) specifies the address of the instruction. The instruction is loaded 1 byte at a time over four cycles from an off-chip memory into the 32-bit instruction register (IR). The `Op` field (bits 31:26 of the instruction) is sent to the controller, which sequences the datapath through the correct operations to execute the instruction. For example, in an `ADD` instruction, the two source registers are read from the register file into temporary registers A and B. On the next cycle, the aludec unit commands the Arithmetic/Logic Unit (ALU) to add the inputs. The result is captured in the `ALUOut` register. On the third cycle, the result is written back to the appropriate destination register in the register file.

The controller contains a finite state machine (FSM) that generates multiplexer select signals and register enables to sequence the datapath. A state transition diagram for the FSM is shown in Figure 1.54. As discussed, the first four states fetch the instruction from

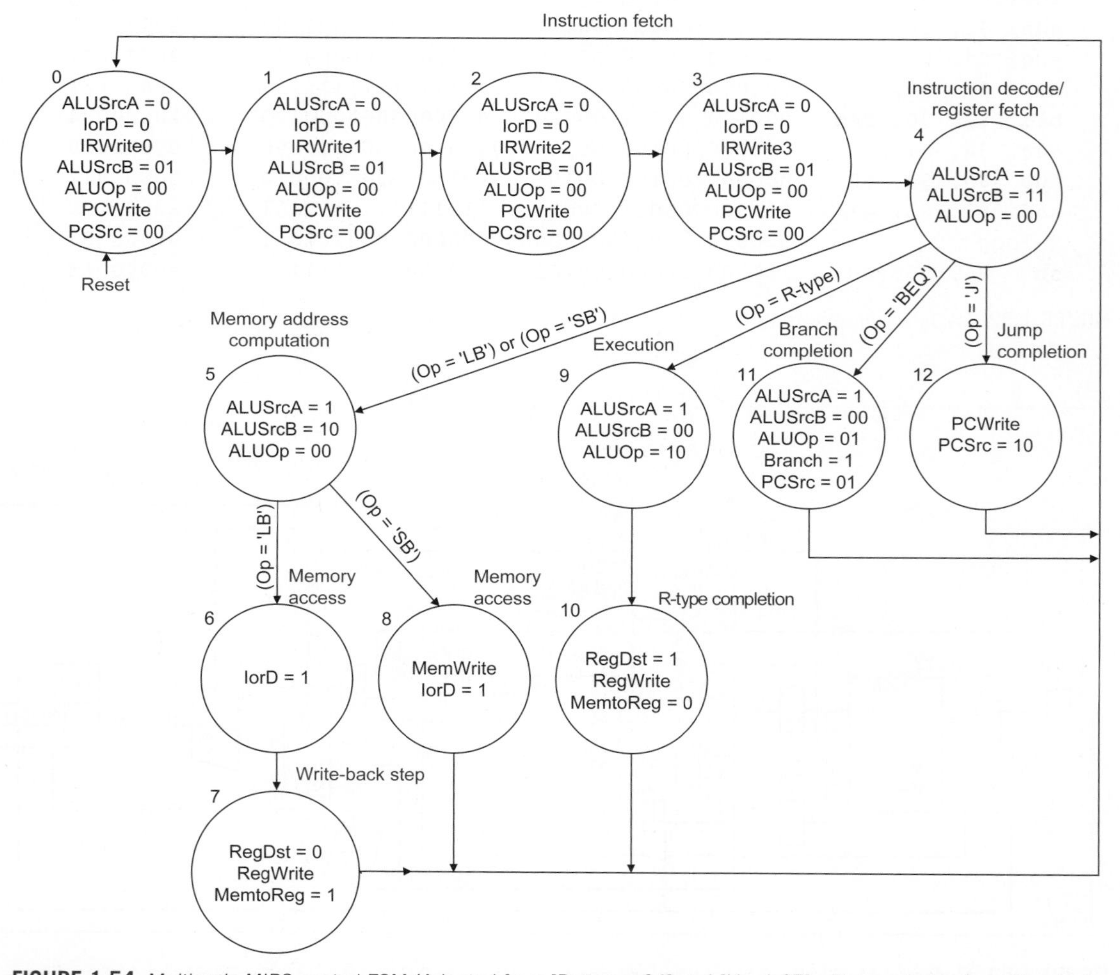

FIGURE 1.54 Multicycle MIPS control FSM (Adapted from [Patterson04] and [Harris07] with permission from Elsevier.)

memory. The FSM then is dispatched based on `Op` to execute the particular instruction. The FSM states for `ADDI` are missing and left as an exercise for the reader.

Observe that the FSM produces a 2-bit `ALUOp` output. The ALU decoder unit in the controller uses combinational logic to compute a 3-bit `ALUControl` signal from the `ALUOp` and `Funct` fields, as specified in Table 1.8. `ALUControl` drives multiplexers in the ALU to select the appropriate computation.

TABLE 1.8 ALUControl determination

ALUOp	Funct	ALUControl	Meaning
00	x	010	ADD
01	x	110	SUB
10	100000	010	ADD
10	100010	110	SUB
10	100100	000	AND
10	100101	001	OR
10	101010	111	SLT
11	x	x	undefined

Example 1.5

Referring to Figures 1.53 and 1.54, explain how the MIPS processor fetches and executes the `SUB` instruction.

SOLUTION: The first step is to fetch the 32-bit instruction. This takes four cycles because the instruction must come over an 8-bit memory interface. On each cycle, we want to fetch a byte from the address in memory specified by the program counter, then increment the program counter by one to point to the next byte.

The fetch is performed by states 0–3 of the FSM in Figure 1.54. Let us start with state 0. The program counter (PC) contains the address of the first byte of the instruction. The controller must select `IorD` = 0 so that the multiplexer sends this address to the memory. `MemRead` must also be asserted so the memory reads the byte onto the `MemData` bus. Finally, `IRWrite0` should be asserted to enable writing `memdata` into the least significant byte of the instruction register (IR).

Meanwhile, we need to increment the program counter. We can do this with the ALU by specifying PC as one input, 1 as the other input, and `ADD` as the operation. To select PC as the first input, `ALUSrcA` = 0. To select 1 as the other input, `ALUSrcB` = 01. To perform an addition, `ALUOp` = 00, according to Table 1.8. To write this result back into the program counter at the end of the cycle, `PCSrc` = 00 and `PCEn` = 1 (done by setting `PCWrite` = 1).

All of these control signals are indicated in state 0 of Figure 1.54. The other register enables are assumed to be 0 if not explicitly asserted and the other multiplexer selects are don't cares. The next three states are identical except that they write bytes 1, 2, and 3 of the IR, respectively.

The next step is to read the source registers, done in state 4. The two source registers are specified in bits 25:21 and 20:16 of the IR. The register file reads these registers and puts the values into the A and B registers. No control signals are necessary for `SUB` (although state 4 performs a branch address computation in case the instruction is `BEQ`).

The next step is to perform the subtraction. Based on the `Op` field (IR bits 31:26), the FSM jumps to state 9 because `SUB` is an R-type instruction. The two source registers are selected as input to the ALU by setting `ALUSrcA` = 1 and `ALUSrcB` = 00. Choosing `ALUOp` = 10 directs the ALU Control decoder to select the `ALUControl` signal as 110, subtraction. Other R-type instructions are executed identically except that the decoder receives a different `Funct` code (IR bits 5:0) and thus generates a different `ALUControl` signal. The result is placed in the ALUOut register.

Finally, the result must be written back to the register file in state 10. The data comes from the ALUOut register so `MemtoReg` = 0. The destination register is specified in bits 15:11 of the instruction so `RegDst` = 1. `RegWrite` must be asserted to perform the write. Then, the control FSM returns to state 0 to fetch the next instruction.

1.8 Logic Design

We begin the logic design by defining the top-level chip interface and block diagram. We then hierarchically decompose the units until we reach leaf cells. We specify the logic with a Hardware Description Language (HDL), which provides a higher level of abstraction than schematics or layout. This code is often called the Register Transfer Level (RTL) description.

1.8.1 Top-Level Interfaces

The top-level inputs and outputs are listed in Table 1.9. This example uses a two-phase clocking system to avoid hold-time problems. `Reset` initializes the PC to 0 and the control FSM to the start state.

TABLE 1.9 Top-level inputs and outputs

Inputs	Outputs
`ph1`	`MemWrite`
`ph2`	`Adr[7:0]`
`reset`	`WriteData[7:0]`
`MemData[7:0]`	

The remainder of the signals are used for an 8-bit memory interface (assuming the memory is located off chip). The processor sends an 8-bit address `Adr` and optionally asserts `MemWrite`. On a read cycle, the memory returns a value on the `MemData` lines while on a write cycle, the memory accepts input from `WriteData`. In many systems, `MemData` and `WriteData` can be combined onto a single bidirectional bus, but for this example we preserve the interface of Figure 1.53. Figure 1.55 shows a simple computer system built from the MIPS processor, external memory, reset switch, and clock generator.

1.8.2 Block Diagrams

The chip is partitioned into two top-level units: the controller and datapath, as shown in the block diagram in Figure 1.56. The controller comprises the control FSM, the ALU decoder, and the two gates used to compute `PCEn`. The ALU decoder consists of combina-

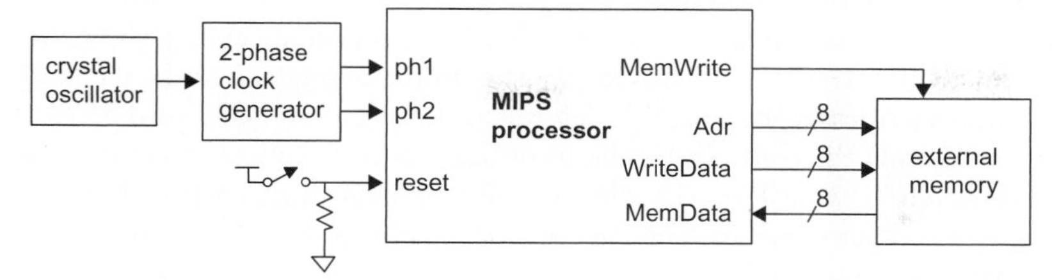

FIGURE 1.55 MIPS computer system

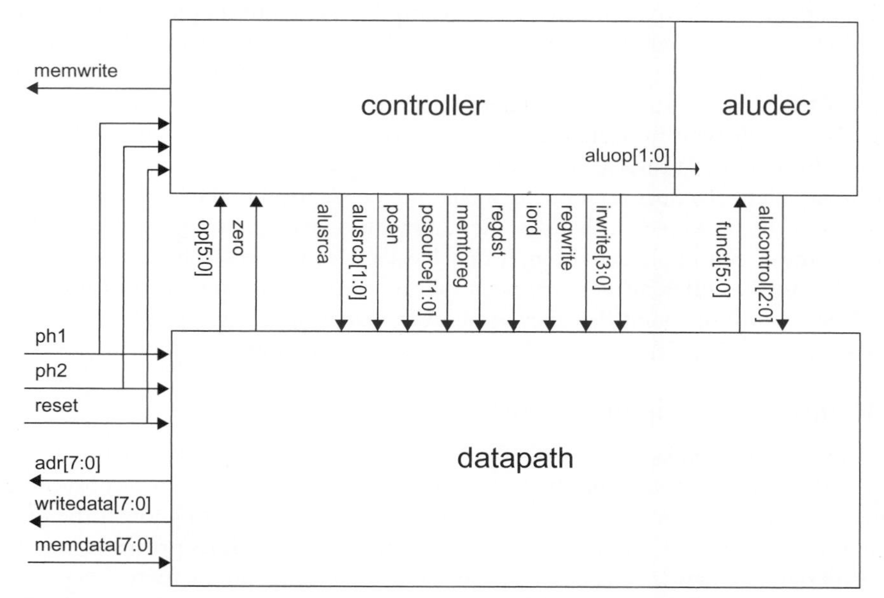

FIGURE 1.56 Top-level MIPS block diagram

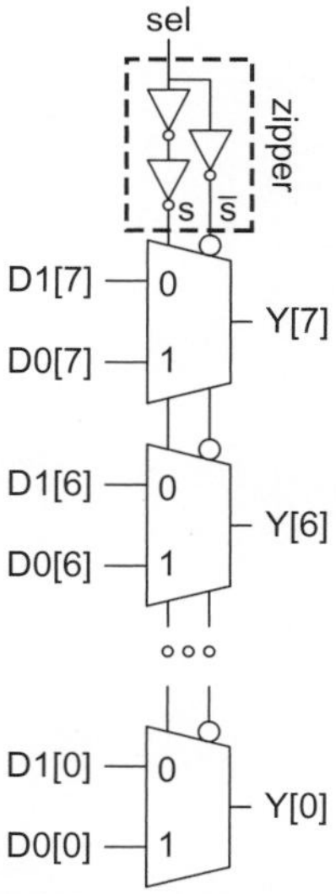

tional logic to determine `ALUControl`. The 8-bit datapath contains the remainder of the chip. It can be viewed as a collection of wordslices or bitslices. A *wordslice* is a column containing an 8-bit flip-flop, adder, multiplexer, or other element. For example, Figure 1.57 shows a wordslice for an 8-bit 2:1 multiplexer. It contains eight individual 2:1 multiplexers, along with a *zipper* containing a buffer and inverter to drive the true and complementary select signals to all eight multiplexers.[6] Factoring these drivers out into the zipper saves space as compared to putting inverters in each multiplexer. Alternatively, the datapath can be viewed as eight rows of *bitslices*. Each bitslice has one bit of each component, along with the horizontal wires connecting the bits together.

The chip partitioning is influenced by the intended physical design. The datapath contains most of the transistors and is very regular in structure. We can achieve high density with moderate design effort by handcrafting each wordslice or bitslice and tiling the

FIGURE 1.57 8-bit 2:1 multiplexer wordslice

[6]In this example, the zipper is shown at the top of the wordslice. In wider datapaths, the zipper is sometimes placed in the middle of the wordslice so that it drives shorter wires. The name comes from the way the layout resembles a plaid sweatshirt with a zipper down the middle.

circuits together. Building datapaths using wordslices is usually easier because certain structures, such as the zero detection circuit in the ALU, are not identical in each bitslice. However, thinking about bitslices is a valuable way to plan the wiring across the datapath.

The controller has much less structure. It is tedious to translate an FSM into gates by hand, and in a new design, the controller is the most likely portion to have bugs and last-minute changes. Therefore, we will specify the controller more abstractly with a hardware description language and automatically generate it using synthesis and place & route tools or a programmable logic array (PLA).

1.8.3 Hierarchy

The best way to design complex systems is to decompose them into simpler pieces. Figure 1.58 shows part of the design hierarchy for the MIPS processor. The controller contains the controller_pla and aludec, which in turn is built from a library of standard cells such as NANDs, NORs, and inverters. The datapath is composed of 8-bit wordslices, each of which also is typically built from standard cells such as adders, register file bits, multiplexers, and flip-flops. Some of these cells are reused in multiple places.

The design hierarchy does not necessarily have to be identical in the logic, circuit, and physical designs. For example, in the logic view, a memory may be best treated as a black box, while in the circuit implementation, it may have a decoder, cell array, column multiplexers, and so forth. Different hierarchies complicate verification, however, because they must be *flattened* until the point that they agree. As a matter of practice, it is best to make logic, circuit, and physical design hierarchies agree as far as possible.

1.8.4 Hardware Description Languages

Designers need rapid feedback on whether a logic design is reasonable. Translating block diagrams and FSM state transition diagrams into circuit schematics is time-consuming and prone to error; before going through this entire process it is wise to know if the top-level design has major bugs that will require complete redesign. HDLs provide a way to specify the design at a higher level of abstraction to raise designer productivity. They were originally intended for documentation and simulation, but are now used to synthesize gates directly from the HDL.

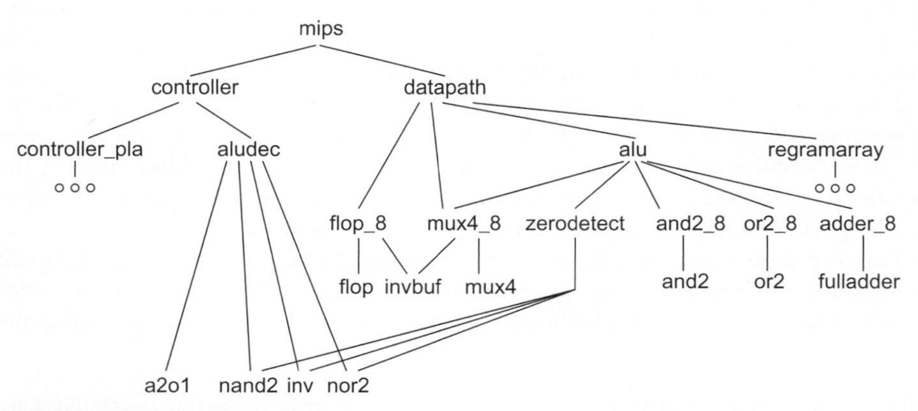

FIGURE 1.58 MIPS design hierarchy

The two most popular HDLs are *Verilog* and *VHDL*. Verilog was developed by Advanced Integrated Design Systems (later renamed Gateway Design Automation) in 1984 and became a *de facto* industry open standard by 1991. In 2005, the SystemVerilog extensions were standardized, and some of these features are used in this book. VHDL, which stands for VHSIC Hardware Description Language, where VHSIC in turn was a Department of Defense project on Very High Speed Integrated Circuits, was developed by committee under government sponsorship. As one might expect from their pedigrees, Verilog is less verbose and closer in syntax to C, while VHDL supports some abstractions useful for large team projects. Many Silicon Valley companies use Verilog while defense and telecommunications companies often use VHDL. Neither language offers a decisive advantage over the other so the industry is saddled with supporting both. Examples in this book are given in Verilog for the sake of brevity.

When coding in an HDL, it is important to remember that you are specifying hardware that operates in parallel rather than software that executes in sequence. There are two general coding styles. *Structural* HDL specifies how a cell is composed of other cells or primitive gates and transistors. *Behavioral* HDL specifies what a cell does.

A *logic simulator* simulates HDL code; it can report whether results match expectations, and can display waveforms to help debug discrepancies. A *logic synthesis* tool is similar to a compiler for hardware: it maps HDL code onto a *library* of gates called *standard cells* to minimize area while meeting some timing constraints. Only a subset of HDL constructs are synthesizable. For example, file I/O commands used in testbenches are obviously not synthesizable. Logic synthesis generally produces circuits that are neither as dense nor as fast as those handcrafted by a skilled designer. Nevertheless, integrated circuit processes are now so advanced that synthesized circuits are good enough for the great majority of application-specific integrated circuits (ASICs) built today. Layout may be automatically generated using place & route tools.

In Verilog, each cell is called a *module*. The inputs and outputs are declared much as in a C program and bit widths are given for busses. Internal signals must also be declared in a way analogous to local variables. The processor is described hierarchically using structural Verilog at the upper levels and behavioral Verilog for the leaf cells. For example, the controller module shows how a finite state machine is specified in behavioral Verilog and the aludec module shows how complex combinational logic is specified. The datapath is specified structurally in terms of wordslices, which are in turn described behaviorally.

For the sake of illustration, the 8-bit adder wordslice could be described structurally as a ripple carry adder composed of eight cascaded full adders. The full adder could be expressed structurally as a sum and a carry subcircuit. In turn, the sum and carry subcircuits could be expressed behaviorally. The full adder block is shown in Figure 1.59 while the carry subcircuit is explored further in Section 1.9.

```
module adder(input  logic [7:0] a, b,
             input  logic       c,
             output logic [7:0] s,
             output logic       cout);

  wire [6:0] carry;
```

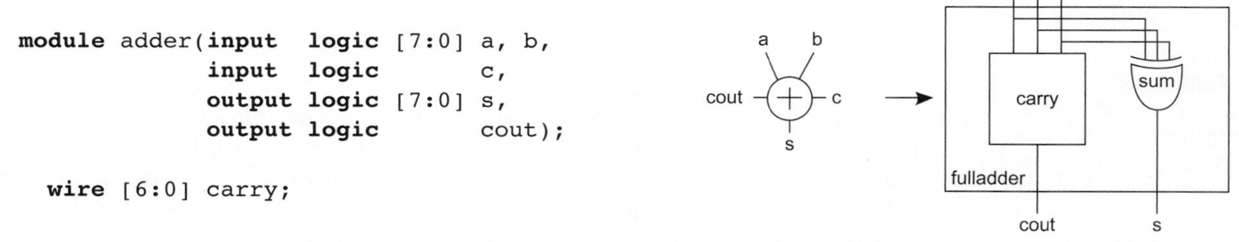

FIGURE 1.59 Full adder

```
    fulladder fa0(a[0], b[0], c,        s[0], carry[0]);
    fulladder fa1(a[1], b[1], carry[0], s[1], carry[1]);
    fulladder fa2(a[2], b[2], carry[1], s[2], carry[2]);
    ...
    fulladder fa7(a[7], b[7], carry[6], s[7], cout);
  endmodule

  module fulladder(input  logic a, b, c,
                   output logic s, cout);

    sum s1(a, b, c, s);
    carry c1(a, b, c, cout);
  endmodule

  module carry(input  logic a, b, c,
               output logic cout);

    assign cout = (a&b) | (a&c) | (b&c);
  endmodule
```

1.9 Circuit Design

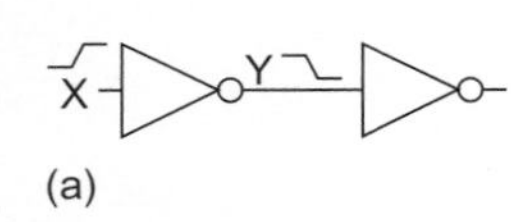

FIGURE 1.60 Circuit delay and power: (a) inverter pair, (b) transistor-level model showing capacitance and current during switching, (c) static leakage current during quiescent operation

Circuit design is concerned with arranging transistors to perform a particular logic function. Given a circuit design, we can estimate the delay and power. The circuit can be represented as a schematic, or in textual form as a netlist. Common transistor level netlist formats include Verilog and SPICE. Verilog netlists are used for functional verification, while SPICE netlists have more detail necessary for delay and power simulations.

Because a transistor gate is a good insulator, it can be modeled as a capacitor, C. When the transistor is ON, some current I flows between source and drain. Both the current and capacitance are proportional to the transistor width.

The delay of a logic gate is determined by the current that it can deliver and the capacitance that it is driving, as shown in Figure 1.60 for one inverter driving another inverter. The capacitance is charged or discharged according to the constitutive equation

$$I = C\frac{dV}{dt}$$

If an average current I is applied, the time t to switch between 0 and V_{DD} is

$$t = \frac{C}{I}V_{DD}$$

Hence, the delay increases with the load capacitance and decreases with the drive current. To make these calculations, we will have to delve below the switch-level model of a transistor. Chapter 2 develops more detailed models of transistors accounting for the current and capacitance. One of the goals of circuit design is to choose transistor widths to meet delay requirements. Methods for doing so are discussed in Chapter 3.

Energy is required to charge and discharge the load capacitance. This is called dynamic power because it is consumed when the circuit is actively switching. The dynamic power consumed when a capacitor is charged and discharged at a frequency f is

$$P_{\text{dynamic}} = CV_{DD}^2 f$$

Even when the gate is not switching, it draws some static power. Because an OFF transistor is leaky, a small amount of current I_{static} flows between power and ground, resulting in a static power dissipation of

$$P_{\text{static}} = I_{\text{static}} V_{DD}$$

Chapter 4 examines power in more detail.

A particular logic function can be implemented in many ways. Should the function be built with ANDs, ORs, NANDs, or NORs? What should be the fan-in and fan-out of each gate? How wide should the transistors be on each gate? Each of these choices influences the capacitance and current and hence the speed and power of the circuit, as well as the area and cost.

As mentioned earlier, in many design methodologies, logic synthesis tools automatically make these choices, searching through the standard cells for the best implementation. For many applications, synthesis is good enough. When a system has critical requirements of high speed or low power or will be manufactured in large enough volume to justify the extra engineering, custom circuit design becomes important for critical portions of the chip.

Circuit designers often draw schematics at the transistor and/or gate level. For example, Figure 1.61 shows two alternative circuit designs for the carry circuit in a full adder. The gate-level design in Figure 1.61(a) requires 26 transistors and four stages of gate delays (recall that ANDs and ORs are built from NANDs and NORs followed by inverters). The transistor-level design in Figure 1.61(b) requires only 12 transistors and two stages of gate delays, illustrating the benefits of optimizing circuit designs to take advantage of CMOS technology.

These schematics are then *netlisted* for simulation and verification. One common netlist format is structural Verilog HDL. The gate-level design can be netlisted as follows:

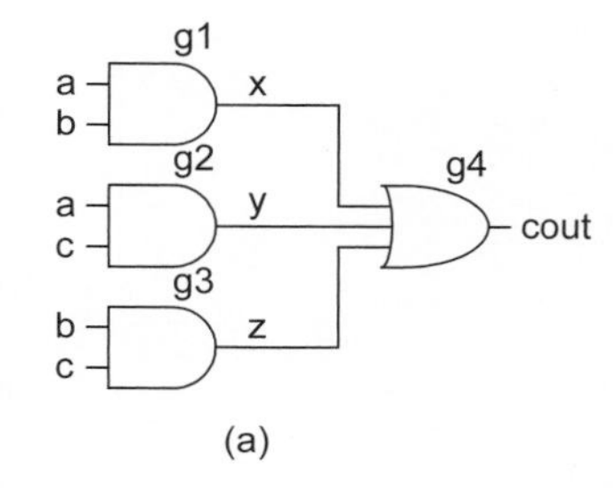

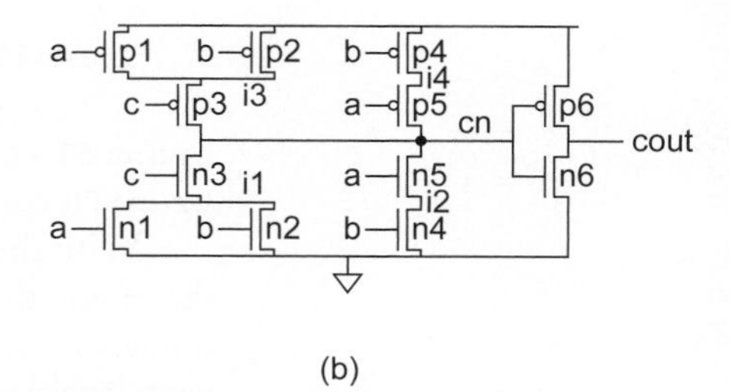

FIGURE 1.61 Carry subcircuit

```
module carry(input  logic a, b, c,
             output logic cout);

  logic x, y, z;

  and g1(x, a, b);
  and g2(y, a, c);
  and g3(z, b, c);
  or  g4(cout, x, y, z);
endmodule
```

This is a technology-independent structural description, because generic gates have been used and the actual gate implementations have not been specified. The transistor-level netlist follows:

```
module carry(input  logic a, b, c,
             output tri   cout);

  tri     i1, i2, i3, i4, cn;
  supply0 gnd;
  supply1 vdd;

  tranif1 n1(i1, gnd, a);
  tranif1 n2(i1, gnd, b);
  tranif1 n3(cn, i1, c);
  tranif1 n4(i2, gnd, b);
  tranif1 n5(cn, i2, a);
  tranif0 p1(i3, vdd, a);
  tranif0 p2(i3, vdd, b);
  tranif0 p3(cn, i3, c);
  tranif0 p4(i4, vdd, b);
  tranif0 p5(cn, i4, a);
  tranif1 n6(cout, gnd, cn);
  tranif0 p6(cout, vdd, cn);
endmodule
```

Transistors are expressed as

```
Transistor-type name(drain, source, gate);
```

`tranif1` corresponds to nMOS transistors that turn ON when the gate is 1 while `tranif0` corresponds to pMOS transistors that turn ON when the gate is 0.

With the description generated so far, we still do not have the information required to determine the speed or power consumption of the gate. We need to specify the size of the transistors and the stray capacitance. Because Verilog was designed as a switch-level and gate-level language, it is poorly suited to structural descriptions at this level of detail. Hence, we turn to another common structural language used by the circuit simulator SPICE. The specification of the transistor-level carry subcircuit at the circuit level might be represented as follows:

```
.SUBCKT CARRY A B C COUT VDD GND
MN1 I1 A GND GND NMOS W=2U L=0.6U AD=1.8P AS=3P
MN2 I1 B GND GND NMOS W=2U L=0.6U AD=1.8P AS=3P
MN3 CN C I1 GND NMOS W=2U L=0.6U AD=3P AS=3P
MN4 I2 B GND GND NMOS W=2U L=0.6U AD=0.9P AS=3P
MN5 CN A I2 GND NMOS W=2U L=0.6U AD=3P AS=0.9P
MP1 I3 A VDD VDD PMOS W=4U L=0.6U AD=3.6P AS=6P
MP2 I3 B VDD VDD PMOS W=4U L=0.6U AD=3.6P AS=6P
MP3 CN C I3 VDD PMOS W=4U L=0.6U AD=6P AS=6P
```

```
MP4 I4 B VDD VDD PMOS W=4U L=0.6U AD=1.8P AS=6P
MP5 CN A I4 VDD PMOS W=4U L=0.6U AD=6P AS=1.8P
MN6 COUT CN GND GND NMOS W=4U L=0.6U AD=6P AS=6P
MP6 COUT CN VDD VDD PMOS W=8U L=0.6U AD=12P AS=12P
CI1 I1 GND 6FF
CI3 I3 GND 9FF
CA A GND 12FF
CB B GND 12FF
CC C GND 6FF
CCN CN GND 12FF
CCOUT COUT GND 6FF
.ENDS
```

Transistors are specified by lines beginning with an M as follows:

```
Mname  drain   gate   source   body   type   W=width  L=length
       AD=drain area   AS=source area
```

Although MOS switches have been masquerading as three terminal devices (gate, source, and drain) until this point, they are in fact four terminal devices with the substrate or well forming the *body* terminal. The body connection was not listed in Verilog but is required for SPICE. The type specifies whether the transistor is a p-device or n-device. The width, length, and area parameters specify physical dimensions of the actual transistors. Units include U (micro, 10^{-6}), P (pico, 10^{-12}), and F (femto, 10^{-15}). Capacitors are specified by lines beginning with C as follows:

```
Cname   node1   node2   value
```

In this description, the MOS model in SPICE calculates the parasitic capacitances inherent in the MOS transistor using the device dimensions specified. The extra capacitance statements in the above description designate additional routing capacitance not inherent to the device structure. This depends on the physical design of the gate. Long wires also contribute resistance, which increases delay. At the circuit level of structural specification, all connections are given that are necessary to fully characterize the carry gate in terms of speed, power, and connectivity. Chapter 7 describes SPICE models in more detail.

1.10 Physical Design

1.10.1 Floorplanning

Physical design begins with a floorplan. The floorplan estimates the area of major units in the chip and defines their relative placements. The floorplan is essential to determine whether a proposed design will fit in the chip area budgeted and to estimate wiring lengths and wiring congestion. An initial floorplan should be prepared as soon as the logic is loosely defined. As usual, this process involves feedback. The floorplan will often suggest changes to the logic (and microarchitecture), which in turn changes the floorplan. For example, suppose microarchitects assume that a cache requires a 2-cycle access latency. If the floorplan shows that the data cache can be placed adjacent to the execution units in the datapath, the cache access time might reduce to a single cycle. This could allow the microarchitects to reduce the cache capacity while providing the same performance. Once the cache shrinks, the floorplan must be reconsidered to take advantage of the newly available space near the datapath. As a complex design begins to stabilize, the floorplan is often hierarchically subdivided to describe the functional blocks within the units.

The challenge of floorplanning is estimating the size of each unit without proceeding through a detailed design of the chip (which would depend on the floorplan and wire lengths). This section assumes that good estimates have been made and describes what a floorplan looks like. The next sections describe each of the types of components that might be in a floorplan and suggests ways to estimate the component sizes.

Figure 1.62 shows the chip floorplan for the MIPS processor including the pad frame. The top-level blocks are the controller and datapath. A wiring channel is located between the two blocks to provide room to route control signals to the datapath. The datapath is further partitioned into wordslices. The *pad frame* includes 40 I/O pads, which are wired to the pins on the chip package. There are 29 pads used for signals; the remainder are V_{DD} and GND.

The floorplan is drawn to scale and annotated with dimensions. The chip is designed in a 0.6 μm process on a 1.5 × 1.5 mm die so the die is 5000 λ on a side. Each pad is 750 λ ×

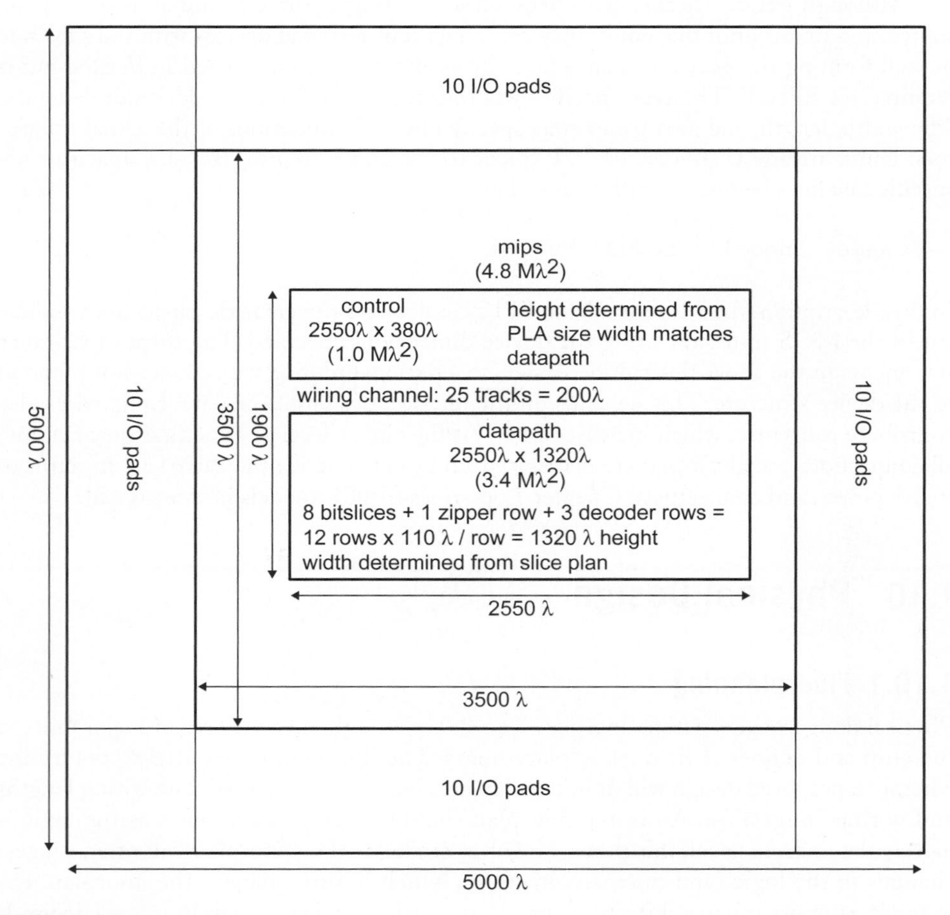

FIGURE 1.62 MIPS floorplan

350 λ, so the maximum possible core area inside the pad frame is 3500 $\lambda \times$ 3500 λ = 12.25 Mλ^2. Due to the wiring channel, the actual core area of 4.8 Mλ^2 is larger than the sum of the block areas. This design is said to be *pad-limited* because the I/O pads set the chip area. Most commercial chips are *core-limited* because the chip area is set by the logic excluding the pads. In general, blocks in a floorplan should be rectangular because it is difficult for a designer to stuff logic into an odd-shaped region (although some CAD tools do so just fine).

Figure 1.63 shows the actual chip layout. Notice the 40 I/O pads around the periphery. Just inside the pad frame are metal2 V_{DD} and GND rings, marked with + and –.

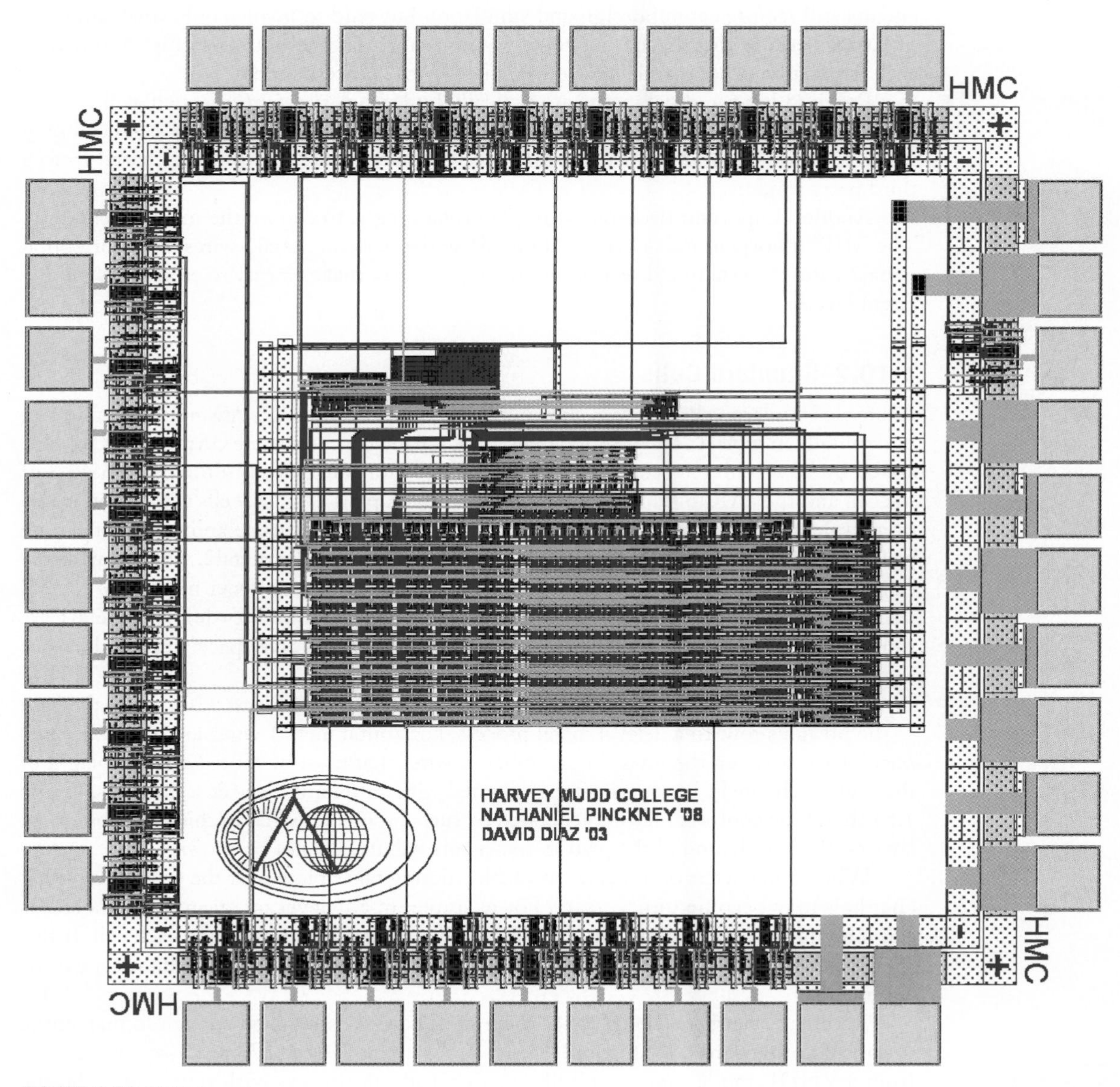

FIGURE 1.63 MIPS layout

On-chip structures can be categorized as *random logic*, *datapaths*, *arrays*, *analog*, and *input/output* (I/O). Random logic, like the aludecoder, has little structure. Datapaths operate on multi-bit data words and perform roughly the same function on each bit so they consist of multiple N-bit wordslices. Arrays, like RAMs, ROMs, and PLAs, consist of identical cells repeated in two dimensions. Productivity is highest if layout can be reused or automatically generated. Datapaths and arrays are good VLSI building blocks because a single carefully crafted cell is reused in one or two dimensions. Automatic layout generators exist for memory arrays and random logic but are not as mature for datapaths. Therefore, many design methodologies ignore the potential structure of datapaths and instead lay them out with random logic tools except when performance or area are vital. Analog circuits still require careful design and simulation but tend to involve only small amounts of layout because they have relatively few transistors. I/O cells are also highly tuned to each fabrication process and are often supplied by the process vendor.

Random logic and datapaths are typically built from *standard cells* such as inverters, NAND gates, and flip-flops. Standard cells increase productivity because each cell only needs to be drawn and verified once. Often, a standard cell library is purchased from a third party vendor.

Another important decision during floorplanning is to choose the metal orientation. The MIPS floorplan uses horizontal metal1 wires, vertical metal2 wires, and horizontal metal3 wires. Alternating directions between each layer makes it easy to cross wires on different layers.

1.10.2 Standard Cells

A simple standard cell library is shown on the inside front cover. Power and ground run horizontally in metal1. These supply rails are 8 λ wide (to carry more current) and are separated by 90 λ center-to-center. The nMOS transistors are placed in the bottom 40 λ of the cell and the pMOS transistors are placed in the top 50 λ. Thus, cells can be connected by abutment with the supply rails and n-well matching up. Substrate and well contacts are placed under the supply rails. Inputs and outputs are provided in metal2, which runs vertically. Each cell is a multiple of 8 λ in width so that it offers an integer number of metal2 tracks. Within the cell, poly is run vertically to form gates and diffusion and metal1 are run horizontally, though metal1 can also be run vertically to save space when it does not interfere with other connections.

Cells are tiled in rows. Each row is separated vertically by at least 110 λ from the base of the previous row. In a 2-level metal process, horizontal metal1 wires are placed in *routing channels* between the rows. The number of wires that must be routed sets the height of the routing channels. Layout is often generated with automatic place & route tools. Figure 1.64 shows the controller layout generated by such a tool. Note that in this and subsequent layouts, the n-well around the pMOS transistors will usually not be shown.

When more layers of metal are available, routing takes place over the cells and routing channels may become unnecessary. For example, in a 3-level metal process, metal3 is run horizontally on a 10 λ pitch. Thus, 11 horizontal tracks can run over each cell. If this is sufficient to accommodate all of the horizontal wires, the routing channels can be eliminated.

Automatic synthesis and place & route tools have become good enough to map entire designs onto standard cells. Figure 1.65 shows the entire 8-bit MIPS processor synthesized from a VHDL model onto a cell library in a 130 nm process with seven metal layers.

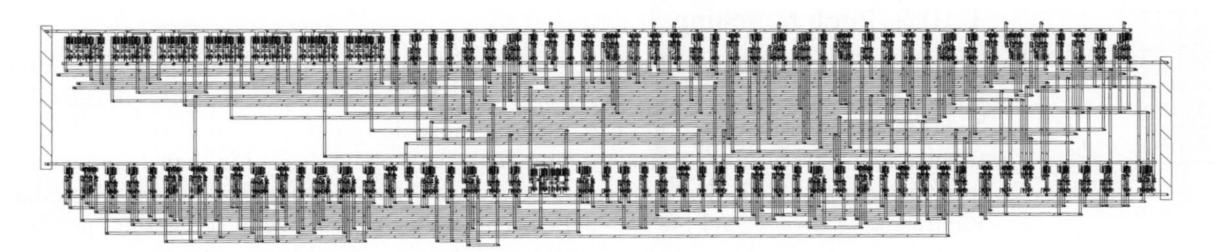

FIGURE 1.64 MIPS controller layout (synthesized)

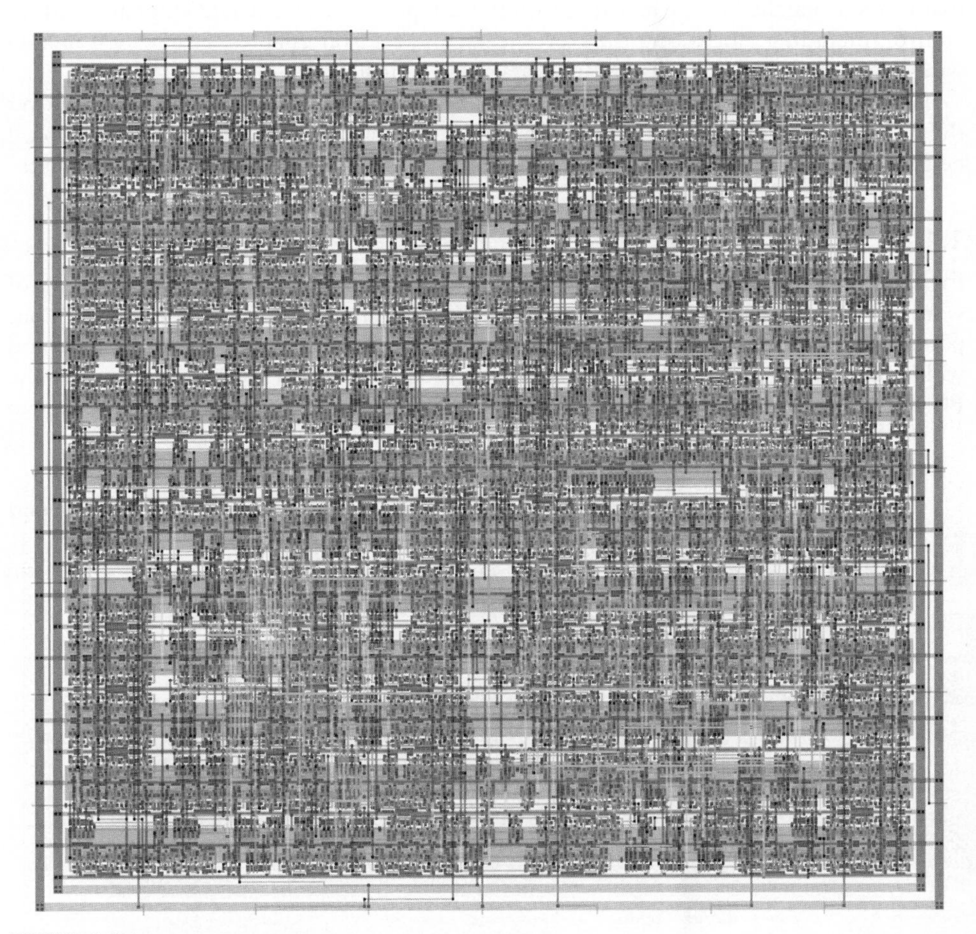

FIGURE 1.65 Synthesized MIPS processor

Compared to Figure 1.63, the synthesized design shows little discernible structure except that 26 rows of standard cells can be identified beneath the wires. The area is approximately 4 $M\lambda^2$. Synthesized designs tend to be somewhat slower than a good custom design, but they also take an order of magnitude less design effort.

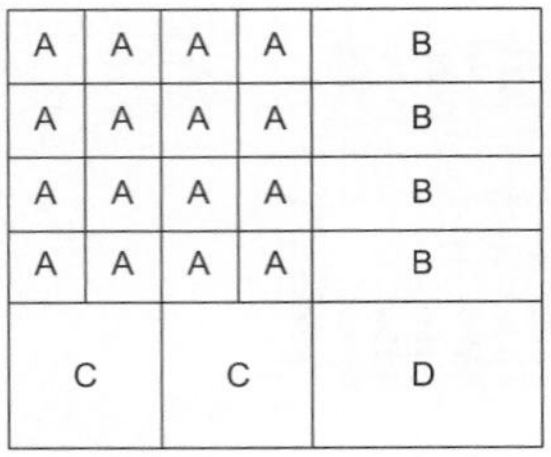

A	A	A	A	B
A	A	A	A	B
A	A	A	A	B
A	A	A	A	B
C		C		D

FIGURE 1.66 Pitch-matching of snap-together cells

1.10.3 Pitch Matching

The area of the controller in Figure 1.64 is dominated by the routing channels. When the logic is more regular, layout density can be improved by including the wires in cells that "snap together." Snap-together cells require more design and layout effort but lead to smaller area and shorter (i.e., faster) wires. The key issue in designing snap-together cells is *pitch-matching*. Cells that connect must have the same size along the connecting edge. Figure 1.66 shows several pitch-matched cells. Reducing the size of cell D does not help the layout area. On the other hand, increasing the size of cell D also affects the area of B and/or C.

Figure 1.67 shows the MIPS datapath in more detail. The eight horizontal bitslices are clearly visible. The zipper at the top of the layout includes three rows for the decoder that is pitch-matched to the register file in the datapath. Vertical metal2 wires are used for control, including clocks, multiplexer selects, and register enables. Horizontal metal3 wires run over the tops of cells to carry data along a bitslice.

The width of the transistors in the cells and the number of wires that must run over the datapath determines the minimum height of the datapath cells. 60–100 λ are typical heights for relatively simple datapaths. The width of the cell depends on the cell contents.

1.10.4 Slice Plans

Figure 1.68 shows a *slice plan* of the datapath. The diagram illustrates the ordering of wordslices and the allocation of wiring tracks within each bitslice. Dots indicate that a bus passes over a cell and is also used in that cell. Each cell is annotated with its type and width (in number of tracks). For example, the program counter (`pc`) is an output of the PC flop and is also used as an input to the srcA and address multiplexers. The slice plan

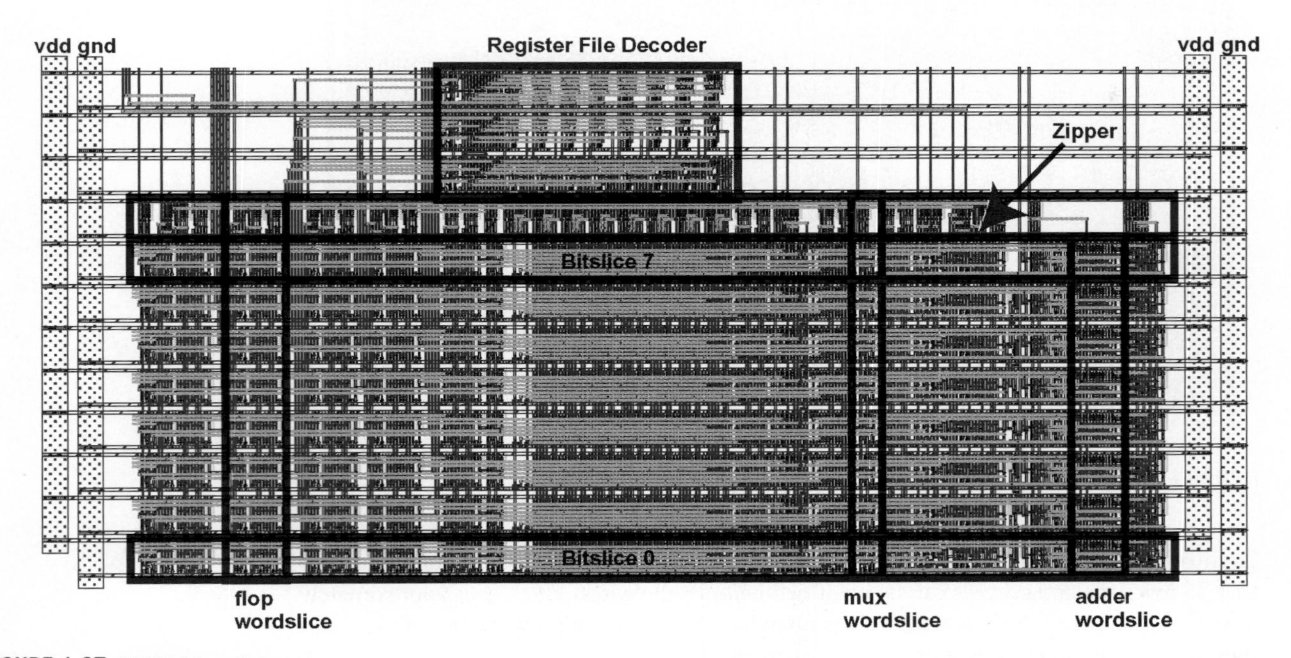

FIGURE 1.67 MIPS datapath layout

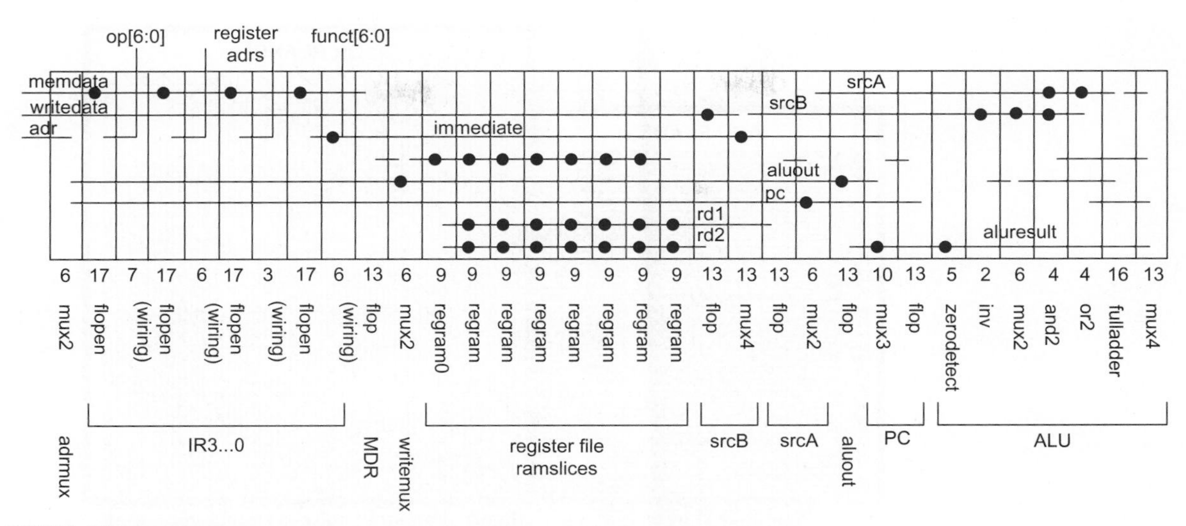

FIGURE 1.68 Datapath slice plan

makes it easy to calculate wire lengths and evaluate wiring congestion before laying out the datapath. In this case, it is evident that the greatest congestion takes place over the register file, where seven wiring tracks are required.

The slice plan is also critical for estimating area of datapaths. Each wordslice is annotated with its width, measured in tracks. This information can be obtained by looking at the cell library layouts. By adding up the widths of each element in the slice plan, we see that the datapath is 319 tracks wide, or 2552 λ wide. There are eight bitslices in the 8-bit datapath. In addition, there is one more row for the zipper and three more for the three register file address decoders, giving a total of 12 rows. At a pitch of 110 λ / row, the datapath is 1320 λ tall. The address decoders only occupy a small fraction of their rows, leaving wasted empty space. In a denser design, the controller could share some of the unused area.

1.10.5 Arrays

Figure 1.69 shows a programmable logic array (PLA) used for the control FSM next state and output logic. A PLA can compute any function expressed in sum of products form. The structure on the left is called the AND plane and the structure on the right is the OR plane. PLAs are discussed further in Section 11.7.

This PLA layout uses 2 vertical tracks for each input and 3 for each output plus about 6 for overhead. It uses 1.5 horizontal tracks for each product or minterm, plus about 14 for overhead. Hence, the size of a PLA is easy to calculate. The total PLA area is 500 $\lambda \times$ 350 λ, plus another 336 $\lambda \times$ 220 λ for the four external flip-flops needed in the control FSM. The height of the controller is dictated by the height of the PLA plus a few wiring tracks to route inputs and outputs. In comparison, the synthesized controller from Figure 1.64 has a size of 1500 $\lambda \times$ 400 λ because the wiring tracks waste so much space.

1.10.6 Area Estimation

A good floorplan depends on reasonable area estimates, which may be difficult to make before logic is finalized. An experienced designer may be able to estimate block area by

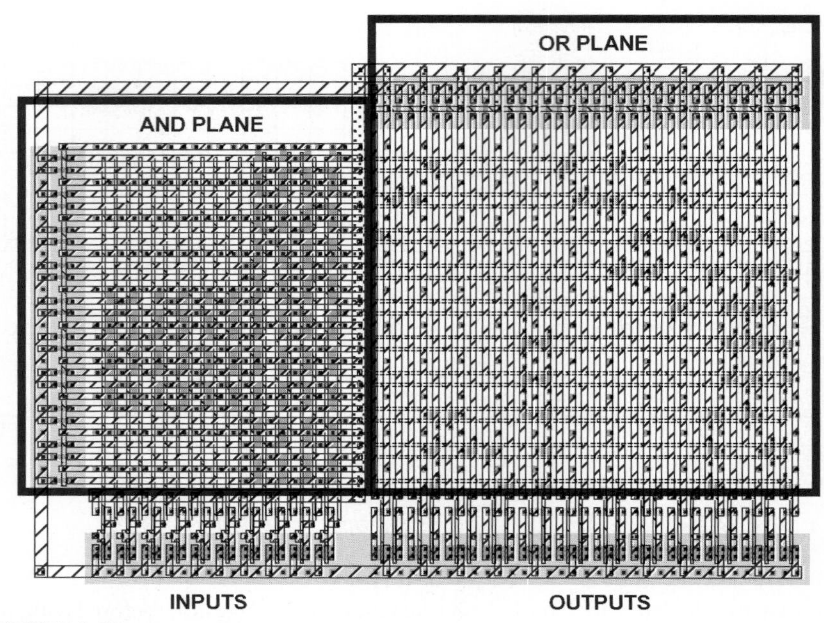

FIGURE 1.69 PLA for control FSM

comparison to the area of a comparable block drawn in the past. In the absence of data for such comparison, Table 1.10 lists some typical numbers. Be certain to account for large wiring channels at a pitch of 8 λ / track. Larger transistors clearly occupy a greater area, so this may be factored into the area estimates as a function of W and L (width and length). For memories, don't forget about the decoders and other periphery circuits, which often take as much area as the memory bits themselves. Your mileage may vary, but datapaths and arrays typically achieve higher densities than standard cells.

TABLE 1.10 Typical layout densities

Element	Area
random logic (2-level metal process)	1000 – 1500 λ^2 / transistor
datapath	250 – 750 λ^2 / transistor or 6 WL + 360 λ^2 / transistor
SRAM	1000 λ^2 / bit
DRAM (in a DRAM process)	100 λ^2 / bit
ROM	100 λ^2 / bit

Given enough time, it is nearly always possible to shave a few lambda here or there from a design. However, such efforts are seldom a good investment unless an element is repeated so often that it accounts for a major fraction of the chip area or if floorplan errors have led to too little space for a block and the block must be shrunk before the chip can be completed. It is wise to make conservative area estimates in floorplans, especially if there is risk that more functionality may be added to a block.

Some cell library vendors specify typical routed standard cell layout densities in kgates / mm^2.[7] Commonly, a gate is defined as a 3-input static CMOS NAND or NOR with six transistors. A 65 nm process ($\lambda \approx 0.03$ μm) with eight metal layers may achieve a density of 160–500 kgates / mm^2 for random logic. This corresponds to about 370–1160 λ^2 / transistor. Processes with many metal layers obtain high density because routing channels are not needed.

1.11 Design Verification

Integrated circuits are complicated enough that if anything can go wrong, it probably will. Design verification is essential to catching the errors before manufacturing and commonly accounts for half or more of the effort devoted to a chip.

As design representations become more detailed, verification time increases. It is not practical to simulate an entire chip in a circuit-level simulator such as SPICE for a large number of cycles to prove that the layout is correct. Instead, the design is usually tested for functionality at the architectural level with a model in a language such as C and at the logic level by simulating the HDL description. Then, the circuits are checked to ensure that they are a faithful representation of the logic and the layout is checked to ensure it is a faithful representation of the circuits, as shown in Figure 1.70. Circuits and layout must meet timing and power specifications as well.

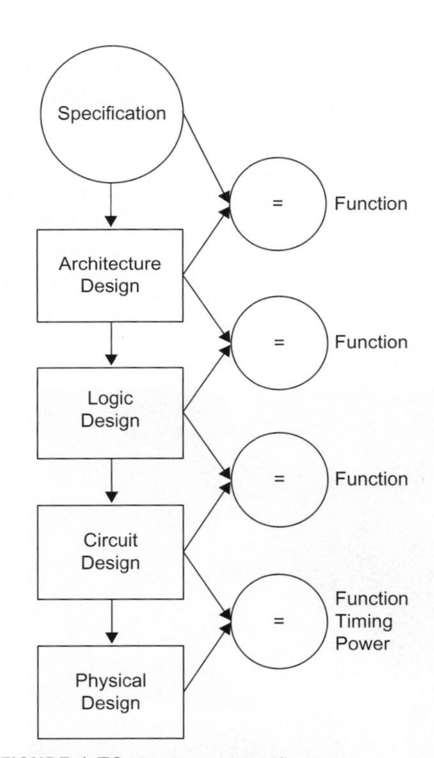

FIGURE 1.70 Design and verification sequence

A *testbench* is used to verify that the logic is correct. The testbench instantiates the logic under test. It reads a file of inputs and expected outputs called *test vectors*, applies them to the module under test, and logs mismatches.

A number of techniques are available for circuit verification. If the logic is synthesized onto a cell library, the postsynthesis gate-level netlist can be expressed in an HDL again and simulated using the same test vectors. Alternatively, a transistor-level netlist can be simulated against the test vector, although this can result in tricky race conditions for sequential circuits. Powerful *formal verification* tools are also available to check that a circuit performs the same Boolean function as the associated logic. Exotic circuits should be simulated thoroughly to ensure that they perform the intended logic function and have adequate noise margins; circuit pitfalls are discussed throughout this book.

Layout vs. Schematic tools (LVS) check that transistors in a layout are connected in the same way as in the circuit schematic. *Design rule checkers* (DRC) verify that the layout satisfies design rules. *Electrical rule checkers* (ERC) scan for other potential problems such as noise or premature wearout; such problems will also be discussed later in the book.

[7] 1 kgate = 1000 gates.

1.12 Fabrication, Packaging, and Testing

Once a chip design is complete, it is taped out for manufacturing. *Tapeout* gets its name from the old practice of writing a specification of masks to magnetic tape; today, the mask descriptions are usually sent to the manufacturer electronically. Two common formats for mask descriptions are the Caltech Interchange Format (CIF) [Mead80] (mainly used in academia) and the Calma GDS II Stream Format (GDS) [Calma84] (used in industry).

Masks are made by etching a pattern of chrome on glass with an electron beam. A set of masks for a nanometer process can be very expensive. For example, masks for a large chip in a 180 nm process may cost on the order of a quarter of a million dollars. In a 65 nm process, the mask set costs about $3 million. The MOSIS service in the United States and its EUROPRACTICE and VDEC counterparts in Europe and Japan make a single set of masks covering multiple small designs from academia and industry to amortize the cost across many customers. With a university discount, the cost for a run of 40 small chips on a multi-project wafer can run about $10,000 in a 130 nm process down to $2000 in a 0.6 μm process. MOSIS offers certain grants to cover fabrication of class project chips.

Integrated circuit fabrication plants (fabs) now cost billions of dollars and become obsolete in a few years. Some large companies still own their own fabs, but an increasing number of fabless semiconductor companies contract out manufacturing to foundries such as TSMC, UMC, and IBM.

Multiple chips are manufactured simultaneously on a single silicon wafer, typically 150–300 mm (6″–12″) in diameter. Fabrication requires many deposition, masking, etching, and implant steps. Most fabrication plants are optimized for wafer throughput rather than latency, leading to turnaround times of up to 10 weeks. Figure 1.71 shows an engineer in a *clean room* holding a completed 300 mm wafer. Clean rooms are filtered to eliminate most dust and other particles that could damage a partially processed wafer. The engineer is wearing a "bunny suit" to avoid contaminating the clean room. Figure 1.72 is a

FIGURE 1.71 Engineer holding processed 12-inch wafer (Photograph courtesy of the Intel Corporation.)

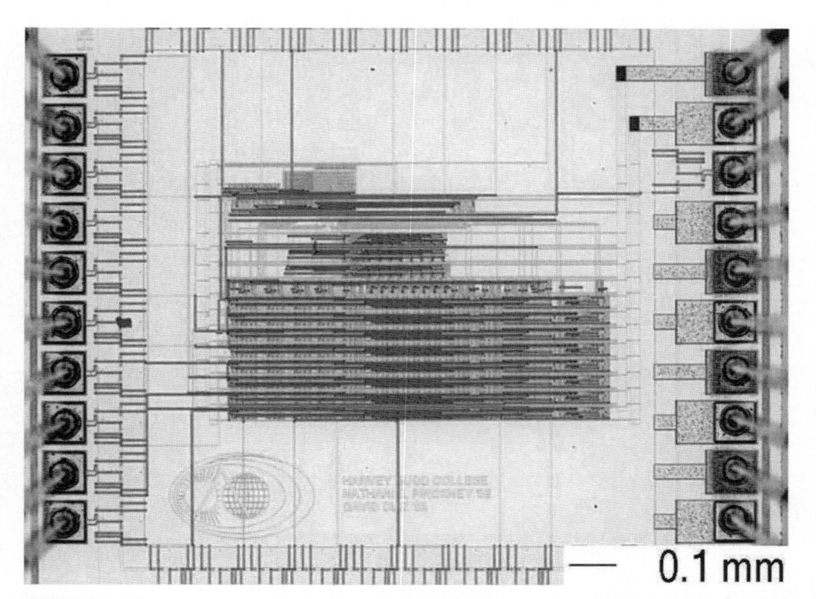

FIGURE 1.72 MIPS processor photomicrograph (only part of pad frame shown)

photomicrograph (a photograph taken under a microscope) of the 8-bit MIPS processor.

Processed wafers are sliced into dice (chips) and packaged. Figure 1.73 shows the 1.5 × 1.5 mm chip in a 40-pin *dual-inline package* (DIP). This *wire-bonded* package uses thin gold wires to connect the pads on the die to the lead frame in the center cavity of the package. These wires are visible on the pads in Figure 1.72. More advanced packages offer different trade-offs between cost, pin count, pin bandwidth, power handling, and reliability, as will be discussed in Section 12.2. Flip-chip technology places small solder balls directly onto the die, eliminating the bond wire inductance and allowing contacts over the entire chip area rather than just at the periphery.

FIGURE 1.73 Chip in a 40-pin dual-inline package

Even tiny defects in a wafer or dust particles can cause a chip to fail. Chips are tested before being sold. Testers capable of handling high-speed chips cost millions of dollars, so many chips use built-in self-test features to reduce the tester time required. Chapter 14 is devoted to design verification and testing.

Summary and a Look Ahead

"If the automobile had followed the same development cycle as the computer, a Rolls-Royce would today cost $100, get one million miles to the gallon, and explode once a year . . ."

—Robert X. Cringely

CMOS technology, driven by Moore's Law, has come to dominate the semiconductor industry. This chapter examined the principles of designing a simple CMOS integrated circuit. MOS transistors can be viewed as electrically controlled switches. Static CMOS gates are built from pull-down networks of nMOS transistors and pull-up networks of pMOS transistors. Transistors and wires are fabricated on silicon wafers using a series of deposition, lithography, and etch steps. These steps are defined by a set of masks drawn as a chip layout. Design rules specify minimum width and spacing between elements in the layout. The chip design process can be divided into architecture, logic, circuit, and physical design. The performance, area, and power of the chip are influenced by interrelated decisions made at each level. Design verification plays an important role in constructing such complex systems; the reliability requirements for hardware are much greater than those typically imposed on software.

Primary design objectives include reliability, performance, power, and cost. Any chip should, with high probability, operate reliably for its intended lifetime. For example, the chip must be designed so that it does not overheat or break down from excessive voltage. Performance is influenced by many factors including clock speed and parallelism. CMOS transistors dissipate power every time they switch, so the dynamic power consumption is related to the number and size of transistors and the rate at which they switch. At feature sizes below 180 nm, transistors also leak a significant amount of current even when they should be OFF. Thus, chips now draw static power even when they are idle. One of the central challenges of VLSI design is making good trade-offs between performance and power for a particular application. The cost of a chip includes nonrecurring engineering (NRE) expenses for the design and masks, along with per-chip manufacturing costs related to the size of the chip. In processes with smaller feature sizes, the per-unit cost goes down because more transistors can be packed into a given area, but the NRE increases. The latest manufacturing processes are only cost-effective for chips that will sell in huge volumes. Nevertheless, plenty of interesting markets exist for chips in mature, inexpensive manufacturing processes.

To quantify how a chip meets these objectives, we must develop and analyze more complete models. The remainder of this book will expand on the material introduced in this chapter. Of course, transistors are not simply switches. Chapter 2 examines the current and capacitance of transistors, which are essential for estimating delay and power. The next several chapters address the fundamental concerns of circuit designers. The models from Chapter 2 are too detailed to apply by hand to large systems, yet not detailed enough to fully capture the complexity of modern transistors. Chapter 3 develops simplified models to estimate the delay of circuits. If modern chips were designed to squeeze out the ultimate possible performance without regard to power, they would burn up. Thus, it is essential to estimate and trade off the power consumption against performance. Moreover, low power consumption is crucial to mobile battery-operated systems. Power is considered in Chapter 4. Wires are as important as transistors in their contribution to overall performance and power, and are discussed in Chapter 5. Chapter 6 addresses design of robust circuits with a high yield and low failure rate.

Simulation is discussed in Chapter 8 and is used to obtain more accurate performance and power predictions as well as to verify the correctness of circuits and logic. Chapter 8 considers combinational circuit design. A whole kit of circuit families are available with different trade-offs in speed, power, complexity, and robustness. Chapter 9 continues with sequential circuit design, including clocking and latching techniques.

The next three chapters delve into CMOS subsystems. Chapter 10 catalogs designs for a host of datapath subsystems including adders, shifters, multipliers, and counters. Chapter 11 similarly describes memory subsystems including SRAMs, DRAMs, CAMs, ROMs, and PLAs. Chapter 12 addresses special-purpose subsystems including power distribution, clocking, and I/O.

The final chapters address practicalities of CMOS system design. Chapter 13 focuses on a range of current design methods, identifying the issues peculiar to CMOS. Testing, design-for-test, and debugging techniques are discussed in Chapter 14. Hardware description languages (HDLs) are used in the design of nearly all digital integrated circuits today. A more detailed description of CMOS processing technology and layout rules is presented in Chapter 15.

A number of sections are marked with an "optional" icon. These sections describe particular subjects in greater detail. You may skip over these sections on a first reading and return to them when they are of practical relevance. A companion text, *Digital VLSI Chip Design with Cadence and Synopsys CAD Tools* [Brunvand09], covers practical details of using the leading industrial CAD tools to build chips.

Exercises

1.1 As the cost of a transistor drops from a microbuck ($\$10^{-6}$) toward a nanobuck, what opportunities can you imagine to change the world with integrated circuits?

1.2 A 3-input majority gate returns a true output if at least two of the inputs are true. A minority gate is its complement. Design a 3-input CMOS minority gate using a single stage of logic.

a) sketch a transistor-level schematic

b) sketch a stick diagram

c) estimate the area from the stick diagram

1.3 Extrapolating the data from Figure 1.4, predict the transistor count of a microprocessor in 2016.

1.4 Consider the design of a CMOS compound OR-AND-INVERT (OAI21) gate computing $F = \overline{(A + B) \cdot C}$.

a) sketch a transistor-level schematic

b) sketch a stick diagram

c) estimate the area from the stick diagram

d) layout your gate with a CAD tool using unit-sized transistors

e) compare the layout size to the estimated area

1.5 Use a combination of CMOS gates (represented by their symbols) to generate the following functions from A, B, and C.

a) $Y = A$ (buffer)

b) $Y = A\overline{B} + \overline{A}B$ (XOR)

c) $Y = \overline{A}\overline{B} + AB$ (XNOR)

d) $Y = AB + BC + AC$ (majority)

1.6 Sketch a transistor-level schematic of a CMOS 3-input XOR gate. You may assume you have both true and complementary versions of the inputs available.

1.7 Sketch a transistor-level schematic for a CMOS 4-input NOR gate.

1.8 Sketch a transistor-level schematic for a compound CMOS logic gate for each of the following functions:

a) $Y = \overline{ABC + D}$

b) $Y = \overline{(AB + C) \cdot D}$

c) $Y = \overline{AB + C \cdot (A + B)}$

1.9 Figure 1.74 shows a stick diagram of a 2-input NAND gate. Sketch a side view (cross-section) of the gate from X to X′.

1.10 Figure 1.75 gives a stick diagram for a level-sensitive latch. Estimate the area of the latch.

1.11 Design a 3-input minority gate using CMOS NANDs, NORs, and inverters. How many transistors are required? How does this compare to a design from Exercise 1.2(a)?

1.12 Sketch a stick diagram for a CMOS 4-input NOR gate from Exercise 1.7.

1.13 Estimate the area of your 4-input NOR gate from Exercise 1.12.

1.14 A carry lookahead adder computes $G = G_3 + P_3(G_2 + P_2(G_1 + P_1G_0))$. Consider designing a compound gate to compute $\overline{G}$.

a) sketch a transistor-level schematic

b) sketch a stick diagram

c) estimate the area from the stick diagram

1.15 `www.cmosvlsi.com` has a series of four labs in which you can learn VLSI design by completing the multicycle MIPS processor described in this chapter. The labs use the open-source Electric CAD tool or commercial tools from Cadence and Synopsys. They cover the following:

a) leaf cells: schematic entry, layout, icons, simulation, DRC, ERC, LVS; hierarchical design

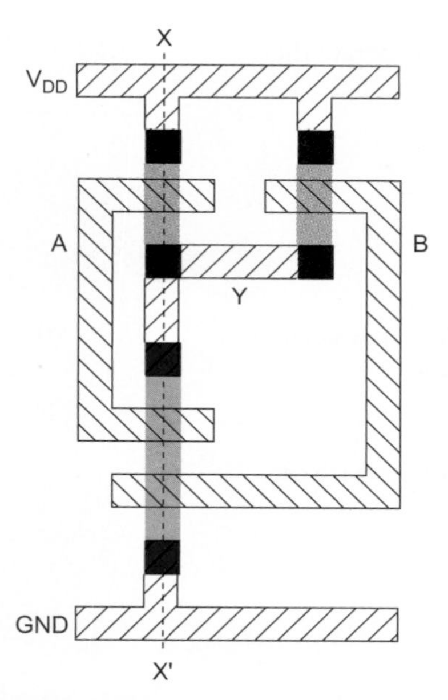

FIGURE 1.74 2-input NAND gate stick diagram

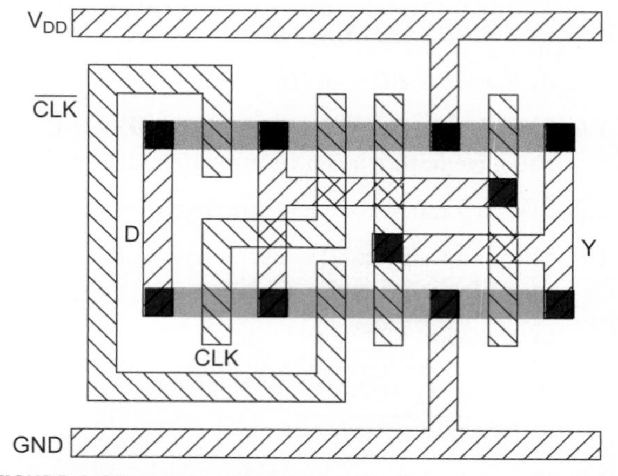

FIGURE 1.75 Level-sensitive latch stick diagram

b) datapath design: wordslices, ALU assembly, datapath routing

c) control design: random logic or PLAs

d) chip assembly, pad frame, global routing, full-chip verification, tapeout

1.16 Using a CAD tool of your choice, layout a 4-input NOR gate. How does its size compare to the prediction from Exercise 1.13?

1.17 Draw a transistor-level schematic for the latch of Figure 1.75. How does the schematic differ from Figure 1.31(b)?

1.18 Read a biography or history about a major event in the development of integrated circuits. For example, see *Crystal Fire* by Lillian Hoddesonor, *Microchip* by Jeffrey Zygmont, or *The Pentium Chronicles* by Robert Colwell. Pick a team or individual that made a major contribution to the field. In your opinion, what were the characteristics that led to success? What traits of the team management would you seek to emulate or avoid in your own professional life?

1.19 Sketch transistor-level schematics for the following logic functions. You may assume you have both true and complementary versions of the inputs available.

a) A 2:4 decoder defined by

$$Y0 = \overline{A0} \cdot \overline{A1}$$
$$Y1 = A0 \cdot A1$$
$$Y2 = \overline{A0} \cdot A1$$
$$Y3 = A0 \cdot A1$$

b) A 3:2 priority encoder defined by

$$Y0 = \overline{A0} \cdot (A1 + \overline{A2})$$
$$Y1 = \overline{A0} \cdot \overline{A1}$$

1.20 Search the Web for transistor counts of Intel's more recent microprocessors. Make a graph of transistor count vs. year of introduction from the Pentium Processor in 1993 to the present on a semilogarithmic scale. How many months pass between doubling of transistor counts?

1.21 Consider the design of a CMOS compound OR-OR-AND-INVERT (OAI22) gate computing $F = \overline{(A + B) \cdot (C + D)}$.

a) sketch a transistor-level schematic

b) sketch a stick diagram

c) estimate the area from the stick diagram

d) layout your gate with a CAD tool using unit-sized transistors

e) compare the layout size to the estimated area

Devices 2

2.1 Introduction

In Chapter 1, the Metal-Oxide-Semiconductor (MOS) transistor was introduced in terms of its operation as an ideal switch. As we saw in Section 1.9, the performance and power of a chip depend on the current and capacitance of the transistors and wires. In this chapter, we will examine the characteristics of MOS transistors in more detail; Chapter 5 addresses wires.

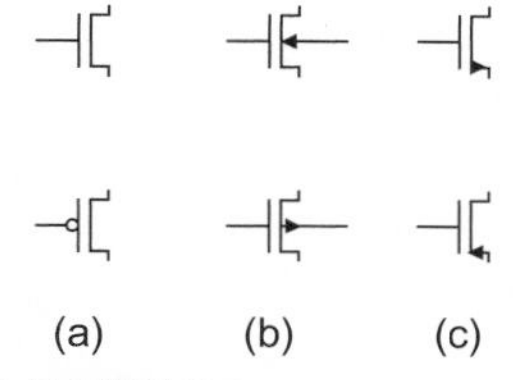

FIGURE 2.1
MOS transistor symbols

Figure 2.1 shows some of the symbols that are commonly used for MOS transistors. The three-terminal symbols in Figure 2.1(a) are used in the great majority of schematics. If the body (substrate or well) connection needs to be shown, the four-terminal symbols in Figure 2.1(b) will be used. Figure 2.1(c) shows an example of other symbols that may be encountered in the literature.

The MOS transistor is a *majority-carrier* device in which the current in a conducting channel between the source and drain is controlled by a voltage applied to the gate. In an nMOS transistor, the majority carriers are electrons; in a pMOS transistor, the majority carriers are holes. The behavior of MOS transistors can be understood by first examining an isolated MOS structure with a gate and body but no source or drain. Figure 2.2 shows a simple MOS structure. The top layer of the structure is a good conductor called the *gate*. Early transistors used metal gates. Transistor gates soon changed to use polysilicon, i.e., silicon formed from many small crystals, although metal gates are making a resurgence at 65 nm and beyond, as will be seen in Section 15.4.1.3. The middle layer is a very thin insulating film of SiO_2 called the *gate oxide*. The bottom layer is the doped silicon body. The figure shows a p-type body in which the carriers are holes. The body is grounded and a voltage is applied to the gate. The gate oxide is a good insulator so almost zero current flows from the gate to the body.[1]

In Figure 2.2(a) , a negative voltage is applied to the gate, so there is negative charge on the gate. The mobile positively charged holes are attracted to the region beneath the gate. This is called the *accumulation* mode. In Figure 2.2(b), a small positive voltage is applied to the gate, resulting in some positive charge on the gate. The holes in the body are repelled from the region directly beneath the gate, resulting in a *depletion* region forming below the gate. In Figure 2.2(c), a higher positive potential exceeding a critical threshold voltage V_t is applied, attracting more positive charge to the gate. The holes are repelled further and some free electrons in the body are attracted to the region beneath the gate. This conductive layer of electrons in the p-type body is called the *inversion* layer. The threshold

[1]Gate oxides are now only a handful of atomic layers thick and carriers sometimes tunnel through the oxide, creating a current through the gate. This effect is explored in Section 2.4.4.2.

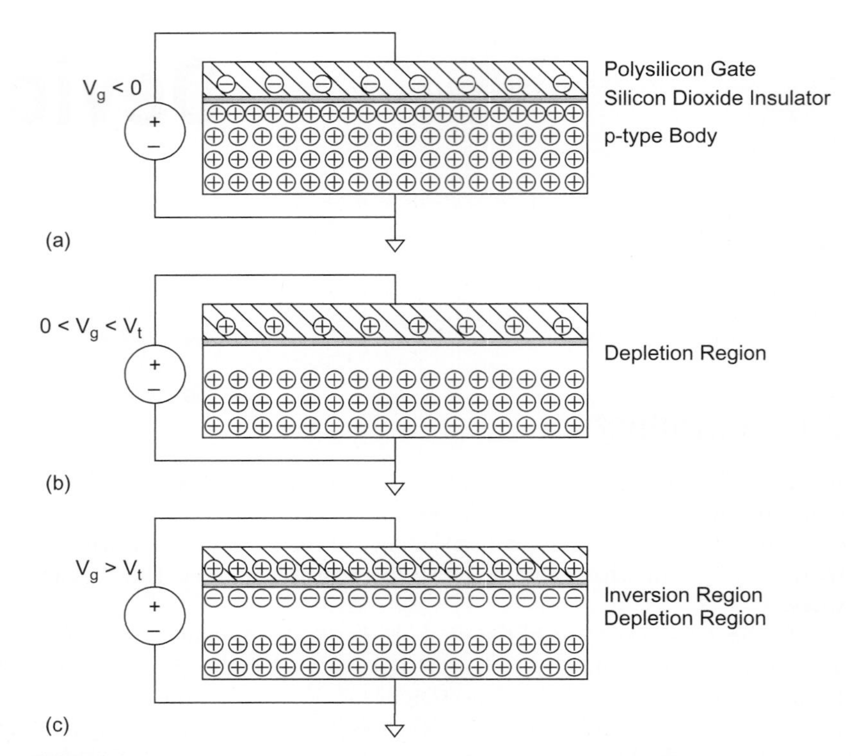

FIGURE 2.2 MOS structure demonstrating (a) accumulation, (b) depletion, and (c) inversion

voltage depends on the number of dopants in the body and the thickness t_{ox} of the oxide. It is usually positive, as shown in this example, but can be engineered to be negative.

Figure 2.3 shows an nMOS transistor. The transistor consists of the MOS stack between two n-type regions called the *source* and *drain*. In Figure 2.3(a), the gate-to-source voltage V_{gs} is less than the threshold voltage. The source and drain have free electrons. The body has free holes but no free electrons. Suppose the source is grounded. The junctions between the body and the source or drain are zero-biased or reverse-biased, so little or no current flows. We say the transistor is OFF, and this mode of operation is called *cutoff*. It is often convenient to approximate the current through an OFF transistor as zero, especially in comparison to the current through an ON transistor. Remember, however, that small amounts of current leaking through OFF transistors can become significant, especially when multiplied by millions or billions of transistors on a chip. In Figure 2.3(b), the gate voltage is greater than the threshold voltage. Now an inversion region of electrons (majority carriers) called the *channel* connects the source and drain, creating a conductive path and turning the transistor ON. The number of carriers and the conductivity increases with the gate voltage. The potential difference between drain and source is $V_{ds} = V_{gs} - V_{gd}$. If $V_{ds} = 0$ (i.e., $V_{gs} = V_{gd}$), there is no electric field tending to push current from drain to source.

When a small positive potential V_{ds} is applied to the drain (Figure 2.3(c)), current I_{ds} flows through the channel from drain to source.[2] This mode of operation is termed *linear*,

[2]The terminology of source and drain might initially seem backward. Recall that the current in an nMOS transistor is carried by moving electrons with a negative charge. Therefore, positive current from drain to source corresponds to electrons flowing from their source to their drain.

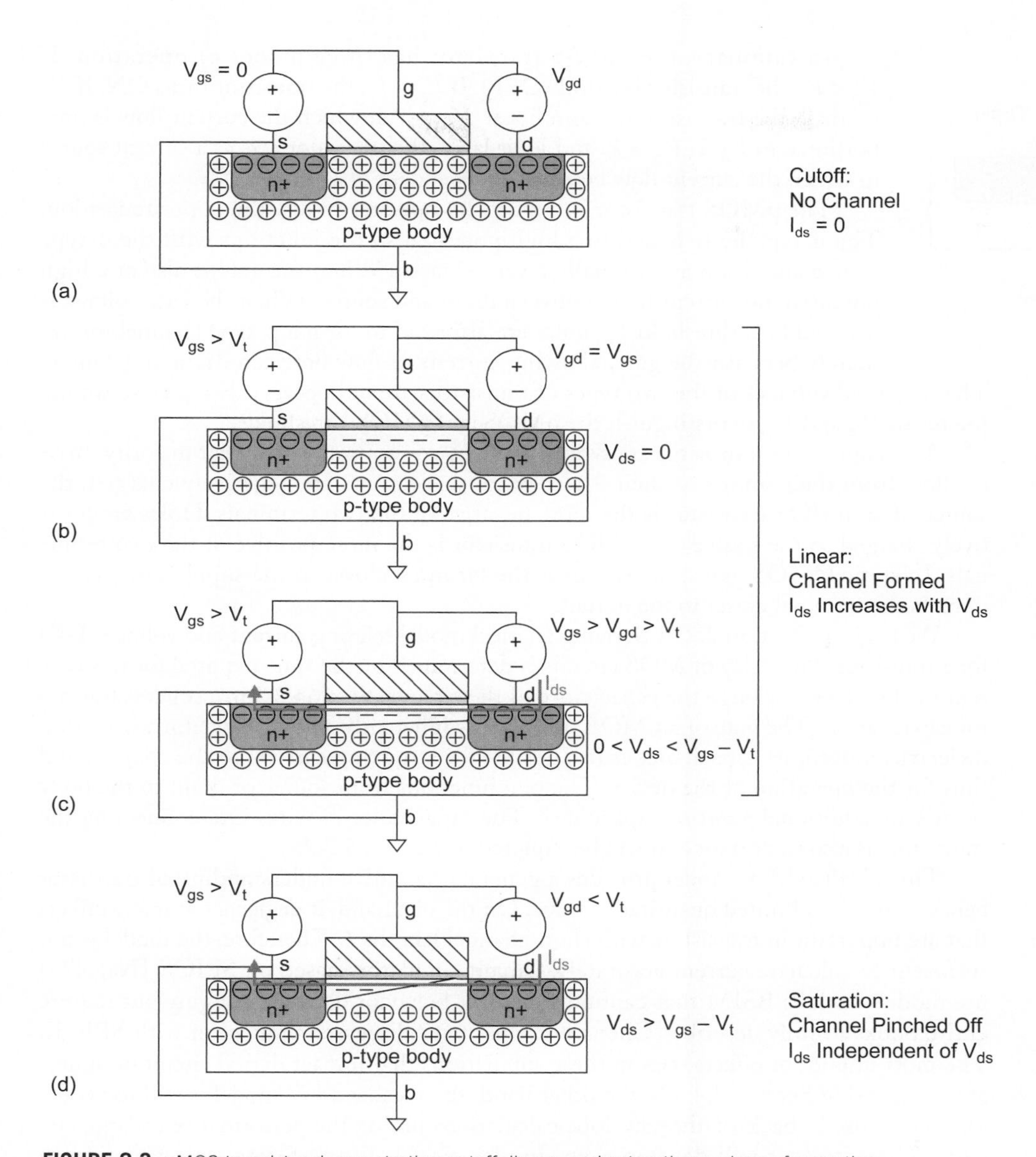

FIGURE 2.3 nMOS transistor demonstrating cutoff, linear, and saturation regions of operation

resistive, *triode*, *nonsaturated*, or *unsaturated*; the current increases with both the drain voltage and gate voltage. If V_{ds} becomes sufficiently large that $V_{gd} < V_t$, the channel is no longer inverted near the drain and becomes *pinched off* (Figure 2.3(d)). However, conduction is still brought about by the drift of electrons under the influence of the positive drain voltage. As electrons reach the end of the channel, they are injected into the depletion region near the drain and accelerated toward the drain. Above this drain voltage the current I_{ds} is controlled only by the gate voltage and ceases to be influenced by the drain. This mode is called *saturation*.

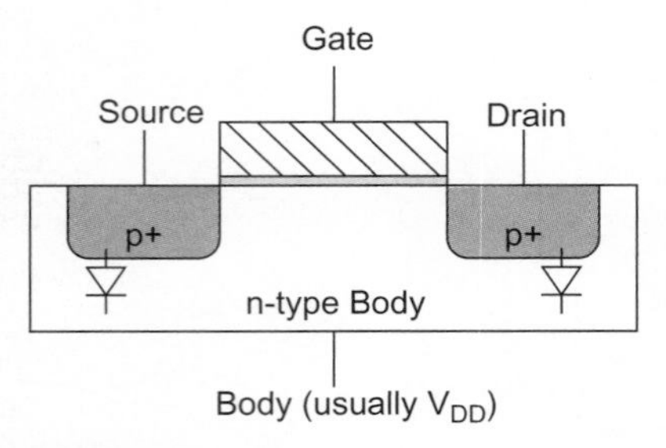

FIGURE 2.4 pMOS transistor

In summary, the nMOS transistor has three modes of operation. If $V_{gs} < V_t$, the transistor is cutoff (OFF). If $V_{gs} > V_t$, the transistor turns ON. If V_{ds} is small, the transistor acts as a linear resistor in which the current flow is proportional to V_{ds}. If $V_{gs} > V_t$ and V_{ds} is large, the transistor acts as a current source in which the current flow becomes independent of V_{ds}.

The pMOS transistor in Figure 2.4 operates in just the opposite fashion. The n-type body is tied to a high potential so the junctions with the p-type source and drain are normally reverse-biased. When the gate is also at a high potential, no current flows between drain and source. When the gate voltage is lowered by a threshold V_t, holes are attracted to form a p-type channel immediately beneath the gate, allowing current to flow between drain and source. The threshold voltages of the two types of transistors are not necessarily equal, so we use the terms V_{tn} and V_{tp} to distinguish the nMOS and pMOS thresholds.

Although MOS transistors are symmetrical, by convention we say that majority carriers flow from their source to their drain. Because electrons are negatively charged, the source of an nMOS transistor is the more negative of the two terminals. Holes are positively charged so the source of a pMOS transistor is the more positive of the two terminals. In static CMOS gates, the source is the terminal closer to the supply rail and the drain is the terminal closer to the output.

We begin in Section 2.2 by deriving an ideal model relating current and voltage (I-V) for a transistor. The delay of MOS circuits is determined by the time required for this current to charge or discharge the capacitance of the circuits. Section 2.3 investigates transistor capacitances. The gate of an MOS transistor is inherently a good capacitor with a thin dielectric; indeed, its capacitance is responsible for attracting carriers to the channel and thus for the operation of the device. The p–n junctions from source or drain to the body contribute additional *parasitic* capacitance. The capacitance of wires interconnecting the transistors is also important and will be explored in Section 5.2.2.

This idealized I-V model provides a general qualitative understanding of transistor behavior but is of limited quantitative value. On the one hand, it neglects too many effects that are important in transistors with short channel lengths L. Therefore, the model is not sufficient to calculate current accurately. Circuit simulators based on SPICE [Nagel75] use models such as BSIM that capture transistor behavior quite thoroughly but require entire books to fully describe [Cheng99]. Chapter 7 discusses simulation with SPICE. The most important effects seen in these simulations that impact digital circuit designers are examined in Section 2.4. On the other hand, the idealized I-V model is still too complicated to use in back-of-the-envelope calculations tuning the performance of large circuits. Therefore, we will develop even simpler models for performance estimation in Chapter 3.

Section 2.5 wraps up this chapter by applying the I-V models to understand the DC transfer characteristics of CMOS gates and pass transistors.

2.2 Long-Channel I-V Characteristics

As stated previously, MOS transistors have three regions of operation:

- Cutoff or subthreshold region
- Linear region
- Saturation region

Let us derive a model [Shockley52, Cobbold70, Sah64] relating the current and voltage (I-V) for an nMOS transistor in each of these regions. The model assumes that the channel length is long enough that the lateral electric field (the field between source and drain) is relatively low, which is no longer the case in nanometer devices. This model is variously known as the *long-channel*, *ideal*, *first-order*, or *Shockley* model. Subsequent sections will refine the model to reflect high fields, leakage, and other nonidealities.

The long-channel model assumes that the current through an OFF transistor is 0. When a transistor turns ON ($V_{gs} > V_t$), the gate attracts carriers (electrons) to form a channel. The electrons drift from source to drain at a rate proportional to the electric field between these regions. Thus, we can compute currents if we know the amount of charge in the channel and the rate at which it moves. We know that the charge on each plate of a capacitor is $Q = CV$. Thus, the charge in the channel Q_{channel} is

$$Q_{\text{channel}} = C_g \left(V_{gc} - V_t\right) \quad \textbf{(2.1)}$$

where C_g is the capacitance of the gate to the channel and $V_{gc} - V_t$ is the amount of voltage attracting charge to the channel beyond the minimum required to invert from p to n. The gate voltage is referenced to the channel, which is not grounded. If the source is at V_s and the drain is at V_d, the average is $V_c = (V_s + V_d)/2 = V_s + V_{ds}/2$. Therefore, the mean difference between the gate and channel potentials V_{gc} is $V_g - V_c = V_{gs} - V_{ds}/2$, as shown in Figure 2.5.

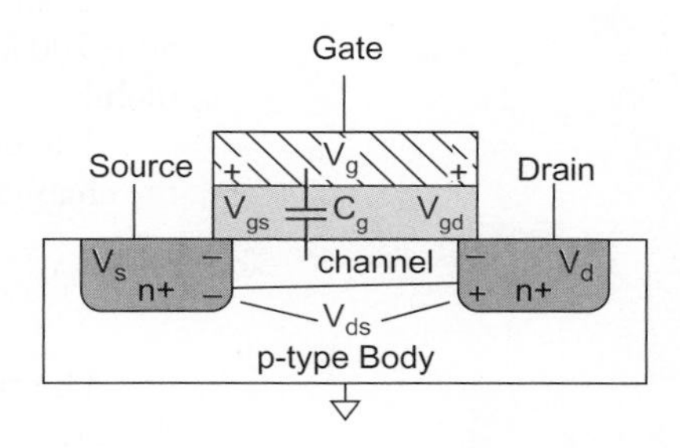

FIGURE 2.5 Average gate to channel voltage

We can model the gate as a parallel plate capacitor with capacitance proportional to area over thickness. If the gate has length L and width W and the oxide thickness is t_{ox}, as shown in Figure 2.6, the capacitance is

$$C_g = k_{\text{ox}}\varepsilon_0 \frac{WL}{t_{\text{ox}}} = \varepsilon_{\text{ox}} \frac{WL}{t_{\text{ox}}} = C_{\text{ox}} WL \quad \textbf{(2.2)}$$

where ε_0 is the permittivity of free space, 8.85×10^{-14} F/cm, and the permittivity of SiO_2 is $k_{\text{ox}} = 3.9$ times as great. Often, the $\varepsilon_{\text{ox}}/t_{\text{ox}}$ term is called C_{ox}, the capacitance per unit area of the gate oxide.

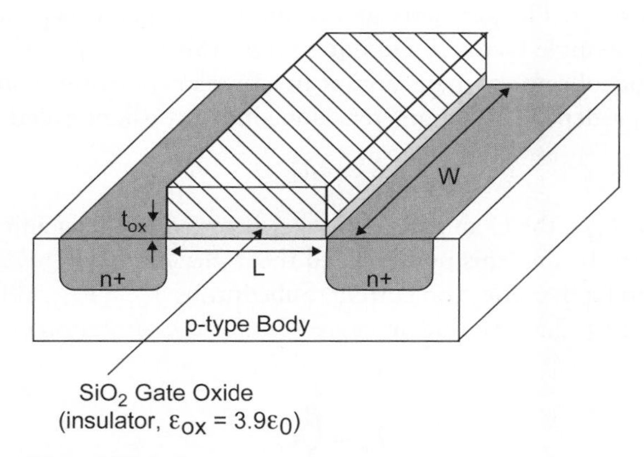

FIGURE 2.6 Transistor dimensions

Some nanometer processes use a different gate dielectric with a higher dielectric constant. In these processes, we call t_{ox} the *equivalent oxide thickness* (EOT), the thickness of a layer of SiO_2 that has the same C_{ox}. In this case, t_{ox} is thinner than the actual dielectric.

Each carrier in the channel is accelerated to an average velocity, v, proportional to the lateral electric field, i.e., the field between source and drain. The constant of proportionality μ is called the *mobility*.

$$v = \mu E \tag{2.3}$$

A typical value of μ for electrons in an nMOS transistor with low electric fields is 500–700 $cm^2/V \cdot s$. However, most transistors today operate at far higher fields where the mobility is severely curtailed (see Section 2.4.1).

The electric field E is the voltage difference between drain and source V_{ds} divided by the channel length

$$E = \frac{V_{ds}}{L} \tag{2.4}$$

The time required for carriers to cross the channel is the channel length divided by the carrier velocity: L/v. Therefore, the current between source and drain is the total amount of charge in the channel divided by the time required to cross

$$\begin{aligned} I_{ds} &= \frac{Q_{\text{channel}}}{L/v} \\ &= \mu C_{ox} \frac{W}{L}\left(V_{gs} - V_t - V_{ds}/2\right)V_{ds} \\ &= \beta\left(V_{GT} - V_{ds}/2\right)V_{ds} \end{aligned} \tag{2.5}$$

where

$$\beta = \mu C_{ox} \frac{W}{L}; \; V_{GT} = V_{gs} - V_t \tag{2.6}$$

The term $V_{gs} - V_t$ arises so often that it is convenient to abbreviate it as V_{GT}. EQ (2.5) describes the linear region of operation, for $V_{gs} > V_t$, but V_{ds} relatively small. It is called *linear* or *resistive* because when $V_{ds} << V_{GT}$, I_{ds} increases almost linearly with V_{ds}, just like an ideal resistor. The geometry and technology-dependent parameters are sometimes merged into a single factor β. Do not confuse this use of β with the same symbol used for the ratio of collector-to-base current in a bipolar transistor. Some texts [Gray01] lump the technology-dependent parameters alone into a constant called "k prime."[3]

$$k' = \mu C_{ox} \tag{2.7}$$

If $V_{ds} > V_{dsat} \equiv V_{GT}$, the channel is no longer inverted in the vicinity of the drain; we say it is pinched off. Beyond this point, called the *drain saturation voltage*, increasing the drain voltage has no further effect on current. Substituting $V_{ds} = V_{dsat}$ at this point of maximum current into EQ (2.5), we find an expression for the saturation current that is independent of V_{ds}.

$$I_{ds} = \frac{\beta}{2} V_{GT}^2 \tag{2.8}$$

[3]Other sources (e.g., MOSIS) define $k' = \frac{\mu C_{ox}}{2}$; check the definition before using quoted data.

This expression is valid for $V_{gs} > V_t$ and $V_{ds} > V_{dsat}$. Thus, long-channel MOS transistors are said to exhibit *square-law behavior* in saturation.

Two key figures of merit for a transistor are I_{on} and I_{off}. I_{on} (also called I_{dsat}) is the ON current, I_{ds}, when $V_{gs} = V_{ds} = V_{DD}$. I_{off} is the OFF current when $V_{gs} = 0$ and $V_{ds} = V_{DD}$. According to the long-channel model, $I_{off} = 0$ and

$$I_{on} = \frac{\beta}{2}\left(V_{DD} - V_t\right) \tag{2.9}$$

EQ (2.10) summarizes the current in the three regions:

$$I_{ds} = \begin{cases} 0 & V_{gs} < V_t & \text{Cutoff} \\ \beta\left(V_{GT} - V_{ds}/2\right)V_{ds} & V_{ds} < V_{dsat} & \text{Linear} \\ \frac{\beta}{2}V_{GT}^2 & V_{ds} > V_{dsat} & \text{Saturation} \end{cases} \tag{2.10}$$

Example 2.1

Consider an nMOS transistor in a 65 nm process with a minimum drawn channel length of 50 nm ($\lambda = 25$ nm). Let $W/L = 4/2\ \lambda$ (i.e., 0.1/0.05 μm). In this process, the gate oxide thickness is 10.5 Å. Estimate the high-field mobility of electrons to be 80 $\text{cm}^2/\text{V}\cdot\text{s}$ at 70 °C. The threshold voltage is 0.3 V. Plot I_{ds} vs. V_{ds} for $V_{gs} = 0$, 0.2, 0.4, 0.6, 0.8, and 1.0 V using the long-channel model.

SOLUTION: We first calculate β.

$$\beta = \mu C_{ox}\frac{W}{L} = \left(80\frac{\text{cm}^2}{\text{V}\cdot\text{s}}\right)\left(\frac{3.9\times 8.85\times 10^{-14}\,\frac{\text{F}}{\text{cm}}}{10.5\times 10^{-8}\,\text{cm}}\right)\left(\frac{W}{L}\right) = 262\frac{W}{L}\frac{\text{A}}{\text{V}^2} \tag{2.11}$$

Figure 2.7(a) shows the I-V characteristics for the transistor. According to the first-order model, the current is zero for gate voltages below V_t. For higher gate voltages, current increases linearly with V_{ds} for small V_{ds}. As V_{ds} reaches the saturation point $V_{dsat} = V_{GT}$, current rolls off and eventually becomes independent of V_{ds} when the transistor is saturated. We will later see that the Shockley model overestimates current at high voltage because it does not account for mobility degradation and velocity saturation caused by the high electric fields.

pMOS transistors behave in the same way, but with the signs of all voltages and currents reversed. The I-V characteristics are in the third quadrant, as shown in Figure 2.7(b). To keep notation simple in this text, we will disregard the signs and just remember that the current flows from source to drain in a pMOS transistor. The mobility of holes in silicon is typically lower than that of electrons. This means that pMOS transistors provide less current than nMOS transistors of comparable size and hence are slower. The symbols μ_n and μ_p are used to distinguish mobility of electrons and of holes in nMOS and pMOS transistors, respectively. The *mobility ratio* μ_n/μ_p is typically 2–3; we will generally use 2 for examples in this book. The pMOS transistor has the same geometry as the nMOS in Figure 2.7(a), but with $\mu_p = 40\ \text{cm}^2/\text{V}\cdot\text{s}$ and $V_{tp} = -0.3$ V. Similarly, β_n, β_p, k'_n, and k'_p are sometimes used to distinguish nMOS and pMOS I-V characteristics.

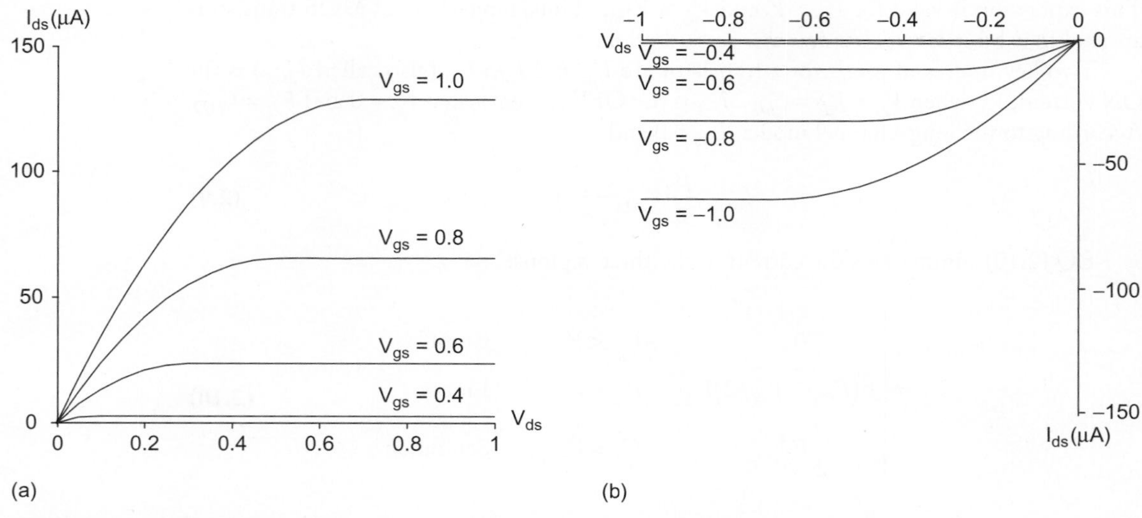

FIGURE 2.7 I-V characteristics of ideal 4/2 λ (a) nMOS and (b) pMOS transistors

2.3 C-V Characteristics

Each terminal of an MOS transistor has capacitance to the other terminals. In general, these capacitances are nonlinear and voltage dependent (C-V); however, they can be approximated as simple capacitors when their behavior is averaged across the switching voltages of a logic gate. This section first presents simple models of each capacitance suitable for estimating delay and power consumption of transistors. It then explores more detailed models used for circuit simulation. The more detailed models may be skipped on a first reading.

2.3.1 Simple MOS Capacitance Models

The gate of an MOS transistor is a good capacitor. Indeed, its capacitance is necessary to attract charge to invert the channel, so high gate capacitance is required to obtain high I_{ds}. As seen in Section 2.2, the gate capacitor can be viewed as a parallel plate capacitor with the gate on top and channel on bottom with the thin oxide dielectric between. Therefore, the capacitance is

$$C_g = C_{ox}WL \tag{2.12}$$

The bottom plate of the capacitor is the channel, which is not one of the transistor's terminals. When the transistor is on, the channel extends from the source (and reaches the drain if the transistor is unsaturated, or stops short in saturation). Thus, we often approximate the gate capacitance as terminating at the source and call the capacitance C_{gs}.

Most transistors used in logic are of minimum manufacturable length because this results in greatest speed and lowest dynamic power consumption.[4] Thus, taking this mini-

[4]Some designs use slightly longer than minimum transistors that have higher thresholds because of the short-channel effect (see Sections 2.4.3.3 and 4.3.3). This avoids the cost of an extra mask step for high-V_t transistors. The change in channel length is small (~5–10%), so the change in gate capacitance is minor.

mum L as a constant for a particular process, we can define

$$C_g = C_{\text{permicron}} \times W \tag{2.13}$$

where

$$C_{\text{permicron}} = C_{\text{ox}} L = \frac{\varepsilon_{\text{ox}}}{t_{\text{ox}}} L \tag{2.14}$$

Notice that if we develop a more advanced manufacturing process in which both the channel length and oxide thickness are reduced by the same factor, $C_{\text{permicron}}$ remains unchanged. This relationship is handy for quick calculations but not exact; $C_{\text{permicron}}$ has fallen from about 2 fF/μm in old processes to about 1 fF/μm at the 90 and 65 nm nodes. Table 7.5 lists gate capacitance for a variety of processes.

In addition to the gate, the source and drain also have capacitances. These capacitances are not fundamental to operation of the devices, but do impact circuit performance and hence are called *parasitic* capacitors. The source and drain capacitances arise from the p–n junctions between the source or drain diffusion and the body and hence are also called *diffusion*[5] capacitance C_{sb} and C_{db}. A *depletion region* with no free carriers forms along the junction. The depletion region acts as an insulator between the conducting p- and n-type regions, creating capacitance across the junction. The capacitance of these junctions depends on the area and perimeter of the source and drain diffusion, the depth of the diffusion, the doping levels, and the voltage. As diffusion has both high capacitance and high resistance, it is generally made as small as possible in the layout. Three types of diffusion regions are frequently seen, illustrated by the two series transistors in Figure 2.8. In Figure

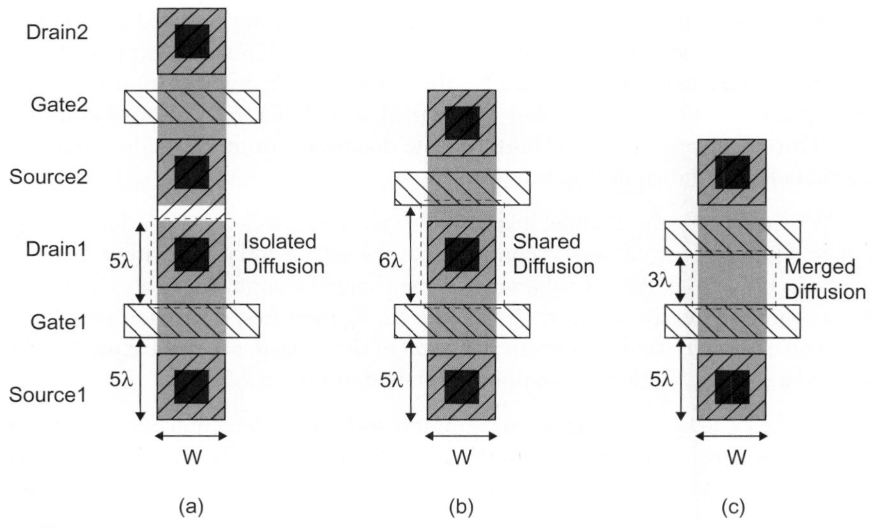

FIGURE 2.8 Diffusion region geometries

[5]Device engineers more properly call this *depletion* capacitance, but the term *diffusion* capacitance is widely used by circuit designers.

2.8(a), each source and drain has its own *isolated* region of contacted diffusion. In Figure 2.8(b), the drain of the bottom transistor and source of the top transistor form a *shared* contacted diffusion region. In Figure 2.8(c), the source and drain are *merged* into an uncontacted region. The average capacitance of each of these types of regions can be calculated or measured from simulation as a transistor switches between V_{DD} and GND. Table 7.5 also lists the capacitance for each scenario for a variety of processes.

For the purposes of hand estimation, you can observe that the diffusion capacitance C_{sb} and C_{db} of contacted source and drain regions is comparable to the gate capacitance (e.g., 1–2 fF/μm of gate width). The diffusion capacitance of the uncontacted source or drain is somewhat less because the area is smaller but the difference is usually unimportant for hand calculations. These values of $C_g = C_{sb} = C_{db} \approx 1\text{fF}/\mu\text{m}$ will be used in examples throughout the text, but you should obtain the appropriate data for your process using methods to be discussed in Section 7.4.

2.3.2 Detailed MOS Gate Capacitance Model

The MOS gate sits above the channel and may partially overlap the source and drain diffusion areas. Therefore, the gate capacitance has two components: the intrinsic capacitance C_{gc} (over the channel) and the overlap capacitances C_{gol} (to the source and drain).

The intrinsic capacitance was approximated as a simple parallel plate in EQ (2.12) with capacitance $C_0 = WLC_{ox}$. However, the bottom plate of the capacitor depends on the mode of operation of the transistor. The intrinsic capacitance has three components representing the different terminals connected to the bottom plate: C_{gb} (gate-to-body), C_{gs} (gate-to-source), and C_{gd} (gate-to-drain). Figure 2.9(a) plots capacitance vs. V_{gs} in the cutoff region and for small V_{ds}, while 2.9(b) plots capacitance vs. V_{ds} in the linear and saturation regions [Dally98].

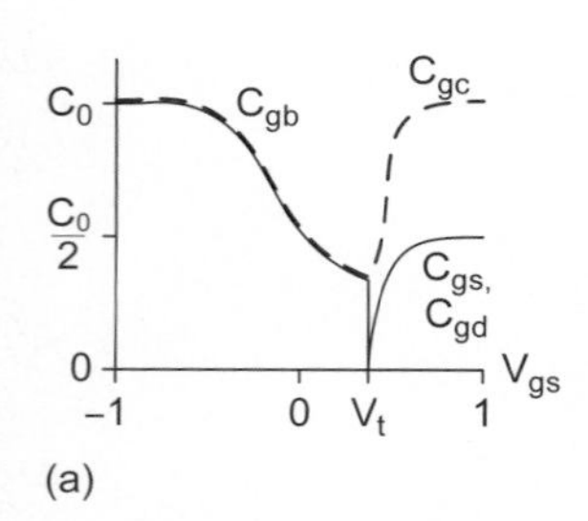

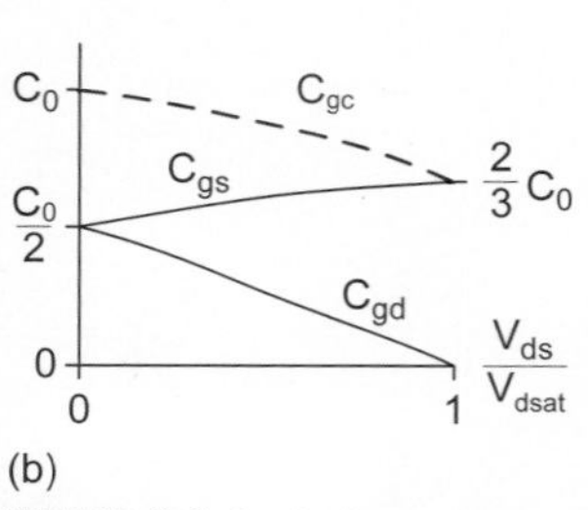

FIGURE 2.9 Intrinsic gate capacitance $C_{gc} = C_{gs} + C_{gd} + C_{gb}$ as a function of (a) V_{gs} and (b) V_{ds}

1. *Cutoff.* When the transistor is OFF ($V_{gs} < V_t$), the channel is not inverted and charge on the gate is matched with opposite charge from the body. This is called C_{gb}, the gate-to-body capacitance. For negative V_{gs}, the transistor is in accumulation and $C_{gb} = C_0$. As V_{gs} increases but remains below a threshold, a depletion region forms at the surface. This effectively moves the bottom plate downward from the oxide, reducing the capacitance, as shown in Figure 2.9(a).
2. *Linear.* When $V_{gs} > V_t$, the channel inverts and again serves as a good conductive bottom plate. However, the channel is connected to the source and drain, rather than the body, so C_{gb} drops to 0. At low values of V_{ds}, the channel charge is roughly shared between source and drain, so $C_{gs} = C_{gd} = C_0/2$. As V_{ds} increases, the region near the drain becomes less inverted, so a greater fraction of the capacitance is attributed to the source and a smaller fraction to the drain, as shown in Figure 2.9(b).
3. *Saturation.* At $V_{ds} > V_{dsat}$, the transistor saturates and the channel pinches off. At this point, all the intrinsic capacitance is to the source, as shown in Figure 2.9(b). Because of pinchoff, the capacitance in saturation reduces to $C_{gs} = 2/3\ C_0$ for an ideal transistor [Gray01].

The behavior in these three regions can be approximated as shown in Table 2.1.

TABLE 2.1 Approximation for intrinsic MOS gate capacitance

Parameter	Cutoff	Linear	Saturation
C_{gb}	$\leq C_0$	0	0
C_{gs}	0	$C_0/2$	$2/3\ C_0$
C_{gd}	0	$C_0/2$	0
$C_g = C_{gs} + C_{gd} + C_{gb}$	C_0	C_0	$2/3\ C_0$

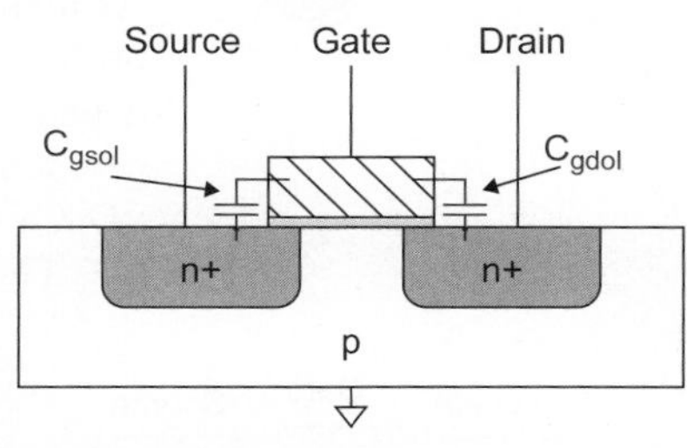

FIGURE 2.10 Overlap capacitance

The gate overlaps the source and drain in a real device and also has fringing fields terminating on the source and drain. This leads to additional overlap capacitances, as shown in Figure 2.10. These capacitances are proportional to the width of the transistor. Typical values are $C_{gsol} = C_{gdol} = 0.2 - 0.4$ fF/μm. They should be added to the intrinsic gate capacitance to find the total.

$$\begin{aligned} C_{gsol(\text{overlap})} &= C_{gsol} W \\ C_{gdol(\text{overlap})} &= C_{gdol} W \end{aligned} \qquad (2.15)$$

It is convenient to view the gate capacitance as a single-terminal capacitor attached to the gate (with the other side not switching). Because the source and drain actually form second terminals, the effective gate capacitance varies with the switching activity of the source and drain. Figure 2.11 shows the effective gate capacitance in a 0.35 μm process for seven different combinations of source and drain behavior [Bailey98].

More accurate modeling of the gate capacitance may be achieved by using a charge-based model [Cheng99]. For the purpose of delay calculation of digital circuits, we usually approximate $C_g = C_{gs} + C_{gd} + C_{gb} \approx C_0 + 2C_{gol}W$ or use an effective capacitance extracted

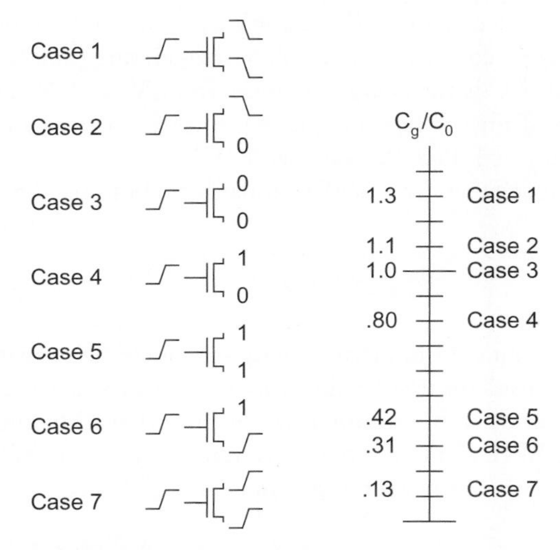

FIGURE 2.11 Data-dependent gate capacitance

from simulation [Nose00b]. It is important to remember that this model significantly overestimates the capacitance of transistors operating just below threshold.

2.3.3 Detailed MOS Diffusion Capacitance Model

As mentioned in Section 2.3.1, the p–n junction between the source diffusion and the body contributes parasitic capacitance across the depletion region. The capacitance depends on both the *area AS* and *sidewall perimeter PS* of the source diffusion region. The geometry is illustrated in Figure 2.12. The area is $AS = WD$. The perimeter is $PS = 2W + 2D$. Of this perimeter, W abuts the channel and the remaining $W + 2D$ does not.

The total source parasitic capacitance is

$$C_{sb} = AS \times C_{jbs} + PS \times C_{jbssw} \tag{2.16}$$

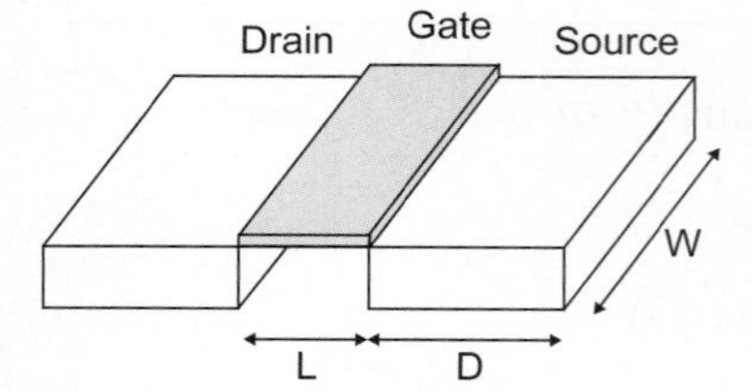

FIGURE 2.12 Diffusion region geometry

where C_{jbs} (the capacitance of the junction between the body and the bottom of the source) has units of capacitance/area and C_{jbssw} (the capacitance of the junction between the body and the side walls of the source) has units of capacitance/length.

Because the depletion region thickness depends on the bias conditions, these parasitics are nonlinear. The area junction capacitance term is [Gray01]

$$C_{jbs} = C_J\left(1 + \frac{V_{sb}}{\psi_0}\right)^{-M_J} \tag{2.17}$$

C_J is the junction capacitance at zero bias and is highly process-dependent. M_J is the *junction grading coefficient*, typically in the range of 0.5 to 0.33 depending on the abruptness of the diffusion junction. ψ_0 is the *built-in potential* that depends on doping levels.

$$\psi_0 = v_T \ln \frac{N_A N_D}{n_i^2} \tag{2.18}$$

v_T is the *thermal voltage* from thermodynamics, not to be confused with the threshold voltage V_t. It has a value equal to kT/q (26 mV at room temperature), where $k = 1.380 \times 10^{-23}$ J/K is Boltzmann's constant, T is absolute temperature (300 K at room temperature), and $q = 1.602 \times 10^{-19}$ C is the charge of an electron. N_A and N_D are the doping levels of the body and source diffusion region. n_i is the intrinsic carrier concentration in undoped silicon and has a value of 1.45×10^{10} cm^{-3} at 300 K.

The sidewall capacitance term is of a similar form but uses different coefficients.

$$C_{jbssw} = C_{JSW}\left(1 + \frac{V_{sb}}{\psi_{SW}}\right)^{-M_{JSW}} \tag{2.19}$$

In processes below about 0.35 μm that employ shallow trench isolation surrounding transistors with an SiO_2 insulator (see Section 15.2.6), the sidewall capacitance along the nonconductive trench tends to be minimal, while the sidewall facing the channel is more significant. In some SPICE models, the capacitance of this sidewall abutting the gate and channel is specified with another set of parameters:

$$C_{jbsswg} = C_{JSW}\left(1 + \frac{V_{sb}}{\psi_{SWG}}\right)^{-M_{JSWG}} \tag{2.20}$$

Section 7.3.4 discusses SPICE perimeter capacitance models further.

The drain diffusion has a similar parasitic capacitance dependent on *AD*, *PD*, and V_{db}. Equivalent relationships hold for pMOS transistors, but doping levels differ. As the capacitances are voltage-dependent, the most useful information to digital designers is the value averaged across a switching transition. This is the C_{sb} or C_{db} value that was presented in Section 2.3.1.

Example 2.2

Calculate the diffusion parasitic C_{db} of the drain of a unit-sized contacted nMOS transistor in a 65 nm process when the drain is at 0 V and again at $V_{DD} = 1.0$ V. Assume the substrate is grounded. The diffusion region conforms to the design rules from Figure 2.8 with $\lambda = 25$ nm. The transistor characteristics are $CJ = 1.2$ fF/μm^2, $MJ = 0.33$, $CJSW = 0.1$ fF/μm, $CJSWG = 0.36$ fF/μm, $MJSW = MJSWG = 0.10$, and $\psi_0 = 0.7$ V at room temperature.

SOLUTION: From Figure 2.8, we find a unit-size diffusion contact is $4 \times 5\ \lambda$, or $0.1 \times 0.125\ \mu$m. The area is 0.0125 μm^2 and perimeter is 0.35 μm plus 0.1 μm along the channel. At zero bias, $C_{jbd} = 1.2$ fF/μm^2, $C_{jbdsw} = 0.1$ fF/μm, and $C_{jbdswg} = 0.36$ fF/μm. Hence, the total capacitance is

$$C_{db}(0\text{ V}) = \left(0.0125\mu\text{m}^2\right)\left(1.2\frac{\text{fF}}{\mu\text{m}^2}\right) + \left(0.35\mu\text{m}\right)\left(0.1\frac{\text{fF}}{\mu\text{m}}\right) + \left(0.1\mu\text{m}\right)\left(0.36\frac{\text{fF}}{\mu\text{m}}\right) = 0.086\text{ fF} \tag{2.21}$$

At a drain voltage of V_{DD}, the capacitance reduces to

$$C_{db}(1\text{ V}) = \left(0.0125\mu\text{m}^2\right)\left(1.2\frac{\text{fF}}{\mu\text{m}^2}\right)\left(1+\frac{1.0}{0.7}\right)^{-0.33} + \left[\left(0.35\mu\text{m}\right)\left(0.1\frac{\text{fF}}{\mu\text{m}}\right) + \left(0.1\mu\text{m}\right)\left(0.36\frac{\text{fF}}{\mu\text{m}}\right)\right]\left(1+\frac{1.0}{0.7}\right)^{-0.10} = 0.076\text{ fF} \tag{2.22}$$

For the purpose of manual performance estimation, this nonlinear capacitance is too much effort. An effective capacitance averaged over the switching range is quite satisfactory for digital applications. In this example, the effective drain capacitance would be approximated as the average of the two extremes, 0.081 fF.

Diffusion regions were historically used for short wires called *runners* in processes with only one or two metal levels. Diffusion capacitance and resistance are large enough that such practice is now discouraged; diffusion regions should be kept as small as possible on nodes that switch.

In summary, an MOS transistor can be viewed as a four-terminal device with capacitances between each terminal pair, as shown in Figure 2.13. The gate capacitance includes an intrinsic component (to the body, source and drain, or source alone, depending on operating regime) and overlap terms with the source and drain. The source and drain have parasitic diffusion capacitance to the body.

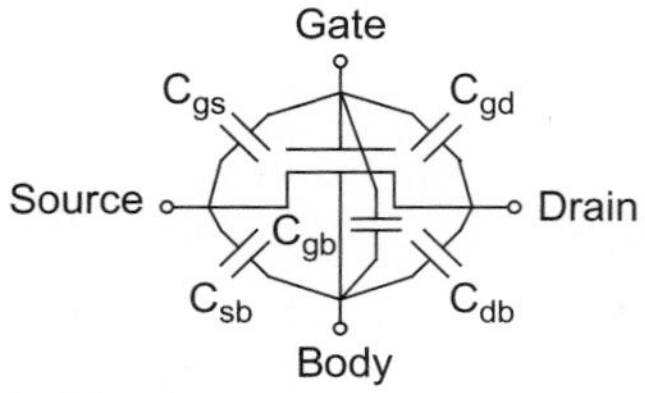

FIGURE 2.13 Capacitance of an MOS transistor

2.4 Nonideal I-V Effects

The long-channel I-V model of EQ (2.10) neglects many effects that are important to devices with channel lengths below 1 micron. This section summarizes the effects of greatest significance to designers, then models each one in more depth.

Figure 2.14 compares the simulated I-V characteristics of a 1-micron wide nMOS transistor in a 65 nm process to the ideal characteristics computed in Section 2.2. The saturation current increases less than quadratically with increasing V_{gs}. This is caused by two effects: velocity saturation and mobility degradation. At high lateral field strengths (V_{ds}/L), carrier velocity ceases to increase linearly with field strength. This is called *velocity saturation* and results in lower I_{ds} than expected at high V_{ds}. At high vertical field strengths (V_{gs}/t_{ox}), the carriers scatter off the oxide interface more often, slowing their progess. This *mobility degradation* effect also leads to less current than expected at high V_{gs}. The saturation current of the nonideal transistor increases somewhat with V_{ds}. This is caused by *channel length modulation*, in which higher V_{ds} increases the size of the depletion region around the drain and thus effectively shortens the channel.

The threshold voltage indicates the gate voltage necessary to invert the channel and is primarily determined by the oxide thickness and channel doping levels. However, other fields in the transistor have some effect on the channel, effectively modifying the threshold voltage. Increasing the potential between the source and body raises the threshold through the *body effect*. Increasing the drain voltage lowers the threshold through *drain-induced barrier lowering*. Increasing the channel length raises the threshold through the *short channel effect*.

Several sources of leakage result in current flow in nominally OFF transistors. When $V_{gs} < V_t$, the current drops off exponentially rather than abruptly becoming zero. This is called *subthreshold conduction*. The current into the gate I_g is ideally 0. However, as the

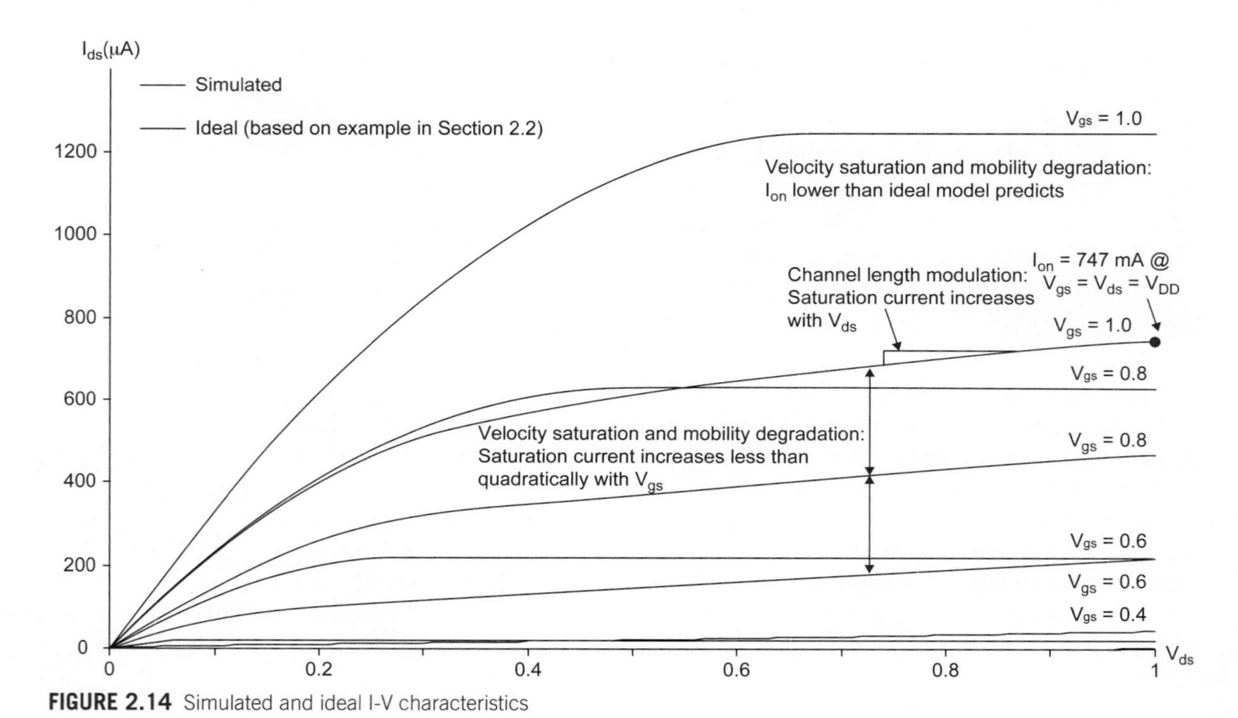

FIGURE 2.14 Simulated and ideal I-V characteristics

thickness of gate oxides reduces to only a small number of atomic layers, electrons *tunnel* through the gate, causing some *gate leakage* current. The source and drain diffusions are typically reverse-biased diodes and also experience *junction leakage* into the substrate or well.

Both mobility and threshold voltage decrease with rising temperature. The mobility effect tends to dominate for strongly ON transistors, resulting in lower I_{ds} at high temperature. The threshold effect is most important for OFF transistors, resulting in higher leakage current at high temperature. In summary, MOS characteristics degrade with temperature.

It is useful to have a qualitative understanding of nonideal effects to predict their impact on circuit behavior and to be able to anticipate how devices will change in future process generations. However, the effects lead to complicated I-V characteristics that are hard to directly apply in hand calculations. Instead, the effects are built into good transistor models and simulated with SPICE or similar software.

2.4.1 Mobility Degradation and Velocity Saturation

Recall from EQ (2.3) that carrier drift velocity, and hence current, is proportional to the lateral electric field $E_{\text{lat}} = V_{ds}/L$ between source and drain. The constant of proportionality is called the carrier mobility, μ. The long-channel model assumed that carrier mobility is independent of the applied fields. This is a good approximation for low fields, but breaks down when strong lateral or vertical fields are applied.

As an analogy, imagine that you have been working all night in the VLSI lab and decide to run down and across the courtyard to the coffee cart.[6] The number of hours you have been up is analogous to the lateral electric field. The longer you have been up, the faster you want to reach coffee: Your speed equals your fatigue times your mobility. There is a strong wind blowing in the courtyard, analogous to the vertical electric field. This wind buffets you against the wall, slowing your progress. In the same way, a high voltage at the gate of the transistor attracts the carriers to the edge of the channel, causing collisions with the oxide interface that slow the carriers. This is called mobility degradation. Moreover, freshman physics is just letting out of the lecture hall. Occasionally, you bounce off a confused freshman, fall down, and have to get up and start running again. This is analogous to carriers scattering off the silicon lattice (technically called collisions with optical phonons). The faster you try to go, the more often you collide. Beyond a certain level of fatigue, you reach a maximum average speed. In the same way, carriers approach a maximum velocity v_{sat} when high fields are applied. This phenomenon is called velocity saturation.[7]

Mobility degradation can be modeled by replacing μ with a smaller μ_{eff} that is a function of V_{gs}. A universal model [Chen96, Chen97] that matches experimental data from multiple processes reasonably well is

$$\mu_{\text{eff}-n} = \frac{540 \frac{\text{cm}^2}{\text{V}\cdot\text{s}}}{1 + \left(\frac{V_{gs} + V_t}{0.54 \frac{\text{V}}{\text{nm}} t_{\text{ox}}} \right)^{1.85}} \qquad \mu_{\text{eff}-p} = \frac{185 \frac{\text{cm}^2}{\text{V}\cdot\text{s}}}{1 + \frac{\left| V_{gs} + 1.5 V_t \right|}{0.338 \frac{\text{V}}{\text{nm}} t_{\text{ox}}}} \tag{2.23}$$

[6]This practice has been observed empirically, but is not recommended. Productivity decreases with fatigue. Beyond a certain point of exhaustion, the net work accomplished per hour becomes negative because so many mistakes are made.

[7]Do not confuse the *saturation* region of transistor operation (where $V_{ds} > V_{gs} - V_t$) with *velocity saturation* (where $E_{\text{lat}} = V_{ds}/L$ approaches E_c). In this text, the word "saturation" alone refers to the operating region while "velocity saturation" refers to the limiting of carrier velocity at high field.

Example 2.3

Compute the effective mobilities for nMOS and pMOS transistors when they are fully ON. Use the physical parameters from Example 2.1.

SOLUTION: Use $V_{gs} = 1.0$ for ON transistors, remembering that we are treating voltages as positive in a pMOS transistor. Substituting $V_t = 0.3$ V and $t_{ox} = 1.05$ nm into EQ (2.23) gives:

$$\mu_{\text{eff-n}}(V_{gs} = 1.0) = 96 \text{ cm}^2/\text{V}, \mu_{\text{eff-p}}(V_{gs} = 1.0) = 36 \text{ cm}^2/\text{V}$$

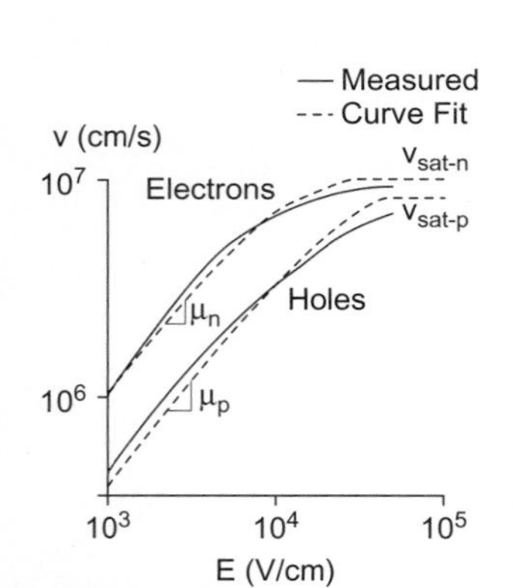

FIGURE 2.15 Carrier velocity vs. electric field at 300 K, adapted from [Jacoboni77]. Velocity saturates at high fields.

Figure 2.15 shows measured data for carrier velocity as a function of the electric field, E, between the drain and source. At low fields, the velocity increases linearly with the field. The slope is the mobility, μ_{eff}. At fields above a critical level, E_c, the velocity levels out at v_{sat}, which is approximately 10^7 cm/s for electrons and 8×10^6 cm/s for holes [Muller03]. As shown in the figure, the velocity can be approximated reasonably well with the following expression [Toh88, Takeuchi94]:

$$v = \begin{cases} \dfrac{\mu_{\text{eff}} E}{1 + \dfrac{E}{E_c}} & E < E_c \\ v_{\text{sat}} & E \geq E_c \end{cases} \tag{2.24}$$

where, by continuity, the *critical electric field* is

$$E_c = \frac{2 v_{\text{sat}}}{\mu_{\text{eff}}} \tag{2.25}$$

The *critical voltage* V_c is the drain-source voltage at which the critical effective field is reached: $V_c = E_c L$.

Example 2.4

Find the critical voltage for fully ON nMOS and pMOS transistors using the effective mobilities from Example 2.3.

SOLUTION: Using EQ (2.25)

$$V_{c-n} = \frac{2\left(10^7 \frac{\text{cm}}{\text{s}}\right)}{96 \frac{\text{cm}^2}{\text{V} \cdot \text{s}}} \left(5 \times 10^{-6} \text{cm}\right) = 1.04 \text{ V}$$

$$V_{c-p} = \frac{2\left(8 \times 10^6 \frac{\text{cm}}{\text{s}}\right)}{36 \frac{\text{cm}^2}{\text{V} \cdot \text{s}}} \left(5 \times 10^{-6} \text{cm}\right) = 2.22 \text{ V}$$

The nMOS transistor is velocity saturated in normal operation because V_{c-n} is comparable to V_{DD}. The pMOS transistor has lower mobility and thus is not as badly velocity saturated.

Using a derivation similar to that of Section 2.2 with the new carrier velocity expression in EQ (2.24) gives modified equations for linear and saturation currents [Sodini84].

$$I_{ds} = \begin{cases} \dfrac{\mu_{\text{eff}}}{1+\dfrac{V_{ds}}{V_c}} C_{\text{ox}} \dfrac{W}{L}\left(V_{GT} - V_{ds}/2\right)V_{ds} & V_{ds} < V_{\text{dsat}} \quad \text{Linear} \\ C_{\text{ox}} W \left(V_{GT} - V_{\text{dsat}}\right) v_{\text{sat}} & V_{ds} > V_{\text{dsat}} \quad \text{Saturation} \end{cases} \tag{2.26}$$

Note that μ_{eff} is a decreasing function of V_{gs} because of mobility degradation. Observe that the current in the linear regime is the same as in EQ (2.5) except that the mobility term is reduced by a factor related to V_{ds}. At sufficiently high lateral fields, the current saturates at some value dependent on the maximum carrier velocity. Equating the two parts of EQ (2.26) at $V_{ds} = V_{\text{dsat}}$ lets us solve for the saturation voltage

$$V_{\text{dsat}} = \frac{V_{GT}V_c}{V_{GT} + V_c} \tag{2.27}$$

Noting that EQ (2.27) is in the same form as a parallel resistor equation, we see that V_{dsat} is less than the smaller of V_{GT} and V_c. Finally, substituting EQ (2.27) into EQ (2.26) gives a simplified expression for saturation current accounting for velocity saturations:

$$I_{\text{dsat}} = WC_{\text{ox}} v_{\text{sat}} \frac{V_{GT}^2}{V_{GT} + V_c} \qquad V_{ds} > V_{\text{dsat}} \tag{2.28}$$

If $V_{GT} << V_c$, velocity saturation effects are negligible and EQ (2.28) reduces to the square-law model. This is also called the *long-channel regime*. But if $V_{GT} >> V_c$, EQ (2.28) approaches the velocity-saturated limit

$$I_{\text{dsat}} \approx WC_{\text{ox}} v_{\text{sat}} V_{GT} \qquad V_{ds} > V_c \tag{2.29}$$

Observe that the drain current is quadratically dependent on voltage in the long-channel regime and linearly dependent when fully velocity saturated. For moderate supply voltages, transistors operate in a region where the velocity neither increases linearly with field, nor is completely saturated. The *α-power law model* given in EQ (2.30) provides a simple approximation to capture this behavior [Sakurai90]. α is called the *velocity saturation index* and is determined by curve fitting measured I-V data. Transistors with long channels or low V_{DD} display quadratic I-V characteristics in saturation and are modeled with $\alpha = 2$. As transistors become more velocity saturated, increasing V_{gs} has less effect on current and α decreases, reaching 1 for transistors that are completely velocity saturated. For simplicity, the model uses a straight line in the linear region. Overall, the model is based on three parameters that can be determined empirically from a curve fit of I-V characteristics: α, βP_c, and P_v.

$$I_{ds} = \begin{cases} 0 & V_{gs} < V_t \quad \text{Cutoff} \\ I_{\text{dsat}} \dfrac{V_{ds}}{V_{\text{dsat}}} & V_{ds} < V_{\text{dsat}} \quad \text{Linear} \\ I_{\text{dsat}} & V_{ds} > V_{\text{dsat}} \quad \text{Saturation} \end{cases} \tag{2.30}$$

where

$$\begin{aligned} I_{\text{dsat}} &= P_c \frac{\beta}{2} V_{GT}^{\alpha} \\ V_{\text{dsat}} &= P_v V_{GT}^{\alpha/2} \end{aligned} \tag{2.31}$$

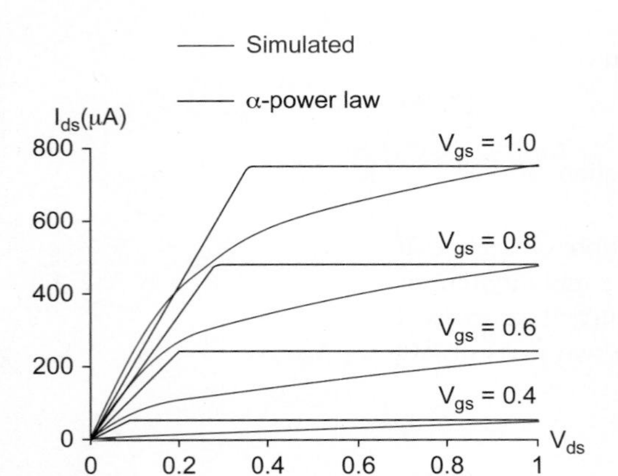

FIGURE 2.16 Comparison of α-power law model with simulated transistor behavior

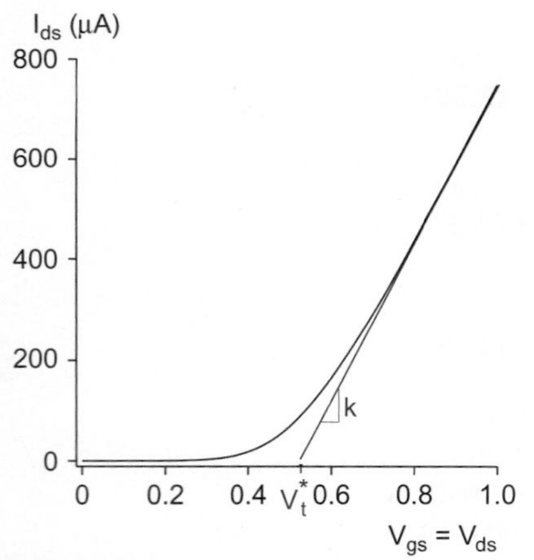

FIGURE 2.17 I_{ds} vs. V_{gs} in saturation, showing good linear fit at high V_{gs}

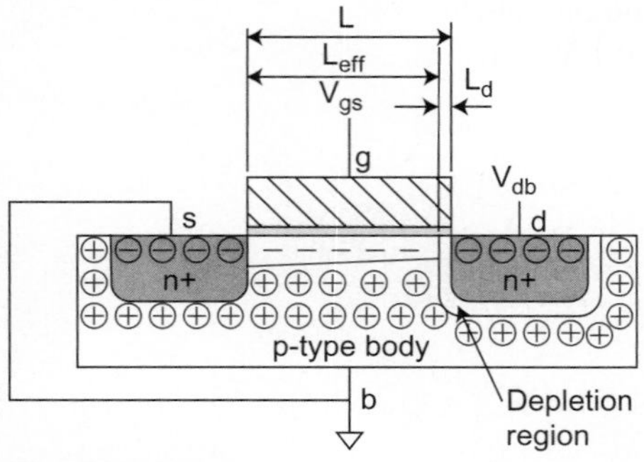

FIGURE 2.18 Depletion region shortens effective channel length

Figure 2.16 compares the α-power law model against simulated results, using $\alpha = 1.3$. The fit is poor at low V_{ds}, but the current at $V_{ds} = V_{DD}$ matches simulation fairly well across the full range of V_{gs}.

The low-field mobility of holes is much lower than that of electrons, so pMOS transistors experience less velocity saturation than nMOS for a given V_{DD}. This shows up as a larger value of α for pMOS than for nMOS transistors.

These models become too complicated to give much insight for hand calculations. A simpler approach is to observe, in velocity-saturated transistors, I_{ds} grows linearly rather than quadratically with V_{gs} when the transistor is strongly ON. Figure 2.17 plots I_{ds} vs. V_{gs} (holding $V_{ds} = V_{gs}$). This is equivalent to plotting I_{on} vs. V_{DD}. For V_{gs} significantly above V_t, I_{ds} fits a straight line quite well. Thus, we can approximate the ON current as

$$I_{ds} = k\left(V_{gs} - V_t^*\right) \qquad \textbf{(2.32)}$$

where V_t^* is the x-intercept.

2.4.2 Channel Length Modulation

Ideally, I_{ds} is independent of V_{ds} for a transistor in saturation, making the transistor a perfect current source. As discussed in Section 2.3.3, the p–n junction between the drain and body forms a depletion region with a width L_d that increases with V_{db}, as shown in Figure 2.18. The depletion region effectively shortens the channel length to

$$L_{\text{eff}} = L - L_d \qquad \textbf{(2.33)}$$

To avoid introducing the body voltage into our calculations, assume the source voltage is close to the body voltage so $V_{db} \approx V_{ds}$. Hence, increasing V_{ds} decreases the effective channel length. Shorter channel length results in higher current; thus, I_{ds} increases with V_{ds} in saturation, as shown in Figure 2.18. This can be crudely modeled by multiplying EQ (2.10) by a factor of $(1 + V_{ds} / V_A)$, where V_A is called the *Early voltage* [Gray01]. In the saturation region, we find

$$I_{ds} = \frac{\beta}{2} V_{\text{GT}}^2 \left(1 + \frac{V_{ds}}{V_A}\right) \qquad \textbf{(2.34)}$$

As channel length gets shorter, the effect of the channel length modulation becomes relatively more important. Hence, V_A is proportional to channel length. This channel length modulation model is a gross oversimplification of nonlinear behavior and is more useful for conceptual understanding than for accurate device modeling.

Channel length modulation is very important to analog designers because it reduces the gain of amplifiers. It is generally unimportant for qualitatively understanding the behavior of digital circuits.

2.4.3 Threshold Voltage Effects

So far, we have treated the threshold voltage as a constant. However, V_t increases with the source voltage, decreases with the body voltage, decreases with the drain voltage, and increases with channel length [Roy03]. This section models each of these effects.

2.4.3.1 Body Effect Until now, we have considered a transistor to be a three-terminal device with gate, source, and drain. However, the body is an implicit fourth terminal. When a voltage V_{sb} is applied between the source and body, it increases the amount of charge required to invert the channel, hence, it increases the threshold voltage. The threshold voltage can be modeled as

$$V_t = V_{t0} + \gamma\left(\sqrt{\phi_s + V_{sb}} - \sqrt{\phi_s}\right) \quad (2.35)$$

where V_{t0} is the threshold voltage when the source is at the body potential, ϕ_s is the *surface potential* at threshold (see a device physics text such as [Tsividis99] for further discussion of surface potential), and γ is the *body effect coefficient*, typically in the range 0.4 to 1 $V^{1/2}$. In turn, these depend on the doping level in the channel, N_A. The body effect further degrades the performance of pass transistors trying to pass the weak value (e.g., nMOS transistors passing a '1'), as we will examine in Section 2.5.4. Section 4.3.4 will describe how a body bias can intentionally be applied to alter the threshold voltage, permitting trade-offs between performance and subthreshold leakage current.

$$\phi_s = 2v_T \ln\frac{N_A}{n_i} \quad (2.36)$$

$$\gamma = \frac{t_{ox}}{\varepsilon_{ox}}\sqrt{2q\varepsilon_{si}N_A} = \frac{\sqrt{2q\varepsilon_{si}N_A}}{C_{ox}} \quad (2.37)$$

For small voltages applied to the source or body, EQ (2.35) can be linearized to

$$V_t = V_{t0} + k_\gamma V_{sb} \quad (2.38)$$

where

$$k_\gamma = \frac{\gamma}{2\sqrt{\phi_s}} = \frac{\sqrt{\dfrac{q\varepsilon_{si}N_A}{v_T \ln\frac{N_A}{n_i}}}}{2C_{ox}} \quad (2.39)$$

Example 2.5

Consider the nMOS transistor in a 65 nm process with a nominal threshold voltage of 0.3 V and a doping level of 8×10^{17} cm^{-3}. The body is tied to ground with a substrate contact. How much does the threshold change at room temperature if the source is at 0.6 V instead of 0?

SOLUTION: At room temperature, the thermal voltage $v_T = kT/q = 26$ mV and $n_i = 1.45 \times 10^{10}$ cm^{-3}. The threshold increases by 0.04 V.

$$\phi_s = 2(0.026\text{ V})\ln\frac{8\times10^{17}\text{ cm}^{-3}}{1.45\times10^{10}\text{cm}^{-3}} = 0.93\text{ V}$$

$$\gamma = \frac{10.5\times10^{-8}\text{cm}}{3.9\times8.85\times10^{-14}\frac{\text{F}}{\text{cm}}}\sqrt{2(1.6\times10^{-19}\text{C})(11.7\times8.85\times10^{-14}\tfrac{\text{F}}{\text{cm}})(8\times10^{17}\text{cm}^{-3})} = 0.16 \quad \textbf{(2.40)}$$

$$V_t = 0.3+\gamma\left(\sqrt{\phi_s+0.6\text{ V}}-\sqrt{\phi_s}\right) = 0.34\text{ V}$$

2.4.3.2 Drain-Induced Barrier Lowering The drain voltage V_{ds} creates an electric field that affects the threshold voltage. This *drain-induced barrier lowering* (DIBL) effect is especially pronounced in short-channel transistors. It can be modeled as

$$V_t = V_{t0} - \eta V_{ds} \quad \textbf{(2.41)}$$

where η is the DIBL coefficient, typically on the order of 0.1 (often expressed as 100 mV/V).

Drain-induced barrier lowering causes I_{ds} to increase with V_{ds} in saturation, in much the same way as channel length modulation does. This effect can be lumped into a smaller Early voltage V_A used in EQ (2.34). Again, this is a bane for analog design but insignificant for most digital circuits. More significantly, DIBL increases subthreshold leakage at high V_{ds}, as we will discuss in Section 2.4.4.

2.4.3.3 Short Channel Effect The threshold voltage typically increases with channel length. This phenomenon is especially pronounced for small L where the source and drain depletion regions extend into a significant portion of the channel, and hence is called the *short channel effect*[8] or V_t *rolloff* [Tsividis99, Cheng99]. In some processes, a *reverse short channel effect* causes V_t to decrease with length.

There is also a *narrow channel effect* in which V_t varies with channel width; this effect tends to be less significant because the minimum width is greater than the minimum length.

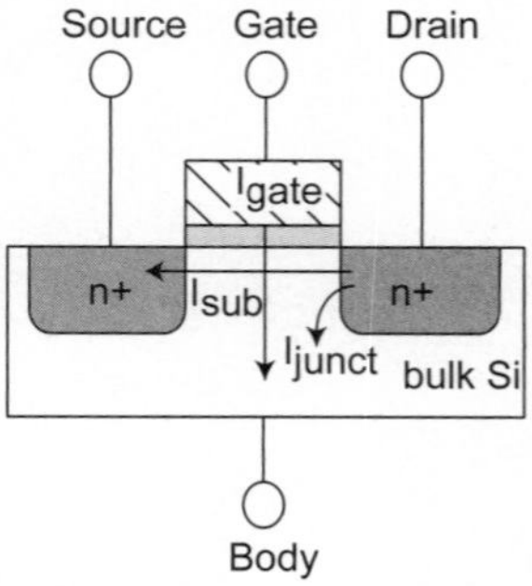

2.4.4 Leakage

Even when transistors are nominally OFF, they leak small amounts of current. Leakage mechanisms include subthreshold conduction between source and drain, gate leakage from the gate to body, and junction leakage from source to body and drain to body, as illustrated in Figure 2.19 [Roy03, Narendra06]. Subthreshold conduction is caused by thermal emission of carriers over the potential barrier set by the threshold. Gate leakage is a quantum-mechanical effect caused by tunneling through the extremely thin gate dielectric. Junction leakage is caused by current through the p-n junction between the source/drain diffusions and the body.

Source Gate Drain
I_{gate}
n+ I_{sub} n+
I_{junct} bulk Si
Body

FIGURE 2.19
Leakage current paths

[8] The term *short-channel effect* is overused in the CMOS literature. Sometimes, it refers to any behavior outside the long-channel models. Other times, it refers to a range of behaviors including DIBL that are most significant for very short channel lengths [Muller03]. In this text, we restrict the term to describe the sensitivity of threshold voltage to channel length.

In processes with feature sizes above 180 nm, leakage was typically insignificant except in very low power applications. In 90 and 65 nm processes, threshold voltage has reduced to the point that subthreshold leakage reaches levels of 1s to 10s of nA per transistor, which is significant when multiplied by millions or billions of transistors on a chip. In 45 nm processes, oxide thickness reduces to the point that gate leakage becomes comparable to subthreshold leakage unless high-k gate dielectrics are employed. Overall, leakage has become an important design consideration in nanometer processes.

2.4.4.1 Subthreshold Leakage The long-channel transistor I-V model assumes current only flows from source to drain when $V_{gs} > V_t$. In real transistors, current does not abruptly cut off below threshold, but rather drops off exponentially, as seen in Figure 2.20. When the gate voltage is high, the transistor is strongly ON. When the gate falls below V_t, the exponential decline in current appears as a straight line on the logarithmic scale. This regime of $V_{gs} < V_t$ is called *weak inversion*. The *subthreshold leakage current* increases significantly with V_{ds} because of drain-induced barrier lowering (see Section 2.4.3.2). There is a lower limit on I_{ds} set by drain junction leakage that is exacerbated by the negative gate voltage (see Section 2.4.4.3).

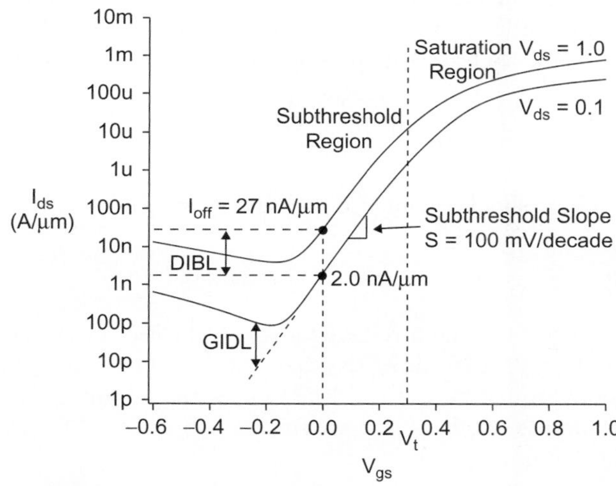

FIGURE 2.20 I-V characteristics of a 65 nm nMOS transistor at 70 °C on a log scale

Subthreshold leakage current is described by EQ (2.42). I_{ds0} is the current at threshold and is dependent on process and device geometry. It is typically extracted from simulation but can also be calculated from EQ (2.43); the $e^{1.8}$ term was found empirically [Sheu87]. n is a process-dependent term affected by the depletion region characteristics and is typically in the range of 1.3–1.7 for CMOS processes. The final term indicates that leakage is 0 if $V_{ds} = 0$, but increases to its full value when V_{ds} is a few multiples of the thermal voltage v_T (e.g., when $V_{ds} > 50$ mV). More significantly, drain-induced barrier lowering effectively reduces the threshold voltage, as indicated by the ηV_{ds} term. This can increase leakage by an order of magnitude for $V_{ds} = V_{DD}$ as compared to small V_{ds}. The body effect also modulates V_t when $V_{sb} \approx 0$.

$$I_{ds} = I_{ds0} e^{\frac{V_{gs} - V_{t0} + \eta V_{ds} - k_\gamma V_{sb}}{n v_T}} \left(1 - e^{\frac{-V_{ds}}{v_T}} \right) \tag{2.42}$$

$$I_{ds0} = \beta v_T^2 e^{1.8} \quad (2.43)$$

Subthreshold conduction is used to advantage in very low-power circuits, as will be explored in Section 8.6. It afflicts dynamic circuits and DRAMs, which depend on the storage of charge on a capacitor. Conduction through an OFF transistor discharges the capacitor unless it is periodically refreshed or a trickle of current is available to counter the leakage. Leakage also contributes to power dissipation in idle circuits. Subthreshold leakage increases exponentially as V_t decreases or as temperature rises, so it is a major problem for chips using low supply and threshold voltages and for chips operating at high temperature.

As shown in Figure 2.20, subthreshold current fits a straight line on a semilog plot. The inverse of the slope of this line is called the *subthreshold slope*, *S*

$$S = \left[\frac{d\left(\log_{10} I_{ds}\right)}{dV_{gs}} \right]^{-1} = nv_T \ln 10 \quad (2.44)$$

The subthreshold slope indicates how much the gate voltage must drop to decrease the leakage current by an order of magnitude. A typical value is 100 mV/decade at room temperature. EQ (2.42) can be rewritten using the subthreshold slope as

$$I_{ds} = I_{\text{off}} 10^{\frac{V_{gs}+\eta\left(V_{ds}-V_{dd}\right)-k_\gamma V_{sb}}{S}} \left(1 - e^{\frac{-V_{ds}}{v_T}}\right) \quad (2.45)$$

where I_{off} is the subthreshold current at $V_{gs} = 0$ and $V_{ds} = V_{DD}$.

Example 2.6

What is the minimum threshold voltage for which the leakage current through an OFF transistor ($V_{gs} = 0$) is 10^3 times less than that of a transistor that is barely ON ($V_{gs} = V_t$) at room temperature if $n = 1.5$? One of the advantages of silicon-on-insulator (SOI) processes is that they have smaller n (see Section 8.5). What threshold is required for SOI if $n = 1.3$?

SOLUTION: $v_T = 26$ mV at room temperature. Assume $V_{ds} >> v_T$ so leakage is significant. We solve

$$\begin{aligned} I_{ds}\left(V_{gs} = 0\right) &= 10^{-3} I_{ds0} = I_{ds0} e^{\frac{-V_t}{nv_T}} \\ V_t &= -nv_T \ln 10^{-3} = 270\text{mV} \end{aligned} \quad (2.46)$$

In the CMOS process, leakage rolls off by a factor of 10 for every 90 mV V_{gs} falls below threshold. This is often quoted as a subthreshold slope of $S = 90$ mV/decade. In the SOI process, the subthreshold slope S is 78 mV/decade, so a threshold of only 234 mV is required.

2.4.4.2 Gate Leakage According to quantum mechanics, the electron cloud surrounding an atom has a probabilistic spatial distribution. For gate oxides thinner than 15–20 Å, there is a nonzero probability that an electron in the gate will find itself on the wrong side of the oxide, where it will get whisked away through the channel. This effect of carriers crossing a thin barrier is called tunneling, and results in leakage current through the gate.

Two physical mechanisms for gate tunneling are called *Fowler-Nordheim* (FN) *tunneling* and *direct tunneling*. FN tunneling is most important at high voltage and moderate oxide thickness and is used to program EEPROM memories (see Section 11.4). Direct tunneling is most important at lower voltage with thin oxides and is the dominant leakage component.

The direct gate tunneling current can be estimated as [Chandrakasan01]

$$I_{\text{gate}} = WA\left(\frac{V_{DD}}{t_{\text{ox}}}\right)^2 e^{-B\frac{t_{\text{ox}}}{V_{DD}}} \quad \textbf{(2.47)}$$

where A and B are technology constants.

Transistors need high C_{ox} to deliver good ON current, driving the decrease in oxide thickness. Tunneling current drops exponentially with the oxide thickness and has only recently become significant. Figure 2.21 plots gate leakage current density (current/area) J_G against voltage for various oxide thicknesses. Gate leakage increases by a factor of 2.7 or more per angstrom reduction in thickness [Rohrer05]. Large tunneling currents impact not only dynamic nodes but also quiescent power consumption and thus limits equivalent oxide thicknesses t_{ox} to at least 10.5 Å to keep gate leakage below 100 A/cm^2. To keep these dimensions in perspective, recall that each atomic layer of SiO_2 is about 3 Å, so such gate oxides are a handful of atomic layers thick. Section 15.4.1.3 describes innovations in gate insulators with higher dielectric constants that offer good C_{ox} while reducing tunneling.

Tunneling current can be an order of magnitude higher for nMOS than pMOS transistors with SiO_2 gate dielectrics because the electrons tunnel from the conduction band while the holes tunnel from the valence band and see a higher barrier [Hamzaoglu02]. Different dielectrics may have different tunneling properties.

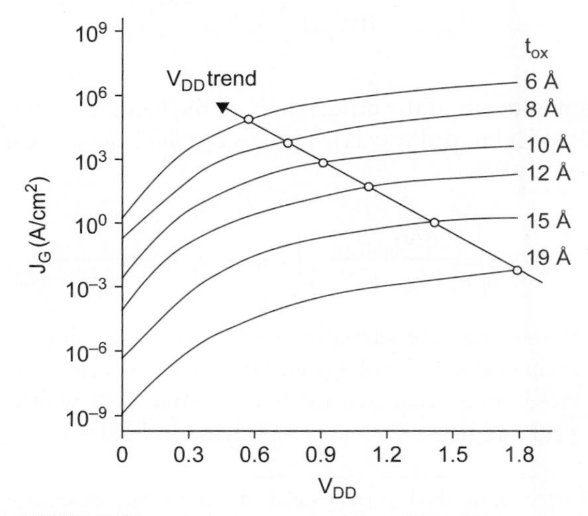

FIGURE 2.21 Gate leakage current from [Song01]

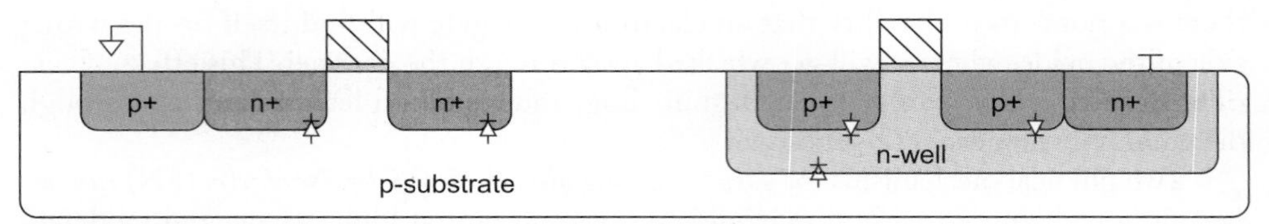

FIGURE 2.22 Substrate to diffusion diodes in CMOS circuits

2.4.4.3 Junction Leakage The p–n junctions between diffusion and the substrate or well form diodes, as shown in Figure 2.22. The well-to-substrate junction is another diode. The substrate and well are tied to GND or V_{DD} to ensure these diodes do not become forward biased in normal operation. However, reverse-biased diodes still conduct a small amount of current I_D.[9]

$$I_D = I_S\left(e^{\frac{V_D}{v_T}} - 1\right) \tag{2.48}$$

where I_S depends on doping levels and on the area and perimeter of the diffusion region and V_D is the diode voltage (e.g., $-V_{sb}$ or $-V_{db}$). When a junction is reverse biased by significantly more than the thermal voltage, the leakage is just $-I_S$, generally in the 0.1–0.01 fA/μm^2 range, which is negligible compared to other leakage mechanisms.

More significantly, heavily doped drains are subject to *band-to-band tunneling* (BTBT) and *gate-induced drain leakage* (GIDL).

BTBT occurs across the junction between the source or drain and the body when the junction is reverse-biased. It is a function of the reverse bias and the doping levels. High halo doping used to increase V_t to alleviate subthreshold leakage instead causes BTBT to grow. The leakage is exacerbated by *trap-assisted tunneling* (TAT) when defects in the silicon lattice called traps reduce the distance that a carrier must tunnel. Most of the leakage occurs along the sidewall closest to the channel where the doping is highest. It can be modeled as

$$I_{BTBT} = WX_j A \frac{E_j}{E_g^{0.5}} V_{dd}\, e^{-B\frac{E_g^{1.5}}{E_j}} \tag{2.49}$$

where X_j is the junction depth of the diffusion, E_g is the bandgap voltage, and A and B are technology constants [Mukhopadhyay05]. The electric field along the junction at a reverse bias of V_{DD} is

$$E_j = \sqrt{\frac{2qN_{halo}N_{sd}}{\varepsilon\left(N_{halo} + N_{sd}\right)}\left(V_{DD} + v_T \ln\frac{N_{halo}N_{sd}}{n_i^2}\right)} \tag{2.50}$$

GIDL occurs where the gate partially overlaps the drain. This effect is most pronounced when the drain is at a high voltage and the gate is at a low voltage. GIDL current is proportional to gate-drain overlap area and hence to transistor width. It is a strong function of the electric field and hence increases rapidly with the drain-to-gate voltage. How-

[9]Beware that I_D and I_S stand for the diode current and diode reverse-biased saturation currents, respectively. The D and S are not related to drain or source.

ever, it is normally insignificant at $|V_{gd}| \le V_{DD}$ [Mukhopadhyay05], only coming into play when the gate is driven outside the rails in an attempt to cut off subthreshold leakage.

2.4.5 Temperature Dependence

Transistor characteristics are influenced by temperature [Cobbold66, Vadasz66, Tsividis99, Gutierrez01]. Carrier mobility decreases with temperature. An approximate relation is

$$\mu(T) = \mu(T_r)\left(\frac{T}{T_r}\right)^{-k_\mu} \tag{2.51}$$

where T is the absolute temperature, T_r is room temperature, and k_μ is a fitting parameter with a typical value of about 1.5. v_{sat} also decreases with temperature, dropping by about 20% from 300 to 400 K.

The magnitude of the threshold voltage decreases nearly linearly with temperature and may be approximated by

$$V_t(T) = V_t(T_r) - k_{vt}(T - T_r) \tag{2.52}$$

where k_{vt} is typically about 1–2 mV/K.

I_{on} at high V_{DD} decreases with temperature. Subthreshold leakage increases exponentially with temperature. BTBT increases slowly with temperature, and gate leakage is almost independent of temperature.

The combined temperature effects are shown in Figure 2.23. At high V_{gs}, the current has a *negative temperature coefficient*; i.e., it decreases with temperature. At low V_{gs}, the current has a positive temperature coefficient. Thus, OFF current increases with temperature. ON current I_{dsat} normally decreases with temperature, as shown in Figure 2.24, so circuit performance is worst at high temperature. However, for systems operating at low V_{DD} (typically < 0.7 – 1.1 V), I_{dsat} increases with temperature [Kumar06].

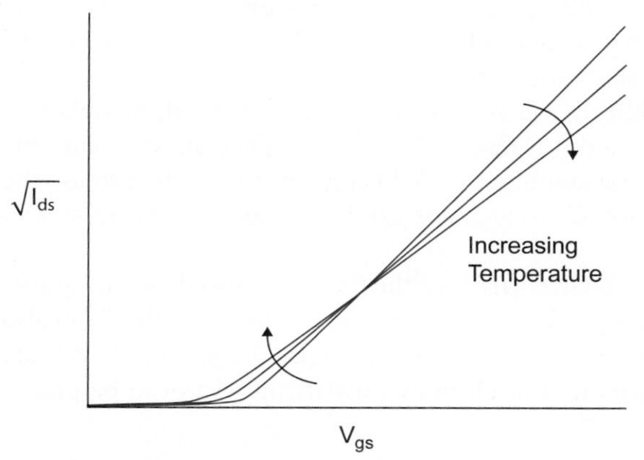

FIGURE 2.23 I–V characteristics of nMOS transistor in saturation at various temperatures

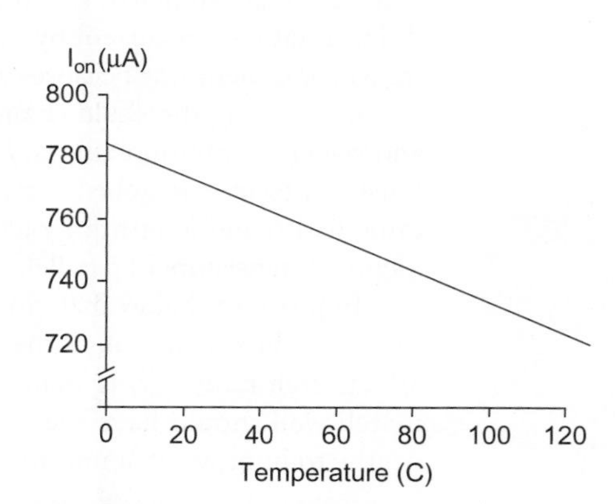

FIGURE 2.24 I_{dsat} vs. temperature

Conversely, circuit performance can be improved by cooling. Most systems use natural convection or fans in conjunction with heat sinks, but water cooling, thin-film refrigerators, or even liquid nitrogen can increase performance if the expense is justified. There are many advantages of operating at low temperature [Keyes70, Sun87]. Subthreshold leakage is exponentially dependent on temperature, so lower threshold voltages can be used. Velocity saturation occurs at higher fields, providing more current. As mobility is also higher, these fields are reached at a lower power supply, saving power. Depletion regions become wider, resulting in less junction capacitance.

Two popular lab tools for determining temperature dependence in circuits are a can of freeze spray and a heat gun. The former can be used to momentarily "freeze" a chip to see whether performance alters and the other, of course, can be used to heat up a chip. Often, these tests are done to quickly determine whether a chip is prone to temperature effects. Be careful—sometimes the sudden temperature change can fracture chips or their packages.

2.4.6 Geometry Dependence

The layout designer draws transistors with width and length W_{drawn} and L_{drawn}. The actual gate dimensions may differ by some factors X_W and X_L. For example, the manufacturer may create masks with narrower polysilicon or may overetch the polysilicon to provide shorter channels (negative X_L) without changing the overall design rules or metal pitch. Moreover, the source and drain tend to diffuse laterally under the gate by L_D, producing a shorter effective channel length that the carriers must traverse between source and drain. Similarly, W_D accounts for other effects that shrink the transistor width. Putting these factors together, we can compute effective transistor lengths and widths that should be used in place of L and W in the current and capacitance equations given elsewhere in the book. The factors of two come from lateral diffusion on both sides of the channel.

$$\begin{aligned} L_{\text{eff}} &= L_{\text{drawn}} + X_L - 2L_D \\ W_{\text{eff}} &= W_{\text{drawn}} + X_W - 2W_D \end{aligned} \tag{2.53}$$

Therefore, a transistor drawn twice as long may have an effective length that is more than twice as great. Similarly, two transistors differing in drawn widths by a factor of two may differ in saturation current by more than a factor of two. Threshold voltages also vary with transistor dimensions because of the short and narrow channel effects.

Combining threshold changes, effective channel lengths, channel length modulation, and velocity saturation effects, I_{dsat} does not scale exactly as $1/L$. In general, when currents must be precisely matched (e.g., in sense amplifiers or A/D converters), it is best to use the same width and length for each device. Current ratios can be produced by tying several identical transistors in parallel.

In processes below 0.25 μm, the effective length of the transistor also depends significantly on the orientation of the transistor. Moreover, the amount of nearby polysilicon also affects etch rates during manufacturing and thus channel length. Transistors that must match well should have the same orientation. Dummy polysilicon wires can be placed nearby to improve etch uniformity.

2.4.7 Summary

Although the physics of nanometer-scale devices is complicated, the impact of nonideal I-V behavior is fairly easy to understand from the designer's viewpoint.

Threshold drops Pass transistors suffer a threshold drop when passing the wrong value: nMOS transistors only pull up to $V_{DD} - V_{tn}$, while pMOS transistors only pull down to $|V_{tp}|$. The magnitude of the threshold drop is increased by the body effect. Therefore, pass transistors do not operate very well in nanometer processes where the threshold voltage is a significant fraction of the supply voltage. Fully complementary transmission gates should be used where both 0s and 1s must be passed well.

Leakage current Ideally, static CMOS gates draw zero current and dissipate zero power when idle. Real gates draw some leakage current. The most important source at this time is subthreshold leakage between source and drain of a transistor that should be cut off. The subthreshold current of an OFF transistor decreases by an order of magnitude for every 60–100 mV that V_{gs} is below V_t. Threshold voltages have been decreasing, so subthreshold leakage has been increasing dramatically. Some processes offer multiple choices of V_t: low-V_t devices are used for high performance in critical circuits, while high-V_t devices are used for low leakage elsewhere.

The transistor gate is a good insulator. However, significant tunneling current flows through very thin gates. This has limited the scaling of gate oxide and led to new high-k gate dielectrics.

Leakage current causes CMOS gates to consume power when idle. It also limits the amount of time that data is retained in dynamic logic, latches, and memory cells. In nanometer processes, dynamic logic and latches require some sort of feedback to prevent data loss from leakage. Leakage increases at high temperature.

V_{DD} Velocity saturation and mobility degradation result in less current than expected at high voltage. This means that there is no point in trying to use a high V_{DD} to achieve fast transistors, so V_{DD} has been decreasing with process generation to reduce power consumption. Moreover, the very short channels and thin gate oxides would be damaged by high V_{DD}.

Delay Transistors in series drop part of the voltage across each transistor and thus experience smaller fields and less velocity saturation than single transistors. Therefore, series transistors tend to be a bit faster than a simple model would predict. For example, two nMOS transistors in series deliver more than half the current of a single nMOS transistor of the same width. This effect is more pronounced for nMOS transistors than pMOS transistors because nMOS transistors have higher mobility to begin with and thus are more velocity saturated.

Matching If two transistors should behave identically, both should have the same dimensions and orientation and be interdigitated if possible.

2.5 DC Transfer Characteristics

Digital circuits are merely analog circuits used over a special portion of their range. The DC transfer characteristics of a circuit relate the output voltage to the input voltage, assuming the input changes slowly enough that capacitances have plenty of time to charge or discharge. Specific ranges of input and output voltages are defined as valid 0 and 1 logic levels. This section explores the DC transfer characteristics of CMOS gates and pass transistors.

2.5.1 Static CMOS Inverter DC Characteristics

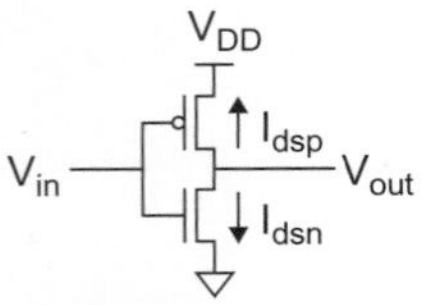

FIGURE 2.25 A CMOS inverter

Let us derive the DC transfer function (V_{out} vs. V_{in}) for the static CMOS inverter shown in Figure 2.25. We begin with Table 2.2, which outlines various regions of operation for the n- and p-transistors. In this table, V_{tn} is the threshold voltage of the n-channel device, and V_{tp} is the threshold voltage of the p-channel device. Note that V_{tp} is negative. The equations are given both in terms of V_{gs}/V_{ds} and V_{in}/V_{out}. As the source of the nMOS transistor is grounded, $V_{gsn} = V_{in}$ and $V_{dsn} = V_{out}$. As the source of the pMOS transistor is tied to V_{DD}, $V_{gsp} = V_{in} - V_{DD}$ and $V_{dsp} = V_{out} - V_{DD}$.

TABLE 2.2 Relationships between voltages for the three regions of operation of a CMOS inverter

	Cutoff	Linear	Saturated
nMOS	$V_{gsn} < V_{tn}$	$V_{gsn} > V_{tn}$	$V_{gsn} > V_{tn}$
	$V_{in} < V_{tn}$	$V_{in} > V_{tn}$	$V_{in} > V_{tn}$
		$V_{dsn} < V_{gsn} - V_{tn}$	$V_{dsn} > V_{gsn} - V_{tn}$
		$V_{out} < V_{in} - V_{tn}$	$V_{out} > V_{in} - V_{tn}$
pMOS	$V_{gsp} > V_{tp}$	$V_{gsp} < V_{tp}$	$V_{gsp} < V_{tp}$
	$V_{in} > V_{tp} + V_{DD}$	$V_{in} < V_{tp} + V_{DD}$	$V_{in} < V_{tp} + V_{DD}$
		$V_{dsp} > V_{gsp} - V_{tp}$	$V_{dsp} < V_{gsp} - V_{tp}$
		$V_{out} > V_{in} - V_{tp}$	$V_{out} < V_{in} - V_{tp}$

The objective is to find the variation in output voltage (V_{out}) as a function of the input voltage (V_{in}). This may be done graphically, analytically (see Exercise 2.16), or through simulation [Carr72]. Given V_{in}, we must find V_{out} subject to the constraint that $I_{dsn} = |I_{dsp}|$. For simplicity, we assume $V_{tp} = -V_{tn}$ and that the pMOS transistor is 2–3 times as wide as the nMOS transistor so $\beta_n = \beta_p$. We relax this assumption in Section 2.5.2.

We commence with the graphical representation of the simple algebraic equations described by EQ (2.10) for the two transistors shown in Figure 2.26(a). The plot shows I_{dsn} and I_{dsp} in terms of V_{dsn} and V_{dsp} for various values of V_{gsn} and V_{gsp}. Figure 2.26(b) shows the same plot of I_{dsn} and $|I_{dsp}|$ now in terms of V_{out} for various values of V_{in}. The possible operating points of the inverter, marked with dots, are the values of V_{out} where $I_{dsn} = |I_{dsp}|$ for a given value of V_{in}. These operating points are plotted on V_{out} vs. V_{in} axes in Figure 2.26(c) to show the inverter DC transfer characteristics. The supply current $I_{DD} = I_{dsn} = |I_{dsp}|$ is also plotted against V_{in} in Figure 2.26(d) showing that both transistors are momentarily ON as V_{in} passes through voltages between GND and V_{DD}, resulting in a pulse of current drawn from the power supply.

The operation of the CMOS inverter can be divided into five regions indicated on Figure 2.26(c). The state of each transistor in each region is shown in Table 2.3. In region *A*, the nMOS transistor is OFF so the pMOS transistor pulls the output to V_{DD}. In region *B*, the nMOS transistor starts to turn ON, pulling the output down. In region *C*, both transistors are in saturation. Notice that ideal transistors are only in region *C* for $V_{in} = V_{DD}/2$ and that the slope of the transfer curve in this example is $-\infty$ in this region, corresponding to infinite gain. Real transistors have finite output resistances on account of channel length modulation, described in Section 2.4.2, and thus have finite slopes over a broader region *C*. In region *D*, the pMOS transistor is partially ON and in region *E*, it is completely

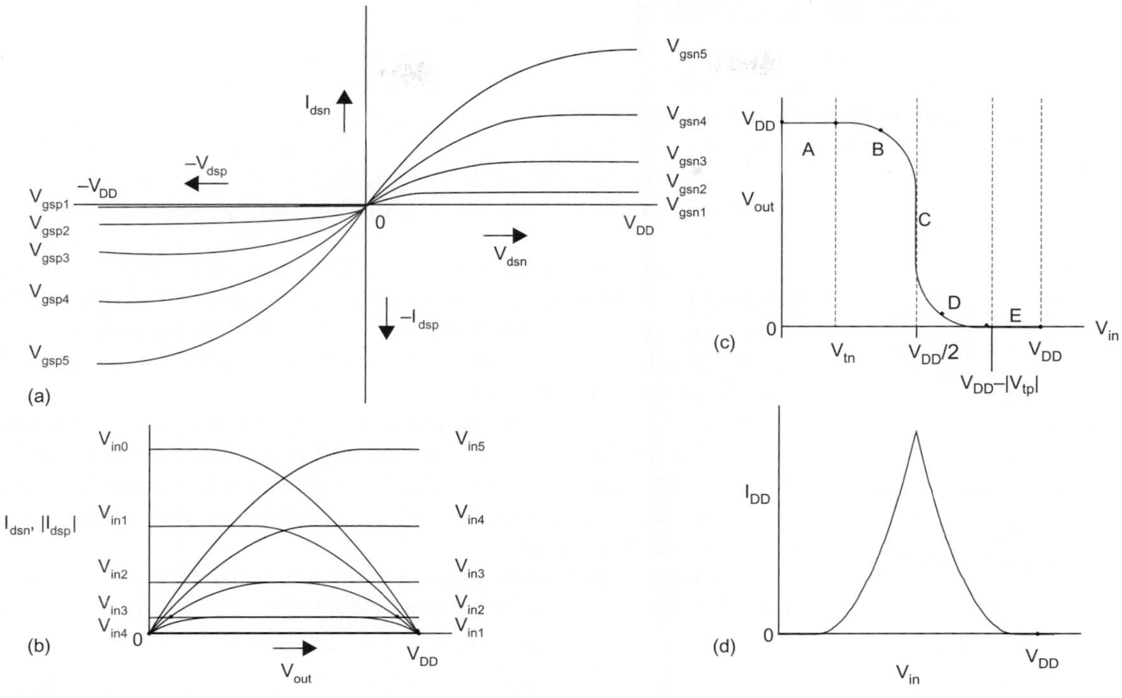

FIGURE 2.26 Graphical derivation of CMOS inverter DC characteristic

OFF, leaving the nMOS transistor to pull the output down to GND. Also notice that the inverter's current consumption is ideally zero, neglecting leakage, when the input is within a threshold voltage of the V_{DD} or GND rails. This feature is important for low-power operation.

TABLE 2.3 Summary of CMOS inverter operation

Region	Condition	p-device	n-device	Output
A	$0 \le V_{in} < V_{tn}$	linear	cutoff	$V_{out} = V_{DD}$
B	$V_{tn} \le V_{in} < V_{DD}/2$	linear	saturated	$V_{out} > V_{DD}/2$
C	$V_{in} = V_{DD}/2$	saturated	saturated	V_{out} drops sharply
D	$V_{DD}/2 < V_{in} \le V_{DD} - \|V_{tp}\|$	saturated	linear	$V_{out} < V_{DD}/2$
E	$V_{in} > V_{DD} - \|V_{tp}\|$	cutoff	linear	$V_{out} = 0$

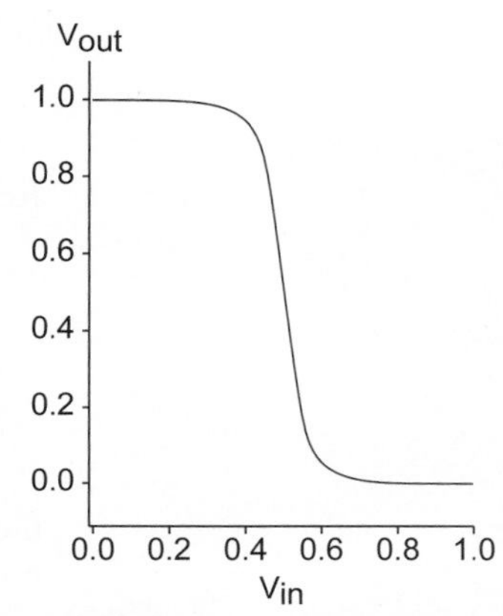

FIGURE 2.27 Simulated CMOS inverter DC characteristic

Figure 2.27 shows simulation results of an inverter from a 65 nm process. The pMOS transistor is twice as wide as the nMOS transistor to achieve approximately equal betas. Simulation matches the simple models reasonably well, although the transition is not quite as steep because transistors are not ideal current sources in saturation.

The crossover point where $V_{inv} = V_{in} = V_{out}$ is called the *input threshold*. Because both mobility and the magnitude of the threshold voltage decrease with temperature for nMOS and pMOS transistors, the input threshold of the gate is only weakly sensitive to temperature.

2.5.2 Beta Ratio Effects

We have seen that for $\beta_p = \beta_n$, the inverter threshold voltage V_{inv} is $V_{DD}/2$. This may be desirable because it maximizes noise margins (see Section 2.5.3) and allows a capacitive load to charge and discharge in equal times by providing equal current source and sink capabilities (see Section 3.2). Inverters with different beta ratios $r = \beta_p / \beta_n$ are called *skewed* inverters [Sutherland99]. If $r > 1$, the inverter is *HI-skewed*. If $r < 1$, the inverter is *LO-skewed*. If $r = 1$, the inverter has normal skew or is *unskewed*.

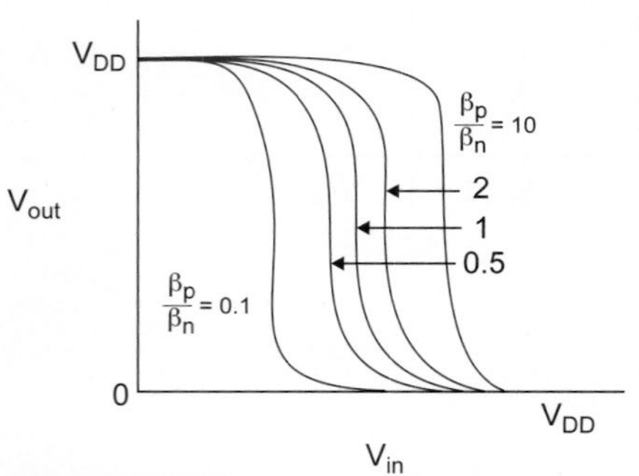

FIGURE 2.28 Transfer characteristics of skewed inverters

A HI-skew inverter has a stronger pMOS transistor. Therefore, if the input is $V_{DD}/2$, we would expect the output will be greater than $V_{DD}/2$. In other words, the input threshold must be higher than for an unskewed inverter. Similarly, a LO-skew inverter has a weaker pMOS transistor and thus a lower switching threshold.

Figure 2.28 explores the impact of skewing the beta ratio on the DC transfer characteristics. As the beta ratio is changed, the switching threshold moves. However, the output voltage transition remains sharp. Gates are usually skewed by adjusting the widths of transistors while maintaining minimum length for speed.

The inverter threshold can also be computed analytically. If the long-channel models of EQ (2.10) for saturated transistors are valid:

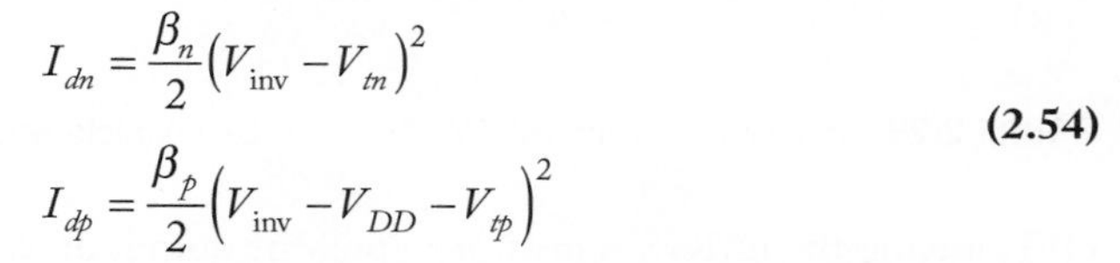

$$I_{dn} = \frac{\beta_n}{2}\left(V_{\text{inv}} - V_{tn}\right)^2$$
$$I_{dp} = \frac{\beta_p}{2}\left(V_{\text{inv}} - V_{DD} - V_{tp}\right)^2 \tag{2.54}$$

By setting the currents to be equal and opposite, we can solve for V_{inv} as a function of r:

$$V_{\text{inv}} = \frac{V_{DD} + V_{tp} + V_{tn}\sqrt{\frac{1}{r}}}{1 + \sqrt{\frac{1}{r}}} \tag{2.55}$$

In the limit that the transistors are fully velocity saturated, EQ (2.29) shows

$$I_{dn} = W_n C_{\text{ox}} v_{\text{sat}-n}(V_{\text{inv}} - V_{tn})$$
$$I_{dp} = W_p C_{\text{ox}} v_{\text{sat}-p}(V_{\text{inv}} - V_{DD} - V_{tp}) \tag{2.56}$$

Redefining $r = W_p v_{\text{sat-}p} / W_n v_{\text{sat-}n}$, we can again find the inverter threshold

$$V_{\text{inv}} = \frac{V_{DD} + V_{tp} + V_{tn}\frac{1}{r}}{1 + \frac{1}{r}} \tag{2.57}$$

In either case, if $V_{tn} = -V_{tp}$ and $r = 1$, $V_{\text{inv}} = V_{DD}/2$ as expected. However, velocity saturated inverters are more sensitive to skewing because their DC transfer characteristics are not as sharp.

DC transfer characteristics of other static CMOS gates can be understood by collapsing the gates into an equivalent inverter. Series transistors can be viewed as a single transistor of greater length. If only one of several parallel transistors is ON, the other transistors can be ignored. If several parallel transistors are ON, the collection can be viewed as a single transistor of greater width.

2.5.3 Noise Margin

Noise margin is closely related to the DC voltage characteristics [Wakerly00]. This parameter allows you to determine the allowable noise voltage on the input of a gate so that the output will not be corrupted. The specification most commonly used to describe noise margin (or *noise immunity*) uses two parameters: the *LOW* noise margin, NM_L, and the *HIGH* noise margin, NM_H. With reference to Figure 2.29, NM_L is defined as the difference in maximum LOW input voltage recognized by the receiving gate and the maximum LOW output voltage produced by the driving gate.

$$NM_L = V_{IL} - V_{OL} \tag{2.58}$$

The value of NM_H is the difference between the minimum HIGH output voltage of the driving gate and the minimum HIGH input voltage recognized by the receiving gate. Thus,

$$NM_H = V_{OH} - V_{IH} \tag{2.59}$$

where

V_{IH} = minimum HIGH input voltage
V_{IL} = maximum LOW input voltage
V_{OH} = minimum HIGH output voltage
V_{OL} = maximum LOW output voltage

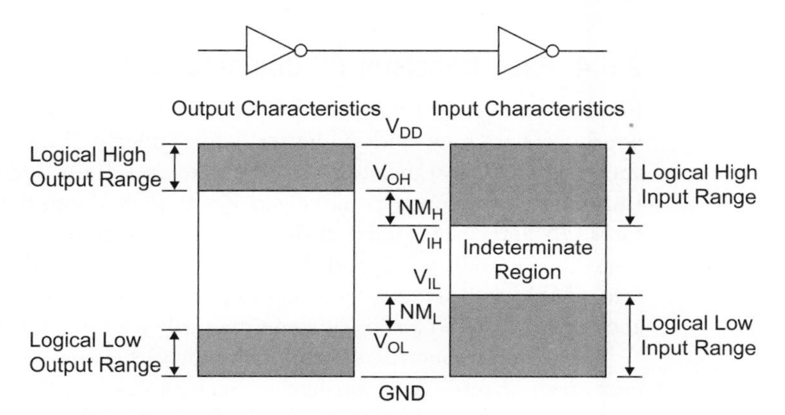

FIGURE 2.29 Noise margin definitions

Inputs between V_{IL} and V_{IH} are said to be in the *indeterminate region* or *forbidden zone* and do not represent legal digital logic levels. Therefore, it is generally desirable to have V_{IH} as close as possible to V_{IL} and for this value to be midway in the "logic swing," V_{OL} to V_{OH}. This implies that the transfer characteristic should switch abruptly; that is, there should be high gain in the transition region. For the purpose of calculating noise margins, the transfer characteristic of the inverter and the definition of voltage levels V_{IL}, V_{OL}, V_{IH}, and V_{OH} are shown in Figure 2.30. Logic levels are defined at the unity gain point where

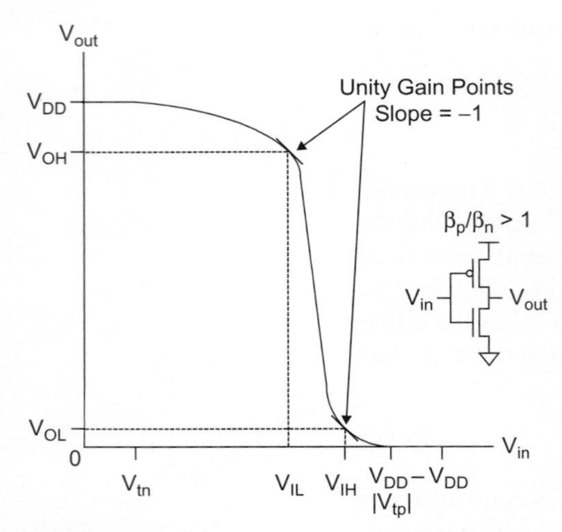

FIGURE 2.30 CMOS inverter noise margins

the slope is −1. This gives a conservative bound on the worst case static noise margin [Hill68, Lohstroh83, Shepard99]. For the inverter shown, the NM_L is 0.46 V_{DD} while the NM_H is 0.13 V_{DD}. Note that the output is slightly degraded when the input is at its worst legal value; this is called *noise feedthrough* or *propagated noise.* The exercises at the end of the chapter examine graphical and analytical approaches of finding the logic levels and noise margins.

If either NM_L or NM_H for a gate are too small, the gate may be disturbed by noise that occurs on the inputs. An unskewed gate has equal noise margins, which maximizes immunity to arbitrary noise sources. If a gate sees more noise in the high or low input state, the gate can be skewed to improve that noise margin at the expense of the other. Note that if $|V_{tp}| = V_{tn}$, then NM_H and NM_L increase as threshold voltages are increased.

Quite often, noise margins are compromised to improve speed. Circuit examples in Chapter 8 will illustrate this trade-off. Noise sources tend to scale with the supply voltage, so noise margins are best given as a fraction of the supply voltage. A noise margin of 0.4 V is quite comfortable in a 1.8 V process, but marginal in a 5 V process.

DC analysis gives us the *static noise margins* specifying the level of noise that a gate may see for an indefinite duration. Larger noise pulses may be acceptable if they are brief; these are described by *dynamic noise margins* specified by a maximum amplitude as a function of the duration [Lohstroh79, Somasekhar00]. Unfortunately, there is no simple amplitude-duration product that conveniently specifies dynamic noise margins.

2.5.4 Pass Transistor DC Characteristics

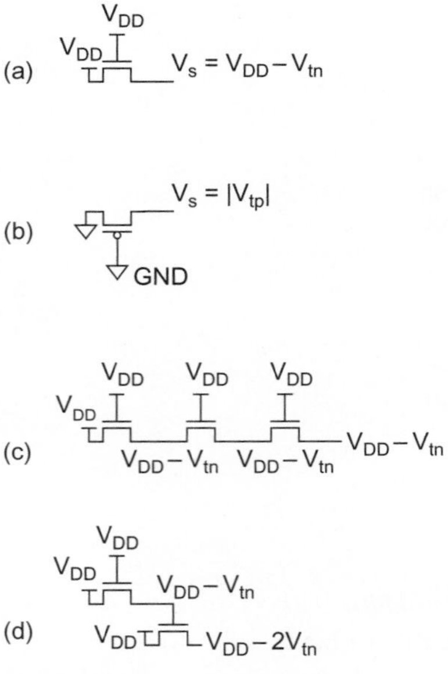

Recall from Section 1.4.6 that nMOS transistors pass '0's well but 1s poorly. We are now ready to better define "poorly." Figure 2.31(a) shows an nMOS transistor with the gate and drain tied to V_{DD}. Imagine that the source is initially at $V_s = 0$. $V_{gs} > V_{tn}$, so the transistor is ON and current flows. If the voltage on the source rises to $V_s = V_{DD} - V_{tn}$, V_{gs} falls to V_{tn} and the transistor cuts itself OFF. Therefore, nMOS transistors attempting to pass a 1 never pull the source above $V_{DD} - V_{tn}$.[10] This loss is sometimes called a *threshold drop.*

Moreover, when the source of the nMOS transistor rises, V_{sb} becomes nonzero. As described in Section 2.4.3.1, this nonzero source to body potential introduces the body effect that increases the threshold voltage. Using the data from the example in that section, a pass transistor driven with V_{DD} = 1 V would produce an output of only 0.65 V, potentially violating the noise margins of the next stage.

Similarly, pMOS transistors pass 1s well but 0s poorly. If the pMOS source drops below $|V_{tp}|$, the transistor cuts off. Hence, pMOS transistors only pull down to within a threshold above GND, as shown in Figure 2.31(b).

(a) $V_s = V_{DD} - V_{tn}$

(b) $V_s = |V_{tp}|$

(c) $V_{DD} - V_{tn}$

(d) $V_{DD} - V_{tn}$, $V_{DD} - 2V_{tn}$

FIGURE 2.31 Pass transistor threshold drops

[10] Technically, the output can rise higher very slowly by means of subthreshold leakage.

As the source can rise to within a threshold voltage of the gate, the output of several transistors in series is no more degraded than that of a single transistor (Figure 2.31(c)). However, if a degraded output drives the gate of another transistor, the second transistor can produce an even further degraded output (Figure 2.31(d)).

If we attempt to use a transistor as a switch, the threshold drop degrades the output voltage. In old processes where the power supply voltage was high and V_t was a small fraction of V_{DD}, the drop was tolerable. In modern processes where V_t is closer to 1/3 of V_{DD}, the threshold drop can produce an invalid or marginal logic level at the output. To solve this problem, CMOS switches are generally built using transmission gates.

Recall from Section 1.4.6 that a transmission gate consists of an nMOS transistor and a pMOS transistor in parallel with gates controlled by complementary signals. When the transmission gate is ON, at least one of the two transistors is ON for any output voltage and hence, the transmission gate passes both 0s and 1s well. The transmission gate is a fundamental and ubiquitous component in MOS logic. It finds use as a multiplexing element, a logic structure, a latch element, and an analog switch. The transmission gate acts as a voltage-controlled switch connecting the input and the output.

2.6 Pitfalls and Fallacies

This section lists a number of pitfalls and fallacies that can deceive the novice (or experienced) designer.

Blindly trusting one's models

Models should be viewed as only approximations to reality, not reality itself, and used within their limitations. In particular, simple models like the Shockley or RC models aren't even close to accurate fits for the I-V characteristics of a modern transistor. They are valuable for the insight they give on trends (i.e., making a transistor wider increases its gate capacitance and decreases its ON resistance), not for the absolute values they predict. Cutting-edge projects often target processes that are still under development, so these models should only be viewed as speculative. Finally, processes may not be fully characterized over all operating regimes; for example, don't assume that your models are accurate in the subthreshold region unless your vendor tells you so. Having said this, modern SPICE models do an extremely good job of predicting performance well into the GHz range for well-characterized processes and models when using proper design practices (such as accounting for temperature, voltage, and process variation).

Using excessively complicated models for manual calculations

Because models cannot be perfectly accurate, there is little value in using excessively complicated models, particularly for hand calculations. Simpler models give more insight on key trade-offs and more rapid feedback during design. Moreover, RC models calibrated against simulated data for a fabrication process can estimate delay just as accurately as elaborate models based on a large number of physical parameters but not calibrated to the process.

Assuming a transistor with twice the drawn length has exactly half the current

To first order, current is proportional to W/L. In modern transistors, the effective transistor length is usually shorter than the drawn length, so doubling the drawn length reduces current by more than a factor of two. Moreover, the threshold voltage tends to increase for longer transistors, resulting in less current. Therefore, it is a poor strategy to try to ratio currents by ratioing transistor lengths.

Assuming two transistors in series deliver exactly half the current of a single transistor
To first order, this would be true. However, each series transistor sees a smaller electric field across the channel and hence are each less velocity saturated. Therefore, two series transistors in a nanometer process will deliver more than half the current of a single transistor. This is more pronounced for nMOS than pMOS transistors because of the higher mobility and the higher degree of velocity saturation of electrons than holes at a given field. Hence, NAND gates perform better than first order estimates might predict.

Ignoring leakage
In contemporary processes, subthreshold and gate leakage can be quite significant. Leakage is exacerbated by high temperature and by random process variations. Undriven nodes will not retain their state for long; they will leak to some new voltage. Leakage power can account for a large fraction of total power, especially in battery-operated devices that are idle most of the time.

Using nMOS pass transistors
nMOS pass transistors only pull up to $V_{DD} - V_t$. This voltage may fall below V_{IH} of a receiver, especially as V_{DD} decreases. For example, one author worked with a scan latch containing an nMOS pass transistor that operated correctly in a 250 nm process at 2.5 V. When the latch was ported to a 180 nm process at 1.8 V, the scan chain stopped working. The problem was traced to the pass transistor and the scan chain was made operational in the lab by raising V_{DD} to 2 V. A better solution is to use transmission gates in place of pass transistors.

Summary

In summary, we have seen that MOS transistors are four-terminal devices with a gate, source, drain, and body. In normal operation, the body is tied to GND or V_{DD} so the transistor can be modeled as a three-terminal device. The transistor behaves as a voltage-controlled switch. An nMOS switch is OFF (no path from source to drain) when the gate voltage is below some threshold V_t. The switch turns ON, forming a channel connecting source to drain, when the gate voltage rises above V_t. This chapter has developed more elaborate models to predict the amount of current that flows when the transistor is ON. The transistor operates in three modes depending on the terminal voltages:

- $V_{gs} < V_t$ Cutoff $I_{ds} \approx 0$
- $V_{gs} > V_t$, $V_{ds} < V_{dsat}$ Linear I_{ds} increases with V_{ds} (like a resistor)
- $V_{gs} > V_t$, $V_{ds} > V_{dsat}$ Saturation I_{ds} constant (like a current source)

In a long-channel transistor, the saturation current depends on V_{GT}^2. pMOS transistors are similar to nMOS transistors, but have the signs reversed and deliver about half the current because of lower mobility.

In a real transistor, the I-V characteristics are more complicated. Modern transistors are extraordinarily small and thus experience enormous electric fields even at low voltage. The high fields cause velocity saturation and mobility degradation that lead to less current than you might otherwise expect. This can be modeled as a saturation current dependent on V_{GT}^{α}, where the velocity saturation index α is less than 2. Moreover, the saturation current does increase slightly with V_{ds} because of channel length modulation. Although simple hand calculations are no longer accurate, the general shape does not change very much and the transfer characteristics can still be derived using graphical or simulation methods.

Even when the gate voltage is low, the transistor is not completely OFF. Subthreshold current through the channel drops off exponentially for $V_{gs} < V_t$, but is nonnegligible for transistors with low thresholds. Junction leakage currents flow through the reverse-biased p–n junctions. Tunneling current flows through the insulating gate when the oxide becomes thin enough.

We can derive the DC transfer characteristics and noise margins of logic gates using either analytical expressions or a graphical load line analysis or simulation. Static CMOS gates have excellent noise margins.

Unlike ideal switches, MOS transistors pass some voltage levels better than others. An nMOS transistor passes 0s well, but only pulls up to $V_{DD} - V_{tn}$ when passing 1s. The pMOS passes 1s well, but only pulls down to $|V_{tp}|$ when passing 0s. This threshold drop is exacerbated by the body effect, which increases the threshold voltage when the source is at a different potential than the body.

There are too many parameters in a modern BSIM model for a designer to deal with intuitively. Instead, CMOS transistors are usually characterized by the following basic figures of merit:

- V_{DD} — Target supply voltage
- $L_{\text{gate}} / L_{\text{poly}}$ — Effective channel length (< feature size)
- t_{ox} — Effective oxide thickness (a.k.a. EOT)
- I_{dsat} — I_{ds} @ $V_{gs} = V_{ds} = V_{DD}$
- I_{off} — I_{ds} @ $V_{gs} = 0$, $V_{ds} = V_{DD}$
- I_g — Gate leakage @ $V_{gs} = V_{DD}$

[Muller03] and [Tsividis99] offer comprehensive treatments of device physics at a more advanced level. [Gray01] describes MOSFET models in more detail from the analog designer's point of view.

Exercises

2.1 Sometimes the substrate is connected to a voltage called the substrate bias to alter the threshold of the nMOS transistors. If the threshold of an nMOS transistor is to be raised, should a positive or negative substrate bias be used?

2.2 Peter Pitfall is offering to license to you his patented noninverting buffer circuit shown in Figure 2.32. Graphically derive the transfer characteristics for this buffer. Assume $\beta_n = \beta_p = \beta$ and $V_{tn} = |V_{tp}| = V_t$. Why is it a bad circuit idea?

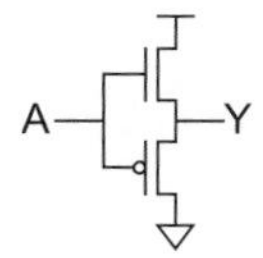

FIGURE 2.32 Noninverting buffer

2.3 Consider an nMOS transistor in a 0.6 μm process with $W/L = 4/2\ \lambda$ (i.e., 1.2/0.6 μm). In this process, the gate oxide thickness is 100 Å and the mobility of electrons is 350 cm^2/V · s. The threshold voltage is 0.7 V. Plot I_{ds} vs. V_{ds} for V_{gs} = 0, 1, 2, 3, 4, and 5 V.

2.4 Show that the current through two transistors in series is equal to the current through a single transistor of twice the length if the transistors are well described by the Shockley model. Specifically, show that $I_{DS1} = I_{DS2}$ in Figure 2.33 when the transistors are in their linear region: $V_{DS} < V_{DD} - V_t$, $V_{DD} > V_t$ (this is also true in saturation). *Hint:* Express the currents of the series transistors in terms of V_1 and solve for V_1.

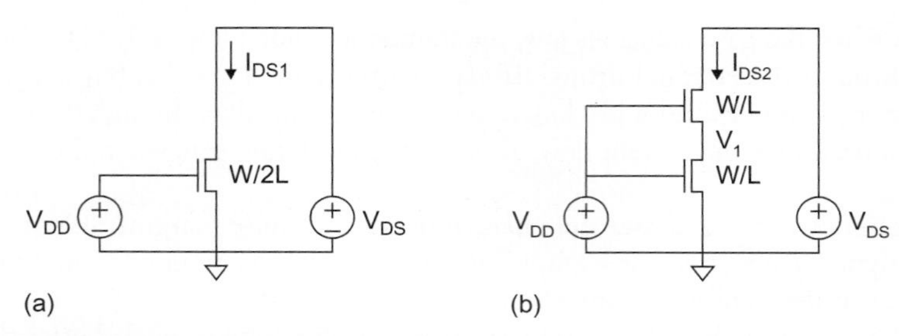

FIGURE 2.33 Current in series transistors

2.5 In Exercise 2.4, the body effect was ignored. If the body effect is considered, will I_{DS2} be equal to, greater than, or less than I_{DS1}? Explain.

2.6 A 90 nm long transistor has a gate oxide thickness of 16 Å. What is its gate capacitance per micron of width?

2.7 Calculate the diffusion parasitic C_{db} of the drain of a unit-sized contacted nMOS transistor in a 0.6 μm process when the drain is at 0 and at $V_{DD} = 5$ V. Assume the substrate is grounded. The transistor characteristics are $CJ = 0.42$ fF/μm^2, $MJ =$ 0.44, $CJSW = 0.33$ fF/μm, $MJSW = 0.12$, and $\psi_0 = 0.98$ V at room temperature.

2.8 Prove EQ (2.27).

2.9 Consider the nMOS transistor in a 0.6 μm process with gate oxide thickness of 100 Å. The doping level is $N_A = 2 \times 10^{17}$ cm^{-3} and the nominal threshold voltage is 0.7 V. The body is tied to ground with a substrate contact. How much does the threshold change at room temperature if the source is at 4 V instead of 0?

2.10 Does the body effect of a process limit the number of transistors that can be placed in series in a CMOS gate at low frequencies?

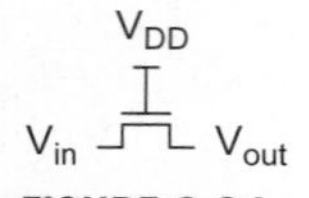

FIGURE 2.34 Single pass transistor

2.11 Suppose $V_{DD} = 1.2$ V and $V_t = 0.4$ V. Determine V_{out} in Figure 2.34 for the following. Neglect the body effect.

a) $V_{in} = 0$ V

b) $V_{in} = 0.6$ V

c) $V_{in} = 0.9$ V

d) $V_{in} = 1.2$ V.

2.12 An nMOS transistor has a threshold voltage of 0.4 V and a supply voltage of $V_{DD} =$ 1.2 V. A circuit designer is evaluating a proposal to reduce V_t by 100 mV to obtain faster transistors.

a) By what factor would the saturation current increase (at $V_{gs} = V_{ds} = V_{DD}$) if the transistor were ideal?

b) By what factor would the subthreshold leakage current increase at room temperature at $V_{gs} = 0$? Assume $n = 1.4$.

c) By what factor would the subthreshold leakage current increase at 120 °C? Assume the threshold voltage is independent of temperature.

2.13 Find the subthreshold leakage current of an inverter at room temperature if the input $A = 0$. Let $\beta_n = 2\beta_p = 1$ mA/V^2, $n = 1.0$, and $|V_t| = 0.4$ V. Assume the body effect and DIBL coefficients are $\gamma = \eta = 0$.

2.14 Repeat Exercise 2.13 for a NAND gate built from unit transistors with inputs $A = B = 0$. Show that the subthreshold leakage current through the series transistors is half that of the inverter if $n = 1$.

2.15 Repeat Exercises 2.13 and 2.14 when $\eta = 0.04$ and $V_{DD} = 1.8$ V, as in the case of a more realistic transistor. γ has a secondary effect, so assume that it is 0. Did the leakage currents go up or down in each case? Is the leakage through the series transistors more than half, exactly half, or less than half of that through the inverter?

2.16 Give an expression for the output voltage for the pass transistor networks shown in Figure 2.35. Neglect the body effect.

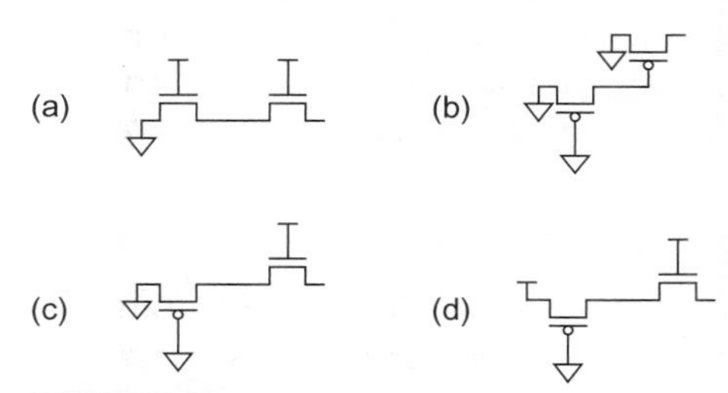

FIGURE 2.35 Pass transistor networks

2.17 A novel inverter has the transfer characteristics shown in Figure 2.36. What are the values of V_{IL}, V_{IH}, V_{OL}, and V_{OH} that give best noise margins? What are these high and low noise margins?

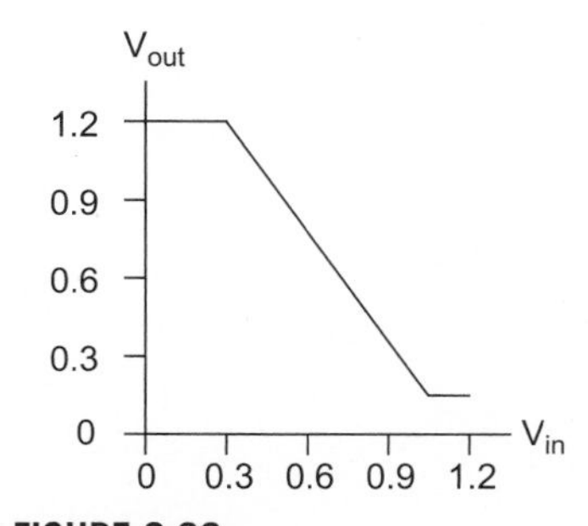

FIGURE 2.36 Transfer characteristics

2.18 Section 2.5.1 graphically determined the transfer characteristics of a static CMOS inverter. Derive analytic expressions for V_{out} as a function of V_{in} for regions B and D of the transfer function. Let $|V_{tp}| = V_{tn}$ and $\beta_p = \beta_n$.

2.19 Using the results from Exercise 2.18, calculate the noise margin for a CMOS inverter operating at 1.0 V with $V_{tn} = |V_{tp}| = 0.35$ V, $\beta_p = \beta_n$.

2.20 Repeat Exercise 2.18 if the thresholds and betas of the two transistors are not necessarily equal. Also solve for the value of V_{in} for region C where both transistors are saturated.

2.21 Using the results from Exercise 2.20, calculate the noise margin for a CMOS inverter operating at 1.0 V with $V_{tn} = |V_{tp}| = 0.35$ V, $\beta_p = 0.5\beta_n$.

Speed 3

3.1 Introduction

In Chapter 1 we learned how to make chips that work. Now we move on to making chips that work *well*. The two most common metrics for a good chip are speed and power, discussed in this chapter and Chapter 5, respectively. Delay and power are influenced as much by the wires as by the transistors, so Chapter 5 delves into interconnect analysis and design. A chip is of no value if it cannot reliably accomplish its function, so Chapter 6 examines how we achieve robustness in designs.

The most obvious way to characterize a circuit is through simulation, and that will be the topic of Chapter 7. Unfortunately, simulations only inform us how a particular circuit behaves, not how to change the circuit to make it better. There are far too many degrees of freedom in chip design to explore each promising path through simulation (although some may try). Moreover, if we don't know approximately what the result of the simulation should be, we are unlikely to catch the inevitable bugs in our simulation model. Mediocre engineers rely entirely on computer tools, but outstanding engineers develop their physical intuition to rapidly predict the behavior of circuits. In this chapter and the next two, we are primarily concerned with the development of simple models that will assist us in understanding system performance.

3.1.1 Definitions

We begin with a few definitions illustrated in Figure 3.1:

- *Propagation delay time*, t_{pd} = maximum time from the input crossing 50% to the output crossing 50%
- *Contamination delay time*, t_{cd} = minimum time from the input crossing 50% to the output crossing 50%
- *Rise time*, t_r = time for a waveform to rise from 20% to 80% of its steady-state value
- *Fall time*, t_f = time for a waveform to fall from 80% to 20% of its steady-state value
- *Edge rate*, $t_{rf} = (t_r + t_f)/2$

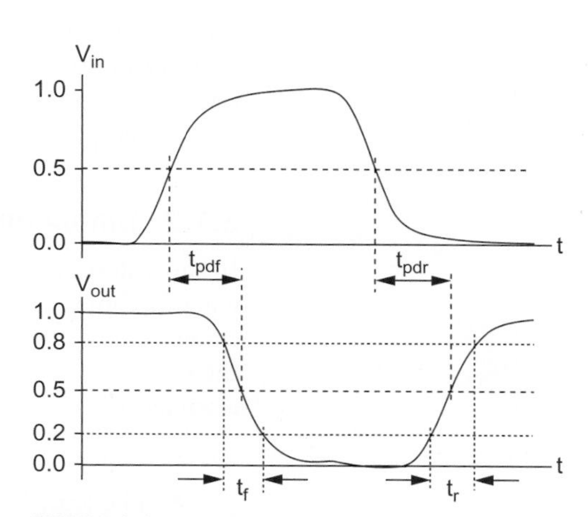

FIGURE 3.1 Propagation delay and rise/fall times

Intuitively, we know that when an input changes, the output will retain its old value for at least the contamination delay and take on its new value in at most the propagation delay. We sometimes differentiate between the delays for the output rising, t_{pdr}/t_{cdr}, and the output falling, t_{pdf}/t_{cdf}. Rise/fall times are also sometimes called *slopes* or *edge rates*. Propagation and contamination delay times are also called *max-time* and *min-time*, respectively. The gate that charges or discharges a node is called the *driver* and the gates and wire being driven are called the *load*. Propagation delay is usually the most relevant value of interest, and is often simply called *delay*.

A *timing analyzer* computes the arrival times, i.e., the latest time at which each node in a block of logic will switch. The nodes are classified as inputs, outputs, and internal nodes. The user must specify the *arrival time* of inputs and the time data is required at the outputs. The arrival time a_i at internal node i depends on the propagation delay of the gate driving i and the arrival times of the inputs to the gate:

$$a_i = \max_{j \in fanin(i)} \left\{ a_j \right\} + t_{pd_i} \tag{3.1}$$

The timing analyzer computes the arrival times at each node and checks that the outputs arrive by their required time. The *slack* is the difference between the required and arrival times. *Positive slack* means that the circuit meets timing. *Negative slack* means that the circuit is not fast enough. Figure 3.2 shows nodes annotated with arrival times. If the outputs are all required at 200 ps, the circuit has 60 ps of slack.

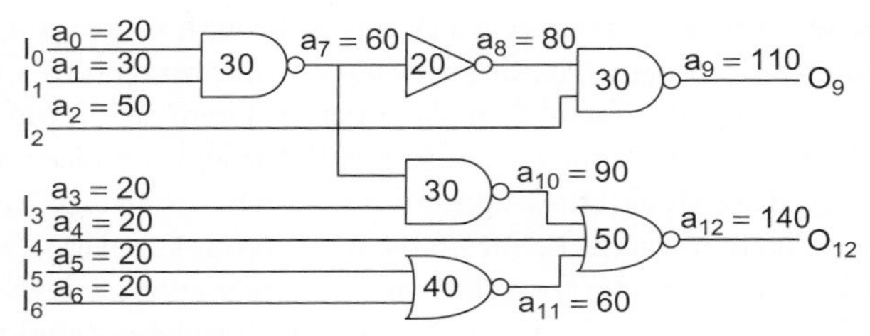

FIGURE 3.2 Arrival time example

A practical timing analyzer extends this arrival time model to account for a number of effects. Arrival times and propagation delays are defined separately for rising and falling transitions. The delay of a gate may be different from different inputs. Earliest arrival times can also be computed based on contamination delays. Considering all of these factors gives a window over which the gate may switch and allows the timing analyzer to verify that setup and hold times are satisfied at each register.

3.1.2 Timing Optimization

In most designs there will be many logic paths that do not require any conscious effort when it comes to speed. These paths are already fast enough for the timing goals of the system. However, there will be a number of *critical paths* that limit the operating speed of the system and require attention to timing details. The critical paths can be affected at four main levels:

- The architectural/microarchitectural level
- The logic level
- The circuit level
- The layout level

The most leverage is achieved with a good microarchitecture. This requires a broad knowledge of both the algorithms that implement the function and the technology being targeted, such as how many gate delays fit in a clock cycle, how quickly addition occurs, how fast memories are accessed, and how long signals take to propagate along a wire. Trade-offs at the microarchitectural level include the number of pipeline stages, the number of execution units (parallelism), and the size of memories.

The next level of timing optimization comes at the logic level. Trade-offs include types of functional blocks (e.g., ripple carry vs. lookahead adders), the number of stages of gates in the clock cycle, and the fan-in and fan-out of the gates. The transformation from function to gates and registers can be done by experience, by experimentation, or, most often, by logic synthesis. Remember, however, that no amount of skillful logic design can overcome a poor microarchitecture.

Once the logic has been selected, the delay can be tuned at the circuit level by choosing transistor sizes or using other styles of CMOS logic. Finally, delay is dependent on the layout. The floorplan (either manually or automatically generated) is of great importance because it determines the wire lengths that can dominate delay. Good cell layouts can also reduce parasitic capacitance.

Many RTL designers never venture below the microarchitectural level. A common design practice is to write RTL code, synthesize it (allowing the synthesizer to do the timing optimizations at the logic, circuit, and placement levels) and check if the results are fast enough. If they are not, the designer recodes the RTL with more parallelism or pipelining, or changes the algorithm and repeats until the timing constraints are satisfied. Timing analyzers are used to check *timing closure*, i.e., whether the circuit meets all of the timing constraints. Without an understanding of the lower levels of abstraction where the synthesizer is working, a designer may have a difficult time achieving timing closure on a challenging system.

This chapter focuses on the logic and circuit optimizations of selecting the number of stages of logic, the types of gates, and the transistor sizes. We begin by examining the transient response of an inverter. Using the device models from Chapter 2, we can write differential equations for voltage as a function of time to calculate delay. Unfortunately, these equations are too complicated to give much insight, yet too simple to give accurate results. This chapter focuses on developing simpler models that offer the designer more intuition. The RC delay model approximates a switching transistor with an effective resistance and provides a way to estimate delay using arithmetic rather than differential equations. The method of Logical Effort simplifies the model even further and is a powerful way to evaluate delay in circuits. The chapter ends with a discussion of other delay models used for timing analysis.

3.2 Transient Response

The most fundamental way to compute delay is to develop a physical model of the circuit of interest, write a differential equation describing the output voltage as a function of input voltage and time, and solve the equation. The solution of the differential equation is called the *transient response*, and the delay is the time when the output reaches $V_{DD}/2$.

The differential equation is based on charging or discharging of the capacitances in the circuit. The circuit takes time to switch because the capacitance cannot change its voltage instantaneously. If capacitance C is charged with a current I, the voltage on the capacitor varies as:

$$I = C\frac{dV}{dt} \tag{3.2}$$

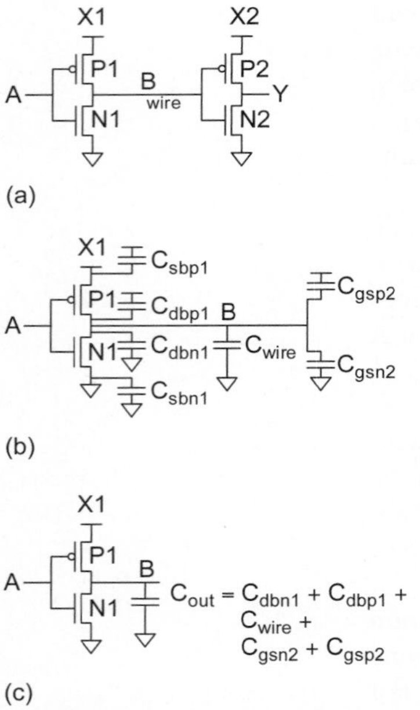

(a)

(b)

(c)

FIGURE 3.3 Capacitances for inverter delay calculations

Every real circuit has some capacitance. In an integrated circuit, it typically consists of the gate capacitance of the load along with the diffusion capacitance of the driver's own transistors, as discussed in Section 2.3. As will be explored further in Section 5.2.2, wires that connect transistors together often contribute the majority of the capacitance. The transistor current depends on the input (gate) and output (source/drain) voltages. To illustrate these points, consider computing the step response of an inverter.

Figure 3.3(a) shows an inverter $X1$ driving another inverter $X2$ at the end of a wire. Suppose a voltage step from 0 to V_{DD} is applied to node A and we wish to compute the propagation delay, t_{pdf}, through $X1$, i.e., the delay from the input step until node B crosses $V_{DD}/2$.

These capacitances are annotated on Figure 3.3(b). There are diffusion capacitances between the drain and body of each transistor and between the source and body of each transistor: C_{db} and C_{sb}. The gate capacitance C_{gs} of the transistors in $X2$ are part of the load. The wire capacitance is also part of the load. The gate capacitance of the transistors in $X1$ and the diffusion capacitance of the transistors in $X2$ do not matter because they do not connect to node B. The source-to-body capacitors C_{sbn1} and C_{sbp1} have both terminals tied to constant voltages and thus do not contribute to the switching capacitance. It is also irrelevant whether the second terminal of each capacitor connects to ground or power because both are constant supplies, so for the sake of simplicity, we can draw all of the capacitors as if they are connected to ground. Figure 3.3(c) shows the equivalent circuit diagram in which all the capacitances are lumped into a single C_{out}.

Before the voltage step is applied, $A = 0$. $N1$ is OFF, $P1$ is ON, and $B = V_{DD}$.

After the step, $A = 1$. $N1$ turns ON and $P1$ turns OFF and B drops toward 0. The rate of change of the voltage V_B at node B depends on the output capacitance and on the current through $N1$:

$$C_{\text{out}} \frac{dV_B}{dt} = -I_{dsn1} \tag{3.3}$$

Suppose the transistors obey the long-channel models. The current depends on whether $N1$ is in the linear or saturation regime. The gate is at V_{DD}, the source is at 0, and the drain is at V_B. Thus, $V_{gs} = V_{DD}$ and $V_{ds} = V_B$. Initially, $V_{ds} = V_{DD} > V_{gs} - V_t$, so $N1$ is in saturation. As V_B falls below $V_{DD} - V_t$, $N1$ enters the linear regime. Substituting EQ (2.10) and rearranging, we find the differential equation governing V_B.

$$\frac{dV_B}{dt} = -\frac{\beta}{C_{\text{out}}} \begin{cases} \dfrac{\left(V_{DD} - V_t\right)^2}{2} & V_B > V_{DD} - V_t \\ \left(V_{DD} - V_t - \dfrac{V_B}{2}\right) V_B & V_B < V_{DD} - V_t \end{cases} \tag{3.4}$$

During saturation, the current is constant and V_B drops linearly until it reaches $V_{DD} - V_t$. Thereafter, the differential equation becomes nonlinear. The response can be computed numerically. The rising output response is computed in an analogous fashion and is symmetric with the falling response if $\beta_p = \beta_n$.

Example 3.1

Plot the response of the inverter to a step input and determine the propagation delay. Assume that the nMOS transistor width is 1 μm and the output capacitance is 20 fF. Use the following long-channel model parameter values for a 65-nm process: $L = 50$ nm, $V_{DD} = 1.0$ V, $V_t = 0.3$ V, $t_{ox} = 10.5$ Å, $\mu = 80$ cm²/V · s.

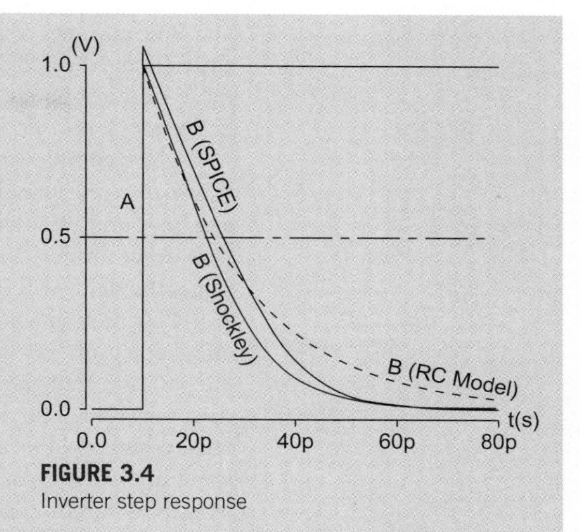

FIGURE 3.4 Inverter step response

SOLUTION: The response is plotted in Figure 3.3. The input, A, rises at 10 ps. The solid blue line indicates the step response predicted by the long-channel model. The output, B, initially follows a straight line, as the saturated nMOS transistor behaves as a constant current source. B eventually curves as it approaches 0 and the nMOS transistor enters the linear regime. The propagation delay is 12.5 ps. The solid black line indicates the step response predicted by SPICE. The propagation delay is 15.8 ps, which is longer because the mobility used in the long-channel model didn't fully account for velocity saturation and mobility degradation effects. SPICE shows that B also initially rises momentarily before falling. This effect is called bootstrapping and will be discussed in Section 3.6.6. The dashed black line shows an RC model that approximates the nMOS transistor as a 1 kΩ resistor when it is ON. The propagation delay predicted by the RC model matches SPICE fairly well, although the fall time is overestimated. RC models will be explored further in Section 3.3.

In a real circuit, the input comes from another gate with a nonzero rise/fall time. This input can be approximated as a ramp with the same rise/fall time. Again, let us consider a rising ramp and a falling output and examine how the nonzero rise time affects the propagation delay.

Assuming $V_{tn} + |V_{tp}| < V_{DD}$, the ramp response includes three phases, as shown in Table 3.1. When A starts to rise, $N1$ remains OFF and B remains at V_{DD}. When A reaches V_{tn}, $N1$ turns ON. It fights $P1$ and starts to gradually pull B down toward an intermediate value predicted by the DC circuit response examined in Section 2.5. When A gets close enough to V_{DD}, $P1$ turns OFF and B falls to 0 unopposed. Thus, we can write the differential equations for V_B in each phase:

$$\begin{aligned} &\text{Phase 1} \quad V_B = V_{DD} \\ &\text{Phase 2} \quad \frac{dV_B}{dt} = \frac{I_{dsp1} - I_{dsn1}}{C_{\text{out}}} \\ &\text{Phase 3} \quad \frac{dV_B}{dt} = \frac{-I_{dsn1}}{C_{\text{out}}} \end{aligned} \quad (3.5)$$

TABLE 3.1 Phases of inverter ramp response

Phase	V_A	$N1$	$P1$	V_B
1	$0 < V_A < V_{tn}$	OFF	ON	V_{DD}
2	$V_{tn} < V_A < V_{DD} - \|V_{tp}\|$	ON	ON	Intermediate
3	$V_{DD} - \|V_{tp}\| < V_A < V_{DD}$	ON	OFF	Falling toward 0

The currents could be estimated using the long-channel model again, but working out the full model is tedious and offers little insight. The key observation is that the propagation delay increases because $N1$ is not fully ON right away and because it must fight $P1$ in Phase 2. Section 3.4.6.1 develops a model for how propagation delay increases with rise time.

More complex gates such as NANDs or NORs have transistors in series. Each series transistor sees a smaller V_{ds} and delivers less current. The current through the transistors can be found by solving the simultaneous nonlinear differential equations, which again is best done numerically. If the transistors have the same dimensions and the load is the same, the delay will increase with the number of series transistors.

This section has shown how to develop a physical model for a circuit, write the differential equation for the model, and solve the equation to compute delay. The physical modeling shows that the delay increases with the output capacitance and decreases with the driver current. The differential equations used the long-channel model for transistor current, which is inaccurate in modern processes. The equations are also too nonlinear to solve in closed form, so they have to be solved numerically and give little insight about delay. Circuit simulators automate this process using more accurate delay equations and give good predictions of delay, but offer even less insight. The rest of this chapter is devoted to developing simpler delay models that offer more insight and tolerable accuracy.

3.3 RC Delay Model

RC delay models approximate the nonlinear transistor I-V and C-V characteristics with an average resistance and capacitance over the switching range of the gate. This approximation works remarkably well for delay estimation despite its obvious limitations in predicting detailed analog behavior.

3.3.1 Effective Resistance

The RC delay model treats a transistor as a switch in series with a resistor. The *effective resistance* is the ratio of V_{ds} to I_{ds} averaged across the switching interval of interest.

A unit nMOS transistor is defined to have effective resistance R. The size of the unit transistor is arbitrary but conventionally refers to a transistor with minimum length and minimum contacted diffusion width (i.e., 4/2 λ). Alternatively, it may refer to the width of the nMOS transistor in a minimum-sized inverter in a standard cell library. An nMOS transistor of k times unit width has resistance R/k because it delivers k times as much current. A unit pMOS transistor has greater resistance, generally in the range of $2R$–$3R$, because of its lower mobility. Throughout this book we will use $2R$ for examples to keep arithmetic simple. R is typically on the order of 10 kΩ for a unit transistor. Sections 3.3.7 and 7.4.5 examine how to determine the effective resistance for transistors in a particular process.

According to the long-channel model, current decreases linearly with channel length and hence resistance is proportional to L. Moreover, the resistance of two transistors in series is the sum of the resistances of each transistor. However, if a transistor is fully velocity-saturated, current and resistance become independent of channel length. Real transistors operate somewhere between these two extremes. This also means that the resistance of transistors in series is somewhat lower than the sum of the resistances, because series transistors see smaller V_{ds} and are less velocity-saturated. The effect is more pronounced for nMOS transistors than pMOS because of the higher mobility and greater degree of velocity saturation. The simplest approach is to neglect velocity-saturation for hand calculations, but recognize that series transistors will be somewhat faster than predicted.

3.3.2 Gate and Diffusion Capacitance

Each transistor also has gate and diffusion capacitance. We define C to be the gate capacitance of a unit transistor of either flavor. A transistor of k times unit width has capacitance kC. Diffusion capacitance depends on the size of the source/drain region. Using the approximations from Section 2.3.1, we assume the contacted source or drain of a unit transistor to also have capacitance of about C. Wider transistors have proportionally greater diffusion capacitance. Increasing channel length increases gate capacitance proportionally but does not affect diffusion capacitance.

Although capacitances have a nonlinear voltage dependence, we use a single average value. As discussed in Section 2.3.1, we roughly estimate C for a minimum length transistor to be 1 fF/μm of width. In a 65 nm process with a unit transistor being 0.1 μm wide, C is thus about 0.1 fF.

3.3.3 Equivalent RC Circuits

Figure 3.5 shows equivalent RC circuit models for nMOS and pMOS transistors of width k with contacted diffusion on both source and drain. The pMOS transistor has approximately twice the resistance of the nMOS transistor because holes have lower mobility than electrons. The pMOS capacitors are shown with V_{DD} as their second terminal because the n-well is usually tied high. However, the behavior of the capacitor from a delay perspective is independent of the second terminal voltage so long as it is constant. Hence, we sometimes draw the second terminal as ground for convenience.

The equivalent circuits for logic gates are assembled from the individual transistors. Figure 3.6 shows the equivalent circuit for a fanout-of-1 inverter with negligible wire capacitance. The unit inverters of Figure 3.6(a) are composed from an nMOS transistor of unit size and a pMOS transistor of twice unit

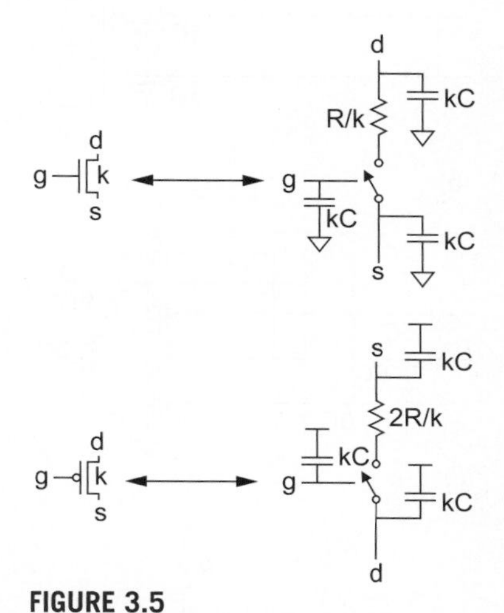

FIGURE 3.5 Equivalent circuits for transistors

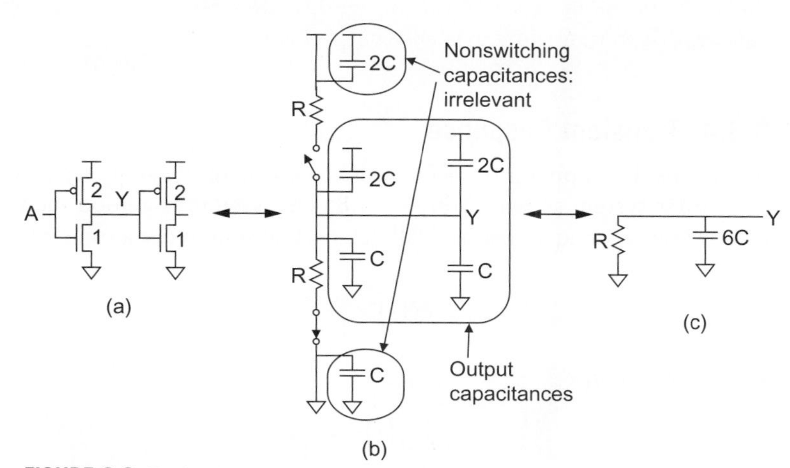

FIGURE 3.6 Equivalent circuit for an inverter

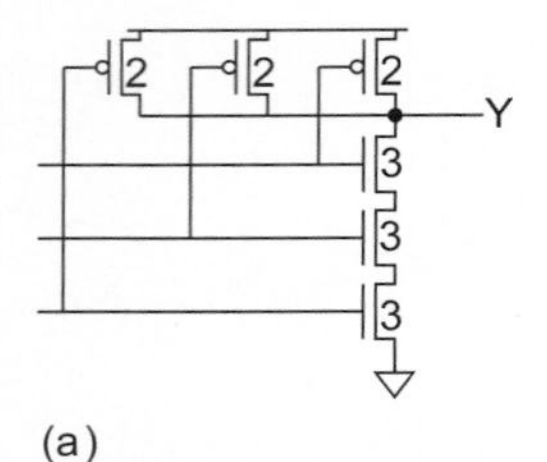

(a)

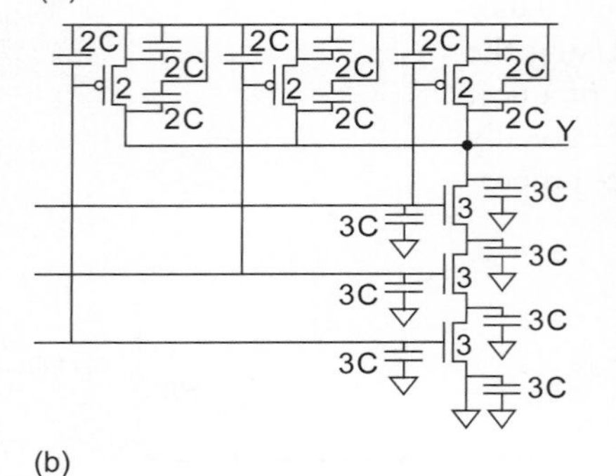

(b)

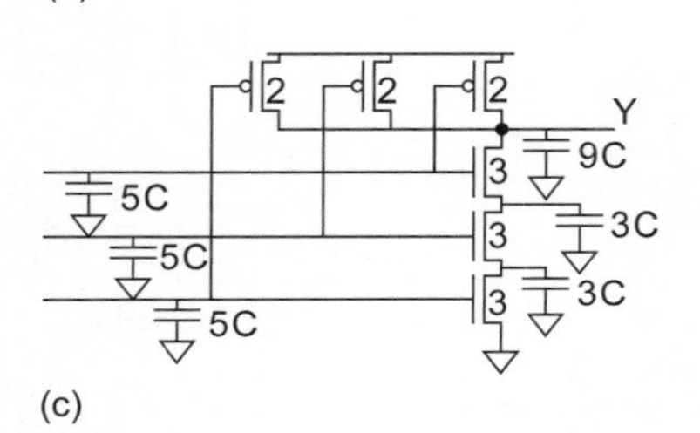

(c)

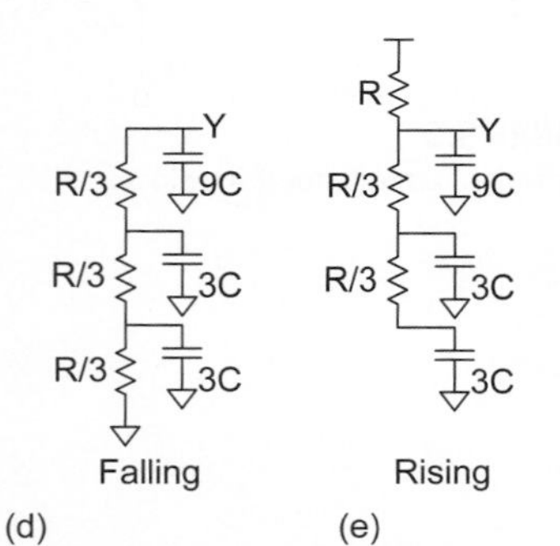

(d) (e)

FIGURE 3.7 Equivalent circuits for a 3-input NAND gate

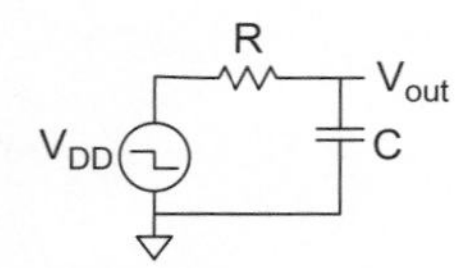

FIGURE 3.8
First-order RC system

width to achieve equal rise and fall resistance. Figure 3.6(b) gives an equivalent circuit, showing the first inverter driving the second inverter's gate. If the input A rises, the nMOS transistor will be ON and the pMOS OFF. Figure 3.6(c) illustrates this case with the switches removed. The capacitors shorted between two constant supplies are also removed because they are not charged or discharged. The total capacitance on the output Y is $6C$.

Example 3.2

Sketch a 3-input NAND gate with transistor widths chosen to achieve effective rise and fall resistance equal to that of a unit inverter (R). Annotate the gate with its gate and diffusion capacitances. Assume all diffusion nodes are contacted. Then sketch equivalent circuits for the falling output transition and for the worst-case rising output transition.

SOLUTION: Figure 3.7(a) shows such a gate. The three nMOS transistors are in series so the resistance is three times that of a single transistor. Therefore, each must be three times unit width to compensate. In other words, each transistor has resistance $R/3$ and the series combination has resistance R. The two pMOS transistors are in parallel. In the worst case (with one of the inputs low), only one of the pMOS transistors is ON. Therefore, each must be twice unit width to have resistance R.

Figure 3.7(b) shows the capacitances. Each input presents five units of gate capacitance to whatever circuit drives that input. Notice that the capacitors on source diffusions attached to the rails have both terminals shorted together so they are irrelevant to circuit operation. Figure 3.7(c) redraws the gate with these capacitances deleted and the remaining capacitances lumped to ground.

Figure 3.7(d) shows the equivalent circuit for the falling output transition. The output pulls down through the three series nMOS transistors. Figure 3.7(e) shows the equivalent circuit for the rising output transition. In the worst case, the upper two inputs are 1 and the bottom one falls to 0. The output pulls up through a single pMOS transistor. The upper two nMOS transistors are still on, so the diffusion capacitance between the series nMOS transistors must also be discharged.

3.3.4 Transient Response

Now, consider applying the RC model to estimate the step response of the first-order system shown in Figure 3.8. This system is a good model of an inverter sized for equal rise and fall delays. The system has a transfer function

$$H(s) = \frac{1}{1 + sRC} \tag{3.6}$$

and a step response

$$V_{out}(t) = V_{DD} e^{-t/\tau} \tag{3.7}$$

where $\tau = RC$. The propagation delay is the time at which V_{out} reaches $V_{DD}/2$, as shown in Figure 3.9.

$$t_{pd} = RC \ln 2 \tag{3.8}$$

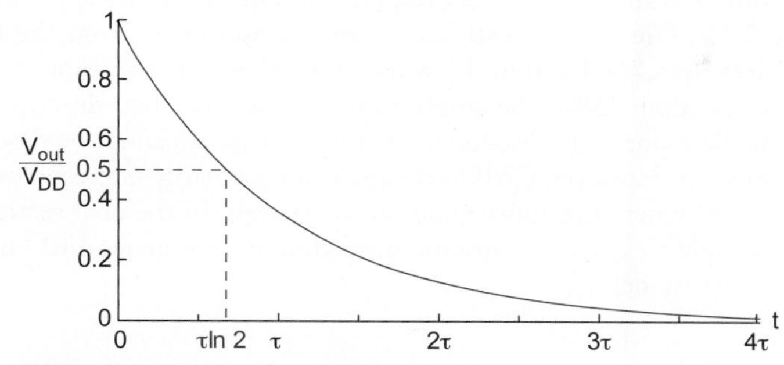

FIGURE 3.9 First-order step response

The factor of ln 2 = 0.69 is cumbersome. The effective resistance R is an empirical parameter anyway, so it is preferable to incorporate the factor of ln 2 to define a new effective resistance $R' = R \ln 2$. Now the propagation delay is simply $R'C$. For the sake of convenience, we usually drop the prime symbols and just write

$$t_{pd} = RC \tag{3.9}$$

where the effective resistance R is chosen to give the correct delay.

Figure 3.10 shows a second-order system. R_1 and R_2 might model the two series nMOS transistors in a NAND gate or an inverter driving a long wire with non-negligible resistance. The transfer function is

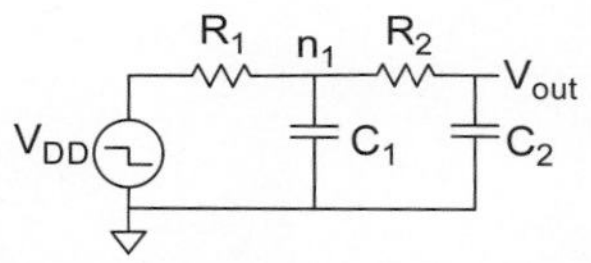

FIGURE 3.10 Second-order RC system

$$H(s) = \frac{1}{1 + s\left[R_1C_1 + (R_1 + R_2)C_2\right] + s^2 R_1C_1R_2C_2} \tag{3.10}$$

The function has two real poles and the step response is

$$V_{out}(t) = V_{DD}\frac{\tau_1 e^{-t/\tau_1} - \tau_2 e^{-t/\tau_2}}{\tau_1 - \tau_2} \tag{3.11}$$

with

$$\tau_{1,2} = \frac{R_1C_1 + (R_1 + R_2)C_2}{2}\left(1 \pm \sqrt{1 - \frac{4R^*C^*}{\left[1 + (1 + R^*)C^*\right]^2}}\right) \tag{3.12}$$
$$R^* = \frac{R_2}{R_1};\ C^* = \frac{C_2}{C_1}$$

EQ (3.12) is so complicated that it defeats the purpose of simplifying a CMOS circuit into an equivalent RC network. However, it can be further approximated as a first-order system with a single time constant:

$$\tau = \tau_1 + \tau_2 = R_1C_1 + (R_1 + R_2)C_2 \quad (3.13)$$

This approximation works best when one time constant is significantly bigger than the other [Horowitz84]. For example, if $R_1 = R_2 = R$ and $C_1 = C_2 = C$, then $\tau_1 = 2.6\ RC$, $\tau_2 = 0.4\ RC$, $\tau = 3\ RC$ and the second-order response and its first-order approximation are shown in Figure 3.11. The error in estimated propagation delay from the first-order approximation is less than 7%. Even in the worst case, where the two time constants are equal, the error is less than 15%. The single time constant is a bad description of the behavior of intermediate nodes. For example, the response at n_1 cannot be described well by a single time constant. However, CMOS designers are primarily interested in the delay to the output of a gate, where the approximation works well. In the next section, we will see how to find a simple single time constant approximation for general RC tree circuits using the Elmore delay model.

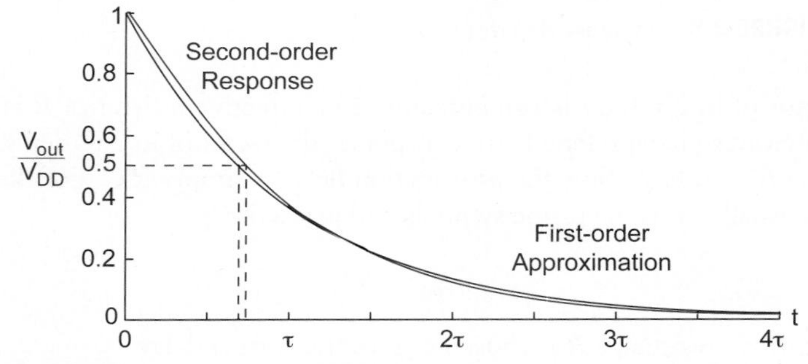

FIGURE 3.11 Comparison of second-order response to first-order approximation

3.3.5 Elmore Delay

In general, most circuits of interest can be represented as an *RC tree*, i.e., an RC circuit with no loops. The root of the tree is the voltage source and the leaves are the capacitors at the ends of the branches. The Elmore delay model [Elmore48] estimates the delay from a source switching to one of the leaf nodes changing as the sum over each node i of the capacitance C_i on the node, multiplied by the effective resistance R_{is} on the shared path from the source to the node and the leaf. Application of Elmore delay is best illustrated through examples.

$$t_{pd} = \sum_i R_{is}C_i \quad (3.14)$$

Example 3.3

Compute the Elmore delay for V_{out} in the 2nd order RC system from Figure 3.10.

SOLUTION: The circuit has a source and two nodes. At node n_1, the capacitance is C_1 and the resistance to the source is R_1. At node V_{out}, the capacitance is C_2 and the resistance to the source is $(R_1 + R_2)$. Hence, the Elmore delay is $t_{pd} = R_1C_1 + (R_1 + R_2)C_2$, just as the single time constant predicted in EQ (3.13). Note that the effective resistances should account for the factor of ln 2.

Example 3.4

Estimate t_{pd} for a unit inverter driving m identical unit inverters.

SOLUTION: Figure 3.12 shows an equivalent circuit for the falling transition. Each load inverter presents $3C$ units of gate capacitance, for a total of $3mC$. The output node also sees a capacitance of $3C$ from the drain diffusions of the driving inverter. This capacitance is called *parasitic* because it is an undesired side-effect of the need to make the drain large enough to contact. The parasitic capacitance is independent of the load that the inverter is driving. Hence, the total capacitance is $(3 + 3m)C$. The resistance is R, so the Elmore delay is $t_{pd} = (3 + 3m)RC$. The equivalent circuit for the rising transition gives the same results.

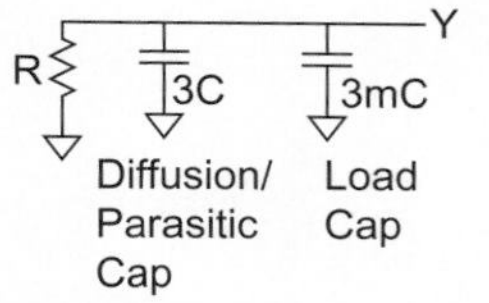

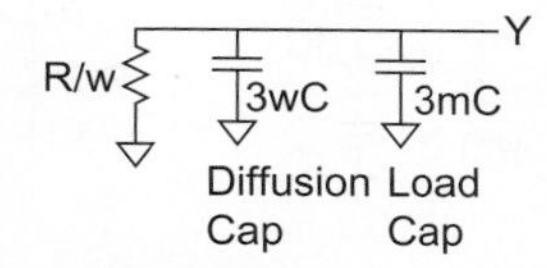

FIGURE 3.12 Equivalent circuit for inverter

Example 3.5

Repeat Example 3.4 if the driver is w times unit size.

SOLUTION: Figure 3.13 shows the equivalent circuit. The driver transistors are w times as wide, so the effective resistance decreases by a factor of w. The diffusion capacitance increases by a factor of w. The Elmore delay is $t_{pd} = ((3w + 3m)C)(R/w) = (3 + 3m/w)RC$.

Define the *fanout* of the gate, h, to be the ratio of the load capacitance to the input capacitance. (Diffusion capacitance is not counted in the fanout.) The load capacitance is $3mC$. The input capacitance is $3wC$. Thus, the inverter has a fanout of $h = m/w$ and the delay can be written as $(3 + 3h)RC$.

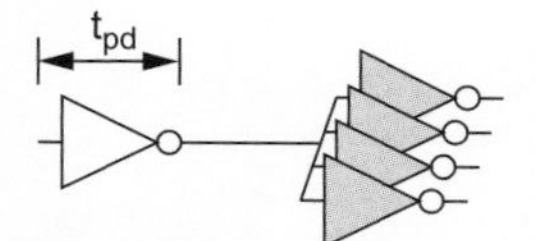

FIGURE 3.13 Equivalent circuit for wider inverter

Example 3.6

If a unit transistor has $R = 10\ \text{k}\Omega$ and $C = 0.1$ fF in a 65 nm process, compute the delay, in picoseconds, of the inverter in Figure 3.14 with a fanout of $h = 4$.

SOLUTION: The RC product in the 65 nm process is $(10\ \text{k}\Omega)(0.1\ \text{fF}) = 1$ ps. For $h = 4$, the delay is $(3 + 3h)(1\ \text{ps}) = 15$ ps. This is called the *fanout-of-4 (FO4) inverter delay* and is representative of gate delays in a typical circuit. Remember that a picosecond is a trillionth of a second. The inverter can switch about 66 billion times per second. This stunning speed partially explains the fantastic capabilities of integrated circuits.

FIGURE 3.14 Fanout-of-4 (FO4) inverter

It is often helpful to express delay in a process-independent form so that circuits can be compared based on topology rather than speed of the manufacturing process. Moreover, with a process-independent measure for delay, knowledge of circuit speeds gained while working in one process can be carried over to a new process. Observe that the delay of an ideal fanout-of-1 inverter with no parasitic capacitance is $\tau = 3RC$[1] [Sutherland99]. We denote the normalized delay d relative to this inverter delay:

$$d = \frac{t_{pd}}{\tau} \tag{3.15}$$

[1]Do not confuse this definition of $\tau = 3RC$, the delay of a parasitic-free fanout-of-1 inverter, with Mead and Conway's definition [Mead80] $\tau = RC$, the delay of an nMOS transistor driving its own gate, or with the use of τ as an arbitrary time constant. For the remainder of this text, $\tau = 3RC$.

Hence, the delay of a fanout-of-h inverter can be written in normalized form as $d = h + 1$, assuming that diffusion capacitance approximately equals gate capacitance. An FO4 inverter has a delay of 5τ. If diffusion capacitance were slightly higher or lower, the FO4 delay would change by only a small amount. Thus, circuit delay measured in FO4 delays is nearly constant from one process to another.[2]

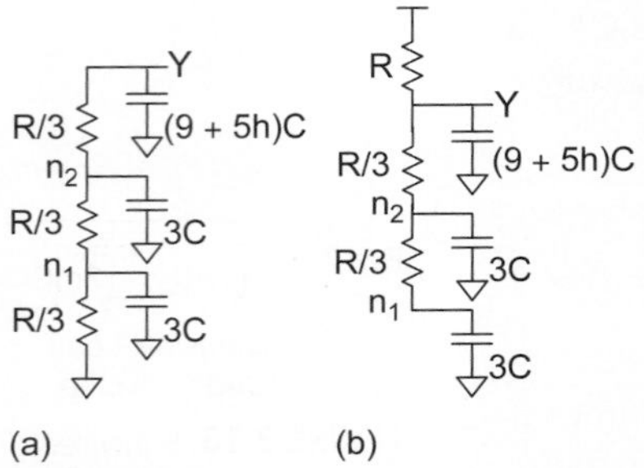

FIGURE 3.15 Equivalent circuits for loaded gate

Example 3.7

Estimate t_{pdf} and t_{pdr} for the 3-input NAND gate from Example 3.2 if the output is loaded with h identical NAND gates.

SOLUTION: Each NAND gate load presents 5 units of capacitance on a given input. Figure 3.15(a) shows the equivalent circuit including the load for the falling transition. Node n_1 has capacitance $3C$ and resistance of $R/3$ to ground. Node n_2 has capacitance $3C$ and resistance $(R/3 + R/3)$ to ground. Node Y has capacitance $(9 + 5h)C$ and resistance $(R/3 + R/3 + R/3)$ to ground. The Elmore delay for the falling output is the sum of these RC products, $t_{pdf} = (3C)(R/3) + (3C)(R/3 + R/3) + ((9 + 5h)C)(R/3 + R/3 + R/3) = (12 + 5h)RC$.

Figure 3.15(b) shows the equivalent circuit for the falling transition. In the worst case, the two inner inputs are 1 and the outer input falls. Y is pulled up to V_{DD} through a single pMOS transistor. The ON nMOS transistors contribute parasitic capacitance that slows the transiton. Node Y has capacitance $(9 + 5h)C$ and resistance R to the V_{DD} supply. Node n_2 has capacitance $3C$. The relevant resistance is only R, not $(R + R/3)$, because the output is being charged only through R. This is what is meant by the resistance on the shared path from the source (V_{DD}) to the node (n_2) and the leaf (Y). Similarly, node n_1 has capacitance $3C$ and resistance R. Hence, the Elmore delay for the rising output is $t_{pdr} = (15 + 5h)RC$. The $R/3$ resistances do not contribute to this delay. Indeed, they shield the diffusion capacitances, which don't have to charge all the way up before Y rises. Hence, the Elmore delay is conservative and the actual delay is somewhat faster.

Although the gate has equal resistance pulling up and down, the delays are not quite equal because of the capacitances on the internal nodes.

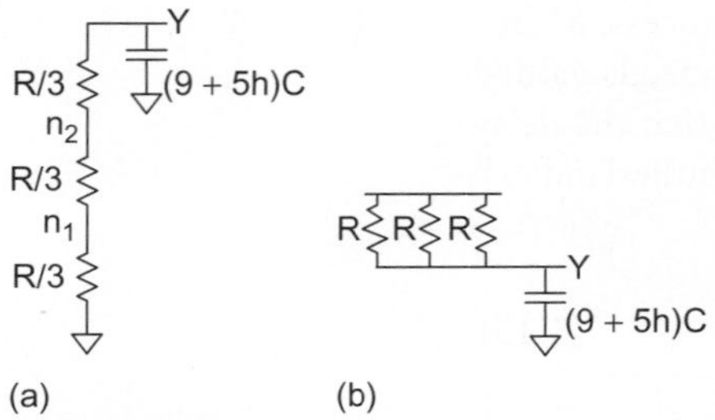

FIGURE 3.16 Equivalent circuits for contamination delay

Example 3.8

Estimate the contamination delays t_{cdf} and t_{cdr} for the 3-input NAND gate from Example 3.2 if the output is loaded with h identical NAND gates.

SOLUTION: The contamination delay is the fastest that the gate might switch. For the falling transition, the best case is that the bottom two nMOS transistors are already ON when the top one turns ON. In such a case, the diffusion capacitances on n_1 and n_2 have already been discharged and do not contribute to the delay. Figure 3.16(a) shows the equivalent circuit and the delay is $t_{cdf} = (9 + 5h)RC$.

[2]This assumes that the circuit is dominated by gate delay. The RC delay of long wires does not track well with the gate delay, as will be explored in Chapter 7.

For the rising transition, the best case is that all three pMOS transistors turn on simultaneously. The nMOS transistors turn OFF, so n_1 and n_2 are not connected to the output and do not contribute to delay. The parallel transistors deliver three times as much current, as shown in Figure 3.16(b), so the delay is $t_{cdr} = (3 + (5/3)h)RC$.

In all of the Examples, the delay consists of two components. The *parasitic delay* is the time for a gate to drive its own internal diffusion capacitance. Boosting the width of the transistors decreases the resistance but increases the capacitance so the parasitic delay is ideally independent of the gate size.[3] The *effort delay* depends on the ratio h of external load capacitance to input capacitance and thus changes with transistor widths. It also depends on the complexity of the gate. The capacitance ratio is called the *fanout* or *electrical effort* and the term indicating gate complexity is called the *logical effort*. For example, an inverter has a delay of $d = h + 1$, so the parasitic delay is 1 and the logical effort is also 1. The NAND3 has a worst case delay of $d = (5/3)h + 5$. Thus, it has a parasitic delay of 5 and a logical effort of 5/3. These delay components will be explored further in Section 3.4.

3.3.6 Layout Dependence of Capacitance

In a good layout, diffusion nodes are shared wherever possible to reduce the diffusion capacitance. Moreover, the uncontacted diffusion nodes between series transistors are usually smaller than those that must be contacted. Such uncontacted nodes have less capacitance (see Sections 2.3.3 and 7.4.4), although we will neglect the difference for hand calculations. A conservative method of estimating capacitances before layout is to assume uncontacted diffusion between series transistors and contacted diffusion on all other nodes. However, a more accurate estimate can be made once the layout is known.

Example 3.9

Figure 3.17(a) shows a layout of the 3-input NAND gate. A single drain diffusion region is shared between two of the pMOS transistors. Estimate the actual diffusion capacitance from the layout.

SOLUTION: Figure 3.17(b) redraws the schematic with these capacitances lumped to ground. The output node has the following diffusion capacitances: $3C$ from the nMOS transistor drain, $2C$ from the isolated pMOS transistor drain, and $2C$ from a pair of pMOS drains that share a contact. Thus, the actual diffusion capacitance on the output is $7C$, rather than $9C$ predicted in Figure 3.15.

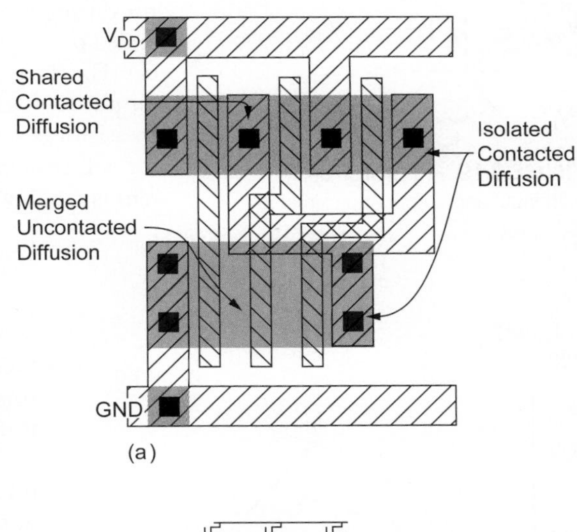

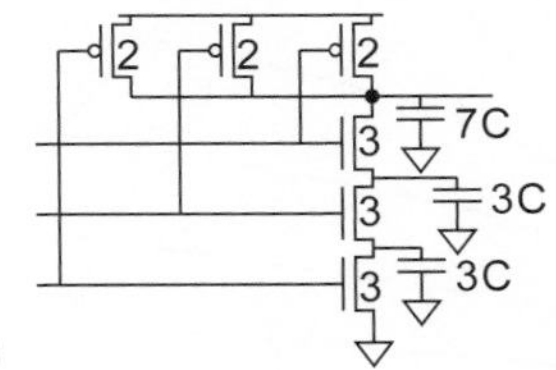

FIGURE 3.17 3-input NAND annotated with diffusion capacitances extracted from the layout

[3] Gates with wider transistors may use layout tricks so the diffusion capacitance increases less than linearly with width, slightly decreasing the parasitic delay of large gates as discussed in Section 3.3.6.

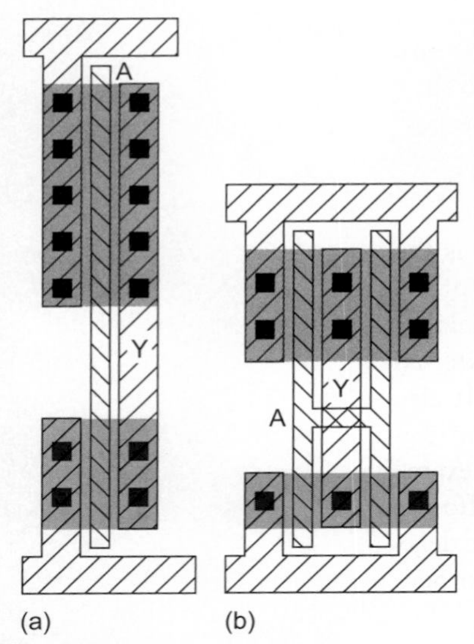

FIGURE 3.18 Layout styles: (a) conventional, (b) folded

The diffusion capacitance can also be decreased by *folding* wide transistors. Figure 3.18(a) shows a conventional layout of a 24/12 λ inverter. Because a unit (4 λ) transistor has diffusion capacitance C, the inverter has a total diffusion capacitance of $9C$. The folded layout in Figure 3.18(b) constructs each transistor from two parallel devices of half the width. Observe that the diffusion area has shrunk by a factor of two, reducing the diffusion capacitance to $4.5C$. In general, folded layouts offer lower parasitic delay than unfolded layouts. The folded layout may also fit better in a standard cell of limited height, and the shorter polysilicon lines have lower resistance. For these reasons, wide transistors are folded whenever possible.

In some nanometer processes (generally 45 nm and below), transistor gates are restricted to a limited choice of pitches to improve manufacturability and reduce variability. For example, the spacing between polysilicon for gates may always be the contacted transistor pitch, even if no contact is required. Moreover, using a single standard transistor width may reduce variability.

3.3.7 Determining Effective Resistance

The effective resistance can be determined through simulation or analysis. Section 7.4.5 explains the simulation technique, which is most accurate. This section, however, offers an analysis that provides more insight into the relationship of resistance to other parameters.

Recall that the effective resistance is the average value of V_{ds} / I_{ds} of a transistor during a switching event. As mentioned in Section 3.3.4, the resistance is scaled by a factor of ln 2 so that propagation delay can be written as an RC product. For the step response of a rising input, we are interested in the time for the output to discharge from V_{DD} to $V_{DD} / 2$ through an nMOS transistor. If the transistor is sufficiently velocity-saturated that $V_{dsat} < V_{DD} / 2$, then the transistor will remain in the saturation region throughout this transition and the current is roughly constant at I_{dsat}. In such a case, the effective resistance is

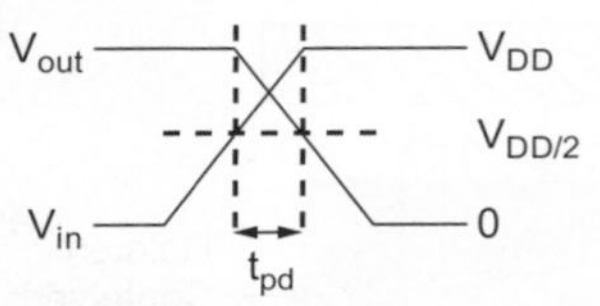

FIGURE 3.19 Propagation delay with input and output approximated as ramps

$$R_{\text{step}} = \frac{\ln 2}{V_{DD}/2} \int_{V_{DD}/2}^{V_{DD}} \frac{V}{I_{\text{dsat}}} dV = \frac{3\ln 2}{4} \frac{V_{DD}}{I_{\text{dsat}}} \approx \frac{V_{DD}}{2I_{\text{dsat}}} \quad (3.16)$$

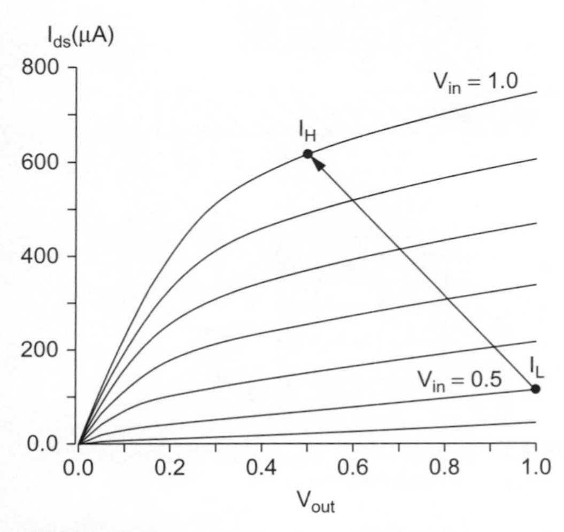

FIGURE 3.20 Approximate switching trajectory

Channel length modulation and DIBL cause the current to decrease somewhat with V_{ds} in a real transistor, slightly increasing the effective resistance.

More importantly, the input has a nonzero rise time and we are interested in the time from when the input rises through $V_{DD} / 2$ until the output falls through $V_{DD} / 2$. Assume that the input and output slopes are equal and that the output starts to fall when the input passes through $V_{DD} / 2$. Then, the output will reach $V_{DD} / 2$ when the input reaches V_{DD}, as shown in Figure 3.19.

Define the transistor current to be I_L at the start of the transition (when $V_{gs} = V_{DD} / 2$, $V_{ds} = V_{DD}$) and I_H at the end of the transition (when $V_{gs} = V_{DD}$, $V_{ds} = V_{DD} / 2$), as shown in Figure 3.20. Then, the transistor can be approximated during the switching event as a current source I_{eff} that is the average of these two extremes [Na02]:

$$I_{\text{eff}} = \frac{I_H + I_L}{2} \quad (3.17)$$

The time for the output to discharge to $V_{DD} / 2$ is thus:

$$t_{pd} = \frac{CV_{DD}}{2I_{\text{eff}}} \tag{3.18}$$

Equating this to $t_{pd} = RC$ gives

$$R = \frac{V_{DD}}{2I_{\text{eff}}} = \frac{V_{DD}}{I_H + I_L} \tag{3.19}$$

3.4 Linear Delay Model

The RC delay model showed that delay is a linear function of the fanout of a gate. Based on this observation, designers further simplify delay analysis by characterizing a gate by the slope and *y*-intercept of this function. In general, the normalized delay of a gate can be expressed in units of τ as

$$d = f + p \tag{3.20}$$

p is the *parasitic delay* inherent to the gate when no load is attached. f is the *effort delay* or *stage effort* that depends on the complexity and fanout of the gate:

$$f = gh \tag{3.21}$$

The complexity is represented by the *logical effort*, g [Sutherland99]. An inverter is defined to have a logical effort of 1. More complex gates have greater logical efforts, indicating that they take longer to drive a given fanout. For example, the logical effort of the 3-input NAND gate from the previous example is 5/3. A gate driving h identical copies of itself is said to have a *fanout* or *electrical effort* of h. If the load does not contain identical copies of the gate, the electrical effort can be computed as

$$h = \frac{C_{\text{out}}}{C_{\text{in}}} \tag{3.22}$$

where C_{out} is the capacitance of the external load being driven and C_{in} is the input capacitance of the gate.[4]

Figure 3.21 plots normalized delay vs. electrical effort for an idealized inverter and 3-input NAND gate. The *y*-intercepts indicate the parasitic delay, i.e., the delay when the gate drives no load. The slope of the lines is the logical effort. The inverter has a slope of 1 by definition. The NAND has a slope of 5/3.

The remainder of this section explores how to estimate the logical effort and parasitic delay and how to use the linear delay model.

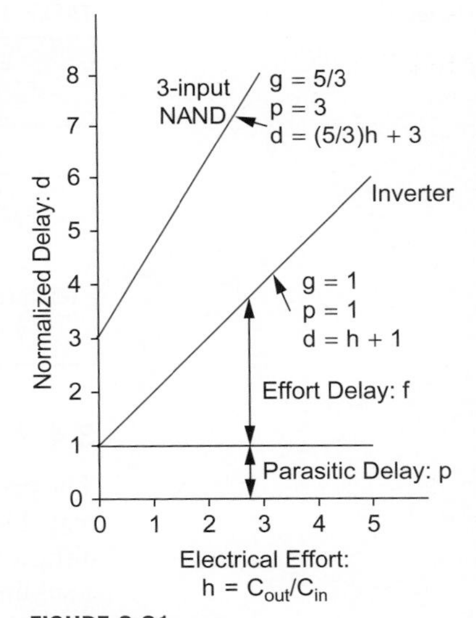

FIGURE 3.21
Normalized delay vs. fanout

[4]Some board-level designers say a device has a fanout of h when it drives h other devices, even if the other devices have different capacitances. This definition would not be useful for calculating delay and is best avoided in VLSI design. The term *electrical effort* avoids this potential confusion and emphasizes the parallels with logical effort.

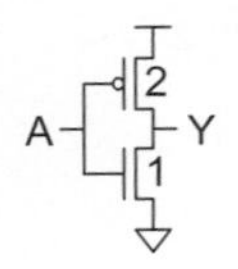

$C_{in} = 3$
(a) g = 3/3

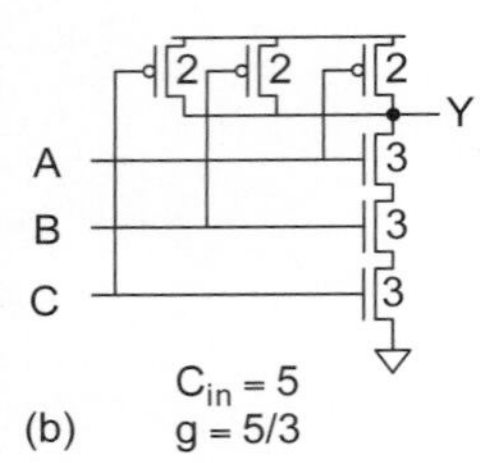

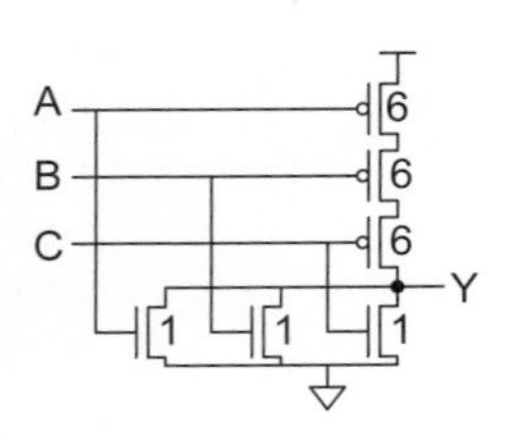

$C_{in} = 7$
(c) g = 7/3

FIGURE 3.22 Logic gates sized for unit resistance

3.4.1 Logical Effort

Logical effort of a gate is defined as *the ratio of the input capacitance of the gate to the input capacitance of an inverter that can deliver the same output current.* Equivalently, logical effort indicates how much worse a gate is at producing output current as compared to an inverter, given that each input of the gate may only present as much input capacitance as the inverter.

Logical effort can be measured in simulation from delay vs. fanout plots as the ratio of the slope of the delay of the gate to the slope of the delay of an inverter, as will be discussed in Section 7.5.3. Alternatively, it can be estimated by sketching gates. Figure 3.22 shows inverter, 3-input NAND, and 3-input NOR gates with transistor widths chosen to achieve unit resistance, assuming pMOS transistors have twice the resistance of nMOS transistors.[5] The inverter presents three units of input capacitance. The NAND presents five units of capacitance on each input, so the logical effort is 5/3. Similarly, the NOR presents seven units of capacitance, so the logical effort is 7/3. This matches our expectation that NANDs are better than NORs because NORs have slow pMOS transistors in series.

Table 3.2 lists the logical effort of common gates. The effort tends to increase with the number of inputs. NAND gates are better than NOR gates because the series transistors are nMOS rather than pMOS. Exclusive-OR gates are particularly costly and have different logical efforts for different inputs. An interesting case is that multiplexers built from ganged tristates, as shown in Figure 1.29(b), have a logical effort of 2 independent of the number of inputs. This might at first seem to imply that very large multiplexers are just as fast as small ones. However, the parasitic delay does increase with multiplexer size; hence, it is generally fastest to construct large multiplexers out of trees of 4-input multiplexers [Sutherland99].

TABLE 3.2 Logical effort of common gates

Gate Type	Number of Inputs				
	1	2	3	4	*n*
inverter	1				
NAND		4/3	5/3	6/3	$(n + 2)/3$
NOR		5/3	7/3	9/3	$(2n + 1)/3$
tristate, multiplexer	2	2	2	2	2
XOR, XNOR		4, 4	6, 12, 6	8, 16, 16, 8	

3.4.2 Parasitic Delay

The parasitic delay of a gate is the delay of the gate when it drives zero load. It can be estimated with RC delay models. A crude method good for hand calculations is to count only diffusion capacitance on the output node. For example, consider the gates in Figure 3.22, assuming each transistor on the output node has its own drain diffusion contact. Transistor widths were chosen to give a resistance of R in each gate. The inverter has three units of diffusion capacitance on the output, so the parasitic delay is $3RC = \tau$. In other words,

[5] This assumption is made throughout the book. Exercises 3.17–3.18 explore the effects of different relative resistances (see also [Sutherland99]). The overall conclusions do not change very much, so the simple model is good enough for most hand estimates. A simulator or static timing analyzer should be used when more accurate results are required.

the normalized parasitic delay is 1. In general, we will call the normalized parasitic delay $p_{inv} \cdot p_{inv}$ is the ratio of diffusion capacitance to gate capacitance in a particular process. It is usually close to 1 and will be considered to be 1 in many examples for simplicity. The 3-input NAND and NOR each have 9 units of diffusion capacitance on the output, so the parasitic delay is three times as great ($3p_{inv}$, or simply 3). Table 3.3 estimates the parasitic delay of common gates. Increasing transistor sizes reduces resistance but increases capacitance correspondingly, so parasitic delay is, on first order, independent of gate size. However, wider transistors can be folded and often see less than linear increases in internal wiring parasitic capacitance, so in practice, larger gates tend to have slightly lower parasitic delay.

TABLE 3.3 Parasitic delay of common gates

Gate Type	Number of Inputs				
	1	2	3	4	n
inverter	1				
NAND		2	3	4	n
NOR		2	3	4	n
tristate, multiplexer	2	4	6	8	$2n$

This method of estimating parasitic delay is obviously crude. More refined estimates use the Elmore delay counting internal parasitics, as in Example 3.7, or extract the delays from simulation. The parasitic delay also depends on the ratio of diffusion capacitance to gate capacitance. For example, in a silicon-on-insulator process in which diffusion capacitance is much less, the parasitic delays will be lower. While knowing the parasitic delay is important for accurately estimating gate delay, we will see in Section 3.5 that the best transistor sizes for a particular circuit are only weakly dependent on parasitic delay. Hence, crude estimates tend to be sufficient to reach a good circuit design.

Nevertheless, it is important to realize that parasitic delay grows more than linearly with the number of inputs in a real NAND or NOR circuit. For example, Figure 3.23 shows a model of an n-input NAND gate in which the upper inputs were all 1 and the bottom input rises. The gate must discharge the diffusion capacitances of all of the internal nodes as well as the output. The Elmore delay is

$$t_{pd} = R(3nC) + \sum_{i=1}^{n-1} \left(\frac{iR}{n}\right)(nC) = \left(\frac{n^2}{2} + \frac{5}{2}n\right) RC \tag{3.23}$$

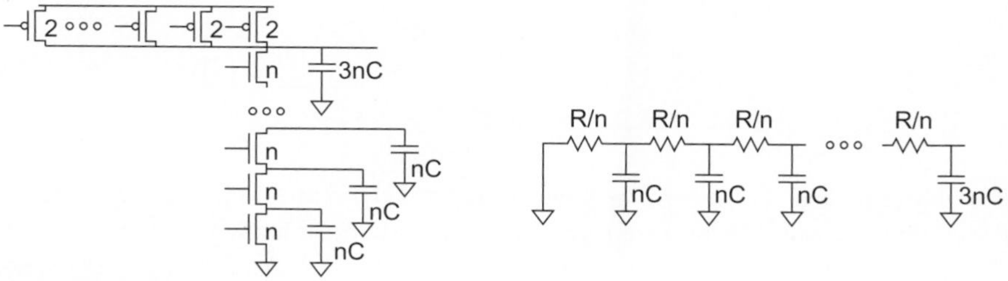

FIGURE 3.23 n-input NAND gate parasitic delay

This delay grows quadratically with the number of series transistors n, indicating that beyond a certain point it is faster to split a large gate into a cascade of two smaller gates. We will see in Section 3.4.6.5 that the coefficient of the n^2 term tends to be even larger in real circuits than in this simple model because of gate-source capacitance. In practice, it is rarely advisable to construct a gate with more than four or possibly five series transistors. When building large fan-in gates, trees of NAND gates are better than NOR gates because the NANDs have lower logical effort.

3.4.3 Delay in a Logic Gate

Consider two examples of applying the linear delay model to logic gates.

Example 3.10

Use the linear delay model to estimate the delay of the fanout-of-4 (FO4) inverter from Example 3.6. Assume the inverter is constructed in a 65 nm process with $\tau = 3$ ps.

SOLUTION: The logical effort of the inverter is $g = 1$, by definition. The electrical effort is 4 because the load is four gates of equal size. The parasitic delay of an inverter is $p_{\text{inv}} \approx 1$. The total delay is $d = gh + p = 1 \times 4 + 1 = 5$ in normalized terms, or $t_{pd} = 15$ ps in absolute terms.

Often path delays are expressed in terms of FO4 inverter delays. While not all designers are familiar with the τ notation, most experienced designers do know the delay of a fanout-of-4 inverter in the process in which they are working. τ can be estimated as 0.2 FO4 inverter delays. Even if the ratio of diffusion capacitance to gate capacitance changes so $p_{\text{inv}} = 0.8$ or 1.2 rather than 1, the FO4 inverter delay only varies from 4.8 to 5.2. Hence, the delay of a gate-dominated logic block expressed in terms of FO4 inverters remains relatively constant from one process to another even if the diffusion capacitance does not.

As a rough rule of thumb, the FO4 delay for a process (in picoseconds) is 1/3 to 1/2 of the drawn channel length (in nanometers). For example, a 65 nm process with a 50 nm channel length may have an FO4 delay of 16–25 ps. Delay is highly sensitive to process, voltage, and temperature variations, as will be examined in Section 6.2. The FO4 delay is usually quoted assuming typical process parameters and worst-case environment (low power supply voltage and high temperature).

Example 3.11

A ring oscillator is constructed from an odd number of inverters, as shown in Figure 3.24. Estimate the frequency of an N-stage ring oscillator.

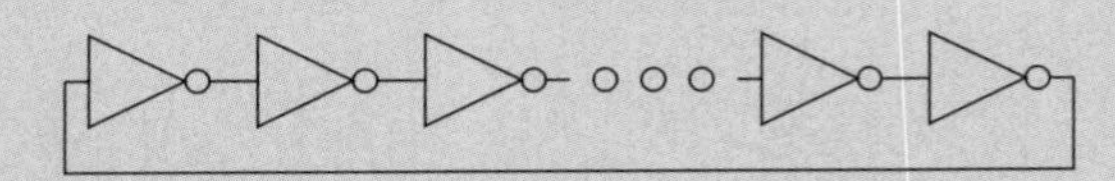

FIGURE 3.24 Ring oscillator

SOLUTION: The logical effort of the inverter is $g = 1$, by definition. The electrical effort of each inverter is also 1 because it drives a single identical load. The parasitic delay is also 1. The delay of each stage is $d = gh + p = 1 \times 1 + 1 = 2$. An N-stage ring oscillator has a period of $2N$ stage delays because a value must propagate twice around the ring to regain the original polarity. Therefore, the period is $T = 2 \times 2N$. The frequency is the reciprocal of the period, $1/4N$.

A 31-stage ring oscillator in a 65 nm process has a frequency of $1/(4 \times 31 \times 3 \text{ ps}) =$ 2.7 GHz.

Note that ring oscillators are often used as process monitors to judge if a particular chip is faster or slower than nominally expected. One of the inverters should be replaced with a NAND gate to turn the ring off when not in use. The output can be routed to an external pad, possibly through a test multiplexer. The oscillation frequency should be low enough (e.g., 100 MHz) that the path to the outside world does not attenuate the signal too badly.

3.4.4 Drive

A good standard cell library contains multiple sizes of each common gate. The sizes are typically labeled with their drive. For example, a unit inverter may be called inv_1x. An inverter of eight times unit size is called inv_8x. A 2-input NAND that delivers the same current as the inverter is called nand2_1x.

It is often more intuitive to characterize gates by their drive, x, rather than their input capacitance. If we redefine a unit inverter to have one unit of input capacitance, then the drive of an arbitrary gate is

$$x = \frac{C_{\text{in}}}{g} \tag{3.24}$$

Delay can be expressed in terms of drive as

$$d = \frac{C_{\text{out}}}{x} + p \tag{3.25}$$

3.4.5 Extracting Logical Effort from Datasheets

When using a standard cell library, you can often extract logical effort of gates directly from the datasheets. For example, Figure 3.25 shows the INV and NAND2 datasheets from the Artisan Components library for the TSMC 180 nm process. The gates in the library come in various drive strengths. INVX1 is the unit inverter; INVX2 has twice the drive. INVXL has the same area as the unit inverter but uses smaller transistors to reduce power consumption on noncritical paths. The X12–X20 inverters are built from three stages of smaller inverters to give high drive strength and low input capacitance at the expense of greater parasitic delay.

From the datasheet, we see the unit inverter has an input capacitance of 3.6 fF. The rising and falling delays are specified separately. We will develop a notation for different delays in Section 8.2.1.5, but will use the average delay for now. The average *intrinsic* or parasitic delay is (25.3 + 14.6)/2 = 20.0 ps. The slope of the delay vs. load capacitance curve is the average of the rising and falling K_{load} values. An inverter driving a fanout of h will thus have a delay of

$$t_{pd} = 20.0 \text{ ps} + \left(3.6 \frac{\text{fF}}{\text{gate}}\right)(h \text{ gates})\left(\frac{4.53 + 2.37}{2} \frac{\text{ns}}{\text{pF}}\right) = (20.0 + 12.4h) \text{ ps} \tag{3.26}$$

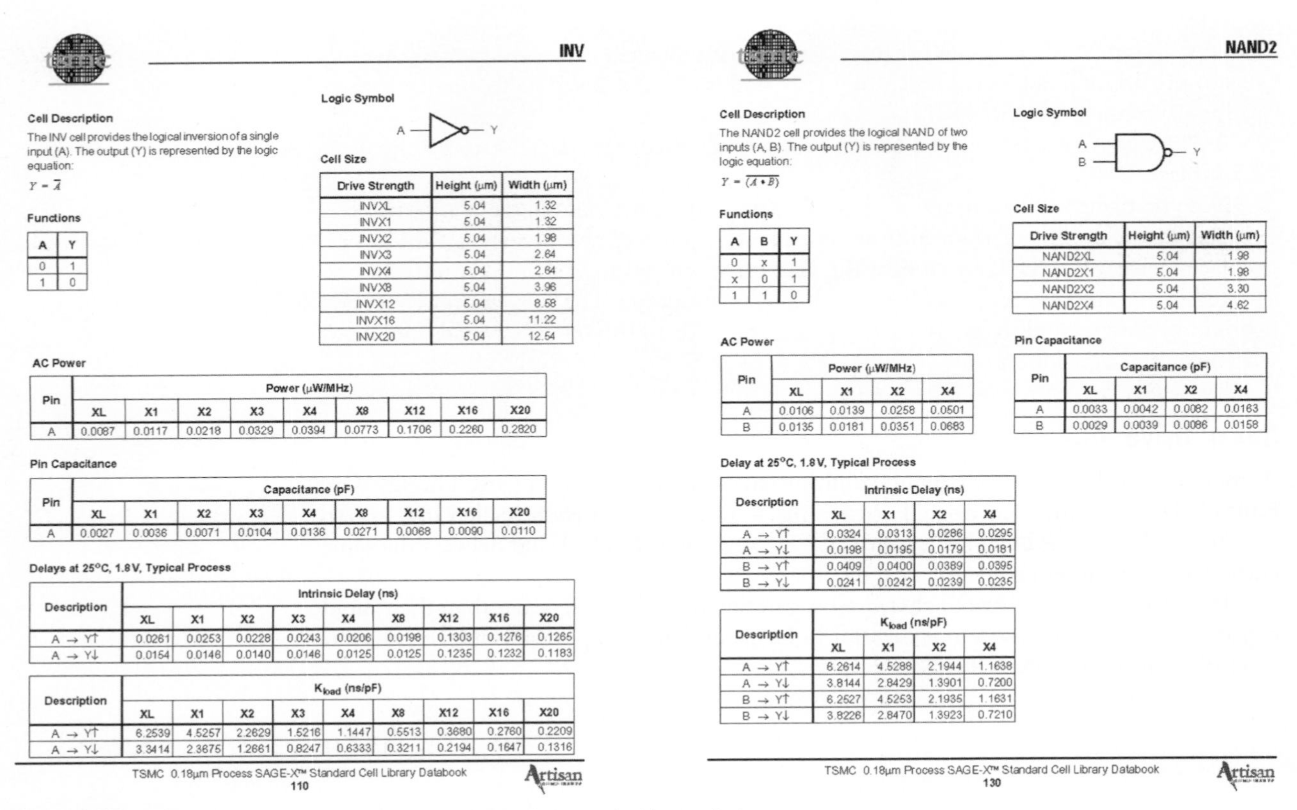

INV

Cell Description

The INV cell provides the logical inversion of a single input (A). The output (Y) is represented by the logic equation:

$Y = \overline{A}$

Functions

A	Y
0	1
1	0

Logic Symbol

Cell Size

Drive Strength	Height (μm)	Width (μm)
INVXL	5.04	1.32
INVX1	5.04	1.32
INVX2	5.04	1.98
INVX3	5.04	2.64
INVX4	5.04	2.64
INVX8	5.04	3.96
INVX12	5.04	8.58
INVX16	5.04	11.22
INVX20	5.04	12.54

AC Power

Pin	Power (μW/MHz)								
	XL	X1	X2	X3	X4	X8	X12	X16	X20
A	0.0087	0.0117	0.0218	0.0329	0.0394	0.0773	0.1706	0.2260	0.2820

Pin Capacitance

Pin	Capacitance (pF)								
	XL	X1	X2	X3	X4	X8	X12	X16	X20
A	0.0027	0.0036	0.0071	0.0104	0.0136	0.0271	0.0068	0.0090	0.0110

Delays at 25°C, 1.8V, Typical Process

Description	Intrinsic Delay (ns)								
	XL	X1	X2	X3	X4	X8	X12	X16	X20
A → Y↑	0.0261	0.0253	0.0228	0.0243	0.0206	0.0198	0.1303	0.1276	0.1265
A → Y↓	0.0154	0.0146	0.0140	0.0146	0.0125	0.0125	0.1235	0.1232	0.1183

Description	K_{load} (ns/pF)								
	XL	X1	X2	X3	X4	X8	X12	X16	X20
A → Y↑	6.2539	4.5257	2.2629	1.5216	1.1447	0.5513	0.3680	0.2760	0.2209
A → Y↓	3.3414	2.3675	1.2661	0.8247	0.6333	0.3211	0.2194	0.1647	0.1316

TSMC 0.18μm Process SAGE-X™ Standard Cell Library Databook
110
Artisan

NAND2

Cell Description

The NAND2 cell provides the logical NAND of two inputs (A, B). The output (Y) is represented by the logic equation:

$Y = \overline{(A \cdot B)}$

Functions

A	B	Y
0	x	1
x	0	1
1	1	0

Logic Symbol

Cell Size

Drive Strength	Height (μm)	Width (μm)
NAND2XL	5.04	1.98
NAND2X1	5.04	1.98
NAND2X2	5.04	3.30
NAND2X4	5.04	4.62

AC Power

Pin	Power (μW/MHz)			
	XL	X1	X2	X4
A	0.0106	0.0139	0.0258	0.0501
B	0.0135	0.0181	0.0351	0.0683

Pin Capacitance

Pin	Capacitance (pF)			
	XL	X1	X2	X4
A	0.0033	0.0042	0.0082	0.0163
B	0.0029	0.0039	0.0086	0.0158

Delay at 25°C, 1.8V, Typical Process

Description	Intrinsic Delay (ns)			
	XL	X1	X2	X4
A → Y↑	0.0324	0.0313	0.0286	0.0295
A → Y↓	0.0198	0.0195	0.0179	0.0181
B → Y↑	0.0409	0.0400	0.0389	0.0395
B → Y↓	0.0241	0.0242	0.0239	0.0235

Description	K_{load} (ns/pF)			
	XL	X1	X2	X4
A → Y↑	6.2614	4.5288	2.1944	1.1638
A → Y↓	3.8144	2.8429	1.3901	0.7200
B → Y↑	6.2527	4.5253	2.1935	1.1631
B → Y↓	3.8226	2.8470	1.3923	0.7210

TSMC 0.18μm Process SAGE-X™ Standard Cell Library Databook
130
Artisan

Figure 3.25 Artisan Components cell library datasheets. Reprinted with permission.

The slope of the delay vs. fanout curve indicates $\tau = 12.4$ ps and the y-intercept indicates $p_{\text{inv}} = 20.0$ ps, or $(20.0/12.4) = 1.61$ in normalized terms. This is larger than the delay of 1 estimated earlier, probably because it includes capacitance of internal wires.

By a similar calculation, we find the X1 2-input NAND gate has an average delay from the inner (A) input of

$$t_{pd} = \left(\frac{31.3 + 19.5}{2}\right)\text{ps} + \left(4.2\frac{\text{fF}}{\text{gate}}\right)(h \text{ gates})\left(\frac{4.53 + 2.84}{2}\frac{\text{ns}}{\text{pF}}\right) = (25.4 + 15.5h)\text{ ps} \quad \textbf{(3.27)}$$

Thus, the parasitic delay is $(25.4/12.4) = 2.05$ and the logical effort is $(15.5/12.4) = 1.25$. The logical effort is slighly better than the theoretical 4/3 value, for reasons to be explored in Section 3.4.6.3. The parasitic delay from the outer (B) input is slightly higher, as expected. The parasitic delay and logical effort of the X2 and X4 gates are similar, confirming our model that logical effort should be independent of gate size for gates of reasonable sizes.

3.4.6 Limitations to the Linear Delay Model

The linear delay model works remarkably well even in advanced technologies; for example, Figure 7.30 shows subpicosecond agreement in a 65 nm process assuming that input and output slopes are matched. Nevertheless, it also has limitations that should be understood when more accuracy is needed.

3.4.6.1 Input and Output Slope The largest source of error in the linear delay model is the input slope effect. Figure 3.26(a) shows a fanout-of-4 inverter driven by ramps with different slopes. Recall that the ON current increases with the gate voltage for an nMOS transistor. We say the transistor is *OFF* for $V_{gs} < V_t$, *fully ON* for $V_{gs} = V_{DD}$, and *partially ON* for intermediate gate voltages. As the rise time of the input increases, the delay also increases because the active transistor is not turned fully ON at once. Figure 3.26(b) plots average inverter propagation delay vs. input rise time. Notice that the delay vs. rise time data fits a straight line quite well [Hedenstierna87].

Accounting for slopes is important for accurate timing analysis (see Section 3.6), but is generally more complex than is worthwhile for hand calculations. Fortunately, we will see in Section 3.5 that circuits are fastest when each gate has the same effort delay and when that delay is roughly 4τ. Because slopes are related to edge rate, fast circuits tend to have relatively consistent slopes. If a cell library is characterized with these slopes, it will tend to be used in the regime in which it most accurately models delay.

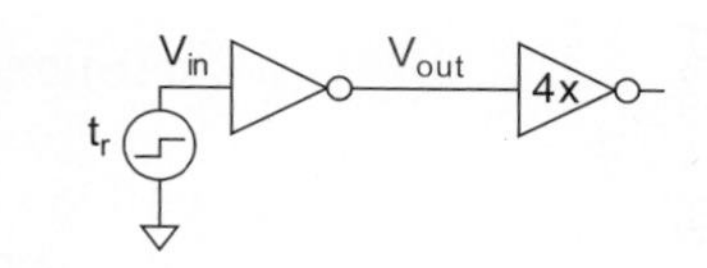

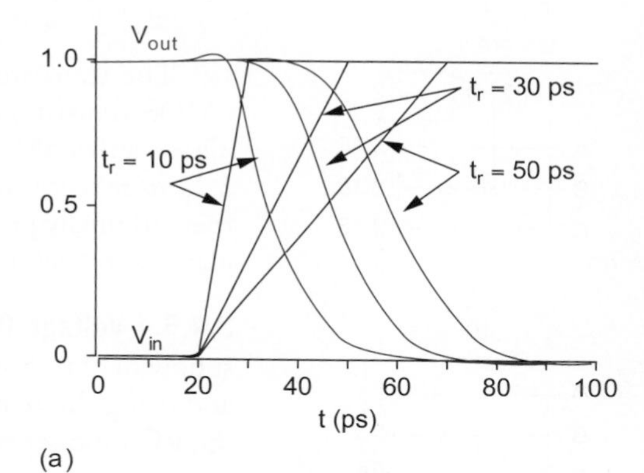

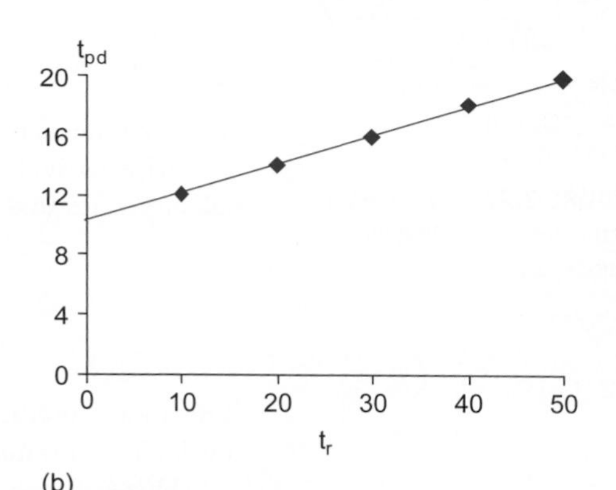

FIGURE 3.26 SPICE simulation of slope effect on CMOS inverter delay

3.4.6.2 Input Arrival Times Another source of error in the linear delay model is the assumption that one input of a multiple-input gate switches while the others are completely stable. When two inputs to a series stack turn ON simultaneously, the delay will be slightly longer than predicted because both transistors are only partially ON during the initial part of the transition. When two inputs to a parallel stack turn ON simultaneously, the delay will be shorter than predicted because both transistors deliver current to the output. The delays are also slightly different depending on which input arrives first, as will be explored in Section 7.5.3.

3.4.6.3 Velocity Saturation The estimated logical efforts assume that N transistors in series must be N times as wide to give equal current. However, as discussed in Section 3.3.1, series transistors see less velocity saturation and hence have lower resistance than we estimated [Sakurai91].

To make a better estimate, observe that N transistors in series are equivalent to one transistor with N times the channel length. Substituting L and NL into EQ (2.28) shows that the ratio of I_{dsat} for two series transistors to that of a single transistor is

$$\frac{I_{dsat-N-series}}{I_{dsat}} = \frac{(V_{DD} - V_t) + V_c}{(V_{DD} - V_t) + NV_c} \quad (3.28)$$

In the limit that the transistors are not at all velocity saturated ($V_c >> V_{DD} - V_t$), the current ratio reduces to $1/N$ as predicted. In the limit that the transistors are completely velocity saturated, the current is independent of the number of series transistors.

Example 3.12

Determine the relative saturation current of 2- and 3-transistor nMOS and pMOS stacks in a 65 nm process. $V_{DD} = 1.0$ V and $V_t = 0.3$ V. Use $V_c = E_c L = 1.04$ V for nMOS devices and 2.22 V for pMOS devices.

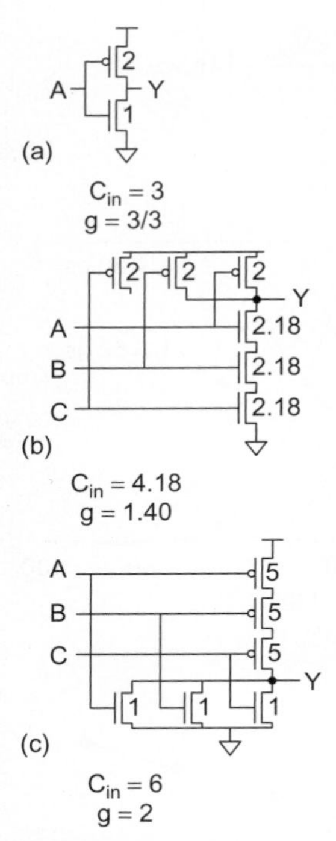

FIGURE 3.27 Logical effort estimates accounting for velocity saturation

SOLUTION: Applying EQ (3.28) gives a ratio of 0.63 for 2 nMOS transistors, 0.46 for 3 nMOS transistors, 0.57 for 2 pMOS transistors, and 0.40 for 3 pMOS transistors. The pMOS are closer to the ideal result of 0.5 and 0.33 because they experience less velocity saturation.

The transistors are scaled to deliver the same current as an inverter. Three series nMOS transistors must be 1/0.46 = 2.18 times as wide, rather than three times as wide. Three series pMOS transistors must be 2.5 times as wide. Figure 3.27 modifies Figure 3.22 to reflect velocity saturation. The logical efforts of the NAND and NOR are lower than originally predicted, and agree with the results obtained by curve-fitting SPICE simulations in Section 7.5.3.

3.4.6.4 Voltage Dependence Designers often need to predict how delay will vary if the supply or threshold voltage is changed. Recalling that delay is proportional to CV_{DD} / I and using the α-power law model of EQ (2.30) for I_{dsat}, we can estimate the scaling of the RC time constant and of gate delay as

$$\tau = k\frac{CV_{DD}}{\left(V_{DD} - V_t\right)^{\alpha}} \tag{3.29}$$

where k reflects process parameters.

Alternatively, using the straight line saturation current model from EQ (2.32) for velocity-saturated transistors, we obtain an even simpler estimate:

$$\tau = k\frac{CV_{DD}}{\left(V_{DD} - V_t^*\right)} = \frac{kC}{1 - \frac{V_t^*}{V_{DD}}} \tag{3.30}$$

This model predicts that the supply voltage can be reduced without changing the delay of a velocity-saturated transistor so long as the threshold is reduced in proportion.

When $V_{DD} < V_t$, delay instead depends on the subthreshold current of EQ (2.45):

$$\tau = k\frac{CV_{DD}}{I_{\text{off}} 10^{\frac{V_{DD}}{S}}} \tag{3.31}$$

3.4.6.5 Gate-Source Capacitance The examples in Section 3.3 assumed that gate capacitance terminates on a fixed supply rail. As discussed in Section 2.3.2, the bottom terminal of the gate oxide capacitor is the channel, which is primarily connected to the source when the transistor is ON. This means that as the source of a transistor changes value, charge is required to change the voltage on C_{gs}, adding to the delay for series stacks.

3.4.6.6 Bootstrapping Transistors also have some capacitance from gate to drain. This capacitance couples the input and output in an effect known as *bootstrapping*, which can be understood by examining Figure 3.28(a). Our models so far have only considered C_{in} (C_{gs}). This figure also considers C_{gd}, the gate to drain capacitance. In the case that the input is rising (the output starts high), the effective input capacitance is $C_{gs} + C_{gd}$. When the output starts to fall, the voltage across C_{gd} changes, requiring the input to supply additional current to charge C_{gd}. In other words, the impact of C_{gd} on gate capacitance is effectively doubled.

To illustrate the effect of the bootstrap capacitance on a circuit, Figure 3.28(b) shows two inverter pairs. The top pair has an extra bit of capacitance between the input and output of the second inverter. The bottom pair has the same amount of extra capacitance from input to ground. When x falls, nodes a and c begin to rise (Figure 3.28(c)). At first, both nodes see approximately the same capacitance, consisting of the two transistors and the extra 3 fF. As node a rises, it initially bumps up b or "lifts b by its own bootstraps." Eventually the nMOS transistors turn ON, pulling down b and d. As b falls, it tugs on a through the capacitor, leading to the slow final transition visible on node a. Also observe that b falls later than d because of the extra charge that must be supplied to discharge the bootstrap capacitor. In summary, the extra capacitance has a greater effect when connected between input and output as compared to when it is connected between input and ground.

Because C_{gd} is fairly small, bootstrapping is only a mild annoyance in digital circuits. However, if the inverter is biased in its linear region near $V_{DD}/2$, the C_{gd} is multiplied by the large gain of the inverter. This is known as the *Miller effect* and is of major importance in analog circuits.

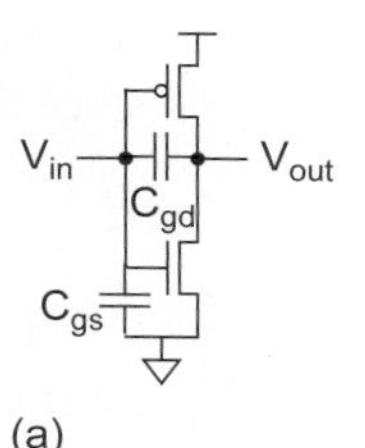

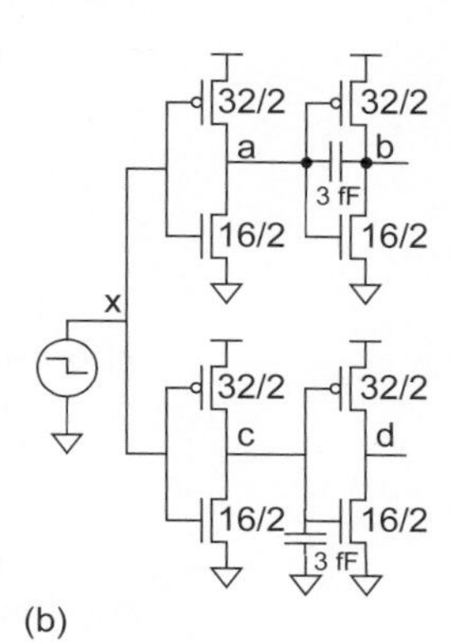

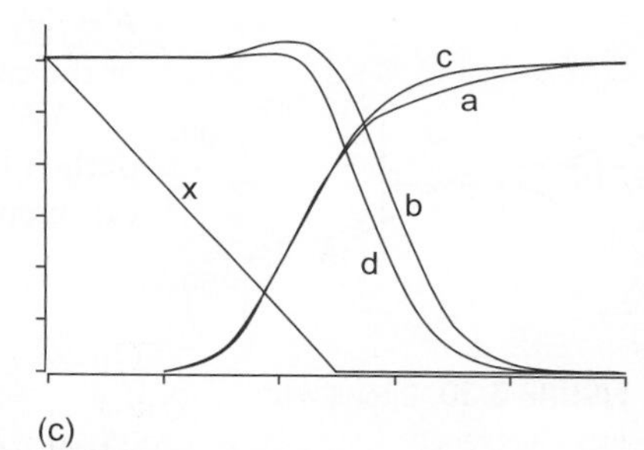

FIGURE 3.28 The effect of bootstrapping on inverter delay and waveform shape

3.5 Logical Effort of Paths

Designers often need to choose the fastest circuit topology and gate sizes for a particular logic function and to estimate the delay of the design. As has been stated, simulation or timing analysis are poor tools for this task because they only determine how fast a particular implementation will operate, not whether the implementation can be modified for better results and if so, what to change. Inexperienced designers often end up in the "simulate and tweak" loop involving minor changes and many fruitless simulations. The method of Logical Effort [Sutherland99] provides a simple method "on the back of an envelope" to choose the best topology and number of stages of logic for a function. Based on the linear delay model, it allows the designer to quickly estimate the best number of stages for a path, the minimum possible delay for the given topology, and the gate sizes that achieve this delay. The techniques of Logical Effort will be revisited throughout this text to understand the delay of many types of circuits.

3.5.1 Delay in Multistage Logic Networks

Figure 3.29 shows the logical and electrical efforts of each stage in a multistage path as a function of the sizes of each stage. The path of interest (the only path in this case) is marked with the dashed blue line. Observe that logical effort is independent of size, while electrical effort depends on sizes. This section develops some metrics for the path as a whole that are independent of sizing decisions.

10   x   y   z   20

$g_1 = 1$, $h_1 = x/10$ $g_2 = 5/3$, $h_2 = y/x$ $g_3 = 4/3$, $h_3 = z/y$ $g_4 = 1$, $h_4 = 20/z$

FIGURE 3.29 Multistage logic network

The *path logical effort* G can be expressed as the products of the logical efforts of each stage along the path.

$$G = \prod g_i \tag{3.32}$$

The *path electrical effort* H can be given as the ratio of the output capacitance the path must drive divided by the input capacitance presented by the path. This is more convenient than defining path electrical effort as the product of stage electrical efforts because we do not know the individual stage electrical efforts until gate sizes are selected.

$$H = \frac{C_{\text{out(path)}}}{C_{\text{in(path)}}} \tag{3.33}$$

The *path effort* F is the product of the stage efforts of each stage. Recall that the stage effort of a single stage is $f = gh$. Can we by analogy state $F = GH$ for a path?

$$F = \prod f_i = \prod g_i h_i \tag{3.34}$$

In paths that branch, $F \neq GH$. This is illustrated in Figure 3.30, a circuit with a two-way branch. Consider a path from the primary input to one of the outputs. The path logical effort is $G = 1 \times 1 = 1$. The path electrical effort is $H = 90/5 = 18$. Thus, $GH = 18$. But $F = f_1 f_2 = g_1 h_1 g_2 h_2 = 1 \times 6 \times 1 \times 6 = 36$. In other words, $F = 2GH$ in this path on account of the two-way branch.

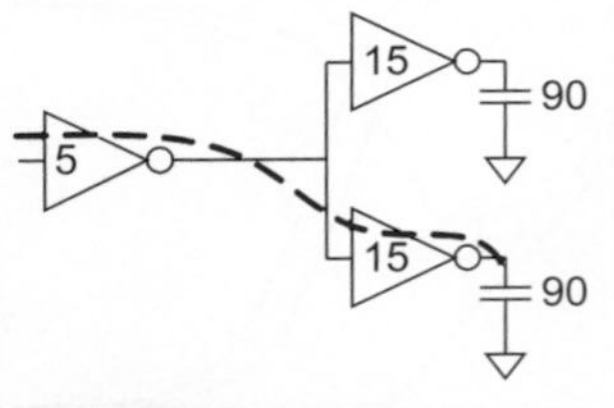

FIGURE 3.30 Circuit with two-way branch

We must introduce a new kind of effort to account for branching between stages of a path. This *branching effort* b is the ratio of the total capacitance seen by a stage to the capacitance on the path; in Figure 3.30 it is $(15 + 15)/15 = 2$.

$$b = \frac{C_{\text{onpath}} + C_{\text{offpath}}}{C_{\text{onpath}}} \tag{3.35}$$

The *path branching effort* B is the product of the branching efforts between stages.

$$B = \prod b_i \tag{3.36}$$

Now we can define the path effort F as the product of the logical, electrical, and branching efforts of the path. Note that the product of the electrical efforts of the stages is actually BH, not just H.

$$F = GBH \tag{3.37}$$

We can now compute the delay of a multistage network. The *path delay* D is the sum of the delays of each stage. It can also be written as the sum of the *path effort delay* D_F and *path parasitic delay* P:

$$\begin{aligned} D &= \sum d_i = D_F + P \\ D_F &= \sum f_i \\ P &= \sum p_i \end{aligned} \tag{3.38}$$

The product of the stage efforts is F, independent of gate sizes. The path effort delay is the sum of the stage efforts. The sum of a set of numbers whose product is constant is minimized by choosing all the numbers to be equal. In other words, the path delay is minimized when each stage bears the same effort. If a path has N stages and each bears the same effort, that effort must be

$$\hat{f} = g_i h_i = F^{1/N} \tag{3.39}$$

Thus, the minimum possible delay of an N-stage path with path effort F and path parasitic delay P is

$$D = NF^{1/N} + P \tag{3.40}$$

This is a key result of Logical Effort. It shows that the minimum delay of the path can be estimated knowing only the number of stages, path effort, and parasitic delays without the need to assign transistor sizes. This is superior to simulation, in which delay depends on sizes and you never achieve certainty that the sizes selected are those that offer minimum delay.

It is also straightforward to select gate sizes to achieve this least delay. Combining EQs (3.21) and (3.22) gives us the *capacitance transformation* formula to find the best input capacitance for a gate given the output capacitance it drives.

$$C_{\text{in}_i} = \frac{C_{\text{out}_i} \times g_i}{\hat{f}} \tag{3.41}$$

Starting with the load at the end of the path, work backward applying the capacitance transformation to determine the size of each stage. Check the arithmetic by verifying that the size of the initial stage matches the specification.

Example 3.13

Estimate the minimum delay of the path from A to B in Figure 3.31 and choose transistor sizes to achieve this delay. The initial NAND2 gate may present a load of 8 λ of transistor width on the input and the output load is equivalent to 45 λ of transistor width.

SOLUTION: The path logical effort is $G = (4/3) \times (5/3) \times (5/3) = 100/27$. The path electrical effort is $H = 45/8$. The path branching effort is $B = 3 \times 2 = 6$. The path effort is $F = GBH = 125$. As there are three stages, the best stage effort is $\hat{f} = \sqrt[3]{125} = 5$. The path parasitic delay is $P = 2 + 3 + 2 = 7$. Hence, the minimum path delay is $D = 3 \times 5 + 7 = 22$ in units of τ, or 4.4 FO4 inverter delays. The gate sizes are computed with the capacitance transformation from EQ (3.41) working backward along the path: $y = 45 \times (5/3)/5 = 15$. $x = (15 + 15) \times (5/3)/5 = 10$. We verify that the initial 2-input NAND gate has the specified size of $(10 + 10 + 10) \times (4/3)/5 = 8$. The transistor sizes in Figure 3.32 are chosen to give the desired amount of input capacitance while achieving equal rise and fall delays. For example, a 2-input NOR gate should have a 4:1 P/N ratio. If the total input capacitance is 15, the pMOS width must be 12 and the nMOS width must be 3 to achieve that ratio.

We can also check that our delay was achieved. The NAND2 gate delay is $d_1 = g_1 h_1 + p_1 = (4/3) \times (10 + 10 + 10)/8 + 2 = 7$. The NAND3

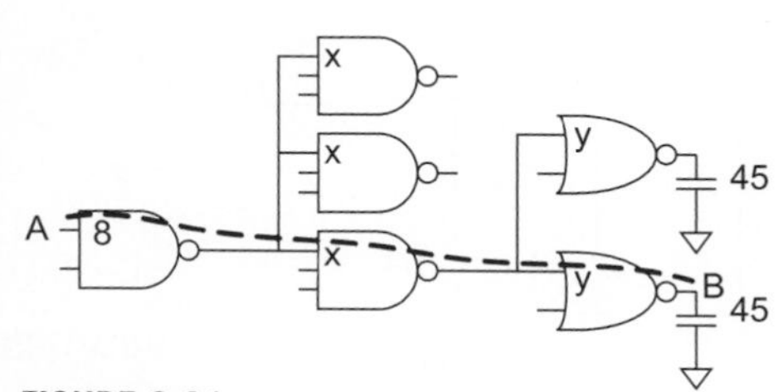

FIGURE 3.31 Example path

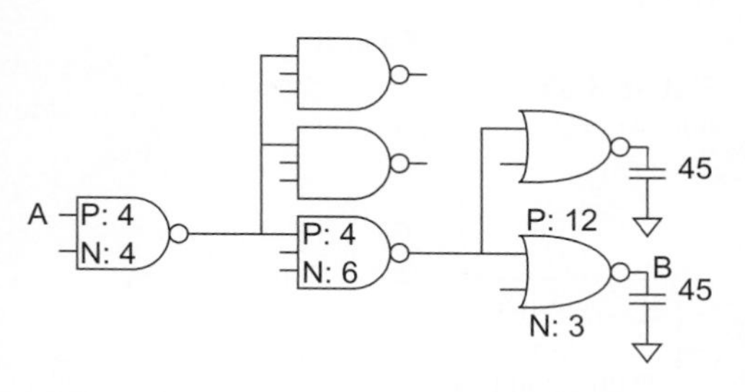

FIGURE 3.32 Example path annotated with transistor sizes

gate delay is $d_2 = g_2h_2 + p_2 = (5/3) \times (15 + 15)/10 + 3 = 8$. The NOR2 gate delay is $d_3 = g_3h_3 + p_3 = (5/3) \times 45/15 + 2 = 7$. Hence, the path delay is 22, as predicted.

Recall that delay is expressed in units of τ. In a 65 nm process with $\tau = 3$ ps, the delay is 66 ps. Alternatively, a fanout-of-4 inverter delay is 5τ, so the path delay is 4.4 FO4s.

Many inexperienced designers know that wider transistors offer more current and thus try to make circuits faster by using bigger gates. Increasing the size of any of the gates except the first one only makes the circuit slower. For example, increasing the size of the NAND3 makes the NAND3 faster but makes the NAND2 slower, resulting in a net speed loss. Increasing the size of the initial NAND2 gate does speed up the circuit under consideration. However, it presents a larger load on the path that computes input A, making that path slower. Hence, it is crucial to have a specification of not only the load the path must drive but also the maximum input capacitance the path may present.

3.5.2 Choosing the Best Number of Stages

Given a specific circuit topology, we now know how to estimate delay and choose gate sizes. However, there are many different topologies that implement a particular logic function. Logical Effort tells us that NANDs are better than NORs and that gates with few inputs are better than gates with many. In this section, we will also use Logical Effort to predict the best number of stages to use.

Logic designers sometimes estimate delay by counting the number of stages of logic, assuming each stage has a constant "gate delay." This is potentially misleading because it implies that the fastest circuits are those that use the fewest stages of logic. Of course, the gate delay actually depends on the electrical effort, so sometimes using fewer stages results in more delay. The following example illustrates this point.

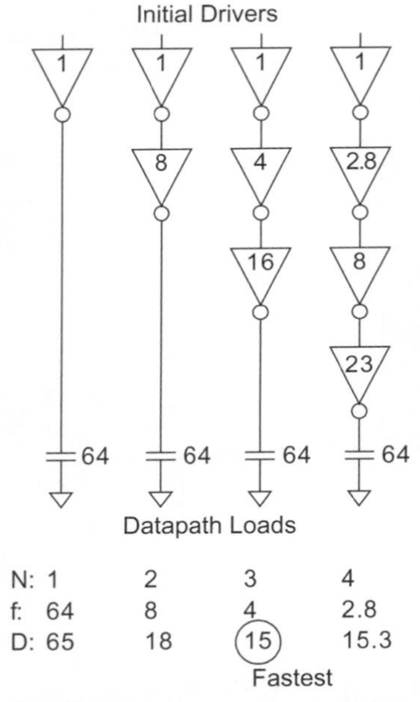

FIGURE 3.33 Comparison of different number of stages of buffers

Example 3.14

A control unit generates a signal from a unit-sized inverter. The signal must drive unit-sized loads in each bitslice of a 64-bit datapath. The designer can add inverters to buffer the signal to drive the large load. Assuming polarity of the signal does not matter, what is the best number of inverters to add and what delay can be achieved?

SOLUTION: Figure 3.33 shows the cases of adding 0, 1, 2, or 3 inverters. The path electrical effort is $H = 64$. The path logical effort is $G = 1$, independent of the number of inverters. Thus, the path effort is $F = 64$. The inverter sizes are chosen to achieve equal stage effort. The total delay is $D = N\sqrt[N]{64} + N$.

The 3-stage design is fastest and far superior to a single stage. If an even number of inversions were required, the two- or four-stage designs are promising. The four-stage design is slightly faster, but the two-stage design requires significantly less area and power.

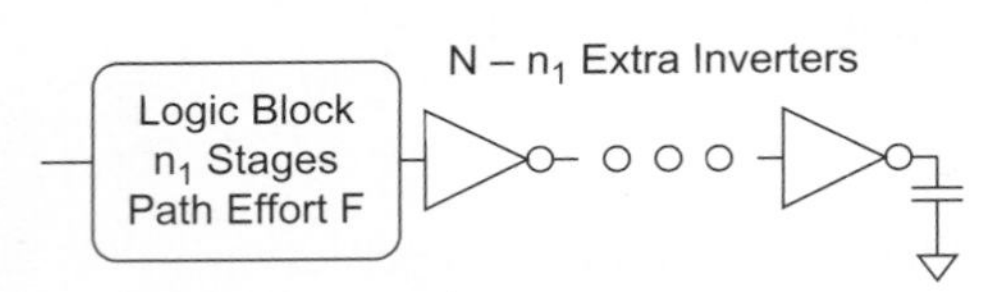

FIGURE 3.34 Logic block with additional inverters

In general, you can always add inverters to the end of a path without changing its function (save possibly for polarity). Let us compute how many should be added for least delay. The logic block shown in Figure 3.34 has n_1 stages and a path effort of F. Consider adding $N - n_1$ inverters to the end to bring the path to N stages. The extra inverters do not change the path logical effort but do add parasitic delay. The delay of the new path is

$$D = NF^{1/N} + \sum_{i=1}^{n_1} p_i + (N - n_1) p_{\text{inv}} \tag{3.42}$$

Differentiating with respect to N and setting to 0 allows us to solve for the best number of stages, which we will call $\hat{N}$. The result can be expressed more compactly by defining

$$\rho = F^{1/\hat{N}}$$

to be the best stage effort.

$$\begin{aligned} \frac{\partial D}{\partial N} &= -F^{1/N} \ln F^{1/N} + F^{1/N} + p_{\text{inv}} = 0 \\ &\Rightarrow p_{\text{inv}} + \rho(1 - \ln \rho) = 0 \end{aligned} \tag{3.43}$$

EQ (3.43) has no closed form solution. Neglecting parasitics (i.e., assuming $p_{\text{inv}} = 0$), we find the classic result that the stage effort $\rho = 2.71828$ (e) [Mead80]. In practice, the parasitic delays mean each inverter is somewhat more costly to add. As a result, it is better to use fewer stages, or equivalently a higher stage effort than e. Solving numerically, when $p_{\text{inv}} = 1$, we find $\rho = 3.59$.

A path achieves least delay by using $\hat{N} = \log_\rho F$ stages. It is important to understand not only the best stage effort and number of stages but also the sensitivity to using a different number of stages. Figure 3.35 plots the delay increase using a particular number of stages against the total number of stages, for $p_{\text{inv}} = 1$. The x-axis plots the ratio of the actual number of stages to the ideal number. The y-axis plots the ratio of the actual delay to the best achievable. The curve is flat around the optimum. The delay is within 15% of the best achievable if the number of stages is within 2/3 to 3/2 times the theoretical best number (i.e., ρ is in the range of 2.4 to 6).

Using a stage effort of 4 is a convenient choice and simplifies mentally choosing the best number of stages. This effort gives delays within 2% of minimum for p_{inv} in the range of 0.7 to 2.5. This further explains why a fanout-of-4 inverter has a "representative" logic gate delay.

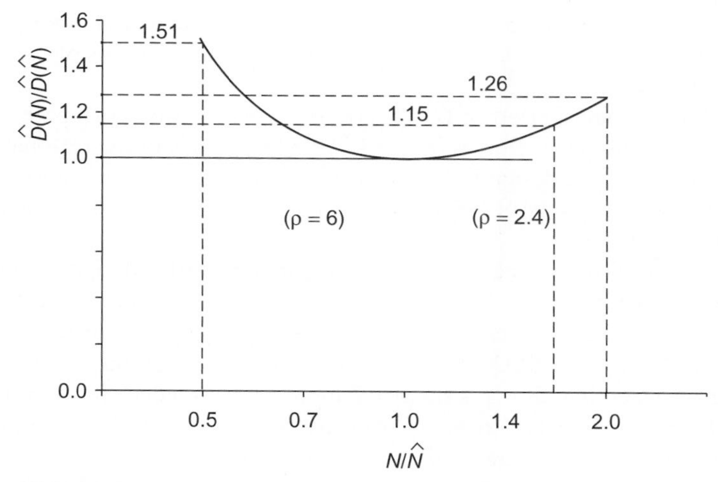

FIGURE 3.35 Sensitivity of delay to number of stages

3.5.3 Example

Consider a larger example to illustrate the application of Logical Effort. Our esteemed colleague Ben Bitdiddle is designing a decoder for a register file in the Motoroil 68W86, an embedded processor for automotive applications. The decoder has the following specifications:

- 16-word register file
- 32-bit words
- Each register bit presents a load of three unit-sized transistors on the word line (two unit-sized access transistors plus some wire capacitance)
- True and complementary versions of the address bits $A[3:0]$ are available
- Each address input can drive 10 unit-sized transistors

As we will see further in Section 11.2.2, a 2^N-word decoder consists of 2^N N-input AND gates. Therefore, the problem is reduced to designing a suitable 4-input AND gate. Let us help Ben determine how many stages to use, how large each gate should be, and how fast the decoder can operate.

The output load on a word line is 32 bits with three units of capacitance each, or 96 units. Therefore, the path electrical effort is $H = 96/10 = 9.6$. Each address is used to compute half of the 16 word lines; its complement is used for the other half. Therefore, a $B = 8$-way branch is required somewhere in the path. Now we are faced with a chicken-and-egg dilemma. We need to know the path logical effort to calculate the path effort and best number of stages. However, without knowing the best number of stages, we cannot sketch a path and determine the logical effort for that path. There are two ways to resolve the dilemma. One is to sketch a path with a random number of stages, determine the path logical effort, and then use that to compute the path effort and the actual number of stages. The path can be redesigned with this number of stages, refining the path logical effort. If the logical effort changes significantly, the process can be repeated. Alternatively, we know that the logic of a decoder is rather simple, so we can ignore the logical effort (assume $G = 1$). Then we can proceed with our design, remembering that the best number of stages is likely slightly higher than predicted because we neglected logical effort.

Taking the second approach, we estimate the path effort is $F = GBH = (1)(8)(9.6) = 76.8$. Targeting a best stage effort of $\rho = 4$, we find the best number of stages is $N = \log_4 76.8 = 3.1$. Let us select a 3-stage design, recalling that a 4-stage design might be a good choice too when logical effort is considered. Figure 3.36 shows a possible 3-stage design (INV-NAND4-INV).

The path has a logical effort of $G = 1 \times (6/3) \times 1 = 2$, so the actual path effort is $F = (2)(8)(9.6) = 154$. The stage effort is $\hat{f} = 154^{1/3} = 5.36$. This is in the reasonable range of 2.4 to 6, so we expect our design to be acceptable. Applying the capacitance transformation, we find gate sizes $z = 96 \times 1/5.36 = 18$ and $y = 18 \times 2 /5.36 = 6.7$. The delay is $3 \times 5.36 + 1 + 4 + 1 = 22.1$.

Logical Effort also allows us to rapidly compare alternative designs using a spreadsheet rather than a schematic editor and a large number of simulations. Table 3.4 compares a number of alternative designs. We find a 4-stage design is somewhat faster, as we suspected. The 4-stage NAND2-INV-NAND2-INV design not only has the theoretical best number of stages, but also uses simpler 2-input gates to reduce the logical effort and parasitic delay to obtain a 12% speedup over the original design. However, the 3-stage design has a smaller total gate area and dissipates less power.

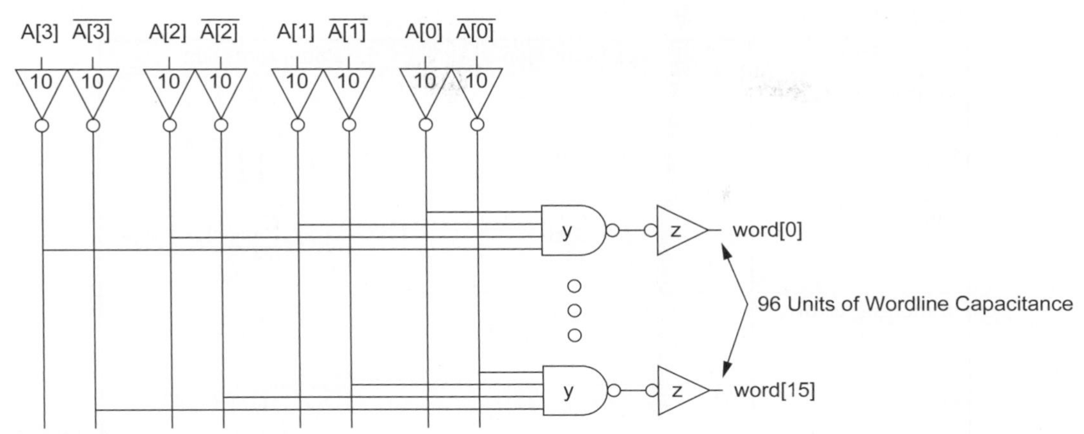

FIGURE 3.36 3-stage decoder design

TABLE 3.4 Spreadsheet comparing decoder designs

Design	Stages *N*	*G*	*P*	*D*
NAND4-INV	2	2	5	29.8
NAND2-NOR2	2	20/9	4	30.1
INV-NAND4-INV	**3**	**2**	**6**	**22.1**
NAND4-INV-INV-INV	4	2	7	21.1
NAND2-NOR2-INV-INV	4	20/9	6	20.5
NAND2-INV-NAND2-INV	4	16/9	6	19.7
INV-NAND2-INV-NAND2-INV	5	16/9	7	20.4
NAND2-INV-NAND2-INV-INV-INV	6	16/9	8	21.6

3.5.4 Summary and Observations

Logical Effort provides an easy way to compare and select circuit topologies, choose the best number of stages for a path, and estimate path delay. The notation takes some time to become natural, but this author has poured through all the letters in the English and Greek alphabets without finding better notation. It may help to remember d for "**d**elay," p for "**p**arasitic," b for "**b**ranching," f for "e**f**fort," g for "lo**g**ical effort" (or perhaps **g**ain), and h as the next letter after "f" and "g." The notation is summarized in Table 3.5 for both stages and paths.

The method of Logical Effort is applied with the following steps:

1. Compute the path effort: $F = GBH$
2. Estimate the best number of stages: $\hat{N} = \log_4 F$
3. Sketch a path using: $\hat{N}$ stages
4. Estimate the minimum delay: $D = \hat{N}F^{1/\hat{N}} + P$
5. Determine the best stage effort: $\hat{f} = F^{1/\hat{N}}$
6. Starting at the end, work backward to find sizes: $C_{\text{in}_i} = \dfrac{C_{\text{out}_i} \times g_i}{\hat{f}}$

TABLE 3.5 Summary of Logical Effort notation

Term	Stage Expression	Path Expression
number of stages	1	N
logical effort	g (see Table 3.2)	$G = \prod g_i$
electrical effort	$h = \frac{C_{\text{out}}}{C_{\text{in}}}$	$H = \frac{C_{\text{out(path)}}}{C_{\text{in(path)}}}$
branching effort	$b = \frac{C_{\text{onpath}} + C_{\text{offpath}}}{C_{\text{onpath}}}$	$B = \prod b_i$
effort	$f = gh$	$F = GBH$
effort delay	f	$D_F = \sum f_i$
parasitic delay	p (see Table 3.3)	$P = \sum p_i$
delay	$d = f + p$	$D = \sum d_i = D_F + P$

CAD tools are very fast and accurate at evaluating complex delay models, so Logical Effort should not be used as a replacement for such tools. Rather, its value arises from "quick and dirty" hand calculations and from the insights it lends to circuit design. Some of the key insights include:

- The idea of a numeric "logical effort" that characterizes the complexity of a logic gate or path allows you to compare alternative circuit topologies and show that some topologies are better than others.
- NAND structures are faster than NOR structures in static CMOS circuits.
- Paths are fastest when the effort delays of each stage are about the same and when these delays are close to four.
- Path delay is insensitive to modest deviations from the optimum. Stage efforts of 2.4–6 give designs within 15% of minimum delay. There is no need to make calculations to more than 1–2 significant figures, so many estimations can be made in your head. There is no need to choose transistor sizes exactly according to theory and there is little benefit in tweaking transistor sizes if the design is reasonable.
- Using stage efforts somewhat greater than 4 reduces area and power consumption at a slight cost in speed. Using efforts greater than 6–8 comes at a significant cost in speed.
- Using fewer stages for "less gate delays" does not make a circuit faster. Making gates larger also does not make a circuit faster; it only increases the area and power consumption.
- The delay of a well-designed path is about $\log_4 F$ fanout-of-4 (FO4) inverter delays. Each quadrupling of the load adds about one FO4 inverter delay to the path. Control signals fanning out to a 64-bit datapath therefore incur an amplification delay of about three FO4 inverters.

- The logical effort of each input of a gate increases through no fault of its own as the number of inputs grows. Considering both logical effort and parasitic delay, we find a practical limit of about four series transistors in logic gates and about four inputs to multiplexers. Beyond this fan-in, it is faster to split gates into multiple stages of skinnier gates.
- Inverters or 2-input NAND gates with low logical efforts are best for driving nodes with a large branching effort. Use small gates after the branches to minimize load on the driving gate.
- When a path forks and one leg is more critical than the others, buffer the noncritical legs to reduce the branching effort on the critical path.

3.5.5 Limitations of Logical Effort

Logical Effort is based on the linear delay model and the simple premise that making the effort delays of each stage equal minimizes path delay. This simplicity is the method's greatest strength, but also results in a number of limitations:

- Logical Effort does not account for interconnect. The effects of nonnegligible wire capacitance and RC delay will be revisited in Chapter 5. Logical Effort is most applicable to high-speed circuits with regular layouts where routing delay does not dominate. Such structures include adders, multipliers, memories, and other datapaths and arrays.
- Logical Effort explains how to design a critical path for maximum speed, but not how to design an entire circuit for minimum area or power given a fixed speed constraint. This problem is addressed in Section 4.2.2.1.
- Paths with nonuniform branching or reconvergent fanout are difficult to analyze by hand.
- The linear delay model fails to capture the effect of input slope. Fortunately, edge rates tend to be about equal in well-designed circuits with equal effort delay per stage.

3.5.6 Iterative Solutions for Sizing

To address the limitations in the previous section, we can write the delay equations for each gate in the system and minimize the latest arrival time. No closed-form solutions exist, but the equations are easy to solve iteratively on a computer and the formulation still gives some insight for the designer. This section examines sizing for minimum delay, while Section 4.2.2.1 examines sizing for minimum energy subject to a delay constraint.

The ith gate is characterized by its logical effort, g_i, parasitic delay, p_i, and drive, x_i. Formally, our goal is to find a nonnegative vector of drives x that minimizes the arrival time of the latest output. This can be done using a commercial optimizer such as MOSEK or, for smaller problems, Microsoft Excel's solver. The arrival time equations are classified as *convex*, which has the pleasant property of having a single optimum; there is no risk of finding a wrong answer. Moreover, they are of a special class of functions called *posynomials*, which allows an especially efficient technique called *geometric programming* to be applied [Fishburn85].

Example 3.15

The circuit in Figure 3.37 has nonuniform branching, reconvergent fanout, and a wire load in the middle of the path, all of which stymie back-of-the-envelope application of Logical Effort. The wire load is given in the same units as the gate capacitances (i.e., multiples of the capacitance of a unit inverter). Assume the inputs arrive at time 0. Write an expression for the arrival time of the output as a function of the gate drives. Determine the sizes to achieve minimum delay.

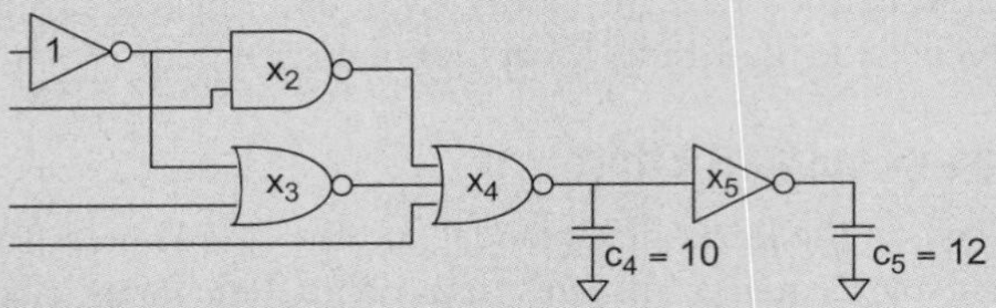

FIGURE 3.37 Example path

SOLUTION: The delay equations for each gate are obtained using EQ (3.25). Note that x indicates drive, not size. According to EQ (3.24), the input capacitance of a gate with logical effort g and drive x is $C_{\text{in}} = gx$.

$$\begin{aligned}
d_1 &= 1 + \tfrac{4}{3}x_2 + \tfrac{5}{3}x_3 \\
d_2 &= 2 + \tfrac{7}{3}\frac{x_4}{x_2} \\
d_3 &= 2 + \tfrac{7}{3}\frac{x_4}{x_3} \\
d_4 &= 3 + \frac{x_5}{x_4} + \frac{10}{x_4} \\
d_5 &= 1 + \frac{12}{x_5}
\end{aligned} \tag{3.44}$$

Write the arrival times using the definitions from EQ (3.1).

$$\begin{aligned}
a_1 &= d_1 \\
a_2 &= a_1 + d_2 \\
a_3 &= a_1 + d_3 \\
a_4 &= \max\{a_2, a_3\} + d_4 \\
a_5 &= a_4 + d_5 = d_1 + \max\{d_2, d_3\} + d_4 + d_5
\end{aligned} \tag{3.45}$$

Use a solver to choose the drives to minimize the latest arrival time. Table 3.6 summarizes the results. The minimum delay is 23.44.

The example leads to several interesting observations:

- In paths that branch, each fork should contribute equal delay. If one fork were faster than the other, it could be downsized to reduce the capacitance it presents to the stage before the branch.

- The stage efforts, f, are equal for each gate in paths with no fixed capacitive loads, but may change after a load.
- To minimize delay, upsize gates on nodes with large fixed capacitances to reduce the effort borne by the gate, while only slightly increasing the effort borne by the predecessor.

TABLE 3.6 Path design for minimum delays

Stage (i)	x_i	f_i	c_{in}	d_i	a_i
1: INV	1	4.85	1	5.85	5.85
2: NAND2	1.62	4.85	2.16	6.85	12.70
3: NOR2	1.62	4.85	2.70	6.85	12.70
4: NOR3	3.37	4.85	7.86	7.85	20.55
5: INV	6.35	1.89	6.35	2.89	23.44

A standard cell library offers a discrete set of sizes. Gate drives must be rounded to the nearest available size. For example, the circuit might use inv_1x, nand2_2x, nor2_2x, nor3_3x, and inv_6x. The delay increases to 23.83, less than a 2% penalty. In general, libraries with a granularity of $\sqrt{2}$ between successive drives are nearly as good as those with continuous sizes, so long as large inverters are available to drive big loads. Even using a granularity of 2 between drives (1x, 2x, 4x, 8x) is sufficient to obtain good results.

Although this section used a linear delay model to build on the insights of Logical Effort, it is also possible to use more elaborate models taking into account sensitivity to edge rate, V_{DD}, and V_t [Patil07]; the extra complexity is not a problem for a numerical solver and the model allows for optimizing supply and threshold voltages as well as sizes. Timing models are discussed further in Section 3.6.

3.6 Timing Analysis Delay Models

To handle a chip with millions of gates, the delay model for a timing analyzer must be easy enough to compute that timing analysis is fast, yet accurate enough to give confidence. This section reviews several delay models for timing analysis that are much faster than SPICE simulations, yet more accurate than the simple linear delay model. Timing (and area, power, and noise) models for each gate in a standard cell library are stored in a .lib file. These models are part of the Liberty standard documented at `www.opensourceliberty.org`. Logical effort parameters for standard cells can be obtained by fitting a straight line to the timing models, assuming equal delays and rise/fall times for the previous stage.

3.6.1 Slope-Based Linear Model

A simple approach is to extend the linear delay model by adding a term reflecting the input slope. Assuming the slope of the input is proportional to the delay of the previous stage, the delays for rising and falling outputs can be expressed as:

```
delay_rise =intrinsic_rise + rise_resistance × capacitance +
            slope_rise × delay_previous
delay_fall =intrinsic_fall + fall_resistance × capacitance +
            slope_fall × delay_previous
```

Linear delay models are not accurate enough to handle the wide range of slopes and loads found in synthesized circuits, so they have largely been superseded by nonlinear delay models.

3.6.2 Nonlinear Delay Model

A nonlinear delay model looks up the delay from a table based on the load capacitance and the input slope. Separate tables are used to lookup rising and falling delays and output slopes. Table 3.7 shows an example of a nonlinear delay model for the falling delay of an inverter. The timing analyzer uses interpolation when a specific load capacitance or slope is not in the table.

TABLE 3.7 Nonlinear Delay Model for inverter t_{pdf} (ps)

C_{out} (fF)	Rise Time (ps)				
	10	20	40	80	160
1	11.5	13.3	17.0	21.2	25.3
2	18.4	20.2	24.1	30.9	38.5
4	32.0	33.8	37.6	43.4	58.5
8	59.2	60.9	65.7	72.3	87.8

Nonlinear delay models are widely used at the time of this writing. However, they do not contain enough information to characterize the delay of a gate driving a complex RC interconnect network with the accuracy desired by some users. They also lack the accuracy to fully characterize noise events. A different model must be created for each voltage and temperature at which the chip might be characterized.

3.6.3 Current Source Model

The limitations of nonlinear delay models have motivated the development of current source models. A *current source model* theoretically should express the output DC current as a nonlinear function of the input and output voltages of the cell. A timing analyzer numerically integrates the output current to find the voltage as a function of time into an arbitrary RC network and to solve for the propagation delay.

The Liberty *Composite Current Source Model* (CCSM) instead stores output current as a function of time for a given input slew rate and output capacitance. The competing *Effective Current Source Model* (ECSM) stores output voltage as a function of time. The two representations are equivalent, and can be synthesized into a true current source model [Chopra06].

3.7 Pitfalls and Fallacies

Defining gate delay for an unloaded gate

When marketing a process, it is common to report gate delay based on an inverter in a ring oscillator (2τ), or even the RC time constant of a transistor charging its own gate capacitance ($1/3\ \tau$). Remember that the delay of a real gate on the critical path should be closer to 5–6τ.

When in doubt, ask how "gate delay" is defined or ask for the FO4 inverter delay.

Trying to increase speed by increasing the size of transistors in a path
Most designers know that increasing the size of a transistor decreases its resistance and thus makes it faster at driving a constant load. Novice designers sometimes forget that increasing the size increases input capacitance and makes the previous stage slower, especially when that previous stage belongs to somebody else's timing budget. The authors have seen this lead to lack of convergence in full-chip timing analysis on a large microprocessor because individual engineers boost the size of their own gates until their path meets timing. Only after the weekly full-chip timing roll-up do they discover that their inputs now arrive later because of the greater load on the previous stage. The solution is to include in the specification of each block not only the arrival time but also the resistance of the driver in the previous block.

Trying to increase speed by using as few stages of logic as possible
Logic designers often count "gate delays" in a path. This is a convenient simplification when used properly. In the hands of an inexperienced engineer who believes each gate contributes a gate delay, it suggests that the delay of a path is minimized by using as few stages of logic as possible, which is clearly untrue.

3.8 Historical Perspective

Figure 1.5 illustrated the exponential increase in microprocessor frequencies over nearly four decades. While much of the improvement comes from the natural improvements in gate delay with feature size, a significant portion is due to better microarchitecture and circuit design with fewer gate delays per cycle. From a circuit perspective, the cycle time is best expressed in FO4 inverter delays.

Figure 3.38 illustrates the historical trends in microprocessor cycle time based on chips reported at the International Solid-State Circuits Conference. Early processors operated at close to 100 FO4 delays per cycle. The Alpha line of microprocessors from Digital Equipment Corporation shocked the staid world of circuit design in the early 1990s by proving that cycle times below 20 FO4 delays were possible. This kicked off a race for higher clock frequencies. By the late 1990s, Intel and AMD marketed processors primarily on frequency. The Pentium II and III reached about 20–24 FO4 delays/cycle. The Pentium 4 drove cycle times down to about 10 FO4 at the expense of a very long

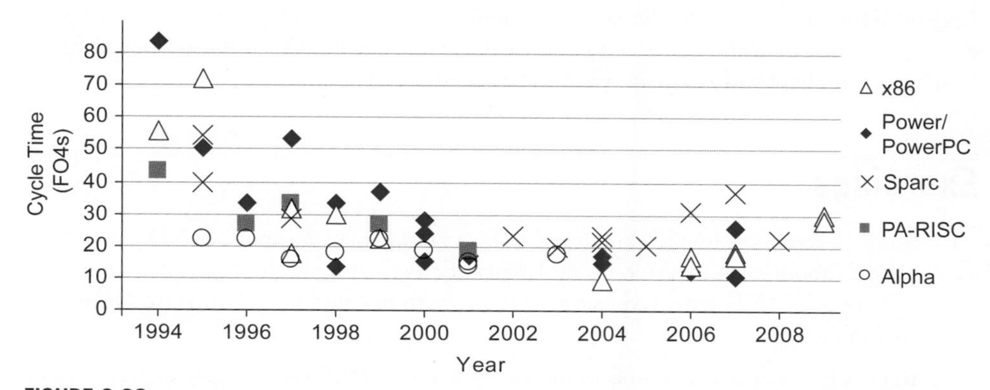

FIGURE 3.38 Microprocessor cylcle time trends. Data has some uncertainty based on estimating FO4 delay as a function of feature size.

pipeline and enormous power consumption. Microarchitects predicted that performance would be maximized at a cycle time of only 8 FO4 delays/cycle [Hrishikesh02].

The short cycle times came at the expense of vast numbers (20–30) of pipeline stages and enormous power consumption (nearly 100 W). As will be seen in the next chapter, power became as important as performance specifications. The number of gates per cycle rebounded to a more power-efficient point. [Srinivasan02] observed that 19–24 FO4 delays per cycle provides a better trade-off between performance and power.

Application-specific integrated circuits have generally operated at much lower frequencies (e.g., 200–400 MHz in nanometer processes) so that they can be designed more easily. Typical ASIC cycle times are 40–100 FO4 delays per cycle [Mai05, Chinnery02], although performance-critical designs sometimes are as fast as 25 FO4s.

Summary

The VLSI designer's challenge is to engineer a system that meets speed requirements while consuming little power or area, operating reliably, and taking little time to design. Circuit simulation is an important tool for calculating delay and will be discussed in depth in Chapter 4, but it takes too long to simulate every possible design; is prone to garbage-in, garbage-out mistakes; and doesn't give insight into why a circuit has a particular delay or how the circuit should be changed to improve delay. The designer must also have simple models to quickly estimate performance by hand and explain why some circuits are better than others.

Although transistors are complicated devices with nonlinear current-voltage and capacitance-voltage relationships, for the purpose of delay estimation in digital circuits, they can be approximated quite well as having constant capacitance and an effective resistance R when ON. Logic gates are thus modeled as RC networks. The Elmore delay model estimates the delay of the network as the sum of each capacitance times the resistance through which it must be charged or discharged. Therefore, the gate delay consists of a parasitic delay (accounting for the gate driving its own internal parasitic capacitance) plus an effort delay (accounting for the gate driving an external load). The effort delay depends on the electrical effort (the ratio of load capacitance to input capacitance, also called *fanout*) and the logical effort (which characterizes the current driving capability of the gate relative to an inverter with equal input capacitance). Even in advanced fabrication processes, the delay vs. electrical effort curve fits a straight line very well. The method of Logical Effort builds on this linear delay model to help us quickly estimate the delay of entire paths based on the effort and parasitic delay of the path. We will use Logical Effort in subsequent chapters to explain what makes circuits fast.

Exercises

3.1 Some designers define a "gate delay" to be a fanout-of-3 2-input NAND gate rather than a fanout-of-4 inverter. Using Logical Effort, estimate the delay of a fanout-of-3 2-input NAND gate. Express your result both in τ and in FO4 inverter delays, assuming $p_{inv} = 1$.

3.2 Repeat Exercise 3.1 in a process with a lower ratio of diffusion to gate capacitance in which $p_{inv} = 0.75$. By what percentage does this change the NAND gate delay, as measured in FO4 inverter delays? What if $p_{inv} = 1.25$?

3.3 Sketch a 2-input NOR gate with transistor widths chosen to achieve effective rise and fall resistances equal to a unit inverter. Compute the rising and falling propagation delays of the NOR gate driving h identical NOR gates using the Elmore delay model. Assume that every source or drain has fully contacted diffusion when making your estimate of capacitance.

3.4 Sketch a stick diagram for the 2-input NOR. Repeat Exercise 3.3 with better capacitance estimates. In particular, if a diffusion node is shared between two parallel transistors, only budget its capacitance once. If a diffusion node is between two series transistors and requires no contacts, only budget half the capacitance because of the smaller diffusion area.

3.5 A 3-stage logic path is designed so that the effort borne by each stage is 12, 6, and 9 delay units, respectively. Can this design be improved? Why? What is the best number of stages for this path? What changes do you recommend to the existing design?

3.6 Suppose a unit inverter with three units of input capacitance has unit drive.

a) What is the drive of a 4x inverter?

b) What is the drive of a 2-input NAND gate with three units of input capacitance?

3.7 Sketch a delay vs. electrical effort graph like that of Figure 3.21 for a 2-input NOR gate using the logical effort and parasitic delay estimated in Section 3.4.2. How does the slope of your graph compare to that of a 2-input NAND? How does the y-intercept compare?

3.8 Let a 4x inverter have transistors four times as wide as those of a unit inverter. If a unit inverter has three units of input capacitance and parasitic delay of p_{inv}, what is the input capacitance of a 4x inverter? What is the logical effort? What is the parasitic delay?

3.9 Design a circuit at the gate level to compute the following function:

```
if (a == b) y = a;
else y = 0;
```

Let a, b, and y be 16-bit busses. Assume the input and output capacitances are each 10 units. Your goal is to make the circuit as fast as possible. Estimate the delay in FO4 inverter delays using Logical Effort if the best gate sizes were used. What sizes do you need to use to achieve this delay?

3.10 An output pad contains a chain of successively larger inverters to drive the (relatively) enormous off-chip capacitance. If the first inverter in the chain has an input capacitance of 20 fF and the off-chip load is 10 pF, how many inverters should be used to drive the load with least delay? Estimate this delay, expressed in FO4 inverter delays.

3.11 The clock buffer in Figure 3.39 can present a maximum input capacitance of 100 fF. Both true and complementary outputs must drive loads of 300 fF. Compute the input capacitance of each inverter to minimize the worst-case delay from input to either output. What is this delay, in τ? Assume the inverter parasitic delay is 1.

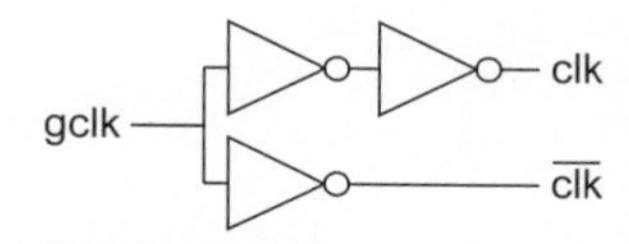

FIGURE 3.39 Clock buffer

3.12 Consider the two designs of a 2-input AND gate shown in Figure 3.40. Give an intuitive argument about which will be faster. Back up your argument with a calculation of the path effort, delay, and input capacitances x and y to achieve this delay.

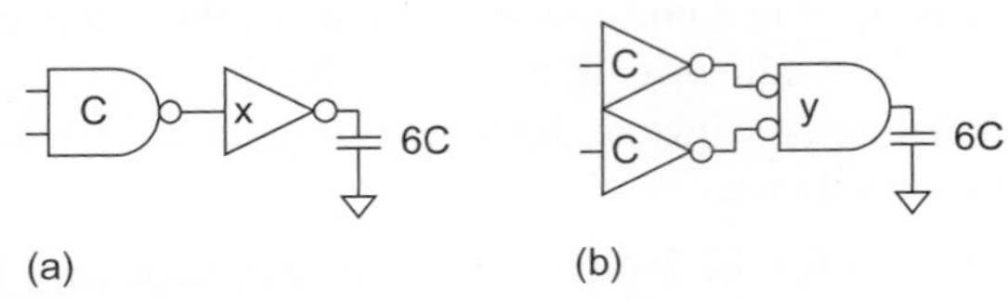

FIGURE 3.40 2-input AND gate

3.13 Consider four designs of a 6-input AND gate shown in Figure 3.41. Develop an expression for the delay of each path if the path electrical effort is H. What design is fastest for $H = 1$? For $H = 5$? For $H = 20$? Explain your conclusions intuitively.

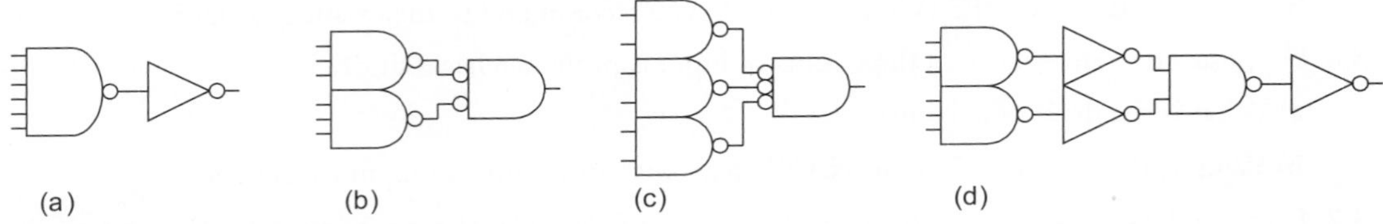

FIGURE 3.41 6-input AND gate

3.14 Find the parasitic delay and logical effort of the X2 and X4 NOR gate A input using Figure 3.42. By what percentage do they differ from that of the X1 gate? What does this imply about our model that parasitic delay and logical effort depend only on gate type and not on transistor sizes?

3.15 Sketch a 4-input NAND gate with transistor widths chosen to achieve equal rise and fall resistance as a unit inverter. Show why the logical effort is 6/3.

3.16 Repeat the decoder design example from Section 3.5.3 for a 32-word register file with 64-bit registers. Determine the fastest decoder design and estimate the delay of the decoder and the transistor widths to achieve this delay.

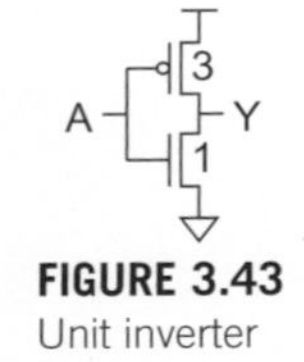

FIGURE 3.43 Unit inverter

3.17 Consider a process in which pMOS transistors have three times the effective resistance as nMOS transistors. A unit inverter with equal rising and falling delays in this process is shown in Figure 3.43. Calculate the logical efforts of a 2-input NAND gate and a 2-input NOR gate if they are designed with equal rising and falling delays.

3.18 Generalize Exercise 3.17 if the pMOS transistors have μ times the effective resistance of nMOS transistors. Find a general expression for the logical efforts of a k-input NAND gate and a k-input NOR gate. As μ increases, comment on the relative desirability of NANDs vs. NORs.

3.19 What are the parasitic delay and logical effort of the X1 NOR gate B input in Figure 3.42? How and why do they differ from the A input?

3.20 Plot the average delay from input A of an FO3 NAND2 gate from the datasheet in Figure 3.25. Why is the delay larger for the XL drive strength than for the other drive strengths?

3.21 Figure 3.42 shows a datasheet for a 2-input NOR gate in the Artisan Components standard cell library for the TSMC 180 nm process. Find the average parasitic delay and logical effort of the X1 NOR gate A input. Use the value of τ from Section 3.4.5.

NOR2

Cell Description

The NOR2 cell provides a logical NOR of two inputs (A, B). The output (Y) is represented by the logic equation:

$Y = \overline{(A+B)}$

Logic Symbol

A, B → Y

Functions

A	B	Y
0	0	1
x	1	0
1	x	0

Cell Size

Drive Strength	Height (μm)	Width (μm)
NOR2XL	5.04	1.98
NOR2X1	5.04	1.98
NOR2X2	5.04	3.30
NOR2X4	5.04	4.62

AC Power

Pin	Power (μW/MHz)			
	XL	X1	X2	X4
A	0.0110	0.0143	0.0275	0.0545
B	0.0139	0.0182	0.0365	0.0728

Pin Capacitance

Pin	Capacitance (pF)			
	XL	X1	X2	X4
A	0.0033	0.0045	0.0083	0.0169
B	0.0029	0.0040	0.0085	0.0160

Delays at 25°C, 1.8V, Typical Process

Description	Intrinsic Delay (ns)			
	XL	X1	X2	X4
A → Y↑	0.0426	0.0403	0.0347	0.0351
A → Y↓	0.0213	0.0196	0.0183	0.0187
B → Y↑	0.0536	0.0519	0.0485	0.0473
B → Y↓	0.0259	0.0244	0.0247	0.0245

Description	K_{load} (ns/pF)			
	XL	X1	X2	X4
A → Y↑	9.4704	6.7329	3.3656	1.6550
A → Y↓	3.5015	2.3672	1.2659	0.6330
B → Y↑	9.4589	6.7278	3.3647	1.6542
B → Y↓	3.5234	2.3779	1.2717	0.6357

TSMC 0.18μm Process SAGE-X™ Standard Cell Library Databook
144
Artisan

FIGURE 3.42 2-input NOR datasheet (Courtesy of Artisan Components.)

3.22 Parasitic delay estimates in Section 3.4.2 are made assuming contacted diffusion on each transistor on the output node and ignoring internal diffusion. Would parasitic delay increase or decrease if you took into account that some parallel transistors on the output node share a single diffusion contact? If you counted internal diffusion capacitance between series transistors? If you counted wire capacitance within the cell?

3.23 Find the rising and falling propagation delays of an unloaded AND-OR-INVERT gate using the Elmore delay model. Estimate the diffusion capacitance based on a stick diagram of the layout.

3.24 The clock buffer from Exercise 3.11 is an example of a *1–2 fork*. In general, if a 1–2 fork has a maximum input capacitance of C_1 and each of the two legs drives a load of C_2, what should the capacitance of each inverter be and how fast will the circuit operate? Express your answer in terms of p_{inv}.

3.25 The 64-bit Naffziger adder [Naffziger96] has a delay of 930 ps in a fast 0.5-μm Hewlett-Packard process with an FO4 inverter delay of about 140 ps. Estimate its delay in a 65 nm process with an FO4 inverter delay of 20 ps.

3.26 Find the worst-case Elmore parasitic delay of an n-input NOR gate.

Power 4

4.1 Introduction

On Earth, apart from nuclear sources, all energy is or has been stockpiled from the sun. In essence, Earth is a huge battery that has been charged up over billions of years via the energy of sunlight in the form of plant growth, which in turn has been turned to carbon and then to oil, gas, coal or other carbon-based fuels. Additionally, in these times, we can harvest energy directly from the sun (solar power), or indirectly from the wind, tides, precipitation (hydro) or geothermal. Energy undergoes transformations. Sunlight to plant growth. Plants to carbon. Carbon to heat. Heat to electricity. Electricity to chemical (battery charging). Chemical to electricity (battery discharging). Electricity to audio (playing an MP3). In the last conversion, some energy is transformed into sound that dissipates into the universe. The rest is turned to heat as the tunes are decoded and played. It is also lost to the universe (perhaps warming our hands slightly on a cold night). So pervasive are energy transformations in everyday life, we are often not at all aware of them. Most times they occur quietly and unnoticed.

Today, we are interested in power from a number of points of view. In portable applications, products normally run off batteries. While battery technology has improved markedly over the years, it remains that a battery of a certain weight and size has a certain energy capacity. For example, a pair of rechargeable AA batteries has an energy capacity of about 7 W-hr, and a good lithium-ion laptop battery has an energy density of about 80 W-hr/lb. Inevitably, the battery runs down and needs recharging or replacement. Product designers are interested in extending the lifetime of the battery while simultaneously adding features and reducing size, so creating low-power IC designs is key. In applications that are permanently connected to a power cord, the ever-present need to reduce dependence on fossil fuels and reduce greenhouse emissions leads us to look for low power solutions to all problems involving electronics. High-performance chips are limited to about 150 W before liquid cooling or other costly heat sinks become necessary. In 2006, data centers and servers in the United States consumed 61 billion kWh of electricity [EPA07]. This represents the output of 15 power plants, costs about $4.5 billion, and amounts to 1.5% of total U.S. energy consumption—more than that consumed by all the television sets in the country. While chip functionality was once limited by area, it is now often constrained by power. High-performance design and energy-efficient design have become synonymous.

In this chapter, we will examine the fundamental theory behind the various sources of power dissipation in a CMOS chip. Next, we will look at methods of estimating and minimizing these sources. Then, some architectural ideas for achieving low power are discussed.

While we concentrate mainly on the methods available as an IC designer to reduce power, it should be remembered that it is the application and architectural level where the major impact on power dissipation can be made. Quite simply stated, the less time you have a circuit turned on, the less power it will dissipate. It is a simple maxim, but drives all of the work on extremely low power circuits. To state this again, you must optimize power in a top-down manner, from the problem definition downward. Do not optimize from the bottom up, i.e., the circuit level; you will be doomed to fail.

4.1.1 Definitions

We have thrown some terms about already including power and energy. It is informative to go back to basics and examine what we mean by these terms and why we are even interested in them.

The *instantaneous power* $P(t)$ consumed or supplied by a circuit element is the product of the current through the element and the voltage across the element

$$P(t) = I(t)V(t) \tag{4.1}$$

The *energy* consumed or supplied over some time interval T is the integral of the instantaneous power

$$E = \int_0^T P(t)dt \tag{4.2}$$

The *average power* over this interval is

$$P_{\text{avg}} = \frac{E}{T} = \frac{1}{T}\int_0^T P(t)dt \tag{4.3}$$

Power is expressed in units of Watts (W). Energy in circuits is usually expressed in Joules (J), where 1 W = 1 J/s. Energy in batteries is often given in W-hr, where 1 W-hr = (1 J/s)(3600 s/hr)(1 hr) = 3600 J.

4.1.2 Examples

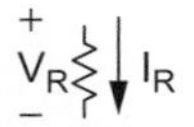

V_R I_R

FIGURE 4.1
Resistor

Figure 4.1 shows a resistor. The voltage and current are related by Ohm's Law, $V = IR$, so the instantaneous power dissipated in the resistor is

$$P_R(t) = \frac{V_R^2(t)}{R} = I_R^2(t)R \tag{4.4}$$

This power is converted from electricity to heat.

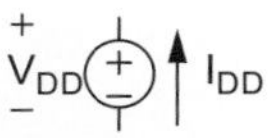

V_{DD} I_{DD}

FIGURE 4.2
Voltage source

Figure 4.2 shows a voltage source V_{DD}. It supplies power proportional to its current

$$P_{VDD}(t) = I_{DD}(t)V_{DD} \tag{4.5}$$

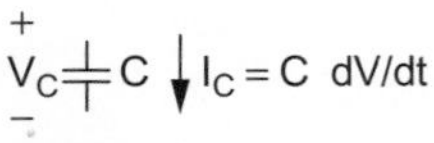

V_C C $I_C = C\ dV/dt$

FIGURE 4.3
Capacitor

Figure 4.3 shows a capacitor. When the capacitor is charged from 0 to V_C, it stores energy E_C

$$E_C = \int_0^\infty I(t)V(t)dt = \int_0^\infty C\frac{dV}{dt}V(t)dt = C\int_0^{V_C} V(t)dV = \tfrac{1}{2}CV_C^2 \tag{4.6}$$

The capacitor releases this energy when it discharges back to 0.

Figure 4.4 shows a CMOS inverter driving a load capacitance. When the input switches from 1 to 0, the pMOS transistor turns ON and charges the load to V_{DD}. According to EQ (4.6), the energy stored in the capacitor is

$$E_C = \tfrac{1}{2} C_L V_{DD}^2 \tag{4.7}$$

The energy delivered from the power supply is

$$E_C = \int_0^{\infty} I(t) V_{DD} dt = \int_0^{\infty} C \frac{dV}{dt} V_{DD} dt = C V_{DD} \int_0^{V_{DD}} dV = C V_{DD}^2 \tag{4.8}$$

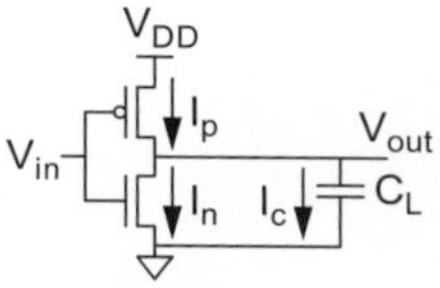

FIGURE 4.4 CMOS inverter

Observe that only half of the energy from the power supply is stored in the capacitor. The other half is dissipated (converted to heat) in the pMOS transistor because the transistor has a voltage across it at the same time a current flows through it. The power dissipated depends only on the load capacitance, not on the size of the transistor or the speed at which the gate switches. Figure 4.5 shows the energy and power of the supply and capacitor as the gate switches.

When the input switches from 0 back to 1, the pMOS transistor turns OFF and the nMOS transistor turns ON, discharging the capacitor. The energy stored in the capacitor is dissipated in the nMOS transistor. No energy is drawn from the power supply during this transition. The same analysis applies for any static CMOS gate driving a capacitive load.

Figure 4.5 shows the waveforms as the inverter drives a 150 fF capacitor at 1 GHz. When V_{in} begins to fall, the pMOS transistor starts to turn ON. It is initially saturated, and the current I_p ramps up and eventually levels out at I_{dsat} as V_{in} falls. Eventually, V_{out} rises to the point that the pMOS shifts to the linear regime. I_p tapers off exponentially, as

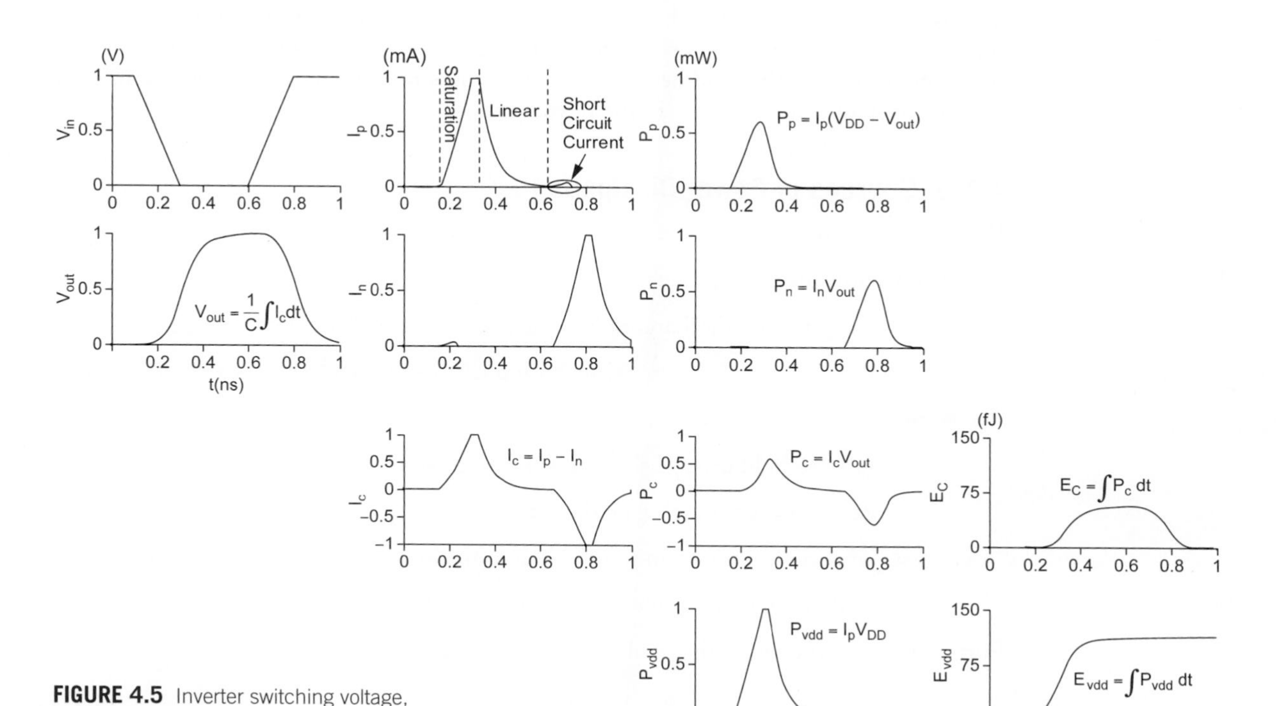

FIGURE 4.5 Inverter switching voltage, current, power, and energy

one would expect charging a capacitor through a linear resistor. When V_{in} rises, the pMOS starts to turn OFF. However, there is a small blip of current while the partially ON pMOS fights against the nMOS. This is called *short-circuit current*. The inverter draws power from V_{DD} as V_{out} rises. Half of the power is dissipated in the pMOS transistor and the other half is delivered to the capacitor. V_{DD} supplies a total of 150 fJ of energy, of which half is stored on the capacitor. The inverter is sized for equal rise/fall times so the falling transition is symmetric. The energy on the capacitor is dumped to GND. The short-circuit current consumes an almost imperceptibly small 2.7 fJ of additional energy from V_{DD} during this transition.

Suppose that the gate switches at some average frequency f_{sw}. Over some interval T, the load will be charged and discharged Tf_{sw} times. Then, according to EQ (4.3), the average power dissipation is

$$P_{switching} = \frac{E}{T} = \frac{Tf_{sw}CV_{DD}^2}{T} = CV_{DD}^2 f_{sw} \tag{4.9}$$

This is called the *dynamic power* because it arises from the switching of the load. Because most gates do not switch every clock cycle, it is often more convenient to express switching frequency f_{sw} as an *activity factor* α times the clock frequency f. Now, the dynamic power dissipation may be rewritten as

$$P_{switching} = \alpha CV_{DD}^2 f \tag{4.10}$$

The activity factor is the probability that the circuit node transitions from 0 to 1, because that is the only time the circuit consumes power. A clock has an activity factor of $\alpha = 1$ because it rises and falls every cycle. Most data has a maximum activity factor of 0.5 because it transitions only once each cycle. Truly random data has an activity factor of 0.25 because it transitions every other cycle. Static CMOS logic has been empirically determined to have activity factors closer to 0.1 because some gates maintain one output state more often than another and because real data inputs to some portions of a system often remain constant from one cycle to the next.

4.1.3 Sources of Power Dissipation

Power dissipation in CMOS circuits comes from two components:

- Dynamic dissipation due to
  - charging and discharging load capacitances as gates switch
  - "short-circuit" current while both pMOS and nMOS stacks are partially ON
- Static dissipation due to
  - subthreshold leakage through OFF transistors
  - gate leakage through gate dielectric
  - junction leakage from source/drain diffusions
  - contention current in ratioed circuits (see Section 8.2.2)

Putting this together gives the total power of a circuit

$$P_{dynamic} = P_{switching} + P_{short\ circuit} \tag{4.11}$$

$$P_{\text{static}} = \left(I_{\text{sub}} + I_{\text{gate}} + I_{\text{junct}} + I_{\text{contention}}\right)V_{DD} \quad (4.12)$$

$$P_{\text{total}} = P_{\text{dynamic}} + P_{\text{static}} \quad (4.13)$$

Power can also be considered in active, standby, and sleep modes. *Active power* is the power consumed while the chip is doing useful work. It is usually dominated by $P_{\text{switching}}$. *Standby power* is the power consumed while the chip is idle. If clocks are stopped and ratioed circuits are disabled, the standby power is set by leakage. In sleep mode, the supplies to unneeded circuits are turned off to eliminate leakage. This drastically reduces the *sleep power* required, but the chip requires time and energy to wake up so sleeping is only viable if the chip will idle for long enough.

[Gonzalez96] found that roughly one-third of microprocessor power is spent on the clock, another third on memories, and the remaining third on logic and wires. In nanometer technologies, nearly one-third of the power is leakage. High-speed I/O contributes a growing component too. For example, Figure 4.6 shows the active power consumption of Sun's 8-core 84 W Niagra2 processor [Nawathe08]. The cores and other components collectively account for clock, logic, and wires.

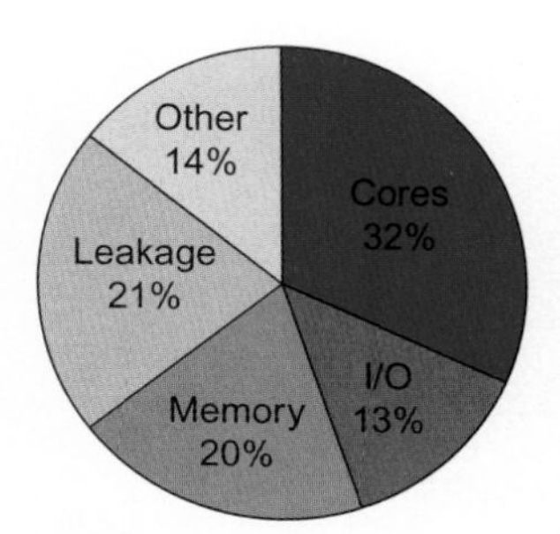

FIGURE 4.6 Power in Niagra2

The next sections investigate how to estimate and minimize each of these components of power. Many tools are available to assist with power estimation; these are discussed further in Sections 7.5.4 and 13.4.1.6.

4.2 Dynamic Power

Dynamic power consists mostly of the switching power, given in EQ (4.10). The supply voltage V_{DD} and frequency f are readily known by the designer. To estimate this power, one can consider each node of the circuit. The capacitance of the node is the sum of the gate, diffusion, and wire capacitances on the node. The activity factor can be estimated using techniques described in Section 4.2.1 or measured from logic simulations. The *effective capacitance* of the node is its true capacitance multiplied by the activity factor. The switching power depends on the sum of the effective capacitances of all the nodes.

Activity factors can be heavily dependent on the particular task being executed. For example, a processor in a cell phone will use more power while running video games than while displaying a calendar. CAD tools do a fine job of power estimation when given a realistic workload. Low power design involves considering and reducing each of the terms in switching power.

As V_{DD} is a quadratic term, it is good to select the minimum V_{DD} that can support the required frequency of operation. Likewise, we choose the lowest frequency of operation that achieves the desired end performance. The activity factor is mainly reduced by putting unused blocks to sleep. Finally, the circuit may be optimized to reduce the overall load capacitance of each section.

Example 4.1

A digital system-on-chip in a 1 V 65 nm process (with 50 nm drawn channel lengths and λ = 25 nm) has 1 billion transistors, of which 50 million are in logic gates and the remainder in memory arrays. The average logic transistor width is 12 λ and the average memory transistor width is 4 λ. The memory arrays are divided into banks and only the necessary bank is activated so the memory activity factor is 0.02. The static CMOS logic gates have an average activity factor of 0.1. Assume each transistor contributes 1 fF/μm of gate capacitance and 0.8 fF/μm of diffusion capacitance. Neglect wire capacitance for now (though it could account for a large fraction of total power). Estimate the switching power when operating at 1 GHz.

SOLUTION: There are (50×10^6 logic transistors)(12 λ)(0.025 μm/λ)((1 + 0.8) fF/μm) = 27 nF of logic transistors and (950×10^6 memory transistors)(4 λ)(0.025 μm/λ)((1 + 0.8) fF/μm) = 171 nF of memory transistors. The switching power consumption is $[(0.1)(27 \times 10^{-9}) + (0.02)(171 \times 10^{-9})](1.0\text{ V})^2(10^9\text{ Hz}) = 6.1\text{ W}$.

Dynamic power also includes a short-circuit power component caused by power rushing from V_{DD} to GND when both the pullup and pulldown networks are partially ON while a transistor switches. This is normally less than 10% of the whole, so it can be conservatively estimated by adding 10% to the switching power.

Switching power is consumed by delivering energy to charge a load capacitance, then dumping this energy to GND. Intuitively, one might expect that power could be saved by shuffling the energy around to where it is needed rather than just dumping it. Resonant circuits, and adiabatic charge-recovering circuits [Maksimovic00, Sathe07] seek to achieve such a goal. Unfortunately, all of these techniques add complexity that detracts from the potential energy savings, and none have found more than niche applications.

4.2.1 Activity Factor

The activity factor is a powerful and easy-to-use lever for reducing power. If a circuit can be turned off entirely, the activity factor and dynamic power go to zero. Blocks are typically turned off by stopping the clock; this is called *clock gating*. When a block is on, the activity factor is 1 for clocks and substantially lower for nodes in logic circuits. The activity factor of a logic gate can be estimated by calculating the switching probability. Glitches can increase the activity factor.

4.2.1.1 Clock Gating Clock gating ANDs a clock signal with an enable to turn off the clock to idle blocks. It is highly effective because the clock has such a high activity factor, and because gating the clock to the input registers of a block prevents the registers from switching and thus stops all the activity in the downstream combinational logic.

Clock gating can be employed on any enabled register. Section 9.3.5 discusses enabled register design. Sometimes the logic to compute the enable signal is easy; for example, a floating-point unit can be turned off when no floating-point instructions are being issued. Often, however, clock gating signals are some of the most critical paths of the chip.

The clock enable must be stable while the clock is active (i.e., 1 for systems using positive edge-triggered flip-flops). Figure 4.7 shows how an enable latch can be used to ensure the enable does not change before the clock falls.

When a large block of logic is turned off, the clock can be gated early in the clock distribution network, turning off not only the registers but also a portion of the global network. The clock network has an activity factor of 1 and a high capacitance, so this saves significant power.

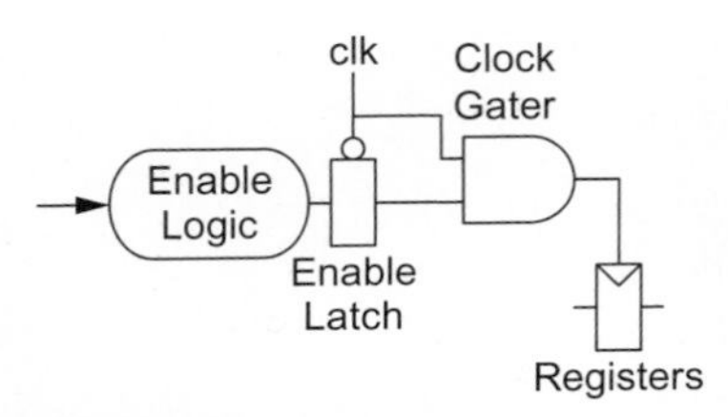

FIGURE 4.7 Clock gating

4.2.1.2 Switching Probability Recall that the activity factor of a node is the probability that it switches from 0 to 1. This probability depends on the logic function. By analyzing the probability that each node is 1, we can estimate the activity factors. Although designers don't manually estimate activity factors very often, the exercise is worth doing here to gain some intuition about switching activity.

Define P_i to be the probability that node i is 1. $\overline{P}_i = 1 - P_i$ is the probability that node i is 0. α_i, the activity factor of node i, is the probability that the node is 0 on one cycle and 1 on the next. If the probability is uncorrelated from cycle to cycle,

$$\alpha_i = \overline{P}_i P_i \tag{4.14}$$

Completely random data has $P = 0.5$ and thus $\alpha = 0.25$. Structured data may have different probabilities. For example, the upper bits of a 64-bit unsigned integer representing a physical quantity such as the intensity of a sound or the amount of money in your bank account are 0 most of the time. The activity factor is lower than 0.25 for such data.

Table 4.1 lists the output probabilities of various gates as a function of their input probabilities, assuming the inputs are uncorrelated. According to EQ (4.14), the activity factor of the output is $\overline{P}_Y P_Y$.

TABLE 4.1 Switching probabilities

Gate	P_Y
AND2	$P_A P_B$
AND3	$P_A P_B P_C$
OR2	$1 - \overline{P}_A \overline{P}_B$
NAND2	$1 - P_A P_B$
NOR2	$\overline{P}_A \overline{P}_B$
XOR2	$P_A \overline{P}_B + \overline{P}_A P_B$

Example 4.2

Figure 4.8 shows a 4-input AND gate built using a tree (a) and a chain (b) of gates. Determine the activity factors at each node in the circuit assuming the input probabilities $P_A = P_B = P_C = P_D = 0.5$.

SOLUTION: Figure 4.9 labels the signal probabilities and the activity factors at each node based on Table 4.1 and EQ (4.14). The chain has a lower activity factor at the intermediate nodes.

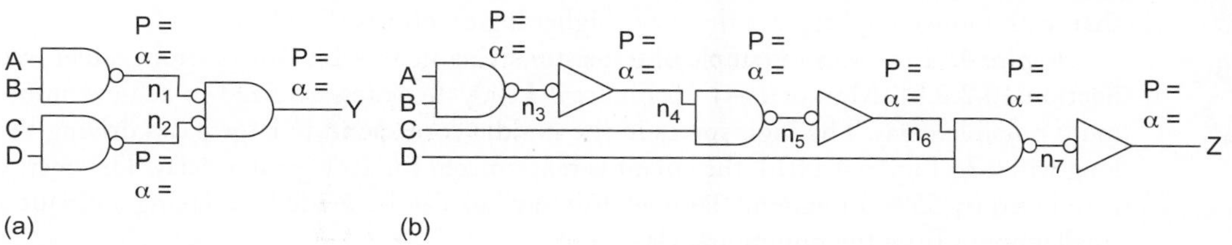

FIGURE 4.8 4-input AND circuits

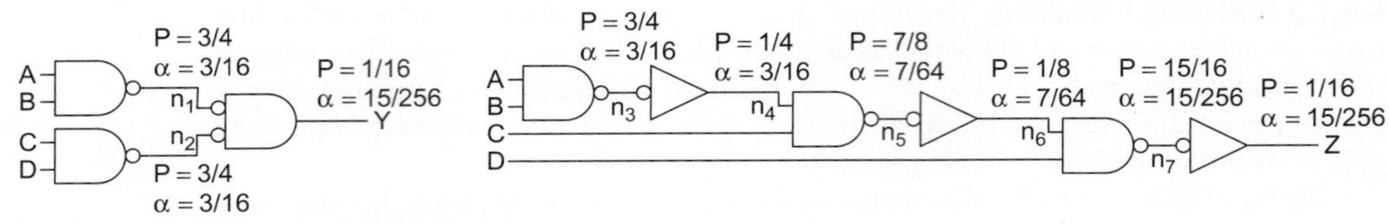

FIGURE 4.9 Signal probabilities and activity factors

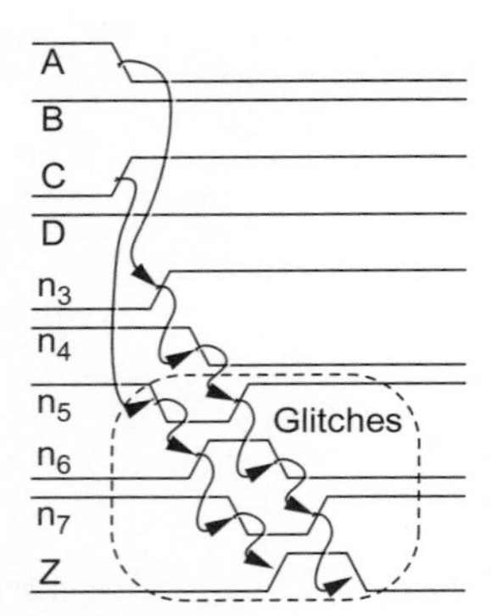

FIGURE 4.10 Glitching in a chain of gates

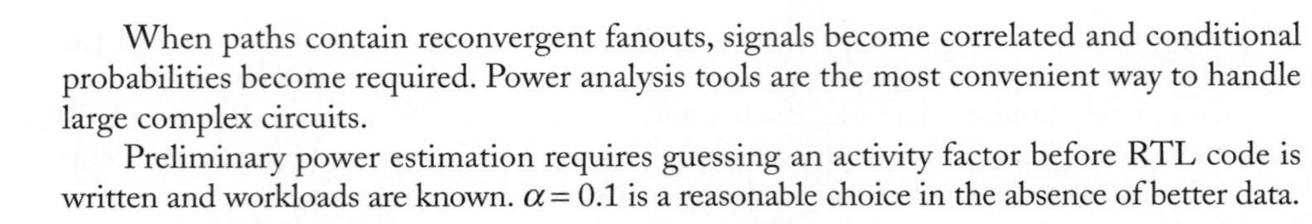

When paths contain reconvergent fanouts, signals become correlated and conditional probabilities become required. Power analysis tools are the most convenient way to handle large complex circuits.

Preliminary power estimation requires guessing an activity factor before RTL code is written and workloads are known. $\alpha = 0.1$ is a reasonable choice in the absence of better data.

4.2.1.3 Glitches The switching probabilities computed in the previous section are only valid if the gates have zero propagation delay. In reality, gates sometimes make spurious transitions called glitches when inputs do not arrive simultaneously. For example, in Figure 4.8(b), suppose *ABCD* changes from 1101 to 0111. Node n_4 was 1 and falls to 0. However, nodes n_5, n_6, n_7, and Z may glitch before n_4 changes, as shown in Figure 4.10. The glitches cause extra power dissipation. Chains of gates are particularly prone to this problem. Glitching can raise the activity factor of a gate above 1 and can account for the majority of power in certain circuits such as ripple carry adders and array multipliers (see Chapter 10). Glitching power can be accurately assessed through simulations accounting for timing.

4.2.2 Capacitance

Switching capacitance comes from the wires and transistors in a circuit. Wire capacitance is minimized through good floorplanning and placement (the locality aspect of structured design). Units that exchange a large amount of data should be placed close to each other to reduce wire lengths.

Device-switching capacitance is reduced by choosing fewer stages of logic and smaller transistors. Minimum-sized gates can be used on non-critical paths. Although Logical Effort finds that the best stage effort is about 4, using a larger stage effort increases delay only slightly and greatly reduces transistor sizes. Therefore, gates that are large or have a high activity factor and thus dominate the power can be downsized with only a small performance impact. For example, buffers driving I/O pads or long wires may use a stage effort of 8–12 to reduce the buffer size. Similarly, registers should use small clocked transistors because their activity factor is an order of magnitude greater than transistors in combinational logic. In Chapter 5, we will see that wire capacitance dominates many circuits. The most energy-efficient way to drive long wires is with inverters or buffers rather than with more complex gates that have higher logical efforts [Stan99].

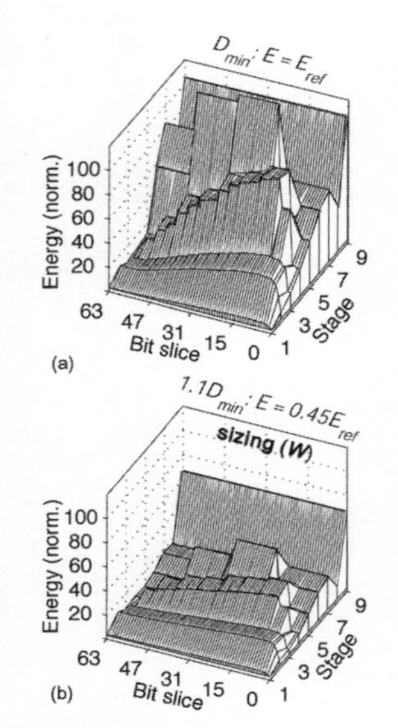

FIGURE 4.11 Adder gate sizing under a delay constraint (Adapted from [Marković04]. © IEEE 2004.)

Figure 4.11 shows an example of transistor sizing in a 64-bit Kogge-Stone adder (see Section 10.2.2.8) [Marković04]. In Figure 4.11(a), the gates are sized to achieve minimum possible delay. The high spikes in the middle correspond to large gates driving the long wires. In Figure 4.11(b), the circuit is reoptimized for 10% greater delay. The energy is reduced by 55%. In general, large energy savings can be made by relaxing a circuit a small amount from the minimum delay point.

Unfortunately, there are no closed-form methods to determine gate sizes that minimize energy under a delay constraint, even for circuits as simple as an inverter chain

[Ma94]. However, it is straightforward to solve the problem numerically, as will be formulated in the next section.

4.2.2.1 Gate Sizing Under a Delay Constraint In Chapter 3, Logical Effort showed us how to minimize delay. In many cases, we are willing to increase delay to save energy. We can extend the iterative technique from Section 3.5.6 to size a circuit for minimum switching energy under a delay constraint.

First, consider a model to compute the energy of a circuit. If a unit inverter has gate capacitance $3C$, then a gate with logical effort g, parasitic delay p, and drive x has gx times as much gate capacitance and px times as much diffusion capacitance. The switching energy of each gate depends on its activity factor, the diffusion capacitance of the gate, the wire capacitance C_{wire}, and the gate capacitance of all the stages it drives. The energy of the entire circuit is the sum of the energies of each gate.

$$\text{Energy} = 3CV_{DD}^2 \sum_{i \in \text{nodes}} \alpha_i \left(\frac{C_{\text{wire}_i}}{3C} + p_i x_i + \sum_{j \in \text{fanout}(i)} g_j x_j \right) \tag{4.15}$$

If wire capacitance is expressed in multiples of the capacitance of a unit inverter as $c = C_{\text{wire}}/3C$ and we normalize energy for the capacitance and voltage of the process, EQ (4.15) becomes the sum of the effective capacitances of the nodes.

$$E = \sum_{i \in \text{nodes}} \alpha_i \left(c_i + p_i x_i + \sum_{j \in \text{fanout}(i)} g_j x_j \right) = \sum_{i \in \text{nodes}} \alpha_i x_i d_i \tag{4.16}$$

Now, we seek to minimize E such that the worst-case arrival time is less than some delay D. The problem is still a posynomial and has a unique solution that can be found quickly by a good optimizer.

Example 4.3

Generate an energy-delay trade-off curve for the circuit from Figure 3.37 as delay varies from the minimum possible ($D_{\text{min}} = 23.44\ \tau$) to 50 τ. Assume that the input probabilities are 0.5.

SOLUTION: Figure 4.12 shows the activity factors of each node. Hence, the energy of this circuit is

$$\begin{aligned} E = &\tfrac{1}{4}\left(1 + \tfrac{4}{3}x_2 + \tfrac{5}{3}x_3\right) + \tfrac{3}{16}\left(2x_2 + \tfrac{7}{3}x_4\right) + \tfrac{3}{16}\left(2x_3 + \tfrac{7}{3}x_4\right) \\ &+ \tfrac{87}{1024}\left(10 + 3x_4 + x_5\right) + \tfrac{87}{1024}\left(12 + x_5\right) \end{aligned} \tag{4.17}$$

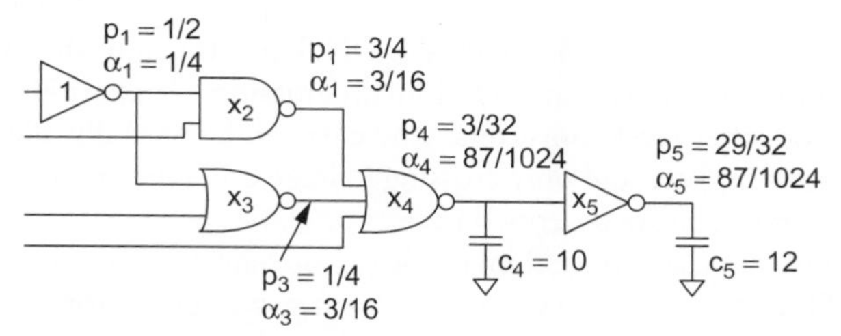

FIGURE 4.12 Activity factors

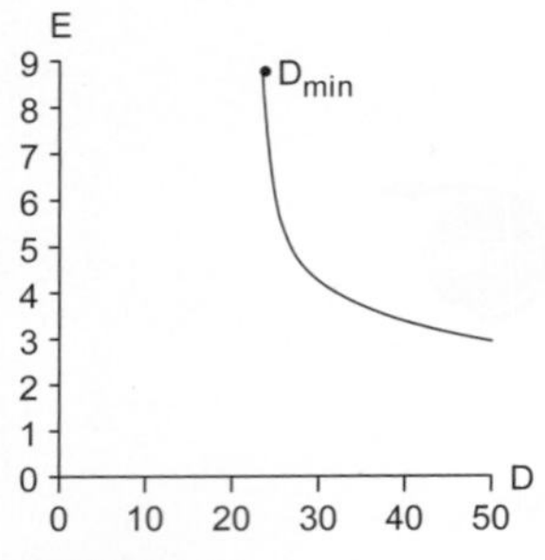

FIGURE 4.13 Energy-delay trade-off curve

Figure 4.13 shows the energy-delay trade-off curve obtained by repeatedly solving for minimum energy under a delay constraint. The curve is steep near D_{min}, indicating that a large amount of energy can be saved for a small increase in delay. The delay cannot be reduced below D_{min} for any amount of energy unless the size of the input inverter is increased (which would increase the delay of the previous circuit).

4.2.3 Voltage

Voltage has a quadratic effect on dynamic power. Therefore, choosing a lower power supply significantly reduces power consumption. As many transistors are operating in a velocity-saturated regime, the lower power supply may not reduce performance as much as long-channel models predict. The chip may be divided into multiple *voltage domains*, where each domain is optimized for the needs of certain circuits. For example, a system-on-chip might use a high supply voltage for memories to ensure cell stability, a medium voltage for a processor, and a low voltage for I/O peripherals running at lower speeds. In Section 4.3.2, we will examine how voltage domains can be turned off entirely to save leakage power during sleep mode.

Voltage also can be adjusted based on operating mode; for example, a laptop processor may operate at high voltage and high speed when plugged into an AC adapter, but at lower voltage and speed when on battery power. If the frequency and voltage scale down in proportion, a cubic reduction in power is achieved. For example, the laptop processor may scale back to 2/3 frequency and voltage to save 70% in power when unplugged.

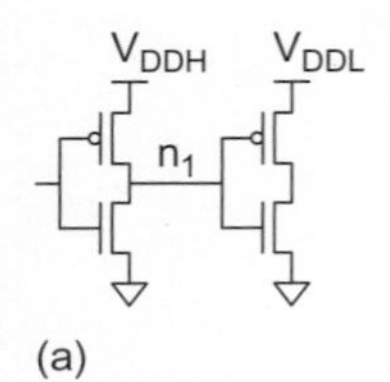

(a)

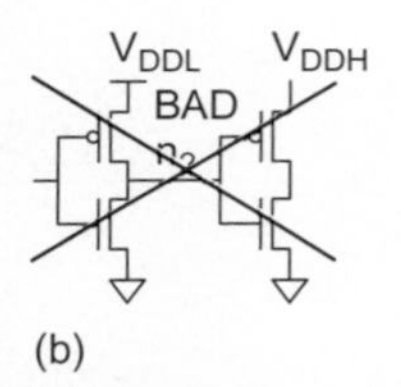

(b)

FIGURE 4.14 Voltage domain crossing

4.2.3.1 Voltage Domains Some of the challenges in using voltage domains include converting voltage levels for signals that cross domains, selecting which circuits belong in which domain, and routing power supplies to multiple domains.

Figure 4.14 shows direct connection of inverters in two domains using high and low supplies, V_{DDH} and V_{DDL}, respectively. A gate in the V_{DDH} domain can directly drive a gate in the V_{DDL} domain. However, the gate in the V_{DDL} domain will switch faster than it would if driven by another V_{DDL} gate. The timing analyzer must consider this when computing the contamination delay, lest a hold time be violated. Unfortunately, the gate in the V_{DDL} domain cannot directly drive a gate in the V_{DDH} domain. When n_2 is at V_{DDL}, the pMOS transistor in the V_{DDH} domain has $V_{gs} = V_{DDH} - V_{DDL}$. If this exceeds V_t, the pMOS will turn ON and burn contention current. Even if the difference is less than V_t, the pMOS will suffer substantially increased leakage. This problem may be alleviated by using a high-V_t pMOS device in the receiver if the voltage difference between domains is small enough [Tawfik09].

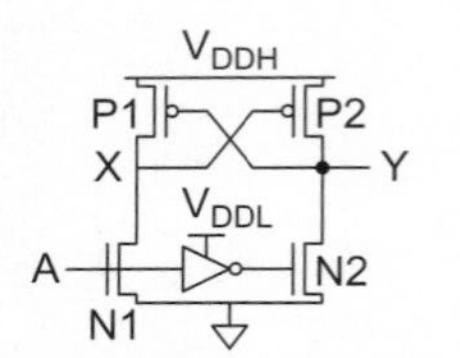

FIGURE 4.15 Level converter

The standard method to handle voltage domain crossings is a *level converter*, shown in Figure 4.15. When $A = 0$, $N1$ is OFF and $N2$ is ON. $N2$ pulls Y down to 0, which turns on $P1$, pulling X up to V_{DDH} and ensuring that $P2$ turns OFF. When $A = 1$, $N1$ is ON and $N2$ is OFF. $N1$ pulls X down to 0, which turns on $P2$, pulling Y up to V_{DDH}. In either case, the level converter behaves as a buffer and properly drives Y between 0 and V_{DDH} without risk of transistors remaining partially ON. Unfortunately, the level converter costs delay (about 2 FO4) and power at each domain crossing. [Kulkarni04] and [Ishihara04] survey a variety of other level converters. The cost can be partially alleviated by building the converter into a register and only crossing voltage domains on clock cycle boundaries. Such level-converter flops are described in Section 9.4.4.

The easiest way to use voltage domains is to associate each domain with a large area of the floorplan. Thus, each domain receives its own power grid. Note that level converters require two power supplies, so they should be placed near the periphery of the domain where necessary for domain crossings.

An alterative approach is called *clustered voltage scaling* (CVS) [Usami95], in which two supply voltages can be used in a single block. Figure 4.16 shows an example of clustered voltage scaling. Gates early in the path use V_{DDH}. Noncritical gates later in the path use V_{DDL}. Voltages are assigned such that a path never crosses from a V_{DDL} gate to a V_{DDH} gate within a block of combinational logic, so level converters are only required at the registers. CVS requires that two power supplies be distributed across the entire block. This can be done by using two power rails. A cell library can have high- and low-voltage versions of each cell, differentiated only by the rail to which the pMOS transistors are connected, so that the flavor of gate can be interchanged. Note that many processes require a large spacing between n-wells at different potentials, which limits the proximity of the V_{DDH} and V_{DDL} gates.

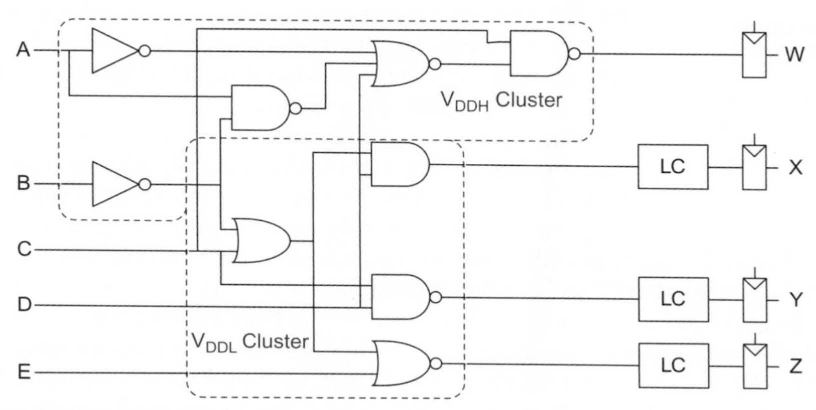

FIGURE 4.16 Clustered voltage scaling

4.2.3.2 Dynamic Voltage Scaling (DVS) Many systems have time-varying performance requirements. For example, a video decoder requires more computation for rapidly moving scenes than for static scenes. A workstation requires more performance when running SPICE than when running Solitaire. Such systems can save large amounts of energy by reducing the clock frequency to the minimum sufficient to complete the task on schedule, then reducing the supply voltage to the minimum necessary to operate at that frequency. This is called *dynamic voltage scaling* (DVS) or *dynamic voltage/frequency scaling* (DVFS) [Burd00]. Figure 4.17 shows a block diagram for a basic DVS system. The DVS controller takes information from the system about the workload and/or the die temperature. It determines the supply voltage and clock frequency sufficient to complete the workload on schedule or to maximize performance without overheating. A switching voltage regulator efficiently steps down V_{in} from a high value to the necessary V_{DD}. The core logic contains a phase-locked loop or other clock synthesizer to generate the specified clock frequency.

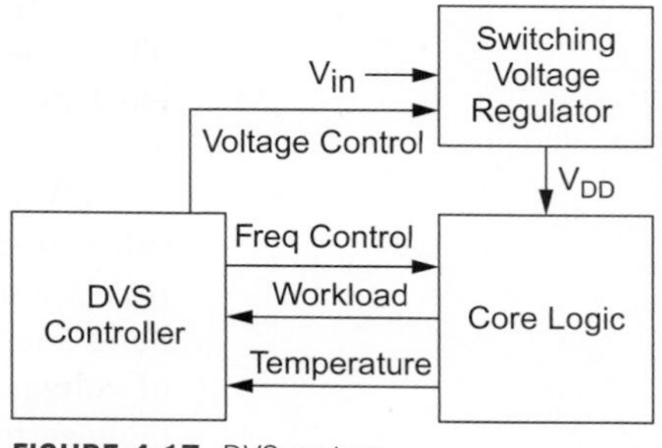

FIGURE 4.17 DVS system

The DVS controller determines the operating frequency, then chooses the lowest supply voltage suitable for that frequency. One method of choosing voltage is with a precharacterized table of voltage vs. frequency. This is inherently conservative because the voltage should be high enough to suffice for even worst-case parts (see Chapter 6 about variability). The quad-core Itanium processor contains a fuse-programmable table that can be tailored to each chip during production [Stackhouse09]. Another method is to use a replica circuit such as a ring oscillator that tracks the performance of the system, as discussed in Section 6.5.3.4.

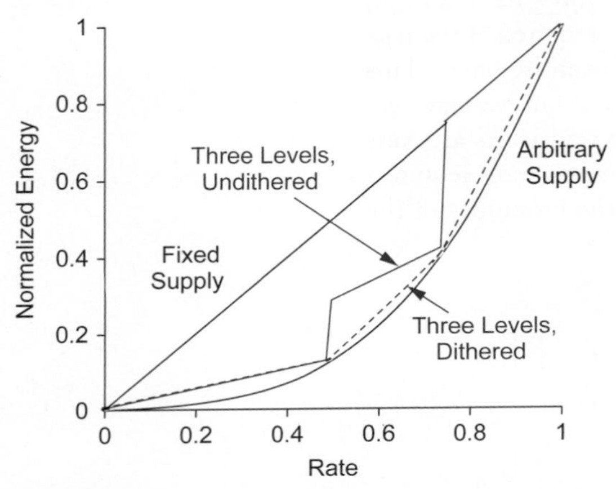

FIGURE 4.18 Energy reduction from DVS

Consider how the energy for a system varies with the workload. Define the *rate* to be the fraction of maximum performance required to complete the workload in a specified amount of time. Figure 4.18 plots energy against rate. If the rate is less than 1, the clock frequency can be adjusted down accordingly, or the system can run at full frequency until the work is done, then stop the clock and go to sleep; this may be simpler than building a continuously adjustable clock. Without DVS, the energy varies linearly with the rate. With ideal DVS, the voltage could also be reduced at lower rates. Assuming a linear relationship between voltage and frequency, the energy is proportional to the rate cubed, giving much greater savings at lower rates. Operating at half the maximum rate costs only one-eighth of the energy.

Such scaling assumes a continuously adjustable supply voltage, which is more expensive than a supply with discrete levels. Characterizing a circuit across a continuous range of voltages and frequencies is also difficult. If the supply voltage is limited to three levels, e.g., 1.0, 0.75, and 0.5 V, and the frequencies limited to three settings as well, much of the benefit of DVS still can be achieved. Better yet, a system can *dither* between these voltages to save even more energy [Gutnik97]. For example, if a rate of 0.6 is required, the system could operate at a rate of 0.75 for 40% of the computation, then switch to a rate of 0.5 for the remaining 60%. Hence, by dithering between three levels, the system can achieve almost as low energy as by using an arbitrary supply voltage. Indeed, dithering between only two supply voltages selected for full and half-rate operation is sufficient to get more than 80% of the benefit of DVS [Aisaka02].

Section 4.3.2 discusses power gating to turn off power to a block during sleep mode. The same mechanism can be used to select from one of several supply voltages for each block during active mode. This allows *local voltage dithering* so that each block can operate at a preferred voltage.

DVS normally operates over a range from the maximum V_{DD} down to about half that value. It can be extended further into the subthreshold regime [Zhai05a, Calhoun06a]; this is sometimes called *ultra-dynamic voltage scaling* (UDVS). It can be challenging to build a replica circuit that tracks the worst case delay on the chip across a very wide range of voltages. DVS is now widely used in systems ranging from consumer electronics to high-performance microprocessors [Keating07, Stackhouse09].

Subthreshold and gate leakage are strongly sensitive to the supply voltage, so DVS also is effective at reducing leakage during periods of low activity.

Operating at varied V_{DD} voltages implies an adjustable voltage regulator that reduces the voltage from a higher supply. Be careful to use a switching type regulator; otherwise, the power will just be dissipated in the regulator.

4.2.4 Frequency

Dynamic power is directly proportional to frequency, so a chip obviously should not run faster than necessary. As mentioned earlier, reducing the frequency also allows downsizing transistors or using a lower supply voltage, which has an even greater impact on power. The performance can be recouped through parallelism (see Section 4.5.2), especially if area is not as important as power.

Even if multiple voltage supplies are not available, a chip may still use multiple frequency domains so that certain portions can run more slowly than others. For example, a microprocessor bus interface usually runs much slower than the core. Low frequency domains can also save energy by using smaller transistors.

Frequency domain crossings are easiest if the frequencies are related by integer multiples and the clocks are synchronized between domains. Section 9.5 discusses synchronization further.

4.2.5 Short-Circuit Current

Short-circuit power dissipation occurs as both pullup and pulldown networks are partially ON while the input switches, as was illustrated in Figure 4.5. It increases as the input edge rates become slower because both networks are ON for more time [Veendrick84]. However, it decreases as load capacitance increases because with large loads the output only switches a small amount during the input transition, leading to a small V_{ds} across the transistor that is causing the short-circuit current. Unless the input edge rate is much slower than the output edge rate, short-circuit current is a small fraction (< 10%) of current to the load and can be ignored in hand calculations. It is good to use relatively crisp edge rates at the inputs to gates with wide transistors to minimize their short-circuit current. This is achieved by keeping the stage effort of the previous stage reasonable, e.g., 4 or less. In general, gates with balanced input and output edge rates have low short-circuit power.

Short-circuit power is strongly sensitive to the ratio $v = V_t / V_{DD}$. In the limit that $v > 0.5$, short-circuit current is eliminated entirely because the pullup and pulldown networks are never simultaneously ON. For $v = 0.3$ or 0.2, short-circuit power is typically about 2% or 10% of switching power, respectively, assuming clean edges [Nose00a]. In nanometer processes, V_t can scarcely fall below 0.3 V without excessive leakage, and V_{DD} is on the order of 1 V, so short-circuit current has become almost negligible.

4.2.6 Resonant Circuits

Resonant circuits seek to reduce switching power consumption by letting energy slosh back and forth between storage elements such as capacitors and inductors rather than dumping the energy to ground. The technique is best suited to applications such as clocks that operate at a constant frequency.

Figure 4.19 shows a model of a resonant clock network [Chan05]. C_{clock} is the capacitance of the clock network. In an ordinary clock circuit, it is driven between V_{DD} and GND by a strong clock buffer. The resonant clock network adds the inductor L_1 and the capacitor C_2, which is approximately $10C_{clock}$. R_{clock} and R_{ind} represent losses in the clock wires and in the inductor that lower the quality of the resonator. In the resonant clock circuit, energy moves back and forth between L_1 and C_{clock}, causing a sinusoidal oscillation at the resonant frequency f. The driver pumps in just enough energy to compensate for the resistive losses. C_2 must be large enough to store excess energy and not interfere with the resonance of the clock capacitance.

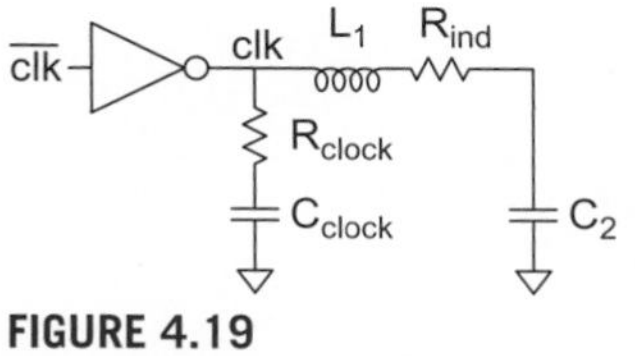

FIGURE 4.19 Resonant clock network

$$f \approx \frac{1}{2\pi\sqrt{L_1 C_{clock}}} \tag{4.18}$$

In a mechanical analogy, inductors represent springs and capacitors represent mass. The clock itself has high capacitance and little inductance, representing a rigid mass suspended on a set of springs corresponding to the inductor L_1. The mass oscillates up and down. The clock driver gives the mass a kick to get it started and compensate for damping in the springs, but little energy is required because the springs do most of the work storing energy on the way down and delivering it back to the mass on the way up.

IBM has demonstrated a resonant global clock distribution system for the Cell processor [Chan09]. At an operating frequency of 4–5 GHz, the system could reduce chip power by 10%. Some of the drawbacks of resonant clocking include the limited range of operating frequencies, the sinusoidal clock output, and the difficulty of building a high-quality inductor in a CMOS process.

4.3 Static Power

Static power is consumed even when a chip is not switching. CMOS has replaced nMOS processes because contention current inherent to nMOS logic limited the number of transistors that could be integrated on one chip. Static CMOS gates have no contention current. Prior to the 90 nm node, leakage power was of concern primarily during sleep mode because it was negligible compared to dynamic power. In nanometer processes with low threshold voltages and thin gate oxides, leakage can account for as much as a third of total active power. Section 2.4.4 introduced leakage current mechanisms. This section briefly reviews each source of static power. It then discusses power gating, which is a key technique to reduce power in sleep mode. Because subthreshold leakage is usually the dominant source of static power, other techniques for leakage reduction are explored, including multiple threshold voltages, variable threshold voltages, and stack forcing.

4.3.1 Static Power Sources

As given in EQ (4.12), static power arises from subthreshold, gate, and junction leakage currents and contention current. Entire books have been written about leakage [Narendra06], but this section summarizes the key effects.

4.3.1.1 Subthreshold Leakage Subthreshold leakage current flows when a transistor is supposed to be OFF. It is given by EQ (2.45). For V_{ds} exceeding a few multiples of the thermal voltage (e.g., $V_{ds} > 50$ mV), it can be simplified to

$$I_{sub} = I_{off} 10^{\frac{V_{gs} + \eta(V_{ds} - V_{DD}) - k_\gamma V_{sb}}{S}} \tag{4.19}$$

where I_{off} is the subthreshold current at $V_{gs} = 0$ and $V_{ds} = V_{DD}$, and S is the subthreshold slope given by EQ (2.44) (about 100 mV/decade). I_{off} is a key process parameter defining the leakage of a single OFF transistor. It ranges from about 100 nA/μm for typical low-V_t devices to below 1 nA/μm for high-V_t devices. η is the DIBL coefficient, typically around 100 mV/V for a 65 nm transistor, and trending upward because the drain exerts an increasing influence on the channel as the geometry shrinks. If V_{ds} is small, I_{sub} may decrease by roughly an order of magnitude from I_{off}. k_γ is the body effect coefficient, which describes how the body effect modulates the threshold voltage. Raising the source voltage or applying a negative body voltage can further decrease leakage.

I_{off} is usually specified at 25 °C and increases exponentially with temperature because V_t decreases with temperature and S is directly proportional to temperature. I_{off} typically increases by one to two orders of magnitude at 125 °C, so limiting die temperature is essential to controlling leakage.

The leakage through two or more series transistors is dramatically reduced on account of the *stack effect* [Ye98, Narendra01]. Figure 4.20 shows two series OFF transistors with gates at 0 volts. The drain of $N2$ is at V_{DD}, so the stack will leak. However, the middle node voltage V_x settles to a point that each transistor has the same current. If V_x is small, $N1$ will see a much smaller DIBL effect and will leak less. As V_x rises, V_{gs} for $N2$ becomes negative, reducing its leakage. Hence, we would expect that the series transistors leak less. This can be demonstrated mathematically by solving for V_x and I_{sub}, assuming that $V_x > 50$ mV.

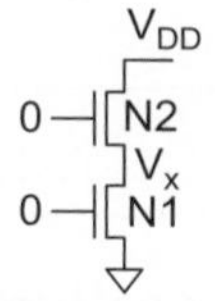

FIGURE 4.20 Series OFF transistors demonstrating the stack effect

$$I_{sub} = \underbrace{I_{off}10^{\frac{\eta(V_x - V_{DD})}{S}}}_{N2} = \underbrace{I_{off}10^{\frac{-V_x+\eta((V_{DD}-V_x)-V_{DD})-k_\gamma V_x}{S}}}_{N1} \tag{4.20}$$

$$V_x = \frac{\eta V_{DD}}{1+2\eta+k_\gamma} \tag{4.21}$$

$$I_{sub} = I_{off}10^{\frac{-\eta V_{DD}\left(\frac{1+\eta+k_\gamma}{1+2\eta+k_\gamma}\right)}{S}} \approx I_{off}10^{\frac{-\eta V_{DD}}{S}} \tag{4.22}$$

Using the typical values above and $V_{DD} = 1.0$ V, we find that the stack effect reduces subthreshold leakage by a factor of about 10. Stacks with three or more OFF transistors have even lower leakage.

Subthreshold leakage cannot be reduced without consideration of other forms of leakage [Mukhopadhyay05]. Raising the halo doping level to raise V_t by controlling DIBL and short-channel effects causes BTBT to increase. Applying a reverse body bias to increase V_t also causes BTBT to increase. Applying a negative gate voltage to turn the transistor OFF more strongly causes GIDL to increase. Figure 4.21 shows how subthreshold leakage dominates in a 50 nm process at low V_t, but how the other sources take over at higher V_t [Agarwal07].

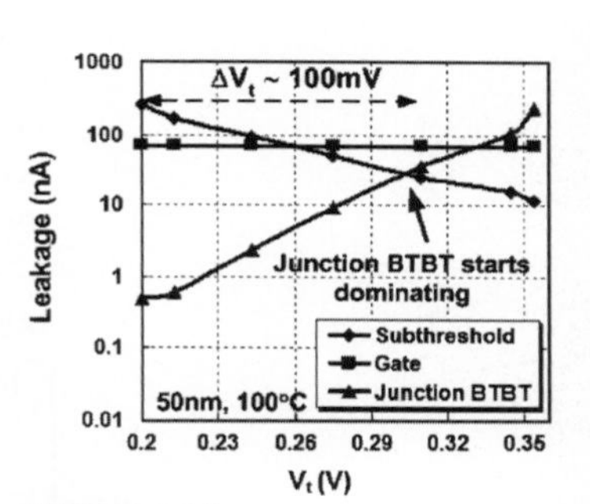

FIGURE 4.21 Leakage as a function of V_t (© IEEE 2007.)

Silicon on Insulator (SOI) circuits are attractive for low-leakage designs because they have a sharper subthreshold current rolloff (smaller n in EQ (2.42)). SOI circuit design will be discussed further in Section 8.4.

4.3.1.2 Gate Leakage Gate leakage occurs when carriers tunnel through a thin gate dielectric when a voltage is applied across the gate (e.g., when the gate is ON). A process usually specifies I_G in nA/μm for a minimum-length gate or in A/mm^2 of transistor gate. Gate leakage is an extremely strong function of the dielectric thickness. It is normally limited to acceptable levels in the process by selection of the dielectric thickness. pMOS gate leakage is an order of magnitude smaller in ordinary SiO_2 gates and can often be ignored, but it can be significant for other gate dielectrics.

Gate leakage also depends on the voltage across the gate. For example, Figure 4.22 shows two series transistors. If $N1$ is ON and $N2$ is OFF, $N1$ has $V_{gs} = V_{DD}$ and experiences full gate leakage. On the other hand, if $N1$ is OFF and $N2$ is on, $N2$ has $V_{gs} = V_t$ and

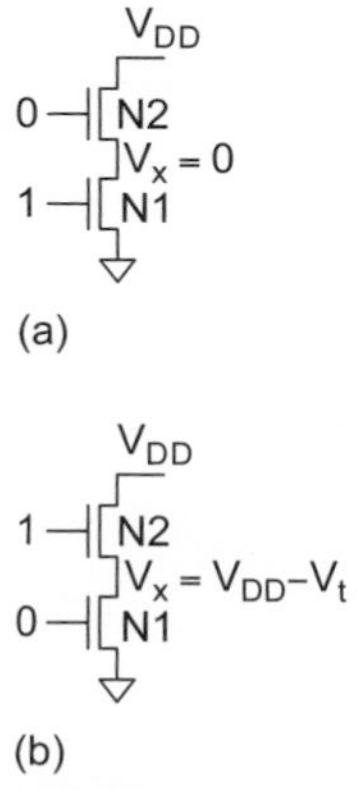

FIGURE 4.22 Gate leakage in series stack

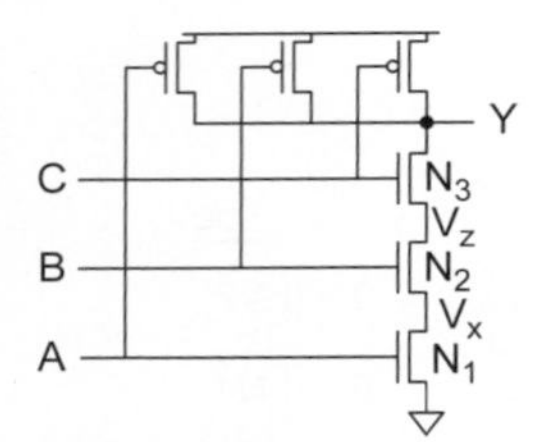

FIGURE 4.23 NAND gate demonstrating pattern dependence of gate and subthreshold leakage

experiences negligible gate leakage [Lee03, Mukhopadhyay03]. In both cases, the OFF transistor has no gate leakage. Thus, gate leakage can be alleviated by stacking transistors such that the OFF transistor is closer to the rail.

Table 4.2 summarizes the combined effects of gate and subthreshold leakage on the 3-input NAND gate shown in Figure 4.23 using data from [Lee03] for a process with 15 Å oxides and 60 nm channel length. The gate leakage through an ON nMOS transistor is 6.3 nA. pMOS gate leakage is negligible. The subthreshold leakage through an nMOS transistor with $V_{ds} = V_{DD}$ is 5.63 nA and the subthreshold leakage through a pMOS transistor with $|V_{ds}| = V_{DD}$ is 9.3 nA.

The NAND3 benefits from the stack effect to reduce subthreshold leakage. In the 000 case, all three nMOS transistors are OFF and the triple stack effect cuts leakage by a factor of 10. Both intermediate nodes drift up to somewhere around 100–200 mV set by the stack effect. In the 001 and 100 cases, two nMOS transistors are OFF and the double stack effect cuts leakage by a factor of 5. In the 110 case, the nMOS stack experiences full subthreshold leakage because only one transistor is OFF and it sees $V_{ds} = V_{DD}$. In the 011 and 101 cases, the single OFF nMOS transistor sees $V_{ds} = V_{DD} - V_t$, so the leakage is partially reduced. In the 111 case, all three parallel pMOS transistors leak.

The NAND3 also sees pattern-dependent gate leakage. In the 000 case, all three nMOS transistors are off, so no gate current flows. In the 001 and 011 cases, the ON transistors see $V_{gs} = V_t$ and thus have little leakage. In the 010 case, gate leakage through $N2$ charges V_x and V_z up to an intermediate voltage until the increase in source/drain voltage reduces the gate current. This raises the source voltage of $N3$, effectively eliminating its subthreshold leakage. In the 101 case, $N1$ sees full gate leakage, while $N3$ has little because V_z is at a high voltage. In the 110 case, $N1$ and $N2$ both see gate leakage, and in the 111 case, all three nMOS transistors leak.

TABLE 4.2 Gate and subthreshold leakage in NAND3 (nA)

Input State (ABC)	I_{sub}	I_{gate}	I_{total}	V_x	V_z
000	0.4	0	0.4	stack effect	stack effect
001	0.7	0	0.7	stack effect	$V_{DD} - V_t$
010	0	1.3	1.3	intermediate	intermediate
011	3.8	0	10.1	$V_{DD} - V_t$	$V_{DD} - V_t$
100	0.7	6.3	7.0	0	stack effect
101	3.8	6.3	10.1	0	$V_{DD} - V_t$
110	5.6	12.6	18.2	0	0
111	28	18.9	46.9	0	0

4.3.1.3 Junction Leakage Junction leakage occurs when a source or drain diffusion region is at a different potential from the substrate. Although the ordinary leakage of reverse-biased diodes is usually negligible, BTBT and GIDL can result in leakage currents that approach subthreshold leakage levels in high-V_t transistors. BTBT is maximum when a strong reverse bias is applied between the drain and body (e.g., $V_{db} = V_{DD}$ for an nMOS transistor). GIDL is maximum when the transistor is OFF and a strong bias is applied to the drain (e.g., $V_{gd} = -V_{DD}$ for an nMOS transistor). Junction leakage is often minor in comparison to the other leakages, but can be expressed in nA/μm of transistor width when it needs to be considered.

4.3.1.4 Contention Current Static CMOS circuits have no contention current. However, certain alternative circuits inherently draw current even while quiescent. For example, pseudo-nMOS gates discussed in Section 8.2.2 experience contention between the nMOS pulldowns and the always-on pMOS pullups when the output is 0. Current-mode logic and many analog circuits also draw static current. Such circuits should be turned OFF in sleep mode by disabling the pullups or current source.

4.3.1.5 Static Power Estimation Static current estimation is a matter of estimating the total width of transistors that are leaking, multiplying by the leakage current per width, and multiplying by the fraction of transistors that are in their leaky state (usually one-half). Add the contention current if applicable. The static power is the supply voltage times the static current.

Example 4.4

Consider the system-on-chip from Example 4.1. Subthreshold leakage for OFF devices is 100 nA/μm for low-threshold devices and 10 nA/μm for high-threshold devices. Gate leakage is 5 nA/μm. Junction leakage is negligible. Memories use low-leakage devices everywhere. Logic uses low-leakage devices in all but 5% of the paths that are most critical for performance. Estimate the static power consumption.

SOLUTION: There are $(50 \times 10^6$ logic transistors$)(0.05)(12\ \lambda)(0.025\ \mu\text{m}/\lambda) = 0.75 \times 10^6$ μm of low-threshold devices and $[(50 \times 10^6$ logic transistors$)(0.95)(12\ \lambda) + (950 \times 10^6$ memory transistors$)(4\ \lambda)](0.025\ \mu\text{m}/\lambda) = 109.25 \times 10^6\ \mu$m of high-threshold devices. Neglecting the benefits of series stacks, half the transistors are OFF and contribute subthreshold leakage. Half the transistors are ON and contribute gate leakage. $I_{sub} = [(0.75 \times 10^6\ \mu\text{m})(100\ \text{nA}/\mu\text{m}) + (109.25 \times 10^6\ \mu\text{m})(10\ \text{nA}/\mu\text{m})]/2 = 584$ mA. $I_{gate} = ((0.75 + 109.25) \times 10^6\ \mu\text{m})(5\ \text{nA}/\mu\text{m})/2 = 275$ mA. $P_{static} = (584\ \text{mA} + 275\ \text{mA})(1\ \text{V}) = 859$ mW. This is 15% of the switching power and is enough to deplete the battery of a hand-held device rapidly.

4.3.2 Power Gating

The easiest way to reduce static current during sleep mode is to turn off the power supply to the sleeping blocks. This technique is called *power gating* and is shown in Figure 4.24. The logic block receives its power from a virtual V_{DD} rail, V_{DDV}. When the block is active, the header switch transistors are ON, connecting V_{DDV} to V_{DD}. When the block goes to sleep, the header switch turns OFF, allowing V_{DDV} to float and gradually sink toward 0. As this occurs, the outputs of the block may take on voltage levels in the forbidden zone. The output isolation gates force the outputs to a valid level during sleep so that they do not cause problems in downstream logic.

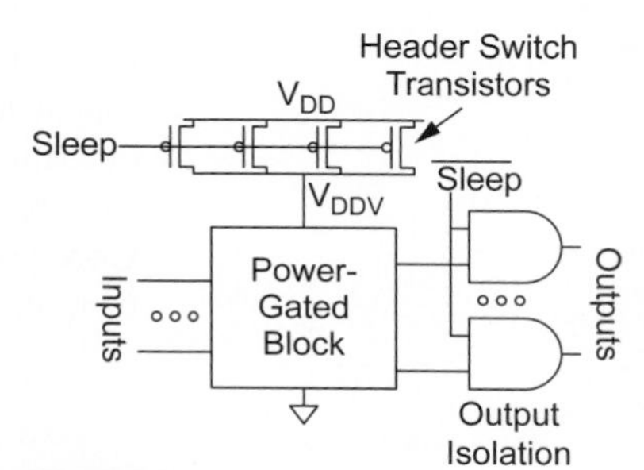

FIGURE 4.24 Power gating

Power gating introduces a number of design issues. The header switch requires careful sizing. It should add minimal delay to the circuit during active operation, and should have low leakage during sleep. The transition between active and sleep modes takes some time and energy, so power gating is only effective when a block is turned off long enough. When a block is gated, the state must either be saved or reset upon power-up. Section 9.4.3 discusses state retention registers that use a second power supply to maintain the state. Alternatively, the important registers can be saved to memory so the entire block can be power-gated. The registers must then be reloaded from memory when power is restored. [Keating07] addresses at length how to use power gating in a standard CAD flow. If power switches are fast enough, they can be used to save leakage power during active mode by powering down clock-gated blocks [Tschanz03, Min06]. If saving or losing the state costs too much overhead, turning the power supply down to the minimum level necessary to retain state (about 300 mV) using DVS is sufficient to eliminate gate leakage and reduce subthreshold leakage energy by more than an order of magnitude [Calhoun04].

Power gating was originally proposed as *Multiple Threshold CMOS* (MTCMOS) [Mutoh95] because it used low-V_t transistors for logic and high-V_t header and footer switches. However, the name is somewhat confusing because a system may use multiple threshold voltages without power gating. Moreover, it is unnecessary to switch both V_{DD} and GND.

4.3.2.1 Power Gate Design Power gating can be done externally with a disable input to a voltage regulator or internally with high-V_t header or footer switches. External power gating completely eliminates leakage during sleep, but it takes a long time and significant energy because the power network may have 100s of nF of decoupling capacitance to discharge.

On-chip power gating can use pMOS header switch transistors or nMOS footer switch transistors. nMOS transistors deliver more current per unit width so they can be smaller. On the other hand, if both internal and external power gating are used, it is more consistent for both methods to cut off V_{DD}. pMOS power gating also is simpler when multiple power supplies are employed. As a practical matter, ensuring that GND is always constant reduces confusion among designers and CAD tools; this alone is enough for many projects to choose pMOS power gating.

Theoretically, it is possible to use *fine-grained power gating* applied to individual logic gates, but placing a switch in every cell has enormous area overhead. Practical designs use *coarse-grained power gating* where the switch is shared across an entire block. The switch has an effective resistance that inevitably causes some voltage droop on V_{DDV} and increases the delay of the block. The switch is commonly sized to keep this delay to 5–10%. One way to achieve this is to calculate or simulate how much voltage droop can occur on V_{DDV} while maintaining acceptable delay. Then the average current of the block is determined through power analysis. The switch width is chosen so that the voltage droop is small enough when the average current flows through the switch. If the block is large enough that switching events are spread over time and has enough capacitance on V_{DDV} to smooth out ripples, this *average current method* [Mutoh99] is satisfactory. Wider switches reduce the droop but have more leakage when OFF and take more energy. For example, 45 nm Core processors use 1.5 meters of low-leakage pMOS power gate transistor per core to turn off the idle cores [Kumar09].

Example 4.5

A cache in a 65 nm process consumes an average power of 2 W. Estimate how wide should the pMOS header switch be if delay should not increase by more than 5%?

SOLUTION: The 65 nm process operates at 1 V, so the average current is 2 W / 1 V = 2 A. The pMOS transistor has an ON resistance of $R = 2\ \text{k}\Omega \cdot \mu\text{m}$. A 5% delay increase corresponds to a droop on V_{DDV} of about 5% (check this using EQ (3.29). Thus, R_{switch} = 0.05 × 1 V / 2 A = 25 mΩ. So the transistor width must be kΩ · μm/25 mΩ = 8× 10^4 μm. The ON resistance at low V_{ds} is lower than R. Circuit simulation shows that a width of 3.7× 10^4 μm suffices to keep droop to 5%.

The power switch is generally made of many transistors in parallel. The length and width of the transistors should be selected to maximize the I_{on} / I_{off} ratio; this is highly process-dependent and generally requires SPICE simulations sweeping L and W. A reverse body bias may be applied to the power switch transistors during sleep mode to improve their I_{on} / I_{off} ratio (see Section 4.3.4). Alternatively, the switch can be overdriven positively or negatively to turn it ON or OFF more effectively so long as the gate oxide is not overstressed [Min06].

When the power switch is turned ON, the sudden inrush of current can cause *IR* and *L di/dt* drop noise (see Section 12.3) and electromigration of the power bus (see Section 6.3.3.1). To alleviate these problems, the switch can be turned on gradually by controlling how many parallel transistors are ON.

4.3.3 Multiple Threshold Voltages and Oxide Thicknesses

Selective application of multiple threshold voltages can maintain performance on critical paths with low-V_t transistors while reducing leakage on other paths with high-V_t transistors.

A multiple-threshold cell library should contain cells that are physically identical save for their thresholds, facilitating easy swapping of thresholds. Good design practice starts with high-V_t devices everywhere and selectively introduces low-V_t devices where necessary.

Using multiple thresholds requires additional implant masks that add to the cost of a CMOS process. Alternatively, designers can increase the channel length, which tends to raise the threshold voltage via the short channel effect. For example, in Intel's 65 nm process, drawing transistors 10% longer reduces I_{on} by 10% but reduces I_{off} by a factor of 3 [Rusu07]. The dual-core Xeon processor uses longer transistors almost exclusively in the caches and in 54% of the core gates.

Most nanometer processes offer a thin oxide for logic transistors and a much thicker oxide for I/O transistors that can withstand higher voltages. The oxide thickness is controlled by another mask step. Gate leakage is negligible in the thick oxide devices, but their performance is inadequate for high speed logic applications. Some processes offer another intermediate oxide thickness to reduce gate leakage.

[Anis03] provides an extensive survey of the applications of multiple thresholds.

4.3.4 Variable Threshold Voltages

Recall from EQ (2.38) that V_{sb} modulates the threshold voltage through the body effect. Another method to achieve high I_{on} in active mode and low I_{off} in sleep mode is to dynamically adjust the threshold voltage of the transistor by applying a body bias. This technique is sometimes called *variable threshold CMOS* (VTCMOS).

For example, low-V_t devices can be used and a *reverse body bias* (RBB) can be applied during sleep mode to reduce leakage [Kuroda96]. Alternatively, higher-V_t devices can be used, and then a *forward body bias* (FBB) can be applied during active mode to increase performance [Narendra03]. Body bias can be applied to the power gating transistors to turn them off more effectively during sleep.

Too much reverse body bias (e.g., < −1.2 V) leads to greater junction leakage through BTBT [Keshavarzi01], while too much forward body bias (> 0.4 V) leads to substantial current through the body to source diodes. According to EQ (2.39), the body effect weakens as t_{ox} becomes thinner, so body biasing offers diminishing returns at 90 nm and below [von Arnim05].

Applying a body bias requires additional power supply rails to distribute the substrate and well voltages. For example, an RBB scheme for a 1.0 V n-well process could bias the p-type substrate at $V_{BBn} = -0.4$ V and the n-well at $V_{BBp} = 1.4$ V. Figure 4.25 shows a schematic and cross-section of an inverter using body bias. In an n-well process, all nMOS transistors share the same p substrate and must use the same V_{BBn}. In a triple-well process, groups of transistors can use different p-wells isolated from the substrate and thus can use different body biases. The well and substrate carry little current, so the bias voltages are relatively easy to generate using a charge pump (see Section 12.3.8).

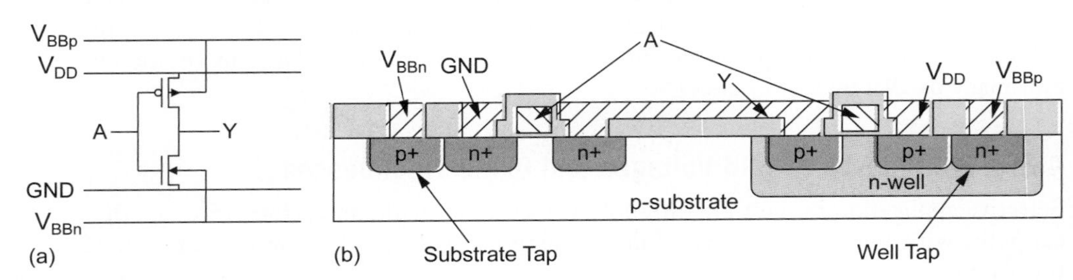

FIGURE 4.25 Body bias

4.3.5 Input Vector Control

As was illustrated in Table 4.2, the stack effect and input ordering cause subthreshold and gate leakage to vary by up to two orders of magnitude between best and worst cases. Therefore, the leakage of a block of logic depends on gate inputs, which in turn depend on the inputs to the block of logic. The idea of *input vector control* is to apply the input pattern that minimizes block leakage when the block is placed in sleep mode [Narendra06, Abdollahi04]. The vector can be applied via set/reset inputs on the registers or via a scan chain. It is hard to control all the gates in a block of logic using only the block inputs, but the best input vectors may save 25–50% of leakage as compared to random vectors. Applying the input vector causes some switching activity, so a block may need to remain in sleep for thousands of cycles to recoup the energy spent entering the sleep state.

4.4 Energy-Delay Optimization

At this point, a natural question is: what is the best choice of V_{DD} and V_t? The answer, of course, depends on the objective. Minimum power by itself is not an interesting objective because it is achieved as the delay for a computation approaches infinity and nothing is accomplished. The time for a computation must be factored into the analysis. Better metrics include minimizing the energy, minimizing the energy-delay product, and minimizing energy under a delay constraint.

4.4.1 Minimum Energy

According to EQ (4.3), the product of the power of an operation and the time for the operation to complete is the energy consumed. Hence, the *power-delay product* (PDP) is simply the energy. The minimum energy point is the least energy that an operation could consume if delay were unimportant. It occurs in subthreshold operation where $V_{DD} < V_t$. The minimum energy point typically consumes an order of magnitude less energy than the conventional operating point, but runs at least three orders of magnitude more slowly [Wang06].

John von Neumann first asserted (without justification) that the "thermodynamic minimum of energy per elementary act of information" was $kT \ln 2$ [von Neumann66]. [Meindl00] proved this result for CMOS by considering the minimum allowable voltage at which an inverter could operate. To achieve nonzero noise margins, an inverter must have a slope steeper than –1 at the switching point, V_{inv}. For an ideal inverter with $n = 1$ in the subthreshold characteristics, this occurs at a minimum operating voltage of

$$V_{min} = 2 \ln 2 v_T = 36 \text{ mV} @ 300 \text{ K} \quad \textbf{(4.23)}$$

The energy stored on the gate capacitance of a single MOSFET is $E = QV_{DD}/2$, where Q is the charge. The minimum possible charge is one electron, q. Substituting V_{min} for V_{DD} gives $E_{min} = kT \ln 2 = 2.9 \times 10^{-21}$ J. In contrast, a unit inverter in a 0.5 μm 5 V process draws about 1.5×10^{-13} J from the supply when switching, and the same inverter in a 65 nm 1 V process draws 3×10^{-16} J.

Inverters have been demonstrated operating with power supplies under 100 mV, but these do not actually minimize energy in a real CMOS process. Although they have extremely low switching energy, they run so slowly that the leakage energy dominates. The true minimum energy point is at a higher voltage that balances switching and leakage energy.

In subthreshold operation, the current drops exponentially as $V_{DD} - V_t$ decreases and thus the delay increases exponentially. The switching energy improves quadratically with V_{DD}. Leakage current improves slowly with V_{DD} because of DIBL, but the leakage energy increases exponentially because the slower gate leaks for a longer time. To achieve minimum energy operation, all transistors should be minimum width. This reduces both switching capacitance and leakage. Gate and junction leakage and short-circuit power are negligible in subthreshold operation, so the total energy is the sum of the switching and leakage energy, which is minimized near the point they crossover, as shown in Figure 4.26.

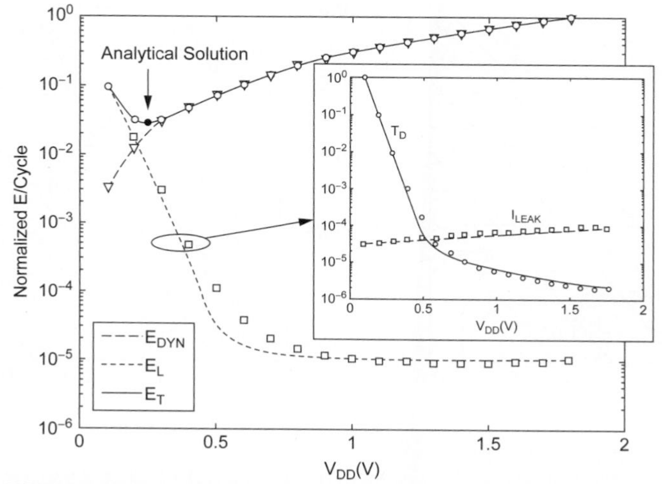

FIGURE 4.26 Minimum energy point (Reprinted from [Calhoun05]. © 2005 IEEE.)

To compute the energy, assume that a circuit has N gates on the critical path, a total effective capacitance C_{eff}, and a total effective width W_{eff} of leaking transistor. The delay of a gate operating subthreshold with a load C_g is given by EQ (3.31). The cycle time is thus

$$D = \frac{NkC_g V_{DD}}{I_{\text{off}} 10^{V_{DD}}} \tag{4.24}$$

The energy consumed in one cycle is

$$\begin{aligned}
E_{\text{switching}} &= C_{\text{eff}} V_{DD}^2 \\
E_{\text{leak}} &= I_{\text{sub}} V_{DD} D = W_{\text{eff}} NkC_g 10^{-V_{DD}} V_{DD}^2 \\
E_{\text{total}} &= E_{\text{switching}} + E_{\text{leak}} = V_{DD}^2 \left(C_{\text{eff}} + W_{\text{eff}} NkC_g 10^{-V_{DD}} \right)
\end{aligned} \tag{4.25}$$

It is possible to differentiate EQ (4.25) with respect to V_{DD} to find the minimum energy point [Calhoun05], but the results are rather messy.

A more intuitive approach is to look at the minimum energy point graphically. Figure 4.27(a) plots the energy and delay contours as a function of V_{DD} and V_t for a ring oscillator in a 180 nm process designed to reflect the behavior of a microprocessor pipeline [Wang02]. As V_{DD} increases or V_t decreases, the operating frequency increases exponentially assuming the circuit is operating at or near threshold. At $V_{DD} = V_t$, the circuit operates at about 10 MHz. The energy contours are normalized to the minimum energy point. This point, marked with a cross, occurs at $V_{DD} = 0.13$ V and $V_t = 0.37$ V. The energy is about 10 times lower than at a typical operating point, but the delay is three to four orders of magnitude greater.

The shape of the curve is only a weak function of process parameters, so it remains valid for nanometer processes. However, the result does depend strongly on the relative switching and leakage energies. Figure 4.27(b) plots the results when the activity factor drops to 0.1, reducing C_{eff}. Switching energy is less important, so the circuit can run at a higher supply voltage. The threshold then increases to cut leakage. The total energy is

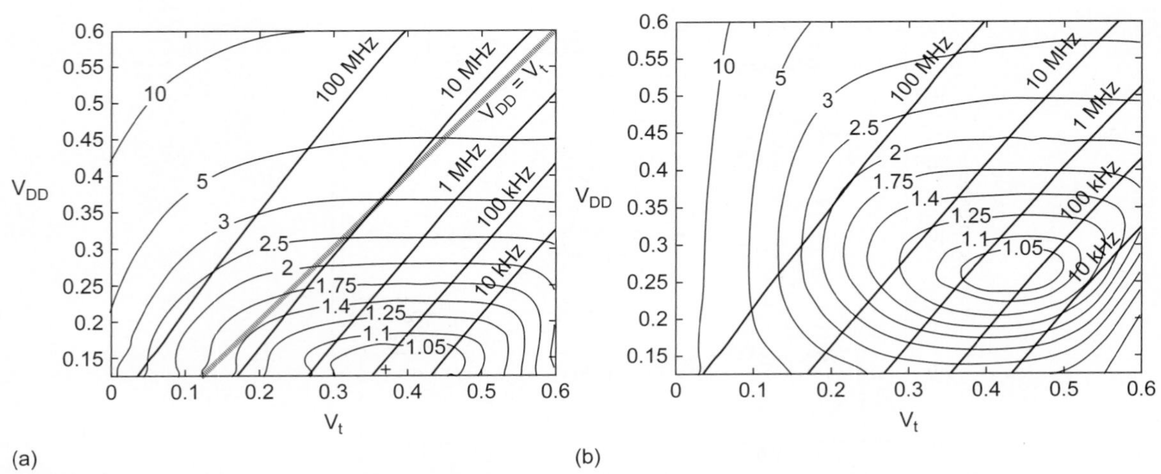

FIGURE 4.27 Contours of energy and delay for ring oscillators with (a) $\alpha = 1$, (b) $\alpha = 0.1$ (Adapted from [Wang02]. © IEEE 2002.)

greatly reduced. The result also depends on temperature: at high temperature, circuits leak more so a higher threshold voltage should be used. Process variation also pushes the best operating point toward higher voltage and energy.

4.4.2 Minimum Energy-Delay Product

The *energy-delay product* (EDP) is a popular metric that balances the importance of energy and delay [Gonzalez97, Stan99, Nose00c]. Neglecting leakage, we can elegantly solve for the supply voltage that minimizes EDP. Considering leakage, the best supply voltage is slightly higher.

First, consider the EDP when leakage is negligible. The energy to charge a load capacitance C_{eff} is given by EQ (4.7). The delay, using an α-power law model, is given by EQ (3.29). Thus, the EDP is

$$\text{EDP} = k\frac{C_{\text{eff}}^2 V_{DD}^3}{\left(V_{DD} - V_t\right)^{\alpha}} \tag{4.26}$$

Differentiating with respect to V_{DD} and setting the result to 0 gives the voltage at which the EDP is minimized

$$V_{DD\text{-opt}} = \frac{3}{3-\alpha}V_t \tag{4.27}$$

Recall that α is between 1 (completely velocity saturated) and 2 (no velocity saturation). For a typical value of α, we come to the interesting conclusion that $V_{DD\text{-opt}} \approx 2V_t$, which is substantially lower than most systems presently run.

EQ (4.26) suggests that the EDP improves as V_t approaches 0, which is obviously not true because leakage power would dominate. When a leakage term is incorporated into EQ (4.27), the results become too messy to reprint here. Figure 4.28 shows contours of EDP and delay as a function of V_{DD} and V_t. EDP is normalized to the best achievable. For typical process parameters, the best V_t is about 100–150 mV and the EDP is about four times better than at a typical operating point of V_{DD} = 1.0 V and V_t = 0.3 V. At the optimum, leakage energy is about half of dynamic energy. The dashed lines indicate contours of equal speed, normalized to the speed at the best EDP point. To operate at higher speed requires increasing the EDP. Section 6.5.3.2 will revisit this analysis considering process variation and show that the minimum EDP point occurs at a higher voltage and threshold when variations are accounted for.

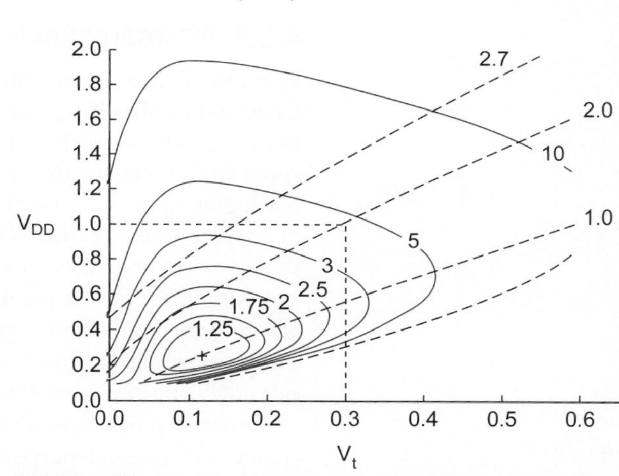

FIGURE 4.28 Contours of energy-delay product (Adapted from [Gonzalez97]. © IEEE 1997.)

4.4.3 Minimum Energy Under a Delay Constraint

In practice, designers generally face the problem of achieving minimum energy under a delay constraint. Equivalently, the power consumption of the system is limited by battery or cooling considerations and the designer seeks to achieve minimum delay under an energy constraint. Figure 4.27(a) showed contours of delay and energy. The best supply voltage and threshold for operation at a given delay is where the delay and energy contours are tangent.

For a given supply voltage and threshold voltage, the designer can make logic and sizing choices that affect delay and energy. Figure 4.13 showed an example of an energy-delay trade-off curve. Such curves can be generated using a logic synthesizer or sizing tool constrained to various delays. The curve becomes steep near the point of minimum delay, so energy-efficient designs should aim to operate at a longer delay.

Energy under a delay constraint is also minimized when leakage is about half of dynamic power [Marković04]. However, the curve is fairly flat around this point, so many designs operate at lower leakage to facilitate power saving during sleep mode.

4.5 Low Power Architectures

VLSI design used to be constrained by the number of transistors that could fit on a chip. Extracting maximum speed from each transistor maximized overall performance. Now that billions of nanometer-scale transistors fit on a chip, many designs have become power constrained and the most energy-efficient design is the highest performer. This is one of the factors that has driven the industry's abrupt shift to multicore processors.

4.5.1 Microarchitecture

Energy-efficient architectures take advantage of the structured design principles of modularity and locality [Horowitz04, Naffziger0b]. [Pollack99] observed that processor performance grows with the square root of the number of transistors. Building complex, sprawling processors to extract the last bit of instruction-level parallelism from a problem is a highly inefficient use of energy. Microarchitectures are moving toward larger numbers of simpler cores seeking to handle task and data-level parallelism. Smaller cores also have shorter wires and faster memory access.

Memories have a much lower power density than logic because their activity factors are miniscule and their regularity simplifies leakage control. If a task can be accelerated using either a faster processor or a larger memory, the memory is often preferable. Memories now comprise more than half the area of many chips.

Special-purpose functional units can offer an order of magnitude better energy efficiency than general-purpose processors. Accelerators for compute-intensive applications such as graphics, networking, and cryptography offload these tasks from the processor. Such heterogeneous architectures, combining regular cores, specialized accelerators, and large amounts of memory, are of growing importance.

Commercial software has historically lagged at least a decade behind hardware advances such as virtual memory, memory protection, 32- and 64-bit datapaths, and robust power-management. Presently, programmers have trouble taking advantage of many cores. Time will tell whether programming practices and tools catch up or whether microarchitectures will have to yield to the needs of programmers.

4.5.2 Parallelism and Pipelining

In the past, *parallelism* and *pipelining* have been effective ways to reduce power consumption, as shown in Figure 4.29 [Chandrakasan92].

Replacing a single functional unit with N parallel units allows each to operate at $1/N$ the frequency. A multiplexer selects between the results. The voltage can be scaled down accordingly, offering quadratic savings in energy at the expense of doubling the area. Replacing a single functional unit with an N-stage pipelined unit also reduces the amount of logic in a clock cycle at the expense of more registers. Again, the voltage can be scaled down. The two techniques can be combined for even better energy efficiency.

When leakage is unimportant, parallelism offers a slight edge because the multiplexer has less overhead than the pipeline registers. Also, perfectly balancing logic across pipeline stages can be difficult. Now that leakage is a substantial fraction of total power, pipelining becomes preferable because the parallel hardware has N times as much leakage [Marković04].

Now that V_{DD} is closer to the best energy-delay point, the potential supply reduction and energy savings are diminishing. Nevertheless, parallelism and pipelining remain primary tools to extract performance from the vast transistor budgets now available.

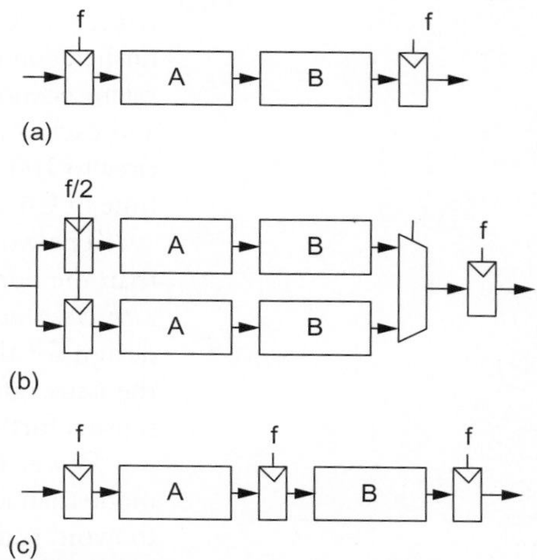

FIGURE 4.29 Functional units: (a) normal, (b) parallel, (c) pipelined

4.5.3 Power Management Modes

As your parents taught you to turn off the lights when you leave a room, chip designers have now learned they must turn off portions of the chip when they are not active by applying clock and power gating. Many chips now employ a variety of power management modes giving a trade-off between power savings and wake-up time.

For example, the Intel Atom processor [Gerosa09] operates at a peak frequency of 2 GHz at 1 V, consuming 2 W. The power management modes are shown in Figure 4.30. In the low frequency mode, the clock drops as slow as 600 MHz while the power supply

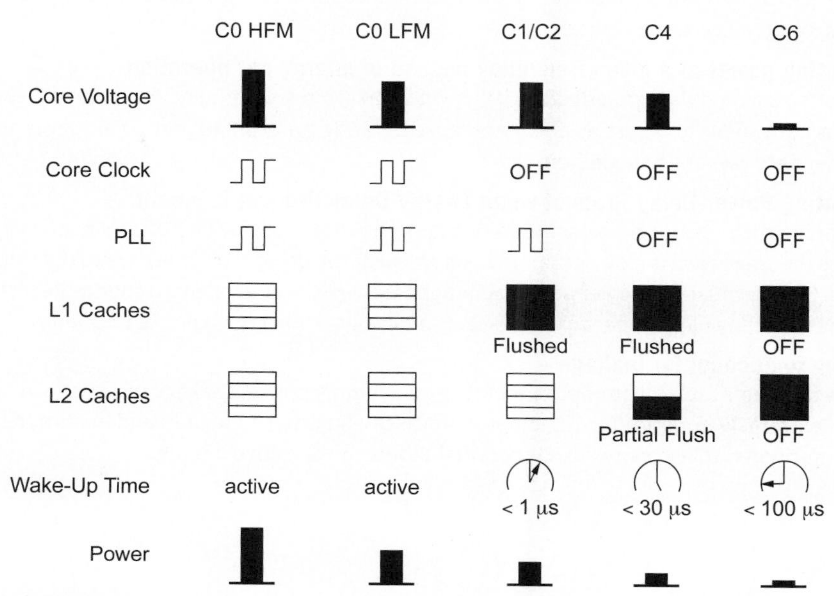

FIGURE 4.30 Atom power management modes (© 2009 IEEE.)

reduces to 0.75 V. In sleep mode C1, the core clock is turned off and the level 1 cache is flushed and power-gated to reduce leakage, but the processor can return to active state in 1 microsecond. In sleep mode C4, the PLL is also turned OFF. In sleep mode C6, the core and caches are all power-gated to reduce power to less than 80 mW, but wake-up time rises to 100 microseconds. For a typical workload, the processor can spend 80–90% of its time in C6 sleep mode, reducing average power to 220 mW.

The worst-case power that a chip may consume can be a factor of two or more greater than the normal power. Code triggering maximal power consumption is sometimes called a *thermal virus* [Naffziger06] because it seeks to burn out the chip. To avoid having to design for this worst case, chips can employ adaptive features, throttling back activity if the issue rate or die temperature becomes too high. Section 12.2.5 discusses temperature sensors further.

Power management results in substantially lower power consumption during idle mode than active mode. The transition between idle and active may require multiple cycles to avoid sudden current spikes that excite power supply resonances and cause excessive supply noise.

4.6 Pitfalls and Fallacies

Oversizing gates

Designers seeking timing closure tend to crank up the size of gates. Doubling the size of all the gates on a gate-dominated path does not improve delay, but doubles the power consumption.

Designing for speed without regard to power

Nanometer processes have reached a point where it is no longer possible to design a large chip for speed without regard to power: the chip will be impossible to cool. Designs must be power efficient. Systems tuned exclusively for speed tend to use large gates and speculative logic that consumes a great deal of power. If a core or processing element can be simplified to offer 80% of the performance at 50% of the power, then two cores in parallel can offer 160% of the throughput at the same power.

Reporting power at a given frequency instead of energy per operation

Sometimes a module is described by its power at an arbitrary frequency (e.g., 10 mW @ 1 GHz). This is equivalent to reporting energy because $E = P/f$ (e.g., 10 pJ). Reporting energy is arguably cleaner because it is a single number.

Reporting Power-Delay Product when Energy-Delay Product is meant

Extending the previous point, sometimes a system is described by its PDP at a given frequency, where the frequency is slower than the reciprocal of the delay. This metric is really a variation of the EDP, because the power at a low enough frequency is equivalent to energy. Reporting the EDP is definitely cleaner because it doesn't involve an arbitrary choice of frequency.

Failing to account for leakage

Many designers are accustomed to focusing on dynamic power. Leakage in all its forms has become extremely important in nanometer processes. Ignoring it not only underestimates power consumption but also can cause functional failures in sensitive circuits.

4.7 Historical Perspective

The history of electronics has been a relentless quest to reduce power so that more capabilities can be provided in a smaller volume.

The Colossus, brought online in 1944, was one of the world's first fully electronic computers. The secret machine was built from 2400 vacuum tubes and consumed 15 kW as it worked day and night decrypting secret German communications. The machine was destroyed after the war, but a functional replica shown in Figure 4.31 was rebuilt in 2007.

Vacuum tube machines filled entire rooms and failed frequently. Imagine the problem of keeping 2400 light bulbs burning simultaneously. By the 1960s, vacuum tubes were surpassed by solid-state transistors that were far smaller and consumed milliwatts rather than watts. Gordon Moore soon issued his famous prophecy about the exponential growth in the number of transistors per chip.

FIGURE 4.31 Reconstructed Colossus Mark 2 (Photograph by Tony Sale. Reprinted with permission.)

MOSFETs entered the scene commercially around 1970. For more than a decade, nMOS technology predominated because it could pack transistors more densely (and hence cheaply) than CMOS. nMOS circuits used *depletion load* (negative-V_t) nMOS pull-ups as resistive loads, so each gate with an output of 0 dissipated contention power. For example, Figure 4.32 shows an nMOS 2-input NOR gate.

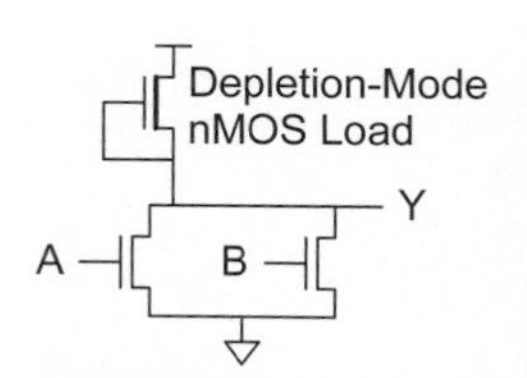

FIGURE 4.32 nMOS NOR gate

CMOS circuits made their debut in watch circuits (pioneered by none other than the Swiss, of course!), where their key ability to draw almost zero power while not switching was critical [Vittoz72]. This use succeeded despite very low circuit densities and low circuit speed of the CMOS technologies of the day. It was not until the mid 1980s that the ever-increasing power dissipation of mainstream circuits such as microprocessors forced a move from nMOS to CMOS technology, again despite density arguments.

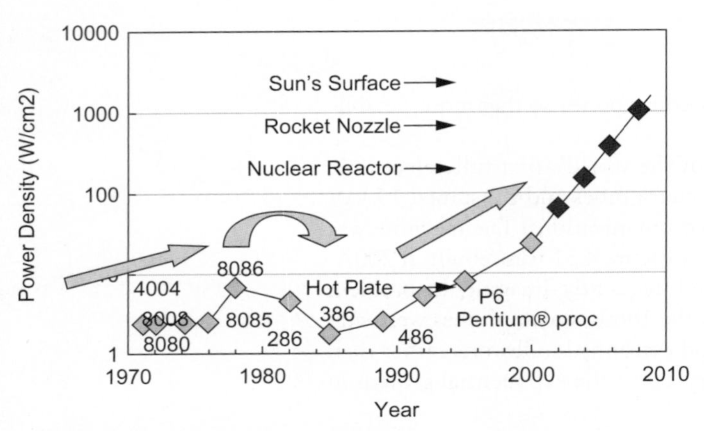

FIGURE 4.33 Microprocessor power density trends as predicted in 2001 (Reprinted with permission of Intel Corporation.)

As engineers became capable of integrating millions (and now billions) of CMOS transistors onto a single chip, power consumption became a concern for CMOS as well. In the 1990s, designers facing the power wall abandoned the long-cherished 5 V standard and began scaling power supplies to reduce dynamic power. Eventually, this forced threshold voltages to decrease until sub-threshold leakage has become an issue. As gate dielectrics have scaled down to a few atoms in thickness, quantum mechanical effects have made gate leakage problematic as well. With supplies at 1 V, there is limited room for further power scaling: we are stuck between a rock and a hot place.

On February 5, 2001, Intel Vice President Patrick Gelsinger gave a keynote speech at the International Solid State Circuits Conference [Gelsinger01]. He showed that microprocessor power consumption has been increasing exponentially (see Figure 4.33) and was forecast to grow even faster in the coming decade. He predicted that "business as usual will not work in the future," and that if scaling continued at this pace, by 2005, high-speed processors would have the power density of a nuclear reactor, by 2010, a rocket nozzle, and by 2015, the surface of the sun! Obviously, business did not proceed as usual and power consumption has leveled out at under 150 W for high-performance processors and much lower for battery-powered systems.

Clock gating was the first widely applied technique for power reduction because it is relatively straightforward. Power gating was initially applied to low-power battery-operated systems to increase the standby lifetime, but is now required for leakage control even in high-performance microprocessors [Rusu10]. Voltage domains are also widely used. Initially, separate supplies were provided for the core and I/O to provide compatibility with legacy I/O standards. The next step was to separate the supplies for the memories from the core logic. Memories arrays often use a constant relatively high supply voltage (as high as the process allows) for reliability. Logic dissipates the bulk of the dynamic power, so it operates at a lower, possibly variable voltage. Sometimes phase-locked loops or sensitive analog circuitry use yet another filtered domain. Dynamic voltage scaling is commonly used to support a range of power/performance trade-offs [Clark01]. For example, laptop processors commonly run at a higher voltage when the system is plugged into wall power.

Body bias has been used for leakage control in applications such as the Intel XScale microprocessor [Clark02], the Transmeta Efficeon microprocessor, and a Toshiba MPEG4 video codec [Takahashi98]. Clustered voltage scaling was also used in the video codec. Both of these techniques introduce overhead routing the bias or voltage lines through a block and controlling noise on these lines. They have not achieved the widespread popularity of other techniques, and the effectiveness of body bias becomes limited below 130 nm because the body effect coefficient decreases along with oxide thickness.

The move to CMOS technology was really the last major movement in mass-market semiconductor technologies. To date, no one has come up with better devices. The hundreds of billions of dollars that have been invested in optimizing CMOS make it a formidable technology to surpass. Rather than looking for a replacement, our best hope is to continue learning to use energy as efficiently as we can.

Summary

The power consumption of a circuit has both dynamic and static components. The dynamic power comes from charging and discharging the load capacitances and depends on the frequency, voltage, capacitance, and activity factor. The static power comes from leakage and from circuits that have an intentional path from V_{DD} to GND. CMOS circuits have historically consumed relatively low power because complementary CMOS gates dissipate almost zero static power when operated at high V_t. However, leakage is increasing as feature size decreases, making static power consumption as great a concern as dynamic power. The best way to control power is to turn off a circuit when it is not in use. The most important techniques are clock gating, which turns off the clock when a unit is idle, and power gating, which turns off the power supply when a unit is in sleep mode.

Exercises

4.1 Derive the switching probabilities in Table 4.1.

4.2 Design an 8-input OR gate with a delay of under 4 FO4 inverters. Each input may present at most 1 unit of capacitance. The load capacitance is 16 units. If the input probabilities are 0.5, compute the switching probability at each node and size the circuit for minimum switching energy.

4.3 Construct a table similar to Table 4.2 for a 2-input NOR gate.

4.4 Design a header switch for a power gating circuit in a 65 nm process. Suppose the pMOS transistor has an ON resistance of about 2.5 k$\Omega \cdot \mu$m. The block being gated has an ON current of 100 mA. How wide must the header transistor be to cause less than a 2% increase in delay?

4.5 You are synthesizing a chip composed of random logic with an average activity factor of 0.1. You are using a standard cell process with an average switching capacitance of 450 pF/mm^2. Estimate the dynamic power consumption of your chip if it has an area of 70 mm^2 and runs at 450 MHz at $V_{DD} = 0.9$ V.

4.6 You are considering lowering V_{DD} to try to save power in a static CMOS gate. You will also scale V_t proportionally to maintain performance. Will dynamic power consumption go up or down? Will static power consumption go up or down?

4.7 The stack effect causes the current through two series OFF transistors to be an order of magnitude less than I_{off} when DIBL is significant. Show that the current is $I_{\text{off}}/2$ when DIBL is insignificant (e.g., $\eta = 0$). Assume $\gamma = 0$, $n = 1$.

4.8 Determine the activity factor for the signal shown in Figure 4.34. The clock rate is 1 GHz.

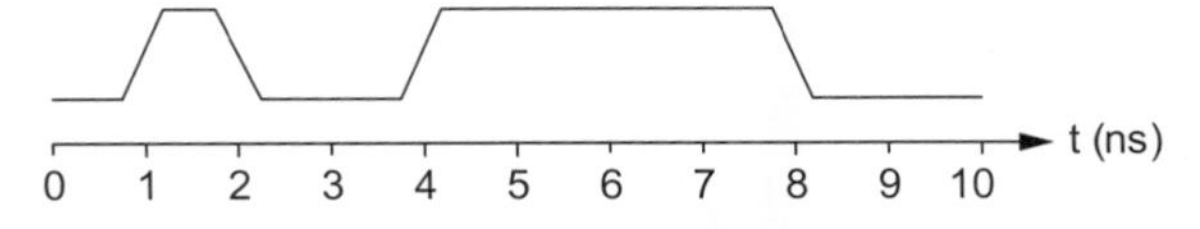

FIGURE 4.34 Signal for Exercise 4.4

4.9 Consider the buffer design problem from Example 3.14. If the delay constraint is 20 τ, how many stages will give the lowest energy, and how should the stages be sized?

4.10 Repeat Exercise 4.9 if the load is 500 rather than 64 and the delay constraint is 30 τ.

Wires 5

5.1 Introduction

The wires linking transistors together are called *interconnect* and play a major role in the performance of modern systems. In the early days of VLSI, transistors were relatively slow. Wires were wide and thick and thus had low resistance. Under those circumstances, wires could be treated as ideal equipotential nodes with lumped capacitance. In modern VLSI processes, transistors switch much faster. Meanwhile, wires have become narrower, driving up their resistance to the point, that in many signal paths, the wire RC delay exceeds gate delay. Moreover, the wires are packed very closely together and thus a large fraction of their capacitance is to their neighbors. When one wire switches, it tends to affect its neighbor through capacitive coupling; this effect is called *crosstalk*. Wires also account for a large portion of the switching energy of a chip. On-chip interconnect inductance had been negligible but is now becoming a factor for systems with fast edge rates and closely packed busses. Considering all of these factors, circuit design is now as much about engineering the wires as the transistors that sit underneath.

The remainder of this section defines the dimensions used to describe interconnect and gives a practical example of wire stacks in nanometer processes. Section 5.2 explores how to model the resistance, capacitance, and inductance of wires. Section 5.3 examines the impact of wires on delay, energy, and noise. Section 5.4 considers the tools at a designer's disposal for improving performance and controlling noise. Section 5.5 extends the method of Logical Effort to give insights about designing paths with interconnect.

5.1.1 Wire Geometry

Figure 5.1 shows a pair of adjacent wires. The wires have width w, length l, thickness t, and spacing of s from their neighbors and have a dielectric of height h between them and the conducting layer below. The sum of width and spacing is called the wire *pitch*. The thickness to width ratio t/w is called the *aspect ratio*.

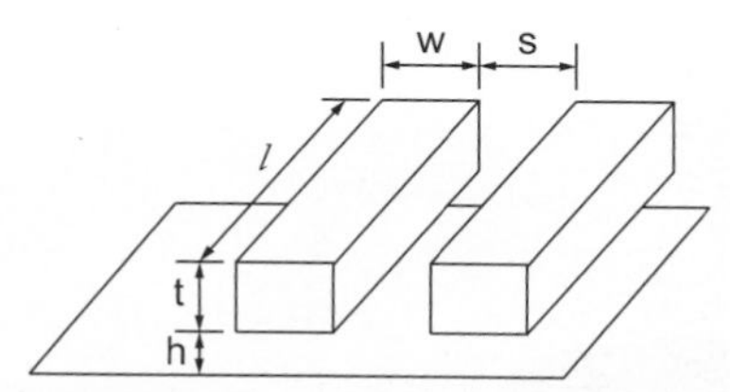

FIGURE 5.1 Interconnect geometry

Early CMOS processes had a single metal layer and until the early 1990s only two or three layers were available, but with advances in chemical-mechanical polishing it became far more practical to manufacture many metal layers. As discussed in Section 15.4.2, aluminum (Al) wires used in older processes gave way to copper (Cu) around the 180 or 130 nm node to reduce resistance. Soon after, manufacturers began replacing the SiO_2 insulator between wires with a succession of materials with lower dielectric constants (*low-k*) to reduce capacitance. A 65 nm process typically has 8–10 metal layers and the layer count has been increasing at a rate of about one layer every process generation or two.

5.1.2 Example: Intel Metal Stacks

Figure 5.2 shows cross-sections of the metal stacks in the Intel 90 and 45 nm processes, shown to scale [Thompson02, Moon08]. The 90 nm process has six metal layers, while the 45 nm process shows the bottom eight metal layers. The transistors are tiny gizmos beneath the vast labyrinth of wire. Metal1 is on the tightest pitch, roughly that of a contacted transistor, to provide dense routing within cells. The upper levels are progressively thicker and on a greater pitch to offer lower-resistance interconnections over progressively longer distances. The wires have a maximum aspect ratio of about 1.8.

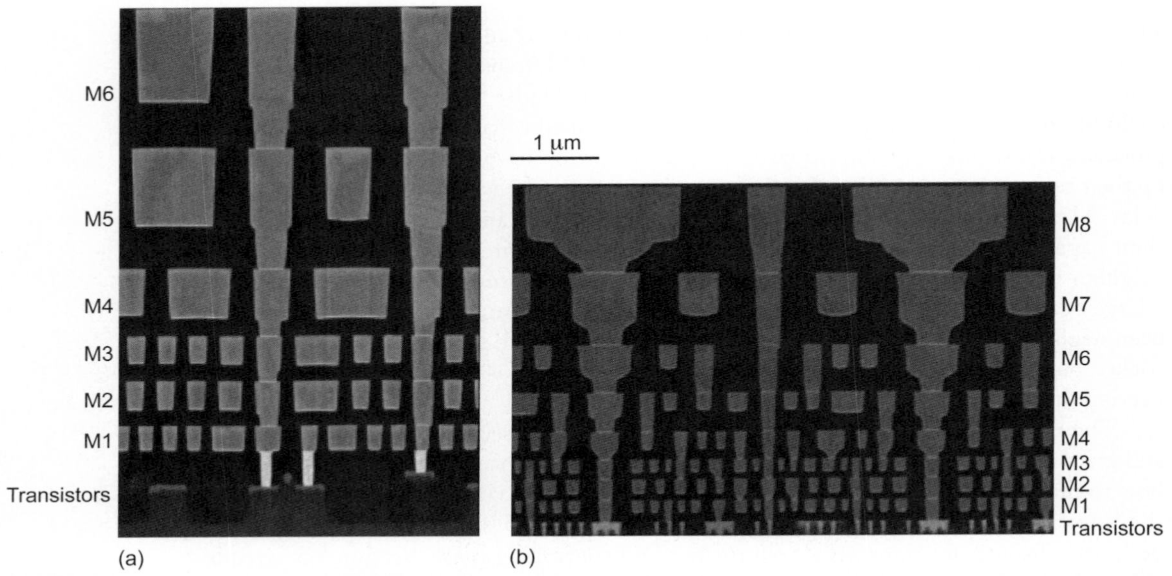

FIGURE 5.2 SEM image of wire cross-sections in Intel's (a) 90 nm and (b) 45 nm processes ((a) From [Thompson02] © 2002 IEEE. (b) From [Moon08] with permission of Intel Corporation.)

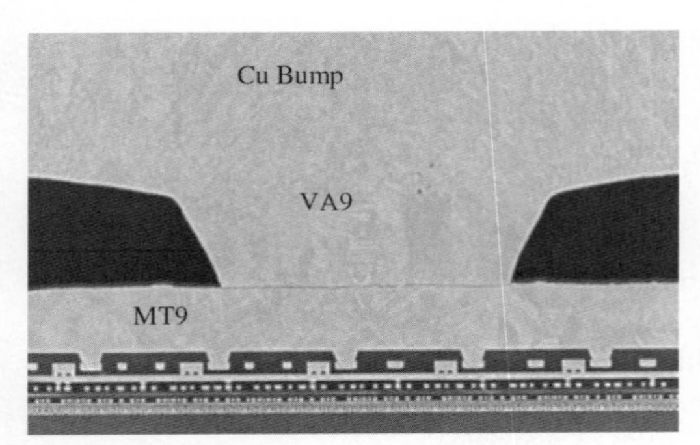

FIGURE 5.3 SEM image of complete cross-section of Intel's 45 nm process including M9 and I/O bump (From [Moon08] with permission of Intel Corporation.)

The top-level metal is usually used for power and clock distribution because it has the lowest resistance. Intel's 45 nm process introduced an unusual extra-thick ninth Cu metal layer used to distribute power to different power-gated domains across the die (see Section 4.3.2). Figure 5.3 shows a full cross-section including this MT9 layer, a Cu bump for connecting to the power or ground network in the package (see Section 12.2.2), and a VA9 via between MT9 and the bump. The lower levels of metal and transistors are scarcely visible beneath these fat top layers. Table 5.1 lists the thickness and minimum pitch for each metal layer.

TABLE 5.1 Intel 45 nm metal stack

Layer	*t* (nm)	*w* (nm)	*s* (nm)	pitch (nm)
M9	7 μm	17.5 μm	13 μm	30.5 μm
M8	720	400	410	810
M7	504	280	280	560
M6	324	180	180	360
M5	252	140	140	280
M4	216	120	120	240
M3	144	80	80	160
M2	144	80	80	160
M1	144	80	80	160

5.2 Interconnect Modeling

A pipe makes a good mechanical analogy for a wire, as shown in Figure 5.4 [Ho07]. The resistance relates to the wire's cross-sectional area. A narrow pipe impedes the flow of current. The capacitance relates to a trough underneath the leaky pipe that must fill up before current passes out the end of the pipe. And the inductance relates to a paddle wheel along the wire with inertia that opposes changes in the rate of flow. Each of these elements is discussed further in this section.

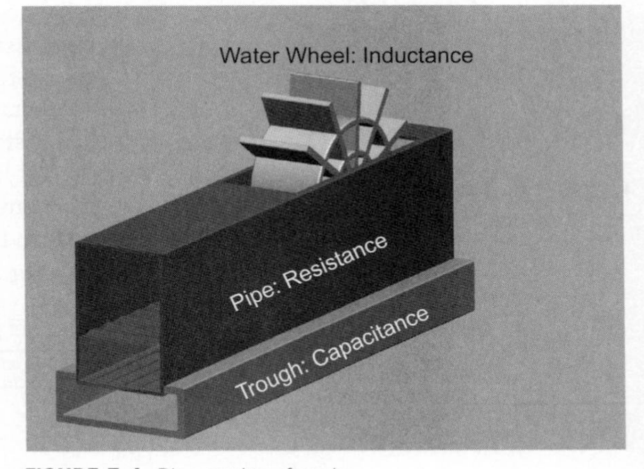

FIGURE 5.4 Pipe analogy for wire

A wire is a distributed circuit with a resistance and capacitance per unit length. Its behavior can be approximated with a number of lumped elements. Three standard approximations are the L-model, π-model, and T-model, so-named because of their shapes. Figure 5.5 shows how a distributed RC circuit is equivalent to N distributed RC segments of proportionally smaller resistance and capacitance, and how these segments can be modeled with lumped elements. As the number of segments approaches infinity, the lumped approximation will converge with the true distributed circuit. The L-model is a poor choice because a large number of segments are required for accurate results. The π-model is much better; three segments are sufficient to give results accurate to 3% [Sakurai83]. The T-model is comparable to the π-model, but produces a circuit with one more node that is slower to solve by hand or with a circuit simulator. Therefore, it is common practice to model long wires with a 3–5 segment π-model for simulation. If inductance is considered, it is placed in series with each resistor. The remainder of this section considers how to compute the resistance, capacitance, and inductance.

N Segments
R C ↔ R/N C/N R/N C/N ∘∘∘ R/N C/N R/N C/N
L-model (R, C) — π-model (R, C/2, C/2) — T-model (R/2, R/2, C)

FIGURE 5.5 Lumped approximation to distributed RC circuit

5.2.1 Resistance

The resistance of a uniform slab of conducting material can be expressed as

$$R = \frac{\rho}{t}\frac{l}{w} \tag{5.1}$$

where ρ is the resistivity.[1] This expression can be rewritten as

$$R = R_{\square}\frac{l}{w} \tag{5.2}$$

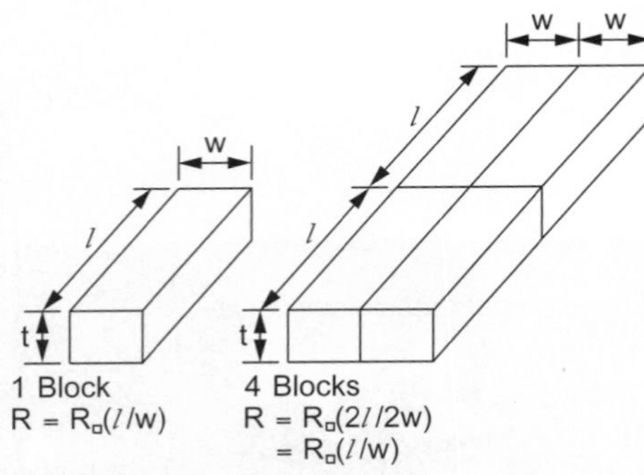

FIGURE 5.6
Two conductors with equal resistance

where $R_{\square} = \rho/t$ is the *sheet resistance* and has units of Ω/square. Note that a square is a dimensionless quantity corresponding to a slab of equal length and width. This is convenient because resistivity and thickness are characteristics of the process outside the control of the circuit designer and can be abstracted away into the single sheet resistance parameter.

To obtain the resistance of a conductor on a layer, multiply the sheet resistance by the ratio of length to width of the conductor. For example, the resistances of the two shapes in Figure 5.6 are equal because the length-to-width ratio is the same even though the sizes are different. Nonrectangular shapes can be decomposed into simpler regions for which the resistance is calculated [Horowitz83].

Table 5.2 shows bulk electrical resistivities of pure metals at room temperature [Bakoglu90]. The resistivity of thin metal films used in wires tends to be higher because of scattering off the surfaces and grain boundaries, e.g., 2.2–2.6 $\mu\Omega \cdot$ cm for Cu and 3.6–4.0 $\mu\Omega \cdot$ cm for Al [Kapur02].

TABLE 5.2 Bulk resistivity of pure metals at 22 °C

Metal	Resistivity ($\mu\Omega \cdot$ cm)
Silver (Ag)	1.6
Copper (Cu)	1.7
Gold (Au)	2.2
Aluminum (Al)	2.8
Tungsten (W)	5.3
Molybdenum (Mo)	5.3
Titanium (Ti)	43.0

FIGURE 5.7 Copper barrier layer and dishing

As shown in Figure 5.7, copper must be surrounded by a lower-conductivity diffusion barrier that effectively reduces the wire cross-sectional area and hence raises the resistance. Moreover, the polishing step can cause *dishing* that thins the metal. Even a 10 nm barrier is quite significant when the wire width is only tens of nanometers. If the average barrier thickness is t_{barrier} and the height is reduced by t_{dish}, the resistance becomes

$$R = \frac{\rho}{\left(t - t_{\text{dish}} - t_{\text{barrier}}\right)}\frac{l}{\left(w - 2t_{\text{barrier}}\right)} \tag{5.3}$$

[1] ρ is used to indicate both resistivity and best stage effort. The meaning should be clear from context.

Example 5.1

Compute the sheet resistance of a 0.22 μm thick Cu wire in a 65 nm process. Find the total resistance if the wire is 0.125 μm wide and 1 mm long. Ignore the barrier layer and dishing.

SOLUTION: The sheet resistance is

$$R_{\square} = \frac{2.2 \times 10^{-8}\ \Omega \cdot \text{m}}{0.22 \times 10^{-6}\ \text{m}} = 0.10\ \Omega/\square \tag{5.4}$$

The total resistance is

$$R = \left(0.10\ \Omega/\square\right)\frac{1000\ \mu\text{m}}{0.125\ \mu\text{m}} = 800\ \Omega \tag{5.5}$$

The resistivity of polysilicon, diffusion, and wells is significantly influenced by the doping levels. Polysilicon and diffusion typically have sheet resistances under 10 Ω/square when silicided and up to several hundred Ω/square when unsilicided. Wells have lower doping and thus even higher sheet resistance. These numbers are highly process-dependent. Large resistors are often made from wells or unsilicided polysilicon.

Contacts and vias also have a resistance, which is dependent on the contacted materials and size of the contact. Typical values are 2–20 Ω. Multiple contacts should be used to form low-resistance connections, as shown in Figure 5.8. When current turns at a right angle or reverses, a square array of contacts is generally required, while fewer contacts can be used when the flow is in the same direction.

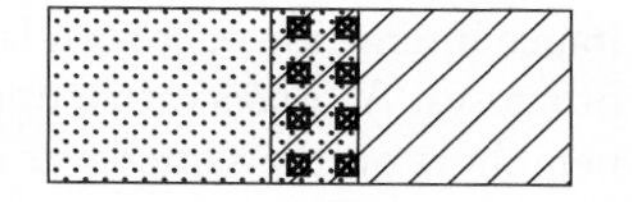

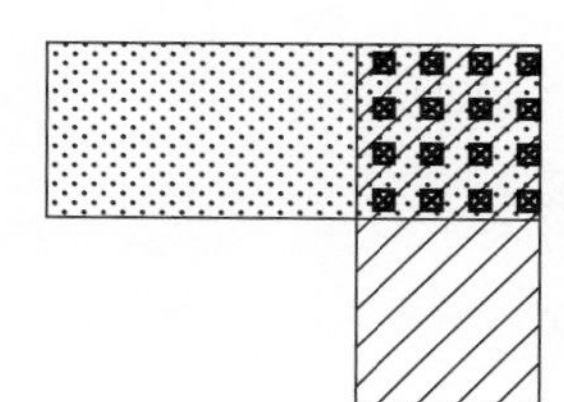

FIGURE 5.8 Multiple vias for low-resistance connections

5.2.2 Capacitance

An isolated wire over the substrate can be modeled as a conductor over a ground plane. The wire capacitance has two major components: the parallel plate capacitance of the bottom of the wire to ground and the fringing capacitance arising from fringing fields along the edge of a conductor with finite thickness. In addition, a wire adjacent to a second wire on the same layer can exhibit capacitance to that neighbor. These effects are illustrated in Figure 5.9. The classic parallel plate capacitance formula is

$$C = \frac{\varepsilon_{ox}}{h} wl \tag{5.6}$$

Note that oxides are often doped with phosphorous to trap ions before they damage transistors; this oxide has $\varepsilon_{ox} \approx k\varepsilon_0$, with $k = 4.1$ as compared to 3.9 for an ideal oxide or lower for low-k dielectrics.

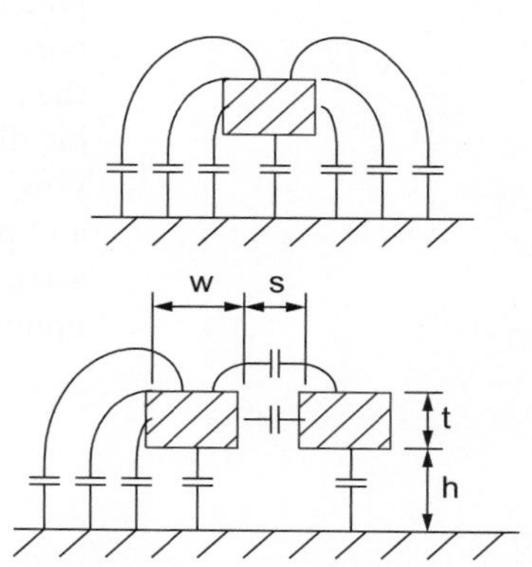

FIGURE 5.9 Effect of fringing fields on capacitance

The fringing capacitance is more complicated to compute and requires a numerical field solver for exact results. A number of authors have proposed approximations to this calculation [Barke88, Ruehli73, Yuan82]. One intuitively

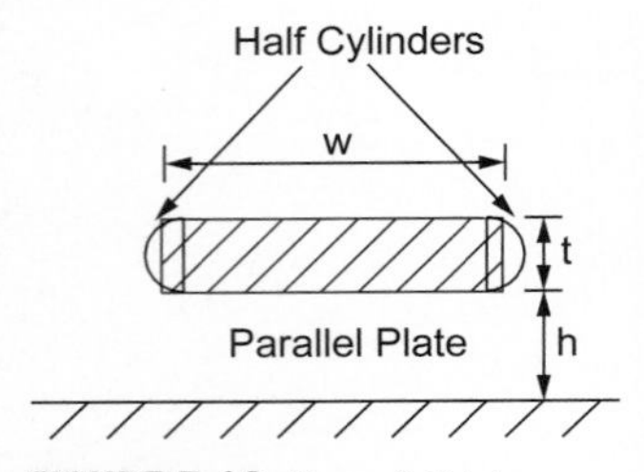

FIGURE 5.10 Yuan & Trick capacitance model including fringing fields

appealing approximation treats a lone conductor above a ground plane as a rectangular middle section with two hemispherical end caps, as shown in Figure 5.10 [Yuan82]. The total capacitance is assumed to be the sum of a parallel plate capacitor of width $w - t/2$ and a cylindrical capacitor of radius $t/2$. This results in an expression for the capacitance that is accurate within 10% for aspect ratios less than 2 and $t \approx h$.

$$C = \varepsilon_{ox} l \left[\frac{w - \frac{t}{2}}{h} + \frac{2\pi}{\ln\left(1 + \frac{2h}{t} + \sqrt{\frac{2h}{t}\left(\frac{2h}{t} + 2\right)}\right)} \right] \quad (5.7)$$

An empirical formula that is computationally efficient and relatively accurate is [Meijs84, Barke88]

$$C = \varepsilon_{ox} l \left[\frac{w}{h} + 0.77 + 1.06\left(\frac{w}{h}\right)^{0.25} + 1.06\left(\frac{t}{h}\right)^{0.5} \right] \quad (5.8)$$

which is good to 6% for aspect ratios less than 3.3.

These formulae do not account for neighbors on the same layer or higher layers. Capacitance interactions between layers can become quite complex in modern multilayer CMOS processes. A conservative upper bound on capacitance can be obtained assuming parallel neighbors on the same layer at minimum spacing and that the layers above and below the conductor of interest are solid ground planes. Similarly, a lower bound can be obtained assuming there are no other conductors in the system except the substrate. The upper bound can be used for propagation delay and power estimation while the lower bound can be used for contamination delay calculations before layout information is available.

A cross-section of the model used for capacitance upper bound calculations is shown in Figure 5.11. The total capacitance of the conductor of interest is the sum of its capacitance to the layer above, the layer below, and the two adjacent conductors. If the layers above and below are not switching,[2] they can be modeled as ground planes and this component of capacitance is called C_{gnd}. Wires do have some capacitance to further neighbors, but this capacitance is generally negligible because most electric fields terminate on the nearest conductors. The dielectrics used between adjacent wires have the lowest possible dielectric constant k_{horiz} to minimize capacitance. The dielectric between layers must provide greater mechanical stability and may have a larger k_{vert}. EQ (5.9) gives a simple and physically intuitive estimate of wire capacitance [Bohr95]. The constant C_{fringe} term accounts for fringing capacitance and gives a better fit for w and s up to several times minimum [Ho01].

$$\begin{aligned} C_{total} &= C_{top} + C_{bot} + 2C_{adj} \\ &\approx \varepsilon_0 l \left[2k_{vert} \frac{w}{h} + 2k_{horiz} \frac{t}{s} \right] + C_{fringe} \end{aligned} \quad (5.9)$$

[2] Or at least consist of a large number of orthogonal conductors that on average cancel each other's switching activities.

The capacitances can be computed by generating a lookup table of data with a field solver such as FastCap [Nabors92] or HSPICE. The table may contain data for different widths and spacings for each metal layer, assuming the layers above and below are occupied or unoccupied. The table should list both C_{adj} and C_{gnd}, because coupling to adjacent lines is of great importance. Figure 5.12 shows representative data for a metal2 wire in a 180 nm process with wire and oxide thicknesses of 0.7 μm. The width and spacing are given in multiples of the 0.32 μm minimum. For an isolated wire above the substrate, the capacitance is strongly influenced by spacing between conductors. For a wire sandwiched between metal1 and metal3 planes, the capacitance is higher and is more sensitive to the width (determining parallel plate capacitance) but less sensitive to spacing once the spacing is significantly greater than the wire thickness. In either case, the y-intercept is greater than zero so doubling the width of a wire results in less than double the total capacitance. The data fits EQ (5.9) with C_{fringe} = 0.05 fF/μm. Tight-pitch metal lines have a capacitance of roughly 0.2 fF/μm.

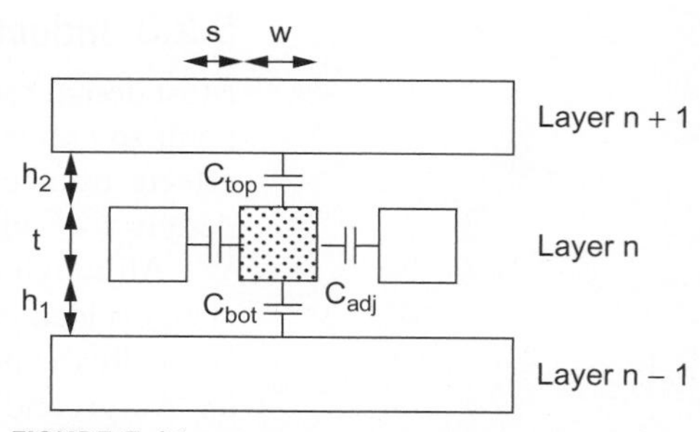

FIGURE 5.11 Multilayer capacitance model

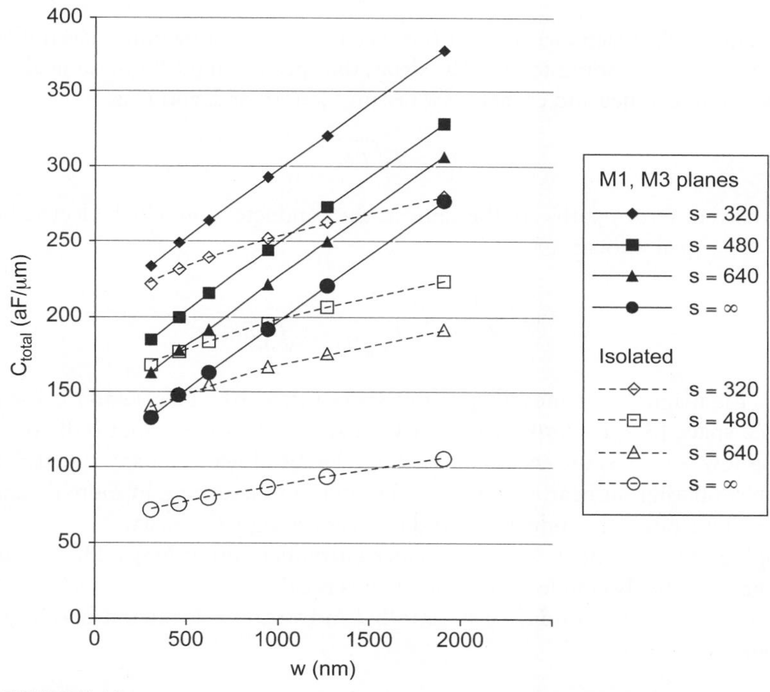

FIGURE 5.12 Capacitance of metal2 line as a function of width and spacing

In practice, the layers above and below the conductor of interest are neither solid planes nor totally empty. One can extract capacitance more accurately by interpolating between these two extremes based on the density of metal on each level. [Chern92] gives formulae for this interpolation accurate to within 10%. However, if the wiring above and below is fairly dense (e.g., a bus on minimum pitch), it is well-approximated as a plane. Dense wire fill is added to many chips for mechanical stability and etch uniformity, making this approximation even more appropriate.

5.2.3 Inductance

Most design tools consider only interconnect resistance and capacitance. Inductance is difficult to extract and model, so engineers prefer to design in such a way that inductive effects are negligible. Nevertheless, inductance needs to be considered in high-speed designs for wide wires such as clocks and power busses.

Although we generally discuss current flowing from a gate output to charge or discharge a load capacitance, current really flows in loops. The return path for a current loop is usually the power or ground network; at the frequencies of interest, the power supply is an "AC ground" because the bypass capacitance forms a low-impedance path between V_{DD} and GND. Currents flowing around a loop generate a magnetic field proportional to the area of the loop and the amount of current. Changing the current requires supplying energy to change the magnetic field. This means that changing currents induce a voltage proportional to the rate of change. The constant of proportionality is called the inductance, L.[3]

$$V = L\frac{dI}{dt} \tag{5.10}$$

Inductance and capacitance also set the speed of light in a medium. Even if the resistance of a wire is zero leading to zero RC delay, the speed of light flight-time along a wire of length with inductance and capacitance per unit length of L and C is

$$t_{pd} = l\sqrt{LC} \tag{5.11}$$

If the current return paths are the same as the conductors on which electric field lines terminate, the signal velocity v is

$$v = \frac{1}{\sqrt{LC}} = \frac{1}{\sqrt{\varepsilon_{ox}\mu_0}} = \frac{c}{\sqrt{3.9}} \tag{5.12}$$

where μ_0 is the magnetic permeability of free space ($4\pi \times 10^{-7}$ H/m) and c is the speed of light in free space (3×10^8 m/s). In other words, signals travel about half the speed of light. Using low-k (< 3.9) dielectrics raises this velocity. However, many signals have electric fields terminating on nearby neighbors, but currents returning in more distant power supply lines. This raises the inductance and reduces the signal velocity.

Changing magnetic fields in turn produce currents in other loops. Hence, signals on one wire can inductively couple onto another; this is called *inductive crosstalk*.

The inductance of a conductor of length l and width w located a height h above a ground plane is approximately

$$L = l\frac{\mu_0}{2\pi}\ln\left(\frac{8h}{w} + \frac{w}{4h}\right) \tag{5.13}$$

assuming $w < h$ and thickness is negligible. Typical on-chip inductance values are in the range of 0.15–1.5 pH/μm depending on the proximity of the power or ground lines. (Wires near their return path have smaller current loops and lower inductance.)

[3] L is used to indicate both inductance and transistor channel length. The meaning should be clear from context.

Extracting inductance in general is a three-dimensional problem and is extremely time-consuming for complex geometries. Inductance depends on the entire loop and therefore cannot be simply decomposed into sections as with capacitance. It is therefore impractical to extract the inductance from a chip layout. Instead, usually inductance is extracted using tools such as FastHenry [Kamon94] for simple test structures intended to capture the worst cases on the chip. This extraction is only possible when the power supply network is highly regular. Power planes are ideal but require a large amount of metal resources. Dense power grids are usually the preferred alternative. Gaps in the power grid force current to flow around the gap, increasing the loop area and greatly increasing inductance. Moreover, large loops couple magnetic fields through other loops formed by conductors at a distance. Therefore, mutual inductive coupling can occur over a long distance, especially when the return path is far from the conductor.

5.2.4 Skin Effect

Current flows along the path of lowest impedance $Z = R + j\omega L$. At high frequency, ω, impedance becomes dominated by inductance. The inductance is minimized if the current flows only near the surface of the conductor closest to the return path(s). This skin effect can reduce the effective cross-sectional area of thick conductors and raise the effective resistance at high frequency. The skin depth for a conductor is

$$\delta = \sqrt{\frac{2\rho}{\omega\mu}} \tag{5.14}$$

where μ is the magnetic permeability of the dielectric (normally the same as in free space, $4\pi \times 10^{-7}$ H/m). The frequency of importance is the highest frequency with significant power in the Fourier transform of the signal. This is not the chip operating frequency, but rather is associated with the faster edges. A sine wave with the same 20–80% rise/fall time as the signal has a period of $8.65t_{rf}$. Therefore, the frequency associated with the edge can be approximated as

$$\omega = \frac{2\pi}{8.65\, t_{rf}} \tag{5.15}$$

where t_{rf} is the average 20–80% rise/fall time.

In a chip with a good power grid, good current return paths are usually available on all sides. Thus, it is a reasonable approximation to assume the current flows in a shell of thickness δ along the four sides of the conductor, as shown in Figure 5.13. If $\min(w, t) > 2\delta$, part of the conductor carries no current and the resistance increases.

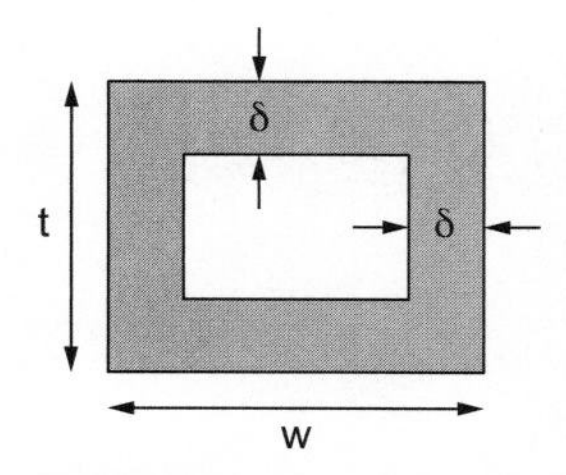

FIGURE 5.13 Current flow in shell determined by skin depth

Example 5.2

Determine the skin depth for a copper wire in a chip with 20 ps edge rates.

SOLUTION: According to EQ (5.15), the maximum frequency of interest is

$$\omega = \frac{2\pi}{8.65 \times 20 \text{ ps}} = 3.6 \times 10^{10} \text{ rad/s} = 5.8 \text{ GHz} \tag{5.16}$$

According to EQ (5.14), the skin depth is thus

$$\delta = \sqrt{\frac{2\left(2.2\times10^{-8}\ \Omega\cdot\text{m}\right)}{\left(3.6\times10^{10}\ \text{rad/s}\right)\left(4\pi\times10^{-7}\ \text{H/m}\right)}} = 0.99\ \mu\text{m} \tag{5.17}$$

This exceeds half the thickness of typical metal layers, so the skin effect is rarely a factor in CMOS circuits.

5.2.5 Temperature Dependence

Interconnect capacitance is independent of temperature, but the resistance varies strongly. The temperature coefficients of copper and aluminum are about 0.4%/°C over the normal operating range of circuits; that is, a 100 °C increase in temperature leads to 40% higher resistance. At liquid nitrogen temperature (77 K), the resistivity of copper drops to 0.22 $\mu\Omega\cdot$cm, an order-of-magnitude improvement. This suggests great advantages for RC-dominated paths in cooled systems.

5.3 Interconnect Impact

Using the lumped models, this section examines the delay, energy, and noise impact of wires.

5.3.1 Delay

Interconnect increases circuit delay for two reasons. First, the wire capacitance adds loading to each gate. Second, long wires have significant resistance that contributes distributed RC delay or *flight time*. It is straightforward to add wire capacitance to the Elmore delay calculations of Section 3.3.5, so in this section we focus on the RC delay.

The Elmore delay of a single-segment L-model is *RC*. As the number of segments of the L-model increases, the Elmore delay decreases toward *RC*/2. The Elmore delay of a π- or T-model is *RC*/2 no matter how many segments are used. Thus, a single-segment π-model is a good approximation for hand calculations.

Example 5.3

A 10x unit-sized inverter drives a 2x inverter at the end of the 1 mm wire from Example 5.1. Suppose that wire capacitance is 0.2 fF/μm and that unit-sized nMOS transistor has R = 10 kΩ and C = 0.1 fF. Estimate the propagation delay using the Elmore delay model; neglect diffusion capacitance.

SOLUTION: The driver has a resistance of 1 kΩ.The receiver has a 2-unit nMOS transistor and a 4-unit pMOS transistor, for a capacitance of 0.6 fF. The wire capacitance is 200 fF.

Figure 5.14 shows an equivalent circuit for the system using a single-segment π-model. The Elmore delay is t_{pd} = (1000 Ω)(100 fF) + (1000 Ω + 800 Ω)(100 fF + 0.6 fF) = 281 ps. The capacitance of the long wire dominates the delay; the capacitance of the 2x inverter is negligible in comparison.

Because both wire resistance and wire capacitance increase with length, wire delay grows quadratically with length. Using thicker and wider wires, lower-resistance metals such as copper, and lower-dielectric constant insulators helps, but long wires nevertheless often have unacceptable delay. Section 5.4.2 describes how repeaters can be used to break a long wire into multiple segments such that the overall delay becomes a linear function of length.

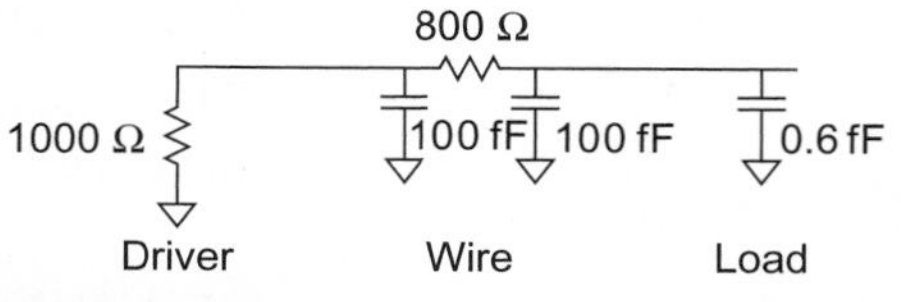

FIGURE 5.14 Equivalent circuit for example

Example 5.4

Find the RC flight time per mm^2 for a wire using the parameters from Example 5.3. Express the result in FO4/mm^2, if the FO4 inverter delay is 15 ps. What is the flight time to cross a 10 mm die?

SOLUTION: $R = 800\ \Omega$/mm. $C = 0.2$ pF/mm. The flight time is $RC/2 = 80$ ps/mm^2, or 5.3 FO4/mm^2. The flight time across a 10 mm die is thus 530 FO4, which is dozens of clock cycles.

Polysilicon and diffusion wires (sometimes called *runners*) have high resistance, even if silicided. Diffusion also has very high capacitance. Do not use diffusion for routing. Use polysilicon sparingly, usually in latches and flip-flops (i.e., do not use for other than intracell routing).

Recall that the Elmore delay model only considers the resistance on the path from the driver to a leaf. Capacitances on other branches are lumped as if they were at the branch point. This gives a conservative result because they are really partially shielded by their resistances.

Example 5.5

Figure 5.15 models a gate driving wires to two destinations. The gate is represented as a voltage source with effective resistance R_1. The two receivers are located at nodes 3 and 4. The wire to node 3 is long enough that it is represented with a pair of π-segments, while the wire to node 4 is represented with a single segment. Find the Elmore delay from input x to each receiver.

SOLUTION: The Elmore delays are

$$\begin{aligned} T_{D_3} &= R_1C_1 + (R_1 + R_2)C_2 + (R_1 + R_2 + R_3)C_3 + R_1C_4 \\ T_{D_4} &= R_1C_1 + R_1(C_2 + C_3) + (R_1 + R_4)C_4 \end{aligned} \tag{5.18}$$

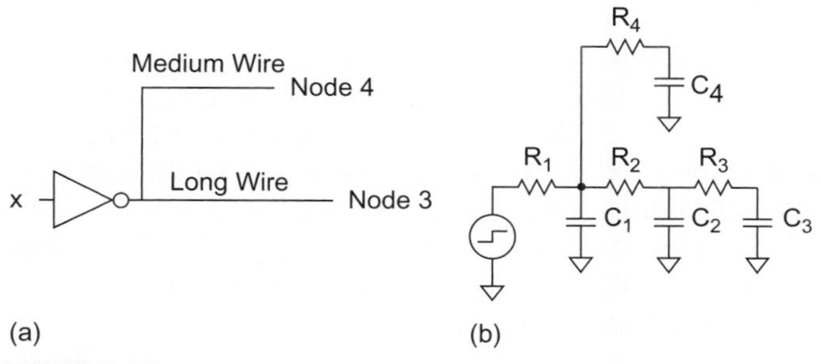

FIGURE 5.15 Interconnect modeling with RC tree

5.3.2 Energy

The switching energy of a wire is set by its capacitance. Long wires have significant capacitance and thus require substantial amounts of energy to switch.

Example 5.6

Estimate the energy per unit length to send a bit of information (one rising and one falling transition) in a CMOS process.

SOLUTION: $E = (0.2 \text{ pF/mm}) (1.0 \text{ V})^2 = 0.2$ pJ/bit/mm. Sometimes energy in a communication link is expressed as power per gigabit per second: 0.2 mW/Gbps.

Example 5.7

Consider a microprocessor on a 20 mm × 20 mm die running at 3 GHz in the 65 nm process. A layer of metal is routed on a 250 nm pitch. Half of the available wire tracks are used. The wires have an average activity factor of 0.1. Determine the power consumed by the layer of metal.

SOLUTION: There are (20 mm) / (250 nm) = 80,000 tracks of metal across the die, of which 40,000 are occupied. The wire capacitance is (0.2 pF/mm)(20 mm)(40,000 tracks) = 160 nF. The power is $(0.1)(160 \text{ nF})(1.0 \text{ V})^2(3 \text{ GHz}) = 48$ W. This is clearly a problem, especially considering that the chip has more than one layer of metal. The activity factor needs to be much lower to keep power under control.

5.3.3 Crosstalk

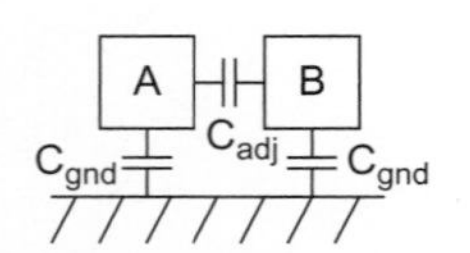

FIGURE 5.16 Capacitances to adjacent neighbor and to ground

As reviewed in Figure 5.16, wires have capacitance to their adjacent neighbors as well as to ground. When wire A switches, it tends to bring its neighbor B along with it on account of capacitive coupling, also called *crosstalk*. If B is supposed to switch simultaneously, this may increase or decrease the switching delay. If B is not supposed to switch, crosstalk causes noise on B. We will see that the impact of crosstalk depends on the ratio of C_{adj} to the total capacitance. Note that the load capacitance is included in the total, so for short wires and large loads, the load capacitance dominates and crosstalk is unimportant. Conversely, crosstalk is very important for long wires.

5.3.3.1 Crosstalk Delay Effects If both a wire and its neighbor are switching, the direction of the switching affects the amount of charge that must be delivered and the delay of the switching. Table 5.3 summarizes this effect. The charge delivered to the coupling capacitor is $Q = C_{adj}\Delta V$, where ΔV is the change in voltage between A and B. If A switches but B does not, $\Delta V = V_{DD}$. The total capacitance effectively seen by A is just the capacitance to ground and to B. If both A and B switch in the same direction, $\Delta V = 0$. Hence, no charge is required and C_{adj} is effectively absent for delay purposes. If A and B switch in the opposite direction, $\Delta V = 2V_{DD}$. Twice as much charge is required. Equivalently, the capacitor can be treated as being effectively twice as large switching through V_{DD}. This is analogous to the Miller effect discussed in Section 3.4.6.6. The *Miller Coupling Factor* (MCF) describes how the capacitance to adjacent wires is multiplied to find the effective capacitance. Some designers use MCF = 1.5 as a statistical compromise when estimating propagation delays before layout information is available.

TABLE 5.3 Dependence of effective capacitance on switching direction

B	ΔV	$C_{\text{eff}(A)}$	MCF
Constant	V_{DD}	$C_{\text{gnd}} + C_{\text{adj}}$	1
Switching same direction as *A*	0	C_{gnd}	0
Switching opposite to *A*	$2V_{DD}$	$C_{\text{gnd}} + 2C_{\text{adj}}$	2

A conservative design methodology assumes neighbors are switching when computing propagation and contamination delays (MCF = 2 and 0, respectively). This leads to a wide variation in the delay of wires. A more aggressive methodology tracks the time window during which each signal can switch. Thus, switching neighbors must be accounted for only if the potential switching windows overlap. Similarly, the direction of switching can be considered. For example, dynamic gates described in Section 8.2.4 precharge high and then fall low during evaluation. Thus, a dynamic bus will never see opposite switching during evaluation.

Example 5.8

Each wire in a pair of 1 mm lines has capacitance of 0.08 fF/μm to ground and 0.12 fF/μm to its neighbor. Each line is driven by an inverter with a 1 kΩ effective resistance. Estimate the contamination and propagation delays of the path. Neglect parasitic capacitance of the inverter and resistance of the wires.

SOLUTION: We find $C_{\text{gnd}} = (0.08\ \text{fF}/\mu\text{m})(1000\ \mu\text{m}) = 80$ fF and $C_{\text{adj}} = 120$ fF. The delay is RC_{eff}. The contamination delay is the minimum possible delay, which occurs when both wires switch in the same direction. In that case, $C_{\text{eff}} = C_{\text{gnd}}$ and the delay is $t_{cd} =$ (1 kΩ)(0.08 pF) = 80 ps. The propagation delay is the maximum possible delay, which occurs when both wires switch in opposite directions. In this case, $C_{\text{eff}} = C_{\text{gnd}} + 2C_{\text{adj}}$ and the delay is $t_{pd} =$ (1 kΩ)(0.32 pF) = 320 ps. This is a factor of four difference between best and worst case.

5.3.3.2 Crosstalk Noise Effects Suppose wire *A* switches while *B* is supposed to remain constant. This introduces noise as *B* partially switches. We call *A* the *aggressor* or *perpetrator* and *B* the *victim*. If the victim is floating, we can model the circuit as a capacitive voltage divider to compute the victim noise, as shown in Figure 5.17. $\Delta V_{\text{aggressor}}$ is normally V_{DD}.

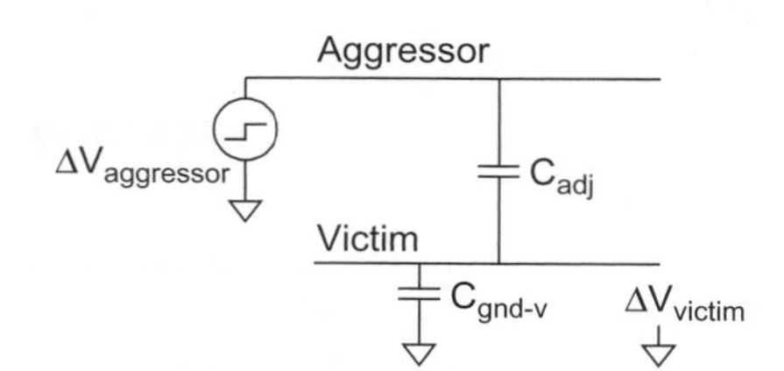

FIGURE 5.17 Coupling to floating victim

$$\Delta V_{\text{victim}} = \frac{C_{\text{adj}}}{C_{\text{gnd}-v} + C_{\text{adj}}} \Delta V_{\text{aggressor}} \tag{5.19}$$

If the victim is actively driven, the driver will supply current to oppose and reduce the victim noise. We model the drivers as resistors, as shown in Figure 5.18. The peak noise becomes dependent on the time constant ratio *k* of the aggressor to the victim [Ho01]:

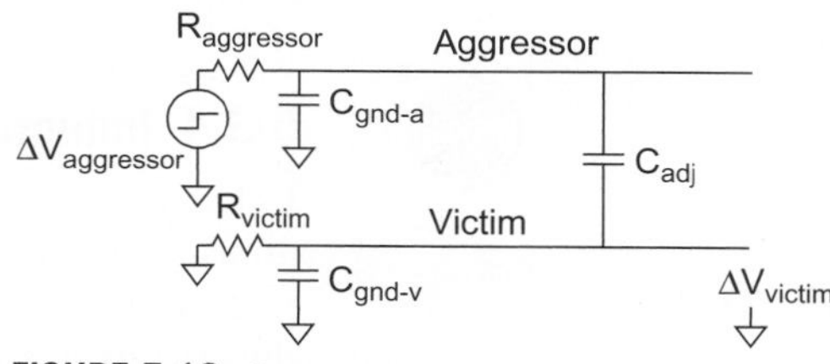

FIGURE 5.18 Coupling to driven victim

$$\Delta V_{\text{victim}} = \frac{C_{\text{adj}}}{C_{\text{gnd}-v} + C_{\text{adj}}} \frac{1}{1+k} \Delta V_{\text{aggressor}} \tag{5.20}$$

where

$$k = \frac{\tau_{\text{aggressor}}}{\tau_{\text{victim}}} = \frac{R_{\text{aggressor}}\left(C_{\text{gnd}-a} + C_{\text{adj}}\right)}{R_{\text{victim}}\left(C_{\text{gnd}-v} + C_{\text{adj}}\right)} \tag{5.21}$$

Figure 5.19 shows simulations of coupling when the aggressor is driven with a unit inverter; the victim is undriven or driven with an inverter of half, equal, or twice the size of the aggressor; and $C_{\text{adj}} = C_{\text{gnd}}$. Observe that when the victim is floating, the noise remains indefinitely. When the victim is driven, the driver restores the victim. Larger (faster) drivers oppose the coupling sooner and result in noise that is a smaller percentage of the supply voltage. Note that during the noise event the victim transistor is in its linear region while the aggressor is in saturation. For equal-sized drivers, this means $R_{\text{aggressor}}$ is two to four times R_{victim}, with greater ratios arising from more velocity saturation [Ho01]. In general, EQ (5.20) is conservative, especially when wire resistance is included [Vittal99]. It is often used to flag nets where coupling can be a problem; then simulations can be performed to calculate the exact coupling noise. Coupling noise is of greatest importance on weakly driven nodes where $k < 1$.

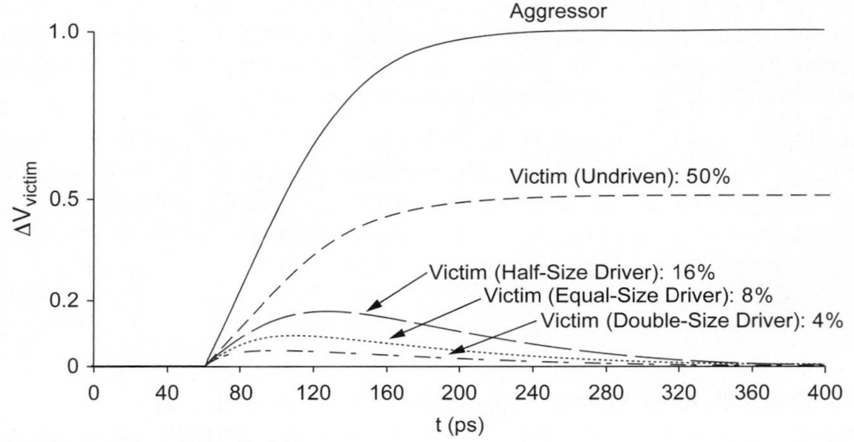

FIGURE 5.19 Waveforms of coupling noise

We have only considered the case of a single neighbor switching. When both neighbors switch, the noise will be twice as great. We have also modeled the layers above and below as AC ground planes, but wires on these layers are likely to be switching. For a long line, you can expect about as many lines switching up and switching down, giving no net contribution to delay or noise. However, a short line running over a 64-bit bus in which all 64 bits are simultaneously switching from 0 to 1 will be strongly influenced by this switching.

5.3.4 Inductive Effects

Inductance has always been important for integrated circuit packages where the physical dimensions are large, as will be discussed in Section 12.2.3. On-chip inductance is important for wires where the speed-of-light flight time is longer than either the rise times of the circuits or the RC delay of the wire. Because speed-of-light flight time increases linearly according to EQ (5.11) and RC delay increases quadratically with length, we can estimate the set of wire lengths for which inductance is relevant [Ismail99].

$$\frac{t_r}{2\sqrt{LC}} < l < \frac{2}{R}\sqrt{\frac{L}{C}} \tag{5.22}$$

Example 5.9

Consider a metal2 signal line with a sheet resistance of 0.10 Ω/□ and a width of 0.125 μm. The capacitance is 0.2 fF/μm and inductance is 0.5 pH/μm. Compute the velocity of signals on the line and plot the range of lengths over which inductance matters as a function of the rise time.

SOLUTION: The velocity is

$$v = \frac{1}{\sqrt{LC}} = \frac{1}{\sqrt{(0.5\text{ pH}/\mu\text{m})(0.2\text{ fF}/\mu\text{m})}} = 10^8\text{ m/s} = \frac{1}{3}c \tag{5.23}$$

Note that this is 100 mm/ns or 1 mm/10 ps. The resistance is (0.1 Ω/□)(1 □/0.125 μm) = 0.8 Ω/μm. Figure 5.20 plots the length of wires for which inductance is relevant against rise times. Above the horizontal line, wires greater than 125 μm are limited by RC delay rather than LC delay. To the right of the diagonal line, rise times are greater than the LC delay. Only in the region between these lines is inductance relevant to delay calculations. This region has very fast edge rates, so inductance is not very important to the delay of highly resistive signal lines.

As the example illustrated, inductance will only be important to the delay of low-resistance signals such as wide clock lines. Inductive crosstalk is also important for wide busses far away from their current return paths. In power distribution networks, inductance means that if one portion of the chip requires a rapidly increasing amount of current, that charge must be delivered from nearby decoupling capacitors or supply pins; portions of the chip further away are unaware of the changing current needs until a speed-of-light flight time has elapsed and hence will not supply current immediately. Adding inductance to the power grid simulation generally reveals greater supply noise than would otherwise be predicted. Power networks will be discussed further in Section 12.3.

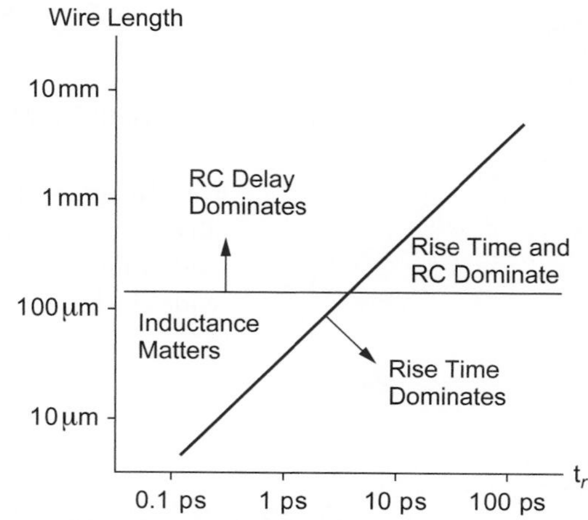

FIGURE 5.20 Wire lengths and edge rates for which inductance impacts delay

In wide, thick, upper-level metal lines, resistance and RC delay may be small. This pushes the horizontal line in Figure 5.20 upward, increasing the range of edge rates for which inductance matters. This is especially common for clock signals. Inductance tends to increase the propagation delay and sharpen the edge rate.

To see the effects of inductance, consider a 5 mm-long clock line above a ground plane driving a 2 pF clock load. If its width is 4.8 μm and thickness is 1.7 μm, it has resistance of 4 Ω/mm, capacitance of 0.4 pF/mm, and inductance of 0.12 nH/mm. Figure 5.21 presents models of the clock line as a 5-stage π-model without (a) and with (b) inductance. Figure 5.21(c) shows the response of each model to an ideal voltage source with 80 ps rise time. The model including inductance shows a greater delay until the clock begins to rise because of the speed-of-light flight time. It also overshoots. However, the rising edge is sharper and the rise time is shorter. In some circumstances when the driver impedance is matched to the characteristic impedance of the wire, the sharper rising edge can actually result in a shorter propagation delay measured at the 50% point.

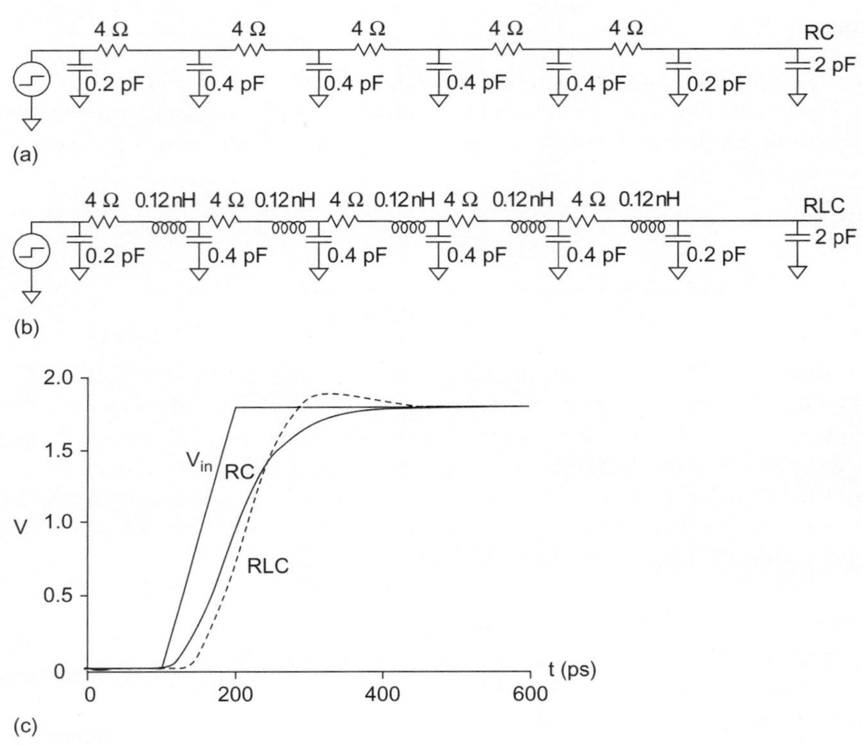

FIGURE 5.21 Wide clock line modeled with and without inductance

To reduce the inductance and the impact of skin effect when no ground plane is available, it is good practice to split wide wires into thinner sections interdigitated with power and ground lines to serve as return paths. For example, Figure 5.22 shows how a 16 μm-wide clock line can be split into four 4 μm lines to reduce the inductance.

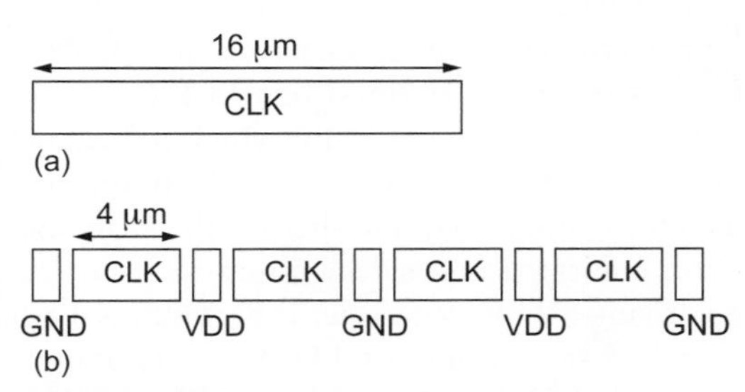

FIGURE 5.22 Wide clock line interdigitated with power and ground lines to reduce inductance

A bus made of closely spaced wires far above a ground plane is particularly susceptible to crosstalk. Figure 5.23 shows the worst case crosstalk scenario. The victim line is in the center. The two adjacent neighbors rise, capacitively coupling the victim upward. The other bus wires fall. Each one creates a loop of current flowing counterclockwise through the wire and back along the ground plane. These loops induce a magnetic field, which in turn induces a current flowing in the other direction in the victim line. This is called mutual inductive coupling and also makes the victim rise. The noise from each aggressor sums on to the victim in much the same way that multiple primary turns in a transformer couple onto a single secondary turn. Computing the inductive crosstalk requires extracting a mutual inductance matrix for the bus and simulating the system. As this is not yet practical for large chips, designers instead either follow design rules that keep the inductive effects small or ignore inductance and hope for the best. The design rules may be of the form that one power or ground wire must be inserted between every N signal lines on each layer. N is called the signal:return (SR) ratio [Morton99]. The returns give an alternative path for current to flow, reducing the mutual inductance. The inductive effects on noise and delay are generally small for $N = 8$ and negligible for $N = 4$ when normal wiring pitches are used [Linderman04]. $N = 2$ means each signal is shielded on one side, also eliminating half the capacitive crosstalk. However, low SR ratios are expensive in terms of metal resources.

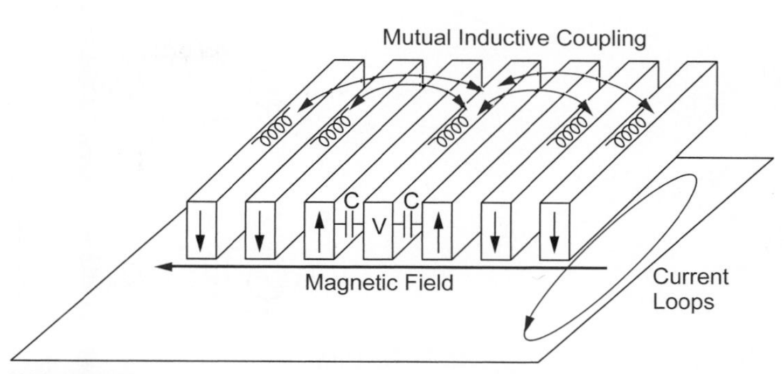

FIGURE 5.23 Inductive and capacitive crosstalk in a bus

In summary, on-chip inductance is difficult to extract. Mutual inductive coupling may occur over a long range, so inductive coupling is difficult to simulate even if accurate values are extracted. Instead, design rules are usually constructed so that inductive effects may be neglected for most structures. The easiest way to do this is to provide a regular power grid in which power and ground are systematically allocated track to keep the SR ratio low. Inductance should be incorporated into simulations of the power and clock networks and into the noise and delay calculations for busses with large SR ratios in high-speed designs.

5.3.5 An Aside on Effective Resistance and Elmore Delay

Recall from Section 3.3.4 that a factor of ln 2 was lumped into the effective resistance of a transistor so that the Elmore delay model predicts propagation delay, yet we have not accounted for the factor in wire resistance. This section examines the discrepancy.

According to the Elmore delay model, a gate with effective resistance R and capacitance C has a propagation delay of RC. A wire with distributed resistance R and capacitance C treated as a single π-segment has propagation delay $RC/2$. Reviewing the properties of RC circuits, we recall that the lumped RC circuit in Figure 5.24(a) has a unit step response of

$$V_{\text{out}}(t) = 1 - e^{\frac{-t}{RC}} \tag{5.24}$$

The propagation delay of this circuit is obtained by solving for t_{pd} when $V_{\text{out}}(t_{pd}) = 1/2$:

$$t_{pd} = R'C \ln 2 = 0.69R'C \tag{5.25}$$

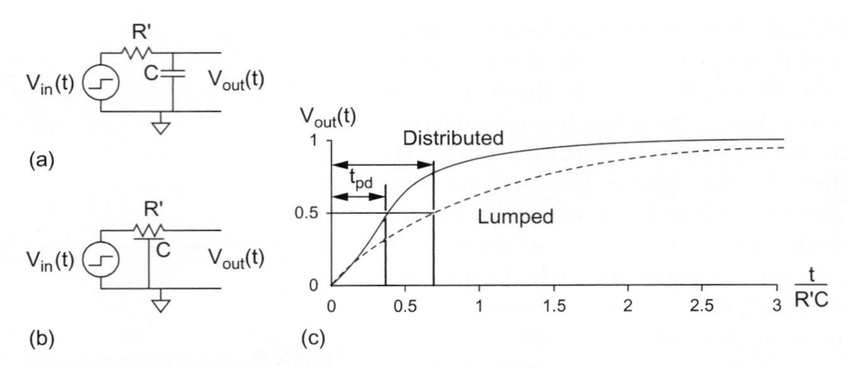

FIGURE 5.24 Lumped and distributed RC circuit response

The distributed RC circuit in Figure 5.24(b) has no closed form time domain response. Because the capacitance is distributed along the circuit rather than all being at the end, you would expect the capacitance to be charged on average through about half the resistance and that the propagation delay should thus be about half as great. A numerical analysis finds that the propagation delay is $0.38R'C$.

To reconcile the Elmore model with the true results for a logic gate, recall that logic gates have complex nonlinear I-V characteristics and are approximated as having an effective resistance. If we characterize that effective resistance as $R = R' \ln 2$, the propagation delay really becomes the product of the effective resistance and the capacitance: $t_{pd} = RC$.

For distributed circuits, observe that

$$0.38R'C \approx \tfrac{1}{2}R'C \ln 2 = \tfrac{1}{2}RC$$

Therefore, the Elmore delay model describes distributed delay well if we use an effective wire resistance scaled by ln 2 from that computed with EQ (5.2). This is somewhat inconvenient. The effective resistance is further complicated by the effect of nonzero rise time on propagation delay. Figure 5.25 shows that the propagation delay depends on the rise time of the input and approaches RC for lumped systems and $RC/2$ for distributed systems when the input is a slow ramp. This suggests that when the input is slow, the effective resistance for delay calculations in a distributed RC circuit is equal to the true resistance. Finally, we note that for many analyses such as repeater insertion calculations in Section 5.4.2, the results are only weakly sensitive to wire resistance, so using the true wire resistance does not introduce great error.

In summary, it is a reasonable practice to estimate the flight time along a wire as $RC/2$ where R is the true resistance of the wire. When more accurate results are needed, it is important to use good transistor models and appropriate input slopes in simulation.

The Elmore delay can be viewed in terms of the first moment of the impulse response of the circuit. CAD tools can obtain greater accuracy by approximating delay based on higher moments using a technique called *moment matching*. *Asymptotic Waveform Evaluation* (AWE) uses moment matching to estimate interconnect delay with better accuracy than the Elmore delay model and faster run times than a full circuit simulation [Celik02].

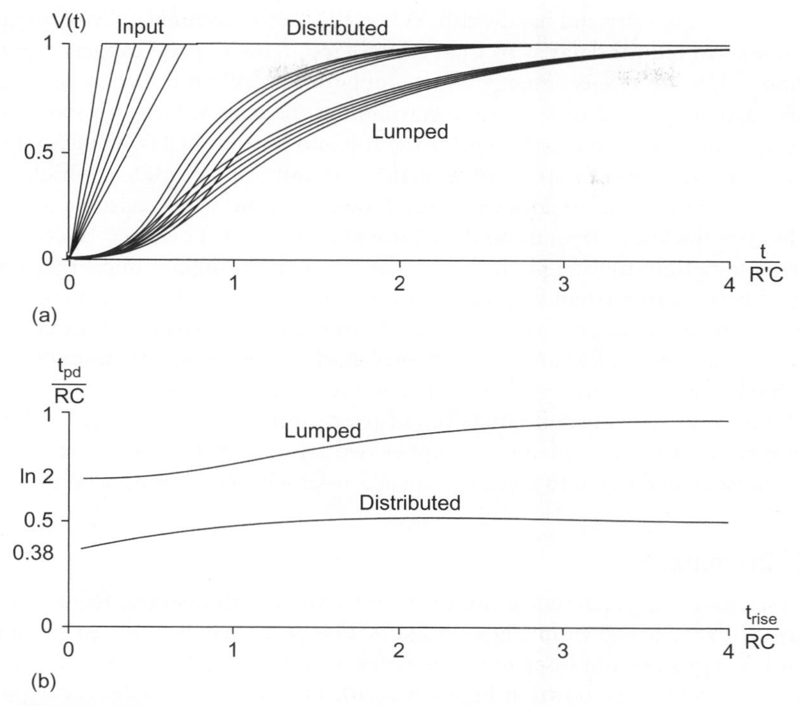

FIGURE 5.25 Effect of rise time on lumped and distributed RC circuit delays

5.4 Interconnect Engineering

As gate delays continue to improve while long wire delays remain constant or even get slower, wire engineering has become a major part of integrated circuit design. It is necessary to develop a floorplan early in the design cycle, identify the long wires, and plan for them. While floorplanning in such a way that critical communicating units are close to one another has the greatest impact on performance, it is inevitable that long wires will still exist. Aspect ratios in old processes were below 1, but are close to 2 in nanometer processes to help the resistance of such narrow lines. This comes at the expense of substantially increased coupling capacitance. The designer has a number of techniques to engineer wires for delay and coupling noise. The width, spacing, and layer usage are all under the designer's control. Shielding can be used to further reduce coupling on critical nets. Repeaters inserted along long wires reduce the delay from a quadratic to a linear function of length. Wire capacitance and resistance complicate the use of Logical Effort in selecting gate sizes.

5.4.1 Width, Spacing, and Layer

The designer selects the wire width, spacing, and layer usage to trade off delay, bandwidth, energy, and noise. By default, minimum pitch wires are preferred for noncritical interconnections for best density and bandwidth. When the load is dominated by wire capacitance, the best way to reduce delay is to increase spacing, reducing the capacitance to nearby neighbors. This also reduces energy and coupling noise. When the delay is dominated by the gate capacitance and wire resistance, widening the wire reduces resistance and delay. However, it increases the capacitance of the top and bottom plates. Widening wires also increases the fraction of capacitance of the top and bottom plates, which somewhat reduces coupling noise from adjacent wires. However, wider wires consume more energy.

The wire thickness depends on the choice of metal layer. The lower layers are thin and optimized for a tight routing pitch. Middle layers are often slightly thicker for lower resistance and better current-handling capability. Upper layers may be even thicker to provide a low-resistance power grid and fast global interconnect. Wiring tracks are a precious resource and are often allocated in the floorplan; the wise designer maintains a reserve of wiring tracks for unanticipated changes late in the design process.

The power grid is usually distributed over multiple layers. Most of the current-handling capability is provided in the upper two layers with lowest resistance. However, the grid must extend down to metal1 or metal2 to provide easy connection to cells.

5.4.2 Repeaters

Both resistance and capacitance increase with wire length l, so the RC delay of a wire increases with l^2, as shown in Figure 5.26(a). The delay may be reduced by splitting the wire into N segments and inserting an inverter or buffer called a *repeater* to actively drive the wire [Glasser85], as shown in Figure 5.26(b). The new wire involves N segments with RC flight time of $(l/N)^2$, for a total delay of l^2/N. If the number of segments is proportional to the length, the overall delay increases only linearly with l.

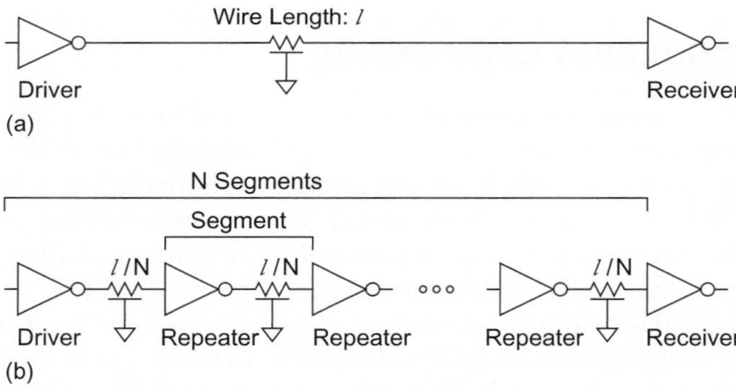

FIGURE 5.26 Wire with and without repeaters

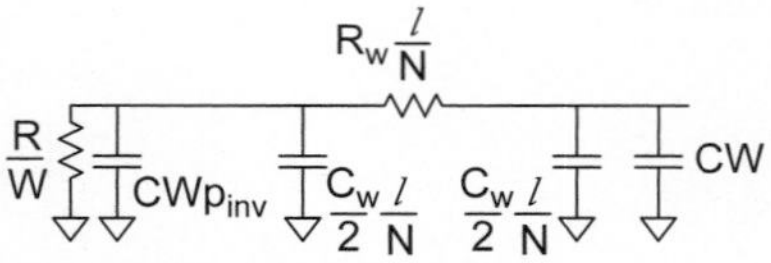

FIGURE 5.27 Equivalent circuit for segment of repeated wire

Using inverters as repeaters gives best performance. Each repeater adds some delay. If the distance is too great between repeaters, the delay will be dominated by the long wires. If the distance is too small, the delay will be dominated by the large number of inverters. As usual, the best distance between repeaters is a compromise between these extremes. Suppose a unit inverter has resistance R, gate capacitance C,[4] and diffusion capacitance Cp_{inv}. A wire has resistance R_w and capacitance C_w per unit length. Consider inserting repeaters of W times unit size.

[4]Note that C now refers to the capacitance of an entire inverter, not a single transistor, so $\tau = RC$.

Figure 5.27 shows a model of one segment. The Elmore delay of the repeated wire is

$$t_{pd} = N\left[\frac{R}{W}\left(C_w\frac{l}{N} + CW\left(1+p_{\text{inv}}\right)\right) + R_w\frac{l}{N}\left(\frac{C_w}{2}\frac{l}{N} + CW\right)\right] \quad (5.26)$$

Differentiating EQ (5.26) with respect to N and W shows that the best length of wire between repeaters is (see Exercise 5.1)

$$\frac{l}{N} = \sqrt{\frac{2RC\left(1+p_{\text{inv}}\right)}{R_w C_w}} \quad (5.27)$$

Recall from Example 3.10 that the delay of an FO4 inverter is $5RC$. Assuming $p_{\text{inv}} \approx 0.5$ using folded transistors, EQ (5.27) simplifies to

$$\frac{l}{N} = 0.77\sqrt{\frac{\text{FO4}}{R_w C_w}} \quad (5.28)$$

The delay per unit length of a properly repeated wire is

$$\frac{t_{pd}}{l} = \left(2+\sqrt{2\left(1+p_{\text{inv}}\right)}\right)\sqrt{RCR_w C_w} \approx 1.67\sqrt{\text{FO4}\; R_w C_w} \quad (5.29)$$

To achieve this delay, the inverters should use an nMOS transistor width of

$$W = \sqrt{\frac{RC_w}{R_w C}} \quad (5.30)$$

The energy per unit length to send a bit depends on the wire and repeater capacitances

$$\frac{E}{l} = C_w + NWC\left(1+p_{\text{inv}}\right) = C_w\left(1+\sqrt{\frac{1+p_{\text{inv}}}{2}}\right)V_{DD}^2 \approx 1.87 C_w V_{DD}^2 \quad (5.31)$$

In other words, repeaters sized for minimum delay add 87% to the energy of an unrepeated wire.

Example 5.10

Compute the delay per mm of a repeated wire in a 65 nm process. Assume the wire is on a middle routing layer and has 2x width, spacing, and height, so its resistance is 200 Ω/mm and capacitance is 0.2 pF/mm. The FO4 inverter delay is 15 ps. Also find the repeater spacing and driver size to achieve this delay and the energy per bit.

SOLUTION: Using EQ (5.29), the delay is

$$t_{pd} = 1.67\sqrt{(15\text{ ps})(200\ \Omega/\text{mm})(0.2\text{ pF/mm})} = 41\text{ ps/mm} \quad (5.32)$$

This delay is achieved using a spacing of 0.45 mm between repeaters and an nMOS driver width of 18 μm (180x unit size). The energy per bit is 0.4 pJ/mm.

As one might expect, the curve of delay vs. distance and driver size is relatively flat near the minimum. Thus, substantial energy can be saved for a small increase in delay. At the minimum EDP point, the segments become 1.7x longer and the drivers are only 0.6x as large. The delay increases by 14% but the repeaters only add 30% to the energy of the unrepeated line [Ho01]. For the parameters in Example 5.10, the minimum EDP can be found numerically at a spacing of about 0.8 mm and a driver width of 11 μm (110x unit size), achieving an energy of 0.26 pJ/mm at a delay of 47 ps/mm. These longer segments are more susceptible to noise.

Unfortunately, inverting repeaters complicate design because you must either ensure an even number of repeaters on each wire or adapt the receiving logic to accept an inverted input. Some designers use inverter pairs (buffers) rather than single inverters to avoid the polarity problem. The pairs contribute more delay. However, the first inverter size W_1 may be smaller, presenting less load on the wire driving it. The second inverter may be larger, driving the next wire more strongly. You can show that the best size of the second inverter is $W_2 = kW_1$, where $k = 2.25$ if $p_{\text{inv}} = 0.5$. The distance between repeaters increases to (see Exercise 5.2)

$$\frac{l}{N} = \sqrt{\frac{2RC\left(k + \frac{1}{k} + 2p_{\text{inv}}\right)}{R_w C_w}} \approx 1.22\sqrt{\frac{\text{FO4}}{R_w C_w}} \qquad \textbf{(5.33)}$$

The delay per unit length becomes

$$\frac{t_{pd}}{l} = 1.81\sqrt{\text{FO4}\, R_w C_w} \qquad \textbf{(5.34)}$$

using transistor widths of

$$W_1 = \frac{W}{\sqrt{k}}, \quad W_2 = W\sqrt{k} \qquad \textbf{(5.35)}$$

and the energy per bit per unit length is

$$\frac{E}{l} \approx 2.2 C_w V_{DD}^2 \qquad \textbf{(5.36)}$$

This typically means that wires driven with noninverting repeaters are only about 8% slower per unit length than those using inverting repeaters. Only about two-thirds as many repeaters are required, simplifying floorplanning. Total repeater area and power increases slightly.

The overall delay is a weak function of the distance between repeaters, so it is reasonable to increase this distance to reduce the difficulty of finding places in the floorplan for repeaters while only slightly increasing delay. Repeaters impose directionality on a wire. Bidirectional busses and distributed tristate busses cannot use simple repeaters and hence are slower; this favors point-to-point unidirectional communications.

5.4.3 Crosstalk Control

Recall from EQ (5.20) that the capacitive crosstalk is proportional to the ratio of coupling capacitance to total capacitance. For modern wires with an aspect ratio (t/w) of 2 or greater, the coupling capacitance can account for 2/3 to 3/4 of the total capacitance and crosstalk can create large amounts of noise and huge data-dependent delay variations. There are several approaches to controlling this crosstalk:

- Increase spacing to adjacent lines
- Shield wires
- Ensure neighbors switch at different times
- Crosstalk cancellation

The easiest approach to fix a minor crosstalk problem is to increase the spacing. If the crosstalk is severe, the spacing may have to be increased by more than one full track. In such a case, it is more efficient to shield critical signals with power or ground wires on one or both sides to eliminate coupling. For example, clock wires are usually shielded so that switching neighbors do not affect the delay of the clock wire and introduce clock jitter. Sensitive analog wires passing near digital signals should also be shielded.

An alternative to shielding is to interleave busses that are guaranteed to switch at different times. For example, if bus *A* switches on the rising edge of the clock and bus *B* switches on the falling edge of the clock, by interleaving the bits of the two busses you can guarantee that both neighbors are constant during a switching event. This avoids the delay impact of coupling; however, you must still ensure that coupling noise does not exceed noise budgets. Figure 5.28 shows wires shielded (a) on one side, (b) on both sides, and (c) interleaved. Critical signals such as clocks or analog voltages can be shielded above and below as well.

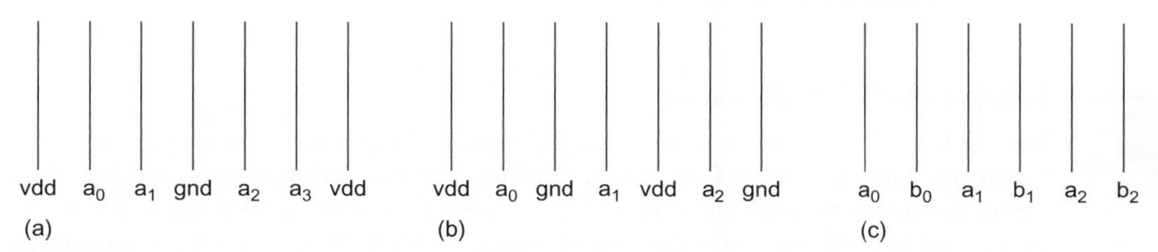

FIGURE 5.28 Wire shielding topologies

Alternatively, wires can be arranged to cancel the effects of crosstalk. Three such methods include *staggered repeaters*, *charge compensation*, and *twisted differential signaling* [Ho03b]. Each technique seeks to cause equal amounts of positive and negative crosstalk on the victim, effectively producing zero net crosstalk.

Figure 5.29(a) shows two wires with staggered repeaters. Each segment of the victim sees half of a rising aggressor segment and half of a falling aggressor segment. Although the cancellation is not perfect because of delays along the segments, staggering is a very effective approach. Figure 5.29(b) shows charge compensation in which an inverter and transistor are added between the aggressor and victim. The transistor is connected to behave as a capacitor. When the aggressor rises and couples the victim upward, the inverter falls and couples the victim downward. By choosing an appropriately sized compensation transistor, most of the noise can be canceled at the expense of the extra circuitry. Figure 5.29(c) shows twisted differential signaling in which each signal is routed differentially. The signals are swapped or *twisted* such that the victim and its complement each see equal coupling from the aggressor and its complement. This approach is expensive in wiring resources, but it effectively eliminates crosstalk. It is widely used in memory designs that are naturally differential, as explored in Section 11.2.3.3.

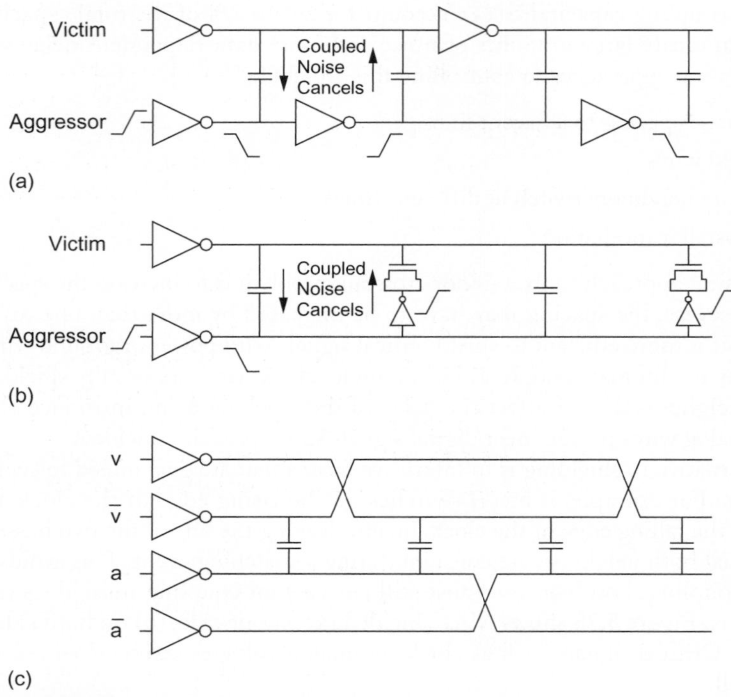

FIGURE 5.29 Crosstalk control schemes

5.4.4 Low-Swing Signaling

Driving long wires is slow because of the RC delay, and expensive in power because of the large capacitance to switch. Low-swing signaling improves performance by sensing when a wire has swung through some small V_{swing} rather than waiting for a full swing. If the driver is turned off after the output has swung sufficiently, the power can be reduced as well. However, the improvements come at the expense of more complicated driver and receiver circuits. Low-swing signaling may also require a twisted differential pair of wires to eliminate common-mode noise that could corrupt the small signal.

The power consumption for low-swing signaling depends on both the driver voltage V_{drive} and the actual voltage swing V_{swing}. Each time the wire is charged and discharged, it consumes $Q = CV_{swing}$. If the effective switching frequency of the wire is αf, the average current is

$$I_{avg} = \frac{1}{T}\int_0^T i_{drive}(t)\,\mathrm{d}t = \alpha f C V_{swing} \quad \textbf{(5.37)}$$

Hence, the dynamic dissipation is

$$P_{dynamic} = I_{avg} V_{drive} = \alpha f C V_{swing} V_{drive} \quad \textbf{(5.38)}$$

In contrast, a rail-to-rail driver uses $V_{drive} = V_{swing} = V_{DD}$ and thus consumes power proportional to V_{DD}^2. V_{swing} must be less than or equal to V_{drive}. By making V_{swing} less than V_{drive}, we speed up the wire because we do not need to wait for a full swing. By making both voltages significantly less than V_{DD}, we can reduce the power by an order of magnitude.

Low-swing signaling involves numerous challenges. A low V_{drive} must be provided to the chip and distributed to low-swing drivers. The signal should be transmitted on differential pairs of wires that are twisted to cancel coupling from neighbors and equalized to prevent interference from the previous data transmitted. The driver must turn on long enough to produce V_{swing} at the far end of the line, then turn off to prevent unnecessary power dissipation. This generally leads to a somewhat larger swing at the near end of the line. The receiver must be clocked at the appropriate time to amplify the differential signal. Distributing a self-timed clock from driver to receiver is difficult because the distances are long, so the time to transmit a full-swing clock exceeds the time for the data to complete its small swing.

Figure 5.30 shows a synchronous low-swing signaling technique using the system clock for both driver and receiver [Ho03a]. During the first half of the cycle, the driver is OFF (high impedance) and the differential wires are equalized to the same voltage. During the second half of the cycle, the drivers turn ON. At the end of the cycle, the receiver senses the differential voltage and amplifies it to full-swing levels. Figure 5.30(a) shows the overall system architecture. Figure 5.30(b) shows the driver for one of the wires. The gates use ordinary V_{DD} while the drive transistors use V_{drive}. Because $V_{\text{drive}} < V_{DD} - V_t$, nMOS transistors are used for both the pullup and pulldown to deliver low effective resistance in their linear regime. A second driver using the complementary input drives the complementary wire. Figure 5.30(c) shows the differential wires with twisting and equalizing. The end of the wire only swings part-way, reducing power consumption. Using medium V_{drive} and small V_{swing} is faster than using a smaller V_{drive} and waiting for the wire to swing all the way. Figure 5.30(d) shows the clocked sense amplifier based on the SA-F/F that will be described further in Section 9.3.8. The sense amplifier uses pMOS input transistors because the small-swing inputs are close to GND and below the threshold of nMOS transistors. Note that the clock period must be long enough to transmit an adequate voltage swing. If the clock period increases, the circuit will actually dissipate more power because the voltage swing will increase to a maximum of V_{drive}.

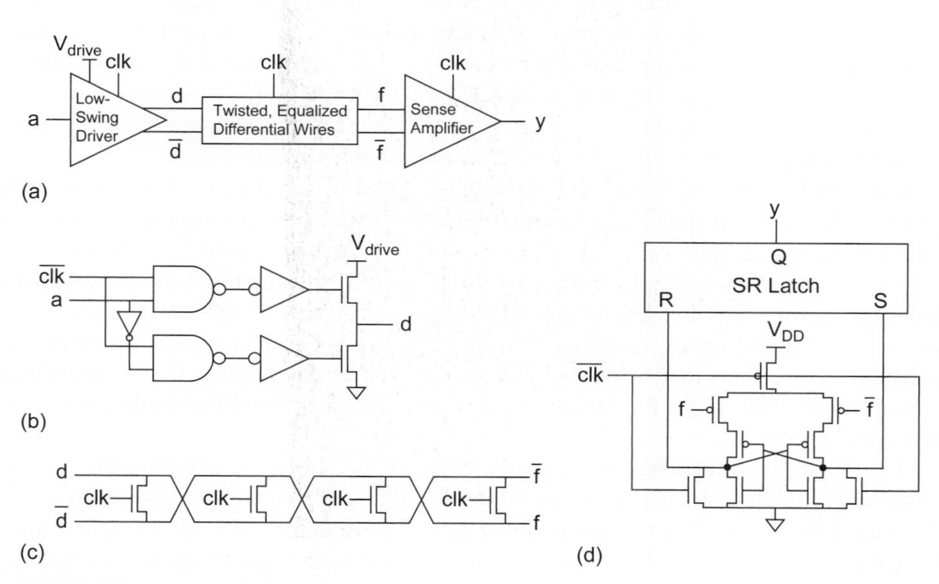

FIGURE 5.30 Low-swing signaling system

5.4.5 Regenerators

Repeaters are placed in series with wires and thus are limited to unidirectional busses. An alternative is to use *regenerators* (also called *boosters*) placed in parallel with wires at periodic intervals, as shown in Figure 5.31. When the wire is initially '0,' the regenerator senses a rising transition and accelerates it. Conversely, when the wire is initially '1,' the regenerator accelerates the falling transition. Regenerators trade off up to 20% better delay or energy for reduced noise margins.

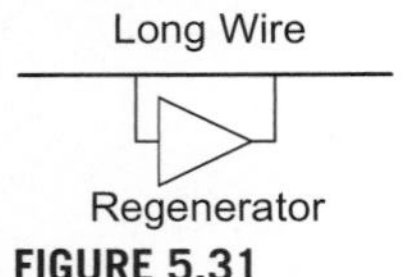

FIGURE 5.31 Regenerator

Regenerators generally use skewed gates to sense a transition. As discussed in Section 8.2.1.5, a HI-skew gate favors the rising output by using a low switching point, and a LO-skew gate does the reverse. Figure 5.32 shows a self-timed regenerator [Dobbalaere95]. When the wire begins to rise, the LO-skewed NAND gate detects the transition midway and turns on the pMOS driver to assist. The normal-skew inverters eventually detect the transition and flip node x, turning off the pMOS driver. When the wire begins to fall, the HI-skewed NOR gate turns on the nMOS to assist. Other regenerator designs include [Nalamalpu02, Singh08].

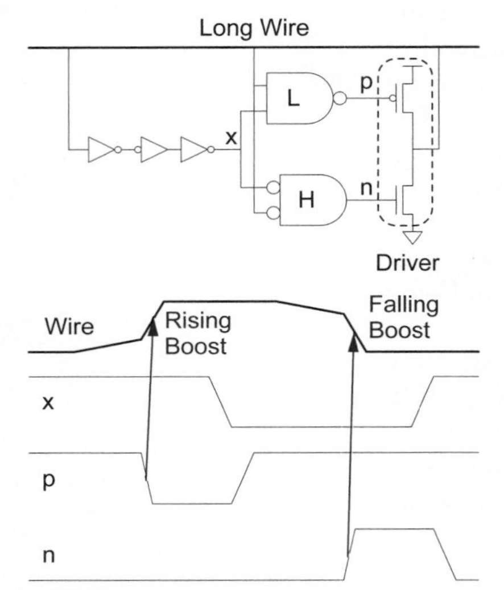

FIGURE 5.32 Regenerator

5.5 Logical Effort with Wires

Interconnect complicates the application of Logical Effort because the wires have a fixed capacitance. The branching effort at a wire with capacitance C_{wire} driving a gate load of C_{gate} is $(C_{gate} + C_{wire}) / C_{gate}$. This branching effort is not constant; it depends on the size of the gate being driven. The simple rule that circuits are fastest when all stages bear equal effort is no longer true when wire capacitance is introduced. If the wire is very short or very long, approximations are possible, but when the wire and gate loads are comparable, there is no simple method to determine the best stage effort.

Every circuit has some interconnect, but when the interconnect is short ($C_{wire} << C_{gate}$), it can be ignored. Alternatively, you can compute the average ratio of wire capacitance to parasitic diffusion capacitance and add this as extra parasitic capacitance when determining parasitic delay. For connections between nearby gates, this generally leads to a best stage effort ρ slightly greater than 4. The path should use fewer stages because each stage contributes wire capacitance. To reduce delay, the gates should be sized larger so that the wire capacitance is a smaller fraction of the whole. However, this comes at the expense of increased energy.

Conversely, when the interconnect is long ($C_{wire} >> C_{gate}$), the gate at the end can be ignored. The path can now be partitioned into two parts. The first part drives the wire while the second receives its input from the wire. The first part is designed to drive the load capacitance of the wire; the extra load of the receiver is negligible. To save energy, the final stage driving the wire should have a low logical effort and a high electrical effort; an inverter is preferred [Stan99]. The size of the receiver is chosen by practical considerations: Larger receivers may be faster, but they also cost area and power. If the wire is long enough that the RC flight time exceeds a few gate delays, it should be broken into segments driven by repeaters.

The most difficult problems occur when $C_{wire} \approx C_{gate}$. These medium-length wires introduce branching efforts that are a strong function of the size of the gates they drive. Writing a delay equation as a function of the gate sizes along the path and the wire capacitance results in an expression that can be differentiated with respect to gate sizes to compute the best sizes. Alternatively, a convex optimizer can be used to minimize delay or generate an energy-delay trade-off curve.

Figure 5.33 shows three stages along a path. By writing the Elmore delay and differentiating with respect to the size of the middle stage, we find the interesting result that the delay caused by the capacitance of a stage should equal the delay caused by the resistance of the stage [Morgenshtein09]:

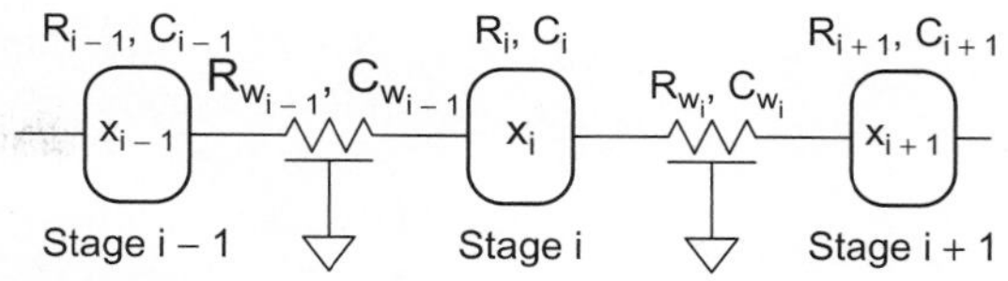

FIGURE 5.33 Path with wires

$$C_i\left(R_{i-1} + R_{w_{i-1}}\right) = R_i\left(C_{i+1} + C_{w_{i+1}}\right) \quad (5.39)$$

Example 5.11

The path in Figure 5.34 contains a medium-length wire modeled as a lumped capacitance. Write an equation for path delay in terms of x and y. How large should the x and y inverters be for shortest path delay? What is the stage effort of each stage?

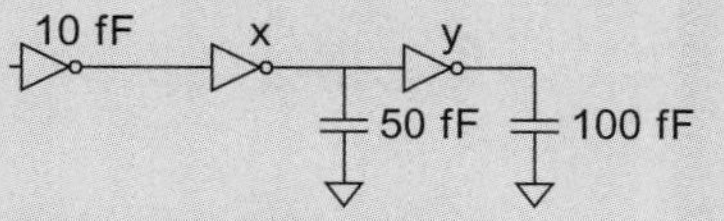

FIGURE 5.34 Path with medium-length wire

SOLUTION: From the Logical Effort delay model, we find the path delay is

$$d = \frac{x}{10} + \frac{y+50}{x} + \frac{100}{y} + P \quad (5.40)$$

Differentiating with respect to each size and setting the results to 0 allows us to solve EQ (5.41) for $x = 33$ fF and $y = 57$ fF.

$$\begin{aligned} \frac{1}{10} - \frac{y+50}{x^2} &= 0 \Rightarrow x^2 = 10y + 500 \\ \frac{1}{x} - \frac{100}{y^2} &= 0 \Rightarrow y^2 = 100x \end{aligned} \quad (5.41)$$

The stage efforts are (33/10) = 3.3, (57 + 50)/33 = 3.2, and (100/57) = 1.8. Notice that the first two stage efforts are equal as usual, but the third stage effort is lower. As x already drives a large wire capacitance, y may be rather large (and will bear a small stage effort) before the incremental increase in delay of x driving y equals the incremental decreases in delay of y driving the output.

5.6 Pitfalls and Fallacies

Designing a large chip without considering the floorplan

In the mid-1990s, designers became accustomed to synthesizing a chip from HDL and "tossing the netlist over the wall" to the vendor who would place & route it and manufacture the chip. Many designers were shielded from considering the physical implementation. Now flight times across the chip are a large portion of the cycle time in slow systems and multiple cycles in faster systems. If the chip is synthesized without a floorplan, some paths with long wires will be discovered to be too slow after layout. This requires resynthesis with new timing constraints to shorten the wires. When the new layout is completed, the long wires simply show up in different paths. The solution to this convergence problem is to make a floorplan early and microarchitect around this floorplan, including budgets for wire flight time between blocks. Algorithms termed *timing directed placement* have alleviated this problem, resulting in place & route tools that converge in one or a few iterations.

Leaving gaps in the power grid

Current always flows in loops. Current flowing along a signal wire must return in the power/ground network. The area of the loop sets the inductance of the signal. A discontinuity in the power grid can force return current to find a path far from the signal wire, greatly increasing the inductance, which increases delay and noise. Because signal inductance is usually not modeled, the delay and noise will not be discovered until after fabrication.

Summary

As feature size decreases, transistors get faster but wires do not. Interconnect delays are now very important. The delay is again estimated using the Elmore delay model based on the resistance and capacitance of the wire and its driver and load. The wire delay grows with the square of its length, so long wires are often broken into shorter segments driven by repeaters. Vast numbers of wires are required to connect all the transistors, so processes provide many layers of interconnect packed closely together. The capacitive coupling between these tightly packed wires can be a major source of noise in a system. These challenges are managed by using many metal layers of various thicknesses to provide high bandwidth for short thin wires and lower delay for longer fat wires. The microarchitecture becomes inherently linked to the floorplan because the design must allocate one or more cycles of pipeline delay for wires that cross the chip.

Exercises

5.1 Derive EQ (5.27)–(5.30). Assume the initial driver and final receiver are of the same size as the repeaters so the total delay is N times the delay of a segment.

5.2 Revisit Exercise 5.1 using a pair of inverters (a noninverting buffer) instead of a single inverter. The first inverter in each pair is $W1$ times unit width. The second is a factor of k larger than the first. Derive EQ (5.33)–(5.36).

5.3 Compute the characteristic velocity (delay per mm) of a repeated metal2 wire in the 180 nm process. A unit nMOS transistor has resistance of 2.5 kΩ and capacitance of 0.7 fF, and the pMOS has twice the resistance. Use the data from Figure 5.12. Consider both minimum pitch and double-pitch (twice minimum width and spacing) wires. Assume solid metal above and below the wires and that the neighbors are not switching.

5.4 Prove EQ (5.39).

5.5 Estimate the resistance per mm of a minimum pitch Cu wire for each layer in the Intel 45 nm process described in Table 5.1. Assume a 10 nm high-resistance barrier layer and negligible dishing.

5.6 Consider a 5 mm-long, 4 λ-wide metal2 wire in a 0.6 μm process. The sheet resistance is 0.08 $\Omega/\square$ and the capacitance is 0.2 fF/μm. Construct a 3-segment π-model for the wire.

5.7 A 10x unit-sized inverter drives a 2x inverter at the end of the 5 mm wire from Exercise 5.6. The gate capacitance is $C = 2$ fF/μm and the effective resistance is $R = 2.5$ k$\Omega \cdot \mu$m for nMOS transistors. Estimate the propagation delay using the Elmore delay model; neglect diffusion capacitance.

5.8 Find the best width and spacing to minimize the RC delay of a metal2 bus in the 180 nm process described in Figure 5.12 if the pitch cannot exceed 960 nm. Minimum width and spacing are 320 nm. First, assume that neither adjacent bit is switching. How does your answer change if the adjacent bits may be switching?

Scaling, Reliability, and Variability

6

6.1 Introduction

A central challenge in building integrated circuits is to get millions or billions of transistors to all function, not just once, but for a quintillion consecutive cycles. Transistors are so small that printing errors below the wavelength of light and variations in the discrete number of dopant atoms have major effects on their performance. Over the course of their operating lives, chips may be subjected to temperatures ranging from freezing to boiling. Intense electric fields gradually break down the gates. Unrelenting currents carry away the atoms of the wires like termites slowly devouring a mansion. Cosmic rays zap the bits stored in tiny memory cells.

Despite these daunting challenges, engineers routinely build robust integrated circuits with lifetimes exceeding ten years of continuous operation. Conventional static CMOS circuits are exceptionally well-suited to the task because they have great noise margins, are minimally sensitive to variations in transistor parameters, and will eventually recover even if a noise event occurs. Fairly simple guidelines on the maximum voltages and currents suffice to ensure long operating life. Fault-tolerant and adaptive architectures can correct for errors and adjust the chip to run at its best despite manufacturing variations and changing operating conditions.

Section 6.2 begins by examining the sources of manufacturing and environmental variations and their effects on a chip. Section 6.3 then discusses reliability, including wearout, soft errors, and catastrophic failures. A good design should work well not only in the current manufacturing process, but also when ported to a more advanced process. Section 6.4 addresses scaling laws to predict how future processes will evolve. Section 6.5 revisits variability with a more mathematical treatment. Section 6.6 examines adaptive and fault-tolerant design techniques to compensate for variations and transient errors.

6.2 Variability

So far, when considering the various aspects of determining a circuit's behavior, we have only alluded to the variations that might occur in this behavior given different operating conditions. In general, there are three different sources of variation—two environmental and one manufacturing:

- Process variation
- Supply voltage
- Operating temperature

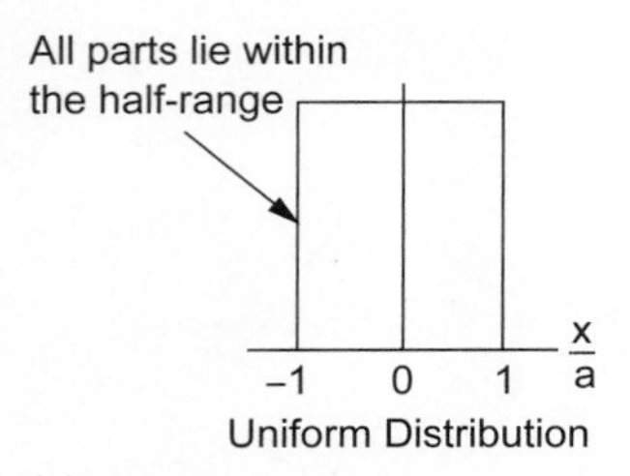

(a)

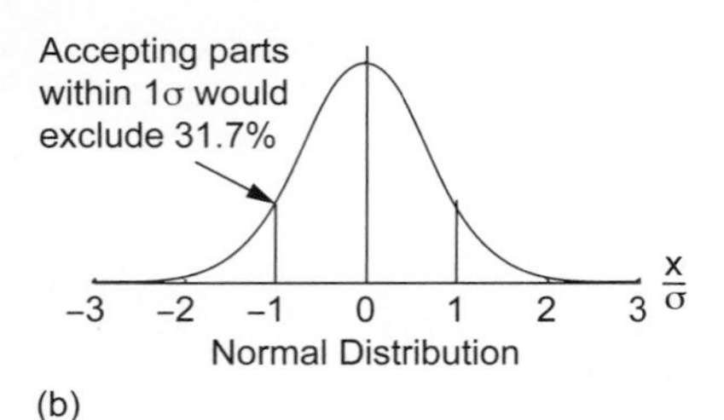

(b)

FIGURE 6.1
Uniform and normal distributions

The variation sources are also known as Process, Voltage, and Temperature (PVT). You must aim to design a circuit that will operate reliably operate over all extremes of these three variables. Failure to do so causes circuit problems, poor yield, and customer dissatisfaction.

Variations are usually modeled with *uniform* or *normal* (Gaussian) statistical distributions, as shown in Figure 6.1. Uniform distributions are specified with a *half-range a*. For good results, accept variations over the entire half-range. For example, a uniform distribution for V_{DD} could be specified at 1.0 V ±10%. This distribution has a 100 mV half-range. All parts should work at any voltage in the range. Normal distributions are specified with a *standard deviation* σ. Processing variations are usually modeled with normal distributions. Retaining parts with a 3σ distribution will result in 0.26% of parts being rejected. A 2σ retention results in 4.56% of parts being rejected, while 1σ results in a 31.74% rejection rate. Obviously, rejecting parts outside 1σ of nominal would waste a large number of parts. A 3σ or 2σ limit is conventional and a manufacturer with a commercially viable CMOS process should be able to supply a set of device parameters describing this range. For components such as memory cells that are replicated millions of times, a 0.26% failure rate is far too high. Such circuits must tolerate 5, 6, or even 7σ of variation. Remember that if only the variations in one direction (e.g., too slow) matter, the reject rate is halved.

6.2.1 Supply Voltage

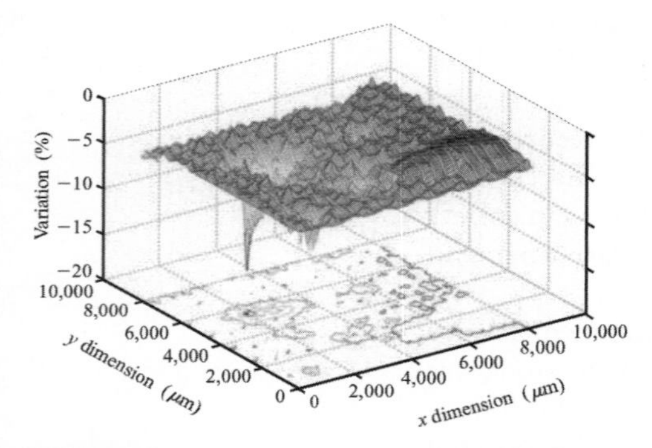

FIGURE 6.2 Voltage droop map (Courtesy of International Business Corporation. Unauthorized use not permitted.)

Systems are designed to operate at a nominal supply voltage, but this voltage may vary for many reasons including tolerances of the voltage regulator, IR drops along supply rails, and di/dt noise. The system designer may trade-off power supply noise against resources devoted to power supply regulation and distribution; typically the supply is specified at ±10% around nominal at each logic gate. The supply varies across the chip as well as in time. For example, Figure 6.2 shows a voltage map indicating the worst case droop as a function of position on a chip [Bernstein06, Su03].

Speed is roughly proportional to V_{DD}, so to first order this leads to ±10% delay variations (check for your process and voltage when this is critical). Power supply variations also appear in noise budgets.

6.2.2 Temperature

Section 2.4.5 showed that as temperature increases, drain current decreases. The junction temperature of a transistor is the sum of the ambient temperature and the temperature rise caused by power dissipation in the package. This rise is determined by the power consumption and the package thermal resistance, as discussed in Section 12.2.4.

Table 6.1 lists the ambient temperature ranges for parts specified to commercial, industrial, and military standards. Parts must function at the bottom end of the ambient range unless they are allowed time to warm up before use. The junction temperature (the temperature at the semiconductor junctions forming the transistors) may significantly exceed the maximum ambient temperature. Commonly commercial parts are verified to operate with junction temperatures up to 125 °C.

TABLE 6.1 Ambient temperature ranges

Standard	Minimum	Maximum
Commercial	0 °C	70 °C
Industrial	–40 °C	85 °C
Military	–55 °C	125 °C

Temperature varies across a die depending on which portions dissipate the most power. The variation is gradual, so all circuits in a given 1 mm diameter see nearly the same temperature. Temperature varies in time on a scale of milliseconds. Figure 6.3 shows a simulated thermal map for the Itanium 2 microprocessor [Harris01b]. The execution core has hot spots exceeding 100 °C, while the caches in the periphery are below 70 °C.

6.2.3 Process Variation

Devices and interconnect have variations in film thickness, lateral dimensions, and doping concentrations [Bernstein99]. These variations can be classified as inter-die (e.g., all the transistors on one die might be shorter than normal because they were etched excessively) and intra-die (e.g., one transistor might have a different threshold voltage than its neighbor because of the random number of dopant atoms implanted).

For devices, the most important variations are channel length L and threshold voltage V_t. Channel length variations are caused by photolithography proximity effects, deviations in the optics, and plasma etch dependencies. Threshold voltages vary because of different doping concentrations and annealing effects, mobile charge in the gate oxide, and discrete dopant variations caused by the small number of dopant atoms in tiny transistors. Threshold voltages gradually change as transistors wear out; such time-dependent variation will be examined in Section 6.3.

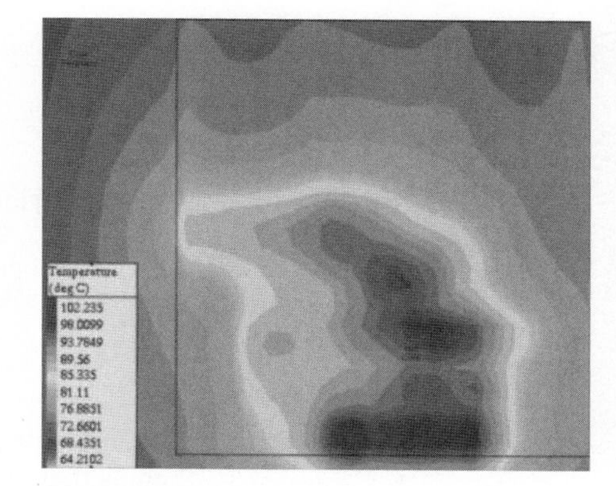

FIGURE 6.3 Thermal map of Itanium 2 (© IEEE 2001.)

For interconnect, the most important variations are line width and spacing, metal and dielectric thickness, and contact resistance. Line width and spacing, like channel length, depend on photolithography and etching proximity effects. Thickness may be influenced by polishing. Contact resistance depends on contact dimensions and the etch and clean steps.

Process variations can be classified as follows:

- Lot-to-lot (*L2L*)
- Wafer-to-wafer (*W2W*)
- Die-to-die (*D2D*), inter-die, or within-wafer (*WIW*)
- Within-die (*WID*) or intra-die

Wafers are processed in batches called *lots*. A lot processed after a furnace has been shut down and cleaned may behave slightly differently than the lot processed earlier. One wafer may be exposed to an ion implanter for a slightly different amount of time than another, causing W2W threshold voltage variation. A die near the edge of the wafer may etch slightly differently than a die in the center, causing D2D channel length variations. For example, Figure 6.4 plots the operating frequency of ring oscillators as a function of their position on the wafer, showing a 20% variation involving both a systematic radial component and a smaller random component. Unless calibrations are made on a per-lot or per-wafer basis, L2L and W2W variations are often lumped into the D2D variations. D2D variations ultimately make one chip faster or slower than another. They can be handled by

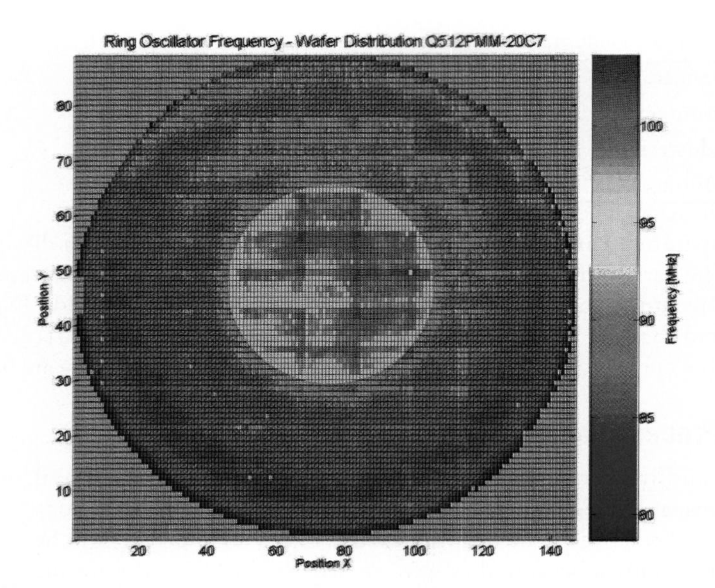

FIGURE 6.4 Wafer map of the frequency distribution of a ring oscillator circuit in 90-nm CMOS technology. From M. Pelgrom, "Nanometer CMOS: An Analog Design Challenge!" *IEEE Distinguished Lecture*, Denver 2006. (Figure courtesy of B. Ljevar (NXP). Reprinted with permission.)

providing enough margin to cover 2 or 3σ of variation and by rejecting the small number of chips that fall outside this bound, as discussed in the next section.

WID variations were once small compared to D2D variations and were largely ignored by digital designers but have become quite important in nanometer processes. Some WID variations are spatially correlated; these are called *process tilt*. For example, an ion implanter might deliver a greater dose near the center of a wafer than near the periphery, causing threshold voltages to tilt radially across the wafer. In summary, transistors on the same die match better than transistors on different dice and adjacent transistors match better than widely separated ones. WID variations are more challenging to manage because some of the millions or billions of transistors on a chip are likely to stray far from typical parameters. Section 6.5 considers the statistics of WID variation.

6.2.4 Design Corners

From the designer's point of view, the collective effects of process and environmental variation can be lumped into their effect on transistors: *typical* (also called *nominal*), *fast*, or *slow*. In CMOS, there are two types of transistors with somewhat independent characteristics, so the speed of each can be characterized. Moreover, interconnect speed may vary independently of devices. When these processing variations are combined with the environmental variations, we define *design* or *process corners*. The term *corner* refers to an imaginary box that surrounds the guaranteed performance of the circuits, as shown in Figure 6.5. The box is not square because some characteristics such as oxide thickness track between devices, making it impossible to find a slow nMOS transistor with thick oxide and a fast pMOS transistor with thin oxide simultaneously.

Table 6.2 lists a number of interesting design corners. The corners are specified with five letters describing the nMOS, pMOS, interconnect, power supply, and temperature, respectively. The letters are F, T, and S, for *fast*, *typical*, and *slow*. The environmental corners for a 1.8 V commercial process are shown in Table 6.3, illustrating that circuits are fastest at high voltage and low temperature. Circuits are most likely to fail at the corners of the design space, so nonstandard circuits should be simulated at all corners to ensure they operate correctly in all cases. Often, integrated circuits are designed to meet a timing specification for typical processing. These parts may be *binned*; faster parts are rated for higher frequency and sold for more money, while slower parts are rated for lower frequency. In any event, the parts must still work in the slowest SSSSS environment. Other integrated circuits are designed to obtain high yield at a relatively low frequency; these parts are simulated for timing in the slow process corner. The fast corner FFFFF has maximum speed. Other corners are used to check for races and ratio problems where the relative strengths and speeds of different transistors or interconnect are important. The FFFFS corner is important for noise because the edge rates are fast, causing more coupling; the threshold voltages are low; and the leakage is high [Shepard99].

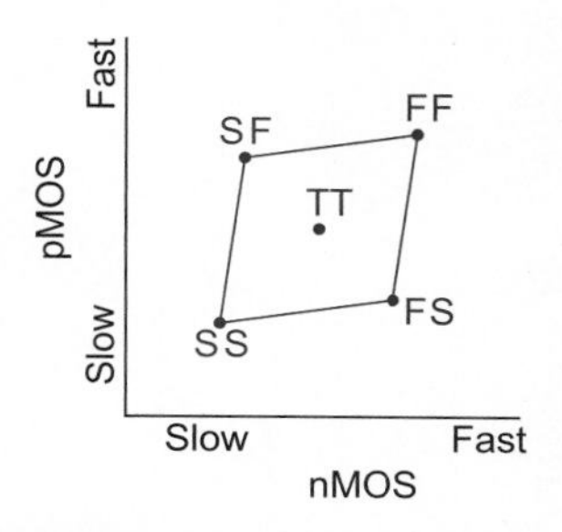

FIGURE 6.5 Design corners

TABLE 6.2 Design corner checks

Corner					Purpose
nMOS	pMOS	Wire	V_{DD}	Temp	
T	T	T	S	S	Timing specifications (binned parts)
S	S	S	S	S	Timing specifications (conservative)
F	F	F	F	F	Race conditions, hold time constraints, pulse collapse, noise
S	S	?	F	S	Dynamic power
F	F	F	F	S	Subthreshold leakage noise and power, overall noise analysis
S	S	F	S	S	Races of gates against wires
F	F	S	F	F	Races of wires against gates
S	F	T	F	F	Pseudo-nMOS and ratioed circuits noise margins, memory read/write, race of pMOS against nMOS
F	S	T	F	F	Ratioed circuits, memory read/write, race of nMOS against pMOS

TABLE 6.3 Environmental corners

Corner	Voltage	Temperature
F	1.98	0 °C
T	1.8	70 °C
S	1.62	125 °C

Often, the corners are abbreviated to fewer letters. For example, two letters generally refer to nMOS and pMOS. Three refer to nMOS, pMOS, and overall environment. Four refer to nMOS, pMOS, voltage, and temperature.

It is important to know the design corner when interpreting delay specifications. For example, the datasheet shown in Figure 3.25 is specified at the 25 °C TTTT corner. The SS corner is 27% slower. The cells are derated at –71% per volt and 0.13%/°C, for additional penalties of 13% each in the low voltage and high temperature corners. These factors are multiplicative, giving SSSS delay of 1.62 times nominal.

[Ho01] and [Chinnery02] find the FO4 inverter delay can be estimated from the effective channel length L_{eff} (also called L_{gate}) as follows:

- $L_{eff} \times$ (0.36 ps/nm) in TTTT corner
- $L_{eff} \times$ (0.50 ps/nm) in TTSS corner
- $L_{eff} \times$ (0.60 ps/nm) in SSSS corner

Note that the effective channel length is aggressively scaled faster than the drawn channel length to improve performance, as shown in Table 15.2. Typically, $L_{eff} = 0.5–0.7\ L_{drawn}$. For example, Intel's 180 nm process was originally manufactured with $L_{eff} = 140$ nm and eventually pushed to $L_{eff} = 100$ nm. This model predicts an FO4 inverter delay of about 50–70 ps in the TTSS corner where design usually takes place. Low-power processes with higher threshold voltages will have longer FO4 delays.

In addition to working at the standard process corners, chips must function in a very high temperature, high voltage *burn-in* corner (e.g., 125 to 140 °C externally, corresponding to an even higher internal temperature, and 1.3–1.7× nominal V_{DD} [Vollertsen99]) during test. While it does not have to run at full speed, it must operate correctly so that all nodes can toggle. The burn-in corner has very high leakage and can dictate the size of keepers and weak feedback on domino gates and static latches.

Processes with multiple threshold voltages and/or multiple oxide thicknesses can see each flavor of transistor independently varying as fast, typical, or slow. This can easily lead to more corners than anyone would care to simulate and raises challenges about identifying what corners must be checked for different types of circuits.

6.3 Reliability

Designing reliable CMOS chips involves understanding and addressing the potential failure modes [Segura04]. This section addresses reliability problems (*hard errors*) that cause integrated circuits to fail permanently, including the following:

- Oxide wearout
- Interconnect wearout
- Overvoltage failure
- Latchup

This section also considers transient failures (*soft errors*) triggered by radiation that cause the system to crash or lose data. Circuit pitfalls and common design errors are discussed in Section 8.3.

6.3.1 Reliability Terminology

A *failure* is a deviation from compliance with the system specification for a given period of time. Failures are caused by *faults*, which are defined as failures of subsystems. Faults have many causes, ranging from design bugs to manufacturing defects to wearout to external disturbances to intentional abuse of a product. Not all faults lead to errors; some are masked. For example, a bug in the multiprocessor interface logic does not cause an error in a single-processor system. A defective via may go unnoticed if it is in parallel with a good one. Studying the underlying faults gives insight into predicting and improving the failure rate of the overall system.

A number of acronyms are commonly used in describing reliability [Tobias95]. MTBF is the *mean time between failures*: (number of devices × hours of operation) / number of failures. FIT is the *failures in time*, the number of failures that would occur every thousand hours per million devices, or equivalently, $10^9 \times$ (failure rate/hour). 1000 FIT is one failure in 10^6 hours = 114 years. This is good for a single chip. However, if a system contains 100 chips each rated at 1000 FIT and a customer purchases 10 systems, the failure rate is $100 \times 1000 \times 10 = 10^6$ FIT, or one failure every 1000 hours (42 days). Reliability targets of less than 100 FIT are desirable.

Most systems exhibit the *bathtub curve* shown in Figure 6.6. Soon after birth, systems with weak or marginal components tend to fail. This period is called *infant mortality*. Reliable systems then enter their *useful operating life*, in which the failure rate is low. Finally, the failure rate increases at the end of life as the system wears out. It is important to age systems past infant mortality before shipping the products. Aging is accelerated by stressing the part through *burn-in* at higher than normal voltage and temperature, as discussed in Section 6.2.4.

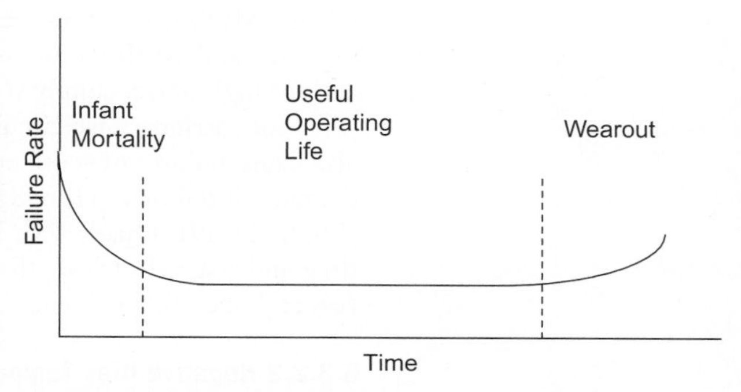

FIGURE 6.6 Reliability bathtub curve

Engineers typically desire product lifetimes exceeding 10 years, but it is clearly impossible to test a product for 10 years before selling it. Fortunately, most wearout mechanisms have been observed to display an exponential relationship with voltage or temperature. Thus, systems are subjected to *accelerated life testing* during burn-in conditions to simulate the aging process and evaluate the time to wearout. The results are extrapolated to normal operating conditions to judge the actual useful operating life. For example, Figure 6.7 shows the measured lifetime of gate oxides in an IBM 32 nm process at elevated voltages [Arnaud08]. The extrapolated results show a lifetime exceeding 10 years at 10% above the nominal 0.9 V V_{DD}.

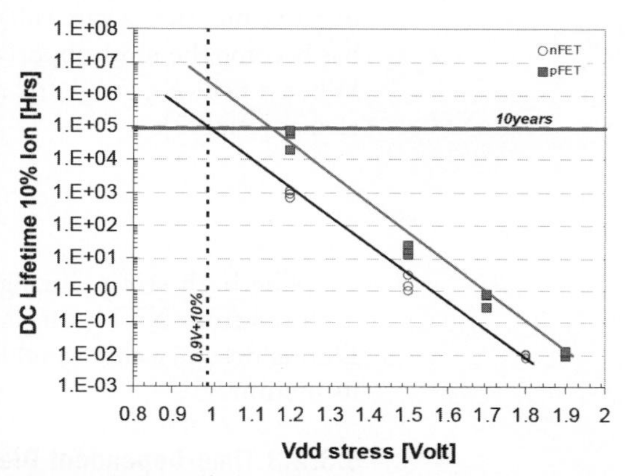

FIGURE 6.7 Accelerated life testing of gate oxides in IBM 32 nm process (© IEEE 2008.)

Life testing is time-consuming and comes right at the end of the project when pressures to get to market are greatest. Part of any high-volume chip design will necessarily include designing a reliability assessment program that consists of burn-in boards deliberately stressing a number of chips over an extended period. Designers have tried to develop reliability simulators to predict lifetime [Hu92, Hsu92], but physical testing remains important. For high-volume parts, the source of failures is tracked and common points of failure can be redesigned and rolled into manufacturing.

6.3.2 Oxide Wearout

As gate oxides are subjected to stress, they gradually wear out, causing the threshold voltage to shift and the gate leakage to increase. Eventually, the circuit fails because transistors become too slow, mismatches become too large, or leakage currents become too great. Processes generally specify a maximum operating voltage to ensure oxide wearout effects are limited during the lifetime of a chip. The primary mechanisms for oxide wearout include the following:

- Hot carriers
- Negative bias temperature instability (NBTI)
- Time-dependent dielectric breakdown (TDDB)

6.3.2.1 Hot Carriers As transistors switch, high-energy ("hot") carriers are occasionally injected into the gate oxide and become trapped there. Electrons have higher mobility and account for most of the hot carriers. The damaged oxide changes the I-V characteristics of the device, reducing current in nMOS transistors and increasing current in pMOS transistors. Damage is maximized when the substrate current I_{sub} is large, which typically occurs when nMOS transistors see a large V_{ds} while ON. Therefore, the problem is worst for inverters and NOR gates with fast rising inputs and heavily loaded outputs [Sakurai86], and for high power supply voltages.

Hot carriers cause circuit wearout as nMOS transistors become too slow. They can also cause failures of sense amplifiers and other matched circuits if matched components degrade differently [Huh98]. Hot electron degradation can be analyzed with simulators [Hu92, Hsu91, Quader94]. The wear is limited by setting maximum values on input rise time and stage electrical effort [Leblebici96]. These maximum values depend on the process and operating voltage.

6.3.2.2 Negative Bias Temperature Instability When an electric field is applied across a gate oxide, dangling bonds called *traps* develop at the Si-SiO_2 interface. The threshold voltage increases as more traps form, reducing the drive current until the circuit fails [Doyle91, Reddy02]. The process is most pronounced for pMOS transistors with a strong negative bias (i.e., a gate voltage of 0 and source voltage of V_{DD}) at elevated temperature. It has become the most important oxide wearout mechanism for many nanometer processes. When a field $E_{ox} = V_{DD}/t_{ox}$ is applied for time t, the threshold voltage shift can be modeled as [Paul07]:

$$\Delta V_t = ke^{\frac{E_{ox}}{E_0}} t^{0.25} \tag{6.1}$$

The high stress during burn-in can lock in most of the threshold voltage shift expected from NBTI; this is good because it allows testing with full NBTI degradation. During design, a chip should be simulated under the worst-case NBTI shift expected over its lifetime.

6.3.2.3 Time-Dependent Dielectric Breakdown As an electric field is applied across the gate oxide, the gate current gradually increases. This phenomenon is called *time-dependent dielectric breakdown* (TDDB) and the elevated gate current is called *stress-induced leakage current* (SILC). The exact physical mechanisms are not fully understood, but TDDB likely results from a combination of charge injection, bulk trap state generation, and trap-assisted conduction [Hicks08]. After sufficient stress, it can result in catastrophic dielectric breakdown that short-circuits the gate.

The failure rate is exponentially dependent on the temperature and oxide thickness [Monsieur01]; for a 10-year life at 125 °C, the field across the gate $E_{ox} = V_{DD}/t_{ox}$ should be kept below about 0.7 V/nm [Moazzami90]. Nanometer processes operate close to this limit. The problem is greatest when voltage overshoots occur; this can be caused by noisy power supplies or reflections at I/O pads. Reliability is improved by lowering the power supply voltage, minimizing power supply noise, and using thicker oxides on the I/O pads.

6.3.3 Interconnect Wearout

High currents flowing through wires eventually can damage the wires. For wires carrying unidirectional (DC) currents, electromigration is the main failure mode. For wires carrying bidirectional (AC) currents, self-heating is the primary concern.

6.3.3.1 Electromigration High current densities lead to an "electron wind" that causes metal atoms to migrate over time. Such *electromigration* causes wearout of metal interconnect through the formation of voids [Hu95]. Figure 6.8 shows a scanning electron micrograph of electromigration failure of a via between M2 and M3 layers [Christiansen06]. Remarkable videos taken under a scanning electron microscope show void formation and migration and wire failure [Meier99]. The problem is especially severe for aluminum wires; it is commonly alleviated with an Al-Cu or Al-Si alloy and is much less important for pure copper wires because of the different grain transport properties. The electromigration properties also depend on the grain structure of the metal film.

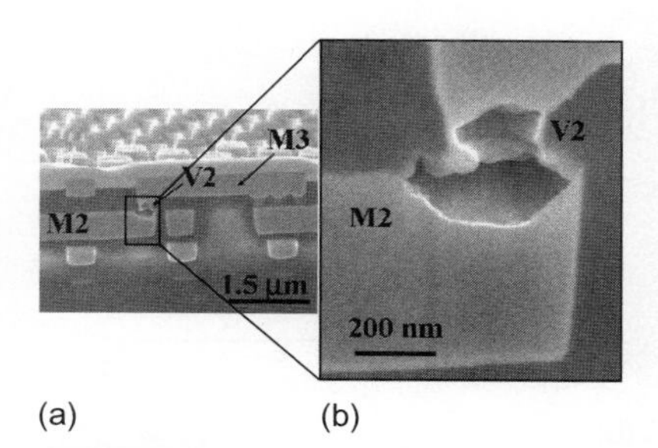

FIGURE 6.8 Electromigration failure of M2-M3 via (© IEEE 2006.)

Electromigration depends on the current density $J = I/wt$. It is more likely to occur for wires carrying a DC current where the electron wind blows in a constant direction than for those with bidirectional currents [Liew90]. Electromigration current limits are usually expressed as a maximum J_{dc}. The mean time to failure (MTTF) also is highly sensitive to operating temperature as given by Black's Equation [Black69]:

$$\text{MTTF} \propto \frac{e^{\frac{E_a}{kT}}}{J_{dc}^n} \tag{6.2}$$

E_a is the activation energy that can be experimentally determined by stress testing at high temperatures and n is typically 2. The electromigration DC current limits vary with materials, processing, and desired MTTF and should be obtained from the fabrication vendor. In the absence of better information, a maximum J_{dc} of 1–2 mA/μm^2 is a conservative limit for aluminum wires at 110 °C [Rzepka98]. Copper is less susceptible to electromigration and may endure current densities of 10 mA/μm^2 or better [Young00]. Current density may be more limited in contact cuts.

Considering the dynamic switching power, we can estimate the maximum switching capacitance that a wire can drive. The current is $I_{dc} = P/V = \alpha CVf$. Thus, for a wire limited to $I_{dc-\max}$, the switching capacitance should be less than

$$C = \frac{I_{dc-\max}}{\alpha V_{DD} f} \tag{6.3}$$

6.3.3.2 Self-Heating While bidirectional wires are less prone to electromigration, their current density is still limited by *self-heating*. High currents dissipate power in the wire. Because the surrounding oxide or low-k dielectric is a thermal insulator, the wire temperature can become significantly greater than the underlying substrate. Hot wires exhibit greater resistance and delay. Electromigration is also highly sensitive to temperature, so self-heating may cause temperature-induced electromigration problems in the bidirectional wires. Brief

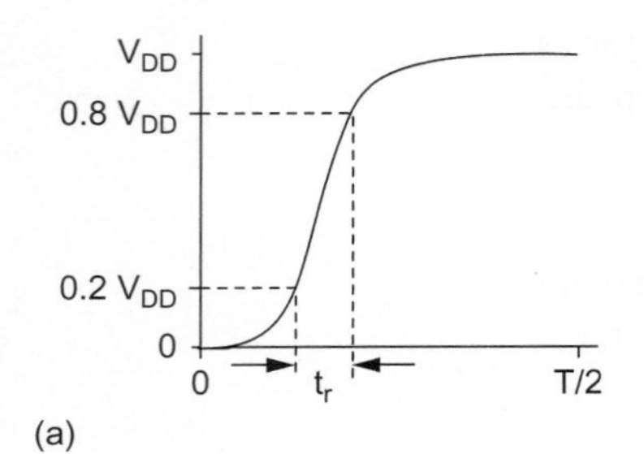

(a)

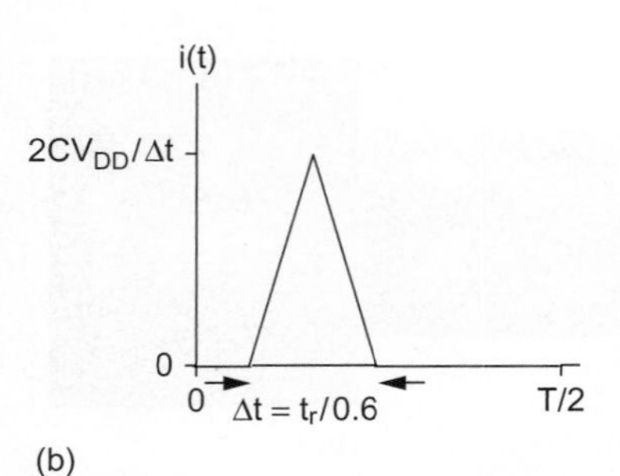

(b)

FIGURE 6.9 Switching waveforms for RMS current estimation

pulses of high peak currents may even melt the interconnect. Self-heating is dependent on the root-mean-square (RMS) current density. This can be measured with a circuit simulator or calculated as

$$I_{rms} = \sqrt{\frac{\int_0^T I(t)^2\, dt}{T}} \tag{6.4}$$

A conservative rule to control reliability problems with self-heating is to keep $J_{rms} < 15$ mA/μm^2 for bidirectional aluminum wires on a silicon substrate [Rzepka98]. The maximum capacitance of the wire can be estimated based on the RMS current. If a signal has symmetric rising and falling edges, we only need to consider half of a period. Figure 6.9(a) shows a signal with a 20–80% rise time t_r and an average period $T = 1/\alpha f$. The switching current $i(t)$ can be approximated as a triangular pulse of duration $\Delta t = t_r / (0.8–0.2)$, as shown in Figure 6.9(b). Then, the RMS current is

$$\begin{aligned} I_{rms} &= \sqrt{\frac{1}{0.5T}\int_0^{0.5T} i^2(t)\,dt} \\ &= \sqrt{\frac{2}{0.5T}\int_0^{\Delta t/2}\left(\frac{2CV_{DD}}{\Delta t}\frac{t}{\Delta t/2}\right)^2 dt} \\ &\approx 1.26CV_{DD}\sqrt{\frac{\alpha f}{t_r}} \end{aligned} \tag{6.5}$$

and, to avoid excessive self-heating, the wire and load capacitance should be less than

$$C = \frac{I_{rms-\max}}{1.26V_{DD}\sqrt{\frac{\alpha f}{t_r}}} \tag{6.6}$$

Example 6.1

A clock signal is routed on the top metal layer using a wire that is 1 μm wide and has a self-heating limit of 10 mA. The wire has a capacitance of 0.4 fF/μm and the load capacitance is 85 fF. The clock switches at 3 GHz and has a 20 ps rise time. How far can the wire run between repeaters without overheating?

SOLUTION: A clock has an activity factor of 1. According to EQ (6.6), the maximum capacitance of the line is

$$C = \frac{10^{-2}\ \text{A}}{1.26(1\ \text{V})\sqrt{\frac{1\cdot 3\times 10^9\ \text{Hz}}{20\times 10^{-12}\ \text{s}}}} = 685\ \text{fF} \tag{6.7}$$

Thus, the maximum wire length is (685 − 85 fF) / (0.4 fF/μm) = 1500 μm.

In summary, electromigration from high DC current densities is primarily a problem in power and ground lines. Self-heating limits the RMS current density in bidirectional signal lines. However, do not overlook the significant unidirectional currents that flow through the wires contacting nMOS and pMOS transistors. For example, Figure 6.10 shows which lines in an inverter are limited by DC and RMS currents. Both problems can be addressed by widening the lines or reducing the transistor sizes (and hence the current).

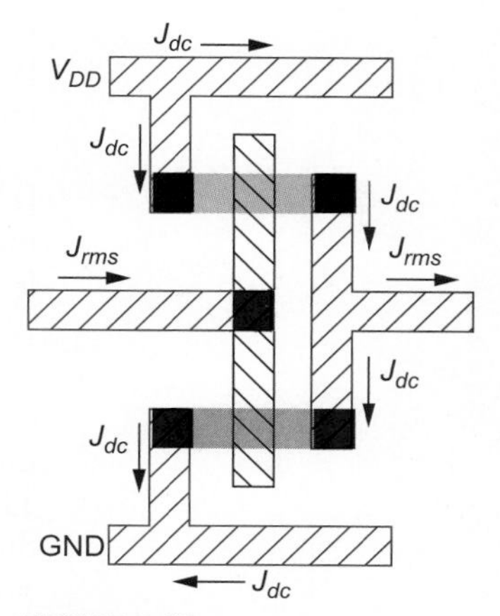

FIGURE 6.10 Current density limits in an inverter

6.3.4 Soft Errors

In the 1970s, as dynamic RAMs (DRAMs) replaced core memories, DRAM vendors were puzzled to find DRAM bits occasionally flipping value spontaneously. At first, the errors were attributed to "system noise," "voltage marginality," "sense amplifiers," or "pattern sensitivity," but the errors were found to be random. When the corrupted bit was rewritten with a new value, it was no more likely than any other bit to experience another error. In a classic paper [May79], Intel identified the source of these soft errors as alpha particle collisions that generate electron-hole pairs in the silicon as the particles lose energy. The excess carriers can be collected into the diffusion terminals of transistors. If the charge collected is comparable to the charge on the node, the voltage can be disturbed.

Soft errors are random nonrecurring errors triggered by radiation striking a chip. Alpha particles, emitted by the decay of trace uranium and thorium impurities in packaging materials, was once the dominant source of soft errors, but they have been greatly reduced by using highly purified materials. Today, high-energy (> 1 MeV) neutrons from cosmic radiation account for most soft errors in many systems [Baumann01, Baumann05]. When a neutron strikes a silicon atom, it can induce fission, shattering the atom into charged fragments that continue traveling through the substrate. These ions leave a trail of electron-hole pairs behind as they travel through the lattice. Figure 6.11 shows the effect of an ion striking a reverse-biased p-n junction [Baumann05]. The ion leaves a cylindrical trail of electrons and holes in its wake, with a radius of less than a micron. Within tens of picoseconds, the electric field at the junction collects the carriers into a funnel-shaped depletion region. Over the subsequent nanoseconds, electrons diffuse into the depletion region. Depending on the type of ion, its energy, its trajectory, and the geometry of the p-n junction, up to several hundred femtocoulombs of charge may be collected onto the junction.

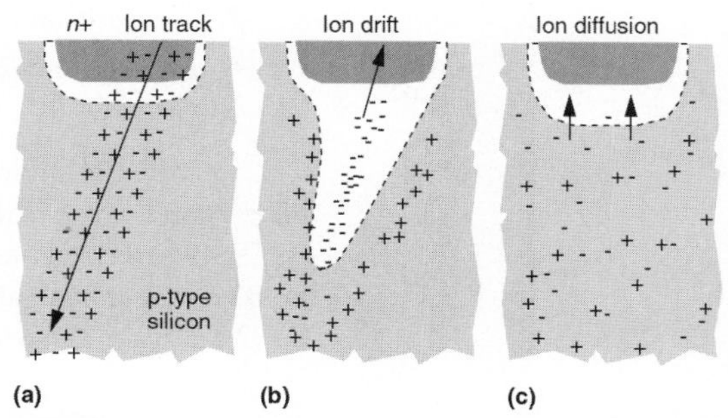

FIGURE 6.11 Generation and collection of carriers after a radiation strike (© IEEE 2005.)

The spike of current is called a *single-event transient* (SET). If the collected charge exceeds a critical amount, Q_{crit}, it may flip the state of the node, causing a fault called a *single-event upset* (SEU). Failures caused by such faults are called *soft errors*. Q_{crit} depends on the capacitance and voltage of the node, and on any feedback devices that may fight against the disturbance. This is a serious challenge because both capacitance and voltage have been decreasing as transistors shrink, reducing Q_{crit}. Fortunately, the amount of charge collected on a smaller junction also decreases, but the net trend has been toward higher soft error rates.

The holes generated by the particle strike flow to a nearby substrate contact where they are collected. The current flowing through the resistive substrate raises the potential of the substrate. This can turn on a parasitic bipolar transistor (see Section 6.3.6) between the source and drain of a nearby nMOS transistor, disturbing that transistor too [Osada04]. Such multinode disturbances can be controlled by using plenty of substrate and well contacts.

At sea level, SRAM generally experiences a *soft error rate* (SER) of 100–2000 FIT/Mb [Hazucha00, Normand96]. The neutron flux from cosmic rays increases by two orders of magnitude at aircraft flight altitudes [Ziegler96] and can cause up to 10^6 FIT/Mb at these levels. Depending on the process and layout, roughly 1% of the soft errors affect multiple nodes [Hazucha04].

Soft errors affect memories, registers, and combinational logic. Memories use error detecting and correcting codes to tolerate soft errors, so these errors rarely turn into failures in a well-designed system. Such codes will be discussed further in Sections 10.7.2 and 11.8.2. Soft errors in registers are becoming much more common as their charge storage diminishes. Radiation-hardening schemes for registers and memory are discussed in Sections 9.3.10 and 11.8.3.

In combinational logic, the collected charge causes a momentary glitch on the output of a gate. This glitch can propagate through downstream logic until it reaches a register. The fault does not necessarily cause a failure. The masking mechanisms include the following:

- *Logical masking*: the SEU may not trigger a sensitized path through the logic. For example, if both inputs to a NAND gate are 0, a SEU on one input does not affect the output.
- *Temporal masking*: The SEU may not reach a register at the time it is sampling.
- *Electrical masking*: The SEU may be attenuated if it is faster than the bandwidth of the gate.

In older technologies, larger gates had more charge, so they were less likely to experience upsets. Even if they did see an upset, it was likely to be attenuated by electrical masking. However, soft errors in combinational logic are a growing problem at 65 nm and below because the gates have less capacitance and higher speed [Mitra05, Rao07]. Section 6.6.2 discusses the use of redundancy to mitigate logic errors.

6.3.5 Overvoltage Failure

Tiny transistors can be easily damaged by relatively low voltages. *Overvoltage* may be triggered by excessive power supply transients or by *electrostatic discharge* (ESD) from static electricity entering the I/O pads, which can cause very large voltage and current transients (see Section 12.6.2).

Overvoltage at the gate node accelerates the oxide wearout. In extreme cases, it can cause *breakdown* and *arcing* across the thin dielectric, destroying the device. The DC oxide breakdown voltage scales with oxide thickness and absolute temperature and can be modeled as [Monsieur01]

$$V_{bd} = at_{\text{ox}} + \frac{b}{T} + V_0 \tag{6.8}$$

with typical values of $a = 1.5$ V/nm, $b = 533$ V · K, and V_0 close to 0. Breakdown occurs around 3 V under worst case (hot) conditions in a 65 nm process.

Higher-than-normal voltages applied between source and drain lead to *punchthrough* when the source/drain depletion regions touch [Tsividis99]. This can lead to abnormally high current flow and ultimately self-destructive overheating.

Both problems lead to a maximum safe voltage that can be applied to transistors. Even when catastrophic failure is avoided, high voltage accelerates the wearout mechanisms. Thus, processes specify a V_{max} for long-term reliable operation. For nanometer processes, this voltage is often much less than the I/O standard voltage, requiring a second type of transistor with thicker oxides and longer channels to endure the higher I/O voltages.

6.3.6 Latchup

Early adoption of CMOS processes was slowed by a curious tendency of CMOS chips to develop low-resistance paths between V_{DD} and GND, causing catastrophic meltdown. The phenomenon, called *latchup*, occurs when parasitic bipolar transistors formed by the substrate, well, and diffusion turn ON. With process advances and proper layout procedures, latchup problems can be easily avoided.

The cause of the latchup effect [Estreich82, Troutman86] can be understood by examining the process cross-section of a CMOS inverter, as shown in Figure 6.12(a), over which is laid an equivalent circuit. In addition to the expected nMOS and pMOS transistors, the schematic depicts a circuit composed of an npn-transistor, a pnp-transistor, and two resistors connected between the power and ground rails (Figure 6.12(b)). The npn-transistor is formed between the grounded n-diffusion source of the nMOS transistor, the p-type substrate, and the n-well. The resistors are due to the resistance through the substrate or well to the nearest substrate and well taps. The cross-coupled transistors form a bistable silicon-controlled rectifier (SCR). Ordinarily, both parasitic bipolar transistors are OFF. Latchup can be triggered when transient currents flow through the substrate during normal chip power-up or when external voltages outside the normal operating range are applied. If substantial current flows in the substrate, V_{sub} will rise, turning ON the npn-transistor. This pulls current through the well resistor, bringing down V_{well} and turning ON the pnp-transistor. The pnp-transistor current in turn raises V_{sub}, initiating a positive feedback loop with a large current flowing between V_{DD} and GND that persists until the power supply is turned off or the power wires melt.

Fortunately, latchup prevention is easily accomplished by minimizing R_{sub} and R_{well}. Some processes use a thin epitaxial layer of lightly doped silicon on top of a heavily doped

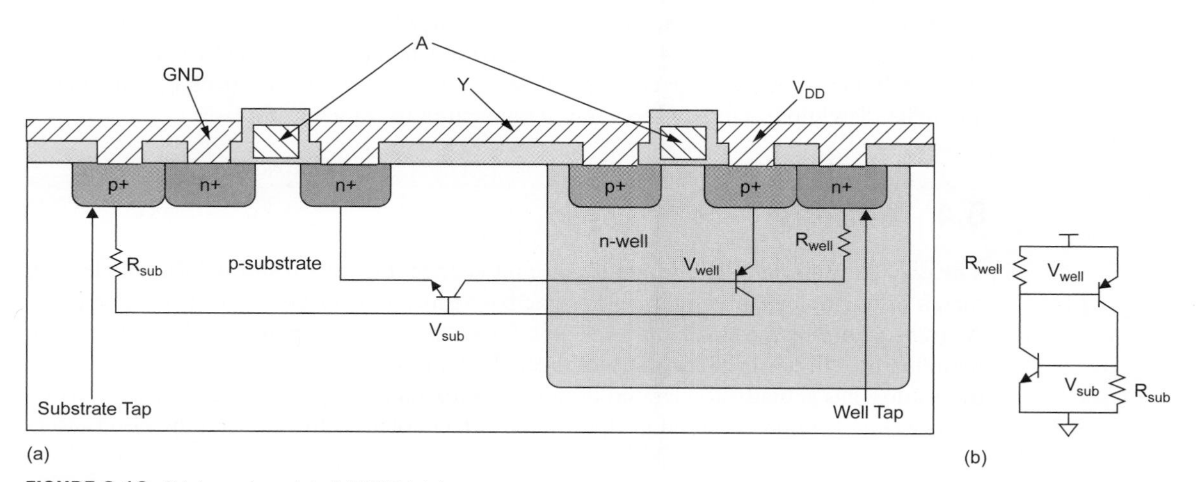

FIGURE 6.12 Origin and model of CMOS latchup

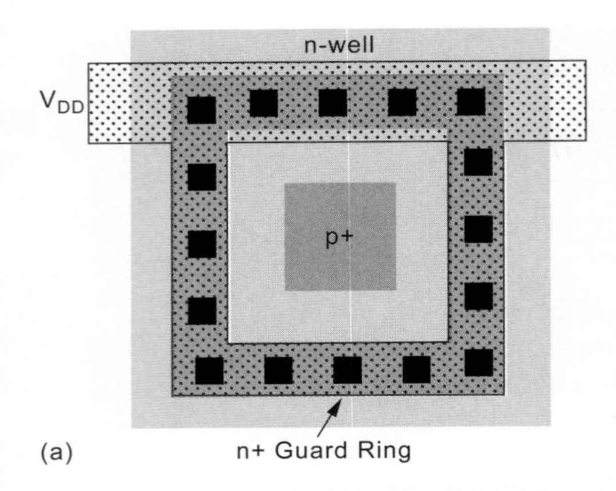

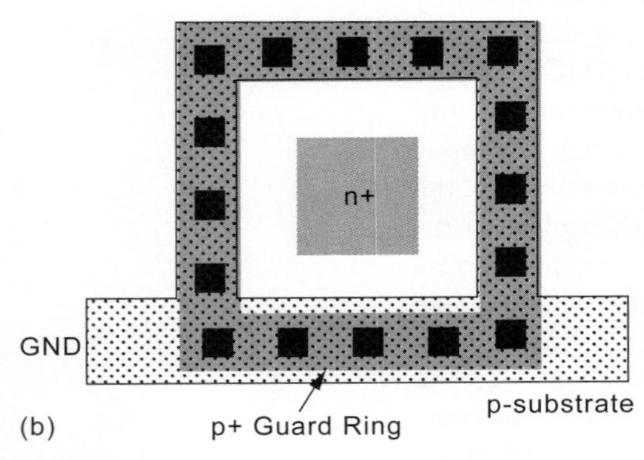

FIGURE 6.13 Guard rings

substrate that offers a low substrate resistance. Most importantly, the designer should place substrate and well taps close to each transistor. A conservative guideline is to place a tap adjacent to every source connected to V_{DD} or GND. If this is not practical, you can obtain more detailed information from the process vendor (they will normally specify a maximum distance for diffusion to substrate/well tap) or try the following guidelines:

- Every well should have at least one tap.
- All substrate and well taps should connect directly to the appropriate supply in metal.
- A tap should be placed for every 5–10 transistors, or more often in sparse areas.
- nMOS transistors should be clustered together near GND and pMOS transistors should be clustered together near V_{DD}, avoiding convoluted structures that intertwine nMOS and pMOS transistors in checkerboard patterns.

I/O pads are especially susceptible to latchup because external voltages can ring below GND or above V_{DD}, forward biasing the junction between the drain and substrate or well and injecting current into the substrate. In such cases, guard rings should be used to collect the current, as shown in Figure 6.13. Guard rings are simply substrate or well taps tied to the proper supply that completely surround the transistor of concern. For example, the n+ diffusion in Figure 6.13(b) can inject electrons into the substrate if it falls a diode drop below 0 volts. The p+ guard ring tied to ground provides a low-resistance path to collect these electrons before they interfere with the operation of other circuits outside the guard ring. *All* diffusion structures in any circuit connected to the external world must be guard ringed; i.e., n+ diffusion by p+ connected to GND or p+ diffusion by n+ connected to V_{DD}. For the ultra-paranoid, double guard rings may be employed; i.e., n+ ringed by p+ to GND, then n+ to V_{DD} or p+ ringed by n+ to V_{DD}, then p+ to GND.

SOI processes avoid latchup entirely because they have no parasitic bipolar structures. Also, processes with $V_{DD} < 1.4–2$ V are immune to latchup because the two parasitic transistors will never have a large enough voltage to sustain positive feedback [Johnston96]. Therefore, latchup has receded to a minor concern in nanometer processes.

6.4 Scaling

The only constant in VLSI design is constant change. Figure 1.6 showed the unrelenting march of technology, in which feature size has reduced by 30% every two to three years. As transistors become smaller, they switch faster, dissipate less power, and are cheaper to manufacture! Since 1995, as the technical challenges have become greater, the pace of innovation has actually accelerated because of ferocious competition across the industry. Such scaling is unprecedented in the history of technology. However, scaling also exacerbates reliability issues, increases complexity, and introduces new problems. Designers need to be able to predict the effect of this feature size scaling on chip performance to plan future products, ensure existing products will scale gracefully to future processes for cost reduction, and anticipate looming design challenges. This section examines how transistors and interconnect scale, and the implications of scaling for design. The Semiconductor Industry Association prepares and maintains an International Technology Roadmap for Semiconductors predicting future scaling. Section 6.8 gives a case study of how scaling has influenced Intel microprocessors over more than three decades.

6.4.1 Transistor Scaling

Dennard's Scaling Law [Dennard74] predicts that the basic operational characteristics of a MOS transistor can be preserved and the performance improved if the critical parameters of a device are scaled by a dimensionless factor S. These parameters include the following:

- All dimensions (in the x, y, and z directions)
- Device voltages
- Doping concentration densities

This approach is also called *constant field* scaling because the electric fields remain the same as both voltage and distance shrink. In contrast, *constant voltage* scaling shrinks the devices but not the power supply. Another approach is *lateral scaling*, in which only the gate length is scaled. This is commonly called a *gate shrink* because it can be done easily to an existing mask database for a design.

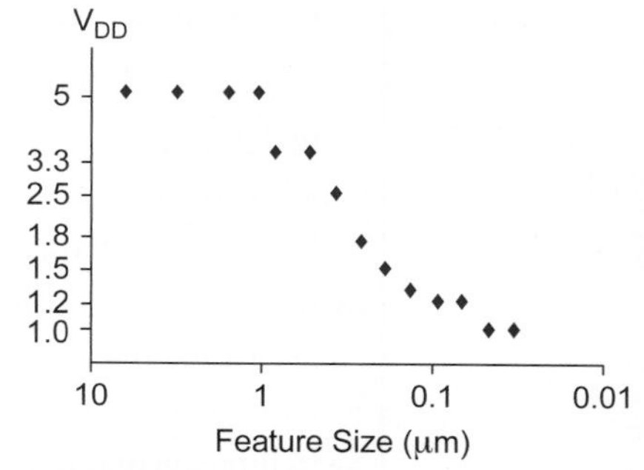

FIGURE 6.14 Voltage scaling with feature size

The effects of these types of scaling are illustrated in Table 6.4 (next page). The industry generally scales process generations with $S = \sqrt{2}$; this is also called a 30% shrink. It reduces the cost (area) of a transistor by a factor of two. A 5% gate shrink ($S = 1.05$) is commonly applied as a process becomes mature to boost the speed of components in that process.

Figure 6.14 shows how voltage has scaled with feature size. Historically, feature sizes were shrunk from 6 μm to 1 μm while maintaining a 5 V supply voltage. This *constant voltage* scaling offered quadratic delay improvement as well as cost reduction. It also maintained continuity in I/O voltage standards. Constant voltage scaling increased the electric fields in devices. By the 1 μm generation, velocity saturation was severe enough that decreasing feature size no longer improved device current. Device breakdown from the high field was another risk. And power consumption became unacceptable. Therefore, Dennard scaling has been the rule since the half-micron node. A 30% shrink with Dennard scaling improves clock frequency by 40% and cuts power consumption per gate by a factor of 2. Maintaining a constant field has the further benefit that many nonlinear factors and wearout mechanisms are essentially unaffected. Unfortunately, voltage scaling has dramatically slowed since the 90 nm generation because of leakage, and this may ultimately limit CMOS scaling.

The FO4 inverter delay will scale as $1/S$ assuming ideal constant-field scaling. As we saw in Section 6.2.4, this delay is commonly 0.5 ps/nm of the effective channel length for typical processing and worst-case environment.

TABLE 6.4 Influence of scaling on MOS device characteristics

Parameter	Sensitivity	Dennard Scaling	Constant Voltage	Lateral Scaling
Scaling Parameters				
Length: L		$1/S$	$1/S$	$1/S$
Width: W		$1/S$	$1/S$	1
Gate oxide thickness: t_{ox}		$1/S$	$1/S$	1
Supply voltage: V_{DD}		$1/S$	1	1
Threshold voltage: V_{tn}, V_{tp}		$1/S$	1	1
Substrate doping: N_A		S	S	1
Device Characteristics				
β	$\frac{W}{L}\frac{1}{t_{ox}}$	S	S	S
Current: I_{ds}	$\beta(V_{DD}-V_t)^2$	$1/S$	S	S
Resistance: R	$\frac{V_{DD}}{I_{ds}}$	1	$1/S$	$1/S$
Gate capacitance: C	$\frac{WL}{t_{ox}}$	$1/S$	$1/S$	$1/S$
Gate delay: τ	RC	$1/S$	$1/S^2$	$1/S^2$
Clock frequency: f	$1/\tau$	S	S^2	S^2
Switching energy (per gate): E	CV_{DD}^2	$1/S^3$	$1/S$	$1/S$
Switching power dissipation (per gate): P	Ef	$1/S^2$	S	S
Area (per gate): A		$1/S^2$	$1/S^2$	1
Switching power density	P/A	1	S^3	S
Switching current density	I_{ds}/A	S	S^3	S

Example 6.2

Nanometer processes have gate capacitance of roughly 1 fF/μm. If the FO4 inverter delay of a process with features size f (in nm) is 0.5 ps $\times f$, estimate the ON resistance of a unit (i.e., 4 λ wide) nMOS transistor.

SOLUTION: An FO4 inverter has a delay of $5\tau = 15RC$. Therefore,

$$RC = \frac{0.5f}{15} = \frac{f}{30}\frac{\text{ps}}{\text{nm}} \tag{6.9}$$

A unit transistor has width $W = 2f$ and thus capacitance of $C = 2f$ fF/μm. Solving for R,

$$R = \left(\frac{f}{30}\frac{\text{ps}}{\text{nm}}\right)\left(\frac{1}{2f}\frac{\mu\text{m}}{\text{fF}}\right) = 16.6\ \text{k}\Omega \tag{6.10}$$

Note that this is independent of feature size. The resistance of a unit transistor is roughly independent of feature size, while the gate capacitance decreases with feature size. Alternatively, the capacitance per micron is roughly independent of feature size while the resistance · micron decreases with feature size.

6.4.2 Interconnect Scaling

Wires also tend to be scaled equally in width and thickness to maintain an aspect ratio close to 2.[1] Table 6.5 shows the resistance, capacitance, and delay per unit length. Wires

TABLE 6.5 Influence of scaling on interconnect characteristics

Parameter	Sensitivity	Scale Factor
Scaling Parameters		
Width: w		$1/S$
Spacing: s		$1/S$
Thickness: t		$1/S$
Interlayer oxide height: h		$1/S$
Die size		D_c
Characteristics per Unit Length		
Wire resistance per unit length: R_w	$\frac{1}{wt}$	S^2
Fringing capacitance per unit length: C_{wf}	$\frac{t}{s}$	1
Parallel plate capacitance per unit length: C_{wp}	$\frac{w}{h}$	1
Total wire capacitance per unit length: C_w	$C_{wf} + C_{wp}$	1
Unrepeated RC constant per unit length: t_{wu}	$R_w C_w$	S^2
Repeated wire RC delay per unit length: t_{wr} (assuming constant field scaling of gates)	$\sqrt{RCR_w C_w}$	$\sqrt{S}$
Crosstalk noise	$\frac{w}{h}$	1
Energy per bit per unit length: E_w	$C_w V_{DD}^2$	$1/S^2$
Local/Semiglobal Interconnect Characteristics		
Length: l		$1/S$
Unrepeated wire RC delay	$l^2 t_{wu}$	1
Repeated wire delay	$l t_{wr}$	$\sqrt{1/S}$
Energy per bit	$l E_w$	$1/S^3$
Global Interconnect Characteristics		
Length: l		D_c
Unrepeated wire RC delay	$l^2 t_{wu}$	$S^2 D_c^2$
Repeated wire delay	$l t_{wr}$	$D_c \sqrt{S}$
Energy per bit	$l E_w$	D_c / S^2

[1]Historically, wires had a lower aspect ratio and could be scaled in width but not thickness. This helped control RC delay. However, coupling capacitance becomes worse at higher aspect ratios and thus crosstalk limits wires to an aspect ratio of 2–3 before the noise is hard to manage.

can be classified as local, semiglobal, and global. *Local* wires run within functional units and use the bottom layers of metal. *Semiglobal* (or *scaled*) wires run across larger blocks or cores, typically using middle layers of metal. Both local and semiglobal wires scale with feature size. *Global* wires run across the entire chip using upper levels of metal. For example, global wires might connect cores to a shared cache. Global wires do not scale with feature size; indeed, they may get longer (by a factor of D_c, on the order of 1.1) because die size has been gradually increasing.

Most local wires are short enough that their resistance does not matter. Like gates, their capacitance per unit length is remaining constant, so their delay is improving just like gates. Semiglobal wires long enough to require repeaters are speeding up, but not as fast as gates. This is a relatively minor problem. Global wires, even with optimal repeaters, are getting slower as technology scales. The time to cross a chip in a nanometer process can be multiple cycles, and this delay must be accounted for in the microarchitecture.

Observe that when wire thickness is scaled, the capacitance per unit length remains constant. Hence, a reasonable initial estimate of the capacitance of a minimum-pitch wire is about 0.2 fF/μm, independent of the process. In other words, wire capacitance is roughly 1/5 of gate capacitance per unit length.

6.4.3 International Technology Roadmap for Semiconductors

The incredible pace of scaling requires cooperation among many companies and researchers both to develop compatible process steps and to anticipate and address future challenges before they hold up production. The Semiconductor Industry Association (SIA) develops and updates the International Technology Roadmap for Semiconductors (ITRS) [SIA07] to forge a consensus so that development efforts are not wasted on incompatible technologies and to predict future needs and direct research efforts. Such an effort to predict the future is inevitably prone to error, and the industry has scaled feature sizes and clock frequencies more rapidly than the roadmap predicted in the late 1990s. Nevertheless, the roadmap offers a more coherent vision than one could obtain by simply interpolating straight lines through historical scaling data.

The ITRS forecasts a major new technology generation, also called *technology node*, approximately every three years. Table 6.6 summarizes some of the predictions, particularly for high-performance microprocessors. However, serious challenges lie ahead, and major breakthroughs will be necessary in many areas to maintain the scaling on the roadmap.

TABLE 6.6 Predictions from the 2007 ITRS

Year	2009	2012	2015	2018	2021
Feature size (nm)	**34**	**24**	**17**	**12**	**8.4**
L_{gate} (nm)	20	14	10	7	5
V_{DD} (V)	1.0	0.9	0.8	0.7	0.65
Billions of transistors/die	1.5	3.1	6.2	12.4	24.7
Wiring levels	12	12	13	14	15
Maximum power (W)	198	198	198	198	198
DRAM capacity (Gb)	2	4	8	16	32
Flash capacity (Gb)	16	32	64	128	256

6.4.4 Impacts on Design

One of the limitations of first-order scaling is that it gives the wrong impression of being able to scale proportionally to zero dimensions and zero voltage. In reality, a number of factors change significantly with scaling. This section attempts to peer into the crystal ball and predict some of the impacts on design for the future. These predictions are notoriously risky because chip designers have had an astonishing history of inventing ingenious solutions to seemingly insurmountable barriers.

6.4.4.1 Improved Performance and Cost The most positive impact of scaling is that performance and cost are steadily improving. System architects need to understand the scaling of CMOS technologies and predict the capabilities of the process several years into the future, when a chip will be completed. Because transistors are becoming cheaper each year, architects particularly need creative ideas of how to exploit growing numbers of transistors to deliver more or better functions. When transistors were first invented, the best predictions of the day suggested that they might eventually approach a fifty-cent manufacturing cost. Figure 6.15 plots the number of transistors and average price per transistor shipped by the semiconductor industry over the past three decades [Moore03]. In 2008, you could buy more than 100,000 transistors for a penny, and the price of a transistor is expected to reach a microcent by 2015 [SIA07].

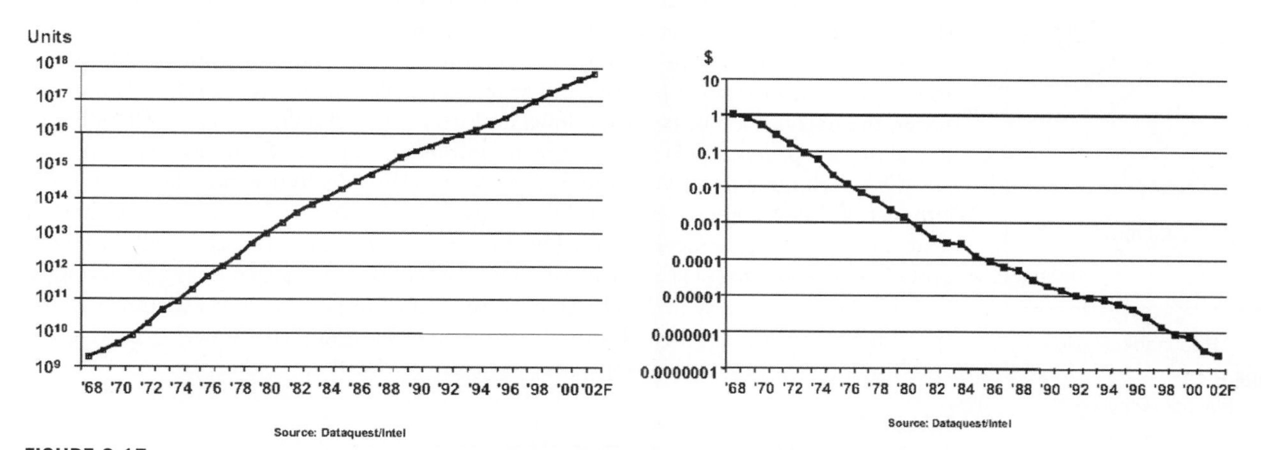

FIGURE 6.15 Transistor shipments and average price (© IEEE 2003.)

6.4.4.2 Interconnect Scaled transistors are steadily improving in delay, but scaled global wires are getting worse. Figure 6.16, taken from the 1997 Semiconductor Industry Association Roadmap [SIA97], forecast the sum of gate and wire bottoming out at the 250 or 180 nm generation and getting worse thereafter. The wire problem motivated a number of papers predicting the demise of conventional wires. However, the plot is misleading in two ways. First, the "gate" delay is shown for a single unloaded transistor (delay = RC) rather than a realistically loaded gate (e.g., an FO4 inverter delay = $15RC$). Second, the wire delays shown are for fixed lengths, but as technology scales, most local wires connecting gates within a unit also become shorter.

In practice, for short wires, such as those inside a logic gate, the wire RC delay is negligible and will remain so for the foreseeable future. However, the long wires present a considerable challenge. It is no longer possible to send a signal from one side of a large,

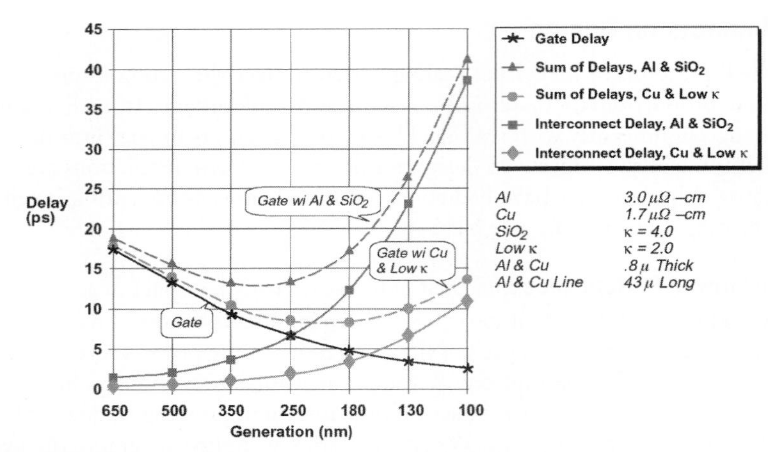

FIGURE 6.16 Gate and wire delay scaling (Reprinted from [SIA97] with permission of the Semiconductor Industry Association.)

high-performance chip to another in a single cycle. Also, the "reachable radius" that a signal can travel in a cycle is steadily getting smaller, as shown in Figure 6.17. This requires that microarchitects understand the floorplan and budget multiple pipeline stages for data to travel long distances across the die.

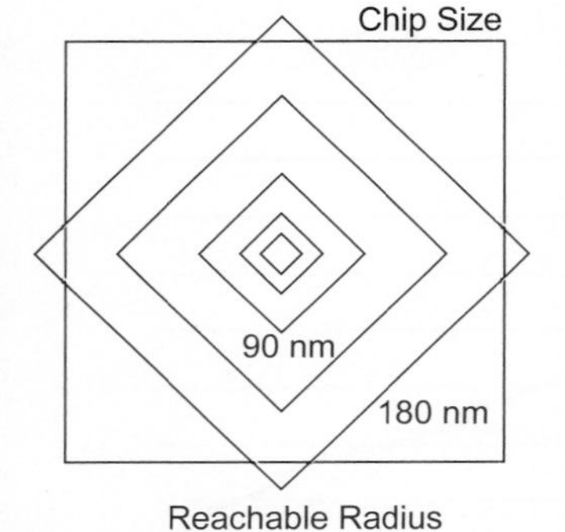

FIGURE 6.17 Reachable radius scaling

Repeaters help somewhat, but even so, interconnect does not keep up. Moreover, the "repeater farms" must be allocated space in the floorplan. As scaled gates become faster, the delay of a repeater goes down and hence, you should expect it will be better to use more repeaters. This means a greater number of repeater farms are required.

One technique to alleviate the interconnect problem is to use more layers of interconnect. Table 6.7 shows the number of layers of interconnect increasing with each generation in TSMC processes. The lower layers of interconnect are classically scaled to provide high-density short connections. The higher layers are scaled less aggressively, or possibly even reverse-scaled to be thicker and wider to provide low-resistance, high-speed interconnect, good clock distribution networks, and a stiff power grid. Copper and low-k dielectrics were also introduced to reduce resistance and capacitance.

TABLE 6.7 Scaling of metal layers in TSMC processes

Process (nm)	Metal Layers
500	3 (Al)
350	4 (Al)
250	5 (Al)
180	6 (Al, low-k)
150	7 (Cu, low-k)
130	8 (Cu, low-k)
90	9 (Cu, low-k)
65	10 (Cu, low-k)
45	10 (Cu, low-k)

Blocks of 50–100 Kgates (1 *Kgate* = 1000 3-input NAND gates or 6000 transistors) will continue to have reasonably short internal wires and acceptably low wire RC delay [Sylvester98]. Therefore, large systems can be partitioned into blocks of roughly this size with repeaters inserted as necessary for communication between blocks.

6.4.4.3 Power In classical constant field scaling, dynamic power density remains constant and overall chip power increases only slowly with die size. In practice, microprocessor power density skyrocketed in the 1990s because extensive pipelining increased clock frequencies much faster than classical scaling would predict and because V_{DD} is somewhat higher than constant field scaling would demand. High-performance microprocessors bumped against the limit of about 150 W that a low-cost fan and heat sink can dissipate. This trend has necessarily ended, and now designers aim for the maximum performance under a power envelope rather than for the maximum clock rate.

Static power is a more serious limitation. Subthreshold leakage power increased exponentially as threshold voltages decreased, and has abruptly changed from being negligible to being a substantial fraction of the total. Section 4.4.2 demonstrated that static power should account for approximately one-third of total power to minimize the energy-delay product. Higher leakage has required the adoption of power gating techniques to control power during sleep mode, especially for battery-powered systems. To limit leakage to −100 nA/μm, V_t has remained fairly constant near 300 mV. The V_{DD}/V_t ratio has dropped from about 5 in older processes toward 3, and EQ (4.27) showed that it may go as low as 2 for best EDP. As the ratio decreased, circuits with threshold drops have ceased to be viable. Performance suffers as the gate overdrive decreases, so V_{DD} scaling has slowed below the 90 nm node. This increases the electric fields, exacerbating velocity saturation and reliability problems. It also raises dynamic power.

Gate leakage current is also important for oxides of less than 15–20 Å, and essentially precludes scaling oxides below 10 Å. If oxides thickness does not scale with the other dimensions, the ratio of ON to OFF current degrades. High-k metal gates solve the problem by offering a lower effective thickness at a higher physical thickness.

Even if power remains constant, lower supply voltage leads to higher current density. This in turn causes higher IR drops and di/dt noise in the supply network (see Sections 8.3.5 and 12.3). These factors lead to more pins and metal resources on a chip being required for the power distribution networks.

All considered, scaling is being squeezed from many directions by power limitations. Some manufacturers are finding that conventional scaling can offer performance or power benefits, but not both [Muller08]. Intel is aggressively introducing new materials such as high-k metal gates and strained silicon to continue to see both performance and power benefits from scaling at the 45 nm node. Even so, the frenetic pace of Moore's Law may begin slowing at last.

6.4.4.4 Variability As transistors shrink, the spread in parameters such as channel length and threshold voltage increases. Variability has moved from being an analog nuisance to becoming a key factor in mainstream digital circuits. Designers are forced to employ wider guard bands to ensure that an acceptable fraction of chips meet specifications. Later sections of this chapter examine variability and variation-tolerant design techniques in more detail.

6.4.4.5 Productivity The number of transistors that fit on a chip is increasing faster than designer productivity (gates/week). This leads to design teams of increasing size, difficulty recruiting enough experienced engineers when the economy is good, and a trend to outsource to locations such as India where more engineering graduates are available. (Bangalore was once considered a low-cost labor market as well, but salaries have been increasing exponentially because of demand and may approach global parity within the decade.) It has driven a search for design methodologies that maximize productivity, even at the expense of performance and area. Now most chips are designed using synthesis and place

& route; the number of situations where custom circuit design is affordable is diminishing. In other words, creativity is shifting from the circuit to the systems level for many designs. On the other hand, performance is still king in the microprocessor world. Design teams in that field are approaching the size of automotive and aerospace teams because the development cost is justified by the size of the market. This drives a need for engineering managers who are skilled in leading such large organizations.

The number of 50–100 Kgate blocks is growing, even in relatively low-end systems. This demands greater attention to floorplanning and placement of the blocks.

One of the key tools to solve the productivity gap is design reuse. Intellectual property (IP) blocks can be purchased and used as black boxes within a system-on-chip (SOC) in much the same way chips are purchased for a board-level design. Early problems with validation of IP blocks have been partially overcome, but the market for IP still lacks transparency.

6.4.4.6 Physical Limits How far will CMOS processes scale? A minimum-sized transistor in a 32 nm process has an effective channel length of less than 100 Si atoms. The gate oxide is only 4 atoms thick. The channel contains approximately 50 dopant atoms. It is clear that scaling cannot continue indefinitely as dimensions reach the atomic scale. Numerous papers have been written forecasting the end of silicon scaling [Svensson03]. For example, in 1972, the limit was placed at the 0.25 μm generation because of tunneling and fluctuations in dopant distributions [Hoeneisen72, Mead80]; at this generation, chips were predicted to operate at 10–30 MHz! In 1999, IBM predicted that scaling would nearly grind to a halt beyond the 100 nm generation in 2004 [Davari99].

In the authors' experience, seemingly insurmountable barriers have seemed to loom about a decade away. Reasons given for these barriers have included the following:

- Subthreshold leakage at low V_{DD} and V_t
- Tunneling current through thin oxides
- Poor I-V characteristics due to DIBL and other short channel effects
- Dynamic power dissipation
- Lithography limitations
- Exponentially increasing costs of fabrication facilities and mask sets
- Electromigration
- Interconnect delay
- Variability

Dennard scaling is beginning to groan under the weight of its own success. At the 32 nm node and beyond, the performance and power benefits of geometrical scaling are starting to diminish as the engineering costs continue to escalate. Nevertheless, scaling still provides a competitive advantage in a cutthroat industry. Improved structures such as copper wires, low-k dielectrics, strained silicon, high-k metal gates, and 3D integration provide benefits independent of reduced feature size. Novel structures are under intensive research. A large number of extremely talented people are continuously pushing the limits and hundreds of billions of dollars are at stake, so we are reluctant to bet against the future of scaling.

6.5 Statistical Analysis of Variability

Variability was introduced in Section 6.2. Die-to-die variability is relatively straightforward to handle with process corners defining the range of acceptable variations (e.g., 3σ); designing to ensure that all chips within the corners meet speed, power, and functionality requirements; and rejecting the few chips that fall outside the corners. Within-die variability is more complicated because a chip has millions or billions of transistors. Even if the die itself is in the TT corner, some transistors are likely to stray at least 5σ from the mean. To achieve acceptable yield, most chips with a few such extreme variations must still be acceptable. Static CMOS gates are so robust that they generally function correctly even when parameters vary enormously. However, their delay and leakage will change, which affects the delay and leakage of the entire chip. Special circuits such as memories, analog circuits, and trickier circuit families may fail entirely under extreme variation.

This section revisits within-die variability from a statistical point of view. It begins with a review of the properties of random variables that are essential for understanding on-chip variability. Then, it examines the sources of variability in more detail. Finally, it considers the impact of variation on circuit delay, energy, and functionality.

6.5.1 Properties of Random Variables

The *probability distribution function* (*PDF*) $f(x)$ specifies the probability that the value of a continuous random variable X falls in a particular interval:

$$P\left[a < X \leq b\right] = \int_a^b f(x)dx \tag{6.11}$$

The *cumulative distribution function* (*CDF*) $F(x)$ specifies the probability that X is less than some value x:

$$F(x) = P\left(X < x\right) = \int_{-\infty}^{x} f(u)du \tag{6.12}$$

Thus, the PDF is the slope of the CDF at any given point.

$$f(x) = \frac{d}{dx}F(x) \tag{6.13}$$

The *mean* or *expected value*, written as $\overline{X}$ or $E[X]$, is the average value of X.

$$\overline{X} = E\left[X\right] = \int_{-\infty}^{\infty} x f(x)dx \tag{6.14}$$

The *standard deviation* $\sigma(X)$ measures the dispersion; i.e., how far X is expected to vary from its mean.

$$\sigma(X) = \sqrt{E\left[\left(x - \overline{X}\right)^2\right]} = \sqrt{\int_{-\infty}^{\infty} \left(x - \overline{X}\right)^2 f(x)dx} \tag{6.15}$$

It is often more convenient to deal with the *variance*, $\sigma^2(X)$, to avoid the square root.

When studying variability in circuits, we are usually interested in the variation from the nominal (mean) value. Thus, a random variable X can be written as $X = \overline{X} + X_v$, where $\overline{X}$ is the mean and X_v is a random variable with zero mean describing the variation. Thus, we will focus on such *zero-mean random variables.*

6.5.1.1 Uniform Random Variables Figure 6.1(a) shows a uniform random variable with zero mean. A uniform random variable distributed between $-a$ and a has the following PDF, CDF, and variance:

$$
\begin{aligned}
f(x) &= \begin{cases} \dfrac{1}{2a} & -a \le x \le a \\ 0 & \text{otherwise} \end{cases} \\
F(x) &= \begin{cases} 0 & x < -a \\ \dfrac{x-a}{2a} & -a \le x \le a \\ 1 & x > a \end{cases} \\
\sigma^2(X) &= \frac{a^2}{3}
\end{aligned}
\tag{6.16}
$$

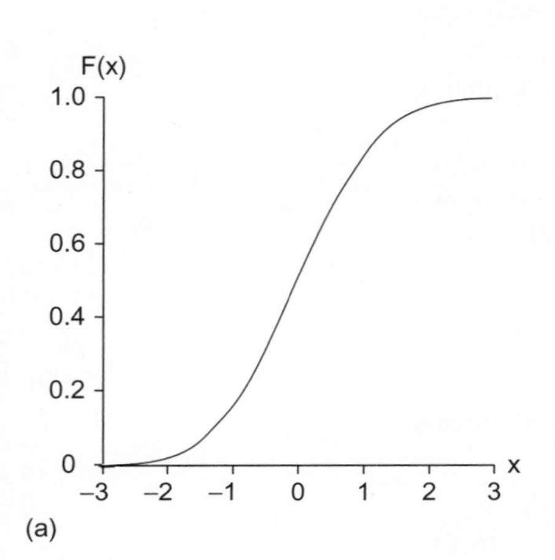

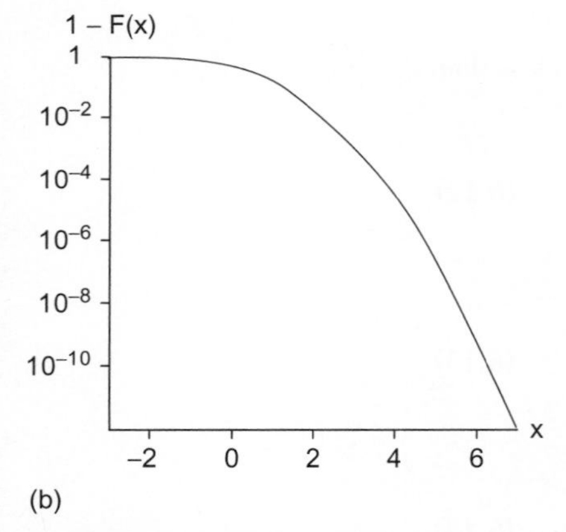

FIGURE 6.18 Cumulative distribution function for a standard normal random variable

6.5.1.2 Normal Random Variables Figure 6.1(b) shows a normal random variable. It is convenient to shift the variable to have zero mean, then scale it to have a standard deviation $\sigma = 1$. The result is called a standard normal distribution and has the following PDF, CDF, and variance:

$$
\begin{aligned}
f(x) &= \frac{1}{\sqrt{2\pi}} e^{-\frac{1}{2}x^2} \\
F(x) &= \tfrac{1}{2}\left[1 + \operatorname{erf}\left(\frac{x}{\sqrt{2}}\right)\right] \\
\sigma^2(X) &= 1
\end{aligned}
\tag{6.17}
$$

where erf(x) is the *error function.*[2] For example, a threshold voltage with a mean of 0.3 V and a standard deviation of 0.025 V can be expressed as $V_t = 0.3 + 0.025\,X$, where X is a standard normal random variable.

A component may fail if a parameter varies too far. The CDF describes the probability that the parameter is less than an upper bound. It is shown in Figure 6.18 and handy values are given in Table 6.8. For example, a chip may be rejected if its delay is more than 3σ above nominal, and this event has a probability of 0.135%.

6.5.1.3 Sums of Random Variables Chip designers are frequently interested in quantities such as path delay that are the sum of independent random variables. The mean is the sum of the means. If the distributions are normal, the sum is another normal distribution with

[2]Microsoft Excel and other spreadsheets define *erf*, which is more convenient than looking it up in a mathematical handbook. In some versions of Excel, you must first select *Add-Ins* from the *Tools* menu and check *Analysis ToolPak* to use the function.

TABLE 6.8 CDF of standard normal random variables

x	*F*(*x*)	1 – *F*(*x*)
1	0.8413	1.59×10^{-1}
2	0.9772	2.28×10^{-2}
3	0.998650	1.35×10^{-3}
4	0.9999683	3.17×10^{-5}
5	0.999999713	2.87×10^{-7}
6	0.999999999013	9.87×10^{-10}

a variance equal to the sum of the variances:

$$\sigma^2 = \sum_i \sigma_i^2 \tag{6.18}$$

Even if the distributions are not normal, the Central Limit Theorem states that EQ (6.18) still holds as the number of variables gets large. Therefore, it is often a reasonable approximation to replace uniformly distributed variables with normal variables that have the same variance.

6.5.1.4 Maximum of Random Variables The cycle time is set by the longest of many possible critical paths that have nearly equal nominal delays. Let M be the maximum of N random variables with independent standard normal distributions. M is *not* normally distributed, but its expected value and standard deviation can be found numerically as a function of N, as given in Table 6.9. As N increases, the expected maximum increases (roughly logarithmically for big N) and its standard deviation decreases. Figure 6.19 shows how the distribution of longest paths change with the number of nearly critical paths. As the number of paths increase, they form a tight wall with an expected worst-case delay that can be significantly longer than nominal [Bowman02]. [Clark61] extends this tabular approach to handle random variables with correlations and unequal standard deviations.

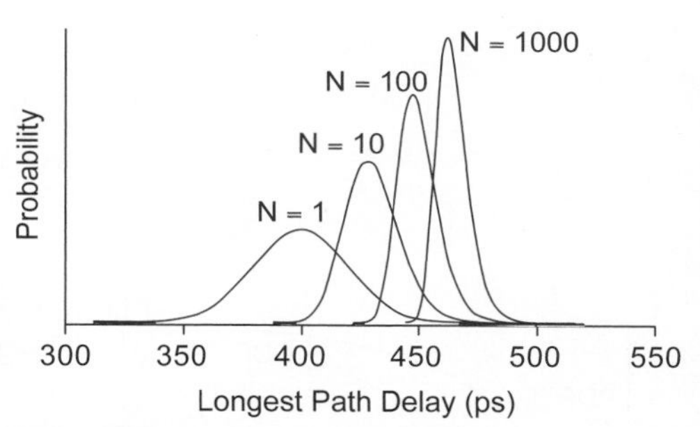

FIGURE 6.19 Delay distributions of typical and longest paths

TABLE 6.9 Behavior of maximum of normal variables

N	*E*[*M*]	*σ*(*M*)
2	0.56	0.82
10	1.54	0.59
100	2.50	0.43
1000	3.24	0.35
10,000	3.85	0.30
100,000	4.40	0.28

Example 6.3

A large chip has 100 paths that are all nearly critical. Each path has a nominal delay of 400 ps and a standard deviation of 20 ps. What is the expected delay of the critical path, and what is the standard deviation in this delay?

SOLUTION: According to Table 6.9, the maximum of 100 standard normal random variables has an expected value of 2.50 and a standard deviation of 0.43. Thus, the expected critical path delay is $400 + 2.50 \times 20 = 450$ ps, and the standard deviation is only $0.43 \times 20 = 9$ ps.

6.5.1.5 Exponential of Normal Random Variables According to EQ (2.42), subthreshold leakage is exponentially related to the threshold voltage. If Y is a normally distributed random variable with mean μ and variance σ^2, then $X = e^Y$ has a *log-normal distribution* with the following properties:

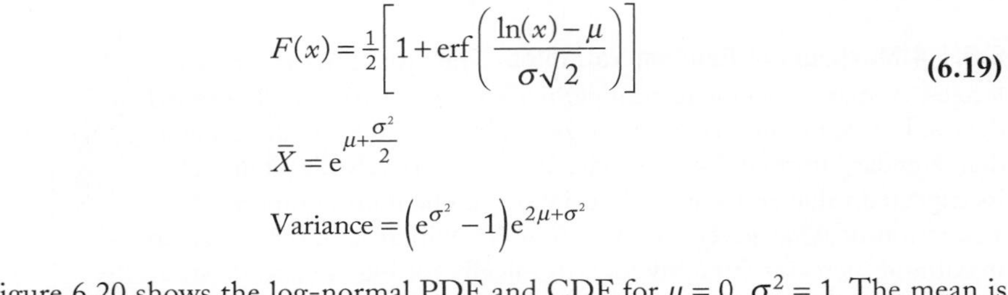

$$
\begin{aligned}
f(x) &= \frac{1}{x\sigma\sqrt{2\pi}} e^{\frac{-(\ln(x)-\mu)^2}{2\sigma^2}} \\
F(x) &= \tfrac{1}{2}\left[1+\mathrm{erf}\left(\frac{\ln(x)-\mu}{\sigma\sqrt{2}}\right)\right] \\
\bar{X} &= e^{\mu+\frac{\sigma^2}{2}} \\
\text{Variance} &= \left(e^{\sigma^2}-1\right)e^{2\mu+\sigma^2}
\end{aligned}
\quad (6.19)
$$

Figure 6.20 shows the log-normal PDF and CDF for $\mu = 0$, $\sigma^2 = 1$. The mean is $x = e^{0.5} = 1.65$ because of the long tail.

FIGURE 6.20 PDF of standard log-normal variable

6.5.1.6 Monte Carlo Simulation For many problems of realistic concern, closed form PDFs do not exist. Monte Carlo simulations are used to evaluate the impact of variations. Such a simulation involves generating N scenarios. In each scenario, each of the variables is given a random value based on its distribution, then the simulation is performed and the characteristics of interest are measured. The collected results of all the scenarios describe the effect of variation on the system. For example, the delay distribution shown in Figure 6.19 can be obtained from the histogram of delays for a large number of simulations of a large number of paths.

6.5.2 Variation Sources

Section 6.2 introduced the major process and environmental variation sources considered when defining design corners. On closer inspection, we can add variations from circuit operation and CAD limitations. Circuit variations include data-dependent crosstalk, simultaneous input switching, and wearout. CAD limitations include imperfect models for SPICE and timing analysis, and approximations made during parasitic extraction.

Variations can be characterized as systematic, random, drift, and jitter. *Systematic* variations have a quantitative relationship with a source. For example, an ion implanter may systematically deliver a different dosage to different regions of a wafer. Similarly, polysilicon gates may systematically be etched narrower in regions of high polysilicon density than low density. Systematic variability can be modeled and nulled out at design time; for example, in principle, you could examine a layout database and calculate the etching variations as a function of nearby layout, then simulate a circuit with suitably adjusted gate lengths. *Random* variations include those that are truly random (such as the number of dopant atoms implanted in a transistor), those whose sources are not fully understood, and those that are too costly to model. Etching variations are usually treated as random because extraction is not worth the effort. Random variations do not change with time, so they can be nulled out by a single calibration step after manufacturing. *Drift*, notably aging and temperature variation, change slowly with time as compared to the operating frequency of the system. Drift can again be nulled by compensation circuits, but such circuits must recalibrate faster than the drift occurs. *Jitter*, often from voltage variations or crosstalk, is the most difficult cause of mismatch. It occurs at frequencies comparable to or faster than the system clock and therefore may not be eliminated through feedback. Systematic and random variations are considered *static*, while drift and jitter are *dynamic*.

The yield is the fraction of manufactured chips that work according to specification. Some chips fail because of gross problems such as open or short circuits caused by contaminants during manufacturing. This is called the *functional* yield. Other operational chips are rejected because they are too slow or consume too much power or have insufficient noise margin. This is called the *parametric yield*. Increasing variability tends to reduce parametric yield, but designers are introducing adaptive techniques to compensate.

According to *Pelgrom's model*, the standard deviation of most random WID variability sources is inversely proportional to the square root of the area (WL) of the transistor [Pelgrom89]. This makes sense intuitively because variations tend to average out over a larger area, and the model is well-supported experimentally.

A good design manual for a nanometer fabrication process will specify the major variation sources and their distributions.

6.5.2.1 Channel Length Channel length varies within-die because of systematic *across-chip linewidth variation* (ACLV) and random *line edge roughness*. ACLV is caused by lithography limitations and by pattern-dependent etch rates.

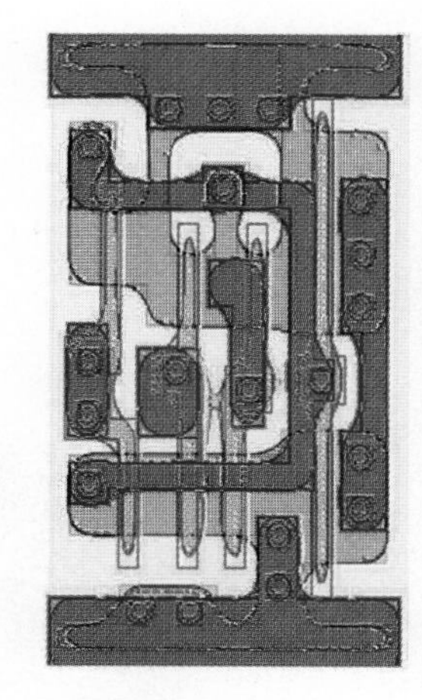

FIGURE 6.21 Discrepancy between drawn and printed layout of NAND gate caused by subwavelength lithography (© 2007 Larry Pileggi, reprinted with permission.)

Figure 6.21 shows the desired layout and actual printed circuit for a NAND gate in a nanometer process. Subwavelength lithography cannot perfectly reproduce the intended polysilicon shapes. The polysilicon tends to be wider near contacts and narrower near its end, causing transistor lengths to deviate from their intended value. In severe cases, the variation can cause shorts between neighboring polysilicon lines, as seen in the center of the gate. Diffusion rounding also changes the transistor widths. Resolution enhancement techniques partially compensate, but some error remains.

The etch rate decreases slightly with the amount of polysilicon that must be etched. *Nested* polysilicon lines are those surrounded by closely spaced parallel lines, while *isolated* lines are those far from other polysilicon. Nested polysilicon tends to be narrower, while isolated lines tend to be wider. Density rules limit the etch rate variation, but again, some remains.

Channel lengths display spatial correlation, called the *proximity effect*. Two adjacent transistors are better matched than two transistors that are hundreds of microns apart. One of the reasons for this is large-scale etch rate variation, where etch rates depend on the average polysilicon density of a large area.

Horizontal polysilicon lines may print differently than vertical lines. This *orientation effect* can be exacerbated when resolution enhancement techniques such as off-axis illumination and double patterning are applied.

Lithography has a shallow depth of focus, leading to variations dependent on the planarity of the underlying wafer. The *topography effect* describes the variation of polysilicon lines dependent on step-height differences between the diffusion and STI regions it crosses.

Many of the factors in ACLV can be controlled by the designer. In nanometer processes, it is good practice to draw gates exclusively in one orientation to avoid variation from the orientation effect. Some processes may require that minimum-width polysilicon run unidirectionally, even where it does not form a gate. In critical circuits such as memories, the density variations are controlled because all the cells are identical. The edge of the array is usually surrounded by one or more dummy rows and columns to provide even more uniformity. Although the remaining variation is systematic and might be predicted by detailed simulation of the lithography and etch effects, it is usually too difficult to model and is thus treated as random. The variance of channel length can be found by summing the variances of the relevant factors.

FIGURE 6.22 SEM of polysilicon showing line edge roughness (Courtesy of Texas Instruments.)

Figure 6.22 shows the line edge roughness (LER) of a polysilicon gate. Roughness, ranging on a scale from atomic to 100 nm, is becoming significant as transistors become so narrow. The standard deviation in channel length caused by LER is inversely proportional to the square root of channel width because variations tend to average out along a wide transistor. [Asenov07] reports variations of about 4 nm in a 35 nm process.

Channel length variation is often expressed as a percentage of the nominal (mean) channel length because delay variations are proportional to this percentage variation. For example, [Misaka96] reported a 0.02 μm standard deviation of channel length in a 0.4 μm process, corresponding to $\sigma/\mu = 5\%$. The amount of variation is highly process-dependent and a foundry should be able to supply detailed variation statistics for processes where it is significant. Controlling variation as a fraction of the nominal value is not getting easier as dimensions shrink. The 2007 International Technology Roadmap for Semiconductors estimates a target $\sigma/\mu = 4\%$.

Corner rounding on the diffusion layer affects the transistor widths. This tends to be a less important effect because the widths are generally longer. For good matching, avoid minimum-width transistors.

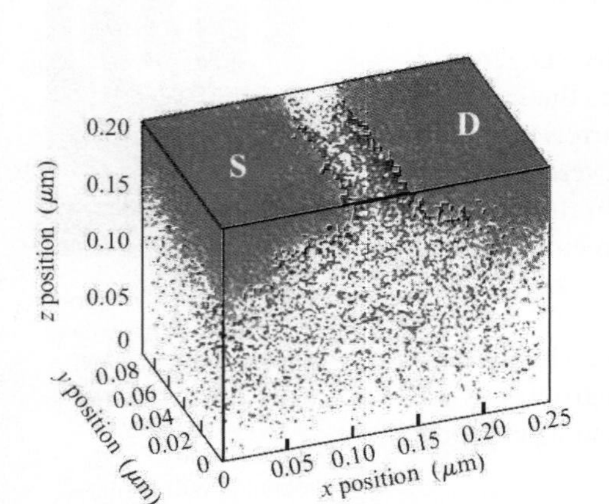

FIGURE 6.23 Random placement of dopant atoms in a 50 nm process. Adapted from [Bernstein06]. (Courtesy of International Business Machines Corporation. Unauthorized use not permitted.)

6.5.2.2 Threshold Voltage The threshold voltage is determined by the number and location of dopant atoms implanted in the channel or halo region. This ion implantation is a stochastic process, leading to random dopant fluctuations (RDF) that cause V_t to vary [Keys75, Tang97]. For example, Figure 6.23 shows the simulated placement of n-type (black) and p-type (blue) dopant atoms along an nMOS transistor in a 50 nm process [Bernstein06]. The variations have become large in nanometer processes because the number of dopant atoms is small.

The standard deviation of V_t caused by RDF can be estimated using [Mizuno94, Stolk98]

$$\sigma_{V_t} = \frac{t_{\text{ox}}}{\varepsilon_{\text{ox}}} \frac{\sqrt[4]{q^3 \varepsilon_{\text{si}} \phi_b N_a}}{\sqrt{2LW}} = \frac{A_{V_t}}{\sqrt{LW}} \tag{6.20}$$

where N_a is the doping level, $\varepsilon_{\text{si}} = 11.8\varepsilon_0$ and ϕ_b is the surface potential. This standard deviation obeys Pelgrom's model that it is inversely proportional to transistor area. High-V_t transistors have higher effective channel doping, so σ_{V_t} increases with V_t [Agarwal07]. Under Dennard scaling, the change in effective oxide thickness cancels the change in square root of area, so the variability scales with the fourth root of the doping level.

[Agarwal07] predicts that the standard deviation in threshold volatage for minimum-sized device is approximately 10 mV in a 90 nm process, 30 mV in a 50 nm process, and 40 mV in a 25 nm process. High-V_t transistors have higher effective channel doping, so the standard deviation increases slightly with V_t. [Bernstein06] reports a standard deviation of 26 mV in an IBM 90 nm process for a minimum-sized transistor. High-k metal gate transistors use the gate work function to control V_t, have a higher dielectric constant, and need a lower halo doping, so they have a smaller threshold variation. [Itoh09] reports A_{vt} of 1.0 – 2.5 mV · μm for 45 nm process with metal gates, and predicts a lower bound of 0.4 mV · μm in future processes.

V_t is also sensitive to the channel length on account of short channel effects. This can be modeled as a threshold variation proportional to the channel length variation. It is important because a systematic decrease in L will cause a systematic decrease in V_t that exponentially increases leakage.

6.5.2.3 Oxide Thickness Average oxide thickness t_{ox} is controlled with remarkable precision, to a fraction of an atomic layer. [Koh01] reports a variation of 0.1 Å in a 10 Å oxide layer. Device variations caused by oxide thickness are presently minor compared to those caused by channel length and threshold voltage. For example, [Bernstein06] finds that they can be accounted for by raising the standard distribution of V_t by 10%.

6.5.2.4 Layout Effects As mentioned in Section 15.2.3, transistors near the edge of a well may have different threshold voltages caused by the well-edge proximity effect. The significance is process-dependent. [Kanamoto07] finds that transistors close to the edge of a well in a 65 nm process may have delays up to 10% higher.

Section 15.4.1.4 described how strain can be used to increase the carrier mobility to improve ON current. Various mechanisms are employed in different processes to create the strain. For example, some processes use the shallow trench isolation (STI) to introduce stress on the transistors. Variations in the layout may change the amount of stress and hence the mobility [Topaloglu07]. This is called *across-chip mobility variation.*

6.5.3 Variation Impacts

Variations affect transistor ON and OFF current, which in turn influence delay and energy. This section offers a first-order analysis of the effects to give some intuition about the effects. More sophisticated analyses to predict parametric yield are given in [Najm07, Agarwal07b]. In practice, Monte Carlo simulations are commonly used to assess the impact of variation.

6.5.3.1 Fundamentals of Yield The yield Y of a manufacturing process is the fraction of products that are operational. Equivalently, it is the probability that a particular product will work. Sometimes it is more convenient to talk about the failure probability $X = 1 - Y$.

If a system is built from N components, each of which must work, then the yield of the system Y_s is the product of the yields Y_c of the components:

$$Y_s = Y_c^N \tag{6.21}$$

Sometimes it is easier to measure the defect density, D, which is the average number of defects per unit area, than the yield of a specific component. If there are M components per unit area and the defects are randomly distributed and uncorrelated, then the average failure rate of a component is $X_c = D/M$. A system with an area A thus has a yield of

$$Y_s = \left(1 - X_c\right)^{MA} = \left[\left(1 - \frac{D}{M}\right)^{M}\right]^{A} \tag{6.22}$$

Taking the limit as M approaches infinity produces a beautiful simplification

$$Y_s = e^{-DA} \tag{6.23}$$

This is called the *Poisson distribution*. Yield drops abruptly for $A > 1/D$.

Section 13.5.2 will discuss defect densities for functional yield. The remainder of this section is concerned with parametric yield.

6.5.3.2 ON and OFF Current The dependence of transistor currents on L and V_t are

$$\begin{aligned} I_{\text{on}} &\propto \frac{W}{L}\left(V_{DD} - V_t\right)^{\alpha} \\ I_{\text{off}} &\propto \frac{W}{L} 10^{-\frac{V_t}{S}} = \frac{W}{L} e^{-\frac{V_t}{nv_T}} \end{aligned} \tag{6.24}$$

Taking partial derivatives with respect to L and V_t and neglecting the dependence of V_t on L, we can estimate the sensitivity to small changes in these parameters

$$\begin{aligned} I_{\text{on}} &= I_{\text{on-nominal}}\left(1 - \frac{\Delta L}{L} - \frac{\alpha}{V_{DD} - V_t}\Delta V_t\right) \\ I_{\text{off}} &= I_{\text{off-nominal}}\left(1 - \frac{\Delta L}{L} - \frac{\Delta V_t}{nv_T}\right) \end{aligned} \tag{6.25}$$

In other words, a 10% change in channel length causes a 10% change in current. If $\alpha = 1.3$, $S = 100$ mV/decade ($n = 1.6$), $V_{DD} = 1.0$ V, and $V_t = 0.3$ V, a 10 mV change in V_t causes a 1.8% change in ON current and a 23% change in OFF current. As one would expect, subthreshold leakage is extremely sensitive to the threshold voltage.

Figure 6.24 shows a scatter plot of I_{on} against I_{off} obtained by a 1500-point Monte Carlo simulation assuming $\sigma/\mu = 0.04$ for L and $\sigma = 25$ mV for V_t. There is a strong positive correlation. However, variation changes OFF current by 6× while changing ON current by only 40%.

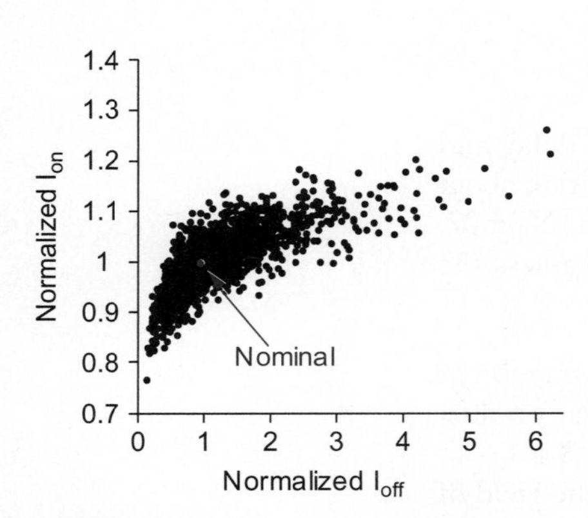

FIGURE 6.24 I_{on} vs. I_{off} with variation

6.5.3.3 Delay A change in ON current changes the delay of an inverter by the same fraction. An M-input gate will have up to M transistors that can vary separately. The delay of an N-stage path is the sum of the delays through each stage. If the variations are completely correlated (e.g., ACLV variation caused by neighboring pattern density), the delay of the path will have the same variance as the delay of a gate. However, if the variations are independent, the variance reduces by a factor of $N \times M$.

Example 6.4

A path contains 16 2-input gates, each of which has a nominal 20 ps delay. Suppose ACLV due to neighboring pattern density causes all of the transistors to experience the same channel length variation, which has a standard deviation of 2% of nominal. Suppose RDF causes a 25 mV standard deviation in each transistor's threshold. Estimate the standard deviation in path delay.

SOLUTION: The nominal path delay is 320 ps. If the path involves two series transistors in half the gates and one parallel transistor in the other half, then 24 transistors are involved in the path. The correlated channel length variation causes a change in I_{on} with a 2% standard deviation, which in turn creates a 2% standard deviation in delay (6.4 ps). We observed below EQ (6.25) that a 10 mV change in V_t causes a 1.8% change in ON current. Thus, a 25 mV standard deviation in V_t causes a $(25/10) \times 1.8\% = 4.6\%$ standard deviation in I_{on}. However, the standard deviation in the delay of the entire path is only $4.6\%/\sqrt{24} = 0.95\%$, or 3.0 ps. The standard deviation considering both effects is the RMS sum, $\sqrt{6.4^2 + 3.0^2} = 7.1$ ps, or 2.2%. Even though the threshold variation accounts for most of the variation in the delay of each individual gate, it adds little to the delay of the path because the chance of all gates seeing worst-case thresholds is miniscule.

As discussed in Section 6.5.1.4, a circuit with many nearly critical paths tends to develop a "wall" of worst case paths 2–3 standard deviations above nominal. Also, paths with fewer gates per pipeline stage suffer more because there is less averaging of random variations.

Example 6.5

A microprocessor in a 0.25 μm process was observed to have an average D2D variation of 8.99% and WID variation of 3% on several critical paths [Bowman02]. If the nominal clock period is T without considering variations and the chip has 1000 nearly critical paths, what clock period should be used to ensure a parametric yield of 97.7%? Neglect clock skew.

SOLUTION: According to Table 6.9, the worst case path due to WID variation has a mean that is $3\% \times 3.24 = 9.7\%$ above nominal and a standard deviation of $3\% \times 0.35 = 1.05\%$ of nominal. The total standard deviation is the RMS sum of the 8.99% and 1.05% D2D and WID components, or 9.05%. According to Table 6.8, 97.7% of chips fall within two standard deviations of the mean. Therefore, the clock period should be increased by $9.7\% + 2 \times 9.05\%$ to $1.28T$ to achieve the desired parametric yield.

6.5.3.4 Energy Variation has a minor impact on dynamic energy, but a major impact on static leakage energy [Rao03]. Variation shifts the minimum energy and EDP operating points toward a higher supply and threshold voltage, and reduces the potential benefits in operating at these points.

Dynamic energy is proportional to the total switching capacitance. Systematic variations affecting the mean channel length or wire widths changes this energy, but the total

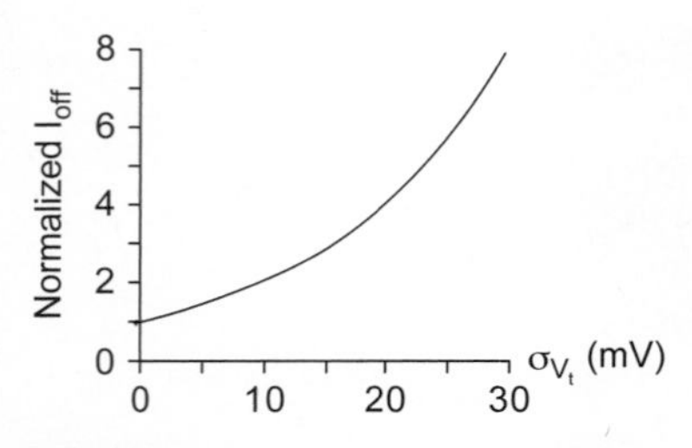

FIGURE 6.25 Impact of systematic threshold variation on worst-case leakage

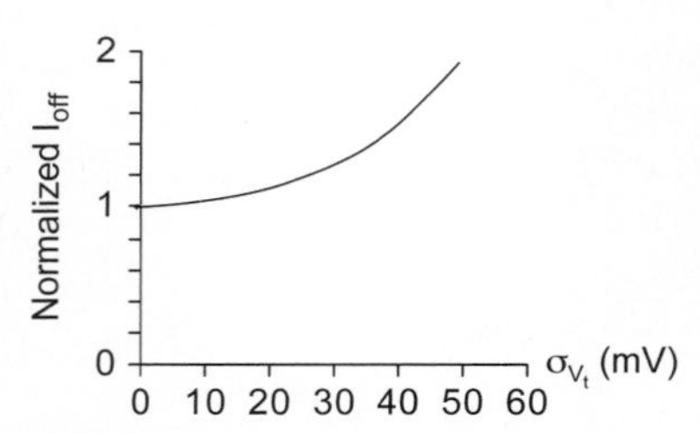

FIGURE 6.26 Impact of random threshold voltage on average leakage

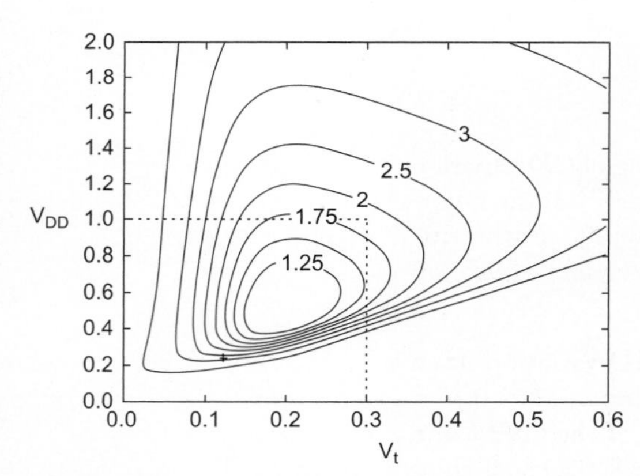

FIGURE 6.27 Contours of equal EDP accounting for variation, adapted from [Gonzalez97] (© IEEE 1997.)

variation is relatively small. Uncorrelated random variations average out over the vast number of circuit elements and have a negligible effect.

Static leakage energy is exponentially sensitive to threshold voltage. Systematic variation in V_t makes a tremendous impact because all transistors are correlated and the exponential has a long tail. Suppose we need to accept all parts with up to 3σ variation. Then, leakage current may be as great as

$$I_{\text{sub}} = I_{\text{off}} e^{\frac{3\sigma_{V_t}}{nv_T}} \tag{6.26}$$

where I_{off} is the nominal leakage. Figure 6.25 shows this exponential dependence of worst-case leakage on systemic threshold voltage variation at room temperature. Systematic threshold voltage variation must be tightly constrained to prevent enormous leakage.

Random dopant fluctuations are uncorrelated, but may have a greater standard deviation, so they can still be important. Leakage variation caused by RDF is averaged across a huge number of gates, so we are interested in the mean of the log-normal leakage distribution. Using EQ (6.19), we compute the expected subthreshold current as follows:

$$I_{\text{sub}} = I_{\text{off}} e^{\frac{1}{2}\left(\frac{\sigma_{V_t}}{nv_T}\right)^2} \tag{6.27}$$

Figure 6.26 shows the impact of random variation on the average leakage.

Figure 6.27 shows contours of equal energy-delay product accounting for temperature and V_t variations [Gonzalez97]. These variations increase the expected leakage. Recall from Section 4.4.2 that the best EDP occurs when leakage is about one third of total energy. Thus, the circuit should operate at a higher V_{DD} and V_t to increase the switching energy and decrease the leakage energy. As compared to the results without variations given in Figure 4.28, the minimum EDP point shifts significantly up and right to a supply of about 500 mV and a threshold of about 200 mV. The relative advantage of operating at the minimum EDP point over the typical point goes down from a factor of 4 to 2. Variation also shifts the minimum energy point to a higher supply voltage and diminishes the relative benefits of operating in the subthreshold regime [Zhai05b].

6.5.3.5 Functionality Variation can cause circuits to malfunction, especially at low voltage. Some of the circuits that are affected include the following:

- Ratioed circuits such as pseudo-nMOS gates and SRAM cells, where one ON device should provide more current than another ON device
- Memories and domino keepers, where one ON device should provide more current than many parallel OFF devices
- Subthreshold circuits, where one not-quite-fully OFF device should provide more current than another OFF device
- Matched circuits such as sense amplifiers that must recognize a small differential voltage

- Circuits with matched delays (see Section 6.5.3.6) that depend on one path being slower than another

These issues will be addressed more closely in subsequent chapters as they arise. In general, using bigger transistors reduces the variability at the expense of greater area and power, which is a good trade-off if only a few circuits are critically sensitive to variation.

Example 6.6

Suppose the offset voltage in a sense amplifier is a normally distributed zero-mean random variable with a standard deviation of 10 mV. If a memory contains 4096 sense amplifiers, how much offset voltage must it tolerate to achieve a 99.9% parametric yield overall?

SOLUTION: Use EQ (6.21) with $Y_s = 0.999$ and $N = 4096$ to solve for $Y_s = 0.99999976$. According to Table 6.8, this requires tolerating about five standard deviations, or 50 mV of amplifier offset.

6.5.3.6 Matched Delays Some circuits rely on *matched delays*. For example, clock-delayed domino needs to provide clocks to gates after their inputs have settled. The clocks must be matched to the gate delay; if they arrive late, the system functions slower, but if they arrive early, the system doesn't work at all. Therefore, it is of great interest to the designer how well two delays can be matched.

The best way to build matched delays is to provide replicas of the gates that are being matched. For example, in a static RAM (see Section 11.2.3.3), replica bitlines are used to determine when the sense amplifier should fire. Any relative variation in wire, diffusion, and gate capacitances happens to both circuits.

In many situations, it is not practical to use replica gates; instead, a chain of inverters can be used. For example, a DVS system may try to set the frequency based on a ring-oscillator intended to run slower than any of the various critical paths [Gutnik97]. Unfortunately, even if there is no within-die process variation, the inverter delay may not exactly track the delay it matches across design corners. For example, if the inverter chain were matching a wire delay in the typical corner, it would be faster than the wire in the FFSFF corner and slower than the wire in the SSFSS corner. This variation requires that the designer provide margin in the typical case so that even in the worst case, the matched delay does not arrive too early [Wei00]. How much margin is necessary?

Figure 6.28 shows how gate delays, measured as a multiple of an FO4 inverter delay, vary with process, design corners, temperature, and voltage. The circuits studied include complementary CMOS NAND and NOR gates, domino AND and OR gates, and a 64-bit domino adder with significant wire RC delay. Figure 6.28(a) shows the gate delay of various circuits in different processes. The adder shows the greatest variation because of its wire-limited paths, but all the circuits track to within 20% across processes. This indicates that if a circuit delay is measured in FO4 inverter delays for one process, it will have a comparable delay in a different process. Figure 6.28(b-c) shows gate delay scaling with power supply voltage and temperature. Figure 6.28(d) shows what combination of design corner, voltage, and temperature gives the largest variation in delay normalized to an FO4 inverter in the same combination in the 0.6 μm process. Observe that the variation is smallest for simple static CMOS gates that most closely resemble inverters and can reach 30% for some gates.

These figures demonstrate that an inverter chain should have a nominal delay about 30% greater than the path it matches so that the inverter output always arrives later than

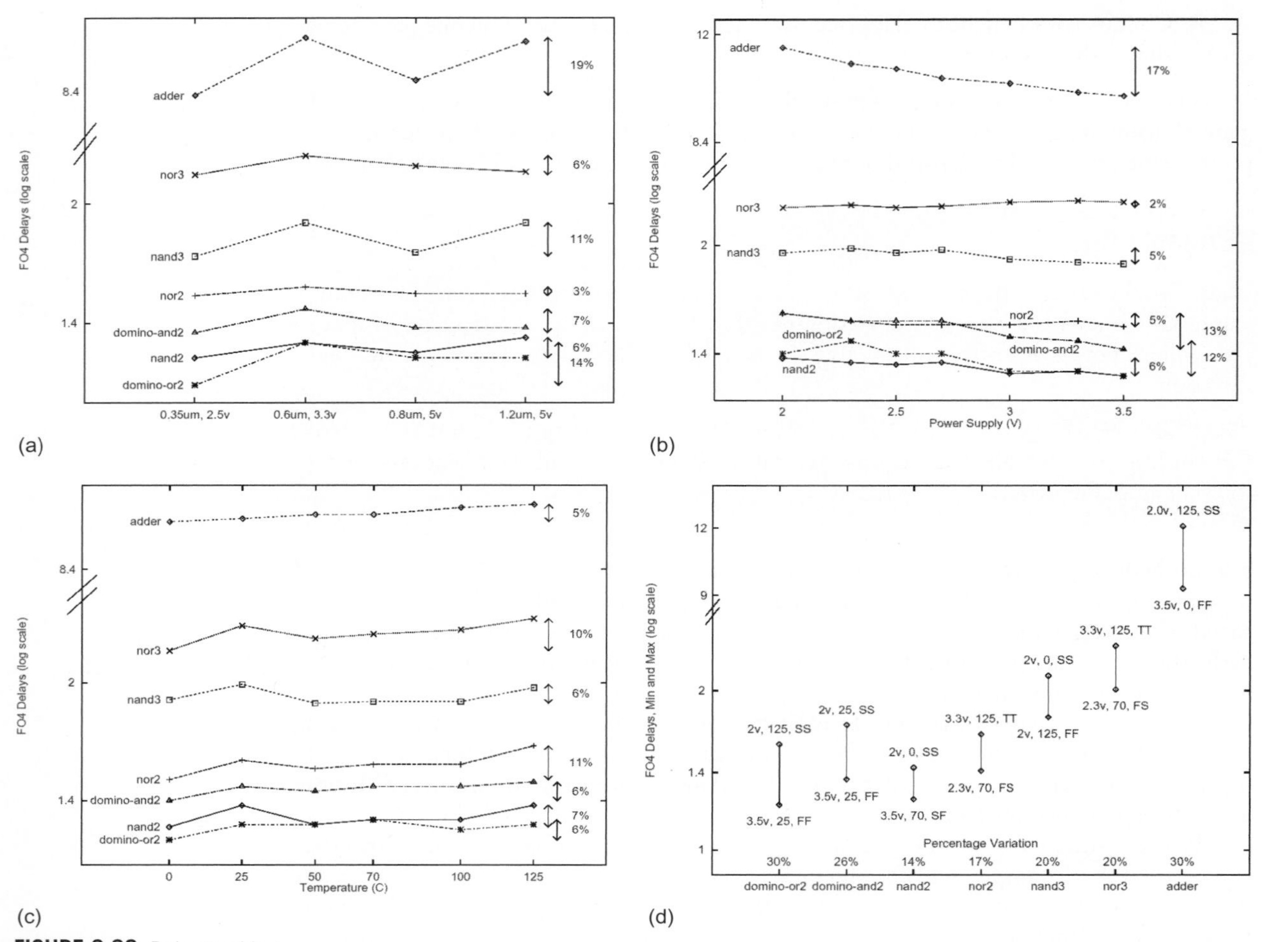

FIGURE 6.28 Delay tracking

the matched path across all combinations of voltage, temperature, and design corners. This is a hefty margin and discourages the casual use of matched delays. Considering within-die variations only makes the margin greater. It is prudent to make the amount of margin adjustable after manufacturing (e.g., via a scan chain or programmable fuse) to avoid extreme conservatism. The Power6 processor reduces the margin using a critical path monitor consisting of several different types of paths (nMOS dominated, pMOS dominated, wire dominated, etc.) and setting the cycle time based on the slowest one [Drake07]. The Montecito Itanium processor used multiple frequency generators distributed across the die to compensate for local voltage variations [Fischer06]. In light of all these issues, circuit designers tend to be moving away from matched delays and instead setting delays based on the clock because failures can be fixed by slowing the clock.

6.6 Variation-Tolerant Design

Variation has traditionally been handled by margining to ensure a good parametric yield. As variability increases, the growing margins severely degrade the performance and power of a chip. Variation-tolerant designs are becoming more important. This section describes methods of using adaptive control and fault tolerance to reduce margins. Chapter 9 addresses skew-tolerant circuits.

6.6.1 Adaptive Control

A chip can measure its operating conditions and adjust parameters such as supply voltage, body bias, frequency, or activity factor on the fly to compensate for variability. This is called *adaptive control* [Wang08a].

Dynamic voltage scaling (DVS) was introduced in Section 4.2.3.2 to save switching energy, and body bias was introduced in Section 4.3.4 to control the threshold voltage. The two techniques can be used together or individually to improve parametric yield [Chen03]. *Adaptive body bias* (ABB) can compensate for systematic die-to-die threshold variations to greatly reduce the spread in leakage and improve performance [Narendra99, Tschanz02]. *Adaptive voltage scaling* (AVS) can trade-off frequency and dynamic energy to compensate for problems in the slow or fast corners. The adjustments tend to be subtle so voltage control requires high resolution (~20 mV) to give significant benefit [Tschanz03b]. If variations are correlated over smaller blocks, the blocks can be individually controlled to run each at its best point [Gregg07].

Chips are usually designed so that worst-case power dissipation remains below a specified level under a worst-case workload. However, in many applications, the chip could work at a higher voltage or frequency if only part of it is active or if the duration is short. For example, a multicore processor running a single-threaded application might benefit from running one core at an accelerated frequency and putting the other cores to sleep.

Adaptive control systems can use one or more temperature sensors (see Section 12.2.5) to monitor die temperature and throttle back voltage or activity when sections of the chip become too hot. For example, the dual-core Itanium processor contains a separate embedded microcontroller that monitors temperature every 20 ms and adjusts core voltage to keep power within limits [McGowen06].

6.6.2 Fault Tolerance

Tolerating occasional faults reduces cost by improving yield and improves performance by reducing the amount of margin necessary. Some techniques include providing spare parts and performing error detection and correction.

Memory designers learned long ago that yield could be improved by providing spare rows and columns of memory cells. If a row or column had a manufacturing error, it could be fixed during manufacturing test by mapping in the spare. This technique will be explored further in Section 11.8.1. This technique generalizes readily to any circuit with multiple identical components. For example, an 8-core processor could be sold as a 6-core model if one or two cores were defective.

If each component has a yield Y_c, the probability P that a system with N components has r defective components is

$$P = \binom{N}{r} Y_c^{N-r} \left(1 - Y_c\right)^r \tag{6.28}$$

where

$$\binom{N}{r} \equiv \frac{(N)(N-1)(N-2)\cdots(N-r+1)}{(r)(r-1)(r-2)\cdots(1)} = \frac{N!}{r!(N-r)!} \tag{6.29}$$

is the number of ways to choose r items from a set of N. Thus, if up to r defects can be repaired with spare components, the system yield improves to

$$Y_s = \sum_{i=0}^{r} \binom{N}{r} Y_c^{N-r} \left(1 - Y_c\right)^r \quad \textbf{(6.30)}$$

If the number of components is large, we may prefer to consider the defect rate per unit area D. Using a limit argument similar to the derivation of EQ (6.23), we obtain an expression based on the Poisson distribution

$$Y_s = e^{-DA} \sum_{i=0}^{r} \frac{(DA)^i}{i!} \quad \textbf{(6.31)}$$

Example 6.7

Suppose each core in a 16-core processor has a yield of 90% and nothing else on the chip fails. What is the yield of the chip? How much better would the yield be if the chip had two spare cores that could replace defective ones?

SOLUTION: If all the cores must work, EQ (6.21) shows that the yield is $(0.9)^{16} = 18.5\%$. If two failures can be replaced, EQ (6.30) predicts that the yield improves to $(0.9)^{16} + 16 \times (0.9)^{15} \times (0.1) + 16 \times 15 / 2 \times (0.9)^{14} \times (0.1)^2 = 78.9\%$.

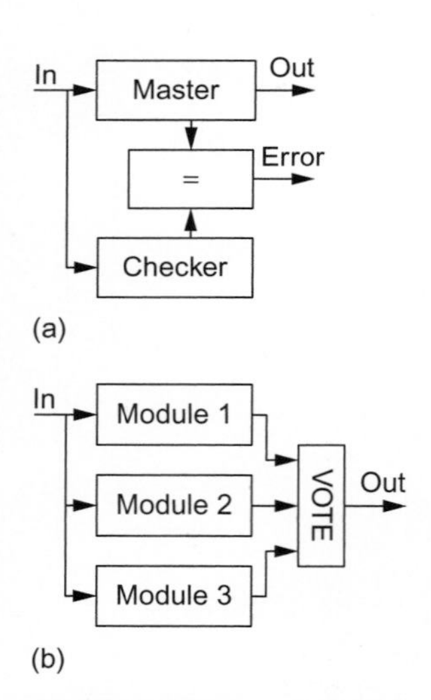

FIGURE 6.29 Master-checker operation and triple-mode redundancy

Memories have also long used error detecting and correcting codes (see Section 11.8.2). The codes are usually used to fix soft errors, but can also fix hard errors. Coding is also common in communication links where noise occasionally flips bits.

Logic fault tolerance is more difficult. Systems that require a high level of dependability (such as life-support) or that are subject to high error rates (such as spacecraft bombarded with cosmic radiation) may use two or three copies of the hardware running in lock-step. In *master-checker* configuration of Figure 6.29(a), the system periodically saves its state to a *checkpoint*. It detects an error when the master and checker differ. The system can then roll back to the last checkpoint and repeat the failed operation. For example, the IBM G5 S/390 mainframe processor contained two identical cores operating in lockstep [Northrop99]. In *triple-mode redundancy* (TMR) shown in Figure 6.29(b), the system uses majority voting to select the correct answer even if one copy of the hardware malfunctions [Lyons62]. This is ideal for real-time systems because the fault is masked and does not slow down operation. In suitably configured systems with many cores, it is possible to lock two or three cores into a fault-tolerant configuration for critical operations.

If a module has a hard failure probability of X_m over its period of service in the field, then the probability that the entire TMR system will fail is the probability that two modules fail plus the probability that three modules fail

$$X_s = \binom{3}{2} X_m^2 \left(1 - X_m\right) + X_m^3 = 3X_m^2 - 2X_m^3 \quad \textbf{(6.32)}$$

Example 6.8

Engineers designing an attitude control computer for a probe traveling to Saturn determine that the computer has a 1% chance of failure from cosmic radiation en route. They choose to use TMR to improve the reliability. What is the new chance of failure?

SOLUTION: Using EQ (6.32), the chance of failure reduces to $3(0.01)^2 - 2(0.01)^3 = 0.0298\%$.

6.7 Pitfalls and Fallacies

Not stating process corner or environment when citing circuit performance

Most products must be guaranteed to work at high temperature, yet many papers are written with transistors operating at room temperature (or lower), giving optimistic performance results. For example, at the International Solid State Circuits Conference, Intel described a Pentium II processor running at a surprisingly high clock rate [Choudhury97], but when asked, the speaker admitted that the measurements were taken while the processor was "colder than an ice cube."

Similarly, the FFFFF design corner is sometimes called the "published paper" corner because delays are reported under these simulation or manufacturing conditions without bothering to state that fact or report the FO4 inverter delay in the same conditions. Circuits in this corner are about twice as fast as in a manufacturable part.

Providing too little margin in matched delays

We have seen that the delay of a chain of inverters can vary by about 30% as compared to the delay of other circuits across design corners, voltage, and temperature. On top of this, you should expect intra-die process variation and errors in modeling and extraction. If a race condition exists where the circuit will fail when the inverter delay is faster than the gate delay, the experienced designer who wishes to sleep well at night provides generous delay margin under nominal conditions. Remember that the consequences of too little margin can be a million dollars in mask costs for another revision of the chip and far more money in the opportunity cost of arriving late to market.

Failing to plan for process scaling

Many products will migrate through multiple process generations. For example, the Intel Pentium Pro was originally designed and manufactured on a 0.6 μm BiCMOS process. The Pentium II is a closely related derivative manufactured in a 0.35 μm process operating at a lower voltage. In the new process, bipolar transistors ceased to offer performance advantages and were removed at considerable design effort. Further derivatives of the same architecture migrated to 0.25 and 0.18 μm processes in which wire delay did not improve at the same rate as gate delay. Interconnect-dominated paths required further redesign to achieve good performance in the new processes. In contrast, the Pentium 4 was designed with process scaling in mind. Knowing that over the lifetime of the product, device performance would improve but wires would not, designers overengineered the interconnect-dominated paths for the original process so that the paths would not limit performance improvement as the process advanced [Deleganes02].

6.8 Historical Perspective

The incredible history of scaling can be seen in the advancement of the microprocessor. The Intel microprocessor line makes a great case study because it spans more than three decades. Table 6.10 summarizes the progression from the first 4-bit microprocessor, the 4004, through the Core i7, courtesy of the Intel Museum. Over the years, feature size has improved more than two orders of magnitude. Transistor budgets multiplied by more than five orders of magnitude and clock frequencies have multiplied more than three orders of magnitude. Even as the challenges have grown in the past decade, scaling has accelerated.

TABLE 6.10 History of Intel microprocessors over three decades

Processor	Year	Feature Size (μm)	Transistors	Frequency (MHz)	Word Size	Power (W)	Cache (L1 / L2 / L3)	Package
4004	1971	10	2.3k	0.75	4	0.5	none	16-pin DIP
8008	1972	10	3.5k	0.5–0.8	8	0.5	none	18-pin DIP
8080	1974	6	6k	2	8	0.5	none	40-pin DIP
8086	1978	3	29k	5–10	16	2	none	40-pin DIP
80286	1982	1.5	134k	6–12	16	3	none	68-pin PGA
Intel386	1985	1.5–1.0	275k	16–25	32	1–1.5	none	100-pin PGA
Intel486	1989	1–0.6	1.2M	25–100	32	0.3–2.5	8K	168-pin PGA
Pentium	1993	0.8–0.35	3.2–4.5M	60–300	32	8–17	16K	296-pin PGA
Pentium Pro	1995	0.6–0.35	5.5M	166–200	32	29–47	16K / 256K+	387-pin MCM PGA
Pentium II	1997	0.35–0.25	7.5M	233–450	32	17–43	32K / 256K+	242-pin SECC
Pentium III	1999	0.25–0.18	9.5–28M	450–1000	32	14–44	32K / 512K	330-pin SECC2
Pentium 4	2000	180–65 nm	42–178M	1400–3800	32/64	21–115	20K+ / 256K+	478-pin PGA
Pentium M	2003	130–90 nm	77–140M	1300–2130	32	5–27	64K / 1M	479-pin FCBGA
Core	2006	65 nm	152M	1000–1860	32	6–31	64K / 2M	479-pin FCBGA
Core 2 Duo	2006	65–45 nm	167–410M	1060–3160	32/64	10–65	64K / 4M+	775-pin LGA
Core i7	2008	45 nm	731M	2660–3330	32/64	45–130	64K / 256K / 8M	1366-pin LGA
Atom	2008	45 nm	47M	800–1860	32/64	1.4–13	56K / 512K+	441-pin FCBGA

Die photos of the microprocessors illustrate the remarkable story of scaling. The 4004 [Faggin96] in Figure 6.30 was handcrafted to pack the transistors onto the tiny die. Observe the 4-bit datapaths and register files. Only a single layer of metal was available, so polysilicon jumpers were required when traces had to cross without touching. The masks were designed with colored pencils and were hand-cut from red plastic rubylith. Observe that diagonal lines were used routinely. The 16 I/O pads and bond wires are clearly visible. The processor was used in the Busicom calculator.

The 80286 [Childs84] shown in Figure 6.31 has a far more regular appearance. It is partitioned into regular datapaths, random control logic, and several arrays. The arrays include the instruction decoder PLA and memory management hardware. At this scale, individual transistors are no longer visible.

The Intel386 (originally 80386, but renamed during an intellectual property battle with AMD because a number cannot be trademarked) shown in Figure 6.32 was Intel's first 32-bit microprocessor. The datapath on the left is clearly recognizable. To the right

FIGURE 6.30 4004 microprocessor (Courtesy of Intel Corporation.)

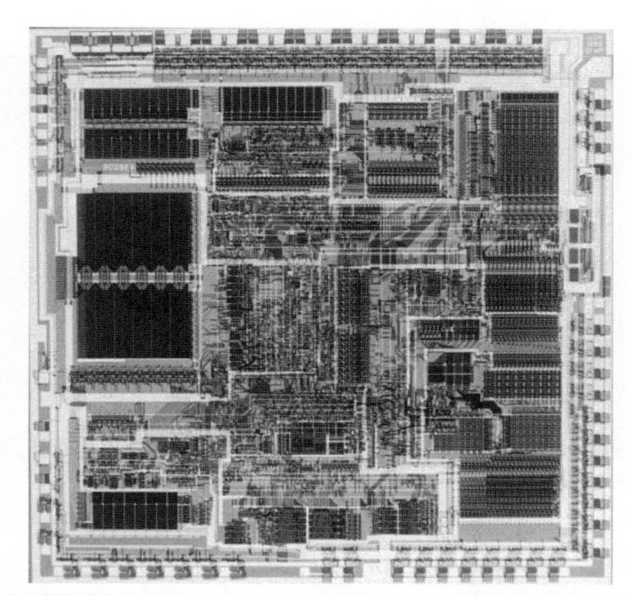

FIGURE 6.31 80286 microprocessor (Courtesy of Intel Corporation.)

are several blocks of synthesized control logic generated with automatic place & route tools. The "more advanced" tools no longer support diagonal interconnect.

The Intel486 integrated an 8 KB cache and floating point unit with a pipelined integer datapath, as shown in Figure 6.33. At this scale, individual gates are not visible. The center row is the 32-bit integer datapath. Above is the cache, divided into four 2 KB subarrays. Observe that the cache involves a significant amount of logic beside the subarrays. The wide datapaths in the upper right form the floating point unit.

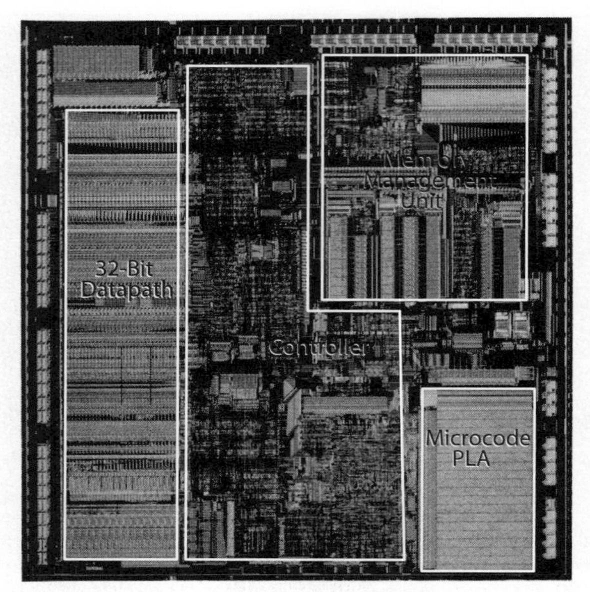

FIGURE 6.32 Intel386 microprocessor (Courtesy of Intel Corporation.)

FIGURE 6.33 Intel486 microprocessor (Courtesy of Intel Corporation.)

The Pentium Processor shown in Figure 6.34 provides a superscalar integer execution unit and separate 8 KB data and instruction caches. The 32-bit datapath and its associated control logic is again visible in the center of the chip, although at this scale, the individual bitslices of the datapath are difficult to resolve. The instruction cache in the upper left feeds the instruction fetch and decode units to its right. The data cache is in the lower left. The bus interface logic sits between the two caches. The pipelined floating point unit, home of the infamous FDIV bug [Price95], is in the lower right. This floorplan is important to minimize wire lengths between units that often communicate, such as the instruction cache and instruction fetch or the data cache and integer datapath. The integer datapath often forms the heart of a microprocessor, and other units surround the datapath to feed it the prodigious quantities of instructions and data that it consumes.

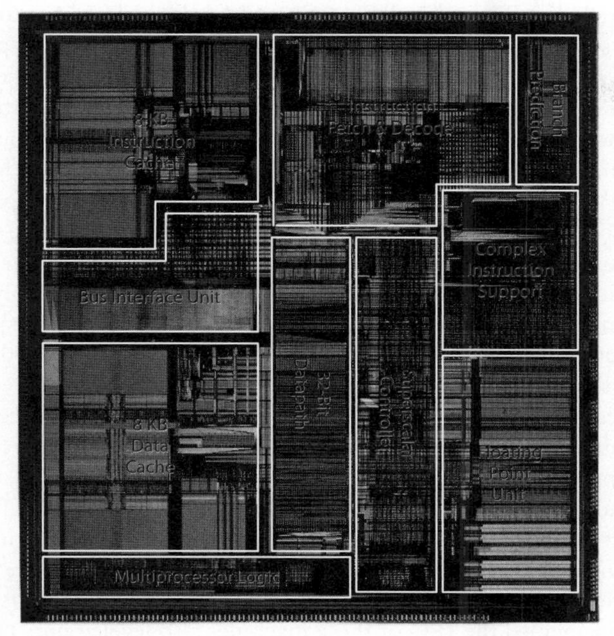

FIGURE 6.34 Pentium microprocessor (Courtesy of Intel Corporation.)

The P6 architecture used in the Pentium Pro, Pentium II, and Pentium III Processors [Colwell95, Choudhury97, Schutz98] converts complex x86 instructions into a sequence of one or more simpler RISC-style "micro-ops." It then issues up to three micro-ops per cycle to an out-of-order pipeline. The Pentium Pro was packaged in an expensive multichip module alongside a level 2 cache chip. The Pentium II and Pentium III reduced the cost by integrating the L2 cache on chip. Figure 6.35 shows the Pentium III Processor. The Integer Execution Unit (IEU) and Floating Point Unit (FPU) datapaths are tiny por-

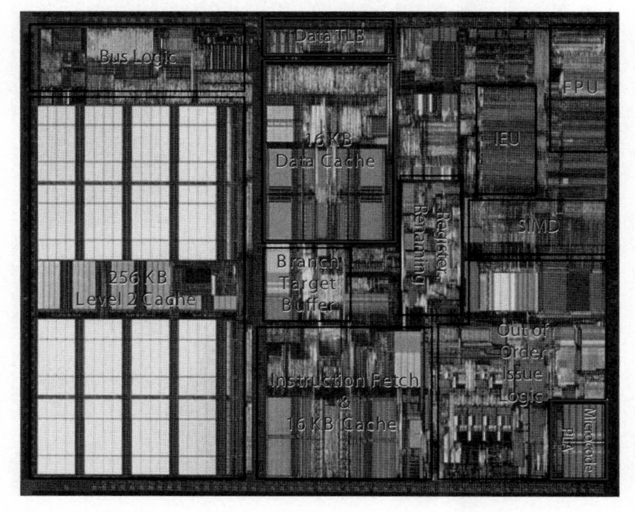

FIGURE 6.35 Pentium III microprocessor (Courtesy of Intel Corporation.)

tions of the overall chip. The entire left portion of the die is dedicated to 256–512 KB of level 2 cache to supplement the 32 KB instruction and data caches. As processor performance outstrips memory bandwidth, the portion of the die devoted to the cache hierarchy continues to grow. *The Pentium Chronicles* [Colwell06] gives a fascinating behind-the-scenes look at the development of the P6 from the perspective of the project leader.

The Pentium 4 Processor [Hinton01, Deleganes02] is shown in Figure 6.36. The complexity of a VLSI system is clear from the enormous number of separate blocks that were each uniquely designed by a team of engineers. Indeed, at this scale, even major functional units become difficult to resolve. The high operating frequency is achieved with a long pipeline using 14 or fewer FO4 inverter delays per cycle. Remarkably, portions of the integer execution unit are "double-pumped" at twice the regular chip frequency. The Pentium 4 was the culmination of the "Megahertz Wars" waged in the 1990s, in which Intel marketed processors based on clock rate rather than performance. Design teams used extreme measures, including 20- to 30-stage pipelines and outlandishly complicated domino circuit techniques to achieve such clock rates.

The Pentium 4's high power consumption was its eventual downfall, especially in laptops where it had to be throttled severely to achieve adequate battery life. In 2004, Intel returned to shorter, simpler pipelines with better energy efficiency, starting with the Pentium M [Gochman03] and continuing with the Core, Core 2, and Core i7 architectures. Clock frequencies leveled out at 2–3 GHz. Adding more execution units and speculation hurts energy efficiency, so the IPC of these machines also leveled out. Thus, these architectures marked the end of the steady advance in single-threaded application performance that had driven microprocessors during the three decades. Instead, the Core line seeks performance through parallelism using 2, 4, 8 [Sakran07, George07, Rusu10], and inevitably more cores. Figure 6.37 shows the Core 2 Duo, in which each core occupies about a quarter of the die and the large cache fills the remainder. The Core i7 appears on the cover of this book. Time will tell if mainstream software uses this parallelism well enough to drive market demand for ever-more cores.

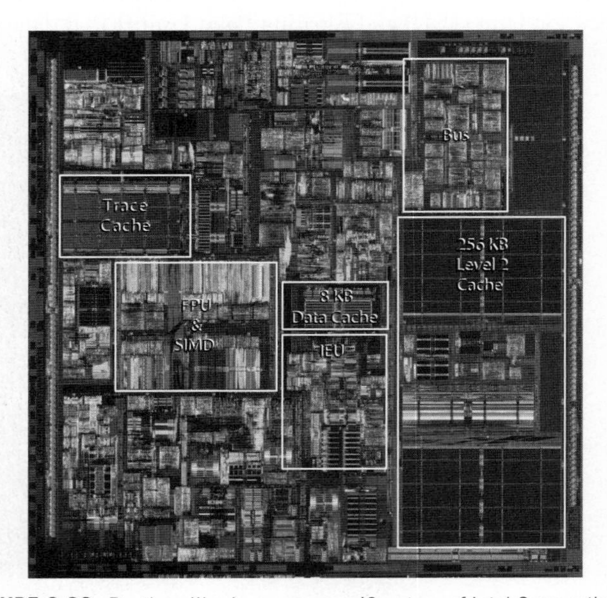

FIGURE 6.36 Pentium III microprocessor (Courtesy of Intel Corporation.)

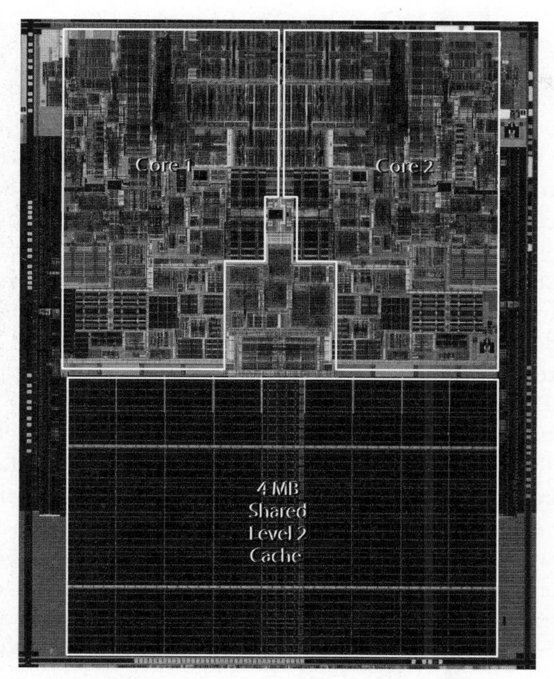

FIGURE 6.37 Core 2 Duo (Courtesy of Intel Corporation.)

It is reasonable to ask if most computer users need the full capability of a multicore CPU operating running at 3 GHz, especially considering that the 66 MHz Pentium was perfectly satisfactory for word processing, e-mail, and Web browsing. The Atom processor, shown in 6.38, is a blast from the past, using an in-order dual-issue pipeline reminiscent of the original Pentium, and achieving 1.86 GHz operation at 2 W and 800 MHz operation at 0.65 W [Gerosa09]. The Atom processor proved to be a stunningly popular CPU for 3-pound *netbooks* offering an all-day battery life and a sale price as low as $300.

FIGURE 6.38 Atom Processor (Courtesy of Intel Corporation.)

Summary

This chapter has covered three main aspects of robust design: managing variability, achieving reliability, and planning for future scaling.

The designer must ensure that the circuit performs correctly across variations in the operating voltage, temperature, and device parameters. Process corners are used to describe the worst-case die-to-die combination of processing and environment for delay, power consumption, and functionality. However, statistical techniques are becoming more important to avoid margining for extremely pessimistic worst cases, especially considering within-die variations. The circuits must also be designed to continue working even as they age or are subject to cosmic rays and electrostatic discharge.

MOS processes have been steadily improving for more than 30 years. A good designer should not only be familiar with the capabilities of current processes, but also be able to predict the capabilities of future processes as feature sizes get progressively smaller. According to Dennard's scaling, all three dimensions should scale equally, and voltage should scale as well. Gate delay improves with scaling. The number of transistors on a chip grows quadratically. The switching energy for each transistor decreases with the cube of channel length, but the dynamic power density remains about the same because chips have more transistors switching at higher rates. Leakage energy goes up as small transistors have exponentially more OFF current. Interconnect capacitance per unit length remains constant, but resistance increases because the wires have a smaller cross-section. Local wires get shorter and have constant delay, while global wires have increasing delay. Since the 90 nm node, Dennard scaling has been suffering from leakage, which is setting lower bounds on threshold voltage and oxide thickness. However, materials innovations have partially compensated and processes continue to improve. VLSI designers increasingly need to understand the effects arising as transistors reach atomic scales. The future of scaling depends on our ability to find innovative solutions to very challenging physical problems and our creativity of using the advanced processes to create compelling new products.

Exercises

6.1 How low can the module yield go before TMR becomes detrimental to system yield?

6.2 A chip contains 100 11-stage ring oscillators. Each inverter has an average delay of 10 ps with a standard deviation of 1 ps, so the average ring oscillator runs at 4.54 GHz. The operating frequency of the chip is defined to be the slowest frequency of any of the oscillators on the chip.

(a) Find the expected operating frequency of a chip.

(b) Find the maximum target operating frequency to achieve 97.7% parametric yield.

6.3 A large chip has a nominal power consumption of 60 W, of which 20 is leakage. The effective channel length is 40 nm, with a 4 nm standard deviation from die to die and a 3 nm standard deviation for uncorrelated random within-die variation. The threshold voltage has a 30 mV standard deviation caused by random dopant fluctuations. It also has a sensitivity to channel length of 2.5 mV/nm caused by short-channel effects. The subthreshold slope is 100 mV/decade. Estimate the maximum power that should be allowed to achieve an 84% parametric yield.

6.4 A circuit is being subjected to accelerated life testing at high voltage. If the measured time to failure is 20 hours at 2 V, 160 hours at 1.8 V, and 1250 hours at 1.6 V, predict the maximum operating voltage for a 10-year lifespan.

6.5 Heavily used subsystems are sometimes designed for "5 9s" yield: 99.999%. How many standard deviations increase must they accept if the parameter leading to failure is normally distributed?

6.6 Design a TMR system that can survive a single-point failure in any component or wire.

6.7 The path from the data cache to the register file of a microprocessor involves 500 ps of gate delay and 500 ps of wire delay along a repeated wire. The chip is scaled using constant field scaling and reduced height wires to a new generation with $S = 2$. Estimate the gate and wire delays of the path. By how much did the overall delay improve?

SPICE 7

7.1 Introduction

Fabricating chips is expensive and time-consuming, so designers need simulation tools to explore the design space and verify designs before they are fabricated. Simulators operate at many levels of abstraction, from process through architecture. *Process simulators* such as SUPREME predict how factors in the process recipe such as time and temperature affect device physical and electrical characteristics. *Circuit simulators* such as SPICE and Spectre use device models and a circuit netlist to predict circuit voltages and currents, which indicate performance and power consumption. *Logic simulators* such as VCS and ModelSim are widely used to verify correct logical operation of designs specified in a hardware description language (HDL). *Architecture simulators*, sometimes offered with a processor's development toolkit, work at the level of instructions and registers to predict throughput and memory access patterns, which influence design decisions such as pipelining and cache memory organization. The various levels of abstraction offer trade-offs between degree of detail and the size of the system that can be simulated. VLSI designers are primarily concerned with circuit and logic simulation. This chapter focuses on circuit simulation with SPICE. Section 14.3 discusses logic simulation.

Is it better to predict circuit behavior using paper-and-pencil analysis, as has been done in the previous chapters, or with simulation? VLSI circuits are complex and modern transistors have nonlinear, nonideal behavior, so simulation is necessary to accurately predict detailed circuit behavior. Even when closed-form solutions exist for delay or transfer characteristics, they are too time-consuming to apply by hand to large numbers of circuits. On the other hand, circuit simulation is notoriously prone to errors: *garbage in, garbage out* (GIGO). The simulator accepts the model of reality provided by the designer, but it is very easy to create a model that is inaccurate or incomplete. Moreover, the simulator only applies the stimulus provided by the designer, and it is common to overlook the worst-case stimulus. In the same way that an experienced programmer doesn't expect a program to operate correctly before debugging, an experienced VLSI designer does not expect that the first run of a simulation will reflect reality. Therefore, the circuit designer needs to have a good intuitive understanding of circuit operation and should be able to predict the expected outcome before simulating. Only when expectation and simulation match can there be confidence in the results. In practice, circuit designers depend on both hand analysis and simulation, or as [Glasser85] puts it, "simulation guided through insight gained from analysis."

This chapter presents a brief SPICE tutorial by example. It then discusses models for transistors and diffusion capacitance. The remainder of the chapter is devoted to simulation techniques to characterize a process and to check performance, power, and correctness of circuits and interconnect.

7.2 A SPICE Tutorial

SPICE (*Simulation Program with Integrated Circuit Emphasis*) was originally developed in the 1970s at Berkeley [Nagel75]. It solves the nonlinear differential equations describing components such as transistors, resistors, capacitors, and voltage sources. SPICE offers many ways to analyze circuits, but digital VLSI designers are primarily interested in *DC* and *transient* analysis that predicts the node voltages given inputs that are fixed or arbitrarily changing in time. SPICE was originally developed in FORTRAN and has some idiosyncrasies, particularly in file formats, related to its heritage. There are free versions of SPICE available on most platforms, but the commercial versions tend to offer more robust numerical convergence. In particular, HSPICE is widely used in industry because it converges well, supports the latest device and interconnect models, and has a large number of enhancements for measuring and optimizing circuits. PSPICE is another commercial version with a free limited student version. LTSpice is a robust free version. The examples throughout this section use HSPICE and generally will not run in ordinary SPICE.

While the details of using SPICE vary with version and platform, all versions of SPICE read an input file and generate a list file with results, warnings, and error messages. The input file is often called a *SPICE deck* and each line a *card* because it was once provided to a mainframe as a deck of punch cards. The input file contains a netlist consisting of components and nodes. It also contains simulation options, analysis commands, and device models. The netlist can be entered by hand or extracted from a circuit schematic or layout in a CAD program.

A good SPICE deck is like a good piece of software. It should be readable, maintainable, and reusable. Comments and white space help make the deck readable. Often, the best way to write a SPICE deck is to start with a good deck that does nearly the right thing and then modify it.

The remainder of this section provides a sequence of examples illustrating the key syntax and capabilities of SPICE for digital VLSI circuits. For more detail, consult the Berkeley SPICE manual [Johnson91], the lengthy HSPICE manual, or any number of textbooks on SPICE (such as [Kielkowski95, Foty96]).

7.2.1 Sources and Passive Components

Suppose we would like to find the response of the RC circuit in Figure 7.1(a) given an input rising from 0 to 1.0 V over 50 ps. Because the RC time constant of 100 fF $\times$ 2 kΩ = 200 ps is much greater than the input rise time, we intuitively expect the output would look like an exponential asymptotically approaching the final value of 1.0 V with a 200 ps time constant. Figure 7.2 gives a SPICE deck for this simulation and Figure 7.1(b) shows the input and output responses.

Lines beginning with * are comments. The first line of a SPICE deck must be a comment, typically indicating the title of the simulation. It is good practice to treat SPICE input files like computer programs and follow similar procedures for commenting the decks. In particular, giving the author, date, and objective of the simulation at the beginning is helpful when the deck must be revisited in the future (e.g., when a chip is in silicon debug and old simulations are being reviewed to track down potential reasons for failure).

Control statements begin with a dot (`.`). The `.option post` statement instructs HSPICE to write the results to a file for use with a waveform viewer. The last statement of a SPICE deck must be `.end`.

Each line in the netlist begins with a letter indicating the type of circuit element. Common elements are given in Table 7.1. In this case, the circuit consists of a voltage source named `Vin`, a resistor named `R1`, and a capacitor named `C1`. The nodes in the circuit are named `in`, `out`, and `gnd`. `gnd` is a special node name defined to be the 0 V reference. The units consist of one or two letters. The first character indicates the order of magnitude, as given in Table 7.2. Take note that mega is `x`, not `m`. The second letter indicates a unit for human convenience (such as `F` for farad or `s` for second) and is ignored by SPICE. For example, the hundred femtofarad capacitor can be expressed as `100fF`, `100f`, or simply `100e–15`. Note that SPICE is case-insensitive but consistent capitalization is good practice nonetheless because the netlist might be parsed by some other tool.

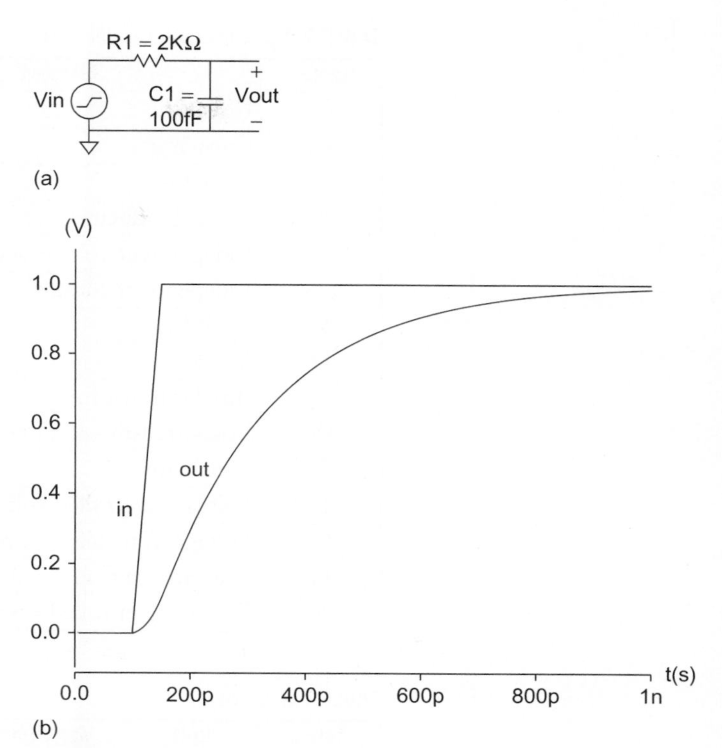

FIGURE 7.1 RC circuit response

```
* rc.sp
* David_Harris@hmc.edu 2/2/03
* Find the response of RC circuit to rising input

*----------------------------------------------------------------
* Parameters and models
*----------------------------------------------------------------
.option post

*----------------------------------------------------------------
* Simulation netlist
*----------------------------------------------------------------
Vin   in    gnd   pwl   0ps 0 100ps 0 150ps 1.0 1ns 1.0
R1    in    out   2k
C1    out   gnd   100f

*----------------------------------------------------------------
* Stimulus
*----------------------------------------------------------------
.tran 20ps 1ns
.plot v(in) v(out)
.end
```

FIGURE 7.2 RC SPICE deck

TABLE 7.1 Common SPICE elements

Letter	Element
R	Resistor
C	Capacitor
L	Inductor
K	Mutual inductor
V	Independent voltage source
I	Independent current source
M	MOSFET
D	Diode
Q	Bipolar transistor
W	Lossy transmission line
X	Subcircuit
E	Voltage-controlled voltage source
G	Voltage-controlled current source
H	Current-controlled voltage source
F	Current-controlled current source

TABLE 7.2 SPICE units

Letter	Unit	Magnitude
a	atto	10^{-18}
f	femto	10^{-15}
p	pico	10^{-12}
n	nano	10^{-9}
u	micro	10^{-6}
m	milli	10^{-3}
k	kilo	10^{3}
x	mega	10^{6}
g	giga	10^{9}

The voltage source is defined as a piecewise linear (PWL) source. The waveform is specified with an arbitrary number of (time, voltage) pairs. Other common sources include DC sources and pulse sources. A DC voltage source named `Vdd` that sets node `vdd` to 2.5 V could be expressed as

```
Vdd   vdd   gnd   2.5
```

Pulse sources are convenient for repetitive signals like clocks. The general form for a pulse source is illustrated in Figure 7.3. For example, a clock with a 1.0 V swing, 800 ps period, 100 ps rise and fall times, and 50% duty cycle (i.e., equal high and low times) would be expressed as

```
Vck   clk   gnd   PULSE 0 1 0ps 100ps 100ps 300ps 800ps
```

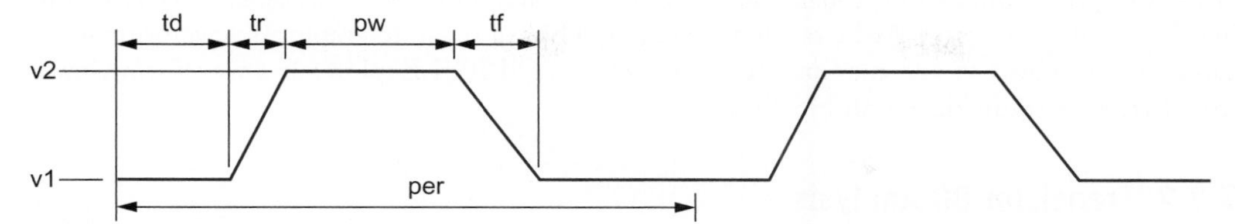

FIGURE 7.3 Pulse waveform

The stimulus specifies that a transient analysis (`.tran`) should be performed using a maximum step size of 20 ps for a duration of 1 ns. When plotting node voltages, the step size determines the spacing between points.

The `.plot` command generates a textual plot of the node variables specified (in this case the voltages at nodes `in` and `out`), as shown in Figure 7.4. Similarly, the `.print` statement prints the results in a multicolumn table. Both commands show the legacy of

```
legend:

 a: v(in)
 b: v(out)

      time         v(in)
  (ab       )   -500.0000m       0.          500.0000m      1.0000      1.5000
                     +             +             +             +             +
      0.        0.  -+------+------2------+------+------+------+------+------+-
   20.0000p     0.   +      +      2      +      +      +      +      +      +
   40.0000p     0.   +      +      2      +      +      +      +      +      +
   60.0000p     0.   +      +      2      +      +      +      +      +      +
   80.0000p     0.   +      +      2      +      +      +      +      +      +
  100.0000p     0.   +      +      2      +      +      +      +      +      +
  120.0000p 400.000m +      +      +b     +   a  +      +      +      +      +
  140.0000p 800.000m +      +      + b    +      +      +a     +      +      +
  160.0000p   1.000  +      +      +   b  +      +      +      a      +      +
  180.0000p   1.000  +      +      +      b      +      +      a      +      +
  200.0000p   1.000 -+------+------+------+-b----+------+------a------+------+-
  220.0000p   1.000  +      +      +      +  b   +      +      a      +      +
  240.0000p   1.000  +      +      +      +    b +      +      a      +      +
  260.0000p   1.000  +      +      +      +      b      +      a      +      +
  280.0000p   1.000  +      +      +      +      +b     +      a      +      +
  300.0000p   1.000  +      +      +      +      + b    +      a      +      +
  320.0000p   1.000  +      +      +      +      +  b   +      a      +      +
  340.0000p   1.000  +      +      +      +      +   b  +      a      +      +
  360.0000p   1.000  +      +      +      +      +    b +      a      +      +
  380.0000p   1.000  +      +      +      +      +     b+      a      +      +
  400.0000p   1.000 -+------+------+------+------+------b------a------+------+-
  420.0000p   1.000  +      +      +      +      +      +b     a      +      +
  440.0000p   1.000  +      +      +      +      +      +b     a      +      +
  460.0000p   1.000  +      +      +      +      +      + b    a      +      +
  480.0000p   1.000  +      +      +      +      +      + b    a      +      +
  500.0000p   1.000  +      +      +      +      +      +  b   a      +      +
  520.0000p   1.000  +      +      +      +      +      +  b   a      +      +
  540.0000p   1.000  +      +      +      +      +      +   b  a      +      +
  560.0000p   1.000  +      +      +      +      +      +   b  a      +      +
  580.0000p   1.000  +      +      +      +      +      +   b  a      +      +
  600.0000p   1.000 -+------+------+------+------+------+---b--a------+------+-
  620.0000p   1.000  +      +      +      +      +      +    b a      +      +
  640.0000p   1.000  +      +      +      +      +      +    b a      +      +
  660.0000p   1.000  +      +      +      +      +      +    b a      +      +
  680.0000p   1.000  +      +      +      +      +      +    b a      +      +
  700.0000p   1.000  +      +      +      +      +      +    b a      +      +
  720.0000p   1.000  +      +      +      +      +      +     ba      +      +
  740.0000p   1.000  +      +      +      +      +      +     ba      +      +
  760.0000p   1.000  +      +      +      +      +      +     ba      +      +
  780.0000p   1.000  +      +      +      +      +      +     ba      +      +
  800.0000p   1.000 -+------+------+------+------+------+-----ba------+------+-
  820.0000p   1.000  +      +      +      +      +      +     ba      +      +
  840.0000p   1.000  +      +      +      +      +      +     ba      +      +
  860.0000p   1.000  +      +      +      +      +      +     ba      +      +
  880.0000p   1.000  +      +      +      +      +      +     ba      +      +
  900.0000p   1.000  +      +      +      +      +      +     ba      +      +
  920.0000p   1.000  +      +      +      +      +      +     ba      +      +
  940.0000p   1.000  +      +      +      +      +      +      2      +      +
  960.0000p   1.000  +      +      +      +      +      +      2      +      +
  980.0000p   1.000  +      +      +      +      +      +      2      +      +
    1.0000n   1.000 -+------+------+------+------+------+------2------+------+-
                     +             +             +             +             +
```

FIGURE 7.4 Textual plot of RC circuit response

FORTRAN and line printers. On modern computers with graphical user interfaces, the `.option post` command is usually preferred. It generates a file (in this case, `rc.tr0`) containing the results of the specified (transient) analysis. Then, a separate graphical waveform viewer can be used to look at and manipulate the waveforms. SPICE Explorer is a waveform viewer from Synopsys compatible with HSPICE.

7.2.2 Transistor DC Analysis

One of the first steps in becoming familiar with a new CMOS process is to look at the I-V characteristics of the transistors. Figure 7.5(a) shows test circuits for a unit (4/2 λ) nMOS transistor in a 65 nm process at V_{DD} = 1.0 V. The I-V characteristics are plotted in Figure 7.5(b) using the SPICE deck in Figure 7.6.

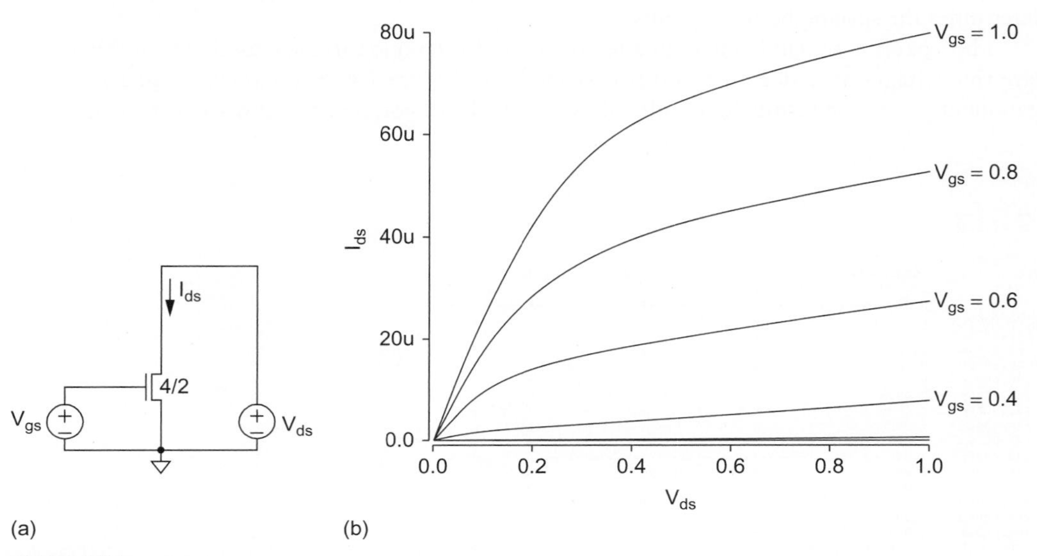

FIGURE 7.5 MOS I-V characteristics. Current in units of microamps (u).

`.include` reads another SPICE file from disk. In this example, it loads device models that will be discussed further in Section 7.3. The circuit uses two independent voltage sources with default values of 0 V; these voltages will be varied by the `.dc` command. The nMOS transistor is defined with the MOSFET element `M` using the syntax

```
Mname   drain   gate   source   body   model   W=<width>   L=<length>
```

Note that this process has λ = 25 nm and a minimum drawn channel length of 50 nm even though it is nominally called a 65 nm process.

The `.dc` command varies the voltage source `Vgs` DC voltage from 0 to 1.0 V in increments of 0.05 V. This is repeated multiple times as `Vgs` is swept from 0 to 1.0 V in 0.2 V increments to compute many I_{ds} vs. V_{ds} curves at different values of V_{gs}.

7.2.3 Inverter Transient Analysis

Figure 7.7 shows the step response of an unloaded unit inverter, annotated with propagation delay and 20–80% rise and fall times. Observe that significant initial overshoot from bootstrapping

```
* mosiv.sp

*----------------------------------------------------------------------
* Parameters and models
*----------------------------------------------------------------------
.include '../models/ibm065/models.sp'
.temp 70
.option post

*----------------------------------------------------------------------
* Simulation netlist
*----------------------------------------------------------------------
*nmos
Vgs    g   gnd    0
Vds    d   gnd    0
M1     d   g      gnd    gnd    NMOS    W=100n   L=50n

*----------------------------------------------------------------------
* Stimulus
*----------------------------------------------------------------------
.dc Vds 0 1.0 0.05 SWEEP Vgs 0 1.0 0.2
.end
```

FIGURE 7.6 MOSIV SPICE deck

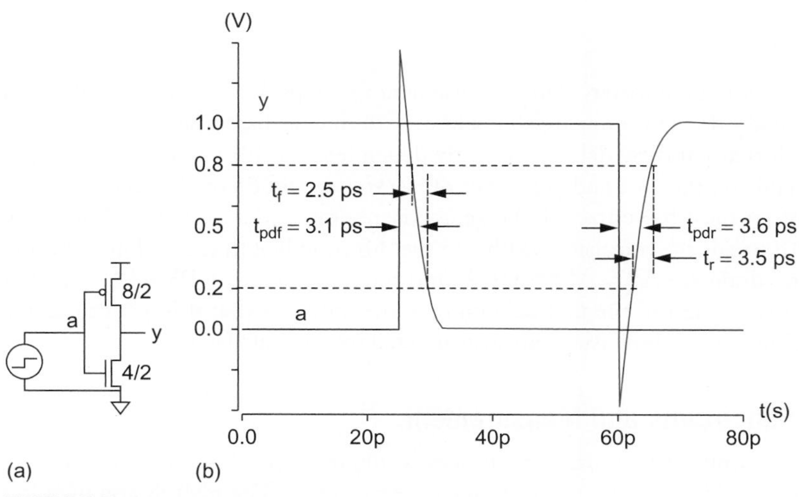

FIGURE 7.7 Unloaded inverter

occurs because there is no load (see Section 3.4.6.6). The SPICE deck for the simulation is shown in Figure 7.8.

This deck introduces the use of parameters and scaling. The `.param` statement defines a parameter named `SUPPLY` to have a value of 1.0. This is then used to set `Vdd` and the amplitude of the input pulse. If we wanted to evaluate the response at a different supply voltage, we would simply need to change the `.param` statement. The `.scale` sets a scale factor for all dimensions that would by default be measured in meters. In this case, it sets the scale to $\lambda = 25$ nm. Now the transistor widths and lengths in the inverter are specified in terms of

```
* inv.sp

*----------------------------------------------------------------------
* Parameters and models
*----------------------------------------------------------------------
.param SUPPLY=1.0
.option scale=25n
.include '../models/ibm065/models.sp'
.temp 70
.option post

*----------------------------------------------------------------------
* Simulation netlist
*----------------------------------------------------------------------
Vdd   vdd   gnd   'SUPPLY'
Vin   a     gnd   PULSE   0 'SUPPLY' 25ps 0ps 0ps 35ps 80ps
M1    y     a     gnd    gnd    NMOS    W=4    L=2
+ AS=20 PS=18 AD=20 PD=18
M2    y     a     vdd    vdd    PMOS    W=8    L=2
+ AS=40 PS=26 AD=40 PD=26

*----------------------------------------------------------------------
* Stimulus
*----------------------------------------------------------------------
.tran 0.1ps 80ps
.end
```

FIGURE 7.8 INV SPICE deck

lambda rather than in meters. This is convenient for chips designed using scalable rules, but is not normally done in commercial processes with micron-based rules.

Recall that parasitic delay is strongly dependent on diffusion capacitance, which in turn depends on the area and perimeter of the source and drain. As each diffusion region in an inverter must be contacted, the geometry resembles that of Figure 2.8(a). The diffusion width equals the transistor width and the diffusion length is 5 λ. Thus, the area of the source and drain are AS = AD = $5W\lambda^2$ and the perimeters are PS = PD = $(2W + 10)\,\lambda$. Note that the + sign in the first column of a line indicates that it is a continuation of the previous line. These dimensions are also affected by the scale factor.

7.2.4 Subcircuits and Measurement

One of the simplest measures of a process's inherent speed is the fanout-of-4 inverter delay. Figure 7.9(a) shows a circuit to measure this delay. The nMOS and pMOS transistor sizes (in multiples of a unit 4/2 λ transistor) are listed below and above each gate, respectively. *X*3 is the inverter under test and *X*4 is its load, which is four times larger than *X*3. To first order, these two inverters would be sufficient. However, the delay of *X*3 also depends on the input slope, as discussed in Section 3.4.6.1. One way to obtain a realistic input slope is to drive node `c` with a pair of FO4 inverters *X*1 and *X*2. Also, as discussed in Section 3.4.6.6, the input capacitance of *X*4 depends not just on its C_{gs} but also on C_{gd}. C_{gd} is Miller-multiplied as node `e` switches and would be effectively doubled if `e` switched instantaneously. When `e` is loaded with *X*5, it switches at a slower, more realistic rate, slightly reducing the effective capacitance presented at node `d` by *X*4. The waveforms in Figure 7.9(b) are annotated with the rising and falling delays.

SPICE decks are easier to read and maintain when common circuit elements are captured as subcircuits. For example, the deck in Figure 7.10 computes the FO4 inverter delay using an inverter subcircuit.

The `.global` statement defines `vdd` and `gnd` as global nodes that can be referenced from within subcircuits. The inverter is declared as a subcircuit with two terminals: `a` and `y`. It also accepts two parameters specifying the width of the nMOS and pMOS transistors; these parameters have default values of 4 and 8, respectively. The source and drain area and perimeter are functions of the transistor widths. HSPICE evaluates functions given inside single quotation marks. The functions can include parameters, constants, parentheses, +, –, *, /, and ** (raised to a power).

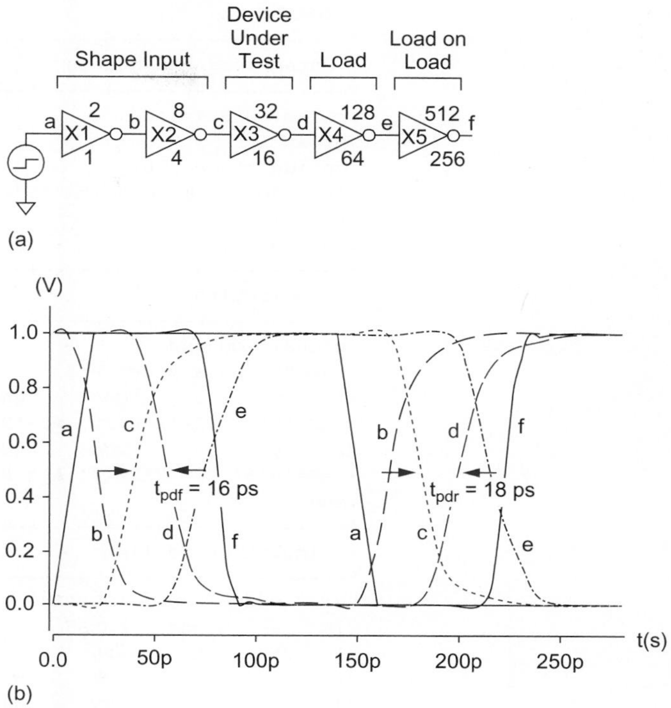

The simulation netlist contains the power supply, input source, and five inverters. Each inverter is a subcircuit (`X`) element. As `N` and `P` are not specified, each uses the default size. The `M` parameter multiplies all the currents in the subcircuit by the factor given, equivalent to `M` elements wired in parallel. In this case, the fanouts are expressed in terms of a parameter `H`. Thus, *X*2 has the capacitance and output current of 4 unit inverters, while *X*3 is equivalent to 16. Another way to model the inverters would have been to use the `N` and `P` parameters:

FIGURE 7.9 Fanout-of-4 inverters

```
X1    a    b    inv    N=4       P=8       * shape input waveform
X2    b    c    inv    N=16      P=32      * reshape input waveform
X3    c    d    inv    N=64      P=128     * device under test
X4    d    e    inv    N=256     P=512     * load
X5    e    f    inv    N=1024    P=2048    * load on load
```

However, a transistor of four times unit width does not have exactly the same input capacitance or output current as four unit inverters tied in parallel, so the `M` parameter is preferred.

In this example, the subcircuit declaration and simulation netlist are part of the SPICE deck. When working with a standard cell library, it is common to keep subcircuit declarations in their own files and reference them with a `.include` statement instead. When the simulation netlist is extracted from a schematic or layout CAD system, it is common to put the netlist in a separate file and `.include` it as well.

The `.measure` statement measures simulation results and prints them in the listing file. The deck measures the rising propagation delay t_{pdr} as the difference between the time that the input `c` first falls through $V_{DD}/2$ and the time that the output `d` first rises through $V_{DD}/2$. `TRIG` and `TARG` indicate the trigger and target events between which delay is measured. The `.measure` statement can also be used to compute functions of other measurements. For example, the average FO4 inverter propagation delay t_{pd} is the mean of t_{pdr} and t_{pdf}, 17 ps. The 20–80% rise time is $t_r = 20$ ps and the fall time is $t_f = 17$ ps.

```
* fo4.sp
*----------------------------------------------------------------------
* Parameters and models
*----------------------------------------------------------------------
.param SUPPLY=1.0
.param H=4
.option scale=25n
.include '../models/ibm065/models.sp'
.temp 70
.option post
*----------------------------------------------------------------------
* Subcircuits
*----------------------------------------------------------------------
.global vdd gnd
.subckt inv a y N=4 P=8
M1     y     a     gnd     gnd     NMOS     W='N'    L=2
+ AS='N*5' PS='2*N+10' AD='N*5' PD='2*N+10'
M2     y     a     vdd     vdd     PMOS     W='P'    L=2
+ AS='P*5' PS='2*P+10' AD='P*5' PD='2*P+10'
.ends
*----------------------------------------------------------------------
* Simulation netlist
*----------------------------------------------------------------------
Vdd   vdd   gnd   'SUPPLY'
Vin   a     gnd   PULSE   0 'SUPPLY' 0ps 20ps 20ps 120ps 280ps
X1    a     b     inv               * shape input waveform
X2    b     c     inv     M='H'     * reshape input waveform
X3    c     d     inv     M='H**2'  * device under test
X4    d     e     inv     M='H**3'  * load
X5    e     f     inv     M='H**4'  * load on load
*----------------------------------------------------------------------
* Stimulus
*----------------------------------------------------------------------
.tran 0.1ps 280ps
.measure tpdr                                   * rising prop delay
+     TRIG v(c)    VAL='SUPPLY/2' FALL=1
+     TARG v(d)    VAL='SUPPLY/2' RISE=1
.measure tpdf                                   * falling prop delay
+     TRIG v(c)    VAL='SUPPLY/2' RISE=1
+     TARG v(d)    VAL='SUPPLY/2' FALL=1
.measure tpd param='(tpdr+tpdf)/2'              * average prop delay
.measure trise                                  * rise time
+     TRIG v(d)    VAL='0.2*SUPPLY' RISE=1
+     TARG v(d)    VAL='0.8*SUPPLY' RISE=1
.measure tfall                                  * fall time
+     TRIG v(d)    VAL='0.8*SUPPLY' FALL=1
+     TARG v(d)    VAL='0.2*SUPPLY' FALL=1
.end
```

FIGURE 7.10 FO4 SPICE deck

7.2.5 Optimization

In many examples, we have assumed that a *P*/*N* ratio of 2:1 gives approximately equal rise and fall delays. The FO4 inverter simulation showed that a ratio of 2:1 gives rising delays that are slower than the falling delays because the pMOS mobility is less than half that of the nMOS. You could repeatedly run simulations with different default values of `P` to find the ratio for equal delay. HSPICE has built-in optimization capabilities that will automatically tweak parameters to achieve some goal and report what parameter value gave the best results. Figure 7.11 shows a modified version of the FO4 inverter simulation using the optimizer.

The subcircuits $X1$–$X5$ override their default pMOS widths to use a width of `P1` instead. In the optimization setup, the difference of t_{pdr} and t_{pdf} is measured. The goal of the optimization will be to drive this difference to 0. To do this, `P1` may be var-

```
* fo4opt.sp
*----------------------------------------------------------------------
* Parameters and models
*----------------------------------------------------------------------
.param SUPPLY=1.0
.option scale=25n
.include '../models/ibm065/models.sp'
.temp 70
.option post
*----------------------------------------------------------------------
* Subcircuits
*----------------------------------------------------------------------
.global vdd gnd
.subckt inv a y N=4 P=8
M1    y    a    gnd    gnd    NMOS    W='N'    L=2
+ AS='N*5' PS='2*N+10' AD='N*5' PD='2*N+10'
M2    y    a    vdd    vdd    PMOS    W='P'    L=2
+ AS='P*5' PS='2*P+10' AD='P*5' PD='2*P+10'
.ends
*----------------------------------------------------------------------
* Simulation netlist
*----------------------------------------------------------------------
Vdd   vdd   gnd   'SUPPLY'
Vin   a     gnd   PULSE     0 'SUPPLY' 0ps 20ps 20ps 120ps 280ps
X1    a     b     inv       P='P1'             * shape input waveform
X2    b     c     inv       P='P1'    M=4      * reshape input waveform
X3    c     d     inv       P='P1'    M=16     * device under test
X4    d     e     inv       P='P1'    M=64     * load
X5    e     f     inv       P='P1'    M=256    * load on load
*----------------------------------------------------------------------
* Optimization setup
*----------------------------------------------------------------------
.param P1=optrange(8,4,16)          * search from 4 to 16, guess 8
.model optmod opt itropt=30         * maximum of 30 iterations
.measure bestratio param='P1/4'     * compute best P/N ratio
*----------------------------------------------------------------------
* Stimulus
*----------------------------------------------------------------------
.tran 0.1ps 280ps SWEEP OPTIMIZE=optrange RESULTS=diff MODEL=optmod
.measure tpdr                              * rising propagation delay
+     TRIG v(c)   VAL='SUPPLY/2' FALL=1
+     TARG v(d)   VAL='SUPPLY/2' RISE=1
.measure tpdf                              * falling propagation delay
+     TRIG v(c)   VAL='SUPPLY/2' RISE=1
+     TARG v(d)   VAL='SUPPLY/2' FALL=1
.measure tpd param='(tpdr+tpdf)/2' goal=0  * average prop delay
.measure diff param='tpdr-tpdf' goal = 0   * diff between delays
.end
```

FIGURE 7.11 FO4OPT SPICE deck

ied from 4 to 16, with an initial guess of 8. The optimizer may use up to 30 iterations to find the best value of `P1`. Because the nMOS width is fixed at 4, the best *P/N* ratio is computed as `P1/4`. The transient analysis includes a `SWEEP` statement containing the parameter to vary, the desired result, and the number of iterations.

HSPICE determines that the *P/N* ratio for equal rise and fall delay is 2.87:1, giving a rising and falling delay of 17.9 ps. This is slower than what the 2:1 ratio provides and requires large, power-hungry pMOS transistors, so such a high ratio is seldom used.

A similar scenario is to find the *P/N* ratio that gives lowest average delay. By changing the `.tran` statement to use `RESULTS=tpd`, we find a best ratio of 1.79:1 with rising, falling, and average propagation delays of 18.8, 15.2, and 17.0 ps, respectively. Whenever you do an optimization, it is important to consider not only the optimum but also the sensitivity to deviations from this point. Further simulation finds that *P/N* ratios of anywhere from 1.5:1 to 2.2:1 all give an average propagation delay of better than 17.2 ps. There is no need to slavishly stick to the 1.79:1 "optimum." The best *P/N* ratio in practice is a compromise between using smaller pMOS devices to save area and power and using larger devices to achieve more nearly equal rise/fall times and avoid the hot electron reliability problems induced by very slow rising edges in circuits with weak pMOS transistors. *P/N* ratios are discussed further in Section 8.2.1.6.

7.2.6 Other HSPICE Commands

The full HSPICE manual fills over 4000 pages and includes many more capabilities than can be described here. A few of the most useful additional commands are covered in this section. Section 7.3 describes transistor models and library calls, and Section 7.6 discusses modeling interconnect with lossy transmission lines.

```
.option accurate
```

Tighten integration tolerances to obtain more accurate results. This is useful for oscillators and high-gain analog circuits or when results seem fishy.

```
.option autostop
```

Conclude simulation when all `.measure` results are obtained rather than continuing for the full duration of the `.tran` statement. This can substantially reduce simulation time.

```
.temp 0 70 125
```

Repeat the simulation three times at temperatures of 0, 70, and 125 °C. Device models may contain information about how changing temperature changes device performance.

```
.op
```

Print the voltages, currents, and transistor bias conditions at the DC operating point.

7.3 Device Models

Most of the examples in Section 7.2 included a file containing transistor models. SPICE provides a wide variety of MOS transistor models with various trade-offs between complexity and accuracy. Level 1 and Level 3 models were historically important, but they are no longer adequate to accurately model very small modern transistors. BSIM models are more accurate and are presently the most widely used. Some companies use their own proprietary models. This section briefly describes the main features of each of these models. It also describes how to model diffusion capacitance and how to run simulations in various process corners. The model descriptions are intended only as an overview of the capabilities and limitations of the models; refer to a SPICE manual for a much more detailed description if one is necessary.

7.3.1 Level 1 Models

The SPICE Level 1, or Shichman-Hodges Model [Shichman68] is closely related to the Shockley model described in EQ (2.10), enhanced with channel length modulation and the body effect. The basic current model is:

$$I_{ds} = \begin{cases} 0 & V_{gs} < V_t & \text{cutoff} \\ \text{KP}\dfrac{W_{\text{eff}}}{L_{\text{eff}}}\left(1+\text{LAMBDA}\times V_{ds}\right)\left(V_{gs}-V_t-\dfrac{V_{ds}}{2}\right)V_{ds} & V_{ds} < V_{gs}-V_t & \text{linear} \\ \dfrac{\text{KP}}{2}\dfrac{W_{\text{eff}}}{L_{\text{eff}}}\left(1+\text{LAMBDA}\times V_{ds}\right)\left(V_{gs}-V_t\right)^2 & V_{ds} > V_{gs}-V_t & \text{saturation} \end{cases} \tag{7.1}$$

The parameters from the SPICE model are given in ALL CAPS. Notice that β is written instead as KP($W_{\text{eff}}/L_{\text{eff}}$), where KP is a model parameter playing the role of k' from EQ (2.7). W_{eff} and L_{eff} are the effective width and length, as described in EQ (2.48). The LAMBDA term (LAMBDA = $1/V_A$) models channel length modulation (see Section 2.4.2).

The threshold voltage is modulated by the source-to-body voltage V_{sb} through the body effect (see Section 2.4.3.1). For nonnegative V_{sb}, the threshold voltage is

$$V_t = \text{VTO} + \text{GAMMA}\left(\sqrt{\text{PHI}+V_{sb}} - \sqrt{\text{PHI}}\right) \tag{7.2}$$

Notice that this is identical to EQ (2.30), where VTO is the "zero-bias" threshold voltage V_{t0}, GAMMA is the body effect coefficient γ, and PHI is the surface potential ϕ_s.

The gate capacitance is calculated from the oxide thickness TOX. The default gate capacitance model in HSPICE is adequate for finding the transient response of digital circuits. More elaborate models exist that capture nonreciprocal effects that are important for analog design.

Level 1 models are useful for teaching because they are easy to correlate with hand analysis, but are too simplistic for modern design. Figure 7.12 gives an example of a Level 1 model illustrating the syntax. The model also includes terms to compute the diffusion capacitance, as described in Section 7.3.4.

```
.model NMOS NMOS (LEVEL=1 TOX=40e-10 KP=155E-6 LAMBDA=0.2
+                VTO=0.4 PHI=0.93 GAMMA=0.6
+                CJ=9.8E-5 PB=0.72 MJ=0.36
+                CJSW=2.2E-10 PHP=7.5 MJSW=0.1)
```

FIGURE 7.12 Sample Level 1 Model

7.3.2 Level 2 and 3 Models

The SPICE Level 2 and 3 models add effects of velocity saturation, mobility degradation, subthreshold conduction, and drain-induced barrier lowering. The Level 2 model is based on the Grove-Frohman equations [Frohman69], while the Level 3 model is based on empirical equations that provide similar accuracy, faster simulation times, and better convergence. However, these models still do not provide good fits to the measured I-V characteristics of modern transistors.

7.3.3 BSIM Models

The Berkeley Short-Channel IGFET[1] Model (BSIM) is a very elaborate model that is now widely used in circuit simulation. The models are derived from the underlying device physics but use an enormous number of parameters to fit the behavior of modern transistors. BSIM versions 1, 2, 3v3, and 4 are implemented as SPICE levels 13, 39, 49, and 54, respectively.

BSIM 3 and 4 require entire books [Cheng99, Dunga07] to describe the models. They include over 100 parameters and the device equations span 27 pages. BSIM is quite good for digital circuit simulation. Features of the model include:

- Continuous and differentiable I-V characteristics across subthreshold, linear, and saturation regions for good convergence
- Sensitivity of parameters such as V_t to transistor length and width
- Detailed threshold voltage model including body effect and drain-induced barrier lowering
- Velocity saturation, mobility degradation, and other short-channel effects
- Multiple gate capacitance models
- Diffusion capacitance and resistance models
- Gate leakage models (in BSIM 4)

Some device parameters such as threshold voltage change significantly with device dimensions. BSIM models can be *binned* with different models covering different ranges of length and width specified by LMIN, LMAX, WMIN, and WMAX parameters. For example, one model might cover transistors with channel lengths from 0.18–0.25 μm, another from 0.25–0.5 μm, and a third from 0.5–5 μm. SPICE will complain if a transistor does not fit in one of the bins.

As the BSIM models are so complicated, it is impractical to derive closed-form equations for propagation delay, switching threshold, noise margins, etc., from the underlying equations. However, it is not difficult to find these properties through circuit simulation. Section 7.4 will show simple simulations to plot the device characteristics over the regions of operation that are interesting to most digital designers and to extract effective capacitance and resistance averaged across the switching transition. The simple RC model continues to give the designer important insight about the characteristics of logic gates.

7.3.4 Diffusion Capacitance Models

The p–n junction between the source or drain diffusion and the body forms a diode. We have seen that the diffusion capacitance determines the parasitic delay of a gate and

[1]IGFET in turn stands for Insulated-Gate Field Effect Transistor, a synonym for MOSFET.

depends on the area and perimeter of the diffusion. HSPICE provides a number of methods to specify this geometry, controlled by the ACM (Area Calculation Method) parameter, which is part of the transistor model. The model must also have values for junction and sidewall diffusion capacitance, as described in Section 2.3.3. The diffusion capacitance model is common across most device models including Levels 1–3 and BSIM.

By default, HSPICE models use ACM = 0. In this method, the designer must specify the area and perimeter of the source and drain of each transistor. For example, the dimensions of each diffusion region from Figure 2.8 are listed in Table 7.3 (in units of λ^2 for area or λ for perimeter). A SPICE description of the shared contacted diffusion case is shown in Figure 7.13, assuming `.option scale` is set to the value of λ.

TABLE 7.3 Diffusion area and perimeter

	AS1 / AD2	PS1 / PD2	AD1 / AS2	PD1 / PS2
(a) Isolated contacted diffusion	$W \times 5$	$2 \times W + 10$	$W \times 5$	$2 \times W + 10$
(b) Shared contacted diffusion	$W \times 5$	$2 \times W + 10$	$W \times 3$	$W + 6$
(c) Merged uncontacted diffusion	$W \times 5$	$2 \times W + 10$	$W \times 1.5$	$W + 3$

```
* Shared contacted diffusion
M1    mid    b    bot    gnd    NMOS    W='w'    L=2
+ AS='w*5' PS='2*w+10' AD='w*3' PD='w+6'
M2    top    a    mid    gnd    NMOS    W='w'    L=2
+ AS='w*3' PS='w+6' AD='w*5' PD='2*w+10'
```

FIGURE 7.13 SPICE model of transistors with shared contacted diffusion

The SPICE models also should contain parameters CJ, CJSW, PB, PHP, MJ, and MJSW. Assuming the diffusion is reverse-biased and the area and perimeter are specified, the diffusion capacitance between source and body is computed as described in Section 2.3.3.

$$C_{sb} = \mathrm{AS} \times \mathrm{CJ} \times \left(1 + \frac{V_{sb}}{\mathrm{PB}}\right)^{-MJ} + \mathrm{PS} \times \mathrm{CJSW} \times \left(1 + \frac{V_{sb}}{\mathrm{PHP}}\right)^{-MJSW} \quad \textbf{(7.3)}$$

The drain equations are analogous, with S replaced by D in the model parameters.

The BSIM3 models offer a similar area calculation model (ACM = 10) that takes into account the different sidewall capacitance on the edge adjacent to the gate. Note that the PHP parameter is renamed to PBSW to be more consistent.

$$C_{sb} = \mathrm{AS} \times \mathrm{CJ} \times \left(1 + \frac{V_{sb}}{\mathrm{PB}}\right)^{-MJ} + (\mathrm{PS} - W) \times \mathrm{CJSW} \times \left(1 + \frac{V_{sb}}{\mathrm{PBSW}}\right)^{-MJSW} + W \times \mathrm{CJSWG} \times \left(1 + \frac{V_{sb}}{\mathrm{PBSWG}}\right)^{-MJSWG} \quad \textbf{(7.4)}$$

If the area and perimeter are not specified, they default to 0 in ACM = 0 or 10, grossly underestimating the parasitic delay of the gate. HSPICE also supports ACM = 1, 2, 3, and 12 that provide nonzero default values when the area and perimeter are not specified. Check your models and read the HSPICE documentation carefully.

The diffusion area and perimeter are also used to compute the junction leakage current. However, this current is generally negligible compared to subthreshold leakage in modern devices.

7.3.5 Design Corners

Engineers often simulate circuits in multiple design corners to verify operation across variations in device characteristics and environment. HSPICE includes the `.lib` statement that makes changing libraries easy. For example, the deck in Figure 7.14 runs three simulations on the step response of an unloaded inverter in the TT, FF, and SS corners.

```
* corner.sp
* Step response of unloaded inverter across process corners

*----------------------------------------------------------------------
* Parameters and models
*----------------------------------------------------------------------
.option scale=25n
.param SUP=1.0 * Must set before calling .lib
.lib '../models/ibm065/opconditions.lib' TT
.option post

*----------------------------------------------------------------------
* Simulation netlist
*----------------------------------------------------------------------
Vdd   vdd   gnd   'SUPPLY'
Vin   a     gnd   PULSE     0 'SUPPLY' 25ps 0ps 0ps 35ps 80ps
M1    y     a     gnd       gnd      NMOS    W=4   L=2
+ AS=20 PS=18 AD=20 PD=18
M2    y     a     vdd       vdd      PMOS    W=8   L=2
+ AS=40 PS=26 AD=40 PD=26

*----------------------------------------------------------------------
* Stimulus
*----------------------------------------------------------------------
.tran 0.1ps 80ps
.alter
.lib '../models/ibm065/opconditions.lib' FF
.alter
.lib '../models/ibm065/opconditions.lib' SS
.end
```

FIGURE 7.14 CORNER SPICE deck

The deck first sets `SUP` to the nominal supply voltage of 1.0 V. It then invokes `.lib` to read in the library specifying the TT conditions. In the stimulus, the `.alter` statement is used to repeat the simulation with changes. In this case, the design corner is changed. Altogether, three simulations are performed and three sets of waveforms are generated for the three design corners.

The library file is given in Figure 7.15. Depending on what library was specified, the temperature is set (in degrees Celsius, with `.temp`) and the V_{DD} value `SUPPLY` is calculated from the nominal `SUP`. The library loads the appropriate nMOS and pMOS transistor models. A fast process file might have lower nominal threshold voltages V_{t0}, greater lateral diffusion L_D, and lower diffusion capacitance values.

```
* opconditions.lib
* For IBM 65 nm process

* TT: Typical nMOS, pMOS, voltage, temperature
.lib TT
.temp 70
.param SUPPLY='SUP'
.include 'modelsTT.sp'
.endl TT

* SS: Slow nMOS, pMOS, low voltage, high temperature
.lib SS
.temp 125
.param SUPPLY='0.9 * SUP'
.include 'modelsSS.sp'
.endl SS

* FF: Fast nMOS, pMOS, high voltage, low temperature
.lib FF
.temp 0
.param SUPPLY='1.1 * SUP'
.include 'modelsFF.sp'
.endl FF

* FS: Fast nMOS, Slow pMOS, typical voltage and temperature
.lib FS
.temp 70
.param SUPPLY='SUP'
.include 'modelsFS.sp'
.endl FS

* SF: Slow nMOS, Fast pMOS, typical voltage and temperature
.lib SF
.temp 70
.param SUPPLY='SUP'
.include 'modelsSF.sp'
.endl SF
```

FIGURE 7.15 OPCONDITIONS library

7.4 Device Characterization

Modern SPICE models have so many parameters that the designer cannot easily read key performance characteristics from the model files. A more convenient approach is to run a set of simulations to extract the effective resistance and capacitance, the fanout-of-4 inverter delay, the I-V characteristics, and other interesting data. This section describes these simulations and compares the results across a variety of CMOS processes.

7.4.1 I-V Characteristics

When familiarizing yourself with a new process, a starting point is to plot the current-voltage (I-V) characteristics. Although digital designers seldom make calculations directly from these plots, it is helpful to know the ON current of nMOS and pMOS transistors, how severely velocity-saturated the process is, how the current rolls off below threshold, how the devices are affected by DIBL and body effect, and so forth. These plots are made with DC sweeps, as discussed in Section 7.2.2. Each transistor is 1 μm wide in a representative 65 nm process at 70 °C with V_{DD} = 1.0 V. Figure 7.16 shows nMOS characteristics and Figure 7.17 shows pMOS characteristics.

Figure 7.16(a) plots I_{ds} vs. V_{ds} at various values of V_{gs}, as was done in Figure 7.5. The saturation current would ideally increase quadratically with $V_{gs} - V_t$, but in this plot it shows closer to a linear dependence, indicating that the nMOS transistor is severely velocity-saturated (α closer to 1 than 2 in the α-power model). The significant increase in saturation current with V_{ds} is caused by channel-length modulation. Figure 7.16(b) makes a similar plot for a device with a drawn channel length of twice minimum. The current drops by less than a factor of two because it experiences less velocity saturation. The current is slightly flatter in saturation because channel-length modulation has less impact at longer channel lengths.

Figure 7.16(c) plots I_{ds} vs. V_{gs} on a semilogarithmic scale for V_{ds} = 0.1 V and 1.0 V. The straight line at low V_{gs} indicates that the current rolls off exponentially below threshold. The difference in subthreshold leakage at the varying drain voltage reflects the effects

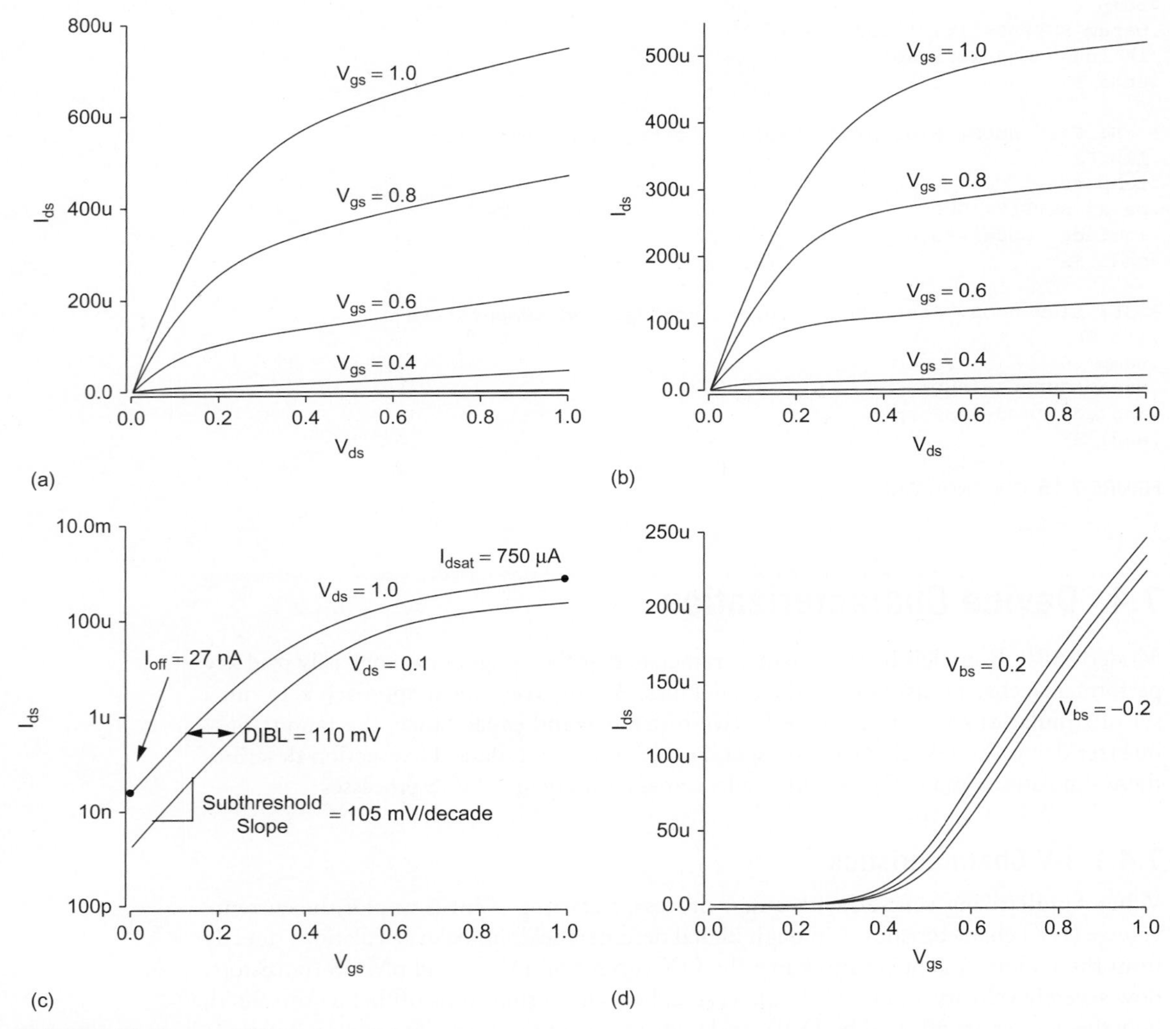

FIGURE 7.16 65 nm nMOS I-V characteristics

of drain-induced barrier lowering (DIBL) effectively reducing V_t at high V_{ds}. The saturation current I_{dsat} is measured at $V_{gs} = V_{ds} = V_{DD}$, while the OFF current I_{off} is measured at $V_{gs} = 0$ and $V_{ds} = V_{DD}$. The subthreshold slope is 105 mV/decade and DIBL reduces the effective threshold voltage by about 110 mV over the range of V_{ds}. The ratio of ON to OFF current is 4–5 orders of magnitude.

Figure 7.16(d) makes a similar plot on a linear scale for $V_{bs} = -0.2$, 0, and 0.2 V. V_{ds} is held constant at 0.1 V. The curves shift horizontally, indicating that the body effect increases the threshold voltage by 125 mV / V as V_{bs} becomes more negative.

Compare the pMOS characteristics in Figure 7.17. The saturation current for a pMOS transistor is lower than for the nMOS (note the different vertical scales), but the device is not as velocity-saturated.

Also compare the 180 nm nMOS characteristics in Figure 7.18. The saturation current is lower in the older technology, leading to lower performance. However, the device characteristics are closer to ideal. The channel-length modulation effect is not as pronounced, though velocity saturation is still severe. The subthreshold slope is 90 nV per decade and DIBL reduces the effective threshold voltage by 40 mV. The ratio of ON to OFF current is 6–7 orders of magnitude.

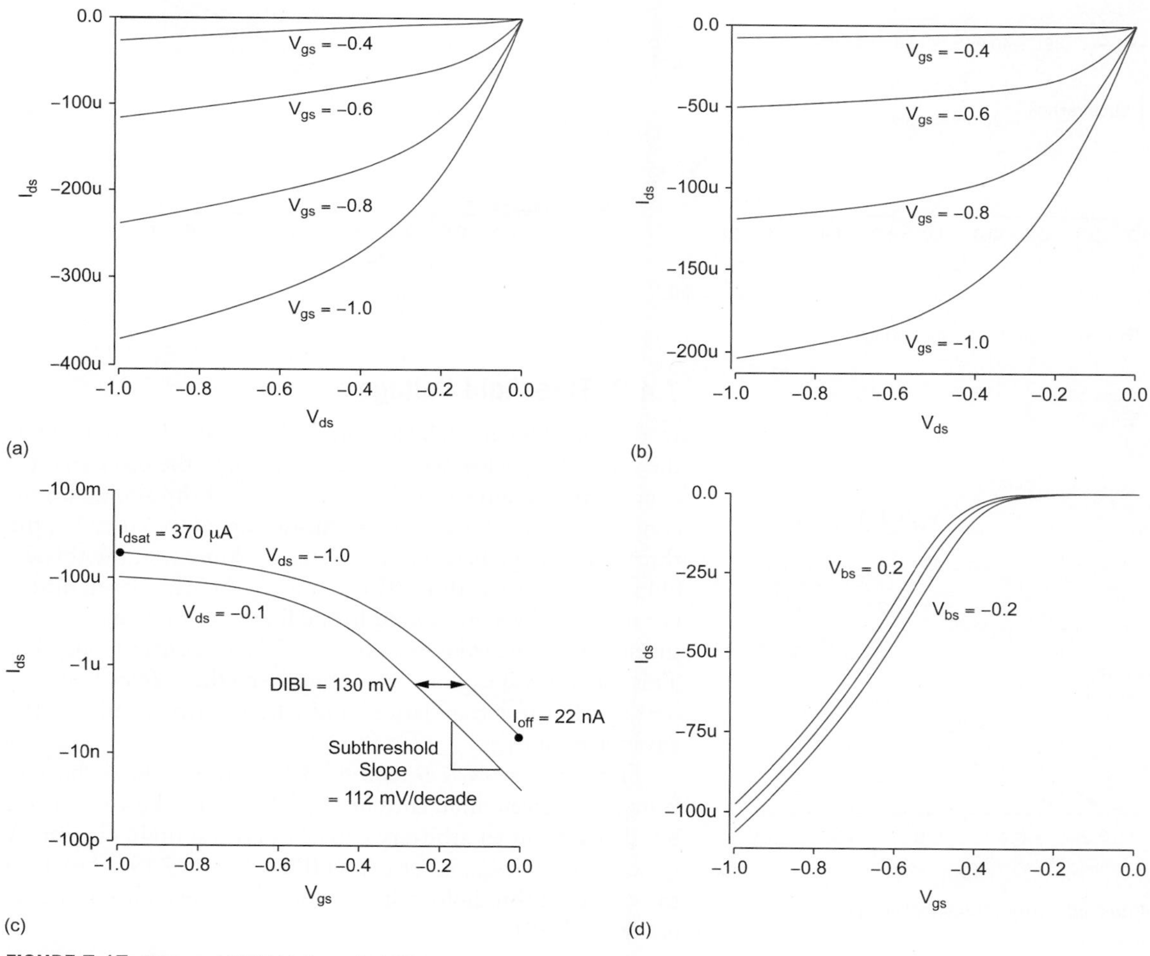

FIGURE 7.17 65 nm pMOS I-V characteristics

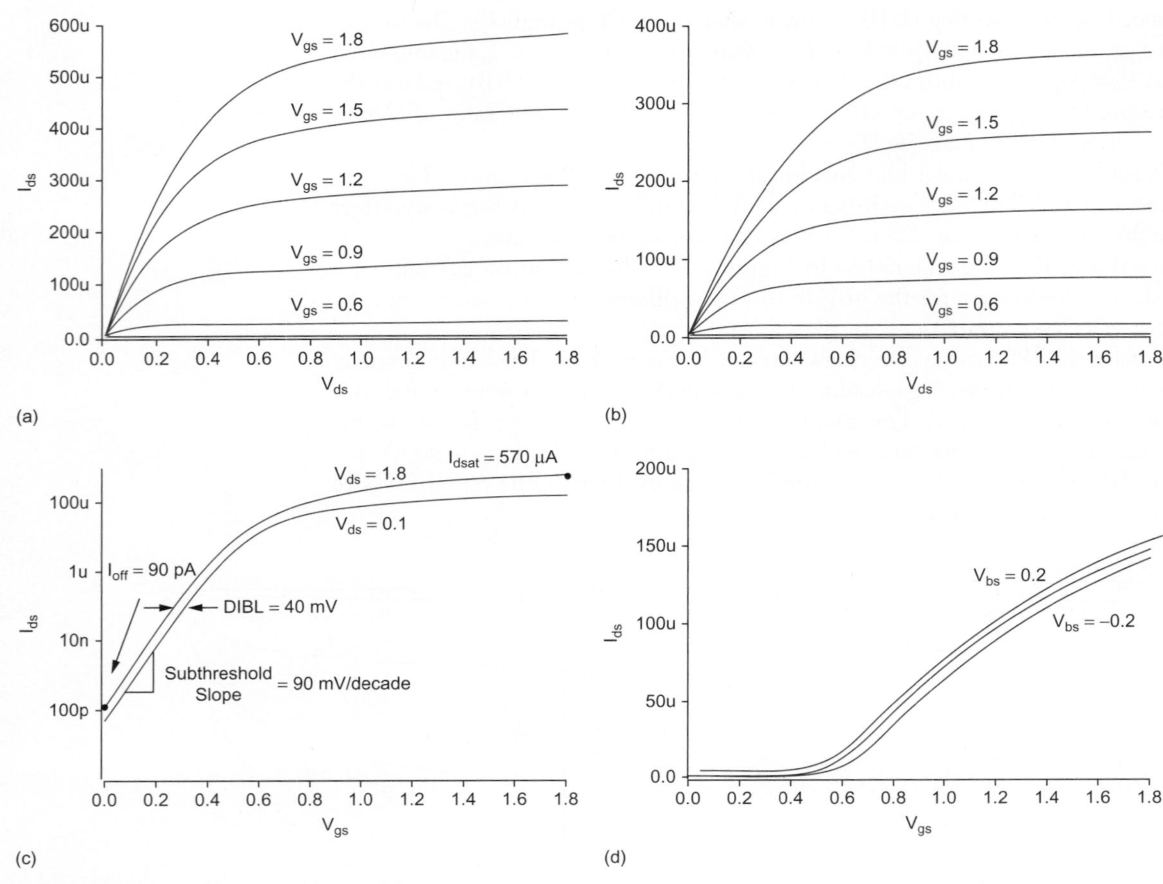

FIGURE 7.18 180 nm nMOS I-V characteristics

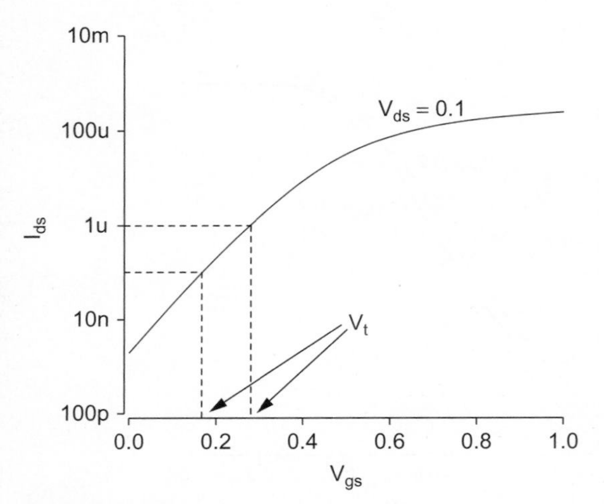

FIGURE 7.19 Constant current threshold voltage extraction method

7.4.2 Threshold Voltage

In the Shockley model, the threshold voltage V_t is defined as the value of V_{gs} below which I_{ds} becomes 0. In the real transistor characteristics shown in Figure 7.16(c), subthreshold current continues to flow for $V_{gs} < V_t$, so measuring or even defining the threshold voltage becomes problematic. Moreover, the threshold voltage varies with L, W, V_{ds}, and V_{bs}. At least eleven different methods have been used in the literature to determine the threshold voltage from measured I_{ds}-V_{gs} data [Ortiz-Conde02]. This section will explore two common methods (constant current and linear extrapolation) and a hybrid that combines the advantages of each.

The *constant current* method defines threshold as the gate voltage at a given drain current I_{crit}. This method is easy to use, but depends on an arbitrary choice of critical drain current. A typical choice of I_{crit} is 0.1 μA $\times$ (W/L). Figure 7.19 shows how the extracted threshold voltage varies with the choice of I_{crit} = 0.1 or 1 μA at V_{ds} = 100 mV.

The *linear extrapolation* (or *maximum-g_m*) method extrapolates the gate voltage from the point of maximum slope on the I_{ds}-V_{gs} characteristics. It is unambiguous but valid only for the linear region of operation (low V_{ds}) because of the series resistance of the source/drain diffusion and because drain-induced barrier lowering effectively reduces the threshold at high V_{ds}. Figure 7.20 shows how the threshold is extracted from measured data using the linear extrapolation method at V_{ds} = 100 mV. Observe that this method can give a significantly different threshold voltage and nonnegligible current at threshold, so it is important to check how the threshold voltage was measured when interpreting threshold voltage specifications. I_{crit} is defined to be the value of I_{ds} at $V_{gs} = V_t$.

[Zhou99] describes a hybrid method of extracting threshold voltage that is valid for all values of V_{ds} and does not depend on an arbitrary choice of critical current. V_t and I_{crit} are found at low V_{ds} (e.g., 100 mV) for a given value of L and W using the linear extrapolation method. For other values of V_{ds}, V_t is defined to be the gate voltage when $I_{ds} = I_{crit}$.

Figure 7.21(a) plots the threshold voltage V_t vs. length for a 16 λ wide device over a variety of design corners and temperatures. The threshold is extracted using the linear extrapolation method and clearly is not constant. It decreases with temperature and is lower in the FF corner than in the SS corner. In an ideal long-channel transistor, the threshold is independent of width and length. In a real device, the geometry sensitivity depends on the particular doping profile of the process. This data shows the threshold decreasing with L, but in many processes, the threshold increases with L. Figure 7.21(b) plots V_t against V_{ds} for 16/2 λ transistors using Zhou's method. The threshold voltage decreases with V_{ds} because of DIBL.

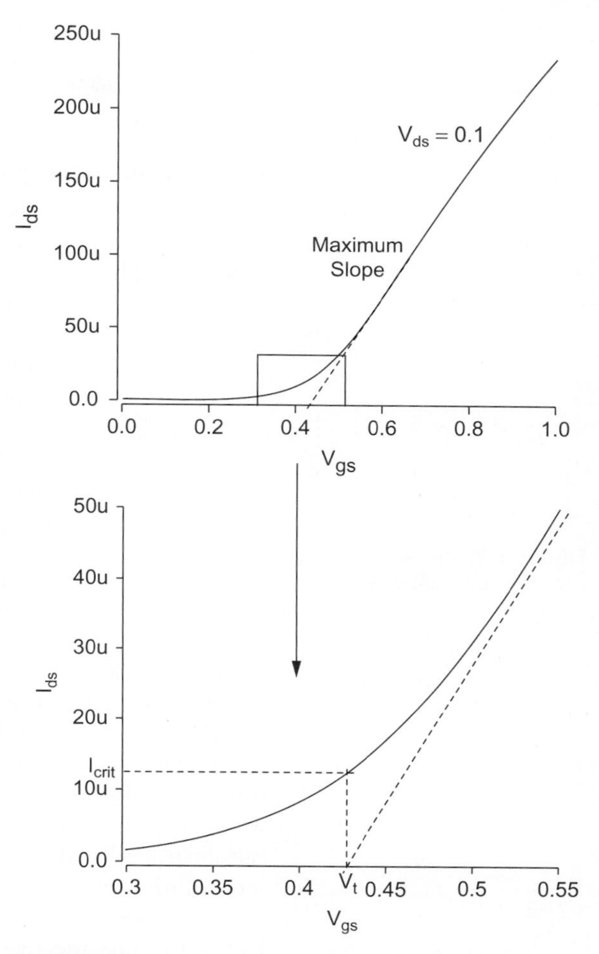

FIGURE 7.20 Linear extrapolation threshold voltage extraction method

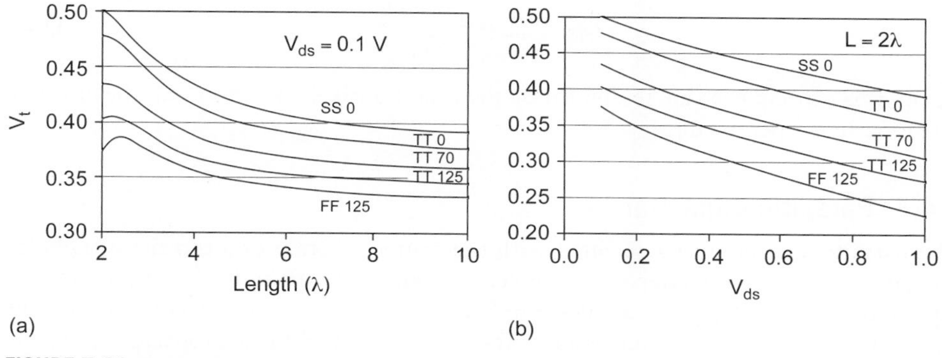

FIGURE 7.21 MOS threshold voltages

The lesson is that V_t depends on length, width, temperature, processing, and how you define it. The current does not abruptly drop to zero at threshold and is significant even for OFF devices in nanometer processes.

7.4.3 Gate Capacitance

When using RC models to estimate gate delay, we need to know the effective gate capacitance for delay purposes. In Section 2.3.2, we saw that the gate capacitance is voltage-dependent. The gate-to-drain component may be effectively doubled when a gate switches because the gate and drain switch in opposite directions. Nevertheless, we can obtain an effective capacitance averaged across the switching time. We use fanout-of-4 inverters to represent gates with "typical" switching times because we know from logical effort that circuits perform well when the stage effort is approximately 4.

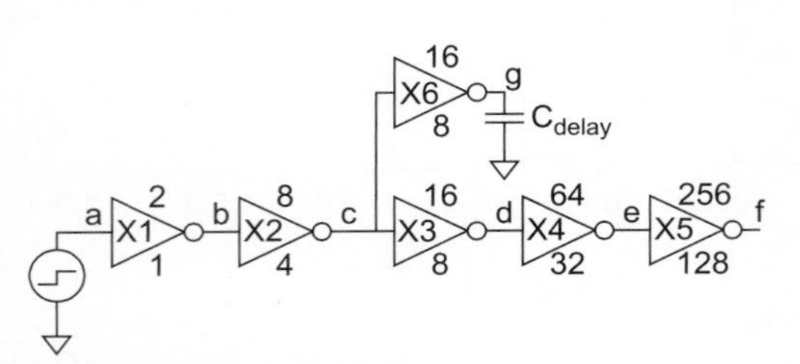

FIGURE 7.22 Circuit for extracting effective gate capacitance for delay estimation

Figure 7.22 shows a circuit for determining the effective gate capacitance of inverter $X4$. The approach is to adjust the capacitance C_{delay} until the average delay from c to g equals the delay from c to d. Because $X6$ and $X3$ have the same input slope and are the same size, when they have the same delay, C_{delay} must equal the effective gate capacitance of $X4$. $X1$ and $X2$ are used to produce a reasonable input slope on node c. A single inverter could suffice, but the inverter pair is even better because it provides a slope on c that is essentially independent of the rise time at a. $X5$ is the load on $X4$ to prevent node e from switching excessively fast, which would overpredict the significance of the gate-to-drain capacitance in $X4$.

Figure 7.23 (on page 309) lists a SPICE deck that uses the optimizer to automatically tune C_{delay} until the delays are equalized. This capacitance is divided by the total gate width (in μm) of $X4$ to obtain the capacitance per micron of gate width $C_{\text{permicron}}$. This capacitance is listed as C_g (delay) in Table 7.5 for a variety of processes. Note that the deck sets diffusion area and perimeter to 0 to measure only the gate capacitance.

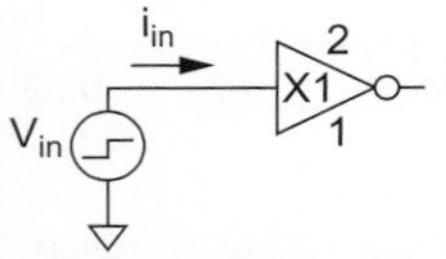

FIGURE 7.24 Circuit for extracting effective gate capacitance for power estimation

Gate capacitance is also important for dynamic power consumption, as was given in EQ (4.10). The effective gate capacitance for power is typically somewhat higher than for delay because C_{gd} is effectively doubled by the Miller effect when we wait long enough for the drain to completely switch. Figure 7.24 shows a circuit for measuring gate capacitance for power purposes. A voltage step is applied to the input, and the current out of the voltage source is integrated. The effective capacitance for dynamic power consumption is:

$$C_{\text{eff-power}} = \frac{\int i_{\text{in}}(t)\,dt}{V_{DD}} \tag{7.5}$$

Again, this capacitance can be divided by the total transistor width to find the effective gate capacitance per micron.

7.4.4 Parasitic Capacitance

The parasitic capacitance associated with the source or drain of a transistor includes the gate-to-diffusion overlap capacitance, C_{gol}, and the diffusion area and perimeter capacitance C_{jb} and C_{jbsw}. As discussed in Section 7.3.4, some models assign a different capacitance C_{jbswg} to the perimeter along the gate side. The diffusion capacitance is voltage-dependent, but as with gate capacitance, we can extract an effective capacitance averaged over the switching transition to use for delay estimation.

```
* capdelay.hsp
* Extract effective gate capacitance for delay estimation.
*----------------------------------------------------------------------
* Parameters and models
*----------------------------------------------------------------------
.option scale=25n
.param SUP=1.0 * Must set before calling .lib
.lib '../models/ibm065/opconditions.lib' TT
.option post
*----------------------------------------------------------------------
* Subcircuits
*----------------------------------------------------------------------
.global  vdd   gnd
.subckt  inv   a     y
M1       y     a     gnd   gnd   NMOS   W=16 L=2 AD=0 AS=0 PD=0 PS=0
M2       y     a     vdd   vdd   PMOS   W=32 L=2 AD=0 AS=0 PD=0 PS=0
.ends
*----------------------------------------------------------------------
* Simulation netlist
*----------------------------------------------------------------------
Vdd      vdd   gnd   'SUPPLY'        * SUPPLY is set by .lib call
Vin      a     gnd   pulse 0 'SUPPLY' 0ps 20ps 20ps 120ps 280ps
X1       a     b     inv             * set appropriate slope
X2       b     c     inv   M=4       * set appropriate slope
X3       c     d     inv   M=8       * drive real load
X4       d     e     inv   M=32      * real load
X5       e     f     inv   M=128     * load on load (important!)
X6       c     g     inv   M=8       * drive linear capacitor
cdelay   g     gnd   'CperMicron*32*(16+32)*25n/1u' * linear capacitor
*----------------------------------------------------------------------
* Optimization setup
*----------------------------------------------------------------------
.measure errorR param='invR - capR' goal=0
.measure errorF param='invF - capF' goal=0
.param CperMicron=optrange(2f, 0.2f, 3.0f)
.model optmod opt itropt=30
.measure CperMic param = 'CperMicron'
*----------------------------------------------------------------------
* Stimulus
*----------------------------------------------------------------------
.tran 1ps 280ns SWEEP OPTIMIZE = optrange
+       RESULTS=errorR,errorF MODEL=optmod
.measure invR
+       TRIG v(c)  VAL='SUPPLY/2' FALL=1
+       TARG v(d)  VAL='SUPPLY/2' RISE=1
.measure  capR
+       TRIG v(c)  VAL='SUPPLY/2' FALL=1
+       TARG v(g)  VAL='SUPPLY/2' RISE=1
.measure invF
+       TRIG v(c)  VAL='SUPPLY/2' RISE=1
+       TARG v(d)  VAL='SUPPLY/2' FALL=1
.measure  capF
+       TRIG v(c)  VAL='SUPPLY/2' RISE=1
+       TARG v(g)  VAL='SUPPLY/2' FALL=1
.end
```

FIGURE 7.23 CAPDELAY SPICE deck

Figure 7.25 shows circuits for extracting these capacitances. They operate in much the same way as the gate capacitance extraction from Section 7.4.3. The first two fanout-of-4 inverters shape the input slope to match a typical gate. *X*3 drives the drain of an OFF

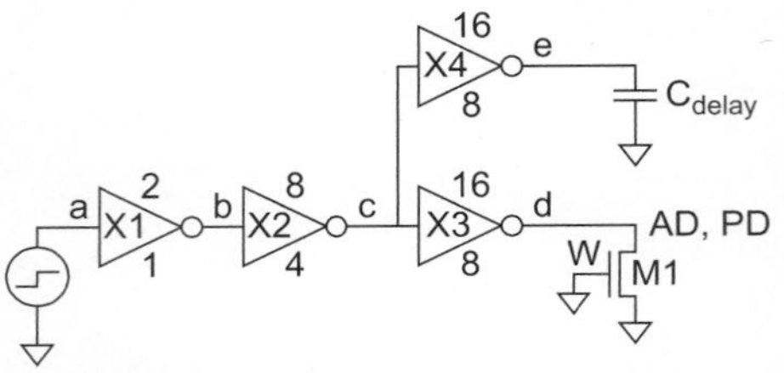

FIGURE 7.25 Circuit for extracting effective parasitic capacitance for delay estimation

transistor $M1$ with specified W, AD, and PD. $X4$ drives a simple capacitor, whose value is optimized so that the delay of $X3$ and $X4$ are equal. This value is the effective capacitance of $M1$'s drain. Similar simulations must be run to find the parasitic capacitances of pMOS transistors.

Table 7.4 lists the appropriate values of W, AD, and PD to extract each of the capacitances. The sizes are chosen such that the gate delays and slope on node d are reasonable when a unit transistor is 16 λ wide (as in Figure 7.23). It also gives values to find the effective capacitance C_d of isolated-contacted, shared-contacted, and merged-uncontacted diffusion regions. The capacitance is found, assuming the transistors are wide enough that the perimeter perpendicular to the polysilicon gate is a negligible fraction of the overall capacitance. The AD and PD dimensions are based on the layouts of Figure 2.8; you should substitute your own design rules. The total capacitance of shared and merged regions should be split between the two transistors sharing the diffusion node. The capacitance can be converted to units per micron (or per micron squared) by normalizing for the value of λ. For example, in our 65 nm process, if C_{delay} is 23 fF for gate overlap, the capacitance per micron is

$$C_{gol} = \frac{23\text{ fF}}{(1600\ \lambda)\left(\frac{0.025\ \mu\text{m}}{\lambda}\right)} = 0.57\frac{\text{fF}}{\mu\text{m}} \tag{7.6}$$

TABLE 7.4 Dimensions for diffusion capacitance extraction

	W (λ)	AD (λ^2)	PD (λ)	To find effective *C* per micron
C_{gol}	1600	0	0	$C_{delay}/1600\lambda$ (per μm)
C_{jb}	0	8000	0	$C_{delay}/8000\lambda^2$ (per μm^2)
C_{jbsw}	0	0	1600	$C_{delay}/1600\lambda$ (per μm)
C_{jbswg}	1600	0	1600	$C_{delay}/1600\lambda - C_{gol}$ (per μm)
C_d (isolated-contacted)	1600	8000	3200	$C_{delay}/1600\lambda$ (per μm of gate width)
C_d (shared-contacted)	3200	9600	3200	$C_{delay}/1600\lambda$ (per μm of gate width)
C_d (merged-uncontacted)	3200	4800	3200	$C_{delay}/1600\lambda$ (per μm of gate width)

7.4.5 Effective Resistance

If a unit transistor has gate capacitance C, parasitic capacitance C_d, and resistance R_n (for nMOS) or R_p (for pMOS), the rising and falling delays of a fanout-of-h inverter with a 2:1 P/N ratio can be found according to Figure 7.26. These delays can readily be measured from the FO4 inverter simulation in Figure 7.10 by changing h.

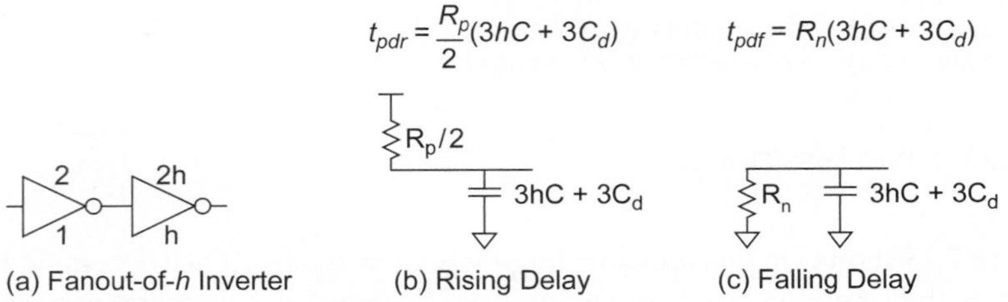

$$t_{pdr} = \frac{R_p}{2}(3hC + 3C_d) \qquad t_{pdf} = R_n(3hC + 3C_d)$$

(a) Fanout-of-*h* Inverter (b) Rising Delay (c) Falling Delay

FIGURE 7.26 RC delay model for fanout-of-*h* inverter

The dependence on parasitics can be removed by calculating the difference between delays at different fanouts. For example, the difference between delays for $h = 3$ and $h = 4$ are

$$\begin{aligned}\Delta t_{pdr} &= \frac{R_p}{2}\left(3\times 4\times C + 3C_d\right) - \frac{R_p}{2}\left(3\times 3\times C + 3C_d\right) = \tfrac{3}{2}R_p C \\ \Delta t_{pdf} &= R_n\left(3\times 4\times C + 3C_d\right) - R_n\left(3\times 3\times C + 3C_d\right) = 3R_n C\end{aligned} \quad (7.7)$$

As C is known from the effective gate capacitance extraction, R_n and R_p are readily calculated. These represent the effective resistance of single nMOS and pMOS transistors for delay estimation.

When two unit transistors are in series, each nominally would have the same effective resistance, giving twice the overall resistance. However, in modern processes where the transistors usually experience some velocity saturation, each transistor sees a smaller V_{ds} and hence less velocity saturation and a lower effective resistance. We can determine this resistance by simulating fanout-of-h tristates in place of inverters, as shown in Figure 7.27. By a similar reasoning, the difference between delays from c to d for $h = 2$ and $h = 3$ is

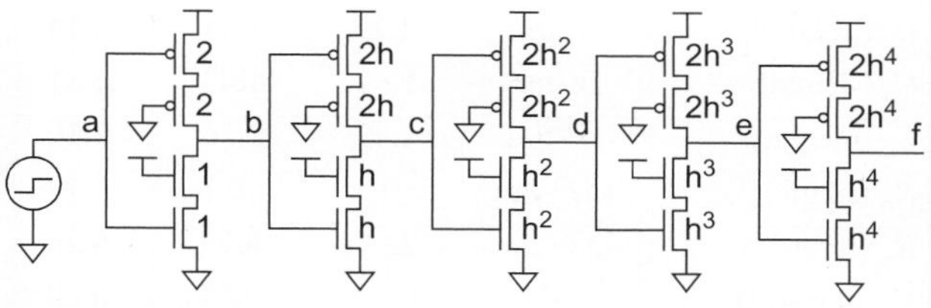

FIGURE 7.27 Circuit for extracting effective series resistance

$$\begin{aligned}\Delta t_{pdr} &= \tfrac{3}{2}\left(2R_{p\text{-series}}\right)C \\ \Delta t_{pdf} &= 3\left(2R_{n\text{-series}}\right)C\end{aligned} \quad (7.8)$$

As C is still known, we can extract the effective resistance of series nMOS and pMOS transistors for delay estimation and should expect this resistance to be smaller than for single transistors.

It is important to use realistic input slopes when extracting effective resistance because the delay varies with input slope. *Realistic* means that the input and output edge rates should be comparable; if a step input is applied, the output will transition faster and the effective resistance will appear to decrease. h was chosen in this section to give stage efforts close to 4.

7.4.6 Comparison of Processes

Table 7.5 compares the characteristics of a variety of CMOS processes with feature sizes ranging from 2 μm down to 65 nm. The older models are obtained from MOSIS wafer test results [Piña02], while the newer models are from IBM or TSMC. The MOSIS models use ACM = 0, so the diffusion sidewall capacitance is treated the same along the gate and the other walls. The 0.6 μm process operates at either V_{DD} = 5 V (for higher speed) or V_{DD} = 3.3 V (for lower power). All characteristics are extracted for TTTT conditions (70 °C) for normal-V_t transistors.

Transistor lengths are usually shorter than the nominal feature size. For example, in the 0.6 μm process, MOSIS preshrinks polysilicon by 0.1 μm before generating masks. In the IBM process, transistors are drawn somewhat shorter than the feature size. Moreover, gates are usually processed such that the effective channel length is even shorter than the drawn channel length. The shorter channels make transistors faster than one might expect simply based on feature size.

TABLE 7.5 Device characteristics for a variety of processes

Vendor		Orbit	HP	AMI	AMI	TSMC	TSMC	TSMC	IBM	IBM	IBM
Model		MOSIS	MOSIS	MOSIS	MOSIS	MOSIS	MOSIS	TSMC	IBM	IBM	IBM
Feature Size f	nm	2000	800	600	600	350	250	180	130	90	65
V_{DD}	V	5	5	5	3.3	3.3	2.5	1.8	1.2	1.0	1.0
Gates											
C_g (delay)	fF/μm	1.77	1.67	1.55	1.48	1.90	2.30	1.67	1.04	0.97	0.80
C_g (power)	fF/μm	2.24	1.70	1.83	1.76	2.20	2.92	2.06	1.34	1.23	1.07
FO4 Inv. Delay	ps	856	297	230	312	210	153	75.6	45.9	37.2	17.2
nMOS											
C_d (isolated)	fF/μm	1.19	1.11	1.14	1.21	1.63	1.88	1.12	0.94	0.89	0.76
C_d (shared)	fF/μm	1.62	1.43	1.41	1.50	2.04	2.60	1.62	1.56	1.60	1.28
C_d (merged)	fF/μm	1.48	1.36	1.19	1.24	1.60	2.16	1.41	1.40	1.51	1.20
R_n (single)	k$\Omega \cdot \mu$m	30.3	10.1	9.19	11.9	5.73	4.02	2.69	2.54	2.35	1.34
R_n (series)	k$\Omega \cdot \mu$m	22.1	6.95	6.28	8.59	4.01	3.10	2.00	1.93	1.81	1.13
V_{tn} (const. I)	V	0.65	0.65	0.70	0.70	0.59	0.48	0.41	0.32	0.32	0.31
V_{tn} (linear ext.)	V	0.65	0.75	0.76	0.76	0.67	0.57	0.53	0.43	0.43	0.43
I_{dsat}	μA/μm	152	380	387	216	450	551	566	478	497	755
I_{off}	pA/μm	2.26	9.36	2.21	1.45	6.57	56.3	93.9	1720	4000	33400
I_{gate}	pA/μm	n/a	n/a	n/a	n/a	n/a	n/a	n/a	1.22	3620	8520
pMOS											
C_d (isolated)	fF/μm	1.42	1.17	1.31	1.42	1.89	2.07	1.24	0.94	0.74	0.73
C_d (shared)	fF/μm	1.92	1.62	1.73	1.86	2.37	2.89	1.79	1.56	1.25	1.25
C_d (merged)	fF/μm	1.52	1.23	1.35	1.43	1.83	2.40	1.56	1.41	1.16	1.18
R_p (single)	k$\Omega \cdot \mu$m	67.1	26.7	19.9	29.6	16.1	8.93	6.51	6.39	5.47	2.87
R_p (series)	k$\Omega \cdot \mu$m	53.9	21.4	15.4	23.6	13.3	6.91	5.41	5.48	4.92	2.42
$\lvert V_{tp} \rvert$ (const. I)	V	0.72	0.91	0.90	0.90	0.83	0.46	0.43	0.33	0.35	0.33
$\lvert V_{tp} \rvert$ (linear ext.)	V	0.71	0.94	0.93	0.93	0.88	0.52	0.51	0.42	0.43	0.42
I_{dsat}	μA/μm	70.5	154	215.3	99.0	181	245	228	177	187	360
I_{off}	pA/μm	2.18	1.57	2.08	1.38	2.06	30.1	25.2	1330	2780	19500
I_{gate}	pA/μm	n/a	n/a	n/a	n/a	n/a	n/a	n/a	0.06	1210	2770

The gate capacitance for delay held steady near 2 fF/μm for many generations, as scaling theory would predict, but abruptly dropped after the 180 nm generation. The gate capacitance for power is slightly higher than that for delay as discussed in Section 7.4.3.

The FO4 inverter delay has steadily improved with feature size as constant field scaling predicts. It fits our rule from Section 3.4.3 of one third to one half of the effective channel length, when delay is measured in picoseconds and length in nanometers.

Diffusion capacitance of an isolated contacted source or drain has been 1–2 fF/μm for both nMOS and pMOS transistors over many generations. The capacitance of a shared contacted diffusion region is slightly higher because it has more area and includes two gate overlaps. The capacitance of the merged diffusion reflects two gate overlaps but a smaller diffusion area. Half the capacitance of the shared and merged diffusions is allocated to each of the transistors connected to the diffusion region.

The effective resistance of a 1 μm wide transistor has decreased with process scaling in proportion to the feature size f. However, the resistance of a unit (4/2 λ) nMOS transistor, $R/2f$, has remained roughly constant around 8 kΩ, as constant field scaling theory would predict. The effective resistance of pMOS transistors is 2–3 times that of nMOS transistors. A pair of nMOS transistors in series each have lower effective resistance than a single device because each has a smaller V_{ds} and thus experiences less velocity saturation. Series pMOS transistors show less pronounced improvement because they were not as velocity-saturated to begin with.

Threshold voltages are reported at V_{ds} = 100 mV for 16/2 λ devices using both the constant current (at $I_{crit} = 0.1(W/L)$ μA for nMOS and $0.06(W/L)$ for pMOS) and linear extrapolation methods. Threshold voltages have generally decreased, but not as fast as channel length or supply voltage (because of subthreshold leakage). Therefore, the V_{DD}/V_t ratio is decreasing and pass transistor circuits with threshold drops do not perform well in modern processes.

Saturation current per micron has increased somewhat through aggressive device design as feature size decreases even though constant field scaling would suggest it should remain constant. OFF current was on the order of a few picoamperes per micron in old processes, but is exponentially increasing in nanometer processes because of subthreshold conduction through devices with low threshold voltages. The current at threshold using the linear extrapolation method is somewhat higher than the constant current I_{crit}, corresponding to the higher threshold voltages found by the linear extrapolation method. Gate leakage has become significant below 90 nm.

7.4.7 Process and Environmental Sensitivity

Table 7.6 shows how the IBM 65 nm process characteristics vary with process corner, voltage, and temperature. The FO4 inverter delay varies by a factor of two between best and worst case. In the TT process, inverter delay varies by about 0.12%/°C and by about 1% for every percent of supply voltage change. These figures agree well with the Artisan library data from Section 6.2.4. Gate and diffusion capacitance change only slightly with process, but effective resistance is inversely proportional to supply voltage and highly sensitive to temperature and device corners. I_{off} subthreshold leakage rises dramatically at high temperature or in the fast corner where threshold voltages are lower.

7.5 Circuit Characterization

The device characterization techniques from the previous section are typically run once by engineers who are familiarizing themselves with a new process. SPICE is used more often to characterize entire circuits. This section gives some pointers on simulating paths and describes how to find the DC transfer characteristics, logical effort, and power consumption of logic gates.

7.5.1 Path Simulations

The delays of most static CMOS circuit paths today are computed with a static timing analyzer (see Sections 3.6 and 13.4.1.4). As long as the noise sources (particularly coupling and power supply noise) are controlled, the circuits will operate correctly and will

TABLE 7.6 Process corners of IBM 65 nm process

nMOS		T	F	S	F	S	T	T	T	T
pMOS		T	F	S	S	F	T	T	T	T
V_{DD}	V	1.0	1.1	0.9	1.0	1.0	1.1	0.9	1.0	1.0
T	°C	70	0	125	70	70	70	70	0	125
Gates										
C_g (delay)	fF/μm	0.80	0.79	0.80	0.82	0.79	0.82	0.78	0.79	0.81
C_g (power)	fF/μm	1.07	1.05	1.07	1.07	1.04	1.07	1.04	1.07	1.06
FO4 Inv. Delay	ps	17.2	12.2	24.4	17.4	17.1	15.1	20.4	16.6	17.5
nMOS										
C_d (isolated)	fF/μm	0.76	0.72	0.79	0.72	0.80	0.75	0.77	0.75	0.76
C_d (shared)	fF/μm	1.28	1.22	1.33	1.22	1.33	1.26	1.29	1.27	1.28
C_d (merged)	fF/μm	1.20	1.15	1.25	1.15	1.25	1.19	1.22	1.20	1.21
R_n (single)	k$\Omega \cdot \mu$m	1.34	0.96	1.92	1.21	1.49	1.16	1.63	1.31	1.37
R_n (series)	k$\Omega \cdot \mu$m	1.13	0.79	1.66	0.96	1.31	0.97	1.39	1.09	1.16
V_{tn} (const. I)	V	0.31	0.32	0.30	0.27	0.34	0.31	0.31	0.36	0.27
V_{tn} (linear ext.)	V	0.43	0.45	0.43	0.40	0.47	0.43	0.43	0.48	0.40
I_{dsat}	μA/μm	755	1094	510	844	672	919	596	793	731
I_{off}	nA/μm	33.4	22.4	38.9	95.0	12.3	41.5	2.7	4.6	120
I_{gate}	nA/μm	8.5	13.6	5.0	8.9	8.1	12.6	5.7	8.1	8.9
pMOS										
C_d (isolated)	fF/μm	0.73	0.69	0.77	0.77	0.70	0.72	0.74	0.72	0.74
C_d (shared)	fF/μm	1.25	1.18	1.32	1.30	1.20	1.23	1.26	1.23	1.26
C_d (merged)	fF/μm	1.18	1.12	1.24	1.23	1.13	1.16	1.20	1.17	1.19
R_p (single)	k$\Omega \cdot \mu$m	2.87	2.09	3.99	3.10	2.65	2.46	3.47	2.82	2.89
R_p (series)	k$\Omega \cdot \mu$m	2.42	1.67	3.47	2.69	2.04	1.16	3.05	2.40	2.35
V_{tn} (const. I)	V	0.33	0.36	0.32	0.37	0.30	0.33	0.33	0.39	0.28
V_{tp} (linear ext.)	V	0.42	0.44	0.41	0.45	0.39	0.42	0.42	0.47	0.39
I_{dsat}	μA/μm	360	517	247	319	407	438	285	373	353
I_{off}	nA/μm	19.5	7.4	27.7	7.0	53.0	24.0	15.7	1.4	86.1
I_{gate}	nA/μm	2.8	4.3	1.7	2.6	2.9	4.0	1.9	2.5	2.9

correlate reasonably well with static timing predictions. However, SPICE-level simulation is important for sensitive circuits such as the clock generator and distribution network, custom memory arrays, and novel circuit techniques.

Most experienced designers begin designing paths based on simple models in order to understand what aspects are most important, evaluate design trade-offs, and obtain a qualitative prediction of the results. The ideal Shockley transistor models, RC delay models, and logical effort are all helpful here because they are simple enough to give insight. When a good first-pass design is ready, the designer simulates the circuit to verify that it operates correctly and meets delay and power specifications. Just as few new software programs run correctly before debugging, the simulation often will be incorrect at first. Unless the designer knows what results to expect, it is tempting to trust the false results that are nicely printed with beguilingly many significant figures. Once the circuit appears to be correct, it should be checked across design corners to verify that it operates in all cases. Section 8.2.4 gives examples of circuits sensitive to various corners.

Simulation is cheap, but silicon revisions are devastating expensive. Therefore, it is important to construct a circuit model that captures all of the relevant conditions, including real input waveforms, appropriate output loading, and adequate interconnect models. When matching is important, you must consider the effects of mismatches that are not given in the corner files (see Section 7.5.5). However, as SPICE decks get more complicated, they run more slowly, accumulate more mistakes, and are more difficult to debug. A good compromise is to start simple and gradually add complexity, ensuring after each step that the results still make sense.

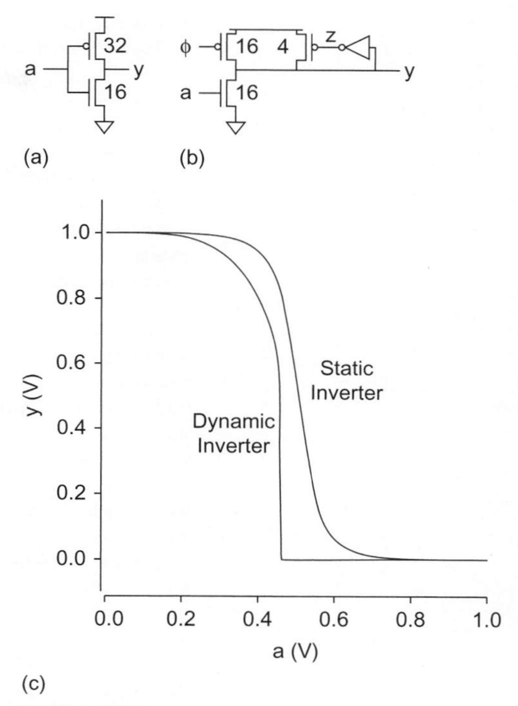

FIGURE 7.28 Circuits for DC transfer analysis

7.5.2 DC Transfer Characteristics

The `.dc` statement is useful for finding the transfer characteristics and noise margins of logic gates. Figure 7.29 shows an example of characterizing static and dynamic inverters (dynamic logic is covered in Section 8.2.4). Figure 7.28(a and b) show the circuit schematics of each gate. Figure 7.28(c) shows the simulation results. The static inverter characteristics are nearly symmetric around $V_{DD}/2$. The dynamic inverter has a lower switching threshold and its output drops abruptly beyond this threshold because positive feedback turns off the keeper.

Note that when the input `a` is 0 and the dynamic inverter is in evaluation ($\phi = 1$), the output would be stable at either 0 or 1. To find the transfer characteristics, we initialize the gate with a 1 output using the `.ic` command.

7.5.3 Logical Effort

The logical effort and parasitic delay of each input of a gate can be measured by fitting a straight line to delay vs. fanout simulation results. As with the FO4 inverter example, it is important to drive the gate with an appropriate input waveform and to provide two stages of loads. Figure 7.30(a) shows an example of a circuit for characterizing the delay of a 2-input NAND gate $X3$ using the `M` parameter to simulate multiple gates in parallel. Figure 7.30(b) plots the delay vs. fanout in a 65 nm process for an inverter and the 2-input NAND. The data is well-fit by a straight line even though the transistors experience all sorts of nonlinear and nonideal effects. This shows that the linear delay model is quite accurate as long as the input and output slopes are consistent.

The `SWEEP` command is convenient to vary the fanout and repeat the transient simulation multiple times. For example, the following statement runs eight simulations varying `H` from 1 to 8 in steps of 1.

```
.tran 1ps 1000ps SWEEP H 1 8 1
```

To characterize an entire library, you can write a script in a language such as Perl or Python that generates the appropriate SPICE decks, invokes the simulator, and post-processes the list files to extract the data and do the curve fit.

Recall that τ is the coefficient of h (i.e., the slope) in a delay vs. fanout plot for an inverter; in this process it is 3.3 ps. The parasitic delay of the inverter is found from the

```
* invdc.sp
* Static and dynamic inverter DC transfer characteristics

*----------------------------------------------------------------------
* Parameters and models
*----------------------------------------------------------------------
.param SUPPLY=1.0
.option scale=25n
.include '../models/ibm065/models.sp'
.temp 70
.option post

*----------------------------------------------------------------------
* Simulation netlist
*----------------------------------------------------------------------
Vdd   vdd   gnd   'SUPPLY'
Va    a     gnd   0
Vclk  clk   gnd   'SUPPLY'
* Static Inverter
M1    y1    a     gnd   gnd   NMOS   W=16   L=2
M2    y1    a     vdd   vdd   PMOS   W=32   L=2
* Dynamic Inverter
M3    y2    a     gnd   gnd   NMOS   W=16   L=2
M4    y2    clk   vdd   vdd   PMOS   W=16   L=2
M5    y2    z     vdd   vdd   PMOS   W=4    L=2
M6    z     y2    gnd   gnd   NMOS   W=4    L=2
M7    z     y2    vdd   vdd   PMOS   W=8    L=2
.ic V(y2) = 'SUPPLY'

*----------------------------------------------------------------------
* Stimulus
*----------------------------------------------------------------------
.dc Va 0 1.0 0.01
.end
```

FIGURE 7.29 INVDC SPICE deck for DC transfer analysis

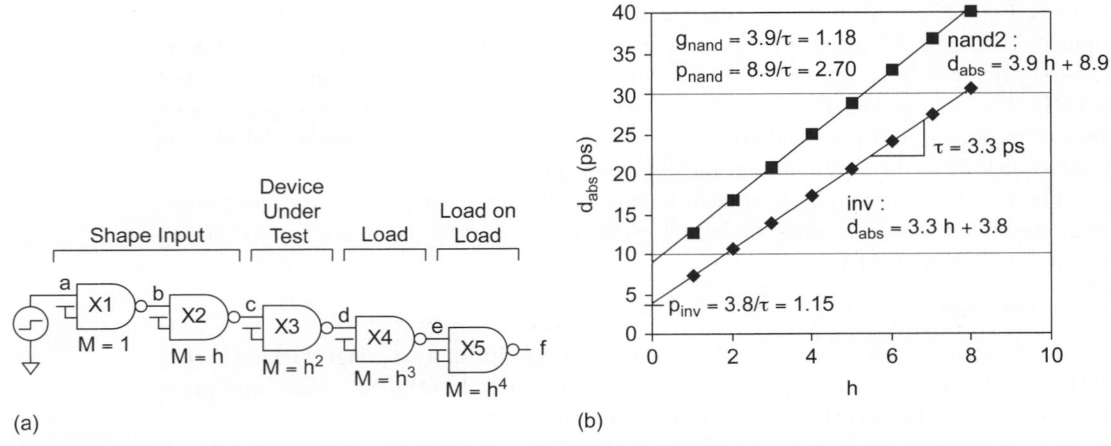

FIGURE 7.30 Logical effort characterization of 2-input NAND gate and inverter

y-intercept of the fit line; it is 3.8 ps, or 1.15 in normalized units. Similarly, the logical effort and parasitic delay of the NAND gate are obtained by normalizing the slope and y-intercept by τ.

Table 7.7 compares the logical effort and parasitic delay of the different inputs of multi-input NAND gates for rising, falling, and average output transitions in the IBM 65 nm process. For rising and falling transitions, we still normalize against the value of τ found from the average delay of an inverter. Input *A* is the outermost (closest to power or ground). As discussed in Section 8.2.1.3, the outer input has higher parasitic delay, but slightly lower logical effort. The rising and falling delays in this process are quite different because pMOS transistors have less than half the mobility of nMOS transistors and because the nMOS transistors are quite velocity-saturated so that series transistors have less resistance than expected.

TABLE 7.7 Logical effort and parasitic delay of different inputs of multi-input NAND gates

# of inputs	Input	Rising Logical Effort g_u	Falling Logical Effort g_u	Average Logical Effort g	Rising Parasitic Delay p_u	Falling Parasitic Delay p_d	Average Parasitic Delay p
2	*A*	1.40	1.12	**1.26**	2.46	2.48	**2.47**
	B	1.31	1.16	**1.24**	1.97	1.82	**1.89**
3	*A*	1.76	1.27	**1.51**	4.77	4.10	**4.44**
	B	1.73	1.32	**1.52**	3.93	3.60	**3.77**
	C	1.59	1.38	**1.48**	3.05	2.43	**2.74**
4	*A*	2.15	1.42	**1.78**	7.63	5.94	**6.79**
	B	2.09	1.48	**1.78**	6.67	5.37	**6.02**
	C	2.08	1.53	**1.80**	5.32	4.51	**4.91**
	D	1.90	1.59	**1.75**	4.04	2.93	**3.49**

Table 7.8 compares the average logical effort and parasitic delay of a variety of gates in many different processes. In each case, the simulations are performed in the TTTT corner for the outer input. For reference, the FO4 inverter delay and τ are given for each process. The logical effort of gates with series transistors is lower than predicted in Section 3.4.1 because one of the transistors is already fully ON and hence has a lower effective resistance than the transistor that is turning ON during the transition. Moreover, the logical effort of NAND gates is even lower because velocity saturation has a smaller effect on series nMOS transistors that see only part of the electric field between drain and source as compared to a single nMOS transistor that experiences the entire field. This effect is less significant for NOR gates because pMOS transistors have lower mobility and thus experience less velocity saturation. The efforts are fairly consistent across process and voltage. In comparison, the velocity-saturated model from Example 3.12 predicts logical efforts of 1.20, 1.39, 1.50, and 2.00 for NAND2, NAND3, NOR2, and NOR3 gates, agreeing reasonably well with the nanometer processes. The parasitic delays show greater spread because of the variation in the relative capacitances of diffusion and gates.

This data includes more detail than the designer typically wants when doing design by hand; the coarse estimates of logical effort from Table 3.2 are generally sufficient for an initial design. However, the accurate delay vs. fanout information, often augmented with input slope dependence, is essential when characterizing a standard cell library to use with

TABLE 7.8 Logical effort and parasitic delay of gates in various processes

Vendor		Orbit	HP	AMI	AMI	TSMC	TSMC	TSMC	IBM	IBM	IBM
Model		MOSIS	MOSIS	MOSIS	MOSIS	MOSIS	MOSIS	TSMC	IBM	IBM	IBM
Feature Size f	nm	2000	800	600	600	350	250	180	130	90	65
V_{DD}	V	5	5	5	3.3	3.3	2.5	1.8	1.2	1.0	1.0
FO4 Delay	ps	856	297	230	312	210	153	75.6	46.0	37.3	17.2
τ	ps	170	59	45	60	40	30	15	9.0	7.4	3.3
Logical Effort											
Inverter		1.00	1.00	1.00	1.00	1.00	1.00	1.00	1.00	1.00	1.00
NAND2		1.13	1.07	1.05	1.08	1.12	1.12	1.14	1.16	1.20	1.26
NAND3		1.32	1.21	1.19	1.24	1.29	1.29	1.31	1.35	1.41	1.51
NAND4		1.53	1.37	1.36	1.42	1.47	1.47	1.50	1.55	1.62	1.78
NOR2		1.57	1.59	1.58	1.60	1.52	1.50	1.50	1.57	1.56	1.50
NOR3		2.16	2.23	2.23	2.30	2.07	2.02	2.00	2.12	2.08	1.96
NOR4		2.76	2.92	2.96	3.09	2.62	2.52	2.53	2.70	2.60	2.43
Parasitic Delay											
Inverter		1.08	1.05	1.18	1.25	1.33	1.18	1.03	1.16	1.07	1.20
NAND2		1.87	1.85	1.92	2.10	2.28	2.07	1.90	2.29	2.25	2.47
NAND3		3.34	3.30	3.40	3.79	4.15	3.65	3.51	4.14	4.10	4.44
NAND4		4.98	5.12	5.22	5.78	6.30	5.47	5.52	6.39	6.39	6.79
NOR2		2.86	2.91	3.29	3.56	3.52	2.95	2.85	3.35	3.01	3.29
NOR3		5.65	6.05	7.02	7.70	6.89	5.61	5.57	6.59	5.76	6.35
NOR4		9.11	10.3	12.4	13.9	11.0	8.76	8.95	10.54	9.11	10.16

a static timing analyzer. The FO4 inverter delays may differ slightly from Table 7.5 because the widths of the transistors are different.

7.5.4 Power and Energy

Recall from Section 4.1 that energy and power are proportional to the supply current. They can be measured based on the current out of the power supply voltage source. For example, the following code uses the `INTEGRAL` command to measure charge and energy delivered to a circuit during the first 10 ns.

```
.measure charge INTEGRAL I(vdd) FROM=0ns TO=10ns
.measure energy param='charge*SUPPLY'
```

Alternatively, HSPICE allows you to directly measure the instantaneous and average power delivered by a voltage source.

```
.print P(vdd)
.measure pwr AVG P(vdd) FROM=0ns TO=10ns
```

Sometimes it is helpful to measure the power consumed by only one gate in a larger circuit. In that case, you can use a separate voltage source for that gate and measure power only from that source. Unfortunately, this means that `vdd` cannot be declared as `.global`.

When the input of a gate switches, it delivers power to the supply through the gate-to-source capacitances. Be careful to differentiate this input power from the power drawn by the gate discharging its internal and load capacitances.

7.5.5 Simulating Mismatches

Many circuits are sensitive to mismatches between nominally identical transistors. For example, sense amplifiers (see Section 11.2.3.3) should respond to a small differential voltage between the inputs. Mismatches between nominally identical transistors add an offset that can significantly increase the required voltage. Merely simulating in different design corners is inadequate because the transistors will still match each other. As discussed in Section 6.5.2, the mismatch between currents in two nominally identical transistors can be primarily attributed to shifts in the threshold voltage and channel length. Figure 7.31 shows an example of simulating this mismatch. Each transistor is replaced by an equivalent circuit with a different channel length and a voltage source modeling the difference in threshold voltage. Note that many binned BSIM models do not allow setting the transistor length shorter than the minimum value supported by the process. Obtaining data on parameter variations was formerly difficult but is now part of the vendor's model guide in nanometer processes.

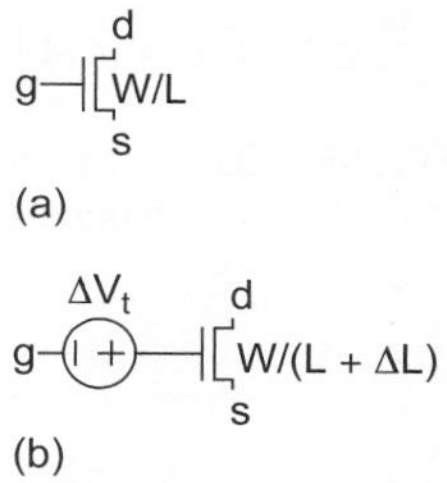

FIGURE 7.31
Modeling mismatch

In many cases, the transistors are not adjacent and may see substantial differences in voltage and temperature. For example, two clock buffers in different corners of the chip that see different environments will cause skew between the two clocks. The voltage difference can be modeled with two different voltage sources. The temperature difference is most easily handled through two separate simulations at different temperatures.

7.5.6 Monte Carlo Simulation

Monte Carlo simulation can be used to find the effects of random variations on a circuit. It consists of running a simulation repeatedly with different randomly chosen parameter offsets. To use Monte Carlo simulation, the statistical distributions of parameters must be part of the model. Manufacturers commonly supply such models for nanometer processes.

For example, consider modifying the FO4 inverter delay simulation from Figure 7.10 to obtain a statistical delay distribution. The transient command must be changed to

```
.tran 1ps 1000ps SWEEP MONTE=30
```

The `.measure` statements report average, minimum, maximum, and standard deviation computed from the 30 repeated simulations. The mean is 17.1 ps and the standard deviation is $\sigma = 0.56$ ps.

Good models will include parameters that the user can set to control whether die-to-die variations, within-die variations, or both are considered. They also may accept information extracted from the layout such as transistor orientation and well edge proximity.

7.6 Interconnect Simulation

Interconnect parasitics can dominate overall delay. When an actual layout is available, the wire geometry can be extracted directly. If only the schematic is available, the designer may need to estimate wire lengths. For small gates, even the capacitances of the wires inside the gate are important. Therefore, some companies use *parasitic estimator* tools to guess wire parasitics in schematics based on the number and size of the transistors. In any case, the designer must explicitly model long wires based on their estimated lengths in the floorplan.

Once wire length and pitch are known or estimated, they can be converted to a wire resistance R and capacitance C using the methods discussed in Section 5.2. A short wire (where wire resistance is much less than gate resistance) can be modeled as a lumped capacitor. A longer wire can be modeled with a multisegment π-model. A four-segment model such as the one shown in Figure 7.32 is generally quite accurate. The model can be readily extended to include coupling between adjacent lines.

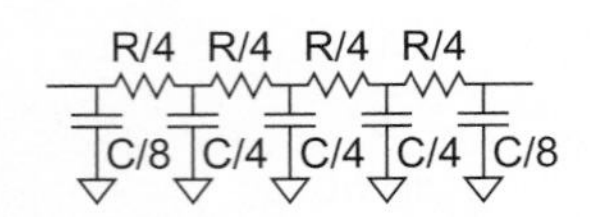

FIGURE 7.32 Four-segment π model for interconnect

In general, interconnect consists of multiple interacting signal and power/ground lines [Young00]. For example, Figure 7.33(a) shows a pair of parallel signals running between a pair of ground wires. Although it is possible to model the ground lines with a resistance and inductance per unit length, it is usually more practical to treat the supply networks as ideal, then account for power supply noise separately in the noise budget. Figure 7.33(b) shows an equivalent circuit using a single π-segment model. Each line has a series resistance and inductance, a capacitance to ground, and mutual capacitance and inductance. The mutual elements describe how a changing voltage or current in one conductor induce a current or voltage in the other.

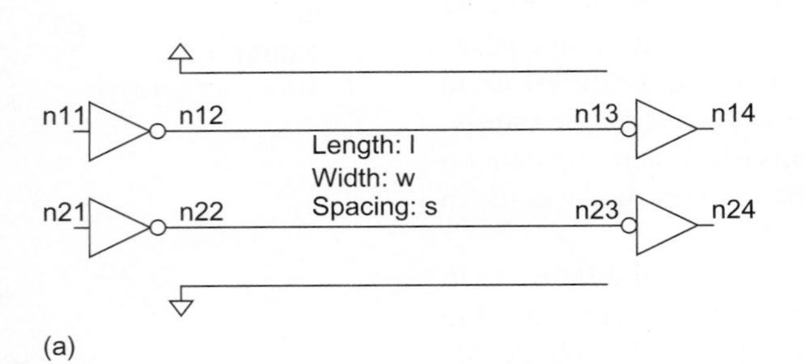

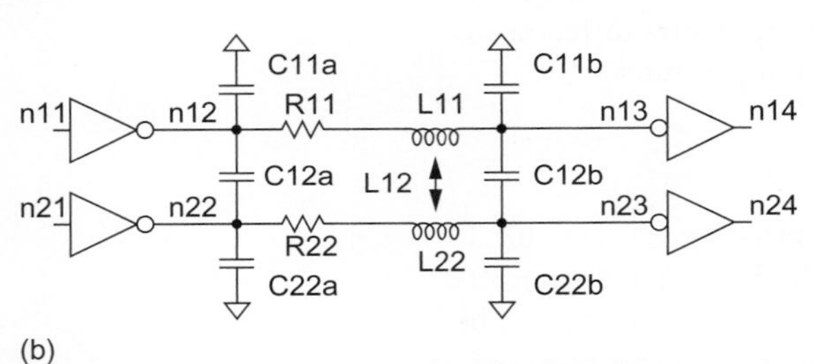

FIGURE 7.33 Lossy multiconductor transmission lines

HSPICE also supports the `W` element that models lossy multiconductor transmission lines. This is more convenient than constructing an enormous π-model with resistance, capacitance, inductance, mutual capacitance, and mutual inductance. Moreover, HSPICE has a built-in two-dimensional field solver that can compute all of the terms from a cross-sectional description of the interconnect. Figure 7.34 gives a SPICE deck that uses the field solver to extract the element values and models the lines with the `W` element.

The deck describes a two-dimensional cross-section of the interconnect that the field solver uses to extract the electrical parameters. The interconnect consists of the two signal traces between two ground wires. Each wire is 2 μm wide and 0.7 μm thick. The copper wires are sandwiched with 0.9 μm of low-k ($\varepsilon = 3.55\varepsilon_0$) dielectric above and below. The $N = 2$ signal traces are spaced 6 μm from the ground lines and 2 μm from each other and have a length of 6 mm. The HSPICE field solver is quite flexible and is fully documented in the HSPICE manual. It generates the transmission line model and writes it to the `coplanar.rlgc` file. The file contains resistance, capacitance, and inductance matrices and is shown in Figure 7.35.

The matrices require a bit of effort to interpret. They are symmetric around the diagonal so only the lower half is printed. The resistances are $R_{11} = R_{22} = 12.4$ Ω/mm. The inductances are $L_{11} = L_{22} = 0.67$ nH/mm and $L_{12} = 0.37$ nH/mm. The capacitance matrix represents coupling capacitances with negative numbers and places the sum of all the capacitances for a wire on the diagonal. Therefore, $C_{11} = C_{22} = 0.0117$ pF/mm and $C_{12} = 0.0137$ pF/mm. In the π-model, half of each of these capacitances is lumped at each end.

Figure 7.36 shows the voltages along the wires. The characteristic velocity of the line is approximately $1/\sqrt{L_{11}(C_{11}+C_{12})} = 2.4\times10^{11}$ mm/s. This is close to the speed of light (3×10^{11} mm/s) because the model assumes air rather than a ground plane outside the dielectric. The flight time down the wire is 6 mm/(2.4×10^{11} mm/s) = 25 ps.

```
* interconnect.sp
*----------------------------------------------------------------------
* Parameters and models
*----------------------------------------------------------------------
.param SUPPLY=1.0
.include '../models/ibm065/models.sp'
.temp 70
.option post
*----------------------------------------------------------------------
* Subcircuits
*----------------------------------------------------------------------
.global vdd gnd
.subckt inv a y N=100nm P=200nm
M1   y   a   gnd   gnd   NMOS   W='N'   L=50nm
+ AS='N*125nm' PS='2*N+250nm' AD='N*125nm' PD='2*N+250nm'
M2   y   a   vdd   vdd   PMOS   W='P'   L=50nm
+ AS='P*125nm' PS='2*P+250nm' AD='P*125nm' PD='2*P+250nm'
.ends
*----------------------------------------------------------------------
* Compute transmission line parameters with field solver
*----------------------------------------------------------------------
.material     oxide          DIELECTRIC          ER=3.55
.material     copper         METAL               CONDUCTIVITY=57.6meg
.layerstack   chipstack      LAYER=(oxide,2.5um)
.fsoptions    opt1           ACCURACY=MEDIUM     PRINTDATA=YES
.shape        widewire       RECTANGLE           WIDTH=2um    HEIGHT=0.7um
.model        coplanar       W                   MODELTYPE=FieldSolver
+ LAYERSTACK=chipstack       FSOPTIONS=opt1      RLGCFILE=coplanar.rlgc
+ CONDUCTOR=(SHAPE=widewire ORIGIN=(0,0.9um)     MATERIAL=copper TYPE=reference)
+ CONDUCTOR=(SHAPE=widewire ORIGIN=(8um,0.9um)   MATERIAL=copper)
+ CONDUCTOR=(SHAPE=widewire ORIGIN=(12um,0.9um)  MATERIAL=copper)
+ CONDUCTOR=(SHAPE=widewire ORIGIN=(20um,0.9um)  MATERIAL=copper TYPE=reference)
*----------------------------------------------------------------------
* Simulation netlist
*----------------------------------------------------------------------
Vdd  vdd   gnd   'SUPPLY'
Vin  n11   gnd   PULSE    0 'SUPPLY' 0ps 20ps 20ps 500ps 1000ps
W1   n12   n22   gnd    n13   n23   gnd   FSmodel=coplanar N=2 l=6mm
X1   n11   n12   inv    M=80
X2   n13   n14   inv    M=40
X3   gnd   n22   inv    M=80
X4   n23   n24   inv    M=40
*----------------------------------------------------------------------
* Stimulus
*----------------------------------------------------------------------
.tran 1ps 250ps
.end
```

FIGURE 7.34 SPICE deck for lossy multiconductor transmission line

```
* L(H/m), C(F/m), Ro(Ohm/m), Go(S/m), Rs(Ohm/(m*sqrt(Hz)), Gd(S/(m*Hz))
.MODEL coplanar W MODELTYPE=RLGC, N=2
+ Lo =  6.68161e-007
+       3.67226e-007  6.68161e-007
+ Co =  2.53841e-011
+       -1.36778e-011  2.53841e-011
+ Ro =  12400.8
+       0  12400.8
+ Go =  0
+       0  0
```

FIGURE 7.35 coplanar.rlgc file

When the input (n11) rises, the near end of the aggressor (n12) begins to fall. n12 levels out for a while at 0.2V as the driver supplies current to charge the rest of the wire. After one flight time (25 ps), the far end of the aggressor (n13) begins to fall. It undershoots to –0.2 V. After a second flight time, n12 levels out near 0. The far end oscillates for a while with a half-period of two flight times (50 ps).

When the aggressor falls, the victim is capacitively coupled down at both ends. The far end (n23) experiences stronger coupling because it is distant from its driver.

The *ringing* can be viewed as either the response of the 2nd order RLC circuit, or as a transmission line reflection. It is visible because the wires are far from their returns (hence having high inductance), are wide and thick enough to have low resistance (that would damp the oscillation), and are driven with an edge much faster than the wire flight time. If the inductance were reduced by moving the ground lines closer to the conductors, the ringing would decrease.

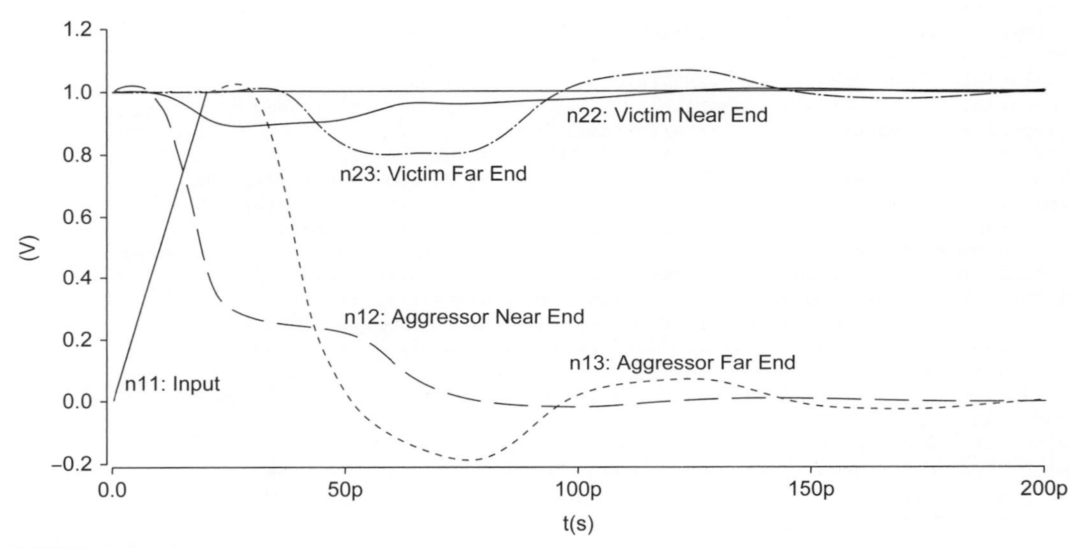

FIGURE 7.36 Transmission line response

7.7 Pitfalls and Fallacies

Failing to estimate diffusion and interconnect parasitics in simulations

The diffusion capacitance can account for 20% of the delay of an FO4 inverter and more than 50% of the delay of a high-fanin, low-fanout gate. Be certain when simulating circuits that the area and perimeter of the source and drain are included in the simulations, or automatically estimated by the models. Interconnect capacitance is also important, but difficult to estimate. For long wires, the capacitance and RC delay represent most of the path delay. A common error is to ignore wires while doing circuit design at the schematic level, and then discover after layout that the wire delay is important enough to demand major circuit changes and complete change of the layout.

Applying inappropriate input waveforms

Gate delay is strongly dependent on the rise/fall time of the input. For example, the propagation delay of an inverter is substantially shorter when a step input is applied than when an input with a realistic rise time is provided.

Applying inappropriate output loading

Gate delay is even more strongly dependent on the output loading. Some engineers, particularly those in the marketing department, report gate delay as the delay of an unloaded inverter. This is about one-fifth of the delay of an FO4 inverter or other gate with "typical" loading. When simulating a critical path, it is important to include the estimated load that the final stage must drive.

Choosing inappropriate transistor sizes

Gate delay also depends on transistor widths. Some papers compare a novel design with carefully selected transistor sizes to a conventional design with poorly selected sizes, and arrive at the misleading conclusion that the novel design is superior.

Identifying the incorrect critical path

During preliminary design, it is much more efficient to compare circuits by modeling only the critical paths rather than the entire circuit. However, this requires that the designer correctly identify the path that will be most critical; sometimes this requires much consideration.

Failing to account for hidden scale factors

Many CAD systems introduce scaling factors. For example, a circuit can be drawn with one set of design rules and automatically scaled to the next process generation. The CAD tools may introduce a scaling factor to reflect this change. Specifying the proper transistor sizes reflecting this scaling is notoriously tricky. Simulation results will look good, but mean nothing if scaling is not accounted for properly.

Blindly trusting results from SPICE

Novice SPICE users often trust the results of simulation far too much. This is exacerbated by the fact that SPICE prints results to many significant figures and generates pretty waveforms. As we have seen, there are a multitude of reasons why simulation results may not reflect the behavior of the real circuit.

When first using a new process or tool set, always predict what the results should be for some simple circuits (e.g., an FO4 inverter) and verify that the simulation matches expectation. It doesn't hurt to be a bit paranoid at first. After proving that the flow is correct, lock down all the models and netlist generation scripts with version control if possible. That way, if any changes are made, a good reason for the change must be evident and the simulations can be revalidated. In general, assume SPICE decks are buggy until proven otherwise. If the simulation does not agree with your expectations, look closely for errors or inadequate modeling in the deck.

Using SPICE in place of thinking

A related error, common among perhaps the majority of circuit designers, is to use SPICE too much and one's brain too little. Circuit simulation should be guided by analysis. In particular, designing to simulation results produced by the optimizer rather than designing based on understanding has led more than one engineer to grief.

Making common SPICE deck errors

Some of the common mistakes in SPICE decks include the following:

- Omitting the comment on the first line
- Omitting the new line at the end of the deck

- Omitting the `.option` post command when using a waveform viewer
- Leaving out diffusion parasitics
- Forgetting to set initial values for dynamic logic or sequential circuits

Using incorrect dimensions when `.option scale` is not set

If `.option scale` is not used, a transistor with W = 4, L = 2 would be interpreted as 4 by 2 meters! This often is outside the legal range of sizes in a BSIM model file, causing SPICE to produce error messages. Similarly, a drain diffusion of $3 \times 0.5\ \mu$m should be specified as PD = 7u AD = 1.5p as opposed to the common mistakes of PD = 7 AD = 1.5 or PD = 7u AD = 1.5u.

Summary

When used properly, SPICE is a powerful tool to characterize the behavior of CMOS circuits. This chapter began with a brief tutorial showing how to perform DC and transient analyses to characterize and optimize simple circuits. SPICE supports many different transistor models. At the time of writing, the BSIM model is most widely used and describes MOSFET behavior quite well for most digital applications. When specifying the MOSFET connection, you must include not only the terminal connections (drain, gate, source, and body) and width and length, but also the area and perimeter of the source and drain that are used to compute parasitic capacitance.

Modern SPICE models have so many parameters that they are intractable for hand calculations. However, the designer can perform some simple simulations to characterize a process. For example, it is helpful to know the effective gate capacitance and resistance, the diffusion capacitance, and the threshold voltage and leakage current. You can also determine the delay of a fanout-of-4 inverter and the logical effort and parasitic delay of a library of gates to make quick estimates of circuit performance.

Most designers use SPICE to characterize real circuits. During preliminary design, you can model the critical path to quickly determine whether a circuit will meet performance requirements. A good model describes not only the circuit itself, but also the input edge rates, the output loading, and parasitics such as diffusion capacitance and interconnect. Most interconnect can be represented with a four-segment π model, although when inductance becomes important, the lossy multiconductor transmission line `W` element is convenient. Novel and "risky" circuits should be simulated in multiple design corners or with Monte Carlo analysis to ensure they will work correctly across variations in processing and environment. As SPICE is prone to garbage-in, garbage-out, it is often best to begin with a simple model and debug until it matches expectations. Then more detail can be added and tested incrementally.

Exercises

Note: This book's Web site at `www.cmosvlsi.com` contains SPICE models and characterization scripts used to generate the data in this chapter. Unless otherwise stated, try the exercises using the mosistsmc180 model file (extracted by MOSIS from test structures manufactured on the TSMC 180 nm process) in TTTT conditions.

7.1 Exercise 3.9 asks you to estimate the delay of a logic function. Simulate your design and compare your results to your estimate. Let one unit of capacitance be a minimum-sized transistor.

7.2 The `charlib.pl` script runs a number of simulations to extract logical effort and parasitic delay of gates in a specified process. Add another column to Table 7.8 for your process.

7.3 Use the `charlib.pl` script to find the logical effort and parasitic delay of a 5-input NAND gate for the outermost input.

7.4 Exercise 3.12 compares two designs of 2-input AND gates. Simulate each design and compare the average delays. What values of x and y give least delay? How much faster is the delay than that achieved using values of x and y suggested from logical effort calculations? How does the best delay compare to estimates using logical effort? Let $C = 10\ \mu$m of gate capacitance.

7.5 Find the average propagation delay of a fanout-of-5 inverter by modifying the SPICE deck shown in Figure 7.10.

7.6 Find the input and output logic levels and high and low noise margins for an inverter with a 3:1 *P/N* ratio.

7.7 What *P/N* ratio maximizes the smaller of the two noise margins for an inverter?

7.8 Generate a set of eight I-V curves like those of Figure 7.16–7.17 for nMOS and pMOS transistors in your process.

7.9 By what percentage does the delay of Exercise 7.5 change if *X5*, the load on the load, is omitted?

7.10 By what percentage does the delay of Exercise 7.5 change if the input is driven by a voltage step rather than a pair of shaping inverters?

7.11 The `char.pl` Perl script runs a number of simulations to characterize a process. Use the script to add another column to Table 7.5 for your process.

Gates 8

8.1 Introduction

Digital logic is divided into combinational and sequential circuits. Combinational circuits are those whose outputs depend only on the present inputs, while sequential circuits have memory. Generally, the building blocks for combinational circuits are logic gates, while the building blocks for sequential circuits are registers and latches. This chapter focuses on combinational logic; Chapter 9 examines sequential logic.

In Chapter 1, we introduced CMOS logic with the assumption that MOS transistors act as simple switches. *Static CMOS* gates used complementary nMOS and pMOS networks to drive 0 and 1 outputs, respectively. In Chapter 3, we used the RC delay model and logical effort to understand the sources of delay in static CMOS logic.

In this chapter, we examine techniques to optimize combinational circuits for lower delay and/or energy. The vast majority of circuits use static CMOS because it is robust, fast, energy-efficient, and easy to design. However, certain circuits have particularly stringent speed, power, or density restrictions that force another solution. Such alternative CMOS logic configurations are called *circuit families*. Section 8.2 examines the most commonly used alternative circuit families: ratioed circuits, dynamic circuits, and pass-transistor circuits. The decade roughly spanning 1994–2004 was the heyday of dynamic circuits, when high-performance microprocessors employed ever-more elaborate structures to squeeze out the highest possible operating frequency. Since then, power, robustness, and design productivity considerations have eliminated dynamic circuits wherever possible, although they remain important for memory arrays where the alternatives are painful. Similarly, other circuit families have been removed or relegated to narrow niches.

Recall from Section 3.3.7 that the delay of a logic gate depends on its output current I, load capacitance C, and output voltage swing ΔV

$$t \propto \frac{C}{I}\Delta V \qquad \textbf{(8.1)}$$

Faster circuit families attempt to reduce one of these three terms. nMOS transistors provide more current than pMOS for the same size and capacitance, so nMOS networks are preferred. Observe that the logical effort is proportional to the C/I term because it is determined by the input capacitance of a gate that can deliver a specified output current.

One drawback of static CMOS is that it requires both nMOS and pMOS transistors on each input. During a falling output transition, the pMOS transistors add significant capacitance without helping the pulldown current; hence, static CMOS has a relatively large logical effort. Many faster circuit families seek to drive only nMOS transistors with the inputs, thus reducing capacitance and logical effort. An alternative mechanism must be provided to pull the output high. Determining when to pull outputs high involves monitoring the inputs, outputs, or some clock signal. Monitoring inputs and outputs inevitably loads the nodes, so clocked circuits are often fastest if the clock can be provided at the ideal time. Another drawback of static CMOS is that all the node voltages must transition between 0 and V_{DD}. Some circuit families use reduced voltage swings to improve propagation delays (and power consumption). This advantage must be weighed against the delay and power of amplifying outputs back to full levels later or the costs of tolerating the reduced swings.

Static CMOS logic is particularly popular because of its robustness. Given the correct inputs, it will eventually produce the correct output so long as there were no errors in logic design or manufacturing. Other circuit families are prone to numerous pathologies examined in Section 8.3, including charge sharing, leakage, threshold drops, and ratioing constraints. When using alternative circuit families, it is vital to understand the failure mechanisms and check that the circuits will work correctly in all design corners.

A host of other circuit families have been proposed, but most have never been used in commercial products and are doomed to reside on dusty library shelves. Every transistor contributes capacitance, so most fast structures are simple. Nevertheless, we will describe some of these circuits as a record of ideas that have been explored. A few hold promise for the future, particularly in specialized applications. Many texts simply catalog these circuit families without making judgments. This book attempts to evaluate the circuit families so that designers can concentrate their efforts on the most promising ones, rather than searching for the "gotchas" that were not mentioned in the original papers. Of course, any such evaluation runs the risk of overlooking advantages or becoming incorrect as technology changes, so you should use your own judgment.

Silicon-on-insulator (SOI) chips eliminate the conductive substrate. They can achieve lower parasitic capacitance and better subthreshold slopes, leading to lower power and/or higher speed, but they have their own special pathologies. Section 8.4 examines considerations for SOI circuits.

CMOS is increasingly applied to ultra-low power systems such as implantable medical devices that require years of operation off of a tiny battery and remote sensors that scavenge their energy from the environment. Static CMOS gates operating in the subthreshold regime can cut the energy per operation by an order of magnitude at the expense of several orders of magnitude performance reduction. Section 8.5 explores design issues for subthreshold circuits.

8.2 Circuit Families

Static CMOS circuits with complementary nMOS pulldown and pMOS pullup networks are used for the vast majority of logic gates in integrated circuits. They have good noise margins, and are fast, low power, insensitive to device variations, easy to design, widely supported by CAD tools, and readily available in standard cell libraries. When noise does exceed the margins, the gate delay increases because of the glitch, but the gate eventually will settle to the correct answer. Most design teams now use static CMOS exclusively for combinational logic. This section begins with a number of techniques for optimizing static CMOS circuits.

Nevertheless, performance or area constraints occasionally dictate the need for other circuit families. The most important alternative is dynamic circuits. However, we begin by considering ratioed circuits, which are simpler and offer a helpful conceptual transition between static and dynamic. We also consider pass transistors, which had their zenith in the 1990s for general-purpose logic and still appear in specialized applications.

8.2.1 Static CMOS

Designers accustomed to AND and OR functions must learn to think in terms of NAND and NOR to take advantage of static CMOS. In manual circuit design, this is often done through bubble pushing. Compound gates are particularly useful to perform complex functions with relatively low logical efforts. When a particular input is known to be latest, the gate can be optimized to favor that input. Similarly, when either the rising or falling edge is known to be more critical, the gate can be optimized to favor that edge. We have focused on building gates with equal rising and falling delays; however, using smaller pMOS transistors can reduce power, area, and delay. In processes with multiple threshold voltages, multiple flavors of gates can be constructed with different speed/leakage power trade-offs.

8.2.1.1 Bubble Pushing CMOS stages are inherently inverting, so AND and OR functions must be built from NAND and NOR gates. DeMorgan's law helps with this conversion:

$$\overline{A \cdot B} = \overline{A} + \overline{B}$$
$$\overline{A + B} = \overline{A} \cdot \overline{B} \qquad (8.2)$$

These relations are illustrated graphically in Figure 8.1. A NAND gate is equivalent to an OR of inverted inputs. A NOR gate is equivalent to an AND of inverted inputs. The same relationship applies to gates with more inputs. Switching between these representations is easy to do on a whiteboard and is often called *bubble pushing*.

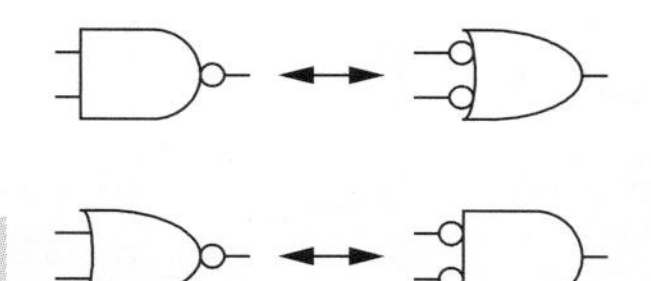

FIGURE 8.1 Bubble pushing with DeMorgan's law

Example 8.1

Design a circuit to compute $F = AB + CD$ using NANDs and NORs.

SOLUTION: By inspection, the circuit consists of two ANDs and an OR, shown in Figure 8.2(a). In Figure 8.2(b), the ANDs and ORs are converted to basic CMOS stages. In Figure 8.2(c and d), bubble pushing is used to simplify the logic to three NANDs.

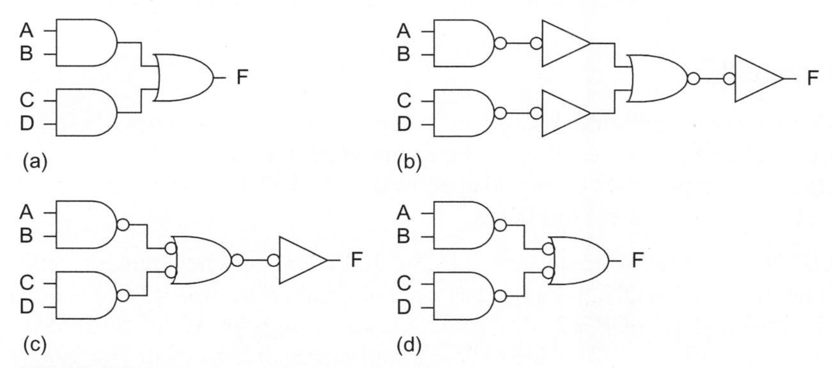

FIGURE 8.2 Bubble pushing to convert ANDs and ORs to NANDs and NORs

8.2.1.2 Compound Gates As described in Section 1.4.5, static CMOS also efficiently handles compound gates computing various inverting combinations of AND/OR functions in a single stage. The function $F = AB + CD$ can be computed with an AND-OR-INVERT-22 (AOI22) gate and an inverter, as shown in Figure 8.3.

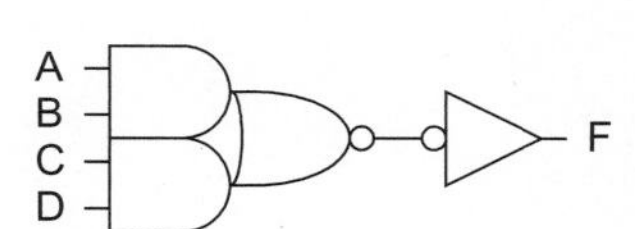

FIGURE 8.3 Logic using AOI22 gate

In general, logical effort of compound gates can be different for different inputs. Figure 8.4 shows how logical efforts can be estimated for the AOI21, AOI22, and a more complex compound AOI gate. The transistor widths are chosen to give the same drive as a unit inverter. The logical effort of each input is the ratio of the input capacitance of that input to the input capacitance of the inverter. For the AOI21 gate, this means the logical effort is slightly lower for the OR terminal (*C*) than for the two AND terminals (*A, B*). The parasitic delay is crudely estimated from the total diffusion capacitance on the output node by summing the sizes of the transistors attached to the output.

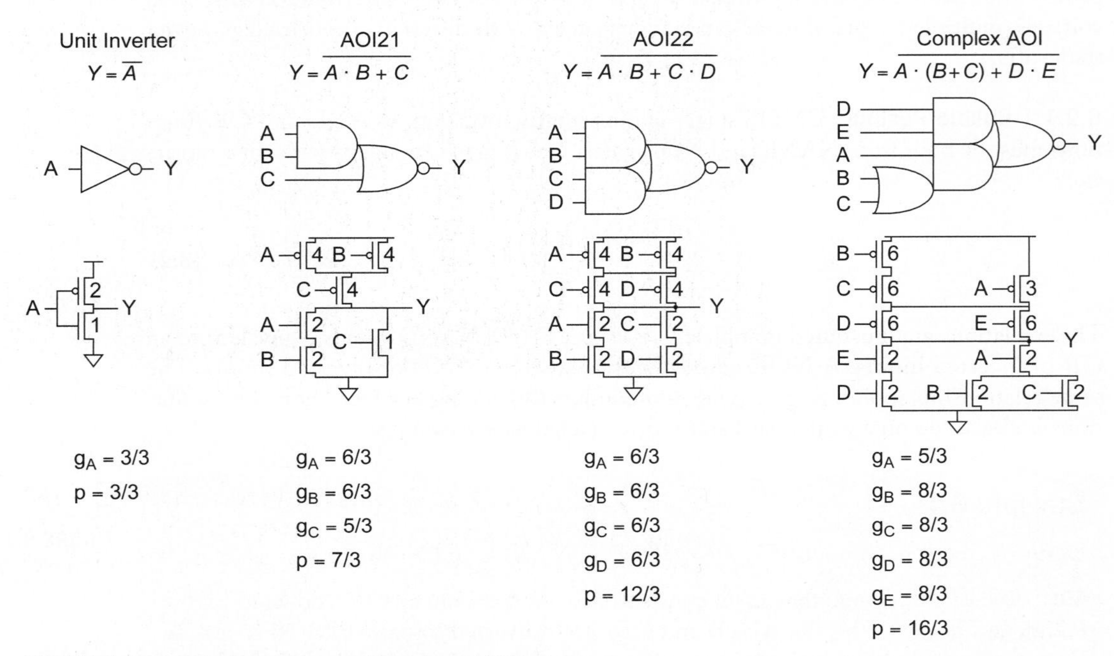

FIGURE 8.4 Logical efforts and parasitic delays of AOI gates

Example 8.2

Calculate the minimum delay, in τ, to compute $F = AB + CD$ using the circuits from Figure 8.2(d) and Figure 8.3. Each input can present a maximum of 20 λ of transistor width. The output must drive a load equivalent to 100 λ of transistor width. Choose transistor sizes to achieve this delay.

SOLUTION: The path electrical effort is $H = 100/20 = 5$ and the branching effort is $B = 1$. The design using NAND gates has a path logical effort of $G = (4/3) \times (4/3) = 16/9$ and parasitic delay of $P = 2 + 2 = 4$. The design using the AOI22 and inverter has a path logical effort of $G = (6/3) \times 1 = 2$ and a parasitic delay of $P = 12/3 + 1 = 5$. Both designs have $N = 2$ stages. The path efforts $F = GBH$ are 80/9 and 10, respectively. The path delays are $NF^{1/N} + P$, or 10.0 τ and 11.3 τ, respectively. Using compound gates does not always result in faster circuits; simple 2-input NAND gates can be quite fast.

To compute the sizes, we determine the best stage efforts, $\hat{f} = F^{1/N} = 3.0$ and 3.2, respectively. These are in the range of 2.4–6 so we know the efforts are reasonable and the design would not improve too much by adding or removing stages. The input capacitance of the second gate is determined by the capacitance transformation

$$C_{\text{in}_i} = \frac{C_{\text{out}_i} \times g_i}{\hat{f}}$$

For the NAND design,

$$C_{\text{in}} = \frac{100\,\lambda \times (4/3)}{3.0} = 44\,\lambda$$

For the AOI22 design,

$$C_{\text{in}} = \frac{100\,\lambda \times (1)}{3.2} = 31\,\lambda$$

The paths are shown in Figure 8.5 with transistor widths rounded to integer values.

8.2.1.3 Input Ordering Delay Effect The logical effort and parasitic delay of different gate inputs are often different. Some logic gates, like the AOI21 in the previous section, are inherently *asymmetric* in that one input sees less capacitance than another. Other gates, like NANDs and NORs, are nominally symmetric but actually have slightly different logical effort and parasitic delays for the different inputs.

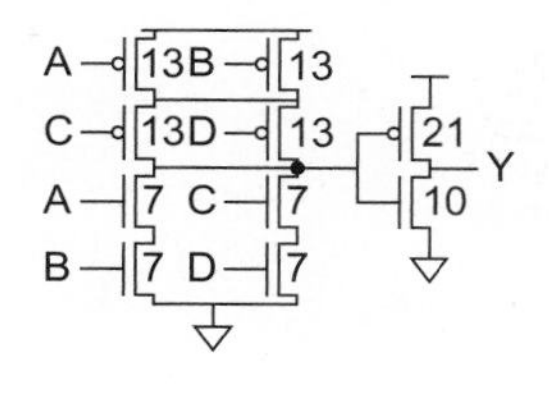

FIGURE 8.5 Paths with transistor widths

Figure 8.6 shows a 2-input NAND gate annotated with diffusion parasitics. Consider the falling output transition occurring when one input held a stable 1 value and the other rises from 0 to 1. If input B rises last, node x will initially be at $V_{DD} - V_t \approx V_{DD}$ because it was pulled up through the nMOS transistor on input A. The Elmore delay is $(R/2)(2C) + R(6C) = 7RC = 2.33\,\tau$.[1] On the other hand, if input A rises last, node x will initially be at 0 V because it was discharged through the nMOS transistor on input B. No charge must be delivered to node x, so the Elmore delay is simply $R(6C) = 6RC = 2\,\tau$.

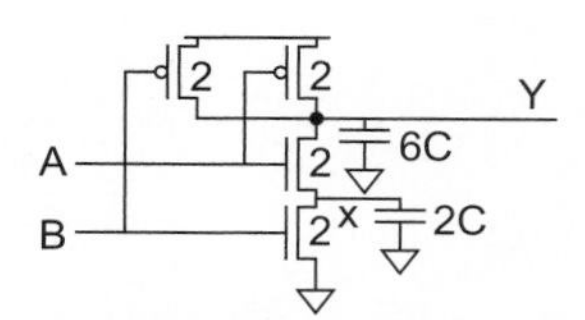

FIGURE 8.6 NAND gate delay estimation

In general, we define the *outer* input to be the input closer to the supply rail (e.g., B) and the *inner* input to be the input closer to the output (e.g., A). The parasitic delay is smallest when the inner input switches last because the intermediate nodes have already been discharged. Therefore, if one signal is known to arrive later than the others, the gate is fastest when that signal is connected to the inner input.

Table 7.7 lists the logical effort and parasitic delay for each input of various NAND gates, confirming that the inner input has a lower parasitic delay. The logical efforts are lower than initial estimates might predict because of velocity saturation. Interestingly, the inner input has a slightly higher logical effort because the intermediate node x tends to rise and cause negative feedback when the inner input turns ON (see Exercise 8.5) [Sutherland99]. This effect is seldom significant to the designer because the inner input remains faster over the range of fanouts used in reasonable circuits.

[1] Recall that $\tau = 3RC$ is the delay of an inverter driving the gate of an identical inverter.

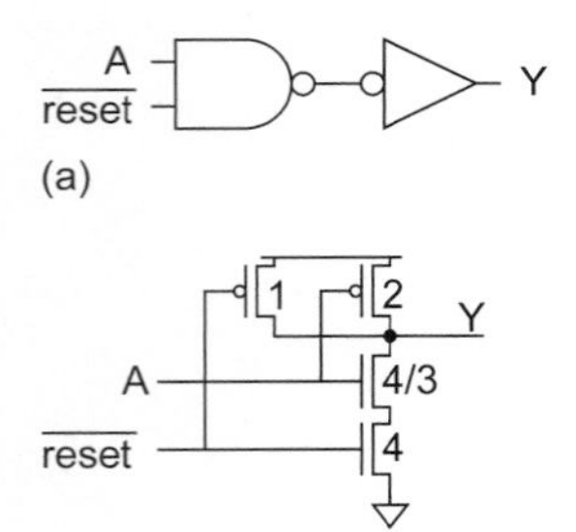

FIGURE 8.7 Resettable buffer optimized for data input

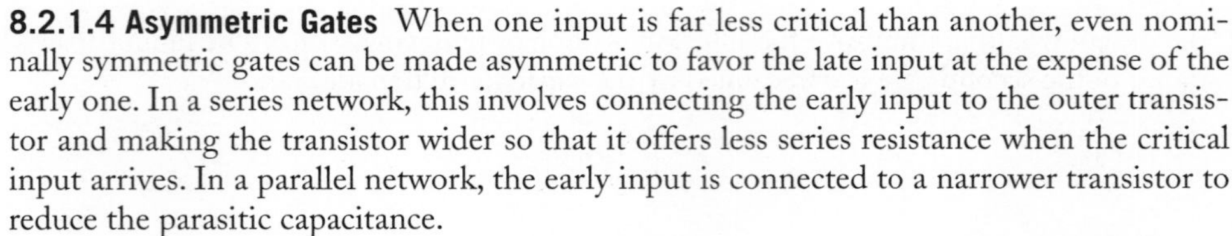

8.2.1.4 Asymmetric Gates When one input is far less critical than another, even nominally symmetric gates can be made asymmetric to favor the late input at the expense of the early one. In a series network, this involves connecting the early input to the outer transistor and making the transistor wider so that it offers less series resistance when the critical input arrives. In a parallel network, the early input is connected to a narrower transistor to reduce the parasitic capacitance.

For example, consider the path in Figure 8.7(a). Under ordinary conditions, the path acts as a buffer between A and Y. When reset is asserted, the path forces the output low. If reset only occurs under exceptional circumstances and can take place slowly, the circuit should be optimized for input-to-output delay at the expense of reset. This can be done with the *asymmetric* NAND gate in Figure 8.7(b). The pulldown resistance is $R/4 + R/(4/3) = R$, so the gate still offers the same driver as a unit inverter. However, the capacitance on input A is only 10/3, so the logical effort is 10/9. This is better than 4/3, which is normally associated with a NAND gate. In the limit of an infinitely large reset transistor and unit-sized nMOS transistor for input A, the logical effort approaches 1, just like an inverter. The improvement in logical effort of input A comes at the cost of much higher effort on the reset input. Note that the pMOS transistor on the reset input is also shrunk. This reduces its diffusion capacitance and parasitic delay at the expense of slower response to reset.

CMOS transistors are usually velocity saturated, and thus series transistors carry more current than the long-channel model would predict. The current can be predicted by collapsing the series stack into an equivalent transistor, as discussed in Section 3.4.6.3. For asymmetric gates, the equivalent width is that of the inner (narrower) transistor. The equivalent length increases by the sum of the reciprocals of the relative widths. The relative current is computed using EQ (3.28), where N is the equivalent length.

Example 8.3

Size the nMOS transistors in the asymmetric NAND gate for unit pulldown current considering velocity saturation. Make the noncritical transistor three times as wide as the critical transistor. Assume $V_{DD} = 1.0$ V and $V_t = 0.3$ V. Use $E_cL = 1.04$ V for nMOS devices. Estimate the logical effort of the gate.

SOLUTION: The equivalent length is $1 + 1/3 = 4/3$ times that of a unit transistor. Applying EQ (3.28) gives a relative current of 0.83. Therefore, the transistors' widths should be 1.20 and 3.60 to deliver unit current. The logical effort is $(1.20 + 2) / 3 = 1.07$, which is even better than predicted without velocity saturation.

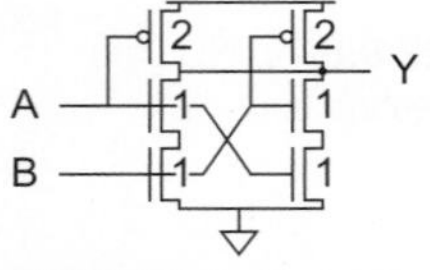

FIGURE 8.8 Perfectly symmetric 2-input NAND gate

In other circuits such as arbiters, we may wish to build gates that are perfectly symmetric so neither input is favored. Figure 8.8 shows how to construct a symmetric NAND gate.

8.2.1.5 Skewed Gates In other cases, one input transition is more important than the other. In Section 2.5.2, we defined *HI-skew* gates to favor the rising output transition and *LO-skew* gates to favor the falling output transition. This favoring can be done by decreasing the size of the noncritical transistor. The logical efforts for the rising (up) and falling (down) transitions are called g_u and g_d, respectively, and are the ratio of the input capacitance of the skewed gate to the input capacitance of an unskewed inverter with equal drive *for that transition*. Figure 8.9(a) shows how a HI-skew inverter is constructed by downsizing the nMOS transistor. This maintains the same effective resistance for the critical transition while reducing the input capacitance relative to the unskewed inverter of Figure 8.9(b), thus reducing the logical effort on that critical transition to $g_u = 2.5/3 = 5/6$. Of course, the improvement comes at the expense of the effort on the noncritical transition. The logical effort for the falling transition is estimated by comparing the inverter to a smaller unskewed inverter with equal pulldown current, shown in Figure 8.9(c), giving a logical effort of $g_d = 2.5/1.5 = 5/3$. The degree of skewing (e.g., the ratio of effective resistance for the fast transition relative to the slow transition) impacts the logical efforts and noise margins; a factor of two is common. Figure 8.10 catalogs HI-skew and LO-skew gates with a skew factor of two. Skewed gates are sometimes denoted with an H or an L on their symbol in a schematic.

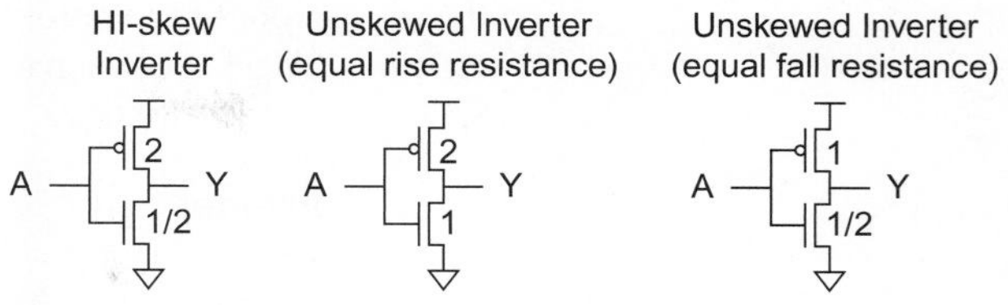

FIGURE 8.9 Logical effort calculation for HI-skew inverter

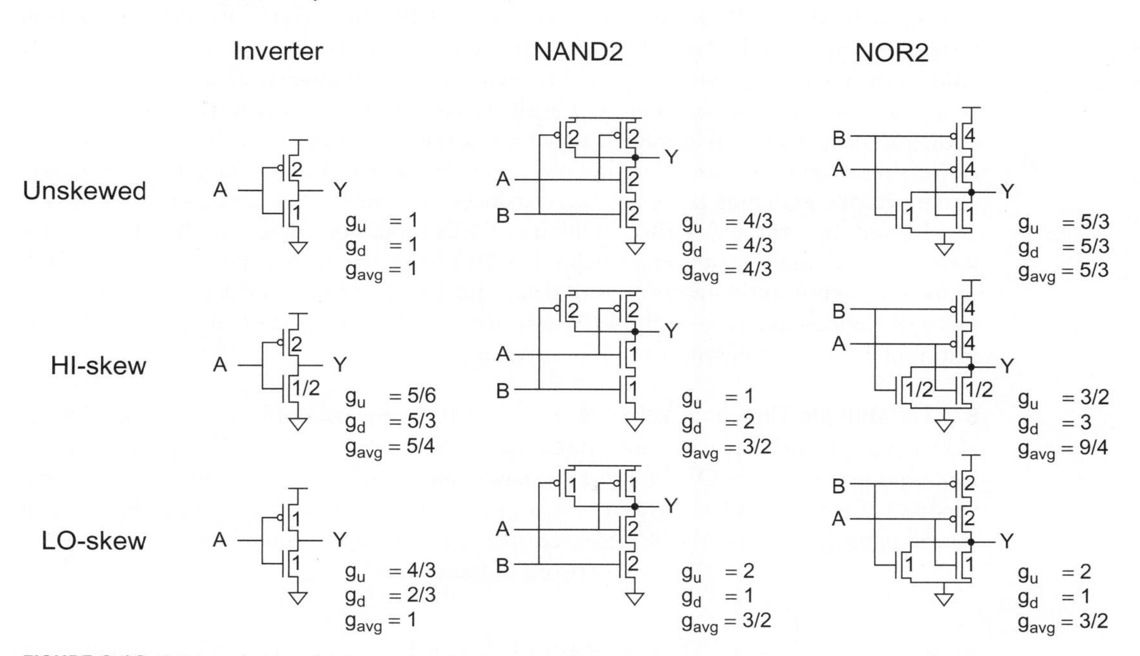

FIGURE 8.10 Catalog of skewed gates

Alternating HI-skew and LO-skew gates can be used when only one transition is important [Solomatnikov00]. Skewed gates work particularly well with dynamic circuits, as we shall see in Section 8.2.4.

8.2.1.6 P/N Ratios Notice in Figure 8.10 that the average logical effort of the LO-skew NOR2 is actually better than that of the unskewed gate. The pMOS transistors in the unskewed gate are enormous in order to provide equal rise delay. They contribute input capacitance for both transitions, while only helping the rising delay. By accepting a slower rise delay, the pMOS transistors can be downsized to reduce input capacitance and average delay significantly.

In general, what is the best P/N ratio for logic gates (i.e., the ratio of pMOS to nMOS transistor width)? You can prove in Exercise 8.13 that the ratio giving lowest average delay is the square root of the ratio that gives equal rise and fall delays. For processes with a mobility ratio of $\mu_n/\mu_p = 2$ as we have generally been assuming, the best ratios are shown in Figure 8.11.

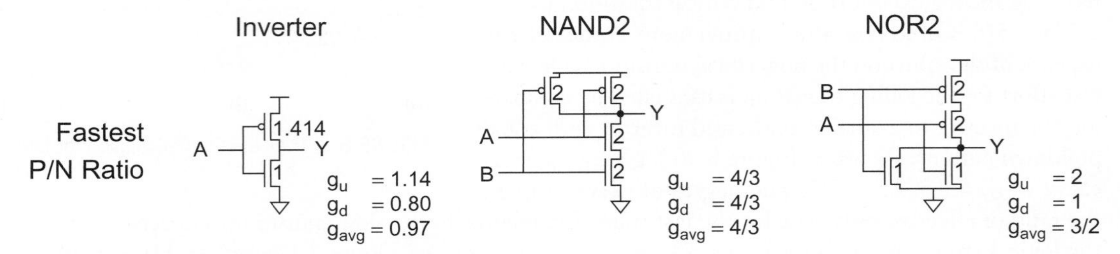

FIGURE 8.11 Gates with *P/N* ratios giving least delay

Reducing the pMOS size from 2 to $\sqrt{2} \approx 1.4$ for the inverter gives the theoretical fastest average delay, but this delay improvement is only 3%. However, this significantly reduces the pMOS transistor area. It also reduces input capacitance, which in turn reduces power consumption. Unfortunately, it leads to unequal delay between the outputs. Some paths can be slower than average if they trigger the worst edge of each gate. Excessively slow rising outputs can also cause hot electron degradation. And reducing the pMOS size also moves the switching point lower and reduces the inverter's noise margin.

In summary, the *P/N* ratio of a library of cells should be chosen on the basis of area, power, and reliability, not average delay. For NOR gates, reducing the size of the pMOS transistors significantly improves both delay and area. In most standard cell libraries, the pitch of the cell determines the *P/N* ratio that can be achieved in any particular gate. Ratios of 1.5–2 are commonly used for inverters.

8.2.1.7 Multiple Threshold Voltages Some CMOS processes offer two or more threshold voltages. Transistors with lower threshold voltages produce more ON current, but also leak exponentially more OFF current. Libraries can provide both high- and low-threshold versions of gates. The low-threshold gates can be used sparingly to reduce the delay of critical paths [Kumar94, Wei98]. Skewed gates can use low-threshold devices on only the critical network of transistors.

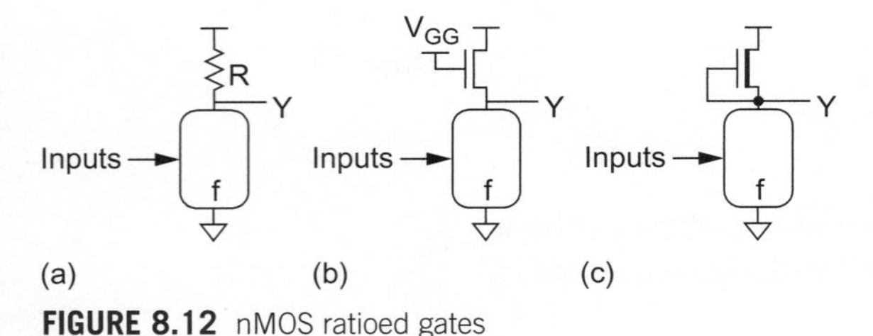

FIGURE 8.12 nMOS ratioed gates

8.2.2 Ratioed Circuits

Ratioed circuits depend on the proper size or resistance of devices for correct operation. For example, in the 1970s and early 1980s before CMOS technologies matured, circuits were often built with only nMOS transistors, as shown in Figure 8.12. Conceptually, the ratioed gate consists of an nMOS pulldown network and some pullup device called the *static load.* When the pulldown network is OFF, the static load pulls the output to 1. When the pulldown network turns ON, it fights the static load. The static load must be weak enough that the output pulls down to an acceptable 0. Hence, there is a ratio constraint between the static load and pulldown network. Stronger static loads produce faster rising outputs, but increase V_{OL}, degrade the noise margin, and burn more static power when the output should be 0. Unlike complementary circuits, the ratio must be chosen so the circuit operates correctly despite any variations from nominal component values that may occur during manufacturing. CMOS logic eventually displaced nMOS logic because the static power became unacceptable as the number of gates increased. However, ratioed circuits are occasionally still useful in special applications.

A resistor is a simple static load, but large resistors consume a large layout area in typical MOS processes. Another technique is to use an nMOS transistor with the gate tied to V_{GG}. If $V_{GG} = V_{DD}$, the nMOS transistor will only pull up to $V_{DD} - V_t$. Worse yet, the threshold is increased by the body effect. Thus, using $V_{GG} > V_{DD}$ was attractive. To eliminate this extra supply voltage, some nMOS processes offered *depletion mode* transistors. These transistors, indicated with the thick bar, are identical to ordinary *enhancement mode* transistors except that an extra ion implantation was performed to create a negative threshold voltage. The depletion mode pullups have their gate wired to the source so $V_{gs} = 0$ and the transistor is always weakly ON.

8.2.2.1 Pseudo-nMOS Figure 8.13(a) shows a *pseudo-nMOS* inverter. Neither high-value resistors nor depletion mode transistors are readily available as static loads in most CMOS

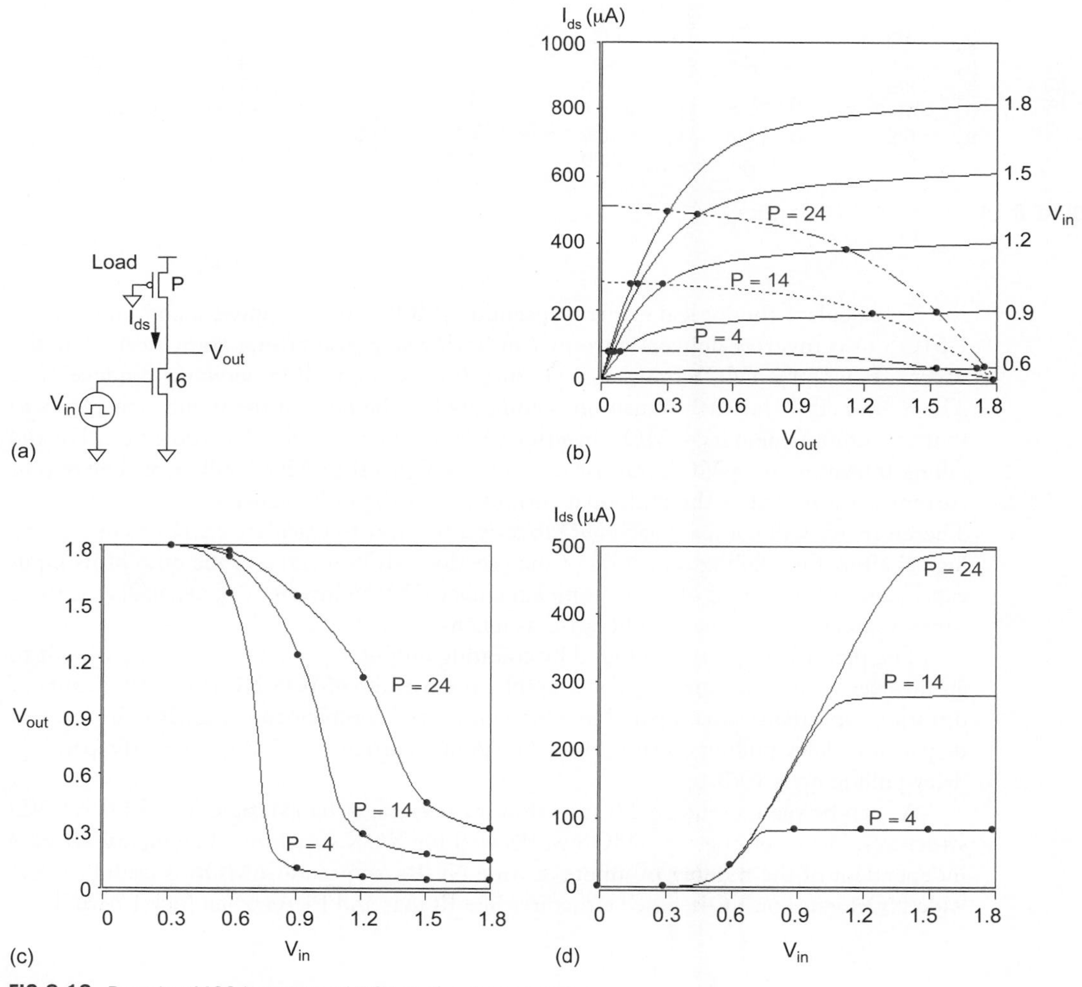

FIG 8.13 Pseudo-nMOS inverter and DC transfer characteristics

processes. Instead, the static load is built from a single pMOS transistor that has its gate grounded so it is always ON. The DC transfer characteristics are derived by finding V_{out} for which $I_{dsn} = |I_{dsp}|$ for a given V_{in}, as shown in Figure 8.13(b–c) for a 180 nm process. The beta ratio affects the shape of the transfer characteristics and the V_{OL} of the inverter. Larger relative pMOS transistor sizes offer faster rise times but less sharp transfer characteristics. Figure 8.13(d) shows that when the nMOS transistor is turned on, a static DC current flows in the circuit.

Figure 8.14 shows several pseudo-nMOS logic gates. The pulldown network is like that of an ordinary static gate, but the pullup network has been replaced with a single pMOS transistor that is grounded so it is always ON. The pMOS transistor widths are selected to be about 1/4 the strength (i.e., 1/2 the effective width) of the nMOS pulldown network as a compromise between noise margin and speed; this best size is process-dependent, but is usually in the range of 1/3 to 1/6.

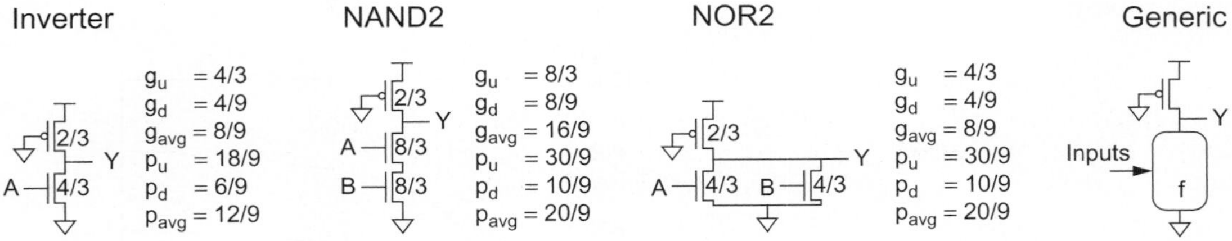

FIGURE 8.14 Pseudo-nMOS logic gates

To calculate the logical effort of pseudo-nMOS gates, suppose a complementary CMOS unit inverter delivers current I in both rising and falling transitions. For the widths shown, the pMOS transistors produce $I/3$ and the nMOS networks produce $4I/3$. The logical effort for each transition is computed as the ratio of the input capacitance to that of a complementary CMOS inverter with equal current for that transition. For the falling transition, the pMOS transistor effectively fights the nMOS pulldown. The output current is estimated as the pulldown current minus the pullup current, $(4I/3 - I/3) = I$. Therefore, we will compare each gate to a unit inverter to calculate g_d. For example, the logical effort for a falling transition of the pseudo-nMOS inverter is the ratio of its input capacitance (4/3) to that of a unit complementary CMOS inverter (3), i.e., 4/9. g_u is three times as great because the current is 1/3 as much.

The parasitic delay is also found by counting output capacitance and comparing it to an inverter with equal current. For example, the pseudo-nMOS NOR has 10/3 units of diffusion capacitance as compared to 3 for a unit-sized complementary CMOS inverter, so its parasitic delay pulling down is 10/9. The pullup current is 1/3 as great, so the parasitic delay pulling up is 10/3.

As can be seen, pseudo-nMOS is slower on average than static CMOS for NAND structures. However, pseudo-nMOS works well for NOR structures. The logical effort is independent of the number of inputs in wide NORs, so pseudo-nMOS is useful for fast wide NOR gates or NOR-based structures like ROMs and PLAs when power permits.

Example 8.4

Design a k-input AND gate with DeMorgan's law using static CMOS inverters followed by a k-input pseudo-nMOS NOR, as shown in Figure 8.15. Let each inverter be unit-sized. If the output load is an inverter of size H, determine the best transistor sizes in the NOR gate and estimate the average delay of the path.

SOLUTION: The path electrical effort is H and the branching effort is $B = 1$. The inverter has a logical effort of 1. The pseudo-nMOS NOR has an average logical effort of 8/9 according to Figure 8.14. The path logical effort is $G = 1 \times (8/9) = 8/9$, so the path effort is $8H/9$. Each stage should bear an effort of $\hat{f} = \sqrt{8H/9}$. Using the capacitance transformation gives NOR pulldown transistor widths of

$$C_{in} = \frac{gC_{out}}{\hat{f}} = \frac{(8/9)H}{\sqrt{8H/9}} = \frac{\sqrt{8H}}{3}$$

unit-sized inverters. As a unit inverter has three units of input capacitance, the NOR transistor nMOS widths should be $\sqrt{8H}$. According to Figure 8.14, the pullup transistor should be half this width. The complete circuit marked with nMOS and pMOS widths is drawn in Figure 8.16.

We estimate the average parasitic delay of a k-input pseudo-nMOS NOR to be $(8k + 4)/9$. The total delay in τ is

$$D = N\hat{f} + P = \frac{4\sqrt{2}}{3}\sqrt{H} + \frac{8k+13}{9}$$

Increasing the number of inputs only impacts the parasitic delay, not the effort delay.

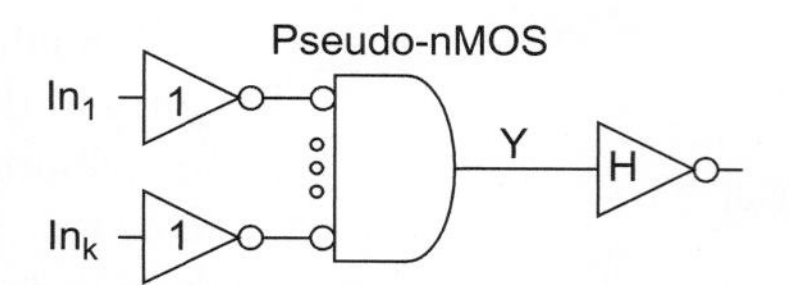

FIGURE 8.15 k-input AND gate driving load of H

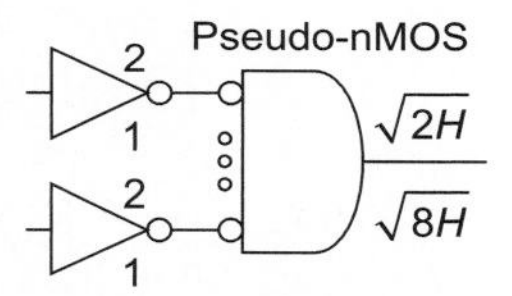

FIGURE 8.16 k-input AND marked with transistor widths

Pseudo-nMOS gates will not operate correctly if $V_{OL} > V_{IL}$ of the receiving gate. This is most likely in the SF design corner where nMOS transistors are weak and pMOS transistors are strong. Designing for acceptable noise margin in the SF corner forces a conservative choice of weak pMOS transistors in the normal corner. A biasing circuit can be used to reduce process sensitivity, as shown in Figure 8.17. The goal of the biasing circuit is to create a V_{bias} that causes $P2$ to deliver 1/3 the current of $N2$, independent of the relative mobilities of the pMOS and nMOS transistors. Transistor $N2$ has width of 3/2 and hence produces current $3I/2$ when ON. Transistor $N1$ is tied ON to act as a current source with 1/3 the current of $N2$, i.e., $I/2$. $P1$ acts as a current mirror using feedback to establish the bias voltage sufficient to provide equal current as $N1$, $I/2$. The size of $P1$ is noncritical so long as it is large enough to produce sufficient current and is equal in size to $P2$. Now, $P2$ ideally also provides $I/2$. In summary, when A is low, the pseudo-nMOS gate pulls up with a current of $I/2$. When A is high, the pseudo-nMOS gate pulls down with an effective current of $(3I/2 - I/2) = I$. To first order, this biasing technique sets the relative currents strictly by transistor widths, independent of relative pMOS and nMOS mobilities.

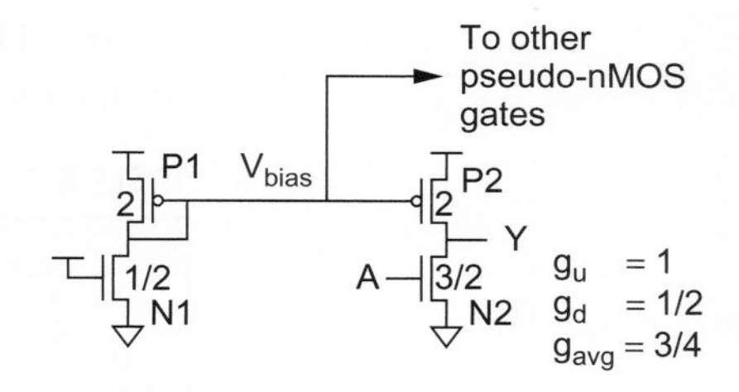

FIGURE 8.17 Replica biasing of pseudo-nMOS gates

Such replica biasing permits the 1/3 current ratio rather than the conservative 1/4 ratio in the previous circuits, resulting in lower logical effort. The bias voltage V_{bias} can be distributed to multiple pseudo-nMOS gates. Ideally, V_{bias} will adjust itself to keep V_{OL} constant across process corners. Unfortunately, the currents through the two pMOS transistors do not exactly match because their drain voltages are unequal, so this technique still has some process sensitivity. Also note that this bias is relative to V_{DD}, so any noise on either the bias voltage line or the V_{DD} supply rail will impact circuit performance.

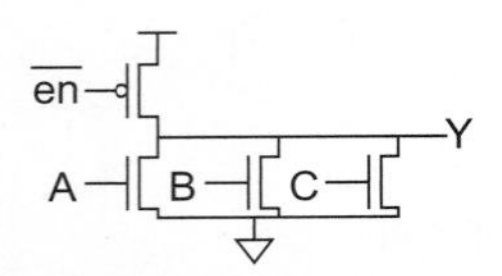

FIGURE 8.18 Pseudo-nMOS gate with enabled pullup

Turning off the pMOS transistor can reduce power when the logic is idle or during IDDQ test mode (see Section 14.6.4), as shown in Figure 8.18.

Example 8.5

Calculate the static power dissipation of a 32-word × 48-bit ROM that contains a 5:32 pseudo-nMOS row decoder and pMOS pullups on the 48-bit lines. The pMOS transistors have an ON current of 360 μA/μm and are minimum width (100 nm). V_{DD} = 1.0 *V*. Assume one of the word lines and 50% of the bitlines are high at any given time.

SOLUTION: Each pMOS transistor dissipates 360 μA/μm × 0.1 μm × 1.0 V = 36 μW of power when the output is low. We expect to see 31 wordlines and 24 bitlines low, so the total static power is 36 μW × (31 + 24) = 1.98 mW.

FIGURE 8.19 Symmetric 2-input NOR gate

8.2.2.2 Ganged CMOS Figure 8.19 illustrates pairs of CMOS inverters ganged together. The truth table is given in Table 8.1, showing that the pair compute the NOR function. Such a circuit is sometimes called a *symmetric*[2] *NOR* [Johnson88], or more generally, *ganged CMOS* [Schultz90]. When one input is 0 and the other 1, the gate can be viewed as a pseudo-nMOS circuit with appropriate ratio constraints. When both inputs are 0, both pMOS transistors turn on in parallel, pulling the output high faster than they would in an ordinary pseudo-nMOS gate. Moreover, when both inputs are 1, both pMOS transistors turn OFF, saving static power dissipation. As in pseudo-nMOS, the transistors are sized so the pMOS are about 1/4 the strength of the nMOS and the pulldown current matches that of a unit inverter. Hence, the symmetric NOR achieves both better performance and lower power dissipation than a 2-input pseudo-nMOS NOR.

TABLE 8.1 Operation of symmetric NOR

A	*B*	*N1*	*P1*	*N2*	*P2*	*Y*
0	0	OFF	ON	OFF	ON	1
0	1	OFF	ON	ON	OFF	~ 0
1	0	ON	OFF	OFF	ON	~ 0
1	1	ON	OFF	ON	OFF	0

Johnson also showed that symmetric structures can be used for NOR gates with more inputs and even for NAND gates (see Exercises 8.23–8.24). The 3-input symmetric NOR also works well, but the logical efforts of the other structures are unattractive.

[2]Do not confuse this use of *symmetric* with the concept of *symmetric* and *asymmetric* gates from Section 8.2.1.4.

8.2.3 Cascode Voltage Switch Logic

Cascode Voltage Switch Logic (CVSL[3]) [Heller84] seeks the benefits of ratioed circuits without the static power consumption. It uses both true and complementary input signals and computes both true and complementary outputs using a pair of nMOS pulldown networks, as shown in Figure 8.20(a). The pulldown network f implements the logic function as in a static CMOS gate, while $\overline{f}$ uses inverted inputs feeding transistors arranged in the conduction complement. For any given input pattern, one of the pulldown networks will be ON and the other OFF. The pulldown network that is ON will pull that output low. This low output turns ON the pMOS transistor to pull the opposite output high. When the opposite output rises, the other pMOS transistor turns OFF so no static power dissipation occurs. Figure 8.20(b) shows a CVSL AND/NAND gate. Observe how the pulldown networks are complementary, with parallel transistors in one and series in the other. Figure 8.20(c) shows a 4-input XOR gate. The pulldown networks share A and $\overline{A}$ transistors to reduce the transistor count by two. Sharing is often possible in complex functions, and systematic methods exist to design shared networks [Chu86].

CVSL has a potential speed advantage because all of the logic is performed with nMOS transistors, thus reducing the input capacitance. As in pseudo-nMOS, the size of the pMOS transistor is important. It fights the pulldown network, so a large pMOS transistor will slow the falling transition. Unlike pseudo-nMOS, the feedback tends to turn off the pMOS, so the outputs will settle eventually to a legal logic level. A small pMOS transistor is slow at pulling the complementary output high. In addition, the CVSL gate requires both the low- and high-going transitions, adding more delay. Contention current during the switching period also increases power consumption.

Pseudo-nMOS worked well for wide NOR structures. Unfortunately, CVSL also requires the complement, a slow tall NAND structure. Therefore, CVSL is poorly suited to general NAND and NOR logic. Even for symmetric structures like XORs, it tends to be slower than static CMOS, as well as more power-hungry [Chu87, Ng96]. However, the ideas behind CVSL help us understand dual-rail domino and complementary pass-transistor logic discussed in later sections.

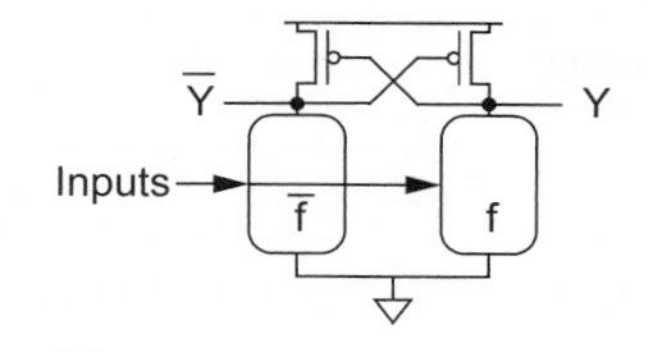

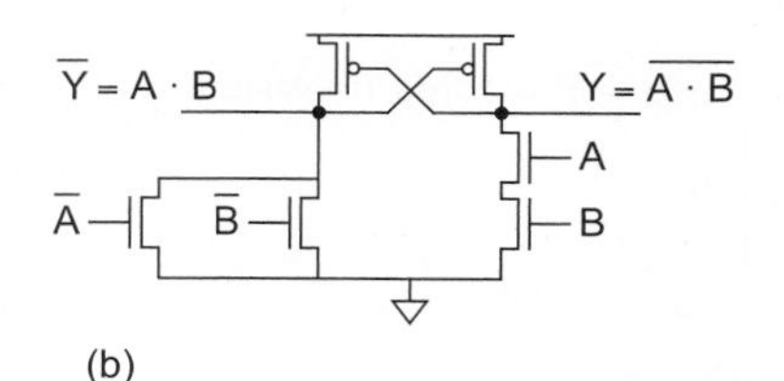

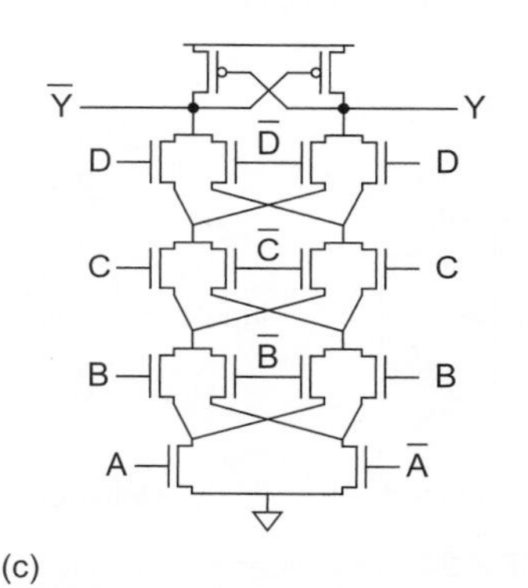

FIGURE 8.20 CVSL gates

8.2.4 Dynamic Circuits

Ratioed circuits reduce the input capacitance by replacing the pMOS transistors connected to the inputs with a single resistive pullup. The drawbacks of ratioed circuits include slow rising transitions, contention on the falling transitions, static power dissipation, and a nonzero V_{OL}. Dynamic circuits circumvent these drawbacks by using a clocked pullup transistor rather than a pMOS that is always ON. Figure 8.21 compares (a) static CMOS, (b) pseudo-nMOS, and (c) dynamic inverters. Dynamic circuit operation is divided into two modes, as shown in Figure 8.22. During *precharge*, the clock ϕ is 0, so the clocked pMOS is ON and initializes the output Y high. During *evaluation*, the clock is 1 and the clocked pMOS turns OFF. The output may remain high or may be discharged low through the pulldown network. Dynamic

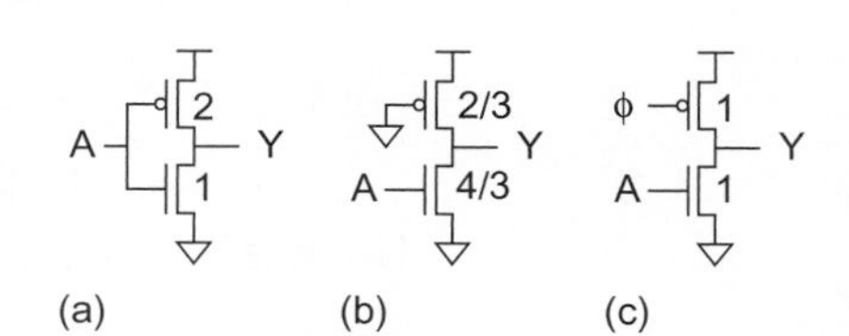

FIGURE 8.21 Comparison of (a) static CMOS, (b) pseudo-nMOS, and (c) dynamic inverters

[3] Many authors call this circuit family *Differential Cascode Voltage Switch Logic* (DCVS [Chu86] or DCVSL [Ng96]). The term *cascode* comes from analog circuits where transistors are placed in series.

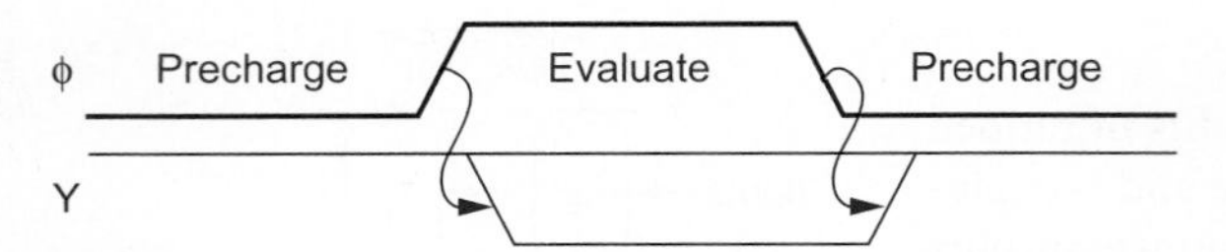

FIGURE 8.22 Precharge and evaluation of dynamic gates

FIGURE 8.23 Footed dynamic inverter

FIGURE 8.24 Generalized footed and unfooted dynamic gates

circuits are the fastest commonly used circuit family because they have lower input capacitance and no contention during switching. They also have zero static power dissipation. However, they require careful clocking, consume significant dynamic power, and are sensitive to noise during evaluation.

In Figure 8.21(c), if the input *A* is 1 during precharge, contention will take place because both the pMOS and nMOS transistors will be ON. When the input cannot be guaranteed to be 0 during precharge, an extra clocked evaluation transistor can be added to the bottom of the nMOS stack to avoid contention as shown in Figure 8.23. The extra transistor is sometimes called a *foot*. Figure 8.24 shows generic *footed* and *unfooted* gates.[4]

Figure 8.25 estimates the falling logical effort of both footed and unfooted dynamic gates. As usual, the pulldown transistors' widths are chosen to give unit resistance. Precharge occurs while the gate is idle and often may take place more slowly. Therefore, the precharge transistor width is chosen for twice unit resistance. This reduces the capacitive load on the clock and the parasitic capacitance at the expense of greater rising delays. We see that the logical efforts are very low. Footed gates have higher logical effort than their unfooted counterparts but are still an improvement over static logic. In practice, the logical effort of footed gates is better than predicted because velocity saturation means series nMOS transistors have less resistance than we have estimated. Moreover, logical efforts are also slightly better than predicted because there is no contention between nMOS and pMOS transistors during the input transition. The size of the foot can be increased relative to the other nMOS transistors to reduce logical effort of the other inputs at the expense of greater clock loading. Like pseudo-nMOS gates, dynamic gates are particularly well suited to wide NOR functions or multiplexers because the logical effort is indepen-

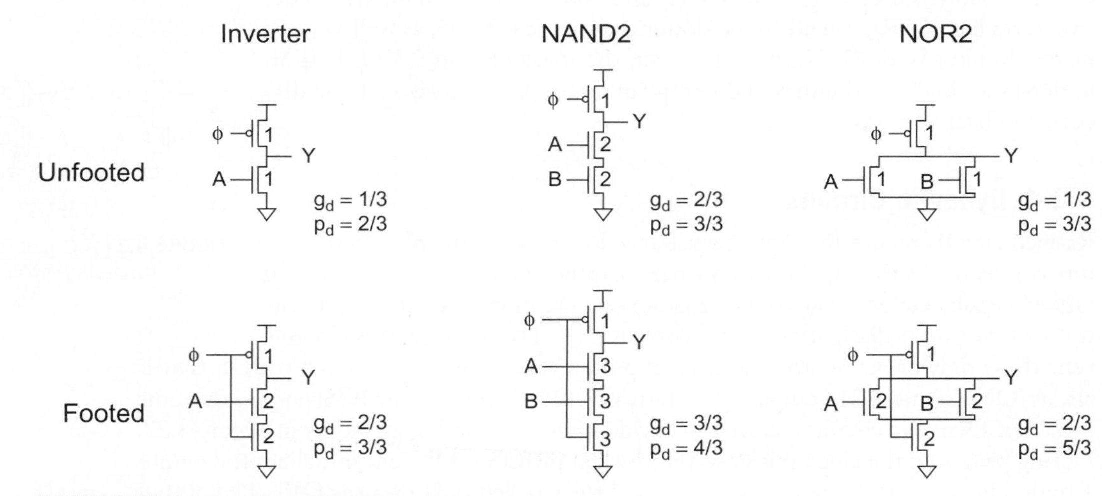

FIGURE 8.25 Catalog of dynamic gates

[4] The footed and unfooted terminology is from IBM [Nowka98]. Intel calls these styles D1 and D2, respectively.

dent of the number of inputs. Of course, the parasitic delay does increase with the number of inputs because there is more diffusion capacitance on the output node. Characterizing the logical effort and parasitic delay of dynamic gates is tricky because the output tends to fall much faster than the input rises, leading to potentially misleading dependence of propagation delay on fanout [Sutherland99].

A fundamental difficulty with dynamic circuits is the *monotonicity* requirement. While a dynamic gate is in evaluation, the inputs must be *monotonically rising*. That is, the input can start LOW and remain LOW, start LOW and rise HIGH, start HIGH and remain HIGH, but not start HIGH and fall LOW. Figure 8.26 shows waveforms for a footed dynamic inverter in which the input violates monotonicity. During precharge, the output is pulled HIGH. When the clock rises, the input is HIGH so the output is discharged LOW through the pulldown network, as you would want to have happen in an inverter. The input later falls LOW, turning off the pulldown network. However, the precharge transistor is also OFF so the output floats, staying LOW rather than rising as it would in a normal inverter. The output will remain low until the next precharge step. In summary, the inputs must be monotonically rising for the dynamic gate to compute the correct function.

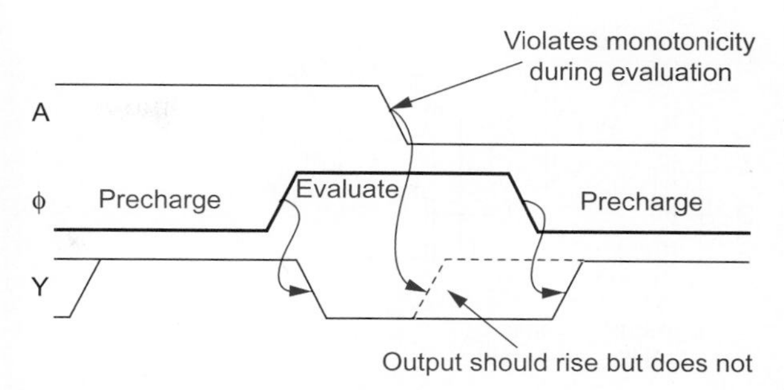

FIGURE 8.26 Monotonicity problem

Unfortunately, the output of a dynamic gate begins HIGH and monotonically falls LOW during evaluation. This monotonically falling output X is not a suitable input to a second dynamic gate expecting monotonically rising signals, as shown in Figure 8.27. Dynamic gates sharing the same clock cannot be directly connected. This problem is often overcome with domino logic, described in the next section.

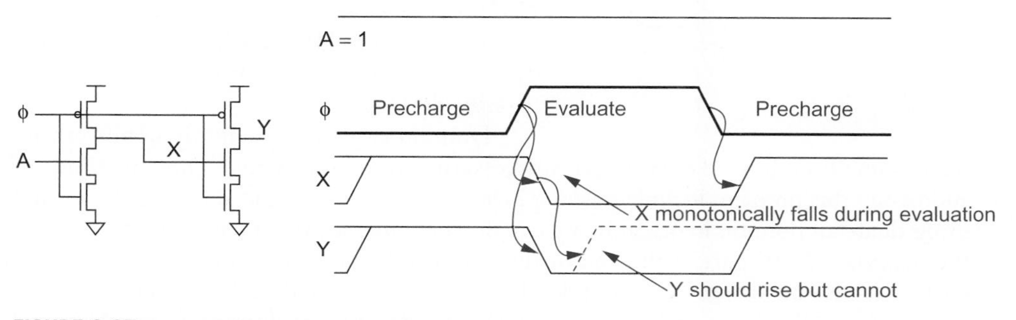

FIGURE 8.27 Incorrect connection of dynamic gates

8.2.4.1 Domino Logic The monotonicity problem can be solved by placing a static CMOS inverter between dynamic gates, as shown in Figure 8.28(a). This converts the monotonically falling output into a monotonically rising signal suitable for the next gate, as shown in Figure 8.28(b). The dynamic-static pair together is called a *domino* gate [Krambeck82] because precharge resembles setting up a chain of dominos and evaluation causes the gates to fire like dominos tipping over, each triggering the next. A single clock can be used to precharge and evaluate all the logic gates within the chain. The dynamic output is monotonically falling during evaluation, so the static inverter output is monotonically rising. Therefore, the static inverter is usually a HI-skew gate to favor this rising output. Observe that precharge occurs in parallel, but evaluation occurs sequentially. This

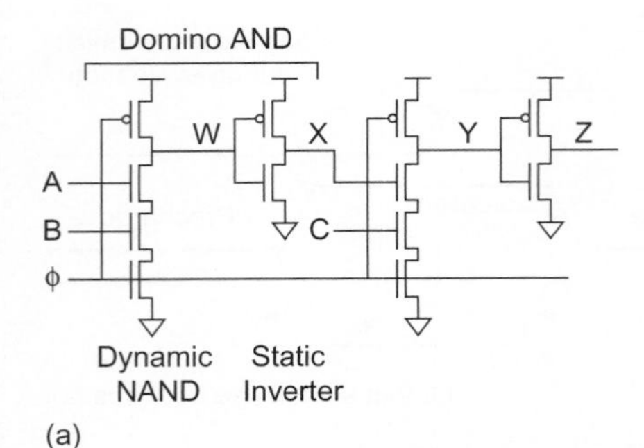

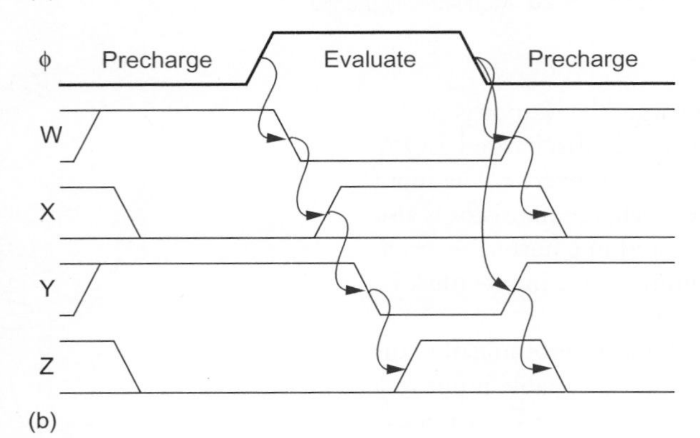

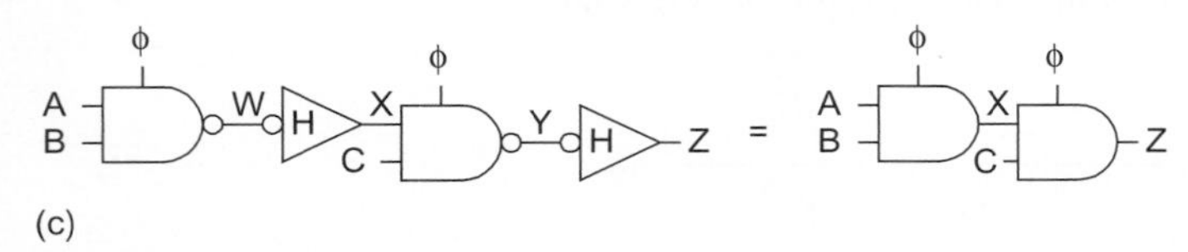

FIGURE 8.28 Domino gates

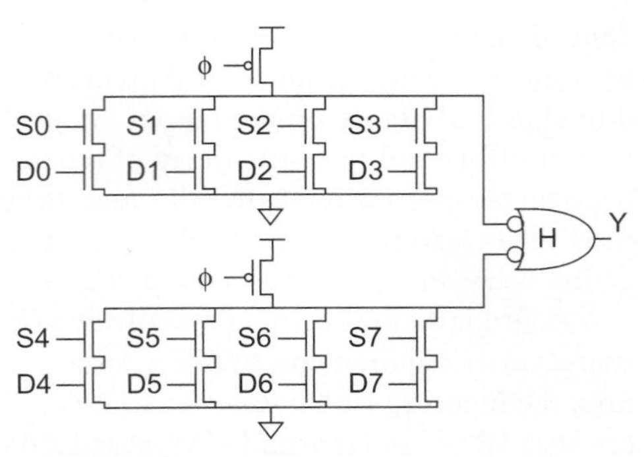

FIGURE 8.29 Domino gate using logic in static CMOS stage

explains why precharge is usually less critical. The symbols for the dynamic NAND, HI-skew inverter, and domino AND are shown in Figure 8.28(c).

In general, more complex inverting static CMOS gates such as NANDs or NORs can be used in place of the inverter [Sutherland99]. This mixture of dynamic and static logic is called *compound domino*. For example, Figure 8.29 shows an 8-input domino multiplexer built from two 4-input dynamic multiplexers and a HI-skew NAND gate. This is often faster than an 8-input dynamic mux and HI-skew inverter because the dynamic stage has less diffusion capacitance and parasitic delay.

Domino gates are inherently noninverting, while some functions like XOR gates necessarily require inversion. Three methods of addressing this problem include pushing inversions into static logic, delaying clocks, and using dual-rail domino logic. In many circuits including arithmetic logic units (ALUs), the necessary XOR gate at the end of the path can be built with a conventional static CMOS XOR gate driven by the last domino circuit. However, the XOR output no longer is monotonically rising and thus cannot directly drive more domino logic. A second approach is to directly cascade dynamic gates without the static CMOS inverter, delaying the clock to the later gates to ensure the inputs are monotonic during evaluation. This is commonly done in content-addressable memories (CAMs) and NOR-NOR PLAs and will be discussed in Section 11.7. The third approach, dual-rail domino logic, is discussed in the next section.

8.2.4.2 Dual-Rail Domino Logic *Dual-rail domino* gates encode each signal with a pair of wires. The input and output signal pairs are denoted with *_h* and *_l*, respectively. Table 8.2 summarizes the encoding. The *_h* wire is asserted to indicate that the output of the gate is "high" or 1. The *_l* wire is asserted to indicate that the output of the gate is "low" or 0. When the gate is precharged, neither *_h* nor *_l* is asserted. The pair of lines should never be both asserted simultaneously during correct operation.

TABLE 8.2 Dual-rail domino signal encoding

sig_h	*sig_l*	**Meaning**
0	0	Precharged
0	1	'0'
1	0	'1'
1	1	Invalid

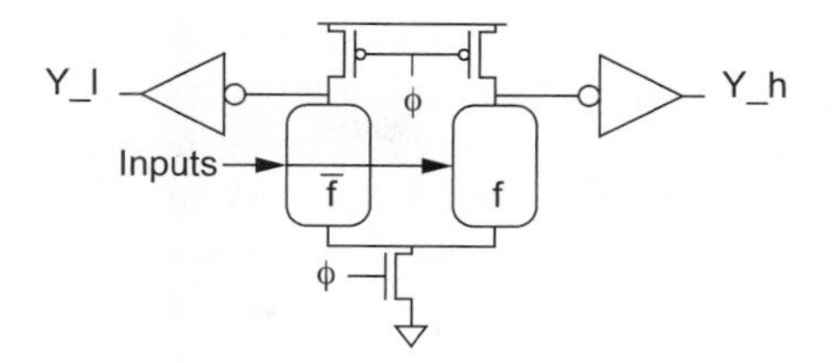

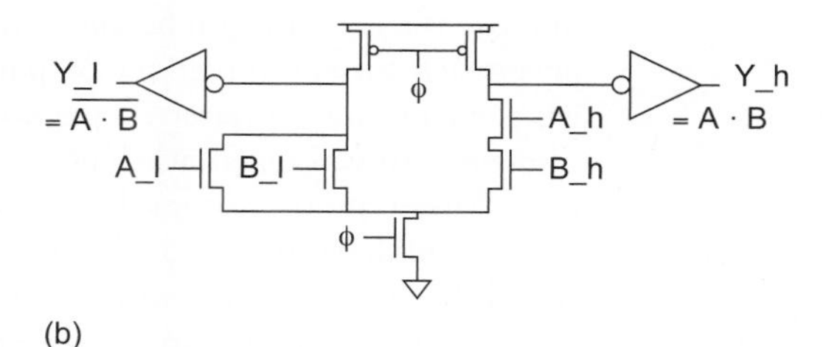

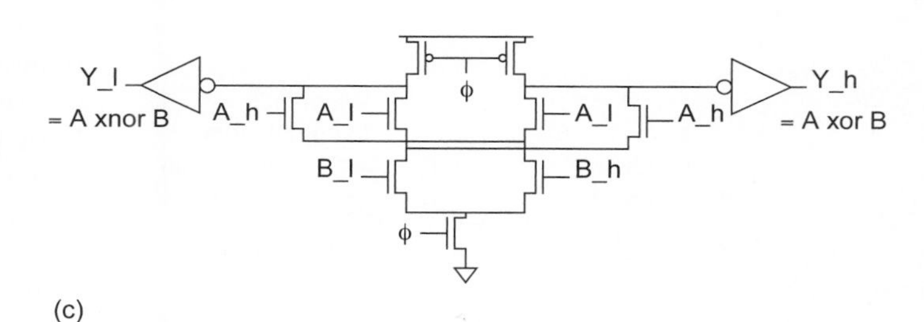

FIGURE 8.30 Dual-rail domino gates

Dual-rail domino gates accept both true and complementary inputs and compute both true and complementary outputs, as shown in Figure 8.30(a). Observe that this is identical to static CVSL circuits from Figure 8.20 except that the cross-coupled pMOS transistors are instead connected to the precharge clock. Therefore, dual-rail domino can be viewed as a dynamic form of CVSL, sometimes called DCVS [Heller84]. Figure 8.30(b) shows a dual-rail AND/NAND gate and Figure 8.30(c) shows a dual-rail XOR/XNOR gate. The gates are shown with clocked evaluation transistors, but can also be unfooted. Dual-rail domino is a *complete* logic family in that it can compute all inverting and noninverting logic functions. However, it requires more area, wiring, and power. Dual-rail structures also lose the efficiency of wide dynamic NOR gates because they require complementary tall dynamic NAND stacks.

Dual-rail domino signals not only the result of a computation but also indicates when the computation is done. Before computation completes, both rails are precharged. When the computation completes, one rail will be asserted. A NAND gate can be used for completion detection, as shown in Figure 8.31. This is particularly useful for asynchronous circuits [Williams91, Sparsø01].

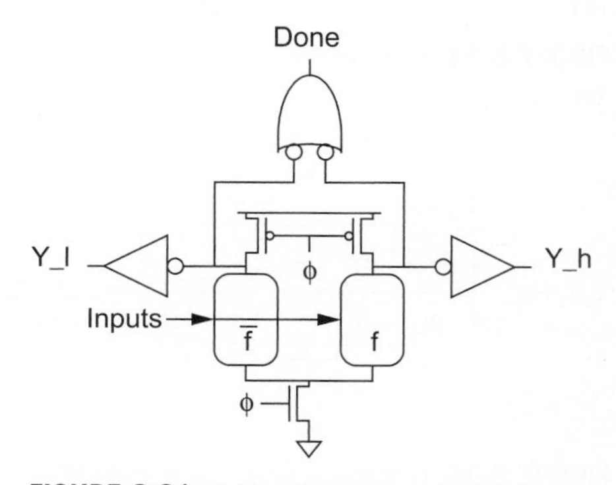

FIGURE 8.31 Dual-rail domino gate with completion detection

Coupling can be reduced in dual-rail signal busses by interdigitating the bits of the bus, as shown in Figure 8.32. Each wire will never see more than one aggressor switching at a time because only one of the two rails switches in each cycle.

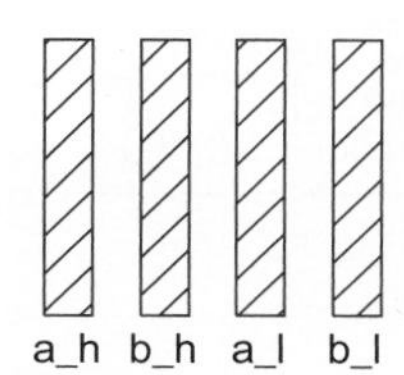

FIGURE 8.32 Reducing coupling noise on dual-rail busses

8.2.4.3 Keepers Dynamic circuits also suffer from charge leakage on the dynamic node. If a dynamic node is precharged high and then left floating, the voltage on the dynamic node will drift over time due to subthreshold, gate, and junction leakage. The time constants tend to be in the millisecond to nanosecond range, depending on process and temperature. This problem is analogous to leakage in dynamic RAMs. Moreover, dynamic circuits have poor input noise margins. If the input rises above V_t while the gate is in evaluation, the input transistors will turn on weakly and can incorrectly discharge the output. Both leakage and noise margin problems can be addressed by adding a *keeper* circuit.

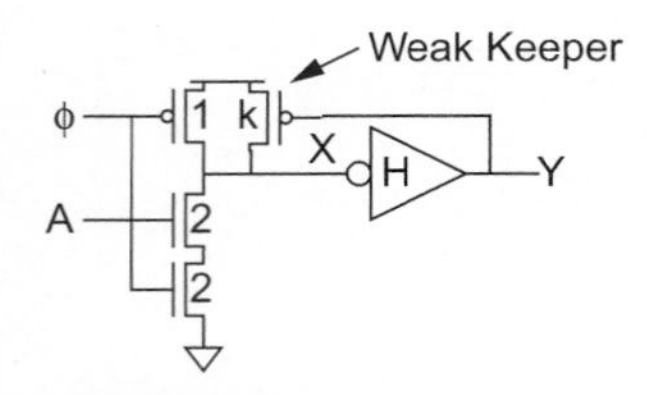

FIGURE 8.33 Conventional keeper

Figure 8.33 shows a conventional keeper on a domino buffer. The keeper is a weak transistor that holds, or *staticizes*, the output at the correct level when it would otherwise float. When the dynamic node X is high, the output Y is low and the keeper is ON to prevent X from floating. When X falls, the keeper initially opposes the transition so it must be much weaker than the pulldown network. Eventually Y rises, turning the keeper OFF and avoiding static power dissipation.

The keeper must be strong (i.e., wide) enough to compensate for any leakage current drawn when the output is floating and the pulldown stack is OFF. Strong keepers also improve the noise margin because when the inputs are slightly above V_t the keeper can supply enough current to hold the output high. Figure 7.28 showed the DC transfer characteristics of a dynamic inverter. As the keeper width k increases, the switching point shifts right. However, strong keepers also increase delay, typically by 5–10%. For example, the 90 nm Itanium Montecito processor selected a pMOS keeper with 6% of the combined width of the leaking pulldown transistors [Naffziger06]. An 8-input NOR with 1 μm wide transistors would thus need a keeper width of 0.48 μm. More advanced processes tend to have greater I_{off}/I_{on} ratios and more variability, so the keepers must be even stronger.

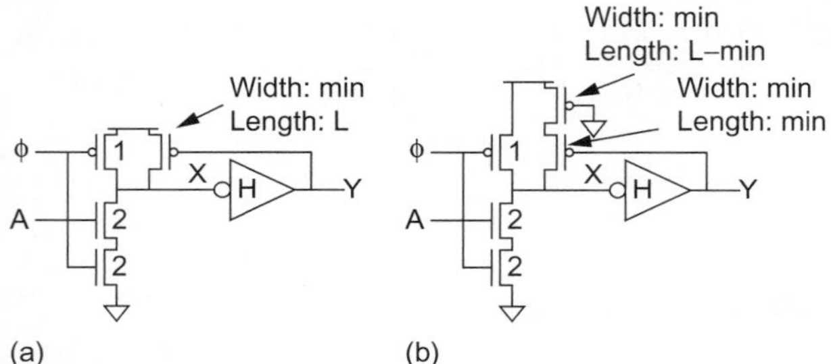

FIGURE 8.34 Weak keeper implementations

For small dynamic gates, the keeper must be weaker than a minimum-sized transistor. This is achieved by increasing the keeper length, as shown in Figure 8.34(a). Long keeper transistors increase the capacitive load on the output Y. This can be avoided by splitting the keeper, as shown in Figure 8.34(b).

Figure 8.35 shows a *differential keeper* for a dual-rail domino buffer. When the gate is precharged, both keeper transistors are OFF and the dynamic outputs float. However, as soon as one of the rails evaluates low, the opposite keeper turns ON. The differential keeper is fast because it does not oppose the falling rail. As long as one of the rails is guaranteed to fall promptly, the keeper on the other rail will turn on before excessive leakage or noise causes failure. Of course, dual-rail domino can also use a pair of conventional keepers.

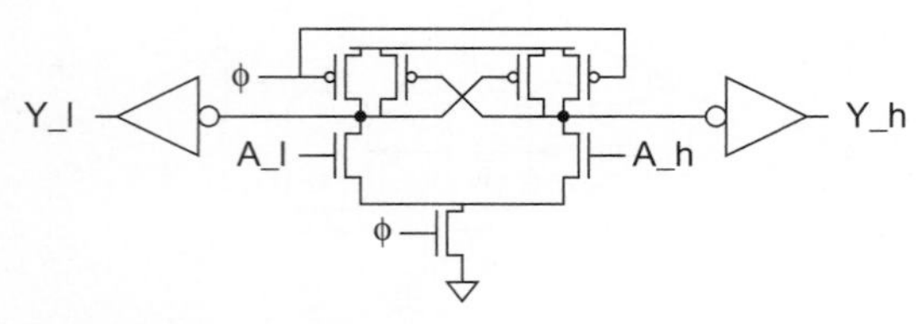

FIGURE 8.35 Differential keeper

During *burn-in*, the chip operates at reduced frequency, but at very high temperature and voltage. This causes severe leakage that can overpower the keeper in wide dynamic NOR gates where many nMOS transistors leak in parallel. Figure 8.36 shows a domino gate with a *burn-in conditional keeper* [Alvandpour02]. The *BI* signal is asserted during burn-in to turn on a second keeper in parallel with the primary keeper. The second keeper slows the gate during burn-in, but provides extra current to fight leakage.

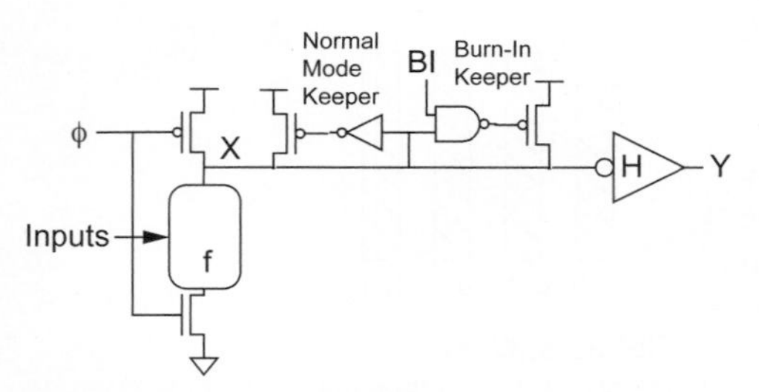

FIGURE 8.36 Burn-in conditional keeper

Noise on the output of the inverter (e.g., from capacitive crosstalk) can reduce the effectiveness of the keeper. In nanometer processes at low voltage where the leakage is high, this effect can significantly increase the required keeper width. Notice how the domino gate in Figure 8.36 used a separate feedback inverter that is not subject to crosstalk noise because it remains inside the cell. This technique is used at Intel even when the burn-in keeper is not employed.

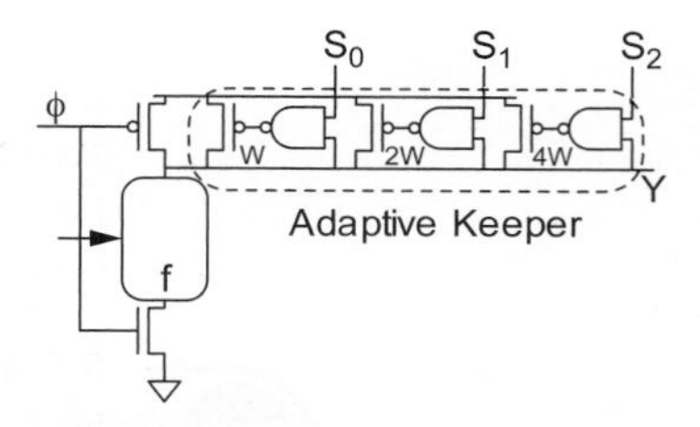

FIGURE 8.37 Adaptive keeper

Like ratioed circuits, domino keepers are afflicted by process variation [Brusamarello08]. The keeper must be wide enough to retain the output in the FS corner. It has the greatest impact on delay in the SF corner. Furthermore, the keeper must be sized to handle roughly 5σ of within-die variation to have negligible impact on yield when the chip has many domino gates. More elaborate keepers can be used to compensate for systemic variations. The *adaptive keeper* of Figure 8.37 has a digitally configurable keeper strength [Kim03]. The *leakage current replica* (LCR) keeper of Figure 8.38 uses a current mirror so that the keeper current tracks the leakage current in a fashion similar to replica biasing of pseudo-nMOS gates [Lih07]. The width of the nMOS transistor in the current mirror is chosen to match the width of the leaking devices. Additional margin is necessary to compensate for noise and random variations.

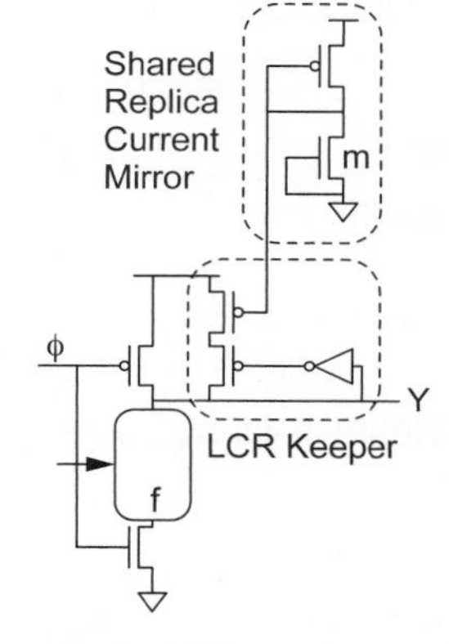

FIGURE 8.38 Leakage current replica keeper

Domino circuits with delayed clocks can use full keepers consisting of cross-coupled inverters to hold the output either high or low.

8.2.4.4 Secondary Precharge Devices Dynamic gates are subject to problems with *charge sharing* [Oklobdzija86]. For example, consider the 2-input dynamic NAND gate in Figure 8.39(a). Suppose the output Y is precharged to V_{DD} and inputs A and B are low. Also suppose that the intermediate node x had a low value from a previous cycle. During evaluation, input A rises, but input B remains low so the output Y should remain high. However, charge is shared between C_x and C_Y, shown in Figure 8.39(b). This behaves as a capacitive voltage divider and the voltages equalize at

$$V_x = V_Y = \frac{C_Y}{C_x + C_Y} V_{DD} \tag{8.3}$$

Charge sharing is most serious when the output is lightly loaded (small C_Y) and the internal capacitance is large. For example, 4-input dynamic NAND gates and complex AOI gates can share charge among multiple nodes. If the charge-sharing noise is small, the keeper will eventually restore the dynamic output to V_{DD}. However, if the charge-sharing noise is large, the output may flip and turn off the keeper, leading to incorrect results.

Charge sharing can be overcome by precharging some or all of the internal nodes with *secondary precharge transistors*, as shown in Figure 8.40. These transistors should be small because they only must charge the small internal capacitances and their diffusion capacitance slows the evaluation. It is often sufficient to precharge every other node in a tall stack. SOI processes are less susceptible to charge sharing in dynamic gates because the diffusion capacitance of the internal nodes is smaller. If some charge sharing is acceptable, a gate can be made faster by predischarging some internal nodes [Ye00].

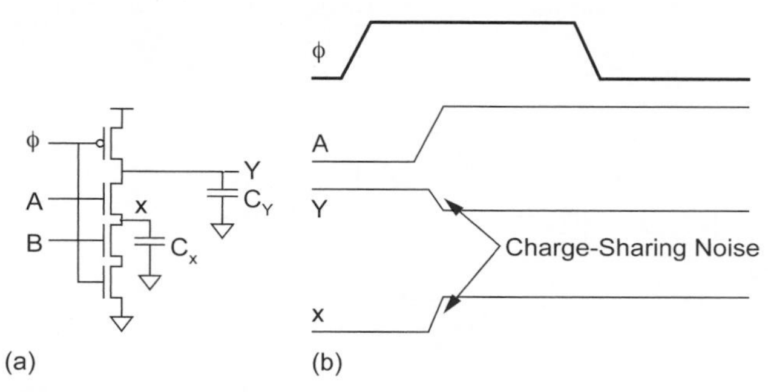

FIGURE 8.39 Charge-sharing noise

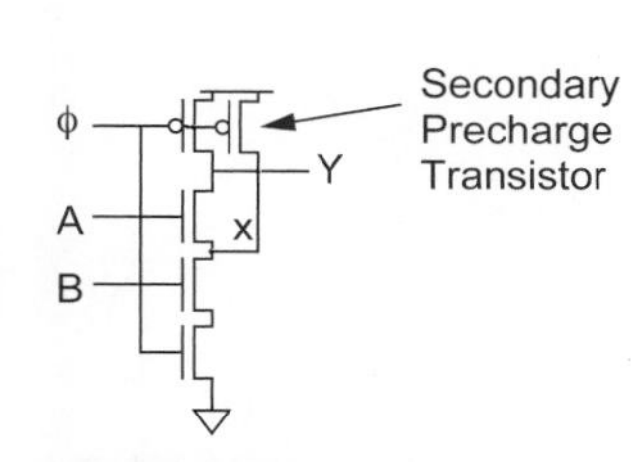

FIGURE 8.40 Secondary precharge transistor

In summary, domino logic was originally proposed as a fast and compact circuit technique. In practice, domino is prized for its speed. However, by the time feet, keepers, and secondary precharge devices are added for robustness, domino is seldom much more compact than static CMOS and it demands a tremendous design effort to ensure robust circuits. When dual-rail domino is required, the area exceeds static CMOS.

8.2.4.5 Logical Effort of Dynamic Paths In Section 3.5.2, we found the best stage effort by hypothetically appending static CMOS inverters onto the end of the path. The best effort depended on the parasitic delay and was 3.59 for $p_{inv} = 1$. When we employ alternative circuit families, the best stage effort may change. For example, with domino circuits, we may consider appending domino buffers onto the end of the path. Figure 8.41 shows that the logical effort of a domino buffer is $G = 5/9$ for footed domino and 5/18 for unfooted domino. Therefore, each buffer appended to a path actually decreases the path effort. Hence, it is better to add more buffers, or equivalently, to target a lower stage effort than you would in a static CMOS design.

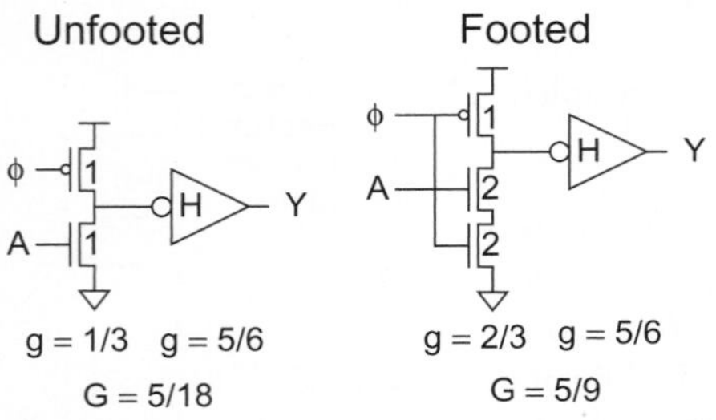

Unfooted: $g = 1/3$ $g = 5/6$, $G = 5/18$

Footed: $g = 2/3$ $g = 5/6$, $G = 5/9$

FIGURE 8.41 Logical efforts of domino buffers

[Sutherland99] showed that the best stage effort is $\rho = 2.76$ for paths with footed domino and 2.0 for paths with unfooted domino. In paths mixing footed and unfooted domino, the best effort is somewhere between these extremes. As a rule of thumb, just as you target a stage effort of 4 for static CMOS paths, you can target a stage effort of 2–3 for domino paths.

We have also seen that it is possible to push logic into the static CMOS stages between dynamic gates. The following example explores under what circumstances this is beneficial.

Example 8.6

Figure 8.42 shows two designs for an 8-input domino AND gate using footed dynamic gates. One uses four stages of logic with static CMOS inverters. The other uses only two stages by employing a HI-skew NOR gate. For what range of path electrical efforts is the 2-stage design faster?

SOLULTION: You might expect that the second design is superior because it scarcely increases the complexity of the static gate and uses half as many stages, but this is only true for low electrical efforts. Figure 8.43 shows the paths annotated with (a) logical effort, (b) parasitic delay, and (c) total delay. The parasitic delays only consider diffusion capacitance on the output node. The delay of each design is plotted against path electrical effort H.[5] For $H > 2.9$, the 4-stage design becomes preferable because the domino gates are effective buffers.

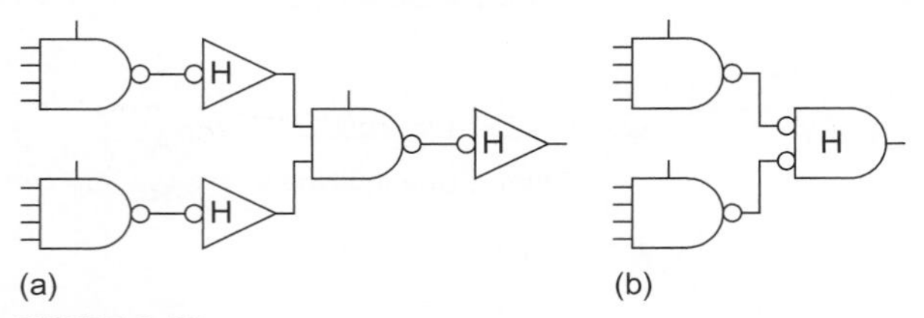

FIGURE 8.42 8-input domino AND gates

[5]Do not confuse the path electrical effort H with the letter H designating the HI-skew static CMOS gates in the schematic.

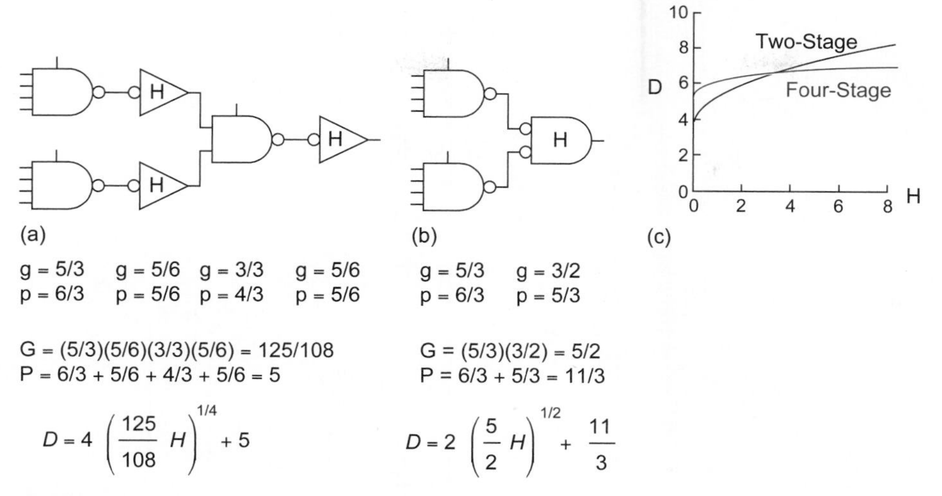

FIGURE 8.43 8-input domino AND delays

In summary, dynamic stages are fast because they build logic using nMOS transistors. Moreover, the low logical efforts suggest that using a relatively large number of stages is beneficial. Pushing logic into the static CMOS stages uses slower pMOS transistors and reduces the number of stages. Thus, it is usually good to use static CMOS gates only on paths with low electrical effort.

8.2.4.6 Multiple-Output Domino Logic (MODL) It is often necessary to compute multiple functions where one is a subfunction of another or shares a subfunction. *Multiple-output domino logic* (MODL) [Hwang89, Wang97] saves area by combining all of the computations into a multiple-output gate.

A popular application is in addition, where the carry-out c_i of each bit of a 4-bit block must be computed, as discussed in Section 10.2.2.2. Each bit position i in the block can either propagate the carry (p_i) or generate a carry (g_i). The carry-out logic is

$$\begin{aligned} c_1 &= g_1 + p_1 c_0 \\ c_2 &= g_2 + p_2\left(g_1 + p_1 c_0\right) \\ c_3 &= g_3 + p_3\left(g_2 + p_2\left(g_1 + p_1 c_0\right)\right) \\ c_4 &= g_4 + p_4\left(g_3 + p_3\left(g_2 + p_2\left(g_1 + p_1 c_0\right)\right)\right) \end{aligned} \tag{8.4}$$

This can be implemented in four compound AOI gates, as shown in Figure 8.44(a). Notice that each output is a function of the less significant outputs. The more compact MODL design shown in Figure 8.44(b) is often called a *Manchester carry chain*. Note that the intermediate outputs require secondary precharge transistors. Also note that care must be taken for certain inputs to be mutually exclusive in order to avoid *sneak paths*. For example, in the adder we must define

$$\begin{aligned} g_i &= a_i b_i \\ p_i &= a_i \oplus b_i \end{aligned} \tag{8.5}$$

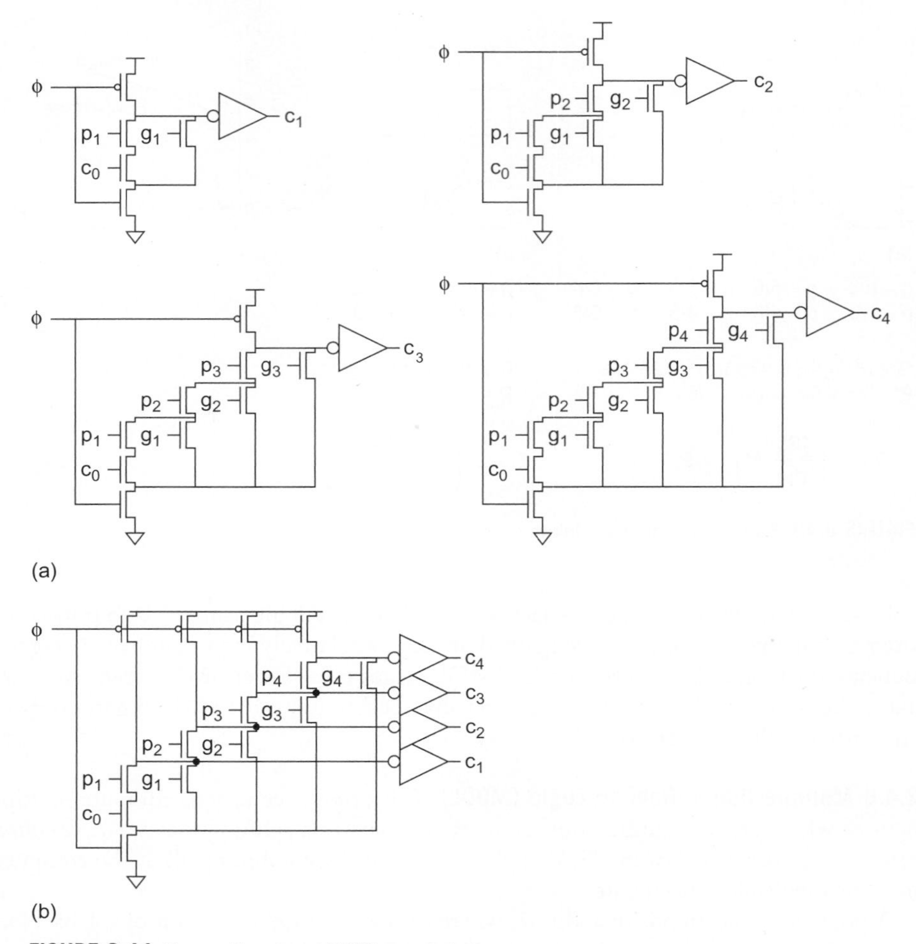

FIGURE 8.44 Conventional and MODL carry chains

If p_i were defined as $a_i + b_i$, a sneak path could exist when a_4 and b_4 are 1 and all other inputs are 0. In that case, $g_4 = p_4 = 1$. c_4 would fire as desired, but c_3 would also fire incorrectly, as shown in Figure 8.45.

8.2.4.7 NP and Zipper Domino Another variation on domino is shown in Figure 8.46(a). The HI-skew inverting static gates are replaced with predischarged dynamic gates using pMOS logic. For example, a footed dynamic p-logic NAND gate is shown in Figure 8.46(b). When ϕ is 0, the first and third stages precharge high while the second stage predischarges low. When ϕ rises, all the stages evaluate. Domino connections are possible, as shown in Figure 8.46(c). The design style is called *NP Domino* or *NORA Domino* (NO RAce) [Gonclaves83, Friedman84].

NORA has two major drawbacks. The logical effort of footed p-logic gates is generally worse than that of HI-skew gates (e.g., 2 vs. 3/2 for NOR2 and 4/3 vs. 1 for NAND2). Secondly, NORA is extremely susceptible to noise. In an ordinary dynamic gate, the input has a low noise margin (about V_t), but is strongly driven by a static CMOS gate. The floating dynamic output is more prone to noise from coupling and charge sharing, but drives another static CMOS gate with a larger noise margin. In NORA, however, the sensitive dynamic inputs are driven by noise-prone dynamic outputs. Given these drawbacks and the extra clock phase required, there is little reason to use NORA.

Zipper domino [Lee86] is a closely related technique that leaves the precharge transistors slightly ON during evaluation by using precharge clocks that swing between 0 and $V_{DD} - |V_{tp}|$ for the pMOS precharge and V_{tn} and V_{DD} for the nMOS precharge. This plays much the same role as a keeper. Zipper never saw widespread use in the industry [Bernstein99].

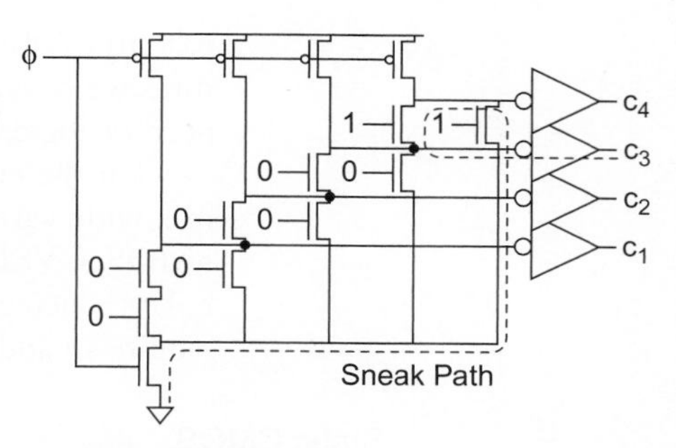

FIGURE 8.45 Sneak path

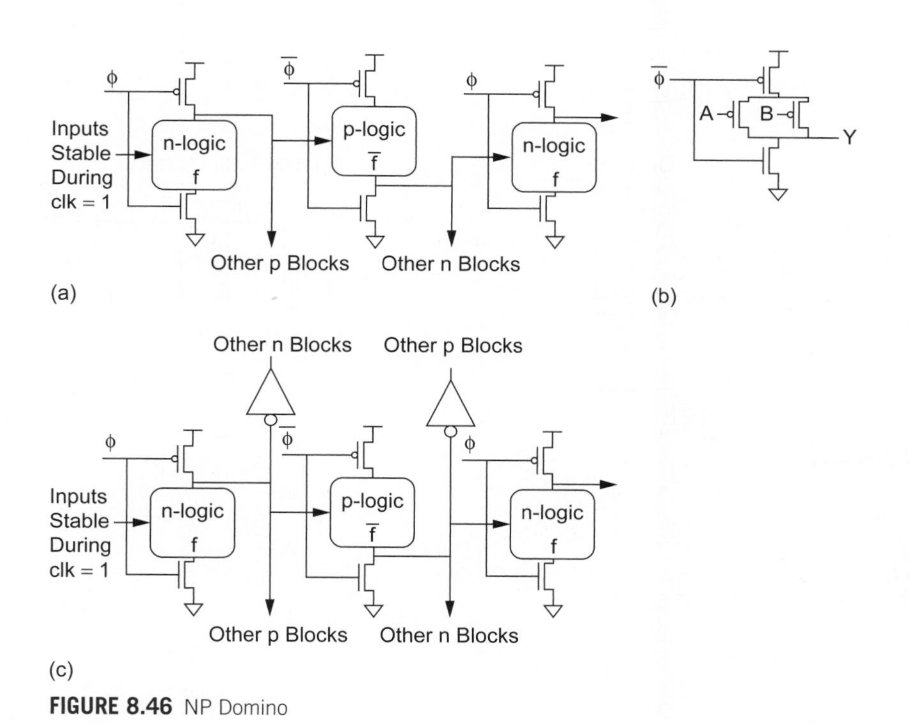

FIGURE 8.46 NP Domino

8.2.5 Pass-Transistor Circuits

In the circuit families we have explored so far, inputs are applied only to the gate terminals of transistors. In pass-transistor circuits, inputs are also applied to the source/drain diffusion terminals. These circuits build switches using either nMOS pass transistors or parallel pairs of nMOS and pMOS transistors called *transmission gates*. Many authors have claimed substantial area, speed, and/or power improvements for pass transistors compared to static CMOS logic. In specialized circumstances this can be true; for example, pass transistors are essential to the design of efficient 6-transistor static RAM cells used in most modern systems (see Section 11.2). Full adders and other circuits rich in XORs also can be efficiently constructed with pass transistors. In certain other cases, we will see that pass-transistor circuits are essentially equivalent ways to draw the fundamental logic structures we have explored before. An independent evaluation finds that for most general-purpose logic, static CMOS is superior in speed, power, and area [Zimmermann97].

For the purpose of comparison, Figure 8.47 shows a 2-input multiplexer constructed in a wide variety of pass-transistor circuit families along with static CMOS, pseudo-nMOS, CVSL, and single- and dual-rail domino. Some of the circuit families are dual-rail, producing both true and complementary outputs, while others are single-rail and may require an additional inversion if the other polarity of output is needed. U XOR V can be

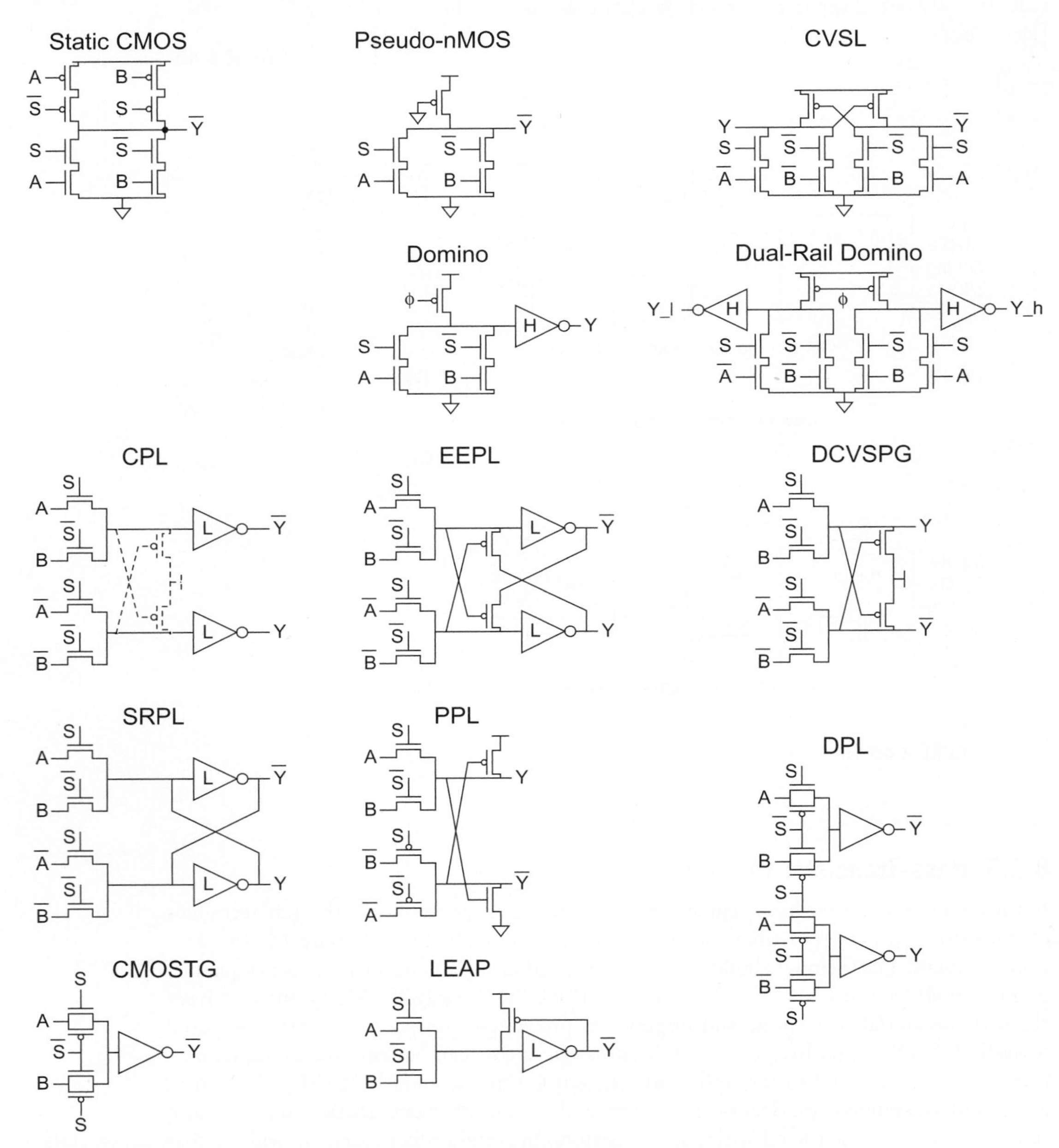

FIGURE 8.47 Comparison of circuit families for 2-input multiplexers

computed with exactly the same logic using $S = U$, $\overline{S} = \overline{U}$, $A = V$, $B = \overline{V}$. This shows that static CMOS is particularly poorly suited to XOR because the complex gate and two additional inverters are required; hence, pass-transistor circuits become attractive. In comparison, static CMOS NAND and NOR gates are relatively efficient and benefit less from pass transistors.

This section first examines mixing CMOS with transmission gates, as is common in multiplexers and latches. It next examines *Complementary Pass-transistor Logic* (CPL), which can work well for XOR-rich circuits like full adders and *LEAn integration with Pass transistors* (LEAP), which illustrates single-ended pass-transistor design. Finally, it catalogs and compares a wide variety of alternative pass-transistor families.

8.2.5.1 CMOS with Transmission Gates Structures such as tristates, latches, and multiplexers are often drawn as transmission gates in conjunction with simple static CMOS logic. For example, Figure 1.28 introduced the transmission gate multiplexer using two transmission gates. The circuit was nonrestoring; i.e., the logic levels on the output are no better than those on the input so a cascade of such circuits may accumulate noise. To buffer the output and restore levels, a static CMOS output inverter can be added, as shown in Figure 8.47 (CMOSTG).

A single nMOS or pMOS pass transistor suffers from a threshold drop. If used alone, additional circuitry may be needed to pull the output to the rail. Transmission gates solve this problem but require two transistors in parallel. The resistance of a unit-sized transmission gate can be estimated as R for the purpose of delay estimation. Current flows through the parallel combination of the nMOS and pMOS transistors. One of the transistors is passing the value well and the other is passing it poorly; for example, a logic 1 is passed well through the pMOS but poorly through the nMOS. Estimate the effective resistance of a unit transistor passing a value in its poor direction as twice the usual value: $2R$ for nMOS and $4R$ for pMOS. Figure 8.48 shows the parallel combination of resistances. When passing a 0, the resistance is $R \,||\, 4R = (4/5)R$. The effective resistance passing a 1 is $2R \,||\, 2R = R$. Hence, a transmission gate made from unit transistors is approximately R in either direction. Note that transmission gates are commonly built using equal-sized nMOS and pMOS transistors. Boosting the size of the pMOS transistor only slightly improves the effective resistance while significantly increasing the capacitance.

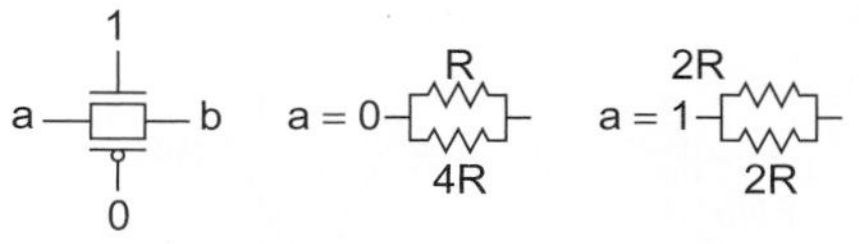

FIGURE 8.48 Effective resistance of a unit transmission gate

At first, CMOS with transmission gates might appear to offer an entirely new range of circuit constructs. A careful examination shows that the topology is actually almost identical to static CMOS. If multiple stages of logic are cascaded, they can be viewed as alternating transmission gates and inverters. Figure 8.49(a) redraws the multiplexer to include the inverters from the previous stage that drive the diffusion inputs but to exclude the output inverter. Figure 8.49(b) shows this multiplexer drawn at the transistor level. Observe that this is identical to the static CMOS multiplexer of Figure 8.47 except that the intermediate nodes in the pullup and pulldown networks are shorted together as $N1$ and $N2$.

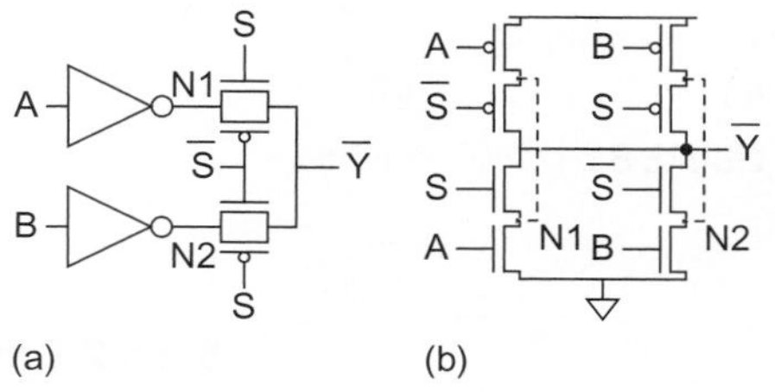

FIGURE 8.49 Alternate representations of CMOSTG in a 2-input inverting multiplexer

The shorting of the intermediate nodes has two effects on delay. The effective resistance decreases somewhat (especially for rising outputs) because the output is pulled up or down through the parallel combination of both pass transistors rather than through a single transistor. However, the effective capacitance increases slightly because of the extra diffusion and wire capacitance required for this shorting. This is apparent from

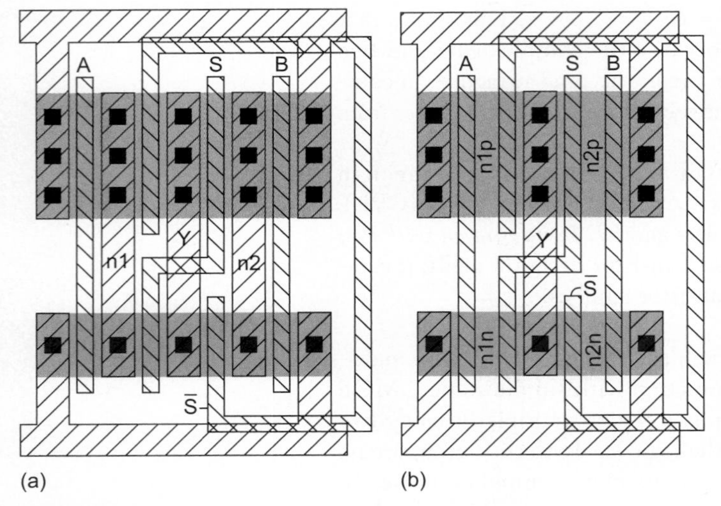

FIGURE 8.50 Multiplexer layout comparison

layouts of the multiplexers; the transmission gate design in Figure 8.50(a) requires contacted diffusion on *N*1 and *N*2 while the static CMOS gate in Figure 8.50(b) does not. In most processes, the improved resistance dominates for gates with moderate fanouts, making shorting generally faster at a small cost in power.

Figure 8.51 shows a similar transformation of a tristate inverter from transmission gate form to conventional static CMOS by unshorting the intermediate node and redrawing the gate. Note that the circuit in Figure 8.51(d) interchanges the *A* and enable terminals. It is logically equivalent, but electrically inferior because if the output is tristated but *A* toggles, charge from the internal nodes may disturb the floating output node. Charge sharing is discussed further in Section 8.3.4.

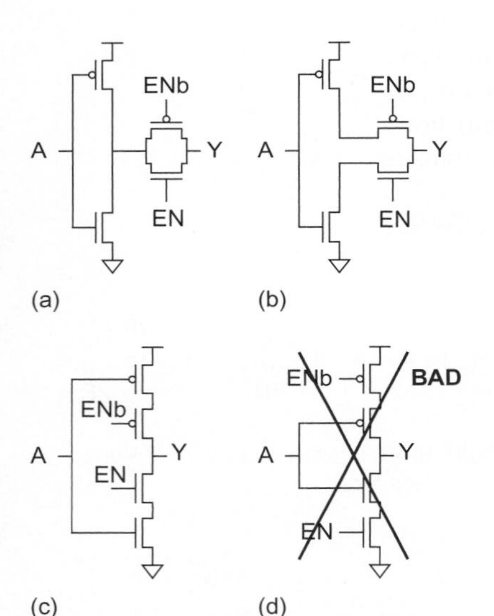

FIGURE 8.51 Tristate inverter

Several factors favor the static CMOS representation over CMOS with transmission gates. If the inverter is on the output rather than the input, the delay of the gate depends on what is driving the input as well as the capacitance driven by the output. This input driver sensitivity makes characterizing the gate more difficult and is incompatible with most timing analysis tools. Novice designers often erroneously characterize transmission gate circuits by applying a voltage source directly to the diffusion input. This makes transmission gate multiplexers look very fast because they only involve one transistor in series rather than two. For accurate characterization, the driver must also be included. A second drawback is that diffusion inputs to tristate inverters are susceptible to noise that may incorrectly turn on the inverter; this is discussed further in Section 8.3.9. Finally, the contacts slightly increase area and their capacitance increases power consumption.

The logical effort of circuits involving transmission gates is computed by drawing stages that begin at gate inputs rather than diffusion inputs, as in Figure 8.52 for a transmission gate multiplexer. The effect of the shorting can be ignored, so the logical effort from either the *A* or *B* terminals is 6/3, just as in a static CMOS multiplexer. Note that the parasitic delay of transmission gate circuits with multiple series transmission gates increases rapidly because of the internal diffusion capacitance, so it is seldom beneficial to use more than two transmission gates in series without buffering.

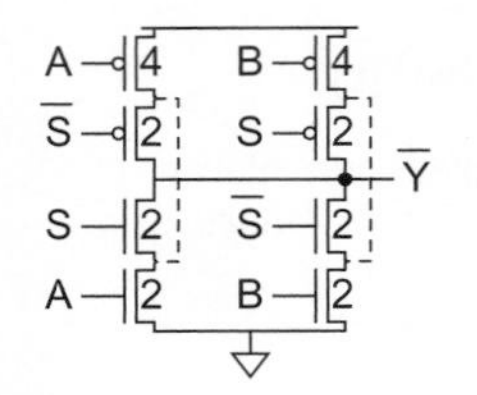

FIGURE 8.52 Logical effort of transmission gate circuit

8.2.5.2 Complementary Pass Transistor Logic (CPL) CPL [Yano90] can be understood as an improvement on CVSL. CVSL is slow because one side of the gate pulls down, and then the cross-coupled pMOS transistor pulls the other side up. The size of the cross-coupled device is an inherent compromise between a large transistor that fights the pulldown excessively and a small transistor that is slow pulling up. CPL resolves this problem by making one half of the gate pull up while the other half pulls down.

Figure 8.53(a) shows the CPL multiplexer from Figure 8.47 rotated sideways. If a path consists of a cascade of CPL gates, the inverters can be viewed equally well as being on the output of one stage or the input of the next. Figure 8.53(b) redraws the mux to

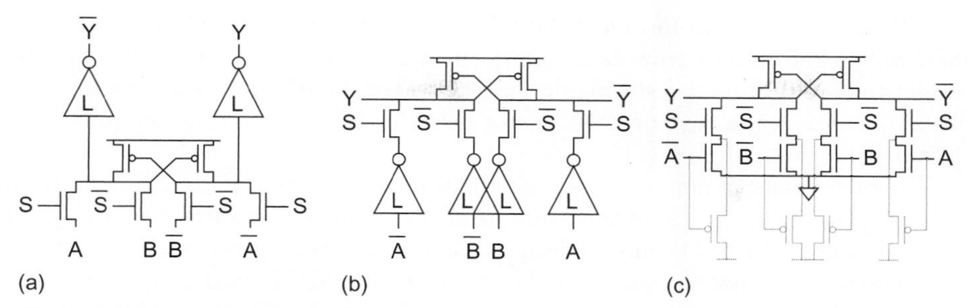

FIGURE 8.53 Alternate representations of CPL

include the inverters from the previous stage that drives the diffusion input, but to exclude the output inverters. Figure 8.53(c) shows the mux drawn at the transistor level. Observe that this is identical to the CVSL gate from Figure 8.47 except that the internal node of the stack can be pulled up through the weak pMOS transistors in the inverters.

When the gate switches, one side pulls down well through its nMOS transistors. The other side pulls up. CPL can be constructed without cross-coupled pMOS transistors, but the outputs would only rise to $V_{DD} - V_t$ (or slightly lower because the nMOS transistors experience the body effect). This costs static power because the output inverter will be turned slightly ON. Adding weak cross-coupled devices helps bring the rising output to the supply rail while only slightly slowing the falling output. The output inverters can be LO-skewed to reduce sensitivity to the slowly rising output.

8.2.5.3 Lean Integration with Pass Transistors (LEAP) Like CPL, LEAP[6] [Yano96] builds logic networks using only fast nMOS transistors, as shown in Figure 8.47. It is a single-ended logic family in that the complementary network is not required, thus saving area and power. The output is buffered with an inverter, which can be LO-skewed to favor the asymmetric response of an nMOS transistor. The nMOS network only pulls up to $V_{DD} - V_t$ so a pMOS feedback transistor is necessary to pull the internal node fully high, avoiding power consumption in the output inverter. The pMOS width is a trade-off between fighting falling transitions and assisting the last part of a rising transition; it generally should be quite weak and the circuit will fail if it is too strong. LEAP can be a good way to build wide 1-of-*N* hot multiplexers with many of the advantages of pseudo-nMOS but without the static power consumption. It was originally proposed for use in a pass transistor logic synthesis system because the cells are compact.

Unlike most circuit families that can operate down to $V_{DD} \geq \max(V_{tn}, |V_{tp}|)$, LEAP is limited to operating at $V_{DD} \geq 2V_t$ because the inverter must flip even when receiving an input degraded by a threshold voltage.

8.2.5.4 Other Pass Transistor Families There have been a host of pass transistor families proposed in the literature, including *Differential Pass Transistor Logic* (DPTL) [Pasternak87, Pasternak91], *Double Pass Transistor Logic* (DPL) [Suzuki93], *Energy Economized Pass Transistor Logic* (EEPL) [Song96], *Push-Pull Pass Transistor Logic* (PPL) [Paik96], *Swing-Restored Pass Transistor Logic* (SRPL) [Parameswar96], and *Differential Cascode Voltage Switch with Pass Gate Logic* (DCVSPG) [Lai97]. All of these are dual-rail families like CPL, as contrasted with the single-rail CMOSTG and LEAP.

[6] The LEAP topology was reinvented under the name *Single Ended Swing Restoring Pass Transistor Logic* [Pihl98].

DPL is a double-rail form of CMOSTG optimized to use single-pass transistors where only a known 0 or 1 needs to be passed. It passes good high and low logic levels without the need for level-restoring devices. However, the pMOS transistors contribute substantial area and capacitance, but do not help the delay much, resulting in large and relatively slow gates.

The other dual-rail families can be viewed as modifications to CPL. EEPL drives the cross-coupled level restoring transistors from the opposite rail rather than V_{DD}. The inventors claimed this led to shorter delay and lower power dissipation than CPL, but the improvements could not be confirmed [Zimmermann97]. SRPL cross-couples the inverters instead of using cross-coupled pMOS pullups. This leads to a ratio problem in which the nMOS transistors in the inverter must be weak enough to be overcome as the pass transistors try to pull up. This tends to require small inverters, which make poor buffers. DCVSPG eliminates the output inverters from CPL. Without these buffers, the output of a DCVSPG gate makes a poor input to the diffusion terminal of another DCVSPG gate because a long unrestored chain of nMOS transistors would be formed, leading to delay and noise problems. PPL also has unbuffered outputs and associated delay and noise issues. DPTL generalizes the output buffer structure to consider alternatives to the cross-coupled pMOS transistors and LO-skewed inverters of CPL. All of the alternatives are slower and larger than CPL.

8.3 Circuit Pitfalls

Circuit designers tend to use simple circuits because they are robust. Elaborate circuits, especially those with more transistors, tend to add more area, more capacitance, and more things that can go wrong. Static CMOS is the most robust circuit family and should be used whenever possible. This section catalogs a variety of circuit pitfalls that can cause chips to fail. They include the following:

- Threshold drops
- Ratio failures
- Leakage
- Charge sharing
- Power supply noise
- Coupling
- Minority carrier injection
- Back-gate coupling
- Diffusion input noise sensitivity
- Race conditions
- Delay matching
- Metastability
- Hot spots
- Soft errors
- Process sensitivity

Capacitive and inductive coupling were discussed in Section 5.3. Sneak paths were discussed in Section 8.2.4.6. Reliability issues such as soft errors impacting circuit design were discussed in Section 6.3. Timing-related problems including race conditions, delay matching, and metastability will be examined in Sections 9.2.3 and 9.5.1. The other pitfalls are described here.

8.3.1 Threshold Drops

Pass transistors are good at pulling in a preferred direction, but only swing to within V_t of the rail in the other direction; this is called a *threshold drop*. For example, Figure 8.54 shows a pass transistor driving a logic 1 into an inverter. The output of the pass transistor only rises to $V_{DD} - V_t$. Worse yet, the body effect increases this threshold voltage because $V_{sb} > 0$ for the pass transistor. The degraded level is insufficient to completely turn off the pMOS transistor in the inverter, resulting in static power dissipation. Indeed, for low V_{DD}, the degraded output can be so poor that the inverter no longer sees a valid input logic level V_{IH}. Finally, the transition becomes lethargic as the output approaches $V_{DD} - V_t$. Threshold drops were sometimes tolerable in older processes where $V_{DD} \approx 5V_t$, but are seldom acceptable in modern processes where the power supply has been scaled down faster than the threshold voltage to $V_{DD} \approx 3V_t$. As a result, pass transistors must be replaced by full transmission gates or may use weak pMOS feedback transistors to pull the output to V_{DD}, as was done in several pass transistor families.

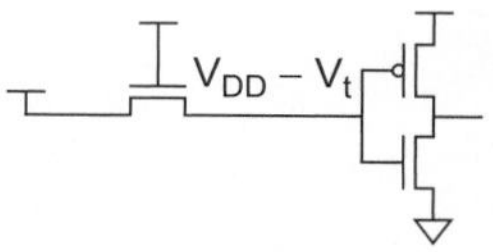

$V_{DD} - V_t$

FIGURE 8.54 Pass transistor with threshold drop

8.3.2 Ratio Failures

Pseudo-nMOS circuits illustrated ratio constraints that occur when a node is simultaneously pulled up and down, typically by strong nMOS transistors and weak pMOS transistors. The weak transistors must be sufficiently small that the output level falls below V_{IL} of the next stage by some noise margin. Ideally, the output should fall below V_t so the next stage does not conduct static power. Ratioed circuits should be checked in the SF and FS corners.

Another example of ratio failures occurs in circuits with feedback. For example, dynamic keepers, level-restoring devices in SRPL and LEAP, and feedback inverters in static latches all have weak feedback transistors that must be ratioed properly.

Ratioing is especially sensitive for diffusion inputs. For example, Figure 8.55(a) shows a static latch with a weak feedback inverter. The feedback inverter must be weak enough to be overcome by the series combination of the pass transistor and the gate driving the D input, as shown in Figure 8.55(b). This cannot be verified by checking the latch alone; it requires a global check of the latch and driver. Worse yet, if the driver is far away, the series wire resistance must also be considered, as shown in Figure 8.55(c).

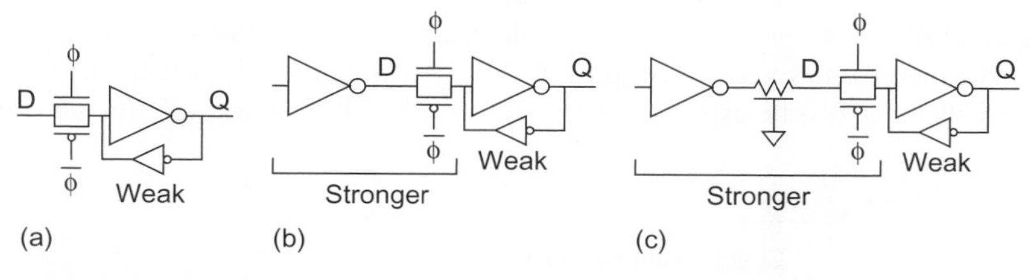

FIGURE 8.55 Ratio constraint on static latch with diffusion input

8.3.3 Leakage

Leakage current is a growing problem as technology scales, especially for dynamic nodes and wide NOR structures. Recall that leakage arises from subthreshold conduction, gate tunneling, and reverse-biased diode leakage. Subthreshold conduction is presently the most important component because V_t is low and getting lower, but gate tunneling will become profoundly important too as oxide thickness diminishes. Besides causing static power dissipation, leakage can result in incorrect values on dynamic or weakly driven nodes. The time required for leakage to disturb a dynamic node by some voltage ΔV is

$$t = \frac{C_{\text{node}} \Delta V}{I_{\text{leak}}} \tag{8.6}$$

Subthreshold leakage gradually discharges dynamic nodes through transistors that are nominally OFF. Fully dynamic gates and latches without keepers are not viable in most modern processes. DRAM refresh times are also set by leakage and DRAM processes must minimize leakage to have satisfactory retention times.

Even when a keeper is used, it must be wide enough. This seems trivial because the keeper is fully ON while leakage takes place through transistors that are supposed to be OFF. However, in wide dynamic NOR structures, many parallel nMOS transistors may be leaking simultaneously. Similar problems apply to wide pseudo-nMOS NOR gates and PLAs. Leakage increases exponentially with temperature, so the problem is especially bad at burn-in. For example, a preliminary version of the Sun UltraSparc V had difficulty with burn-in because of excess leakage.

Subthreshold leakage is much lower through two OFF transistors in series than through a single transistor because the outer transistor has a lower drain voltage and sees a much lower effect from DIBL. Multiple threshold voltages are also frequently used to achieve high performance in critical paths and lower leakage in other paths.

8.3.4 Charge Sharing

Charge sharing was introduced in Section 8.2.4.4 in the context of a dynamic gate. Charge sharing can also occur when dynamic gates drive pass transistors. For example, Figure 8.56 shows a dynamic inverter driving a transmission gate. Suppose the dynamic gate has been precharged and the output is floating high. Further suppose the transmission gate is OFF and $Y = 0$. If the transmission gate turns on, charge will be shared between X and Y, disturbing the dynamic output.

8.3.5 Power Supply Noise

V_{DD} and GND are not constant across a large chip. Both are subject to *power supply noise* caused by IR drops and di/dt noise. IR drops occur across the resistance R of the power supply grid between the supply pins and a block drawing a current I, as shown in Figure 8.57. di/dt noise occurs across the power supply inductance L as the current rapidly changes. di/dt noise can be especially important for blocks that are idle for several cycles

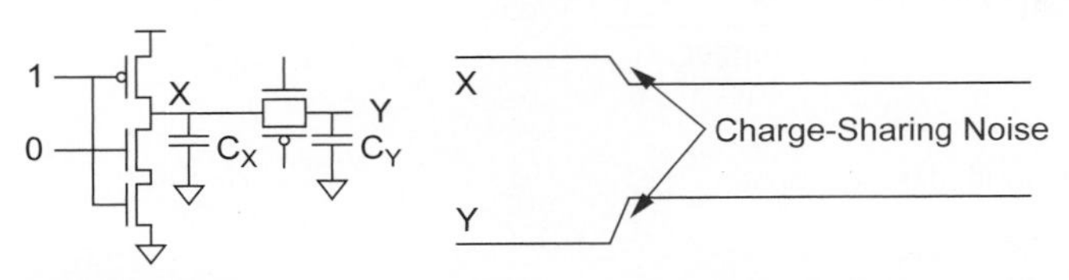

FIGURE 8.56 Charge sharing on dynamic gate driving pass transistor

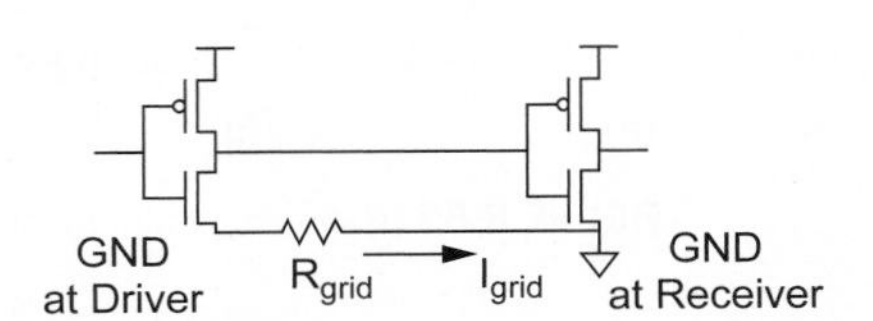

FIGURE 8.57 Power supply IR drops

and then begin switching. Power supply noise hurts performance and can degrade noise margins. Typical targets are for power supply noise on the order of 5–10% of V_{DD}. Power supply noise causes both noise margin problems and delay variations. The noise margin issues can be managed by placing sensitive circuits near each other and having them share a common low-resistance power wire.

Power supply noise can be estimated from simulations of the chip power grid, bypass capacitance, and packaging, as discussed in Section 12.3. Figure 6.2 shows a map of power supply noise across a chip.

8.3.6 Hot Spots

Transistor performance degrades with temperature, so care must be taken to avoid excessively *hot spots*. These can be caused by nonuniform power dissipation even when the overall power consumption is within budget. The nonuniform temperature distribution leads to variation in delay between gates across the chip. Full-chip temperature plots can be generated through electrothermal simulation [Petegem94, Cheng00]; this can begin when the floorplan and preliminary power estimates for each unit are available. Figure 6.3 shows a thermal map of the Itanium 2. A particularly localized form of hot spots is self-heating in resistive wires, described in Section 6.3.3.2.

8.3.7 Minority Carrier Injection

It is sometimes possible to drive a signal momentarily outside the rails, either through capacitive coupling or through inductive ringing on I/O drivers. In such a case, the junctions between drain and body may momentarily become forward-biased, causing current to flow into the substrate. This effect is called *minority carrier injection* [Chandrakasan01]. For example, in Figure 8.58, the drain of an nMOS transistor is driven below GND, injecting electrons into the p-type substrate. These can be collected on a nearby transistor

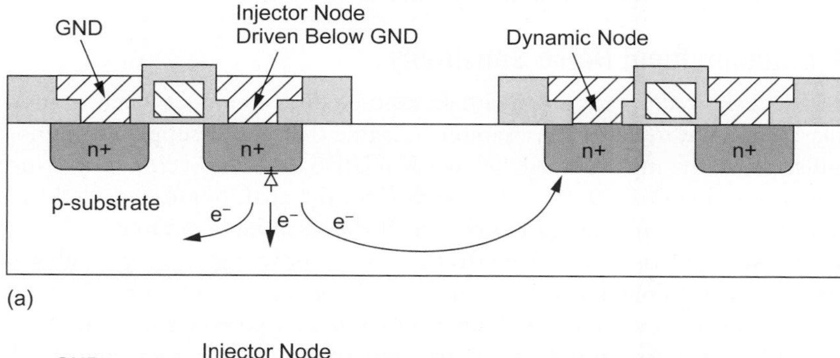

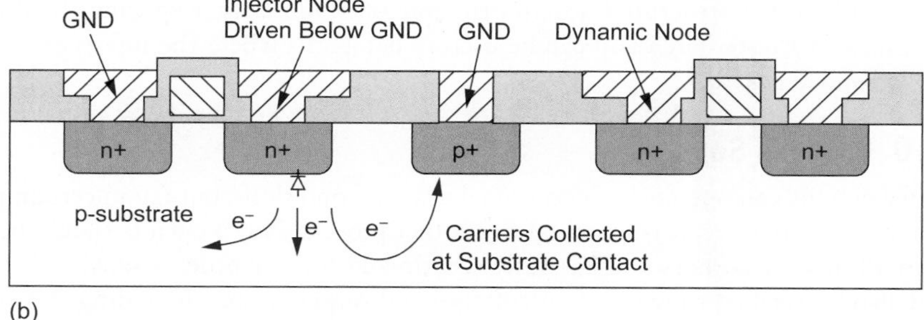

FIGURE 8.58 Minority carrier injection and collection

diffusion node (Figure 8.58(a)), disturbing a high voltage on the node. This is a particular problem for dynamic nodes and sensitive analog circuits.

Minority carrier injection problems are avoided by keeping injection sources away from sensitive nodes. In particular, I/O pads should not be located near sensitive nodes. Noise tools can identify potential coupling problems so the layout can be modified to reduce coupling. Alternatively, the sensitive node can be protected by an intermediate substrate or well contact. For example in Figure 8.58(b), most of the injected electrons will be collected into the substrate contact before reaching the dynamic node. In I/O pads, it is common to build *guard rings* of substrate/well contacts around the output transistors. Guard rings were illustrated in Figure 6.13.

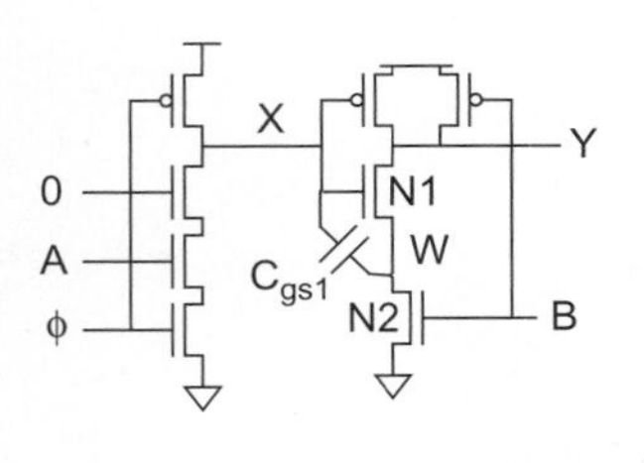

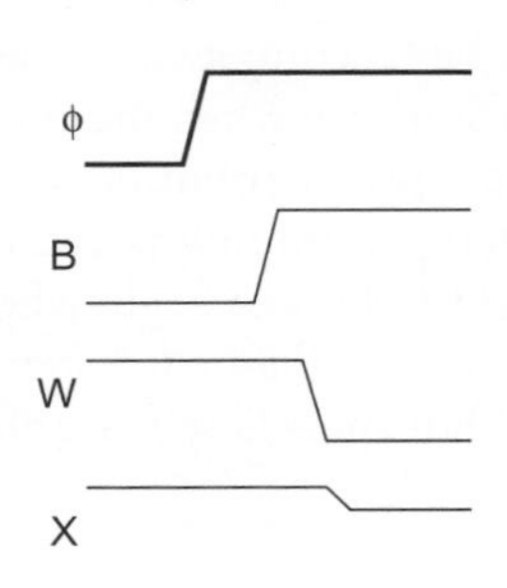

FIGURE 8.59 Back-gate coupling

8.3.8 Back-Gate Coupling

Dynamic gates driving multiple-input static CMOS gates are susceptible to the *back-gate coupling* effect [Chandrakasan01] illustrated in Figure 8.59. In this example, a dynamic NAND gate drives a static NAND gate. The gate-to-source capacitance C_{gs1} of $N1$ is shown explicitly. Suppose that the dynamic gate is in evaluation and its output X is floating high. The other input B to the static NAND gate is initially low. Therefore, the NAND output Y is high and the internal node W is charged up to $V_{DD} - V_t$. At some time B rises, discharging Y and W through transistor $N2$. The source of $N1$ falls. This tends to bring the gate along for the ride because of the C_{gs1} capacitance, resulting in a droop on the dynamic node X. As with charge sharing, the magnitude of the droop depends on the ratio of C_{gs1} to the total capacitance on node X.

Back-gate coupling is eliminated by driving the input closer to the rail. For example, if X drove $N2$ instead of $N1$, the problem would be avoided. Otherwise, the back-gate coupling noise must be included in the dynamic noise budget.

8.3.9 Diffusion Input Noise Sensitivity

Coupling & Supply Noise
0
D
V_{DD}
Q
0
$-2V_t$
V_{DD}
Weak

FIGURE 8.60 Noise on diffusion input of latch

Figure 8.55(a) showed a static latch with an exposed diffusion input. Such an input is also particularly sensitive to noise. For example, imagine that power supply noise and/or coupling noise drove the input voltage below $-V_t$ relative to GND seen by the transmission gate, as shown in Figure 8.60. V_{gs} now exceeds V_t for the nMOS transistor in the transmission gate, so the transmission gate turns on. If the latch had contained a 1, it could be incorrectly discharged to 0. A similar effect can occur for voltage excursions above V_{DD}.

For this reason, along with the ratio issues discussed in Section 8.3.2, standard cell latches are usually built with buffered inputs rather than exposed diffusion nodes. This is a good example of the structured design principle of modularity. Exposing the diffusion input results in a faster latch and can be used in datapaths where the inputs are carefully controlled and checked.

8.3.10 Process Sensitivity

Marginal circuits can operate under nominal process conditions, but fail in certain process corners or when the circuit is migrated to another process. Novel circuits should be simulated in all process corners and carefully scrutinized for any process sensitivities. They should also be verified to work at all voltages and temperatures, including the elevated voltages and temperatures used during burn-in and the lower voltage that might be used for low-power versions of a part.

When a design is likely to be migrated to another process for cost-reduction, circuits should be designed to facilitate this migration. You can expect that leakage will increase, threshold drops will become a greater fraction of the supply voltage, wire delay will become a greater portion of the cycle time, and coupling may get worse as aspect ratios of wires increase. For example, the Pentium 4 processor was originally fabricated in a 180 nm process. Designers placed repeaters closer than was optimal for that process because they knew the best repeater spacing would become smaller as transistor dimensions were reduced later in the product's life [Kumar01].

8.3.11 Example: Domino Noise Budgets

Domino logic requires careful verification because it is sensitive to noise. Noise in static CMOS gates usually results in greater delay, but noise in domino logic can produce incorrect results. This section reviews the various noise sources that can affect domino gates and presents a sample noise budget.

Dynamic outputs are especially susceptible to noise when they float high, held only by a weak keeper. Dynamic inputs have low noise margins (approximately V_t). Noise issues that should be considered include [Chandrakasan01]:

- **Charge leakage** Subthreshold leakage on the dynamic node is presently most important, but gate leakage will become important, too. Subthreshold leakage is worst for wide NOR structures at high temperature (especially during burn-in). Keepers must be sized appropriately to compensate for leakage.
- **Charge sharing** Charge sharing can take place between the dynamic output node and the nodes within the dynamic gate. Secondary precharge transistors should be added when the charge sharing could be excessive. Do not drive dynamic nodes directly into transmission gates because charge sharing can occur when the transmission gate turns ON.
- **Capacitive coupling** Capacitive coupling can occur on both the input and output. The inputs of dynamic gates have the lowest noise margin, but are actively driven by a static gate, which fights coupling noise. The dynamic outputs have more noise tolerance, but are weakly driven. Coupling is minimized by keeping wires short and increasing the spacing to neighbors or shielding the lines. Coupling can be extremely bad in processes below 250 nm because the wires have such high aspect ratios.
- **Back-gate coupling** Dynamic gates connected to multiple-input CMOS gates should drive the outer input when possible. This is not a factor for dynamic gates driving inverters.
- **Minority carrier injection** Dynamic nodes should be protected from nodes that can inject minority carriers. These include I/O circuits and nodes that can be coupled far outside the supply rails. Substrate/well contacts and guard rings can be added to protect dynamic nodes from potential injectors.
- **Power supply noise** Static gates should be located close to the dynamic gates they drive to minimize the amount of power supply noise seen.
- **Soft errors** Alpha particles and cosmic rays can disturb dynamic nodes. The probability of failure is reduced through large node capacitance and strong keepers.

- **Noise feedthrough** Noise that pushes the input of a previous stage to near its noise margin will cause the output to be slightly degraded, as shown in Figure 2.30.
- **Process corner effects** Noise margins are degraded in certain process corners. Dynamic gates have the smallest noise margin in the FS corner where the nMOS transistors have a low threshold and the pMOS keepers are weak. HI-skew static gates have the smallest noise margins in the SF corner where the gates are most skewed.

In a domino gate, the noise-prone dynamic output drives a static gate with a reasonable noise margin. The noise-sensitive dynamic gate is strongly driven by a noise-resistant static gate. In an NP domino gate or clock-delayed domino gate, the noise-prone dynamic output directly drives a noise-sensitive dynamic input, making such circuits particularly risky.

Consider a noise budget for a 3.3 V process [Harris01a]. A HI-skew inverter in this process has $V_{IH} = 2.08$ V, resulting in $NM_H = 37\%$ of V_{DD} if $V_{OH} = V_{DD}$ A dynamic gate with a small keeper has $V_{IL} = 0.63$ V, resulting in $NM_L = 19\%$ of V_{DD}. Table 8.3 allocates these margins to the primary noise sources. In a full design methodology, different margins can be used for different gates. For example, wide NOR structures have no charge-sharing noise, but may see significant leakage instead. More coupling noise could be tolerated if other noise sources are known to be smaller. Noise analysis tools are discussed further in Section 13.4.2.6.

TABLE 8.3 Sample domino noise budget

Source	Dynamic Output	Dynamic Input
Charge sharing	10	n/a
Coupling	17	7
Supply noise	5	5
Feedthrough noise	5	7
Total	37%	19%

8.4 Silicon-On-Insulator Circuit Design

Silicon-on-Insulator (SOI) technology has been a subject of research for decades, but has become commercially important since it was adopted by IBM for PowerPC microprocessors in 1998 [Shahidi02]. SOI is attractive because it offers potential for higher performance and lower power consumption, but also has a higher manufacturing cost and some unusual transistor behavior that complicates circuit design.

The fundamental difference between SOI and conventional bulk CMOS technology is that the transistor source, drain, and body are surrounded by insulating oxide rather than the conductive substrate or well (called the *bulk*). Using an insulator eliminates most of the parasitic capacitance of the diffusion regions. However, it means that the body is no longer tied to GND or V_{DD} through the substrate or well. Any change in body voltage modulates V_t, leading to both advantages and complications in design.

Figure 8.61 shows a cross-section of an inverter in a SOI process. The process is similar to standard CMOS, but starts with a wafer containing a thin layer of SiO_2 buried beneath a thin single-crystal silicon layer. Section 15.4.1.2 discusses several ways to form this buried oxide. Shallow trench isolation is used to surround each transistor by an oxide insulator. Figure 8.62 shows a scanning electron micrograph of a 6-transistor static RAM cell in a 0.22 μm IBM SOI process.

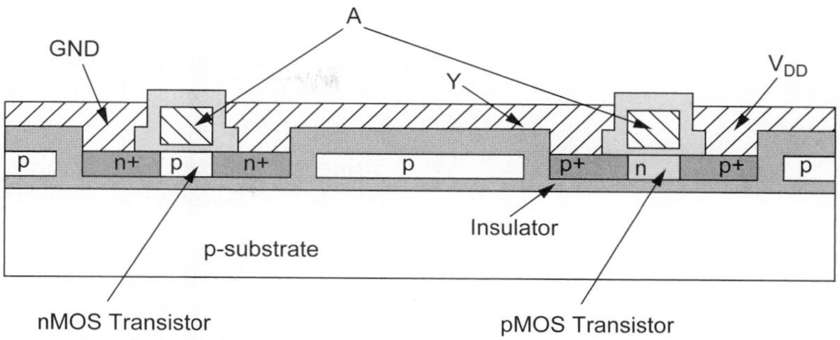

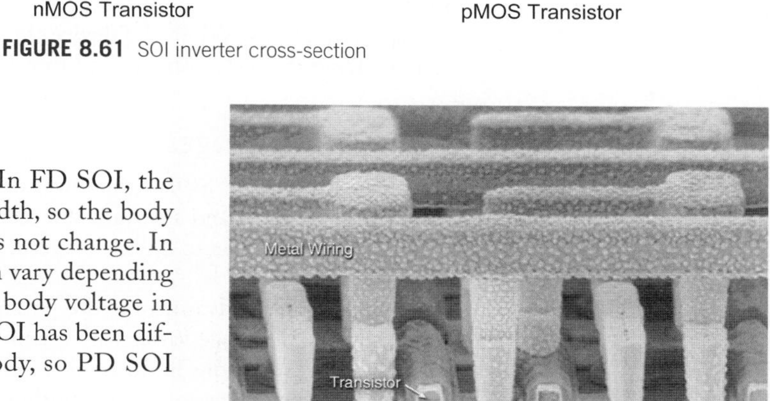

FIGURE 8.61 SOI inverter cross-section

SOI devices are categorized as partially depleted (PD) or fully depleted (FD). A depletion region empty of free carriers forms in the body beneath the gate. In FD SOI, the body is thinner than the channel depletion width, so the body charge is fixed and thus the body voltage does not change. In PD SOI, the body is thicker and its voltage can vary depending on how much charge is present. This varying body voltage in turn changes V_t through the body effect. FD SOI has been difficult to manufacture because of the thin body, so PD SOI appears to be the most promising technology.

Throughout this section we will concentrate on nMOS transistors. pMOS transistors have analogous behaviors.

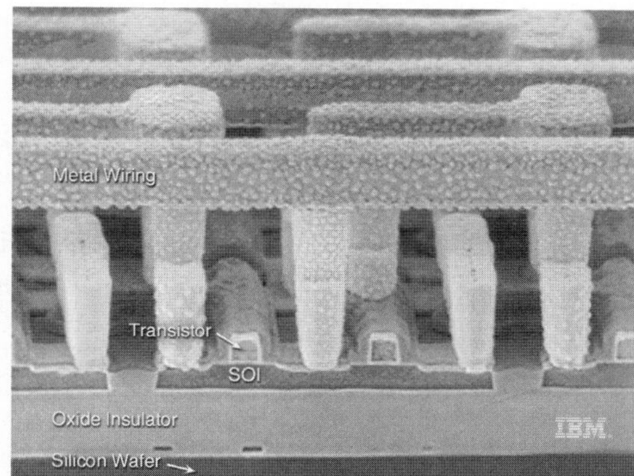

FIGURE 8.62 IBM SOI process electron micrograph (Courtesy of International Business Machines Corporation. Unauthorized use not permitted.)

8.4.1 Floating Body Voltage

The key to understanding PD SOI is to follow the body voltage. If the body voltage were constant, the threshold voltage would be constant as well and the transistor would behave much like a conventional bulk device except that the diffusion capacitance is lower.

In PD SOI, the floating body voltage varies as it charges or discharges. Figure 8.63 illustrates the mechanisms by which charges enter into or exit from the body [Bernstein00]. There are two paths through which charge can slowly build up in the body:

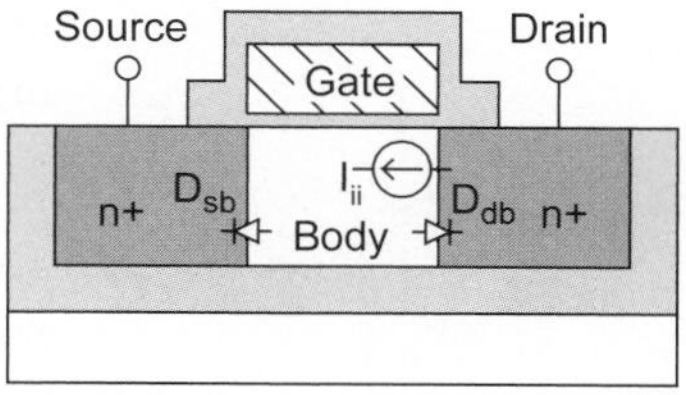

- Reverse-biased drain-to-body D_{db} and possibly source-to-body D_{sb} junctions carry small diode leakage currents into the body.
- High-energy carriers cause impact ionization, creating electron-hole pairs. Some of these electrons are injected into the gate or gate oxide. (This is the mechanism for hot-electron wearout described in Section 6.3.2.1.) The corresponding holes accumulate in the body. This effect is most pronounced at V_{DS} above the intended operating point of devices and is relatively unimportant during normal operation. The impact ionization current into the body is modeled as a current source I_{ii}.

FIGURE 8.63 Charge paths to/from floating body

The charge can exit the body through two other paths:

- As the body voltage increases, the source-to-body D_{sb} junction becomes slightly forward-biased. Eventually, the charge exiting from this junction equals the charge leaking in from the drain-to-body D_{db} junction.
- A rising gate or drain capacitively couples the body upward, too. This may strongly forward-bias the source-to-body D_{sb} junction and rapidly spill charge out of the body.

In summary, when a device is idle long enough (on the order of microseconds), the body voltage will reach equilibrium when based on the leakage currents through the source and drain junctions. When the device then begins switching, the charge may spill off the body, shifting the body voltage (and threshold voltage) significantly.

8.4.2 SOI Advantages

A major advantage of SOI is the lower diffusion capacitance. The source and drain abut oxide on the bottom and sidewalls not facing the channel, essentially eliminating the parasitic capacitance of these sides. This results in a smaller parasitic delay and lower dynamic power consumption.

A more subtle advantage is the potential for lower threshold voltages. In bulk processes, threshold voltage varies with channel length. Hence, variations in polysilicon etching show up as variations in threshold voltage. The threshold voltage must be high enough in the worst (lowest) case to limit subthreshold leakage, so the nominal threshold voltage must be higher. In SOI processes, the threshold variations tend to be smaller. Hence, the nominal V_t can be closer to worst-case. Lower nominal V_t results in faster transistors, especially at low V_{DD}.

According to EQ (2.44), CMOS devices have a subthreshold slope of $nv_T \ln 10$, where $v_T = kT/q$ is the thermal voltage (26 mV at room temperature) and n is process-dependent. Bulk CMOS processes typically have $n \approx 1.5$, corresponding to a subthreshold slope of 90 mV/decade. In other words, for each 90 mV decrease in V_{gs} below V_t, the subthreshold leakage current reduces by an order of magnitude. Misleading claims have been made suggesting SOI has $n = 1$ and thus an ideal subthreshold slope of only 60 mV/decade. IBM has found that real SOI devices actually have subthreshold slopes of 75–85 mV/decade. This is better than bulk, but not as good as the hype would suggest. FinFETs discussed in Section 15.4.4 are variations on SOI transistors that offer lower subthreshold slopes because the gate surrounds the channel on more sides and thus turns the transistor off more abruptly.

Finally, SOI is immune to latchup because the insulating oxide eliminates the parasitic bipolar devices that could trigger latchup.

8.4.3 SOI Disadvantages

PD SOI suffers from the *history effect*. Changes in the body voltage modulate the threshold voltage and thus adjust gate delay. The body voltage depends on whether the device has been idle or switching, so gate delay is a function of the switching history. Overall, the elevated body voltage reduces the threshold and makes the gates faster, but the uncertainty makes circuit design more challenging. The history effect can be modeled in a simplified way by assigning different propagation and contamination delays to each gate. IBM found the history effect tends to result in about an 8% variation in gate delay, which is modest compared to the combined effects of manufacturing and environmental variations [Shahidi02].

Unfortunately, the history effect causes significant mismatches between nominally identical transistors. For example, if a sense amplifier has repeatedly read a particular input value, the threshold voltages of the differential pair will be different, introducing an offset voltage in the sense amplifier. This problem can be circumvented by adding a contact to tie the body to ground or to the source for sensitive analog circuits.

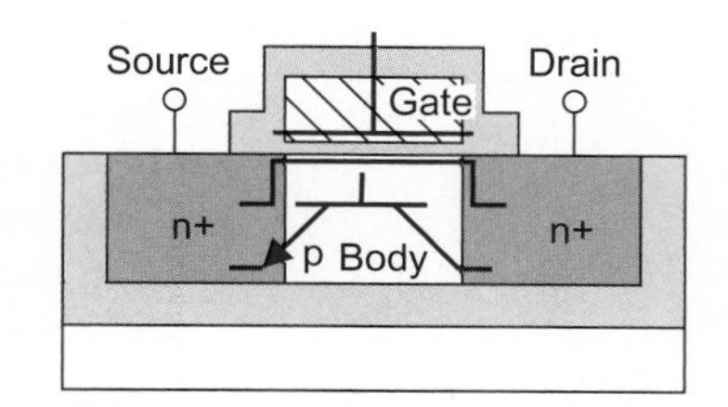

FIGURE 8.64 Parasitic bipolar transistor in PD SOI

Another PD SOI problem is the presence of a parasitic bipolar transistor within each transistor. As shown in Figure 8.64, the source, body, and drain form an emitter, base, and collector of an npn bipolar transistor. In an ordinary transistor, the body is tied to a supply, but in SOI, the body/base floats. If the source and drain are both held high for an extended period of time while the gate is low, the base will float high as well through diode leakage. If the source should then be pulled low, the npn transistor will turn ON. A current I_B flows from body/base to source/emitter. This causes βI_B to flow from the drain/collector to source/emitter. The bipolar transistor gain β depends on the channel length and doping levels but can be greater than 1. Hence, a significant pulse of current can flow from drain to source when the source is pulled low even though the transistor should be OFF.

This pulse of current is sometimes called *pass-gate leakage* because it commonly happens to OFF pass transistors where the source and drain are initially high and then pulled low. It is not a major problem for static circuits because the ON transistors oppose the glitch. However, it can cause malfunctions in dynamic latches and logic. Thus, dynamic nodes should use strong keepers to hold the node steady.

A third problem common to all SOI circuits is self-heating. The oxide is a good thermal insulator as well as an electrical insulator. Thus, heat dissipated in switching transistors tends to accumulate in the transistor rather than spreading rapidly into the substrate. Individual transistors dissipating large amounts of power may become substantially warmer than the die as a whole. At higher temperature they deliver less current and hence are slower. Self-heating can raise the temperature by 10–15 °C for clock buffer and I/O transistors, although the effects tend to be smaller for logic transistors.

8.4.4 Implications for Circuit Styles

In summary, SOI is attractive for fast CMOS logic. The smaller diffusion capacitance offers a lower parasitic delay. Lower threshold voltages offer better drive current and lower gate delays. Moreover, SOI is also attractive for low-power design. The smaller diffusion capacitance reduces dynamic power consumption. The speed improvements can be traded for lower supply voltage to reduce dynamic power further. Sharper subthreshold slopes offer the opportunity for reduced static leakage current, especially in FinFETs.

Complementary static CMOS gates in PD SOI behave much like their bulk counterparts except for the delay improvement. The history effect also causes pattern-dependent variation in the gate delay.

Circuits with dynamic nodes must cope with a new noise source from pass gate leakage. In particular, dynamic latches and dynamic gates can lose the charge on the dynamic node. Figure 8.65 shows the pass gate leakage mechanism. In each case, the dynamic node X is initially high and the transistor connected to the node is OFF. The source of this transistor starts high and pulls

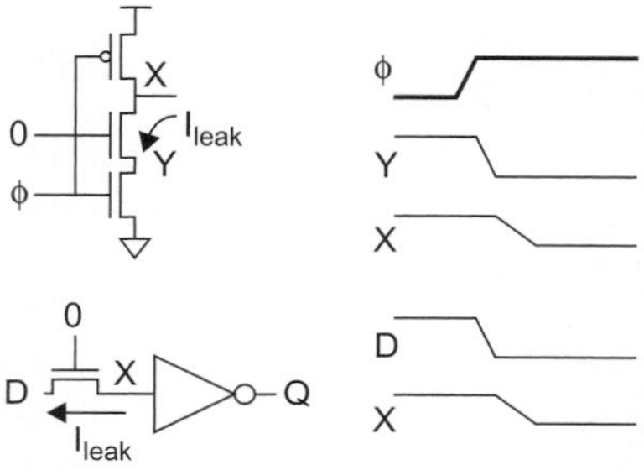

FIGURE 8.65 Pass gate leakage in dynamic latches and gates

low, turning on the parasitic bipolar transistor and partially discharging *X*. To overcome pass gate leakage, *X* should be staticized with a cross-coupled inverter pair for latches or a pMOS keeper for dynamic gates. The staticizing transistors must be relatively strong (e.g., 1/4 as strong as the normal path) to fight the leakage. The gates are slower because they must overcome the strong keepers. Dynamic gates may predischarge the internal nodes to prevent pass gate leakage, but then must deal with charge sharing onto those internal nodes.

Analog circuits, sense amplifiers, and other circuits that depend on matching between transistors suffer from major threshold voltage mismatches caused by the history of the floating body. They require body contacts to eliminate the mismatches by holding the body at a constant voltage. Gated clocks also have greater clock skew because the history effect makes the clock switch more slowly on the first active cycle after the clock has been disabled for an extended time

8.4.5 Summary

In summary, Silicon-on-Insulator is attractive because it greatly reduces the source/drain diffusion capacitance, resulting in faster and power-efficient transistors. It also is immune to latchup. Partially depleted SOI is the most practical technology and also boosts drive current because the floating body leads to lower threshold voltages.

SOI design is more challenging because of the floating body effects. Gate delay becomes history-dependent because the voltage of the body depends on the previous state of the device. This complicates device modeling and delay estimation. It also contributes to mismatches between devices. In specialized applications like sense amplifiers, a body contact may be added to create a fully depleted device.

A second challenge with SOI design is pass-gate leakage. Dynamic nodes may be discharged from this leakage even when connected to OFF transistors. Strong keepers can fight the leakage to prevent errors.

Finally, the oxide surrounding SOI devices is a good thermal insulator. This leads to greater self-heating. Thus, the operating temperature of individual transistors may be up to 10–15 °C higher than that of the substrate. Self-heating reduces ON current and makes modeling more difficult.

This section only scratches the surface of a subject worthy of entire books. In particular, SOI static RAMs require special care because of pass gate leakage and floating bodies. [Bernstein00] offers a definitive treatment of partially depleted SOI circuit design and [Kuo01] surveys the literature of SOI circuits.

8.5 Subthreshold Circuit Design

In a growing body of applications, performance requirements are minimal and battery life is paramount. For example, a pacemaker would ideally last for the life of the patient because surgery to replace the battery carries significant risk and expense. In other applications, the battery can be eliminated entirely if the system can scavenge enough energy from the environment. For example, a tire pressure sensor could obtain its energy from the vibration of the rolling tire. Such applications demand the lowest possible energy consumption.

As discussed in Section 4.4.1, the minimum energy point typically occurs at $V_{DD} < V_t$, which is called the subthreshold regime. All the transistors in the circuit are

OFF, but some are more OFF than others. According to EQ (2.45), subthreshold leakage increases exponentially with V_{gs}. Assuming a subthreshold slope of $S = 100$ mV, a transistor with $V_{gs} = 0.3$ will nominally leak 1000 times more current than a transistor with $V_{gs} = 0$. This difference is sufficient to perform logic, albeit slowly. Gate leakage and junction leakage drop off rapidly with V_{DD}, so they are negligible compared to subthreshold leakage.

In the subthreshold regime, delay increases exponentially as the supply voltage decreases. Reducing the supply voltage reduces the switching energy but causes the OFF transistors to leak for a longer time, increasing the leakage energy. The minimum energy point is where the sum of dynamic and leakage energies is smallest. This point is typically at a supply close to 300–500 mV; a somewhat higher voltage is preferable when leakage dominates (e.g., at low activity factor or high temperature). At this voltage, static CMOS logic operates at kHz or low MHz frequencies and consumes an order of magnitude lower energy per operation than at typical voltages. The power consumption is many orders of magnitude lower because the operating frequency is so slow. It is possible to operate at a voltage and frequency below the minimum energy point to reduce power further at the expense of increased energy per operation. However, if system considerations permit, the average power is even lower if the system operates at the minimum energy point, then turns off its power supply until the next operation is required.

This section outlines the key points, including transistor sizing, DC transfer characteristics, and gate selection. Section 11.2.6.3 examines subthreshold memories. [Wang06] devotes an entire book to subthreshold circuit design and [Hanson06] explores design issues at the minimum energy point. One of the earliest applications of subthreshold circuits was in a frequency divider for a wristwatch [Vittoz72]. More recently, [Hanson09] and [Kwong09] have demonstrated experimental microcontrollers achieving power as low as nanowatts in active operation and picowatts in sleep.

8.5.1 Sizing

Transistor sizing offers at best a linear performance benefit, while supply voltage offers an exponential performance benefit. As a general rule, minimum energy under a performance constraint is thus achieved by using minimum width transistors and raising the supply voltage if necessary from the minimum energy point until the performance is achieved (assuming the performance requirement is low enough that the circuit remains in the subthreshold regime) [Calhoun05].

If V_t variations from random dopant fluctuations are extremely high, wider transistors might become advantageous to reduce the variability and its attendant risk of high leakage [Kwong06]. Also, if one path through a circuit is far more critical than the others, upsizing the transistors in that path for speed might be better than raising the supply voltage to the entire circuit.

When minimum-width transistors are employed, wires are likely to contribute the majority of the switching capacitance. To shorten wires, subthreshold cells should be as small as possible; the cell height is generally set by the minimum height of a flip-flop. Good floorplanning and placement is essential.

8.5.2 Gate Selection

A logic gate must have a slope steeper than –1 in its DC transfer characteristics to achieve restoring behavior and maintain noise margins. Decades ago, static CMOS logic was shown to have good transfer characteristics at supply voltages as low as 100 mV

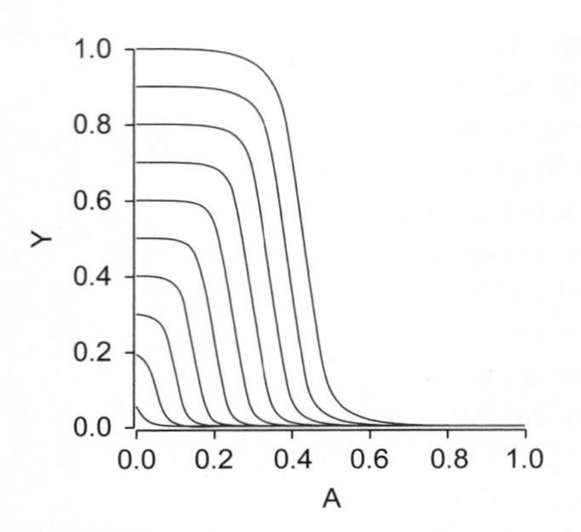

FIGURE 8.66 Inverter DC transfer characteristics at low voltage

[Swanson72]. Figure 8.66 shows the typical characteristics as the supply voltage varies in a 65 nm process using minimum-width transistors. The switching point is skewed because the pMOS and nMOS thresholds are unequal and the gate is not designed for equal rise/fall currents, but the behavior still looks good to 300 mV and is tolerable at 200 mV.

Unfortunately, process variation degrades the switching characteristics. In the worst case corners (usually SF or FS), the supply voltage may need to be 300 mV, or higher for complex gates, to guarantee proper operation. Gates with multiple series and parallel transistors require a higher supply voltage to ensure the ON current through the series stack exceeds the OFF current through all of the parallel transistors. Moreover, the stack effect degrades the ON current and speed for the series transistors. Thus, subthreshold circuits should use simple gates (e.g., no more complicated than an AOI22 or NAND3).

Static structures with many parallel transistors such as wide multiplexers do not work well at low voltage because the leakage through the OFF transistors can exceed the current through the ON transistor, especially considering variation. This is an important consideration for subthreshold RAM design.

Ratioed circuits do not work well at low voltage because exponential sensitivity to variation makes it difficult to ensure that the proper transistor is stronger. Latches and registers with weak feedback devices should thus be avoided. The conventional register shown in Figure 9.19(b) works well in subthreshold.

Additionally, dynamic circuits are not robust in subthreshold operation because leakage easily disturbs the dynamic node. Keepers present a ratioing problem that is difficult to resolve across the range of process variations.

Subthreshold circuits can be synthesized using commercially available low-power standard cell libraries by excluding all the cells that are too complex or that exceed that smallest available size.

8.6 Pitfalls and Fallacies

Failing to plan for advances in technology

There are many advances in technology that change the relative merits of different circuit techniques. For example, interconnect delays are not improving as rapidly as gate delays, threshold drops are becoming a greater portion of the supply voltage, and leakage currents are increasing. Failing to anticipate these changes leads to inventions whose usefulness is short-lived.

A salient example is the rise and fall of BiCMOS circuits. Bipolar transistors have a higher current output per unit input capacitance (i.e., a lower logical effort) than CMOS circuits in the 0.8 μm generation, so they became popular, particularly for driving large loads. In the early 1990s, hundreds of papers were written on the subject. The Pentium and Pentium Pro processors were built using BiCMOS processes. Investors poured at least $40 million into a startup company called Exponential, which sought to build a fast PowerPC processor in a BiCMOS process.

Unfortunately, technology scaling works against BiCMOS because of the faster CMOS transistors, lower supply voltages, and larger numbers of transistors on a chip. The relative benefit of bipolar transistors over fine-geometry CMOS decreased. The V_{be} drop became an unacceptable fraction of the power supply. Finally, the static power consumption caused by bipolar base currents limits the number of bipolar transistors that can be used.

The Pentium II was based on the Pentium Pro design, but the bipolar transistors had to be removed because they no longer provided advantages in the 0.35 μm generation. Despite a talented engineering team, Exponential failed entirely, ultimately producing a processor that lacked compelling performance advantages and dissipated far more power than anything else on the market [Maier97].

Comparing a well-tuned new circuit to a poor example of existing practice
A time-honored way to make a new invention look good is to tune it as well as possible and compare it to an untuned strawman held up as an example of "existing practice." For example, [Zimmermann97] points out that most papers finding pass-transistor adders faster than static CMOS adders use 40-transistor static adder cells rather than the faster and smaller 28-transistor cells (Figure 10.4).

Ignoring driver resistance when characterizing pass-transistor circuits
Another way to make pass-transistor circuit families look about twice as fast as they really are is to drive diffusion inputs with a voltage source rather than with the output stage of the previous gate.

Reporting only part of the delay of a circuit
Clocked circuits all have a setup time and a clock-to-output delay. A good way to make clocked circuits look fast is to only report the clock-to-output delay. This is particularly common for the sense-amplifier logic families.

Making outrageous claims about performance
Many published papers have made outrageous performance claims. For example, while comparing full adder designs, some authors have found that DSL and dual-rail domino are 8–10× faster than static CMOS. Neither statement is anywhere close to what designers see in practice; for example, [Ng96] finds that an 8 × 8 multiplier built from DSL is 1.5× faster and one built from dual-rail domino is 2× faster than static CMOS.

In general, "there ain't no such thing as a free lunch" in circuit design. CMOS design is a fairly mature field and designers are not stupid (or at least not all designers are stupid all the time), so if some new invention seems too good to be true, it probably is. Beware of papers that push the advantages of a new invention without disclosing the inevitable trade-offs. The trade-offs may be acceptable, but they must be understood.

Building circuits without adequate verification tools
It is impractical to manually verify circuits on chips that have many millions (soon billions) of transistors. Automated verification tools should check for any pitfalls common to widely used circuit families. If you cannot afford to buy or write appropriate tools, stick with robust static CMOS logic.

Sizing subthreshold circuits for speed
The purpose of operating in the subthreshold regime is to minimize energy. A number of papers have proposed using wide transistors to achieve higher speed. Given the exponential relationship between voltage and speed, the same speed could have been achieved at a lower energy by increasing the supply voltage slightly.

8.7 Historical Perspective

Ratioed and dynamic circuits predate the widespread use of CMOS. In an nMOS process, pMOS transistors were not available to build complementary gates. One strategy was to build ratioed gates, which consume static power whenever the outputs are low. The speed

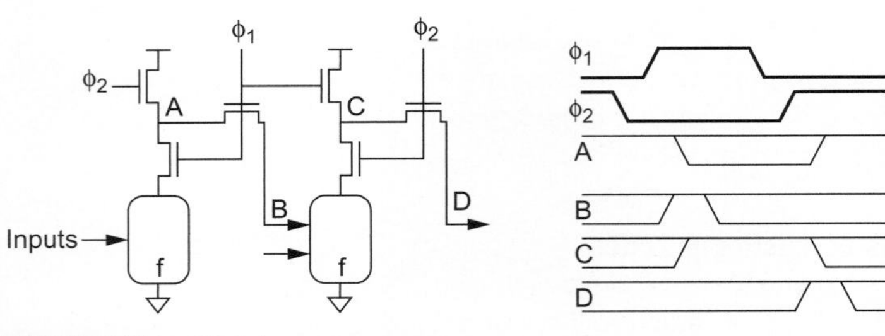

FIGURE 8.67 nMOS dynamic gates

is proportional to the RC product, so fast gates need low-resistance pullups, exacerbating the power problem. An alternative was to use dynamic gates. The classic MOS textbook of the early 1970s [Penney72] devotes 29 pages to describing a multitude of dynamic gate configurations. Unfortunately, dynamic gates suffer from the monotonicity problem, so each phase of logic may contain only one gate. Phases were separated using nMOS pass transistors that behaved as dynamic latches. Figure 8.67 shows an approach using two-phase nonoverlapping clocks. Each gate precharges in one phase while the subsequent latch is opaque. It then evaluates while making the latch transparent. This approach is prone to charge-sharing noise when the latch opens and precharge only rises to $V_{DD} - V_t$. Numerous four-phase clocking techniques were also developed.

With the advent of CMOS technology, dynamic logic lost its advantage of power consumption. However, chip space was at a premium and dynamic gates could eliminate most of the pMOS transistors to save area. Domino gates were developed at Bell Labs for a 32-bit adder in the BELLMAC-32A microprocessor to solve problems of both area and speed [Krambeck82, Shoji82]. Domino allows multiple noninverting gates to be cascaded in a single phase.

Some older domino designs leave out the keeper to save area and gain a slight performance advantage. This has become more difficult as leakage and coupling noise have increased with process scaling. The 0.35 μm Alpha 21164 was one of the last designs to have no keeper (and to use dynamic latches). Its fully dynamic operation gave advantages in both speed and area, but during test it had a minimum operating frequency of 20 MHz to retain state. In the Alpha 21264, leakage current had increased to the point that keepers were essential. Modern designs always need keepers. As an interesting aside, the Alpha microprocessors also did not use scan latches because scan cost area and a small amount of performance. This proved unfortunate on the Alpha 21264, which was difficult to debug because of the limited observability into the processor state. Now virtually all design methodologies require scan capability in the latches or registers, as discussed in Section 14.6.

High-performance microprocessors have boosted clock speeds faster than simple process improvement would allow, so the number of gate delays per cycle has shrunk. The DEC Alpha microprocessors pioneered this trend through the 1990s [Gronowski98] and most other CPUs have followed. During the "MHz Wars" from about 1994 through 2004 when microprocessors were marketed primarily on clock frequency, the number of FO4 inverter delays per cycle dropped from more than 24 down to only 10–12. Domino circuits became crucial to achieving these fast cycle times. Intel moved domino gates with overlapping clocks in the Pentium Pro / II / III [Colwell95, Choudhury97] and Itanium series [Naffziger02]. The initial 180 nm "Willamette" Pentium 4 adopted even more elaborate self-resetting domino and double-pumped the integer execution unit at twice the core frequency [Hinton01]. The 90 nm "Prescott" Pentium 4 moved to the extraordinarily complex LVS logic family with long chains of nMOS transistors connected to sense amplifiers [Deleganes04, Deleganes05]. The integer core required painstaking custom design of 6.8M transistors by a team of circuit wizards.

Unfortunately, the low-swing logic did not scale well as supply voltages decreased and variability and coupling increased. Moreover, dynamic circuits have a high activity factor and thus consume a great deal of power, which makes them unsuited to power-constrained designs. Tricky circuit techniques have often been the cause of problems during silicon debug [Josephson02]. A six-month delay can cost hundreds of millions of dollars in a competitive market and a year-long delay can kill a product entirely, giving designers yet another reason to be conservative. The "Tejas" team was in the midst of stripping out the hard-won LVS logic when the project was canceled in 2004. Intel moved to the Core architecture with longer cycle times and better power efficiency. Dynamic logic continues to be essential for dense memory arrays, but it has largely been eliminated from datapaths.

Pass-transistor logic families enjoyed a period of intense popularity in Japan in the 1990s. Advocates claimed speed or power advantages, though these claims have been disputed, as discussed in Section 8.2.5. They suffer from a lack of modularity: the delay driving a diffusion input depends on the previous stage as well as the current stage. This is an obstacle for conventional static timing analysis. The effort to build cell libraries is another drawback. Given the marginal benefits and clear costs, pass transistor logic families have faded from commercial application.

IBM is notable for having always relied on static CMOS logic and fast time to market in cutting-edge SOI processes [Curran02]. For example, the POWER6 can operate up to 5 GHz without needing dynamic logic in the datapaths [Stolt08].

For many years, inventing a circuit family, giving it a three- or four-letter acronym, and publishing it in the IEEE *Journal of Solid-State Circuits* was seemingly grounds to claim a Ph.D. degree. This intensive research led to an enormous proliferation of circuit families, of which only a miniscule proportion have ever seen commercial application. Today, even the few circuit families that were used have been largely removed in favor of static CMOS circuits that are robust, perform quite well, and offer the fastest design and debug time. Circuit innovation has moved on to more rewarding areas such as low-voltage memories, high-speed I/O, phase-locked loops, and analog and RF circuits.

Summary

Circuit delay is related to the $(C/I)\Delta V$ product of gates. This chapter explored alternative combinational circuit structures to improve the C/I ratio or respond to smaller voltage swings. Many of these techniques trade higher power consumption and/or lower noise margins for better delay. While complementary CMOS circuits are quite robust, the alternative circuit families have pitfalls that must be understood and managed.

Most logic outside arrays now uses static CMOS. Many techniques exist for optimizing static CMOS logic, including gate selection and sizing, input ordering, asymmetric and skewed gates, and multiple threshold voltages. Silicon-on-insulator processes reduce the parasitic capacitance and improve leakage, allowing lower power or higher performance. Operating circuits in the subthreshold region at a supply voltage of 300–500 mV can save an order of magnitude of energy when performance is not important.

Three of the historically important alternatives to complementary CMOS are domino, pseudo-nMOS, and pass transistor logic. Each attempts to reduce the input capacitance by performing logic mostly through nMOS transistors. Power, robustness, and productivity issues have largely eliminated these techniques from datapaths and random logic, though niche applications still exist, especially in arrays.

Pseudo-nMOS replaces the pMOS pullup network with a single weak pMOS transistor that is always ON. The pMOS transistor dissipates static power when the output is low. If it is too weak, the rising transition is slow. If it is too strong, V_{OL} is too high and the power consumption increases. When the static power consumption is tolerable, pseudo-nMOS gates work well for wide NOR functions.

Dynamic gates resemble pseudo-nMOS, but use a clocked pMOS transistor in place of the weak pullup. When the clock is low, the gates precharge high. When the clock rises, the gates evaluate, pulling the output low or leaving it floating high. The input of a dynamic gate must be monotonically rising while the gate is in evaluation, but the output monotonically falls. Domino gates consist of a dynamic gate followed by an inverting static gate and produce monotonically rising outputs. Therefore, domino gates can be cascaded, but only compute noninverting functions. Dual-rail domino accepts true and complementary inputs and produces true and complementary outputs to provide any logic function at the expense of larger gates and twice as many wires. Dynamic gates are also sensitive to noise because V_{IL} is close to the threshold voltage V_t and the output floats. Major noise sources include charge sharing, leakage, and coupling. Therefore, domino circuits typically use secondary precharge transistors, keepers, and shielded or carefully routed interconnect. The high-activity factors of the clock and dynamic node make domino power hungry. Despite all of these challenges, domino offers a 1.5–2× speedup over static CMOS, giving it a compelling advantage for the critical paths of high-performance systems.

Pass-transistor circuits use inputs that drive the diffusion inputs as well as the gates of transistors. Many pass-transistor techniques have been explored and Complementary Pass Transistor logic has proven to be one of the most effective. This dual-rail technique uses networks of nMOS transistors to compute true and complementary logic functions. The nMOS transistors only pull up to $V_{DD} - V_t$, so cross-coupled pMOS transistors boost the output to full-rail levels. Some designers find that pass-transistor circuits are faster and smaller for functions such as XOR, full adders, and multiplexers that are clumsy to implement in static CMOS. Because of the threshold drop, the circuits do not scale well as V_{DD}/V_t decreases.

Exercises

8.1 Design a fast 6-input OR gate in each of the following circuit families. Sketch an implementation using two stages of logic (e.g., NOR6 + INV, NOR3 + NAND2, etc.). Label each gate with the width of the pMOS and nMOS transistors. Each input can drive no more than 30 λ of transistor width. The output must drive a 60/30 inverter (i.e., an inverter with a 60 λ wide pMOS and 30 λ wide nMOS transistor). Use logical effort to choose the topology and size for least average delay. Estimate this delay using logical effort. When estimating parasitic delays, count only the diffusion capacitance on the output node.

a) static CMOS

b) pseudo-nMOS with pMOS transistors 1/4 the strength of the pulldown stack

c) domino (a footed dynamic gate followed by a HI-skew inverter); only optimize the delay from rising input to rising output

8.2 Simulate each gate you designed in Exercise 8.1. Determine the average delay (or rising delay for the domino design). Logical effort is only an approximation. Tweak the transistor sizes to improve the delay. How much improvement can you obtain?

8.3 Sketch a schematic for a 12-input OR gate built from NANDs and NORs of no more than three inputs each.

8.4 Design a static CMOS circuit to compute $F = (A + B)(C + D)$ with least delay. Each input can present a maximum of 30 λ of transistor width. The output must drive a load equivalent to 500 λ of transistor width. Choose transistor sizes to achieve least delay and estimate this delay in τ.

8.5 Figure 8.68 shows two series transistors modeling the pulldown network of 2-input NAND gate.

a) Plot I vs. A using long-channel transistor models for $0 \leq A \leq 1$, $B = Y = 1$, $V_t = 0$, $\beta = 1$. On the same axes, plot I vs. B for $0 \leq B \leq 1$, $A = 1$. *Hint*: You will need to solve for x; this can be done numerically.

b) Using your results from (a), explain why the inner input of a 2-input NAND gate has a slightly greater logical effort than the outer input.

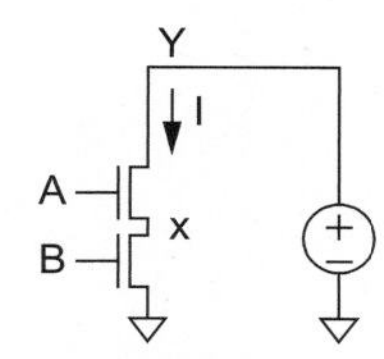

FIGURE 8.68 Current in series transistors

8.6 What is the logical effort of an OR-AND-INVERT gate at either of the OR terminals? At the AND terminal? What is the parasitic delay if only diffusion capacitance on the output is counted?

8.7 Simulate a 3-input NOR gate in your process. Determine the logical effort and parasitic delay from each input.

8.8 Using the datasheet from Figure 3.25, find the rising and falling logical effort and parasitic delay of the $X1$ 2-input NAND gate from the A input.

8.9 Repeat Exercise 8.8 for the B input. Explain why the results are different for the different inputs.

8.10 Sketch HI-skew and LO-skew 3-input NAND and NOR gates. What are the logical efforts of each gate on its critical transition?

8.11 Derive a formula for g_u, g_d, and g_{avg} for HI-skew and LO-skew k-input NAND gates with a skew factor of $s < 1$ (i.e., the noncritical transistor is s times normal size) as a function of s and k.

8.12 Design an asymmetric 3-input NOR gate that favors a critical input over the other two. Choose transistor sizes so the logical effort on the critical input is 1.5. What is the logical effort of the noncritical inputs?

8.13 Prove that the P/N ratio that gives lowest average delay in a logic gate is the square root of the ratio that gives equal rise and fall delays.

8.14 Let $\rho(g, p)$ be the best stage effort of a path if one is free to add extra buffers with a parasitic delay p and logical effort g. For example, Section 3.5.2 shows that $\rho(1, 1) = 3.59$. It is easy to make a plot of $\rho(1, p)$ by solving EQ (3.19) numerically; this gives the best stage effort of static CMOS circuits where the inverter has a parasitic delay of p. Prove the following result, which is useful for determining the best stage effort of domino circuits where buffers have lower logical efforts:

$$\rho(g, p) = g\rho(1, \frac{p}{g})$$

8.15 Simulate a fanout-of-4 inverter. Use a unit-sized nMOS transistor. How wide must the pMOS transistor be to achieve equal rising and falling delays? What is the delay? How wide must the pMOS transistor be to achieve minimum average delay? What is the delay? How much faster is the average delay?

8.16 Many standard cell libraries choose a *P/N* ratio for an inverter in between that which would give equal rising and falling delays and that which would give minimum average delay. Why is this done?

8.17 A static CMOS NOR gate uses four transistors, while a pseudo-nMOS NOR gate uses only three. Unfortunately, the pseudo-nMOS output does not swing rail to rail. If both the inputs and their complements are available, it is possible to build a 3-transistor NOR that swings rail to rail without using any dynamic nodes. Show how to do it. Explain any drawbacks of your circuit.

8.18 Sketch pseudo-nMOS 3-input NAND and NOR gates. Label the transistor widths. What are the rising, falling, and average logical efforts of each gate?

8.19 Sketch a pseudo-nMOS gate that implements the function

$$F = \overline{A\left(B+C+D\right)+E\cdot F\cdot G}$$

8.20 Design an 8-input AND gate with an electrical effort of six using pseudo-nMOS logic. If the parasitic delay of an n-input pseudo-nMOS NOR gate is $(4n + 2)/9$, what is the path delay?

8.21 Simulate a pseudo-nMOS inverter in which the pMOS transistor is half the width of the nMOS transistor. What are the rising, falling, and average logical efforts? What is V_{OL}?

8.22 Repeat Exercise 8.21 in the FS and SF process corners.

8.23 Sketch a 3-input symmetric NOR gate. Size the inverters so that the pulldown is four times as strong as the net worst-case pullup. Label the transistor widths. Estimate the rising, falling, and average logical efforts. How do they compare to a static CMOS 3-input NOR gate?

8.24 Sketch a 2-input symmetric NAND gate. Size the inverters so that the pullup is four times as strong as the net worst-case pulldown. Label the transistor widths. Estimate the rising, falling, and average logical efforts. How do they compare to a static CMOS 2-input NAND gate?

8.25 Compare the average delays of a 2, 4, 8, and 16-input pseudo-NMOS and SFPL NOR gate driving a fanout of four identical gates.

8.26 Sketch a 3-input CVSL OR/NOR gate.

8.27 Sketch dynamic footed and unfooted 3-input NAND and NOR gates. Label the transistor widths. What is the logical effort of each gate?

8.28 Sketch a 3-input dual-rail domino OR/NOR gate.

8.29 Sketch a 3-input dual-rail domino majority/minority gate. This is often used in domino full adder cells. Recall that the majority function is true if more than half of the inputs are true.

8.30 Compare a standard keeper with the noise tolerant precharge device. Larger pMOS transistors result in a higher V_{IL} (and thus better noise margins) but more delay. Simulate a 2-input footed NAND gate and plot V_{IL} vs. delay for various sizes of keepers and noise tolerant precharge transistors.

8.31 Design a 4-input footed dynamic NAND gate driving an electrical effort of 1. Estimate the worst charge-sharing noise as a fraction of V_{DD} assuming that diffusion capacitance on uncontacted nodes is about half of gate capacitance and on contacted nodes it equals gate capacitance.

8.32 Repeat Exercise 8.31, generating a graph of charge-sharing noise vs. electrical effort for $h = 0, 1, 2, 4$, and 8.

8.33 Repeat Exercise 8.31 if a small secondary precharge transistor is added on one of the internal nodes.

8.34 Perform a simulation of your circuits from Exercise 8.31. Explain any discrepancies.

8.35 Design a domino circuit to compute $F = (A + B)(C + D)$ as fast as possible. Each input may present a maximum of 30 λ of transistor width. The output must drive a load equivalent to 500 λ of transistor width. Choose transistor sizes to achieve least delay and estimate this delay in τ.

8.36 Redesign the memory decoder from Section 3.5.3 using footed domino logic. You can assume you have both true and complementary monotonic inputs available, each capable of driving 10 unit transistors. Label gate sizes and estimate the delay.

8.37 Sketch an NP Domino 8-input AND circuit.

8.38 Sketch a 4:1 multiplexer. You are given four data signals *D0*, *D1*, *D2*, and *D3*, and two select signals, *S0* and *S1*. How many transistors does each design require?

a) Use only static CMOS logic gates.

b) Use a combination of logic gates and transmission gates.

8.39 Sketch 3-input XOR functions using each of the following circuit techniques:

a) Static CMOS

b) Pseudo-nMOS

c) Dual-rail domino

d) CPL

e) EEPL

f) DCVSPG

g) SRPL

h) PPL

i) DPL

j) LEAP

8.40 Repeat Exercise 8.39 for a 2-input NAND gate.

8.41 Design sense-amplifier gates using each of the following circuit families to compute an 8-input XOR function in a single gate: SSDL, ECDL, LCDL, DCSL1,

DCSL2, DCSL3. Each true or complementary input can drive no more than 24 λ of transistor width. Each output must drive a 32/16 λ inverter. Simulate each circuit to determine the setup time and clock-to-out delays.

8.42 Figure 8.69 shows a Switched Output Differential Structure (SODS) gate. Explain how the gate operates and sketch waveforms for the gate acting as an inverter/buffer.

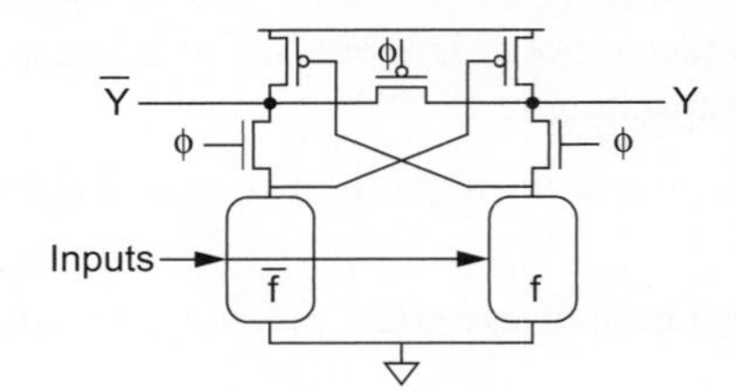

FIGURE 8.69 SODS

Comment on the strengths and weaknesses of the circuit family.

8.43 Choose one of the circuit families (besides SODS, Exercise 8.42) mentioned in a a recent published paper. Critically evaluate the original paper in which the circuit was proposed. Sketch an inverter or buffer and explain how it operates, including appropriate waveforms. What are the strengths of the circuit family? If you were the circuit manager choosing design styles for a large chip, what concerns might you have about the circuit family?

8.44 Derive V_{out} using the long-channel models for the pseudo-nMOS inverter from Figure 8.13 with $V_{in} = V_{DD}$ as a function of the threshold voltages and beta values of the two transistors. Assume $V_{out} < |V_{tp}|$.

9 Sequencing

9.1 Introduction

Chapter 8 addressed *combinational* circuits in which the output is a function of the current inputs. This chapter discusses *sequential* circuits in which the output depends on previous as well as current inputs; such circuits are said to have *state*. Finite state machines and pipelines are two important examples of sequential circuits.

Sequential circuits are usually designed with flip-flops or latches, which are sometimes called *memory elements*, that hold data called *tokens*. The purpose of these elements is not really memory; instead, it is to enforce sequence, to distinguish the *current* token from the *previous* or *next* token. Therefore, we will call them *sequencing elements* [Harris01a]. Without sequencing elements, the next token might catch up with the previous token, garbling both. Sequencing elements delay tokens that arrive too early, preventing them from catching up with previous tokens. Unfortunately, they inevitably add some delay to tokens that are already critical, decreasing the performance of the system. This extra delay is called *sequencing overhead*.

This chapter considers sequencing for both static and dynamic circuits. *Static circuits* refer to gates that have no clock input, such as complementary CMOS, pseudo-nMOS, or pass transistor logic. *Dynamic circuits* refer to gates that have a clock input, especially domino logic. To complicate terminology, sequencing elements themselves can be either static or dynamic. A sequencing element with *static storage* employs some sort of feedback to retain its output value indefinitely. An element with *dynamic storage* generally maintains its value as charge on a capacitor that will leak away if not refreshed for a long period of time. The choices of static or dynamic for gates and for sequencing elements can be independent.

Sections 9.2–9.4 explore sequencing elements for static circuits, particularly flip-flops, 2-phase transparent latches, and pulsed latches. A periodic clock is commonly used to indicate the timing of a sequence. Section 9.5 describes how external signals can be synchronized to the clock and analyzes the risks of synchronizer failure. Wave pipelining is discussed in Section 9.6. Clock generation and distribution will be examined further in Section 12.4.

The choice of sequencing strategy is intimately tied to the design flow that is being used by an organization. Thus, it is important before departing on a design direction to ensure that all phases of design capture, synthesis, and verification can be accommodated. This includes such aspects as cell libraries (are the latch or flip-flop circuits and models available?); tools such as timing analyzers (can timing closure be achieved easily?); and automatic test generation (can self-test elements be inserted easily?).

9.2 Sequencing Static Circuits

Recall from Section 1.4.9 that *latches* and *flip-flops* are the two most commonly used sequencing elements. Both have three terminals: data input (D), clock (*clk*), and data output (Q). The latch is transparent when the clock is high and opaque when the clock is low; in other words, when the clock is high, D flows through to Q as if the latch were just a buffer, but when the clock is low, the latch holds its present Q output even if D changes. The flip-flop is an edge-triggered device that copies D to Q on the rising edge of the clock and ignores D at all other times. These are illustrated in Figure 9.1. The unknown state of Q before the first rising clock edge is indicated by the pair of lines at both low and high levels.

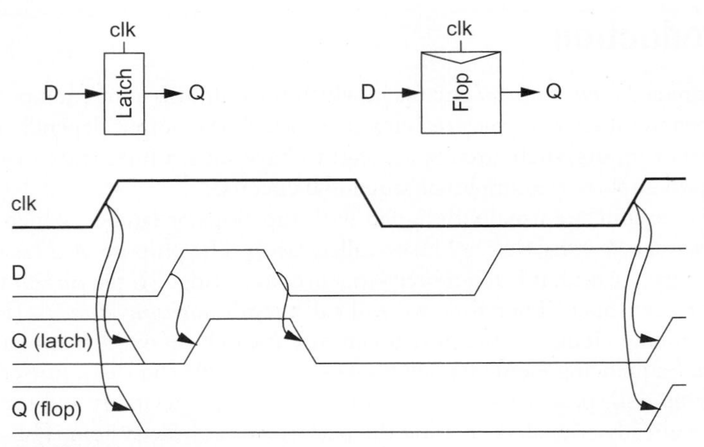

FIGURE 9.1 Latches and flip-flops

This section explores the three most widely used methods of sequencing static circuits with these elements: flip-flops, 2-phase transparent latches, and pulsed latches [Unger86]. An ideal sequencing methodology would introduce no sequencing overhead, allow sequencing elements back-to-back with no logic in between, grant the designer flexibility in balancing the amount of logic in each clock cycle, tolerate moderate amounts of clock skew without degrading performance, and consume zero area and power. We will compare these methods and explore the trade-offs they offer. We will also examine a number of transistor-level circuit implementations of each element.

9.2.1 Sequencing Methods

Figure 9.2 illustrates three methods of sequencing blocks of combinational logic. In each case, the clock waveforms, sequencing elements, and combinational logic are shown. The horizontal axis corresponds to the time at which a token reaches a point in the circuit. For example, the token is captured in the first flip-flop on the first rising edge of the clock. It propagates through the combinational logic and reaches the second flip-flop on the second rising edge of the clock. The dashed vertical lines indicate the boundary between one clock cycle and the next. The clock period is T_c. In a 2-phase system, the phases may be separated by $t_{\text{nonoverlap}}$. In a pulsed system, the pulse width is t_{pw}.

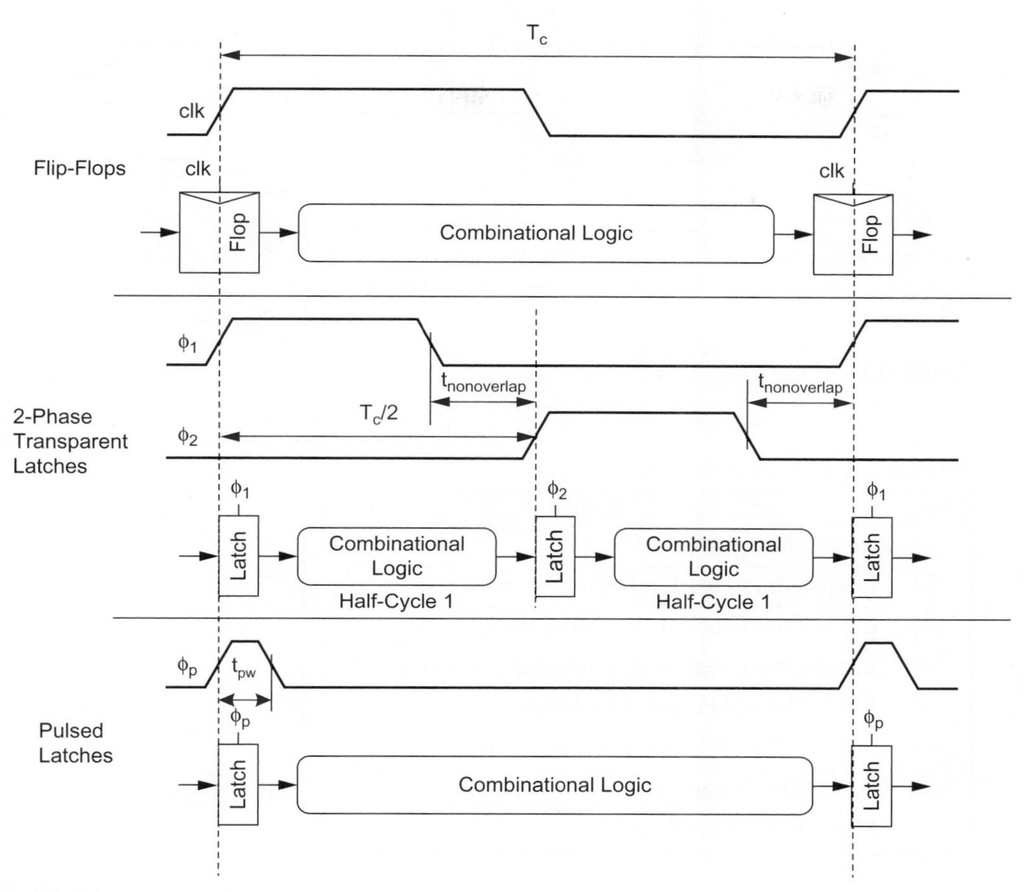

FIGURE 9.2 Static sequencing methods

Flip-flop-based systems use one flip-flop on each cycle boundary. Tokens advance from one cycle to the next on the rising edge. If a token arrives too early, it waits at the flip-flop until the next cycle. Recall that the flip-flop can be viewed as a pair of back-to-back latches using *clk* and its complement, as shown in Figure 9.3. If we separate the latches, we can divide the full cycle of combinational logic into two phases, sometimes called *half-cycles*. The two latch clocks are often called ϕ_1 and ϕ_2. They may correspond to *clk* and its complement $\overline{clk}$ or may be nonoverlapping ($t_{\text{nonoverlap}} > 0$). At any given time, at least one clock is low and the corresponding latch is opaque, preventing one token from catching up with another. The two latches behave in much the same manner as two water-tight gates in a canal lock [Mead80]. Pulsed latch systems eliminate one of the latches from each cycle and apply a brief pulse to the remaining latch. If the pulse is shorter than the delay through the combinational logic, we can still expect that a token will only advance through one clock cycle on each pulse.

Table 9.1 defines the delays and timing constraints of the combinational logic and sequencing elements. These delays may differ significantly for rising and falling transitions and can be distinguished with an *r* or *f* suffix. For brevity, we will use the overall maximum and minimum.

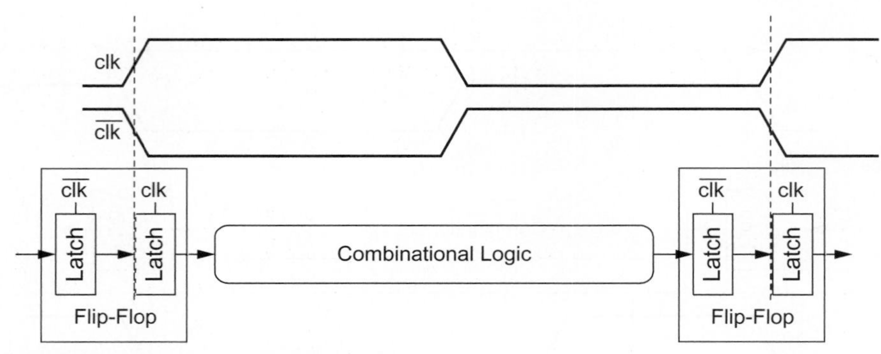

FIGURE 9.3 Flip-flop viewed as back-to-back latch pair

TABLE 9.1 Sequencing element timing notation

Term	Name
t_{pd}	Logic Propagation Delay
t_{cd}	Logic Contamination Delay
t_{pcq}	Latch/Flop Clock-to-Q Propagation Delay
t_{ccq}	Latch/Flop Clock-to-Q Contamination Delay
t_{pdq}	Latch D-to-Q Propagation Delay
t_{cdq}	Latch D-to-Q Contamination Delay
t_{setup}	Latch/Flop Setup Time
t_{hold}	Latch/Flop Hold Time

Figure 9.4 illustrates these delays in a *timing diagram*. In a timing diagram, the horizontal axis indicates time and the vertical axis indicates logic level. A single line indicates that a signal is high or low at that time. A pair of lines indicates that a signal is stable but that we don't care about its value. Criss-crossed lines indicate that the signal might change at that time. A pair of lines with cross-hatching indicates that the signal may change once or more over an interval of time.

Figure 9.4(a) shows the response of combinational logic to the input A changing from one arbitrary value to another. The output Y cannot change instantaneously. After the contamination delay t_{cd}, Y may begin to change or *glitch*. After the propagation delay t_{pd}, Y must have settled to a final value. The contamination delay and propagation delay may be very different because of multiple paths through the combinational logic. Figure 9.4(b) shows the response of a flip-flop. The data input must be stable for some window around the rising edge of the flop if it is to be reliably sampled. Specifically, the input D must have settled by some *setup time* t_{setup} before the rising edge of *clk* and should not change again until a *hold time* t_{hold} after the clock edge. The output begins to change after a *clock-to-Q contamination delay* t_{ccq} and completely settles after a *clock-to-Q propagation delay* t_{pcq}. Figure 9.4(c) shows the response of a latch. Now the input D must set up and hold around the falling edge that defines the end of the sampling period. The output initially changes t_{ccq} after the latch becomes transparent on the rising edge of the clock and settles by t_{pcq}. While the latch is transparent, the output will continue to track the input after some

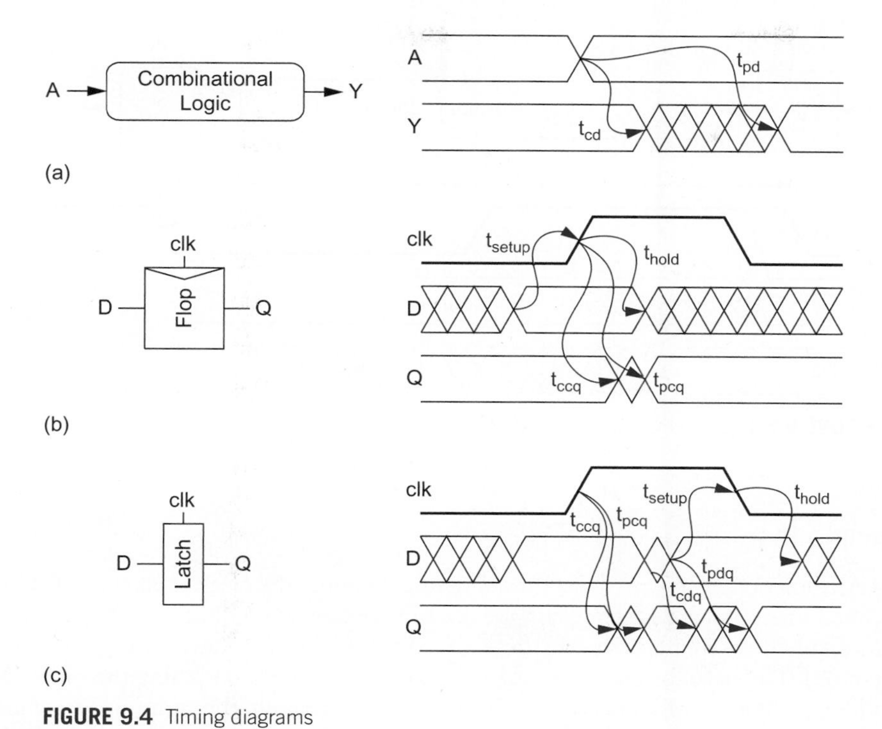

FIGURE 9.4 Timing diagrams

D-to-*Q* delay t_{cdq} and t_{pdq}. Section 9.4.2 discusses how to measure the setup and hold times and propagation delays in simulation.

9.2.2 Max-Delay Constraints

Ideally, the entire clock cycle would be available for computations in the combinational logic. Of course, the sequencing overhead of the latches or flip-flops cuts into this time. If the combinational logic delay is too great, the receiving element will miss its setup time and sample the wrong value. This is called a *setup time failure* or *max-delay failure*. It can be solved by redesigning the logic to be faster or by increasing the clock period. This section computes the actual time available for logic and the sequencing overhead of each of our favorite sequencing elements: flip-flops, two-phase latches, and pulsed latches.

Figure 9.5 shows the max-delay timing constraints on a path from one flip-flop to the next, assuming ideal clocks with no skew. The path begins with the rising edge of the clock triggering $F1$. The data must propagate to the output of the flip-flop $Q1$ and through the combinational logic to $D2$, setting up at $F2$ before the next rising clock edge. This implies that the clock period must be at least

$$T_c \geq t_{pcq} + t_{pd} + t_{\text{setup}} \tag{9.1}$$

Alternatively, we can solve for the maximum allowable logic delay, which is simply the cycle time less the sequencing overhead introduced by the propagation delay and setup time of the flip-flop.

$$t_{pd} \leq T_c - \underbrace{\left(t_{\text{setup}} + t_{pcq}\right)}_{\text{sequencing overhead}} \tag{9.2}$$

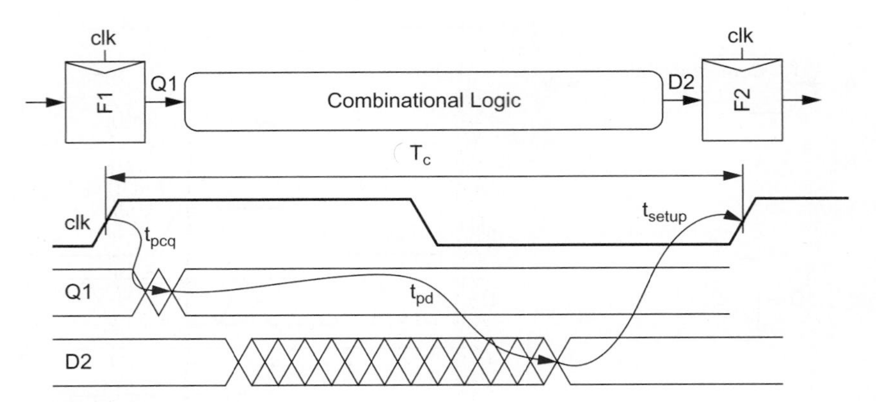

FIGURE 9.5 Flip-flop max-delay constraint

Example 9.1

The Arithmetic/Logic Unit (ALU) *self-bypass path* limits the clock frequency of some pipelined microprocessors. For example, the Integer Execution Unit (IEU) of the Itanium 2 contains self-bypass paths for six separate ALUs, as shown in Figure 9.6(a) [Fetzer02]. The path for one of the ALUs begins at registers containing the inputs to an adder, as shown in Figure 9.6(b). The adder must compute the sum (or difference, for subtraction). A *result multiplexer* chooses between this sum, the output of the logic unit, and the output of the shifter. Then a series of *bypass multiplexers* selects the inputs to the ALU for the next cycle. The early bypass multiplexer chooses among results of ALUs from previous cycles and is not on the critical path. The 8:1 middle bypass multiplexer chooses a result from any of the six ALUs, the early bypass mux, or the register file. The 4:1 late bypass multiplexer chooses a result from either of two results returning from the data cache, the middle bypass mux result, or the immediate operand specified by the next instruction. The late bypass mux output is driven back to the ALU to use on the next cycle. Because the six ALUs and the bypass multiplexers occupy a significant amount of area, the critical path also involves 2 mm wires from the result mux to middle bypass mux and from the middle bypass mux back to the late bypass mux. (*Note*: In the Itanium 2, the ALU self-bypass path is built from four-phase skew-tolerant domino circuits. For the purposes of these examples, we will hypothesize instead that it is built from static logic and flip-flops or latches.)

For our example, the propagation delays and contamination delays of the path are given in Table 9.2. Suppose the registers are built from flip-flops with a setup time of 62 ps, hold time of −10 ps, propagation delay of 90 ps, and contamination delay of 75 ps. Calculate the minimum cycle time T_c at which the ALU self-bypass path will operate correctly.

SOLUTION: The critical path involves propagation delays through the adder (590 ps), result mux (60 ps), middle bypass mux (80 ps), late bypass mux (70 ps), and two 2-mm wires (100 ps each), for a total of $t_{pd} = 1000$ ps. According to EQ (9.1), the cycle time T_c must be at least $90 + 1000 + 62 = 1152$ ps.

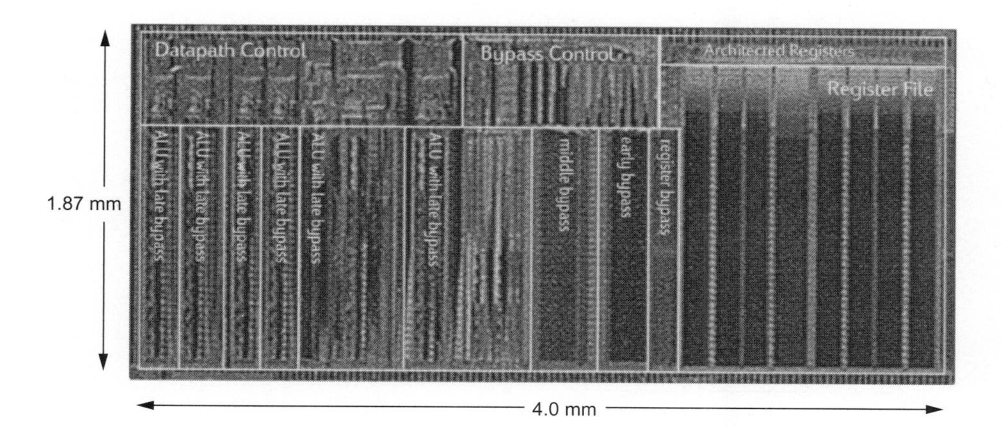

(a)

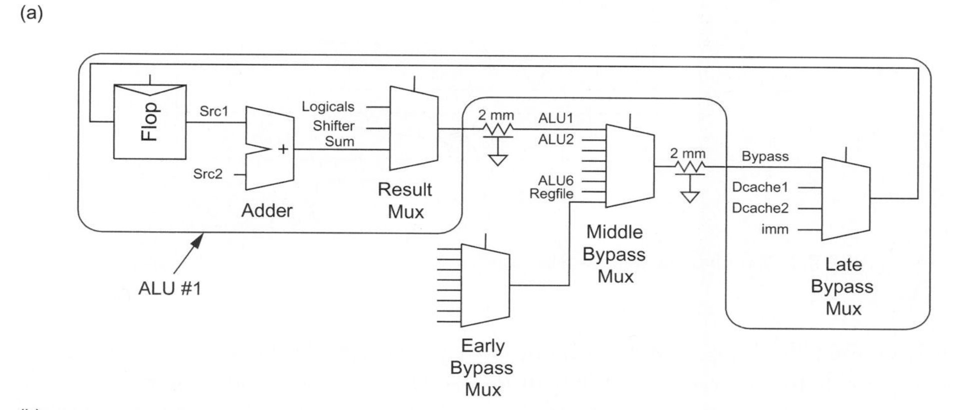

(b)

FIGURE 9.6 Itanium 2 ALU self-bypass path ((a) © IEEE 2002.)

TABLE 9.2 Combinational logic delays

Element	Propagation Delay	Contamination Delay
Adder	590 ps	100 ps
Result Mux	60 ps	35 ps
Early Bypass Mux	110 ps	95 ps
Middle Bypass Mux	80 ps	55 ps
Late Bypass Mux	70 ps	45 ps
2-mm Wire	100 ps	65 ps

Figure 9.7 shows the analogous constraints on a path using two-phase transparent latches. Let us assume that data $D1$ arrives at $L1$ while the latch is transparent (ϕ_1 high). The data propagates through $L1$, the first block of combinational logic, $L2$, and the second block of combinational logic. Technically, $D3$ could arrive as late as a setup time before the falling edge of ϕ_1 and still be captured correctly by $L3$. To be fair, we will insist

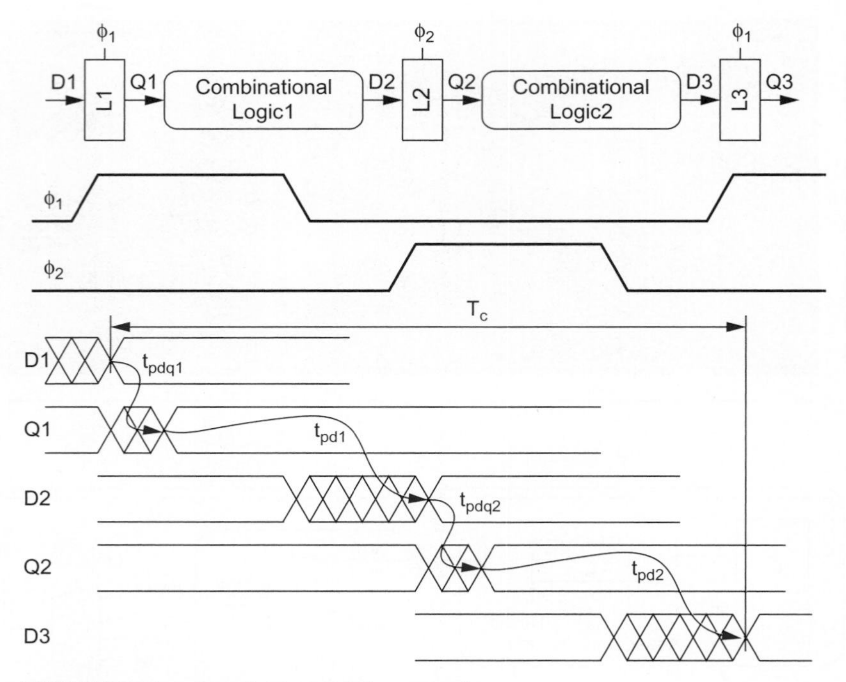

FIGURE 9.7 Two-phase latch max-delay constraint

that $D3$ nominally arrive no more than one clock period after $D1$ because, in the long run, it is impossible for every single-cycle path in a design to consume more than a full clock period. Certain paths may take longer if other paths take less time; this technique is called *time borrowing* and will be addressed in Section 9.2.4. Assuming the path takes no more than a cycle, we see the cycle time must be

$$T_c \geq t_{pdq1} + t_{pd1} + t_{pdq2} + t_{pd2} \tag{9.3}$$

Once again, we can solve for the maximum logic delay, which is the sum of the logic delays through each of the two phases. The sequencing overhead consists of the two latch propagation delays. Notice that the nonoverlap between clocks does not degrade performance in the latch-based system because data continues to propagate through the combinational logic between latches even while both clocks are low. Realizing that a flip-flop can be made from two latches whose delays determine the flop propagation delay and setup time, we see EQ (9.4) is closely analogous to EQ (9.2).

$$t_{pd} = t_{pd1} + t_{pd2} \leq T_c - \underbrace{\left(2t_{pdq}\right)}_{\text{sequencing overhead}} \tag{9.4}$$

The max-delay constraint for pulsed latches is similar to that of two-phase latches except that only one latch is in the critical path, as shown in Figure 9.8(a). However, if the pulse is narrower than the setup time, the data must set up before the pulse rises, as shown in Figure 9.8(b). Combining these two cases gives

$$T_c \geq \max\left(t_{pdq} + t_{pd},\ t_{pcq} + t_{pd} + t_{\text{setup}} - t_{pw}\right) \tag{9.5}$$

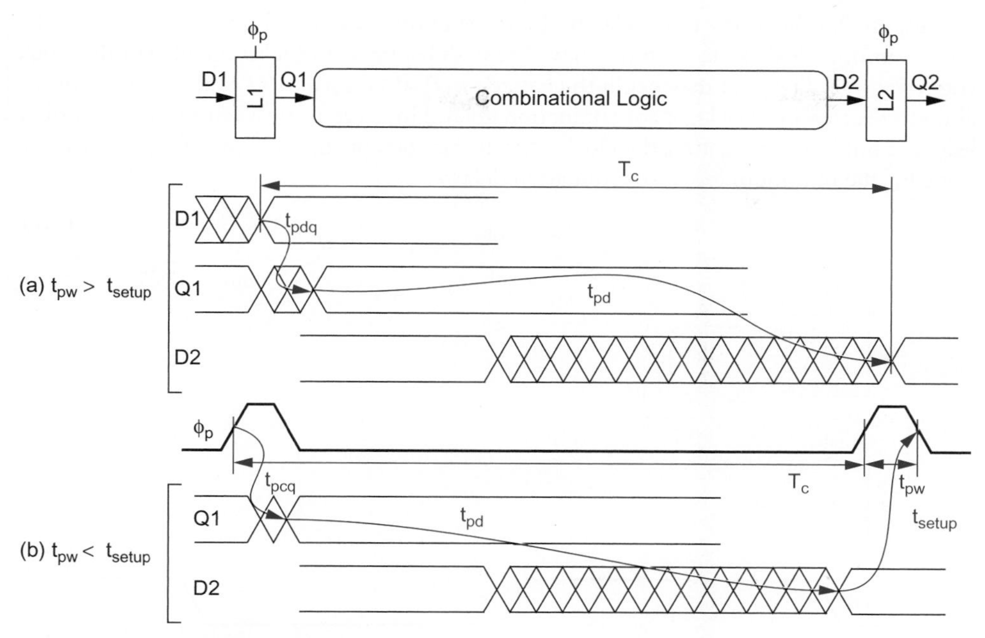

FIGURE 9.8 Pulsed latch max-delay constraint

Solving for the maximum logic delay shows that the sequencing overhead is just one latch delay if the pulse is wide enough to hide the setup time

$$t_{pd} \leq T_c - \underbrace{\max\left(t_{pdq}, t_{pcq} + t_{\text{setup}} - t_{pw}\right)}_{\text{sequencing overhead}} \tag{9.6}$$

Example 9.2

Recompute the ALU self-bypass path cycle time if the flip-flop is replaced with a pulsed latch. The pulsed latch has a pulse width of 150 ps, a setup time of 40 ps, a hold time of 5 ps, a *clk*-to-*Q* propagation delay of 82 ps and contamination delay of 52 ps, and a *D*-to-*Q* propagation delay of 92 ps.

SOLUTION: t_{pd} is still 1000 ps. According to EQ (9.5), the cycle time must be at least 92 + 1000 = 1092 ps.

9.2.3 Min-Delay Constraints

Ideally, sequencing elements can be placed back-to-back without intervening combinational logic and still function correctly. For example, a pipeline can use back-to-back registers to sequence along an instruction opcode without modifying it. However, if the hold time is large and the contamination delay is small, data can incorrectly propagate through two successive elements on one clock edge, corrupting the state of the system. This is called a *race condition*, *hold-time failure*, or *min-delay failure*. It can only be fixed by redesigning the logic, not by slowing the clock. Therefore, designers should be very conservative in avoiding such failures because modifying and refabricating a chip is catastrophically expensive and time-consuming.

Figure 9.9 shows the min-delay timing constraints on a path from one flip-flop to the next assuming ideal clocks with no skew. The path begins with the rising edge of the clock triggering $F1$. The data may begin to change at $Q1$ after a *clk*-to-Q contamination delay, and at $D2$ after another logic contamination delay. However, it must not reach $D2$ until at least the hold time t_{hold} after the clock edge, lest it corrupt the contents of $F2$. Hence, we solve for the minimum logic contamination delay:

$$t_{cd} \geq t_{\text{hold}} - t_{ccq} \tag{9.7}$$

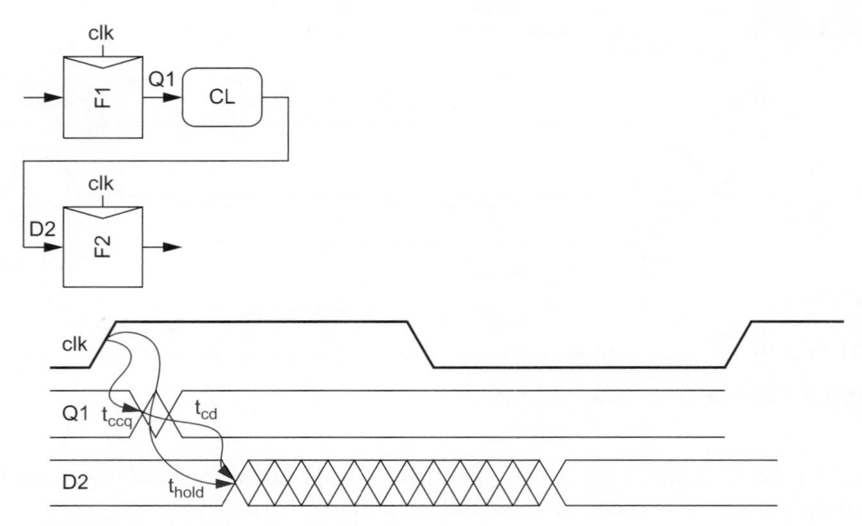

FIGURE 9.9 Flip-flop latch min-delay constraint

Example 9.3

In the ALU self-bypass example with flip-flops from Figure 9.6, the earliest input to the late bypass multiplexer is the *imm* value coming from another flip-flop. Will this path experience any hold-time failures?

SOLUTION: No. The late bypass mux has $t_{cd} = 45$ ps. The flip-flops have $t_{\text{hold}} = -10$ ps and $t_{ccq} = 75$ ps. Hence, EQ (9.7) is easily satisfied.

If the contamination delay through the flip-flop exceeds the hold time, you can safely use back-to-back flip-flops. If not, you must explicitly add delay between the flip-flops (e.g., with a buffer) or use special slow flip-flops with greater than normal contamination delay on paths that require back-to-back flops. Scan chains are a common example of paths with back-to-back flops.

Figure 9.10 shows the min-delay timing constraints on a path from one transparent latch to the next. The path begins with data passing through $L1$ on the rising edge of ϕ_1. It must not reach $L2$ until a hold time after the previous falling edge of ϕ_2 because $L2$ should have become safely opaque before $L1$ becomes transparent. As the edges are separated by $t_{\text{nonoverlap}}$, the minimum logic contamination delay through each phase of logic is

$$t_{cd1}, t_{cd2} \geq t_{\text{hold}} - t_{ccq} - t_{\text{nonoverlap}} \tag{9.8}$$

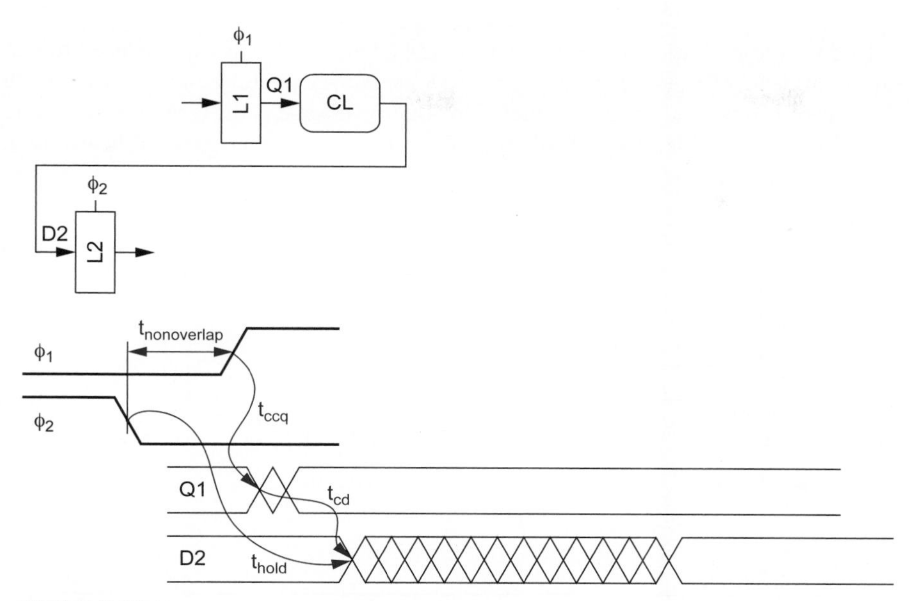

FIGURE 9.10 Two-phase latch min-delay constraint

(Note that our derivation found the minimum delay through the first half-cycle, but that the second half-cycle has the same constraint.)

This result shows that by making $t_{nonoverlap}$ sufficiently large, hold-time failure can be avoided entirely. However, generating and distributing nonoverlapping clocks is challenging at high speeds. Therefore, most commercial transparent latch-based systems use the clock and its complement. In this case, $t_{nonoverlap} = 0$ and the contamination delay constraint is the same between the latches and flip-flops.

This leads to an apparent paradox: The contamination delay constraint applies to each phase of logic for latch-based systems, but to the entire cycle of logic for flip-flops. Therefore, latches seem to require twice the overall logic contamination delay as compared to flip-flops. Yet flip-flops can be built from a pair of latches! The paradox is resolved by observing that a flip-flop has an internal race condition between the two latches. The flip-flop must be carefully designed so that it always operates reliably.

Figure 9.11 shows the min-delay timing constraints on a path from one pulsed latch to the next. Now data departs on the rising edge of the pulse but must hold until after the falling edge of the pulse. Therefore, the pulse width effectively increases the hold time of the pulsed latch as compared to a flip-flop.

$$t_{cd} \geq t_{hold} - t_{ccq} + t_{pw} \tag{9.9}$$

Example 9.4

If the ALU self-bypass path uses pulsed latches in place of flip-flops, will it have any hold-time problems?

SOLUTION: Yes. The late bypass mux has $t_{cd} = 45$ ps. The pulsed latches have $t_{pw} = 150$ ps, $t_{hold} = 5$ ps, and $t_{ccq} = 52$ ps. Hence, EQ (9.9) is badly violated. *Src*1 may receive *imm* from the next instruction rather than the current instruction. The problem could be solved by adding buffers after the *imm*-pulsed latch. The buffers would need to add a minimum delay of $t_{\text{hold}} - t_{ccq} + t_{pw} - t_{cd} = 58$ ps. Alternatively, the *imm*-pulsed latch could be replaced with a flip-flop without slowing the critical path. If the flip-flop were designed with a very long (> 110 ps) contamination delay, the race would be avoided.

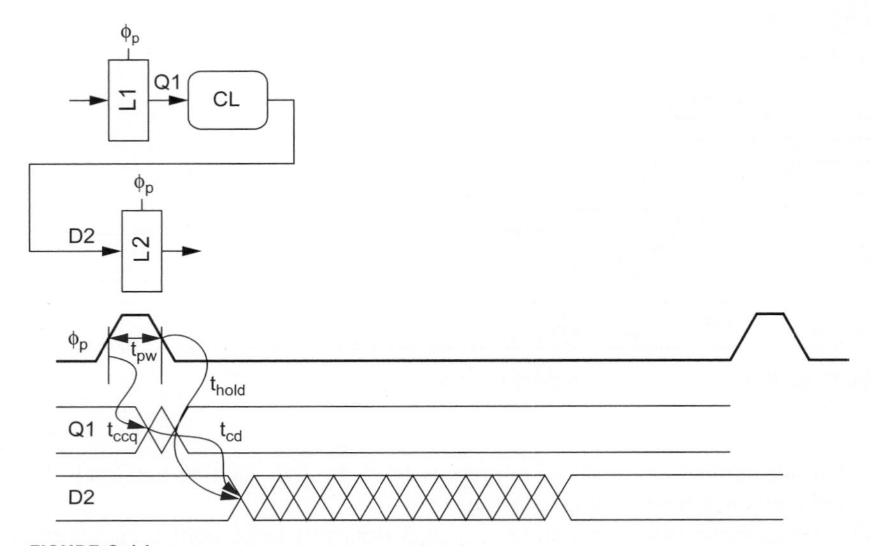

FIGURE 9.11 Pulsed latch min-delay constraint

9.2.4 Time Borrowing

In a system using flip-flops, data departs the first flop on the rising edge of the clock and must set up at the second flop before the next rising edge of the clock. If the data arrives late, the circuit produces the wrong result. If the data arrives early, it is blocked until the clock edge, and the remaining time goes unused. Therefore, we say the clock imposes a *hard edge* because it sharply delineates the cycles.

In contrast, when a system uses transparent latches, the data can depart the first latch on the rising edge of the clock, but does not have to set up until the falling edge of the clock on the receiving latch. If one half-cycle or stage of a pipeline has too much logic, it can borrow time into the next half-cycle or stage, as illustrated in Figure 9.12(a) [Bernstein99]. *Time borrowing* can accumulate across multiple cycles. However, in systems with feedback, the long delays must be balanced by shorter delays so that the overall loop completes in the time available. For example, Figure 9.12(b) shows a single-cycle self-bypass loop in which time borrowing occurs across half-cycles, but the entire path must fit in one cycle. A typical example of a self-bypass loop is the execution stage of a pipelined processor in which an ALU must complete an operation and bypass the result back for use in the ALU on a dependent instruction. Most critical paths in digital systems occur in self-bypass loops because otherwise latency does not matter.

Figure 9.13 illustrates the maximum amount of time that a two-phase latch-based system can borrow (beyond the $T_c/2 - t_{pdq}$ nominally available to each half-cycle of logic).

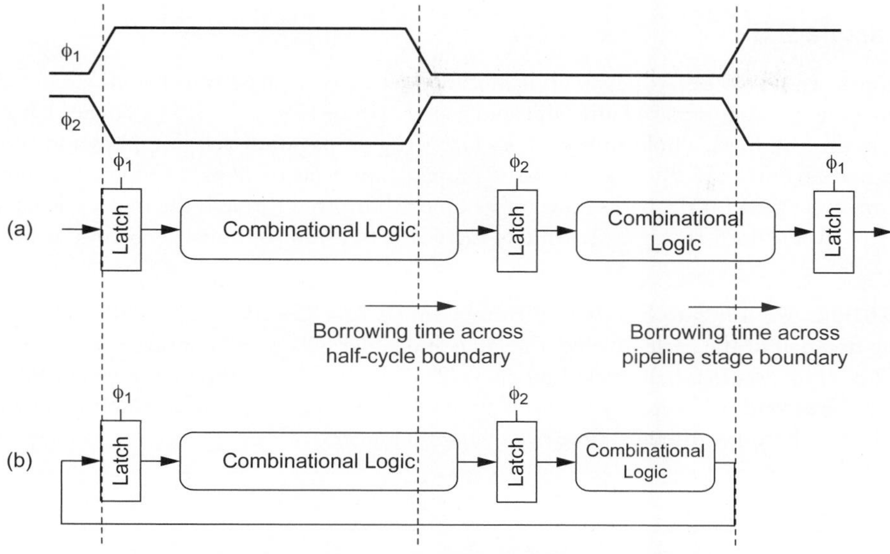

FIGURE 9.12 Time borrowing

Because data does not have to set up until the falling edge of the receiving latch's clock, one phase can borrow up to half a cycle of time from the next (less setup time and nonoverlap):

$$t_{\text{borrow}} \leq \frac{T_c}{2} - \left(t_{\text{setup}} + t_{\text{nonoverlap}}\right) \tag{9.10}$$

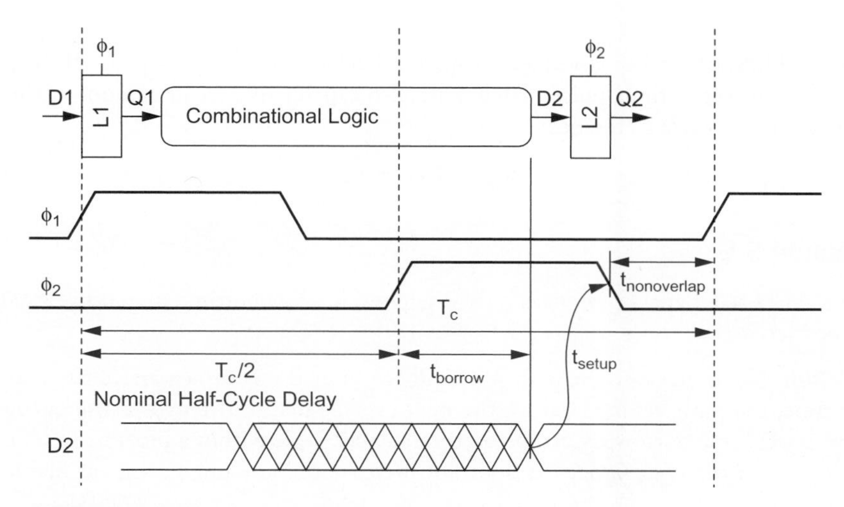

FIGURE 9.13 Maximum amount of time borrowing

Example 9.5

Suppose the ALU self-bypass path is modified to use two-phase transparent latches. A mid-cycle ϕ_2 latch is placed after the adder, as shown in Figure 9.14. The latches have a setup time of 40 ps, a hold time of 5 ps, a *clk*-to-*Q* propagation delay of 82 ps and contamination delay of 52 ps, and a *D*-to-*Q* propagation delay of 82 ps. Compute the minimum cycle time for the path. How much time is borrowed through the mid-cycle latch at this cycle time? If the cycle time is increased to 2000 ps, how much time is borrowed?

SOLUTION: According to EQ (9.3), the cycle time is $T_c = 82 + 590 + 82 + 410 = 1164$ ps. The first half of the cycle involves the latch and adder delays and consumes $82 + 590 = 672$ ps. The nominal half-cycle time is $T_c/2 = 582$ ps. Hence, the path borrows 90 ps from the second half-cycle. If the cycle time increases to 2000 ps and the nominal half-cycle time becomes 1000 ps, time borrowing no longer occurs.

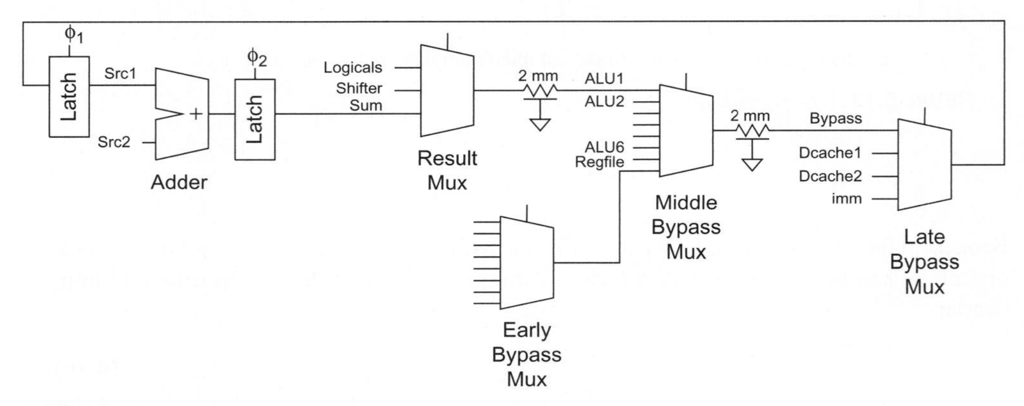

FIGURE 9.14 ALU self-bypass path with two-phase latches

Pulsed latches can be viewed as transparent latches with a narrow pulse. If the pulse is wider than the setup time, pulsed latches are also capable of a small amount of time borrowing from one cycle to the next.

$$t_{\text{borrow}} \leq t_{pw} - t_{\text{setup}} \quad (9.11)$$

Example 9.6

If the ALU self-bypass path uses pulsed latches, how much time may it borrow from the next cycle?

SOLUTION: None. Because the path is a feedback loop, if its outputs arrive late and borrow time, the path begins later on the next cycle. This in turn causes the outputs to arrive later. Time borrowing can be used to balance logic within a pipeline but, despite the wishes of many designers, it does not increase the amount of time available in a clock cycle.

Time borrowing has two benefits for the system designer. The most obvious is *intentional time borrowing*, in which the designer can more easily balance logic between half-cycles and pipeline stages. This leads to potentially shorter design time because the balancing can take place during circuit design rather than requiring changes to the microarchitecture to explicitly move functions from one stage to another. The other benefit is *opportunistic time borrowing*. Even if the designer carefully equalizes the delay in each stage at design time, the delays will differ from one stage to another in the fabricated chip because of process and environmental variations and inaccuracies in the timing model used by the CAD system. In a system with hard edges, the longest cycle sets the minimum clock period. In a system capable of time borrowing, the slow cycles can opportunistically borrow time from faster ones and average out some of the variation.

Some experienced design managers forbid the use of intentional time borrowing until the chip approaches tapeout. Otherwise designers are overly prone to assuming that their pipeline stage can borrow time from adjacent stages. When many designers make this same assumption, all of the paths become excessively long. Worse yet, the problem may be hidden until full-chip timing analysis begins, at which time it is too late to redesign so many paths. Another solution is to do full-chip timing analysis starting early in the design process.

9.2.5 Clock Skew

The analysis so far has assumed ideal clocks with zero skew. In reality, clocks have some uncertainty in their arrival times that can cut into the time available for useful computation, as shown in Figure 9.15(a). The bold *clk* line indicates the latest possible clock arrival time. The hashed lines show that the clock might arrive over a range of earlier times because of skew. The worst scenario for max delay in a flip-flop-based system is that the launching flop receives its clock late and the receiving flop receives its clock early. In this case, the clock skew is subtracted from the time available for useful computation and appears as sequencing overhead. The worst scenario for min delay is that the launching flop receives its clock early and the receiving clock receives its clock late, as shown in Figure 9.15(b). In this case, the clock skew effectively increases the hold time of the system.

$$t_{pd} \le T_c - \underbrace{\left(t_{pcq} + t_{\text{setup}} + t_{\text{skew}}\right)}_{\text{sequencing overhead}} \tag{9.12}$$

$$t_{cd} \ge t_{\text{hold}} - t_{ccq} + t_{\text{skew}} \tag{9.13}$$

In the system using transparent latches, clock skew does not degrade performance. Figure 9.16 shows how the full cycle (less two latch delays) is available for computation even when the clocks are skewed because the data can still arrive at the latches while they are transparent. Therefore, we say that transparent latch-based systems are *skew-tolerant*. However, skew still effectively increases the hold time in each half-cycle. It also cuts into the window available for time borrowing.

$$t_{pd} \le T_c - \underbrace{\left(2t_{pdq}\right)}_{\text{sequencing overhead}} \tag{9.14}$$

$$t_{cd1}, t_{cd2} \ge t_{\text{hold}} - t_{ccq} - t_{\text{nonoverlap}} + t_{\text{skew}} \tag{9.15}$$

$$t_{\text{borrow}} \le \frac{T_c}{2} - \left(t_{\text{setup}} + t_{\text{nonoverlap}} + t_{\text{skew}}\right) \tag{9.16}$$

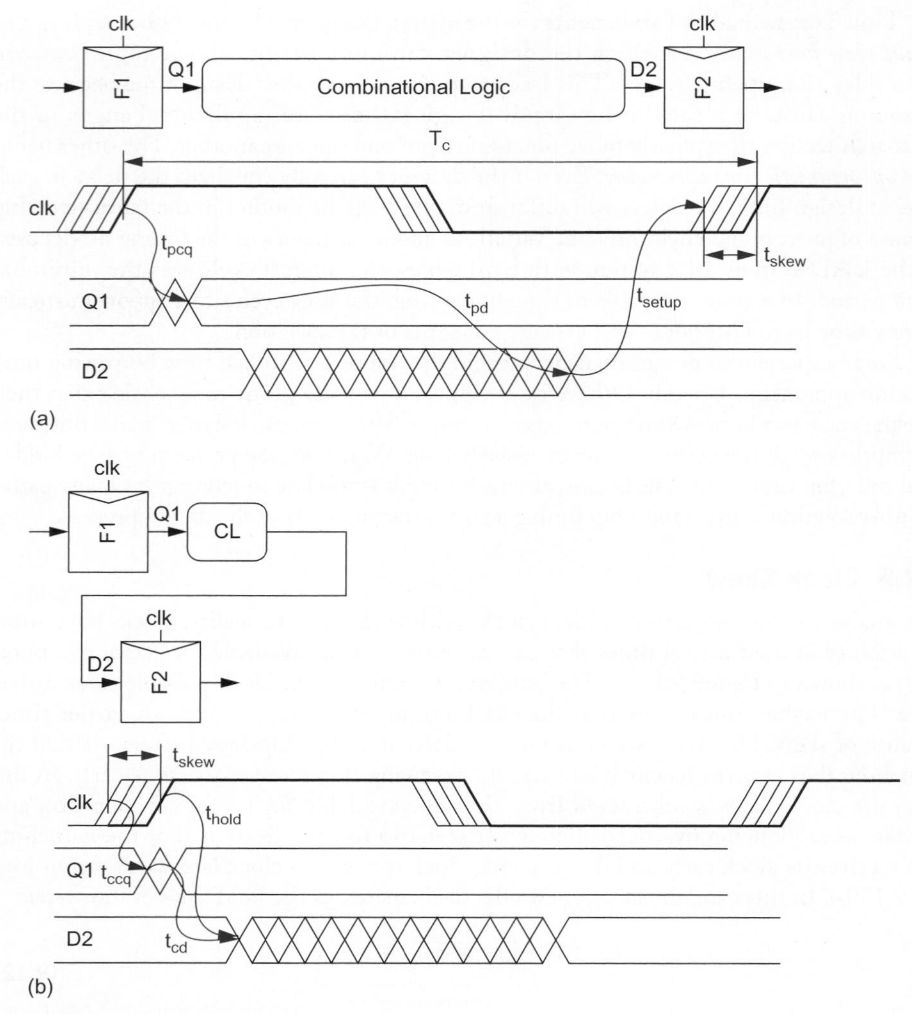

FIGURE 9.15 Clock skew and flip-flops

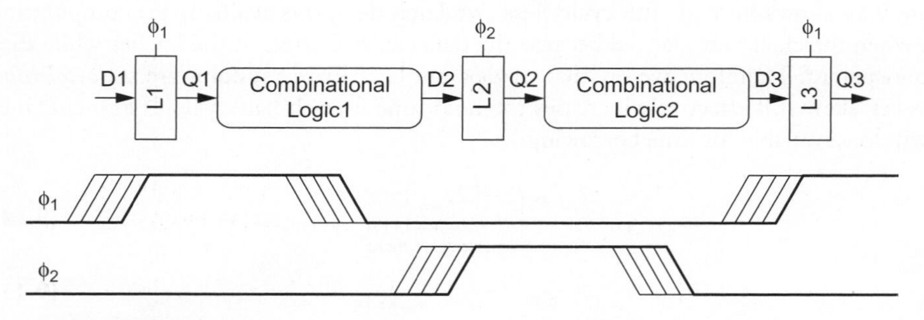

FIGURE 9.16 Clock skew and transparent latches

Example 9.7

If the ALU self-bypass path from Figure 9.6 can experience 50 ps of skew from one cycle to the next between flip-flops in the various ALUs, what is the minimum cycle time of the system? How much clock skew can the system have before hold-time failures occur?

SOLUTION: According to EQ (9.12), the cycle time should increase by 50 ps to 1202 ps. The maximum skew for which the system can operate correctly at any cycle time is $t_{cd} - t_{\text{hold}} + t_{ccq} = 45 - (-10) + 75 = 130$ ps.

Pulsed latches can tolerate an amount of skew proportional to the pulse width. If the pulse is wide enough, the skew will not increase the sequencing overhead because the data can arrive while the latch is transparent. If the pulse is narrow, skew can degrade performance. Again, skew effectively increases the hold time and reduces the amount of time available for borrowing (see Exercise 9.9).

$$t_{pd} \le T_c - \underbrace{\max\left(t_{pdq},\ t_{pcq} + t_{\text{setup}} - t_{pw} + t_{\text{skew}}\right)}_{\text{sequencing overhead}} \quad (9.17)$$

$$t_{cd} \ge t_{\text{hold}} + t_{pw} - t_{ccq} + t_{\text{skew}} \quad (9.18)$$

$$t_{\text{borrow}} \le t_{pw} - \left(t_{\text{setup}} + t_{\text{skew}}\right) \quad (9.19)$$

In summary, systems with hard edges (e.g., flip-flops) subtract clock skew from the time available for useful computation. Systems with softer edges (e.g., latches) take advantage of the window of transparency to tolerate some clock skew without increasing the sequencing overhead. Clock skew will be addressed further in Section 12.4. In particular, different amounts of skew can be budgeted for min-delay and max-delay checks. Moreover, nearby sequential elements are likely to see less skew than elements on opposite corners of the chip. Current automated place & route tools spend considerable effort to model clock delays and insert buffer elements to minimize clock skew, but skew is a growing problem for systems with aggressive cycle times.

9.3 Circuit Design of Latches and Flip-Flops

Conventional CMOS latches are built using pass transistors or tristate buffers to pass the data while the latch is transparent and feedback to hold the data while the latch is opaque. We begin by exploring circuit designs for basic latches, then build on them to produce flip-flops and pulsed latches. Many latches accept reset and/or enable inputs. It is also possible to build logic functions into the latches to reduce the sequencing overhead.

A number of alternative latch and flip-flop structures have been used in commercial designs. The True Single Phase Clocking (TSPC) technique uses a single clock with no inversions to simplify clock distribution. The Klass Semidynamic Flip-Flop (SDFF) is a fast flip-flop using a domino-style input stage. Differential flip-flops are good for certain applications. Each of these alternatives are described and compared.

9.3.1 Conventional CMOS Latches

Figure 9.17(a) shows a very simple transparent latch built from a single transistor. It is compact and fast but suffers four limitations. The output does not swing from *rail-to-rail* (i.e., from GND to V_{DD}); it never rises above $V_{DD} - V_t$. The output is also *dynamic*; in other words, the output floats when the latch is opaque. If it floats long enough, it can be disturbed by leakage (see Section 8.3.3). D drives the *diffusion input* of a pass transistor directly, leading to potential noise issues (see Section 8.3.9) and making the delay harder to model with static timing analyzers. Finally, the state node is *exposed*, so noise on the output can corrupt the state. The remainder of the figures illustrate improved latches using more transistors to achieve more robust operation.

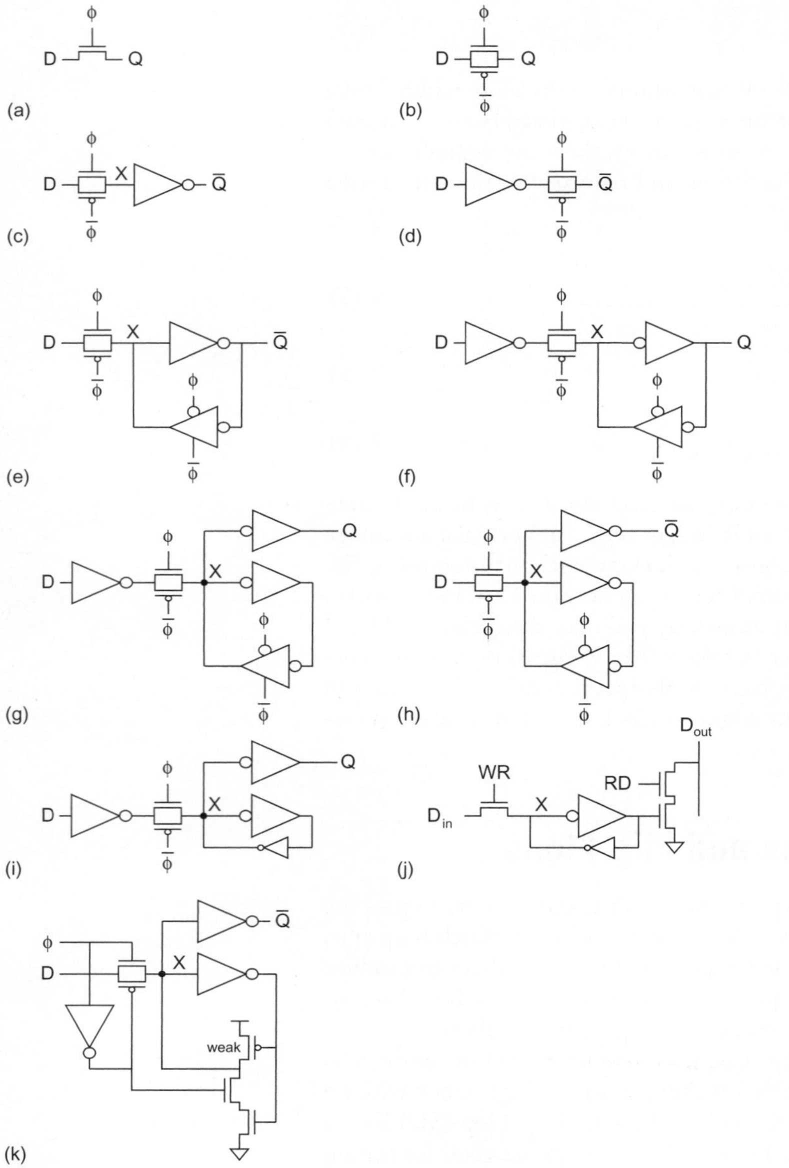

FIGURE 9.17 Transparent latches

Figure 9.17(b) uses a CMOS transmission gate in place of the single nMOS pass transistor to offer rail-to-rail output swings. It requires a complementary clock $\overline{\phi}$, which can be provided as an additional input or locally generated from ϕ through an inverter. Figure 9.17(c) adds an output inverter so that the state node X is isolated from noise on the output. Of course, this creates an inverting latch. Figure 9.17(d) also behaves as an inverting latch with a buffered input but unbuffered output. As discussed in Section 8.2.5.1, the inverter followed by a transmission gate is essentially equivalent to a tristate inverter but has a slightly lower logical effort because the output is driven by both transistors of the transmission gate in parallel. Figure 9.17(c) and (d) are both fast dynamic latches.

In modern processes, subthreshold leakage is large enough that dynamic nodes retain their values for only a short time, especially at the high temperature and voltage encountered during burn-in test. Therefore, practical latches need to be staticized, adding feedback to prevent the output from floating, as shown in Figure 9.17(e). When the clock is 1, the input transmission gate is ON, the feedback tristate is OFF, and the latch is transparent. When the clock is 0, the input transmission gate turns OFF. However, the feedback tristate turns ON, holding X at the correct level. Figure 9.17(f) adds an input inverter so the input is a transistor gate rather than unbuffered diffusion. Unfortunately, both (e) and (f) reintroduced output noise sensitivity: A large noise spike on the output can propagate backward through the feedback gates and corrupt the state node X. Figure 9.17(g) is a robust transparent latch that addresses all of the deficiencies mentioned so far: The latch is static, all nodes swing rail-to-rail, the state noise is isolated from output noise, and the input drives transistor gates rather than diffusion. Such a latch is widely used in standard cell applications including the Artisan standard cell library [Artisan02]. It is recommended for all but the most performance- or area-critical designs.

In semicustom datapath applications where input noise can be better controlled, the inverting latch of Figure 9.17(h) may be preferable because it is faster and more compact. Intel uses this as a standard datapath latch [Karnik01]. Figure 9.17(i) shows the *jamb latch*, a variation of Figure 9.17(g) that reduces the clock load and saves two transistors by using a weak feedback inverter in place of the tristate. This requires careful circuit design to ensure that the tristate is strong enough to overpower the feedback inverter in all process corners. Figure 9.17(j) shows another jamb latch commonly used in register files and Field Programmable Gate Array (FPGA) cells. Many such latches read out onto a single D_{out} wire and only one latch is enabled at any given time with its RD signal. The Itanium 2 processor uses the latch shown in Figure 9.17(k) [Naffziger02]. In the static feedback, the pulldown stack is clocked, but the pullup is a weak pMOS transistor. Therefore, the gate driving the input must be strong enough to overcome the feedback. The Itanium 2 cell library also contains a similar latch with an additional input inverter to buffer the input when the previous gate is too weak or far away. With the input inverter, the latch can be viewed as a cross between the designs shown in (g) and (i). Some latches add one more inverter to provide both true and complementary outputs.

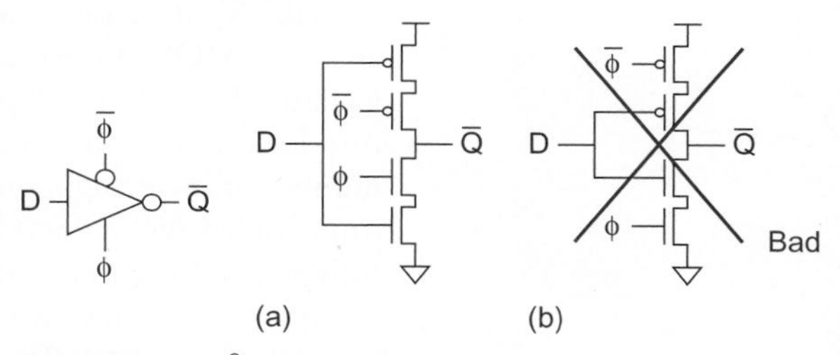

FIGURE 9.18 C²MOS Latch

The dynamic latch of Figure 9.17(d) can also be drawn as a clocked tristate, as shown in Figure 9.18(a). Such a form is sometimes called *clocked CMOS* (C^2MOS) [Suzuki73]. The conventional form using the inverter and transmission gate is slightly faster because the output is driven through the nMOS and pMOS working in parallel. C^2MOS is slightly smaller because it eliminates two contacts. Figure 9.18(b) shows another form of the tristate that swaps the data and clock terminals. It is logically equivalent but electrically inferior because toggling D while the latch is opaque can cause charge-sharing noise on the output node [Suzuki73].

All of the latches shown so far are transparent while ϕ is high. They can be converted to active-low latches by swapping ϕ and $\overline{\phi}$.

9.3.2 Conventional CMOS Flip-Flops

Figure 9.19(a) shows a dynamic inverting flip-flop built from a pair of back-to-back dynamic latches [Suzuki73]. Either the first or the last inverter can be removed to reduce delay at the expense of greater noise sensitivity on the unbuffered input or output. Figure 9.19(b) adds feedback

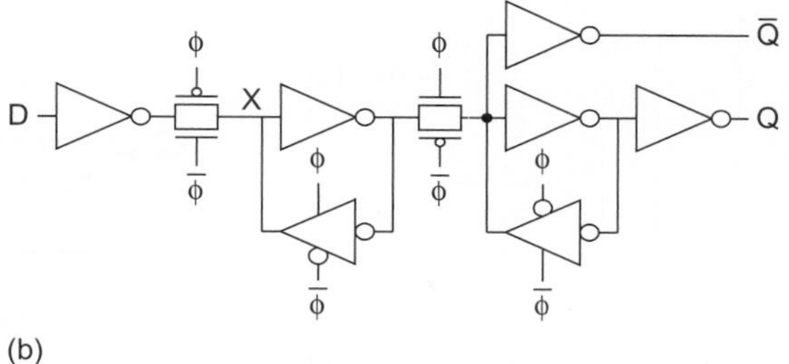

FIGURE 9.19 Flip-flops

and another inverter to produce a noninverting static flip-flop. The PowerPC 603 microprocessor datapath used this flip-flop design without the input inverter or $\overline{Q}$ output [Gerosa94]. Most standard cell libraries employ this design because it is simple, robust, compact, and energy-efficient [Stojanovic99]. However, some of the alternatives described later are faster.

Flip-flops usually take a single clock signal ϕ and locally generate its complement $\overline{\phi}$. If the clock rise/fall time is very slow, it is possible that both the clock and its complement will simultaneously be at intermediate voltages, making both latches transparent and increasing the flip-flop hold time. In ASIC standard cell libraries (such as the Artisan library), the clock is both complemented and buffered in the flip-flop cell to sharpen up the edge rates at the expense of more inverters and clock loading. However, the clock load should be kept as small as possible because it has an activity factor of 1 and thus accounts for much of the power consumption in the flip-flop.

Recall that the flip-flop also has a potential internal race condition between the two latches. This race can be exacerbated by skew between the clock and its complement caused by the delay of the inverter. Figure 9.20(a) redraws Figure 9.19(a) with a built-in clock inverter. When ϕ falls, both the clock and its complement are momentarily low as shown in Figure 9.20(b), turning on the clocked pMOS transistors in both transmission gates. If the skew (i.e., inverter delay) is too large, the data can sneak through both latches on the falling clock edge, leading to incorrect operation. Figure 9.20(c) shows a C^2MOS dynamic flip-flop built using C^2MOS latches rather than inverters and transmission gates [Suzuki73]. Because each stage inverts, data passes through the nMOS stack of one latch and the pMOS of the other, so skew that turns on both clocked pMOS transistors is not a hazard. However, the flip-flop is still susceptible to failure from very slow edge rates that turn both transistors partially ON. The same skew advantages apply even when an even number of inverting logic stages are placed between the latches; this technique is sometimes called *NO RAce* (NORA) [Gonclaves83]. In practice, most flip-flop designs carefully control the delay of the clock inverter so the transmission gate design is safe and slightly faster than C^2MOS [Chao89].

All of these flip-flop designs still present potential min-delay problems between flip-flops, especially when there is little or no logic between flops and the clock skew is large or

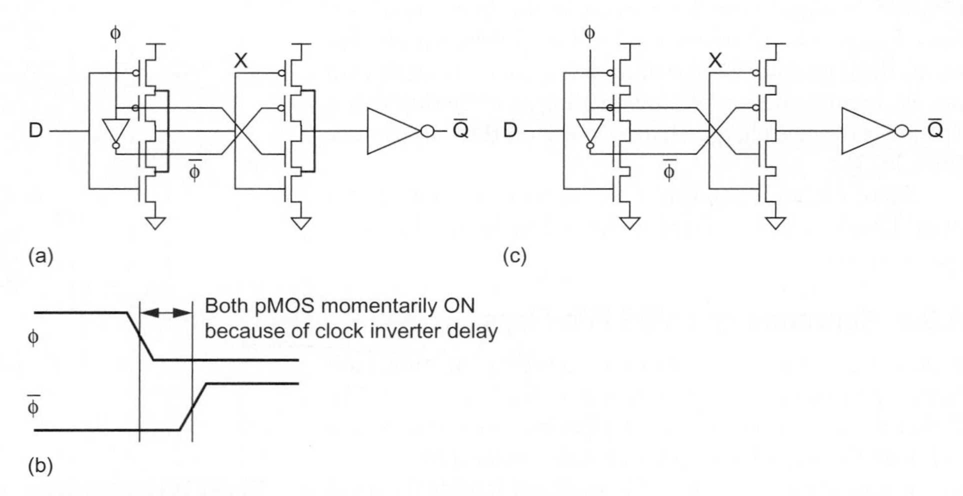

FIGURE 9.20 Transmission gate and NORA dynamic flip-flops

poorly analyzed. For VLSI class projects where careful clock skew analysis is too much work and performance is less important, a reasonable alternative is to use a pair of two-phase nonoverlapping clocks instead of the clock and its complement, as shown in Figure 9.21. The flip-flop captures its input on the rising edge of ϕ_1. By making the nonoverlap large enough, the circuit will work despite large skews. However, the nonoverlap time is not used by logic, so it directly increases the setup time and sequencing overhead of the flip-flop (see Exercise 9.10). The layout for the flip-flop is shown on the inside front cover and is readily adapted to use a single clock. Observe how diffusion nodes are shared to reduce parasitic capacitance.

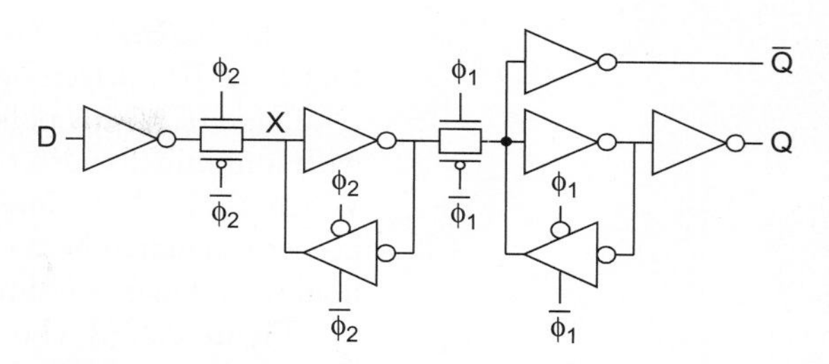

FIGURE 9.21 Flip-flop with two-phase nonoverlapping clocks

9.3.3 Pulsed Latches

A pulsed latch can be built from a conventional CMOS transparent latch driven by a brief clock pulse. Figure 9.22(a) shows a simple pulse generator, sometimes called a *clock chopper* or *one-shot* [Harris01a]. The pulsed latch is faster than a regular flip-flop because it involves a single latch rather than two and because it allows time borrowing. It can also consume less energy, although the pulse generator adds to the energy consumption (and is ideally shared across multiple pulsed latches for energy and area efficiency). The drawback is the increased hold time.

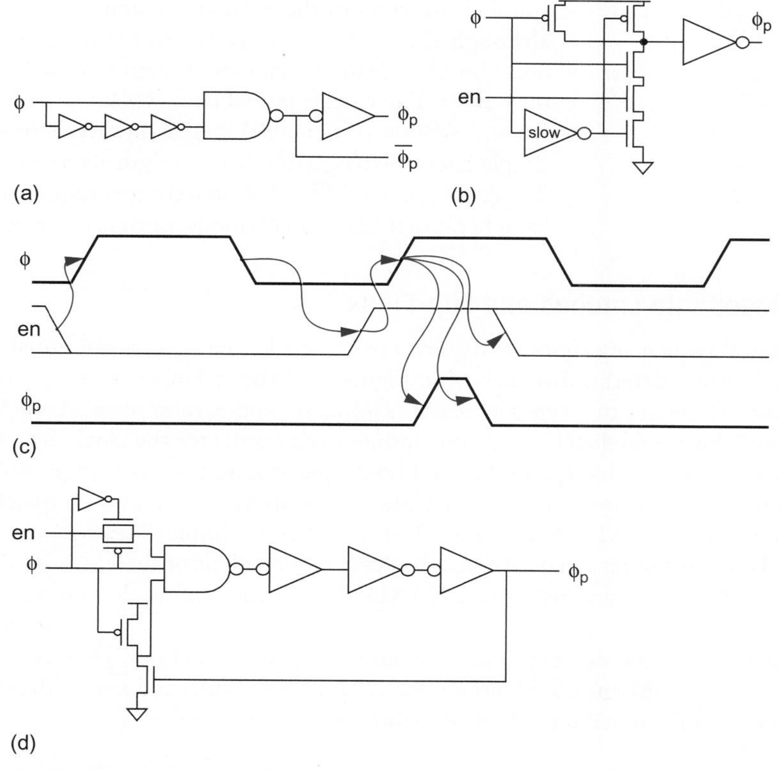

FIGURE 9.22 Pulse generators

The *Naffziger pulsed latch* used on the Itanium 2 processor consists of the latch from Figure 9.17(k) driven by even shorter pulses produced by the generator of Figure 9.22(b) [Naffziger02]. This pulse generator uses a fairly slow (weak) inverter to produce a pulse with a nominal width of about one-sixth of the cycle (125 ps for 1.2 GHz operation). When disabled, the internal node of the pulse generator floats high momentarily, but no keeper is required because the duration is short. Of course, the enable signal has setup and hold requirements around the rising edge of the clock, as shown in Figure 9.22(c).

Figure 9.22(d) shows yet another pulse generator used on an NEC RISC processor [Kozu96] to produce substantially longer pulses. It includes a built-in dynamic transmission-gate latch to prevent the enable from glitching during the pulse.

Many designers consider short pulses risky. The pulse generator should be carefully simulated across process corners and possible RC loads to ensure the pulse is not degraded too badly by process variation or routing. However, the Itanium 2 team found that the pulses could be used just as regular clocks as long as the pulse generator had adequate drive. The quad-core Itanium pulse generator selects between 1- and 3-inverter delay chains using a transmission gate multiplexer [Stackhouse09]. The wider pulse offers more robust latch operation across process and environmental variability and permits more time borrowing, but increases the hold time. The multiplexer select is software-programmable to fix problems discovered after fabrication.

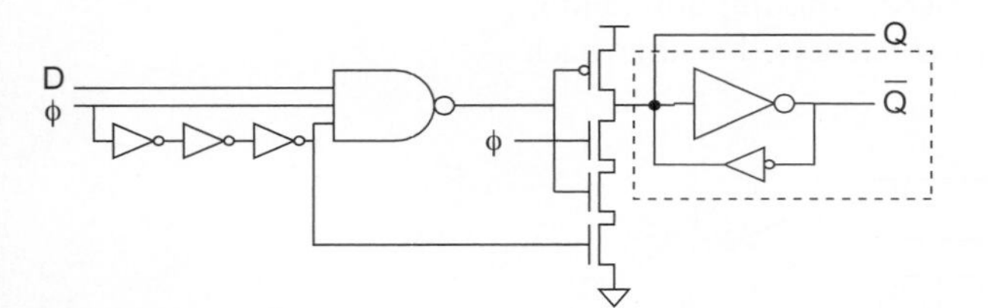

FIGURE 9.23 Partovi pulsed latch

The *Partovi pulsed latch* in Figure 9.23 eliminates the need to distribute the pulse by building the pulse generator into the latch itself [Partovi96, Draper97]. The weak cross-coupled inverters in the dashed box staticize the circuit, although the latch is susceptible to back-driven output noise on Q or $\overline{Q}$ unless an extra inverter is used to buffer the output. The Partovi pulsed latch was used on the AMD K6 and Athlon [Golden99], but is slightly slower than a simple latch [Naffziger02]. It was originally called an *Edge Triggered Latch* (ETL), but strictly speaking is a pulsed latch because it has a brief window of transparency.

9.3.4 Resettable Latches and Flip-Flops

Most practical sequencing elements require a reset signal to enter a known initial state on startup and ensure deterministic behavior. Figure 9.24 shows latches and flip-flops with reset inputs. There are two types of reset: *synchronous* and *asynchronous*. Asynchronous reset forces Q low immediately, while synchronous reset waits for the clock. Synchronous reset signals must be stable for a setup and hold time around the clock edge while asynchronous reset is characterized by a propagation delay from reset to output. Synchronous reset simply requires ANDing the input D with $\overline{reset}$. Asynchronous reset requires gating both the data and the feedback to force the reset independent of the clock. The tristate NAND gate can be constructed from a NAND gate in series with a clocked transmission gate.

Settable latches and flip-flops force the output high instead of low. They are similar to resettable elements of Figure 9.24 but replace NAND with NOR and $\overline{reset}$ with *set*. Figure 9.25 shows a flip-flop combining both asynchronous set and reset.

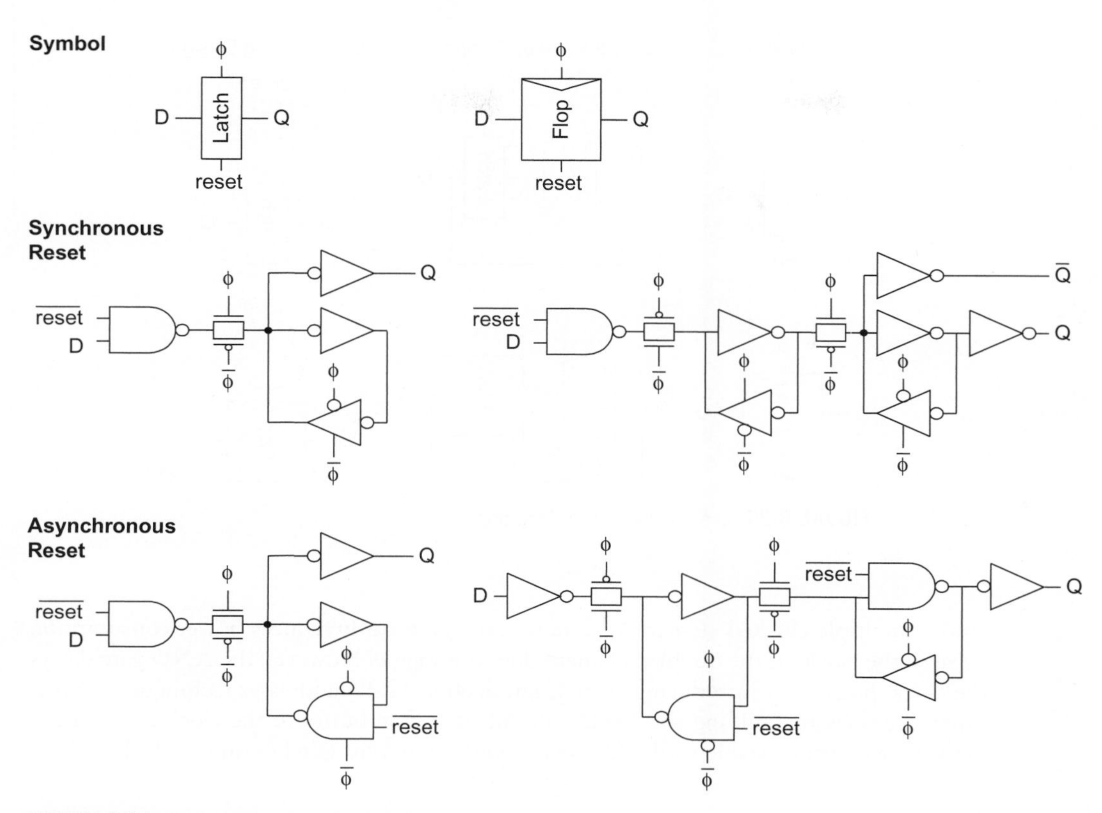

FIGURE 9.24 Resettable latches and flip-flops

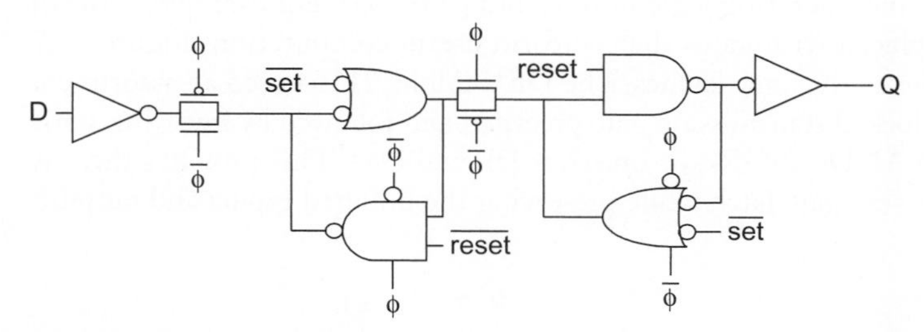

FIGURE 9.25 Flip-flop with asynchronous set and reset

9.3.5 Enabled Latches and Flip-Flops

Sequencing elements also often accept an enable input. When enable *en* is low, the element retains its state independently of the clock. The enable can be performed with an input multiplexer or clock *gating*, as shown in Figure 9.26. The input multiplexer feeds back the old state when the element is disabled. The multiplexer adds area and delay. Clock gating does not affect delay from the data input and the AND gate can be shared

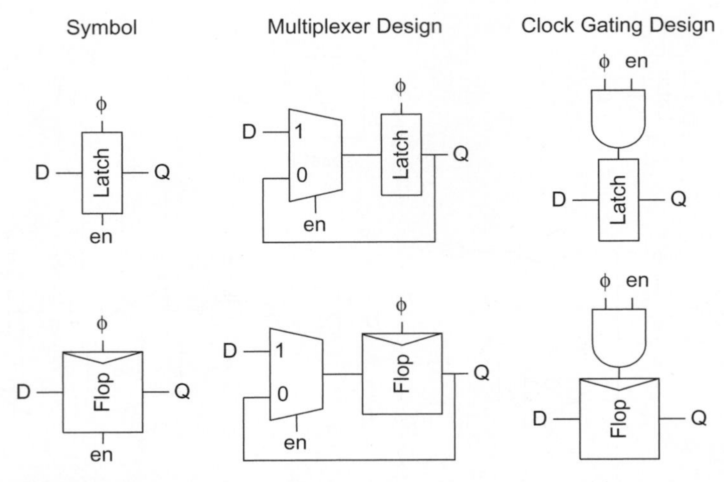

FIGURE 9.26 Enabled latches and flip-flops

among multiple clocked elements. Moreover, it significantly reduces power consumption because the clock on the disabled element does not toggle. However, the AND gate delays the clock, potentially introducing clock skew. Section 12.4.5 addresses techniques to minimize the skew by building the AND gate into the final buffer of the clock distribution network. *en* must be stable while the clock is high to prevent glitches on the clock.

9.3.6 Incorporating Logic into Latches

Since the early days of computing, engineers have recognized that they can reduce sequencing overhead by incorporating logic into latches [Earle65]. For example, some of the inverters can be replaced with gates that perform useful computation. Figure 9.27 shows two ways to do this in dynamic latches. The DEC Alpha 21164 used an assortment of latches built from a clocked transmission gate preceded and followed by inverting static CMOS gates such as NANDs, NORs, or inverters [Bowhill95]. This provides the low overhead of the transmission gate latch while preserving the buffered inputs and outputs.

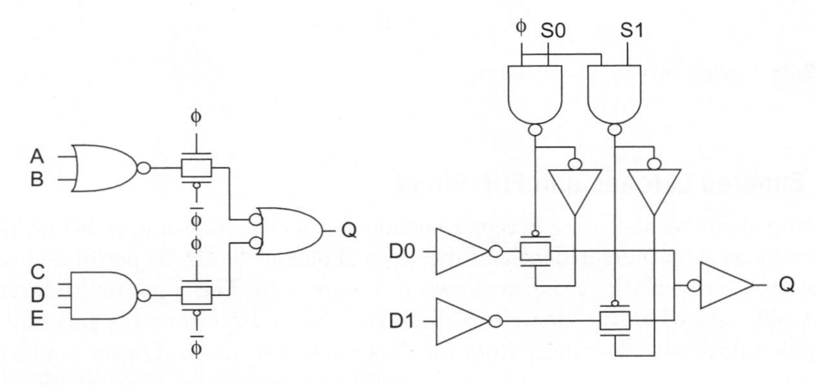

FIGURE 9.27 Combining logic and latches

The *mux-latch* consists of two transmission gates in parallel controlled by clocks gated with the corresponding select signals. It integrates the multiplexer function with no extra delay from the D inputs to the Q outputs except the small amount of extra diffusion capacitance on the state node. Note that the setup time on the select inputs is relatively high. The clock gating will introduce skew unless the clocking methodology systematically plans to gate all clocks. The same principles extend to static latches and flip-flops.

9.3.7 Klass Semidynamic Flip-Flop (SDFF)

The *Klass semidynamic flip-flop* (SDFF) [Klass99] shown in Figure 9.28 is a cross between a pulsed latch and a flip-flop. Like the Partovi pulsed latch, it operates on the principle of intersecting pulses. However, it uses a dynamic NAND gate in place of the static NAND. While the clock is low, X precharges high and Q holds its old state. When the clock rises, the dynamic NAND evaluates. If D is 0, X remains high and the top nMOS transistor turns OFF. If D is 1 and X starts to fall low, the transistor remains ON to finish the transition. This allows for a short pulse and hold time. The dynamic front end serves as the master latch, while the second stage serves as the slave. The weak cross-coupled inverters staticize the flip-flop and the final inverter buffers the output node.

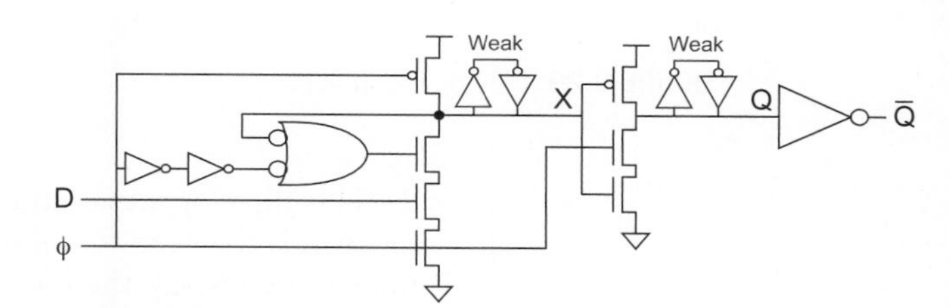

FIGURE 9.28 Klass semidynamic flip-flop

Like a pulsed latch, the SDFF accepts rising inputs slightly after the rising clock edge. Like a flip-flop, falling inputs must set up before the rising clock edge. It is called *semidynamic* because it combines the dynamic input stage with static operation. The SDFF is slightly faster than the Partovi pulsed latch but loses the skew tolerance and time borrowing capability. It also has a higher energy consumption because of the large number of nodes with high activity factors.

The Sun UltraSparc III built logic into the SDFF very efficiently by replacing the single transistor connected to D with a collection of transistors performing the OR or multiplexer functions [Heald00]. The Cell processor similarly employed dynamic mux-latches with up to 4 inputs (plus a fifth input for scan) [Warnock06].

9.3.8 Differential Flip-Flops

Differential flip-flops accept true and complementary inputs and produce true and complementary outputs. They are built from a clocked sense amplifier so that they can rapidly respond to small differential input voltages. While they are larger than an ordinary single-ended flip-flop—having an extra inverter to produce the complementary output—they work well with low-swing inputs such as register file bitlines (Section 11.2.3.3) and low-swing busses (Section 5.4.4).

Figure 9.29(a) shows a differential *sense-amplifier flip-flop* (SA-F/F) receiving differential inputs and producing a differential output [Matsui94]. When the clock is low, the internal nodes X and $\overline{X}$ precharge. When the clock rises, one of the two nodes is pulled down, while the cross-coupled pMOS transistors act as a keeper for the other node. The SR latch formed by the cross-coupled NAND gates behaves as a slave stage, capturing the output and holding it through precharge. The flip-flop can amplify and respond to small differential input voltages, or it can use an inverter to derive the complementary input

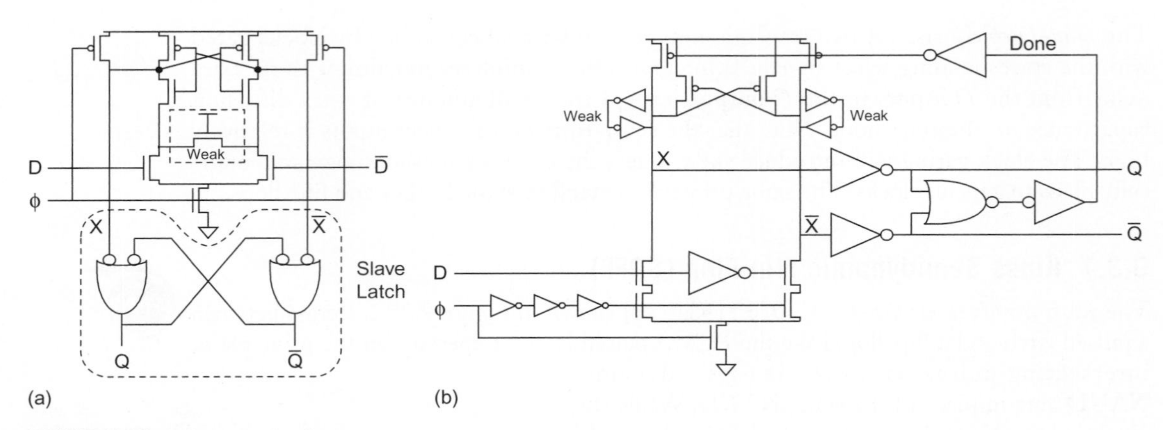

FIGURE 9.29 Differential flip-flops

from *D*. This flip-flop was used in the Alpha 21264 [Gronowski98]. It has a small clock load and avoids the need for an inverted clock. However, the structure is fairly large and consumes more energy than a conventional flip-flop. If the two input transistors are replaced by true and complementary nMOS logic networks, the SA-F/F can also perform logic functions at the expense of greater setup time [Klass99].

The original SA-F/F suffers from the possibility that one of the internal nodes will float low if the inputs switch while the clock is high. The StrongArm 110 processor [Montanaro96] adds the weak nMOS transistor shown in Figure 9.29(a) to fully staticize the flip-flop at the expense of a small amount more internal loading and delay.

Although the sense amplifier stage is fast, the propagation delay through the two cross-coupled NAND gates hurts performance. The NAND gates serve as a slave SR latch and are only necessary to convert the monotonically falling pulsed *X* signals to static *Q* outputs; they can be replaced by HI-skew inverters when *Q* drives domino gates. [Nikolić00], [Kim00], and [Strollo05] all propose alternative slave latch designs that are faster but use more transistors.

The AMD K6 used another differential flip-flop shown in Figure 9.29(b) at the interface from static to self-resetting domino logic [Draper97]. The master stage consists of a self-resetting dual-rail domino gate. Assume the internal nodes are initially precharged. On the rising edge of the clock, one of the two will pull down and drive the corresponding output high. The OR gate detects this and produces a *done* signal that precharges the internal nodes and resets the outputs. Therefore, the flip-flop produces pulsed outputs primarily suitable for use in subsequent self-resetting domino gates. The cross-coupled pMOS transistors improve the noise immunity while the cross-coupled inverters staticize the internal nodes.

9.3.9 Dual Edge-Triggered Flip-Flops

Many researchers have proposed flip-flops that sample data on both the rising and falling edges of the clock to save energy by operating at half the clock frequency. A major drawback is sensitivity to duty cycle variation that increases the skew of the falling clock edge. (The skew from rising edge to rising edge tends to be smaller than the skew from rising edge to falling edge because it involves the same transitions and thus matches better in the face of variation.) To first order, a dual edge-triggered (DET) flip-flop has half the clock frequency and twice the activity factor, so the energy consumed in the flip-flop is unchanged. However, the energy in the global clock distribution network is cut by a factor of two from the reduced frequency. In a well-designed system, the energy is usually dominated by the registers and not by the clock distribution. Moreover, the DET flip-flop tends to have some overhead in area, delay, and energy. The extra skew caused by duty cycle variation further increases the sequencing overhead. By the time the path is modified to recover the extra delay, the net energy savings may be small or negative. Even if the savings are real, DET flip-flops require modifications to timing analysis and other CAD flows. For all these reasons, DET flip-flops have yet to find widespread use in commercial systems.

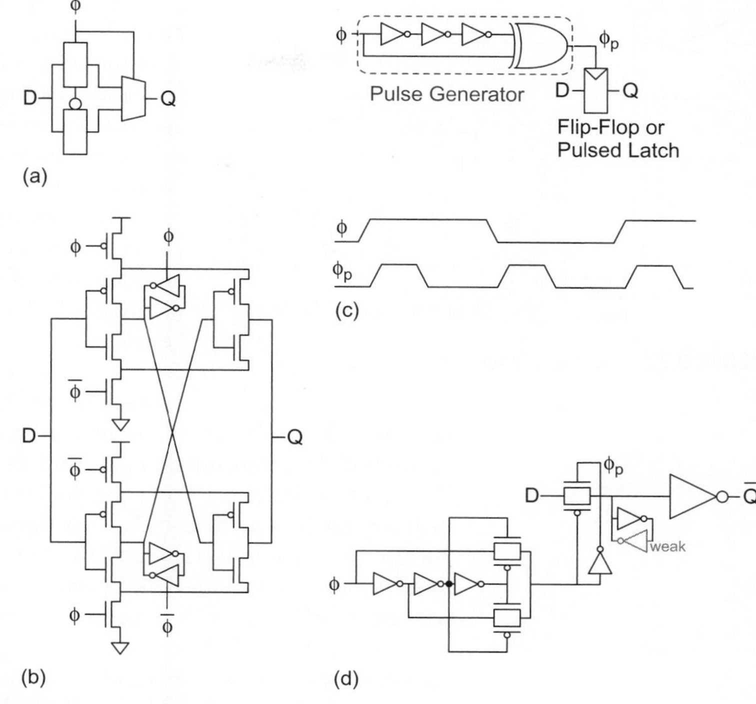

FIGURE 9.30 DET flip-flops

Two conceptual designs for DET flip-flops are shown in Figure 9.30 along with circuit realizations [Tschanz01, Gago93]. In the master-slave design of Figure 9.30(a), two separate master latches operate on opposite phases of the clock. The multiplexer, serving in place of the slave latch, selects the result of the opaque master. Figure 9.30(b) shows a transistor-level implementation of this design. In the pulsed design of Figure 9.30(c), a pulse generator produces a pulse on both edges of the clock. This pulse serves as the clock to an ordinary flip-flop or pulsed latch. Figure 9.30(d) shows a transistor-level design using a pulsed latch and an efficient pulse generator.

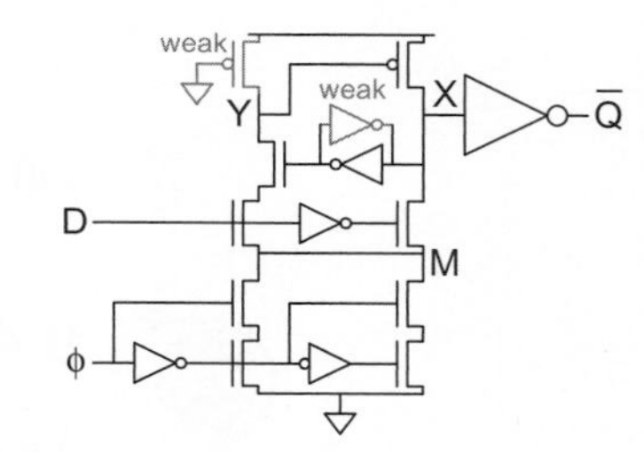

FIGURE 9.31 Zhao implicitly pulsed DET flip-flop

Figure 9.31 shows the Zhao implicitly pulsed DET flip-flop [Zhao07]. In contrast to the explicit pulse generator in Figure 9.30(c), the bottom two pairs of nMOS transistors act as an implicit pulse generator, pulling down node M for a brief interval on the rising and falling edges of the clock. During these intervals, if D is 0, X gets pulled down to 0. If D is 1 and X is 0, Y is briefly pulled down to 0, causing X to rise to 1. For the remainder of the cycle, Y is held at 1 by the weak pMOS transistor and X is held at its current value by the weak inverter. Note that there is a severe ratio constraint: the weak transistors must be overcome by up to four series nMOS transistors.

9.3.10 Radiation-Hardened Flip-Flops

Soft errors caused by alpha particles or cosmic rays were once of primary concern in memories because RAM cells have the smallest node capacitance and weakest feedback, so they are easily disturbed, as discussed in Section 6.3.4. As transistors have scaled, soft error rates for flip-flops have increased to the point that they are important for high-reliability systems. *Radiation-hardened* flip-flops are designed to resist such errors. They are also critically important for space applications where the cosmic ray flux is much greater.

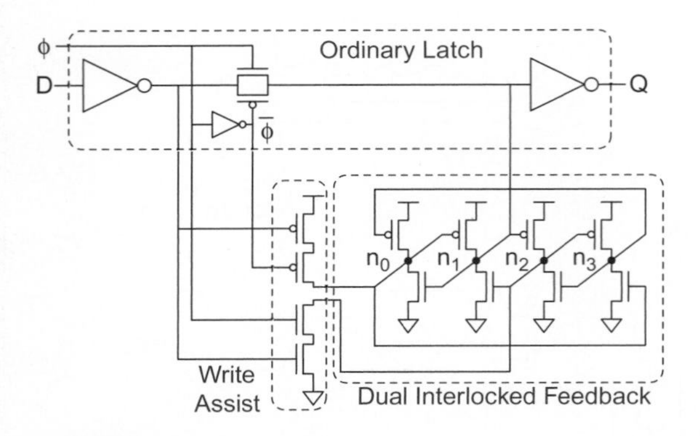

FIGURE 9.32 Radiation-hardened latch

The simplest way to minimize soft errors is to use a storage node holding enough charge that a particle strike is unlikely to flip the state. This has become difficult in nanometer processes because scaling reduces both the capacitance and voltage, greatly decreasing the charge. An unusually large storage node can still reduce the probability of disturbance, but it comes at a cost in performance, energy, and area.

Another option is to use triple-mode redundancy with three registers per bit, and to use majority voting to tolerate an upset in one of the bits (see Section 6.6.2). This is clearly even more costly, but is an effective way of protecting critical state elements.

Figure 9.32 shows a radiation-hardened latch [Stackhouse09, Hazucha04] used on the quad-core Itanium processor. The soft-error resistance is based on the *dual interlocked cell* (DICE) principle [Calin96]. The transmission gate and three inverters at the top form an ordinary latch. The latch is staticized using the dual interlocked feedback circuitry underneath. In an ordinary latch, a particle strike that flipped the state of one of the internal nodes would corrupt the value in the latch. In the DICE approach, nodes n_0 and n_2 normally have the same value as Q. n_1 and n_3 also normally have the complementary value. When the cell is written, n_1 is driven to $\overline{D}$. To prevent contention, the nMOS and pMOS feedback transistors driving n_1 should be turned off during the write. This is performed by the write assist circuit, which ensures $n_2 = 0$ and $n_0 = 1$ during writes. If one of the four state nodes n_0–n_3 is disturbed by a soft error, the interlocked feedback will correct the value. The latch is still vulnerable to radiation strikes that disturb two nodes. Separating the nodes in the cell layout reduces this risk. The quad-core Itanium found that the latch reduced soft errors by two orders of magnitude with no delay penalty at a cost of 34% in area and 25% in power.

The Razor latch discussed in Section 9.4.5 uses a redundant storage node to detect soft errors. In combination with a replay mechanism, it can eliminate these errors.

9.4 Static Sequencing Element Methodology

This section examines a number of issues designers must address when selecting a sequencing element methodology. We begin with general issues, and then proceed to techniques specific to flip-flops, pulsed latches, and transparent latches.

Until the 0.5 μm generation, leakage was relatively low and thus dynamic latches held their state for acceptably long times. The DEC Alpha 21164 was one of the last major microprocessors to use a dynamic latching methodology in a 0.35 μm process in the mid-1990s. It required a minimum operating frequency of 1/10th full speed to retain state, even during testing. Modern systems generally require static sequencing elements to hold state when clocks are gated or the system is tested at a moderate frequency. Leakage is usually worst during burn-in testing at elevated temperature and voltage, where the chip must still function correctly to ensure good toggle coverage. Static elements are larger and somewhat slower than their dynamic counterparts.

Similarly, the growing difficulty and cost of debugging and testing has induced engineers to build design-for-test (DFT) features into the sequencing elements. The most important feature is *scan*, a special mode in which the latches or flip-flops can be chained together into a large shift register so that they can be read and written under external control during testing. This technique is discussed further in Section 14.6.2. Scan has become particularly important because chips have so many metal layers that most internal signals cannot be directly reached with probes. Moreover, some *flip-chips* are mounted upside down, making physical access even more difficult. Scan can dramatically decrease the time required to debug a chip and reduce the cost of testing, so most design methodologies dictate that all sequencing elements must be scannable despite the extra area this entails. The Alpha 21264 did not support full scan and was very difficult to debug, leading to a later-than-desired release.

Clock distribution is another key challenge. As we will see in Section 12.4, it is very difficult to distribute a single clock across a large die in a fashion that gets it to all sequencing elements at nearly the same time. Controlling the clock skew on more than one clock is even more difficult, so almost all modern designs distribute a single high-speed clock in any given region. Other signals such as complementary clocks, pulses, and delayed clocks are generated locally where they are needed. The clock edge rates must be relatively sharp to avoid races in which both the master and slave latches are partially on simultaneously. The global clock may have slow edge rates after propagating along long wires, so it is typically buffered locally (either in each sequencing element or in a buffer cell serving a bank of elements) to sharpen the edge rates. Clock power, from the clock distribution network and the clocked loads, typically accounts for one third to one half of the total chip power consumption. Therefore, clocks are often gated with an AND gate in the local clock buffer to turn off the sequencing elements for inactive units of the chip.

All bistable elements are subject to soft errors from alpha particles or cosmic rays striking the circuits and injecting charge onto sensitive nodes (see Section 6.3.4). Sequencing elements require relatively high capacitance on the state node to achieve low soft error rates. This can set a lower bound on the minimum transistor sizes on that node.

9.4.1 Choice of Elements

Flip-flops, pulsed latches, and transparent latches offer trade-offs in sequencing overhead, skew tolerance, and simplicity.

9.4.1.1 Flip-Flops As we have seen, flip-flops have fairly high sequencing overhead but are popular because they are so simple. Nearly all engineers understand how flip-flops work. Some synthesis tools and timing analyzers handle flip-flops much more gracefully than transparent latches. Most ASIC methodologies use flip-flops exclusively for pipelines and state machines. If performance requirements are not near the cutting edge of a process, flip-flops are clearly the right choice in today's CAD flows.

9.4.1.2 Pulsed Latches Pulsed latches are faster than flip-flops and offer some time- borrowing capability at the expense of greater hold times. They have fewer clocked transistors and hence lower power consumption. If intentional time borrowing is not necessary, you can model a pulsed latch as a flip-flop triggered on the rising edge of the pulse with a lower delay but a lengthy hold time. This makes pulsed latches relatively easy to integrate into flip-flop-based CAD flows. Moreover, the pulsed latches still offer opportunistic time borrowing to compensate for modeling inaccuracies even if the intentional time borrowing is not used. Pulsed latches are used in some microprocessors where their performance justifies the effort managing hold times.

The long hold times make pulsed latches unsuitable for use in pipelines with no logic between pipeline stages. One solution is to use ordinary flip-flops in place of the pulsed latches in these circumstances where speed is not important. Unfortunately, some pulsed latches fan out to multiple paths, some of which are short and others long. The Itanium 2 processor used the *clocked deracer* in conjunction with Naffziger pulsed latches, as shown in Figure 9.33 [Naffziger02]. These were placed before the receiving latches on short paths and block incoming paths while the receiving latch is transparent. They automatically adapt to pulse with variation and hence have a shorter nominal propagation delay than buffers, but also consume more power than buffers because of the clock loading [Rusu03].

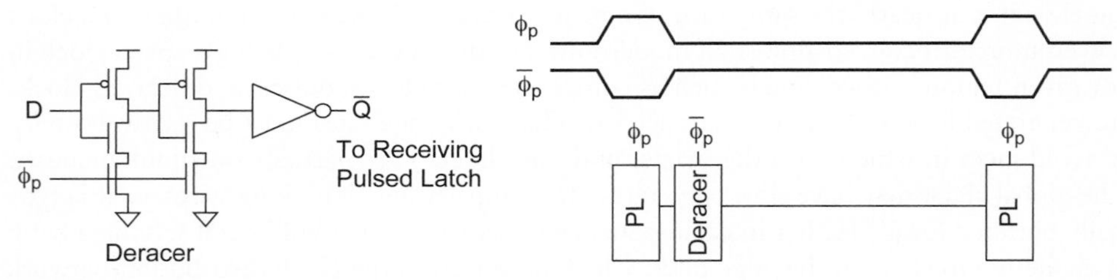

FIGURE 9.33 Clocked deracer

9.4.1.3 Transparent Latches Transparent latches also have lower sequencing overhead than flip-flops and are attractive because they permit nearly half a cycle of time borrowing. One latch must be placed in each half-cycle. Data can arrive at the latch any time the latch is transparent. A convenient design approach is to nominally place the latch at the beginning of each half-cycle. Then time borrowing occurs when the logic in one half-cycle is longer than nominal and data does not arrive at the next latch until some time into the next half-cycle.

Figure 9.34 illustrates pipeline timing for short and long logic paths between latches. When the path is short (a), the data arrives at the second latch early and is delayed until the rising edge of ϕ_2. Therefore, it is natural to consider latches residing at the beginning of their half-cycle because short paths automatically adjust to operate this way. When the

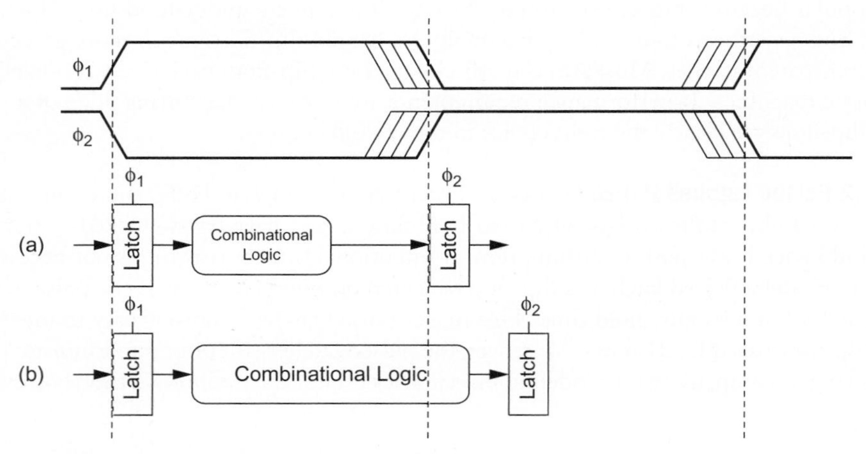

FIGURE 9.34 Latch placement and time borrowing

path is longer (b), it borrows time from the first half-cycle into the second. Notice how clock skew does not slow long paths because the data does not arrive at the latch until after the latest skewed rising edge.

Logic blocks involving multiple signals must ensure that each signal path passes through two latches in each cycle. Signals can be classified as Phase 1 or Phase 2 and logic gates must receive all their inputs from the same phase.

9.4.2 Characterizing Sequencing Element Delays

Previous sections have derived sequencing element performance in terms of the setup and hold times and propagation and contamination delays. These delays are interrelated and are used for budgeting purposes. For example, a flip-flop might still capture its input properly if the data changes slightly less than a setup time before the clock edge. However, the clock-to-*Q* delay might be quite long in this situation. The best way to define these timing parameters is to minimize the overall *D*-to-*Q* delay from when the data must set up until the output is stable. If we call t_{DC} the time that the data actually sets up before the clock edge and t_{CQ} the actual delay from clock to *Q*, we could define t_{setup} as the smallest value of t_{DC} such that $t_{CQ} \leq t_{pcq}$. Moreover, we could choose t_{pcq} to minimize the sequencing overhead $t_{setup} + t_{pcq}$. In this section we will explore how to characterize these delays through simulation.

Figure 9.35 shows the timing of a conventional static edge-triggered flip-flop from Figure 9.19(b). Delays are normalized to an FO4 inverter. The actual *clk*-to-*Q* (t_{CQ}) and *D*-to-*Q* (t_{DQ}) delays for a rising input are plotted against the *D*-to-*clk* (t_{DC}) delay, i.e., how long the data arrived before the clock rises. If the data arrives long before the clock, t_{CQ} is short and essentially independent of t_{DC} delay. $t_{DQ} = t_{DC} + t_{CQ}$, so it increases linearly as data arrives earlier because the data is blocked and waits for the clock before proceeding. As the data arrives closer to the clock, t_{CQ} begins to rise. However, t_{DQ} initially decreases and reaches a minimum when t_{CQ} has a slope of -1 (note that the axes are not to scale).

Therefore, let us define the setup time t_{setup} as t_{DC} at which this minimum t_{DQ} occurs and the propagation delay t_{pcq} as t_{CQ} at this time. The contamination delay t_{ccq} is the minimum t_{CQ} that occurs when the input arrives early. The hold time is the minimum delay from clock to *D* changing such that the $t_{CQ} \leq t_{pcq}$.

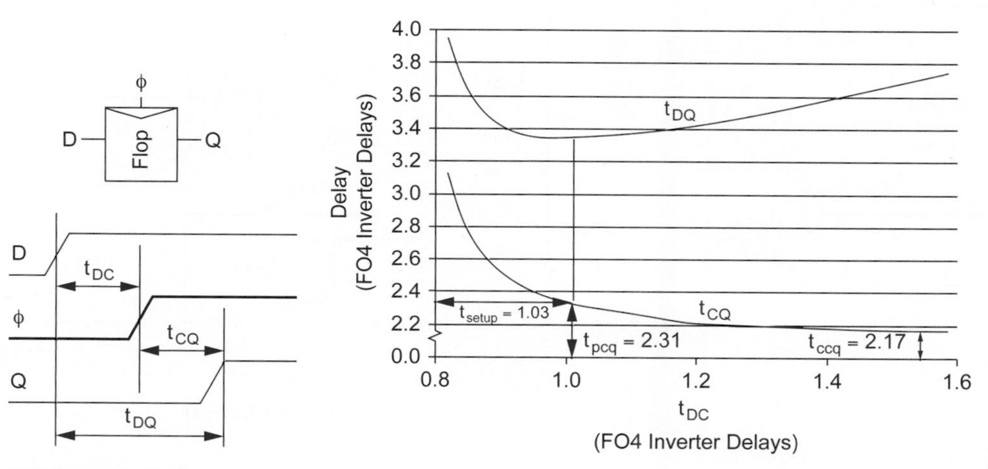

FIGURE 9.35 Flip-flop delay vs. data arrival time

In general, the delays will differ for inputs of 0 and 1. Figure 9.36 plots t_{CQ} vs. t_{DC} for the four combinations of rising and falling D and Q. The setup times t_{setup0} and t_{setup1} are the times that D must fall or rise, respectively, before the clock so that the data is properly captured with the least possible t_{DQ}. Observe that this flip-flop has a longer setup time but shorter propagation delay for low inputs than high inputs. The hold times t_{hold0} and t_{hold1} are the times that D must rise or fall, respectively, after the clock so that the old value of 0 or 1 is captured instead of the new value. Observe that the hold times are typically negative. The contamination delay $t_{ccq0/1}$ again is the lowest possible t_{CQ} and occurs when the input changes well before the clock edge. When only one delay is quoted for a flip-flop timing parameter, it is customarily the worst of the 0 and 1 delays.

The *aperture width* t_a is the width of the window around the clock edge during which the data must not transition if the flip-flop is to produce the correct output with a propagation delay less than t_{pcq}. The aperture times for rising and falling inputs are

$$
\begin{aligned}
t_{ar} &= t_{\text{setup1}} + t_{\text{hold0}} \\
t_{af} &= t_{\text{setup0}} + t_{\text{hold1}}
\end{aligned}
\tag{9.20}
$$

If the data transitions within the aperture, Q can become metastable and take an unbounded amount of time to settle. Metastability is discussed further in Section 9.5.1.

If D is a very short pulse, the flip-flop may fail to capture it even if D is stable during the setup and hold times around the rising clock edge. Similarly, if the clock pulse is too short, the flip-flop may fail to capture stable data. Well-characterized libraries sometimes

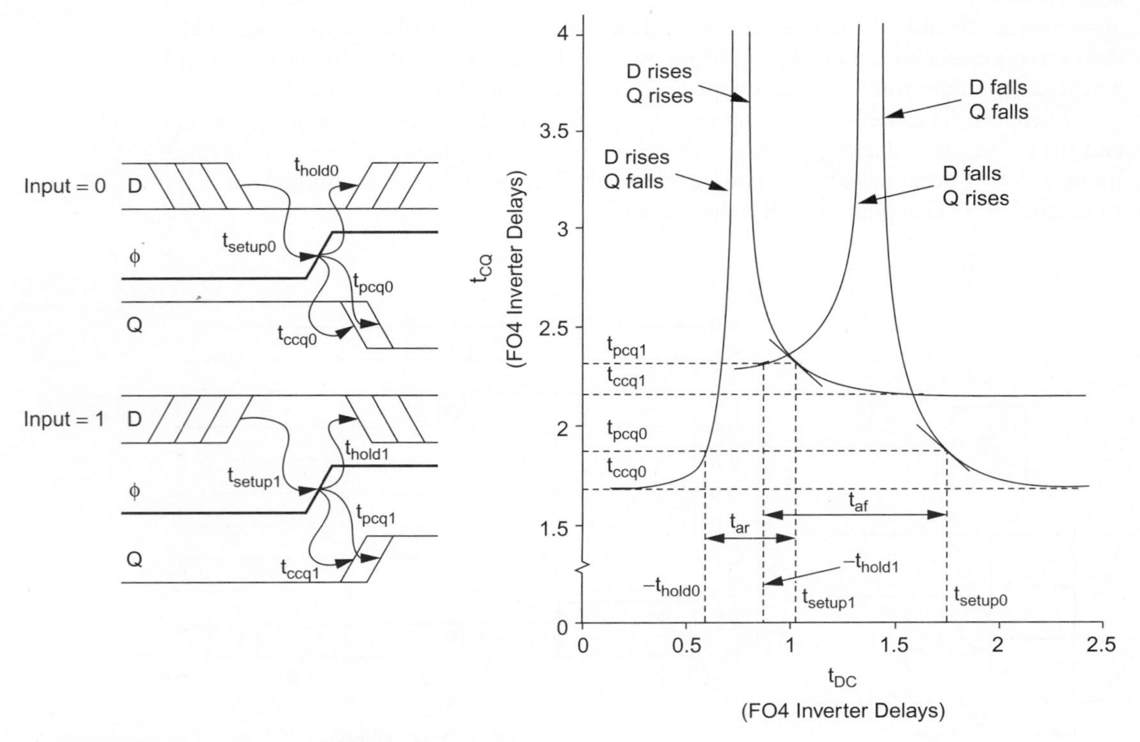

FIGURE 9.36 Flip-flop setup and hold times

specify minimum pulse widths for the clock and/or data as well as setup and hold times.

Level-sensitive latches have somewhat different timing constraints because of their transparency, as shown in Figure 9.37 for a conventional static latch from Figure 9.17(g) using a pulse width of 4 FO4 inverter delays. As with an edge-triggered flip-flop, if the data arrives before the clock rises ($t_{DCr} > 0$), it must wait for the clock. In this region, the clock-to-Q t_{CrQ} delay is nearly constant and t_{DQ} increases as the data arrives earlier. If the data arrives after the clock rises while the latch is transparent, t_{DQ} is essentially independent of the arrival time. The data must set up before the falling edge of the clock. The second set of labels on the X-axis indicates the D-to-clk fall time t_{DCf}. As the data arrives too close to the falling edge, t_{DQ} increases. Now, to achieve low t_{DQ}, we choose the setup time before the knee of the curve, e.g., 5% greater than its minimum value. The setup time is measured relative to the falling edge of the clock. If the data changes less than a hold time after the falling edge of the clock, Q may momentarily glitch. Thus, the hold time t_{hold} for a latch is defined to be $-t_{DCf}$ for which Q displays a negligible glitch.

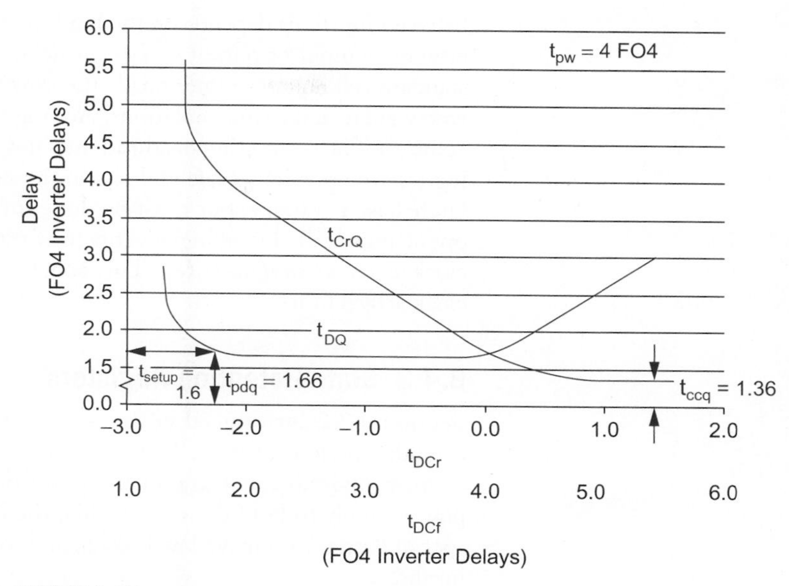

FIGURE 9.37 Latch delay vs. data arrival time

Pulsed latches have setup and hold times measured around the falling edge of the clock. However, designers often wish to treat pulsed latches as edge-triggered flip-flops from the perspective of timing analysis. Therefore, we can define "virtual" setup and hold times relative to the rising clock edge [Stojanovic99]. For example, the pulsed latch in Figure 9.37 has $t_{\text{setup-virtual}} = t_{\text{setup}} - t_{pw} = -2.4$ FO4 but $t_{pcq\text{-virtual}} = t_{pdq} + (t_{pw} - t_{\text{setup}}) =$ 4.06 FO4, so the total sequencing overhead of $t_{pdq} = t_{\text{setup-virtual}} + t_{pcq\text{-virtual}}$ is unaffected by the change of reference or pulse width. The virtual hold time is now $t_{\text{hold-virtual}} = t_{\text{hold}} + t_{pw} = 2.6$ FO4, which is positive as one should expect because the input must hold long after the rising edge of the clock.

The delays vary with input slope, voltage, and temperature. The contamination delay should be measured in the environment where it is shortest while the setup and hold times and propagation delay should be measured in the environment where they are longest.

The designer can trade off setup time, hold time, and propagation delay. Figure 9.38 shows the effects of adding delay t_{buf} to the clock, D, or Q terminals of a flip-flop. Recall that the sequencing overhead depends on the sum of the setup time and propagation delay while the minimum delay

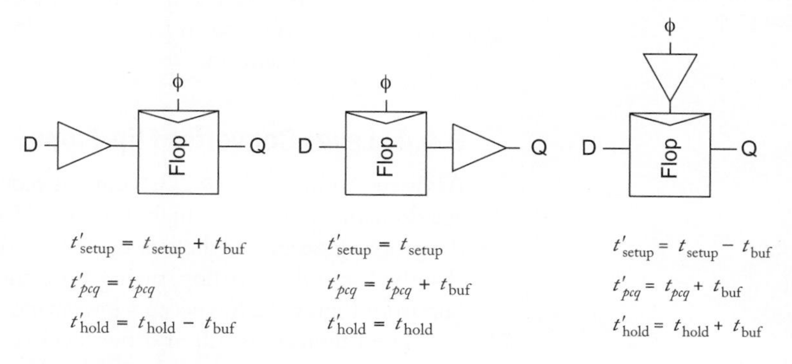

FIGURE 9.38 Delay trade-offs

between flip-flops depends on the hold time less the contamination delay. Adding delay on either the input or output eases min-delay at the expense of sequencing overhead. Many standard cell libraries intentionally use slow flip-flops so that logic designers do not have to worry about hold-time violations. Adding delay on the clock simply shifts when the flop activates. The sequencing overhead does not change, but the system can accommodate more logic in the previous cycle and less in the next cycle. This is similar to time borrowing in latch-based systems, but must be done intentionally by adjusting the clock rather than opportunistically by taking advantage of transparency. Some authors refer to delaying the clock as *intentional clock skew*. This book reserves the term *clock skew* for uncertainty in the clock arrival times.

9.4.3 State Retention Registers

Section 4.3.2 introduced power gating to save leakage power while a unit is idle for extended periods of time. The unit must either reinitialize itself when power is reapplied or must maintain its state during powerdown. *State retention registers* receive a second power supply to hold their state while the rest of the unit is powered down. They require special design to achieve low leakage and to prevent corruption when their inputs become invalid.

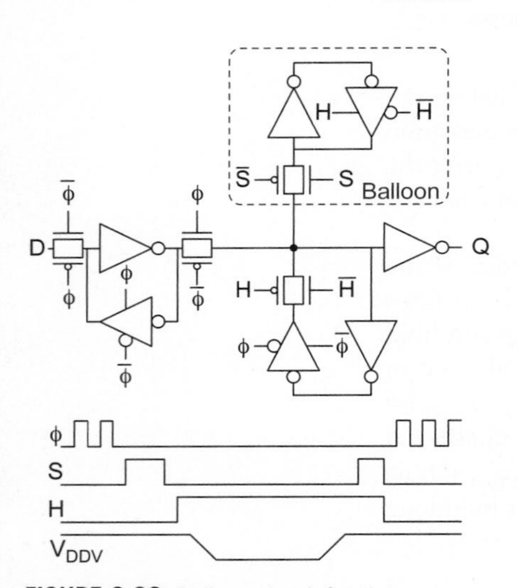

FIGURE 9.39 Balloon circuit for state retention

Figure 9.39 shows a flip-flop with a balloon circuit for state retention [Shigematsu97]. The cross-coupled inverters in the balloon circuit use low-leakage transistors connected to a separate power supply to hold state while the power is gated to the remainder of the flip-flop. The balloon circuit typically uses minimum-sized high-V_t and/or thick oxide transistors to minimize leakage; for example, I/O transistors typically have these properties and are available at no extra cost. The control signals SAMPLE (S) and HOLD (H) are 0 during normal operation, as shown in the timing diagram. When the unit is about to be power-gated, ϕ stops low with the slave latch opaque. SAMPLE is pulsed for long enough to write the state into the balloon (potentially a long time if the transistors are particularly slow). Then HOLD is asserted to retain the state. Now, the virtual power rail can be deactivated to power down the unit. Even if the clock or other latch control signals such as reset or data toggle during powerdown, the copy of the state will be safely stored in the balloon. When the unit powers back up, SAMPLE is pulsed again to copy the state from the balloon back to the slave latch. Then HOLD is deasserted and finally the unit can restart ϕ and resume normal operation. The same balloon circuit could be attached to the state node of a transparent latch or pulsed latch for state retention.

9.4.4 Level-Converter Flip-Flops

As discussed in Section 4.2.3.1, circuits require level converters when crossing between voltage domains from low to high. Figure 4.15 showed a standard differential level converter. If the crossing occurs on a clock cycle boundary, the overhead of the level converter can be absorbed into the flip-flop, saving time and energy. For example, the sense-amplifier flip-flop from Figure 9.29(a) accepts low-swing inputs.

The literature is full of other level-converter flip-flops. The general principle is that the low-swing inputs should only drive nMOS transistors or pass transistors because they cannot fully turn OFF pMOS transistors connected to V_{DDH}. Figure 9.40 shows an assortment of approaches. The blue inverters and tristates use V_{DDL}; the other gates use V_{DDH}. Both D and ϕ may use V_{DDL} levels. Figure 9.40(a) shows a flip-flop with a pair of slave latches connected to a differential level converter [Hamada98]. The cross-coupled nMOS transistors serve to staticize the slave latches. Figure 9.40(b) shows a simple latch level converter [Usami95]. The cross-coupled inverters perform level restoration as well as staticizing the latch. They must be weak enough to be overcome by the nMOS pulldown stacks. Figure 9.40(c) shows Zhao's implicitly pulsed level converter [Zhao09]. It is similar to the implicitly pulsed DET flip-flop from Figure 9.31. [Zhao09] and [Ishihara04] survey a variety of other designs. However, commercial designs still tend to use standard flip-flops and differential level converters.

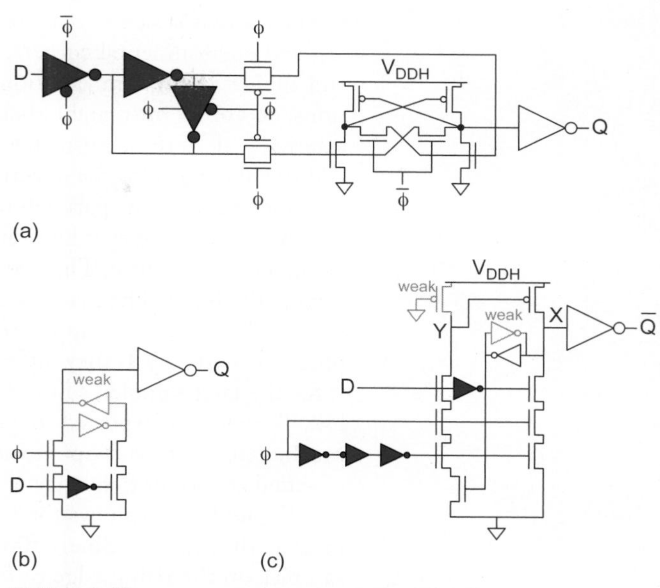

FIGURE 9.40 Level-converter flip-flops and latches

9.4.5 Design Margin and Adaptive Sequential Elements

Sequential circuits require some margin in voltage or frequency to ensure that they work reliably despite variations. All considered, the margin forces designers to derate performance or power by 30% or more from what could be achieved under TT processing and nominal operating conditions.[1] *Adaptive* (or *variation-tolerant*) *sequential elements* seek to reduce this margin by measuring and compensating for the variation.

Dynamic voltage scaling is a particularly good application for adaptive sequential elements because the voltage-frequency trade-off must be made at multiple operating points. The problem can be viewed as selecting the minimum voltages necessary to achieve each of several frequency targets, although an equivalent dual problem is selecting the maximum frequencies the part can work at each of several voltage points. The simplest approach is to precharacterize the chip and create a table of voltage-frequency pairs that are guaranteed to work even under worst case variation. This is a common technique in commercial microprocessors because it is simple to build and easy to test, but it requires the most conservative margins [Stackhouse09]. By measuring the temperature, voltage droop, and/or supply current and providing these to the lookup table, the margins can be relaxed somewhat [Tschanz07].

An adaptive approach introduced in Section 6.5.3.6 is to build a delay chain that mimics the worst case path on the chip and to use that delay to set the operating frequency. This is called a *canary circuit*: in the same way that miners sent a canary into the tunnel to see if the air is safe to breathe, the chip uses the canary circuit to determine the

[1]For example, some PC enthusiasts enjoy trying to recoup some of this performance by overclocking their CPUs, taking advantage of the fact that the processing is likely better than worst case. They often use a fancy heat sink to keep the operating temperature below worst case, then crank up the supply voltage to achieve even higher performance. And occasionally they burn out their CPUs by overstressing them at high voltage and/or temperature.

frequency that is safe to operate [Calhoun04]. The canary circuit tracks with the processing and environmental corners, so some of the margin can be eliminated. However, it is still subject to random variations, process tilt, within-die voltage and temperature variations, and other mismatches between the canary circuit and the true critical paths. Characterizing all of these mismatch sources is difficult, so a conservative designer will provide additional margin for the uncertainty. Better yet, the amount of margin can be adjusted at runtime to ensure the part will function at some speed.

A fascinating recent innovation is to let the circuits themselves indicate when they are at the edge of failure. This can be done by modifying sequential elements to double-sample the input. The main path through the sequential element is unchanged, but a secondary checking path samples the input slightly later. If the two results agree, the circuit is operating correctly. If they differ, the data missed its setup time at the main path but made it for the later sampler, so the frequency is slightly too high or the voltage is slightly too low. This error is reported to a system controller. If the system is designed with a replay mechanism to repeat operations from a last known good state, the operation can be repeated at a lower frequency or higher voltage where it works correctly.

Figure 9.41(a) shows the basic concept of the *Razor flip-flop* [Ernst03, Das06]. The main path uses an ordinary flip-flop, while the checking path uses a latch. The flip-flop samples on the rising edge of ϕ_p, while the latch samples some time later on the falling edge of ϕ_p. Figure 9.41(b) illustrates the operation of the circuit. If the data arrives at least a setup time before the rising edge of ϕ_p, both elements sample the same value. If the data arrives late, the flip-flop misses the data and the XOR generates an *ERR* signal. The *ERR* signals from all the flip-flops in the system (or at least those on potentially critical paths) are ORed together to indicate an error and trigger the replay mechanism.

The operating voltage and frequency are adjusted until the system is barely working so that very little margin is provided: the circuit is functioning "on the razor's edge." Variations such as power supply noise, unusually large crosstalk, or even activation of a rarely

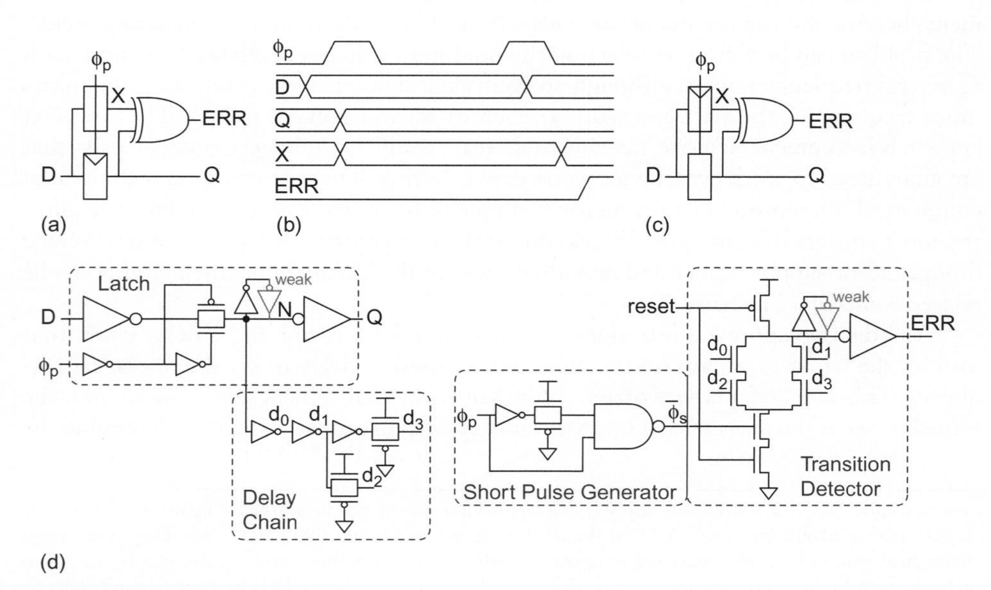

FIGURE 9.41 Adaptive sequencing elements

triggered critical path, are sufficient to delay the arrival of *D* and cause an occasional error. The width of the clock pulse presents a trade-off between error detection and hold time. Wider pulses allow later inputs to be detected as errors, which increases the allowable difference between typical and worst-case delay. However, the hold time increases with the pulse width, just like a pulsed latch. Managing long hold times is difficult, so a relatively narrow pulse (e.g., < 3 FO4 delays) is preferable.

The Razor circuit has the drawback that the flip-flop may become metastable if *D* changes during the aperture. If *Q* resolves to the same value as the latch, no error will be flagged, but the propagation through the flip-flop can increase by an unbounded amount of time. [Ernst03] suggests adding a metastability detector, which significantly increases the overhead of the circuit.

Figure 9.41(c) shows an improved structure called *Double Sampling with Time Borrowing* (DSTB) that moves metastability out of the data path and onto the error path [Bowman09]. If the data arrives slightly late, the pulsed latch will still capture it correctly. The flip-flop will either miss it, causing *ERR* to rise and signaling that the system is near the edge of failure, or will become metastable. Assuming that the error path has plenty of slack, the metastability can resolve before *ERR* is sampled.

Figure 9.41(d) shows the *Razor II* pulsed latch [Das09], which consists of an ordinary pulsed latch, a short pulse generator, and a transition detector. The short pulse generator produces a brief downgoing pulse when the latch becomes transparent. The transition detector signals an error if any changes are observed outside this brief pulse. The transition detector uses a dynamic XOR structure precharged by the *reset* signal, which must be reapplied after each error is detected. The short pulse width sets the time borrowing, the long pulse width sets the hold time, and the difference sets the detection window during which delay errors can be detected.

In addition to detecting late data, these adaptive sequencing elements can detect soft errors. A particle strike that corrupts the latch or flip-flop will trigger the *ERR* signal. A particle strike that induces a glitch in the combinational logic is only significant if it causes the sequential element to capture the wrong value. As long as the detection window is longer than the glitch, *ERR* will also rise. The replay mechanism can then be used to recompute the result correctly.

9.5 Synchronizers

Sequencing elements are characterized by their setup and hold time. If the data input changes before the setup time, the output reflects the new value after a bounded propagation delay. If the data changes after the hold time, the output reflects the old value after a bounded propagation delay. If the data changes during the *aperture* between the setup and hold times, the output may be unpredictable and the time for the output to settle to a good logic level may be unbounded. Properly designed synchronous circuits guarantee the data is stable during the aperture. However, many interesting systems must interface with data coming from sources that are not synchronized to the same clock. For example, the user may press a key at any time and data coming over a network may be aligned with a clock of differing phase or frequency.

A *synchronizer* is a circuit that accepts an input that can change at arbitrary times and produces an output aligned to the synchronizer's clock. Because the input can change during the synchronizer's aperture, the synchronizer has a nonzero probability of producing a *metastable* output [Chaney73]. This section first examines the response of a latch to an analog voltage that can change near the sampling clock edge. The latch can enter a metastable state for some amount of time that is unbounded, although the probability of remaining metastable drops off exponentially with time. Therefore, you can build a simple synchronizer by sampling a signal, waiting until the probability of metastability is acceptably low, then sampling again. In certain circumstances, the relationship of the data and clock timing is more predictable, permitting faster and more reliable synchronizers.

9.5.1 Metastability

A latch is a bistable device; i.e., it has two stable states (0 and 1). Under the right conditions, that latch can enter a metastable state in which the output is at an indeterminate level between 0 and 1. For example, Figure 9.42 shows a simple model for a static latch consisting of two switches (probably transmission gates in practice) and two inverters. While the latch is transparent, the sample switch is closed and the hold switch open (Figure 9.42(a)). When the latch goes opaque, the sample switch opens and the hold switch closes (Figure 9.42(b)). Figure 9.42(c) shows the DC transfer characteristics of the two inverters. Because $A = B$ when the latch is opaque, the stable states are $A = B = 0$ and $A = B = V_{DD}$. The metastable state is $A = B = V_m$, where V_m is an invalid logic level. This point is called *metastable* because the voltages are self-consistent and can remain there indefinitely. However, any noise or other disturbance will cause A and B to switch to one

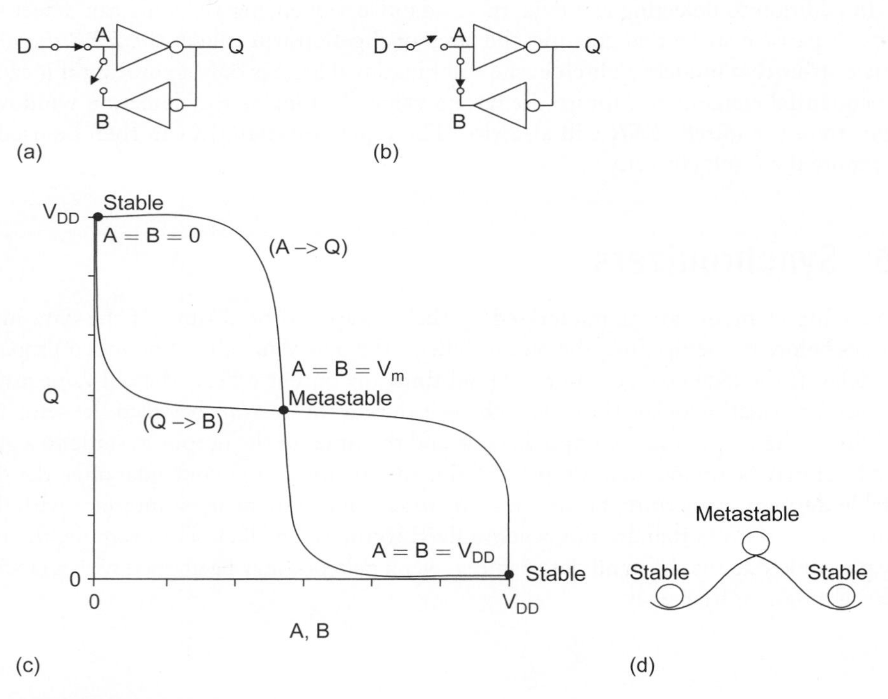

FIGURE 9.42 Metastable state in static latch

of the two stable states. Figure 9.42(d) shows an analogy of a ball delicately balanced on a hill. The top of the hill is a metastable state. Any disturbance will cause the ball to roll down to one of the two stable states on the left or right side of the hill.

Figure 9.43(a) plots the output of the latch from Figure 9.17(g) as the data transitions near the falling clock edge. If the data changes at just the wrong time t_m within the aperture, the output can remain at the metastable point for some time before settling to a valid logic level. Figure 9.43(b) plots t_{DQ} vs. $t_{DC} - t_m$ on a semilogarithmic scale for a rising input and output. The delay is less than or equal to t_{pdq} for inputs that meet the setup time and increases for inputs that arrive too close to t_m. The points marked on the graph will be used in the example at the end of this section.

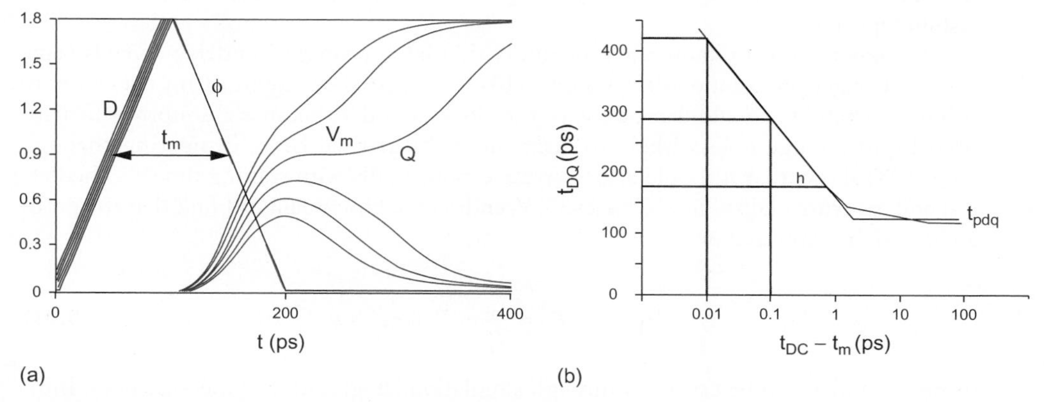

FIGURE 9.43 Metastable transients and propagation delay

The cross-coupled inverters behave like a linear amplifier with gain G when A is near the metastable voltage V_m. The inverter delay can be modeled with an output resistance R and load capacitance C. We can predict the behavior in metastability by assuming that the initial voltage on node A when the latch becomes opaque at time $t = 0$ is

$$A(0) = V_m + a(0) \tag{9.21}$$

where $a(0)$ is a small signal offset from the metastable point. Figure 9.44 shows a small-signal model for $a(t)$. The behavior after time 0 is given by the first-order differential equation

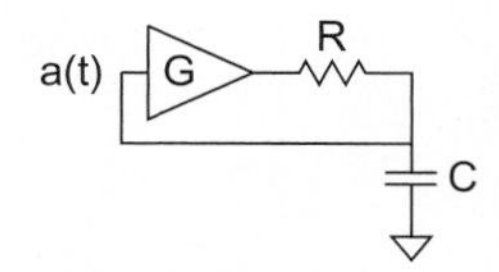

FIGURE 9.44 Small signal model of bistable element in metastability

$$\frac{Ga(t) - a(t)}{R} = C\frac{da(t)}{dt} \tag{9.22}$$

Solving this equation shows that the positive feedback drives $a(t)$ exponentially away from the metastable point with a time constant determined by the gain and RC delay of the cross-coupled inverter loop.

$$a(t) = a(0)e^{\frac{t}{\tau_s}}; \; \tau_s = \frac{RC}{G-1} \tag{9.23}$$

Suppose the node is defined to reach a legal logic level when $|a(t)|$ exceeds some deviation ΔV. The time to reach this level is

$$t_{DQ} = \tau_s \left[\ln \Delta V - \ln a(0) \right] \quad \textbf{(9.24)}$$

This shows that the latch propagation delay increases as $A(0)$ approaches the metastable point and $a(0)$ approaches 0. The delay approaches infinity if $a(0)$ is precisely 0, but this can never physically happen because of noise. However, there is no upper bound on the possible waiting time t required for the signal to become valid. If the input $A(t)$ is a ramp that passes through V_m at time t_m, $a(0)$ is proportional to $t_{DC} - t_m$. Observe that EQ (9.24) is a good fit to the log-linear portion of Figure 9.43(b). The time constant τ_s is essentially the reciprocal of the gain-bandwidth product [Flannagan85]. Therefore, the feedback loop in a latch should have a high gain-bandwidth product to resolve from metastability quickly.

Designers need to know the probability that latch propagation delay exceeds some time t'. Longer propagation delays are less likely because they require $a(0)$ to be closer to 0. This probability should decrease with the clock period T_c because a uniformly distributed input change is less likely to occur near the critical time. Projecting through EQ (9.24) shows that it should also decrease exponentially with waiting time t'. Theoretical and experimental studies [Chaney83, Veendrick80, Horstmann89] find that the probability can be expressed as

$$P\left(t_{DQ} > t'\right) = \frac{T_0}{T_c} e^{-\frac{t'}{\tau_s}} \quad \text{for } t' > h \quad \textbf{(9.25)}$$

where T_0 and τ_s can be extracted through simulation [Baghini02] or measurement. Intuitively, T_0/T_c describes the probability that the input would change during the aperture, causing metastability, and the exponential term describes the probability that the output has not resolved after t' if it did enter metastability. The model is only valid for sufficiently long propagation delays (h significantly greater than t_{pdq}).

Example 9.8

Find τ_s, T_0, and h for the latch using the data in Figure 9.43.

SOLUTION: h is the propagation delay above which the data fits a good straight line on a log-linear scale. In Figure 9.43, this appears to be approximately 175 ps. The probability that the delay exceeds some t' is the chance that the input changing at a random time falls within the small aperture that leads to the high delay. We can choose two points on the linear portion of the plot and solve for the two unknowns. For example, choosing (0.1 ps, 290 ps) and (0.01 ps, 415 ps), we solve

$$\begin{aligned} P\left(t_{DQ} > 290 \text{ ps}\right) &= \frac{0.1 \text{ ps}}{T_c} = \frac{T_0}{T_c} e^{-\frac{290 \text{ ps}}{\tau_s}} \\ P\left(t_{DQ} > 415 \text{ ps}\right) &= \frac{0.01 \text{ ps}}{T_c} = \frac{T_0}{T_c} e^{-\frac{415 \text{ ps}}{\tau_s}} \end{aligned} \quad \textbf{(9.26)}$$

T_c drops out of the equations and we find $\tau_s = 54$ ps and $T_0 = 21$ ps. Recall that this data was taken for a rising input. A conservative design should also consider the falling input and take data in the slow rather than typical environment.

We have seen that a good synchronizer latch should have a feedback loop with a high-gain-bandwidth product. Conventional latches have data and clock transistors in series, increasing the delay (i.e., reducing the bandwidth). Figure 9.45 shows a synchronizer flip-flop in which the feedback loops simplify to cross-coupled inverter pairs [Dike99]. Furthermore, the flip-flop is reset to 0, and then is only set to 1 if $D = 1$ to minimize loading on the feedback loop.

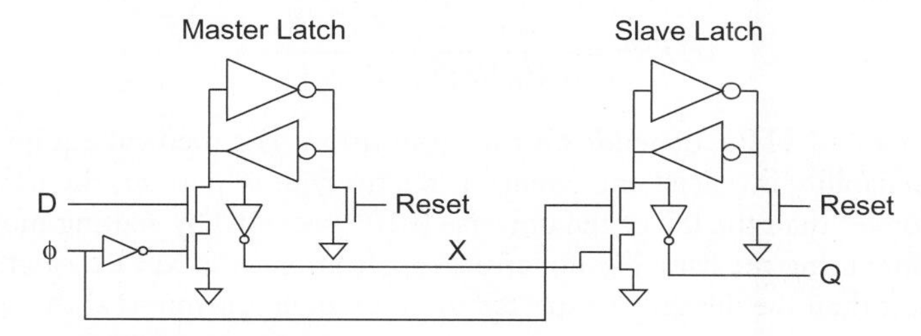

FIGURE 9.45 Fast synchronizer flip-flop

The flip-flop consists of master and slave jamb latches. Each latch is reset to 0 while $D = 0$. When D rises before ϕ, the master output X is driven high. This in turn drives the slave output Q high when ϕ rises. The pulldown transistors are just large enough to overpower the cross-coupled inverters, but should add as little stray capacitance to the feedback loops as possible. X and Q are buffered with small inverters so they do not load the feedback loops.

9.5.2 A Simple Synchronizer

A synchronizer accepts an input D and a clock ϕ. It produces an output Q that ought to be valid some bounded delay after the clock. The synchronizer has an aperture defined by a setup and hold time around the rising edge of the clock. If the data is stable during the aperture, Q should equal D. If the data changes during the aperture, Q can be chosen arbitrarily. Unfortunately, it is impossible to build a perfect synchronizer because the duration of metastability can be unbounded. We define synchronizer failure as occurring if the output has not settled to a valid logic level after some time t'.

Figure 9.46 shows a simple synchronizer built from a pair of flip-flops. $F1$ samples the asynchronous input D. The output X may be metastable for some time, but will settle to a good level with high probability if we wait long enough. $F2$ samples X and produces an output Q that should be a valid logic level and be aligned with the clock. The synchronizer has a latency of one clock cycle, T_c. It can fail if X has not settled to a valid level by a setup time before the second clock edge.

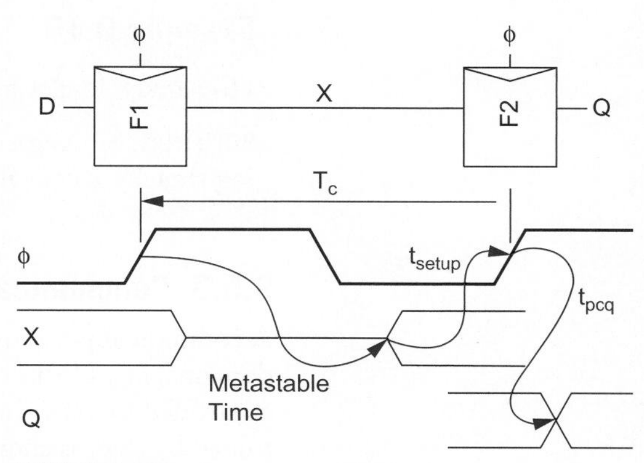

FIGURE 9.46 Simple synchronizer

Each flip-flop samples on the rising clock edge when the master latch becomes opaque. The slave latch merely passes along the contents of the master and does not significantly affect the probability of metastability. If the synchronizer receives an average of N asynchronous input changes at D each second, the probability of synchronizer failure in any given second is

$$P(\text{failure}) = N\frac{T_0}{T_c}e^{\frac{-(T_c - t_{\text{setup}})}{\tau_s}} \tag{9.27}$$

and the mean time between failures increases exponentially with cycle time

$$MTBF = \frac{1}{P(\text{failure})} = \frac{T_c e^{\frac{T_c - t_{\text{setup}}}{\tau_s}}}{NT_0} \tag{9.28}$$

The acceptable MTBF depends on the application. For medical equipment where synchronizer reliability is crucial and latency is relatively unimportant, the MTBF can be chosen to be longer than the life of the universe (~10^{19} seconds) by waiting more than one clock cycle before using the data. For noncritical applications, the MTBF can be chosen to be merely longer than the designer's expected duration of employment at the company!

Example 9.9

A particular synchronizer flip-flop in a 0.25 μm process has $\tau_s = 20$ ps and $T_0 = 15$ ps [Dike99]. Assuming the input toggles at $N = 50$ MHz and the setup time is negligible, what is the minimum clock period T_c for which the MTBF exceeds one year?

SOLUTION: 1 year $\approx \pi \times 10^7$ seconds. Thus, we must solve

$$\pi \times 10^7 = \frac{T_c e^{\frac{T_c}{20 \times 10^{-12}}}}{(5 \times 10^7)(15 \times 10^{-12})} \tag{9.29}$$

numerically for a minimum clock period of 625 ps (1.6 GHz).

Example 9.10

How much longer must we wait for a 1000-year MTBF?

SOLUTION: Solving an equation similar to EQ (9.29) gives 760 ps. Increasing the waiting time by 135 ps improved MTBF by a factor of 1000.

9.5.3 Communicating Between Asynchronous Clock Domains

A common application of synchronizers is in communication between asynchronous clock domains, i.e., blocks of circuits that do not share a common clock. Suppose System A is controlled by *clkA* that needs to transmit N-bit data words to System B, which is controlled by *clkB*, as shown in Figure 9.47. The systems can represent separate chips or separate units within a chip using unrelated clocks. Each word should be received by system B exactly once. System A must guarantee that the data is stable while the flip-flops in System B sample the word. It indicates when new data is valid by using a request signal (*Req*), so System B receives the word exactly once rather than zero or multiple times. System B replies with an acknowledge signal (*Ack*) when it has sampled the data so System A knows when the data can safely be changed. If the relationship between *clkA* and *clkB* is completely unknown, a synchronizer is required at the interface.

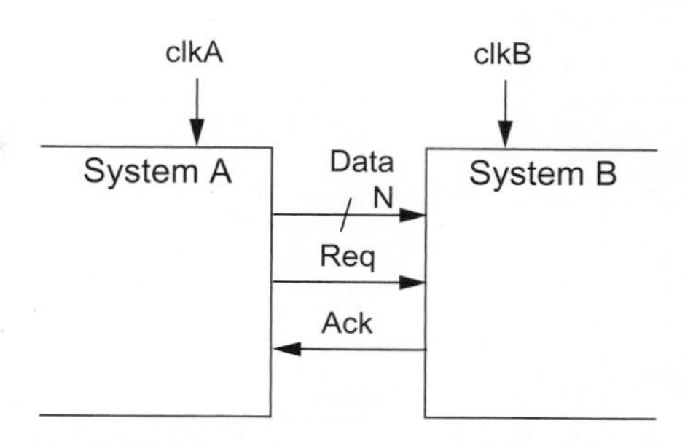

FIGURE 9.47 Communication between asynchronous systems

The request and acknowledge signals are called *handshaking* lines. Figure 9.48 illustrates two-phase and four-phase handshaking protocols. The four-phase handshake is level-sensitive while the two-phase handshake is edge-triggered. In the four-phase handshake, system A places data on the bus. It then raises *Req* to indicate that the data is valid. System B samples the data when it sees a high value on *Req* and raises *Ack* to indicate that the data has been captured. System A lowers *Req*, then system B lowers *Ack*. This protocol requires four transitions of the handshake lines. In the two-phase handshake, system A places data on the bus. Then it changes *Req* (low to high or high to low) to indicate that the data is valid. System B samples the data when it detects a change in the level of *Req* and toggles *Ack* to indicate that the data has been captured. This protocol uses fewer transitions (and thus possibly less time and energy), but requires circuitry that responds to edges rather than levels.

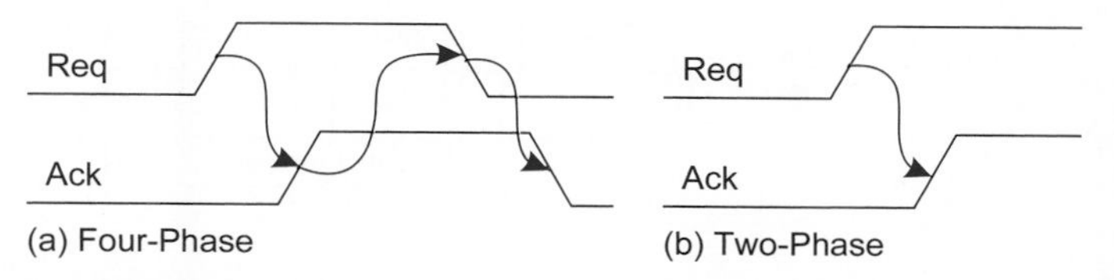

FIGURE 9.48 Four-phase and two-phase handshake protocols

Req is not synchronized to *clkB*. If it changes at the same time *clkB* rises, System *B* may receive a metastable value. Thus, System *B* needs a synchronizer on the *Req* input. If the synchronizer waits long enough, the request will resolve to a valid logic level with very high probability. The synchronizer may resolve high or low. If it resolves high, the rising request was detected and System *B* can sample the data. If it resolves low, the rising request was just missed. However, it will be detected on the next cycle of *clkB*, just as it would have been if the rising request occurred just slightly later. *Ack* is not synchronized to *clkA*, so it also requires a synchronizer.

Figure 9.49 shows a typical two-phase handshaking system [Crews03]. *clkA* and *clkB* operate at unrelated frequencies and each system may not know the frequency of its counterpart. Each system contains a synchronizer, a level-to-pulse converter, and a pulse-to-level converter. System A asserts *ReqA* for one cycle when *DataA* is ready. We will refer to this as a *pulse*. The XOR and flip-flop form a pulse-to-level converter that toggles the level of *Req*. This level is synchronized to *clkB*. When an edge is detected, the level-to-pulse converter produces a pulse on *ReqB*. This pulse in turn toggles *Ack*. The acknowledge level is synchronized to *clkA* and converted back to a pulse on *AckA*. The synchronizers add significant latency so the throughput of asynchronous communication can be much lower than that of synchronous communication.

9.5.4 Common Synchronizer Mistakes

Although a synchronizer is a simple circuit, it is notoriously easy to misuse. For example, the AMD 9513 system timing controller, AMD 9519 interrupt controller, Zilog Z-80 Serial I/O interface, Intel 8048 microprocessor, and AMD 29000 microprocessor are all

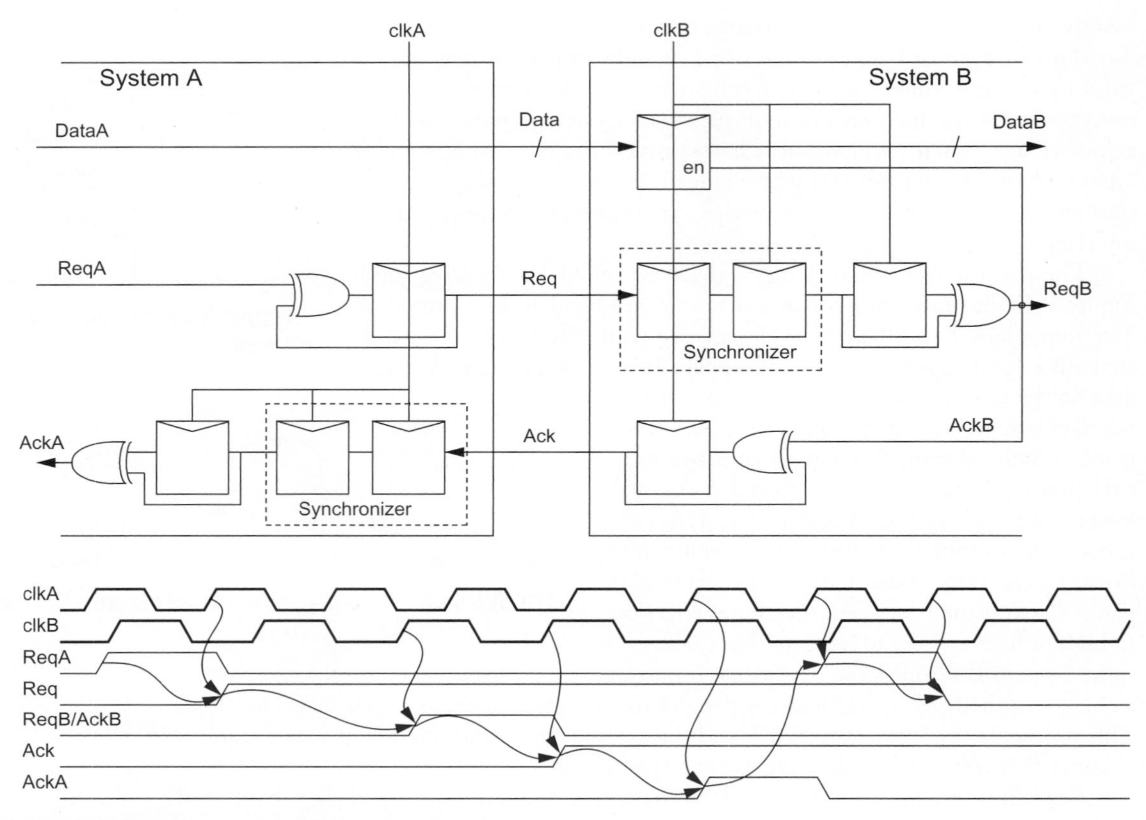

FIGURE 9.49 Two-phase handshake circuitry with synchronizers

said to have suffered from metastability problems [Wakerly00]. [Ginosar03] has even written a paper on *Fourteen Ways to Fool Your Synchronizer* illustrating overly imaginative designs.

One way to build a bad synchronizer is to use a bad latch or flip-flop. The synchronizer depends on positive feedback to drive the output to a good logic level. Therefore, dynamic latches without feedback such as Figure 9.17(a–d) do not work. The probability of failure grows exponentially with the time constant of the feedback loop. Therefore, the loop should be lightly loaded. The latch from Figure 9.17(f) is a poor choice because a large capacitive load on the output will increase the time constant; Figure 9.17(g) is a much better choice.

Another error is to capture inconsistent data. For example, Figure 9.50(a) shows a single signal driving two synchronizers (each consisting of a pair of back-to-back flip-flops). If the signal is stable through the aperture, *Q*1 and *Q*2 will be the same. However, if the signal changes during the aperture, *Q*1 and *Q*2 might resolve to different values. If the system requires that *Q*1 and *Q*2 be identical representations of the data input, they must come from a single synchronizer.

Another example is to synchronize a multibit word where more than one bit might be changing at a time. For example, if the word in Figure 9.50(b) is transitioning from 0000 to 1111, the synchronizer might produce a value such as 0101 that is neither the old nor the new data word. For this reason, the system in Figure 9.49 synchronized only the *Req/*

Ack signals and used them to indicate that data was stable to sample or finished being sampled. *Gray codes* (see Section 10.7.3) are also useful for counters whose outputs must be synchronized because exactly one bit changes on each count so that the synchronizer is guaranteed to find either the old or the new data value.

In general, synchronizer bugs are intermittent and notoriously difficult to locate and diagnose. For this reason, asynchronous interfaces should be reviewed closely.

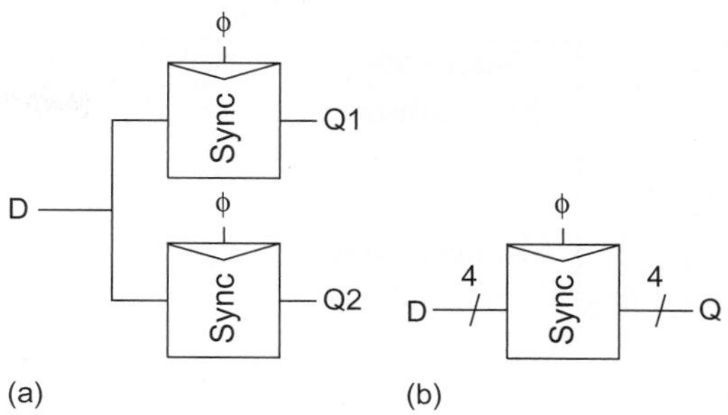

FIGURE 9.50 Bad synchronizer designs

9.5.5 Arbiters

The *arbiter* of Figure 9.51(a) is closely related to the synchronizer. It determines which of two inputs arrived first. If the spacing between the inputs exceeds some aperture time, the first input should be acknowledged. If the spacing is smaller, exactly one of the two inputs should be acknowledged, but the choice is arbitrary. For example, in a television game show, two contestants may pound buttons to answer a question. If one presses the button first, she should be acknowledged. If both press the button at times too close to distinguish, the host may choose one of the two contestants arbitrarily (but must not lock up or catch on fire).

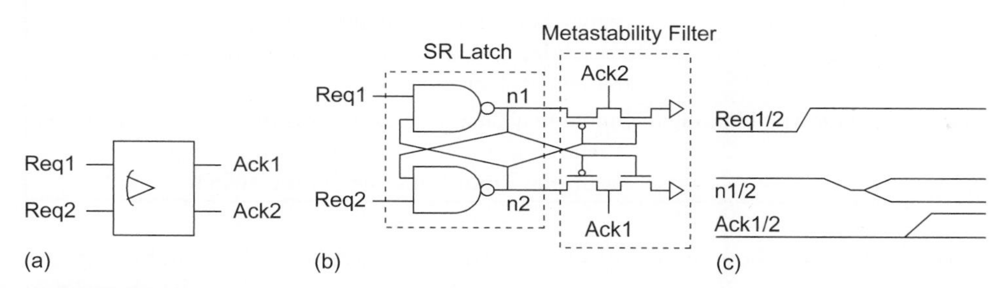

FIGURE 9.51 Arbiter

Figure 9.51(b) shows an arbiter built from an SR latch and a four-transistor metastability filter. If one of the request inputs arrives well before the other, the latch will respond appropriately. However, if they arrive at nearly the same time, the latch may be driven into metastability, as shown in Figure 9.51(c). The filter keeps both acknowledge signals low until the voltage difference between the internal nodes n_1 and n_2 exceeds V_t, indicating that a decision has been made. Such an asynchronous arbiter will never produce metastable outputs. However, the time required to make the decision can be unbounded, so the acknowledge signals must be synchronized before they are used in a clocked system.

Arbiters can be generalized to select 1-of-*N* or *M*-of-*N* inputs. However, such arbiters have multiple metastable states and require careful design [van Berkel99].

9.5.6 Degrees of Synchrony

The simple synchronizer from Section 9.5.2 accepts inputs that can change at any time, but has two-cycle latency and a nonzero probability of failure. In practice, many inputs may not be aligned to a single system clock, but they may still be predictable. Table 9.3 provides a classification of degrees of synchrony between input signals and the receiver system clock [Messerschmitt90] based on the difference in phase $\Delta\phi$ and frequency Δf.

TABLE 9.3 Degrees of synchrony

Classification	Periodic	$\Delta\phi$	Δf	Description
Synchronous	Yes	0	0	Signal has same frequency and phase as clock. Safe to sample signal directly with the clock. **Example:** Flip-flop to flip-flop on chip.
Mesochronous	Yes	Constant	0	Signal has same frequency, but is out of phase with the clock. Safe to sample signal if it is delayed by a constant amount to fall outside aperture. **Example:** Chip-to-chip where chips use same clock signal, but might have arbitrarily large skews.
Plesiochronous	Yes	Varies slowly	Small	Signal has nearly the same frequency. Phase drifts slowly over time. Safe to sample signal if it is delayed by a variable but predictable amount. Difference in frequency can lead to dropped or duplicated data. **Example:** Board-to-board where boards use clock crystals with small mismatches in nominally identical rates.
Periodic	Yes	Varies rapidly	Large	Signal is periodic at an arbitrary frequency. Periodic nature can be exploited to predict and delay accordingly when data will change during aperture. **Example:** Board-to-board where boards use different frequency clocks.
Asynchronous	No	Unknown	Unknown	Signal may change at arbitrary times. Full synchronizer is required. **Example:** Input from pushbutton switch.

[Dally98] describes a number of synchronizers that have zero failure probability and possibly lower latency when the input is predictable. They are based on the observation that either the signal or a copy of the signal delayed by t_a will be stable throughout the aperture. Hence, a synchronizer that can predict the input arrival time can choose the signal or its delayed counterpart to safely sample. Mesochronous signals are synchronized by measuring the phase difference and delaying the input enough to ensure it falls outside the aperture. Plesiochronous signals can be synchronized in a similar fashion, but the phase difference slowly varies, so the delay must be occasionally adjusted. Because the frequencies differ, the synchronizer requires some control flow to handle the missing or extra data items. Periodic signals also require control flow and use a clock predictor to calculate where the next clock edge will occur and whether the signal must be delayed to avoid falling in the aperture.

9.6 Wave Pipelining

Recall that sequencing elements are used in pipelined systems to prevent the current token from overtaking the next token or from being overtaken by the previous token in the pipeline. If the elements propagate through the pipeline at a fairly constant rate, explicit sequencing elements may not be necessary to maintain sequence. As an analogy, fiber optic cables carry data as a series of light pulses. Many pulses enter the cable before the first one reaches the end, yet the cable does not need internal latches to keep the pulses separated because they propagate along the cable at a well-controlled velocity. The maximum data rate is limited by the dispersion along the line that causes pulses to smear over time and blur into one another if they become too short.

Figure 9.52 compares traditional pipelining with wave pipelining. In both cases, the pipeline contains combinational logic separated by registers (Figure 9.52(a)). The registers $F1$ and $F2$ receive clocks *clk*1 and *clk*2 that are nominally identical, but might experience skew. Figure 9.52(b) shows traditional pipelining. The data is launched on the rising edge of *clk1*. Its propagation is indicated by the hashed cone. $D2$ becomes stable somewhere between the contamination and propagation delays after the clock edge (neglecting the flip-flop *clk*-to-Q delay). $D2$ must not change during the setup and hold aperture around *clk*2, marked with the blue box. The figure shows two successive cycles in which tokens i and $i + 1$ move through the pipeline. Each token passes through the combinational logic in a single cycle. Figure 9.52(c) shows wave pipelining with a clock of twice the frequency. Token i enters the combinational logic, but takes two cycles to reach $F2$. Meanwhile, token $i + 1$ enters the logic a cycle later. As long as each token is stable to sample at $F2$ and the cones do not overlap, the pipeline will operate correctly with the same latency but twice the throughput.

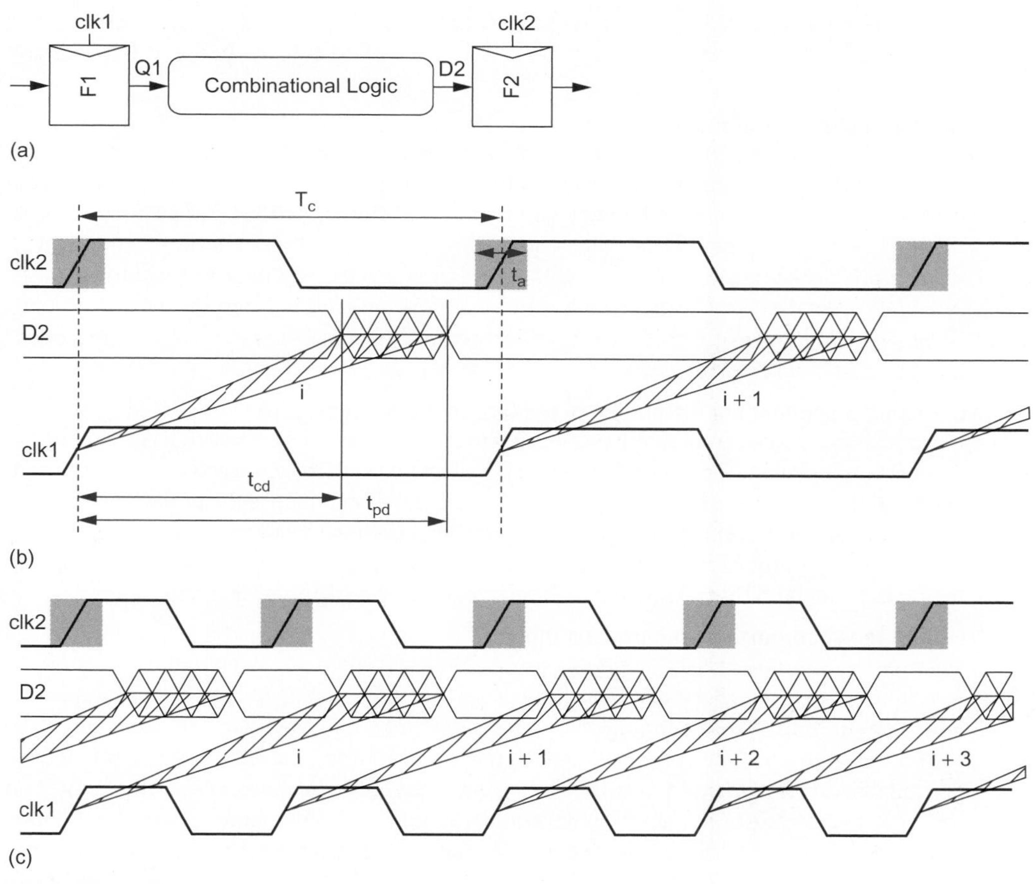

FIGURE 9.52 Wave pipelining

[Burleson98] gives a tutorial on wave pipelining and derives the timing constraints. In general, a wave pipeline can contain N tokens between each pair of registers. The maximum value of N is limited by the ratio of propagation delay to dispersion of the logic cones:

$$N < \frac{t_{pd}}{t_{pd} - t_{cd}} \tag{9.30}$$

If the contamination and propagation delays are nearly equal, the combinational logic can contain many tokens simultaneously. In practice, the delays tend to be widely variable because of voltage, temperature, and processing as well as differences in path lengths through the logic. Clock skew and sequencing overhead also eat into the timing budgets. In practice, even achieving $N = 2$ simultaneous tokens can be difficult and wave pipelining has not achieved widespread popularity for general-purpose logic.

9.7 Pitfalls and Fallacies

Incompletely reporting flip-flop delay

The effective delay of a flip-flop is its minimum *D*-to-*Q* time. This is the sum of the setup time t_{setup} and the *clk*-to-*Q* delay t_{pdq} if these delays are defined to minimize the sum. Some engineers focus on only the *clk*-to-*Q* delay or define setup and *clk*-to-*Q* delays in a way that does not minimize the sum.

Failing to check hold times

One of the leading reasons that chips fail to operate even though they appear to simulate correctly is hold-time violations, especially violations caused by unexpected clock skew. Unless a design uses two-phase nonoverlapping clocks, the clock skew should be carefully modeled and the hold times should be checked with a static timing analyzer. These checks should happen as soon as a block is designed so that errors can be corrected immediately. For example, a large microprocessor used a wide assortment of delayed clocks to solve setup time problems on long paths. Hold times were not checked until shortly before tapeout, leading to a significant schedule slip when many violations were found.

Choosing a sequencing methodology too late in the design cycle

Designers may choose from many sequencing methodologies, each of which has trade-offs. The best methodology for a particular application is very debatable, and engineers love a good debate. If the sequencing methodology is not settled at the beginning of the project, experience shows that engineers will waste tremendous amounts of time redoing work as the method changes, or supporting and verifying multiple methodologies. Projects need a strong technical manager to demand that a team choose one method at the beginning and stick with it.

Failing to synchronize asynchronous inputs

Unsynchronized inputs can cause strange and wonderful sporadic system failures that are very difficult to locate. For example, a finite state machine running off one clock received a *READY* input from a *UART* running on another clock when the *UART* had data available, as shown in Figure 9.53. The designer reasoned that synchronizing the *READY* signal was unimportant because if it changed near the clock edge of the FSM, she did not care whether it was detected in one cycle or the next. Moreover, the clock was so slow that metastability would have time to resolve. However, the FSM occasionally failed by jumping to seemingly random

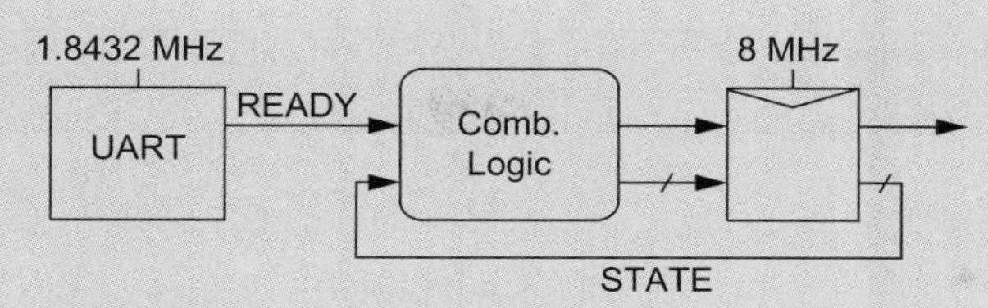

FIGURE 9.53 Unsynchronized input

states that could never legally occur. After two months of debugging, she realized that the problem was triggered if the asynchronous *READY* signal was asserted a few gate delays before the FSM clock edge. The propagation delay through the combinational logic was different for various bits of the next state logic. Some bits had changed to their new values while others were still at their old values, so the FSM could jump to an undefined state. Registering the *READY* signal with the FSM clock before it drove the combinational logic solved the problem.

Building faulty synchronizers

Designers have found many ways to build faulty synchronizers. For example, if an asynchronous input drives more than one synchronizer, the two synchronizers can resolve to different values. If they must produce consistent outputs, only one synchronizer should be used. In another example, synchronizers must not accept multibit inputs where more than one of the bits can change simultaneously. This would pose the risk that some of the bits resolve as changed while others resolve in their old state, resulting in an invalid pattern that is neither the old nor the new input word. In yet another example, synchronizers with poorly designed feedback loops can be much slower than expected and can have exponentially worse mean time between failures.

Summary

This chapter has examined the trade-offs of sequencing with flip-flops, two-phase transparent latches, and pulsed latches. Minimizing sequencing overhead is critical in these high-performance systems. Flip-flops are the simplest, but have the greatest sequencing overhead. Transparent latches are most tolerant of skew and allow the most time borrowing, but require greater design effort to partition logic into half-cycles instead of cycles. Pulsed latches have the lowest sequencing overhead, but are most susceptible to min-delay problems. Table 9.4 compares the sequencing overhead, minimum delay constraint, and time borrowing capability of each technique. All of the techniques are used in commercial products, and the designer's choice depends on the design constraints and CAD tools.

TABLE 9.4 Comparison of sequencing elements

	Sequencing Overhead ($T_c - t_{pd}$)	**Minimum Logic Delay** t_{cd}	**Time Borrowing** t_{borrow}
Flip-Flops	$t_{pcq} + t_{setup} + t_{skew}$	$t_{hold} - t_{ccq} + t_{skew}$	0
Two-Phase Transparent Latches	$2t_{pdq}$	$t_{hold} - t_{ccq} + t_{nonoverlap} + t_{skew}$ in each half-cycle	$T_c/2 - (t_{setup} + t_{nonoverlap} + t_{skew})$
Pulsed Latches	$\max(t_{pdq}, t_{pcq} + t_{setup} - t_{pw} + t_{skew})$	$t_{hold} - t_{ccq} + t_{pw} + t_{skew}$	$t_{pw} - (t_{setup} + t_{skew})$

In class projects for introductory VLSI classes, timing analysis is often rudimentary or nonexistent. Using two-phase nonoverlapping clocks generated off chip is attractive because you can guarantee the chip will have no max-delay or min-delay failures if the clock period and nonoverlap are sufficiently large. However, it is not practical to generate and distribute two nonoverlapping phases on a large, high-performance commercial chip.

The great majority of low- and mid-performance designs and some high-speed designs use flip-flops. Flip-flops are easy to use and are well understood by most designers. Even more importantly, they are handled well by synthesis tools and timing analyzers. Unfortunately, in systems with few gate delays per cycle, the sequencing overhead can consume a large fraction of the cycle. Moreover, many standard cell flip-flops are intentionally rather slow to prevent hold-time violations at the expense of greater sequencing overhead.

Most two-phase latch systems distribute a single clock and locally invert it to drive the second latch. These systems tolerate significant amounts of clock skew without loss of performance and can borrow time to balance delay intentionally or opportunistically. However, the systems require more effort to understand because time borrowing distributes the timing constraints across many stages of a pipeline rather than isolating them at each stage. Not all timing analyzers handle latches gracefully, especially when there are different amounts of clock skew between different clocks [Harris99]. Two-phase latches have been used in the Alpha 21064 and 21164 [Gronowski98] and a variety of older chips, but are rarely used today.

Pulsed latches have low sequencing overhead. They present a trade-off when choosing pulse width: A wide pulse permits more time borrowing and skew tolerance, but makes min-delay constraints harder to meet. Pulsed latches are also popular because they can be modeled as fast flip-flops with a lousy hold time from the point of view of a timing analyzer (or novice designer) if intentional time borrowing is not permitted. The min-delay problems can be largely overcome by mixing pulsed latches for long paths and flip-flops for short paths. Unfortunately, many real designs have paths in which the propagation delay is very long but the contamination delay is very short, making robust design more challenging. Pulsed latches have been used on Itanium 2 [Naffziger02], Pentium 4 [Kurd01], Athlon [Draper97], and CRAY 1 [Unger86]. However, they can wreak havoc with conventional commercially available design flows and are best avoided unless the performance requirements are extreme.

When inputs to a system arrive asynchronously, they cannot be guaranteed to meet setup or hold times at clocked elements. Even if we do not care whether an input arrived in one cycle or the next, we must ensure that the clocked element produces a valid logic level. Unfortunately, if the element samples a changing input at just the wrong time, it may produce a metastable output that remains invalid for an unbounded amount of time. The probability of metastability drops off exponentially with time. Systems use synchronizers to sample the asynchronous input and hold it long enough to resolve to a valid logic level with very high probability before passing it onward.

Most synchronous VLSI systems use opaque sequencing elements to separate one token from the next. In contrast, many optical systems transmit data as pulses separated in time. As long as the propagation medium does not disperse the pulses too badly, they can be recovered at a receiver. Similarly, if a VLSI system has low dispersion, i.e., nearly equal contamination and propagation delays, it can send more than one wave of data without explicit latching. Such wave pipelining offers the potential of high throughput and low sequencing overhead. However, it is difficult to perform in practice because of the variability of data delay.

Exercises

Use the timing parameters in Table 9.5 for the following exercises.

TABLE 9.5 Sequencing element parameters

	Setup Time	*clk*-to-*Q* Delay	*D*-to-*Q* Delay	Contamination Delay	Hold Time
Flip-Flops	65 ps	50 ps	n/a	35 ps	30 ps
Latches	25 ps	50 ps	40 ps	35 ps	30 ps

9.1 A synchronizer uses a flip-flop with $\tau_s = 54$ ps and $T_0 = 21$ ps. Assuming the input toggles at 10 MHz and the setup time is negligible, what is the minimum clock period for which the mean time between failures exceeds 100 years?

9.2 Simulate the synchronizer flip-flop of Figure 9.45 and make a plot analogous to Figure 9.43. From your plot, find Δ_{DQ}, h, τ, and T_0.

9.3 For each of the following sequencing styles, determine the maximum logic propagation delay available within a 500 ps clock cycle. Assume there is zero clock skew and no time borrowing takes place.

9.4 Repeat Exercise 9.3 if the clock skew between any two elements can be up to 50 ps.

a) Flip-flops

b) Two-phase transparent latches

c) Pulsed latches with 80 ps pulse width

9.5 For each of the following sequencing styles, determine the minimum logic contamination delay in each clock cycle (or half-cycle, for two-phase latches). Assume there is zero clock skew.

a) Flip-flops

b) Two-phase transparent latches with 50% duty cycle clocks

c) Two-phase transparent latches with 60 ps of nonoverlap between phases

d) Pulsed latches with 80 ps pulse width

9.6 Repeat Exercise 9.5 if the clock skew between any two elements can be up to 50 ps.

9.7 Suppose one cycle of logic is particularly critical and the next cycle is nearly empty. Determine the maximum amount of time the first cycle can borrow into the second for each of the following sequencing styles. Assume there is zero clock skew and that the cycle time is 500 ps.

a) Flip-flops

b) Two-phase transparent latches with 50% duty cycle clocks

c) Two-phase transparent latches with 60 ps of nonoverlap between phases

d) Pulsed latches with 80 ps pulse width

9.8 Repeat Exercise 9.7 if the clock skew between any two elements can be up to 50 ps.

9.9 Prove EQ (9.17).

9.10 Consider a flip-flop built from a pair of transparent latches using nonoverlapping clocks. Express the setup time, hold time, and clock-to-Q delay of the flip-flop in terms of the latch timing parameters and $t_{nonoverlap}$, relative to the rising edge of ϕ_1.

9.11 For the path in Figure 9.54, determine which latches borrow time and if any setup time violations occur. Repeat for cycle times of 1200, 1000, and 800 ps. Assume there is zero clock skew and that the latch delays are accounted for in the propagation delay

a) $\Delta1 = 550$ ps; $\Delta2 = 580$ ps; $\Delta3 = 450$ ps; $\Delta4 = 200$ ps

b) $\Delta1 = 300$ ps; $\Delta2 = 600$ ps; $\Delta3 = 400$ ps; $\Delta4 = 550$ ps

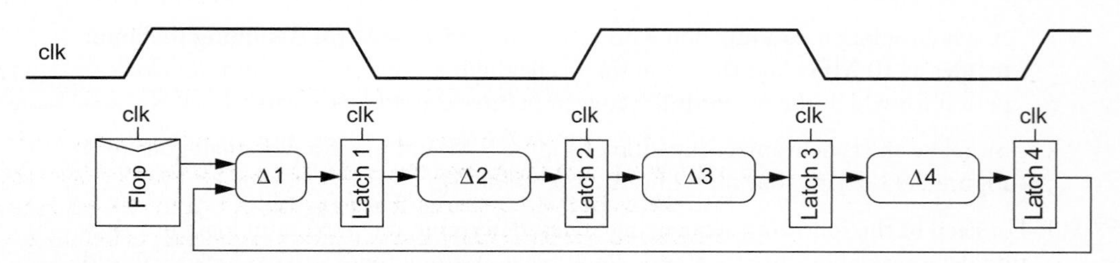

FIGURE 9.54 Example path

9.12 Determine the minimum clock period at which the circuit in Figure 9.55 will operate correctly for each of the following logic delays. Assume there is zero clock skew and that the latch delays are accounted for in the propagation delay

a) $\Delta1 = 300$ ps; $\Delta2 = 400$ ps; $\Delta3 = 200$ ps; $\Delta4 = 350$ ps

b) $\Delta1 = 300$ ps; $\Delta2 = 400$ ps; $\Delta3 = 400$ ps; $\Delta4 = 550$ ps

c) $\Delta1 = 300$ ps; $\Delta2 = 900$ ps; $\Delta3 = 200$ ps; $\Delta4 = 350$ ps

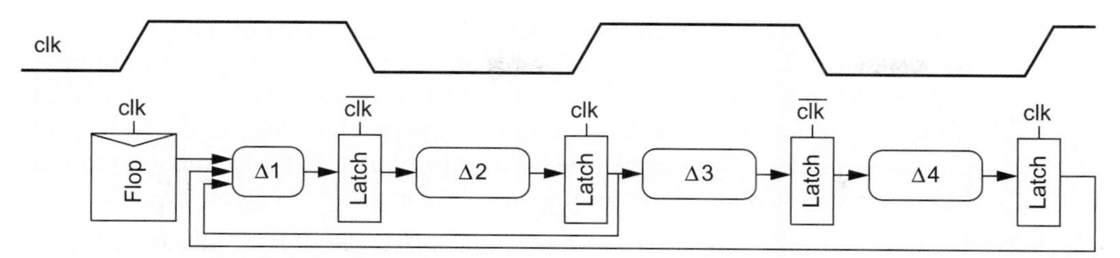

FIGURE 9.55 Another example path

9.13 Repeat Exercise 9.12 if the clock skew is 100 ps.

9.14 Label the timing types of each signal in the circuit from Figure 9.54. The flip-flop is constructed with back-to-back transparent latches—the first controlled by *clk_b* and the second by *clk*.

9.15 InferiorCircuits, Inc., wants to sell you a perfect synchronizer that they claim never produces a metastable output. The synchronizer consists of a regular flip-flop followed by a high-gain comparator that produces a high output for inputs above $V_{DD}/4$ and a low output for inputs below that point. The VP of marketing argues that even if the flip-flop enters metastability, its output will hover near $V_{DD}/2$ so the synchronizer will produce a good high output after the comparator. Why wouldn't you buy this synchronizer?

9.16 Using a simulator, find the setup and hold times of a TSPC latch under the assumptions of Exercise 9.25.

9.17 Determine the maximum logic propagation delay available in a cycle for a traditional domino pipeline using a 500 ps clock cycle. Assume there is zero clock skew.

9.18 Repeat Exercise 9.17 if the clock skew between any two elements can reach 50 ps.

9.19 Determine the maximum logic propagation delay available in a cycle for a four-phase skew-tolerant domino pipeline using a 500 ps clock cycle. Assume there is zero clock skew.

9.20 Repeat Exercise 9.19 if the clock skew between any two elements can be up to 50 ps.

9.21 How much time can one phase borrow into the next in Exercise 9.20 if the clocks each have a 50% duty cycle? Assume $t_{\text{hold}} = 0$.

9.22 Repeat Exercise 9.20 if the clocks have a 65% duty cycle.

9.23 Design a fast pulsed latch. Make the gate capacitance on the clock and data inputs equal. Let the latch drive an output load of four identical latches. Simulate your latch and find the setup and hold times and clock-to-*Q* propagation and contamination delays. Express your results in FO4 inverter delays.

9.24 Simulate the worst-case propagation delay of an 8-input dynamic NOR gate driving a fanout of 4. Report the delay in all 16 design corners (voltage, temperature, nMOS, pMOS). Also determine the delay of a fanout-of-4 inverter in each of these corners. By what percentage does the absolute propagation delay of the NOR gate vary across corners? By what percentage does its normalized delay vary (in terms of FO4 inverters)? Comment on the implications for circuits using matched delays.

9.25 Using a simulator, compare the D-to-Q propagation delays of a conventional dynamic latch from Figure 9.17(d) and a TSPC latch from Section 9.3.11. Assume each latch is loaded with a fanout of 4. Use 4 λ-wide clocked transistors and tune the other transistor sizes for least propagation delay.

Datapaths 10

10.1 Introduction

Chip functions generally can be divided into the following categories:

- Datapath operators
- Memory elements
- Control structures
- Special-purpose cells
  - I/O
  - Power distribution
  - Clock generation and distribution
  - Analog and RF

CMOS system design consists of partitioning the system into subsystems of the types listed above. Many options exist that make trade-offs between speed, density, programmability, ease of design, and other variables. This chapter addresses design options for common datapath operators. The next chapter addresses arrays, especially those used for memory. Control structures are most commonly coded in a hardware description language and synthesized. Special-purpose subsystems are considered in Chapter 12.

As introduced in Chapter 1, datapath operators benefit from the structured design principles of hierarchy, regularity, modularity, and locality. They may use N identical circuits to process N-bit data. Related data operators are placed physically adjacent to each other to reduce wire length and delay. Generally, data is arranged to flow in one direction, while control signals are introduced in a direction orthogonal to the dataflow.

Common datapath operators considered in this chapter include adders, one/zero detectors, comparators, counters, Boolean logic units, error-correcting code blocks, shifters, and multipliers.

10.2 Addition/Subtraction

"Multitudes of contrivances were designed, and almost endless drawings made, for the purpose of economizing the time and simplifying the mechanism of carriage."

—Charles Babbage, on Difference Engine No. 1, 1864 [Morrison61]

Addition forms the basis for many processing operations, from ALUs to address generation to multiplication to filtering. As a result, adder circuits that add two binary numbers are of great interest to digital system designers. An extensive, almost endless, assortment of adder architectures serve different speed/power/area requirements. This section begins with half adders and full adders for single-bit addition. It then considers a plethora of carry-propagate adders (CPAs) for the addition of multibit words. Finally, related structures such as subtracters and multiple-input adders are discussed.

10.2.1 Single-Bit Addition

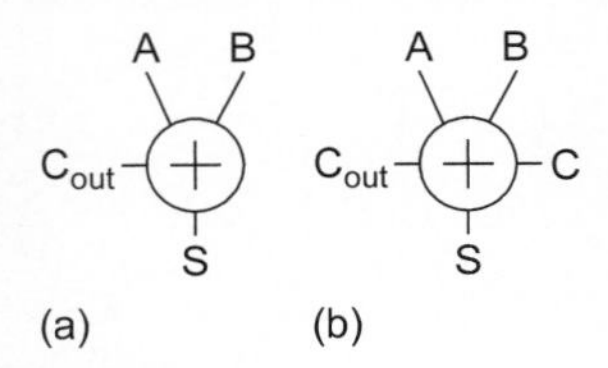

FIGURE 10.1
Half and full adders

The *half adder* of Figure 10.1(a) adds two single-bit inputs, A and B. The result is 0, 1, or 2, so two bits are required to represent the value; they are called the sum S and carry-out C_{out}. The carry-out is equivalent to a carry-in to the next more significant column of a multibit adder, so it can be described as having double the *weight* of the other bits. If multiple adders are to be cascaded, each must be able to receive the carry-in. Such a *full adder* as shown in Figure 10.1(b) has a third input called C or C_{in}.

The truth tables for the half adder and full adder are given in Tables 10.1 and 10.2. For a full adder, it is sometimes useful to define *Generate* (G), *Propagate* (P), and *Kill* (K) signals. The adder generates a carry when C_{out} is true independent of C_{in}, so $G = A \cdot B$. The adder kills a carry when C_{out} is false independent of C_{in}, so $K = \overline{A} \cdot \overline{B} = \overline{A + B}$. The adder propagates a carry; i.e., it produces a carry-out if and only if it receives a carry-in, when exactly one input is true: $P = A \oplus B$.

TABLE 10.1 Truth table for half adder

A	B	C_{out}	S
0	0	0	0
0	1	0	1
1	0	0	1
1	1	1	0

TABLE 10.2 Truth table for full adder

A	B	C	G	P	K	C_{out}	S
0	0	0	0	0	1	0	0
		1				0	1
0	1	0	0	1	0	0	1
		1				1	0
1	0	0	0	1	0	0	1
		1				1	0
1	1	0	1	0	0	1	0
		1				1	1

From the truth table, the half adder logic is

$$\begin{aligned} S &= A \oplus B \\ C_{\text{out}} &= A \cdot B \end{aligned} \tag{10.1}$$

and the full adder logic is

$$
\begin{aligned}
S &= A\bar{B}\bar{C} + \bar{A}B\bar{C} + \bar{A}\bar{B}C + ABC \\
&= (A \oplus B) \oplus C = P \oplus C \\
C_{\text{out}} &= AB + AC + BC \\
&= AB + C(A + B) \\
&= \overline{\bar{A}\bar{B} + \bar{C}(\bar{A} + \bar{B})} \\
&= \text{MAJ}(A,\ B,\ C)
\end{aligned}
\qquad \textbf{(10.2)}
$$

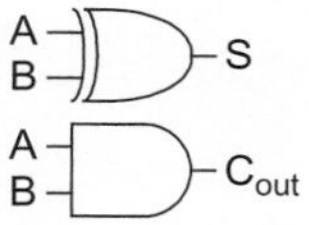

FIGURE 10.2 Half adder design

The most straightforward approach to designing an adder is with logic gates. Figure 10.2 shows a half adder. Figure 10.3 shows a full adder at the gate (a) and transistor (b) levels. The carry gate is also called a *majority* gate because it produces a 1 if at least two of the three inputs are 1. Full adders are used most often, so they will receive the attention of the remainder of this section.

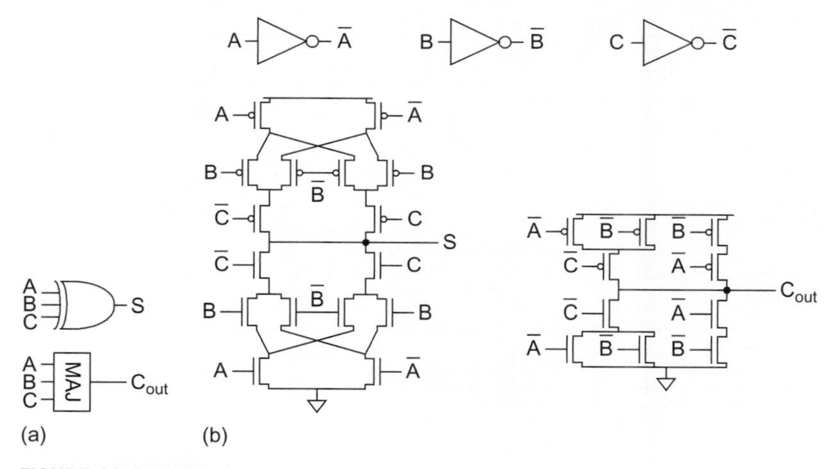

FIGURE 10.3 Full adder design

The full adder of Figure 10.3(b) employs 32 transistors (6 for the inverters, 10 for the majority gate, and 16 for the 3-input XOR). A more compact design is based on the observation that S can be factored to reuse the C_{out} term as follows:

$$
S = ABC + (A + B + C)\bar{C}_{\text{out}} \qquad \textbf{(10.3)}
$$

Such a design is shown at the gate (a) and transistor (b) levels in Figure 10.4 and uses only 28 transistors. Note that the pMOS network is identical to the nMOS network rather than being the conduction complement, so the topology is called a *mirror adder*. This simplification reduces the number of series transistors and makes the layout more uniform. It is possible because the addition function is *symmetric*; i.e., the function of complemented inputs is the complement of the function.

The mirror adder has a greater delay to compute S than C_{out}. In carry-ripple adders (Section 10.2.2.1), the critical path goes from C to C_{out} through many full adders, so the

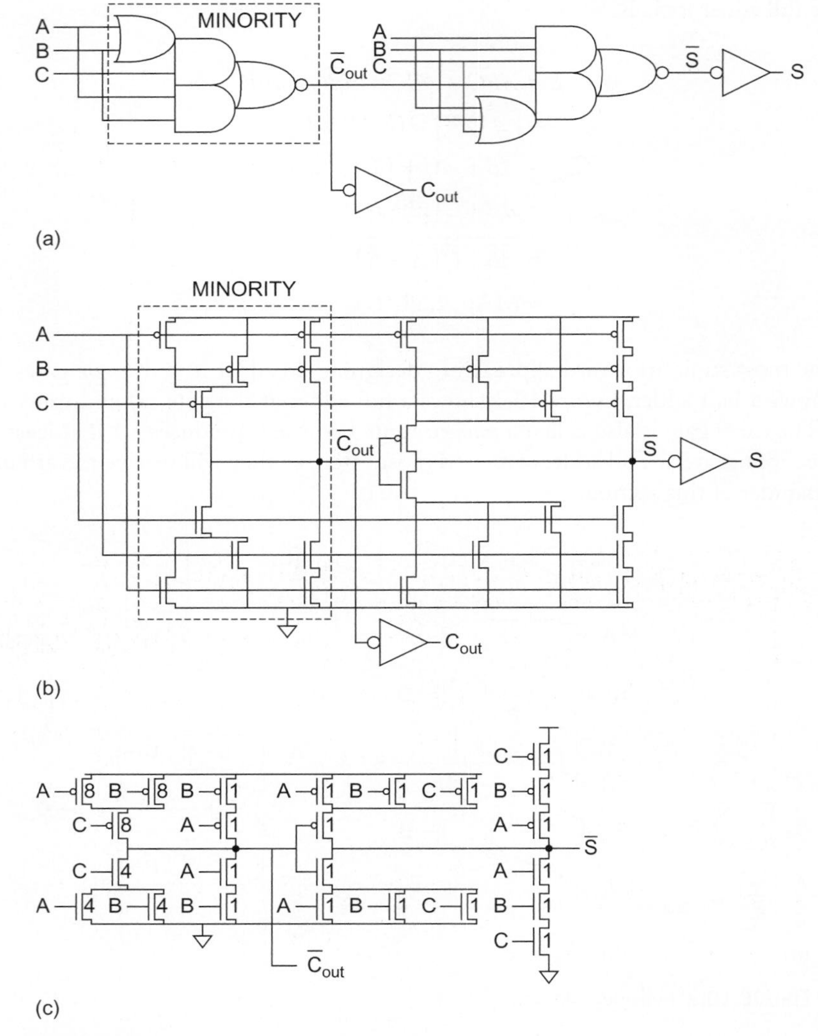

FIGURE 10.4 Full adder for carry-ripple operation

extra delay computing S is unimportant. Figure 10.4(c) shows the adder with transistor sizes optimized to favor the critical path using a number of techniques:

- Feed the carry-in signal (C) to the inner inputs so the internal capacitance is already discharged.
- Make all transistors in the sum logic whose gate signals are connected to the carry-in and carry logic minimum size (1 unit, e.g., 4 λ). This minimizes the branching effort on the critical path. Keep routing on this signal as short as possible to reduce interconnect capacitance.
- Determine widths of series transistors by logical effort and simulation. Build an asymmetric gate that reduces the logical effort from C to C_{out} at the expense of effort to S.

- Use relatively large transistors on the critical path so that stray wiring capacitance is a small fraction of the overall capacitance.
- Remove the output inverters and alternate positive and negative logic to reduce delay and transistor count to 24 (see Section 10.2.2.1).

Figure 10.5 shows two layouts of the adder (see also the inside front cover). The choice of the aspect ratio depends on the application. In a standard-cell environment, the layout of Figure 10.5(a) might be appropriate when a single row of nMOS and pMOS transistors is used. The routing for the *A*, *B*, and *C* inputs is shown inside the cell, although it could be placed outside the cell because external routing tracks have to be assigned to these signals anyway. Figure 10.5(b) shows a layout that might be appropriate for a dense datapath (if horizontal polysilicon is legal). Here, the transistors are rotated and all of the wiring is completed in polysilicon and metal1. This allows metal2 bus lines to pass over the cell horizontally. Moreover, the widths of the transistors can increase

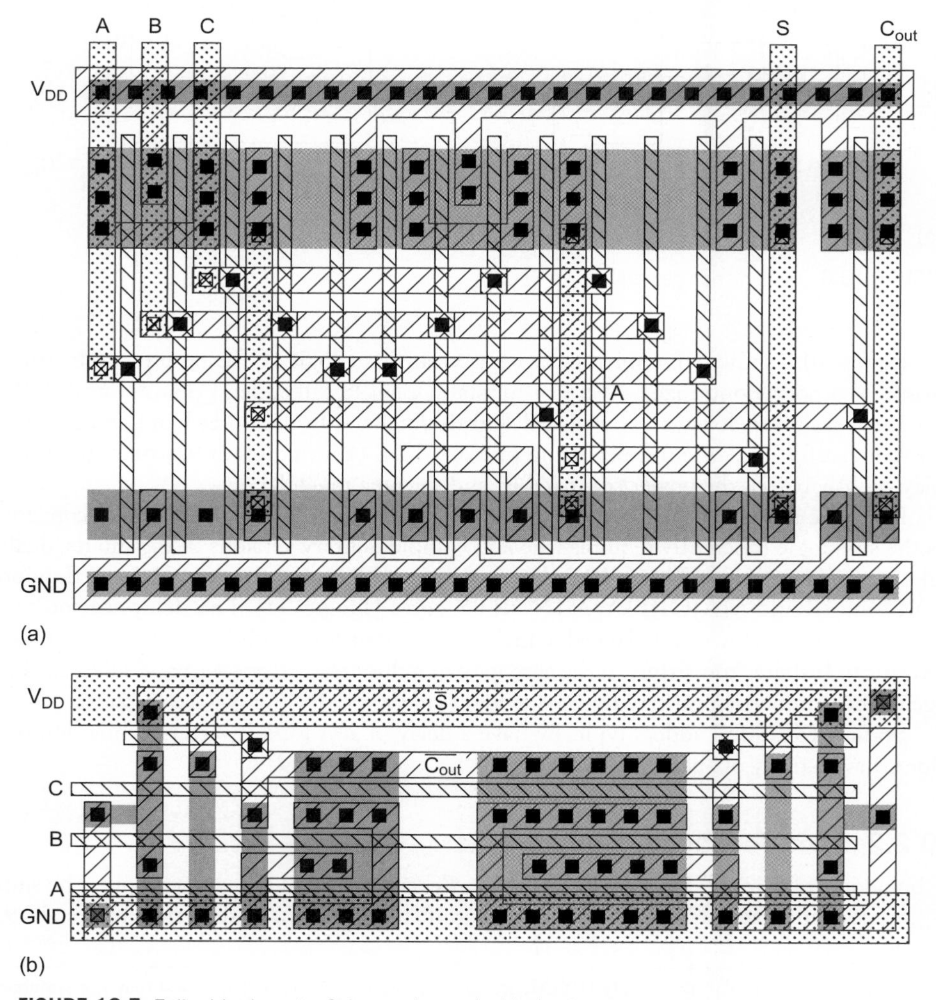

FIGURE 10.5 Full adder layouts. Color version on inside front cover.

without impacting the bit-pitch (height) of the datapath. In this case, the widths are selected to reduce the C_{in} to C_{out} delay that is on the critical path of a carry-ripple adder.

A rather different full adder design uses transmission gates to form multiplexers and XORs. Figure 10.6(a) shows the transistor-level schematic using 24 transistors and providing buffered outputs of the proper polarity with equal delay. The design can be understood by parsing the transmission gate structures into multiplexers and an "invertible inverter" XOR structure (see Section 10.7.4), as drawn in Figure 10.6(b).[1] Note that the multiplexer choosing S is configured to compute $P \oplus C$, as given in EQ (10.2).

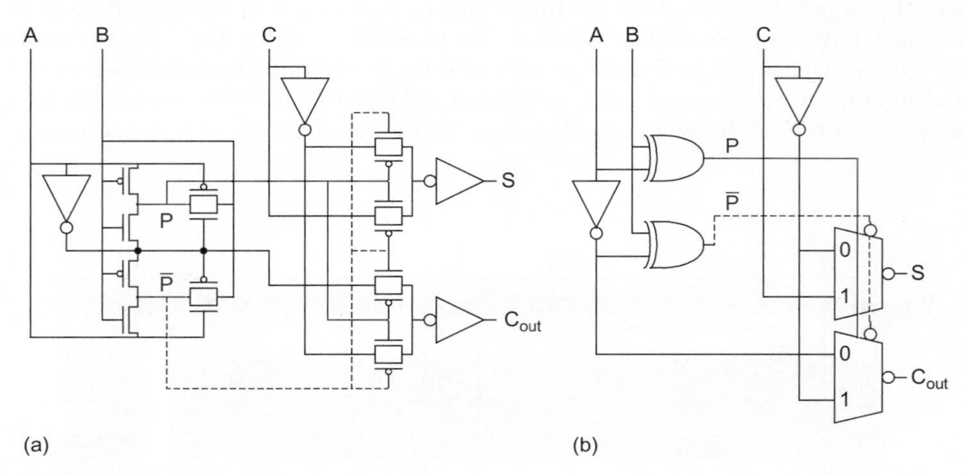

FIGURE 10.6 Transmission gate full adder

Figure 10.7 shows a complementary pass-transistor logic (CPL) approach. In comparison to a poorly optimized 40-transistor static CMOS full adder, [Yano90] finds CPL is twice as fast, 30% lower in power, and slightly smaller. On the other hand, in comparison to a careful implementation of the mirror adder, [Zimmermann97] finds the CPL delay slightly better, the power comparable, and the area much larger.

Dynamic full adders are widely used in fast multipliers when power is not a concern. As the sum logic inherently requires true and complementary versions of the inputs, dual-rail domino is necessary. Figure 10.8 shows such an adder using footless dual-rail domino XOR/XNOR and MAJORITY/MINORTY gates [Heikes94]. The delays to the two outputs are reasonably well balanced, which is important for multipliers where both paths are critical. It shares transistors in the sum gate to reduce transistor count and takes advantage of the symmetric property to provide identical layouts for the two carry gates.

Static CMOS full adders typically have a delay of 2–3 FO4 inverters, while domino adders have a delay of about 1.5.

10.2.2 Carry-Propagate Addition

N-bit adders take inputs $\{A_N, \ldots, A_1\}$, $\{B_N, \ldots, B_1\}$, and carry-in C_{in}, and compute the sum $\{S_N, \ldots, S_1\}$ and the carry-out of the most significant bit C_{out}, as shown in Figure 10.9.

[1] Some switch-level simulators, notably IRSIM, are confused by this XOR structure and may not simulate it correctly.

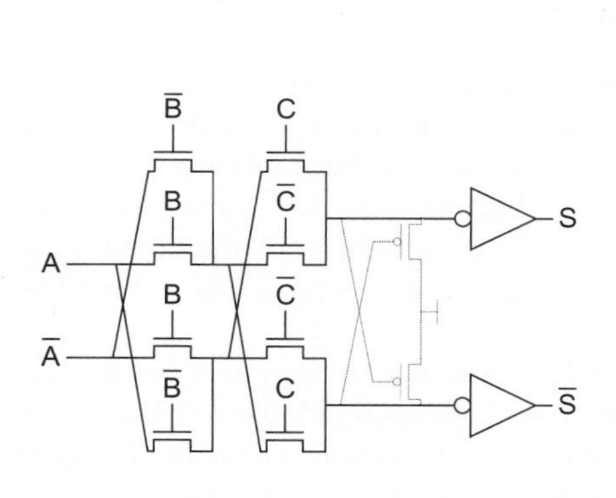

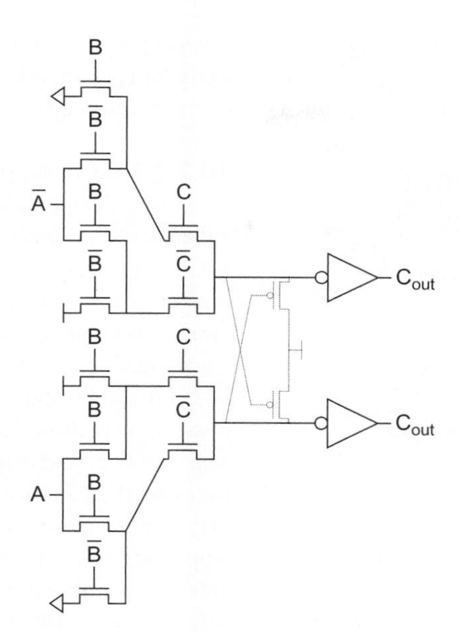

FIGURE 10.7 CPL full adder

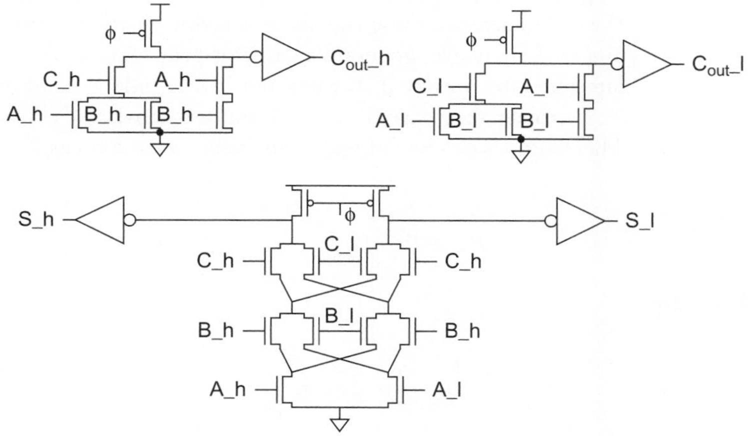

FIGURE 10.8 Dual-rail domino full

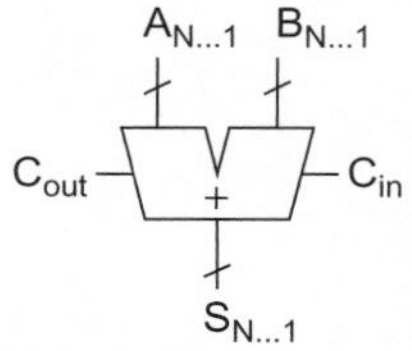

FIGURE 10.9 Carry-propagate adder

(Ordinarily, this text calls the least significant bit A_0 rather than A_1. However, for adders, the notation developed on subsequent pages is more graceful if column 0 is reserved to handle the carry.) They are called *carry-propagate adders* (CPAs) because the carry into each bit can influence the carry into all subsequent bits. For example, Figure 10.10 shows the addition $1111_2 + 0000_2 + 0/1$, in which each of the sum and carry bits is influenced by C_{in}. The simplest design is the carry-ripple adder in which the carry-out of one bit is simply connected as the carry-in to the next. Faster adders look ahead to predict the carry-out of a multibit group. This is usually done by computing group PG signals to indicate

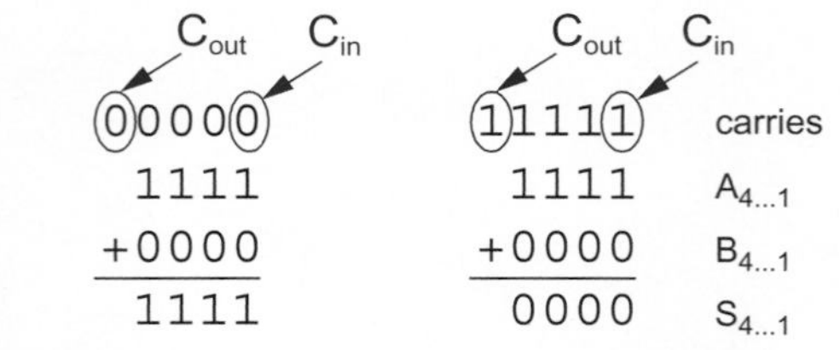

FIGURE 10.10 Example of carry propagation

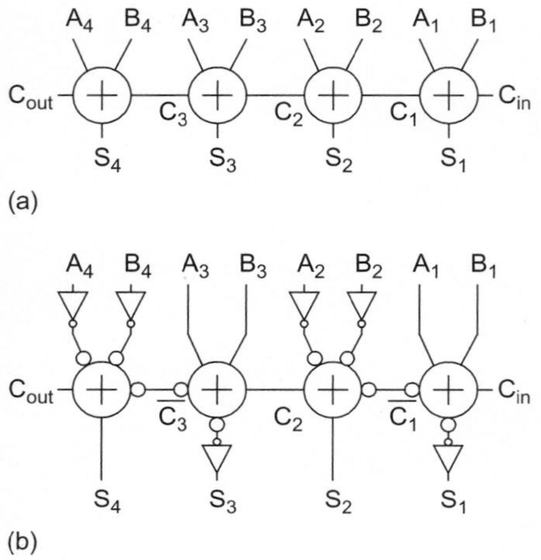

FIGURE 10.11 4-bit carry-ripple adder

whether the multibit group will propagate a carry-in or will generate a carry-out. Long adders use multiple levels of lookahead structures for even more speed.

10.2.2.1 Carry-Ripple Adder An N-bit adder can be constructed by cascading N full adders, as shown in Figure 10.11(a) for $N = 4$. This is called a *carry-ripple adder* (or *ripple-carry adder*). The carry-out of bit i, C_i, is the carry-in to bit $i + 1$. This carry is said to have twice the *weight* of the sum S_i. The delay of the adder is set by the time for the carries to ripple through the N stages, so the $t_{C \to Cout}$ delay should be minimized.

This delay can be reduced by omitting the inverters on the outputs, as was done in Figure 10.4(c). Because addition is a *self-dual function* (i.e., the function of complementary inputs is the complement of the function), an inverting full adder receiving complementary inputs produces true outputs. Figure 10.11(b) shows a carry-ripple adder built from inverting full adders. Every other stage operates on complementary data. The delay inverting the adder inputs or sum outputs is off the critical ripple-carry path.

10.2.2.2 Carry Generation and Propagation This section introduces notation commonly used in describing faster adders. Recall that the P (*propagate*) and G (*generate*) signals were defined in Section 10.2.1. We can generalize these signals to describe whether a group spanning bits $i...j$, inclusive, generate a carry or propagate a carry. A group of bits generates a carry if its carry-out is true independent of the carry-in; it propagates a carry if its carry-out is true when there is a carry-in. These signals can be defined recursively for $i \geq k > j$ as

$$\begin{aligned} G_{i:j} &= G_{i:k} + P_{i:k} \cdot G_{k-1:j} \\ P_{i:j} &= P_{i:k} \cdot P_{k-1:j} \end{aligned} \tag{10.4}$$

with the base case

$$\begin{aligned} G_{i:i} &\equiv G_i = A_i \cdot B_i \\ P_{i:i} &\equiv P_i = A_i \oplus B_i \end{aligned} \tag{10.5}$$

In other words, a group generates a carry if the upper (more significant) or the lower portion generates and the upper portion propagates that carry. The group propagates a carry if both the upper and lower portions propagate the carry.[2]

The carry-in must be treated specially. Let us define $C_0 = C_{in}$ and $C_N = C_{out}$. Then we can define generate and propagate signals for bit 0 as

$$\begin{aligned} G_{0:0} &= C_{in} \\ P_{0:0} &= 0 \end{aligned} \tag{10.6}$$

[2]Alternatively, many adders use $\overline{K}_i = A_i + B_i$ in place of P_i because OR is faster than XOR. The group logic uses the same gates: $G_{i:j} = G_{i:k} + \overline{K}_{i:k} \cdot G_{k-1:j}$ and $\overline{K}_{i:j} = \overline{K}_{i:k} \cdot \overline{K}_{k-1:j}$. However, $P_i = A_i \oplus B_i$ is still required in EQ (10.7) to compute the final sum. It is sometimes renamed X_i or T_i to avoid ambiguity.

Observe that the carry into bit i is the carry-out of bit i–1 and is $C_{i-1} = G_{i-1:0}$. This is an important relationship; *group generate* signals and *carries* will be used synonymously in the subsequent sections. We can thus compute the sum for bit i using EQ (10.2) as

$$S_i = P_i \oplus G_{i-1:0} \qquad \textbf{(10.7)}$$

Hence, addition can be reduced to a three-step process:

1. Computing bitwise generate and propagate signals using EQs (10.5) and (10.6)
2. Combining PG signals to determine group generates $G_{i-1:0}$ for all $N \geq i \geq 1$ using EQ (10.4)
3. Calculating the sums using EQ (10.7)

These steps are illustrated in Figure 10.12. The first and third steps are routine, so most of the attention in the remainder of this section is devoted to alternatives for the group PG logic with different trade-offs between speed, area, and complexity. Some of the hardware can be shared in the bitwise PG logic, as shown in Figure 10.13.

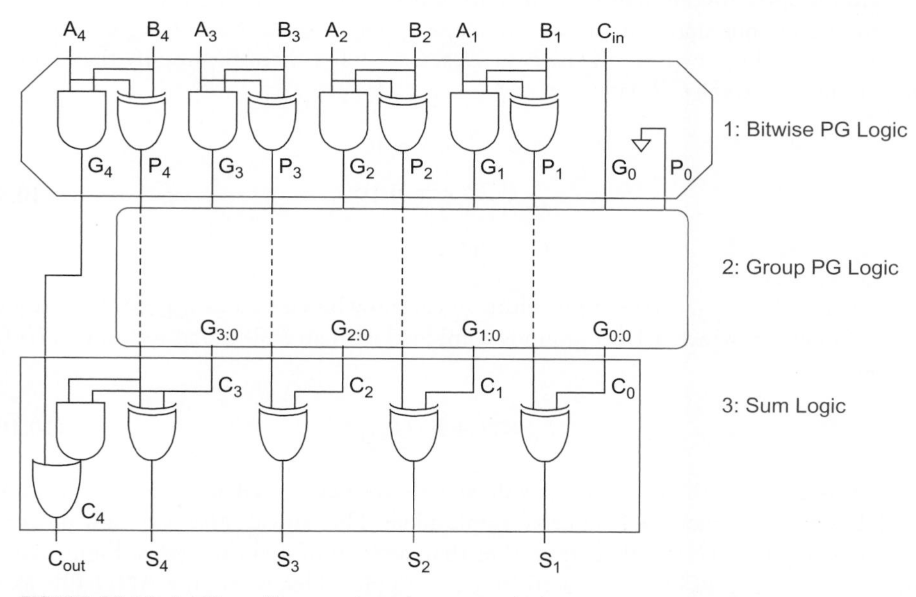

FIGURE 10.12 Addition with generate and propagate logic

Many notations are used in the literature to describe the group PG logic. In general, PG logic is an example of a *prefix* computation [Leighton92]. It accepts inputs $\{P_{N:N}, \ldots, P_{0:0}\}$ and $\{G_{N:N}, \ldots, G_{0:0}\}$ and computes the *prefixes* $\{G_{N:0}, \ldots, G_{0:0}\}$ using the relationship given in EQ (10.4). This relationship is given many names in the literature including the *delta operator*, *fundamental carry operator*, and *prefix operator*. Many other problems such as priority encoding can be posed as prefix computations and all the techniques used to build fast group PG logic will apply, as we will explore in Section 10.10.

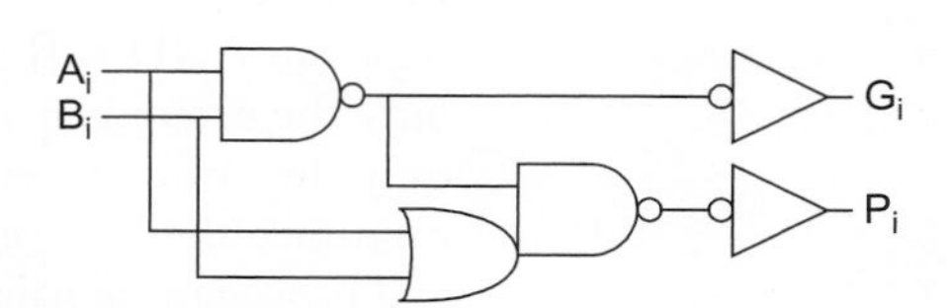

FIGURE 10.13 Shared bitwise PG logic

EQ (10.4) defines *valency*-2 (also called *radix*-2) group PG logic because it combines pairs of smaller groups. It is also possible to define higher-valency group logic to use fewer stages of more complex gates [Beaumont-Smith99], as shown in EQ (10.8) and later in Figure 10.16(c). For example, in valency-4 group logic, a group propagates the carry if all four portions propagate. A group generates a carry if the upper portion generates, the second portion generates and the upper propagates, the third generates and the upper two propagate, or the lower generates and the upper three propagate.

$$\left.\begin{aligned} G_{i:j} &= G_{i:k} + P_{i:k} \cdot G_{k-1:l} + P_{i:k} \cdot P_{k-1:l} \cdot G_{l-1:m} + P_{i:k} \cdot P_{k-1:l} \cdot P_{l-1:m} \cdot G_{m-1:j} \\ &= G_{i:k} + P_{i:k}\left(G_{k-1:l} + P_{k-1:l}\left(G_{l-1:m} + P_{l-1:m} G_{m-1:j}\right)\right) \\ P_{i:j} &= P_{i:k} \cdot P_{k-1:l} \cdot P_{l-1:m} \cdot P_{m-1:j} \end{aligned}\right\} \quad (i \geq k > l > m > j) \qquad \textbf{(10.8)}$$

Logical Effort teaches us that the best stage effort is about 4. Therefore, it is not necessarily better to build fewer stages of higher-valency gates; simulations or calculations should be done to compare the alternatives for a given process technology and circuit family.

10.2.2.3 PG Carry-Ripple Addition The critical path of the carry-ripple adder passes from carry-in to carry-out along the carry chain majority gates. As the P and G signals will have already stabilized by the time the carry arrives, we can use them to simplify the majority function into an AND-OR gate:[3]

$$\begin{aligned} C_i &= A_i B_i + \left(A_i + B_i\right) C_{i-1} \\ &= A_i B_i + \left(A_i \oplus B_i\right) C_{i-1} \\ &= G_i + P_i C_{i-1} \end{aligned} \qquad \textbf{(10.9)}$$

Because $C_i = G_{i:0}$, carry-ripple addition can now be viewed as the extreme case of group PG logic in which a 1-bit group is combined with an i-bit group to form an $(i+1)$-bit group

$$G_{i:0} = G_i + P_i \cdot G_{i-1:0} \qquad \textbf{(10.10)}$$

In this extreme, the group propagate signals are never used and need not be computed. Figure 10.14 shows a 4-bit carry-ripple adder. The critical carry path now proceeds through a chain of AND-OR gates rather than a chain of majority gates. Figure 10.15 illustrates the group PG logic for a 16-bit carry-ripple adder, where the AND-OR gates in the group PG network are represented with gray cells.

Diagrams like these will be used to compare a variety of adder architectures in subsequent sections. The diagrams use black cells, gray cells, and white buffers defined in Figure 10.16(a) for valency-2 cells. Black cells contain the group generate and propagate logic (an AND-OR gate and an AND gate) defined in EQ (10.4). Gray cells containing only the group generate logic are used at the final cell position in each column because only the group generate signal is required to compute the sums. Buffers can be used to minimize the load on critical paths. Each line represents a *bundle* of the group generate and propagate signals (propagate signals are omitted after gray cells). The bitwise PG and

[3]Whenever positive logic such as AND-OR is described, you can also use an AOI gate and alternate positive and negative polarity stages as was done in Figure 10.11(b) to save area and delay.

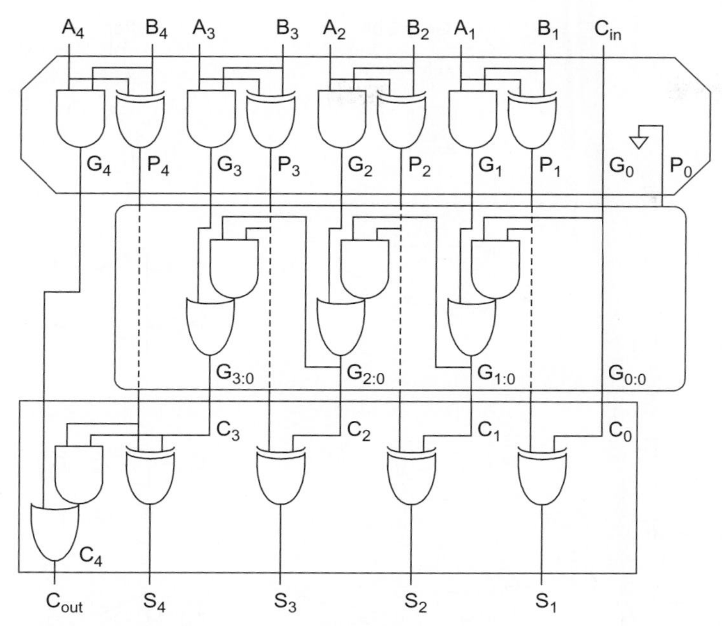

FIGURE 10.14 4-bit carry-ripple adder using PG logic

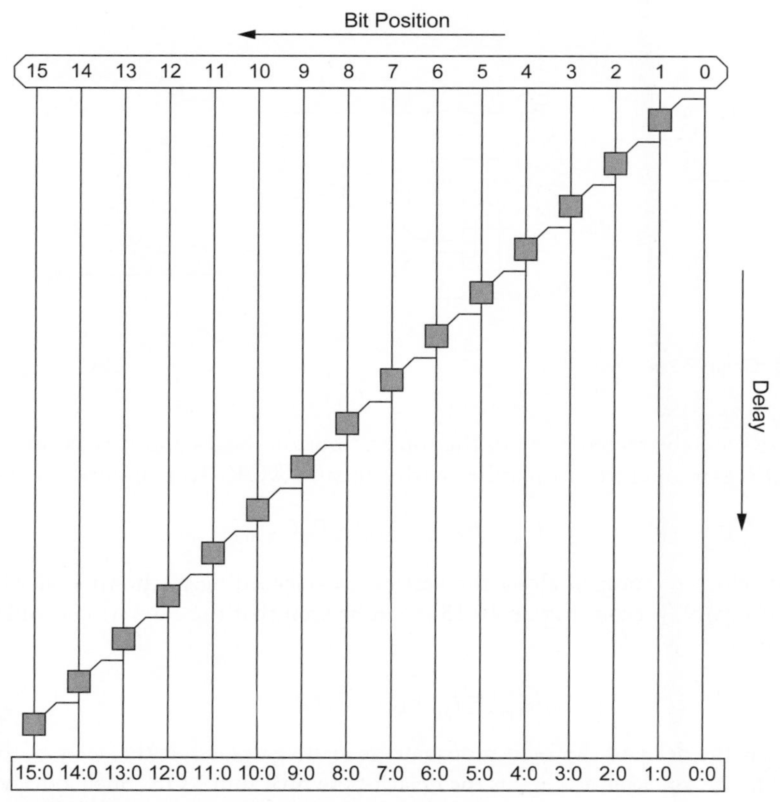

FIGURE 10.15 Carry-ripple adder group PG network

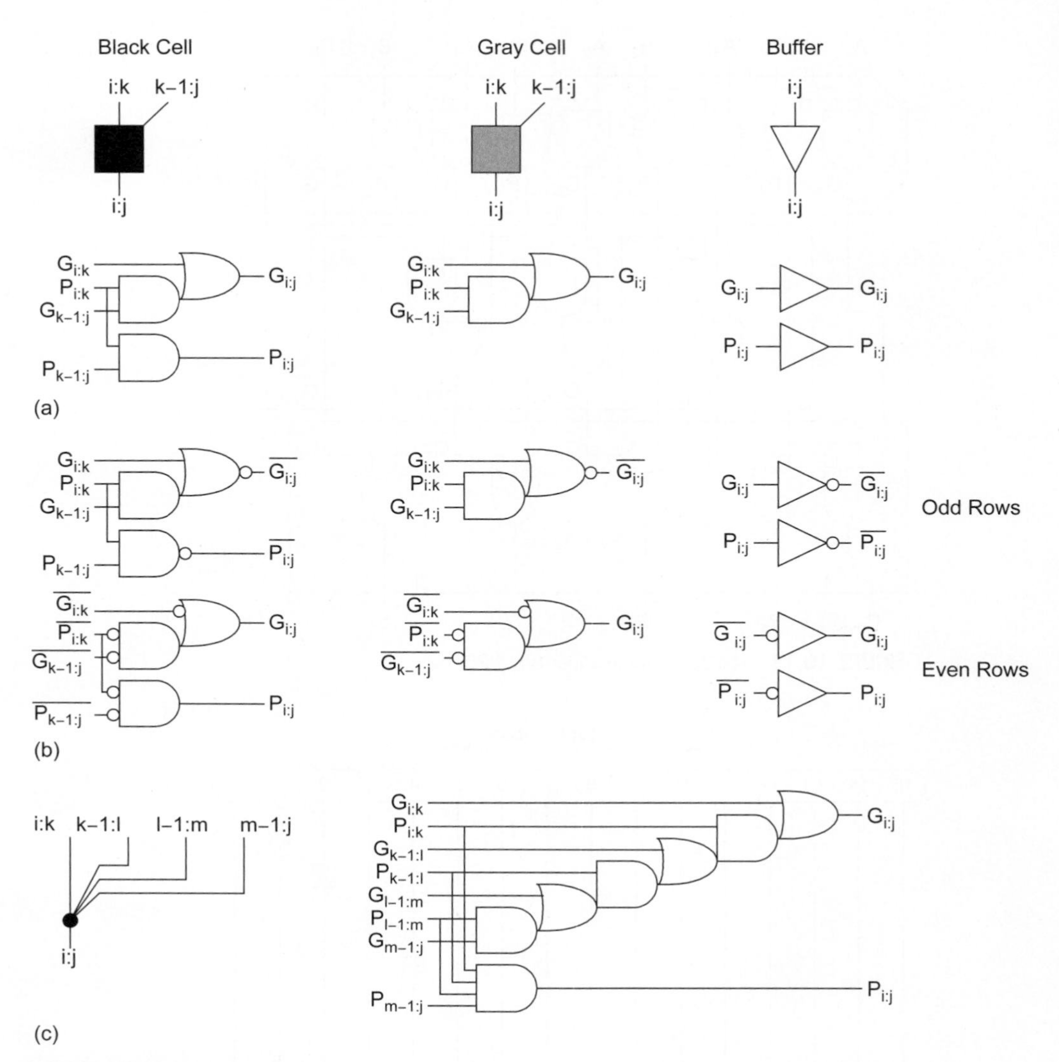

FIGURE 10.16 Group PG cells

sum XORs are abstracted away in the top and bottom boxes and it is assumed that an AND-OR gate operates in parallel with the sum XORs to compute the carry-out:

$$C_{\text{out}} = G_{N:0} = G_N + P_N G_{N-1:0} \tag{10.11}$$

The cells are arranged along the vertical axis according to the time at which they operate [Guyot97]. From Figure 10.15 it can be seen that the carry-ripple adder critical path delay is

$$t_{\text{ripple}} = t_{pg} + (N-1)\, t_{AO} + t_{\text{xor}} \tag{10.12}$$

where t_{pg} is the delay of the 1-bit propagate/generate gates, t_{AO} is the delay of the AND-OR gate in the gray cell, and t_{xor} is the delay of the final sum XOR. Such a delay estimate is only qualitative because it does not account for fanout or sizing.

Often, using noninverting gates leads to more stages of logic than are necessary. Figure 10.16(b) shows how to alternate two types of inverting stages on alternate rows of the group PG network to remove extraneous inverters. For best performance, $G_{k-1:j}$ should drive the inner transistor in the series stack. You can also reduce the number of stages by using higher-valency cells, as shown in Figure 10.16(c) for a valency-4 black cell.

10.2.2.4 Carry-Skip Adder The critical path of CPAs considered so far involves a gate or transistor for each bit of the adder, which can be slow for large adders. The *carry-skip* (also called *carry-bypass*) adder, first proposed by Charles Babbage in the nineteenth century and used for many years in mechanical calculators, shortens the critical path by computing the group propagate signals for each carry chain and using this to skip over long carry ripples [Morgan59, Lehman61]. Figure 10.17 shows a carry skip adder built from 4-bit groups. The rectangles compute the bitwise propagate and generate signals (as in Figure 10.15), and also contain a 4-input AND gate for the propagate signal of the 4-bit group. The skip multiplexer selects the group carry-in if the group propagate is true or the ripple adder carry-out otherwise.

The critical path through Figure 10.17 begins with generating a carry from bit 1, and then propagating it through the remainder of the adder. The carry must ripple through the next three bits, but then may skip across the next two 4-bit blocks. Finally, it must ripple

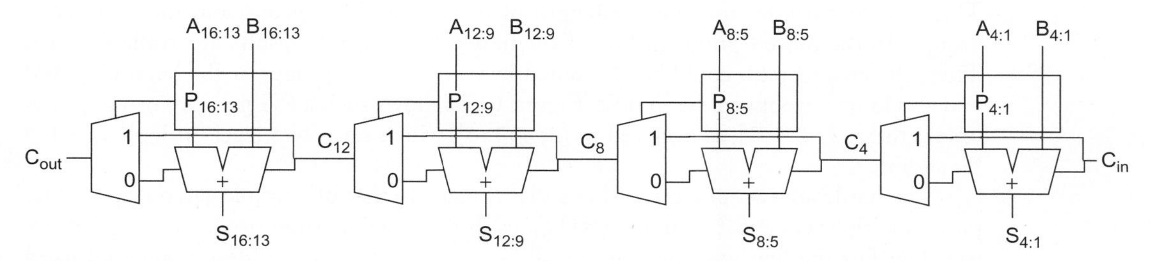

FIGURE 10.17 Carry-skip adder

through the final 4-bit block to produce the sums. This is illustrated in Figure 10.18. The 4-bit ripple chains at the top of the diagram determine if each group generates a carry. The carry skip chain in the middle of the diagram skips across 4-bit blocks. Finally, the 4-bit ripple chains with the blue lines represent *the same adders* that can produce a carry-out when a carry-in is bypassed to them. Note that the final AND-OR and column 16 are not strictly necessary because C_{out} can be computed in parallel with the sum XORs using EQ (10.11).

The critical path of the adder from Figures 10.17 and 10.18 involves the initial PG logic producing a carry out of bit 1, three AND-OR gates rippling it to bit 4, three multiplexers bypassing it to C_{12}, 3 AND-OR gates rippling through bit 15, and a final XOR to produce S_{16}. The multiplexer is an AND22-OR function, so it is slightly slower than the AND-OR function. In general, an N-bit carry-skip adder using k n-bit groups ($N = n \times k$) has a delay of

$$t_{\text{skip}} = t_{pg} + 2(n-1)t_{AO} + (k-1)t_{\text{mux}} + t_{\text{xor}} \qquad \textbf{(10.13)}$$

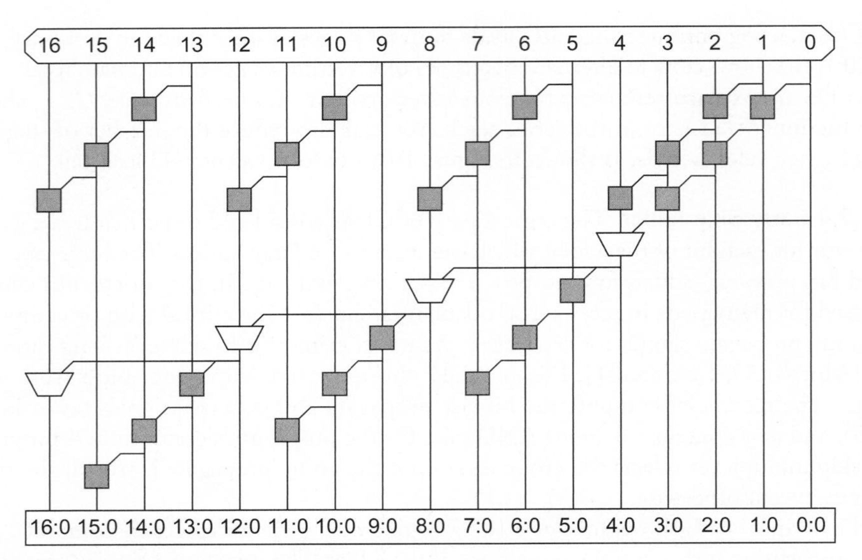

FIGURE 10.18 Carry-skip adder PG network

This critical path depends on the length of the first and last group and the number of groups. In the more significant bits of the network, the ripple results are available early. Thus, the critical path could be shortened by using shorter groups at the beginning and end and longer groups in the middle. Figure 10.19 shows such a PG network using groups of length [2, 3, 4, 4, 3], as opposed to [4, 4, 4, 4], which saves two levels of logic in a 16-bit adder.

The hardware cost of a carry-skip adder is equal to that of a simple carry-ripple adder plus k multiplexers and k n-input AND gates. It is attractive when ripple-carry adders are too slow, but the hardware cost must still be kept low. For long adders, you could use a multilevel skip approach to skip across the skips. A great deal of research has gone into choosing the best group size and number of levels [Majerski67, Oklobdzija85, Guyot87, Chan90, Kantabutra91], although now, parallel prefix adders are generally used for long adders instead.

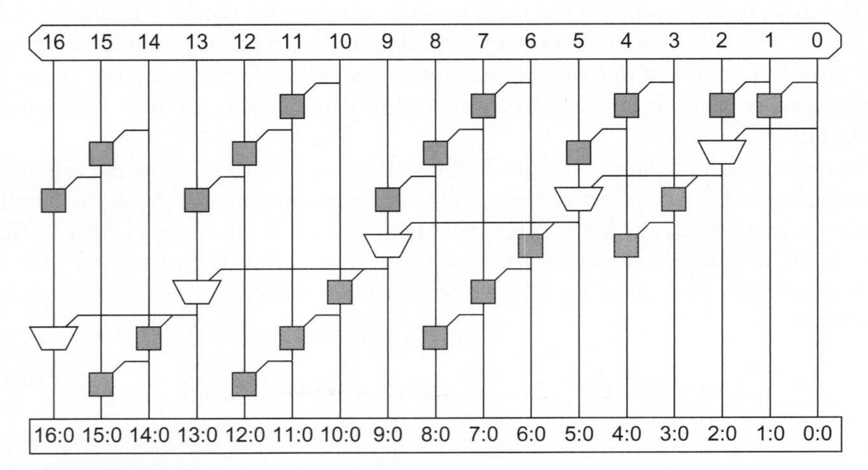

FIGURE 10.19 Variable group size carry-skip adder PG network

It might be tempting to replace each skip multiplexer in Figures 10.17 and 10.18 with an AND-OR gate combining the carry-out of the n-bit adder or the group carry-in and group propagate. Indeed, this works for domino-carry skip adders in which the carry out is precharged each cycle; it also works for carry-lookahead adders and carry-select adders covered in the subsequent section. However, it introduces a sneaky long critical path into an ordinary carry-skip adder. Imagine summing 111...111 + 000...000 + C_{in}. All of the group propagate signals are true. If C_{in} = 1, every 4-bit block will produce a carry-out. When C_{in} falls, the falling carry signal must ripple through all N bits because of the path through the carry out of each n-bit adder. Domino-carry skip adders avoid this path because all of the carries are forced low during precharge, so they can use AND-OR gates.

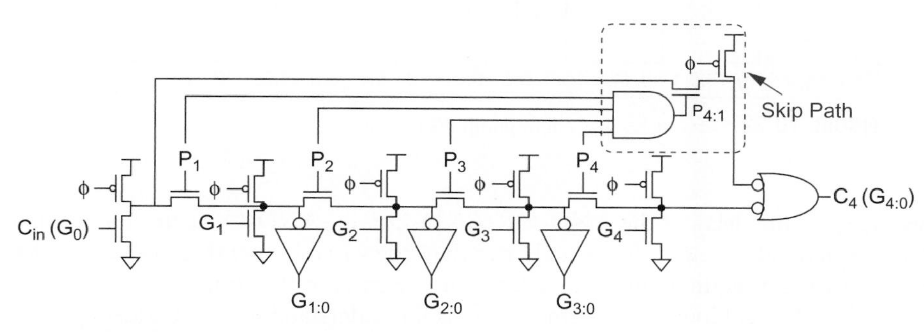

FIGURE 10.20 Carry-skip adder Manchester stage

Figure 10.20 shows how a Manchester carry chain can be modified to perform carry skip [Chan90]. A valency-5 chain is used to skip across groups of 4 bits at a time.

10.2.2.5 Carry-Lookahead Adder The *carry-lookahead adder* (CLA) [Weinberger58] is similar to the carry-skip adder, but computes group generate signals as well as group propagate signals to avoid waiting for a ripple to determine if the first group generates a carry. Such an adder is shown in Figure 10.21 and its PG network is shown in Figure 10.22 using valency-4 black cells to compute 4-bit group PG signals.

In general, a CLA using k groups of n bits each has a delay of

$$t_{\text{cla}} = t_{pg} + t_{pg(n)} + \left[(n-1)+(k-1)\right] t_{AO} + t_{\text{xor}} \tag{10.14}$$

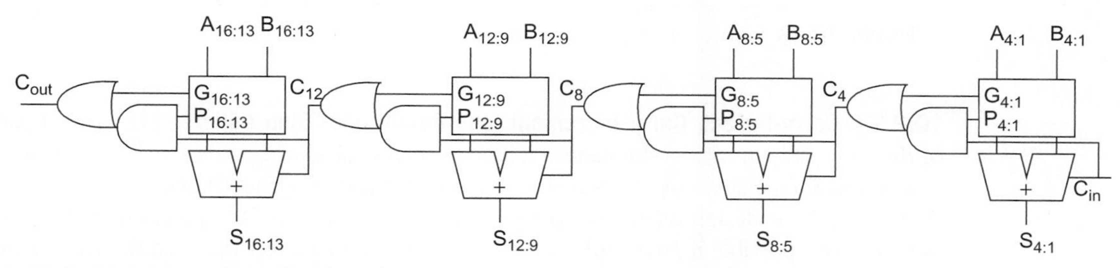

FIGURE 10.21 Carry-lookahead adder

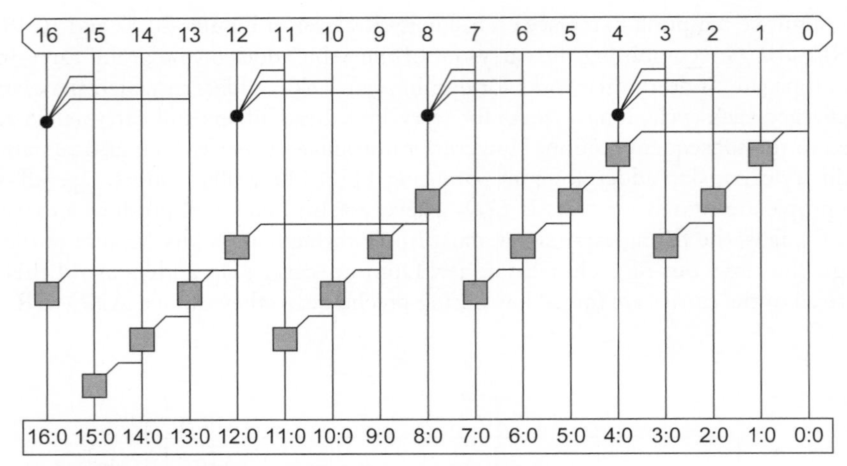

FIGURE 10.22 Carry-lookahead adder group PG network

where $t_{pg(n)}$ is the delay of the AND-OR-AND-OR-...-AND-OR gate computing the valency-n generate signal. This is no better than the variable-length carry-skip adder in Figure 10.19 and requires the extra n-bit generate gate, so the simple CLA is seldom a good design choice. However, it forms the basis for understanding faster adders presented in the subsequent sections.

CLAs often use higher-valency cells to reduce the delay of the n-bit additions by computing the carries in parallel. Figure 10.23 shows such a CLA in which the 4-bit adders are built using Manchester carry chains or multiple static gates operating in parallel.

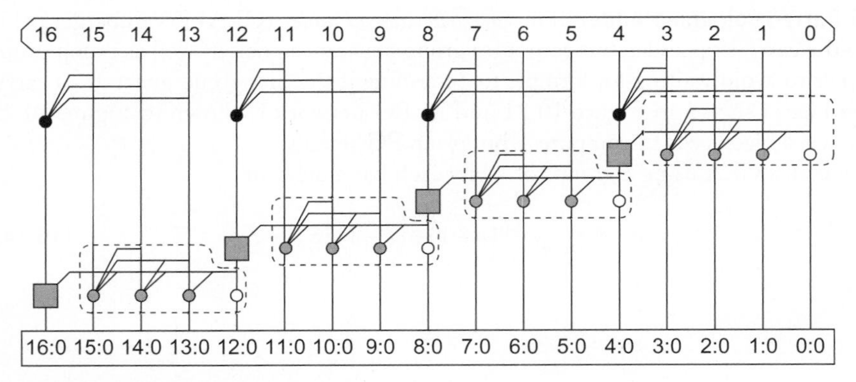

FIGURE 10.23 Improved CLA group PG network

10.2.2.6 Carry-Select, Carry-Increment, and Conditional-Sum Adders The critical path of the carry-skip and carry-lookahead adders involves calculating the carry into each n-bit group, and then calculating the sums for each bit within the group based on the carry-in. A standard logic design technique to accelerate the critical path is to precompute the outputs for both possible inputs, and then use a multiplexer to select between the two output choices. The *carry-select adder* [Bedrij62] shown in Figure 10.24 does this with a pair of

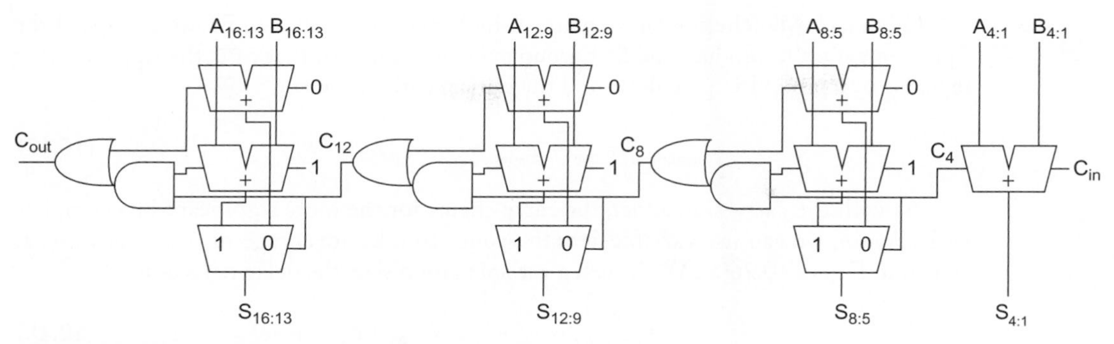

FIGURE 10.24 Carry-select adder

n-bit adders in each group. One adder calculates the sums assuming a carry-in of 0 while the other calculates the sums assuming a carry-in of 1. The actual carry triggers a multiplexer that chooses the appropriate sum. The critical path delay is

$$t_{\text{select}} = t_{pg} + \left[n + (k-2)\right] t_{AO} + t_{\text{mux}} \tag{10.15}$$

The two n-bit adders are redundant in that both contain the initial PG logic and final sum XOR. [Tyagi93] reduces the size by factoring out the common logic and simplifying the multiplexer to a gray cell, as shown in Figure 10.25. This is sometimes called a *carry-increment* adder [Zimmermann96]. It uses a short ripple chain of black cells to compute the PG signals for bits within a group. The bits spanned by each group are annotated on the diagram. When the carry-out from the previous group becomes available, the final gray cells in each column determine the carry-out, which is true if the group generates a carry or if the group propagates a carry and the previous group generated a carry. The carry-increment adder has about twice as many cells in the PG network as a carry-ripple adder. The critical path delay is about the same as that of a carry-select adder because a mux and XOR are comparable, but the area is smaller.

$$t_{\text{increment}} = t_{pg} + \left[(n-1) + (k-1)\right] t_{AO} + t_{\text{xor}} \tag{10.16}$$

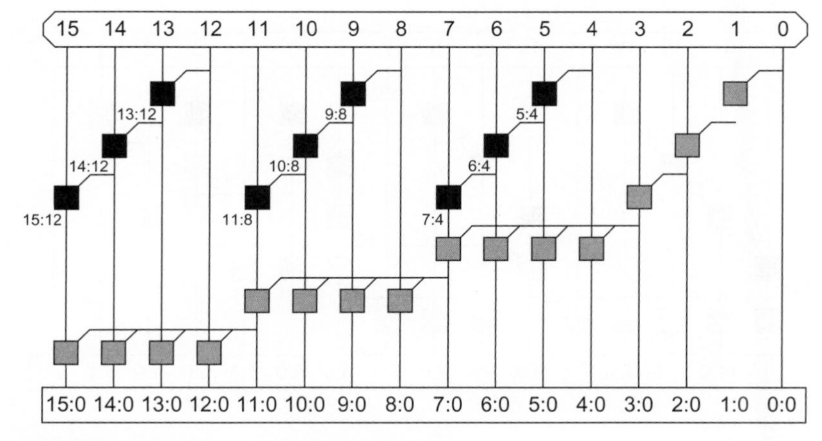

FIGURE 10.25 Carry-increment adder PG network

Of course, Manchester carry chains or higher-valency cells can be used to speed the ripple operation to produce the first group generate signal. In that case, the ripple delay is replaced by a group PG gate delay and the critical path becomes

$$t_{\text{increment}} = t_{pg} + t_{pg(n)} + \left[k-1\right] t_{AO} + t_{\text{xor}} \tag{10.17}$$

As with the carry-skip adder, the carry chains for the more significant bits complete early. Again, we can use variable-length groups to take advantage of the extra time, as shown in Figure 10.26(a). With such a variable group size, the delay reduces to

$$t_{\text{increment}} \approx t_{pg} + \sqrt{2N}\, t_{AO} + t_{\text{xor}} \tag{10.18}$$

The delay equations do not account for the fanout that each stage must drive. The fanouts in a variable-length group can become large enough to require buffering between stages. Figure 10.26(b) shows how buffers can be inserted to reduce the branching effort while not impeding the critical lookahead path; this is a useful technique in many other applications.

In wide adders, we can recursively apply multiple levels of carry-select or carry-increment. For example, a 64-bit carry-select adder can be built from four 16-bit carry-

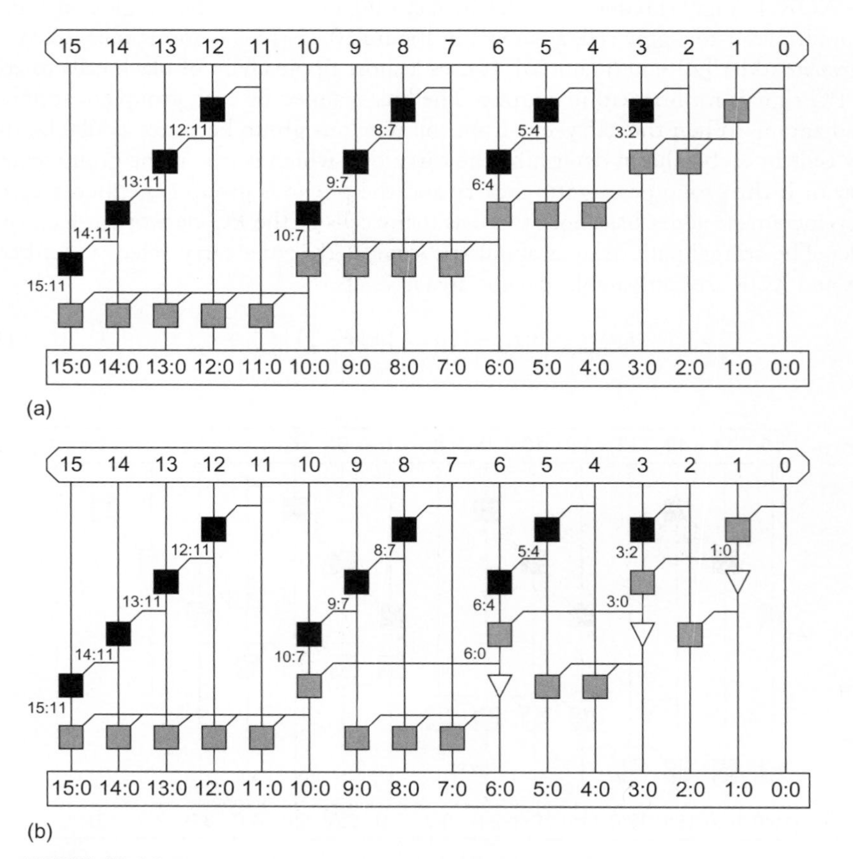

FIGURE 10.26 Variable-length carry-increment adder

select adders, each of which selects the carry-in to the next 16-bit group. Taking this to the limit, we obtain the *conditional-sum* adder [Sklansky60] that performs carry-select starting with groups of 1 bit and recursively doubling to $N/2$ bits. Figure 10.27 shows a 16-bit conditional-sum adder. In the first two rows, full adders compute the sum and carry-out for each bit assuming carries-in of 0 and 1, respectively. In the next two rows, multiplexer pairs select the sum and carry-out of the upper bit of each block of two, again assuming carries-in of 0 and 1. In the next two rows, multiplexers select the sum and carry-out of the upper two bits of each block of four, and so forth.

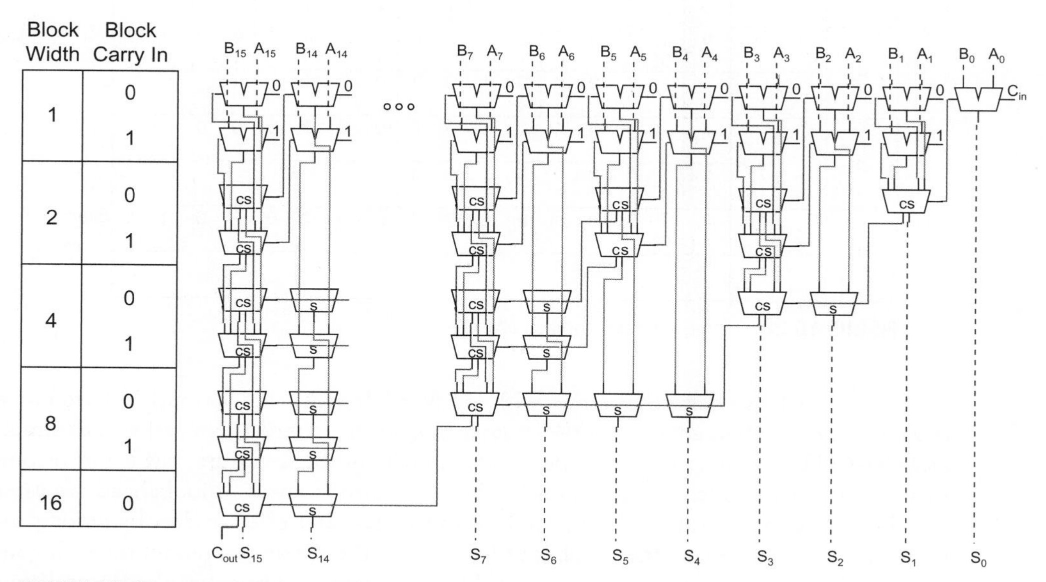

FIGURE 10.27 Conditional-sum adder

Figure 10.28 shows the operation of a conditional-sum adder in action for $N = 16$ with $C_{in} = 0$. In the block width 1 row, a pair of full adders compute the sum and carry-out for each column. One adder operates assuming the carry-in to that column is 0, while the other assumes it is 1. In the block width 2 row, the adder selects the sum for the upper half of each block (the even-numbered columns) based on the carry-out of the lower half. It also computes the carry-out of the pair of bits. Again, this is done twice, for both possibilities of carry-in to the block. In the block width 4 row, the adder again selects the sum for the upper half based on the carry-out of the lower half and finds the carry-out of the entire block. This process is repeated in subsequent rows until the 16-bit sum and the final carry-out are selected.

The conditional-sum adder involves nearly $2N$ full adders and $2N\log_2 N$ multiplexers. As with carry-select, the conditional-sum adder can be improved by factoring out the sum XORs and using AND-OR gates in place of multiplexers. This leads us to the Sklansky tree adder discussed in the next section.

10.2.2.7 Tree Adders For wide adders (roughly, $N > 16$ bits), the delay of carry-lookahead (or carry-skip or carry-select) adders becomes dominated by the delay of passing the carry through the lookahead stages. This delay can be reduced by looking ahead across the lookahead blocks [Weinberger58]. In general, you can construct a multilevel tree of look-ahead

Block Width	Block Carry In		16	15	14	13	12	11	10	9	8	7	6	5	4	3	2	1	C_{in}	
		a	1	0	1	1	1	0	1	1	0	1	1	0	1	1	0	1	0	
		b	0	0	0	1	1	0	0	1	1	0	1	1	0	1	1	0		
			Block Sum and Carry Out																	
1	0	s	1	0	1	0	0	0	1	0	1	1	0	1	1	0	1	1		
		c	0	0	0	1	1	0	0	1	0	0	1	0	0	1	0	0		
	1	s	0	1	0	1	1	1	0	1	0	0	1	0	0	1	0			
		c	1	0	1	1	1	0	1	1	1	1	1	1	1	1	1			
2	0	s	1	0	0	0	0	0	0	0	1	1	0	1	0	0	1	1		
		c	0		1		1		1		0		1		1		0			
	1	s	1	1	0	1	0	1	0	1	0	0	1	0	0	1				
		c	0		1		1		1		1		1		1					
4	0	s	1	1	0	0	0	1	0	0	0	0	0	1	0	0	1	1		
		c	0				1				1				1					
	1	s	1	1	0	1	0	1	0	1	0	0	1	0						
		c	0				1				1									
8	0	s	1	1	0	1	0	1	0	0	0	0	1	0	0	0	1	1		
		c	0								1									
	1	s	1	1	0	1	0	1	0	1										
		c	0																	
16	0	s	1	1	0	1	0	1	0	1	0	0	1	0	0	0	1	1		Sum
		c	0																	C_{out}
	1	s																		
		c																		

FIGURE 10.28 Conditional-sum addition example

structures to achieve delay that grows with log N. Such adders are variously referred to as *tree* adders, *logarithmic* adders, *multilevel-lookahead* adders, *parallel-prefix* adders, or simply *lookahead* adders. The last name appears occasionally in the literature, but is not recommended because it does not distinguish whether multiple levels of lookahead are used.

There are many ways to build the lookahead tree that offer trade-offs among the number of stages of logic, the number of logic gates, the maximum fanout on each gate, and the amount of wiring between stages. Three fundamental trees are the Brent-Kung, Sklansky, and Kogge-Stone architectures. We begin by examining each in the valency-2 case that combines pairs of groups at each stage.

The *Brent-Kung* tree [Brent82] (Figure 10.29(a)) computes prefixes for 2-bit groups. These are used to find prefixes for 4-bit groups, which in turn are used to find prefixes for 8-bit groups, and so forth. The prefixes then fan back down to compute the carries-in to each bit. The tree requires $2\log_2 N - 1$ stages. The fanout is limited to 2 at each stage. The diagram shows buffers used to minimize the fanout and loading on the gates, but in practice, the buffers are generally omitted.

The *Sklansky* or *divide-and-conquer* tree [Sklansky60] (Figure 10.29(b)) reduces the delay to $\log_2 N$ stages by computing intermediate prefixes along with the large group prefixes. This comes at the expense of fanouts that double at each level: The gates fanout to [8, 4, 2, 1] other columns. These high fanouts cause poor performance on wide adders unless the high fanout gates are appropriately sized or the critical signals are buffered before being used for the intermediate prefixes. Transistor sizing can cut into the regularity of the layout because multiple sizes of each cell are required, although the larger gates can spread into adjacent columns. Note that the recursive doubling in the Sklansky tree is analogous to the conditional-sum adder of Figure 10.27. With appropriate buffering, the fanouts can be reduced to [8, 1, 1, 1], as explored in Exercise 10.23.

The *Kogge-Stone* tree [Kogge73] (Figure 10.29(c)) achieves both $\log_2 N$ stages and fanout of 2 at each stage. This comes at the cost of many long wires that must be routed between stages. The tree also contains more PG cells; while this may not impact the area if

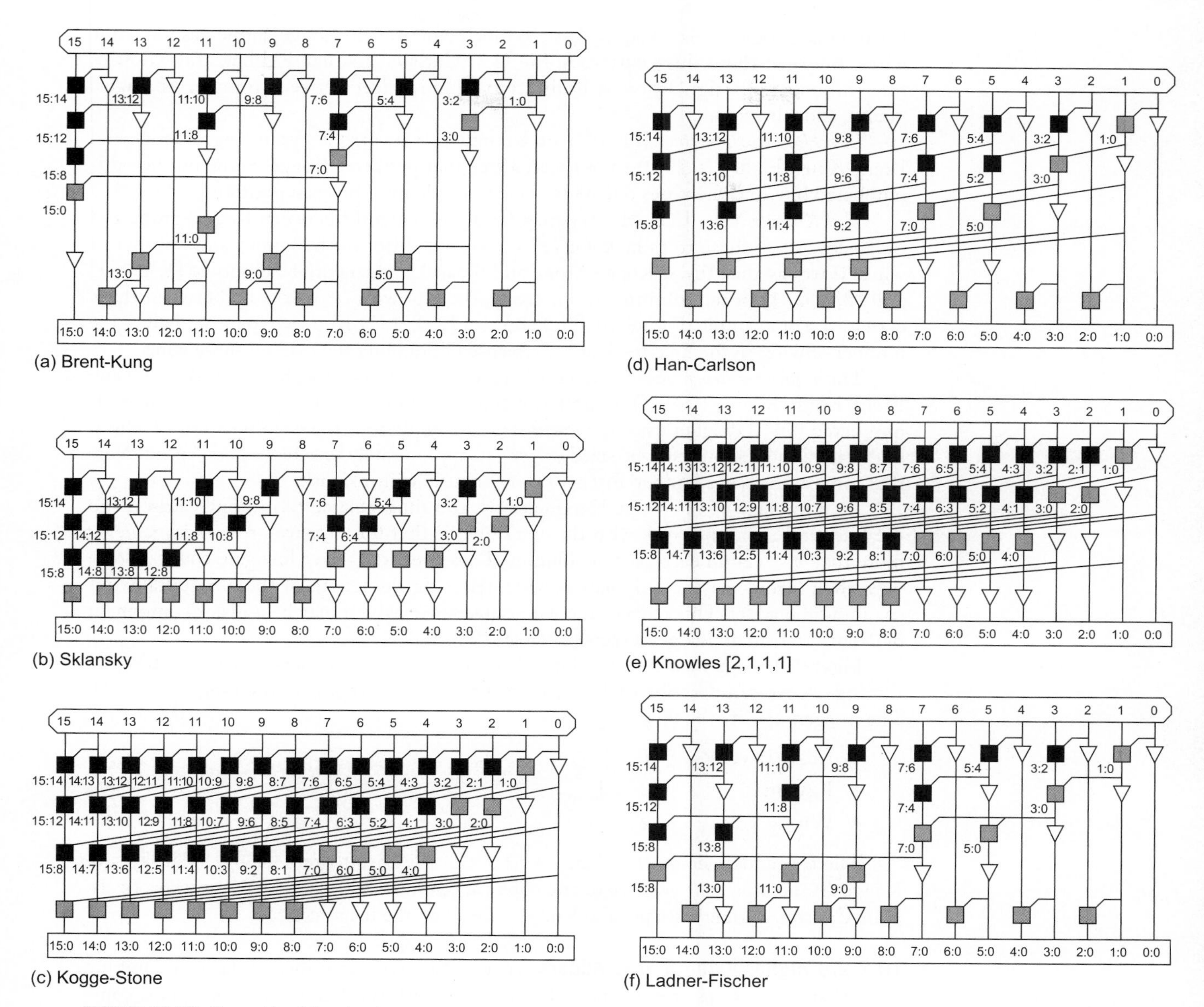

FIGURE 10.29 Tree adder PG networks

the adder layout is on a regular grid, it will increase power consumption. Despite these costs, the Kogge-Stone tree is widely used in high-performance 32-bit and 64-bit adders.

In summary, a Sklansky or Kogge-Stone tree adder reduces the critical path to

$$t_{\text{tree}} \approx t_{pg} + \lceil \log_2 N \rceil t_{AO} + t_{\text{xor}} \tag{10.19}$$

An ideal tree adder would have $\log_2 N$ levels of logic, fanout never exceeding 2, and no more than 1 wiring track ($G_{i:j}$ and $P_{i:j}$ bundle) between each row. The basic tree architectures represent cases that approach the ideal, but each differ in one respect. Brent-Kung has too many logic levels. Sklansky has too much fanout. And Kogge-Stone has too many wires. Between these three extremes, the Han-Carlson, Ladner-Fischer, and Knowles trees fill out the design space with different compromises between number of stages, fanout, and wire count.

The *Han-Carlson* trees [Han87] are a family of networks between Kogge-Stone and Brent-Kung. Figure 10.29(d) shows such a tree that performs Kogge-Stone on the odd-numbered bits, and then uses one more stage to ripple into the even positions.

The *Knowles* trees [Knowles01] are a family of networks between Kogge-Stone and Sklansky. All of these trees have $\log_2 N$ stages, but differ in the fanout and number of wires. If we say that 16-bit Kogge-Stone and Sklansky adders drive fanouts of [1, 1, 1, 1] and [8, 4, 2, 1] other columns, respectively, the Knowles networks lie between these extremes. For example, Figure 10.29(e) shows a [2, 1, 1, 1] Knowles tree that halves the number of wires in the final track at the expense of doubling the load on those wires.

The *Ladner-Fischer* trees [Ladner80] are a family of networks between Sklansky and Brent-Kung. Figure 10.29(f) is similar to Sklansky, but computes prefixes for the odd-numbered bits and again uses one more stage to ripple into the even positions. Cells at high-fanout nodes must still be sized or ganged appropriately to achieve good speed. Note that some authors use Ladner-Fischer synonymously with Sklansky.

An advantage of the Brent-Kung network and those related to it (Han-Carlson and the Ladner-Fischer network with the extra row) is that for any given row, there is never more than one cell in each pair of columns. These networks have low gate count. Moreover, their layout may be only half as wide, reducing the length of the horizontal wires spanning the adder. This reduces the wire capacitance, which may be a major component of delay in 64-bit and larger adders [Huang00].

Figure 10.30 shows a 3-dimensional taxonomy of the tree adders [Harris03]. If we let $L = \log_2 N$, we can describe each tree with three integers (l, f, t) in the range $[0, L-1]$. The integers specify the following:

- Logic Levels: $L + l$
- Fanout: $2^f + 1$
- Wiring Tracks: 2^t

The tree adders lie on the plane $l + f + t = L - 1$. 16-bit Brent-Kung, Sklansky, and Kogge-Stone represent vertices of the cube (3, 0, 0), (0, 3, 0) and (0, 0, 3), respectively. Han-Carlson, Ladner-Fischer, and Knowles lie along the diagonals.

10.2.2.8 Higher-Valency Tree Adders Any of the trees described so far can combine more than two groups at each stage [Beaumont-Smith01]. The number of groups combined in each gate is called the *valency* or *radix* of the cell. For example, Figure 10.31 shows 27-bit valency-3 Brent-Kung, Sklansky, Kogge-Stone, and Han-Carlson trees. The rounded boxes mark valency-3 carry chains (that could be constructed using a Manchester carry chain, multiple-output domino gate, or several discrete gates). The trapezoids mark carry-increment operations. The higher-valency designs use fewer stages of logic, but each stage has greater delay. This tends to be a poor trade-off in static CMOS circuits because the stage efforts become much larger than 4, but is good in domino because the logical efforts are much smaller so fewer stages are necessary.

Nodes with large fanouts or long wires can use buffers. The prefix trees can also be internally pipelined for extremely high-throughput operation. Some higher-valency designs combine the initial PG stage with the first level of PG merge. For example, the

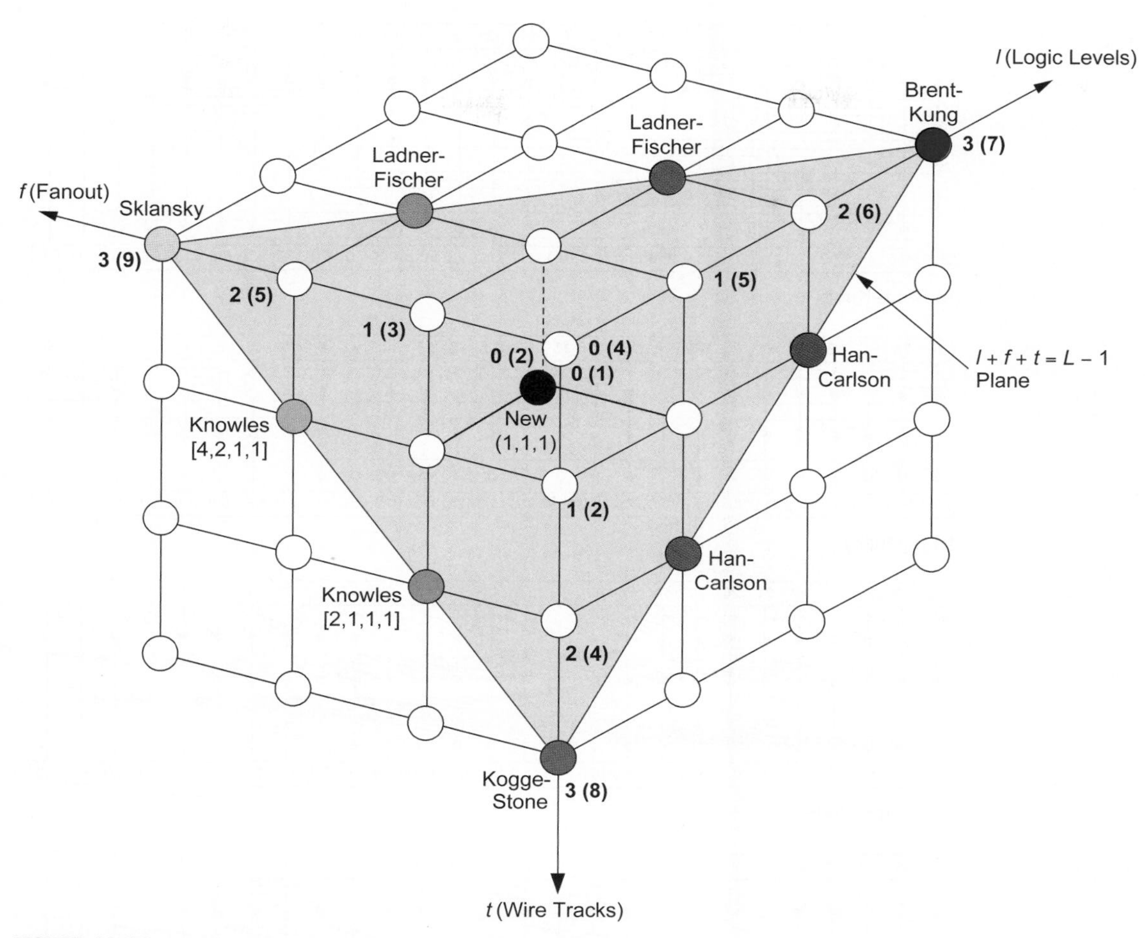

FIGURE 10.30 Taxonomy of prefix networks

Ling adder described in Section 10.2.2.10 computes generate and propagate for up to 4-bit groups from the primary inputs in a single stage.

Higher valency (v) adders can still be described in a 3-dimensional taxonomy with $L = \log_v N$ and $l + f + t = L - 1$. There are $L + l$ logic levels, a maximum fanout of $(v - 1)v^f + 1$, and $(v - 1)v^t$ wiring tracks at the worst level.

10.2.2.9 Sparse Tree Adders Building a prefix tree to compute carries in to every bit is expensive in terms of power. An alternative is to only compute carries into short groups (e.g., $s = 2, 4, 8$, or 16 bits). Meanwhile, pairs of s-bit adders precompute the sums assuming both carries-in of 0 and 1 to each group. A multiplexer selects the correct sum for each group based on the carries from the prefix tree. The group length can be balanced such that the carry-in and precomputed sums become available at about the same time. Such a hybrid between a prefix adder and a carry select adder is called a *sparse tree*. s is the *sparseness* of the tree.

The *spanning-tree adder* [Lynch92] is a sparse tree adder based on a higher-valency Brent-Kung tree of Figure 10.31(a). Figure 10.32 shows a simple valency-3 version that

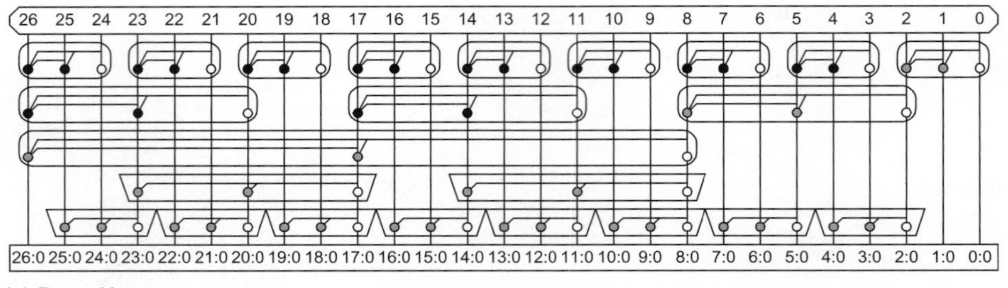

(a) Brent-Kung

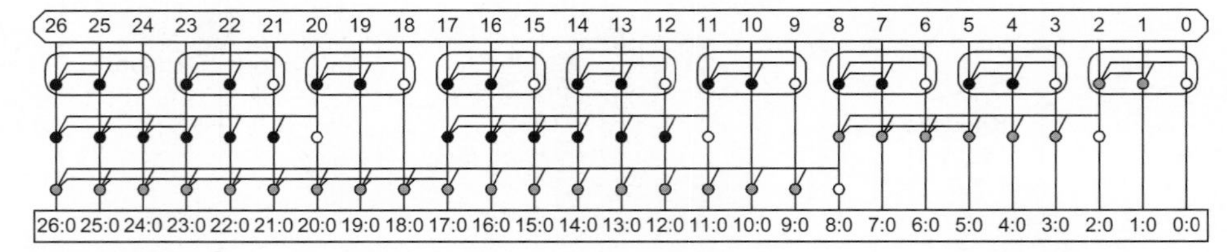

(b) Sklansky

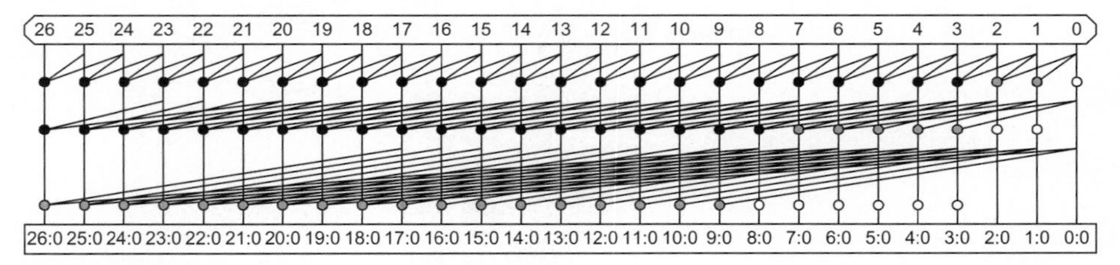

(c) Kogge-Stone

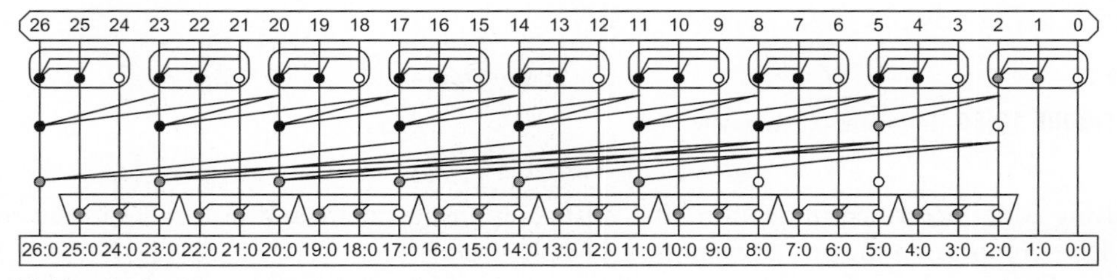

(d) Han-Carlson

FIGURE 10.31 Higher-valency tree adders

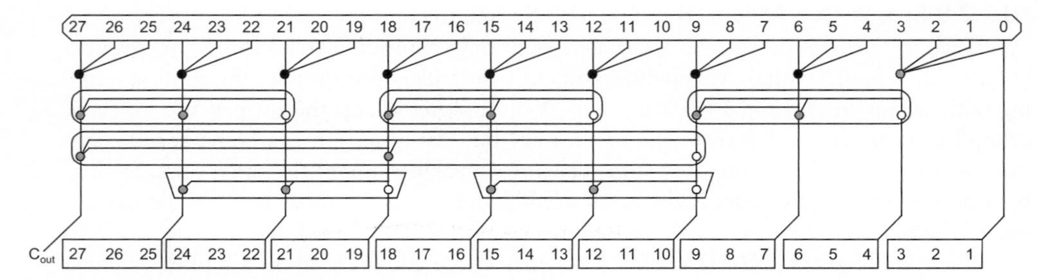

FIGURE 10.32 Valency-3 Brent-Kung sparse tree adder with $s = 3$

precomputes sums for $s = 3$-bit groups and saves one logic level by selecting the output based on the carries into each group. The carry-out (C_{out}) is explicitly shown. Note that the least significant group requires a valency-4 gray cell to compute $G_{3:0}$, the carry-in to the second select block.

[Lynch92] describes a 56-bit spanning-tree design from the AMD AM29050 floating-point unit using valency-4 stages and 8-bit carry select groups. [Kantabutra93] and [Blackburn96] describe optimizing the spanning-tree adder by using variable-length carry-select stages and appropriately selecting transistor sizes.

A carry-select box spanning bits $i \ldots j$ is shown in Figure 10.33(a). It uses short carry-ripple adders to precompute the sums assuming carry-in of 0 and 1 to the group, and then selects between them with a multiplexer, as shown in Figure 10.33(b). The adders can be simplified somewhat because the carry-ins are constant, as shown in Figure 10.33(c) for a 4-bit group.

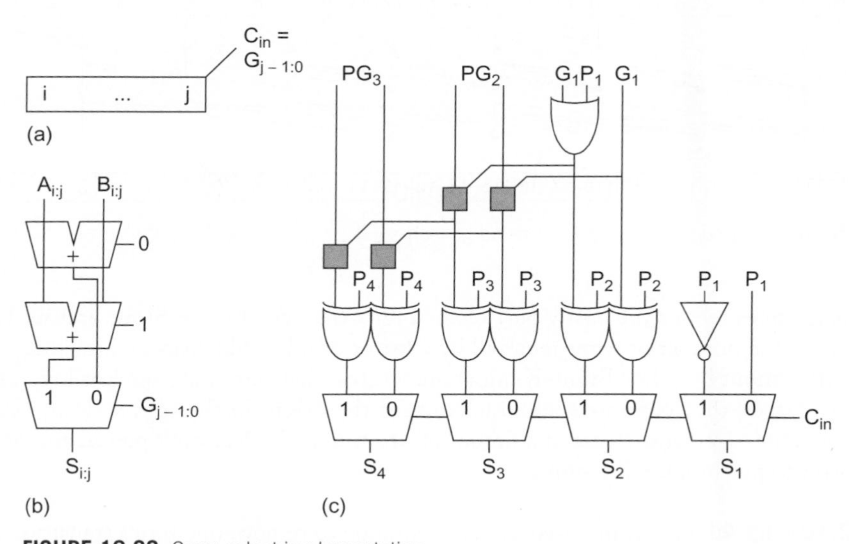

FIGURE 10.33 Carry-select implementation

[Mathew03] describes a 32-bit *sparse-tree adder* using a valency-2 tree similar to Sklansky to compute only the carries into each 4-bit group, as shown in Figure 10.34. This reduces the gate count and power consumption in the tree. The tree can also be viewed as a (2, 2, 0) Ladner-Fischer tree with the final two tree levels and XOR replaced by the select multiplexer. The adder assumes the carry-in is 0 and does not produce a carry-out, saving one input to the least-significant gray box and eliminating the prefix logic in the four most significant columns.

These sparse tree approaches are widely used in high-performance 32–64-bit higher-valency adders because they offer the small number of logic levels of higher-valency trees while reducing the gate count and power consumption in the tree. Figure 10.35 shows a 27-bit valency-3 Kogge-Stone design with carry-select on 3-bit groups. Observe how the number of gates in the tree is reduced threefold. Moreover, because the number of wires is also reduced, the extra area can be used for shielding to reduce path delay. This design can also be viewed as the Han-Carlson adder of Figure 10.31(d) with the last logic level replaced by a carry-select multiplexer.

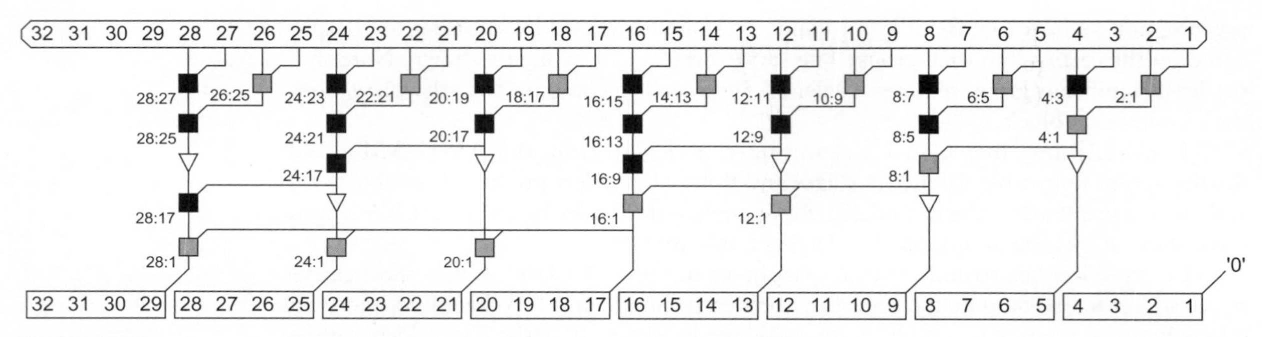

FIGURE 10.34 Intel valency-2 Sklansky sparse tree adder with $s = 4$

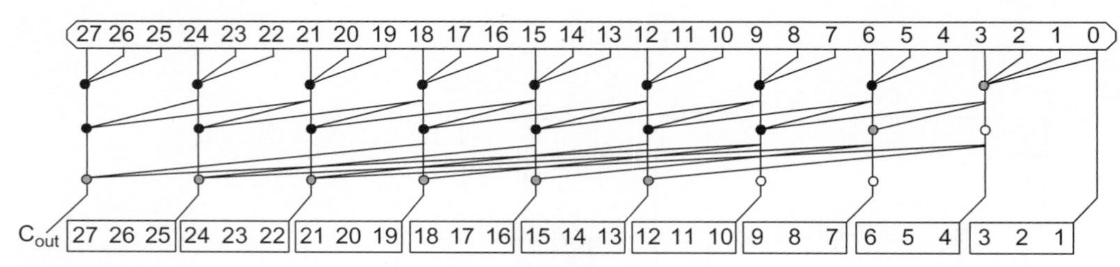

FIGURE 10.35 Valency-3 Kogge-Stone sparse tree adder with $s = 3$

Sparse trees reduce the costly part of the prefix tree. For Kogge-Stone architectures, they reduce the number of wires required by a factor of s. For Sklansky architectures, they reduce the fanout by s. For Brent-Kung architectures, they eliminate the last $\log_v s$ logic levels. In effect, they can move an adder toward the origin in the (l, f, t) design space. These benefits come at the cost of a fanout of s to the final select multiplexer, and of area and power to precompute the sums.

10.2.2.10 Ling Adders Ling discovered a technique to remove one series transistor from the critical group generate path through an adder at the expense of another XOR gate in the sum precomputation [Ling81, Doran88, Bewick94]. The technique depends on using $\overline{K}$ in place of P in the prefix network, and on the observation that $G_i\overline{K}_i = (A_iB_i)(A_i + B_i) = G_i$.

Define a *pseudogenerate* (sometimes called *pseudo-carry*) signal $H_{i:j} = G_i + G_{i-1:j}$. This is simpler than $G_{i:j} = G_i + P_iG_{i-1:j}$. $G_{i:j}$ can be obtained later from $H_{i:j}$ with an AND operation when it is needed:

$$\overline{K}_i H_{i:j} = \overline{K}_i G_i + \overline{K}_i G_{i-1:j} = G_i + \overline{K}_i G_{i-1:j} = G_{i:j} \quad \textbf{(10.20)}$$

The advantage of pseudogenerate signals over regular generate is that the first row in the prefix network is easier to compute.

Also define a *pseudopropagate* signal I that is simply a shifted version of propagate: $I_{i:j} = \overline{K}_{i-1:j-1}$. Group pseudogenerate and pseudopropagate signals are combined using the same black or gray cells as ordinary group generate and propagate signals, as you may show in Exercise 10.17.

$$\begin{aligned} H_{i:j} &= H_{i:k} + I_{i:k} H_{k-1:j} \\ I_{i:j} &= I_{i:k} I_{k-1:j} \end{aligned} \quad \textbf{(10.21)}$$

The true group generate signals are formed from the pseudogenerates using EQ (10.20). These signals can be used to compute the sums with the usual XOR: $S_i = P_i \oplus G_{i-1:0} = P_i \oplus (\overline{K}_{i-1}H_{i-1:0})$. To avoid introducing an AND gate back onto the critical path, we expand S_i in terms of $H_{i-1:0}$

$$S_i = H_{i-1:0}\left[P_i \oplus \overline{K}_{i-1}\right] + \overline{H}_{i-1:0}\left[P_i\right] \quad (10.22)$$

Thus, sum selection can be performed with a multiplexer choosing either $P_i \oplus \overline{K}_{i-1}$ or P_i based on $H_{i-1:0}$.

The Ling adder technique can be used with any form of adder that uses black and gray cells in a prefix network. It works with any valency and for both domino and static designs. The initial PG stage and the first levels of the prefix network are replaced by a cell that computes the group H and I signals directly. The middle of the prefix network is identical to an ordinary prefix adder but operates on H and I instead of G and P. The sum-selection logic uses the multiplexer from EQ (10.22) rather than an XOR. In sparse trees, the sum out of s-bit blocks is selected directly based on the H signals.

For a valency-v adder, the Ling technique converts a generate gate with v series nMOS transistors and v series pMOS transistors to a pseudogenerate gate with $v - 1$ series nMOS but still v series pMOS. For example, in valency 2, the AOI gate becomes a NOR2 gate. This is not particularly helpful for static logic, but is beneficial for domino implementations because the series pMOS are eliminated and the nMOS stacks are shortened.

Another advantage of the Ling technique is that it allows the first level pseudogenerate and pseudopropagate signals to be computed directly from the A_i and B_i inputs rather than based on G_i and K_i gates. For example, Figure 10.36 compares static gates that compute $G_{2:1}$ and $H_{2:1}$ directly from $A_{2:1}$ and $B_{2:1}$. The H gate has one fewer series transistor and much less parasitic capacitance. $H_{3:1}$ can also be computed directly from $A_{3:1}$ and $B_{3:1}$ using the complex static CMOS gate shown in Figure 10.37(a) [Quach92]. Similarly, Figure 10.37(b) shows a compound domino gate that directly computes $H_{4:1}$ from A and B using only four series transistors rather than the five required for $G_{4:1}$ [Naffziger96, Naffziger98].

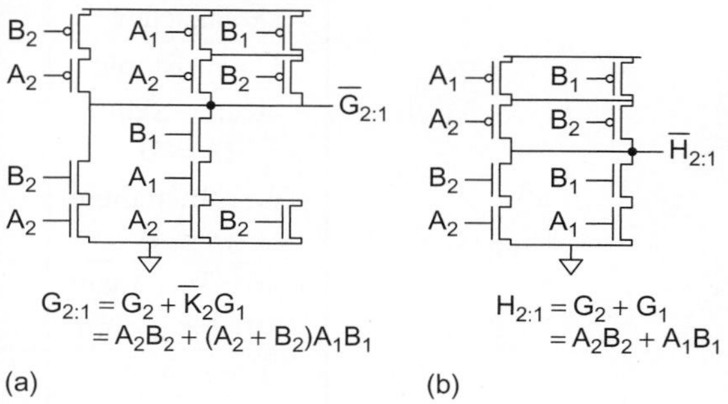

$G_{2:1} = G_2 + \overline{K}_2G_1 = A_2B_2 + (A_2 + B_2)A_1B_1$ (a)

$H_{2:1} = G_2 + G_1 = A_2B_2 + A_1B_1$ (b)

FIGURE 10.36 2-bit generate and pseudogenerate gates using primary inputs

$H_{3:1} = G_3 + G_2 + \overline{K}_2G_1 = A_3B_3 + A_2B_2 + (A_2 + B_2)A_1B_1$ (a)

$H_{4:1} = G_4 + G_3 + \overline{K}_3(G_2 + \overline{K}_2G_1) = A_4B_4 + A_3B_3 + (A_3 + B_3)(A_2B_2 + (A_2 + B_2)A_1B_1)$ (b)

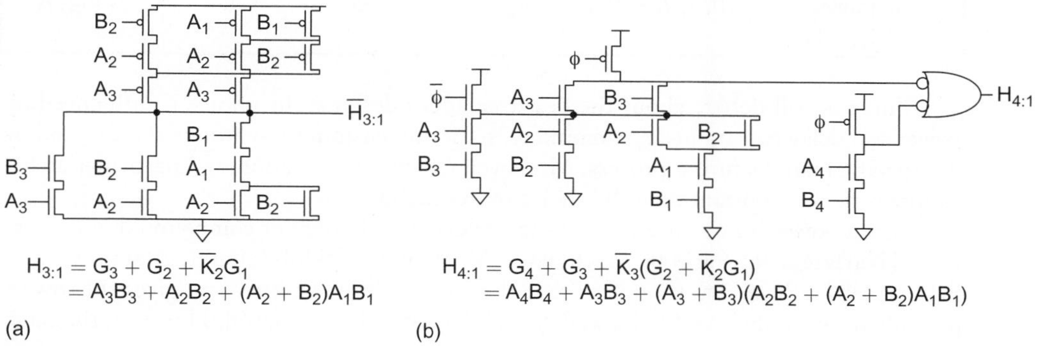

FIGURE 10.37 3-bit and 4-bit pseudogenerate gates using primary inputs

[Jackson04] proposed applying the Ling method recursively to factor out the $\overline{K}$ signal elsewhere in the adder tree. [Burgess09] showed that this recursive Ling technique opens up a new design space containing faster and smaller adders.

10.2.2.11 Summary Having examined so many adders, you probably want to know which adder should be used in which application. Table 10.3 compares the various adder architectures that have been illustrated with valency-2 prefix networks. The category "logic levels" gives the number of AND-OR gates in the critical path, excluding the initial PG logic and final XOR. Of course, the delay depends on the fanout and wire loads as well as the number of logic levels. The category "cells" refers to the approximate number of gray and black cells in the network. Carry-lookahead is not shown because it uses higher-valency cells. Carry-select is also not shown because it is larger than carry-increment for the same performance.

In general, carry-ripple adders should be used when they meet timing constraints because they use the least energy and are easy to build. When faster adders are required, carry-increment and carry-skip architectures work well for 8–16 bit lengths. Hybrids combining these techniques are also popular. At word lengths of 32 and especially 64 bits, tree adders are distinctly faster.

TABLE 10.3 Comparison of adder architectures

Architecture	Classification	Logic Levels	Max Fanout	Tracks	Cells
Carry-Ripple		$N-1$	1	1	N
Carry-Skip ($n = 4$)		$N/4 + 5$	2	1	$1.25N$
Carry-Increment ($n = 4$)		$N/4 + 2$	4	1	$2N$
Carry-Increment (variable group)		$\sqrt{2N}$	$\sqrt{2N}$	1	$2N$
Brent-Kung	$(L-1, 0, 0)$	$2\log_2 N-1$	2	1	$2N$
Sklansky	$(0, L-1, 0)$	$\log_2 N$	$N/2 + 1$	1	$0.5\, N\log_2 N$
Kogge-Stone	$(0, 0, L-1)$	$\log_2 N$	2	$N/2$	$N\log_2 N$
Han-Carlson	$(1, 0, L-2)$	$\log_2 N+1$	2	$N/4$	$0.5\, N\log_2 N$
Ladner Fischer ($l = 1$)	$(1, L-2, 0)$	$\log_2 N+1$	$N/4 + 1$	1	$0.25\, N\log_2 N$
Knowles [2,1,...,1]	$(0, 1, L-2)$	$\log_2 N$	3	$N/4$	$N\log_2 N$

There is still debate about the best tree adder designs; the choice is influenced by power and delay constraints, by domino vs. static and custom vs. synthesis choices, and by the specific manufacturing process. Moreover, careful optimization of a particular architecture is more important than the choice of tree architecture.

When power is no concern, the fastest adders use domino or compound domino circuits [Naffziger96, Park00, Mathew03, Mathew05, Oklobdzija05, Zlatanovici09, Wijeratne07]. Several authors find that the Kogge-Stone architecture gives the lowest possible delay [Silberman98, Park00, Oklobdzija05, Zlatanovici09]. However, the large number of long wires consume significant energy and require large drivers for speed. Other architectures such as Sklansky [Mathew03] or Han-Carlson [Vangal02] offer better energy efficiency because they have fewer long wires. Valency-4 dynamic gates followed by inverters tend to give a slight speed advantage [Naffziger96, Park00, Zlatanovici09, Harris04, Oklobdzija05], but compound domino implementations using valency-2 dynamic gates followed by valency-2 HI-skew static gates are also used [Mathew03]. Sparse trees save energy in domino adders with little effect on performance [Naffziger96, Mathew03, Zlatanovici09]. The Ling optimization is not used universally, but several studies have found it to be beneficial [Quach92, Naffziger96, Zlatanovici09, Grad04]. The UltraSparc III used a dual-rail domino Kogge-Stone adder [Heald00]. The Itanium 2 and Hewlett Packard PA-RISC lines of 64-bit microprocessors used a dual-rail domino sparse tree Ling adder [Naffziger96, Fetzer02]. The 65 nm Pentium 4 uses a compound domino radix-2 Sklansky sparse tree [Wijeratne07]. A good 64-bit domino adder takes 7–9 FO4 delays and has an area of 4–12 Mλ^2 [Naffziger96, Zlatanovici09, Mathew05].

Power-constrained designs use static adders, which consume one third to one tenth the energy of dynamic adders and have a delay of about 13 FO4 [Oklobdzija05, Harris03, Zlatanovici09]. For example, the CELL processor floating point unit uses a valency-2 static Kogge-Stone adder [Oh06].

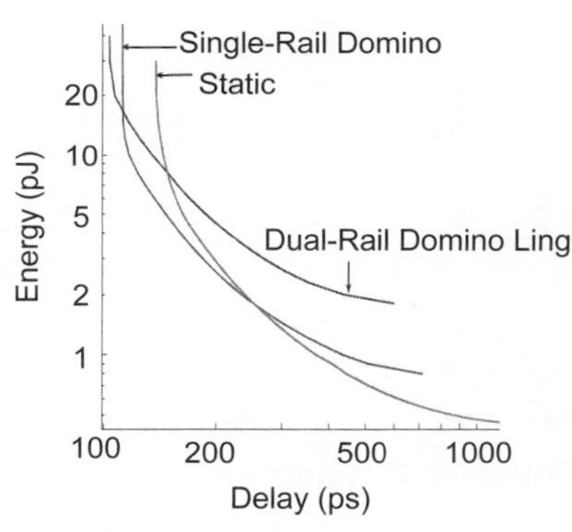

FIGURE 10.38 Energy-delay trade-offs for 90 nm 32-bit Sklansky static, domino, and dual-rail domino adders. FO4 inverter delay in this process at 1.0 V and nominal V_t is 31 ps. (© IEEE 2007.)

[Patil07] presents a comprehensive study of energy-delay design space for adders. The paper concludes that the Sklansky architecture is most energy efficient for any delay requirement because it avoids the large number of power-hungry wires in Kogge-Stone and the excessive number of logic levels in Brent-Kung. The high-fanout gates in the Sklansky tree are upsized to maintain a reasonable logical effort. Static adders are most efficient using valency-2 cells, which provide a stage effort of about 4. Domino adders are most efficient alternating valency-4 dynamic gates with static inverters. The sum precomputation logic in a static sparse tree adder costs more energy than it saves from the prefix network. In a domino adder, a sparseness of 2 does save energy because the sum precomputation can be performed with static gates. Figure 10.38 shows some results, finding that static adders are most energy-efficient for slow adders, while domino become better at high speed requirements and dual-rail domino Ling adders are preferable only for the very fastest and most energy-hungry adders. The very fast delays are achieved using a higher V_{DD} and lower V_t. [Zlatanovici09] explores the energy-delay space for 64-bit domino adders and came to the contradictory conclusion that Kogge-Stone is superior. Again, alternating valency-4 dynamic gates with static inverters and using a sparseness of 2 gave the best results, as shown in Figure 10.39. Other reasonable adders are almost as good in the energy-delay space, so there is not a compelling reason to choose one topology over another and the debate about the "best" adder will doubtlessly rage into the future.

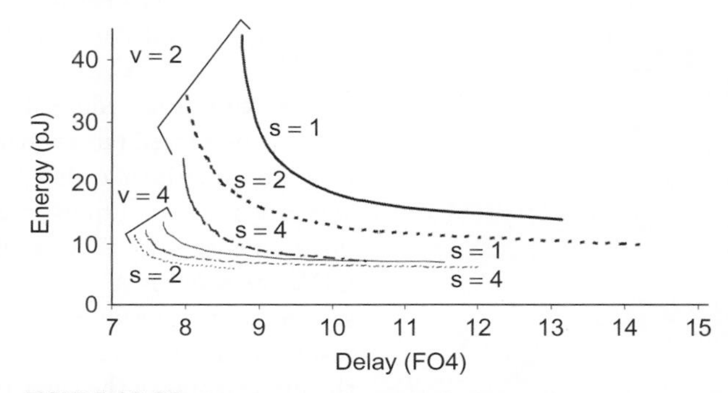

FIGURE 10.39 Energy-delay trade-offs for 90 nm 64-bit domino Kogge-Stone Ling adders as a function of valency (v) and sparseness (s). (© IEEE 2009.)

Good logic synthesis tools automatically map the "+" operator onto an appropriate adder to meet timing constraints while minimizing area. For example, the Synopsys DesignWare libraries contain carry-ripple adders, carry-select adders, carry-lookahead adders, and a variety of prefix adders. Figure 10.40 shows the results of synthesizing

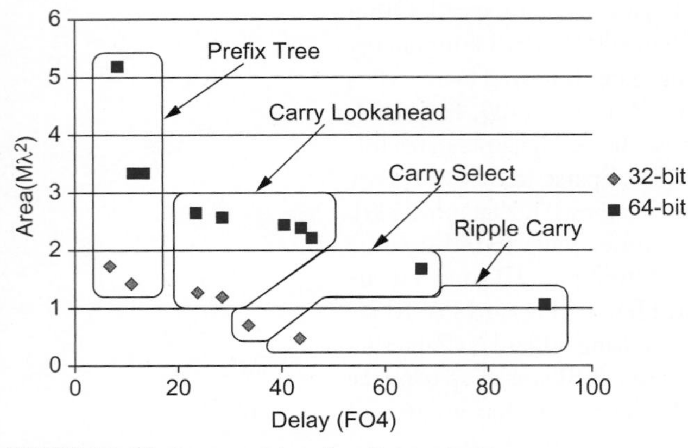

FIGURE 10.40 Area vs. delay of synthesized adders

32-bit and 64-bit adders under different timing constraints. As the latency decreases, synthesis selects more elaborate adders with greater area. The results are for a 0.18 μm commercial cell library with an FO4 inverter delay of 89 ps in the TTTT corner and the area includes estimated interconnect as well as gates. The fastest designs use tree adders and achieve implausibly fast prelayout delays of 7.0 and 8.5 FO4 for 32-bit and 64-bit adders, respectively, by creating nonuniform designs with side loads carefully buffered off the critical path. The carry-select adders achieve an interesting area/delay trade-off by using carry-ripple for the lower three-fourths of the bits and carry-select only on the upper fourth. The results will be somewhat slower when wire parasitics are included.

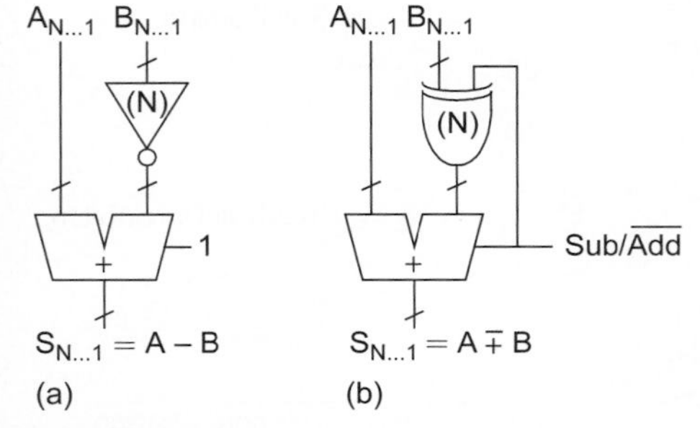

FIGURE 10.41 Subtracters

10.2.3 Subtraction

An N-bit subtracter uses the two's complement relationship

$$A - B = A + \overline{B} + 1 \tag{10.23}$$

This involves inverting one operand to an N-bit CPA and adding 1 via the carry input, as shown in Figure 10.41(a). An adder/subtracter uses XOR gates to conditionally invert B, as shown in Figure 10.41(b). In prefix adders, the XOR gates on the B inputs are sometimes merged into the bitwise PG circuitry.

10.2.4 Multiple-Input Addition

The most obvious method of adding k N-bit words is with $k - 1$ cascaded CPAs as illustrated in Figure 10.42(a) for 0001 + 0111 + 1101 + 0010. This approach consumes a large amount of hardware and is slow. A better technique is to note that a full adder sums three inputs of unit weight and produces a sum output of unit weight and a carry output of double weight. If N full adders are used in parallel, they can accept three N-bit input words $X_{N...1}$, $Y_{N...1}$, and $Z_{N...1}$, and produce two N-bit output words $S_{N...1}$ and $C_{N...1}$, satisfying $X + Y + Z = S + 2C$, as shown in Figure 10.42(b). The results correspond to the sums and carries-out of each adder. This is called *carry-save redundant format* because the carry outputs are preserved rather than propagated along the adder. The full adders in this application are sometimes called *[3:2] carry-save adder* (CSA) because they accept three inputs and produce two outputs in carry-save form. When the carry word C is shifted left by one position (because it has double weight) and added to the sum word S with an ordinary CPA, the result is $X + Y + Z$. Alternatively, a fourth input word can be added to the carry-save redundant result with another row of CSAs, again resulting in a carry-save redundant result. Such carry-save addition of four numbers is illustrated in Figure 10.42(c), where the underscores in the carry outputs serve as reminders that the carries must be shifted left one column on account of their greater weight.

The critical path through a [3:2] adder is for the sum computation, which involves one 3-input XOR, or two levels of XOR2. This is much faster than a CPA. In general, k

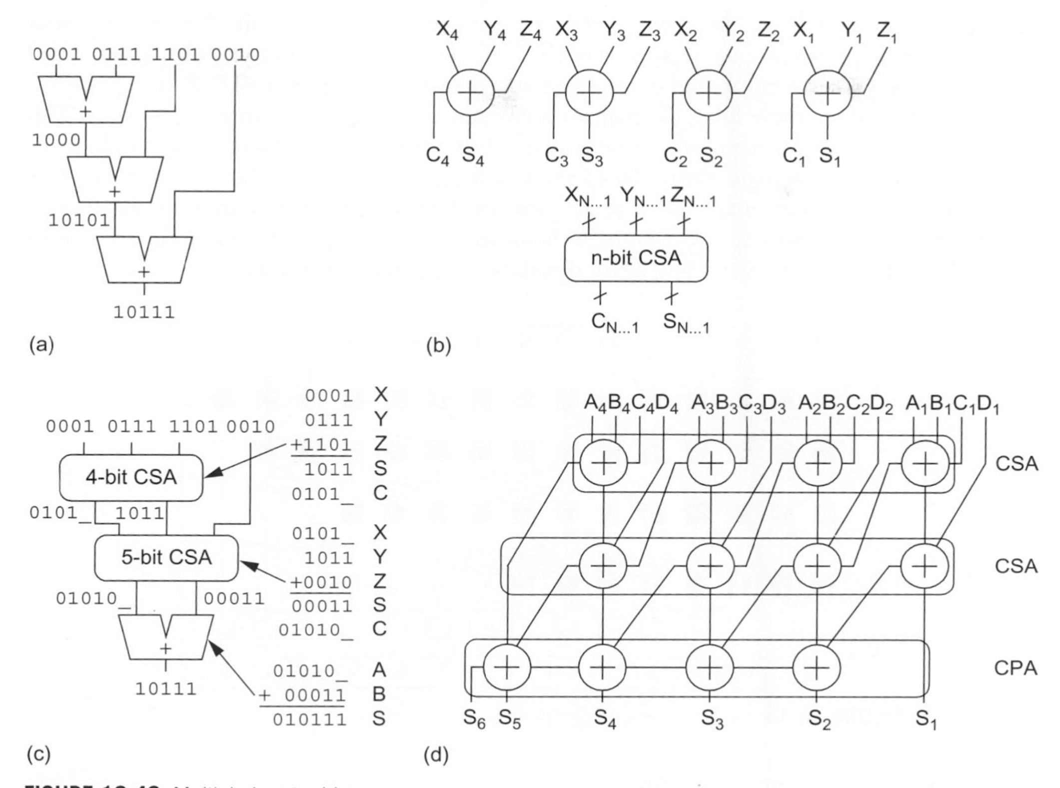

FIGURE 10.42 Multiple-input adders

numbers can be summed with $k - 2$ [3:2] CSAs and only one CPA. This approach will be exploited in Section 10.9 to add many partial products in a multiplier rapidly. The technique dates back to von Neumann's early computer [Burks46].

When one of the inputs to a CSA is a constant, the hardware can be reduced further. If a bit of the input is 0, the CSA column reduces to a half-adder. If the bit is 1, the CSA column simplifies to $S = \overline{A \oplus B}$ and $C = A + B$.

10.2.5 Flagged Prefix Adders

Sometimes it is necessary to compute either $A + B$, and then, depending on a late-arriving control signal, adding 1. Some applications include calculating $A + B \bmod 2^n{-}1$ for cryptography and Reed-Solomon coding, computing the absolute difference $|A - B|$, doing addition/subtraction of sign-magnitude numbers, and performing rounding in certain floating-point adders [Beaumont-Smith99]. A straightforward approach is to build two adders, provide a carry to one, and select between the results. [Burgess02] describes a clever alternative called a *flagged prefix adder* that uses much less hardware.

A flagged prefix adder receives A, B, and a control signal, *inc*, and computes $A + B + inc$. Recall that an ordinary adder computes the prefixes $G_{i-1:0}$ as the carries into each column i, then computes the sum $S_i = P_i \oplus G_{i-1:0}$. In this situation, there is no C_{in} and hence column 0 is omitted; $G_{i-1:1}$ is used instead. The goal of the flagged prefix adder is to adjust these carries when *inc* is asserted. A flagged prefix adder instead uses

$G'_{i-1:1} = G_{i-1:1} + P_{i-1:1} \cdot inc$. Thus, if *inc* is true, it generates a carry into all of the low order bits whose group propagate signals are TRUE. The modified prefixes, $G'_{i-1:1}$, are called *flags*. The sums are computed in the same way with an XOR gate: $S_i = P_i \oplus G'_{i-1:1}$.

To produce these flags, the flagged prefix adder uses one more row of gray cells. This requires that the former bottom row of gray cells be converted to black cells to produce the group propagate signals. Figure 10.43 shows a flagged prefix Kogge-Stone adder. The new row, shown in blue, is appended to perform the late increment. Column 0 is eliminated because there is no C_{in}, but column 16 is provided because applications of flagged adders will need the generate and propagate signals spanning the entire n bits.

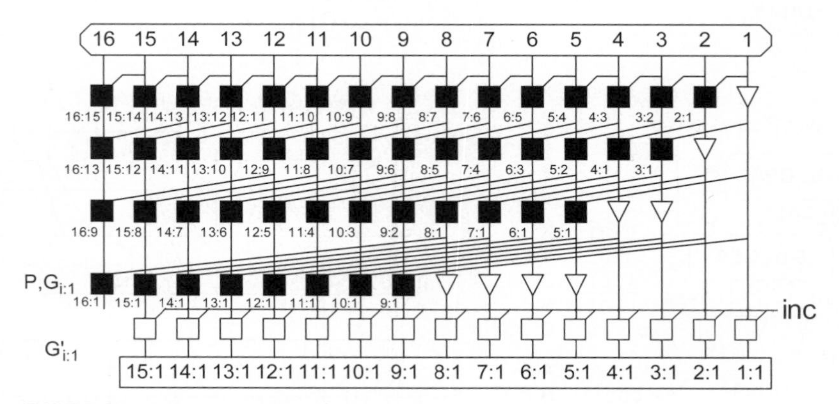

FIGURE 10.43 Flagged prefix Kogge-Stone adder

10.2.5.1 Modulo $2^n - 1$ Addition To compute $A + B \bmod 2^n - 1$ for unsigned operands, an adder should first compute $A + B$. If the sum is greater than or equal to $2^n - 1$, the result should be incremented and truncated back to n bits. $G_{n:1}$ is TRUE if the adder will overflow; i.e., the result is greater than $2^n - 1$. $P_{n:1}$ is TRUE if all columns propagate, which only occurs when the sum equals $2^n - 1$. Hence, modular addition can done with a flagged prefix adder using $inc = G_{n:1} + P_{n:1}$.

Compared to ordinary addition, modular addition requires one more row of black cells, an OR gate to compute *inc*, and a buffer to drive *inc* across all n bits.

10.2.5.2 Absolute Difference $|A - B|$ is called the *absolute difference* and is commonly used in applications such as video compression. The most straightforward approach is to compute both $A - B$ and $B - A$, then select the positive result. A more efficient technique is to compute $A + \overline{B}$ and look at the sign, indicated by $\overline{G}_{n:1}$. If the result is negative, it should be inverted to obtain $B - A$. If the result is positive, it should be incremented to obtain $A - B$.

All of these operations can be performed using a flagged prefix adder enhanced to invert the result conditionally. Modify the sum logic to calculate $S_i = (P_i \oplus inv) \oplus G'_{i-1:1}$. Choose $inv = \overline{G}_{n:1}$ and $inc = G_{n:1}$.

Compared to ordinary addition, absolute difference requires a bank of inverters to obtain $\overline{B}$, one more row of black cells, buffers to drive inv and inc across all n bits, and a row of XORs to invert the result conditionally. Note that $(P_i \oplus inv)$ can be precomputed so this does not affect the critical path.

10.2.5.3 Sign-Magnitude Arithmetic Addition of sign-magnitude numbers involves examining the signs of the operands. If the signs agree, the magnitudes are added and the sign is unchanged. If the signs differ, the absolute difference of the magnitudes must be computed. This can be done using the flagged carry adder described in the previous section. The sign of the result is sign(A) ⊕ $G_{n:1}$.

Subtraction is identical except that the sign of B is first flipped.

10.3 One/Zero Detectors

Detecting all ones or zeros on wide N-bit words requires large fan-in AND or NOR gates. Recall that by DeMorgan's law, AND, OR, NAND, and NOR are fundamentally the same operation except for possible inversions of the inputs and/or outputs. You can build a tree of AND gates, as shown in Figure 10.44(a). Here, alternate NAND and NOR gates have been used. The path has log N stages. In general, the minimum logical effort is achieved with a tree alternating NAND gates and inverters and the path logical effort is

$$G_{\text{and}}(N) = \left(\frac{4}{3}\right)^{\log_2 N} = N^{\log_2 \frac{4}{3}} = N^{0.415} \quad \textbf{(10.24)}$$

A rough estimate of the path delay driving a path electrical effort of H using static CMOS gates is

$$D \approx (\log_4 F)\, t_{FO4} = (\log_4 H + 0.415 \log_4 N)\, t_{FO4} \quad \textbf{(10.25)}$$

where t_{FO4} is the fanout-of-4 inverter delay.

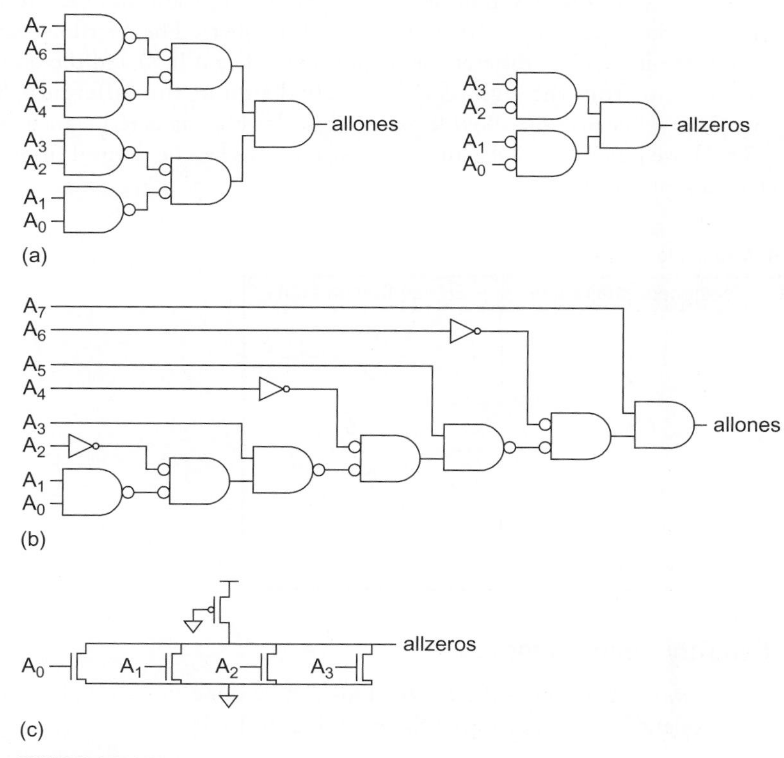

FIGURE 10.44 One/zero detectors

If the word being checked has a natural skew in the arrival time of the bits (such as at the output of a ripple adder), the designer might consider an asymmetric design that favors the late-arriving inputs, as shown in Figure 10.44(b). Here, the delay from the latest bit A_7 is a single gate.

Another fast detector uses a pseudo-nMOS or dynamic NOR structure to perform the "wired-OR," as shown in Figure 10.44(c). This works well for words up to about 16 bits; for larger words, the gates can be split into 8–16-bit chunks to reduce the parasitic delay and avoid problems with subthreshold leakage.

10.4 Comparators

10.4.1 Magnitude Comparator

A *magnitude comparator* determines the larger of two binary numbers. To compare two unsigned numbers A and B, compute $B - A = B + \overline{A} + 1$. If there is a carry-out, $A \leq B$; otherwise, $A > B$. A zero detector indicates that the numbers are equal. Figure 10.45 shows a 4-bit unsigned comparator built from a carry-ripple adder and two's complementer. The relative magnitude is determined from the carry-out (C) and zero (Z) signals according to Table 10.4. For wider inputs, any of the faster adder architectures can be used.

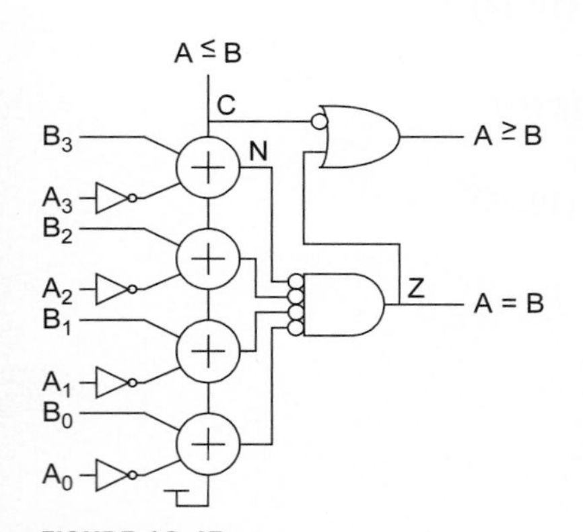

FIGURE 10.45
Unsigned magnitude comparator

Comparing signed two's complement numbers is slightly more complicated because of the possibility of overflow when subtracting two numbers with different signs. Instead of simply examining the carry-out, we must determine if the result is negative (N, indicated by the most significant bit of the result) and if it overflows the range of possible signed numbers. The overflow signal V is true if the inputs had different signs (most significant bits) and the output sign is different from the sign of B. The actual sign of the difference $B - A$ is $S = N \oplus V$ because overflow flips the sign. If this corrected sign is negative ($S = 1$), we know $A > B$. Again, the other relations can be derived from the corrected sign and the Z signal.

TABLE 10.4 Magnitude comparison

Relation	Unsigned Comparison	Signed Comparison
$A = B$	Z	Z
$A \neq B$	$\overline{Z}$	$\overline{Z}$
$A < B$	$C \cdot \overline{Z}$	$\overline{S} \cdot \overline{Z}$
$A > B$	C	S
$A \leq B$	C	$\overline{S}$
$A \geq B$	$\overline{C} + Z$	$S + Z$

10.4.2 Equality Comparator

An *equality comparator* determines if ($A = B$). This can be done more simply and rapidly with XNOR gates and a ones detector, as shown in Figure 10.46.

10.4.3 $K = A + B$ Comparator

Sometimes it is necessary to determine if ($A + B = K$). For example, the sum-addressed memory [Heald98] described in Section 11.2.2.4 contains a decoder that must match against the sum of two numbers, such as a register base address and an immediate offset. Remarkably, this comparison can be done faster than computing $A + B$ because no carry propagation is necessary. The key is that if you know A and B, you also know what the carry into each bit must be if $K = A + B$ [Cortadella92]. Therefore, you only need to check adjacent pairs of bits to verify that the previous bit produces the carry required by the current bit, and then use a ones detector to check that the condition is true for all N pairs. Specifically, if $K = A + B$, Table 10.5 lists what the carry-in c_{i-1} must have been for this to be true and what the carry-out c_i will be for each bit position i.

OPTIONAL

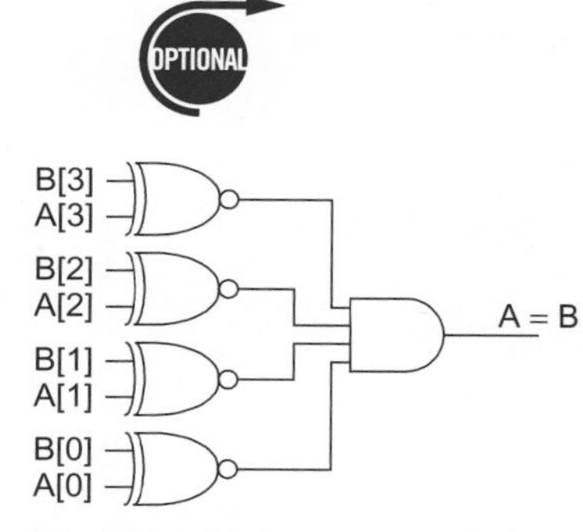

FIGURE 10.46 Equality comparator

TABLE 10.5 Required and generated carries if $K = A + B$

A_i	B_i	K_i	c_{i-1} (required)	c_i (produced)
0	0	0	0	0
0	0	1	1	0
0	1	0	1	1
0	1	1	0	0
1	0	0	1	1
1	0	1	0	0
1	1	0	0	1
1	1	1	1	1

From this table, you can see that the required c_{i-1} for bit i is

$$c_{i-1} = A_i \oplus B_i \oplus K_i \qquad \textbf{(10.26)}$$

and the c_{i-1} produced by bit $i - 1$ is

$$c_{i-1} = \left(A_{i-1} \oplus B_{i-1}\right)\overline{K}_{i-1} + A_{i-1} \cdot B_{i-1} \qquad \textbf{(10.27)}$$

Figure 10.47 shows one bitslice of a circuit to perform this operation. The XNOR gate is used to make sure that the required carry matches the produced carry at each bit position; then the AND gate checks that the condition is satisfied for all bits.

10.5 Counters

Two commonly used types of counters are *binary counters* and *linear-feedback shift registers*. An N-bit binary counter sequences through 2^N outputs in binary order. Simple designs have a minimum cycle time that increases with N, but faster designs operate in constant time. An N-bit linear-feedback shift register sequences through up to $2^N - 1$ outputs in pseudo-random order. It has a short minimum cycle time independent of N, so it is useful for extremely fast counters as well as pseudo-random number generation.

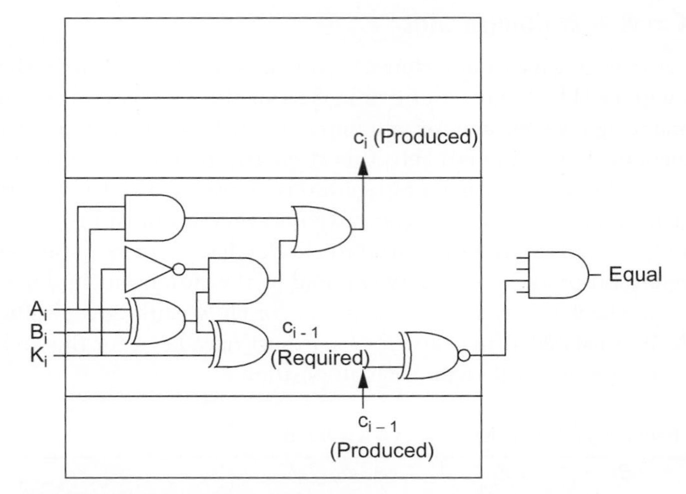

FIGURE 10.47 $A + B = K$ comparator

Some of the common features of counters include the following:

- *Resettable*: counter value is reset to 0 when *RESET* is asserted (essential for testing)
- *Loadable*: counter value is loaded with *N*-bit value when *LOAD* is asserted
- *Enabled*: counter counts only on clock cycles when *EN* is asserted
- *Reversible*: counter increments or decrements based on $\overline{UP}/DOWN$ input
- *Terminal Count: TC* output asserted when counter overflows (when counting up) or underflows (when counting down)

In general, divide-by-*M* counters ($M < 2^N$) can be built using an ordinary *N*-bit counter and circuitry to reset the counter upon reaching *M*. *M* can be a programmable input if an equality comparator is used. Alternatively, a loadable counter can be used to restart at $N - M$ whenever *TC* indicates that the counter overflowed.

10.5.1 Binary Counters

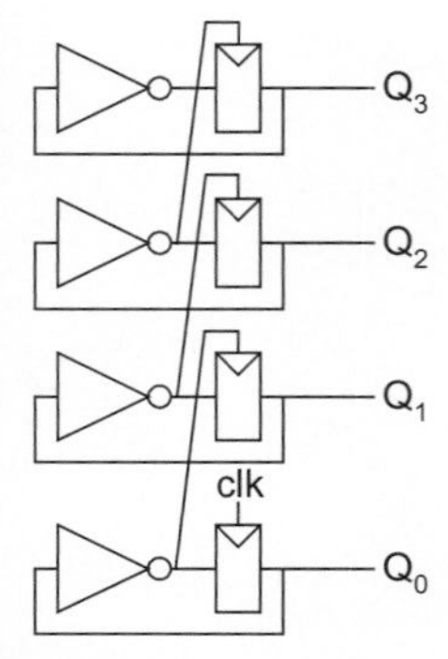

FIGURE 10.48 Asynchronous ripple-carry counter

The simplest binary counter is the *asynchronous ripple-carry counter*, as shown in Figure 10.48. It is composed of *N* registers connected in toggle configuration, where the falling transition of each register clocks the subsequent register. Therefore, the delay can be quite long. It has no reset signal, making it difficult to test. In general, asynchronous circuits introduce a whole assortment of problems, so the ripple-carry counter is shown mainly for historical interest and is not recommended for commercial designs.

A general *synchronous up/down counter* is shown in Figure 10.49(a). It uses a resettable register and full adder for each bit position. The cycle time is limited by the ripple-carry delay. While a faster adder could be used, the next section describes a better way to build fast counters. If only an *up counter* (also called an *incrementer*) is required, the full adder degenerates into a half adder, as shown in Figure 10.49(b). Including an input multiplexer allows the counter to load an initialization value. A clock enable is also often provided to each register for conditional counting. The terminal count (*TC*) output indicates that the counter has overflowed or underflowed. Figure 10.50 shows a fully featured resettable loadable enabled synchronous up/down counter.

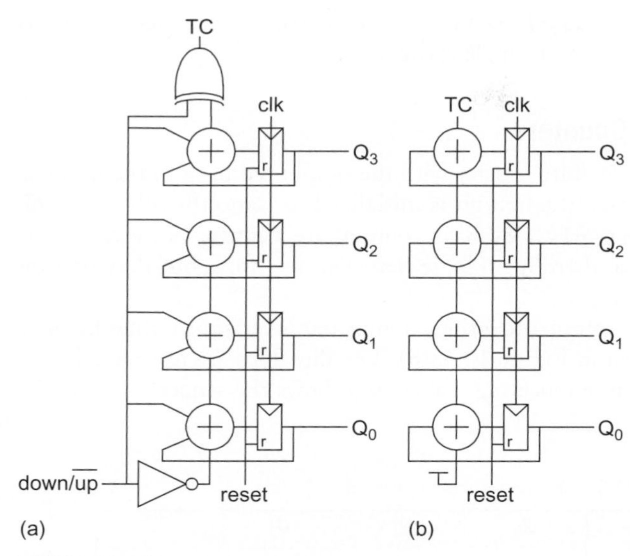

FIGURE 10.49 Synchronous counters

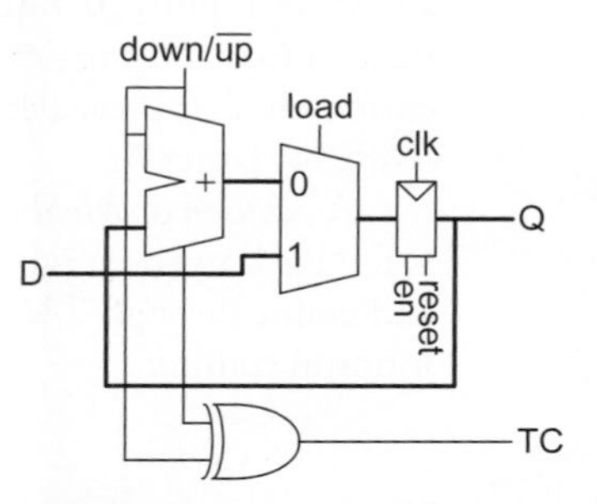

FIGURE 10.50 Synchronous up/down counter with reset, load, and enable

10.5.2 Fast Binary Counters

The speed of the counter in Figure 10.49 is limited by the adder. This can be overcome by dividing the counter into two or more segments [Ercegovac89]. For example, a 32-bit counter could be constructed from a 4-bit *prescalar* counter and a 28-bit counter, as shown in Figure 10.51. The *TC* output of the prescalar enables counting on the more significant segment. Now, the cycle time is limited only by the prescalar speed because the 28-bit adder has 2^4 cycles to produce a result. By using more segments, a counter of arbitrary length can run at the speed of a 1- or 2-bit counter.

Prescaling does not suffice for up/down counters because the more significant segment may have only a single cycle to respond when the counter changes direction. To solve this, a *shadow register* can be used on the more significant segments to hold the previous value that should be used when the direction changes [Stan98]. Figure 10.52 shows the more significant segment for a fast up/down counter. On reset (not shown in the figure), the *dir* register is set to 0, Q to 0, and *shadow* to −1. When $\overline{UP}/DOWN$ changes, *swap* is

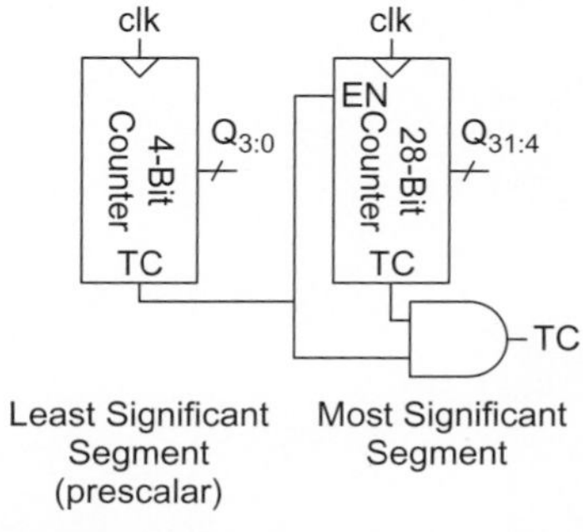

FIGURE 10.51 Fast binary counter

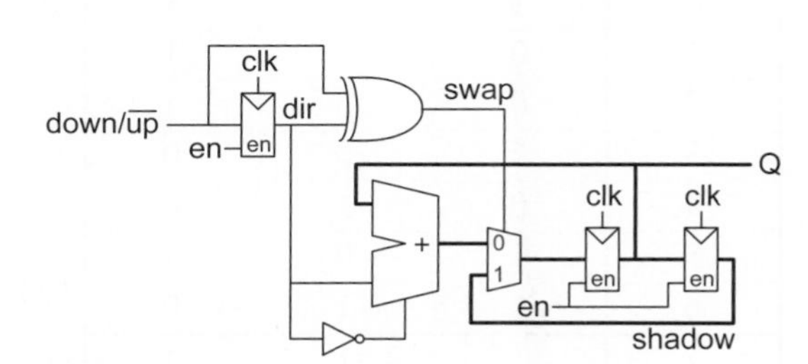

FIGURE 10.52 Fast binary up/down counter (most significant segment)

asserted for a cycle to load the new *count* from the *shadow* register rather than the adder (which may not have had enough time to ripple carries).

10.5.3 Ring and Johnson Counters

A *ring counter* consists of an M-bit shift register with the output fed back to the input, as shown in Figure 10.53(a). On reset, the first bit is initialized to 1 and the others are initialized to 0. TC toggles once every M cycles. Ring counters are a convenient way to build extremely fast prescalars because there is no logic between flip-flops, but they become costly for larger M.

A *Johnson* or *Mobius counter* is similar to a ring counter, but inverts the output before it is fed back to the input, as shown in Figure 10.53(b). The flip-flops are reset to all zeros and count through $2M$ states before repeating. Table 10.6 shows the sequence for a 3-bit Johnson counter.

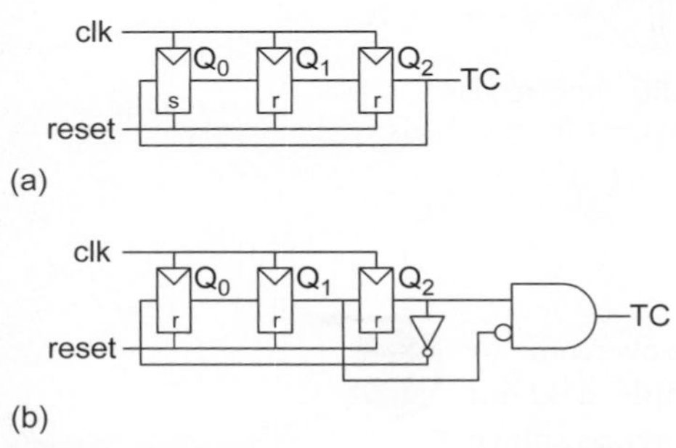

FIGURE 10.53 3-bit ring and Johnson counters

TABLE 10.6 Johnson counter sequence

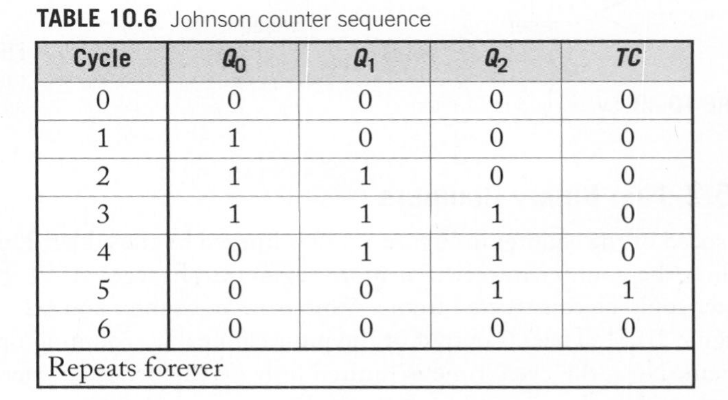

Cycle	Q_0	Q_1	Q_2	TC
0	0	0	0	0
1	1	0	0	0
2	1	1	0	0
3	1	1	1	0
4	0	1	1	0
5	0	0	1	1
6	0	0	0	0
Repeats forever				

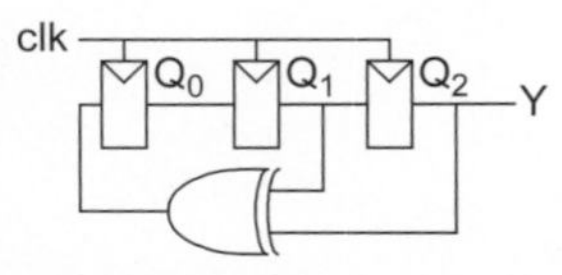

FIGURE 10.54 3-bit LFSR

10.5.4 Linerar-Feedback Shift Registers

A linear-feedback shift register (LFSR) consists of N registers configured as a shift register. The input to the shift register comes from the XOR of particular bits of the register, as shown in Figure 10.54 for a 3-bit LFSR. On reset, the registers must be initialized to a nonzero value (e.g., all 1s). The pattern of outputs for the LFSR is shown in Table 10.7.

TABLE 10.7 LFSR sequence

Cycle	Q_0	Q_1	Q_2 / Y
0	1	1	1
1	0	1	1
2	0	0	1
3	1	0	0
4	0	1	0
5	1	0	1
6	1	1	0
7	1	1	1
Repeats forever			

This LFSR is an example of a *maximal-length* shift register because its output sequences through all $2^n - 1$ combinations (excluding all 0s). The inputs fed to the XOR are called the *tap sequence* and are often specified with a *characteristic polynomial*. For example, this 3-bit LFSR has the characteristic polynomial $1 + x^2 + x^3$ because the taps come after the second and third registers.

The output *Y* follows the 7-bit sequence [1110010]. This is an example of a *pseudo-random bit sequence* (PRBS). LFSRs are used for high-speed counters and pseudo-random number generators. The pseudo-random sequences are handy for built-in self-test and bit-error-rate testing in communications links. They are also used in many spread-spectrum communications systems such as GPS and CDMA where their correlation properties make other users look like uncorrelated noise.

Table 10.8 lists characteristic polynomials for some commonly used maximal-length LFSRs. For certain lengths, *N*, more than two taps may be required. For many values of *N*, there are multiple polynomials resulting in different maximal-length LFSRs. Observe that the cycle time is set by the register and a small number of XOR delays. [Golomb81] offers the definitive treatment on linear-feedback shift registers.

TABLE 10.8 Characteristic polynomials

N	Polynomial
3	$1 + x^2 + x^3$
4	$1 + x^3 + x^4$
5	$1 + x^3 + x^5$
6	$1 + x^5 + x^6$
7	$1 + x^6 + x^7$
8	$1 + x^1 + x^6 + x^7 + x^8$
9	$1 + x^5 + x^9$
15	$1 + x^{14} + x^{15}$
16	$1 + x^4 + x^{13} + x^{15} + x^{16}$
23	$1 + x^{18} + x^{23}$
24	$1 + x^{17} + x^{22} + x^{23} + x^{24}$
31	$1 + x^{28} + x^{31}$
32	$1 + x^{10} + x^{30} + x^{31} + x^{32}$

Example 10.1

Sketch an 8-bit linear-feedback shift register. How long is the pseudo-random bit sequence that it produces?

SOLUTION: Figure 10.55 shows an 8-bit LFSR using the four taps after the 1st, 6th, 7th, and 8th bits, as given in Table 10.7. It produces a sequence of $2^8 - 1 = 255$ bits before repeating.

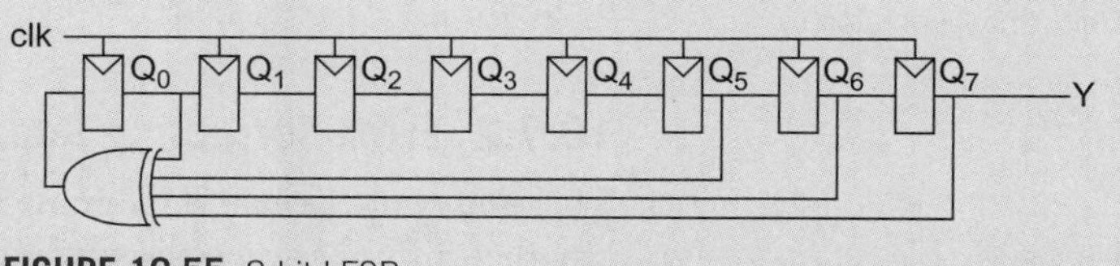

FIGURE 10.55 8-bit LFSR

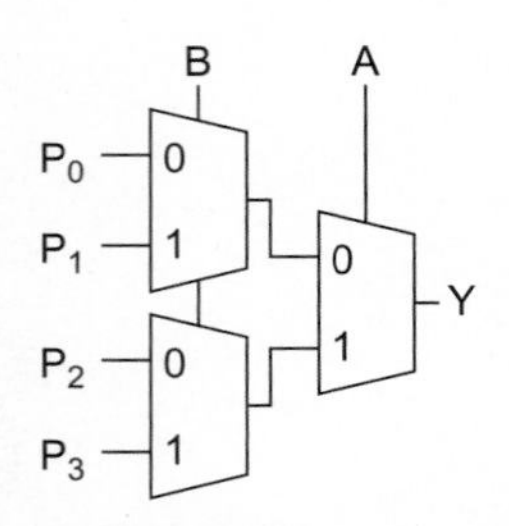

FIGURE 10.56
Boolean logical unit

10.6 Boolean Logical Operations

Boolean logical operations are easily accomplished using a multiplexer-based circuit, as shown in Figure 10.56. Table 10.9 shows how the inputs are assigned to perform different logical functions. By providing different P values, the unit can perform other operations such as XNOR(A, B) or NOT(A). An *Arithmetic Logic Unit* (ALU) requires both arithmetic (add, subtract) and Boolean logical operations.

TABLE 10.9 Functions implemented by Boolean unit

Operation	P_0	P_1	P_2	P_3
AND(A, B)	0	0	0	1
OR(A, B)	0	1	1	1
XOR(A, B)	0	1	1	0
NAND(A, B)	1	1	1	0
NOR(A, B)	1	0	0	0

10.7 Coding

Error-detecting and error-correcting codes are used to increase system reliability. Memory arrays are particularly susceptible to soft errors caused by alpha particles or cosmic rays flipping a bit. Such errors can be detected or even corrected by adding a few extra *check bits* to each word in the array. Codes are also used to reduce the bit error rate in communication links.

The simplest form of error-detecting code is *parity*, which detects single-bit errors. More elaborate *error-correcting codes* (ECC) are capable of single-error correcting and double-error detecting (SEC-DED). Gray codes are another useful alternative to the standard binary codes. All of the codes are heavily based on the XOR function, so we will examine a variety of CMOS XOR designs.

10.7.1 Parity

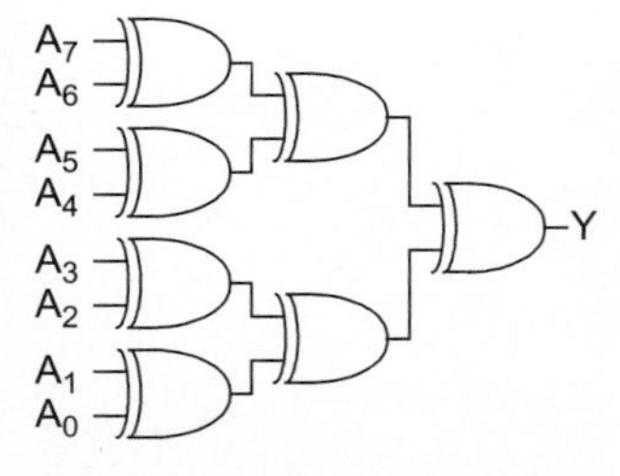

FIGURE 10.57
8-bit parity generator

A parity bit can be added to an N-bit word to indicate whether the number of 1s in the word is even or odd. In *even parity*, the extra bit is the XOR of the other N bits, which ensures the $(N + 1)$-bit coded word has an even number of 1s:

$$A_n = \text{PARITY} = A_0 \oplus A_1 \oplus A_2 \oplus \ldots \oplus A_{n-1} \quad \textbf{(10.28)}$$

Figure 10.57 shows a conventional implementation. Multi-input XOR gates can also be used.

10.7.2 Error-Correcting Codes

The *Hamming distance* [Hamming50] between a pair of binary numbers is the number of bits that differ between the two numbers. A single-bit error transforms a data word into another word separated by a Hamming distance of 1. Error-correcting codes add check bits to the data word so that the minimum Hamming distance between valid words increases. Parity is an example of a code with a single check bit and a Hamming distance of 2 between valid words, so that single-bit errors lead to invalid words and hence are detectable. If more check bits are added so that the minimum distance between valid words is 3, a single-bit error can be corrected because there will be only one valid word within a distance of 1. If the minimum distance between valid words is 4, a single-bit error can be corrected and an error corrupting two bits can be detected (but not corrected). If the probability of bit errors is low and uncorrelated from one bit to another, such single error-correcting, double error-detecting (SEC-DED) codes greatly reduce the overall error rate of the system. Larger Hamming distances improve the error rate further at the expense of more check bits.

In general, you can construct a distance-3 Hamming code of length up to $2^c - 1$ with c check bits and $N = 2^c - c - 1$ data bits using a simple procedure [Wakerly00]. If the bits are numbered from 1 to $2^c - 1$, each bit in a position that is a power of 2 serves as a check bit. The value of the check bit is chosen to obtain even parity for all bits with a 1 in the same position as the check bit, as illustrated in Figure 10.58(a) for a 7-bit code with 4 data bits and 3 check bits. The bits are traditionally reorganized into contiguous data and check bits, as shown in Figure 10.58(b). The structure is called a *parity-check matrix* and each check bit can be computed as the XOR of the highlighted data bits:

$$\begin{aligned} C_0 &= D_3 \oplus D_1 \oplus D_0 \\ C_1 &= D_3 \oplus D_2 \oplus D_0 \\ C_2 &= D_3 \oplus D_2 \oplus D_1 \end{aligned} \qquad \textbf{(10.29)}$$

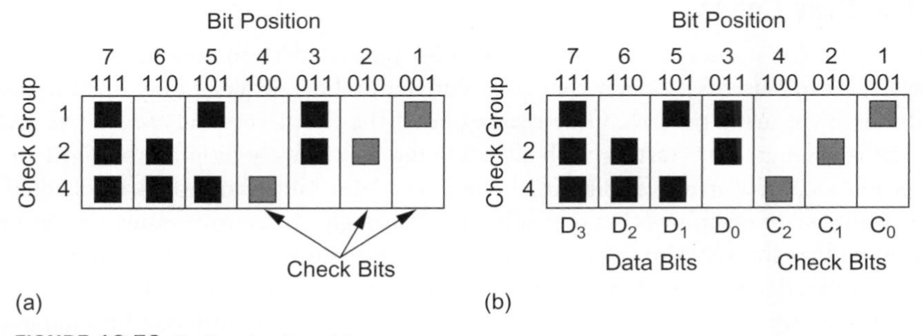

FIGURE 10.58 Parity-check matrix

The error-correcting decoder examines the check bits. If they all have even parity, the word is considered to be correct. If one or more groups have odd parity, an error has occurred. The pattern of check bits that have the wrong parity is called the *syndrome* and corresponds to the bit position that is incorrect. The decoder must flip this bit to recover the correct result.

Example 10.2

Suppose the data value 1001 were to be transmitted using a distance-3 Hamming code. What are the check bits? If the data bits were garbled into 1101 during transmission, explain what the syndrome would be and how the data would be corrected.

SOLUTION: According to EQ (10.29), the check bits should be 100, corresponding to a transmitted word of 1001100. The received word is 1101100. The syndrome is 110, i.e., odd parity on check bits C_2 and C_1, which indicates an error in bit position 110 = 6. This position is flipped to produce a corrected word of 1001100 and the check bits are discarded, leaving the proper data value of 1001.

A SEC-DED distance-4 Hamming code can be constructed from a distance-3 code by adding one more parity bit for the entire word. If there is a single-bit error, parity will fail and the check bits will indicate how to correct the data. If there is a double-bit error, the check bits will indicate an error, but parity will pass, indicating a detectable but uncorrectable double-bit error.

The parity check matrix determines the number of XORs required in the encoding and decoding logic. A SEC-DED Hamming code for a 64-bit data word has 8 check bits. It requires 296 XOR gates. The parity logic for the entire word has 72 inputs. The Hsiao SEC-DED achieves the same function with the same number of data and check bits but is ingeniously designed to minimize the cost, using only 216 XOR gates and parity logic with a maximum of 27 inputs. [Hsiao70] shows parity-check matrices for 16, 32, and 64-bit data words with 6, 7, and 8 check bits.

As the data length and allowable decoder complexity increase, other codes become efficient. These include Reed-Solomon, BCH, and Turbo codes. [Lin83, Sweeney02, Sklar01, Fujiwara06] and many other texts provide extensive information on a variety of error-correcting codes.

10.7.3 Gray Codes

The *Gray codes*, named for Frank Gray, who patented their use on shaft encoders [Gray53], have a useful property that consecutive numbers differ in only one bit position. While there are many possible Gray codes, one of the simplest is the *binary-reflected Gray code* that is generated by starting with all bits 0 and successively flipping the right-most bit that produces a new string. Table 10.10 compares 3-bit binary and binary-reflected Gray codes. Finite state machines that typically move through consecutive states can save power by Gray-coding the states to reduce the number of transitions. When a counter value must be synchronized across clock domains, it can be Gray-coded so that the synchronizer is certain to receive either the current or previous value because only one bit changes each cycle.

TABLE 10.10 3-bit Gray code

Number	Binary	Gray Code
0	000	000
1	001	001
2	010	011
3	011	010
4	100	110
5	101	111
6	110	101
7	111	100

Converting between N-bit binary B and binary-reflected Gray code G representations is remarkably simple.

$$
\begin{array}{lll}
\text{Binary} \rightarrow \text{Gray} & \text{Gray} \rightarrow \text{Binary} & \\
G_{N-1} = B_{N-1} & B_{N-1} = G_{N-1} & \\
G_i = B_{i+1} \oplus B_i & B_i = B_{i+1} \oplus G_i & N-1 > i \geq 0
\end{array} \tag{10.30}
$$

10.7.4 XOR/XNOR Circuit Forms

One of the chronic difficulties in CMOS circuit design is to construct a fast, compact, low-power XOR or XNOR gate. Figure 10.59 shows a number of common static single-rail 2-input XOR designs; XNOR designs are similar. Figure 10.59(a) and Figure 10.59(b) show gate-level implementations; the first is cute, but the second is slightly more efficient. Figure 10.59(c) shows a complementary CMOS gate. Figure 10.59(d) improves the gate by optimizing out two contacts and is a commonly used standard cell design. Figure 10.59(e) shows a transmission gate design. Figure 10.59(f) is the 6-transistor "invertible inverter" design. When A is 0, the transmission gate turns on and B is passed to the output. When A is 1, the A input powers a pair of transistors that invert B. It is compact, but nonrestoring. Some switch-level simulators such as IRSIM cannot handle this unconventional design. Figure 10.59(g) [Wang94] is a compact and fast 4-transistor pass-gate design, but does not swing rail to rail.

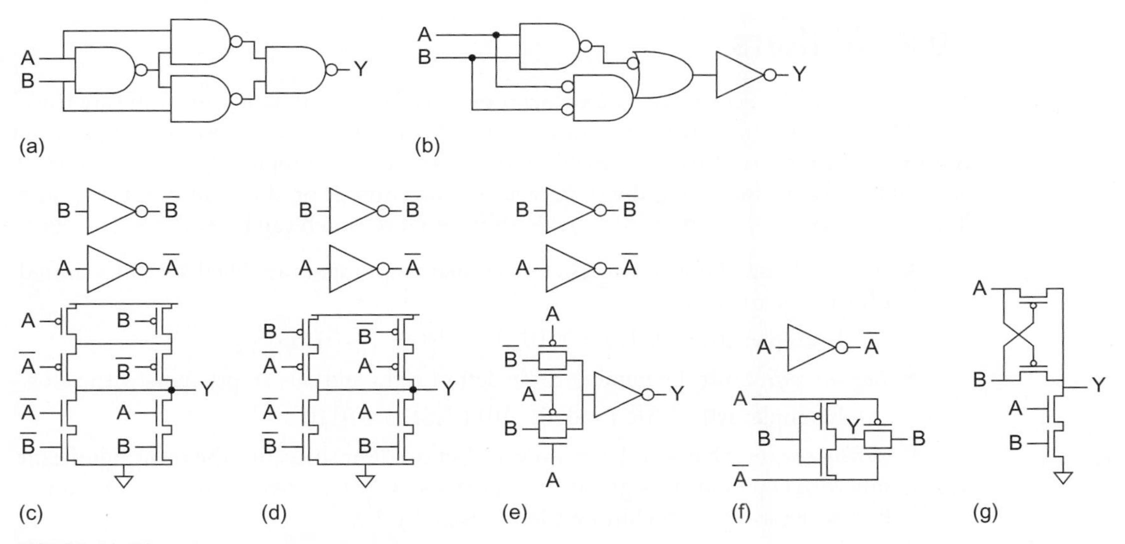

FIGURE 10.59 Static 2-input XOR designs

XOR gates with 3 or 4 inputs can be more compact, although not necessarily faster than a cascade of 2-input gates. Figure 10.60(a) is a 4-input static CMOS XOR [Griffin83] and Figure 10.60(b) is a 4-input CPL XOR/XNOR, while Figure 8.20(c) showed a 4-input CVSL XOR/XNOR. Observe that the true and complementary trees share most of the transistors. As mentioned in Chapter 8, CPL does not perform well at low voltage.

Dynamic XORs pose a problem because both true and complementary inputs are required, violating the monotonicity rule. The common solutions mentioned in Section 10.2.2.10 are to either push the XOR to the end of a chain of domino logic and build it

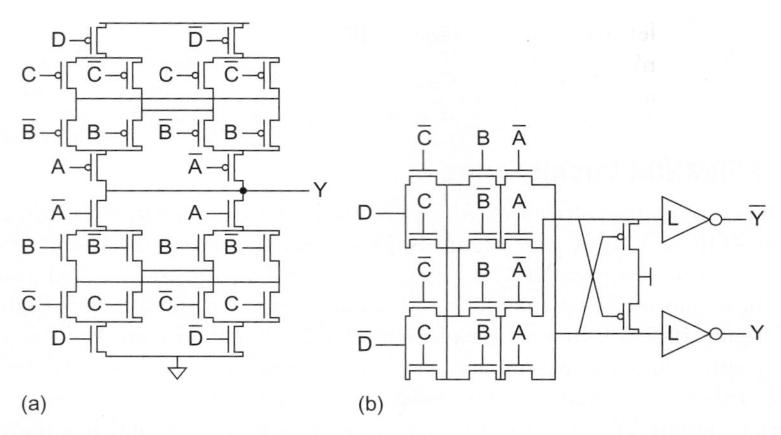

FIGURE 10.60 4-input XOR designs

with static CMOS or to construct a dual-rail domino structure. A dual-rail domino 2-input XOR was shown in Figure 8.30(c).

10.8 Shifters

Shifts can either be performed by a constant or variable amount. Constant shifts are trivial in hardware, requiring only wires. They are also an efficient way to perform multiplication or division by powers of two. A variable shifter takes an N-bit input, A, a shift amount, k, and control signals indicating the shift type and direction. It produces an N-bit output, Y. There are three common types of variable shifts, each of which can be to the left or right:

- *Rotate*: Rotate numbers in a circle such that empty spots are filled with bits shifted off the other end
  - Example: 1011 ROR 1 = 1101; 1011 ROL 1 = 0111
- *Logical shift*: Shift the number to the left or right and fills empty spots with zeros.
  - Example: 1011 LSR 1 = 0101; 1011 LSL 1 = 0110
- *Arithmetic shift*: Same as logical shifter, but on right shifts fills the most significant bits with copies of the sign bit (to properly sign, extend two's complement numbers when using right shift by k for division by 2^k).
  - Example: 1011 ASR 1 = 1101; 1011 ASL 1 = 0110

Conceptually, rotation involves an array of N N-input multiplexers to select each of the outputs from each of the possible input positions. This is called an *array shifter*. The array shifter requires a decoder to produce the 1-of-N-hot shift amount. In practice, multiplexers with more than 4–8 inputs have excessive parasitic capacitance, so they are faster to construct from $\log_v N$ levels of v-input multiplexers. This is called a *logarithmic shifter*. For example, in a radix-2 logarithmic shifter, the first level shifts by $N/2$, the second by $N/4$, and so forth until the final level shifts by 1. In a logarithmic shifter, no decoder is necessary. The CMOS transmission gate multiplexer of Figure 8.47 is especially well-suited to logarithmic shifters because the hefty wire capacitance is driven directly by an inverter rather than through a pair of series transistors. 4:1 or 8:1 transmission gate multiplexers reduce the number of levels by a factor of 2 or 3 at the expense of more wiring and fanout. Pairs or triplets of the shift amount are decoded to drive one-hot mux selects at each level. [Tharakan92] describes a domino logarithmic shifter using 3:1 multiplexers to reduce the number of logic levels.

A left rotate by k bits is equivalent to a right rotate by $N - k$ bits. Computing $N - k$ requires a subtracter in the critical path. Taking advantage of two's complement arithmetic and the fact that rotation is cyclic modulo N, $N - k = N + \bar{k} + 1 = \bar{k} + 1$. Thus, the left rotate can be performed by preshifting right by 1, then doing a right rotate by the complemented shift amount.

Logical and arithmetic shifts are similar to rotates, but must replace bits at one end or the other with a *kill value* (either 0 or the sign bit). The two major shifter architectures are funnel shifters and barrel shifters. In a *funnel shifter*, the kill values are incorporated at the beginning, while in a *barrel shifter*, the kill values are chosen at the end. Each of these architectures is described below. Both barrel and funnel shifters can use array or logarithmic implementations. [Huntzicker08] examines the energy-delay trade-offs in static shifters. For general-purpose shifting, both architectures are comparable in energy and delay. Given typical parasitics capacitances, the theory of Logical Effort shows that logarithmic structure using 4:1 multiplexers is most efficient. If only shift operations (but not rotates) are required, the funnel architecture is simpler, while if only rotates (but not shifts) are required, the barrel is simpler.

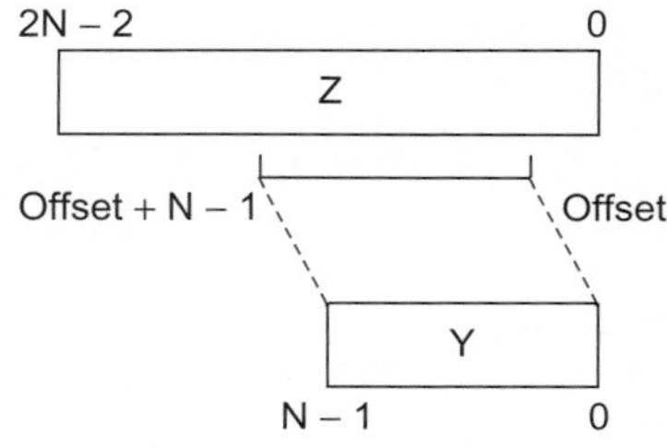

FIGURE 10.61 Funnel shifter function

10.8.1 Funnel Shifter

The *funnel shifter* creates a $2N - 1$-bit input word Z from A and/or the kill values, then selects an N-bit field from this input word, as shown in Figure 10.61. It gets its name from the way the wide word funnels down to a narrower one. Table 10.11 shows how Z is formed for each type of shift. Z incorporates the 1-bit preshift for left shifts.

TABLE 10.11 Funnel shifter source generator

Shift Type	$Z_{2N-2:N}$	Z_{N-1}	$Z_{N-2:0}$	Offset
Logical Right	$A_{N-2:0}$	A_{N-1}	$A_{N-2:0}$	k
Arithmetic Right	0	A_{N-1}	$A_{N-2:0}$	k
Rotate Right	sign	A_{N-1}	$A_{N-2:0}$	k
Logical/Arithmetic Left	$A_{N-1:1}$	A_0	$A_{N-1:1}$	$\bar{k}$
Rotate Left	$A_{N-1:1}$	A_0	0	$\bar{k}$

The simplest funnel shifter design consists of an array of N N-input multiplexers accepting 1-of-N-hot select signals (one multiplexer for each output bit). Such an array shifter is shown in Figure 10.62 using nMOS pass transistors for a 4-bit shifter. The shift amount is conditionally inverted and decoded into select signals that are fed vertically across the array. The outputs are taken horizontally. Each row of transistors attached to an output forms one of the multiplexers. The $2N - 1$ inputs run diagonally to the appropriate mux inputs. Figure 10.63 shows a stick diagram for one of the N^2 transistors in the array. nMOS pass transistors suffer a threshold drop, but the problem can be solved by precharging the outputs (done in the Alpha 21164 [Gronowski96]) or by using full CMOS transmission gates.

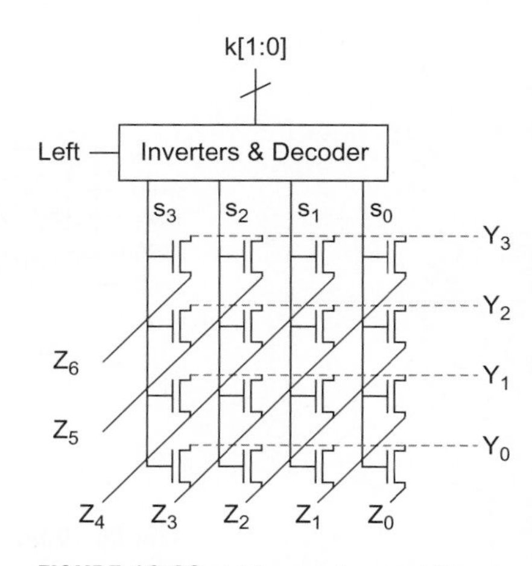

FIGURE 10.62 4-bit array funnel shifter

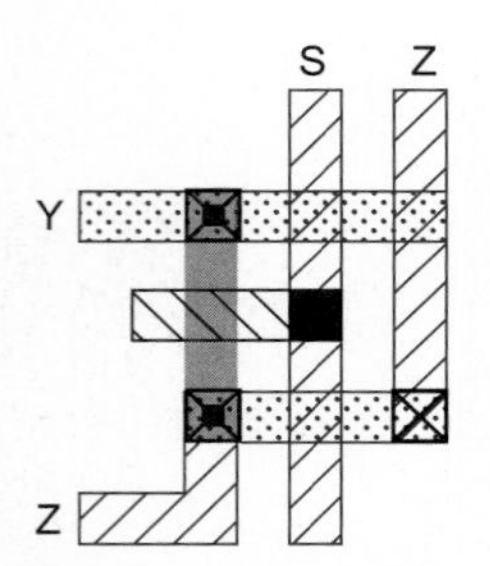

FIGURE 10.63 Array funnel shifter cell stick diagram

The array shifter works well for small shifters in transistor-level designs, but has high parasitic capacitance in larger shifters, leading to excessive delay and energy. Moreover, array shifters are not amenable to standard cell designs. Figure 10.64 shows a 4-bit logarithmic shifter based on multiple levels of 2:1 multiplexers (which, of course, can be transmission gates) [Lim72]. The XOR gates on the control inputs conditionally invert the shift amount for left shifts.

Figure 10.65 shows a 32-bit funnel shifter using a 4:1 multiplexer followed by an 8:1 multiplexer [Huntzicker08]. The source generator selects the 63-bit *Z*. The first stage performs a coarse shift right by 0, 8, 16, or 24 bits. The second stage performs a fine shift right by 0–7 bits. The mux decode block conditionally inverts *k* for left shifts, computes the 1-hot selects, and buffers them to drive the wide multiplexers.

Conceptually, the source generator consists of a $2N-1$-bit 5:1 multiplexer controlled by the shift type and direction. Figure 10.66 shows how the source generator logic can be simplified. The horizontal control lines need to be buffered to drive the high fanout and they are on the critical path. Even if they are available early, the sign bit is still critical. If only certain types of shifts or rotates are supported, the logic can be optimized down further.

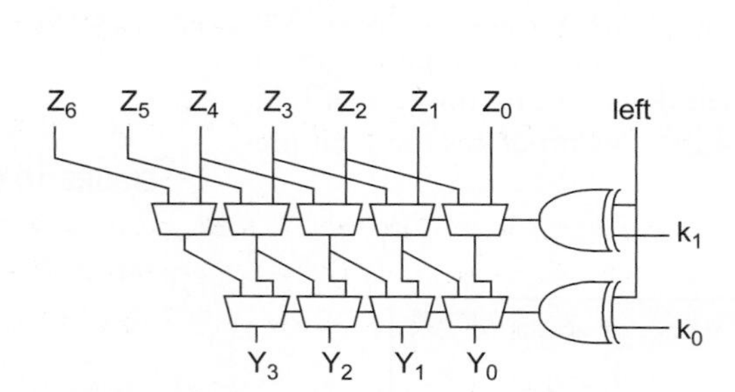

FIGURE 10.64 4-bit logarithmic funnel shifter

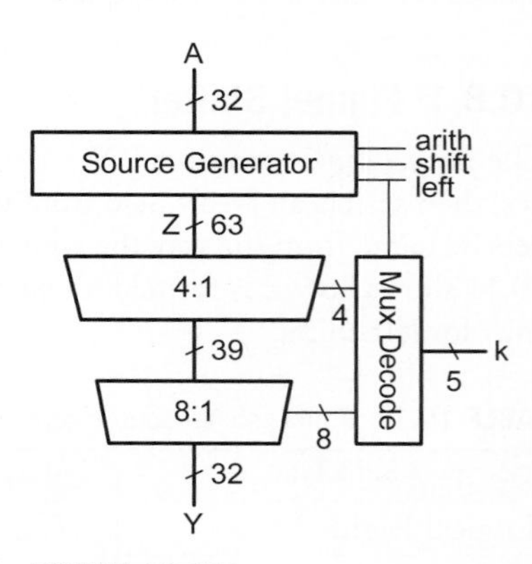

FIGURE 10.65 32-bit logarithmic funnel shifter

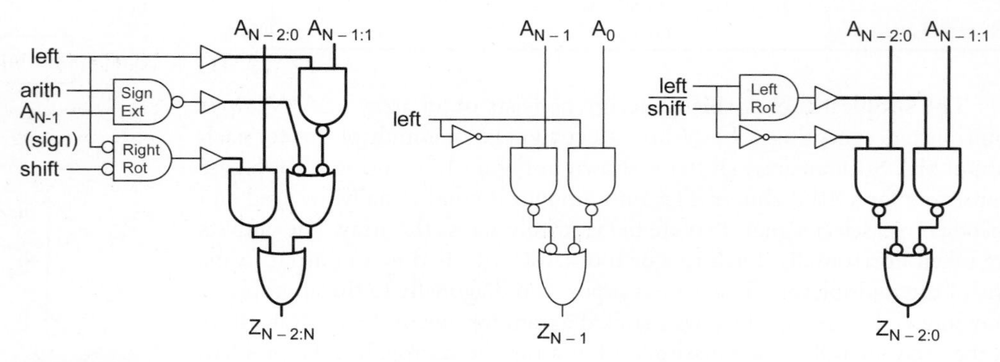

FIGURE 10.66 Optimized source generator logic

The funnel shifter presents a layout problem because the source generator and early stages of multiplexers are wider than the rest of the datapath. Figure 10.67 shows a floorplan in which the source generator is folded to fit the datapath. Such folding also reduces wire lengths, saving energy. Depending on the layout constraints, the extra seven most significant bits of the first-level multiplexer may be folded into another row or incorporated into the zipper.

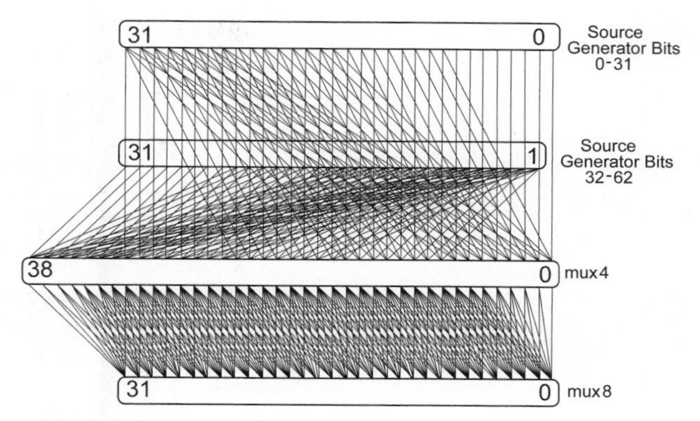

FIGURE 10.67 Funnel shifter floorplans

10.8.2 Barrel Shifter

A *barrel shifter* performs a right rotate operation [Davis69]. As mentioned earlier, it handles left rotations using the complementary shift amount. Barrel shifters can also perform shifts when suitable masking hardware is included. Barrel shifters come in array and logarithmic forms; we focus on logarithmic barrel shifters because they are better suited for large shifts.

Figure 10.68(a) shows a simple 4-bit barrel shifter that performs right rotations. Notice how, unlike funnel shifters, barrel shifters contain long wrap-around wires. In a large shifter, it is beneficial to upsize or buffer the drivers for these wires. Figure 10.68(b) shows an enhanced version that can rotate left by prerotating right by 1, then rotating right by $\bar{k}$. Performing logical or arithmetic shifts on a barrel shifter requires a way to mask out the bits that are rotated off the end of the shifter, as shown in Figure 10.68(c).

Figure 10.69 shows a 32-bit barrel shifter using a 5:1 multiplexer and an 8:1 multiplexer. The first stage rotates right by 0, 1, 2, 3, or 4 bits to handle a prerotate of 1 bit and a fine rotate of up to 3 bits combined into one stage. The second stage rotates right by 0, 4, 8, 12, 16, 20, 24, or 28 bits. The critical path starts with decoding the shift amount for the first stage. If the shift amount is available early, the delay from A to Y improves substantially.

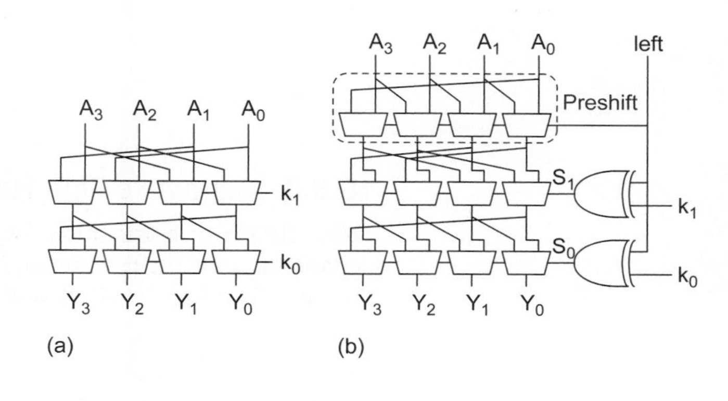

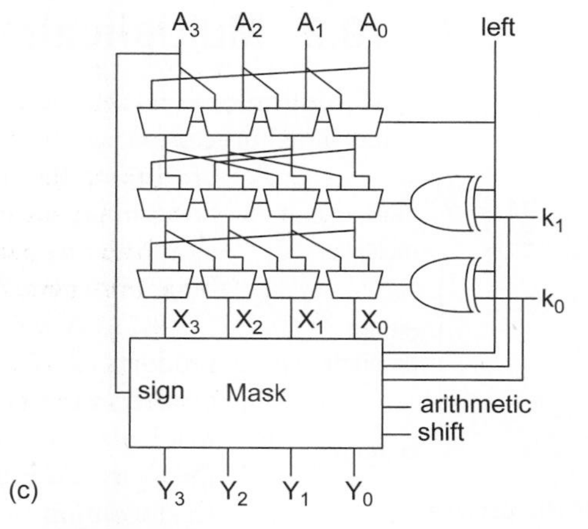

FIGURE 10.68 Barrel shifters: (a) rotate right, (b) rotate left or right, (c) rotates and shifts

While the rotation is taking place, the masking unit generates an N-bit mask with ones where the kill value should be inserted for right shifts. For a right shift by m, the m most significant bits are ones. This is called a thermometer code and the logic to compute it is described in Section 10.10. When the rotation result X is complete, the masking unit replaces the masked bits with the kill value. For left shifts, the mask is reversed. Figure 10.70 shows masking logic. If only certain shifts are supported, the unit can be simplified, and if only rotates are supported, the masking unit can be eliminated, saving substantial hardware, power, and delay.

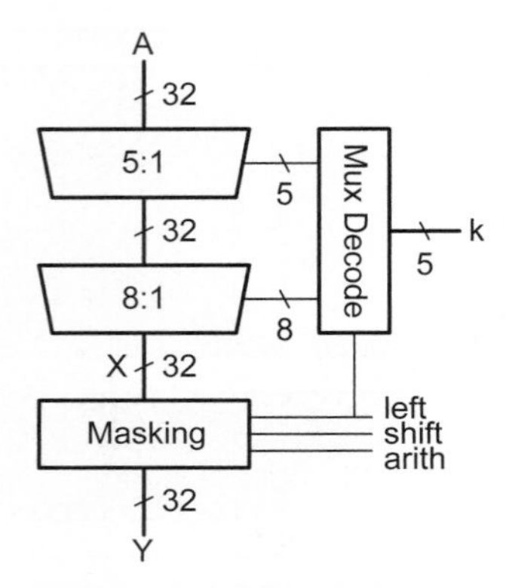

FIGURE 10.69 32-bit logarithmic barrel shifter

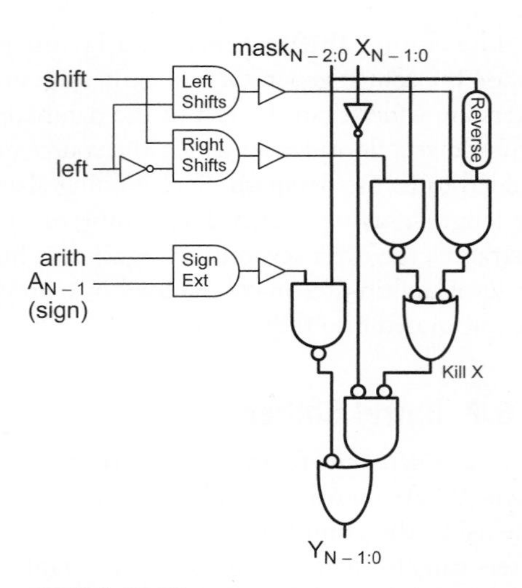

FIGURE 10.70 Barrel shifter masking logic

10.8.3 Alternative Shift Functions

Other flavors of shifts, including shuffles, bit-reversals, interchanges, extraction, and deposit, are sometimes required, especially for cryptographic and multimedia applications [Hilewitz04, Hilewitz07]. These are also built from appropriate combinations of multiplexers.

10.9 Multiplication

Multiplication is less common than addition, but is still essential for microprocessors, digital signal processors, and graphics engines. The most basic form of multiplication consists of forming the product of two unsigned (positive) binary numbers. This can be accomplished through the traditional technique taught in primary school, simplified to base 2. For example, the multiplication of two positive 6-bit binary integers, 25_{10} and 39_{10}, proceeds as shown in Figure 10.71.

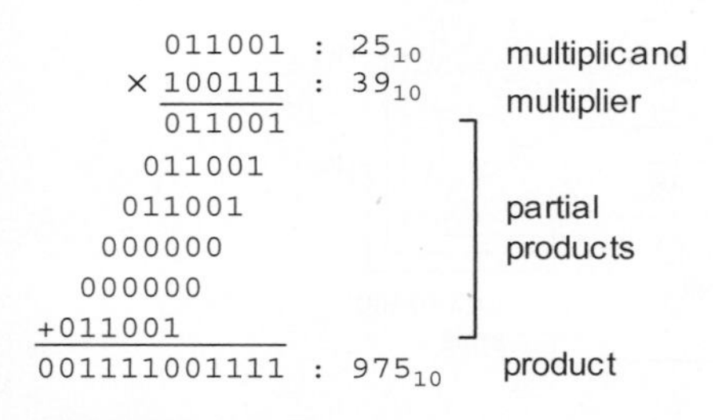

FIGURE 10.71 Multiplication example

$M \times N$-bit multiplication $P = Y \times X$ can be viewed as forming N partial products of M bits each, and then summing the appropriately shifted partial products to produce an $M + N$-bit result P. Binary multiplication is equivalent to a logical AND operation. Therefore, generating partial products consists of the logical ANDing of the appropriate bits of the multiplier and multiplicand. Each column of partial products must then be added and, if necessary, any carry values passed to the next column. We denote the multiplicand as $Y = \{y_{M-1}, y_{M-2}, \ldots, y_1, y_0\}$ and the multiplier as $X = \{x_{N-1}, x_{N-2}, \ldots, x_1, x_0\}$. For unsigned multiplication, the product is given in EQ (10.31). Figure 10.72 illustrates the generation, shifting, and summing of partial products in a 6×6-bit multiplier.

$$P = \left(\sum_{j=0}^{M-1} y_j 2^j \right) \left(\sum_{i=0}^{N-1} x_i 2^i \right) = \sum_{i=0}^{N-1} \sum_{j=0}^{M-1} x_i y_j 2^{i+j} \tag{10.31}$$

						y_5	y_4	y_3	y_2	y_1	y_0	Multiplicand
						x_5	x_4	x_3	x_2	x_1	x_0	Multiplier
						x_0y_5	x_0y_4	x_0y_3	x_0y_2	x_0y_1	x_0y_0	Partial Products
					x_1y_5	x_1y_4	x_1y_3	x_1y_2	x_1y_1	x_1y_0		
				x_2y_5	x_2y_4	x_2y_3	x_2y_2	x_2y_1	x_2y_0			
			x_3y_5	x_3y_4	x_3y_3	x_3y_2	x_3y_1	x_3y_0				
		x_4y_5	x_4y_4	x_4y_3	x_4y_2	x_4y_1	x_4y_0					
	x_5y_5	x_5y_4	x_5y_3	x_5y_2	x_5y_1	x_5y_0						
p_{11}	p_{10}	p_9	p_8	p_7	p_6	p_5	p_4	p_3	p_2	p_1	p_0	Product

FIGURE 10.72 Partial products

Large multiplications can be more conveniently illustrated using *dot diagrams*. Figure 10.73 shows a dot diagram for a simple 16 × 16 multiplier. Each dot represents a placeholder for a single bit that can be a 0 or 1. The partial products are represented by a horizontal boxed row of dots, shifted according to their weight. The multiplier bits used to generate the partial products are shown on the right.

There are a number of techniques that can be used to perform multiplication. In general, the choice is based upon factors such as latency, throughput, energy, area, and design complexity. An obvious approach is to use an $M + 1$-bit carry-propagate adder (CPA) to add the first two partial products, then another CPA to add the third partial product to the running sum, and so forth. Such an approach requires $N - 1$ CPAs and is slow, even if a fast CPA is employed. More efficient parallel approaches use some sort of array or tree of full adders to sum the partial products. We begin with a simple array for unsigned multipliers, and then modify the array to handle signed two's complement numbers using the Baugh-Wooley algorithm. The number of partial products to sum can be reduced using Booth encoding and the number of logic levels required to perform the summation can be reduced with Wallace trees. Unfortunately, Wallace trees are complex to lay out and have long, irregular wires, so hybrid array/tree structures may be more attractive. For completeness, we consider a serial multiplier architecture. This was once popular when gates were relatively expensive, but is now rarely necessary.

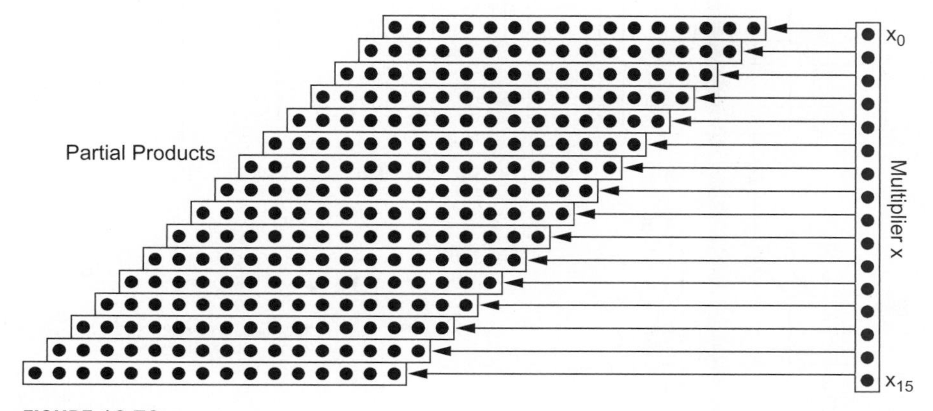

FIGURE 10.73 Dot diagram

10.9.1 Unsigned Array Multiplication

Fast multipliers use carry-save adders (CSAs, see Section 10.2.4) to sum the partial products. A CSA typically has a delay of 1.5–2 FO4 inverters independent of the width of the partial product, while a carry-propagate adder (CPA) tends to have a delay of 4–15+ FO4 inverters depending on the width, architecture, and circuit family. Figure 10.74 shows a 4×4 array multiplier for unsigned numbers using an array of CSAs. Each cell contains a 2-input AND gate that forms a partial product and a full adder (CSA) to add the partial product into the running sum. The first row converts the first partial product into carry-save redundant form. Each later row uses the CSA to add the corresponding partial product to the carry-save redundant result of the previous row and generate a carry-save redundant result. The least significant N output bits are available as sum outputs directly from CSAs. The most significant output bits arrive in carry-save redundant form and require an M-bit carry-propagate adder to convert into regular binary form. In Figure 10.74, the CPA is implemented as a carry-ripple adder. The array is regular in structure and uses a single type of cell, so it is easy to design and lay out. Assuming the carry output is faster than the sum output in a CSA, the critical path through the array is marked on the figure with a dashed line. The adder can easily be pipelined with the placement of registers between rows. In practice, circuits are assigned rectangular blocks in the floorplan so the parallelogram shape wastes space. Figure 10.75 shows the same adder squashed to fit a rectangular block.

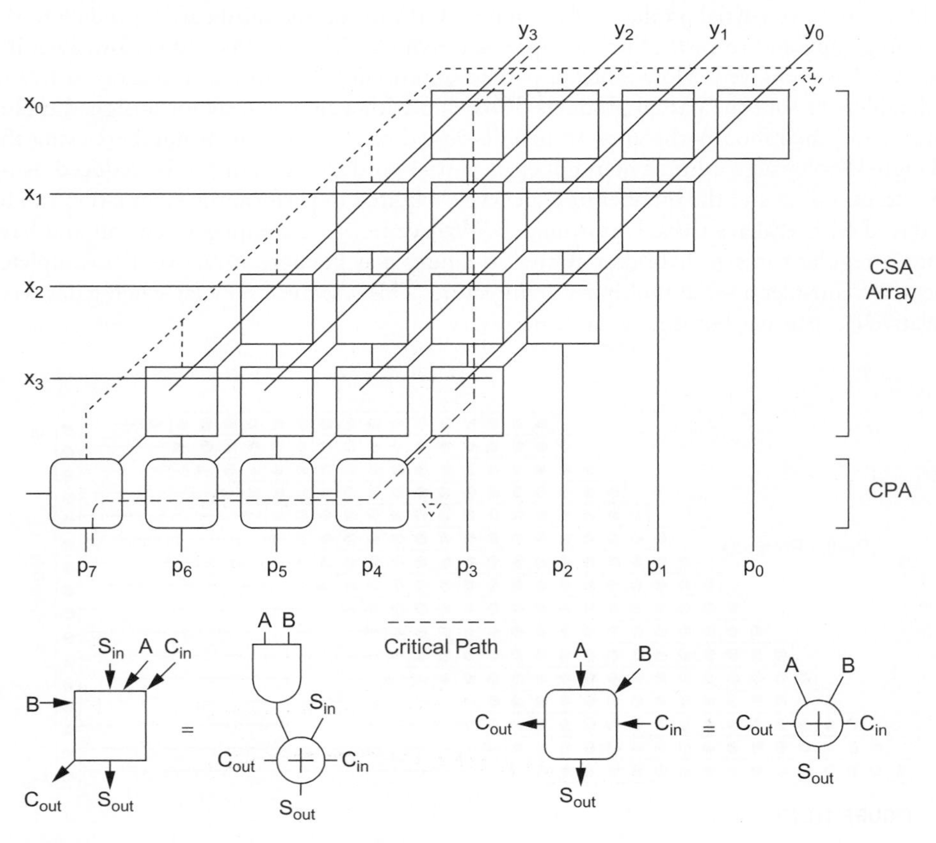

FIGURE 10.74 Array multiplier

A key element of the design is a compact CSA. This not only benefits area but also helps performance because it leads to short wires with low wire capacitance. An ideal CSA design has approximately equal sum and carry delays because the greater of these two delays limits performance. The mirror adder from Figure 10.4 is commonly used for its compact layout even though the sum delay exceeds the carry delay. The sum output can be connected to the faster carry input to partially compensate [Sutherland99, Hsu06a].

Note that the first row of CSAs adds the first partial product to a pair of 0s. This leads to a regular structure, but is inefficient. At a slight cost to regularity, the first row of CSAs can be used to add the first three partial products together. This reduces the number of rows by two and correspondingly reduces the adder propagation delay. Yet another way to improve the multiplier array performance is to replace the bottom row with a faster CPA such as a lookahead or tree adder. In summary, the critical path of an array multiplier involves N–2 CSAs and a CPA.

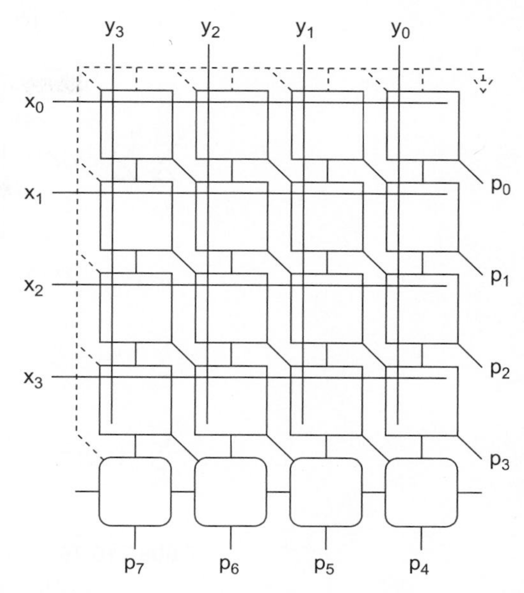

FIGURE 10.75 Rectangular array multiplier

10.9.2 Two's Complement Array Multiplication

Multiplication of two's complement numbers at first might seem more difficult because some partial products are negative and must be subtracted. Recall that the most significant bit of a two's complement number has a negative weight. Hence, the product is

$$
\begin{aligned}
P &= \left(-y_{M-1}2^{M-1} + \sum_{j=0}^{M-2} y_j 2^j\right)\left(-x_{n-1}2^{N-1} + \sum_{i=0}^{N-2} x_i 2^i\right) \\
&= \sum_{i=0}^{N-2}\sum_{j=0}^{M-2} x_i y_j 2^{i+j} + x_{N-1}y_{M-1}2^{M+N-2} - \left(\sum_{i=0}^{N-2} x_i y_{M-1} 2^{i+M-1} + \sum_{j=0}^{M-2} x_{N-1} y_j 2^{j+N-1}\right)
\end{aligned}
\quad \textbf{(10.32)}
$$

In EQ (10.32), two of the partial products have negative weight and thus should be subtracted rather than added. The *Baugh-Wooley* [Baugh73] multiplier algorithm handles subtraction by taking the two's complement of the terms to be subtracted (i.e., inverting the bits and adding one). Figure 10.76 shows the partial products that must be summed. The upper parallelogram represents the unsigned multiplication of all but the most significant bits of the inputs. The next row is a single bit corresponding to the product of the most significant bits. The next two pairs of rows are the inversions of the terms to be subtracted. Each term has implicit leading and trailing zeros, which are inverted to leading and trailing ones. Extra ones must be added in the least significant column when taking the two's complement.

The multiplier delay depends on the number of partial product rows to be summed. The *modified Baugh-Wooley multiplier* [Hatamian86] reduces this number of partial products by precomputing the sums of the constant ones and pushing some of the terms upward into extra columns. Figure 10.77 shows such an arrangement. The parallelogram-shaped array can again be squashed into a rectangle, as shown in Figure 10.78, giving a design almost identical to the unsigned multiplier of Figure 10.75. The AND gates are replaced by NAND gates in the hatched cells and 1s are added in place of 0s at two of the

	p_{11}	p_{10}	p_9	p_8	p_7	p_6	p_5	p_4	p_3	p_2	p_1	p_0
							y_5	y_4	y_3	y_2	y_1	y_0
							x_5	x_4	x_3	x_2	x_1	x_0
$\sum_{i=0}^{N-2}\sum_{j=0}^{M-2} x_i y_j 2^{i+j}$								x_0y_4	x_0y_3	x_0y_2	x_0y_1	x_0y_0
							x_1y_4	x_1y_3	x_1y_2	x_1y_1	x_1y_0	
						x_2y_4	x_2y_3	x_2y_2	x_2y_1	x_2y_0		
					x_3y_4	x_3y_3	x_3y_2	x_3y_1	x_3y_0			
				x_4y_4	x_4y_3	x_4y_2	x_4y_1	x_4y_0				
$x_{N-1} y_{M-1} 2^{M+N-2}$		x_5y_5										
$-\sum_{i=0}^{N-2} x_i y_{M-1} 2^{i+M-1}$	1	1	$\overline{x_4y_5}$	$\overline{x_3y_5}$	$\overline{x_2y_5}$	$\overline{x_1y_5}$	$\overline{x_0y_5}$	1	1	1	1	1
												1
$-\sum_{j=0}^{M-2} x_{N-1} y_j 2^{j+N-1}$	1	1	$\overline{x_5y_4}$	$\overline{x_5y_3}$	$\overline{x_5y_2}$	$\overline{x_5y_1}$	$\overline{x_5y_0}$	1	1	1	1	1
												1
	p_{11}	p_{10}	p_9	p_8	p_7	p_6	p_5	p_4	p_3	p_2	p_1	p_0

FIGURE 10.76 Partial products for two's complement multiplier

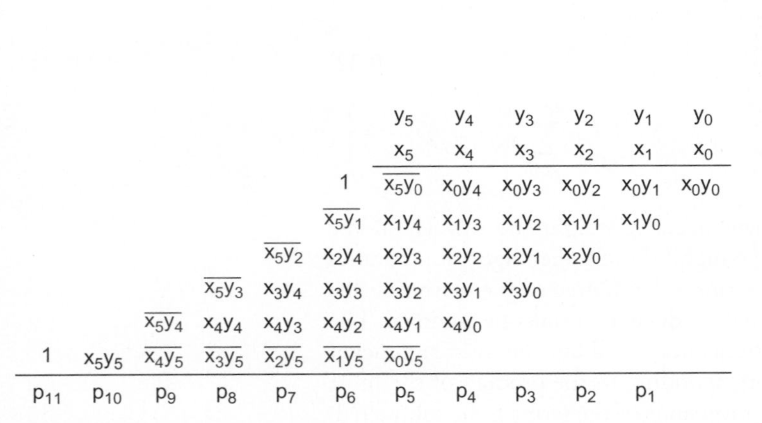

p_{11}	p_{10}	p_9	p_8	p_7	p_6	p_5	p_4	p_3	p_2	p_1
					y_5	y_4	y_3	y_2	y_1	y_0
					x_5	x_4	x_3	x_2	x_1	x_0
				1	$\overline{x_5y_0}$	x_0y_4	x_0y_3	x_0y_2	x_0y_1	x_0y_0
				$\overline{x_5y_1}$	x_1y_4	x_1y_3	x_1y_2	x_1y_1	x_1y_0	
			$\overline{x_5y_2}$	x_2y_4	x_2y_3	x_2y_2	x_2y_1	x_2y_0		
		$\overline{x_5y_3}$	x_3y_4	x_3y_3	x_3y_2	x_3y_1	x_3y_0			
		$\overline{x_5y_4}$	x_4y_4	x_4y_3	x_4y_2	x_4y_1	x_4y_0			
1	x_5y_5	$\overline{x_4y_5}$	$\overline{x_3y_5}$	$\overline{x_2y_5}$	$\overline{x_1y_5}$	$\overline{x_0y_5}$				
p_{11}	p_{10}	p_9	p_8	p_7	p_6	p_5	p_4	p_3	p_2	p_1

FIGURE 10.77 Simplified partial products for two's complement multiplier

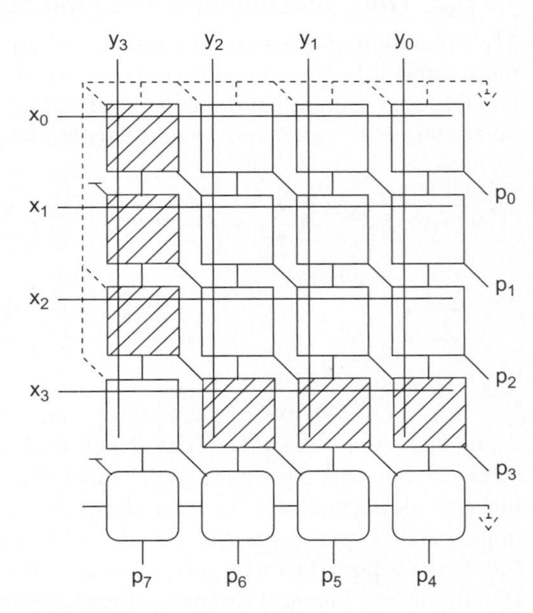

FIGURE 10.78 Modified Baugh-Wooley two's complement multiplier

unused inputs. The signed and unsigned arrays are so similar that a single array can be used for both purposes if XOR gates are used to conditionally invert some of the terms depending on the mode.

10.9.3 Booth Encoding

The array multipliers in the previous sections compute the partial products in a radix-2 manner; i.e., by observing one bit of the multiplier at a time. Radix 2^r multipliers produce

N/r partial products, each of which depend on r bits of the multiplier. Fewer partial products leads to a smaller and faster CSA array. For example, a radix-4 multiplier produces $N/2$ partial products. Each partial product is 0, Y, $2Y$, or $3Y$, depending on a pair of bits of X. Computing $2Y$ is a simple shift, but $3Y$ is a *hard multiple* requiring a slow carry-propagate addition of $Y + 2Y$ before partial product generation begins.

Booth encoding was originally proposed to accelerate serial multiplication [Booth51]. *Modified Booth encoding* [MacSorley61] allows higher radix parallel operation without generating the hard $3Y$ multiple by instead using negative partial products. Observe that $3Y = 4Y - Y$ and $2Y = 4Y - 2Y$. However, $4Y$ in a radix-4 multiplier array is equivalent to Y in the next row of the array that carries four times the weight. Hence, partial products are chosen by considering a pair of bits along with the most significant bit from the previous pair. If the most significant bit from the previous pair is true, Y must be added to the current partial product. If the most significant bit of the current pair is true, the current partial product is selected to be negative and the next partial product is incremented.

Table 10.12 shows how the partial products are selected, based on bits of the multiplier. Negative partial products are generated by taking the two's complement of the multiplicand (possibly left-shifted by one column for $-2Y$). An unsigned radix-4 Booth-encoded multiplier requires $\lceil (N+1)/2 \rceil$ partial products rather than N. Each partial product is $M + 1$ bits to accommodate the $2Y$ and $-2Y$ multiples. Even though X and Y are unsigned, the partial products can be negative and must be sign extended properly. The Booth selects will be discussed further after an example.

TABLE 10.12 Radix-4 modified Booth encoding values

Inputs			Partial Product	Booth Selects		
x_{2i+1}	x_{2i}	x_{2i-1}	PP_i	SINGLE_i	DOUBLE_i	NEG_i
0	0	0	0	0	0	0
0	0	1	Y	1	0	0
0	1	0	Y	1	0	0
0	1	1	$2Y$	0	1	0
1	0	0	$-2Y$	0	1	1
1	0	1	$-Y$	1	0	1
1	1	0	$-Y$	1	0	1
1	1	1	-0 $(= 0)$	0	0	1

Example 10.3

Repeat the multiplication of $P = Y \times X = 011001_2 \times 100111_2$ from Figure 10.71, applying Booth encoding to reduce the number of partial products.

SOLUTION: Figure 10.79 shows the multiplication. X is written vertically and the bits are used to select the four partial products. Each partial product is shifted two columns left of the previous one because it has four times the weight. The upper bits are sign-extended with 1s for negative partial products and 0s for positive partial products. The partial products are added to obtain the result.

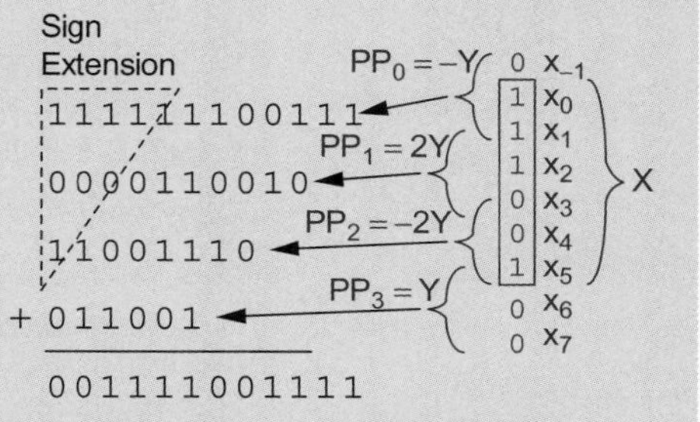

FIGURE 10.79 Booth-encoded example

In a typical radix-4 Booth-encoded multiplier design, each group of 3 bits (a pair, along with the most significant bit of the previous pair) is encoded into several select lines ($SINGLE_i$, $DOUBLE_i$, and NEG_i, given in the rightmost columns of Table 10.12) and driven across the partial product row as shown in Figure 10.80. The multiplier Y is distributed to all the rows. The select lines control Booth selectors that choose the appropriate multiple of Y for each partial product. The Booth selectors substitute for the AND gates of a simple array multiplier to determine the ith partial product. Figure 10.80 shows a conventional Booth encoder and selector design [Goto92]. Y is zero-extended to $M + 1$ bits. Depending on $SINGLE_i$ and $DOUBLE_i$, the A22OI gate selects either 0, Y, or $2Y$. Negative partial products should be two's-complemented (i.e., invert and add 1). If NEG_i is asserted, the partial product is inverted. The extra 1 can be added in the least significant column of the next row to avoid needing a CPA.

Even in an unsigned multiplier, negative partial products must be sign-extended to be summed correctly. Figure 10.81 shows a 16-bit radix-4 Booth partial product array for an unsigned multiplier using the dot diagram notation. Each dot in the Booth-encoded mul-

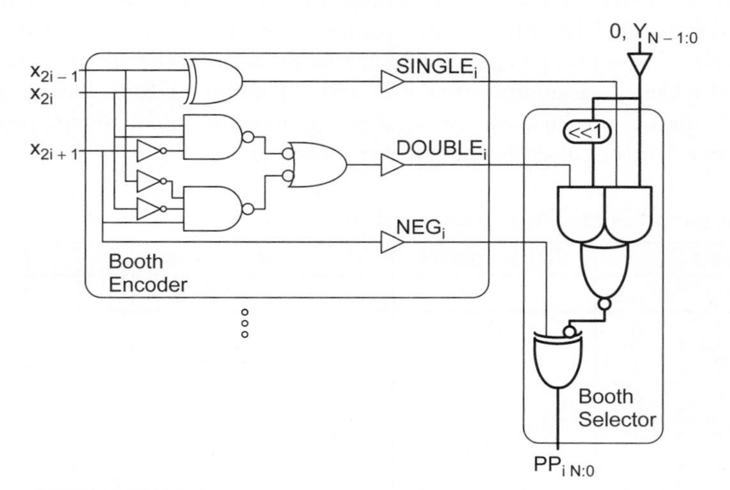

FIGURE 10.80 Radix-4 Booth encoder and selector

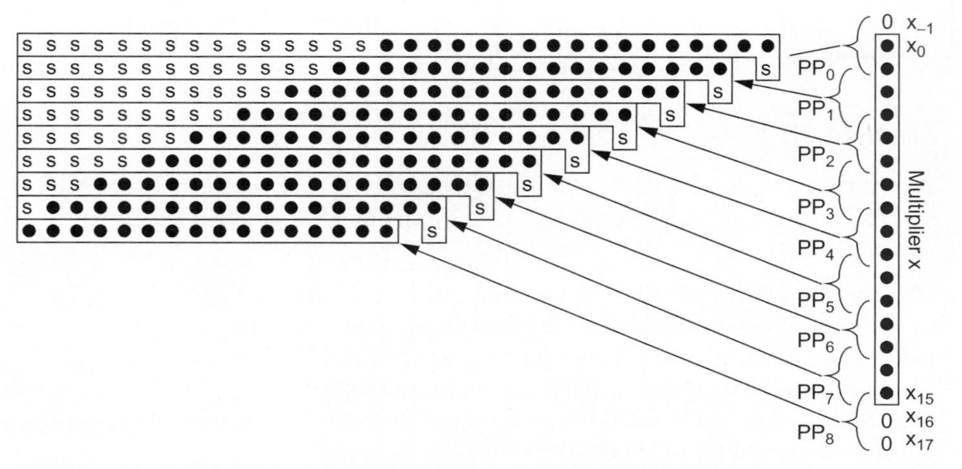

FIGURE 10.81 Radix-4 Booth-encoded partial products with sign extension

tiplier is produced by a Booth selector rather than a simple AND gate. Partial products 0–7 are 17 bits. Each partial product i is sign extended with $s_i = NEG_i = x_{2i+1}$, which is 1 for negative multiples (those in the bottom half of Table 10.12) or 0 for positive multiples. Observe how an extra 1 is added to the least significant bit in the next row to form the 2's complement of negative multiples. Inverting the implicit leading zeros generates leading ones on negative multiples. The extra terms increase the size of the multiplier. PP_8 is required in case PP_7 is negative; this partial product is always 0 or Y because x_{16} and x_{17} are 0. Hence, partial product 8 is only 16 bits.

Observe that the sign extension bits are all either 1s or 0s. If a single 1 is added to the least significant position in a string of 1s, the result is a string of 0s plus a carry-out the top bit that may be discarded. Therefore, the large number of s bits in each partial product can be replaced by an equal number of constant 1s plus the inverse of s added to the least significant position, as shown in Figure 10.82(a). These constants mostly can be optimized out of the array by precomputing their sum. The simplified result is shown in Figure 10.82(b). As usual, it can be squashed to fit a rectangular floorplan.

The critical path of the multiplier involves the Booth decoder, the select line drivers, the Booth selector, approximately $N/2$ CSAs, and a final CPA. Each partial product fills about $M + 5$ columns. 54 × 54-bit radix-4 Booth multipliers for IEEE double-precision floating-point units are typically 20–50% smaller (and arguably up to 20% faster) than nonencoded counterparts, so the technique is widely used. The multiplier requires $M \times N/2$ Booth selectors.

Because the selectors account for a substantial portion of the area and only a small fraction of the critical path, they should be optimized for size over speed. For example,

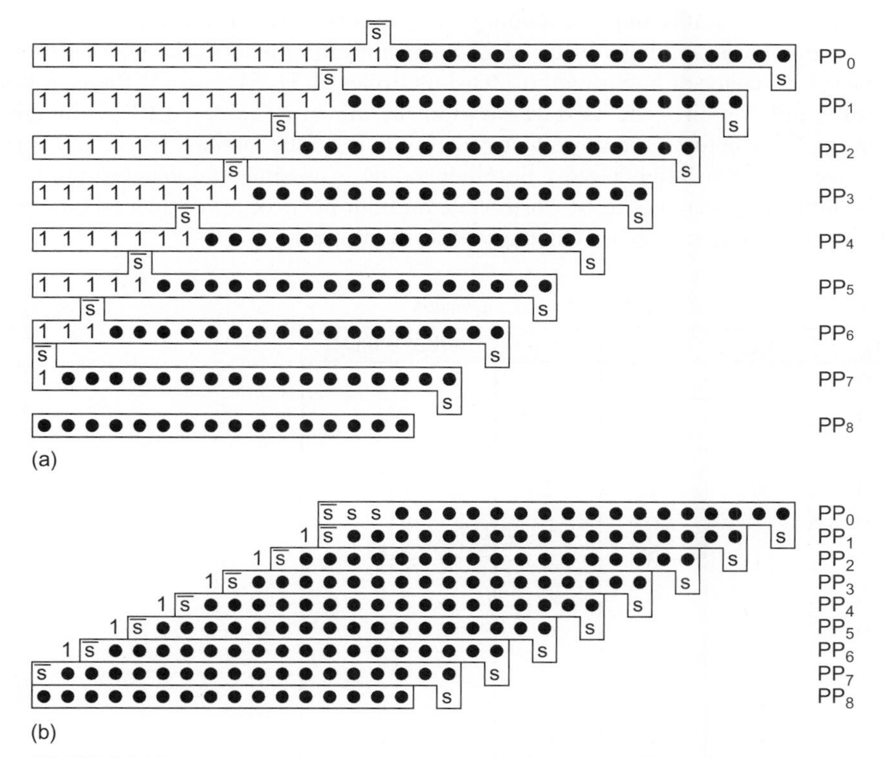

FIGURE 10.82 Radix-4 Booth-encoded partial products with simplified sign extension

[Goto97] describes a *sign select* Booth encoder and selector that uses only 10 transistors per selector bit at the expense of a more complex encoder. [Hsu06a] presents a *one-hot* Booth encoder and selector that chooses one of the six possible partial products using a transmission gate multiplexer. Exercise 10.2 explores yet another encoding.

10.9.3.1 Booth Encoding Signed Multipliers Signed two's complement multiplication is similar, but the multiplicand may have been negative so sign extension must be done based on the sign bit of the partial product, PP_{iM} [Bewick94]. Figure 10.83 shows such an array, where the sign extension bit is $e_i = PP_{iM}$. Also notice that PP_8, which was either Y or 0 for unsigned multiplication, is always 0 and can be omitted for signed multiplication because the multiplier x is sign-extended such that $x_{17} = x_{16} = x_{15}$. The same Booth selector and encoder can be employed (see Figure 10.80), but Y should be sign-extended rather than zero-extended to $M + 1$ bits.

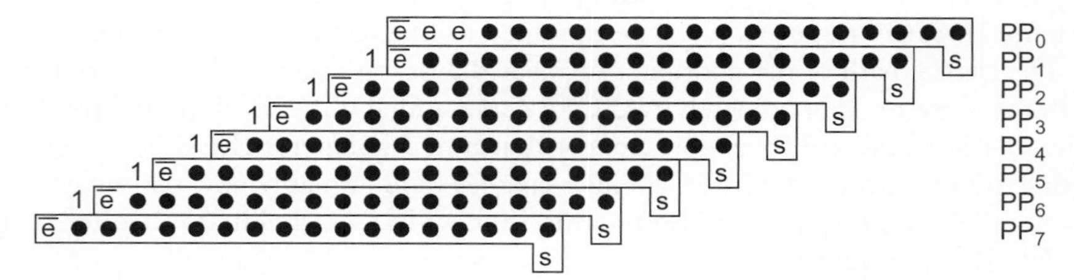

FIGURE 10.83 Radix-4 Booth-encoded partial products for signed multiplication

10.9.3.2 Higher Radix Booth Encoding Large multipliers can use Booth encoding of higher radix. For example, ordinary radix-8 multiplication reduces the number of partial products by a factor of 3, but requires hard multiples of $3Y$, $5Y$, and $7Y$. Radix-8 Booth-encoding only requires the hard $3Y$ multiple, as shown in Table 10.13. Although this requires a CPA before partial product generation, it can be justified by the reduction in array size and delay. Higher-radix Booth encoding is possible, but generating the other hard multiples appears not to be worthwhile for multipliers of fewer than 64 bits. Similar techniques apply to sign-extending higher-radix multipliers.

TABLE 10.13 Radix-8 modified Booth encoding values

x_{i+2}	x_{i+1}	x_i	x_{i-1}	Partial Product
0	0	0	0	0
0	0	0	1	Y
0	0	1	0	Y
0	0	1	1	$2Y$
0	1	0	0	$2Y$
0	1	0	1	$3Y$
0	1	1	0	$3Y$
0	1	1	1	$4Y$
1	0	0	0	$-4Y$
1	0	0	1	$-3Y$
1	0	1	0	$-3Y$

continues

TABLE 10.13 Radix-8 modified Booth encoding values (continued)

1	0	1	1	−2Y
1	1	0	0	−2Y
1	1	0	1	−Y
1	1	1	0	−Y
1	1	1	1	−0

10.9.4 Column Addition

The critical path in a multiplier involves summing the dots in each column. Observe that a CSA is effectively a "ones counter" that adds the number of 1s on the A, B, and C inputs and encodes them on the sum and carry outputs, as summarized in Table 10.14. A CSA is therefore also known as a *(3,2) counter* because it converts three inputs into a count encoded in two outputs [Dadda65]. The carry-out is passed to the next more significant column, while a corresponding carry-in is received from the previous column. This is called a *horizontal path* because it crosses columns. For simplicity, a carry is represented as being passed directly down the column. Figure 10.84 shows a dot diagram of an array multiplier column that sums N partial products sequentially using N–2 CSAs. For example, the 16 × 16 Booth-encoded multiplier from Figure 10.82(b) sums nine partial products with seven levels of CSAs. The output is produced in carry-save redundant form suitable for the final CPA.

TABLE 10.14 An adder as a ones counter

A	*B*	*C*	Carry	Sum	Number of 1s
0	0	0	0	0	0
0	0	1	0	1	1
0	1	0	0	1	1
0	1	1	1	0	2
1	0	0	0	1	1
1	0	1	1	0	2
1	1	0	1	0	2
1	1	1	1	1	3

FIGURE 10.84 Dot diagram for array multiplier

The column addition is slow because only one CSA is active at a time. Another way to speed the column addition is to sum partial products in parallel rather than sequentially. Figure 10.85 shows a *Wallace tree* using this approach [Wallace64]. The Wallace tree requires

$$\left\lceil \log_{3/2}\left(N/2\right)\right\rceil$$

levels of (3,2) counters to reduce N inputs down to two carry-save redundant form outputs.

Even though the CSAs in the Wallace tree are shown in two dimensions, they are logically packed into a single column of the multiplier. This leads to long and irregular wires along the column to connect the CSAs. The wire capacitance increases the delay and energy of multiplier, and the wires can be difficult to lay out.

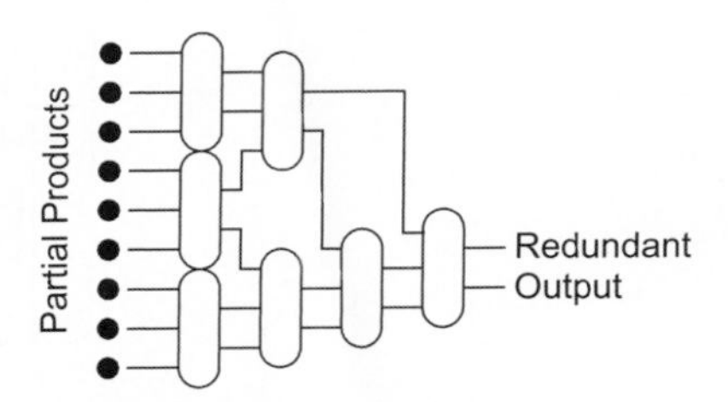

FIGURE 10.85 Dot diagram for Wallace tree multiplier

10.9.4.1 [4:2] Compressor Trees

[4:2] compressors can be used in a binary tree to produce a more regular layout, as shown in Figure 10.86 [Weinberger81, Santoro89]. A [4:2] compressor takes four inputs of equal weight and produces two outputs. It can be constructed from two (3,2) counters as shown in Figure 10.87. Along the way, it generates an intermediate carry, t_i, into the next column and accepts a carry, t_{i-1}, from the previous column, so it may more aptly be called a *(5,3) counter*. This horizontal path does not impact the delay because the output of the top CSA in one column is the input of the bottom CSA in the next column. The [4:2] CSA symbol emphasizes only the primary inputs and outputs to emphasize the main function of reducing four inputs to two outputs. Only

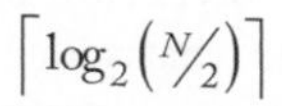

$$\left\lceil \log_2\left(N/2\right) \right\rceil$$

levels of [4:2] compressors are required, although each has greater delay than a CSA. The regular layout and routing also make the binary tree attractive.

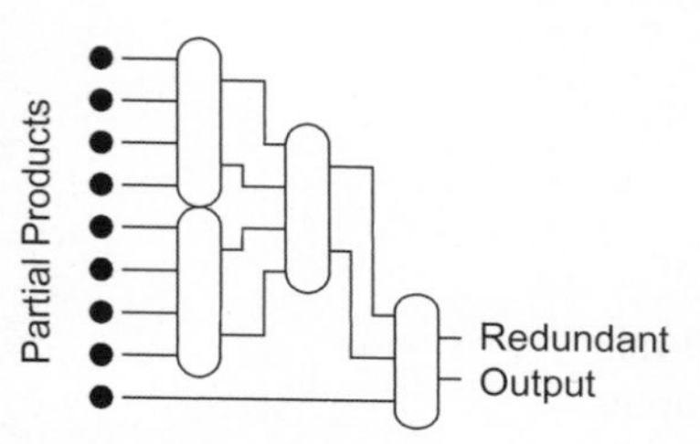

FIGURE 10.86 Dot diagram for [4:2] tree multiplier

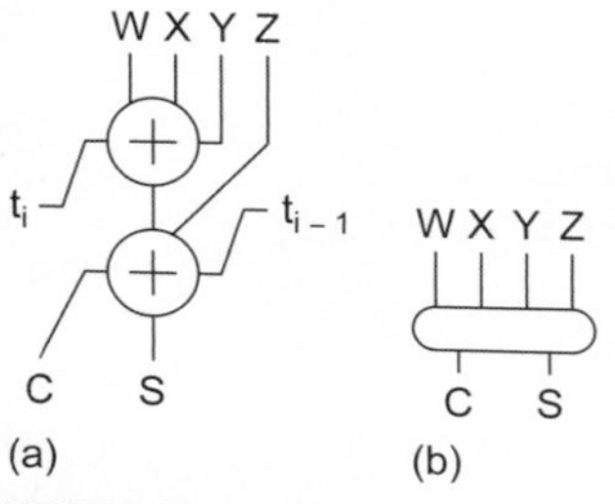

FIGURE 10.87 [4:2] compressor (a) implementation with two CSAs (b) symbol

To see the benefits of a [4:2] compressor, we introduce the notion of fast and slow inputs and outputs. Figure 10.88 shows a simple gate-level CSA design. The longest path through the CSA involves two levels of XOR2 to compute the sum. X is called a *fast input*, while Y and Z are *slow inputs* because they pass through a second level of XOR. C is the *fast output* because it involves a single gate delay, while S is the *slow output* because it involves two gate delays. A [4:2] compressor might be expected to use four levels of XOR2s. Figure 10.89 shows various [4:2] compressor designs that reduce the critical path to only 3 XOR2s. In Figure 10.89(a), the slow output of the first CSA is connected to the fast input of the second. In Figure 10.89(b), the [4:2] compressor has been munged into a single cell,

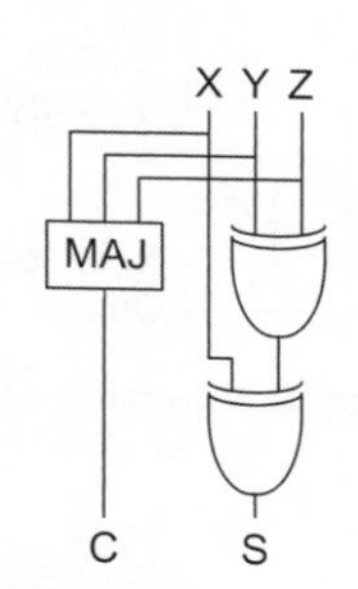

FIGURE 10.88 Gate-level carry-save adder

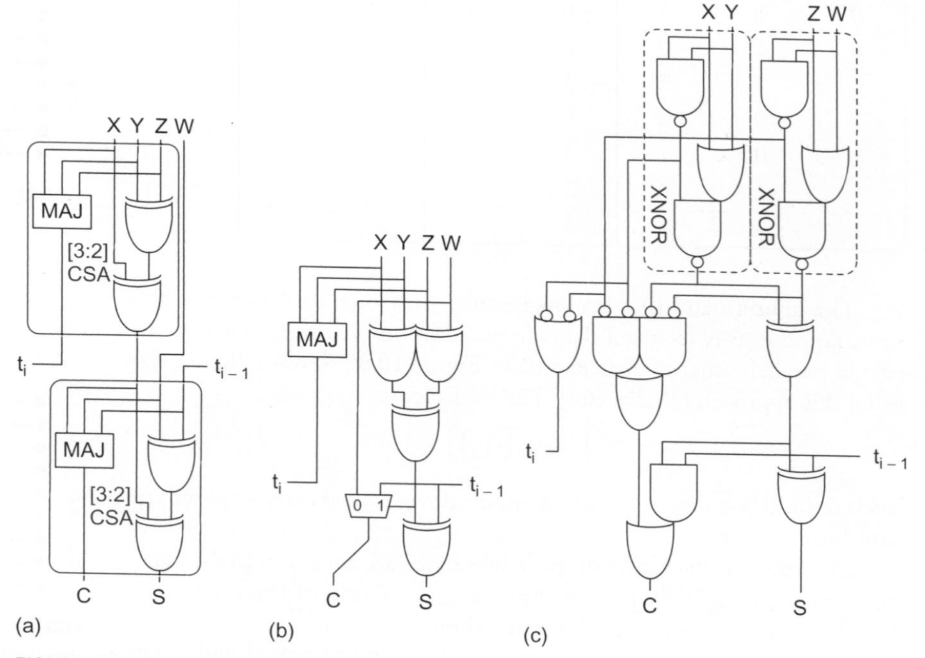

FIGURE 10.89 [4:2] compressors

allowing a majority gate to be replaced with a multiplexer. In Figure 10.89(c), the initial XORs have been replaced with 2-level XNOR circuits that allow some sharing of subfunctions, reducing the transistor count [Goto92].

Figure 10.90 shows a transmission gate implementation of a [4:2] compressor from [Goto97]. It uses only 48 transistors, allowing for a smaller multiplier array with shorter wires. Note that it uses three distinct XNOR circuit forms and two transmission gate multiplexers.

Figure 10.91 compares floorplans of the 16 × 16 Booth-encoded array multiplier from Figure 10.84, the Wallace tree from Figure 10.85, and the [4:2] tree from Figure 10.86. Each row represents a horizontal slice of the multiplier containing a Booth selector or a CSA. Vertical busses connect CSAs. The Wallace tree has the most irregular and lengthy wiring. In practice, the parallelogram may be squashed into a rectangular form to make better use of the space. [Itoh01n] and [Huang05] describes floorplanning issues in tree multipliers.

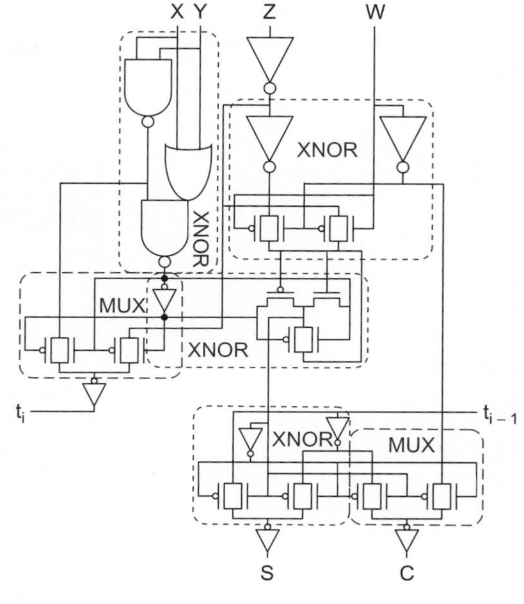

FIGURE 10.90 Transmission gate [4:2] compressor

10.9.4.2 Three-Dimensional Method

The notion of connecting slow outputs to fast inputs generalizes to compressors with more than four inputs. By examining the entire partial product array at once, one can construct trees for each column that sum all of the partial products in the shortest possible time. This approach is called the *three-dimensional method* (TDM) because it considers the arrival time as a third dimension along with rows and columns [Oklobdzija96, Stelling98].

Figure 10.92 shows an example of a 16 × 16 multiplier. The parallelogram at the top shows the dot diagram from Figure 10.82(b) containing nine partial product rows obtained through Booth encoding. The partial products in each of the 32 columns must be summed to produce the 32-bit result. As we have seen, this is done with a compressor to produce a pair of outputs, followed by a final CPA.

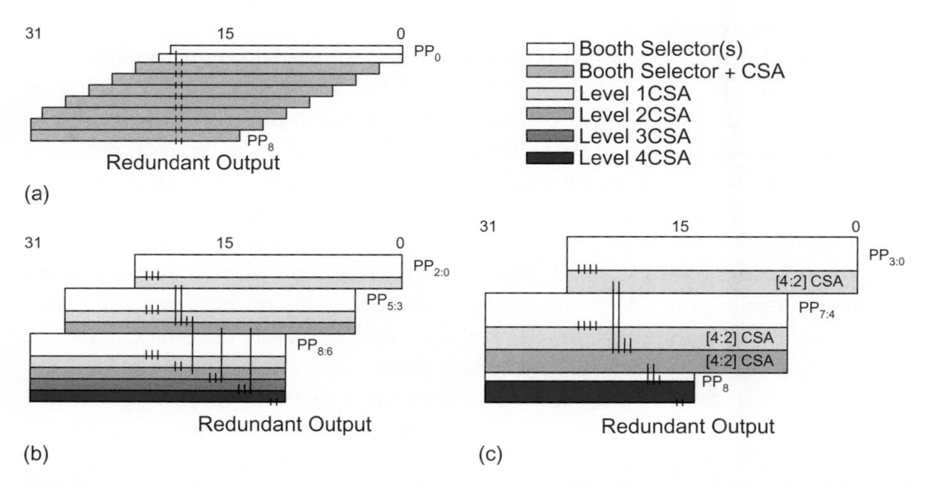

FIGURE 10.91 16 × 16 Booth-encoded multiplier floorplans: (a) array, (b) Wallace tree, (c) [4:2] tree

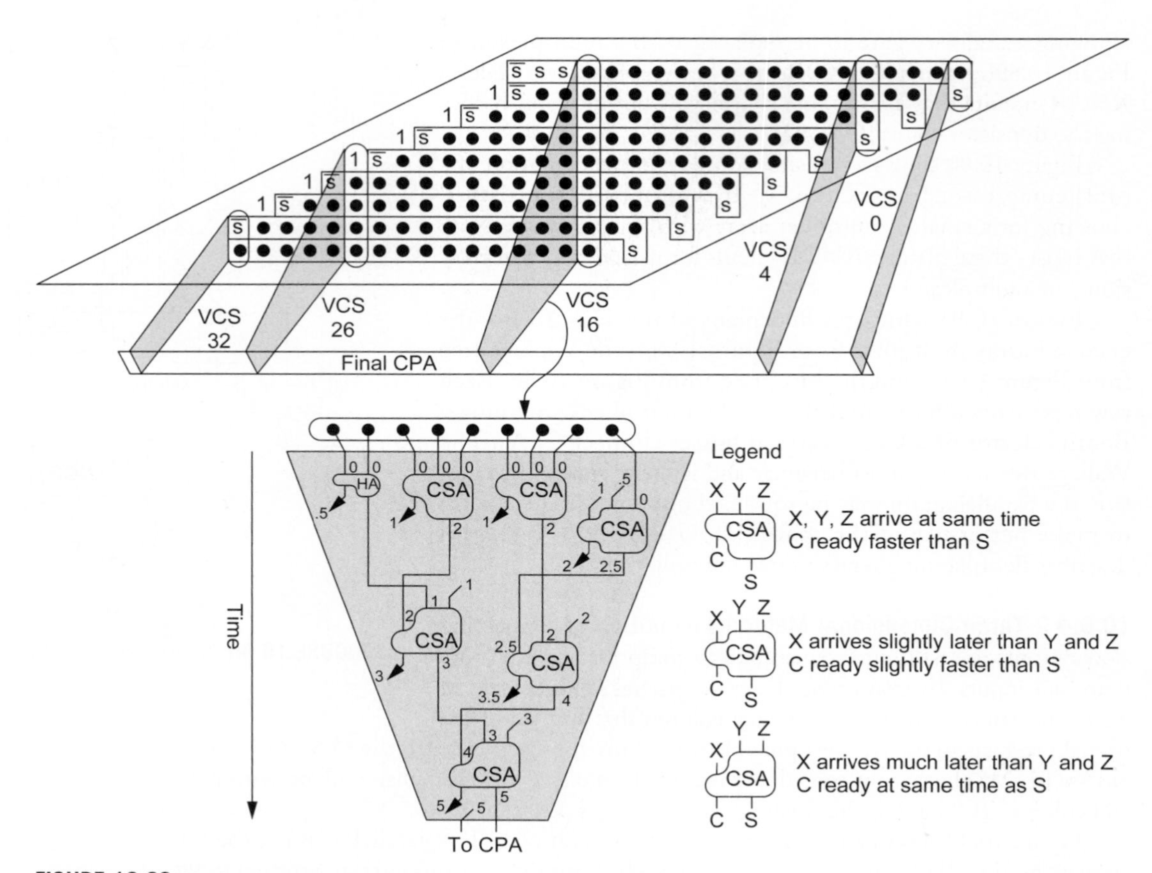

FIGURE 10.92 Vertical compressor slices in a TDM multiplier

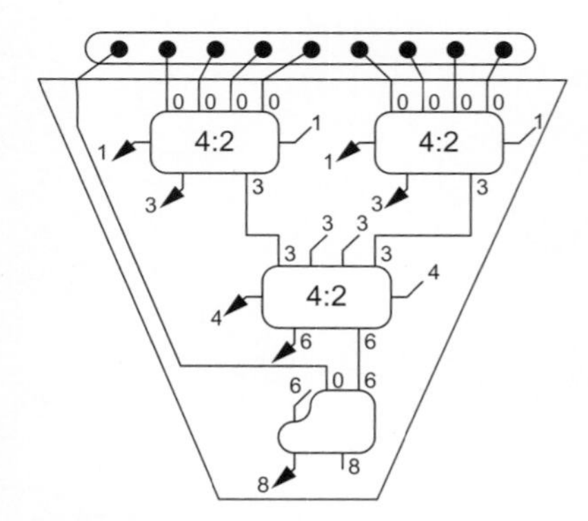

FIGURE 10.93 Vertical compressor slice using [4:2] compressors

In the three-dimensional method, each column is summed with a *vertical compressor slice* (VCS) made of CSAs. In Figure 10.92, VCS 16 adds nine partial products. In this diagram, the horizontal carries between compressor slices are shown explicitly.

Each wire is labeled with its arrival time. All partial product inputs arrive at time 0. The diagram assumes that an XOR2 and a majority gate each have unit delay. Thus, a path through a CSA from any input to *C* or from *X* to *S* takes one unit delay, and that a path from *Y* or *Z* to *S* takes two unit delays. A half adder is assumed to have half the delay. Horizontal carries are represented by diagonal lines coming from behind the slice or pointing out of the slice. VCS 16 receives five horizontal carries in from VCS 15 and produces six horizontal carries out to VCS 17. The final carry out is also shifted by one column before driving the CPA. The inputs to the CSAs are arranged based on their arrival times to minimize the delay of the multiplier. Note how the CSA shape is drawn to emphasize the asymmetric delays. Also, note that VCS 16 is not the slowest; some of the subsequent slices have one unit more delay because the horizontal carries arrive later. [Oklobdzija96] describes an algorithm for choosing the fastest arrangement of CSAs in each VCS given arbitrary CSA delays. In comparison, Figure 10.93 shows the same VCS 16 using [4:2] CSAs; more XOR levels are required but the wiring is more regular.

Table 10.15 lists the number of XOR levels on the critical path for various numbers of partial products. [4:2] trees offer a substantial improvement over Wallace trees in logic levels as well as wiring complexity. TDM generally saves one level of XOR over [4:2] trees, or more for very large multiplies. This savings comes at the cost of irregular wiring, so [4:2] trees and variants thereof remain popular.

TABLE 10.15 Comparison of XOR levels in multiplier trees

# Partial Products	Wallace Tree	4:2 Tree	TDM
8	8	6	5
9	8	8	6
16	12	9	8
24	14	11	10
32	16	12	11
64	20	15	14

10.9.4.3 Hybrid Multiplication Arrays offer regular layout, but many levels of CSAs. Trees offer fewer levels of CSAs, but less regular layout and some long wires. A number of hybrids have been proposed that offer trade-offs between these two extremes. These include *odd/even arrays* [Hennessy90], *arrays of arrays* [Dhanesha95], *balanced delay trees* [Zuras86], *overturned-staircase trees* [Mou90], and *upper/lower left-to-right leapfrog* (ULLRF) trees [Huang05]. They can achieve nearly as few levels of logic as the Wallace tree while offering more regular (and faster) wiring. None have caught on as distinctly better than [4:2] trees.

10.9.5 Final Addition

The output of the partial product array or tree is an $M + N$-bit number in carry-save redundant form. A CPA performs the final addition to convert the result back to nonredundant form.

The inputs to the CPA have nonuniform arrival times. As Figure 10.91 illustrated, the partial products form a parallelogram, with the middle columns having more partial products than the left or right columns. Hence, the middle columns arrive at the CPA later than the others. This can be exploited to simplify the CPA [Zimmermann96, Oklobdzija96]. Figure 10.94 shows an example of a 32-bit prefix network that takes advantage of nonuniform arrival times out of a 16 × 16-bit multiplier. The initial and final stages to compute bitwise PG signals and the sums are not shown. The path from the latest middle inputs to the output involves only four levels of cells. The total number of cells

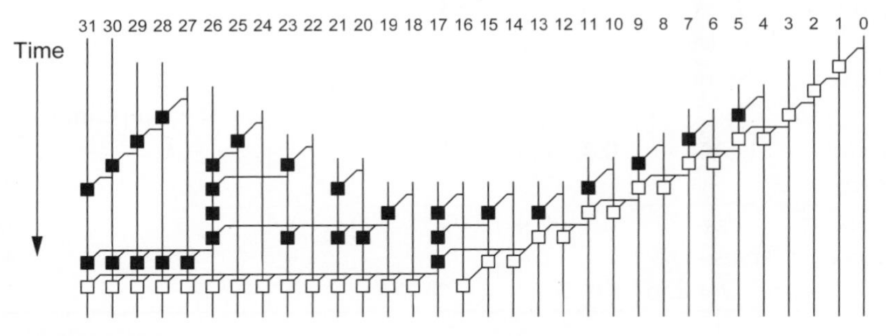

FIGURE 10.94 CPA prefix network with nonuniform input arrival times

and the energy consumption is much less than that of a conventional Kogge-Stone or Sklansky CPA.

10.9.6 Fused Multiply-Add

Many algorithms, particularly in digital signal processing, require computing $P = X \times Y + Z$. While this can be done with a multiplier and adder, it is much faster to use a *fused multiply-add* unit, which is simply an ordinary multiplier modified to accept another input Z that is summed just like the other partial products [Montoye90]. The extra partial product increases the delay of an array multiplier by just one extra CSA.

10.9.7 Summary

The three steps of multiplication are partial product generation, partial product reduction, and carry propagate addition. A simple $M \times N$ multiplier generates N partial products using AND gates. For multipliers of 16 or more bits, radix-4 Booth encoding is typically used to cut the number of partial products in two, saving substantial area and power. Some implementations find Booth encoding is faster, while others find it has little speed benefit. The partial products are then reduced to a pair of numbers in carry-save redundant form using an array or tree of CSAs. Trees have fewer levels of logic, but longer and less regular wiring; nevertheless most large multipliers use trees or hybrid structures. Pass transistor Booth selectors and CSAs were popular in the 1990s, but the trend is toward static CMOS as supply voltage scales. Finally, a CPA converts the result to nonredundant form. The CPA can be simplified based on the nonuniform arrival times of the bits.

Table 10.16 compares reported implementations of 54×54-bit multipliers for double-precision floating point arithmetic. All of the implementations use radix-4 Booth encoding.

TABLE 10.16 54×54-bit multipliers

Design	Process (μm)	PP Reduction	Circuits	Area (mm × mm)	Area ($M\lambda^2$)	Transistors	Latency (ns)	Power (mW)
[Mori91]	0.5	4:2 tree	Pass Transistor XOR	3.6 × 3.5	200	82k	10	870
[Goto92]	0.8	4:2 tree	Static	3.4 × 3.9	80	83k	13	875
[Heikes94]	0.8	array	Dual-Rail Domino	2.1 × 2.2	28		20 (2-stage pipeline)	
[Ohkubo95]	0.25	4:2 tree	Pass Transistors	3.7 × 3.4	805	100k	4.4	
[Goto97]	0.25	4:2 tree	Pass Transistors	1.0 × 1.3	84	61k	4.1	
[Itoh01]	0.18	4:2 tree	Static	1 × 1	100		3.2 (2-stage pipeline)	
[Belluomini05]	90 nm	3:2 and 4:2 tree	LSDL	0.4 × 0.3	61			1800 @ 8 GHz
[Kuang05]	90 nm	3:2 and 4:2 tree	Pass Transistor and Domino	0.5 × 0.4	94			426 @ 4 GHz

10.10 Parallel-Prefix Computations

Many datapath operations involve calculating a set of outputs from a set of inputs in which each output bit depends on all the previous input bits. Addition of two N-bit inputs $A_N \ldots A_1$ and $B_N \ldots B_1$ to produce a sum output $Y_N \ldots Y_1$ is a classic example; each output Y_i depends on a carry-in c_{i-1} from the previous bit, which in turn depends on a carry-in c_{i-2} from the bit before that, and so forth. At first, this dependency chain might seem to suggest that the delay must involve about N stages of logic, as in a carry-ripple adder. However, we have seen that by looking ahead across progressively larger blocks, we can construct adders that involve only log N stages. Section 10.2.2.2 introduced the notion of addition as a prefix computation that involves a bitwise precomputation, a tree of group logic to form the prefixes, and a final output stage, shown in Figure 10.12. In this section, we will extend the same techniques to other prefix computations with associative group logic functions.

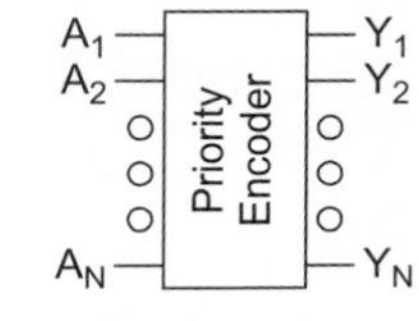

FIGURE 10.95
Priority encoder

Let us begin with the *priority encoder* shown in Figure 10.95. A common application of a priority encoder circuit is to arbitrate among N units that are all requesting access to a shared resource. Each unit i sends a bit A_i indicating a request and receives a bit Y_i indicating that it was granted access; access should only be granted to a single unit with highest priority. If the least significant bit of the input corresponds to the highest priority, the logic can be expressed as follows:

$$\begin{aligned}
Y_1 &= A_1 \\
Y_2 &= A_2 \cdot \overline{A_1} \\
Y_3 &= A_3 \cdot \overline{A_2} \cdot \overline{A_1} \\
&\ldots \\
Y_N &= A_N \cdot \overline{A_{N-1}} \cdot \ldots \cdot \overline{A_1}
\end{aligned} \tag{10.33}$$

We can express priority encoding as a prefix operation by defining a prefix $X_{i:j}$ indicating that none of the inputs $A_i \ldots A_j$ are asserted. Then, priority encoding can be defined with bitwise precomputation, group logic, and output logic with $i \geq k > j$:

$$\begin{aligned}
X_{i:i} &= \overline{A}_i && \text{bitwise precomputation} \\
X_{i:j} &= X_{i:k} \cdot X_{k-1:j} && \text{group logic} \\
Y_i &= A_i \cdot X_{i-1:1} && \text{output logic}
\end{aligned} \tag{10.34}$$

Any of the group networks (e.g., ripple, skip, lookahead, select, increment, tree) discussed in the addition section can be used to build the group logic to calculate the $X_{i:0}$ prefixes. Short priority encoders use the ripple structure. Medium-length encoders may use a skip, lookahead, select, or increment structure. Long encoders use prefix trees to obtain log N delay. Figure 10.96 shows four 8-bit priority encoders illustrating the different group logic. Each design uses an initial row of inverters for the $X_{i:i}$ precomputation and a final row of AND gates for the Y_i output logic. In between, ripple, lookahead, increment, and Sklansky networks form the prefixes with various trade-offs between gate count and delay. Compare these trees to Figure 10.15, Figure 10.22, Figure 10.25, and Figure 10.29(b), respectively. [Wang00, Delgado-Frias00, Huang02] describe a variety of priority encoder implementations.

An *incrementer* can be constructed in a similar way. Adding 1 to an input word consists of finding the least significant 0 in the word and inverting all the bits up to this point. The X prefix plays the role of the propagate signal in an adder. Again, any of the prefix networks can be used with varying area-speed trade-offs.

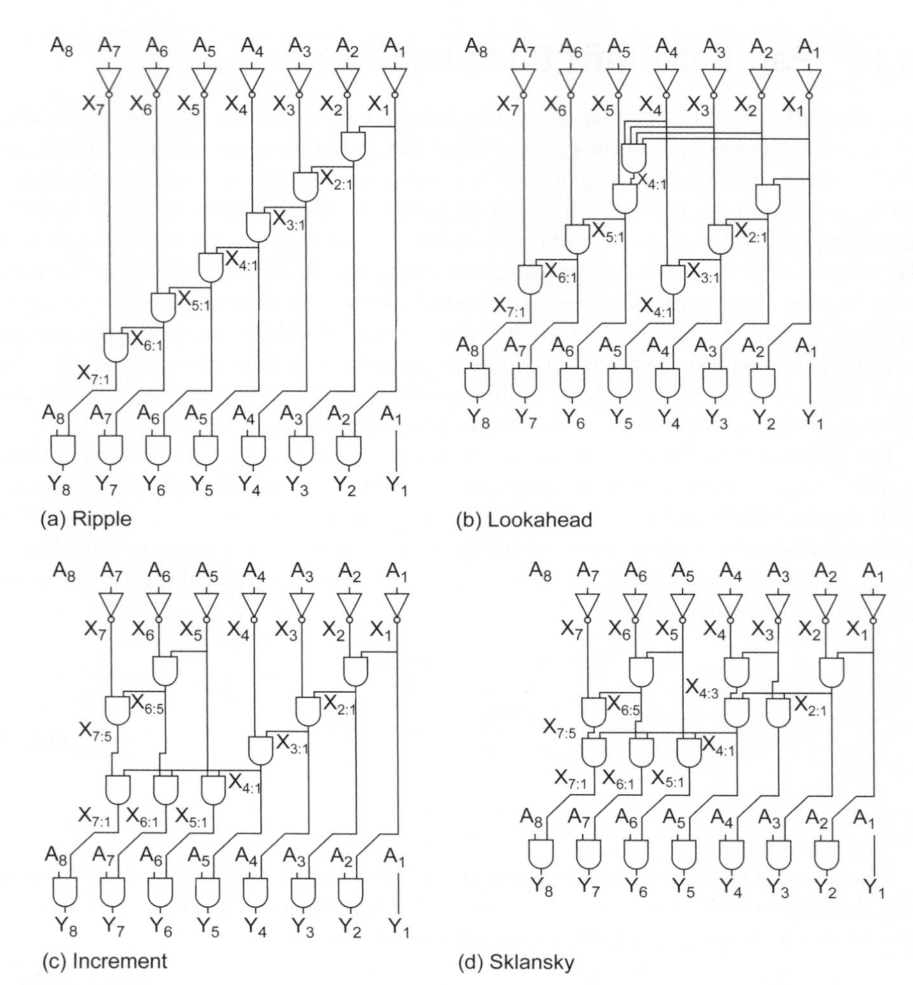

FIGURE 10.96 Priority encoder trees

$$
\begin{aligned}
X_{i:i} &= A_i && \text{bitwise precomputation} \\
X_{i:j} &= X_{i:k} \cdot X_{k-1:j} && \text{group logic} \\
Y_i &= A_i \oplus X_{i-1:1} && \text{output logic}
\end{aligned}
\tag{10.35}
$$

Decrementers and *two's complement* circuits are also similar [Hashemian92]. The decrementer finds the least significant 1 and inverts all the bits up to this point. The two's complement circuit negates a signed number by inverting all the bits above the least significant 1.

A binary-to-thermometer decoder is another application of a prefix computation. The input B is a k-bit representation of the number M. The output Y is a 2^k-bit number with the M most significant bits set to 1, as given in Table 10.17. A simple approach is to use an ordinary k:2^k decoder to produce a one-hot 2^k-bit word A. Then, the following prefix computation can be applied:

$$
\begin{aligned}
X_{i:i} &= A_{N-i} && \text{bitwise precomputation} \\
X_{i:j} &= X_{i:k} + X_{k-1:j} && \text{group logic} \\
Y_i &= X_{i:0} && \text{output logic}
\end{aligned}
\tag{10.36}
$$

TABLE 10.17 Binary to thermometer decoder

B	*Y*
000	00000000
001	10000000
010	11000000
011	11100000
100	11110000
101	11111000
110	11111100
111	11111110

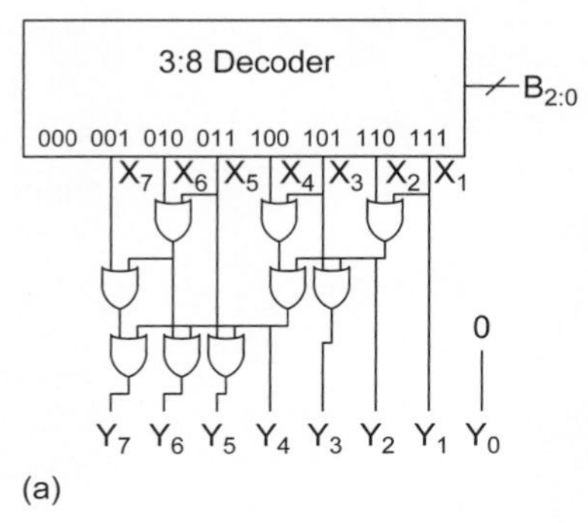

(a)

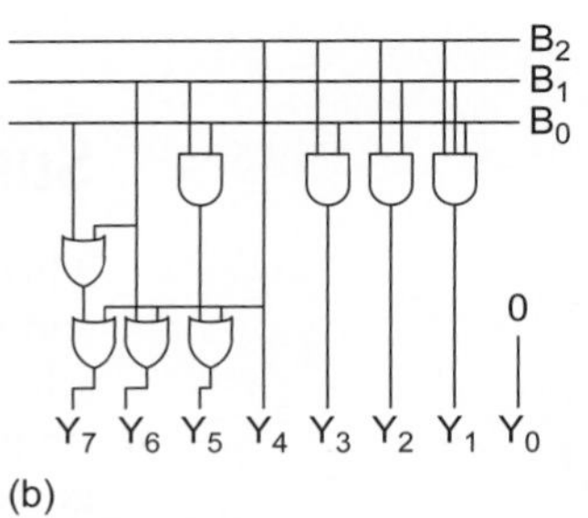

(b)

FIGURE 10.97
Binary-to-thermometer decoders

Figure 10.97(a) shows an 8-bit binary-to-thermometer decoder using a Sklansky tree. The 3:8 decoder contains eight 3-input AND gates operating on true and complementary versions of the input. However, the logic can be significantly simplified by eliminating the complemented AND inputs, as shown in Figure 10.97(b)

In a slightly more complicated example, consider a modified priority encoder that finds the first two 1s in a string of binary numbers. This might be useful in a cache with two write ports that needs to find the first two free words in the cache. We will use two prefixes: X and W. Again, $X_{i:j}$ indicates that none of the inputs $A_i \ldots A_j$ are asserted. $W_{i:j}$ indicates exactly one of the inputs $A_i \ldots A_j$ is asserted. We will produce two 1-hot outputs, Y and Z, indicating the first two 1s.

$$
\begin{aligned}
X_{i:i} &= \overline{A_i} \\
W_{i:i} &= A_i && \text{bitwise precomputation} \\
X_{i:j} &= X_{i:k} \cdot X_{k-1:j} \\
W_{i:j} &= W_{i:k} \cdot X_{k-1:j} + X_{i:k} \cdot W_{k-1:j} && \text{group logic} \\
Y_i &= A_i \cdot X_{i-1:1} \\
Z_i &= A_i \cdot W_{i-1:1} && \text{output logic}
\end{aligned}
\tag{10.37}
$$

10.11 Pitfalls and Fallacies

Equating logic levels and delay

Comparing a novel design with the best existing design is difficult. Some engineers cut corners by merely comparing logic levels. Unfortunately, delay depends strongly on the logical effort of each stage, the fanout it must drive, and the wiring capacitance. For example, [Srinivas92] claims that a novel adder is 20–28% faster than the fastest known binary lookahead adder, but does not present simulation results. Moreover, it reports some of the speed advantages to three or four significant figures. On closer examination [Dobson95], the adder proves to just be a hybrid tree/carry-select design with some unnecessary precomputation.

Designing circuits with threshold drops

In modern processes, single-pass transistors that pull an output to $V_{DD} - V_t$ are generally unacceptable because the threshold drop (amplified by the body effect) results in an output with too little noise margin. Moreover, when they drive the gate terminals of a subsequent stage, the stage turns partially ON and consumes static power. Many 10-transistor full-adder cells have been proposed that suffer from such a threshold drop problem.

Reinventing adders

There is an enormous body of literature on adders with various trade-offs among speed, area, and power consumption. The design space has been explored fairly well and many designers (one of the authors included) have spent quite a bit of time developing a "new" adder, only to find that it is only a minor variation on an existing theme. Similarly, a number of recent publications on priority encoders reinvent prefix network techniques that have already been explored in the context of addition.

Summary

This chapter has presented a range of datapath subsystems. How one goes about designing and implementing a given CMOS chip is largely affected by the availability of tools, the schedule, the complexity of the system, and the final cost goals of the chip. In general, the simplest and least expensive (in terms of time and money) approach that meets the target goals should be chosen. For many systems, this means that synthesis and place & route is good enough. Modern synthesis tools draw on a good library of adders and multipliers with various area/speed trade-offs that are sufficient to cover a wide range of applications. For systems with the most stringent requirements on performance or density, custom design at the schematic level still provides an advantage. Domino parallel-prefix trees provide the fastest adders when the high power consumption can be tolerated. Domino CSAs are also used in fast multipliers. However, in multiplier design, the wiring capacitance is paramount and a multiplier with compact cells and short wires can be fast as well as small and low in power.

Exercises

10.1 Design a Gray-coded counter in which only one bit changes on each cycle.

10.2 Table 10.12 and Figure 10.80 illustrated radix-4 Booth encoding using *SINGLE*, *DOUBLE*, and *NEG*. An alternative encoding is to use *POS*, *NEG*, and *DOUBLE*. *POS* is true for the multiples Y and $2Y$. *NEG* is true for the multiples $-Y$ and $-2Y$. *DOUBLE* is true for the multiples $2Y$ and $-2Y$. Design a Booth encoder and selector using this encoding.

10.3 Adapt the priority encoder logic of EQ (10.37) to produce three 1-hot outputs corresponding to the first three 1s in an input string.

10.4 Sketch a 16-bit priority encoder using a Kogge-Stone prefix network.

10.5 Use Logical Effort to estimate the delay of the priority encoder from Exercise 10.4. Assume the path electrical effort is 1.

10.6 Write equations for a prefix computation that determines the second location in which the pattern 10 appears in an N-bit input string. For example, 010010 should return 010000.

10.7 Design a fast 8-bit adder. The inputs may drive no more than 30 λ of transistor width each and the output must drive a 20/10 λ inverter. Simulate the adder and determine its delay.

10.8 When adding two unsigned numbers, a carry-out of the final stage indicates an overflow. When adding two signed numbers in two's complement format, overflow detection is slightly more complex. Develop a Boolean equation for overflow as a function of the most significant bits of the two inputs and the output.

10.9 Repeat Exercise 10.8 for a signed add/subtract unit like that shown in Figure 10.41(b). Your overflow output should be a function of the subsignal and the most significant bits of the two inputs and the output.

10.10 Develop equations for the logical effort and parasitic delay with respect to the C_0 input of an n-stage Manchester carry chain computing $C_1 \ldots C_n$. Consider all of the internal diffusion capacitances when deriving the parasitic delay. Use the transistor widths shown in Figure 10.98 and assume the P_i and G_i transistors of each stage share a single diffusion contact.

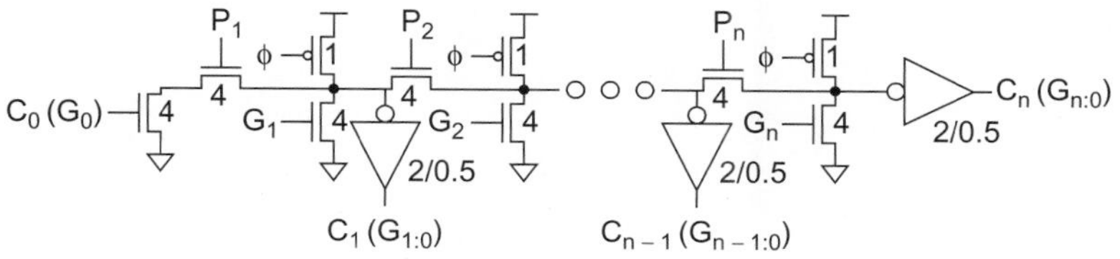

FIGURE 10.98 Manchester carry chain

10.11 Using the results of Exercise 10.10, what Manchester carry chain length gives the least delay for a long adder?

10.12 The carry increment adder in Figure 10.26(b) with variable block size requires five stages of valency-2 group PG cells for 16-bit addition. How many stages are required for 32-bit addition? For 64-bit addition?

10.13 [Jackson04] proposes an extension of the Ling adder formulation to simplify cells later in the prefix network. Design a 16-bit adder using this technique and compare it to a conventional 16-bit Ling adder.

10.14 Figure 10.29 shows PG networks for various 16-bit adders and Figure 10.30 illustrates how these networks can be classified as the intersection of the $l + f + t = 3$

plane with the face of a cube. The plane also intersects one point inside the cube at $(l, f, t) = (1, 1, 1)$ [Harris03]. Sketch the PG network for this 16-bit adder.

10.15 Sketch a diagram of the group PG tree for a 32-bit Ladner-Fischer adder.

10.16 Write a Boolean expression for C_{out} in the circuit shown in Figure 10.6(b). Simplify the equation to prove that the pass-transistor circuits do indeed compute the majority function.

10.17 Prove EQ (10.21).

10.18 Sketch a design for a comparator computing $A - B = k$.

10.19 Show how the layout of the parity generator of Figure 10.57 can be designed as a linear column of XOR gates with a tree-routing channel.

10.20 Design an ECC decoder for distance-3 Hamming codes with $c = 3$. Your circuit should accept a 7-bit received word and produce a 4-bit corrected data word. Sketch a gate-level implementation.

10.21 How many check bits are required for a distance-3 Hamming code for 8-bit data words? Sketch a parity-check matrix and write the equations to compute each of the check bits.

10.22 Find the 4-bit binary-reflected Gray code values for the numbers 0–15.

10.23 Sketch the PG network for a modified 16-bit Sklansky adder with fanout of [8, 1, 1, 1] rather than [8, 4, 2, 1]. Use buffers to prevent the less-significant bits from loading the critical path.

Memories 11

11.1 Introduction

Memory arrays often account for the majority of transistors in a CMOS system-on-chip. Arrays may be divided into categories as shown in Figure 11.1. *Programmable Logic Arrays* (PLAs) perform logic rather than storage functions, but are also discussed in this chapter.

Random access memory is accessed with an *address* and has a latency independent of the address. In contrast, *serial access memories* are accessed sequentially so no address is necessary. *Content addressable memories* determine which address(es) contain data that matches a specified *key*.

Random access memory is commonly classified as *read-only memory* (ROM) or *read/write memory* (confusingly called RAM). Even the term ROM is misleading because many ROMs can be written as well. A more useful classification is *volatile* vs. *nonvolatile* memory. Volatile memory retains its data as long as power is applied, while nonvolatile memory will hold data indefinitely. RAM is synonymous with volatile memory, while ROM is synonymous with nonvolatile memory.

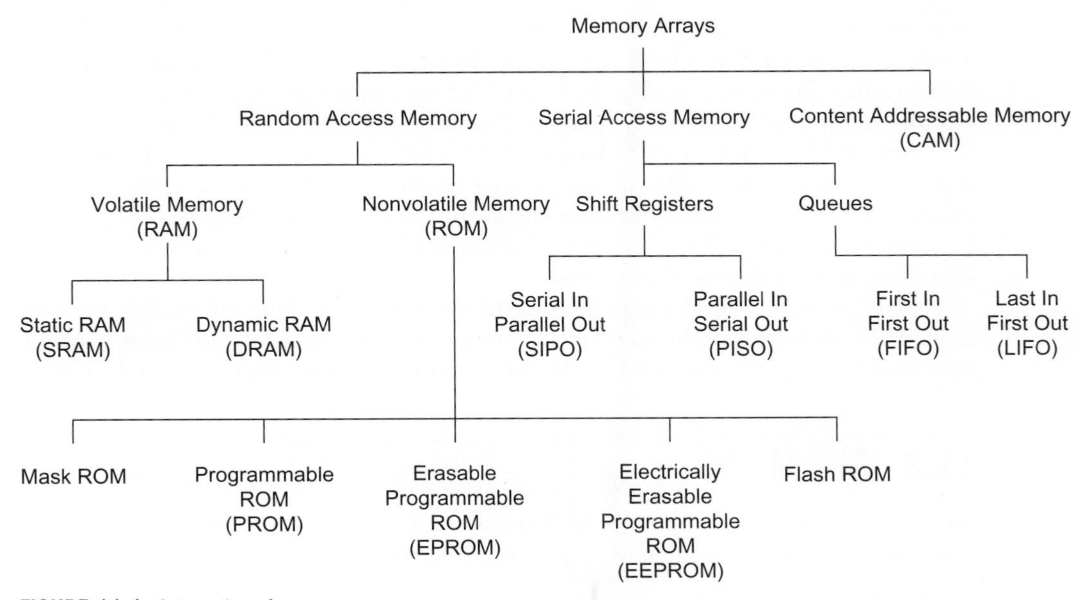

FIGURE 11.1 Categories of memory arrays

Like sequencing elements, the memory cells used in volatile memories can further be divided into *static* structures and *dynamic* structures. Static cells use some form of feedback to maintain their state, while dynamic cells use charge stored on a floating capacitor through an access transistor. Charge will leak away through the access transistor even while the transistor is OFF, so dynamic cells must be periodically read and rewritten to refresh their state. Static RAMs (SRAMs) are faster and less troublesome, but require more area per bit than their dynamic counterparts (DRAMs).

Some nonvolatile memories are indeed read-only. The contents of a mask ROM are hardwired during fabrication and cannot be changed. But many nonvolatile memories can be written, albeit more slowly than their volatile counterparts. A *programmable* ROM (PROM) can be programmed once after fabrication by blowing on-chip fuses with a special high programming voltage. An *erasable programmable* ROM (EPROM) is programmed by storing charge on a floating gate. It can be erased by exposure to ultraviolet (UV) light for several minutes to knock the charge off the gate. Then the EPROM can be reprogrammed. *Electrically erasable programmable* ROMs (EEPROMs) are similar, but can be erased in microseconds with on-chip circuitry. *Flash* memories are a variant of EEPROM that erases entire blocks rather than individual bits. Sharing the erase circuitry across larger blocks reduces the area per bit. Because of their good density and easy in-system reprogrammability, Flash memories have replaced other nonvolatile memories in most modern CMOS systems.

Memory cells can have one or more *ports* for access. On a read/write memory, each port can be read-only, write-only, or capable of both read and write.

A memory array contains 2^n *words* of 2^m bits each. Each bit is stored in a memory cell. Figure 11.2 shows the organization of a small memory array containing 16 4-bit words ($n = 4$, $m = 2$). Figure 11.2(a) shows the simplest design with one row per word and one column per bit. The row decoder uses the address to activate one of the rows by asserting the wordline. During a read operation, the cells on this wordline drive the *bitlines*, which may have been conditioned to a known value in advance of the memory access. The column circuitry may contain amplifiers or buffers to sense the data. A typical memory array may have thousands or millions of words of only 8–64 bits each, which would lead to a tall, skinny layout that is hard to fit in the chip floorplan and slow because of the long vertical wires. Therefore, the array is often folded into fewer rows of more columns. After folding, each row of the memory contains 2^k words, so the array is physically organized as 2^{n-k} *rows* of 2^{m+k} *columns* or bits. Figure 11.2(b) shows a two-way fold ($k = 1$) with eight rows and eight columns. The column decoder controls a multiplexer in the column circuitry to select 2^m bits from the row as the data to access. Larger memories are generally built from multiple smaller subarrays so that the wordlines and bitlines remain reasonably short, fast, and low in power dissipation.

We begin in Section 11.2 with SRAM, the most widely used form of on-chip memory. SRAM also illustrates all the issues of cell design, decoding, and column circuitry design. Subsequent sections address DRAMs, ROMs, serial access memories, CAMs, and PLAs.

11.2 SRAM

Static RAMs use a memory cell with internal feedback that retains its value as long as power is applied. It has the following attractive properties:

- Denser than flip-flops
- Compatible with standard CMOS processes

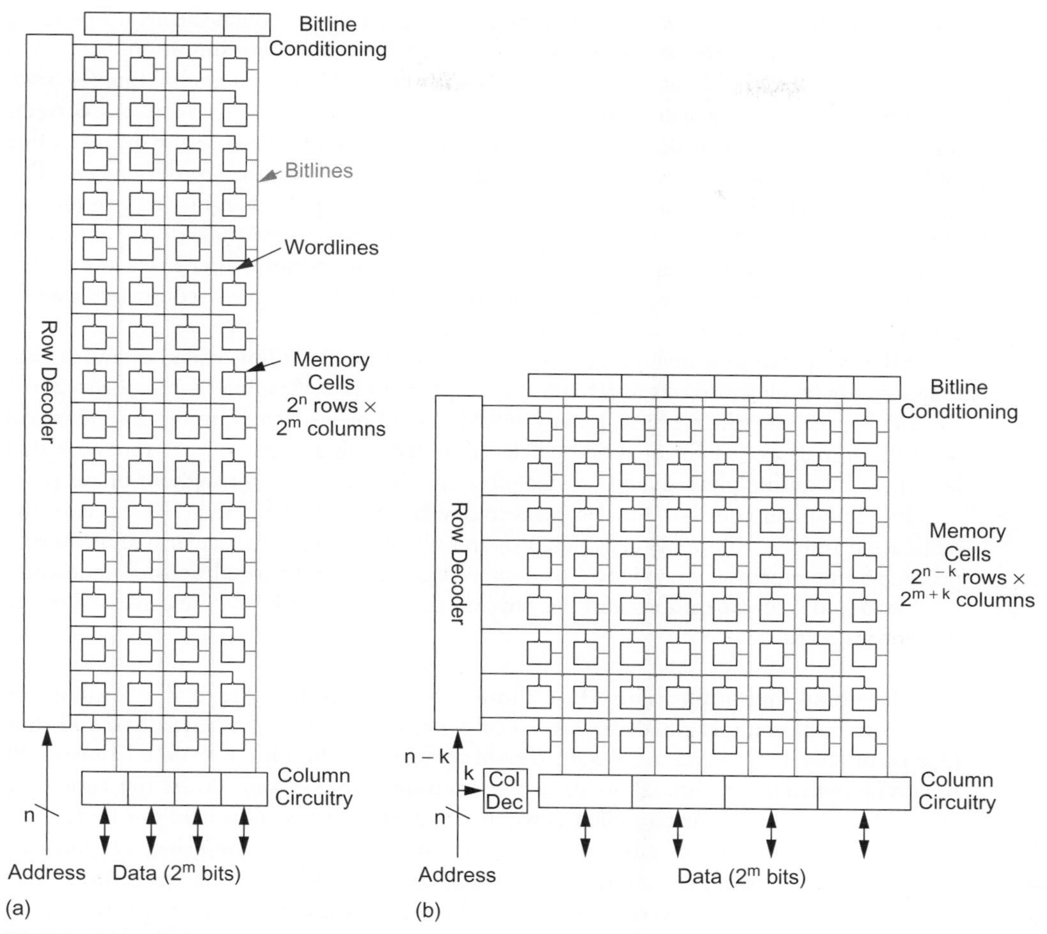

FIGURE 11.2 Memory array architecture

- Faster than DRAM
- Easier to use than DRAM

For these reasons, SRAMs are widely used in applications from caches to register files to tables to scratchpad buffers. The SRAM consists of an array of memory cells along with the row and column circuitry. This section begins by examining the design and operation of each of these components. It then considers important special cases of SRAMs, including multiported register files, large SRAMs and subthreshold SRAMs.

11.2.1 SRAM Cells

A SRAM cell needs to be able to read and write data and to hold the data as long as the power is applied. An ordinary flip-flop could accomplish this requirement, but the size is quite large. Figure 11.3 shows a standard 6-transistor (6T) SRAM cell that can be an order of magnitude smaller than a flip-flop. The 6T cell achieves its compactness at the expense of more complex peripheral circuitry for reading and writing the cells. This is a

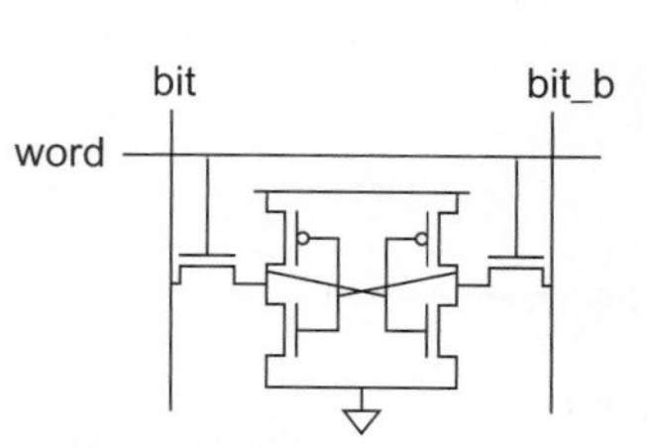

FIGURE 11.3 6T SRAM cell

good trade-off in large RAM arrays where the memory cells dominate the area. The small cell size also offers shorter wires and hence lower dynamic power consumption.

The 6T SRAM cell contains a pair of weak cross-coupled inverters holding the state and a pair of access transistors to read or write the state. The positive feedback corrects disturbances caused by leakage or noise. The cell is written by driving the desired value and its complement onto the bitlines, *bit* and *bit_b*, then raising the wordline, *word*. The new data overpowers the cross-coupled inverters. It is read by precharging the two bitlines high, then allowing them to float. When *word* is raised, *bit* or *bit_b* pulls down, indicating the data value. The central challenges in SRAM design are minimizing its size and ensuring that the circuitry holding the state is weak enough to be overpowered during a write, yet strong enough not to be disturbed during a read.

SRAM operation is divided into two phases. The phases will be called ϕ_1 and ϕ_2, but may actually be generated from *clk* and its complement *clkb*. Assume that in phase 2, the SRAM is precharged. In phase 1, the SRAM is read or written. Timing diagrams will label the signals as _q1 for qualified clocks (ϕ_1 gated with an enable), _v1 for those that become valid during phase 1, and _s1 for those that remain stable throughout phase 1.

It is no longer common for designers to develop their own SRAM cells. Usually, the fabrication vendor will supply cells that are carefully tuned to the particular manufacturing process. Some processes provide two or more cells with different speed/density trade-offs.

Read and write operations and the physical design of the SRAM are discussed in the subsequent sections.

11.2.1.1 Read Operation Figure 11.4 shows a SRAM cell being read. The bitlines are both initially floating high. Without loss of generality, assume Q is initially 0 and thus Q_b is initially 1. Q_b and *bit_b* both should remain 1. When the wordline is raised, *bit* should be pulled down through *driver* and *access* transistors $D1$ and $A1$. At the same time *bit* is being pulled down, node Q tends to rise. Q is held low by $D1$, but raised by current flowing in from $A1$. Hence, the driver $D1$ must be stronger than the access transistor $A1$. Specifically, the transistors must be ratioed such that node Q remains below the switching threshold of the *P2*/*D2* inverter. This constraint is called *read stability.* Waveforms for the read operation are shown in Figure 11.4(b) as a 0 is read onto *bit*. Observe that Q momentarily rises, but does not glitch badly enough to flip the cell.

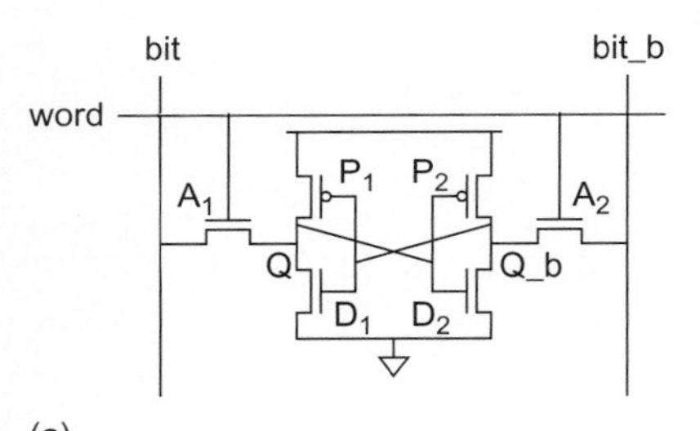

(a)

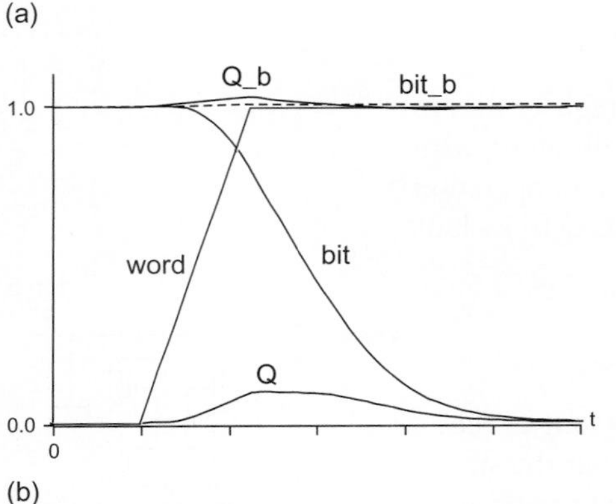

(b)

FIGURE 11.4 Read operation for 6T SRAM cell

Figure 11.5 shows the same cell in the context of a full column from the SRAM. During phase 2, the bitlines are precharged high. The wordline only rises during phase 1; hence, it can be viewed as a _q1 qualified clock. Many SRAM cells share the same bitline pair, which acts as a distributed dual-rail footless dynamic multiplexer. The capacitance of the entire bitline must be discharged through the access transistor. The output can be sensed by a pair of HI-skew inverters. By raising the switching threshold of the sense inverters, delay can be reduced at the expense of noise margin. The outputs are dual-rail monotonically rising signals, just as in a domino gate.

11.2.1.2 Write Operation Figure 11.6 shows the SRAM cell being written. Again, assume Q is initially 0 and that we wish to write a 1 into the cell. *bit* is precharged high and left floating. *bit_b* is pulled low by a write driver. We know on account of the read stability constraint that *bit* will be unable to force Q high through $A1$. Hence, the cell must be written by forcing Q_b low through $A2$. $P2$ opposes this operation; thus, $P2$ must be weaker than $A2$ so that Q_b can be pulled low enough. This constraint is called *writability*. Once Q_b falls low, $D1$ turns OFF and $P1$ turns ON, pulling Q high as desired.

Figure 11.7(a) again shows the cell in the context of a full column from the SRAM. During phase 2, the bitlines are precharged high. Write drivers pull the bitline or its complement low during phase 1 to write the cell. The write drivers can consist of a pair of transistors on each bitline for the data and the write enable, or a single transistor driven by the appropriate combination of signals (Figure 11.7(b)). In either case, the series resistance of the write driver, bitline wire, and access transistor must be low enough to overpower the pMOS transistor in the SRAM cell. Some arrays use tristate write drivers to improve writability by actively driving one bitline high while the other is pulled low.

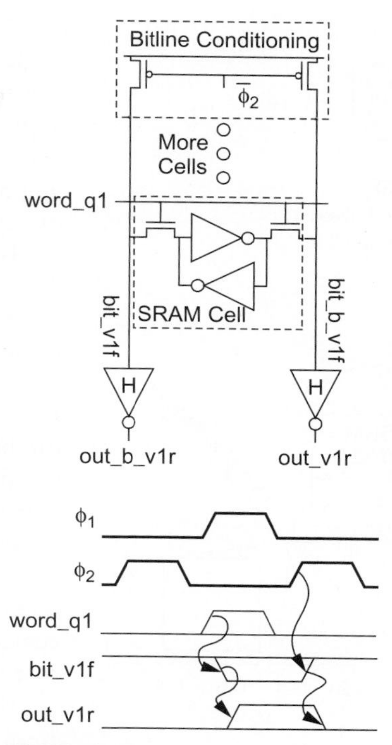

FIGURE 11.5 SRAM column read

11.2.1.3 Cell Stability To ensure both read stability and writability, the transistors must satisfy ratio constraints. The nMOS pulldown transistor in the cross-coupled inverters must be strongest. The access transistors are of intermediate strength, and the pMOS pullup transistors must be weak. To achieve good layout density, all of the transistors must be relatively small. For example, the pulldowns could be 8/2 λ, the access transistors 4/2, and the pullups 3/3. The SRAM cells must operate correctly at all voltages and temperatures despite process variation.

The stability and writability of the cell are quantified by the hold margin, the read margin, and the write margin, which are determined by the static noise margin of the cell in its various modes of operation. A cell should have two stable states during hold and read operation, and only one stable state during write. The *static noise margin* (SNM) measures how much noise can be applied to the inputs of the two cross-coupled inverters before a stable state is lost (during hold or read) or a second stable state is created (during write).

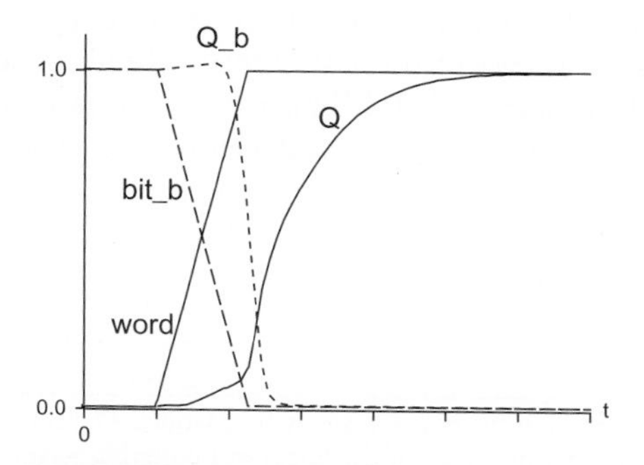

FIGURE 11.6 Write operation for 6T SRAM cell

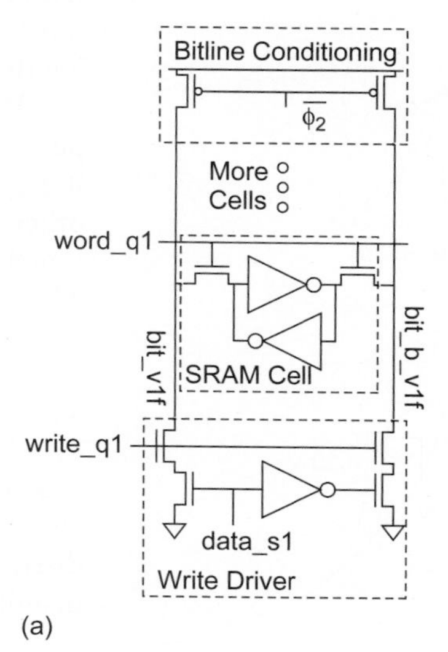

(a)

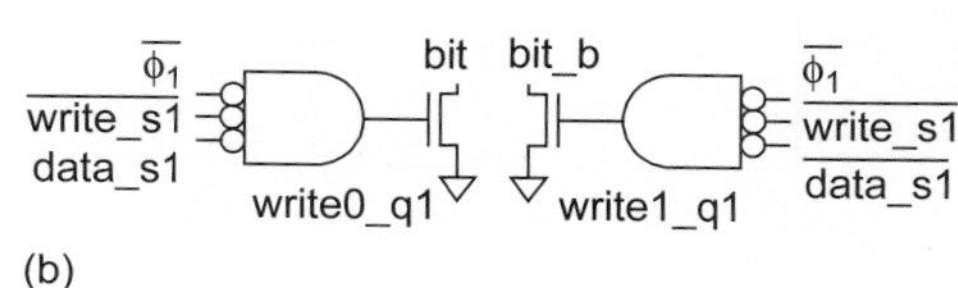

(b)

FIGURE 11.7 SRAM column write

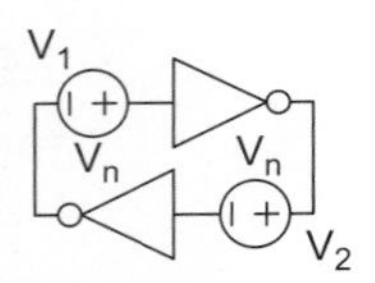

FIGURE 11.8 Cross-coupled inverters with noise sources for hold margin

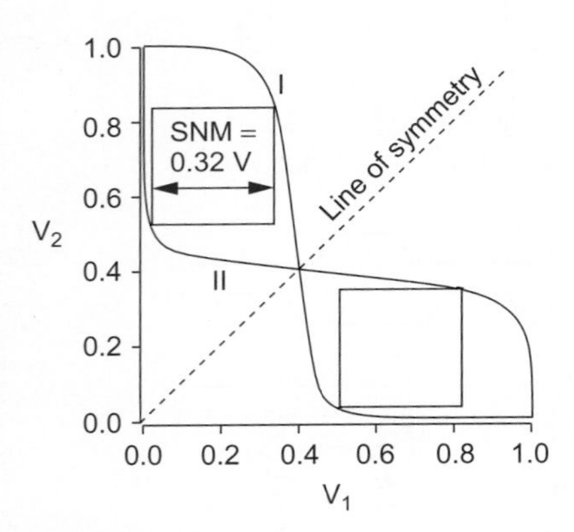

FIGURE 11.9 Butterfly diagram indicating hold margin

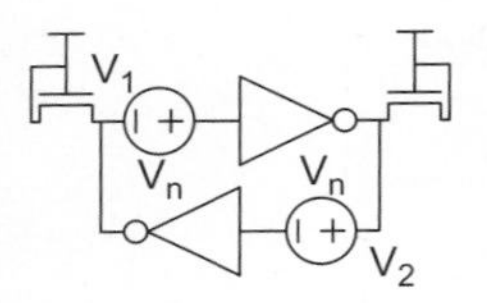

FIGURE 11.10 Read margin circuit

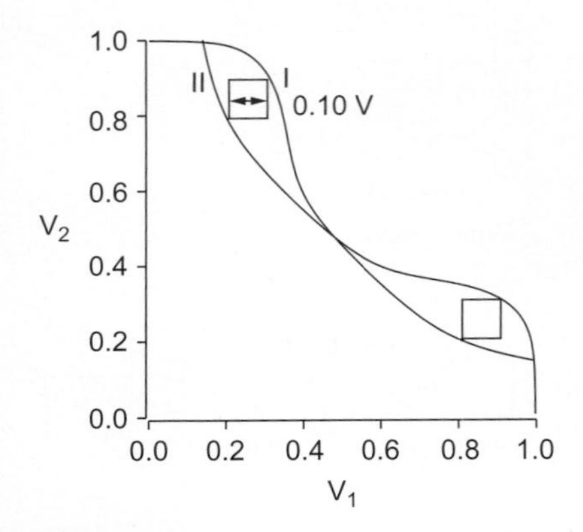

FIGURE 11.11 Read margin

Figure 11.8 shows the test circuit for determining the *hold margin* (i.e., the static noise margin while the cell is holding its state and being neither read nor written; this is *unrelated* to the hold time of flip-flop). A noise source V_n is applied to each of the cross-coupled inverters. The access transistors are OFF and do not affect the circuit behavior. The static noise margin can be determined graphically from a *butterfly diagram* shown in Figure 11.9. The plot is generated by setting $V_n = 0$ and plotting V_2 against V_1 (curve I) and V_1 against V_2 (curve II). If the inverters are identical, the DC transfer curves are mirrored across the line of $V_1 = V_2$. The butterfly plot shows two stable states (with one output low and the other high) and one metastable state (with $V_1 = V_2$). A positive value of noise shifts curve I left and curve II up. Excessive noise eliminates the stable state of $V_1 = 0$ and $V_2 = V_{DD}$, forcing the cell into the opposite state. The static noise margin is determined by the length of the side of the largest square that can be inscribed between the curves [Lohstroh83, Seevinck87]. If the inverters are identical, the butterfly diagram is symmetric, so the high and low static noise margins are equal.[1] If the inverters are not identical, the static noise margin is the lesser of the two cases. The noise margin increases with V_{DD} and V_t.

When the cell is being read, the bitlines are initially precharged and the access transistor tends to pull the low node up. This distorts the voltage transfer characteristics. The static noise margin under these circumstances is called the *read margin* and is smaller than the hold margin. It can be obtained by performing the same simulation on the circuit in Figure 11.10 with the bitlines tied to V_{DD}. Figure 11.11 shows the results. The read margin depends on the relative strength of the pulldown transistor D to the access transistor A. The ratio of these two transistors' widths is called the *beta ratio* or *cell ratio*. A higher beta ratio increases the read margin but takes more area to build the wide pulldown transistors. The read margin also improves by increasing V_{DD} or V_t or by reducing the wordline voltage relative to V_{DD}.

When the cell is being written, the access transistor A must overpower the pullup P to create a single stable state. The *write margin* is determined by a similar simulation as read margin, with one access transistor pulling to 0 and the other to 1. If $|V_n|$ is too large, a second stable state will exist, preventing the function of writes. Figure 11.12 shows the characteristics while bit is held at 0. The write margin is the size of the smallest square inscribed between the two curves [Bhavnagarwala05]. The write margin improves as the access transistor becomes stronger, the pullup becomes weaker, or the word line voltage increases. These trends are in conflict with improving the read margin.

Threshold voltage mismatch caused by random dopant fluctuations is a particular problem in nanometer processes because of the vast number of cells on a chip and the increasing variability [Bhavnagarwala01]. This variation creates a distribution of read, write, and hold margins. If any cell develops a negative margin, it is inoperable.

[1] In contrast, the unity gain noise margins defined in Section 2.5.3 may be unequal. The static noise margin found by the butterfly diagram sacrifices part of the larger noise margin to improve the smaller one.

Example 11.1

Suppose the cells in a 64 Mb SRAM have normally distributed read margins with 15 mV standard deviations. Assume the array is unreliable if any cell has a negative read margin (this is optimistic; some margin should be budgeted for noise). What must the mean read margin be to achieve 90% parametric yield for the array?

SOLUTION: Using EQ (6.21), each cell must have a failure probability of

$$X_c = 1 - \sqrt[N]{Y} = 1 - \sqrt[2^{26}]{0.9} = 1.6 \times 10^{-9} \quad \textbf{(11.1)}$$

According to Table 6.8, this means that nearly 6σ of Gaussian variation must be accepted. Thus, the read margin should be at least 90 mV.

This analysis should be taken with several caveats. The calculation of X_c assumes that the cell failure probabilities are independent (though not necessarily Gaussian). The distribution of read margins is not necessarily Gaussian and a distribution with a differently shaped tail will require a different amount of margin to achieve X_c. The failure criteria of zero read margin does not account for noise that might disturb the cell. The choice of 90% parametric yield is arbitrary and possibly misleading. If the memory were a small part of a larger chip, its parametric yield would have to be larger to achieve good parametric yield for the whole chip. And point defects that cause functional failure have not been considered.

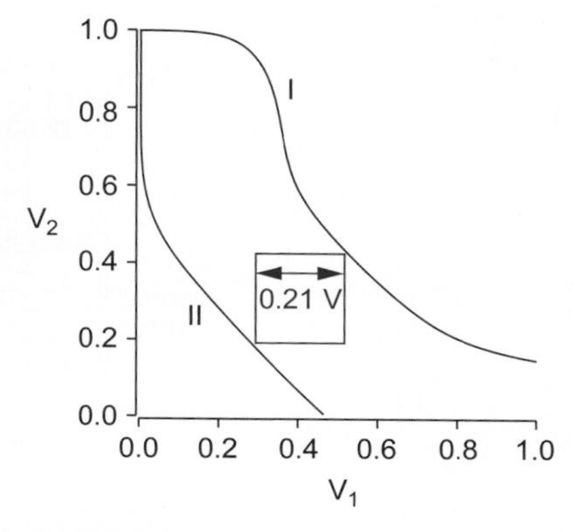

FIGURE 11.12 Write margin

Verifying such failure rates through brute force Monte Carlo simulation requires billions of simulations, which becomes impractical. However, the tails of the static noise margins have been found empirically to follow normal distributions [Calhoun06b]. Therefore, a smaller number of Monte Carlo parameters can be used to fit a model, which in turn is used to predict the behavior of the long tails. This should be done with caution because if the tail distribution does not closely match the model, the results can be seriously inaccurate. Alternatively, a technique called *importance sampling* performs simulations using random values near the point of failure. The samples are then weighted to produce the corrected probability of failure [Kanj06].

Because the static noise margins depend on V_{DD}, SRAMs have a minimum voltage at which they can reliably operate. This voltage is called $V_{\min}$ and is typically on the order of 0.7–1.0 V when 6T cells are employed. $V_{\min}$ presents an obstacle to continued voltage scaling. Section 11.2.6.1 investigates alternatives for low-voltage SRAM design.

Static noise margins are conservative because they assume DC operation: noise sources are constant, access transistors are ON indefinitely, and bitlines remain at their full precharged level. These assumptions can be relaxed to define larger *dynamic noise margins* [Khalil08, Sharifkhani09].

11.2.1.4 Physical Design SRAM cells require clever layout to achieve good density. A traditional design was used until the 90 nm generation, and a lithographically friendly design has been used since.

Figure 11.13(a) shows a stick diagram of a traditional 6T cell. The cell is designed to be mirrored and overlapped to share V_{DD} and GND lines between adjacent cells along the cell boundary, as shown in Figure 11.13(b). Note how a single diffusion contact to the bit-

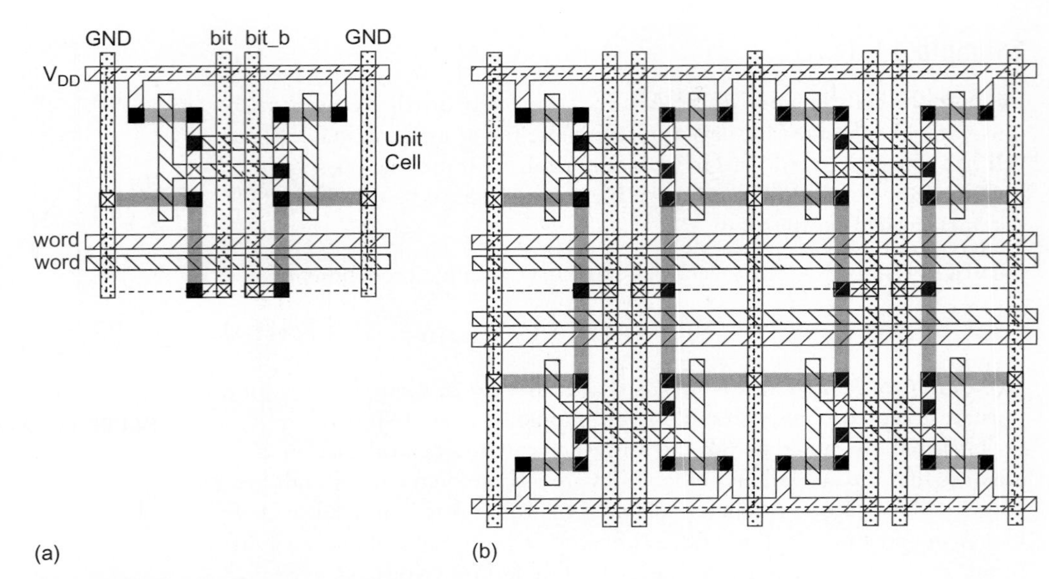

FIGURE 11.13 Stick diagram of 6T SRAM cell

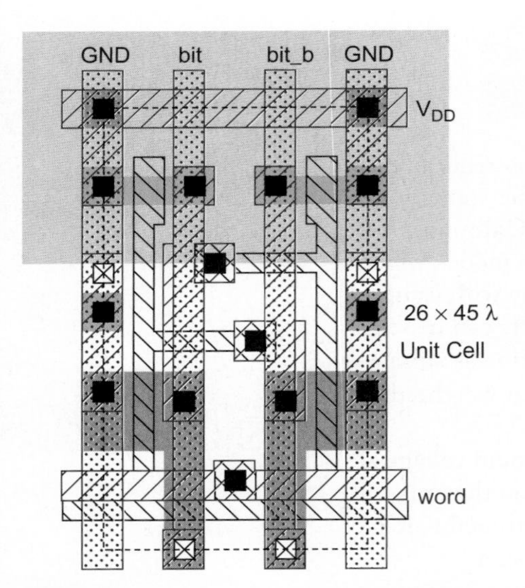

FIGURE 11.14 Layout of 6T SRAM cell. Color version on inside front cover.

line is shared between a pair of cells. This halves the diffusion capacitance, and hence reduces the delay discharging the bitline during a read access. The wordline is run in both metal1 and polysilicon; the two layers must occasionally be strapped (e.g., every four or eight cells). Figure 11.14 shows a conservative cell of $26 \times 45\ \lambda$, obeying the MOSIS submicron design rules. In this layout, the metal1 and polysilicon wordlines are contacted in each cell. The substrate and well are also contacted in each cell.

The bends in polysilicon and diffusion are difficult to precisely fabricate when the feature size is smaller than the wavelength of light. Moreover, mask misalignments in the traditional cell further increase the variability. Thus, nanometer processes now use the *lithographically friendly 6T cell* shown in Figure 11.15 [Osada01]. Diffusion runs strictly in the vertical direction and polysilicon runs strictly in the horizontal direction. The cell is long and skinny, reducing the critical bitline capacitance at the expense of longer wordlines. It is thus sometimes called a *thin cell* [Khare02]. The layout occupies two horizontal metal1 tracks and six vertical metal2 tracks. It uses local interconnect or *trench contacts* to bridge between the pMOS drain and the nMOS transistors and polysilicon routing. Again, substrate and well contacts are shared between multiple cells.

The nMOS diffusion is of unequal width to achieve a beta ratio greater than 1. The notch tends to round out because of lithography limitations. Thus, misalignment of the polysilicon to the diffusion can change the effective width of the access transistor. An alternative layout uses minimum-width diffusion for both nMOS transistor and a beta ratio of 1. This is called a *rectangular-diffusion* [Yamaoka04] or *diffusion-notch-free* [Khellah09] cell. The layout reduces the nominal read margin but reduces the variability of the cell.

Figure 11.16 shows how SRAM cell size has scaled over five process generations. The micrographs show the diffusion and polysilicon regions. Observe the transition from the traditional cell to the thin cell. Figure 11.17 plots cell size vs. feature size. The cell size has scaled well despite the growing challenges of lithography and variability. SRAM is so important that design rules are scrutinized and bent where possible to minimize cell area in commercial processes. The substrate and well contacts are shared among multiple cells to save area at the expense of regularity. Figure 15.12 shows another micrograph of a traditional 6T SRAM cell that uses local interconnect in place of metal1 to connect the nMOS and pMOS transistors.

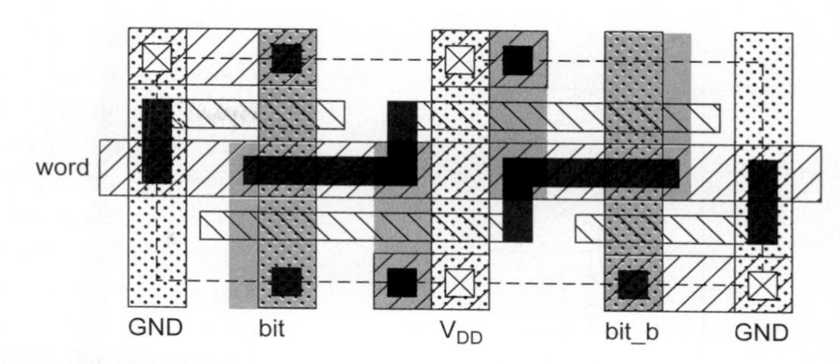

FIGURE 11.15 Lithographically friendly 6T SRAM cell

11.2.1.5 Alternative Cells Figure 11.18 shows a *dual-port* SRAM cell using eight transistors to provide independent read and write ports. For a write, the data and its complement are applied to the *wbl* and *wbl_b* bitlines and the *wwl* wordline is asserted. For a read, the *rbl* bitline is precharged, then the *rwl* wordline is asserted. Notice that read operation does not backdrive the state nodes through the access transistor, so read margin is as good as hold margin. Multiported cells are discussed further in Section 11.2.4

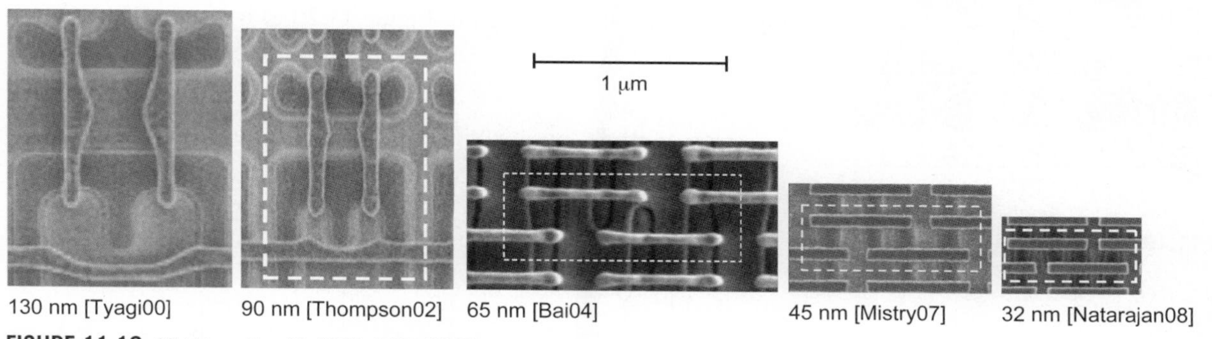

FIGURE 11.16 SRAM scaling (© 2000–2008 IEEE.)

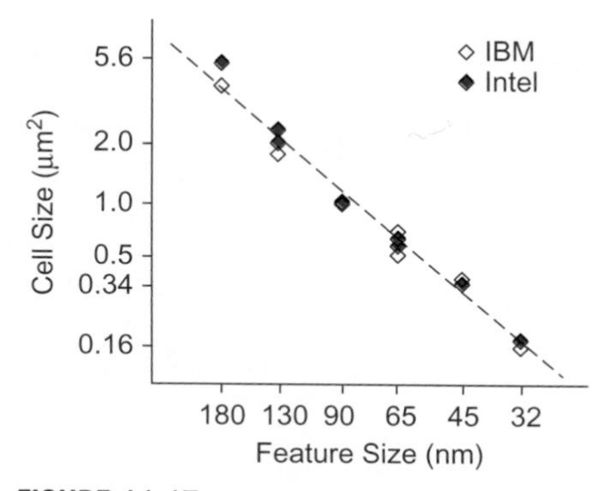

FIGURE 11.17 SRAM cell size vs. feature size

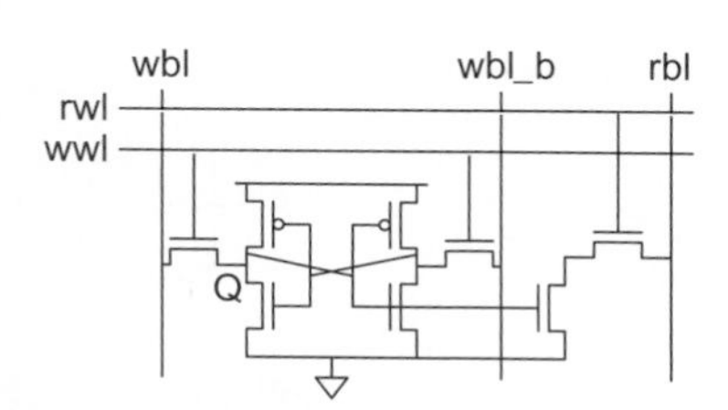

FIGURE 11.18 8T dual-port SRAM cell

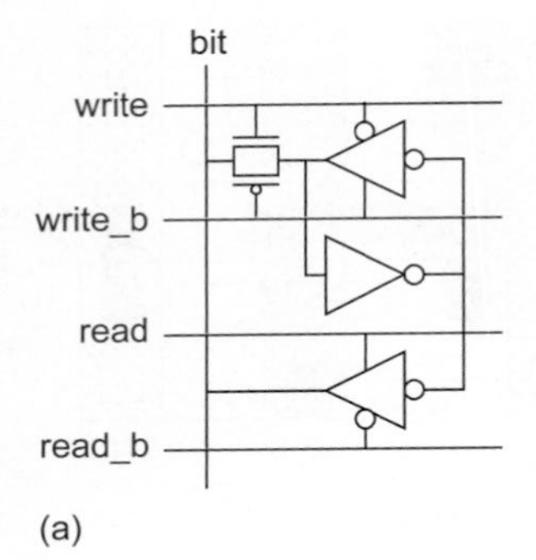

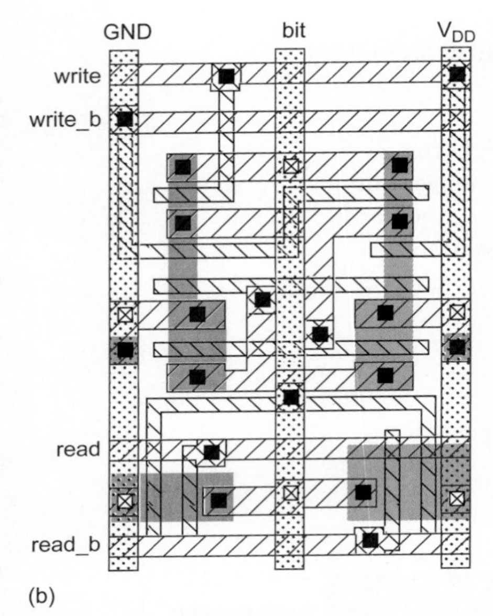

FIGURE 11.19 12T SRAM cell

The trade-off between read margin, write margin, transistor sizes, and operating voltage limits the minimum operating voltage of a compact 6T cell. Using an 8T dual-port cell for single-ported operation circumvents these trade-offs and allows lower-voltage operation [Chang08]. Intel switched from 6T to 8T cells within the cores for its 45 nm line of Core processors [Kumar09].

SRAMs require careful design to ensure that the ratio constraints are met and to protect the dynamic bitline from leakage and noise. For small memories, a static design may be preferable. Figure 11.19(a) shows a 12-transistor SRAM cell built from a simple static latch and tristate inverter. The cell has a single bitline. True and complementary read and write signals are used in place of a single wordline. A representative layout in Figure 11.19(b) has an area of $46 \times 75\ \lambda$. The power and ground lines can be shared between mirrored adjacent cells, but the area is still limited by the wires. This cell is well-suited to low-voltage operation, to small register files (< 32 entries), and to class projects where design time is more important than density.

11.2.2 Row Circuitry

The row circuitry consists of the decoder and word line drivers. The simplest decoder is a collection of AND gates using true and complementary versions of the address bits. Figure 11.20 shows several straightforward implementations. The design in Figure 11.20(a) is a static NAND gate

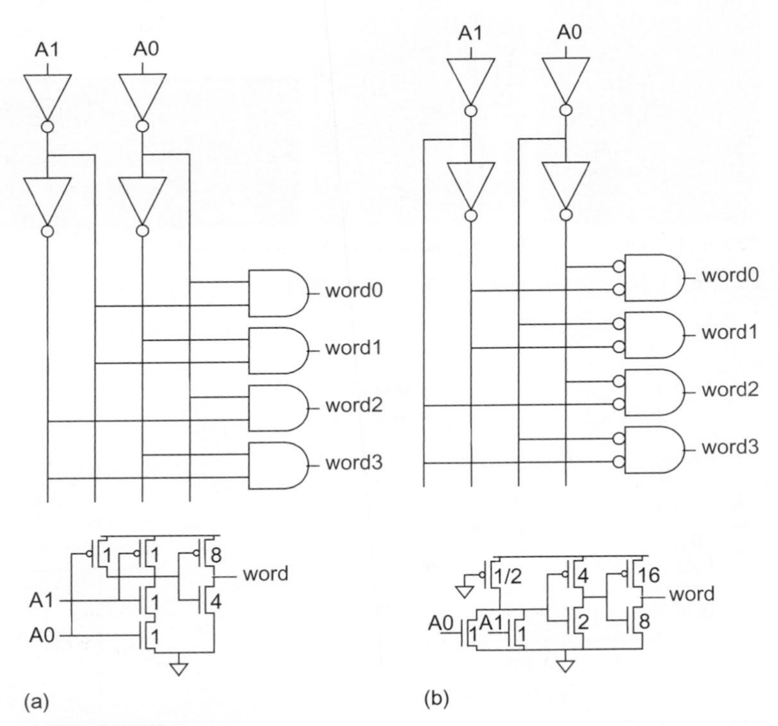

FIGURE 11.20 Decoders

followed by an inverter. This structure is useful for up to 5–6 inputs or more if speed is not critical. The NAND transistors are usually made minimum size to reduce the load on the buffered address lines because there are 2^{n-k} transistors on each true and complementary address line in the row decoder. The design in Figure 11.20(b) uses a pseudo-nMOS NOR gate buffered with two inverters. The NOR gate transistors can be made minimum size and the inverters can be scaled appropriately to drive the wordline. This design is easy to build but requires verifying the ratio constraints and consumes too much power to use in a large array.

The wordline generally must be qualified with the clock for proper bitline timing. This is often performed with another AND gate after the decoder or with an extra clk input to the final stage of decoding. The clock qualification behaves like a static-to-domino interface so the address must setup long enough before the clock edge. Figure 11.21 shows how to take advantage of the 1-hot nature of decoder outputs to share the clocked nMOS transistor across multiple final 2-input AND gates, reducing wordline clock power [Hsu06b]. Similarly, the wordline driver inverters are large and contribute a significant amount of leakage current. At most one driver produces a 1 output at a time. The figure also shows a fine-grained sleep transistor that cuts off leakage for the drivers in the 0 state when the array is inactive [Kitsukawa93, Gerosa09]. The sleep transistor only needs to be wide enough to supply current to a single inverter.

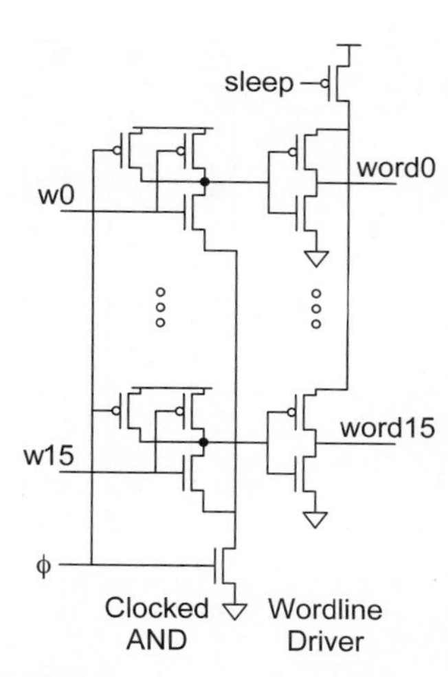

FIGURE 11.21 Shared clock and transistors in wordline driver

The layout of the decoder must be pitch-matched to the memory array; i.e., the height of each decoder gate must match the height of the row it drives. This can be tricky for SRAM and even harder for ROMs and other arrays with small memory cells. Figure 11.22(a) shows a layout of a conventional standard-cell style approach. The minimum-sized transistors in the NAND gate drive a larger buffer inverter. The decoder height grows with the number of inputs. The AND gates are easily programmed by connecting the polysilicon inputs to the appropriate address inputs. Figure 11.22(b) shows a layout on a pitch that is tighter and independent of the number of inputs. The decoder is programmed by placement of transistors and metal straps; this is best done with scripting software that generates layout. The polysilicon address lines should be strapped with metal2 to reduce their resistance, but the metal2 is left out of the figure for readability. The decoder pitch is 5 tracks or 40 λ. If every other row is mirrored to share V_{DD} and GND, the pitch can be reduced to 4 tracks or 32 λ.

11.2.2.1 Predecoding Decoders typically have high electrical and branching effort. Therefore, they need many stages, so the fastest design is the one that minimizes the logical effort. A tree of 2- and 3-input NAND gates and inverters offers the lowest logical effort to build high fan-in gates in static CMOS [Sutherland99]. For example, Figure 11.23(a) shows a 16-word decoder in which the 4-input AND function is built from a pair of 2-input NANDs followed by a 2-input NOR.

Many NAND gates share exactly the same inputs and are thus redundant. The decoder area can be improved by factoring these common NANDs out, as shown in Figure 11.23(b). This technique is called *predecoding*. It does not change the path effort of the decoder, but does improve area. In general, blocks of p address bits can be predecoded into 1-of-2^p-hot predecoded lines that serve as inputs to the final stage decoder. For example, Figure 11.23(b) shows a $p = 2$-bit design that decodes each pair of address bits into a 1-of-4-hot code.

The wordline is a large capacitive load. When the decoder is designed for minimum delay, the NAND gates tend to be large to drive this load. Placing a buffer between the decoder and wordline saves a large amount of dynamic power at a small cost in delay.

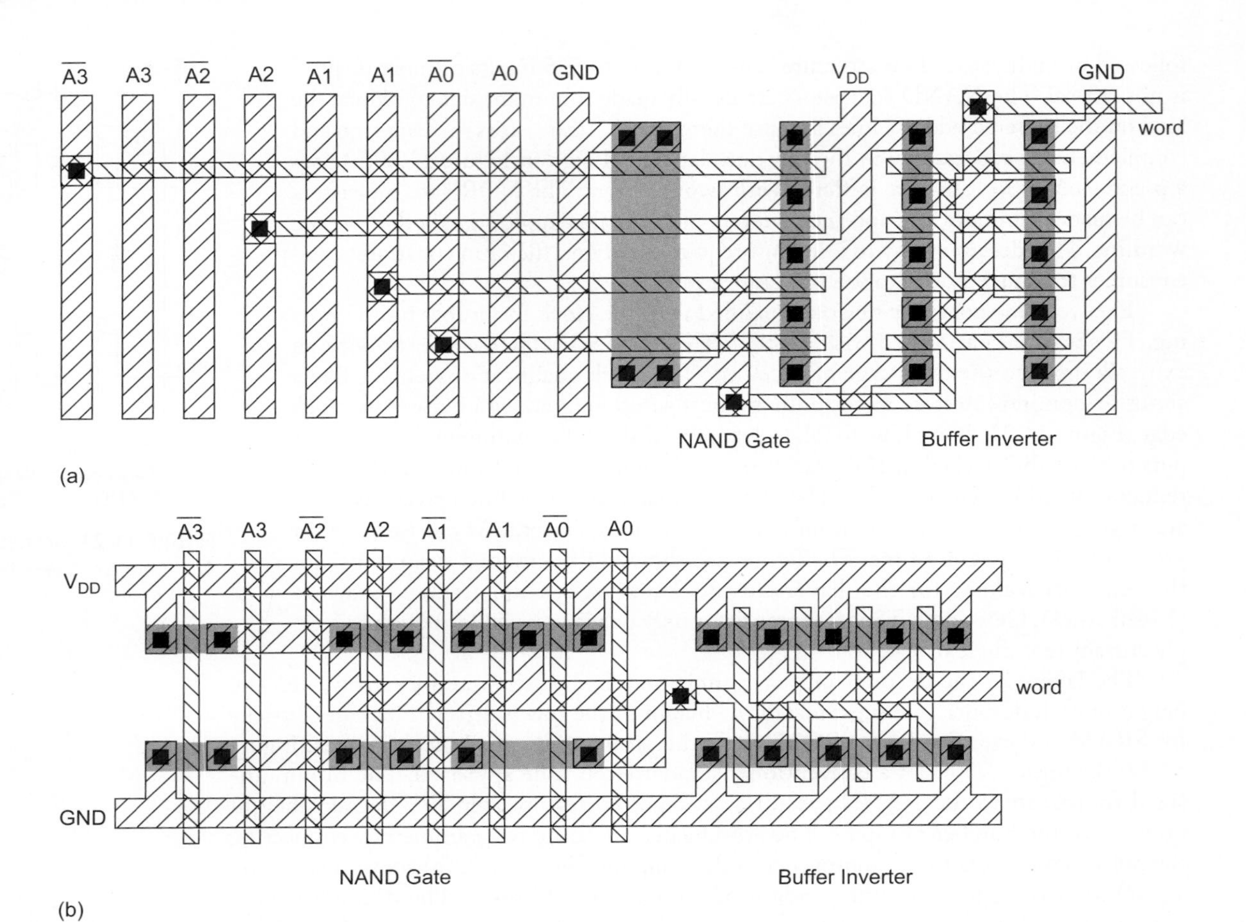

FIGURE 11.22 Stick diagrams of two decoder layouts

11.2.2.2 Hierarchical Wordlines The wordline is heavily loaded. It also has a high resistance because it is constructed from a narrow lower-level metal wire. This leads to a long RC flight time for large arrays. An alternative is to divide the wordline into global and local segments with one more level of distributed decoding, as shown in Figure 11.24 [Yoshimoto83, Itoh97]. These are also called *hierarchical* or *divided wordlines*. The *local wordlines* (*lwl*) are shorter and each drive a smaller group of cells. The *global wordlines* (*gwl*) are still long, but have lighter loads and can be constructed with a wider and thicker level of metal. The arrangement also saves energy because only those bitlines activated by the local wordline will switch.

11.2.2.3 Dynamic Decoders Dynamic gates are attractive for fast decoders because they have lower logical effort. A major problem with traditional domino decoders is the high power consumption. For example, even though only one of the 256 wordlines in the previous example will rise on each cycle, all 256 AND gates must precharge so the clock load is extremely large. A much lower-power approach is to use self-resetting domino gates that only precharge the wordline that evaluated. [Amrutur01] shows some variations that work with long input pulses. Self-resetting domino has essentially the same performance as

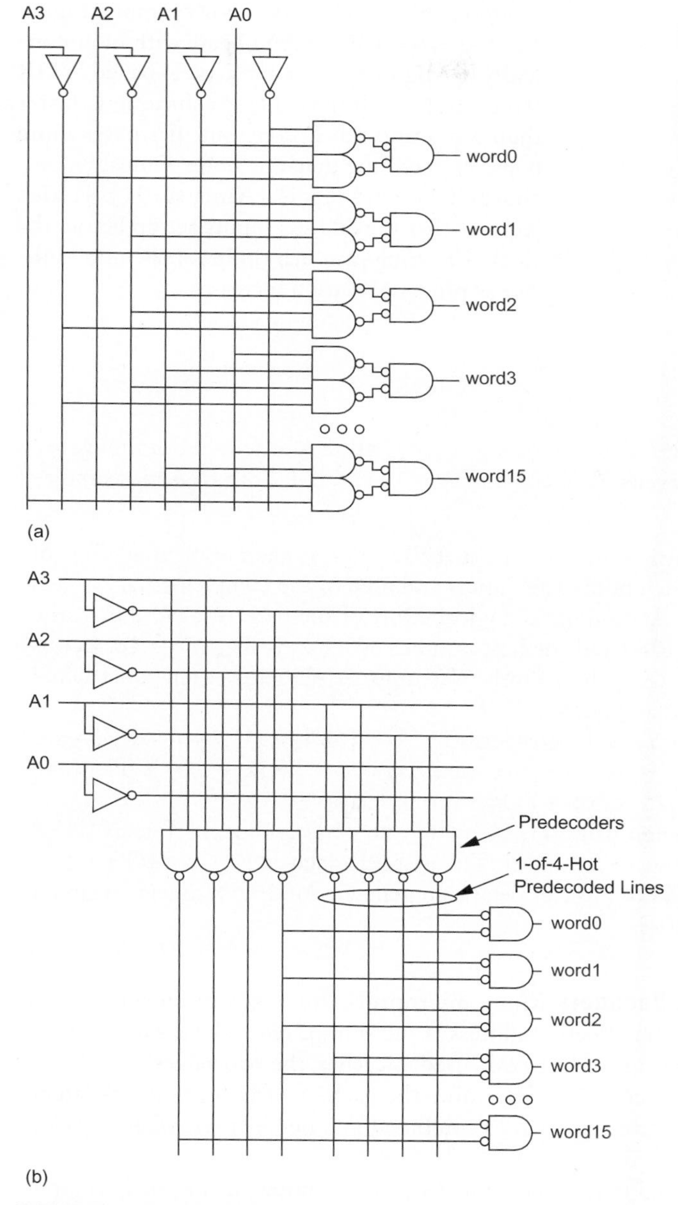

FIGURE 11.23 Ordinary and predecoding circuits

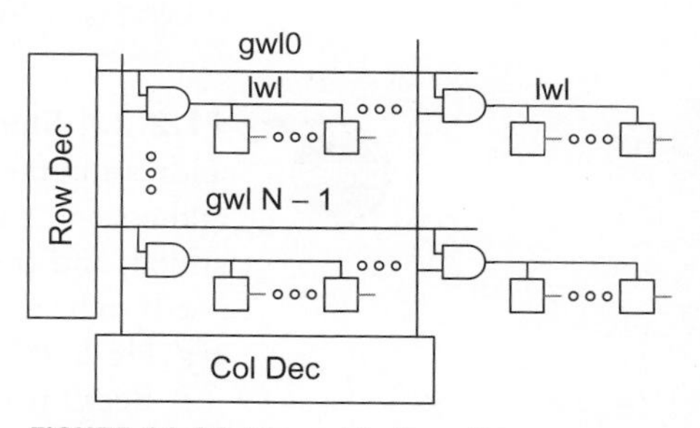

FIGURE 11.24 Hierarchical wordlines

traditional domino because it uses the same basic gates. The pulses create timing races that lead to chip failure if designed incorrectly or subjected to excessive variation. [Samson08] describes another domino decoder in which each gate triggers precharge of its successor to save energy.

Yet another approach for dynamic decoders is to use wide NOR structures in which N–1 of the N outputs discharge on each cycle. As most memories require monotonically rising outputs but the NORs are monotonically falling, such decoders require race-based

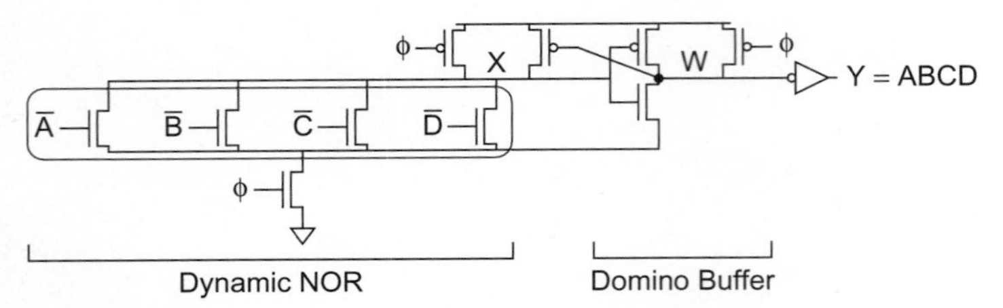

FIGURE 11.25 4-input AND using race-based NOR

nonmonotonic techniques. For example, Figure 11.25 shows a 4-input AND gate with monotonically rising output using a race-based NOR structure [Nambu98]. This technique is faster than a domino AND tree, but dissipates more power because the dynamic node X must be precharged on each cycle [Amrutur01]. It also requires that the address inputs set up before the clock. Ensuring race margin becomes more difficult as process variation increases.

Example 11.2

Estimate the delays of 8:256 decoders using static CMOS and footed domino gates. Assume the decoder has an electrical effort of $H = 10$ and that both true and complementary inputs are available.

SOLUTION: The decoder consists of 256 8-input AND gates. It has a branching effort of $B = 256/2 = 128$ because each of the true inputs and each of the complementary inputs are used by half the gates. Assuming the logical effort of the path G is close to 1, the path effort is $F = GBH = 1280$ and the best number of stages is $\log_4 F = 5.16$. Let us consider a 6-stage design using three levels of 2-input AND gates, each constructed from a 2-input NAND and an inverter.

The static CMOS design has a logical effort of $G = [(4/3) \times (1)]^3 = 64/27$. Therefore, the stage effort is $F = 3034$. The parasitic delay is $P = 3 \times (2 + 1) = 9$. The total delay is $D = NF^{1/N} + P = 31.8\ \tau$ or 6.4 FO4 inverter delays.

The footed domino design using HI-skew inverters has a logical effort of $[(1) \times (5/6)]^3 = 125/256$ and a stage effort of 625. The parasitic delay is $P = 3 \times (4/3 + 5/6) = 6.5$. The total delay is 4.8 FO4 inverter delays. In general, domino decoders are about 33% faster than static CMOS.

11.2.2.4 Sum-Addressed Decoders Many microprocessor instruction sets include addressing modes in which the effective address is the sum of two values, such as a base address and an offset. In conventional SRAMs used as caches, the two values must first be added, and then the result decoded to determine the cache wordline. If access latency needs to be minimized, these two steps can be combined into one in a *sum-addressed memory* [Heald98].

Recall from Section 10.4.3 that checking if $A + B = K$ is faster than actually computing $A + B$ because no carry propagation need occur. A *sum-addressed decoder* for an N-word memory accepts two inputs, A and B. In a simple form, it contains N comparators driving the N wordlines. The first checks if $A + B = 0$. The second checks if $A + B = 1$, and so forth. The comparators contain redundant logic repeated across wordlines. [Heald98] shows how to reduce the area by factoring out common terms in a predecoder.

11.2.3 Column Circuitry

The column circuitry consists of the bitline conditioning circuitry, the write driver, the bitline sensing circuitry, and the column multiplexers. Figures 11.5 and 11.7 showed simple column circuitry with no column multiplexing. The bitlines are initially precharged. During a write, the write driver pulls down one of the bitlines. During a read, data is sensed with a high-skew inverter. The dynamic bitline is connected to many transistors in parallel, so leakage can be a serious problem. As discussed in Section 8.2.4.3, the bitline may require a strong keeper, especially during burn-in. Moreover, the parasitic delay of the bitline contributes a major portion of the read time.

Example 11.3

A subarray of a large memory is organized as 256 words × 136 bits. Estimate the parasitic delay of the bitline. Assume the driver and access transistors are unit-sized and that wire capacitance is comparable to diffusion capacitance.

SOLUTION: The bitline has 256 cells attached, but pairs of cells are mirrored to share a bitline, so the diffusion capacitance is $128C$. Wire capacitance is comparable, so the total capacitance is $256C$. The bitline is pulled down through the driver and access transistors in series, with a total resistance of $2R$. Therefore, the delay is $512RC$, or 34.1 FO4 inverter delays. This is unacceptably large for many applications.

Bitline sensing can be classified as large-signal or small-signal. In *large-signal* or *single-ended sensing*, a bitline swings between V_{DD} and GND just like an ordinary digital signal. The high-skew inverter is an example of large-signal sensing. To reduce the parasitic delay, the bitline can be hierarchically divided into multiple local bitlines, then combined to drive a global wordline. In *small-signal* or *differential sensing*, one of the two bitlines changes by a small amount. A sense amplifier detects the small difference and produces a digital output. This saves the delay of waiting for a full bitline swing and also reduces energy consumption if the bitline swing is terminated after sensing. However, the array requires a timing circuit to indicate when the sense amplifier should fire, and if the time is too short, the wrong answer may be sensed. Process variation leads to offsets in the sense amplifier that increase the required bitline swing. Historically, small SRAM arrays such as register files used large-signal sensing while big SRAM and DRAM arrays used small-signal sensing to improve speed and power, but the trend is toward large-signal sensing in nanometer processes.

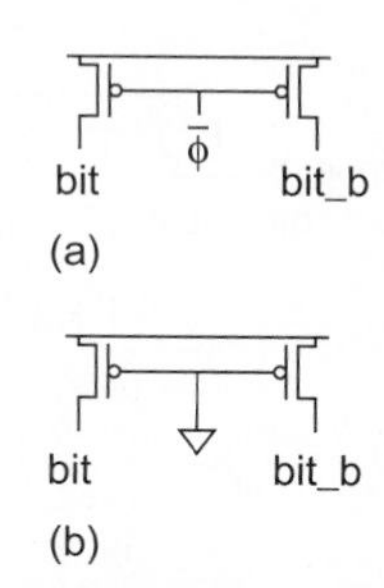

FIGURE 11.26 Bitline conditioning circuits

11.2.3.1 Bitline Conditioning The bitline conditioning circuitry is used to precharge the bitlines high before operation. A simple conditioner consists of a pair of pMOS transistors, as shown in Figure 11.26(a). It is also possible to construct pseudo-nMOS SRAMs with weak pullup transistors in place of the precharge transistors (Figure 11.26(b)) where no clock is available. The contention slows the read and creates a ratio constraint, so it is not suitable to low-voltage operation.

11.2.3.2 Large-Signal Sensing The bitline delay is proportional to the number of words attached to the bitline. Small memories (e.g., up to 16–32 words) may be fast enough with a simple inverter sensing the bitline. Larger memories can read onto *hierarchical* or *divided bitlines*, as shown in Figure 11.27. Small groups of cells are attached to *local bitlines* (*lbl*). Pairs of local bitlines are combined with a HI-skew NAND gate, which in turn can pull down the dynamic *global bitline* (*gbl*). The local bitline can be viewed as an unfooted domino multiplexer comprised of the access and driver transistors for each cell. Recall that a dynamic multiplexer has a constant logical effort but a parasitic delay proportional to the

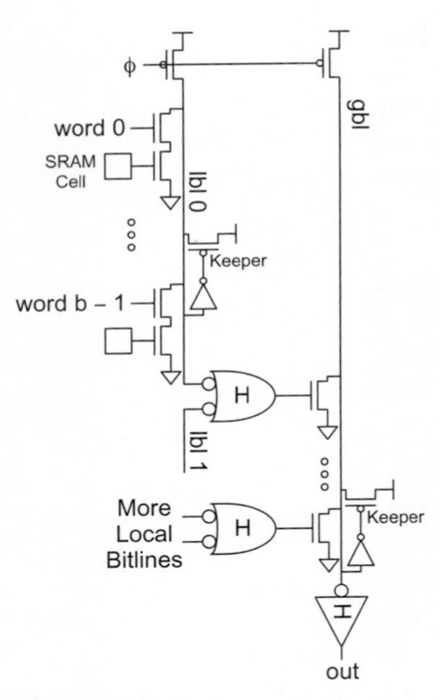

FIGURE 11.27 Hierarchical bitlines

number of inputs (i.e., words on the local bitline), so local bitlines become quite slow for more than 32 words. The global bitline can be viewed as an unfooted domino OR gate. The global bitline drivers are interspersed between the groups of cells. They use larger transistors to drive the long global bitline. The global bitline typically runs over the top of the cell using a higher level of metal (e.g., metal3 or metal4) so that it does not increase the area of the array.

The maximum number of transistors connected to each bitline may be limited by leakage. The worst case occurs when the cell being read contains a 0 and all the others contain a 1. The local bitline should remain at 1 but subthreshold leakage from all the unaccessed cells tends to pull the bitline down. Section 8.2.4.3 described conditional and adaptive keepers to fight leakage when many cells share the same bitline. The data read out must be latched before feeding static logic so that it is not lost during precharge. Examples of large-signal sensing include the Power6 SRAM arrays [Stolt08] and the Itanium register file [Fetzer06].

11.2.3.3 Small-Signal Sensing In a small-signal sensing scheme, the access transistors are activated long enough to swing the bitlines by a small amount (e.g., 100–300 mV), then the differential bitline voltage is sensed. The wordline is turned OFF when sensing occurs to avoid the bitline swinging further and consuming more power. Many *sense amplifiers* have been invented to provide faster sensing by responding to a small voltage swing.

The differential sense amplifier in Figure 11.28(a) is based on an analog differential pair and requires no clock. However, the circuit consumes a significant amount of DC power. It is also difficult to bias at low voltage to keep all the transistors in saturation.

The clocked sense amplifier in Figure 11.28(b) consumes power only while activated, but requires a timing chain to activate at the proper time. When the sense clock is low, the amplifier is inactive. When the sense amplifier rises, it effectively turns on the cross-coupled inverter pair, which pulls one output low and the other high through *regenerative feedback*. The *isolation transistors* speed up the response by disconnecting the outputs from the highly capacitive bitlines during sensing. The sense amplifier flip-flop from Figure 9.29(a) is also commonly used because it inherently isolates the sensing nodes from the bitline [Hart06].

Power dissipation can be reduced for read operations by turning off the wordlines once sufficient differential voltage has been achieved on the bitlines. This reduces the bitline swing and hence the charge required to restore the bitlines to V_{DD} after sensing.

Sense amplifiers are highly susceptible to differential noise on the bitlines because they detect small voltage differences. If bitlines are not precharged long enough, residual voltages on the lines from the previous read may cause pattern-dependent failure. An equalizer transistor (Figure 11.29(a)) can be added to the bitline conditioning circuits to reduce the required precharge time by ensuring that *bit* and *bit_b* are at nearly equal voltage levels even if they have not precharged quite all the way to

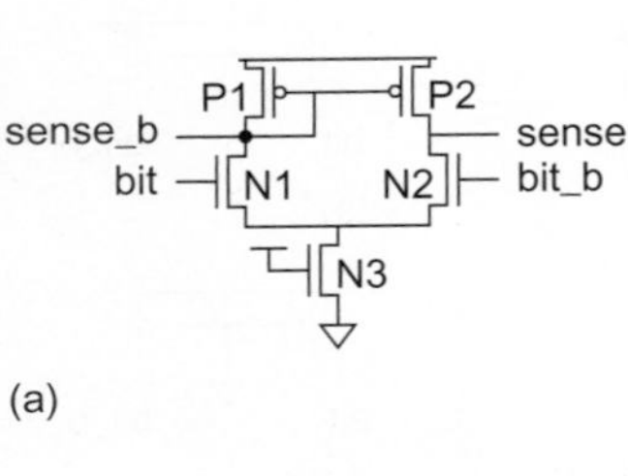

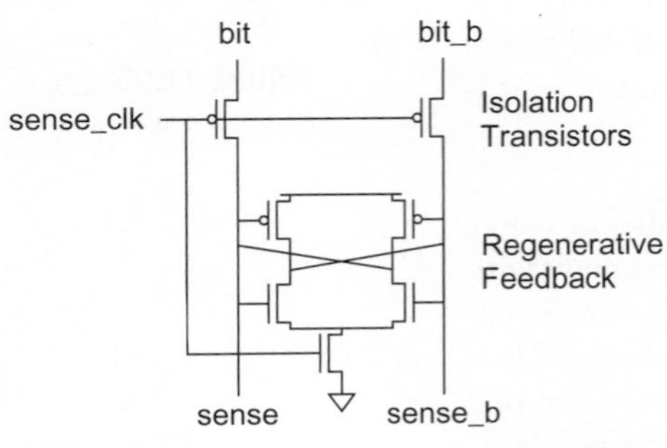

FIGURE 11.28 Sense amplifiers

V_{DD}. Coupling from transitioning bitlines in neighboring cells may also introduce noise. The bitlines can be *twisted* or *transposed* to cause equal coupling onto both the bitline and its complement, as shown in Figure 11.29(b). For example, careful inspection shows that $b1$ couples to $b0\_b$ for the first quarter of its length, $b2$ for the next quarter, $b2\_b$ for the third quarter, and $b0$ for the final quarter. $b1\_b$ also couples to each of these four aggressors for a quarter of its length, so the coupling will be the same onto both lines.

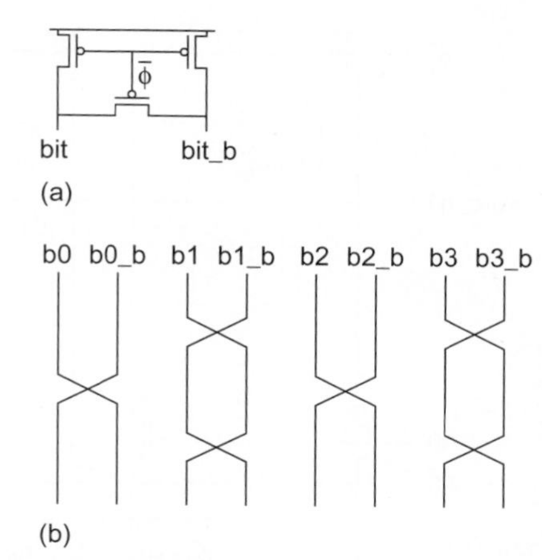

FIGURE 11.29 Bitline noise reduction through equalizers and twisting

The sense amplifier offset voltage is the differential input voltage ($bit - bit\_b$) necessary to produce zero differential output voltage ($sense - sense\_b$). If $N1$ is identical to $N2$ and $P1$ to $P2$, the sense amplifier will ideally have zero offset voltage. In practice, the offset voltage is nonzero because of statistical dopant fluctuations and NBTI degradation that affect V_t. The differential input must substantially exceed the offset voltage to be sensed reliably. A typical budget for offset voltage is 50 mV [Amrutur00]. Unfortunately, the threshold variations and offset voltage are not changing very much with technology scaling, so the offset voltage is becoming a larger fraction of the supply voltage, making sense amplifiers less effective [Mizuno94].

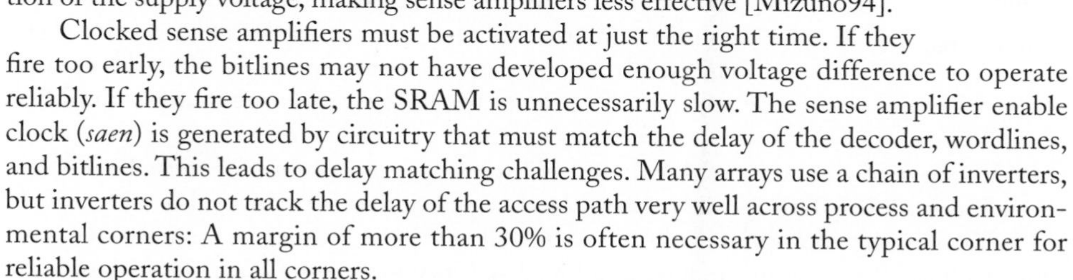

Clocked sense amplifiers must be activated at just the right time. If they fire too early, the bitlines may not have developed enough voltage difference to operate reliably. If they fire too late, the SRAM is unnecessarily slow. The sense amplifier enable clock (*saen*) is generated by circuitry that must match the delay of the decoder, wordlines, and bitlines. This leads to delay matching challenges. Many arrays use a chain of inverters, but inverters do not track the delay of the access path very well across process and environmental corners: A margin of more than 30% is often necessary in the typical corner for reliable operation in all corners.

Alternatively, the array may use *replica* cells and bitlines to more closely track the access path, as shown in Figure 11.30 [Amrutur98]. The block decoder determines that a particular memory block is selected (*bs*). The appropriate local wordline (*lwl*) is activated, turning on a SRAM cell in a column and causing the *bit* or *bit_b* to begin discharging. Meanwhile, the block select signal also activates one cell in the replica column. The replica column has only $1/r$ as many cells connected to the bitline (e.g., $r = 10$), so it discharges r times faster. When the replica bitline (*rbl*) falls low, a reset signal is generated to start deactivating the block. Meanwhile, the signal is buffered to drive the sense amplifiers. By the time *saen* is enabled, the bitline swing will be approximately V_{DD}/r. Thus, r can be selected to obtain the desired bitline swing. Because the replica path involves most of the same elements as the real path, its delay tracks fairly well with PVT variations, reducing the amount of margin required on *saen*. Nevertheless, providing a degree of tunability is prudent so that the nominal margin can be reasonably aggressive, yet the margin can be increased if variation is greater than expected and the circuit malfunctions.

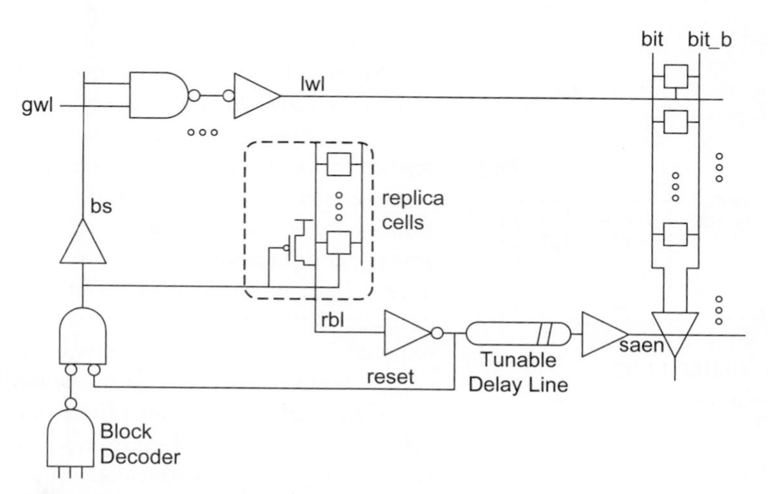

FIGURE 11.30 Replica delay for sense amplifier enable

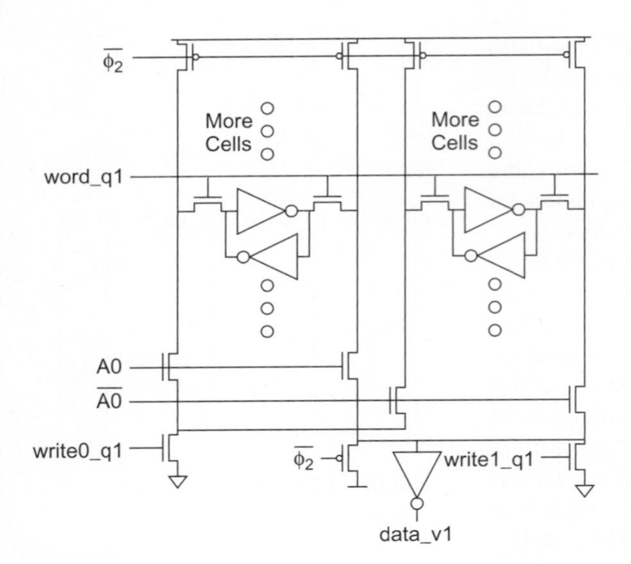

FIGURE 11.31 Complete pair of columns for two-way multiplexed SRAM

11.2.3.4 Column Multiplexing In general, 2^k:1 column multiplexers may be required to extract 2^m bits from the 2^{m+k} bits of each row. The column decoding takes place in parallel with row decoding so it does not impact the critical path. Figure 11.31 shows two-way column multiplexing with large-signal sensing using nMOS pass transistor multiplexers. The output of the multiplexer is precharged high. Both the write drivers and the read sensing inverter are connected to the multiplexer outputs.

In small-signal sensing, the bitlines voltages are close to V_{DD}, so pMOS pass transistors are required. Thus, the array may use transmission gates, or may use separate nMOS transistors in the write path and pMOS transistors in the read path.

Column multiplexing is also helpful because the bit pitch of each column is so narrow that it can be difficult to lay out a sense amplifier for each column. After multiplexing, multiple columns are available for the remainder of the column circuitry. Moreover, placing sense amplifiers after the column multiplexers reduces the number of power-hungry amplifiers required in the array.

When writing an array with column multiplexing, only a subset of the cells in a row should be modified. This is called a *partial write* operation. It is performed by only driving the bitlines in the appropriate columns, while allowing the bitlines in the unwritten columns to float. Partial writes require good read stability so that the unwritten columns are not disturbed; this can be a challenge at low voltage [Chang08].

11.2.4 Multi-Ported SRAM and Register Files

Register files are generally fast SRAMs with multiple read and write ports. They are used in many tables and buffers beyond simply holding the architectural registers; for example, the Core 2 has 54 different register files in each core [George07]. Data caches in superscalar microprocessors often require multiple ports to handle multiple simultaneous loads and stores.

Figure 11.18 showed a conventional 8T dual-ported SRAM cell. An alternative 6T dual-ported SRAM adds a second wordline, as shown in Figure 11.32 [Horowitz87]. Such a *split-wordline cell* can perform two reads or one write in each cycle. The reads are performed by independently selecting different words with the two wordlines. Read becomes a single-ended operation; one read appears on *bit*, while the other appears in complementary form on *bit_b*. For example, asserting *wordA*[7] and *wordB*[3] reads the third word onto *bit* and the complement of the seventh onto *bit_b*. Write still requires both *bit* and *bit_b*, so only a single write can occur. With careful timing, accesses can be performed each half-cycle, permitting two reads in the first phase and a write in the second phase, as commonly required for a register file in a single-issue RISC processor. This cell is used in dual-ported caches in the UltraSPARC [Konstadinidis09] and Power6 [Plass07].

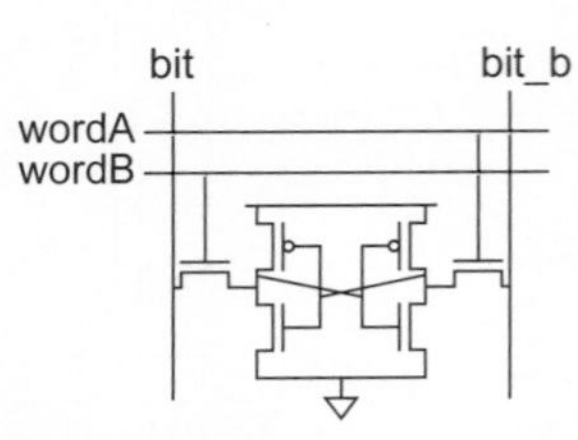

FIGURE 11.32 Simple dual-ported SRAM

Cells with multiple read ports need to isolate the read ports from the state nodes to achieve reasonable read margin, as was done with the 8T cell. Each additional single-ended read port can be provided at the cost of a read wordline, a read bitline, and two read transistors. Differential read ports double the number of read bitlines and transistors.

Cells with multiple write ports simply attach the ports to the state node. External logic should ensure that two ports do not attempt to simultaneously write different values to the register. Each additional write port can be provided at the cost of a write wordline, true and complementary write bitlines, and two access transistors. For cells with many ports, the area of the wires dwarfs the area of the transistors. To save space, the complementary write bitline can be eliminated by adding a transistor or inverter within the cell, as shown in Figure 11.33. The inverter approach requires one more transistor but improves the writability.

This style of cell readily extends to any number of ports by adding one wordline and one bitline for each port. Figure 11.34 shows a SRAM cell with three write ports and four read ports.

Register files for superscalar processors often require an enormous number of ports. For example, the Itanium 2 processor issues up to six integer instructions in a cycle, each of which requires two source registers and a destination. The register file requires four more write ports for late cache data returns, leading to a total of 12 read ports and 10 write ports [Fetzer06]. The area of the large register file is dominated by the mesh of wordlines and bitlines. A rough rule for estimating multiport SRAM cell area is to count the number of tracks for the wordlines and bitlines and then add three in each dimension for internal wiring. The area of a 22-ported register file is enormous, leading to excessive delay and power driving the lengthy wordlines and bitlines.

Two techniques exist for reducing the register file area: time-multiplexing and multiple banks. These techniques can be applied individually or in tandem. As mentioned earlier for the 6T two-ported cell, a register file can be *time-multiplexed* or *double-pumped* by reading in one half of the cycle and writing in the other half. The Itanium 2 register file adopts this technique to cut the number of wordlines to 11. Alternatively, each read and write port can be used twice per cycle. These approaches involve pulsed wordlines and bitlines. In a *multiple bank* design, a register file with R read ports and W write ports is divided into two banks, each with $R/2$ read ports and W write ports. Writes always update both banks so they contain identical data. Reads then can take place from either bank. This technique generalizes to larger numbers of banks. For example, a single-ended register file with 16 read ports and four write ports has a cell size of 23×23 tracks, or about $184 \times 184\ \lambda = 33856\ \lambda^2$. The area can be improved by partitioning the register file into two banks, each with eight read ports and four write ports. The cell size is now 15×15 tracks with an area of $14400\ \lambda^2$ per file, or $28800\ \lambda^2$ all together. The partitioned register file is not only smaller but also faster because of the shorter bitlines and wordlines.

[Golden99, Hart06, and Warnock06] show other designs for the large register files of the AMD Athlon, Sun UltraSparc IV+, and IBM/Sony/Toshiba Cell processors, respectively.

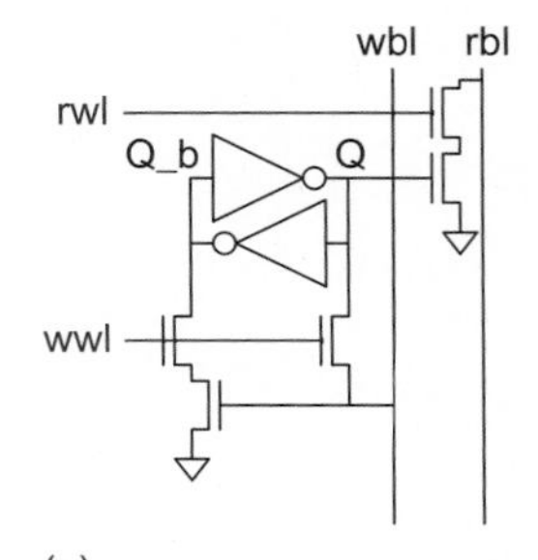

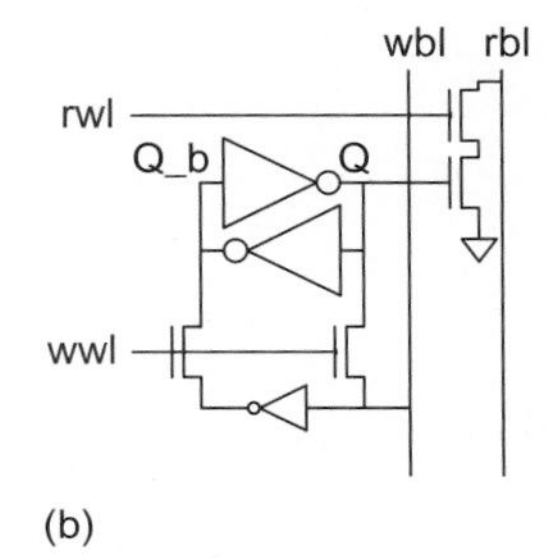

FIGURE 11.33 Register cell with single-ended write port

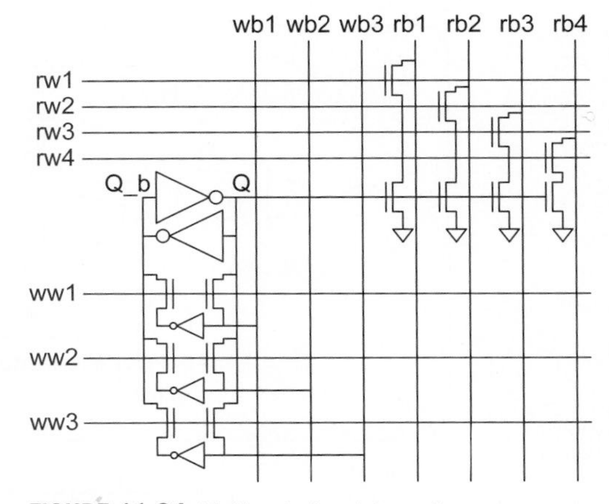

FIGURE 11.34 Multiported register cell

11.2.5 Large SRAMs

The critical path in a static RAM read cycle includes the clock to address delay time, the row address driver time, row decode time, bitline sense time, and the setup time to any data register. The write operation is usually faster than the read cycle because the bitlines are actively driven by large transistors. However, the bitlines may have to recover to their quiescent values before the next read cycle takes place.

If the memory array becomes large, the wordlines and bitlines become rather long. The long lines have high capacitance, leading to long delay and high power consumption. Thus, large memories are partitioned into multiple smaller memory arrays called *banks* or *subarrays*. Each subarray presents some area overhead for its periphery circuitry, so the size of the subarrays represents a trade-off between area and speed.

The delay of the bitline is proportional to the number of cells and the bitline swing. Large SRAMs use hierarchical bitlines or sense amplifiers for speed. Typical subarrays accommodate 128 or 256 words per bitline. The wordline presents an RC delay from the resistance of the wire and gate capacitance of the transistors it drives. This increases with the square of the number of bits on a wordline. Typical subarrays also use 128 or 256 bits on each wordline.

Figure 11.35 shows a typical 16 KB subarray for a large SRAM. The subarray is divided into four 4 KB *banks* or *blocks* of 256 words by 128 bits each. The word line decoders and column circuits are shared between banks to reduce the layout area. Each wordline decoder block performs predecoding and then regular decoding to create a 1-hot 256-bit signal, which in turn is gated with the clock and bank select signals and buffered to drive the wordline of the appropriate bank. The column circuitry includes 4:1 column multiplexers and the sense amplifier and write driver for each group of columns. The timing circuitry generates the sense amplifier enable signal and any other required timing pulses.

Information is carried to and from the subarrays on *datalines*. The large SRAM requires repeaters for the datalines and another decoder to select the appropriate subarray. The clock for inactive subarrays is gated to save power.

Figure 11.36 shows a 512 KB L2 cache from a 130 nm UltraSparc Gemini processor [Shin05]. It is built from four 128 KB arrays, each of which contains sixteen 8 KB banks organized as 256 rows by 256 columns. The data arrays have an area efficiency of about

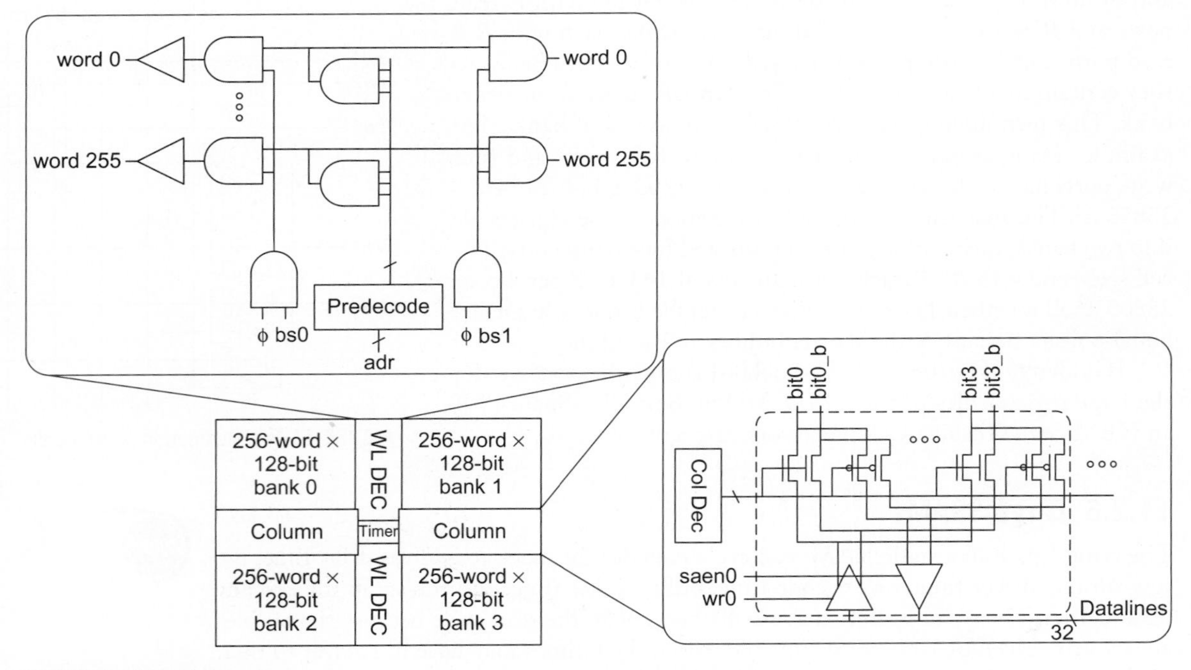

FIGURE 11.35 16 KB subarray

60%, while the overall cache has an area efficiency of 40% because of the other control and routing blocks. See [Chappell91, Weiss02, Shin05, Zhang05, Warnock06, Chang07, Plass07, Hamzaoglu09] for more examples of large embedded SRAMs.

The array efficiency of a memory is the fraction of the area occupied by memory cells. Large SRAM arrays typically achieve an efficiency of 70–75% [Lu08], although faster memories tend to have lower efficiency.

Large memories with multiple subarrays can simulate more than one access port even if each subarray is single-ported. For example, in a system with two subarrays, even-numbered words could be stored in one subarray while odd-numbered words are stored in the other. Two accesses could occur simultaneously if one addresses an even word and another an odd word. If both address an even word, we encounter a *bank conflict* and one access must wait. Increasing the number of banks offers more parallelism and lower probability of bank conflicts.

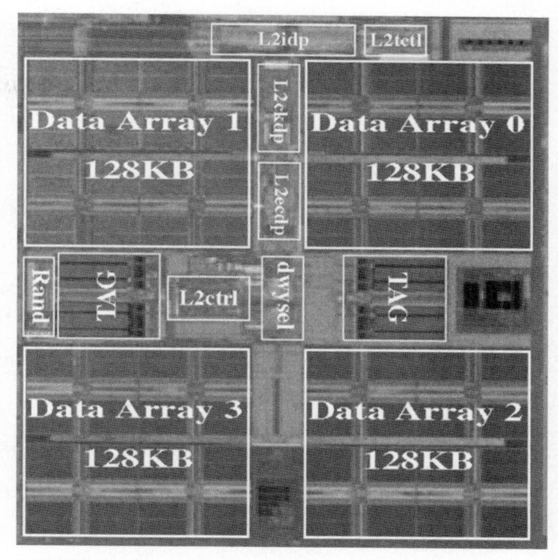

FIGURE 11.36 512 KB cache array (© 2005 IEEE.)

11.2.6 Low-Power SRAMs

SRAM occupies a large fraction of the area of most nanometer chips and consumes a significant part of the dynamic and leakage power. For example, in the dual-core Xeon processor with a 16 MB L3 cache [Rusu07, Chang07], the 6T cells in the various caches account for 77% of the 1.3 billion total transistors and about half of the chip area. The dynamic power is minimized by activating only 0.8% of the L3 cache for an access, and the leakage is minimized by keeping the remainder of the cache in sleep mode. Nevertheless, the L3 cache consumes about 14 W out of a 110 W typical total for the chip, and about half of this cache power is leakage.

This section explores the challenges of low-power SRAM design. The general principles are to turn only the necessary subarrays to minimize dynamic power, to keep the other subarrays in a sleep mode to minimize leakage, and to run at as low a voltage as possible to minimize total power. Maintaining read and write margins at low voltage in the face of process variation can be difficult. Many techniques are used for leakage reduction. When minimum energy is the goal, modified SRAMs can operate subthreshold.

11.2.6.1 Low Voltage Operation The minimum operating voltage, V_{min}, for RAMs is set by the read stability and writability constraints. As discussed in Section 11.2.1.3, within-die variability results in a distribution of read and write margins. The nominal margin required to obtain a satisfactory yield increases with the standard deviation of the margins and the number of cells, both of which are rising with technology scaling. V_{min} for a standard 6T SRAM is around 0.7 V in a 90 nm process [Calhoun07] and is forecast to increase with process scaling [Itoh09]. SRAM cells tend to use high threshold transistors to reduce leakage, leading to slow operation at low voltage.

SRAM transistors with nearly minimum-sized transistors achieve better density but have worse read/write margins and greater variability, increasing V_{min}. For example, the Intel 65 nm process has a high-performance SRAM cell with V_{min} = 0.7 V during operation and 0.6 V during standby (when it retains state but cannot read or write). It also provides a high-density SRAM cell that packs 44% more memory into a given area but is limited to 1.1/1.0 V operation [Khellah07].

Dynamic voltage scaling conflicts with the V_{min} constraint. For example, the 65 nm quad-core Itanium operates at a core supply of 0.9–1.2 V as the frequency varies from 1.2 to 2.4 GHz [Stackhouse09]. However, the chip uses the high density SRAM cell to build a 30 MB cache. The simplest approach to solving this problem is to use a fixed, relatively high 1.1 V supply for the memories and to perform level conversion at the interface [Khellah07].

V_{min} can be reduced with external circuitry to assist the read and write operations. Examples of *read assist* techniques to improve read stability include the following:

- Pulsing the wordline or bitline briefly to exploit dynamic noise margins that are larger than the static noise margins [Khellah06]
- Lowering the wordline voltage [Ohbayashi07, Yabuuchi07]
- Raising the cell V_{DD} during reads [Zhang06, Bhavnagarwala04]

Examples of *write assist* techniques to improve writability include the following:

- Driving the bitline to a negative voltage
- Raising the wordline voltage [Morita06]
- Floating the cell GND during writes [Yamaoka04b]
- Floating the cell V_{DD} during writes [Yamaoka06]
- Lowering the cell V_{DD} during writes [Zhang06, Ohbayashi07]

A simpler approach is to avoid the problematic 6T cell altogether at low voltage. The 8T dual-ported cell of Figure 11.18 solves the read stability problem and thus can operate at lower voltage [Chang08]. The cell area increases by about 30%. Intel switched to an 8T cell in the processor cores of the 45 nm Core family to support dynamic voltage scaling down to 0.7 V [Kumar09]. However, the L3 cache that accounts for much of the die size still uses the denser 6T cell operating at a higher voltage.

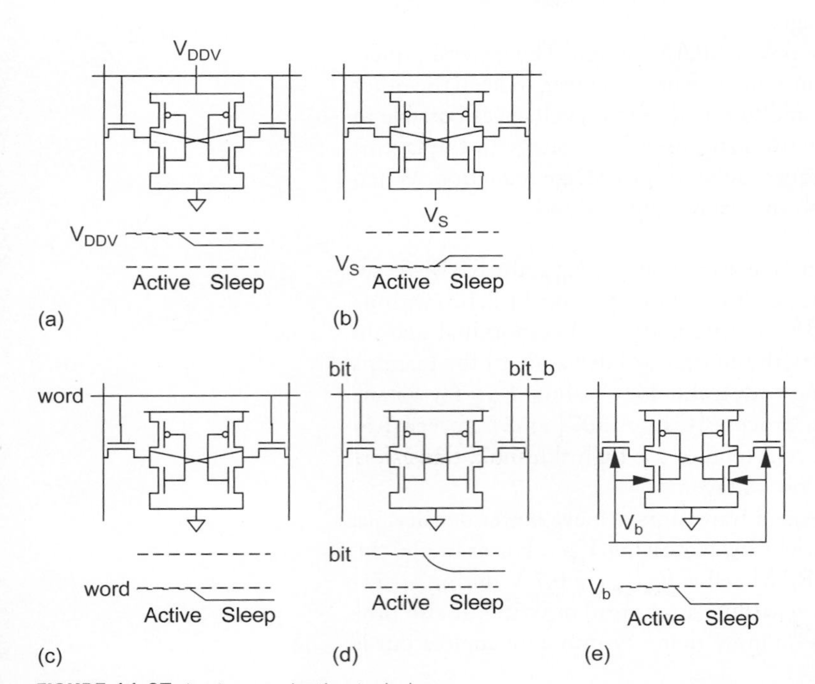

FIGURE 11.37 Leakage reduction techniques

11.2.6.2 Leakage Control Most of the subarrays in a large memory are inactive at any given time, so minimizing leakage in this state is critical. Leakage influences the selection of threshold voltage and oxide thickness for large memories. The three general ways to control leakage dynamically are to reduce V_{ds}, provide a negative V_{gs}, or provide a negative V_{bs} [Nakagome03]. Figure 11.37 illustrates these approaches [Kim05].

The supply voltage necessary to hold a cell's state is lower than that necessary for operation. Reducing the voltage across the transistors reduces the DIBL effect and thus decreases subthreshold leakage. Moreover, it greatly decreases gate leakage and BTBT junction leakage. Hence, this is a common technique for cutting the overall leakage power. It can be done with power switches that permit V_{DD} to droop [Kanda02, Nii04] (Figure 11.37(a)) or GND to rise [Zhang05] (Figure 11.37(b)) by a controlled amount during sleep. The soft error rate increases in this state, so ECC is essential to protect the data.

Example 11.4

Consider a process with a subthreshold slope of 100 mV/decade and a DIBL coefficient of 0.15. How far must the power supply droop to cut subthreshold leakage by a factor of 2?

SOLUTION: According to EQ (2.45), if the voltage across the cell droops by ΔV, the subthreshold leakage becomes

$$I_{sub} = I_{off} 10^{-\frac{\eta \Delta V}{S}} \tag{11.2}$$

Solving for $I_{sub} = I_{off}/2$ gives

$$\Delta V = \frac{S}{\eta} \log_{10} \frac{I_{off}}{I_{sub}} = 200 \text{ mV} \tag{11.3}$$

Figure 11.38 shows an example of partial power gating during sleep [Gerosa09, Hamzaoglu09]. The technique is similar to full power gating described in Section 4.3.2, but the supply collapse must be limited so that the memory retains its state. When the subarray is about to be accessed, a wide power gating transistor activates to connect the array's V_{DDV} to V_{DD}. When the subarray enters sleep mode, the power gating transistor shuts OFF but an adjustable sleep transistor turns ON. The sleep current is set to a level such that V_{DD} droops to the minimum retention voltage. When the subarray is completely disabled, the sleep transistor is also turned OFF. The transition from sleep to active mode requires some time (e.g., two cycles) and energy, so unnecessary transitions should be avoided. The turn-on process can begin as soon as the subarray to be accessed is known; this is usually before row decoding completes. The subarray may remain ON for several cycles after the access in case it is accessed again soon. Several options are available to adjust the sleep transistor [Khellah07]. Closed-loop control involves measuring V_{DDV} and adjusting a control voltage accordingly. Alternatively, the sleep transistor can be built from multiple smaller devices. After manufacturing, a chip calibration step can determine how many should be ON during sleep and this value can be programmed into a set of fuses.

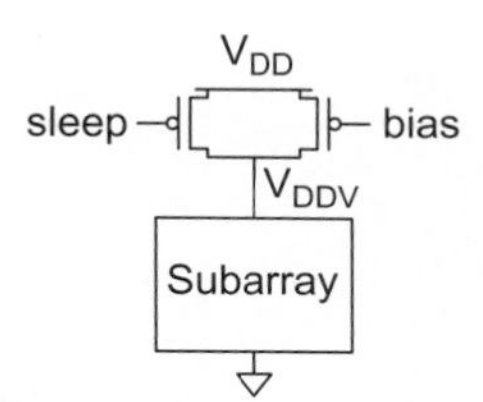

FIGURE 11.38 Partial power gating

Leakage through the access transistors can be reduced by driving inactive wordlines to a negative voltage (Figure 11.37(c)) [Itoh96, Wang07]. Beware: in some processes, the increased gate-induced leakage overwhelms the savings in subthreshold leakage. Reduced leakage increases the number of cells that can be connected to the bitline. During standby, the bitlines can be floated to reduce the access transistor leakage as well (Figure 11.37(d)) [Heo02]. As mentioned in Section 4.3.4, body bias is another way to reduce subthreshold leakage in sleep mode (Figure 11.37(e)) or increase speed in active mode.

11.2.6.3 Subthreshold Memories Conventional 6T SRAM cells do not function reliably in the subthreshold regime because the ratio constraints for read stability and writability cannot be guaranteed, especially in light of threshold variations [Calhoun06b, Chen07]. Moreover, the poor ratio of I_{on} to I_{off} limits the number of cells that can be connected to a local bitline.

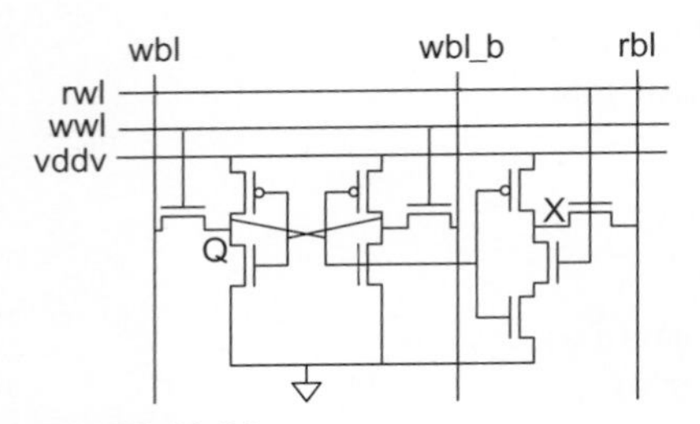

FIGURE 11.39
10T subthreshold memory cell

The 12T cell from Figure 11.19 operates correctly down to voltages as low as static CMOS registers because it has the same circuit form and eliminates any ratio constraints. However, the 12T cell is three times larger than a 6T cell. Moreover, the number of cells sharing a bitline is small because of leakage. The 8T dual-ported cell is dense and can operate at a lower voltage than a 6T cell, but it becomes unwritable near threshold when the access transistor can't be assured of overpowering the pMOS pullup.

The 10T cell of Figure 11.39 is designed specifically for subthreshold operation [Calhoun07]. It looks much like the 8T cell, but adds two transistors to reduce read port leakage and substitutes a virtual supply line to improve writability. The read bitline *rbl* is precharged to V_{DD}. When *rwl* is 0, *rbl* is isolated from GND through two series transistors. Because of the stack effect, leakage is reduced by an order of magnitude. The pMOS transistor connected to node X is optional and serves to further reduce leakage. When it is ON, it pulls X up to V_{DD}. Even when it is OFF, its leakage pulls X to an intermediate voltage above GND. In either case, the nMOS transistor connected to *rbl* will see a negative V_{gs}, further reducing its leakage. Leakage is low enough to allow hundreds of cells to share a common bitline. During write operations, the virtual supply line V_{DDV} is floated. This eliminates contention with the pMOS pullup, allowing the access transistors to flip the state of the cell. V_{DDV} is the restored to V_{DD} to stabilize the cell before the write operation concludes.

The literature is full of other subthreshold memory cells such as [Chen06, Zhai08, Kim09]. Some of these cells only work properly in processes with specific characteristics such as a strong reverse short channel effect, so check the read and write margins carefully in your process while considering variability. Even using specialized cells, subthreshold memories tend to have lower yields than memories operating at higher voltage.

11.2.7 Area, Delay, and Power of RAMs and Register Files

11.2.7.1 Area The area of a memory containing N bits can be predicted as

$$A = \frac{NA_{\text{bit}}}{E} \tag{11.4}$$

where A_{bit} is the area of a memory cell, and E is the array efficiency. Cell areas for 6T SRAM cells were shown in Figure 11.17. A_{bit} is about 600 λ^2 using industrial layouts or 1200 λ^2 using MOSIS design rules. According to Section 11.2.4, a p-ported register file in the MOSIS rules has an area of approximately $64(p + 3)^2 \lambda^2$; industrial layouts may be tighter depending on the pitch of metal3 and metal4 used for the wordlines and bitlines. An array efficiency of 0.7 is a reasonable target. Peripheral circuitry such as a cache controller are not considered in this model.

11.2.7.2 Delay The method of Logical Effort is helpful to estimate the delay of a static RAM or register file. The critical read path for a small single-ported RAM with no column multiplexing involves the decoder to drive the wordline and the SRAM cell that pulls down the bitline. Figure 11.40 highlights this path for a 2^n word by 2^m-bit memory with total storage of $N = 2^{n+m}$ bits.

The decoder is modeled as an n-input AND gate taking some combination of true and complemented address inputs. It has a logical effort of $(n + 2)/3$ and parasitic delay of n according to Tables 3.2 and 3.3. The bitline is discharged in the SRAM cell through two series transistors that behave like a dynamic multiplexer. Suppose each cell has two unit-sized access transistors and stray wire capacitance approximately equal to another unit-sized transistor, for a total capacitance of $3C$ presented by each cell to the wordline. Because there are two transistors in series, the cell delivers about half the current of a unit inverter with input capacitance $3C$. Hence, the logical effort is 2 because the cell delivers half the current of an inverter with the same input capacitance. Suppose each cell presents $1C$ of diffusion capacitance on the bitline, so the total bitline capacitance is $2^n C$. The cell has an effective resistance of $2R$ discharging the bitline through two series unit transistors. Hence, the bitline has a parasitic delay of $2^{n+1}RC$. Normalized by $\tau = 3RC$, this gives $p = 2^{n+1}/3$.

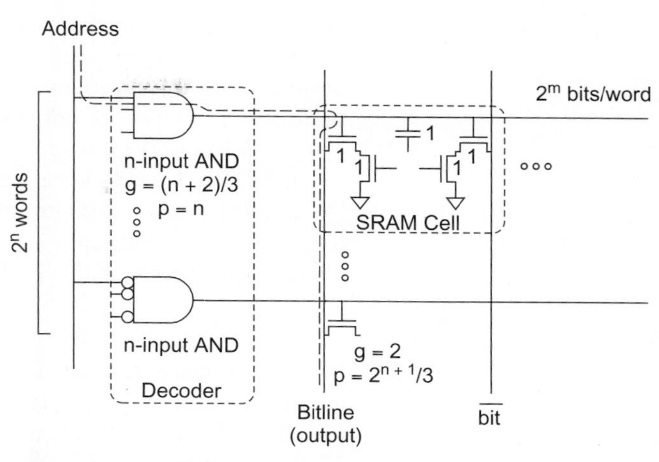

FIGURE 11.40 Critical path for read of small SRAM

Putting these two stages together, the path logical effort is $G = (n + 2)/3 \times 2$. If the true and complementary bitline outputs each drive capacitance equal to half that seen by the address inputs, the path electrical effort is $H = 1/2$. Within the path are a 2^n-way branch as each address bit is needed by each wordline decoder and another 2^m-way branch as each wordline drives all the bits on that word. Hence, the branching effort is $B = N$. The path effort delay is $F = GBH = N(n + 2)/3$. The parasitic delay is $P = n + 2^{n+1}/3$. The best number of stages is approximately $\log_4 F = (m + n)/2 + \log_4 [(n + 2)/3]$. These stages would include buffers in the address driver, multiple levels of gates in the decoder, buffers to drive the wordline, and an inverter on the bitline output. The path delay is

$$D = 4\log_4 F + P = 2(m+n) + 4\log_4[(n+2)/3] + n + 2^{n+1}/3 \quad \textbf{(11.5)}$$

For a 32-word × 32-bit register file, $n = 5$, $N = 2^{10}$, and $D = 48.8\tau = 9.8$ FO4 inverter delays.

This model is clearly an oversimplification valid only for subarray. The n-input AND gate is usually constructed out of a chain of low fan-in gates, but this only slightly improves its logical effort. We also neglect the effort of the clock gating to drive the wordlines on the clock edge. We assume the RAM is small enough that sense amplifiers are not used and neglect the wire resistance and capacitance. The pulldown transistor inside the SRAM cell may be larger than the access transistor. Nevertheless, the model offers insights into the number of stages that the memory should use and its approximate delay. For example, it shows that, without sense amplifiers, putting too many words on a bitline causes excessive parasitic delay.

[Amrutur00] models the delay of large SRAMs using Logical Effort in substantially more detail than can be repeated here. The overall delay includes components contributed by both the gates and the wire RC. In a well-designed N-bit SRAM ($N \geq 2^{16}$) using static CMOS decoders, the gate delay component is approximately

$$D = 1.2\log_2 N - 4 \quad \textbf{(11.6)}$$

FO4 inverter delays. More aggressive decoders using domino or race-based NOR techniques from Section 11.2.2 can reduce this delay by about 15% [Amrutur01]. Wire delay becomes important for RAMs beyond the 1 Mbit capacity. A lower bound for wire delay is set by the speed of light at about 1.75 FO4 for 4-Mbit memories. This delay doubles for each quadrupling in memory size. In practice, the wire delay depends on the wire width and thickness and repeater strategy, but can be several times this lower bound. In processes beyond the 100 nm generation, sense amplifiers will need larger bitline swings because their offset voltages are not scaling with the supply voltage. This will add several FO4 inverter delays to the bitline-sensing time.

CACTI (Cache Access and Cycle Time) is another model for cache delay [Wilton96]. [Agarwal01] extends this model to account for process scaling of wires and transistors. For caches up to 256 KB, the model predicts an access time of a single-ported direct-mapped cache with a 32-byte block size in a 50 nm process of roughly

$$D = 1.5\sqrt{C} + 13 \tag{11.7}$$

FO4 delays, where C is the capacity in KB. For example, the access time for a 16 KB cache is approximately 19 FO4 delays. The model also predicts the delay of a six-ported register file with 64-bit words to vary from 12–16 FO4 delays as the capacity increases from 32–256 registers.

11.2.7.3 Power Memory power has dynamic and leakage components. The dynamic power is proportional to the number of cells in a bank and the number of banks that are activated (typically 1). For large caches, the dynamic power of the datalines to route the data out of the cache is also significant. This power grows with the wire length, which depends on the square root of the capacity. The leakage power is proportional to the total number of cells in the memory. Dynamic and leakage power both grow linearly with the number of ports. [Evans95] describes SRAM power modeling further.

11.3 DRAM

Dynamic RAMs (DRAMs) store their contents as charge on a capacitor rather than in a feedback loop. Thus, the basic cell is substantially smaller than SRAM, but the cell must be periodically read and refreshed so that its contents do not leak away. Commercial DRAMs are built in specialized processes optimized for dense capacitor structures. They offer a factor of 10–20 greater density (bits/cm^2) than high-performance SRAM built in a standard logic process [Nakagome03], but they also have much higher latency. DRAM circuit design is a specialized art and is the topic of excellent books such as [Keeth07]. This section provides an overview of the general issues.

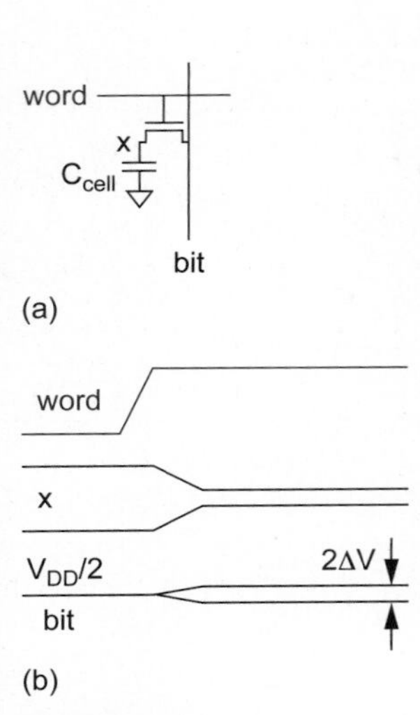

FIGURE 11.41
1T DRAM cell read operation

A 1-transistor (1T) dynamic RAM cell consists of a transistor and a capacitor, as shown in Figure 11.41(a). Like SRAM, the cell is accessed by asserting the wordline to connect the capacitor to the bitline. On a read, the bitline is first precharged to $V_{DD}/2$. When the wordline rises, the capacitor shares its charge with the bitline, causing a voltage change ΔV that can be sensed, as shown in Figure 11.41(b). The read disturbs the cell contents at x, so the cell must be rewritten after each read. On a write, the bitline is driven high or low and the voltage is forced onto the capacitor. Some DRAMs drive the wordline to $V_{DDP} = V_{DD} + V_t$ to avoid a degraded level when writing a '1.'

The DRAM capacitor C_{cell} must be as physically small as possible to achieve good density. However, the bitline is contacted to many DRAM cells and has a relatively large capacitance C_{bit}. Therefore, the cell capacitance is typically much smaller than the bitline capacitance. According to the charge-sharing equation, the voltage swing on the bitline during readout is

$$\Delta V = \frac{V_{DD}}{2} \frac{C_{\text{cell}}}{C_{\text{cell}} + C_{\text{bit}}} \quad (11.8)$$

We see that a large cell capacitance is important to provide a reasonable voltage swing. It also is necessary to retain the contents of the cell for an acceptably long time and to minimize soft errors. For example, 30 fF is a typical target. The most compact way to build such a high capacitance is to extend into the third dimension. For example, Figure 11.42 shows a cross-section and SEM image of *trench capacitors* etched under the source of the transistor. The walls of the trench are lined with an oxide-nitride-oxide dielectric. The trench is then filled with a polysilicon conductor that serves as one terminal of the capacitor attached to the transistor drain, while the heavily doped substrate serves as the other terminal. A variety of three-dimensional capacitor structures have been used in specialized DRAM processes that are not available in conventional CMOS processes.

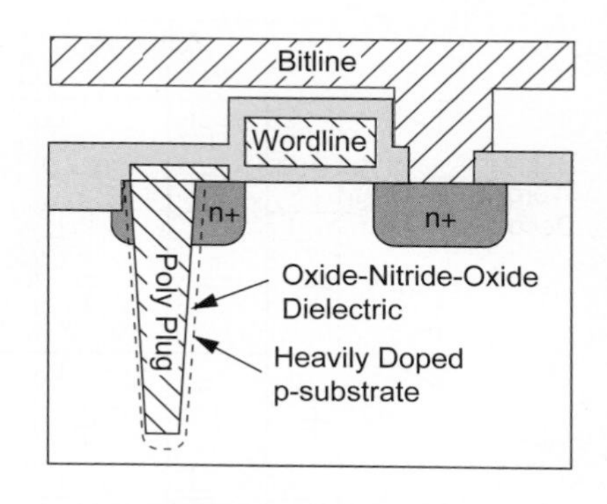

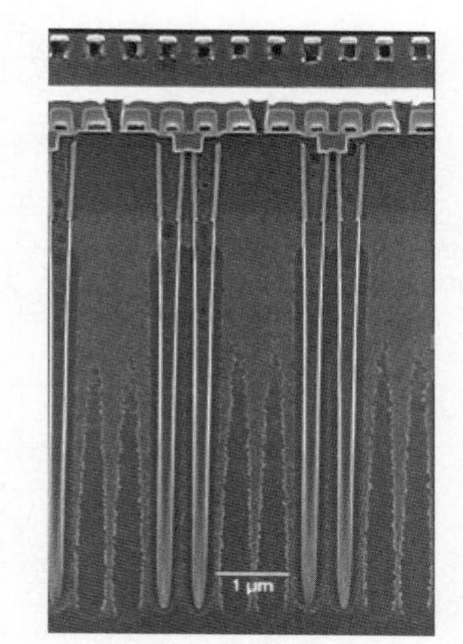

FIGURE 11.42 Trench capacitor

11.3.1 Subarray Architectures

Like SRAMs described in Section 11.2.5, large DRAMs are divided into multiple subarrays. The subarray size represents a trade-off between density and performance. Larger subarrays amortize the decoders and sense amplifiers across more cells and thus achieve better array efficiency. But they also are slow and have small bitline swings because of the high wordline and bitline capacitance. A typical subarray size is 256 words by 512 bits, as shown in Figure 11.43. Array efficiencies are typically 50–60%.

A subarray of this size has an order of magnitude higher capacitance on the bitline than in the cell, so the bitline voltage swing ΔV during a read is tiny. The array uses a sense amplifier to compare the bitline voltage to that of an idle bitline (precharged to $V_{DD}/2$). The sense amplifier must also be compact to fit the tight pitch of the array. The low-swing bitlines are sensitive to noise. Three

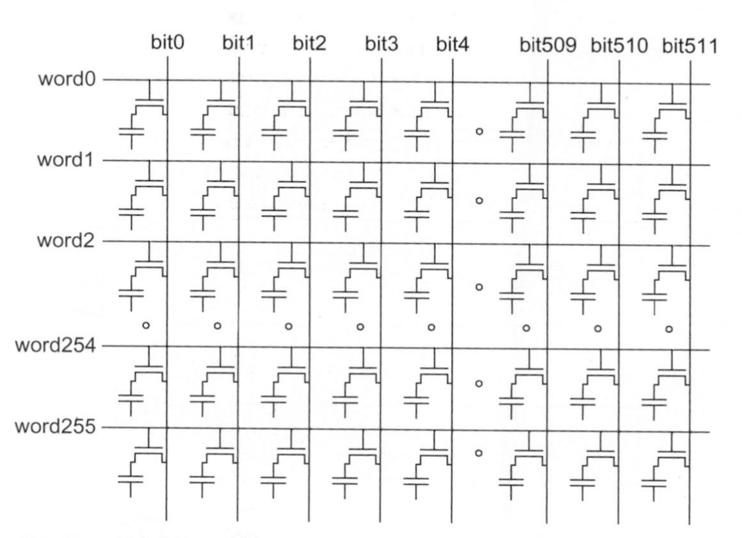

FIGURE 11.43 DRAM subarray

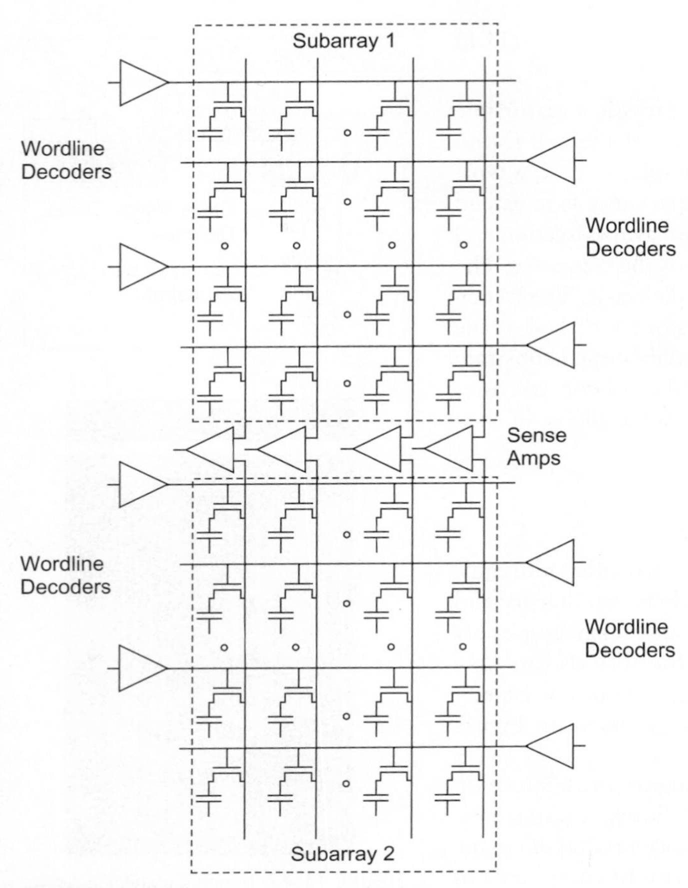

FIGURE 11.44 Open bitlines

bitline architectures, *open*, *folded*, and *twisted*, offer different compromises between noise and area.

Early DRAMs (until the 64-kbit generation) used the open bitline architecture shown in Figure 11.44. In this architecture, the sense amplifier receives one bitline from each of two subarrays. The wordline is only asserted in one array, leaving the bitlines in the other array floating at the reference voltage. The arrays are very dense. However, any noise that affects one array more than the other will appear as differential noise at the sense amplifier. Thus, open bitlines have unacceptably low signal-to-noise ratios for high-capacity DRAM.

The folded bitline architecture is shown in Figure 11.45. In this architecture, each bitline connects to only half as many cells. Adjacent bitlines are organized in pairs as inputs to the sense amplifiers. When a wordline is asserted, one bitline will switch while its neighbor serves as the quiet reference. Many noise sources will couple equally onto the two adjacent bitlines so they tend to appear as common mode noise that is rejected by the sense amplifier. This noise advantage comes at the expense of greater layout area. Figure 11.46 shows a clever layout for a 6×8 folded bitline subarray that is only 33% larger than an open bitline layout. Observe how DRAM processes push the design rules and use diagonal polysilicon to reduce area. Notice how pairs of cells in the layout share a single bitline contact to minimize the bitline capacitance.

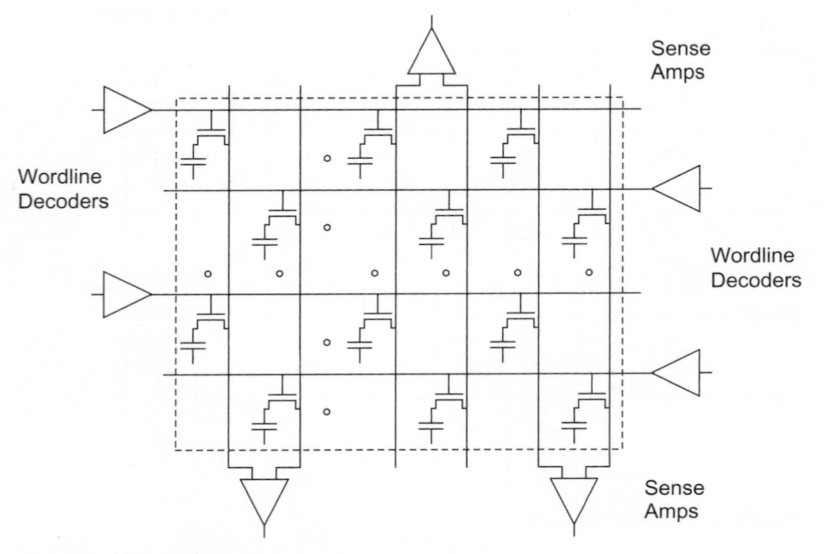

FIGURE 11.45 Folded bitlines

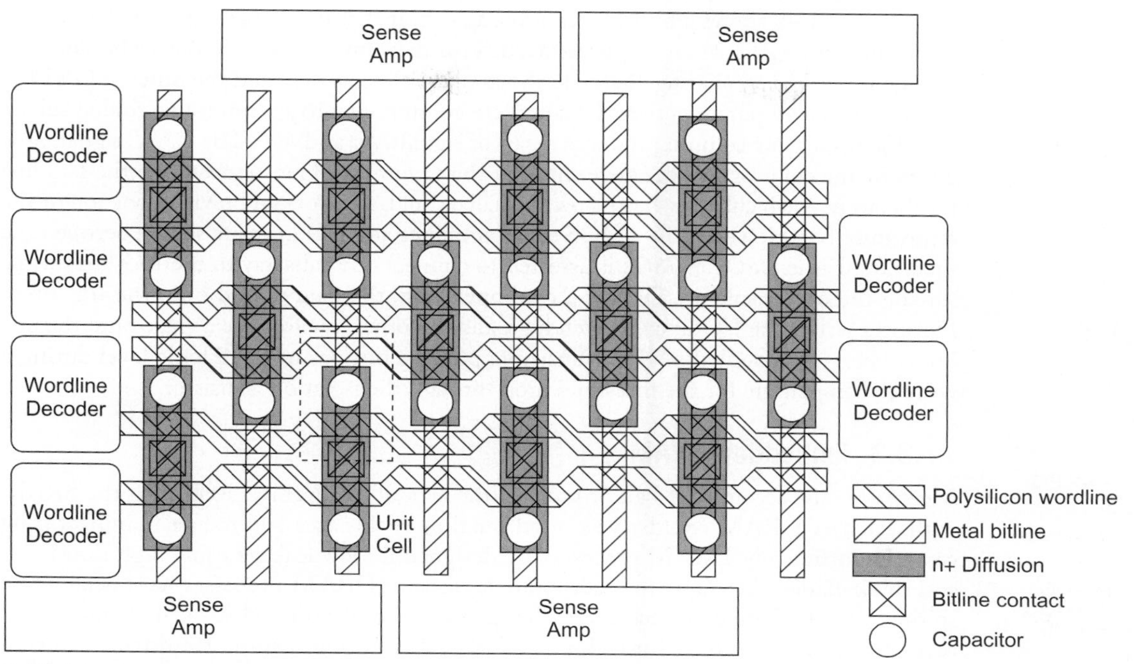

FIGURE 11.46 Layout of folded bitline subarray

Unfortunately, the folded bitline architecture is still susceptible to noise from a neighboring switching bitline that capacitively couples more strongly onto one of the bitlines in the pair. Capacitive coupling is very significant in modern processes. The twisted bitline architecture [Hidaka89] solves this problem by swapping the positions of the folded bitlines part way along the array in much the same way as SRAM bitlines were twisted in Figure 11.29(b). The twists cost a small amount of extra area within the array.

11.3.2 Column Circuitry

The column circuitry in a DRAM includes the sense amplifiers, write drivers, column multiplexing, and bitline conditioning circuits. In a folded or twisted bitline architecture, the column circuitry is placed on both sides of the array so that it can be laid out on four times the pitch of a single column, as shown in Figure 11.46. Part of the circuitry can be shared between two adjacent subarrays.

Figure 11.47(a) shows a sense amplifier built from cross-coupled inverters with supplies tied to control voltages. Initially, the two bitlines *bit* and *bit** are precharged to $V_{DD}/2$, the bottom voltage V_n is at $V_{DD}/2$, and the top voltage V_p is at 0 so all of the transistors in the amplifier are OFF. During a read, one of the bitlines will change by a small amount while the other floats at $V_{DD}/2$. V_n is then pulled low. As it falls to a threshold voltage below the higher of the two bitline voltages, the cross-coupled nMOS transistors will begin to pull the lower bitline voltage down to 0. After a small delay, V_p is pulled high. The cross-coupled pMOS transistors pull the higher bitline voltage up to V_{DD}. For example, Figure 11.47(b) shows the waveforms while reading a '0' on *bit* while using *bit** as a reference. Driving the active bitline to one of the rails has the side effect of rewriting the cell with the value that was just read.

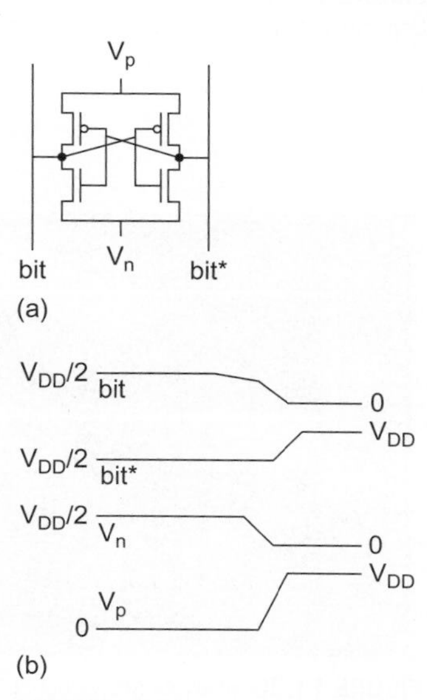

FIGURE 11.47 Sense amplifier

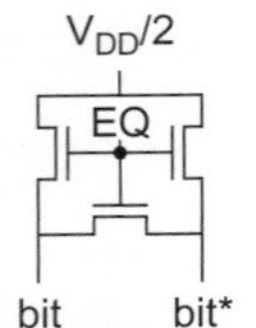

FIGURE 11.48 Bitline conditioning

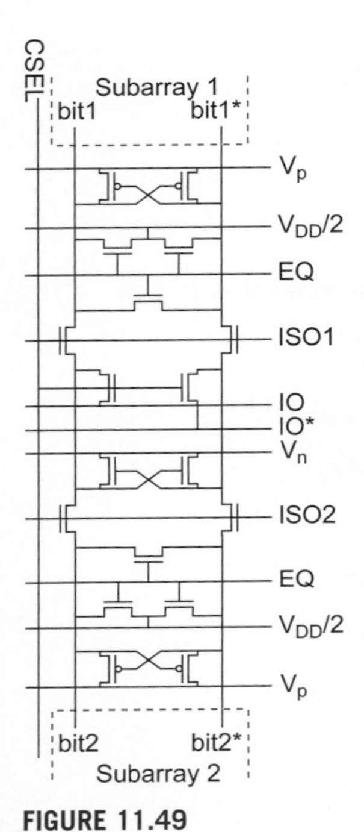

FIGURE 11.49 Column circuitry

Figure 11.48 shows a bitline conditioning circuit that precharges and equalizes a pair of bitlines to $V_{DD}/2$ when EQ is asserted. This consumes very little power because the voltage is reached by sharing charge between one bitline at V_{DD} and the other at GND.

Figure 11.49 puts together the complete column circuitry serving two folded subarrays. Each subarray column produces a pair of signals, *bit* and *bit**. The *CSEL* signal, produced by the column decoder, determines if this column will be connected to the I/O line for the array. Each subarray has its own equalization transistors and pMOS portion of the sense amplifier. However, the nMOS sense amplifier and I/O lines are shared between the subarrays. Either *ISO*1 or *ISO*2 is asserted to connect one subarray to the I/O lines while leaving the other isolated. During a read operation, the data is read onto the I/O lines. During a write, one I/O line is driven high and the other low to force a value onto the bitlines. The cross-coupled pMOS transistors pull the bitlines to a full logic level during a write to compensate for the threshold drop through the isolation transistor.

11.3.3 Embedded DRAM

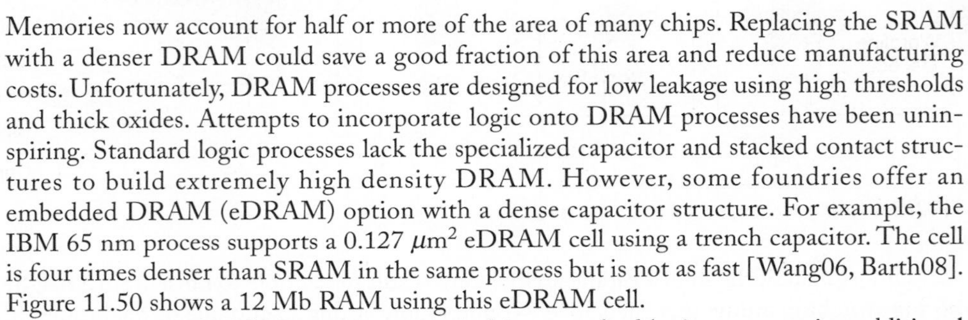

Memories now account for half or more of the area of many chips. Replacing the SRAM with a denser DRAM could save a good fraction of this area and reduce manufacturing costs. Unfortunately, DRAM processes are designed for low leakage using high thresholds and thick oxides. Attempts to incorporate logic onto DRAM processes have been uninspiring. Standard logic processes lack the specialized capacitor and stacked contact structures to build extremely high density DRAM. However, some foundries offer an embedded DRAM (eDRAM) option with a dense capacitor structure. For example, the IBM 65 nm process supports a 0.127 μm^2 eDRAM cell using a trench capacitor. The cell is four times denser than SRAM in the same process but is not as fast [Wang06, Barth08]. Figure 11.50 shows a 12 Mb RAM using this eDRAM cell.

Alternatively, DRAM can be constructed in a standard logic process using additional transistors in place of the capacitor. Figure 11.51 shows some 3T and 4T DRAM *gain cells* [Nakagome03]. These cells store a value on the gate capacitance of a transistor. The read operation involves an active transistor rather than simple charge sharing, so they can produce a stronger signal. Early DRAMs used these cells, but they were superseded by Dennard's invention of the 1T cell at IBM in 1968 [Dennard68]. They might become relevant again as technology and power supplies continue to scale.

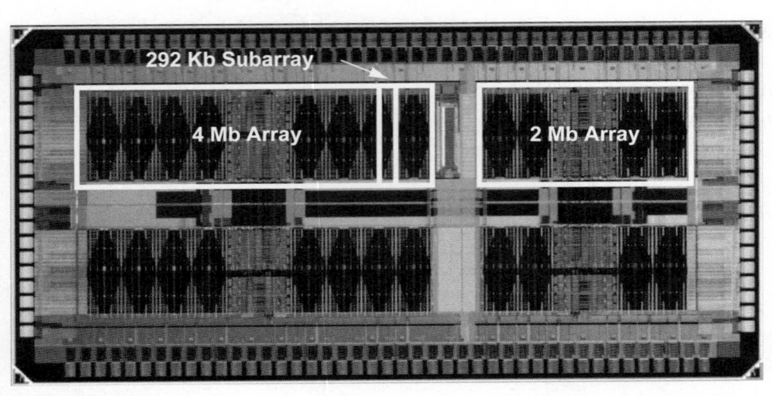

FIGURE 11.50 eDRAM arrays (© 2008 IEEE.)

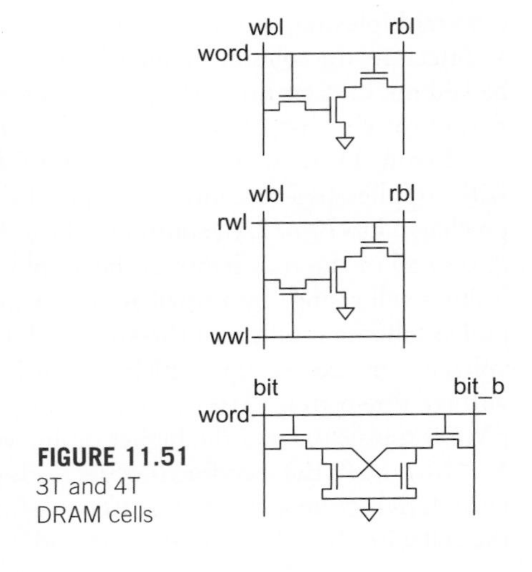

FIGURE 11.51 3T and 4T DRAM cells

11.4 Read-Only Memory

Read-Only Memory (ROM) cells can be built with only one transistor per bit of storage. A ROM is a nonvolatile memory structure in that the state is retained indefinitely—even without power. A ROM array is commonly implemented as a single-ended NOR array. Commercial ROMs are normally dynamic, although pseudo-nMOS is simple and suffices for small structures. As in SRAM cells and other footless dynamic gates, the wordline input must be low during precharge on dynamic NOR gates. In situations where DC power dissipation is acceptable and the speed is sufficient, the pseudo-nMOS ROM is the easiest to design, requiring no timing. The DC power dissipation can be significantly reduced in multiplexed ROMs by placing the pullup transistors after the column multiplexer.

Figure 11.52 shows a 4-word by 6-bit ROM using pseudo-nMOS pullups with the following contents:

```
word0: 010101
word1: 011001
word2: 100101
word3: 101010
```

The contents of the ROM can be symbolically represented with a dot diagram in which dots indicate the presence of 1s, as shown in Figure 11.53. The dots correspond to nMOS pulldown transistors connected to the bitlines, but the outputs are inverted.

Mask-programmed ROMs can be configured by the presence or absence of a transistor or contact, or by a threshold implant that turns a transistor permanently OFF where it is not needed. Omitting transistors has the advantage of reducing capacitance on the wordlines and power consumption. Programming with metal contacts was once popular because such ROMs could be completely manufactured except for the metal layer, and then programmed according to customer requirements through a metallization step. The advent of EEPROM and Flash memory chips has reduced demand for such mask-programmed ROMs. Figure 11.54 shows a layout for the 4-word by 6-bit ROM array.

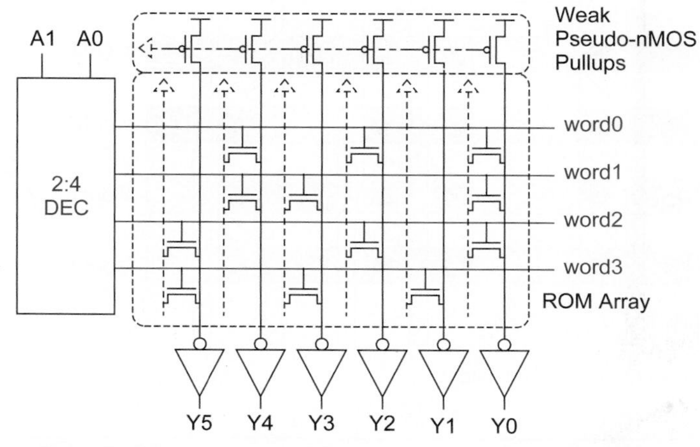

FIGURE 11.52 Pseudo-nMOS ROM

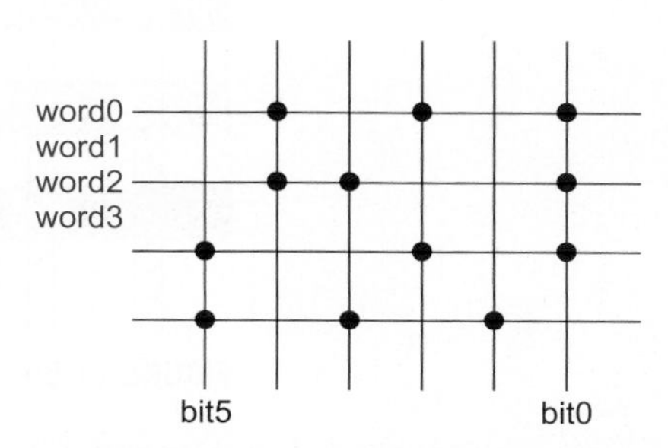

FIGURE 11.53 Dot diagram representation of ROM

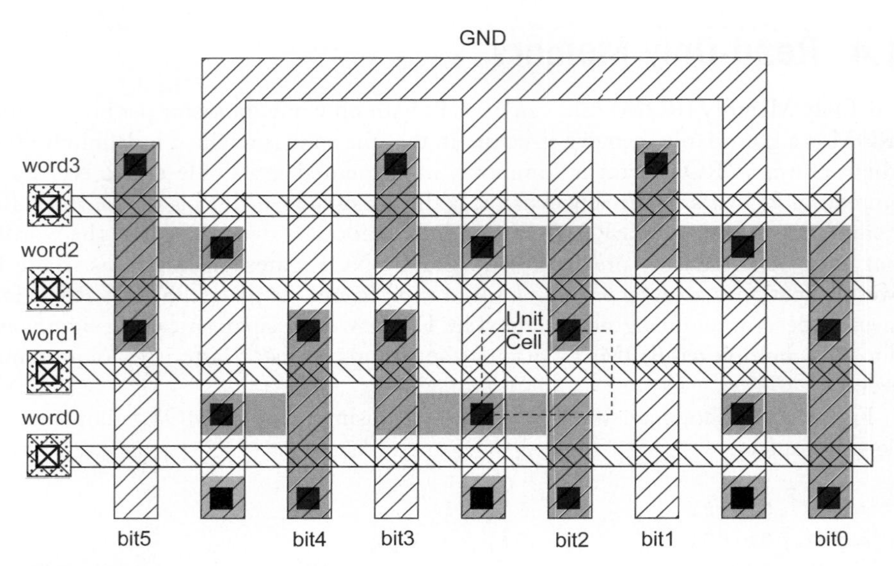

FIGURE 11.54 ROM array layout

The wordlines run horizontally in polysilicon, while the bitlines and grounds run vertically in metal1. Notice how each ground is shared between a pair of cells. Each bit of the ROM occupies a $12 \times 8\ \lambda$ cell[2]. Polysilicon wordlines are only appropriate for small or slow ROMs. A larger ROM can run metal2 straps over the polysilicon and contact the two periodically (e.g., every eight columns). Occasional substrate contacts are also required.

Row decoders for ROMS are similar to those for RAMs except that they are usually tightly constrained by the ROM wordline pitch. Figure 11.55 shows how each output of a 2:4 decoder can be shoehorned into a single horizontal track using vertical polysilicon true

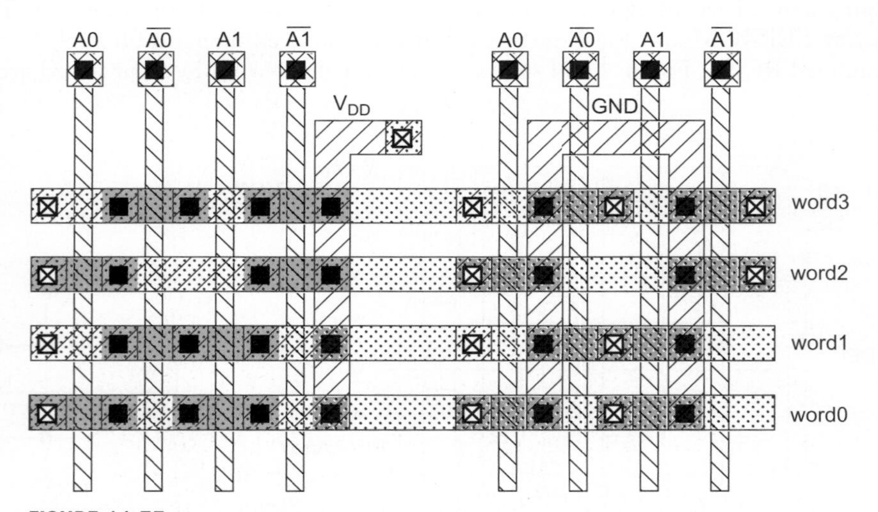

FIGURE 11.55 Row decoder layout on tight pitch

[2]The cell can be reduced to $11 \times 7\ \lambda$ by running the ground line in diffusion and by reducing the width and spacing to $3\ \lambda$.

and complementary address lines and metal supply lines. Column decoders for ROMs are usually simpler than those for RAMs because single-ended sensing is commonly employed.

Figure 11.56 shows a complete pseudo-nMOS ROM including row decoder, cell array, pMOS pullups, and output inverters.

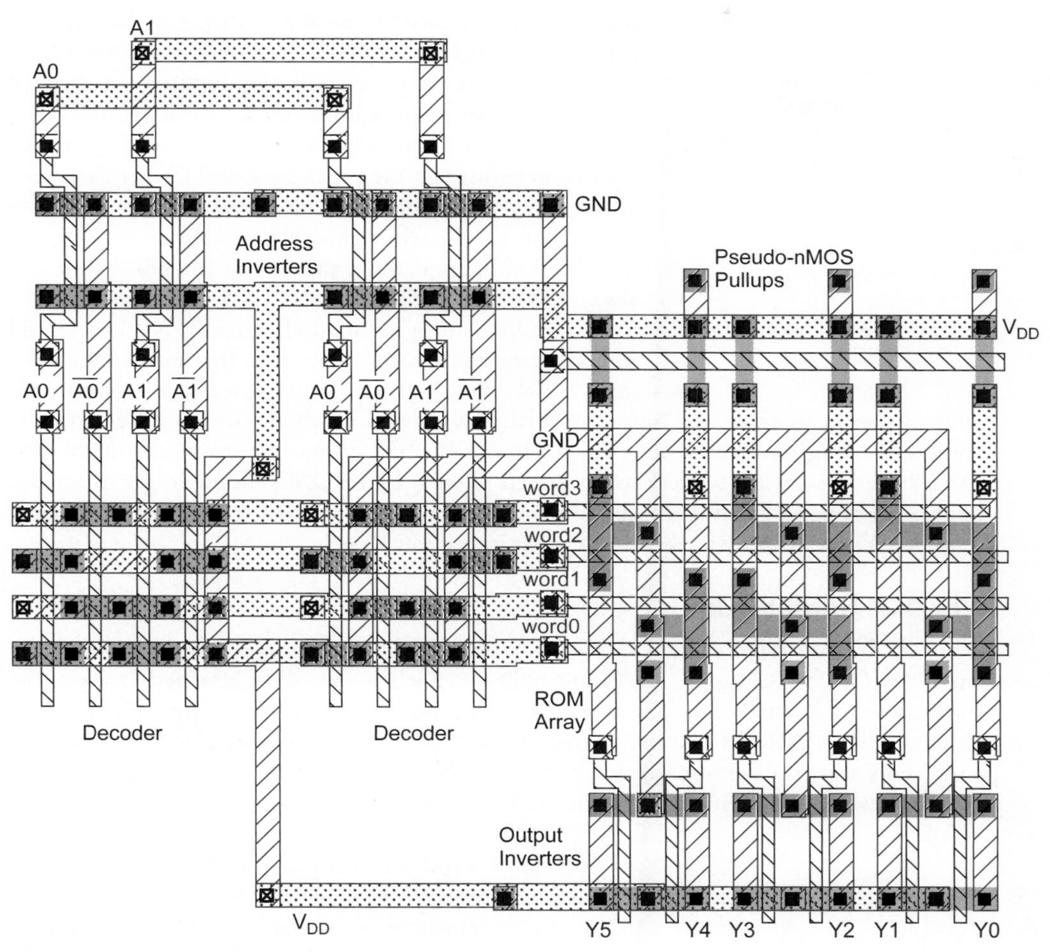

FIGURE 11.56 Complete ROM layout

11.4.1 Progammable ROMs

It is often desirable for the user to be able to program or reprogram a ROM after it is manufactured. Programming/writing speeds are generally slower than read speeds for ROMs. Four types of nonvolatile memories include *Programmable* ROMs (PROMs), *Erasable Programmable* ROMs (EPROMs), *Electrically Erasable Programmable* ROMs (EEPROMs), and *Flash* memories. All of these memories require some enhancements to a standard CMOS process: PROMs use fuses while EPROMs, EEPROMs, and Flash use charge stored on a floating gate.

Programmable ROMs can be fabricated as ordinary ROMs fully populated with pulldown transistors in every position. Each transistor is placed in series with a fuse made of polysilicon, nichrome, or some other conductor that can be burned out by applying a high current. The user typically configures the ROM in a specialized PROM programmer before putting it in the system. As there is no way to repair a blown fuse, PROMs are also referred to as *one-time programmable* memories.

As technology has improved, reprogrammable nonvolatile memory has largely displaced PROMs. These memories, including EPROM, EEPROM, and Flash, use a second layer of polysilicon to form a floating gate between the control gate and the channel, as shown in Figure 11.57. The floating gate is a good conductor, but it is not attached to anything. Applying a high voltage to the control gate causes electrons to jump through the thin oxide onto the floating gate through the processes called *Fowler-Nordheim (FN) tunneling*. Injecting the electrons induces a negative voltage on the floating gate, effectively increasing the threshold voltage of the transistor to the point that it is always OFF.

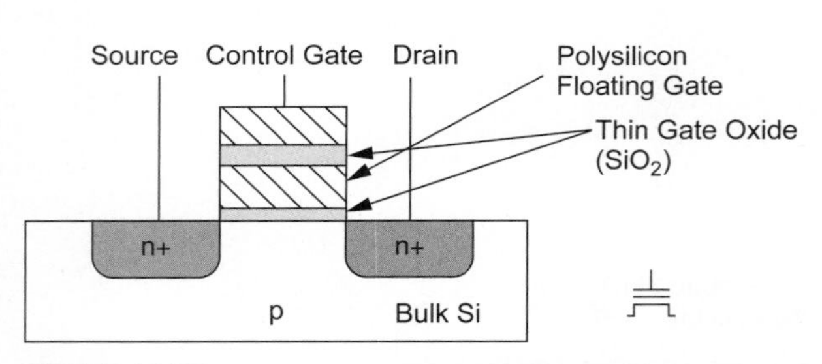

FIGURE 11.57 Cross-section of floating gate nMOS transistor

EPROM is programmed electrically, but it is erased through exposure to ultraviolet light that knocks the electrons off the floating gate. It offers a dense cell, but it is inconvenient to erase and reprogram. EEPROM and Flash can be erased electrically without being removed from the system. EEPROM offers fine-grained control over which bits are erased, while Flash is erased in bulk. EEPROM cells are larger to achieve this versatility, so Flash has become the most economical form of convenient nonvolatile storage. Flash memory is discussed further in Section 11.4.3.

11.4.2 NAND ROMs

The ROM from Figure 11.52 is called a NOR ROM because each of the bitlines is just a pseudo-nMOS NOR gate. The bitline pulls down when a wordline attached to any of the transistors is asserted high. The size of the cell is limited by the ground line. Figure 11.58 shows a NAND ROM that uses active-low wordlines. Transistors are placed in series and the transistors on the nonselected rows are ON. If no transistor is associated with the selected word, the bitline will pull down. If a transistor is present, the bitline will remain high.

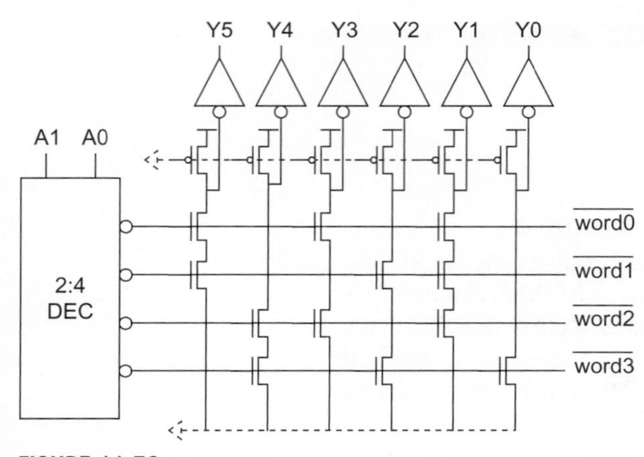

FIGURE 11.58 Pseudo-nMOS NAND ROM

Figure 11.59(a) shows a layout of the NAND ROM. The cell size is only 7×8 λ. The contents are specified by using either a transistor or a metal jumper in each bit position. The contacts limit the cell size. Figure 11.59(b) shows an even smaller layout in which transistors are located at every position. In this design, an extra implantation step can be used to create a negative threshold voltage, turning certain transistors permanently ON where they are not needed. In such a process, the cell size reduces to only 6×5 λ, assuming that the decoder and bitline circuitry can be built on such a tight pitch.

A disadvantage of the NAND ROM is that the delay grows quadratically with the number of series transistors discharging the bitline. NAND structures with more than 8–16 series transistors become extremely slow, so NAND

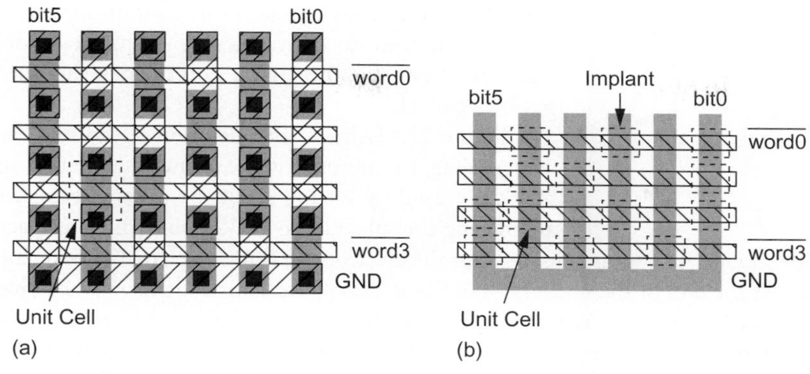

FIGURE 11.59 NAND ROM array layouts

ROMs are often broken into multiple small banks with a limited number of series transistors. Nevertheless, these NAND structures are attractive for Flash memories in which density and cost are more important than access time.

11.4.3 Flash

Flash memory was invented by Fujio Masuoka and colleagues at Toshiba in 1984 [Masuoka84]. Masuoka coined the name because blocks of memory were erased all at once "in a flash." By 1988, the long-term reliability had been proven and volume production began with 256 KB parts [Kynett88]. Meanwhile, Masuoka developed the NAND architecture that cut the area per bit by 30% [Masuoka87]. Flash memory has become tremendously popular because of its nonvolatile storage and exceptionally low cost per bit. For example, Flash memory cards are widely used in digital cameras to store hundreds of high-resolution images. Flash is also useful for firmware or configuration data because it can be rewritten to upgrade a system in the field without opening the case or removing parts. Most of the Flash market has become a commodity business driven almost entirely by cost, with performance and even reliability being secondary considerations. This section summarizes the principles of Flash operation. [Brewer08] describes the many flavors of Flash in great detail.

Most stand-alone Flash memory uses the NAND architecture to minimize bit cell size and cost. NAND Flash memories are divided into *blocks*, which in turn are made of *pages*. The memory is written one page at a time and erased one block at a time. For example, a conventional NAND flash memory might be made of 8 KB (64 Kb) blocks, each of which contain sixteen 512 B (4 Kb) pages.

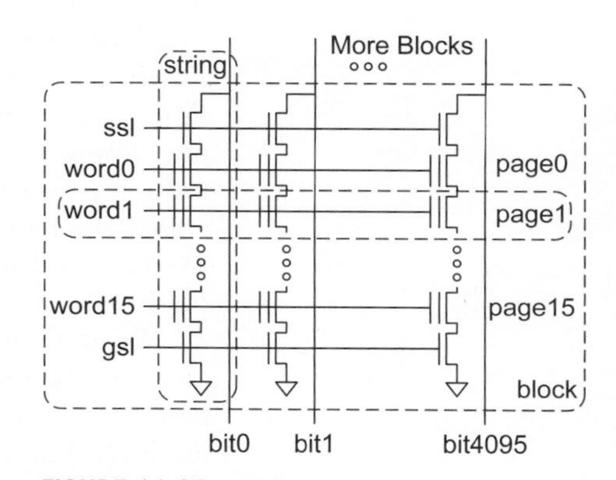

FIGURE 11.60 NAND Flash string

Recall that Flash uses floating gate transistors as memory cells. The charge on the floating gate determines the threshold of the transistor and indicates the state of the cell. A negative threshold represents a logic '1' and a positive threshold represents a logic 0.'

In NAND Flash, the floating gate transistors are connected in series to form *strings*. Figure 11.60 shows the organization of a string, page, and block in a simple Flash memory. Each string consists of 16 cells, a *string select* transistor, and a *ground select* transistor all connected in series and attached to the bitline. The control gate of each cell is connected to

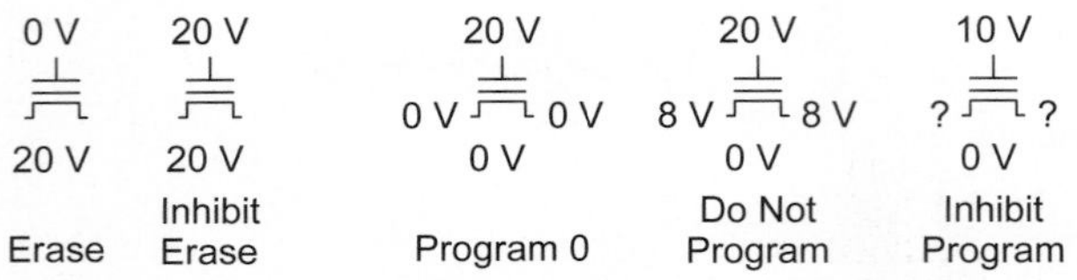

FIGURE 11.61 Erase and program operations

a wordline. The array contains one column for each bit in a page. Each column contains one string per block. The number of cells in the string determines the number of pages per block.

Figure 11.61 shows the operation of the Flash memory using voltages representative of a multimegabit design. The block is erased by setting all of the control gates to GND and raising the substrate to 20 V. The high voltage across the gate oxide induces FN tunneling, causing the electrons to flow from the floating gate to the substrate. At the end of the erase step, all the floating gate transistors have a negative V_t and thus represent 1. Tunneling is a slow process, so block erase takes on the order of a millisecond. The wordlines for other blocks on the chip are set to the same voltage as the substrate to inhibit erasing. An on-chip charge pump (see Section 12.3.7) is used to generate the high voltages.

A cell is programmed (written) to 0 by tunneling electrons onto the floating gate. The programming cannot restore 1 values, so the block must be erased before any cell is reprogrammed. An entire page is programmed at once. To program a page, the bitlines are driven with the data values: 0 V for a logic 0 and 8 V for a logic 1. The substrate is held at ground. The wordline is set to 20 V for the page being programmed and 10 V for the other pages in the block. The ground select line (*gsl*) is left OFF but the string select line (*ssl*) for the block is turned ON, passing the voltage on the bitline to the channels of all the transistors being programmed. Thus, cells being programmed to 0 see 20 V on the control gate and 0 V on the channel. This high voltage difference induces FN tunneling that drives electrons onto the floating gate, raising V_t to a positive voltage. The other cells see a smaller voltage that is insufficient to cause tunneling.

A page is read in a similar fashion to a conventional NAND ROM. The bitlines are precharged. *ssl* and *gsl* are both set to 3.3 V to activate the selected block. The active-low wordline for the selected page is set to 0 V and the wordlines for all the other pages in the block are set to 4.5 V, which is much higher than V_t. Thus, all the transistors in the stack are ON except possibly the one corresponding to the selected page. If the cell being read has a negative V_t, it turns ON too and the bitline discharges. If the cell being read has a positive V_t, it remains OFF and the bitline does not switch.

To achieve higher densities, *multilevel* Flash cells store more than one bit on a transistor by programming the threshold to one of several levels. The threshold can be sensed by adjusting the voltage on the selected wordline. The number of bits that can be stored depends on how accurately the threshold can be programmed and sensed.

Two reliability metrics for Flash memories are retention time and endurance. The *retention time* is the duration for which a Flash cell will hold its value. Under normal conditions, the charge on the floating gate would take thousands or millions of years to leak off. However, defects in the oxide may increase leakage for some cells. Manufacturers typically specify a 10 year retention time. *Endurance* is the number of times that a cell can be erased and reprogrammed. The high voltages stress the oxide and can eventually cause it to wear out. Endurance of 100,000 erase-program cycles are typical, but some multilevel Flash cells have endurance as low as 5000 cycles.

Some foundries offer an embedded Flash option, in which extra masks and process steps are used to create the floating gate transistors. The embedded Flash is commonly used for code storage in applications such as microcontrollers. These applications typically use NOR Flash instead of NAND because they need fast access to individual words rather than slow access to entire pages.

Figure 11.62 shows a die photograph of a 64 Gb NAND Flash chip from Toshiba and SanDisk built in a 43 nm process with 3 metal layers [Trinh09]. The chip uses a 16-level cell to store 4 bits per transistor. The memory is divided into two 32 Gb (4 GB) panes that can operate independently to double the throughput. Each pane has 64K columns. Hence, each page is 64 Kb (8 KB). Each string contains 64 series transistors. Thus, each block holds (64 transistors/string) × (4 bits/transistor) = 256 pages, or 2 MB of data. Each pane has 2K blocks. The chip operates at 3.3 V and has a programming bandwidth of 5.6 MB/s.

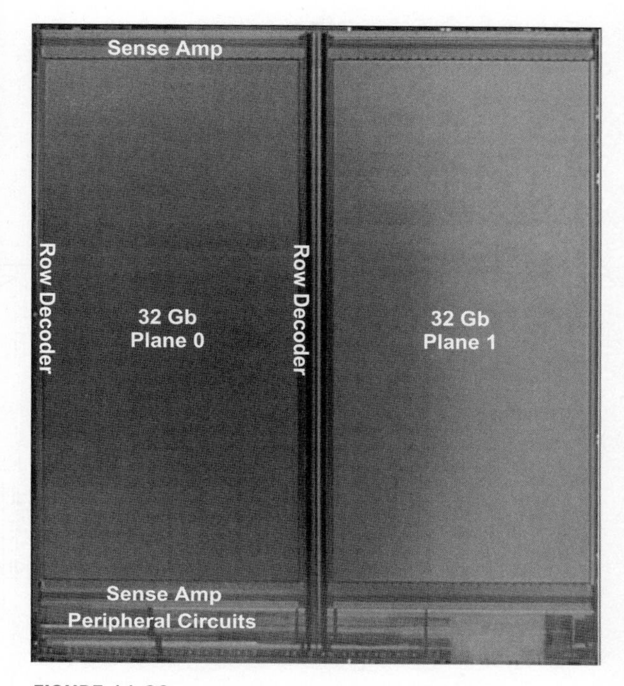

FIGURE 11.62 64 Gb NAND Flash (© 2009 IEEE.)

11.5 Serial Access Memories

Using the basic SRAM cell and/or registers, we can construct a variety of serial access memories including shift registers and queues. These memories avoid the need for external logic to track addresses for reading or writing.

11.5.1 Shift Registers

A *shift register* is commonly used in signal-processing applications to store and delay data. Figure 11.63(a) shows a simple 4-stage 8-bit shift register constructed from 32 flip-flops. As there is no logic between the registers, particular care must be taken that hold times are satisfied. Flip-flops are rather big, so large, dense shift registers use dual-port RAMs instead. The RAM is configured as a circular buffer with a pair of counters specifying where the data is read and written. The read counter is initialized to the first entry and the write counter to the last entry on reset, as shown in Figure 11.63(b). Alternately, the counters in an N-stage shift register can use two 1-of-N hot registers to track which entries should be read and written. Again, one is initialized to point to the first entry and the other to the last entry. These registers can drive the wordlines directly without the need for a separate decoder, as shown in Figure 11.63(c).

The *tapped delay line* is a shift register variant that offers a variable number of stages of delay. Figure 11.64 shows a 64-stage tapped delay line that could be used in a video processing system. Delay blocks are built from 32-, 16-, 8-, 4-, 2-, and 1-stage shift registers. Multiplexers control pass-around of the delay blocks to provide the appropriate total delay.

Another variant is a serial/parallel memory. Figure 11.65(a) shows a 4-stage Serial In Parallel Out (SIPO) memory and Figure 11.65(b) shows a 4-stage Parallel In Serial Out (PISO) memory. These are also often useful in signal processing and communications systems.

11.5.2 Queues (FIFO, LIFO)

Queues allow data to be read and written at different rates. Figure 11.66 shows an interface to a queue. The read and write operations each are controlled by their own clocks that may be asynchronous. The queue asserts the *FULL* flag when there is no room remaining to write data and the *EMPTY* flag when there is no data to read. Because of other system

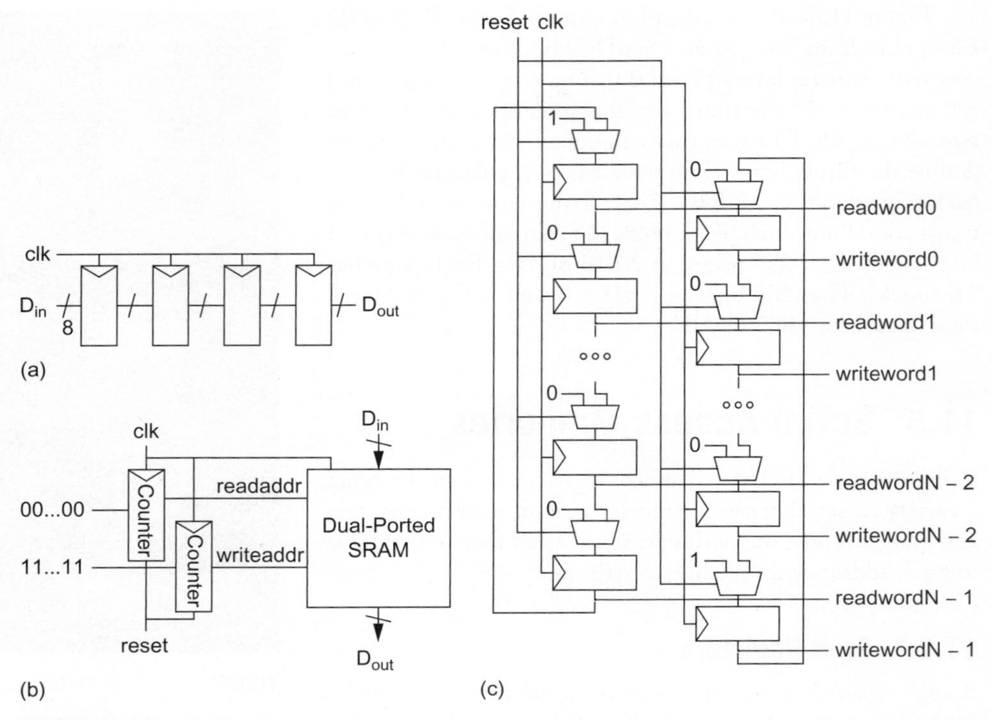

FIGURE 11.63 Shift registers

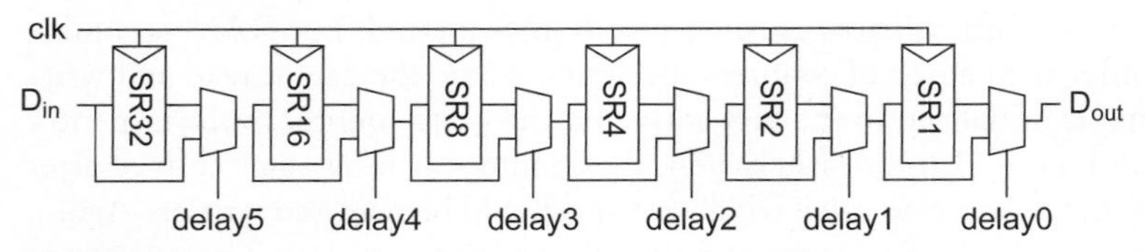

FIGURE 11.64 Tapped delay line

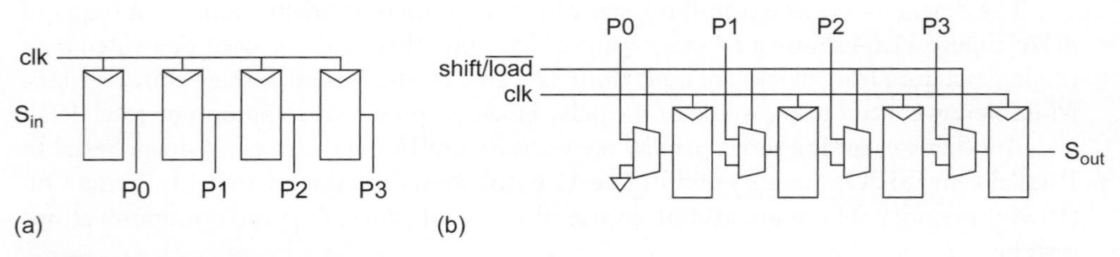

FIGURE 11.65 Serial/parallel memories

delays, some queues also provide ALMOST-FULL and ALMOST-EMPTY flags to communicate the impending state and halt write or read requests. The queue internally maintains read and write pointers indicating which data should be accessed next. As with a shift register, the pointers can be counters or 1-of-N hot registers.

First In First Out (*FIFO*) queues are commonly used to buffer data between two asynchronous streams. Like a shift register, the FIFO is organized as a circular buffer. On reset, the read and write pointers are both initialized to the first element and the FIFO is EMPTY. On a write, the write pointer advances to the next element. If it is about to catch the read pointer, the FIFO is FULL. On a read, the read pointer advances to the next element. If it catches the write pointer, the FIFO is EMPTY again.

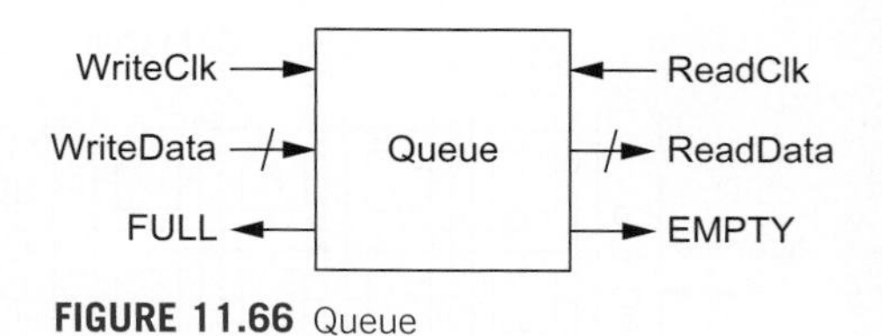

FIGURE 11.66 Queue

Last In First Out (*LIFO*) queues, also known as *stacks*, are used in applications such as subroutine or interrupt stacks in microcontrollers. The LIFO uses a single pointer for both read and write. On reset, the pointer is initialized to the first element and the LIFO is EMPTY. On a write, the pointer is incremented. If it reaches the last element, the LIFO is FULL. On a read, the pointer is decremented. If it reaches the first element, the LIFO is EMPTY again.

11.6 Content-Addressable Memory

Figure 11.67 shows the symbol for a content-addressable memory (CAM). The CAM acts as an ordinary SRAM that can be read or written given *adr* and *data*, but also performs *matching* operations. Matching asserts a *matchline* output for each word of the CAM that contains a specified *key*.

A common application of CAMs is translation lookaside buffers (TLBs) in microprocessors supporting virtual memory. The virtual address is given as the key to the TLB CAM. If this address is in the CAM, the corresponding matchline is asserted. This matchline can serve as the wordline to access a RAM containing the associated physical address, as shown in Figure 11.68. A NOR gate processing all of the matchlines generates a *miss* signal for the CAM. Note that the *read*, *write*, and *adr* lines for updating the TLB entries are not drawn.

Figure 11.69 shows a 10T CAM cell consisting of a normal SRAM cell with additional transistors to perform the match. Multiple CAM cells in the same word are tied to the same matchline. The matchline is either precharged or pulled high as a distributed pseudo-nMOS gate. The key is placed on the bitlines. If the key and the value stored in the cell differ, the matchline will be pulled down. Only if all of the key bits match all of the bits stored in the word of memory will the matchline for that word remain high. The key can contain a "don't care" by setting both *bit* and *bit_b* low. The inside front cover shows a layout of this cell in a $56 \times 43\ \lambda$ area; CAMs generally have about twice the area of SRAM

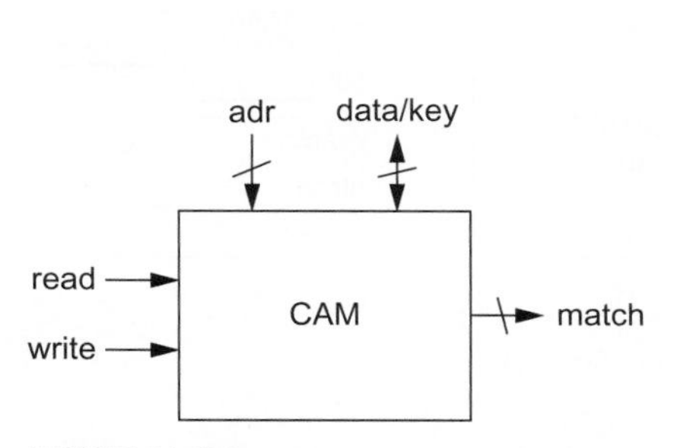

FIGURE 11.67 Content-addressable memory

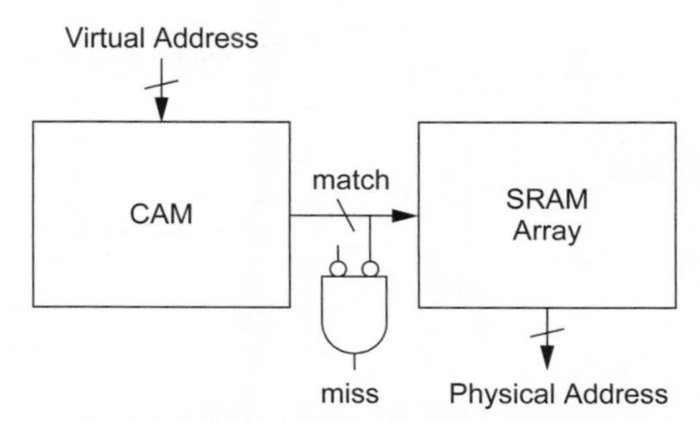

FIGURE 11.68 Translation Lookaside Buffer (TLB) using CAM

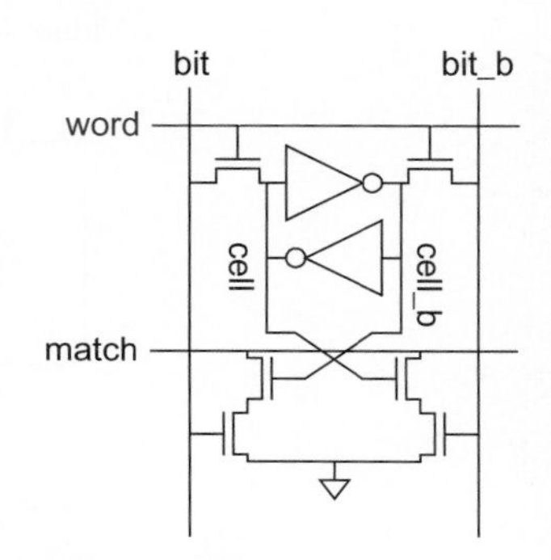

FIGURE 11.69 CAM cell implementation

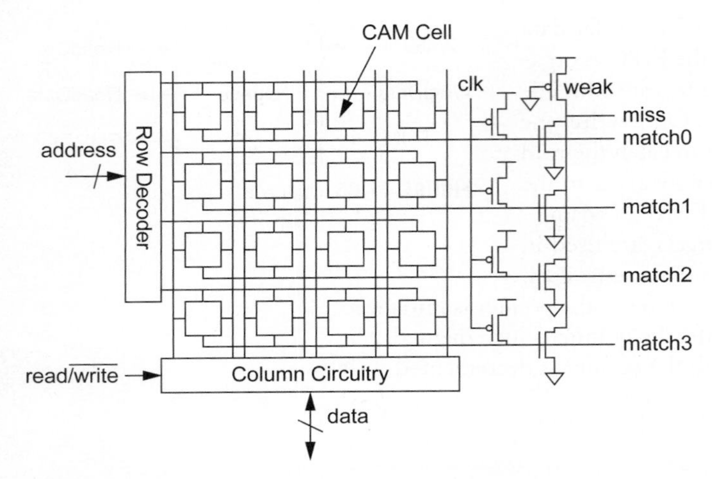

FIGURE 11.70 4 × 4 CAM array

cells. Sometimes the key is provided on separate *searchlines* rather than on the bitlines to reduce the capacitance and power consumption of a search.

Figure 11.70 shows a complete 4 × 4 CAM array. Like an SRAM, it consists of an array of cells, a decoder, and column circuitry. However, each row also produces a dynamic matchline. The matchlines are precharged with the clocked pMOS transistors. The *miss* signal is produced with a distributed pseudo-nMOS NOR.

When the matchlines are used to access a RAM, the monotonicity problem must be considered. Initially, all the matchlines are high. During CAM operation, the lines pull down, leaving at most one line asserted to indicate which row contains the key. However, the RAM requires a monotonically rising wordline. Figure 11.71 refines Figure 11.68 with strobed AND gates driving the wordlines as early as possible after the matchlines have settled. The strobe can be timed with an inverter chain or replica delay line in much the same way that the sense amplifier clock for an SRAM was generated in Section 11.2.3.3. As usual, self-timing margin must be provided so the circuit operates correctly across all design corners. The strobe must be deasserted before the match lines precharge.

In some applications, a CAM doesn't care about the value of certain bits. For example, a CAM used in a network router may not care about the subnet address when it is seeking to route data to the correct continent. A *ternary CAM* (TCAM) can store X (don't care) values as well as 0 and 1 bits. Figure 11.72 shows a TCAM cell using two bits of state to store the three values. This cell also illustrates separating the search lines from the bitlines. When $A = 1$ and $B = 0$, the cell matches a 0. When $A = 0$ and $B = 1$, the cell matches a 1. When $A = 0$ and $B = 0$, the cell matches both 0 and 1.

Large CAMs can use many of the same techniques as large RAMs, including sense amplifiers and multiple subarrays. They tend to consume relatively large amounts of power because the matchlines are heavily loaded and have an activity factor close to 1. [Pagiamtzis06] surveys many alternative CAM structures such as NAND architectures. [Agrawal08] offers a power and delay model.

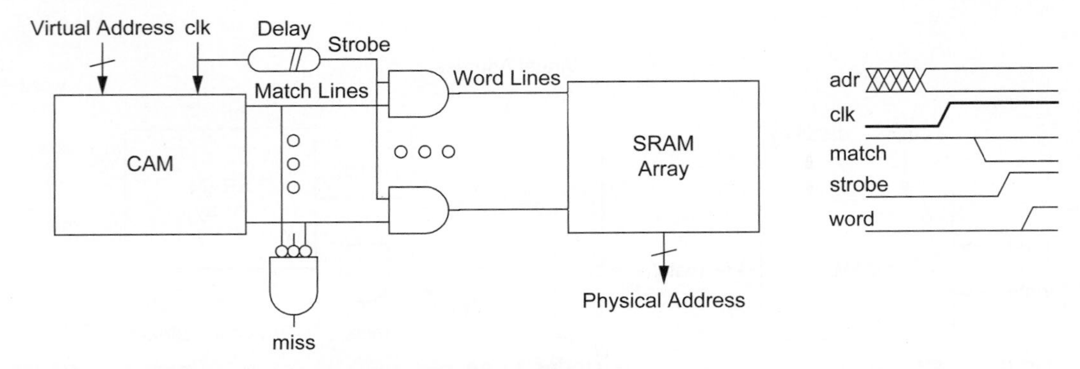

FIGURE 11.71 Refined TLB path with monotonic wordlines

11.7 Programmable Logic Arrays

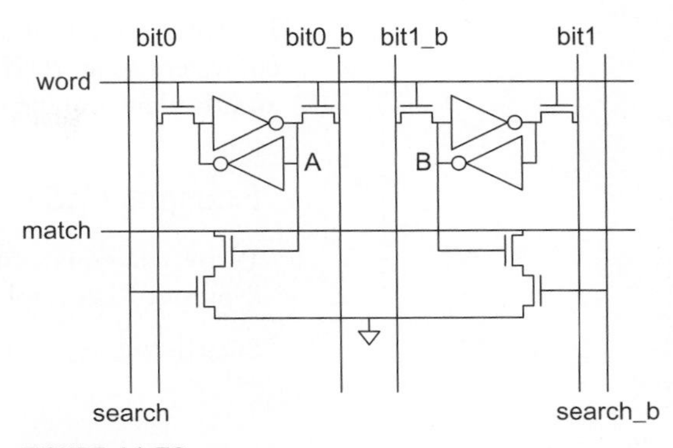

FIGURE 11.72 TCAM

A *programmable logic array* (PLA) provides a regular structure for implementing combinational logic specified in *sum-of-products canonical form*. If outputs are fed back to inputs through registers, PLAs also can form finite state machines. PLAs were most popular in the early days of VLSI when two-level logic minimization was well understood, but multilevel logic optimizers were still immature. They are dense and fast ways to implement simple functions, and with suitable CAD support, are easy to change when logic bugs are discovered. Logic synthesis tools have greatly improved and now control logic is usually synthesized instead. Moreover, pseudo-nMOS PLAs dissipate static power, while dynamic PLAs require careful design of timing chains. Nevertheless, the Cell processor used 27 dynamic PLAs in each core to calculate control signals where static logic would not meet timing [Warnock06].

Any logic function can be expressed in sum-of-products form; i.e., where each output is the OR (sum) of the ANDs (products) of true and complementary inputs. The inputs and their complements are called *literals*. The AND of a set of literals is called a *product* or *minterm*. The outputs are ORs of minterms. The PLA consists of an *AND plane* to compute the minterms and an *OR plane* to compute the outputs.

NOR gates are particularly efficient in pseudo-nMOS and dynamic logic because they use only parallel, never series, transistors. Hence, we use DeMorgan's law to replace the AND and OR gates with NORs after inverting inputs and outputs, as shown in Figure 11.73. For brevity, we often represent the PLA with a dot diagram, shown in Figure 11.74. Experienced designers often add a few unused rows and columns to their PLAs to accommodate last-minute design changes without changing the overall footprint of the

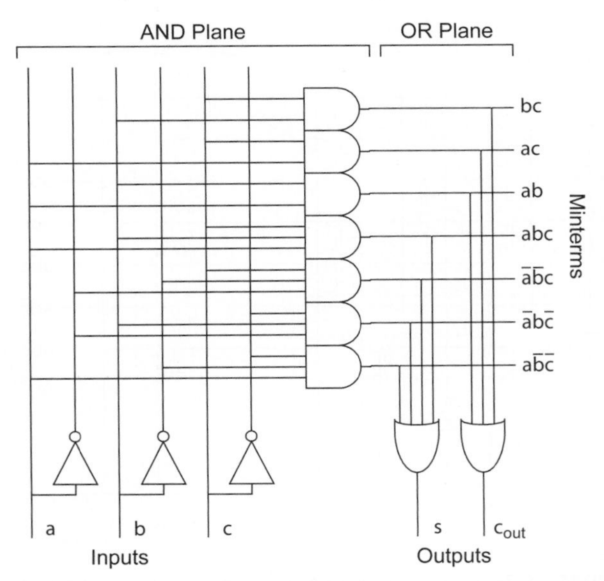

FIGURE 11.73 OR/NOR representation of PLA

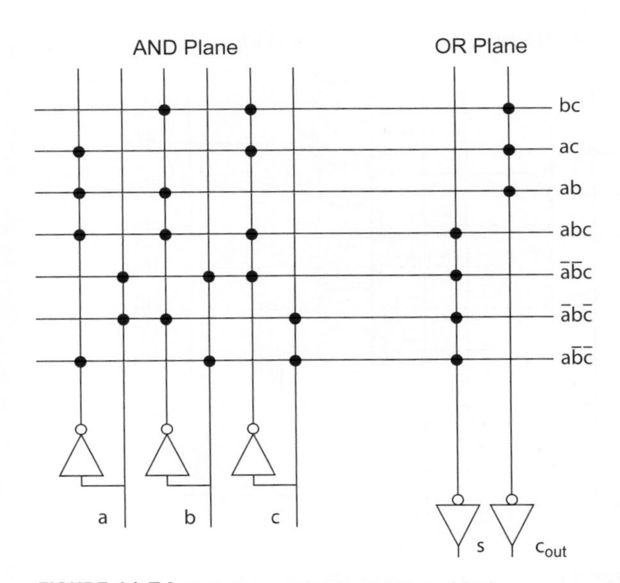

FIGURE 11.74 Dot diagram representation of PLA

PLA. Observe that a ROM and a PLA are very similar in form. The ROM decoder is equivalent to an AND plane generating all 2^n minterms. The ROM array corresponds to an OR plane producing the outputs.

Example 11.5

Write the equations for a full adder in sum-of-products form. Sketch a 3-input, 2-output PLA implementing this logic.

SOLUTION: Figure 11.75 shows the PLA. The logic equations are

$$\begin{aligned} s &= a\overline{b}\overline{c} + \overline{a}b\overline{c} + \overline{a}\overline{b}c + abc \\ c_{out} &= ab + bc + ac \end{aligned} \tag{11.9}$$

The most straightforward design for a small PLA uses a pseudo-nMOS NOR gate. Figure 11.76 shows the circuit diagram for the full adder PLA. Advantages of this PLA include simplicity and small size. Disadvantages include the static power dissipation of the NOR gates, the slow pullup response, and the fact that they don't fit into a conventional logic synthesis flow today. Figure 11.77 shows a layout for the pseudo-nMOS PLA. The transistor gates are run in polysilicon and could be strapped with metal2. Observe how ground lines can be shared between pairs of minterms and outputs so that each minterm and output can be placed on a 1.5 track pitch. The inverters require careful layout to fit the tight pitch. The pMOS pullups may be tied to an enable instead of GND so that the static current can be turned OFF when the PLA is not in use.

Dynamic PLAs eliminate the contention current and are faster than their pseudo-nMOS counterparts. Figure 11.78(a) shows a PLA using footed dynamic NORs for both

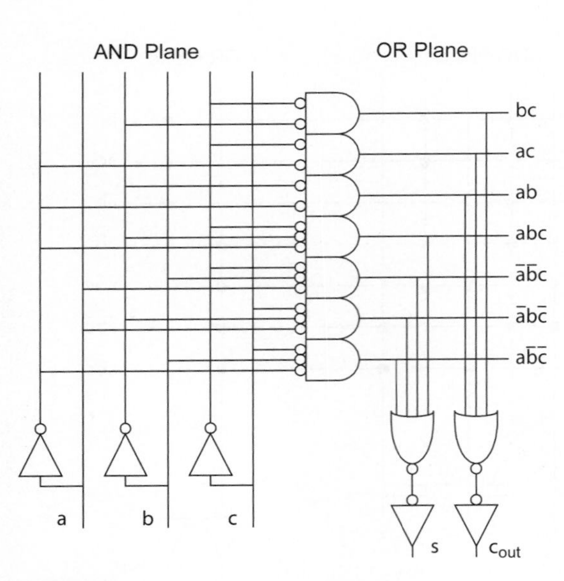

FIGURE 11.75 AND/OR representation of PLA

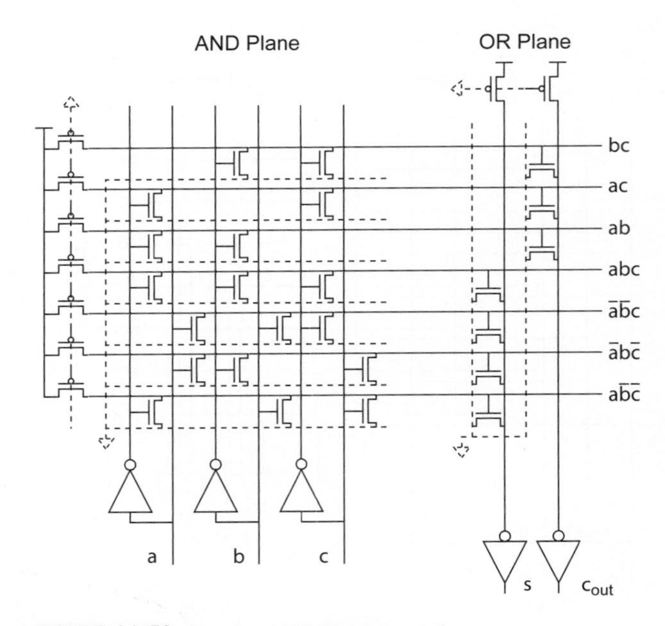

FIGURE 11.76 Pseudo-nMOS PLA schematic

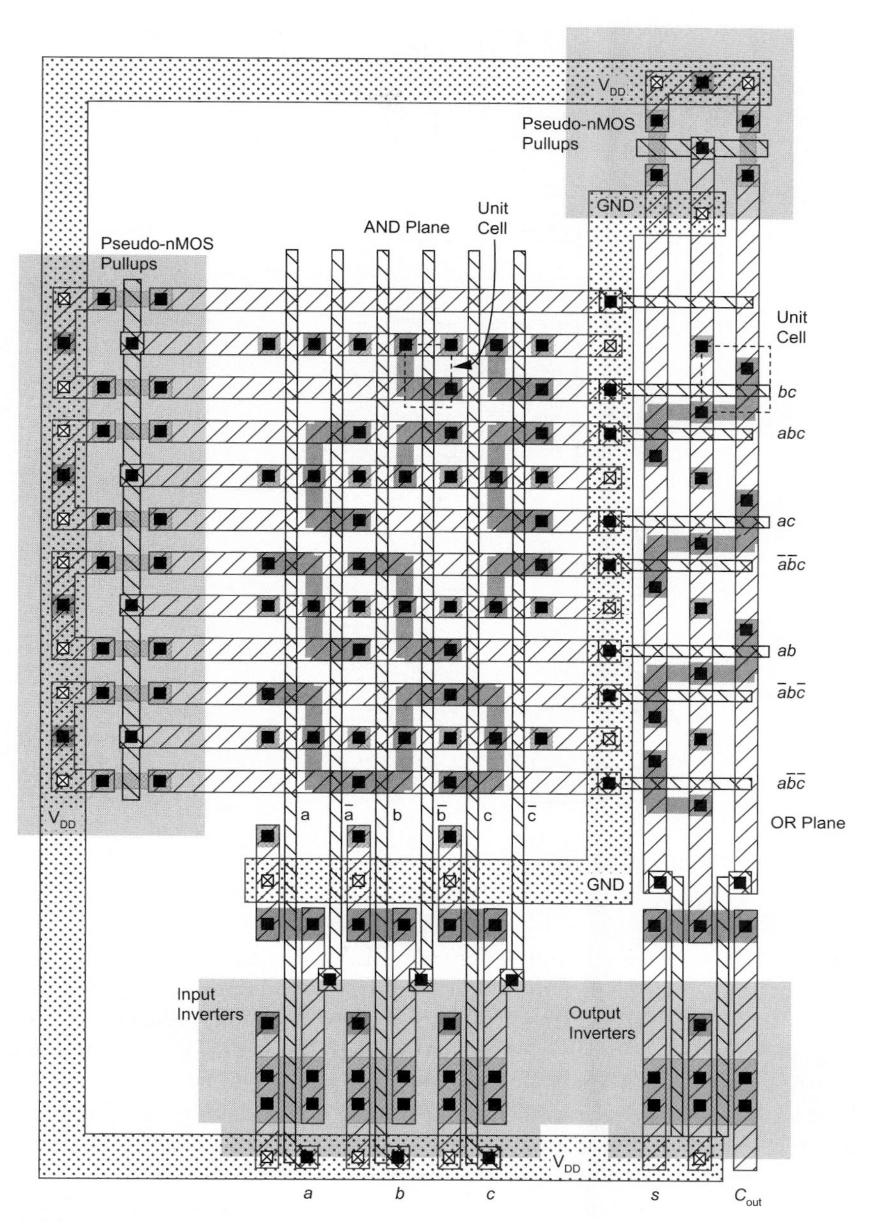

FIGURE 11.77 Pseudo-nMOS PLA layout

the AND and OR planes. Unfortunately, the AND plane must drive the OR plane directly, violating monotonicity. The OR plane must take a clock phase that is delayed until the minterms adequately discharge (to below V_t). This clock is often generated with a replica delay line that is guaranteed to be no faster than the slowest minterm in the AND plane. Moreover, the OR plane outputs must be captured before the AND plane precharges so that the results are not corrupted. To accomplish this, the PLA may be supplied by clocks similar to those shown in Figure 11.78(b). The dynamic power is high because the activity factor on the heavily loaded minterm lines is close to 1.

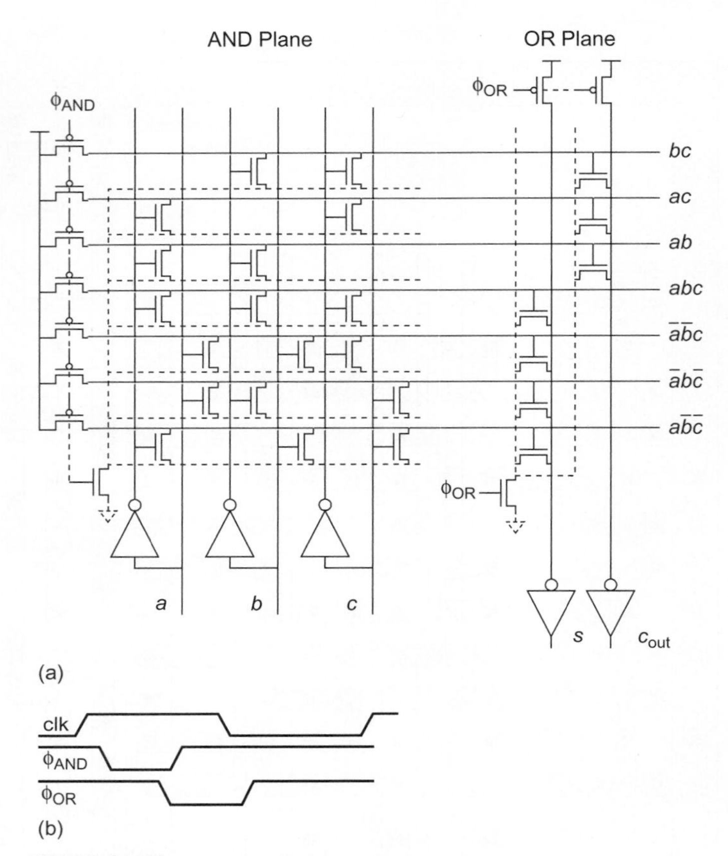

FIGURE 11.78 Dynamic PLA schematic

Figure 11.79 shows a self-timed dynamic PLA using two dummy rows as replica delay lines. Assume that the inputs arrive from flip-flops and settle shortly after the rising edge of the clock. The clocked circuitry acts as a pulse generator, producing a low-going precharge pulse on ϕ_{AND} shortly after the clock edge. The width of the pulse is equal to the delay of dummy AND row 1 plus two inverters and should be great enough to fully precharge all of the real AND rows. Thus, the loading on the dummy AND row is chosen to equal or exceed the worst loading of any real row. This worst loading consists of one nMOS drain for each input and one gate for each output. In this figure, the size of the inverter loading the AND line can be selected to contribute the desired gate load. Once the AND plane enters evaluation, the second dummy AND row starts to discharge through a single transistor. Again, this row is loaded to equal or exceed the delay of the worst real AND row. The three inverters provide some self-timing margin to ensure that ϕ_{OR} will not rise until the AND plane has fully evaluated. The output of the OR plane can be sampled into flip-flops on the next rising edge of the clock.

[Wang01] surveys a variety of other PLA designs. [Samson09] describes a NAND-NOR architecture in which the AND plane is constructed with domino AND gates. This approach is monotonic and thus avoids the race condition. However, performance degrades when the number of series transistors becomes large.

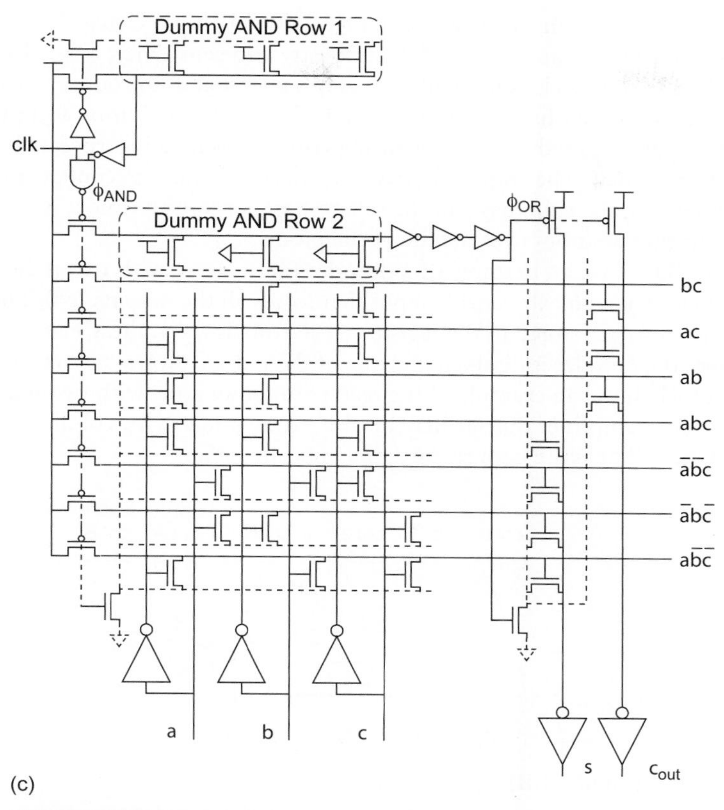

FIGURE 11.79 Dynamic PLA schematic

11.8 Robust Memory Design

Because arrays occupy a large fraction of the die area of many system-on-chip and microprocessor designs, they strongly influence the overall chip yield and reliability. Fortunately, their regular structure makes it easy to enhance the design for better yield and reliability. Redundant rows, columns, and even subarrays are used to fix defective memories. Error correcting codes are used to correct soft errors. Radiation-hardened cells reduce the soft error rate. This section also examines wearout mechanisms.

11.8.1 Redundancy

A single defect in logic circuits will usually render the entire chip useless. Memory yield is improved by providing spare parts that can replace defective elements. Each subarray is equipped with extra rows and columns to fix bad cells. Extra subarrays can be used to replace subarrays that are beyond repair. Alternatively, if the exact memory capacity is unimportant, the defective subarrays can be disabled and the chip can be sold anyway. The challenge in redundancy is to minimize the overhead of the replacement logic.

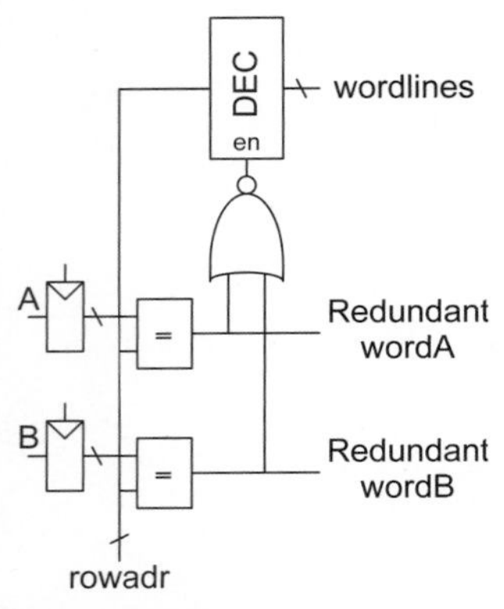

FIGURE 11.80
Row redundancy

Spare rows and columns date back to 64 Kb DRAM chips [Cenker79]. The number of spares depends on the anticipated defect density and sensitivity; a small number can make a big difference in yield. Originally, the chip was tested in the factory to identify bad cells, then a laser zapped links to disable the bad rows and columns and program the address of the spares. To reduce manufacturing cost, arrays now incorporate built-in self-test (see Section 14.6.3). The chip itself may blow electronic fuses to configure the replacements. Alternatively, the chip may run the self-test sequence each time it is reset so that it can detect cells that wear out over the life of the product.

Figure 11.80 shows an example of a decoder for an array with two redundant rows. The row decoder takes an extra enable input that forces all the outputs to 0. The addresses of the defective rows are stored in the registers at startup. If the address matches one of the defective rows, the decoder is disabled and one of the redundant rows is activated instead.

Figure 11.81 shows an example of the read path for an array with two redundant columns. Each sense amplifier output may be shifted by one or two columns to skip over defective columns. The write path requires the opposite logic.

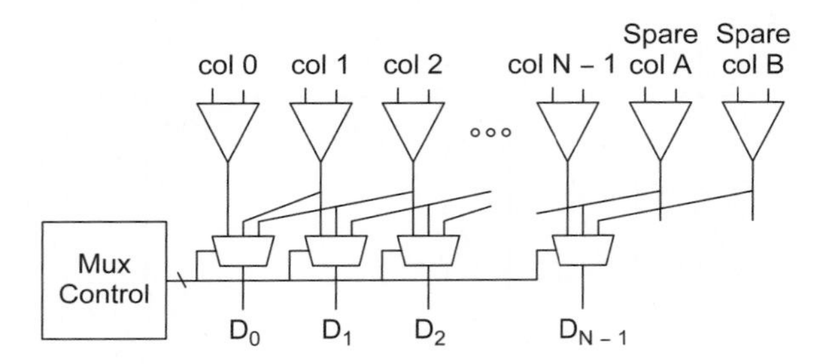

FIGURE 11.81 Column redundancy

Example 11.6

Example 11.1 considered a 64 Mb SRAM assuming read margins were normally distributed with a standard deviation of 15 mV. To achieve 90% parametric yield in the absence of repair mechanisms, the mean read margin had to exceed 6 standard deviations (90 mV). Assume that the array is divided into $S = 2048$ subarrays of $M = 32$ kb each. How does the result improve in each of the following scenarios?

(a) The array can repair two defective cells per subarray.
(b) The array can replace two defective subarrays.

SOLUTION: (a) Using EQ (6.21), the probability of a subarray failing must be less than

$$X_{\text{sub}} = 1 - \sqrt[S]{0.9} = 5.1 \times 10^{-5} \tag{11.10}$$

According to EQ (6.30), if a cell has a probability X_c of failure, the probability that a subarray has more than two failures and is unrepairable is

$$X_{\text{sub}} = 1 - \left[(1 - X_c)^M + M(1 - X_c)^{M-1} X_c + \frac{M(M-1)}{2}(1 - X_c)^{M-2} X_c^2 \right] \tag{11.11}$$

Solving numerically finds $X_c = 2.1 \times 10^{-6}$ to achieve the required level of subarray reliability. By interpolating Table 6.8 or solving the CDF in EQ (6.17), we find this corresponds to 4.6 standard deviations or 69 mV of read margin.

(b) It is arguably more convenient to work in terms of yield $Y = 1 - X$. According to EQ (6.30), the array yield is

$$Y = Y_{\text{sub}}^{S} + SY_{\text{sub}}^{S-1}\left(1 - Y_{\text{sub}}\right) + \frac{S(S-1)}{2} Y_{\text{sub}}^{S-2}\left(1 - Y_{\text{sub}}\right)^{2} \quad \textbf{(11.12)}$$

Solving numerically for $Y = 0.9$ gives $Y_{\text{sub}} = 1 - 5.4 \times 10^{-4}$. The cell yield must be

$$Y_c = \sqrt[M]{Y_{\text{sub}}} = 1 - 1.6 \times 10^{-8} \quad \textbf{(11.13)}$$

This corresponds to 5.6 standard deviations or 84 mV of read margin.

Photolithography and etch problems occur most often along the perimeter of large repetitive structures. Memory yield improves if a *dummy* row and column is placed on each edge. These dummy cells are never activated. For example, the Sun Niagra processor used spares to repair large caches and dummies to improve the yield on unrepairable register files [Leon07]. Similarly, a NAND Flash string adds a dummy bit at each end of the string [Trinh09].

Replacing defective subarrays requires remapping addresses to subarrays. This is easiest in associative structures where a level of address indirection is already built into the system. For example, the Itanium 2 contained 9 MB of memory organized as an 18-way set associative L3 cache [Chang05]. The cache provides six spare 48 kB subarrays for repairs. If this is insufficient to fix a problem, one or more defective ways can be disabled with a fuse bit. A die with at least 12 functional ways can be sold as a product with a 6 MB cache.

NAND Flash memories also tolerate high defect levels to minimize cell size. If a block has too many defects to fix, it is marked as bad. The Flash memory controller performs a mapping of logical addresses to physical blocks in much the same way as a hard disk controller. The controller simply avoids using bad blocks. Blocks that wear out because of their finite endurance are added to the bad block list. Good controllers also perform *wear-leveling* by shuffling the mapping each time a block is erased so no block sees an unusually high number of program-erase cycles.

11.8.2 Error Correcting Codes (ECC)

RAMs are prone to soft errors that spontaneously flip a bit stored in one of the cells, as discussed in Section 6.3.4. Scaling trends are increasing the soft error rate because of the smaller charge on the cell and the larger number of bits on a chip. Soft errors also increase if the power supply is lowered to reduce leakage during sleep mode [Degalahal05].

Error correcting codes (ECC) are commonly used to recover from soft errors, as discussed in Section 10.7.2. For example, adding 8 check bits to a 64-bit word in a memory is sufficient to correct any error in the word and detect any pair of errors. ECC supplements redundancy to dramatically improve yield as well. ECC increases the delay and area of a memory, so it is best suited to large memories where the delay and area are already large.

11.8.3 Radiation Hardening

Radiation hardening is used to reduce soft error susceptibility in aviation and space applications when the flux of radiation is much higher and in high-reliability terrestrial applications such as mainframes or medical devices. The same dual-interlocked cell (DICE)

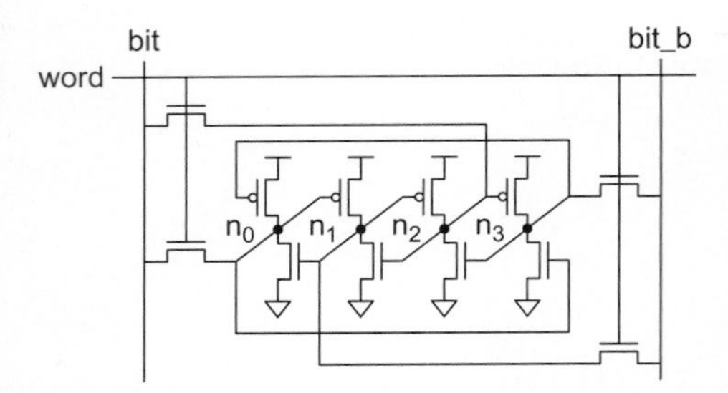

FIGURE 11.82 Radiation-hardened SRAM cell

technique used in registers in Section 9.3.10 also works for SRAM cells [Calin96]. Figure 11.82 shows such a 12T radiation-hardened cell. An upset on any single node is corrected by the feedback. The 12T cell has approximately twice the area of an ordinary 6T cell.

A particle strike may disturb not only the node it hits, but also adjacent nodes due to the parasitic bipolar effect. This can flip bits in adjacent SRAM cells. While ECC is effective at correcting a single error per word, complicated and lengthy codes are required to correct multiple bit errors. The number of adjacent cells that can be affected depends on the layout. Upsets to cells in adjacent rows does not matter because the other rows are independently protected by ECC. Column multiplexing is an effective way to protect cells in the same row because it spreads out the cells that represent a word. For example, a memory may store four 64-bit words (plus 8-bit ECC for each word) in a 288-bit row using 4:1 column multiplexing. Every fourth bit belongs to the same word. Hence, a strike that impacts four adjacent bits in the same row only corrupts one bit in each of the four words and is correctable. In the thin-cell layout, the n-well provides isolation so strikes rarely disturb more than two adjacent cells in the same row. Hence, using 2:1 column multiplexing provides effective resistance to uncorrectable multibit errors [Osada06].

11.9 Historical Perspective

MOS memory made a splash in 1970 when Intel announced sales of the first 1103 1 kb DRAM chip and IBM replaced magnetic core memories with semiconductor memories in its 370-series mainframe computers. Since then, DRAM has become a commodity business characterized by ferocious price competition (and occasional price fixing) among a rather small number of manufacturers. Indeed, in 1986, Intel left what was then its core business when the market was flooded by cheap chips during a cyclical downturn. DRAM capacity per chip has increased by 60% per year and cost per bit has decreased by 27% per year. Feature size improvement accounts for part but not all of the capacity gains. The area per bit has shrunk faster than feature size because of clever cell designs such as the 1T DRAM cell, innovative layout, and three-dimensional capacitor structures. Larger dice have become economical because of manufacturing yield improvements. Growing DRAM capacity has benefited system designers as much as the advances in processor performance.

DRAM density quadrupled approximately every three years for the first three decades of its development. More recently, the trend has slowed toward doubling roughly every three years. Table 11.1 lists some of the innovations at each DRAM generation [Itoh01k, Isaac08]. Early DRAMs were built in nMOS processes requiring high supply voltages. V_{DD} standardized at 5 V through the 1980s and 1990s and CMOS peripheral circuitry was eventually adopted to save power. Other improvements addressed the signal-to-noise ratio, bandwidth and latency, power consumption, and test time.

TABLE 11.1 DRAM generations

Capacity	Years of Volume Shipment	Power Supply (V)	Memory Cell	Circuit Innovations
1 kb	1970s	> 12	3T or 4T	MOS technology Differential sensing
4 kb	1970s	> 12	3T or 4T	Multiplexed addresses
16 kb	1978–1984	12	1T	Dynamic amplifier Dynamic driver
64 kb	1981–1987	5	1T	Folded bitline Word bootstrapping Substrate bias generator
256 kb	1984–1992	5	1T	Shared amplifier Metal-strapped wordline Redundancy
1 Mb	1987–1997	5	1T	CMOS peripheral circuits Half-V_{DD} precharge Multidivided data line BIST
4 Mb	1991–2000	5	1T	3-D capacitor structure
16 Mb	1994–2003	5	1T	On-chip voltage converter Twisted bitlines
64 Mb	1997–2006	3.3	1T	Synchronous small-signal I/O Multidivided wordlines
256 Mb	2001–	1.8–3.3	1T	Double data rate interface
512 Mb	2004–	2.5	1T	DDR2 interface
1 Gb	2007–	1.25–1.8	1T	DDR3 interface
2 Gb	2010–	1.2–1.5	1T	

Summary

Arrays repeat a basic cell in two dimensions. The cell is carefully optimized to provide very high density. For performance or density reasons, the nodes within the array do not always swing from rail to rail. Periphery circuitry restores the output swings to full digital logic levels.

The static RAM is very widely used in CMOS systems. The ubiquitous 6T cell consists of a cross-coupled inverter pair to hold the state and two access transistors for differential reads and writes. The bitlines are first preconditioned to a known value. A decoder asserts one of the wordlines. That word is read onto the bitlines and sensed. A column multiplexer may select only a subset of the bits as outputs. SRAMs are used in caches and other embedded memories. Multiported SRAMs are used in register files.

Content-addressable memories are similar to SRAMs. However, they also provide a lookup mode in which a key is placed on the bitlines and each word that contains that key asserts its matchline. CAMs are important for looking up addresses in translation look-aside buffers and network routers.

Dynamic RAMs store information on a capacitor using a single access transistor. With specialized process steps to build compact capacitors, they offer an order of magnitude higher density of data storage than SRAM. However, the data gradually leaks off the capacitors, so DRAMs must be periodically refreshed to maintain their state. DRAMs are usually built in specialized processes on dedicated chips, but potentially may be useful for high-capacity embedded memories on digital CMOS processes.

Read-only memories also use a single access transistor, but their contents are wired to a constant value. They are commonly used to store code and are convenient because they can be easily changed late in the design process to correct bugs or add features. Flash memories are electronically programmable and erasable and provide extremely high storage density.

A ROM can also be viewed as a lookup table. In general, a ROM of 2^x words by y bits can serve as a lookup table to perform any function of x inputs and y outputs. If a function is written in sum-of-products form, the ROM decoder performs the AND operation while the ROM array performs the OR. Many functions are relatively sparse. A programmable logic array optimizes out the unnecessary entries by replacing the decoder with an AND plane. In some cases, PLAs are smaller than ROMs, yet provide the same flexibility of easy changes late in the design cycle. PLAs were commonly used for microcoded finite state machines in the 1980s. They are still occasionally used, but good logic synthesis tools now deliver the same ease of change for random logic while avoiding the complicated circuit design needed for an efficient PLA.

A good design flow should provide automatic generators for simple SRAMs and ROMs. The designer should be comfortable with using these arrays where they are appropriate. High-performance designs need more elaborate multiported SRAM, large memory arrays, and CAMs. Most of these arrays demand skilled circuit design and thorough simulation.

Exercises

11.1 Sketch a schematic for an 8-word × 2-bit NAND ROM that serves as a lookup table to implement a full adder.

11.2 Explain the advantages and disadvantages of NAND ROMs as compared to NOR ROMs.

11.3 Develop a model for the read time of a ROM with 2^n rows and 2^m columns analogous to that of the SRAM from Section 11.2.6. Assume the wire capacitance in the ROM array is negligible compared to the gate and diffusion capacitance. Assume the ROM cells are laid out such that two cells share a single diffusion contact and hence each contributes only $C/2$ of diffusion capacitance.

11.4 Estimate the dimensions of the SRAM array in Exercise 11.13 using a 1.3 × 1.44 μm SRAM cell, assuming periphery circuitry adds 10% to each dimension of the core.

11.5 Sketch designs for a 6:64 decoder with and without predecoding. Comment on the pros and cons of predecoding.

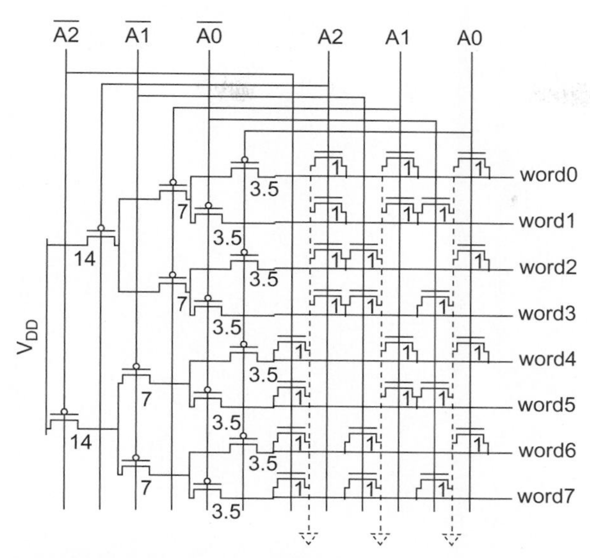

FIGURE 11.83 Lyon-Schediwy decoder

11.6 Figure 11.83 shows a 3:8 decoder [Lyon87]. How does the logical effort of each input compare to an ordinary decoder made of 3-input NORs? Does the decoder have any performance drawbacks?

11.7 Estimate the minimum delay of a 10:1024 decoder driving an electrical effort of $H = 20$ using

(a) static CMOS gates
(b) footless domino gates

11.8 Design the footless domino decoder from Exercise 11.7(b) using self-resetting domino gates. Assume the inputs are available in true and complementary form as pulses with a duration of 3 FO4 inverters and can each drive 48 λ of gate width. Indicate transistor sizes and estimate the delay of the decoder.

11.9 Develop a model of wordline decoder delay for a RAM with 2^n rows and 2^m columns. Assume true and complementary inputs are available and that the input capacitance equals the capacitance of one of the columns so $H = 2^m$. Use static CMOS gates and express your result in terms of n and m.

11.10 Explain the trade-offs between open, closed, and twisted bitlines in a dynamic RAM array.

11.11 Sketch a dot diagram for a 2-input XOR using a ROM.

11.12 Sketch a dot diagram for a 2-input XOR using a PLA.

11.13 An embedded SRAM contains 2048 8-bit words. If it is physically arranged in a square fashion, how many inputs does each column multiplexer require?

Packaging, Power, Clock, I/O

12

Coauthored by Dr. Jaeha Kim

12.1 Introduction

This chapter describes a variety of special-purpose subsystems that a digital designer may encounter. These subsystems are usually designed by a specialist or obtained from a third-party vendor, and each is the subject of entire books. However, the skilled digital designer should be conversant in each area in order to understand the impact of the other subsystems on a core digital design.

The chapter begins with packaging because the package strongly influences other elements of the system. It continues with the power and clock distribution subsystems. Phase-locked loops (PLLs) receive special attention because they are critical to high performance systems. Input/Output (I/O) subsystems connect the chip to the package to receive power, clock, and data. The chapter concludes with a handful of random circuits.

12.2 Packaging and Cooling

The chip package provides a mechanical and electrical connection between the chip and a circuit board. It is no longer possible to separate the design of a high-performance integrated circuit from the design of its package. An ideal package has the following properties:

- Connects signals and power between the chip and board with little delay or distortion
- Removes heat produced by the chip
- Protects the chip from mechanical damage and thermal expansion stress
- Is inexpensive to manufacture and test

To provide good signal and power connections, the package must offer short wires with low resistance and inductance. The impacts of the package on the power supply and I/O are discussed further in Sections 12.3 and 12.6, respectively. The remainder of this section describes some of the types of packages commonly available and how they remove heat from the chip.

12.2.1 Package Options

Table 12.1 lists a variety of common integrated circuit packages. Figure 12.1 shows photographs of these packages. The I/O count includes connections for both signals and power.

TABLE 12.1 Package options

Package	# I/Os	Description
Dual Inline Package (DIP)	8, 14, 16, 20, 28, 40, 64	Two rows of through-hole pins on 100 mil centers. Low cost. Long wires between chip and corner pins.
Pin Grid Array (PGA)	65–391+	Array of through-hole pins on 100 mil centers. Low thermal resistance and high pin counts.
Small Outline IC (SOIC)	8,10, 14, 16, 20, 24, 28	Two rows of SMT pins on 50 mil centers. Low cost, good for low-power parts with small pin counts.
Thin Small Outline Package (TSOP)	28–86+	Two rows of SMT pins on 0.5 or 0.8 mm centers in a thin package. Commonly used for DRAMs.
Plastic Leadless Chip Carrier (PLCC)	20, 28, 44, 68, 84	J-shaped SMT pins on all four sides on 50 mil centers. Sturdy leads are convenient for socketing.
Quad Flat Pack (QFP)	44–240	SMT pins on all four sides on 15.7–50 mil centers. High density of I/Os. Available in thin (TQFP) and very thin (VQFP) forms as thin as 1.6 mm.
Ball Grid Array (BGA)	49–2000+	Array of SMT solder balls on underside of package on 15.7–50 mil centers. Extremely high density of I/Os with low parasitics. Requires specialized assembly and inspection equipment to blindly attach to array of pads on printed circuit board.
Land Grid Array (LGA)	Many	Similar to BGA, but with gold-plated pads rather than solder balls. Connects to a socket or pads on the PCB.
Chip Scale Packaging (CSP)	Variable	SMT package no larger than 1.2× the die size. A common form of CSP is the flip-chip, which directly connects to a printed circuit board through solder balls on top metal layer of chip. Even higher I/O density and lower parasitics than BGA. Popular for mobile devices.

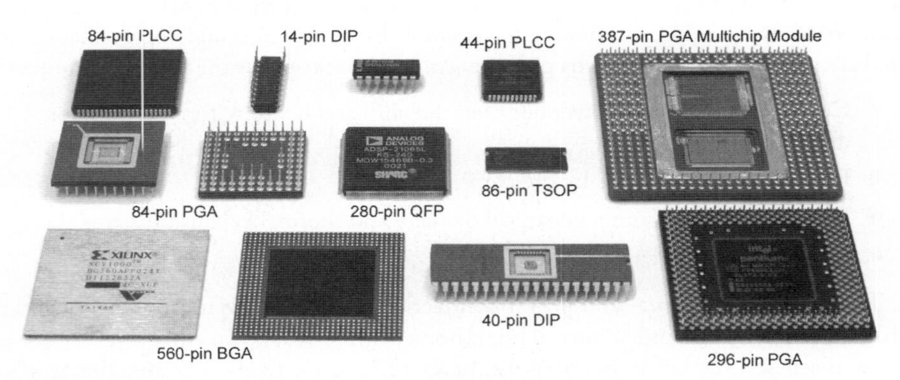

FIGURE 12.1 Integrated circuit packages (© 2003 Harvey Mudd College. Reprinted with permission.)

I/O spacing is typically specified in the archaic unit of mils (1 mil = 0.001 inch = 25.4 μm). Packages come in both ceramic and plastic varieties; plastic is cheaper, but cannot remove as much heat. Older DIP and PGA packages used *through-hole* pins, which pass through holes in a printed circuit board and are soldered from below. The pins contribute inductance, and the size of the holes limits the density of the pins. *Surface mount* (SMT) packages are soldered to the surface of a printed circuit board to alleviate these problems. DIP, PGA, PLCC, and LGA packages are easy to insert into low-cost sockets, so they are convenient for components that might be removed for reprogramming or replacement. Ball Grid Array (BGA) packages and their offshoots become the preferred approach for parts that require a large number of high-bandwidth signals in a compact form factor. Package design is a rapidly advancing field and new packages are being adopted each year.

12.2.2 Chip-to-Package Connections

Conventionally, chips have been connected to their packages through thin (25 μm) gold wires bonded to metal pads. The pads are organized into a ring around the periphery of the chip called a *pad frame*. The minimum pitch of the pads is limited by the bonding machine to approximately 100–200 μm. Thus, a 1-cm^2 chip is limited to several hundred I/Os. Chips with large numbers of I/Os sometimes are *pad-limited*, meaning that the chip size is determined by the pad frame rather than by the logic within the chip. Figure 1.63 showed an example of a pad-limited chip in a 40-pin pad frame. Some chips have used a second ring of pads, but this approach results in longer bond wires and greater risk that the wires will accidentally touch.

The bond wires connect to a metal lead frame in the package. This lead frame distributes the I/Os to the periphery of the package and is bent to form the pins of the package. Many packages also include a heat spreader to help distribute the heat from the die across the package and ultimately out to the heat sink. Figure 12.2 shows a cutaway of a dual-in-line package showing a corner of the chip with bond wires connecting to the lead frame [Mahalingam85]. The metal leads contribute parasitic inductance and coupling capacitance to their neighbors. More advanced packages internally resemble printed circuit boards, using multiple layers of signals and power/ground planes to distribute the I/Os on controlled-impedance transmission lines.

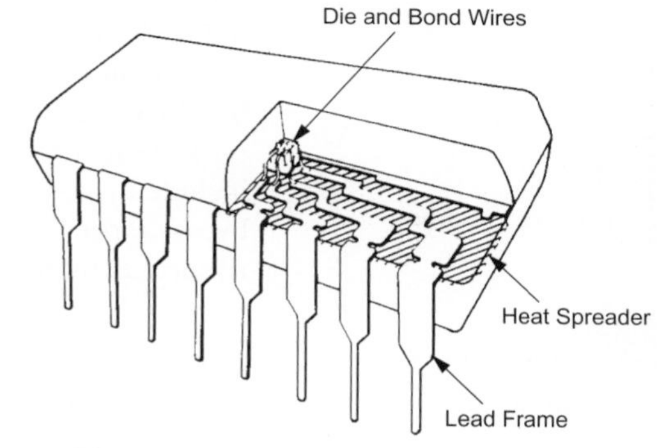

FIGURE 12.2 Cutaway view of dual-in-line package (© IEEE 1985.)

Since the late 1990s, many manufacturers have adopted *flip-chip* connections. This technology, also called *Controlled Collapse Chip Connection* (C4), was developed by IBM in the 1960s and has been used on their mainframes for decades. In a flip-chip design, the surface of the chip is covered with an array of pads on the top level of metal. Lead solder balls are bonded to these pads in a final process step called *wafer bumping*. The chip is flipped upside down and connected to the package by heating the balls until they melt. The bonding requires careful alignment, but surface tension from the solder helps pull the chip into place. The chip is in nearly direct contact with the package, eliminating the inductance associated with the bond wires. The bumps can be placed on a pitch of 150 μm or less, offering thousands of connections between the die and package. For example, a Xeon processor has 13,164 solder bumps, most of which are dedicated to power and ground to bring 120 A of current onto the chip [Rusu07]. Flip-chip technology introduces new testing problems because the top-level metal wires are no longer accessible for probing during debug.

Figure 12.3 shows a Core i7 microprocessor in an LGA package. The LGA substrate can be viewed as a small circuit board. The image on the left is a top view of the bare die flip-chip mounted onto the LGA substrate. Solder balls form connections between top-level metal pads on the die and matching pads on the substrate. The image in the middle shows the LGA after the integrated heat spreader has been attached. The heat spreader

FIGURE 12.3 LGA package (Courtesy of Intel Corporation.)

provides a good thermal path to the heat sink. The image on the right shows the bottom view. The substrate has 1366 gold-plated pads that connect to a socket on the motherboard. Notice the array of bypass capacitors in the center. These provide a low-inductance connection to the die on the opposite side.

12.2.3 Package Parasitics

Figure 12.4 shows a model of an integrated circuit package. The bond wires and lead frame contribute parasitic inductance to the signal traces. They also have some mutual inductive and capacitive coupling to nearby signal traces, potentially causing crosstalk when multiple signals switch. The V_{DD} and GND wires also have inductance from both bond wires and the lead frame. Moreover, they have nonzero resistance, which becomes important for chips drawing large supply current. High-performance packages often include bypass capacitors between V_{DD} and GND. As we will see in Section 12.3.5, the bypass capacitors have their own parasitic resistance and inductance that limit their effectiveness at high frequencies.

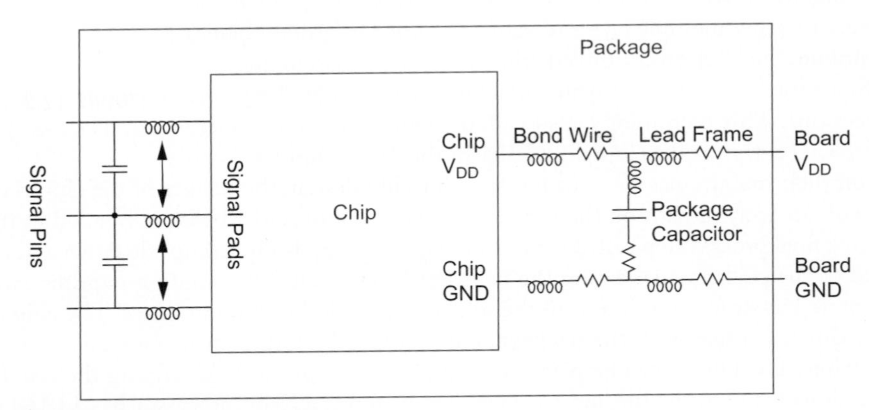

FIGURE 12.4 Package parasitics

12.2.4 Heat Dissipation

A 60-watt light bulb has a surface area of about 120 cm^2 and is too hot to touch. In comparison, a high-performance microprocessor dissipates 150 W on a 1.6 cm^2 die, resulting in a power density 180 times as great! Clearly, removing heat from chips is a major challenge for the package.

The heat generated by a chip flows from the transistor junctions where it is generated through the substrate and package. It can be spread across a heat sink, and then carried away through the air by means of convection. Just as current flow is determined by voltage difference and electrical resistance, the heat flow is determined by temperature difference and thermal resistance. Thus, the temperature difference ΔT between the transistor junctions and the ambient air is

$$\Delta T = \Theta_{ja} P \quad \textbf{(12.1)}$$

where Θ_{ja} is the thermal resistance (in °C/W) between the junction and ambient and P is the power consumption of the chip. The thermal resistance in turn can be modeled as the series resistance from the die to the package Θ_{jp} and from the package to the air Θ_{pa}.

$$\Theta_{ja} = \Theta_{jp} + \Theta_{pa} \quad \textbf{(12.2)}$$

For most low-cost packages, Θ_{pa} dominates the resistance. Still air can transfer about 0.001 W/(cm^2 °C) [Glasser85]. Thus, a package with a surface area of 10 cm^2 has a thermal resistance of about Θ_{pa} = 100 °C/W. Such a package cannot handle chips dissipating more than about 1 watt. Forced air transfers 0.01–0.03 W/(cm^2 °C). High-power chips add a large heat sink and a fan to the package to reduce the thermal resistance. For example, a 72-pin ceramic PGA package has a thermal resistance Θ_{pa} of 34 °C/W in still air, 18 °C/W in 400 feet/minute airflow, and 10 °C/W in 400 feet/minute airflow with a good heat sink. Liquid cooling is costly but highly effective, offering thermal resistance as low as 0.3 °C/W. MEMS microchannels and microfluidics offer the potential for extremely low thermal resistance cooling integrated directly into the die or package [Paik08].

Example 12.1

You are planning to package an ASIC in a ball grid array package with a passive heat sink. The system box contains a large fan providing 250 linear feet/minute (LFM) of airflow. The package vendor specs the thermal resistance from the junction to package at 0.9 °C/W. The heat sink vendor specs the thermal resistance from the package to ambient for this airflow at 4.0 °C/W for the heat sink plus 0.1 °C/W for the heat sink adhesive between the package and heat sink. The system box ambient temperature may reach 55 °C. What is the maximum power dissipation of your ASIC if its junction temperature is not to exceed 100 °C?

SOLUTION: The thermal resistance is Θ_{ja} = 0.9 + 0.1 + 4.0 = 5 °C/W. The temperature difference between the junction and ambient must not exceed ΔT = 100 − 55 = 45 °C. Therefore, the maximum power dissipation is $P = \Delta T / \Theta_{ja}$ = 9 W.

Advances in heat sinks, fans, and packages have raised the practical limit for heat removal from about 8 W in 1985 to about 130 W in 2008 for low-cost packaging. Forced-air cooling appears to be reaching its limits, setting a cap on the power consumption of chips.

12.2.5 Temperature Sensors

If a cooling fan motor fails or air intake vents become clogged, a chip may rapidly overheat to the point of self-destruction. Moreover, chips are normally designed to function correctly in the worst-case environment (e.g., 70 °C inside the system box), so they could operate at a higher power and performance level at room temperature. Most high-performance microprocessors now include one or more temperature sensors placed at hot spots on the die for adaptive control. Based on the temperature, the chip performs dynamic voltage scaling or throttles activity to avoid overheating [McGowen06, Pham06, Sakran07].

The most common method of sensing temperature on-chip is based on the relationship between absolute temperature T, collector current I_c, and base-emitter voltage V_{BE} for a bipolar transistor [Pertijs06]:

$$I_c = I_s e^{\frac{qV_{BE}}{kT}} \tag{12.3}$$

In this equation, q is the charge on an electron and k is the Boltzmann constant. I_s is a function of the transistor geometry and processing, and is also highly sensitive to temperature. Solving EQ (12.3) for V_{BE} gives

$$V_{BE} = \frac{kT}{q} \ln \frac{I_c}{I_s} \tag{12.4}$$

Unfortunately, this base-emitter voltage is a complex function of temperature because of the I_s dependence. However, the difference between base-emitter voltages of two identical transistors operating at different collector currents I_{c1} and I_{c2} eliminates the I_s term and becomes directly *proportional to absolute temperature* (PTAT).

$$\Delta V_{BE} = V_{BE1} - V_{BE2} = \frac{kT}{q}\left(\ln \frac{I_{c1}}{I_s} - \ln \frac{I_{c2}}{I_s} \right) = \frac{kT}{q}\left(\ln \frac{I_{c1}}{I_{c2}} \right) \tag{12.5}$$

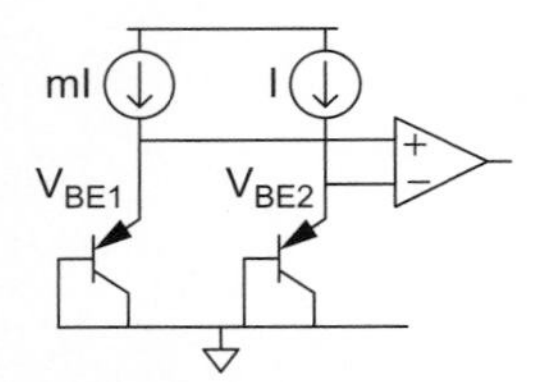

FIGURE 12.5
Temperature sensor

As shown in Section 15.4.3.5, an ordinary CMOS process contains a vertical pnp bipolar transistor formed by p-diffusion, an n-well, and the p-substrate. This structure is exploited to build temperature sensors without costly process modifications.[1] Figure 12.5 shows an implementation of a simple temperature sensor with a current ratio of m. The output voltage could be measured with an A/D converter to produce a digital representation of temperature or simply compared with a reference voltage to generate an over-temperature warning signal. The reference current I is typically on the order of 1 μA to avoid non idealities from low- or high-injection.

Example 12.2

Estimate the temperature coefficient of a temperature sensor if the collector current ratio is 10.

SOLUTION: The temperature coefficient is

$$\frac{\Delta V_{BE}}{T} = \frac{k}{q}\left(\ln \frac{I_{c1}}{I_{c2}} \right) = \frac{1.38 \times 10^{-23} \frac{J}{K}}{1.602 \times 10^{-19} C} \ln 10 = 198 \frac{\mu V}{K} \tag{12.6}$$

[1] A diode has a similar temperature dependence to the I-V characteristics and one might wonder why it couldn't be an even simpler sensor element. The trouble is that diodes suffer from recombination of electron-hole pairs in the depletion region, which introduces inaccuracies in the measurement.

In practice, the relationship of ΔV_{BE} to temperature is not perfectly linear, introducing measurement error. The accuracy is greatly improved by calibrating the sensor at a known temperature. Such calibration involves placing the chip or wafer in a thermal chamber and allowing time for temperature to equilibrate. The increased test time is expensive in a high-volume manufacturing environment. If thermal calibration is limited to 1 second, inaccuracies of about 0.5 °C can be achieved. Two-point calibration produces better results, but is impractically time-consuming.

12.3 Power Distribution

The power distribution subsystem of a chip consists of metal wires or planes on the chip, in the package, and on the printed circuit board. It also includes bypass capacitors to supply the instantaneous current requirements of the system. An ideal power distribution network has the following properties:

- Maintains a stable voltage with little noise
- Provides average and peak power demands
- Provides current return paths for signals
- Avoids wearout from electromigration and self-heating
- Consumes little chip area and wiring
- Easy to lay out

Real networks must balance these competing demands, meeting targets of noise and reliability as inexpensively as possible. The noise goal is typically ±10%; for example, a system with nominal V_{DD} = 1.0 V may guarantee the actual supply remains within 0.9–1.1 V. Reliability goals demand enough vias and metal cross-sectional area to carry the supply current, as was discussed in Section 6.3.3. The two fundamental sources of power supply noise are IR drops and L di/dt noise.

Figure 12.6 plots the power consumption versus time for a microprocessor [Gauthier02]. The power varies on a number of time scales. While the processor is active, the power depends on the operations and data. It also spikes near the clock edges when the large clock loads switch. When the processor becomes idle, clock gating turns off the

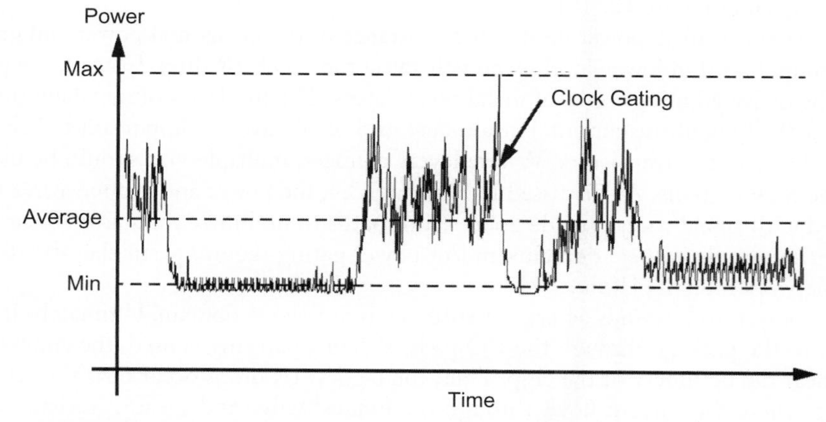

FIGURE 12.6 Time-dependent power consumption of microprocessor (Reprinted with permission of Sun Microsystems.)

clock to unused units, driving the power down significantly. As the supply voltage is nearly constant, the supply current I (also called I_{DD}) is proportional to the instantaneous power demand. As this current flows through the resistance R of the power distribution network, it causes a voltage droop proportional to IR. Moreover, as the changing current flows through the inductance of the printed circuit board and package, it also causes a voltage drop proportional to the rate of change: L di/dt.

This section begins by examining the physical design of a power distribution network. It then discusses IR drops and L di/dt noise. The key to controlling noise from current spikes is to provide adequate bypass capacitance on and off the chip to provide low supply impedance at all frequencies. The power network is complicated enough that manual analysis is inadequate; instead, it typically must be modeled in a finite element simulation. The power network also provides return paths for current flowing in signal wires. The geometry of the network affects the inductance of on-chip signals. Some critical circuits such as phase-locked loops and analog blocks require a quiet supply for good performance. RC filters can reduce much of the supply noise. In sensitive circuits, noise carried through the substrate is also important.

12.3.1 On-Chip Power Distribution Network

The on-chip power distribution network consists of power and ground wires within the cells and more wires connecting the cells together. Most cells contain internal power and ground busses routed on metal1 or metal2. These wires are typically wider than minimum to provide lower resistance and better electromigration immunity. For example, the cells on the inside front cover use 8 λ metal1 power/ground busses. These wires are normally connected between adjacent cells by abutment. Standard cell designs and datapaths both can use rows of cells sharing common power and ground lines.

In a small, low-power design, these rows can be strapped together with even wider vertical metal wires. Figure 12.7(a) shows an abstract diagram of this strapping. Figure 1.64 showed a standard cell design strapped with power on the left and ground on the right. In this example, the nMOS and pMOS transistors in adjacent rows are separated by a routing channel, so spacing between the wells is not a problem. In modern processes, the routing is typically done over the cell in upper-level metal. Therefore, the rows of cells can be packed more closely together and well spacing limits the packing density. Alternatively, every other row can be *mirrored* (flipped upside down) so that the wells of adjacent rows abut, as shown in Figure 12.7(b).

In a larger or high-power design, the resistance of the horizontal power and ground busses routed on thin lower-level metal will cause too much IR drop. Instead, the power should be delivered using a grid of metal on all layers. The top levels of metal are thickest and carry the bulk of the current, but a robust grid on all layers is important to bring the current down to the transistors. Where layers connect, multiple vias should be used to carry the high currents. As discussed in Section 5.3.4, the power and ground wires interdigitated with signal wires provide good return paths to minimize inductive effects. Systems with multiple voltage domains and/or power gating require particular attention to power network integrity [Kanno07].

The power grid extends across the entire chip or voltage domain. Ultimately, it must connect to the package through the I/O pads. When a pad ring is used, the connections are all near the periphery of the chip. Thus, the biggest IR drops occur near the center of the chip where the current flows through the longest wires and greatest resistance. C4 solder bumps distributed across the die are much better for power distribution because

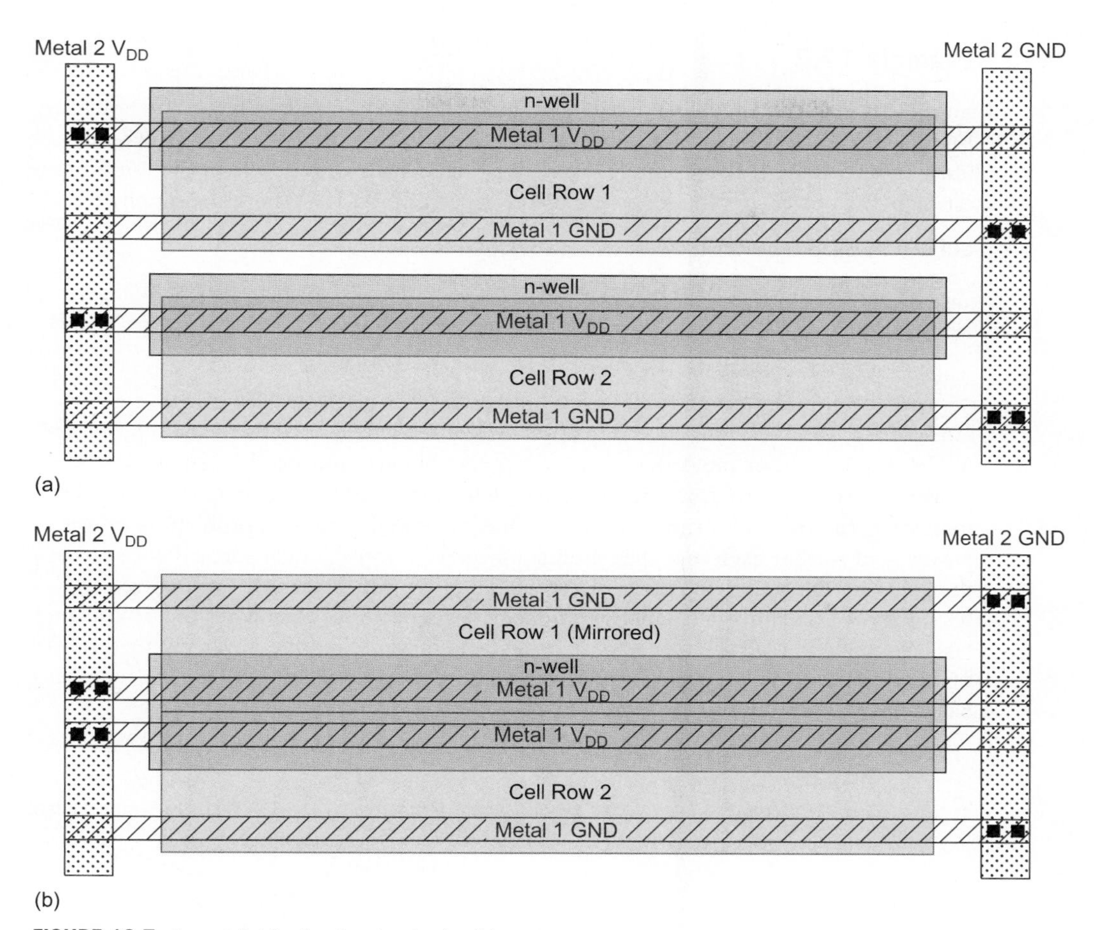

FIGURE 12.7 Power distribution for standard cell layout

they can deliver the current from the low-resistance power plane in the package directly to the area of the chip where the current is needed. Thus, less on-chip metal resources are needed for power distribution.

12.3.2 IR Drops

The resistance of the power supply network includes the resistance of the on-chip wires and vias, the resistance of the bond wires or solder bumps to the package, the resistance of the package planes or traces, and the resistance of the printed circuit board planes. Because the package and printed circuit board typically use copper that is much thicker and wider than on-chip wires, the on-chip network dominates the resistive drop.

IR drops arise from both average and instantaneous current requirements. The instantaneous current may be much larger than the average drop because current draw tends to locally spike near the clock edge when many registers and gates switch simultaneously. Bypass capacitance near the switching gates can supply much of this instantaneous current, so a well-bypassed power supply network only needs low enough resistance to deliver the average current demand, not necessarily the peak.

Example 12.3

Suppose a row of 64 repeaters share a common metal2 power bus like that shown in Figure 12.7(a). The bus is 320 μm long and 1 μm wide. The metal2 has a sheet resistance of 0.05 Ω/□. If the repeaters drive 0.4 pF wire loads with 200 ps transition times, estimate the power supply droop seen by the repeater for a 1.8 V nominal supply.

SOLUTION: Each repeater draws a current of approximately

$$I = C\frac{\Delta V}{\Delta t} = (0.4 \text{ pF})\frac{1.8 \text{ V}}{200 \text{ ps}} = 3.6 \text{ mA} \tag{12.7}$$

The power and ground busses each have a length of 320 squares and thus a resistance of $R = 16\ \Omega$. The supply droop at the end of the wire caused by the 64 repeaters is $64\ IR/2 = 1.85$ V, or more than V_{DD}, which is obviously impossible. Instead, as the power supply begins to droop, the repeaters deliver less current, reducing the droop, but increasing the transition time and delay. One way to alleviate this problem is to use a power grid so that each repeater obtains its current from its own vertical wire rather than sharing the single horizontal wire with all of the simultaneously switching neighbors. Figure 12.8 shows a simulation of one of the repeaters. It compares the two power bus layouts. When all the repeaters share a single power wire, the power supply droops by nearly 30% and the propagation delay is more than doubled. When each repeater has its own power wire so the supply noise is negligible, the output is crisper.

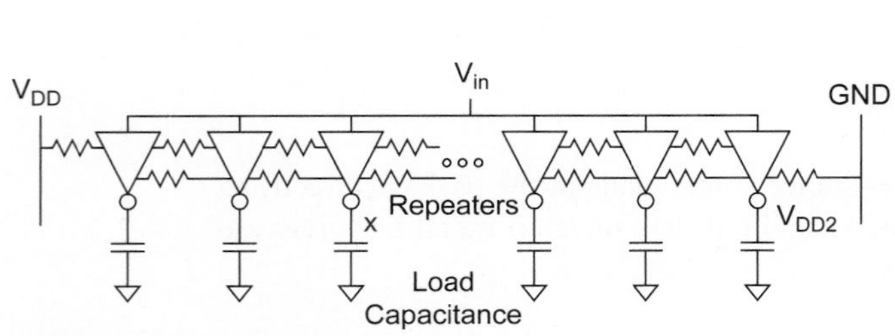

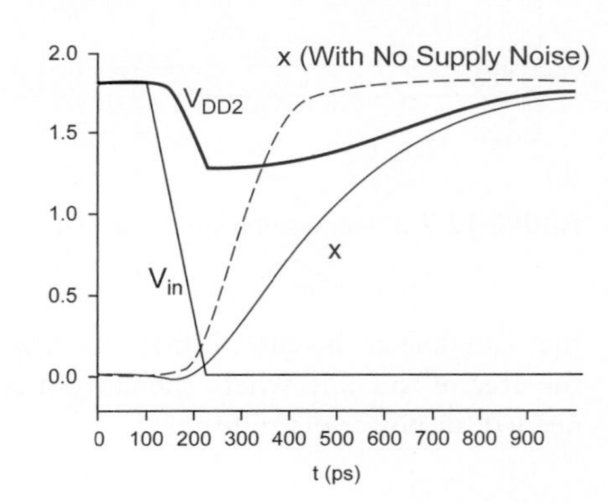

FIGURE 12.8 Power supply droop

12.3.3 L di/dt Noise

The inductance of the power supply is typically dominated by the inductance of the bond wires or C4 bumps connecting the die to the package. A typical bond wire has an inductance of about 1 nH/mm, while a C4 ball is on the order of 100 pH. Recall that the inductance of multiple inductors in parallel is reduced. Modern packages devote many (often 50% or more) of their pins or bumps to power and ground to minimize supply inductance. The two largest sources of current transients are switching I/O signals and changes between idle and active mode in the chip core.

Example 12.4

A 1 GHz chip transitions from idle to full power operation in a single cycle. The idle mode draws 20 A and the full power mode draws 60 A. If the power supply has 20 pH of series inductance, estimate the power supply noise caused by this transition if the chip has no internal bypass capacitance.

SOLUTION: The current transient is

$$\frac{\Delta I}{\Delta t} = \frac{(60\text{ A} - 20\text{ A})}{1\text{ ns}} = 40\text{ GA/s} \quad \textbf{(12.8)}$$

The inductive noise is $L\ \Delta I/\Delta t = 0.8$ V. This is clearly unacceptable in a low-voltage system. Once again, the chip needs internal bypass capacitance to supply the instantaneous current, reducing the transient seen by the I/O pins.

L di/dt noise is becoming enough of a problem that some high-power systems must resort to microarchitectural solutions that prevent the chip from transitioning between minimum and maximum power in a single cycle. For example, a pipeline may enter or exit idle mode one stage at a time rather than all at once to spread the current change over many cycles.

12.3.4 On-Chip Bypass Capacitance

As we have seen, chips need a substantial amount of capacitance between power and ground to provide the instantaneous current demands of the chip. This is called *bypass* or *decoupling* capacitance. The bypass capacitance is distributed across the chip so that a local spike in current can be supplied from nearby bypass capacitance rather than through the resistance of the overall power grid. It also greatly reduces the di/dt drawn from the package.

Example 12.5

How much bypass capacitance is needed to supply a sudden current spike of 40 A for 1 ns with no more than a 200 mV supply droop?

SOLUTION: We solve

$$I = C\frac{\Delta V}{\Delta t} \Rightarrow C = \frac{(40\text{ A})(1\text{ ns})}{0.2\text{ V}} = 200\text{ nF} \quad \textbf{(12.9)}$$

Fortunately, the inherent gate capacitance of quiescent transistors provides a significant amount of *symbiotic* bypass capacitance [Dally98]. For example, Figure 12.9 shows one inverter driving another. The gate-to-source capacitances of the load inverter are shown explicitly. When $A = 1$ and $B = 0$, $M1$ is ON, charging up C_{gs4}. Similarly, when $A = 0$ and $B = 1$, $M2$ is ON, charging up C_{gs3}. The charged capacitor stores energy that can be released to supply sudden current demands. At any given time, approximately half of the gate capacitance of any quiescent circuit behaves as symbiotic bypass capacitance. Moreover, because only a small fraction of the gates are likely to be switching at any given time, nearly half of the entire gate capacitance on the chip will serve as bypass capacitance.

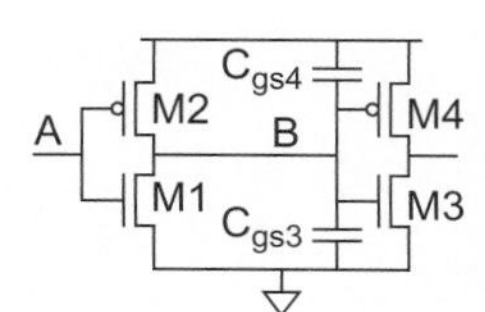

FIGURE 12.9 Symbiotic bypass capacitance

Example 12.6

Estimate the symbiotic bypass capacitance per square millimeter for a chip with feature size f if gate capacitance is 1 fF/μm of transistor width and transistor gates occupy 9% of chip area.

SOLUTION: The capacitance density of a 1 μm wide transistor of length f is $1/f$ fF/μm^2. At 9% utilization, this corresponds to $0.09/f$ nF/mm^2. Half of that, or $0.045/f$ nF/mm^2, serves as symbiotic bypass capacitance on average. In an $f = 65$ nm process, this means the symbiotic bypass capacitance is approximately 0.7 nF/mm^2.

In most low- and medium-power chips, this symbiotic capacitance provides adequate bypassing to filter instantaneous IR drops and L di/dt noise. In high-power chips, additional explicit capacitance is necessary. For example, the Sun Niagra2 processor added 700 nF of on-chip decoupling capacitance [Nawathe08]. The only dielectric available in a standard CMOS process to build compact high-capacitance structures is gate oxide, so the extra bypass capacitance is commonly built with an nMOS transistor with the gate tied to V_{DD} and the source and drain tied to GND. Decoupling capacitor layout should maximize the capacitance per unit area. [Meng08] describes bypass capacitor layout techniques.

In some nanometer processes, gate leakage is significant for thin-oxide transistors. Thicker-oxide transistors may be preferable to save leakage at the cost of lower capacitance density. Sun used thick-oxide transistors in the Rock processor with a 20% loss in capacitance density [Konstadinidis09].

12.3.5 Power Network Modeling

Figure 12.10 shows a lumped model of the power distribution network for a system including the voltage regulator, the printed circuit board planes, the package, and the chip. The network also includes bypass capacitors near the voltage regulator, near the chip package, possibly inside the chip package, and definitely on chip. The external capacitors are modeled as an ideal capacitor with an effective series resistance (ESR) and effective series inductance (ESL) representing the parasitics of the capacitor package. Larger capacitors have bigger effective series inductances.

The voltage regulator seeks to produce a constant output voltage independent of the load current. It is modeled as an ideal voltage source in series with a small resistance and the inductance of its pins. Near the regulator is a large bulk capacitor (typically electrolytic or tantalum). Power and ground planes on the printed circuit board carry the supply

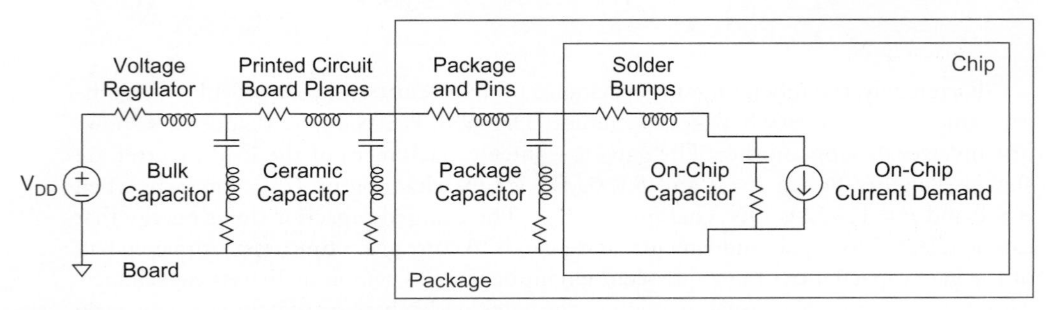

FIGURE 12.10 Power distribution system model

current to the package, contributing some resistance and inductance. Typically, the board designer places several small ceramic capacitors near the package. The package and its pins again contribute resistance and inductance. High-frequency packages often contain small capacitors inside the package for further decoupling. Finally, the chip connects to the package through solder bumps or bond wires with additional resistance and inductance. The dynamic and static current demands of the chip are modeled as a variable current source with a waveform that might resemble Figure 12.6. The on-chip bypass capacitance consists of the symbiotic capacitance and possibly some explicit decoupling capacitance. It typically has negligible inductance because it is located so close to the switching loads.

As one moves from the chip toward the voltage regulator, each capacitor typically increases by about an order of magnitude. However, each series inductance increases by a similar amount. [Budnik06] illustrates a representative power delivery network for a high-performance 90 nm microprocessor. The capacitance is on the order of 1 μF on the die, 10's of μF in the package, and 100's of μF on the board, and 1 mF at the voltage regulator. The inductance is on the order of 1 pH between the die and package, 10 pH between the package and board, and 100 pH along the board to the voltage regulator. The resistance is a fraction of an mΩ at each link.

12.3.5.1 Power Supply Impedance A good power distribution network should offer a low impedance at all frequencies of interest so that the supply voltage remains steady independent of the changing chip current demands. If the system draws P watts of power and the maximum allowable power supply ripple is $r \times V_{DD}$ (e.g., $r = 0.1$ for 10% supply noise), then the supply impedance must be less than

$$Z = \frac{rV_{DD}^2}{P} \tag{12.10}$$

This relationship shows that required supply impedance is dropping quadratically with voltage scaling. It is also dropping as power consumption increases. This impedance requirement has driven the adoption of improved packages and flip-chip bonding with solder bumps instead of bond wires. It means chips need to use more metal and on-chip bypass capacitance. For example, a 1.0 V system dissipating 100 W of power draws 100 A. To keep supply noise down to 10% of V_{DD}, the power supply impedance must be 1 mΩ.

If the system had no bypass capacitance, the distribution network would consist of only the resistance and inductance, so it would have an impedance of $Z = R + j\omega L$. This impedance increases with frequency ω and becomes unacceptably high for most systems by about 1 MHz.

The bypass capacitors in parallel with the supply provide an alternative low-impedance path at higher frequencies. An ideal capacitor has impedance that decreases with frequency as $Z = 1/j\omega C$. Unfortunately, the effective series inductance of the capacitors limits the useful frequency range of the real capacitor. The impedance of a capacitor C with effective series resistance R and inductance L is

$$Z = \frac{1}{j\omega C} + R + j\omega L \tag{12.11}$$

This impedance has a minimum of $Z = R$ at the self-resonant frequency of

$$f_{\text{resonant}} = \frac{\omega_{\text{resonant}}}{2\pi} = \frac{1}{2\pi\sqrt{LC}} \tag{12.12}$$

Figure 12.11 plots the magnitude of the impedance of a 1 μF capacitor with 0.25 nH of series inductance and 0.03 Ω of series resistance. The capacitor has low impedance near its resonant frequency of 10 MHz, but higher impedance elsewhere.

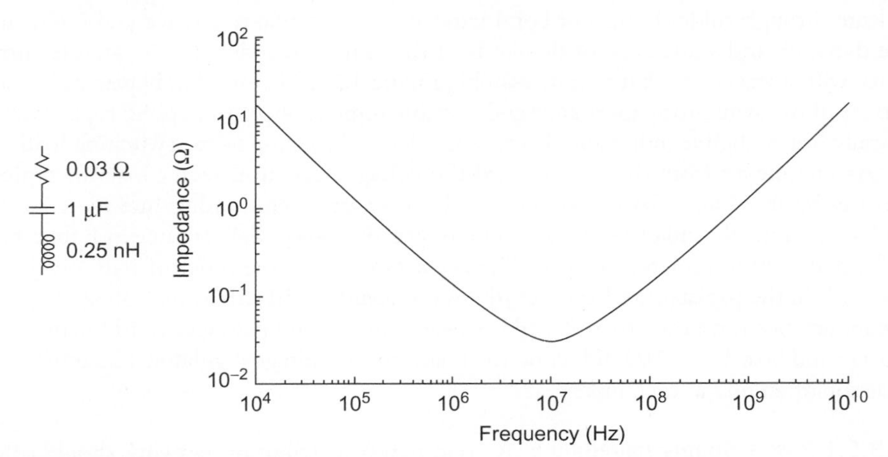

FIGURE 12.11 Impedance of bypass capacitor

Larger capacitors tend to have higher effective series inductances and therefore have lower self-resonant frequencies beyond which they are not useful. Thus, the system uses many capacitors of different sizes to provide low impedance over all the frequencies of interest. Also, capacitors closer to the chip are more useful at high frequencies because they have less inductance in the board and package between them and the chip. The bulk and ceramic capacitors are most effective over the 1–10 MHz range. Capacitors in the package tend to be useful in the 10–200 MHz range. Above a few hundred MHz, the inductance of the solder bumps or bond wires renders all but the on-chip decoupling capacitors ineffective.

Figure 12.12 shows the simulated impedance of the Pentium 4 power distribution network as a function of frequency, illustrating the resonances caused by the package, socket, board, and regulator [Xu08]. Note the large increase in impedance near 100 MHz caused by the package.

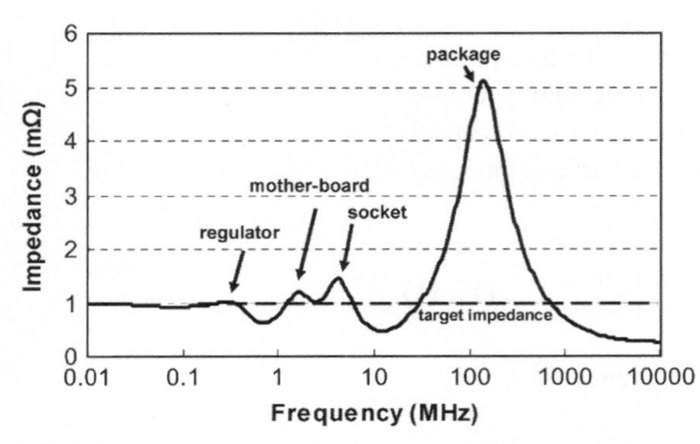

FIGURE 12.12 Pentium 4 simulated power supply impedance (© 2008 IEEE.)

12.3.5.2 Power Supply Step Response Another way to think about the need for nearby bypass capacitance is to imagine a sudden step in current on the chip. Some round-trip propagation delay must occur before the spike reaches the power supply, the supply adjusts the current it is delivering, and that current returns to the chip. A lower bound on this delay is the speed of light. Therefore, when a gate switches, the voltage regulator will not know about the event until sometime after the transition has completed. In the meantime, the charge must be drawn from the bypass capacitors. This results in a series of droops as each capacitor becomes depleted before the next one kicks in. Yet another perspective is to remember that an inductor does not like to change its current instantaneously. Thus, larger inductors introduce a longer lag.

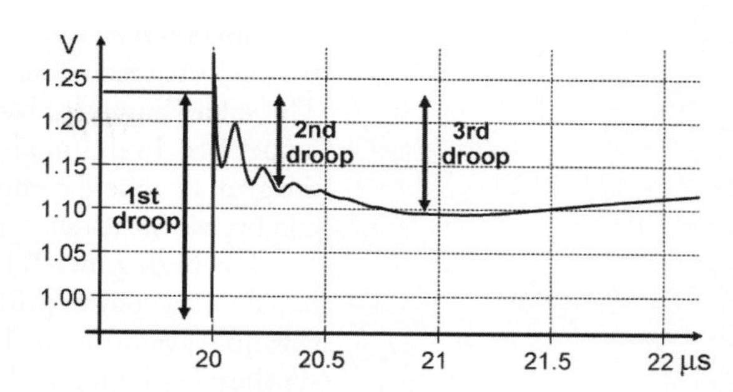

FIGURE 12.13 Pentium 4 simulated voltage droops (© 2006 IEEE.)

Figure 12.13 shows the simulated response to an abrupt increase in current demand on the Pentium 4 illustrating a sequence of three droops characteristic of power distribution networks [Wong06]. Before the step, the voltage regulator delivers some amount of current sufficient to meet the average needs of the chip. When the current demand suddenly increases, the extra charge is initially drawn out of the on-chip bypass capacitors. As these capacitors discharge, the supply voltage drops precipitously. This is called the *first droop*. Soon the current through the solder bumps increases to recharge the on-chip capacitors. The delay depends on the inductance of the bumps. Moreover, this inductance may cause the supply voltage to overshoot and oscillate. Meanwhile, the capacitors in the package supplying this current start to discharge and the voltage droops again. This *second droop* occurs on a longer time scale determined by the package capacitance. Eventually, the current through the package pins and socket increases to begin recharging the package capacitors. Meanwhile, the capacitors on the printed circuit board discharge, leading to a *third droop* before the voltage regulator catches up with the increased current demand. The second and third droops are minimized by providing an adequate number of high-quality, low ESL capacitors at each stage in the power distribution network [Smith99].

Designers typically assume that adding on-chip bypass capacitance to reduce supply droop improves operating frequency. While more capacitance certainly does reduce the droop, the frequency does not necessarily improve. In a striking experiment, [Wong06] fabricated several wafers of Pentium 4 processors with and without decoupling capacitors. Without capacitors, the first droop increased by 8% of V_{DD}, but the operating frequency only slowed by 1%. The anomaly was explained by showing that under certain conditions the noise modulates the clock period in a way that tracks the critical path delay.

12.3.5.3 Distributed Power Supply Models The model presented so far is a lumped approximation that is convenient for analysis and facilitates gaining intuition about chip behavior. Chip designers also are concerned about the variation in supply voltage across the chip. This requires a distributed model, which we can approximate with a mesh of small elements as shown in Figure 12.14. The mesh represents the resistance and inductance of the

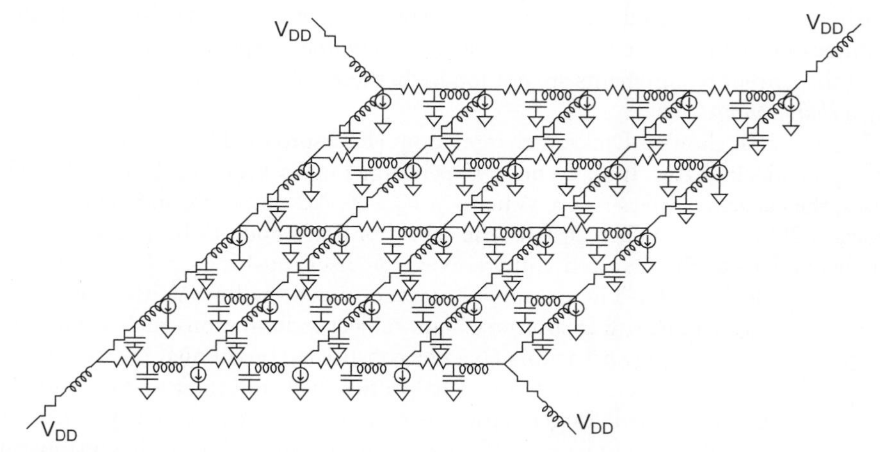

FIGURE 12.14 Impedance of bypass capacitor

on-chip power supply grid. Symbiotic or explicit decoupling capacitors are distributed across the chip. At each node, a current source represents the local current demand of the circuitry. The solder bumps or bond wires to the package are modeled with additional resistance and inductance. In this model, the package is treated as a perfect V_{DD} connected to the corners of the grid. In a more complex model, you also could add the distributed resistance, inductance, and bypass capacitance of the package itself.

For high-power chips, the designer can extract a mesh model as a SPICE netlist based on the power grid wiring and the amount of local decoupling capacitance. Different current waveforms can be applied at different nodes; for example, the current signatures of synthesized logic, SRAM, repeater banks, and domino logic are all quite different. The full-chip power grid simulation often takes many days to run and results in a map of voltage vs. time for the current pattern applied. Figure 6.2 shows a snapshot of the voltage droop on the Itanium 2 microprocessor. The droop was greatest in the integer execution unit, where several power-hungry domino adders all contribute to the IR drop.

12.3.6 Power Supply Filtering

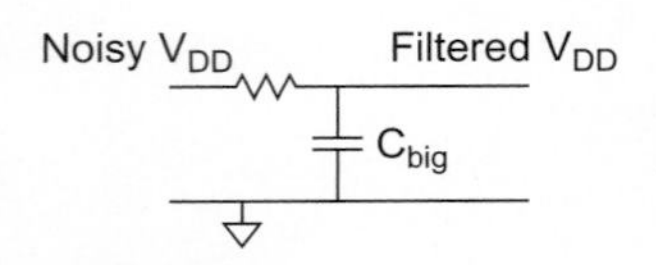

FIGURE 12.15 Power supply filter

Certain structures such as the phase-locked loop (PLL), clock buffers, and analog circuits are particularly sensitive to power supply noise. For example, supply noise on the clock buffers can directly increase clock jitter. Figure 12.15 shows an RC power supply filter circuit that eliminates the high-frequency noise on the local supply. The local filtered power supply is typically connected to the power grid through a single wire or solder bump. The resistance of this wire must be low enough to carry the current demand of the local circuitry without excessive IR drop, yet low enough to produce an RC time constant that will filter noise at frequencies of interest. Typically, this requires a huge filter capacitor as well, making power supply filtering expensive in terms of chip area.

For example, the Pentium 4 uses a power supply filter on the clock buffers to reduce clock jitter [Kurd01]. The filter attenuates typical supply noise from 10 to 2% of V_{DD} using a pMOS transistor as the resistor. It has an RC time constant of 2.5 ns with an IR drop of 70 mV.

12.3.7 Charge Pumps

Many circuits require a positive voltage exceeding V_{DD} or a negative voltage. For example, a Flash memory may require 20 V to erase floating-gate transistors. Reverse body bias techniques need a negative voltage. Extra external voltage regulators add to the system cost. If the current requirements are not too high, these voltages can be generated on-chip using a *charge pump*.

Figure 12.16 shows a Dickson charge pump [Dickson76]. The pump uses two nonoverlapping clock phases. Initially, node V_1 is charged up to $V_{DD} - V_t$ through $N1$. When ϕ_1 rises, the capacitor drives V_1 up. When $V_1 - V_2 > V_t$, $N2$ turns ON and begins charging V_2 toward $2(V_{DD} - V_t)$. When ϕ_1 falls, the capacitor drags node V_1 back down. $N2$ turns OFF, leaving V_2 at the elevated voltage. Next, ϕ_2 rises, pushing up V_2 and V_3 toward $3(V_{DD} - V_t)$. The pumping continues down the line. With enough stages, V_{out} can be driven arbitrary high, subject to limitations such as breakdown. The pumping capacitors C can be constructed out of nMOS transistors with their source and drain connected to the clock and the gate tied to the node being pumped. Larger capacitors pumped at higher frequency f increase the available output current. If the each of the pumped nodes has a stray capac-

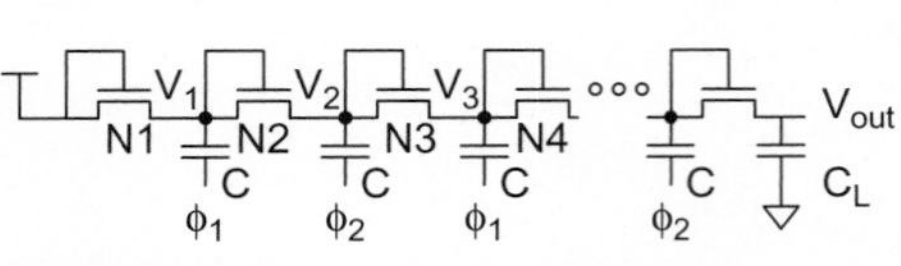

FIGURE 12.16 Dickson charge pump

itance C_s (such as the gate and diffusion capacitance of the transistors to the right and left) then the output voltage is approximately

$$V_{out} = N\left[\frac{CV_{DD} - \frac{I_{out}}{f}}{C + C_s} - V_t\right] \tag{12.13}$$

A large load capacitor C_L helps smooth out the ripple on V_{out}.

Figure 12.17 shows a charge pump for negative voltages. The pump operates in a similar fashion, but the pMOS transistors pull the voltage down on the falling transition of the clock. The pMOS bodies can be tied to GND to reduce the body effect.

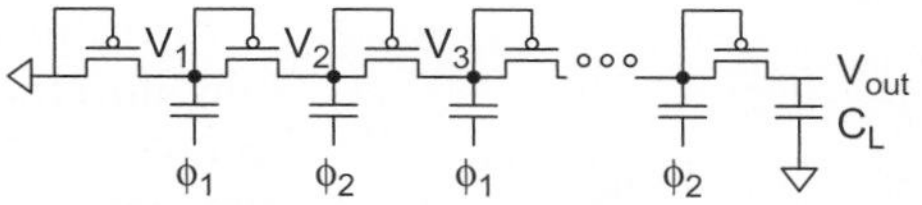

FIGURE 12.17 Negative charge pump

12.3.8 Substrate Noise

The body terminal of a bulk CMOS transistor is connected to the substrate or well. The p-type substrate for an nMOS transistor is normally connected to GND and the n-well for a pMOS transistor is normally connected to V_{DD}. The connection is made through a relatively high-resistance substrate or well contact. Current flow in the substrate causes noise on the body terminal. This current may come from capacitive coupling through the reverse-biased source/drain to substrate diodes or from impact ionization as current flows through an ON transistor. The substrate noise modulates threshold voltages by means of the body effect.

Substrate noise is also a problem for *mixed-signal* designs where separate power supplies are used for noisy digital circuits and quiet analog circuits. The large number of rapidly switching digital circuits creates noise on the digital ground that propagates to the sensitive analog circuitry via the common substrate.

The substrate and well should use plenty of contacts to guarantee a low-resistance path to the power network. Guard rings, described in Section 6.3.6, provide some protection against noise caused by currents in the nearby substrate. Analog circuits should be physically separated from digital circuits and protected by guard rings connected to a quiet analog supply. Twin-tub or triple-well processes and SOI also experience much less substrate coupling because transistors are isolated in their own wells.

Modeling and analyzing substrate noise is beyond the scope of this book. See [Donnay03] for extensive coverage of the subject.

12.3.9 Energy Scavenging

Energy sources are a chronic challenge for portable systems. Most systems use batteries, which eventually require replacement. This ranges from annoying (remembering to change the battery in your fire alarm each year) to downright difficult (changing the battery in an implanted pacemaker). *Energy scavenging* is an emerging field with tremendous promise for ultra-low power systems. The idea is to extract enough energy from the environment to operate the device. The technique is particularly attractive when combined with subthreshold circuits operating at microwatt or nanowatt average power levels. The power demand typically varies with time, so the energy may be stored in a capacitor or *microbattery* until it is needed.

Micropower generators can take advantage of many sources of energy. *Solar cells* are the best known [Guilar09]; solar calculators are among the oldest and best-known applications of energy scavenging. *Thermoelectric* microgenerators use thermocouples to produce a voltage proportional to the temperature difference across the elements [Lhermet08]. *Piezoelectric* microgenerators convert mechanical vibration into electricity [Le06, Ramadass10]. Radio-frequency identification (RFID) tags use a coil to collect RF energy radiated by the reader, then broadcast their ID back. Power output for these sources depends on the amount of energy available for scavenging and the size of the generator, but tens to hundreds of microwatts per square centimeter are commonly achieved.

Microbattery manufacturing is also evolving. Microbatteries are made from layers of thin films that can be deposited on top of an integrated circuit after the standard steps are completed. A 10-μm thick lithium-based battery presently achieves an energy density of 100 μW-hr/cm^2 [Lhermet08].

12.4 Clocks

Synchronous systems use a clock to distinguish one step in a computation from the previous or next step. Ideally, this clock should arrive at all clocked elements in the system simultaneously so that the system shares a common time reference. These elements include latches and flip-flops, memories, and dynamic gates. In practice, the arrival time differs somewhat from one point to another; this difference is called *clock skew.* The central challenge in clock system design is to deliver the clock to all the clocked elements on the chip while finding an acceptable compromise among skew, power consumption, metal resource usage, and design effort.

12.4.1 Definitions

A system is designed to use one or more *logical clocks.* The logical clocks are idealized signals with no skew used by the logic designer when describing the system with a hardware description language. For example, a system with flip-flops requires a single logical clock, usually called *clk.* A system using two-phase transparent latches requires two logical clocks ϕ_1 and ϕ_2 (or `ph1` and `ph2` in a hardware description language). Unfortunately, mismatched clock network paths and processing and environmental variations make it impossible for all clocks to arrive at their ideal times, so the designer must settle for actually receiving a multitude of skewed *physical clocks.*

Distributing a single clock across the entire chip in a low-skew fashion is challenging. Distributing more than one is nearly impossible. Therefore, most systems distribute a single *global clock* even though they may need multiple logical clocks. *Local clock gaters* located near the clocked elements produce the physical clocks and drive them to the elements over short wires. Examples of clock gaters include buffers, AND gates to stop the clock to unused units, inverters to produce complementary clocks, and pulse generators for pulsed latches.

The term *clock skew* has been used informally in many ways. We define skew as the *difference between the nominal and actual interarrival time of a pair of physical clocks.* For example, Figure 12.18(a) shows a system with two flip-flops. Both should receive the logical clock *clk* with zero interarrival time, but they actually receive physical clocks clk_1 and clk_2. Because of differences in the delay of the clock distribution wires and the local clock buffers, clk_1 arrives 25 ps before clk_2. Therefore, we say the clock skew is 25 ps. Figure 12.18(b) shows a system with three transparent latches. The latches use complementary logical clocks ϕ_1 and ϕ_2 with a nominal interarrival time of $T_c/2$ between rising edges.

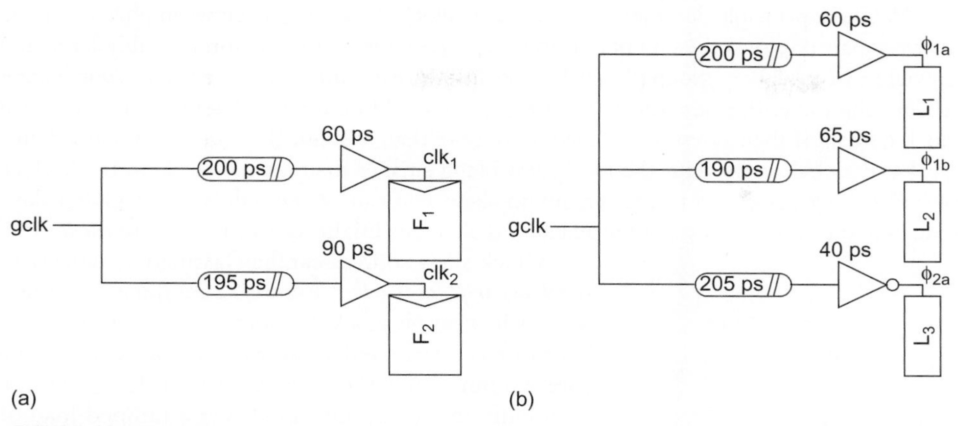

FIGURE 12.18 Clock skew example

They actually receive physical clocks ϕ_{1a}, ϕ_{1b}, and ϕ_{2a}. We see the clock skews are

$$t_{\text{skew}}^{\phi_{1a},\phi_{1b}} = 5 \text{ ps}; \ t_{\text{skew}}^{\phi_{1a},\phi_{2a}} = 15 \text{ ps}; \ t_{\text{skew}}^{\phi_{1b},\phi_{2a}} = 10 \text{ ps}$$

Sometimes designers intentionally delay clocks to solve setup or hold time problems. For example, suppose that a critical path existed between F_1 and F_2 in Figure 12.18(a). The designer might intentionally delay the clock to F_2 by 30 ps to give the path more time by using the slower local clock buffer on clk_2. In this case, the nominal interarrival time of clk_1 and clk_2 is 30 ps. The actual interarrival time is 25 ps, so the clock skew is 5 ps. Some designers call this 30 ps delay *intentional skew*. We prefer to call it *intentional delay* and reserve the term *clock skew* to account for unintentional differences in clock arrival times.

Clock skew can also be measured between different edges of the clock or between different cycles. For example, Figure 12.19 shows two physical clock waveforms in which the edges differ from their nominal timing. The clock skews are defined based on the edge (*r*ising/*f*alling) and the number of intervening cycles as well as the physical clock:

$$t_{\text{skew}}^{clk_1,clk_1,(r,r,0)} = 0 \text{ ps}; \ t_{\text{skew}}^{clk_1,clk_1,(r,f,0)} = 30 \text{ ps}; \ t_{\text{skew}}^{clk_1,clk_1,(r,r,1)} = 70 \text{ ps}$$

$$t_{\text{skew}}^{clk_1,clk_2,(r,r,0)} = 0 \text{ ps}; \ t_{\text{skew}}^{clk_1,clk_2,(r,f,0)} = 0 \text{ ps}; \ t_{\text{skew}}^{clk_1,clk_2,(r,r,1)} = 40 \text{ ps}$$

For a path between two flip-flops, the hold time constraint depends on the skew between the same rising edges of both physical clocks. The setup time constraint depends on the skew between the rising edge of one physical clock and the subsequent rising edge of the other. We will see that clock distribution networks tend to introduce more skew from one cycle to the next so setup and hold time constraints can budget different amounts of skew.

The actual clock skew between two clocked elements varies with time and is different from one chip to another. Moreover, it is unknowable at design time. From the engineering perspective, a more useful parameter is the *clock skew budget*. The clock skew budget should be larger than the actual skew encountered on any long or short path on any working chip, yet no larger than necessary lest the chip be overdesigned.

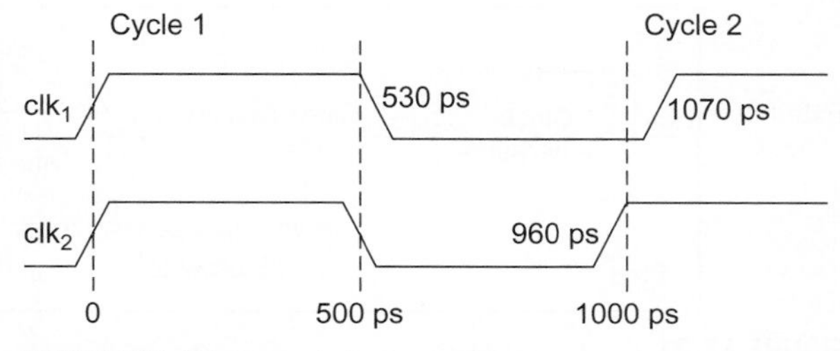

FIGURE 12.19 Skewed clock waveforms

While in principle designers could tabulate clock skew budgets between physical clocks at every pair of clocked elements on the chip, the table would be unreasonably large and unwieldy. Instead, they group physical clocks into *clock domains* and use a single skew budget to describe the entire domain. For example, you could define two latches to be in a local clock domain if their physical distance is no more than 500 μm. Then you could just define local and global skews, with the local skew being smaller than the global skew. If the clock period is long compared to the maximum skew, you can define only a single global skew budget and pessimistically assume all clocked elements might see this worst-case skew.

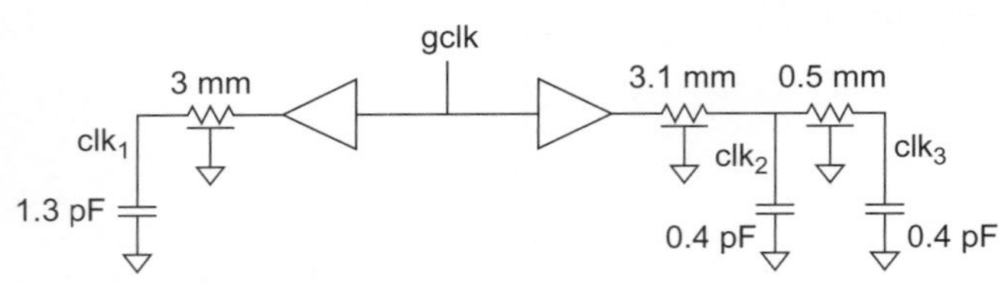

FIGURE 12.20 Simple clock distribution network

Clock skew sources can be classified as *systematic*, *random*, *drift*, and *jitter*. Figure 12.20 illustrates these sources in a simple clock distribution network. The global clock is distributed along wires to two gaters. One wire is 3 mm, while the other is 3.1 mm. The gaters are nominally identical, but one drives a lumped load of 1.3 pF while the other drives a load of 0.8 pF distributed along a 0.5 mm wire. The *systematic* clock skew is the portion that exists even under nominal conditions; this component can be predicted by simulation. By adjusting the size of one of the gaters, the systematic skew between clk_1 and clk_2 could be driven to zero. However, some systematic skew will always exist between clk_2 and clk_3 because of the flight time along the wire after the gater.

The *random* component of skew is caused by manufacturing variations that could affect the wire width, thickness, or spacing and the transistor channel length, threshold voltage, or oxide thickness. These cause unpredictable changes in resistance, capacitance, and transistor current, introducing additional skew. In principle, the actual random skew could be measured during chip test or on startup, and adjustable delay elements could be calibrated to compensate for the random skew.

Drift is caused by time-dependent environmental variations that occur relatively slowly. For example, after the chip turns on, it will heat up. The temperature affects gate and wire delay differently. Also, a temperature gradient across the chip leads to skew. Drift can also be nulled out with adjustable delay elements. Unlike random skew, compensating for drift must take place periodically rather than just once at startup. The frequency of calibration depends on the thermal time constant of the chip.

Jitter is caused by high-frequency environmental variation, particularly power supply noise. This noise leads to delay variation in the clock buffers and gaters in both time and space. Jitter is particularly insidious because it occurs too rapidly for compensation circuits to be able to counter it.

Some engineers do not report jitter as part of the skew. In such a case, they must include both jitter and skew in the setup and hold time budgets.

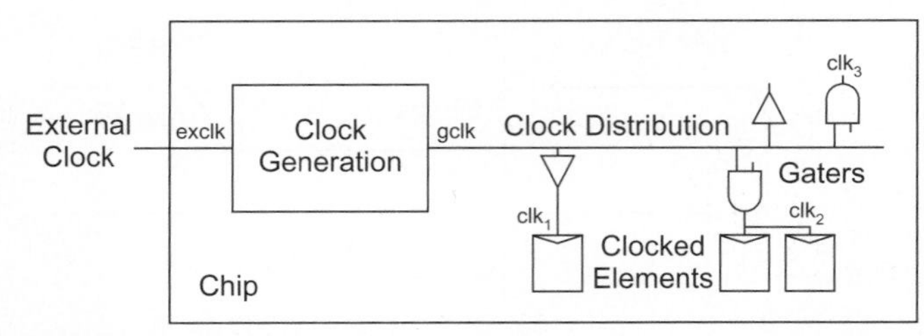

FIGURE 12.21 Clock subsystem

12.4.2 Clock System Architecture

Figure 12.21 shows an overview of a typical clock subsystem. The chip receives an external clock signal through the I/O pads. The clock generation unit may include a *phase-locked loop* (PLL) or *delay-locked loop* (DLL) to adjust the frequency or phase of the global clock, as shall be discussed in Section 12.5. This global clock is then distributed across the chip to points near all of the clocked elements. The clock distribution network must be carefully designed to minimize clock skew. Local clock gaters receive this global clock and drive the physical clock signals along short wires to small groups of clocked elements.

12.4.3 Global Clock Generation

The global clock generator receives an external clock signal and produces the global clock that will be distributed across the die. In the simplest case, the clock generator is simply a chain of buffers to drive the large capacitance load on the clock distribution network. However, such a simple clock generator may suffer from a number of issues.

First, the input pad, buffers, wires, and clock gaters add significant delay that can cause a large delay (often 0.5 to 1 ns across a large chip) between the external clock and the internal clocks distributed to the clocked elements on the chip. This delay can also vary with process variation and environment conditions, and fluctuate rapidly over time due to supply or substrate noise present on the chip. Due to this uncontrolled amount of skew and jitter, the clock domains inside the chip become unsynchronized with the external clock domain, making reliable communication difficult. This is particularly problematic at high frequencies where the skew becomes a significant portion of the clock period.

To mitigate these issues, more sophisticated clock generators use either phase-locked loops (PLLs) or delay-locked loops (DLLs) to regulate the delay to a constant value in the presence of variation and noise. Note that if this delay is equal to an integer multiple of the clock period, the delayed clock is indistinguishable from the original clock with no delay. This way, the external and internal clock domains remain synchronized in spite of the delays introduced by the additional elements in the clock distribution network. For this reason, the PLLs and DLLs used in this purpose are often called *zero-delay buffers*.

Figure 12.22 illustrates the use of a PLL or a DLL to compensate for the on-chip clock delays. The circuits contain a phase detector (PD) that produces a signal proportional to the phase difference between the input and output clocks. The loop filter (LF)

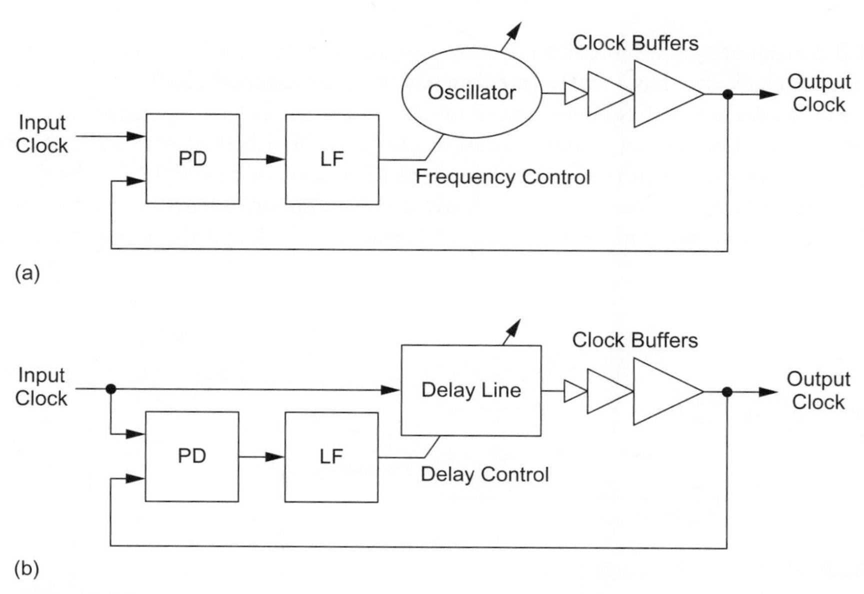

FIGURE 12.22 Zero-delay buffers using (a) a PLL, (b) a DLL

converts this phase error into a control signal adjusting the frequency of an oscillator or the delay of a delay line. Section 12.5 examines the design of each of these elements. The output is then buffered to drive the large output clock load. PLLs and DLLs share the same principle of *feedback control*; they both monitor the distributed clocks and correct them if they are in misalignment with the external input clock. The only difference is that upon the detection of misalignment, a PLL adjusts the frequency of the clock (subsequently its phase) while a DLL adjusts the delay of the clock. Nonetheless, both types of feedback loops strive to distribute the clocks whose edge positions are aligned with those of the external clock.

12.4.3.1 PLLs vs. DLLs The main difference between a PLL and a DLL is that the PLL uses an oscillator that *creates* a new clock whereas the DLL uses a variable delay line that simply *delays* the input clock. While both can serve as the actuator element that adjusts the edge position of the clock, the oscillator is more versatile in a sense that it can also vary the frequency of the clock. This property makes it easy for PLLs to multiply the clock frequency by an integer or even by a fractional amount when desired. However, a PLL loop filter is generally more complicated than the DLL counterpart because it has to control two quantities (i.e., the frequency and phase of the oscillator clock) instead of just one (i.e., the delay).

12.4.3.2 Bandwidth and Stability A key metric for feedback loops is how quickly they can respond to various disturbances and adjust the output clock. For example, if the disturbance is supply noise, we would want the PLL or DLL to counteract the disturbance as soon as possible. However, if the disturbance is the input clock jitter, we may want the loop to respond slowly, so that the output clock will track the average position of the input clock and thus have lower jitter. The most used quantity that describes this promptness in the response is *bandwidth*. Another critical metric is *stability*, which describes how reliably the feedback loop converges to the locked condition. Generally, PLLs require more attention than DLLs in order to achieve good stability.

12.4.3.3 Frequency Multiplication In some applications, it may be necessary to generate an on-chip clock that has a different frequency than the external clock. For example, one may want to use a low-frequency quartz clock source that is less expensive than a high-frequency one. The frequency multiplication can be easily achieved with PLLs by inserting a frequency divider in the feedback path, as illustrated in Figure 12.23. As the phase detector now compares the output clock divided by a factor N with the input reference clock divided by a factor of M, when the PLL reaches a lock and those two clocks are in

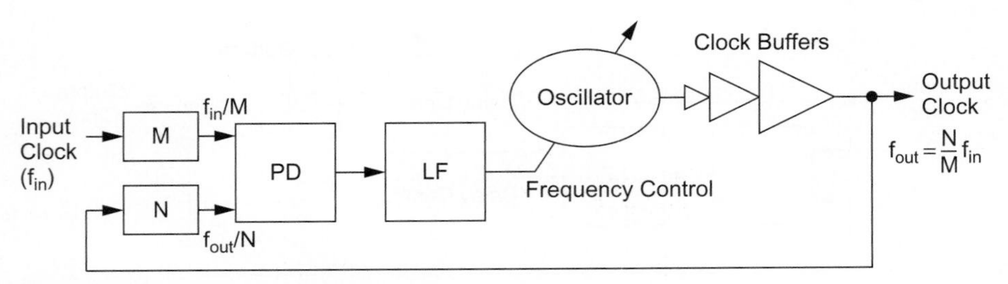

FIGURE 12.23 Frequency multiplication using a PLL

alignment, the output clock from the oscillator should have a frequency that is *N/M* times the input clock frequency. Thus, the PLL can produce an output that is any rational multiple of the input frequency.

12.4.4 Global Clock Distribution

The global clock must be distributed across the chip in a way that reaches all of the clocked elements at nearly the same time. In antiquated processes with slow transistors and fast wires, the clock wire had negligible delay and any convenient routing plan could be used to distribute the clock. In nanometer processes, the RC delay of the resistive clock wire driving its own capacitance and the clock load capacitance tends to be close to 1 ns for a well-designed distribution network covering a 15 mm square die. If the clock were routed randomly, this would lead to a clock skew of about 1 ns between physical clocks near and far from the clock generator. This could be several times the cycle time of the system. Thus, the clock distribution system must be carefully designed to equalize the *flight time* between the clock generator and the clocked receivers. Global clock distribution networks can be classified as *grids*, *H-trees*, *spines*, *ad hoc*, or *hybrid* [Restle98].

Random skew, drift, and jitter from the clock distribution network are proportional to the delay through the network because they are caused by process or environmental variations in the distribution elements. Therefore, the designer should try to keep this distribution delay low. Unfortunately, as chips are getting larger, wires are getting slower, and clock loads are increasing, the distribution delay tends to go up even as cycle times are going down. In the past, systematic clock skew was the dominant component. Now, good clock distribution networks achieve low systematic skews, but the random, drift, and jitter components are becoming an increasing fraction of the cycle time.

12.4.4.1 Grids A clock grid is a mesh of horizontal and vertical wires driven from the middle or edges. The mesh is fine enough to deliver the clock to points nearby every clocked element. The resistance is low between any two nearby points in the mesh so the skew is also low between nearby clocked elements. This reduces the chance of hold-time problems because such problems tend to occur between nearby elements where the propagation delay between elements is also small. Grids also compensate for much of the random skew because shorting the clock together makes variations in delays irrelevant. The grids can be routed early in the design without detailed knowledge of latch placement. However, grids do have significant systematic skew between the points closest to the drivers and the points farthest away. They also consume a large amount of metal resources and hence have a high switching capacitance and power consumption.

12.5.4.2 H-Trees An H-tree is a fractal structure built by drawing an H shape, then recursively drawing H shapes on each of the vertices, as shown in Figure 12.24. With enough recursions, the H-tree can distribute a clock from the center to within an arbitrarily short distance of every point on the chip while maintaining exactly equal wire lengths. Buffers are added as necessary to serve as repeaters. If the clock loads were uniformly distributed around the chip, the H-tree would have zero systematic skew. Moreover, the trees tend to use less wire and thus have lower capacitance than grids [Restle98].

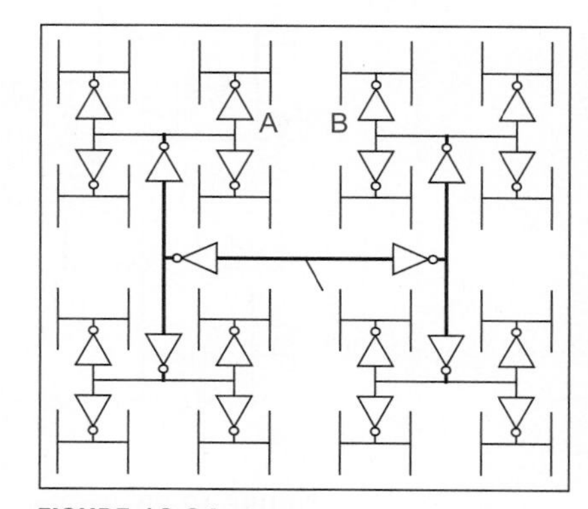

FIGURE 12.24 H-tree

In practice, the H-tree still shows some skew because the clock loads are not uniform, loading some leaves of the tree more than others. Moreover, the tree often must be routed around obstructions such as memory arrays. The leaves of the H do not reach every point on the chip, so some short physical clock wires are required after the local clock gater. Nevertheless, with careful tapering of the wires and sizing of the clock gaters, H-trees can deliver nearly zero systematic skew. A drawback of H-trees is that they may have high random skew, drift, and jitter between two nearby points that are leaves of different legs of the tree. For example, the points A and B in Figure 12.24 might experience large skews. As the points are close, this is a particular problem for hold times.

Figure 12.25 shows a modified H-tree used on the Itanium 2. The primary clock driver in the center of the chip sends a differential output to four differential repeaters on the leaves of the H. These repeaters drive a somewhat irregular pattern of wiring to second-level clock buffers (SLCBs) serving units all across the chip. The wiring and SLCB placement is determined by the nonuniform clock loads and obstructions on the chip. A custom clock router automatically generated the tree based on the actual clock loads so that the tree could be easily rerouted when loads change late in the design process. The SLCBs drive local clock gaters, producing the multitude of clock waveforms used on the microprocessor. Some of these waveforms were shown in Section 9.8.2.

Figure 12.26 shows the differential driver used as a primary clock buffer and repeater on the Itanium 2 [Anderson02]. The input stage is a differential amplifier sensitive to the point where the differential inputs cross over. The repeater pulses either p_1 or n_1 and p_2 or n_2 to switch the internal nodes y and $\bar{y}$. The small tristate keeper prevents these nodes from floating after the pulse terminates. The SLCB uses the same structure, but produces only a single-ended output. It also provides a current-starved adjustable delay line to compensate for systematic skew and to help locate critical paths during debug. The repeater provides a

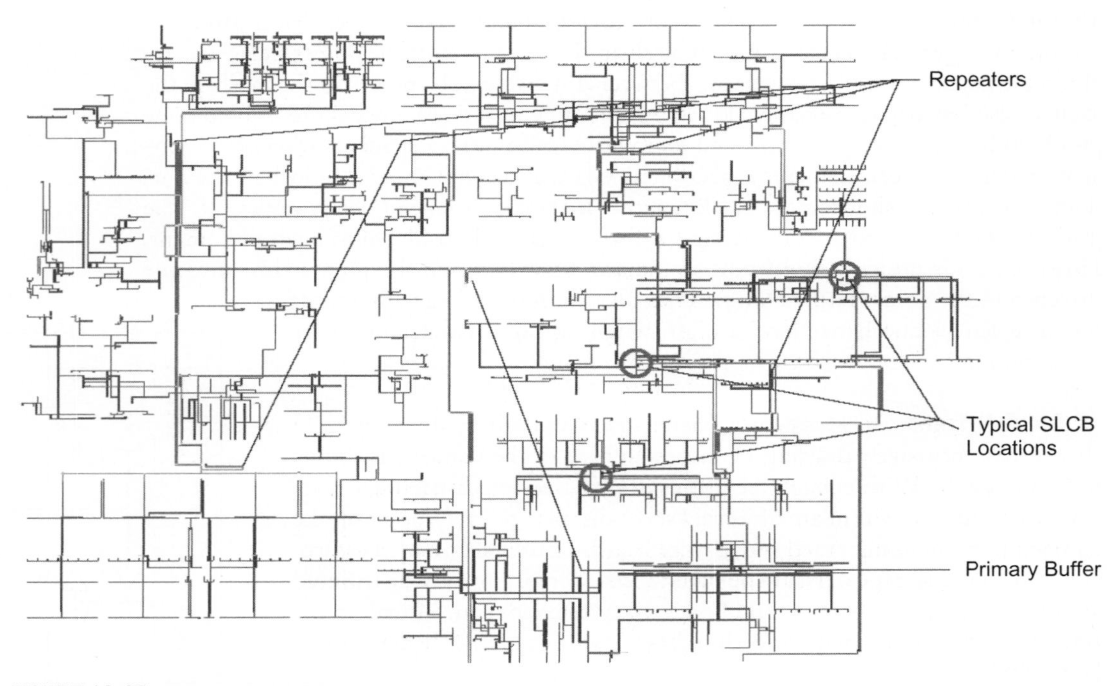

FIGURE 12.25 Itanium 2 modified H-tree

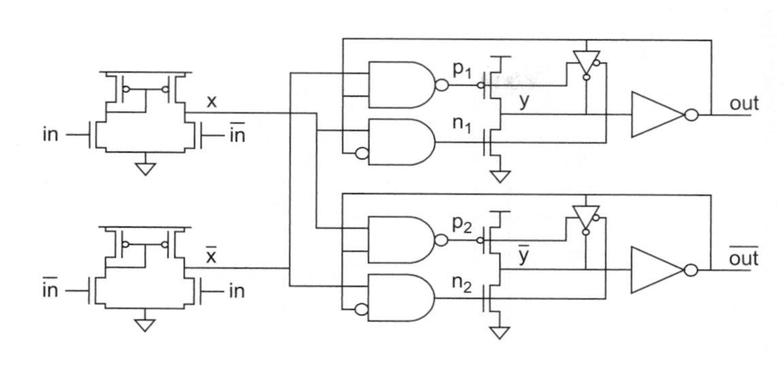

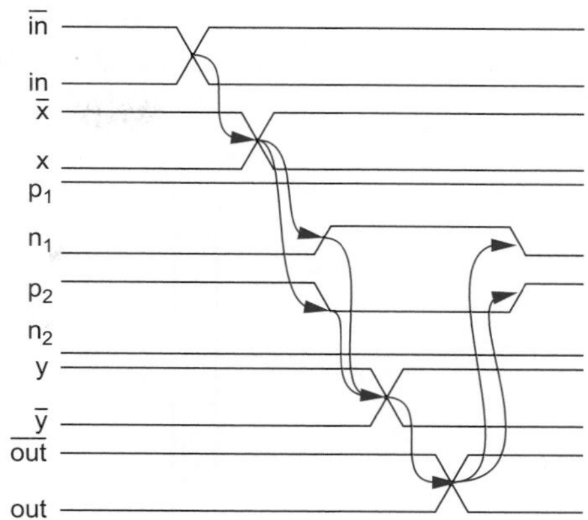

FIGURE 12.26 Itanium 2 repeater

high drive capability with a low input capacitance. Thus, few stages of clock buffering are needed in the network. With so few repeaters, the area overhead of providing a filtered power supply is modest. Although the repeaters are relatively slow, their jitter is controlled with supply filtering.

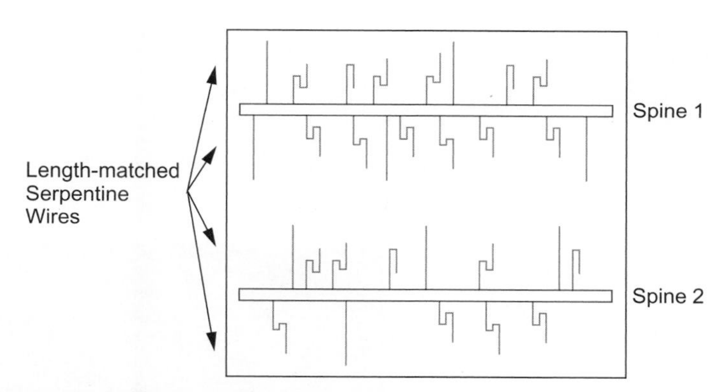

FIGURE 12.27 Clock spines with serpentine routing

12.4.4.3 Spines Figure 12.27 shows a clock distribution scheme using a pair of spines. As with the grid, the clock buffers are located in a few rows across the chip. However, instead of driving a single clock grid across the entire die, the spines drive length-matched serpentine wires to each small group of clocked elements. If the loads are uniform, the spine avoids the systematic skew of the grid by matching the length of the clock wires. Each serpentine is driven individually so gaters can be used to save power by not switching certain wires. The serpentine is also easy to design and each load can be tuned individually. However, a system with many clocked elements may require a large number of serpentine routes, leading to high area and capacitance for the clock network. Like trees, spines also may have large local skews between nearby elements driven by different serpentines.

The Pentium II and III use a pair of clock spines [Geannopoulos98]. The Pentium 4 adds a third clock spine to reduce the length of the final clock wires [Kurd01]. Figure 12.28(a) shows the global clock buffers distributing the clock to the three spines on the Pentium 4 with zero systematic skew while Figure 12.28(b) shows a photograph of the chip annotated with the clock spine locations. The spines drive 47 independent clock domains, each of which can be gated individually. The clock domain gaters also contain adjustable delay buffers used to null out systematic and random skew and even to force deliberate clock delay to improve performance.

12.4.4.4 Ad Hoc Many ASICs running at relatively low frequencies (hundreds of MHz) still get away with an ad hoc clock distribution network in which the clock is routed haphazardly with some attempt to equalize wire lengths or add buffers to equalize delay.

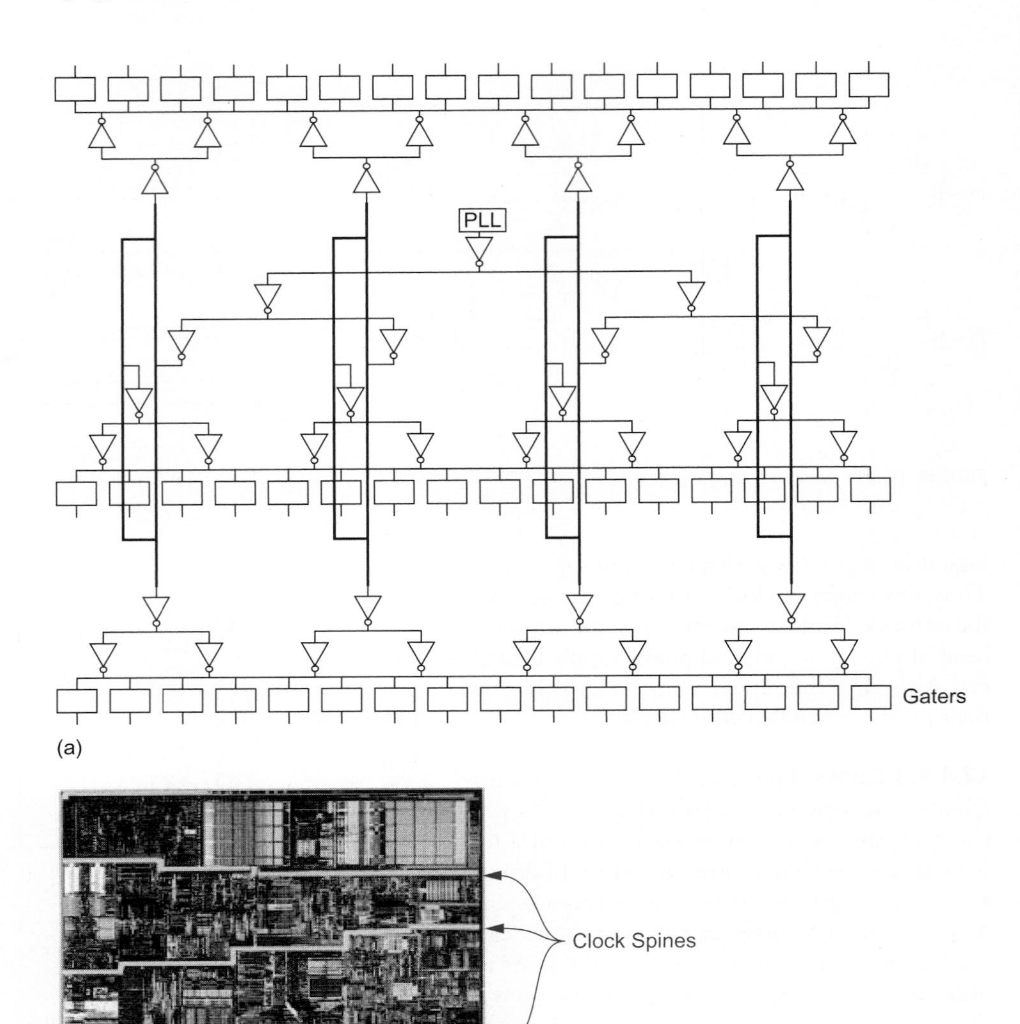

FIGURE 12.28 Pentium 4 clock spines (© 2001 IEEE.)

Such ad hoc networks can have reasonably low systematic skews because the buffer sizes can be adjusted until the nominal delays are nearly equal. However, they are subject to severe random skew when process variations affect wire and gate delays differently. This is the level that most commonly available tools support. Most design teams using ad hoc clock networks also lack the resources to do a careful analysis of random skew, jitter, and drift. Therefore, they should be conservative in defining a skew budget and must be careful about hold time violations.

12.4.4.5 Hybrid A hybrid combination of the H-tree and grid offers lower skew than either an H-tree or grid alone. In the hybrid approach, an H-tree is used to distribute the clock to a large number of points across the die. A grid shorts these points together. Compared to a simple grid, the hybrid approach has lower systematic skew because the grid is driven from many points instead of just the middle or edge. Compared to an H-tree, the hybrid approach is less susceptible to skew from nonuniform load distributions. The grid also reduces local skew and brings the clock near every location where it is needed. Finally, the hybrid approach is regular, making layout of well-controlled transmission line structures easier.

IBM has used such a hybrid distribution network on a variety of microprocessors including the Power4, PowerPC, and S/390 [Restle01]. A primary buffered H-tree drives 16–64 sector buffers arranged across the chip. Each sector buffer drives a smaller tree network. Each tree can be tuned to accommodate nonuniform load capacitance by adjusting the wire widths. Together, the tunable trees drive the global clock grid at up to 1024 points. IBM uses a specialized tool to perform the tuning.

12.4.4.6 Layout Issues High-speed clock distribution networks require careful layout to minimize skew. The two guiding principles are that the network should be as uniform and as fast as possible. In a uniform network, chip-wide process or environmental variations should affect all clock paths identically. In a fast network, localized variations that cause a fractional difference between two clock path delays lead only to modest amounts of skew. For example, voltage noise that causes a 10% delay variation between two paths through an H-tree will lead to 80 ps of jitter if the tree delay is 800 ps, but 160 ps of jitter if the tree delay is 1600 ps.

Building a fast clock network requires low-resistance global clock wires with proper repeater insertion. The thick, top-level metal layer is well-suited to clock distribution. The wide wires should be shielded on both sides with V_{DD} or GND lines to prevent capacitive coupling between the clock and signal lines. The clock can even be shielded on a lower metal layer to form a microstrip waveguide [Anderson02].

Wide, low-resistance wires also have significant inductive effects, including faster than expected edge rates and overshoot. The fast edges are desirable, but overshoot should be minimized to prevent overvoltage damage. High-performance clock networks must be extracted with a field solver and modeled as transmission lines [Huang03]. Uniformity is again important: Even if the RC delays appear to be matched in a nonuniform layout, the RLC delays can be significantly different. As discussed in Section 6.3.4, wide wires should be split into multiple narrower traces interdigitated with V_{DD}/GND wires that provide a low-inductance current return path and minimize skin effect.

12.4.5 Local Clock Gaters

Local clock gaters receive the global clock and produce the physical clocks required by the clocked elements. The output of the gaters typically run a short distance (< 1 mm) to the clocked elements. Clock gaters are often used to stop or *gate* the clock to unused blocks of logic to save power. As discussed in Chapter 9, they can produce a variety of modified clock waveforms including pulsed clocks, delayed clocks, stretched clocks, nonoverlapping clocks, and double-frequency pulsed clocks. When used to modify the clock edges, they are sometimes called *clock choppers* or *clock stretchers*. Figure 12.29 shows a variety of clock gaters.

Most systems require a large number of clock gaters, so it is impractical to filter the power supply at every one. Variations in clock gater delay caused by voltage noise, cross-die process variation, and nonuniform temperature distribution cause skew between clocks produced by different gaters. The best way to limit this skew is to make the gater delay as

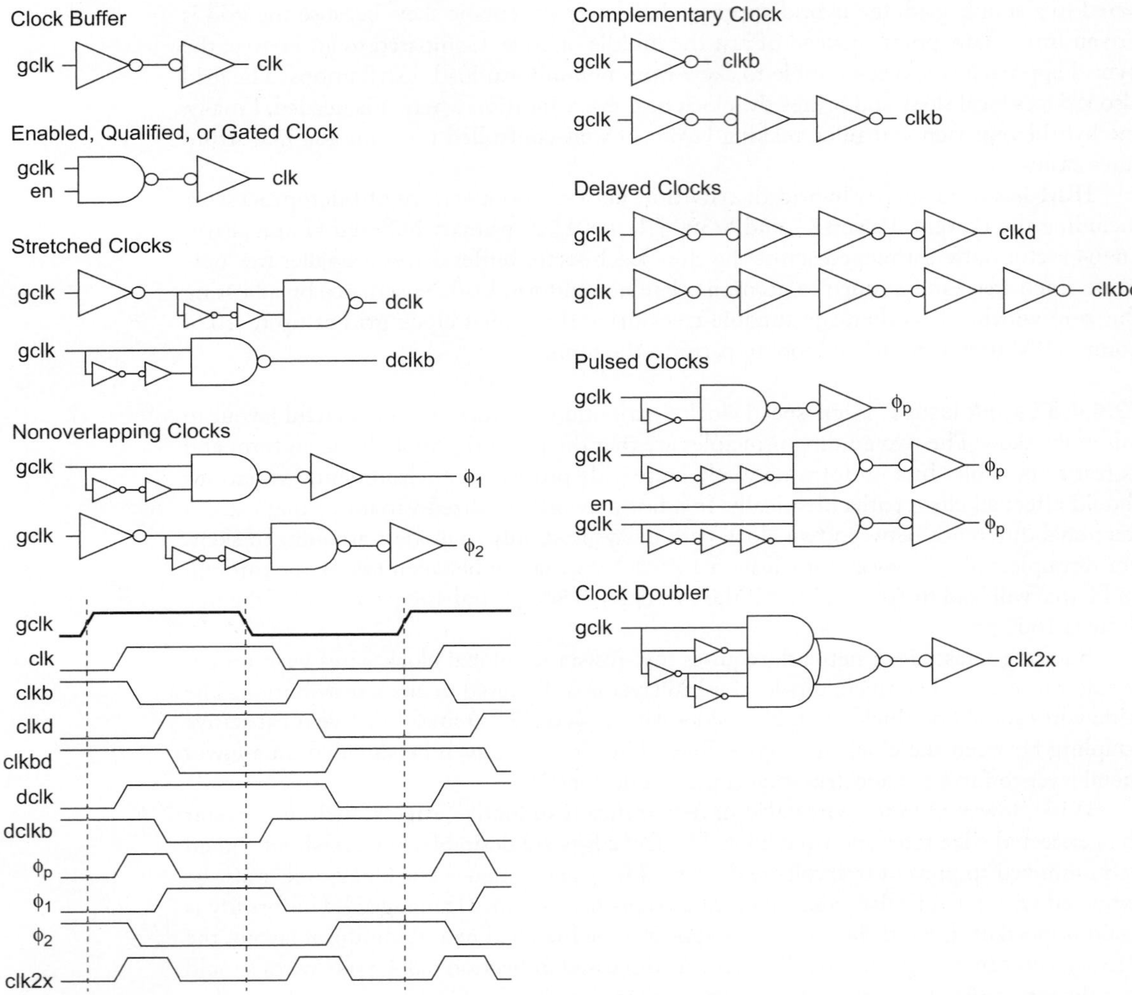

FIGURE 12.29 Clock gaters

short as possible. Variations in the input threshold of the clocked elements also causes skew. The best way to limit this skew is to produce crisp rise/fall times at the clock gaters. The final stage should have a fanout of no more than about 4.

Clock gaters may introduce some systematic delay between phases. For example, if *clkb* is produced with three inverters while *clk* is produced with only two, *clkb* may be delayed slightly from *clk*. The designer can either choose to carefully size the inverters such that the net delay is equal or accept that the delays are unequal and simply roll the systematic difference into timing analysis.

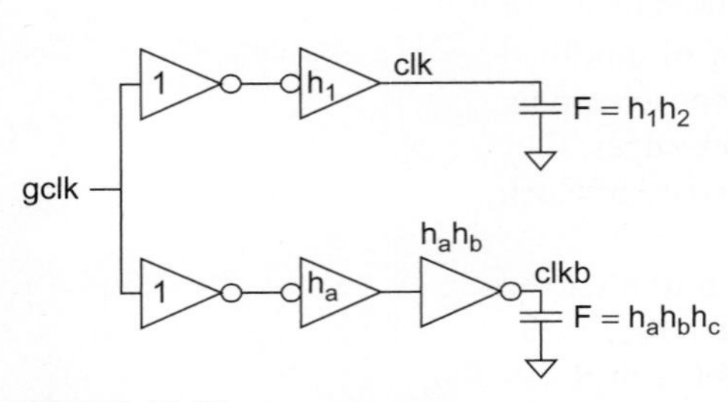

FIGURE 12.30 2- and 3-inverter paths

Figure 12.30 shows a circuit in which the delay of two inverters is matched against the delay of three when driving a fanout of F. The inverters are annotated with their size. The two inverters have electrical efforts of h_1 and h_2, respectively, while the three inverters have electrical efforts of h_a, h_b, and h_c. The electrical efforts should be chosen so that the delays of the chains are equal:

$$D = h_1 + h_2 + 2p_{\text{inv}} = h_a + h_b + h_c + 3p_{\text{inv}} \tag{12.14}$$

Even if the inverters have equal rise and fall delays in the TT corner, they will have unequal delays in the FS or SF corner. This can lead to skew between *clk* and *clkb* in these corners. However, if the delay of the second inverter in each chain is equal ($h_2 = h_b$), the two gaters will have equal delay in all process corners [Shoji86].

We can solve for the best electrical efforts that satisfy this constraint while giving least delay through the path. Recall that a path has least delay when its stage efforts are equal. Thus, choose $h_a = h_c = h^*$. This implies $h_1 = h^{*2}$. The delay of the first inverter in the *clk* path must equal the sum of the delays of the first and third inverter in the *clkb* path:

$$h^{*2} + p_{\text{inv}} = 2h^* + 2p_{\text{inv}} \tag{12.15}$$

This gives a quadratic equation that can be solved for h^*:

$$h^* = 1 + \sqrt{1 + p_{\text{inv}}} \tag{12.16}$$

For $p_{\text{inv}} = 1$, this implies the best stage efforts are

$$\begin{aligned} h_1 &= 5.8 \quad h_2 = \frac{F}{5.8} \\ h_a &= 2.4 \quad h_b = \frac{F}{5.8} \quad h_c = 2.4 \end{aligned} \tag{12.17}$$

In this case, the rise/fall times of the different stages may be rather different, so the Logical Effort delay model is not especially accurate. These efforts make a good starting point, but further tuning should be done with a circuit simulator. The same approach can be used when the gater uses a NAND gate in place of one of the inverters.

Another approach is to try to match the delay of two inverters against one inverter and a transmission gate, as shown in Figure 12.31. This matching will not be perfect across all process corners. However, the gater may have less overall delay and hence produce less jitter from power supply noise.

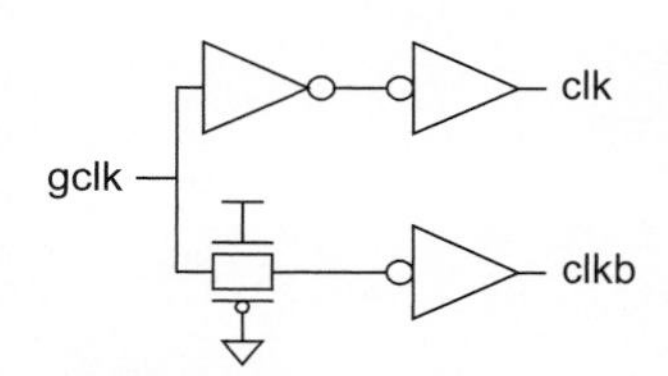

FIGURE 12.31
2- and 1-inverter paths

12.4.6 Clock Skew Budgets

Developing an appropriate clock skew budget for design is a tricky process. The designer has a number of choices, including ignoring clock skew, budgeting worst-case clock skew everywhere, or budgeting different amounts of clock skew between different clock domains. Ultimately, the designer's objective is to build a system that achieves performance targets and has no hold time failures while consuming as little area, power, and design effort as possible. The performance target can be a fixed number set by a standard or can simply be "as fast as possible."

It is possible to ignore clock skew if you are conservative about hold times and simply want the system to run as fast as possible. You must take reasonable care in the clock distribution network so that the skew between back-to-back flip-flops is unlikely to be too large. Many ASIC and FPGA flip-flops are designed with long contamination delays so they can tolerate significant skew before violating hold times. Build the system to run as fast as possible. When it is manufactured, clock skew will cause it to run slower than expected. The advantage of this methodology is that designers can be more productive because they do not need to think about clock skew. A disadvantage is that it uses slow flip-flops. Another drawback is that some paths really will have more skew than others. If all paths are designed to have equal delay, the paths with more skew will limit performance, while the other paths will be overdesigned and will consume more area and power than necessary. Moreover, if skew-tolerant circuit techniques are used in some places but not others, the nontolerant circuits will tend to form the critical paths.

A related approach is to estimate the worst-case clock skew and budget it everywhere. In systems using only flip-flops, this can be done by designing to a shorter clock period. For example, if an ASIC must meet a 4 ns clock period and is predicted to have 500 ps of skew, it can be designed to meet a 3.5 ns clock period with no skew. This method requires work on the part of the clock designer to predict the clock skew, but still protects most of the designers from worrying about skew.

As cycle times get shorter than about 25 FO4 inverter delays, budgeting worst-case skew everywhere makes design impossible. Instead, multiple skew budgets must be developed that reflect smaller amounts of skew between elements in a local clock domain. This method entails more thought on the part of designers to take advantage of locality and requires a static timing analyzer that applies the appropriate skew. A good timing analyzer also properly handles skew-tolerant techniques such as transparent latches and domino gates with overlapping clocks [Harris99].

12.4.6.1 Clock Skew Sources As discussed earlier, clock skew comes from many sources. The output of the phase-locked loop has some jitter because of noise in the PLL and jitter in the external clock source. The clock distribution network introduces more skew from variations in the buffers and wire. The buffers may have different delays because of differences in V_{DD} and temperature, as well as random variations in their channel length and threshold voltages. The wire length and loading between buffers may not be perfectly matched. Each gater drives a physical clock along a wire, so clocked elements at different ends of the wire will see different RC delays. As mentioned in Section 2.3.2, the effective gate capacitance of the clocked loads depends on the switching activity of the source and drain. For some clocked elements, this causes significant data dependence in the clocked capacitance and the local wire delay.

For hold time checks, we are concerned with the skew between two consecutive clocked elements at a particular moment in time. For setup time checks, we are concerned with the skew between elements from one cycle to the next. Jitter in the clock distribution network can affect the instantaneous clock period, so setup time skew budgets must include the cycle-to-cycle jitter of the entire clock distribution system even for elements in the same local clock domain. Hence, we can define separate clock skew budgets for setup time and hold time analyses.

The sources can be categorized as systematic, random, drift, and jitter. Recall that systematic skews can be modeled as extra delay and taken out of the skew budget if you are willing to do the modeling. Good clock distribution networks have close to zero systematic skew. Systematic and random skews can also be eliminated by calibrating delay lines, as will be discussed in Section 12.4.7. Drift occurs slowly enough that it can be eliminated by periodic recalibration of the delay lines. Ultimately, jitter is the most serious source of skew because it changes too rapidly to predict and counteract.

12.4.6.2 Statistical Clock Skew Budgeting The most conservative approach to estimating clock skew is to find the worst-case value of each skew source and sum these values. A real chip is unlikely to simultaneously see all of these worst cases, so such a sum is pessimistic and makes design of high-speed chips nearly impossible.

Most skew sources do not have Gaussian distributions, so taking the root sum square of the sources is inappropriate. A better approach is to perform a Monte Carlo simulation of the different skew sources to find the likely distribution of skews. The skew budget is selected at some point in this distribution. For hold times, the skew must be budgeted conservatively because the chip will not work if a hold time is violated. For example, the hold time skew budget can be selected so that 95–99% of chips will have no hold time violations.

If the goal is to build a chip that operates as fast as possible, any fixed amount of skew that affects all paths equally is irrelevant to the designer because there is nothing to do about it from the point of view of meeting setup times. However, if different paths experience different amounts of skew, a path that sees less skew can contain more logic than a path that sees a larger skew. Moreover, a path using skew-tolerant sequencing elements can contain more logic than a path between flip-flops. Hence, it is useful to predict the median skew seen in various clock domains for the purpose of setup time analysis.

As the systematic clock skew tends to be low, most clock skew sources occur from random process variations and noise. However, critical paths also experience random process variation and noise, so some will be slower than simulation predicts while others will be faster. If the chip is tuned until many critical paths have nearly the same cycle time in simulation, it is likely that a few paths will be slower than expected in the fabricated part and will limit the chip speed. It is improbable that the paths with worst-case variations in data delay are also those affected by the worst clock skew. Hence, a Monte Carlo simulation considering both variations in delay of the data paths and clock network will predict a smaller and more realistic clock skew budget [Harris01b]. [Agarwal04] describes an efficient method of directly determining the probabilistic skew.

Overall, choosing the appropriate clock skew budget is an ongoing source of research and debate among designers. In practice, many design teams seem to perform some calculations, and then fudge the numbers until the clock skew budget is about 10% of the cycle time. This strategy has historically led to functional chips most of the time, but becomes more risky as cycle times decrease. Measured clock skew numbers reported in publications are notoriously optimistic; for example, [Mule02] finds an average reported skew of 3.2% of the cycle time in recent microprocessors. Part of the reason is that measuring the worst case skew is difficult. Measurements tend to be made at only a few clocked elements for a small number of clock cycles, while the chip must be designed to operate correctly for the largest skew seen anywhere on the chip anytime during its ~10^{17} cycle life span.

12.4.7 Adaptive Deskewing

Just as a PLL or a DLL can compensate for the overall clock distribution delay, additional adjustable delay buffers can compensate for mismatches in clock distribution delay along various paths. For example, the Pentium II and 4 use such buffers at the leaves of the clock spine to eliminate systematic and random variations in the clock distribution network. Figure 12.32 shows an example of a digitally adjustable delay line with eight levels of adjustment. The select signals use a *thermometer code*[2] to produce a monotonically decreasing propagation delay as more pass transistors are turned on.

In the Pentium II, a phase comparator checks the arrival times of the physical clocks and adjusts the digitally controlled delay lines to make all clocks arrive simultaneously.

[2]In an N-bit thermometer code, a number $n \in [0, N]$ is represented with n 1s in the least significant positions. For example, the number 3 is represented in an 8-bit thermometer code as 00000111.

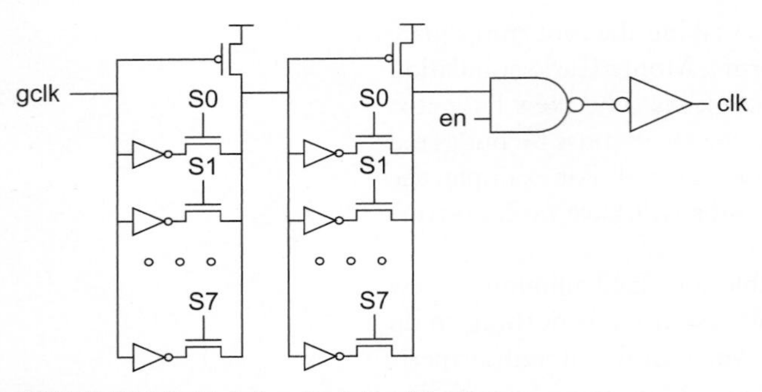

FIGURE 12.32 Digitally adjustable delay line

The loop bandwidth is low enough to ignore jitter, but high enough to compensate for temperature drift. This technique is sometimes called *adaptive deskewing* [Geannopoulos98]. In the Pentium 4, the delay line is adjusted using a scan chain through the boundary scan test access port. 46 phase comparators measure the phase of the clock gaters. Their results can be shifted out through the TAP. The delay lines can be adjusted to reduce systematic and random skew to ±8 ps, as compared to approximately 64 ps before adjustment. The delay lines can also deliberately delay certain clocks to improve performance or assist with debug [Kurd01]. The Itanium series of microprocessors uses similar deskew techniques [Tam00, Anderson02, Stinson03, Tam04]. In the 1.5 GHz Itanium 2, deskew takes place during manufacturing test; on-chip fuses are blown to eliminate the systematic and random skew without needing calibration upon reset or during normal operation.

A drawback of adaptive deskewing is that the buffers introduce extra delay. Voltage noise on the buffers appears as jitter. Unless all of the deskew buffers use well-filtered power supplies, the extra jitter from the deskew buffers can overwhelm the improvement in systematic and random skew.

12.5 PLLs and DLLs

As introduced in Section 12.4.3, phase-locked loops and delay-locked loops are widely used in clock generation and in clock-data recovery for high-speed I/O. A PLL adjusts an oscillator until it produces an output clock matching the frequency and phase of an input clock. A DLL adjusts a delay line until it produces an output clock delayed by the desired amount (typically one cycle) from the input clock. This section examines the operating principles of the PLL and DLL in further detail. We explore circuit designs and linear system models for each component.

12.5.1 PLLs

A phase-locked loop is a dynamical system that produces an output clock in response to the frequency and phase of the input clock. To understand its characteristic behaviors such as bandwidth and stability, it is a common practice to build a simple linear continuous-time system model for the PLL. The model describes the deviations from the lock point.

We can model clocks as ideal square waves alternating between 0 and 1. The key to analyzing PLLs is learning to think about variables representing *phase* rather than voltage. Each clock is described by its phase $\Phi(t)$

$$\text{clk} = \begin{cases} 1 & \Phi(t) \bmod 2\pi < \pi \\ 0 & \Phi(t) \bmod 2\pi \geq \pi \end{cases} \tag{12.18}$$

The phase is the integral of the instantaneous frequency $f(t)$

$$\Phi(t) = 2\pi \int_0^t f(t)\,dt \tag{12.19}$$

If the frequency is constant, the phase is a linear ramp and the clock is periodic as shown in Figure 12.33 for a 250 MHz clock. However, if the clock has jitter, the instantaneous frequency will vary and the phase will cease to be a straight line.

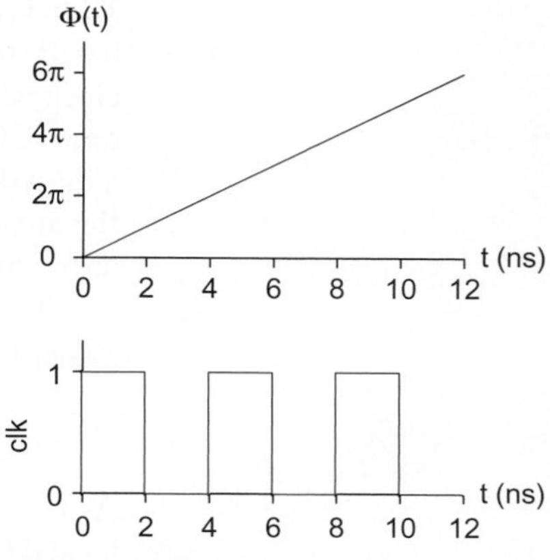

FIGURE 12.33 250 MHz clock phase and waveform

Suppose a multiply-by-N PLL receives an input clock with a nominal phase $\Phi(t)$. The actual clock may have some jitter, causing a small time-varying change in phase $\Delta\Phi_{in}(t)$. When the PLL is locked, the output clock should oscillate N times as fast. However, it may also have some phase offset $\Delta\Phi_{out}(t)$. Thus, the actual input and output clock phases can be written as

$$\begin{aligned}\Phi_{in}(t) &= \Phi(t) + \Delta\Phi_{in}(t)\\ \Phi_{out}(t) &= N\Phi(t) + \Delta\Phi_{out}(t)\end{aligned} \tag{12.20}$$

Figure 12.34 shows a linear system model for a multiply-by-N PLL under these assumptions. The model describes the time-varying phase offsets from the nominal locked operating point. The input and output variables are $\Delta\Phi_{in}$ and $\Delta\Phi_{out}$, the small changes in the input and output clock phases from their nominal values, respectively. The variables are expressed in the s-domain (i.e., after Laplace transformation) rather than the time domain to compactly express operations such as differentiation (multiply by s) and integration (divide by s).

Be sure to remember the assumptions that underlie such a linear system model:

- A linear system model describes how the PLL responds to a small change in the input clock phase ($\Delta\Phi_{in}$) when the PLL is near the locked condition. The response is also expressed by the small change in the output clock phase ($\Delta\Phi_{out}$) from the nominal locked position.
- A PLL may exhibit highly nonlinear behavior when it is far from the locked condition. This lock-acquisition behavior cannot be explained by a linear system model and special attention is required to ensure that the PLL can always reach the desired locked condition (see Section 12.5.3).
- PLLs are typically discrete-time systems that perform phase detection once per cycle. However, we assume that the bandwidth is sufficiently low compared to the input frequency (e.g., < 1/10 of the input frequency) so that the PLL can be well approximated as a continuous-time system. If the bandwidth is too high, the phase detection delay may destabilize the feedback loop.

The remainder of this section discusses each component's function and CMOS implementation.

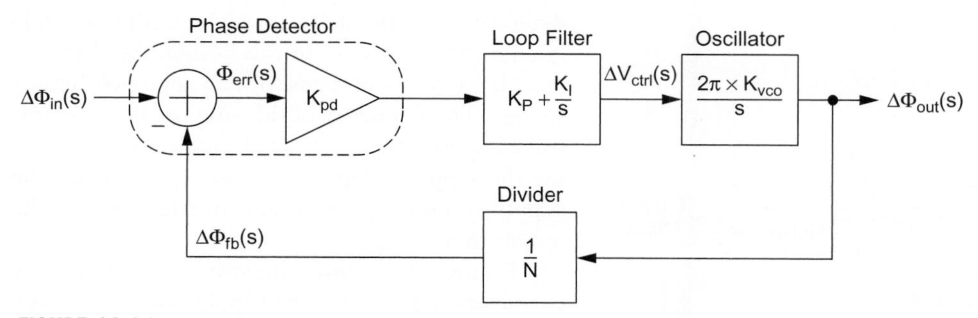

FIGURE 12.34 Linear system model of a PLL

12.5.1.1 Oscillator The oscillator in a PLL generates a clock whose frequency is adjusted based on a control input. For example, a *voltage-controlled oscillator* (VCO) generates a clock whose frequency varies with an input voltage. There are also *current-controlled oscillators* (ICOs) and *digitally controlled oscillators* (DCOs) whose control inputs are a current or a digital number, respectively. We will consider the case of a VCO in this discussion but the analyses and models for other types of oscillators are essentially the same except the different units for the control input.

The VCO control voltage V_{ctrl} can be written as the sum of the value during lock, V_{ctrl0}, and some small offset ΔV_{ctrl}.

$$V_{\text{ctrl}}(t) = V_{\text{ctrl0}} + \Delta V_{\text{ctrl}}(t) \quad (12.21)$$

As the VCO's clock frequency f_{out} changes with the control voltage, the offset from the locked frequency is

$$\frac{\Delta f_{\text{out}}}{\Delta V_{\text{ctrl}}} = K_{vco} \quad (12.22)$$

This small-deviation assumption allows us to express their relationship with a single gain factor, K_{vco}, which is often referred to as the *VCO gain*. When f_{out} is expressed in Hertz and V_{ctrl} is in Volts, the VCO gain has a unit of Hz/V. The above equation also assumes that the frequency responds to the input change almost instantaneously, which is the case for most practical VCO implementations and is also the requirement for the PLL to be stable. Because phase is the integral of frequency, the resulting change in the output clock phase $\Delta\Phi_{\text{out}}$ can be expressed in the *s*-domain:

$$\frac{\Delta\Phi_{\text{out}}(s)}{\Delta V_{\text{ctrl}}(s)} = \frac{2\pi K_{vco}}{s} \quad (12.23)$$

Acute readers may notice that the change in the control voltage does not immediately shift the clock phase of a VCO. The phase rather changes with the time-integration of the control voltage. In other words, *it takes time* to change the phase of a VCO. This characteristic leads to an often-cited phenomenon called *jitter accumulation*. That is, phase error in a PLL does not get corrected immediately after it has been detected by the phase detector and acted on by the loop filter. For a short duration, the phase error may even keep growing! For the same reason, PLLs are also more sensitive to stability issues than DLLs.

Figure 12.35 shows an example circuit implementation of a VCO using a *ring oscillator*. Recall that a ring oscillator consists of an odd number of inverting stages. The clock period is determined by the delay for a clock edge to circle around the ring twice. In this design, the *delay element* is a CMOS inverter with an adjustable supply voltage. The frequency of this ring oscillator is controlled by varying the delay of each stage by adjusting the supply voltage. A voltage regulator sets the supply voltage V_{reg}. A level converter restores the output to full-swing levels.

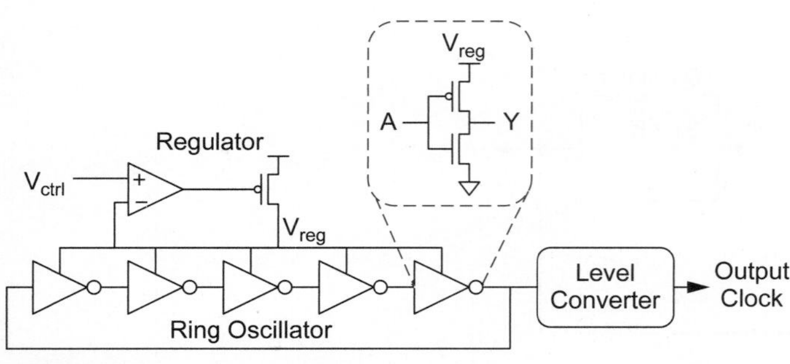

FIGURE 12.35 Voltage-controlled oscillator

Figure 12.36 plots the voltage-to-frequency characteristics of a 9-stage supply-regulated VCO.

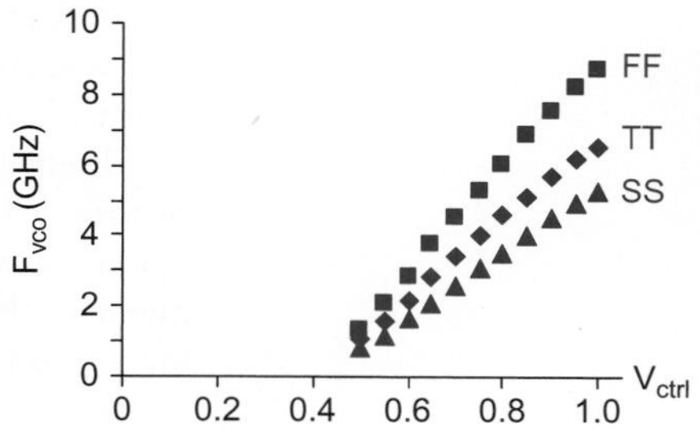

For example, under TT conditions, the VCO gain $K_{vco} = 12$ GHz / V around a 3 GHz operating point. Observe that the curve is not a straight line across the entire range, implying that K_{vco} varies depending on the locked frequency. Moreover, the curve shifts significantly with process and temperature variations. Therefore, one of the difficulties in VCO design is to achieve a wide enough tuning range that can always encompass the desired frequency range across all possible process/environment conditions. Another difficulty lies with the variation in the VCO gain value, which makes it hard to achieve predictable PLL bandwidth and stability.

FIGURE 12.36 VCO voltage-to-frequency characteristics over different process conditions

Of course, there are alternative ways of varying the delay of the stages. Since delay is a function of the load capacitance and drive resistance, it can be varied by adjusting either one. Figure 12.37 illustrates these options, using either a control voltage, a control current, or a digital control value. The adjustable resistance method is called a *current-starved inverter*. These methods tend to provide a smaller range of achievable delays than the adjustable power supply of Figure 12.35.

Some oscillators are based on resonant structures such as inductor-capacitor (LC) tanks and quartz crystals rather than rings oscillators [Razavi03]. While resonance-based oscillators have superior noise performance, ring oscillators are still popular choices for many practical applications because of their wide tuning ranges and ease of integration with other digital CMOS circuits.

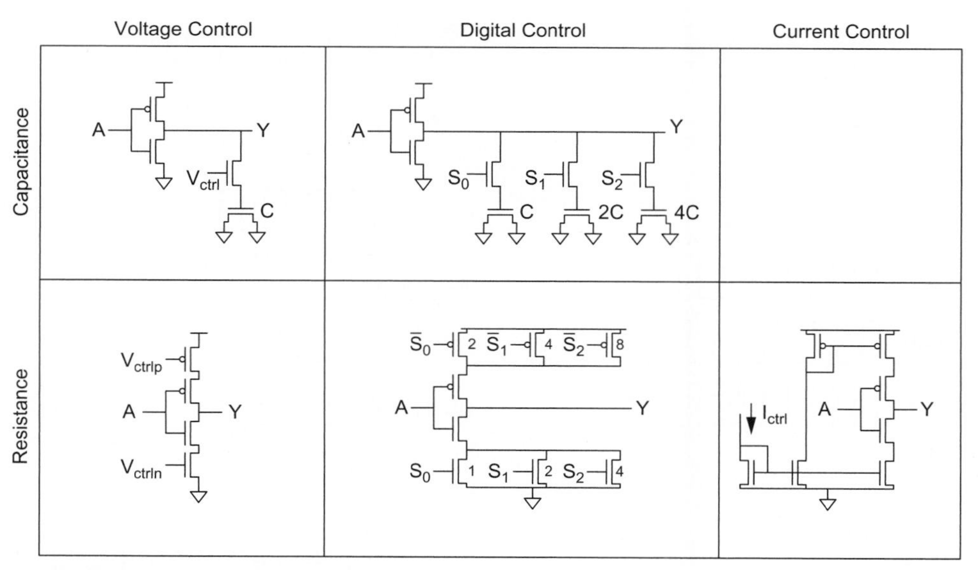

FIGURE 12.37 Alternative delay elements

12.5.1.2 Divider PLLs that produce clocks with the different frequencies than the input clock may have a frequency divider in their feedback paths, as was shown in Figure 12.23. The frequency divider simply divides its input frequency and phase by a factor N:

$$\Delta f_{\text{fb}} = \frac{\Delta f_{\text{out}}}{N}$$
$$\Delta \Phi_{\text{fb}} = \frac{\Delta \Phi_{\text{out}}}{N} \tag{12.24}$$

where N is the division ratio which also corresponds to the frequency multiplication factor. Δf_{fb} and $\Delta \Phi_{fb}$ denote the changes in the frequency and phase of the clock that is fed back to the phase detector, respectively.

Frequency dividers are most commonly realized as modulo-N counters as described in Section 10.5. It is important to keep in mind that the frequency divider has to correctly operate at well beyond the nominal frequency because the VCO may produce higher frequencies during its start-up transients. Otherwise, the PLL may be trapped into a deadlocked condition. See Section 12.5.3 for more details on this pitfall.

12.5.1.3 Phase Detector A phase detector (PD) measures the phase difference between two clocks. In a PLL, it compares the input clock against the feedback clock. The phase error is $\Phi_{err} = \Delta \Phi_{in} - \Delta \Phi_{fb}$.

Although numerous phase detectors have been invented, the two most common are the XOR phase detector and the phase-frequency detector (PFD), shown in Figure 12.38(a, d). These phase detectors produce an output with a duty cycle proportional to the phase difference. If the loop filter bandwidth is much lower than the input clock frequency, the phase detector output can be treated as the average value.

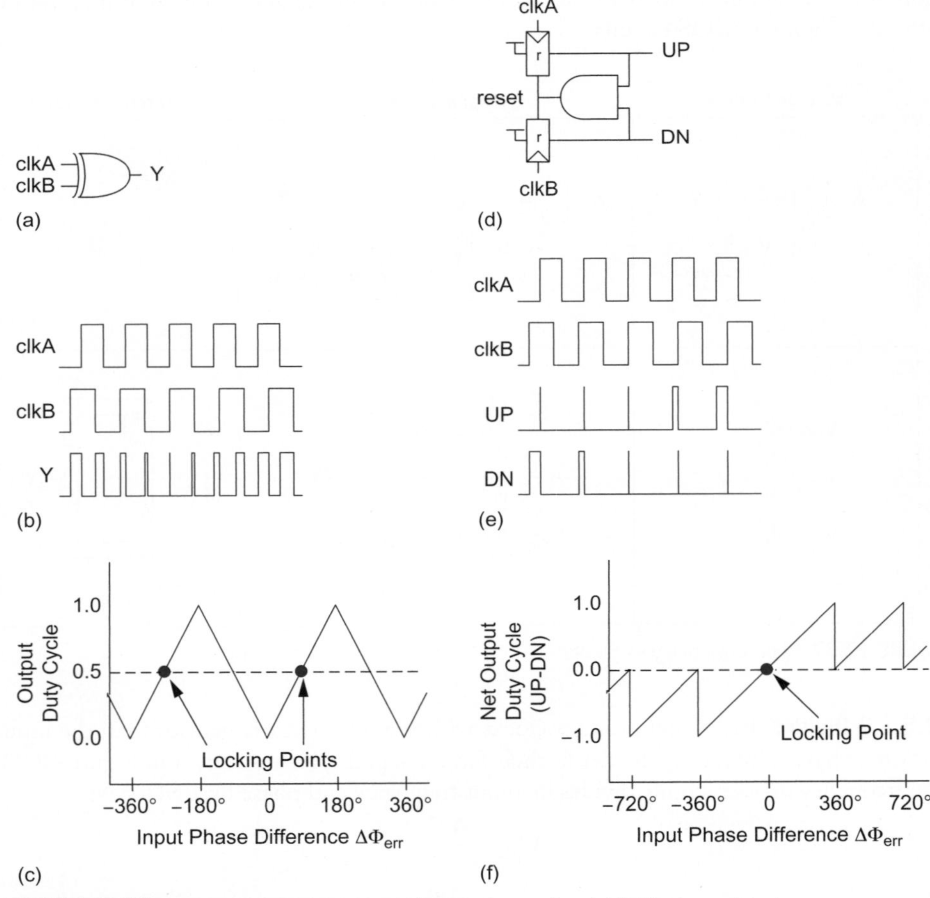

FIGURE 12.38 Phase detector implementations and operation (a) XOR phase detector, (d) phase-frequency detector

The XOR PD produces a high output whenever the two input clocks are at different levels. An example output pulse waveform for various input phase differences is plotted in Figure 12.38(b). A common way to describe the characteristic response of a PD is to plot the output duty cycle (the fraction of the time the output is 1) as a function of the input phase difference, as shown in Figure 12.38(c). Assuming both clocks have 50% duty cycles,[3] the PD produces a full low-pulse (0% duty cycle, interpreted as −1) when the input clocks have identical pulses and a full high-pulse (100% duty cycle, interpreted as +1) when they are out of phase (π radians apart). The duty cycle varies linearly between the two points, crossing 50% (interpreted as 0) for input phase differences at $\pi(1/2 + n)$ for any integer n. However, notice that the PD has positive gains for a half of those zero-level points while it has negative gains for the other half. A PLL can converge only to the points where the PD gain results in the negative feedback. If such PD gain is positive, then an XOR PD can be said to have locking points at $\pi(1/2 + 2n)$. The XOR PD produces an average voltage output

$$\frac{V_{pd}(s)}{\Phi_{\text{err}}(s)} = \frac{V_{DD}}{\pi} = K_{pd} \tag{12.25}$$

The PFD in Figure 12.38(d) belongs to a class of sequential PDs with internal state. The waveforms in Figure 12.38(e) illustrate the operation of this PD. Sequential PDs may produce different outputs for the same input phase difference depending on the past history, which can help extend the linear range in the characteristic curve as plotted in Figure 12.38(f). Assume that initially both outputs of the PD, UP and DN, are at 0s. When the reference clock rises first, the flip-flop triggered by the clock asserts UP high. When the feedback clock rises later, the other flip-flop asserts DN as well. But then, the AND logic connected to the asynchronous reset input of the flip-flops deasserts both UP and DN signals to 0 as soon as they both reach 1s, returning the PD to the original state. The resulting difference in the UP and DN pulse widths corresponds to the timing difference between the two clocks' rising edges.

A PFD typically uses a charge pump to convert the UP and DN pulses into a current output, as shown in Figure 12.39. Near the point of lock, the PFD and charge pump together have a transfer function

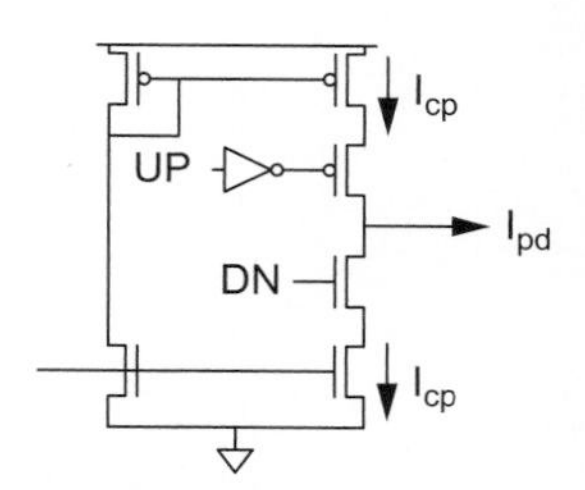

FIGURE 12.39 Charge pump

$$\frac{I_{pd}(s)}{\Phi_{\text{err}}(s)} = \frac{I_{cp}}{2\pi} = K_{pd} \tag{12.26}$$

Sequential PDs have a number of advantages over combinational PDs such as the XOR PD. First, they can be insensitive to the variations in the input clock duty cycles, by being triggered by either the rising edges or falling edges of the input clocks, but not by both. Second, notice that the characteristic curve in Figure 12.38(f) does not alternate its sign every π radians as it did in Figure 12.38(c). Rather, it maintains its sign to indicate the correct polarity of the phase difference. This property makes the PFD serve as a frequency detector as well, when the two input clocks have sizeable frequency difference. If the PLL starts up at the wrong frequency, the PFD will adjust the frequency up or down as required. PFDs are preferred in clock generation PLLs because they help PLLs acquire locks reliably and quickly. However, misuse of PFDs in DLLs may result in intermittent

[3]One problem with the XOR PD is that the output duty cycle may vary depending on the duty cycles of the input clocks.

dead-lock problems (see Section 12.5.3). Moreover, clock-data recovery (CDR) circuits require XOR-based PDs for reasons discussed in Section 12.7.6.

12.5.1.4 Loop Filter A loop filter (LF) is the central element of any PLLs because it determines how much adjustment should be made on the VCO control voltage based on the phase error. Understanding the loop filter dynamics is the key to designing a high-performance PLL.

A typical loop filter produces a control voltage that is proportional to both the phase error and the integral of the phase error. Assuming a PFD and charge pump producing a current output, this can be expressed as

$$\frac{V_{\text{ctrl}}(s)}{I_{pd}(s)} = \frac{K_I}{s} + K_P \tag{12.27}$$

where K_I/s term implies the time integration of the phase error. In essence, the integral control term adjusts V_{ctrl0} so that the VCO oscillates at the desired frequency when the phase error is zero. If V_{ctrl0} is at a wrong value, then the nonzero phase error will shift V_{ctrl0} toward the direction to reduce the error. The integral term will settle to a final value only when the phase error becomes zero.

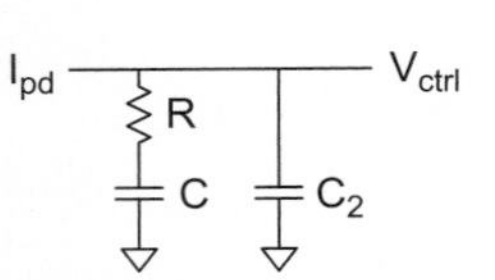

FIGURE 12.40 Loop filter implementation based on a charge pump followed by an RC filter

In conventional analog PLLs with PFDs, this LF is usually implemented with an RC filter, as shown in Figure 12.40. C_2 is much smaller than C and can be disregarded for initial analysis. The RC filter converts the current to the voltage V_{ctrl}:

$$\frac{V_{\text{ctrl}}(s)}{I_{pd}(s)} = \frac{1}{sC} + R \tag{12.28}$$

Any low-frequency phase error produces a current that is integrated on the capacitor C until V_{ctrl} reaches V_{ctrl0} such that the PLL is in lock with no phase error. If high-frequency noise introduces a phase error disturbing the lock, the resistor R produces a control voltage proportional to the error to correct for the noise.

A realistic loop filter has some additional capacitance C_2 between V_{ctrl} and GND due to parasitics and the load presented by the VCO. This capacitance smooths out ripples on V_{ctrl} caused by the charge pump turning ON and OFF, reducing jitter. However, it can destabilize the loop if it is too large. Typically, C is selected to be at least an order of magnitude larger than C_2 so that C_2 can be ignored.

12.5.1.5 Loop Dynamics Now that we have analyzed the behaviors of the individual components in the PLL, we can discuss how the overall PLL will respond to the input clock phase when we close the feedback loop. Specifically, the linear system analysis using the models derived in the previous subsections will help us understand how the key PLL characteristics such as bandwidth and stability are determined by the component parameters such as VCO gain (K_{vco}), charge pump current (I_{cp}), and loop filter resistance (R) and capacitance (C). Some backgrounds on linear systems and control theory may be required to fully understand the material in this subsection.

The response of the PLL's output clock phase $\Delta\Phi_{\text{out}}$ to the input reference clock phase $\Delta\Phi_{\text{in}}$ is given by the closed-loop transfer function of the PLL:

$$H(s) = \frac{\Delta\Phi_{\text{out}}(s)}{\Delta\Phi_{\text{in}}(s)} = \frac{K_{pd}\left(R + \frac{1}{sC}\right)\frac{2\pi K_{vco}}{s}}{1 + \frac{1}{N}K_{pd}\left(R + \frac{1}{sC}\right)\frac{2\pi K_{vco}}{s}} \tag{12.29}$$

The transfer function can be rewritten as a standard second-order system with a natural frequency ω_n and a damping factor ζ. The gain is N, corresponding to frequency multiplication by a factor of N.

$$H(s) = N\frac{2\zeta\omega_n s + \omega_n^2}{s^2 + 2\zeta\omega_n s + \omega_n^2} \tag{12.30}$$

$$\omega_n = \sqrt{\frac{I_{cp}K_{vco}}{NC}} \tag{12.31}$$

$$\zeta = \frac{\omega_n}{2}RC \tag{12.32}$$

The natural frequency is a measure of the loop bandwidth. Loops with greater bandwidth track input changes more rapidly. The bandwidth is typically selected to minimize output clock jitter. If the output jitter is dominated by on-chip noise disturbing V_{ctrl}, high bandwidth is desirable to rapidly correct the control voltage. However, if output jitter is dominated by input clock jitter, then low bandwidth is preferable to reject the input clock noise. In any event, the natural frequency should be at least an order of magnitude below the input clock frequency so that the continuous-time model is valid.

The damping factor is a measure of the loop stability. If the damping factor is less than $1/\sqrt{2}$, the PLL will ring in response to a step change in phase. This is often considered undesirable because it can increase jitter, so ζ is usually selected in the range of 0.7–1.

12.5.1.6 Validation The second-order analysis in the previous section is only an approximation of the behavior of the nonlinear system. The nonlinearities can lead to locking problems. Moreover, lag in the response can lead to instability. After drafting a reasonable paper design, simulation is essential to ensure the loop locks and is stable in all process corners.

Designers typically simulate the closed-loop response of the PLL to a known set of input patterns in SPICE. Popular choices of those input patterns are steps, impulses, or sinusoids, with which one can estimate the closed-loop transfer function $H(s)$ and subsequently evaluate the bandwidth and stability. A clever strategy is to use the small-signal AC analysis capabilities of SPICE to analyze the response in the phase domain, enabling direct characterization of the transfer function [Kim07].

12.5.1.7 Advanced PLL Architectures PVT variations make it difficult to design a stable PLL that meets performance requirements with good yield. Moreover, the loop bandwidth that minimizes jitter depends on the operating frequency. *Self-biased PLLs* adjust parameters such as charge pump current and loop filter resistance to track operating frequency and compensate for process variations [Maneatis03, Kim03b].

Analog components are troublesome to build in nanometer CMOS processes. *All-digital PLLs* (ADPLLs) are a growing field of interest. A typical approach uses a DCO and a digital loop filter [Tierno08].

12.5.2 DLLs

A delay-locked loop aims at the same goal of aligning the output clock to the input reference clock but operates on a slightly different principle. It adjusts the delay of a buffer chain instead of the frequency of an oscillator. As stated earlier, this difference makes the loop filter design for DLLs simpler and less prone to stability problems than in PLLs. This section explores the components of a DLL and the loop characteristics.

Recall that Figure 12.22(b) showed the architecture of a DLL. The input clock is fed into a variable delay line which also includes the buffers to drive the on-chip load. The output clock distributed to the final load is compared back to the input clock. If their edges are not aligned, the phase detector generates error information upon which the loop filter makes appropriate actions to the delay line to reduce the error.

Figure 12.41 shows a linear system model. Compare and contrast this diagram with the PLL in Figure 12.34. Now the state variables are time (T) rather than phase (Φ). The input is ideally periodic with a period T_c. When the DLL is locked, the output is delayed by exactly T_c. The model again describes the effect of small variations ΔT from the operating point for the input cycle time and output delay. The same caveats apply that the linear model is only valid for small deviations from lock and when the bandwidth is less than 1/10 of the input clock frequency. The DLL uses a delay line in place of a VCO and an integrator in place of a PI loop filter. The DLL is a first-order system, so it avoids many of the stability risks of the second-order PLL.

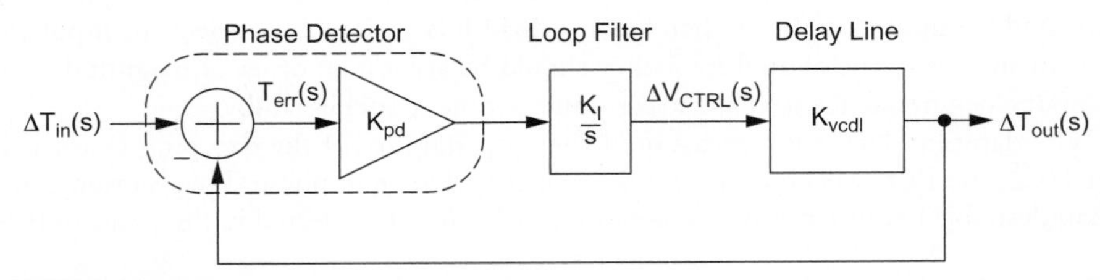

FIGURE 12.41 DLL linear system model

A DLL can produce multiple clock outputs with known phase relationships by tapping from several points along the delay line. For example, if the delay line has eight stages, tapping every other stage yields clocks delayed by 1/4, 1/2, and 3/4 of a cycle as well as a full cycle.

12.5.2.1 Delay Line The variable delay line adjusts the delay between its input and output clocks as directed by the control input. The control input may be a voltage, current, digital number, etc. A *voltage-controlled delay line* (VCDL) is commonly used. For VCDLs, the voltage-to-delay characteristics can be modeled by the following linear equation between the small deviations in the delay and the control voltage (V_{ctrl}) from their respective locked values:

$$\frac{\Delta T_{\text{out}}(s)}{\Delta V_{\text{ctrl}}(s)} = K_{vcdl} \tag{12.33}$$

As in the VCO case, the conversion factor K_{vcdl} is called the *VCDL gain* and has a unit of seconds/V. Unlike VCOs that adjust the clock timing via the time integration of the control input, VCDLs can shift the clock timing almost instantaneously by changing the control voltage. Therefore, DLLs do not typically exhibit jitter accumulation.

Any of the variable delay elements discussed in Section 12.5.1 can be used for a VCDL as well. For example, the delay line in Figure 12.42(a) is built from four stages of current-starved inverters. The bias voltage varies the current and therefore the delay.

Figure 12.42(b) plots the voltage-to-delay characteristic curves for a 16-stage line under various process conditions. The delay tuning range must be wide enough for the delay line to provide the delay shift that can align the clocks for all possible conditions. However, the wide tuning range of a VCDL may make a DLL vulnerable to false locking problems, to be discussed in Section 15.5.3.

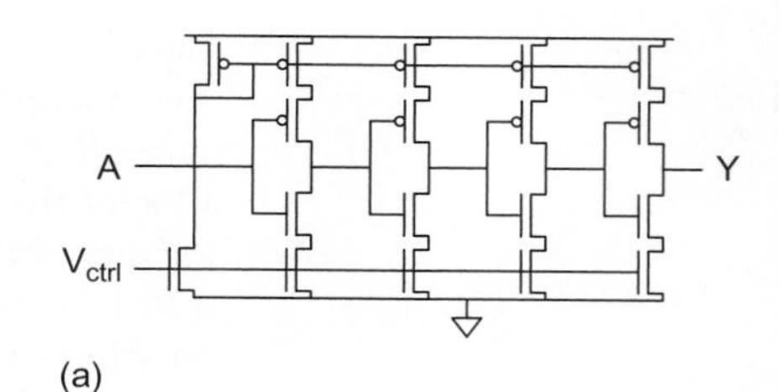

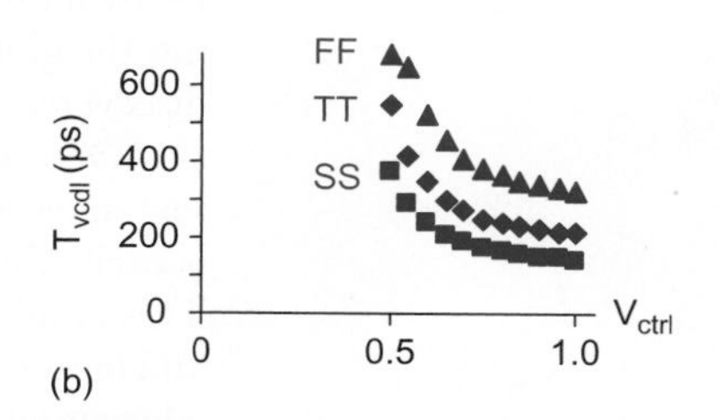

FIGURE 12.42 An example of a voltage-controlled delay line (VCDL) (a) a current-starved inverter chain, (b) its voltage-to-delay characteristics for various process conditions

12.5.2.2 Phase Detector A DLL can use the same types of phase detectors as a PLL. A PFD followed by a charge pump is a common option. It produces an output current with the following transfer function

$$\frac{I_{pd}(s)}{T_{\text{err}}(s)} = \frac{I_{cp}}{T_c} \tag{12.34}$$

12.5.2.3 Loop Filter The loop filter for a DLL has the similar role to that of a PLL, controlling the delay based on the detected phase error. The loop filter design for DLLs is simpler as an integral control alone is typically sufficient to stabilize the feedback loop. Figure 12.43 shows a loop filter consisting of a single capacitor that integrates the current out of the phase detector.

The integral control adjusts the control voltage until the phase error reaches 0. As discussed in the case for PLLs, this integral control is essential in maintaining a low skew between the external and internal clocks in the presence of process and environmental variations. Expressed in the s-domain, the loop filter behavior can be modeled as

$$\frac{\Delta V_{\text{ctrl}}(s)}{I_{pd}(s)} = \frac{K_I}{s} = \frac{1}{sC} \tag{12.35}$$

FIGURE 12.43 Charge-pump based loop filter implementation for a DLL

12.5.2.4 Loop Dynamics The DLL has a closed loop transfer function of

$$H(s) = \frac{\Delta T_{\text{out}}(s)}{\Delta T_{\text{in}}(s)} = \frac{1}{s\tau + 1} \tag{12.36}$$

with a time constant

$$\tau = \frac{1}{K_{pd} K_I K_{vcdl}} = \frac{C T_c}{I_{cp} K_{vcdl}} \tag{12.37}$$

Observe that the transfer function has a magnitude of 1 at low frequencies, indicating that the DLL tracks changes in the input cycle time. The time constant τ indicates how long the DLL needs to respond to abrupt changes in frequency. τ should be at least $10T_c$ so that the continuous-time approximation is valid.

Note that the DLL simply delays the input clock. Any jitter propagates directly to the output. If the input is noisy, a PLL is a better way to filter the noise.

12.5.3 Pitfalls

So far we have used linear system analysis to understand how PLLs and DLLs react to input changes and how the design parameters such as charge pump current or loop filter capacitance influence the key loop dynamics such as bandwidth and stability. While this is the most prevailing methodology to design PLL/DLLs today, it is important to keep in mind that the linear system analysis relies on the assumptions stated in Section 12.5.1. One of them is that the linear system model describes the system behavior only at the vicinity of its locked condition. In other words, even if the linear system analysis says that a PLL is stable, it cannot guarantee that the PLL will always converge when it starts from an arbitrary condition far from the desired locking point. Many of the design pitfalls can be attributed as convergence failures. Unfortunately, there is no systematic way of validating the global convergence yet. The best practice is to try not to repeat the bugs that are discovered so far. A few representative cases are listed in this subsection.

One failure example for a PLL is when its frequency divider does not have an operating range as wide as the VCO. Suppose that the PLL starts in a condition where the VCO is oscillating at a frequency higher than the maximum operating frequency of the divider. This condition is difficult to avoid unless the circuits are checked for all possible global and local variations. In this case, the usual response of the divider is that it misses the clock edges intermittently. As a net result, the divider produces a lower-frequency clock than it is supposed to. When the phase detector compares this clock to the reference clock, it can erroneously determine that the VCO frequency is too low and direct the loop filter to increase it even higher. The PLL cannot escape from this dead-lock condition because all the forces in the feedback loop are toward the wrong directions. A possible fix is to reset the initial value of the VCO control voltage so that the VCO can be guaranteed to start at a low enough frequency for the divider to operate correctly.

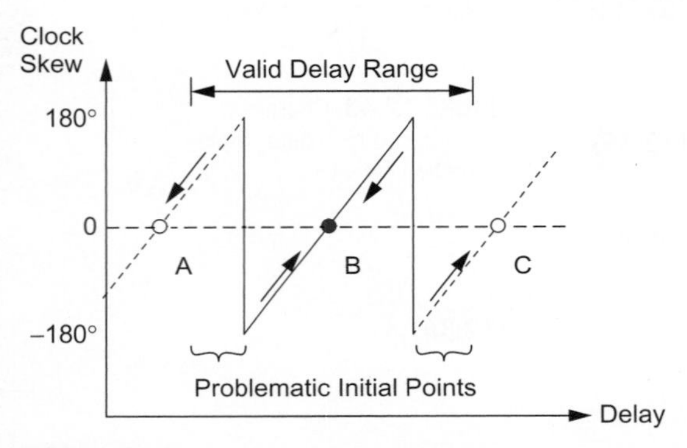

FIGURE 12.44 Illustration of convergence failure examples in a DLL

A DLL may also have a convergence failure even though it does not have a frequency divider. The DLL tries to lock its delay to an integer multiple of the clock period so that the external and internal clock edges become aligned. A problem is that the DLL does not care which integer multiple it is to lock to. Therefore, the DLL has potentially more than one locking point. If the DLL locks to a delay of more than one cycle, it will have more jitter. A more serious problem may occur because the delay line has a finite delay range. This case is illustrated in Figure 12.44. The points A, B, and C are potential lock points while A and C are not within the delay range and therefore cannot be realized. However, depending on the initial condition of the DLL, the phase detector may drive the delay toward A or C, putting the DLL into a dead-lock state chasing a fictitious locking point.

As discussed in Section 12.5.1.3, phase-frequency detectors (PFDs) have certain advantages over phase-only detectors when used for PLLs. However, for DLLs, PFDs can be detrimental. PFDs have internal states that enable them to distinguish 0° from 360 or 720° and DLLs with PFDs can lock only at one particular locking point out of all the possibilities. If the internal states are not properly initialized, the PFD may direct the DLL to lock to a point outside the delay range, forcing it to a dead-locked condition.

12.6 I/O

The input/output (I/O) subsystem is responsible for communicating data between the chip and the external world. A good I/O subsystem has the following properties:

- Drives large capacitances typical of off-chip signals
- Operates at voltage levels compatible with other chips
- Provides adequate bandwidth
- Limits slew rates to control high-frequency noise
- Protects chip against damage from electrostatic discharge (ESD)
- Protects against over-voltage damage
- Has a small number of pins (low cost)

I/O pad design requires specialized analog expertise and knowledge of process-specific ESD structures. Process and library vendors normally supply well-characterized pad libraries tailored to a given manufacturing process. This section summarizes some of the basic design options in I/O subsystems.

A pad consists of a square of top-level metal of approximately 100 μm on a side that is either soldered to a bond wire connecting to the package or coated with a lead solder ball. The term *pad* sometimes refers to just the metal square and other times to the complete cell containing the metal, ESD protection circuitry, and I/O transistors. Input and output pads usually contain built-in receiver and driver circuits to perform level conversion and amplification.

12.6.1 Basic I/O Pad Circuits

Basic I/O pads include V_{DD} and GND, digital input, output, and bidirectional pads, and analog pads.

12.6.1.1 V_{DD} and GND Pads Power and ground pads are simply squares of metal connected to the package and the on-chip power grid. Most high-performance chips devote about half of their pins to power and ground. This large number of pins is required to carry the high current and to provide low supply inductance.

One of the largest sources of noise in many chips is the ground bounce caused when output pads switch. The pads must rapidly charge the large external capacitive loads, causing a big current spike and high L di/dt noise. The problem is especially bad when many pins switch simultaneously, as could be the case in a 64-bit off-chip data bus. Such busses should be interdigitated with many power and ground pins to supply the output current through a low-inductance path. In many designs, the *dirty* power and ground lines serving the output pads are separated from the main power grid to reduce the coupling of I/O-related noise into the core.

Many chips use separate pads for the I/O power supply and for the core. This is essential if the I/O runs at a different voltage than the core, but it also serves to isolate the noisy I/O power from the quieter core.

12.6.1.2 Output Pads First and foremost, an output pad must have sufficient drive capability to deliver adequate rise and fall times into a given capacitive load. If the pad drives resistive loads, it must also deliver enough current to meet the required DC transfer characteristics. Given a load capacitance (typically 2–50 pF) and a rise/fall time specification, the output transistor widths can be calculated or determined through simulation. Typically, these transistors must be very wide and are folded into many legs.

Output pads generally contain additional buffering to reduce the load seen by the on-chip circuitry driving the pad. The method of Logical Effort tells us that the fastest buffers

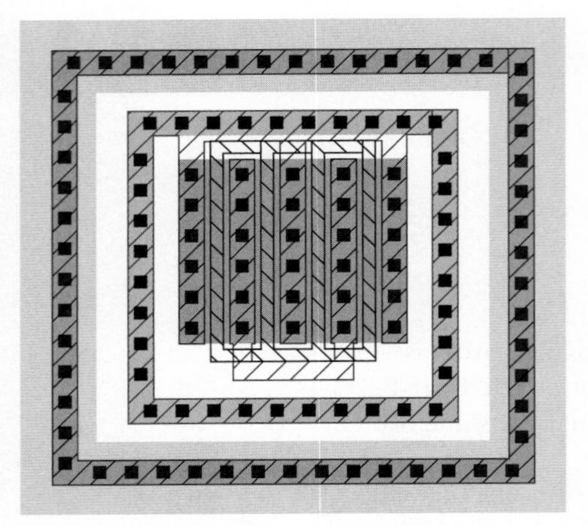

FIGURE 12.45 Double guard rings around folded nMOS output transistor

are built from strings of inverters with fanouts of about 4. In practice, a higher fanout (e.g., 6–8) gives nearly as good delay while reducing the area and power consumption of the buffer. The final stage may have an especially high fanout because the edge rates in the external world are normally an order of magnitude longer than those on chip. However, the final stage must be large enough to source or sink reasonable amounts of current with a small voltage drop.

Latchup, introduced in Section 6.3.6, is a particular problem near output pads, especially when the pads experience voltage transients above V_{DD} or below GND. These transients are likely to occur because of ringing from the bond wire inductance and/or from driving improperly terminated transmission lines. These transients cause the drain-to-body diodes to become forward-biased, forcing current to flow into the substrate or well and potentially causing latchup.

To avoid latchup, the nMOS and pMOS transistors should be separated by substantial distances and surrounded by guard rings. If possible, the output transistors (i.e., those whose drains connect directly to external circuitry) should be doubly guard-ringed, as shown in Figure 12.45. This means that an n-transistor should be encircled with p+ substrate contacts connected to GND, and then further encircled with n+ well contacts in an n-well connected to V_{DD}. The rings should be continuous in diffusion with frequent contacts to metal. Furthermore, dummy collectors consisting of p+ connections to GND and n+ in n-well connections to V_{DD} should be placed between the output transistors and any internal circuitry. These dummy collectors and guard rings serve to capture most of the stray carriers injected into the substrate when the diodes are forward-biased.

The output transistors also often have gates longer than normal to prevent avalanche breakdown damage when overvoltage is applied to the drains. Nonsilicided gates are also preferable because the polysilicon gate resistance better distributes overvoltage across the legs of the output transistor, preventing damage.

12.6.1.3 Input Pads Input pads also contain an inverter or buffer to convert the signal from the noisy external world into a valid logic level for the core circuitry. The input pad also contains electrostatic discharge protection circuitry, described in Section 12.6.2. The buffer may perform level conversion, as will be discussed in Section 12.6.4. In a high-speed system, the buffer typically drives a clocked input register. Section 12.7.4 discusses the timing in depth. Pads can include pullup or pulldown resistors to place an unconnected pad in a known state.

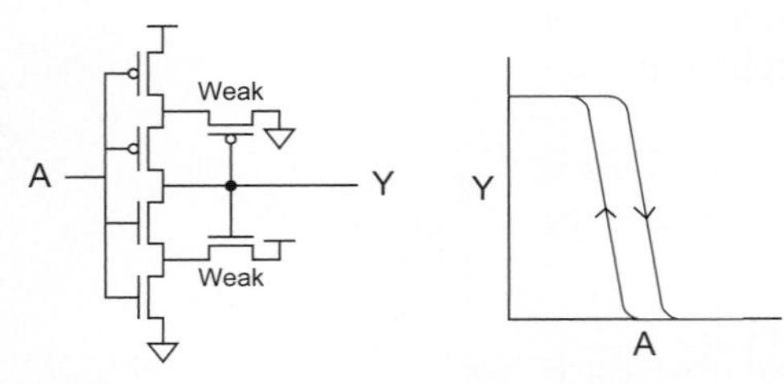

FIGURE 12.46 Schmitt trigger

Some input pads also contain *Schmitt triggers*, as shown in Figure 12.46 [Schmitt38]. A Schmitt trigger has hysteresis that raises the switching point when the input is low and lowers the switching point when the input is high. This helps filter out glitches that might occur if the input rises slowly or is rather noisy.

12.6.1.4 Bidirectional Pads Figure 12.47 shows a bidirectional pad with an output driver that can be tristated and an input receiver. The output driver consists of independently controlled nMOS and pMOS transistors. When the enable is 1, one of the two transistors turns ON. When the enable is 0, both transistors are OFF so the pad is tristated. This design is preferable to the four-transistor "totem pole" tristate from Section 1.4.7 when

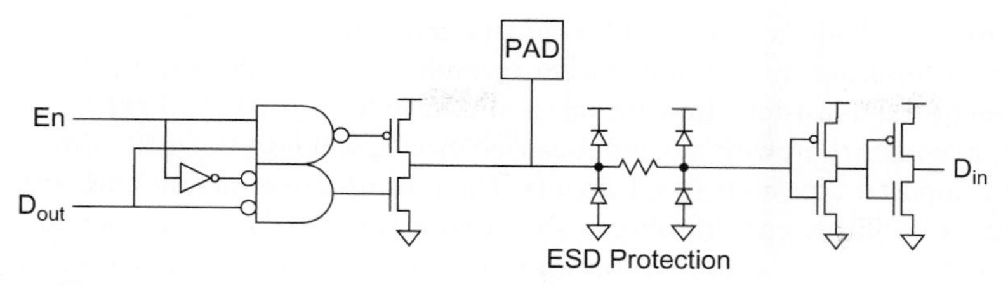

FIGURE 12.47 Bidirectional pad circuitry

driving large capacitances because it has only two rather than four huge transistors in the final stage and the transistors need only be half as wide. Figure 12.48 shows a clever variation on this design in which the NAND and NOR are merged together into a single six-transistor network with two outputs. Such a tristate buffer is smaller and presents less input capacitance on the D_{out} terminal.

Many pad libraries provide only a bidirectional pad. By hardwiring the enable signal to 1 or 0, the pad can be used as an output or input.

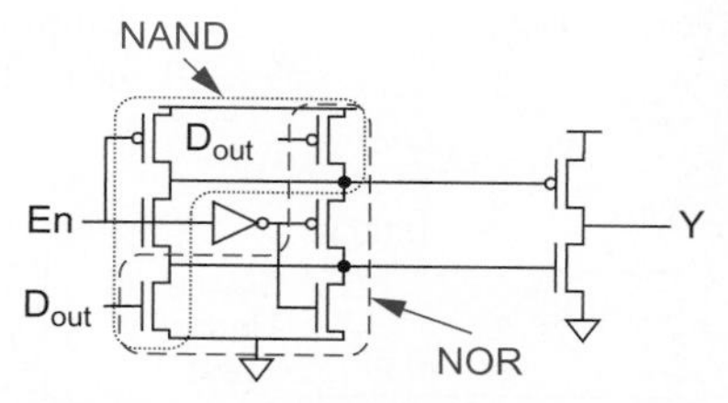

FIGURE 12.48 Improved tristate buffer

12.4.1.5 Analog Pads Analog inputs and outputs connect to simple metal pads and then directly to the on-chip analog circuitry without any digital buffer or driver. Analog pads still require ESD protection circuitry.

12.6.2 Electrostatic Discharge Protection

On a dry day, you have probably experienced a shock when you walk across a carpet and then touch a metal object because you have built up so much charge on your body. Such shocks can destroy integrated circuits. Input pads have transistor gates connected directly to the external world. These gates are subject to damage from electrostatic discharge (ESD) that can puncture and break down the oxide. The breakdown voltage was 40–100 V for older processes with thick (> 100 Å) oxides but now is 5 V or less for modern thin oxides. High ESD voltage on transistor drains can also cause *punchthrough*, in which the source and drain depletion regions meet, allowing large amounts of current to flow through an OFF transistor until overheating and permanent damage occur. ESD voltage outside the power rails also raises the risk of latchup. ESD events cause billions of dollars of losses in the semiconductor industry annually.

The essence of ESD protection is to provide a controlled path to discharge high voltages without damaging the gate oxides [Dabral98]. The path consists of extra circuit elements that clamp the I/O pins to safe levels. The elements are divided into *breakdown* and *nonbreakdown* devices. Nonbreakdown devices are diodes, MOSFETs, and bipolar transistors operating in conventional ways. Breakdown devices include silicon-controlled rectifiers (SCRs), thick field oxide (TFO) transistors, spark gaps, and other devices that break down before the I/O transistors. Breakdown devices are smaller to provide the same level of protection, but are much more difficult to model and design. Therefore, nonbreakdown protection devices are used when possible.

Figure 12.49 shows a typical ESD input protection circuit consisting of diode clamps and a current-limiting resistor. The primary diode clamps turn on if the pad voltage becomes greater than about V_{DD} + 0.7 V or less than –0.7 V, shunting ESD current into the robust V_{DD} or GND networks. A good protection diode has an ON resistance of approximately 1 Ω. A large ESD event may result in 10–20 Å of current flowing, producing a

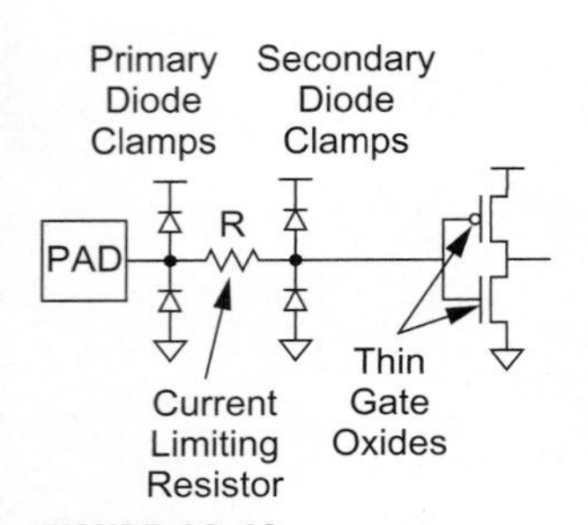

FIGURE 12.49 Input protection circuitry

voltage across the diode large enough to damage transistors. Thus, the protection circuit adds a current limiting resistor and smaller secondary diode clamps to further limit the voltage seen by the transistors. Resistor values anywhere from 100 Ω to 3 kΩ are used. This resistance, in conjunction with any input capacitance *C*, will lead to an RC time constant that can be important for high-speed circuits. The resistors are sometimes made from several squares of unsilicided p+ diffusion in an n-well. Clamping diodes are formed using n+ diffusion to the substrate and p+ diffusion to n-wells. As with output transistors, these diodes and resistors should be double guard-ringed so that they do not inject charge into the substrate and cause latchup.

ESD protection circuits are tested by zapping the pin with an external high voltage. Engineers use standard test circuits shown in Figure 12.50 to characterize ESD robustness. The capacitor is charged to a high voltage, then a switch is closed to connect the capacitor to the pin through a resistor and/or inductor. The *human body model* (HBM) represents the discharge that takes place when an ungrounded person touches a pin of the chip. The *charged device model* (CDM) represents the pin *triboelectrically* charging during manufacturing (i.e., charging through contact with a different material) and then rapidly discharging when it comes in contact with a grounded conductor. The CDM zap is more difficult to protect against, but is also more difficult to perform precisely in the lab. The ESD robustness of the pad is measured as the maximum voltage that the pad can endure. For example, ±15 kV is good for parts such as serial port transceivers that might be exposed to ESD by an end user handling a cable. Parts in an enclosed system are only subject to damage during assembly and can allow limits in the 2–4 kV range.

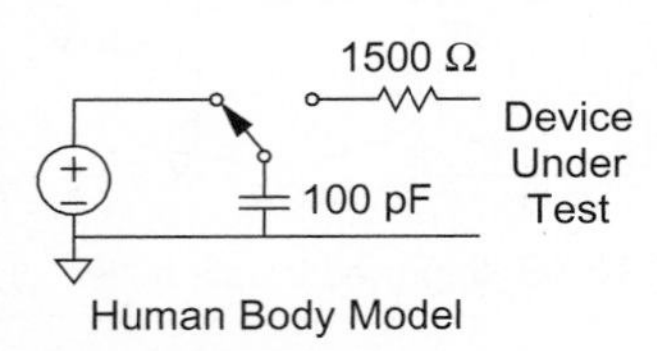

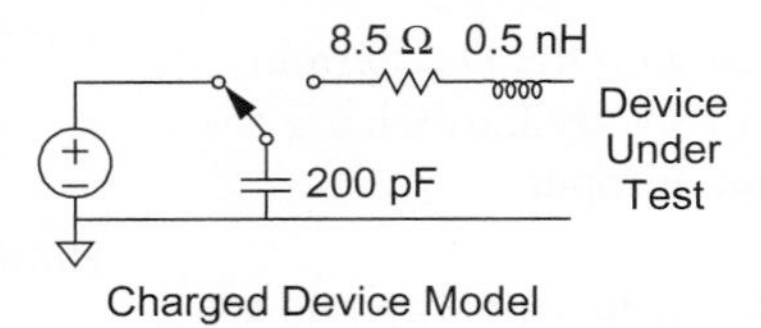

FIGURE 12.50 ESD test circuits

Analog pad protection circuitry must be carefully designed so it does not degrade the bandwidth or signal integrity of the analog components. This is achieved by minimizing the protection diode area. RF pads are extremely demanding because any extra load can compromise performance.

12.6.3 Example: MOSIS I/O Pads

Figure 12.51 shows a layout of a bidirectional pad from the MOSIS service for a 1.6 μm two-metal layer process illustrating the general principles of pad design (see also the inside front cover). The overall cell is about 200 μm on a side. The pad is the large (100 × 75 μm) rectangle consisting of a sandwich of metal1 and metal2 connected with many vias. The SiO_2 overglass covering the metal2 is etched away over the pad so the bond wire can be connected directly to the pad. Two large metal2 rectangles cover most of the pad. The upper one with the legs sticking up is GND, while the lower is V_{DD}.

The bidirectional pad schematic is shown in Figure 12.52. The input protection circuitry consists of some resistance, a thick oxide transistor, and the drain diffusion diodes of the wide output transistors. The resistors are n+ and p+ diffusion wires, each 3.5 squares long. They have nominal sheet resistances of 53 and 75 Ω/□, so the parallel combination of resistance is 109 Ω. To the left and right of the metal pad are thick oxide nMOS transistors consisting of interdigitated fingers. They consist of a source and drain separated by 3 λ, but have no gate. They help protect the pad from ESD because high voltages will punch through the channel and dissipate. The effectiveness of thick oxide transistors is

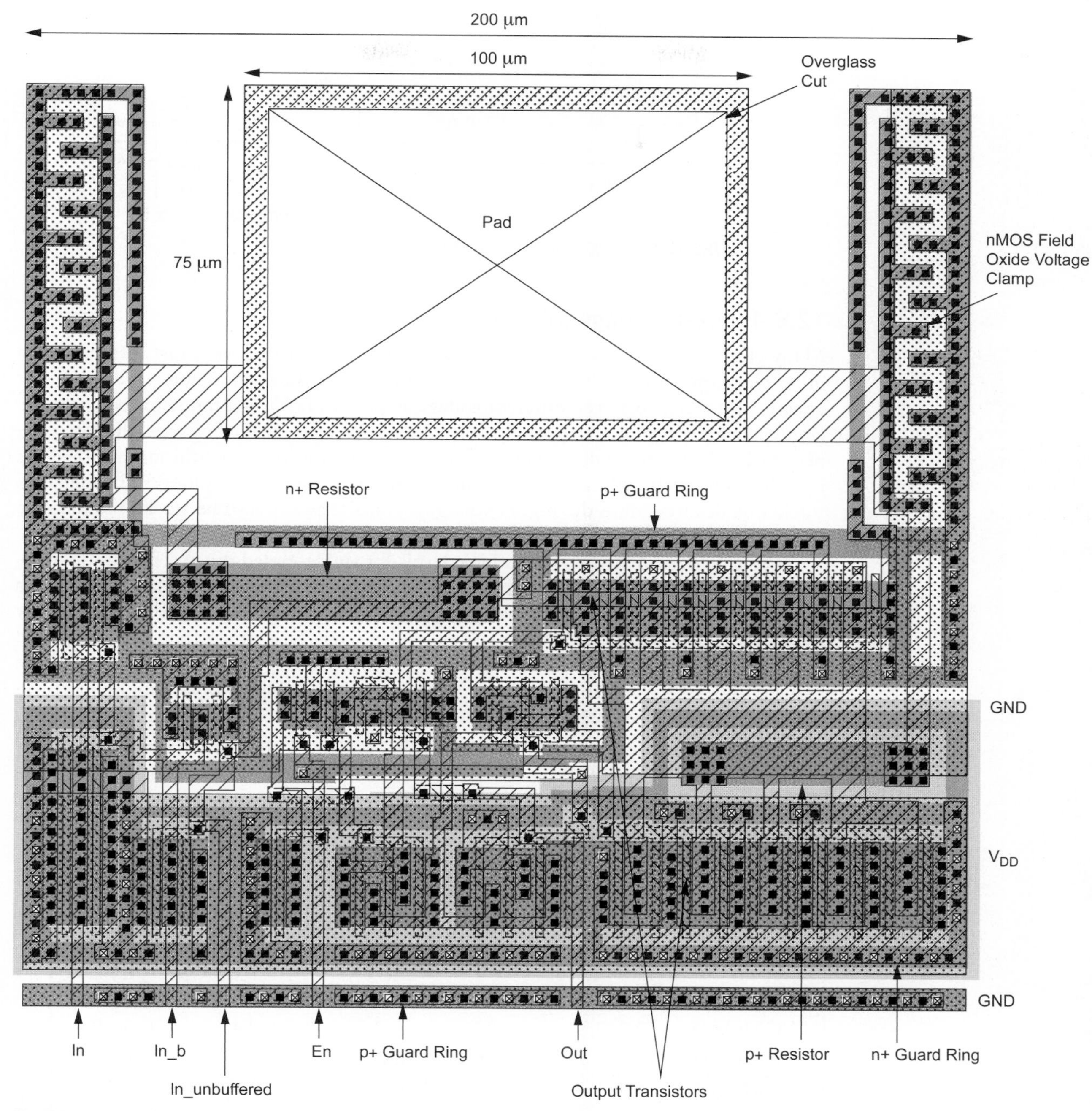

FIGURE 12.51 MOSIS 1.6 μm bidirectional pad. Color version on inside front cover.

process-dependent. The pad uses many substrate/well contacts and is surrounded by double guard rings to prevent latchup during ESD events. The tristate driver and receiver use extensively folded transistors to fit in the space available.

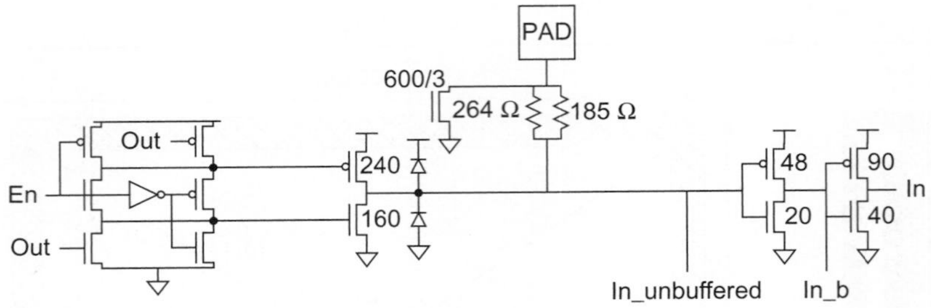

FIGURE 12.52 MOSIS bidirectional pad schematic

12.6.4 Mixed-Voltage I/O

Many chips require a low core voltage for the logic transistors, yet must interface with other chips operating at higher voltages. The I/O pads thus can include level converter circuits to translate between different voltage standards. If V_{ds} of a transistor becomes too large, punchthrough occurs, possibly causing excessive current flowing until the interconnect melts. Transistors with smaller dimensions have a lower punchthrough voltage. As introduced in Section 15.2.7, I/O circuits often use transistors with longer channels and thicker oxides to endure the higher voltages. Transistors can also be stacked to increase their voltage tolerance.

Table 12.2 summarizes typical logic levels for single-ended drivers. Beware that the logic levels definitions vary somewhat between vendors. The popular 74-series logic gates of the 1970s and 1980s used the 5 V *transistor-transistor logic* (TTL) standard with highly asymmetric logic levels because outputs are pulled down by a strong transistor but pulled up by a weaker resistor. The 5 V CMOS standard was more symmetric. In the 1990s, low-voltage (3.3 V) flavors of TTL and CMOS were introduced. Bipolar circuits perform poorly below 3.3 V, so CMOS standards prevailed as voltage continued to decrease. The 5 V CMOS and TTL standards are now completely obsolete, but 3.3 V LVCMOS is still widely supported for compatibility even when the core operates at a much lower voltage. Section 12.7.3 describes differential signaling.

TABLE 12.2 Single-Ended I/O Standards

Standard	V_{DD}	V_{IL}	V_{IH}	V_{OL}	V_{OH}
TTL	4.75–5.25	0.8	2.0	0.4	2.4
CMOS	4.5–6	1.35	3.15	0.33	3.84
LVTTL	3.0–3.6	0.8	2.0	0.4	2.4
LVCMOS33	3.0–3.6	0.8	2.0	0.36	2.7
LVCMOS25	2.3–2.7	0.7	1.7	0.4	V_{DD}–0.4
LVCMOS18	1.65–1.95	0.35 V_{DD}	0.65 V_{DD}	0.45	V_{DD}–0.45
LVCMOS15	1.4–1.6	0.35 V_{DD}	0.65 V_{DD}	0.25 V_{DD}	0.75 V_{DD}
LVCMOS12	1.1–1.3	0.35 V_{DD}	0.65 V_{DD}	0.25 V_{DD}	0.75 V_{DD}

Figure 12.53 shows some simple level converters for chips using a low V_{DDL} core voltage and higher V_{DDH} I/O voltage. Figure 12.53(a) is an output driver that takes a low-swing input voltage and produces a higher-swing output voltage. It uses a CVSL structure consisting of four high-voltage transistors indicated in bold. The inverter uses low-voltage

transistors and the low-voltage power supply. The output Y can be followed by a high-voltage inverter or buffer to deliver more uniform rise/fall times. Figure 12.53(b) is an input receiver that takes a high-swing input voltage and produces a lower-swing voltage for core circuits. It consists of a simple inverter using high-voltage transistors that can withstand the large gate voltages.

To avoid the need for high-voltage transistors, some output drivers use stacked transistors. For example, Figure 12.54 shows a *cascoded* driver for a 3.3 V output in a 2.5 V process [Greenhill97]. The inner (cascode) transistors are tied to supplies in such a way that V_{gs} and V_{ds} across an individual transistor never exceed 2.5 V even though the output has a larger swing. If the voltages on the cascode transistors are provided externally rather than generated internally, the system must apply them in the proper sequence to avoid momentarily exposing the I/O circuitry to damaging electric fields.

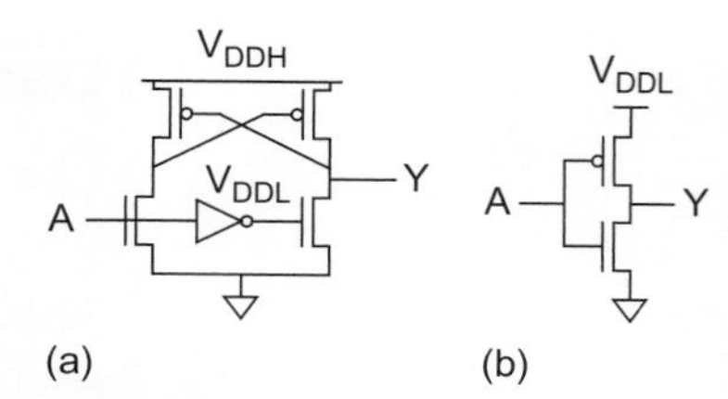

FIGURE 12.53 Level converters

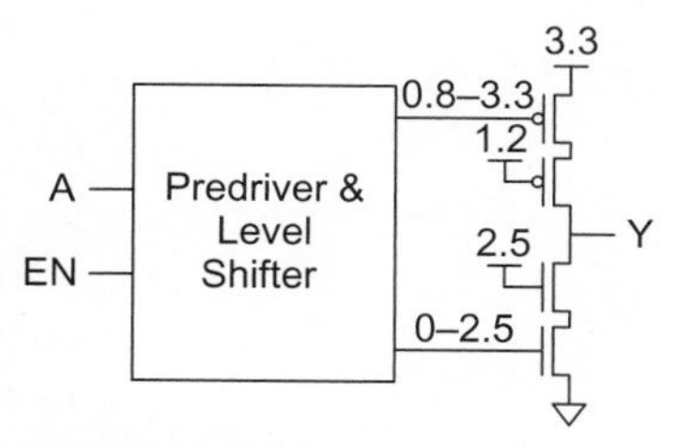

FIGURE 12.54 Cascoded high voltage output driver

12.7 High-Speed Links

As chips integrate more functions on a single die and process more data, the demand for high communication bandwidth between chips continues to rise. While adding more pins is a simple way to increase the I/O bandwidth, it may increase the package cost and chip area significantly. An alternative is to increase the speed of communication per pin. This section discusses the fundamentals of high-speed I/O design.

The basic digital I/O described in Section 12.6 faces a number of challenges as one tries to increase the rate at which the bits are transmitted. The following subsections will discuss these challenges and address the currently established solutions that enable high-speed I/O operation. The challenges are namely:

- Designing high-speed circuits that can generate fast pulses and reliably detect them as digital 1s and 0s
- Propagating signals through a lossy, finite-latency medium (referred to as *transmission lines*)
- Distinguishing one bit from another when they are transmitted successively

12.7.1 High-Speed I/O Channels

In a basic I/O configuration shown in Figure 12.55, a transmitter (or driver) sends an electrical signal to a receiver via a conducting wire. At low transmission speeds, this conductor acts as an ideal wire (or at worst, a resistance in series) that keeps the voltage potentials on both of its ends equal. For example, when the transmitter generates a 1 V signal to represent a Boolean symbol of 1, the same voltage appears on the other side and the receiver interprets it as 1.

At high frequencies, however, the conductor can no longer be treated as an ideal wire. Instead, it acts as a *transmission line* along which the voltage and current propagate as *waves*. A conductor should be treated as a transmission line rather than as an *equipotential net* when the propagation delay along the conductor becomes comparable to the signal rise/fall times.

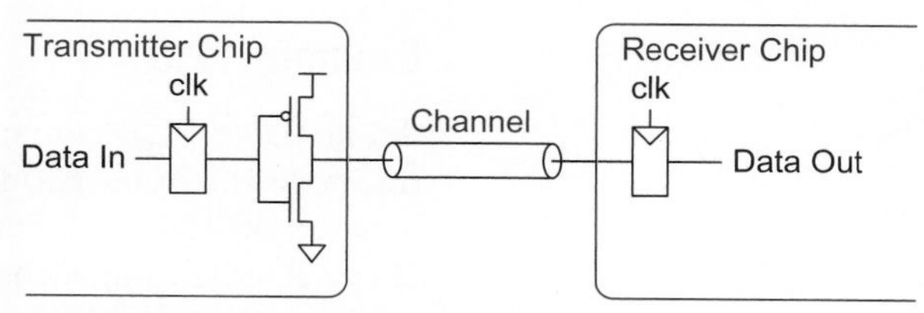

FIGURE 12.55 Basic digital I/O

Example 12.7

Above what frequency must a 10 cm trace on a printed circuit board be treated as a transmission line?

SOLUTION: A typical PCB consists of copper wires embedded in a flame-retardant epoxy material called FR4. FR4 has a dielectric constant of approximately $\varepsilon = 4.35\varepsilon_0$, so signals propagate at a velocity of

$$v = \frac{c}{\sqrt{4.35}} = \frac{3 \times 10^8 \frac{\text{m}}{\text{s}}}{2.086} = 14.4 \frac{\text{cm}}{\text{ns}} \tag{12.38}$$

Thus, the signal takes 700 ps to propagate along the trace. The rise/fall time of a signal should be no more than about one-quarter of the cycle time so that the high and low states are recognizable. Thus, if signals have a period of less than 2.8 ns (i.e., a frequency exceeding 350 MHz), they should be modeled as waves propagating along a transmission line.

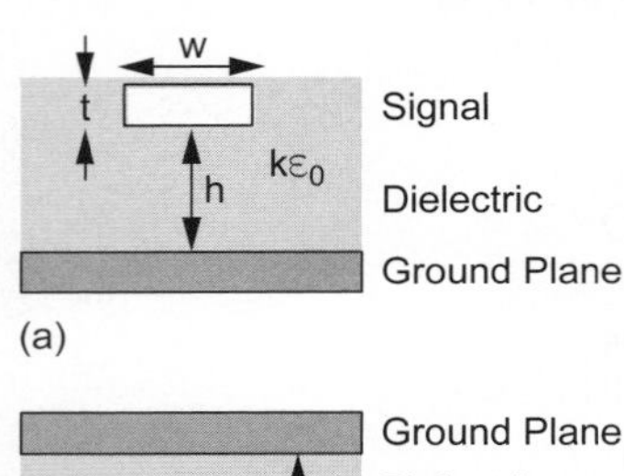

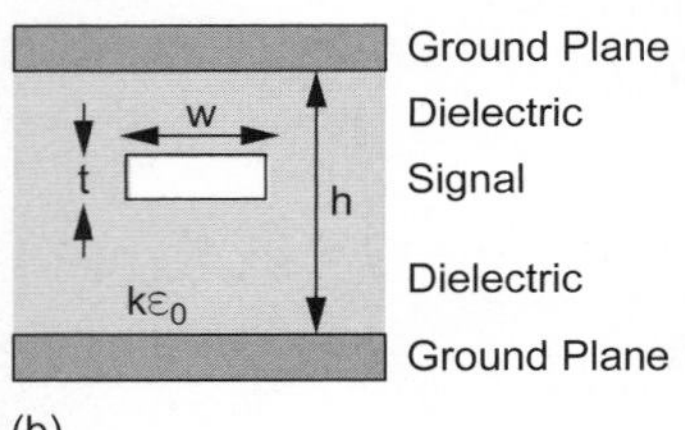

FIGURE 12.56 Transmission lines (a) microstrip, (b) stripline

Another implication of the finite propagation time is that the transmitter cannot see what is connected at the receiving end at the time it launches a pulse down the conducting channel. Instead, it only sees the load impedance presented by the channel itself. This impedance is called *characteristic impedance*, Z_0, of the channel and typical values are around 50 Ω. The initial pulse amplitude is thus determined by the characteristic impedance, not by the load impedance at the receiving end. The characteristic impedance also indicates the ratio between the voltage and current waves that travel down the channel.

To obtain well-controlled impedance and predictable current return paths, high-speed printed circuit boards normally allocate half of the metal layers to power or ground planes. Figure 12.56 shows two common ways in which signals are routed on a PCB. A signal running on an outer layer is called a *microstrip*. It sees a ground plane on one side and free space on the other. The characteristic impedance of a microstrip is approximately [Mears96]

$$Z_0 = \frac{60}{\sqrt{0.475\text{k} + 0.67}} \ln \frac{4h}{0.67(0.8w + t)} \tag{12.39}$$

A signal running on an inner layer between planes is called a *stripline* and has a characteristic impedance of approximately

$$Z_0 = \frac{60}{\sqrt{\text{k}}} \ln \frac{4h}{0.67\pi(0.8w + t)} \tag{12.40}$$

Example 12.8

A four-layer PCB contains power and ground planes on the inner layers and signal traces on the outer two layers. The layers use 1 ounce copper.[4] The FR4 dielectric between the layers is 8.7 mils thick. How wide should the signal traces be to achieve 50 Ω characteristic impedance?

[4]Printed circuit boards describe copper thickness in the obscure unit of ounces, describing the weight of a 1 foot square sheet of metal foil of a particular thickness. 1 ounce Cu is 1.4 mils thick. 1 mil = 10^{-3} inches.

SOLUTION: Solve EQ(12.39) numerically with $h = 8.7$ mils, $t = 1.4$ mils for $w = 15$ mils. This is relatively wide compared to the typical minimum trace width of 6–7 mils. The width can be reduced by selecting a thinner dielectric.

If the waves propagate through the channel according to its characteristic impedance, what happens when they reach the end and find that the final load impedance is actually different from Z_0? The energy that has been traveling down the line cannot be fully absorbed or dissipated by the final load. This is called *impedance mismatch*. If the energy cannot be fully absorbed at the receiving end, the remaining energy must go back toward the transmitter. In other words, the waves are reflected. The *reflection coefficient* Γ is the ratio of the incident to the reflected wave. According to transmission line theory [Hall00], the reflection coefficient can be expressed in terms of the load impedance Z_L and the characteristic impedance Z_0:

$$\Gamma = \frac{Z_L - Z_0}{Z_L + Z_0} \quad (12.41)$$

Reflections are undesirable for several reasons. First, the receiver does not receive the full energy of the signal sent by the transmitter. In other words, the reflected energy is simply wasted. Second, the reflected waves can interfere with other signals that are later sent by the transmitter. The phenomenon of one signal energy spilling over into other signals' energy is in general referred to as *inter-symbol interference* (ISI).

Therefore, in order to suppress such reflections, high-speed I/Os use channels that are properly "terminated" at either end of the channel, as illustrated in Figure 12.57. Terminating a channel means matching the load impedance to the characteristic impedance, therefore achieving zero reflection according to EQ(12.41). As we will see in Section 12.7.3, the channel can be terminated either at the transmitter or at the receiver. However, many industrial standards require both ends be terminated because some unwanted signals may get coupled into the middle of the channel and reflected from the unterminated end to interfere with the desired signal. Terminating both ends reduces the voltage swing by 50% for the same amount of drive current because the equivalent load resistance is $Z_0/2$ rather than Z_0.

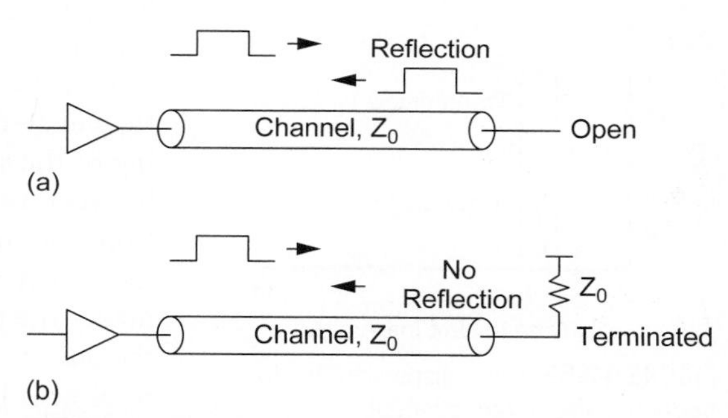

FIGURE 12.57 Transmission line reflections and termination

Notice that with properly terminated channels, the transmitter can send the next bit before the current bit reaches the receiver because the bits propagate through the channel at the same speed and do not interfere with each other. Without terminations, the transmitter would need to wait until the reflections caused by the current bit transmission disappear, which can take multiple round-trip times of the channel. For example, if the bits are transmitted at 100 ps intervals (i.e., 10 Gb/s) via a 10 cm FR4 trace which has one-way propagation delay of 700 ps, seven bits concurrently propagate along the channel at any given time. On the other hand, if the reflections are severe and settle only after two round-trips (2.8ns), then the maximum bit rate would be limited to 350 Mb/s. A properly terminated channel that avoids reflections is therefore the first requisite for high-speed I/O operation.

Device I/Os can be connected in various configurations and the bus and point-to-point configurations shown in Figure 12.58 are the representative examples. While the multidrop bus in Figure 12.58(a) has been the popular choice for low data rates as it

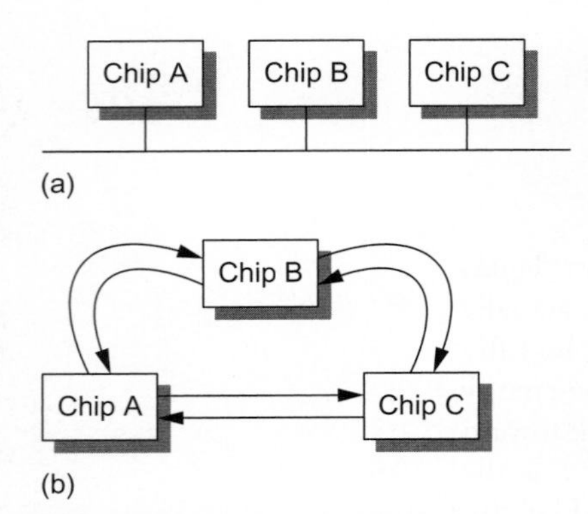

FIGURE 12.58 (a) multidrop bus vs. (b) point-to-point links

requires fewer wires to be routed, the point-to-point links in Figure 12.58(b) are finding widespread use in high-speed applications because they can connect two points without any splitting junctions in the middle. The splitting junctions in the bus configurations cause discontinuities in the characteristic impedance, resulting in reflections. In comparison, point-to-point links are much easier to engineer for minimal reflections.

12.7.2 Channel Noise and Interference

In the previous section, we discussed reflection as one cause for inter-symbol interference limiting reliable data transmission. This section discusses other types of noise and interference that may corrupt the signal propagating through the channel, including dispersion, crosstalk, and return path noise.

12.7.2.1 Dispersion The channel may attenuate certain parts of the signal energy due to resistance along the conductor and dielectric loss through an imperfect insulator. The attenuation is frequency-dependent, generally with low-pass behavior. For example, Section 5.2.4 describes skin effect, where the conductor loss increases with frequency as the current crowds toward the surface of the conductor. The dielectric loss also increases with frequency. This frequency-dependent attenuation causes *dispersion*; i.e., distortion and widening of the signal shape. Suppose that a transmitter sends a lone one-bit pulse between the strings of 0s. As illustrated in Figure 12.59, the pulse emerges from the transmission line with lower amplitude and greater width such that its energy extends beyond its assigned bit period. The smaller amplitude makes the pulse harder to detect. Worse yet, the 0-bit immediately following the one-bit experiences the remnant of the energy from the previous bit, so the receiver is less certain about it being a 0. Therefore, dispersion leads to *inter-symbol interference* (ISI). In Section 12.7.3.3, we will discuss how *equalizers* are used to undo such dispersion.

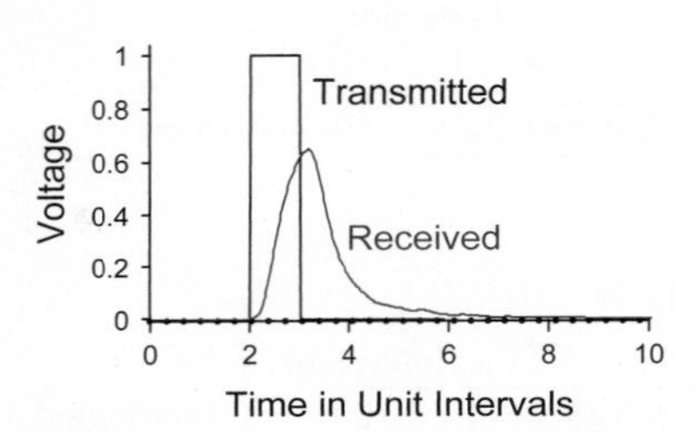

FIGURE 12.59 Pulse dispersion due to frequency-dependent attenuation

12.7.2.2 Crosstalk Capacitive or inductive coupling causes interference called crosstalk between nearby I/O channels in which energy from one channel propagates into another, as discussed in Section 5.3.3. Crosstalk is more challenging than dispersion because the effects cannot be undone unless the coupling mechanisms and the aggressors' bit patterns are known to the victims [Zerbe01]. Instead, crosstalk is usually suppressed by designing the channels to minimize coupling. For example, shielding the channels from one another with ground lines is one approach. Using differential signaling also helps if the aggressor signals affect both the lines equally. While not prevalent in high-speed I/Os, advanced digital communication systems may use error-correcting/detecting codes, sequence detection, or multi-input/multioutput (MIMO) estimation to detect the digital bits reliably in the presence of crosstalk [Barry03, Proakis08].

12.7.2.3 Return Path Effects Figure 12.57 is sometimes misleading because it does not show the path through which the current returns from the receiver back to the transmitter. Conservation of charge dictates that any current leaving a system must come back. Providing a good return path is as important as a good signal path; in fact, many of the signal integrity problems stem from overlooking the return paths. Any voltage drop across the return path due to its finite impedance will appear as additional noise to the transmitted signal. If the return path impedance changes with frequency, then the resulting noise will also vary with frequency.

An example of the return path problem is *ground bounce*. In the single-ended link example shown in Figure 12.60(a), the return paths are the ground nodes shared by the two chips. The transmitter generates the signal voltage in reference to its local ground, but the receiver reads the arrived voltage in reference to its local ground. While these two local grounds should nominally be at the same potentials, the return current may cause temporary difference between them. For example, if the return path is inductive (e.g., due to bonding wires or package leads that connect the grounds of the chip to the die and circuit board), then the voltage difference will vary with the time-derivative of the current. Therefore, the resulting noise that the signal experiences is frequency-dependent and can cause another form of ISI.

When multiple I/O links share a common return path (e.g., the same ground nodes) as illustrated in Figure 12.60(b), the return current from one I/O link can develop a voltage difference between the two ground levels which can interfere with all the other I/O link operations. This is called *simultaneous switching noise* (SSN). Differential signaling, described in Section 12.7.3.2, can be regarded as a way of providing a dedicated return path to each signal path; hence, alleviating many of the SSN and ground bounce issues.

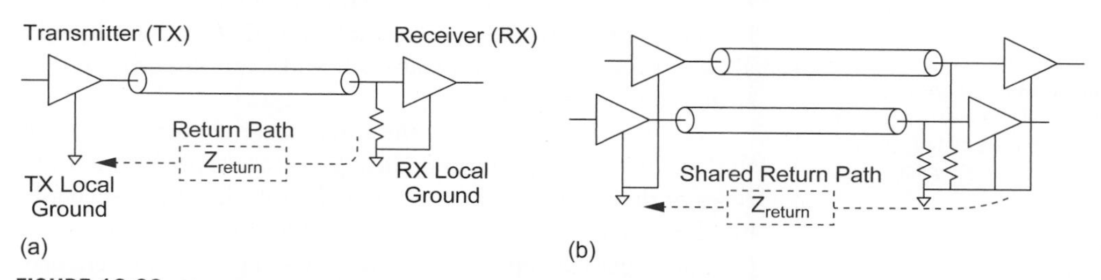

FIGURE 12.60 Simultaneous switching noise mechanism (a) ISI, (b) crosstalk

12.7.3 High-Speed Transmitters and Receivers

Besides channels that can propagate signals with minimum reflection and interference, high-speed I/O requires transmitters and receivers that can generate and detect signal pulses at very high rates. This subsection explores the circuit issues for building such high-speed transmitters and receivers. Recall that the simple I/O link in Figure 12.55 uses a CMOS inverter as the transmitter and a flip-flop as the receiver. As we seek higher data rates, we face various challenges with this basic link. This subsection focuses particularly on the issues related to the inverter as the transmitter. The next subsection will focus on how to maintain the correct timing to trigger the receiver flip-flop.

12.7.3.1 Single-Ended Transmitters The basic problem with the CMOS inverter as a high-speed transmitter is that its output impedance can vary significantly across its output range. When the output voltage is near the supply or ground, either its pMOS or nMOS operates in linear region, making the output impedance low. On the other hand, when the output voltage is in the middle between the supply and ground, both transistors are in saturation and have high output impedance. Due to this wide variation, one can never design an inverter whose output impedance is matched to the channel's characteristic impedance.

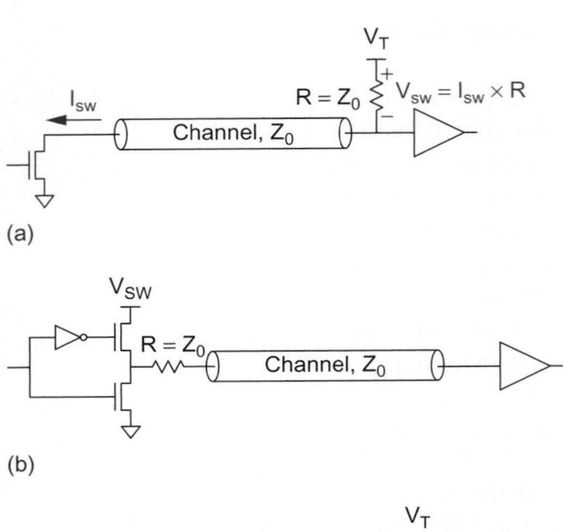

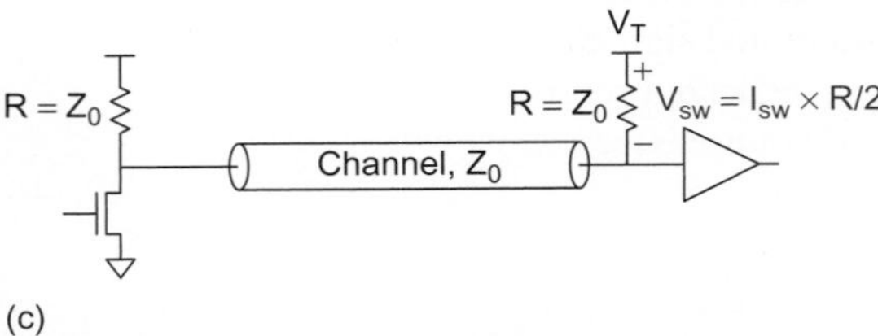

FIGURE 12.61 Transmission line drivers (a) current-mode driver (parallel termination at the receiving end), (b) voltage-mode driver (series termination at the transmitting end), (c) double-terminated driver (distinction between voltage and current is irrelevant)

Without a proper termination at the receiving end, the signal waves can be reflected back at the transmitter side and cause inter-symbol interference.

Figure 12.61 shows several methods of building a single-ended transmitter with more uniform impedance than a simple inverter. The current-mode driver in Figure 12.61(a) uses an open-drain transistor operated in saturation with a high output impedance. The parallel termination at the far end of the transmission line converts the current to voltage. *Gunning Transceiver Logic* (GTL) [Gunning92] uses this style of driver, with $V_T = 1.2$ V and a low output of 0.4 V. It employs a differential receiver (see Figure 11.28(a)) to compare the output against a 0.8 V reference. The voltage-mode driver in Figure 12.61(b) uses wide transistors operated in their linear regime with low output impedance. It adds a series resistor to match the channel impedance. Building a precise resistor is difficult in CMOS because of process variation. An alternative, called *digitally controlled impedance*, builds the driver out of multiple parallel transistors of binary-weighted widths and turns on the proper set to achieve the desired output impedance [Gabara92, DeHon93]. In Figure 12.61(c), the line is parallel terminated at both ends. This eliminates reflections at both ends, but cuts the output swing by a factor of two.

Another way to classify the transmitter circuits is to see if the driver is a *push-pull* type or a *pull-only* type. While both types of drivers generate binary signals, a push-pull type creates bipolar signals centered around 0 and a pull-only type uses 0 (i.e., no signal) as one of the signal levels. The transmitters previously shown in Figures 12.61(a) and (b) are examples of the pull-only and push-pull drivers, respectively. While pull-only drivers are in general a bit easier to design (fewer active switches), push-pull drivers may consume less power for the same voltage/current swing because it uses half the current of the pull-only driver.

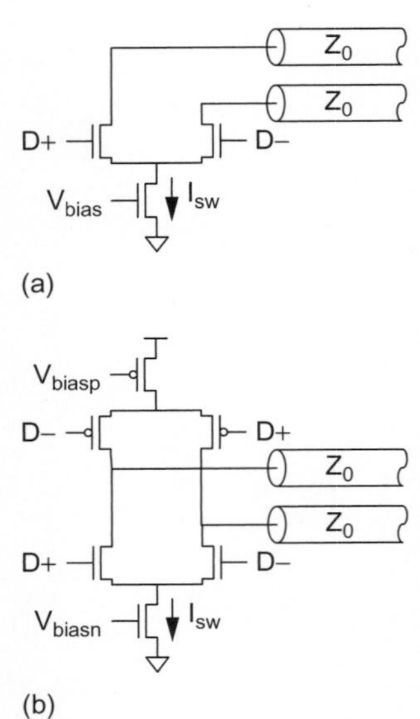

FIGURE 12.62 Differential drivers (a) current-mode logic (CML) (b) low-voltage differential signaling (LVDS)

12.7.3.2 Differential Transmitters *Differential signaling* is a widely adopted way of improving the noise immunity by representing the signal with a difference between two voltages or currents. Even in the presence of external noise or interference, the difference is unaffected as long as the disturbance influences both the signals equally. Two differential transmitter circuits are illustrated in Figure 12.62. As with single-ended drivers, differential drivers can be either voltage- or current-mode and either push-pull or pull-only type. Most differential drivers are made of differential pairs, which steer the current between two outputs while keeping their sum nearly constant. The driver circuit in Figure 12.62(a) generates pull-down currents only while the one in Figure 12.62(b) uses two differential pairs to generate both pull-up and pull-down currents.

Low-voltage differential signaling (LVDS) [National08] switches a 3.5 mA current into a 100 Ω load providing a differential termination between the two transmission lines. Thus, it produces a 350 mV output swing that is detected by a differential receiver. It is suitable for operation up to 3.125 Gb/s and is popular because of the low power consumption. Current mode logic (CML) is not a formal standard; the switching current and voltage levels vary widely. Using higher currents and wider swings, CML can operate beyond

10 Gb/s at the expense of more power. Low-voltage positive-emitter-coupled logic (LVPECL) is a closely related system with similar trade-offs.

12.7.3.3 Transmitter Variations Some applications may require additional features from the transmitter, such as AC coupling, slew rate control, or programmable swing. AC coupling (or DC blocking), as shown in Figure 12.63, is a convenient way to connect a transmitter and a receiver that have different signal ranges. An example is a receiver that operates with multiple signaling standards. A series capacitor inserted in the channel blocks the DC content and propagates only the high-frequency content of the signal. Since the capacitor turns the channel into an open circuit at DC, the signal ranges can be set independently at the transmitter and the receiver sides. However, one must ensure that no data is lost by these DC blocking capacitors. One way is to encode the data with redundancy so that they contain no information in low-frequency spectrums. 8b/10b [Widmer83] encoding is widely used. It recodes 8-bit bytes into 10-bit symbols such that no more than five consecutive 0s and 1s appear and the number of 0s and 1s are roughly balanced. 64b/66b codes are used in 10 Gb Ethernet because they have a lower overhead.

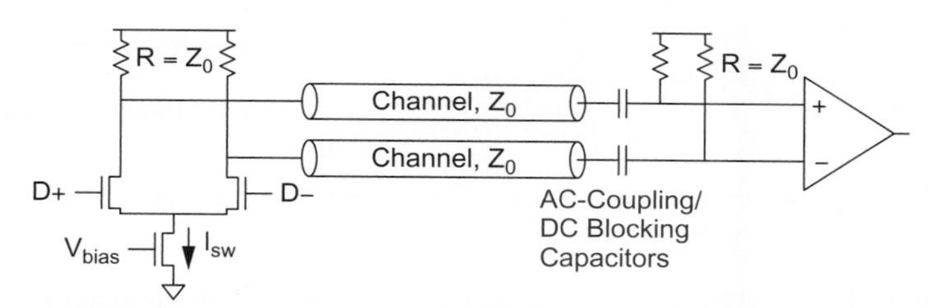

FIGURE 12.63 AC coupling

In many cases, it is desired that the transmitter swing be programmable. Figure 12.64 illustrates a transmitter with segmented driver devices designed for this purpose. The select signals determine how many devices turn on to pull the currents and hence how large the signal swing is at the output.

In addition to the swing, drivers may control how fast the signal transitions (i.e., slew rate). While ideal pulses may have infinitely sharp edges, such sharp edges may have adverse effects in real applications. For example, signals with sharp edges can cause more severe crosstalk, suffer more from reflections, and excite ringing due to parasitic resonance in packages or connectors. If the transmitter creates too fast a signal, one can deliberately slow its transitions down by first dividing the driver device into multiple pieces and then turning them on sequentially, as shown in Figure 12.65. The rate at which these devices are switched on determines the slew rate of the transmitted signal.

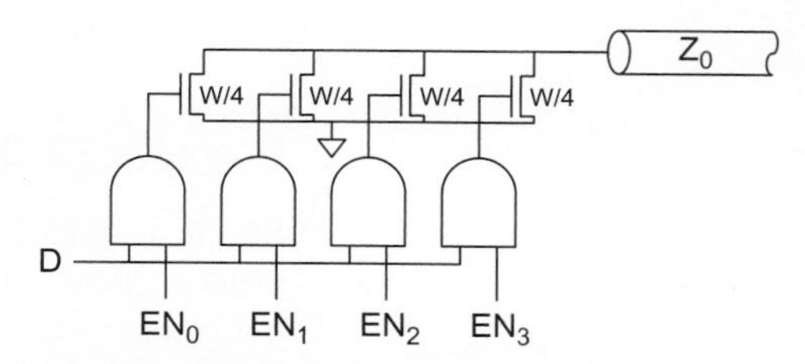

FIGURE 12.64 Programmable drive current

Section 12.7.2 discussed how channels may have frequency-dependent attenuation that causes dispersion and intersymbol interference. Equalizers are circuits that can compensate such undesirable effects. Equalizers are basically filtering circuits that try to make the combined channel response "flat" over the entire frequency range. There are two ways

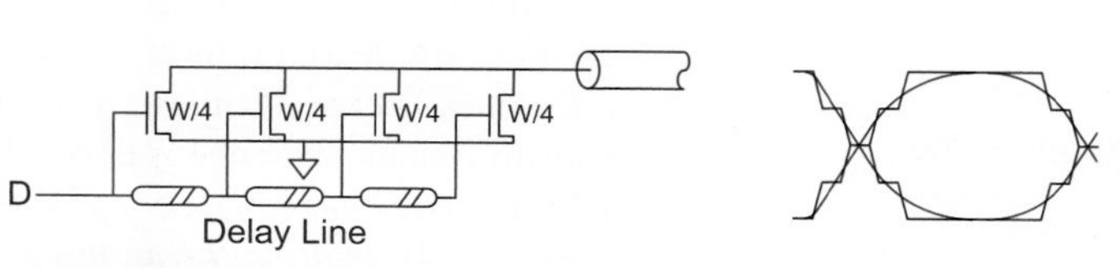

FIGURE 12.65 Slew rate control

that one can achieve this: either amplifying the signal spectrum being attenuated by the channel, or attenuating the other parts of the signal spectrum so that the whole spectrum sees the same level of attenuation. While the former should sound like a better idea, many high-speed I/O circuits adopted the latter mostly because it is easier to implement. Figure 12.66 depicts a so-called *de-emphasizing* transmitter that is commonly used for this purpose. This transmitter is a combination of two sub-transmitters: one for the main data pulses and the other for the inverted, scaled-down pulses of the same data delayed by one bit period. In essence, this transmitter generates the smaller swings for the bits that repeat the preceding ones and larger swings for those that change. It is equivalent to a high-pass filter that counteracts the low-pass responses of the I/O channels.

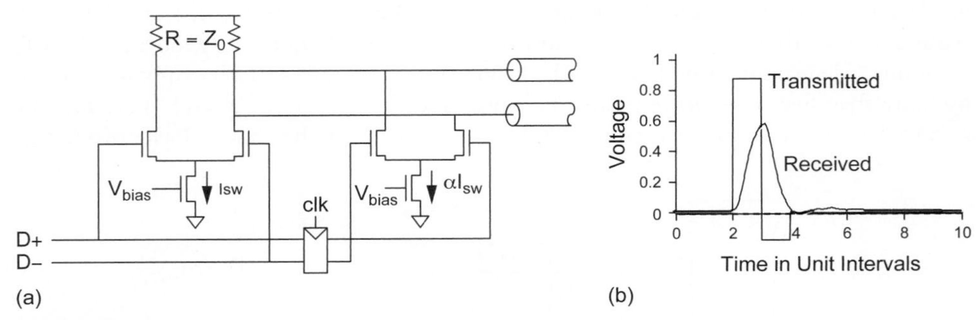

FIGURE 12.66 De-emphasizing transmitter (a) circuit, (b) de-emphasized pulses

12.7.3.4 Higher Data Rates One may wonder what determines the maximum speed of the transmitter circuits described so far. It turns out that the transmitter itself is not the major limiter for the speed. Most high-speed I/O circuits rely on precise clock signals to generate the data pulses at constant intervals and the maximum data rate is often dictated by the highest clock frequency that can be propagated on the chip. The shortest clock period can be estimated as 8 times the delay of each clock buffer stage, which gives rise and fall times each occupying about 25% of the period. Pushing for higher frequency results in clock waveforms that do not reach full swings.

Example 12.9

Suppose clock buffers are built from FO4 inverters with a delay of 15 ps in a 65 nm process. What is the maximum rate at which data can be transmitted if one bit is sent per clock cycle?

SOLUTION: 8 FO4 inverter delays is 120 ps, corresponding to a maximum data rate of 8.3 Gb/s.

It is possible to achieve higher data rates using *time interleaving* or *multilevel signaling*. In time-interleaved transceivers (Figure 12.67), *N* drivers connected in parallel can generate a data stream *N* times greater than that of a single driver. The timing to select each transmitter in sequence is derived from different phases of the clock. Most high-speed I/Os use two-way interleaving because it requires only two clock phases (true and complementary). In *multilevel transmitters* (Figure 12.68), more than two levels may be used to

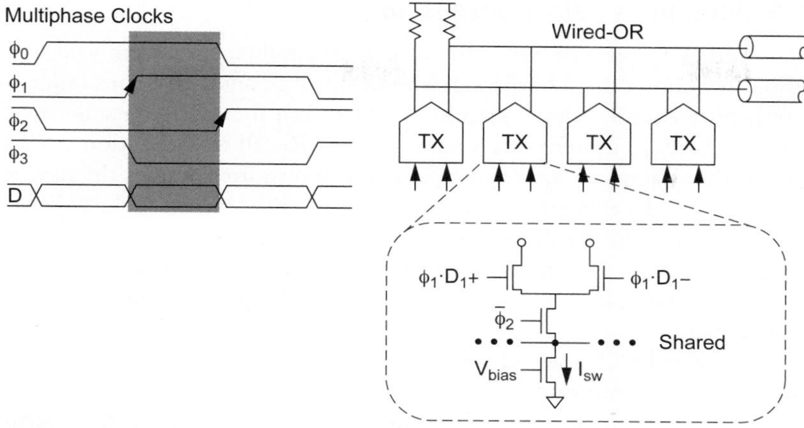

FIGURE 12.67 Time-interleaved transmitter

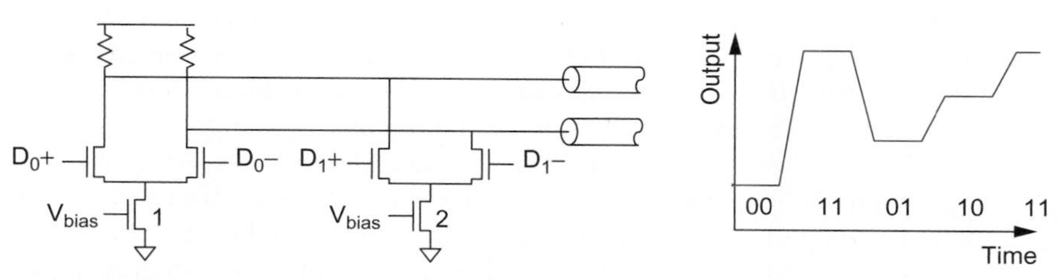

FIGURE 12.68 Multilevel transmitter

represent more than one bit. Multilevel transmitters rely on greater precision in voltage rather than time. The effectiveness of these options generally depend on the attenuation and noise of the channel. For example, time-interleaving is preferred to multilevel signaling when the attenuation in the channel is benign.

12.7.3.5 Receivers The receiver is typically a simple flip-flop that samples the data at the correct time. For differential signaling, a differential flip-flop such as the SA-F/F from Figure 9.29 is required to detect the small swing signal. Time-interleaved signaling uses multiple receivers activated at staggered times. The timing typically comes from a PLL or DLL with multiple outputs tapped from the VCO or VCDL. Multilevel signaling uses a small A/D converter in the receiver. The central challenge of receiver design is to sample the data at the correct time; various solutions are discussed in Sections 12.7.4–12.7.7.

12.7.3.6 Bit Error Rate The main performance metric for any I/O link is *bit error rate* (BER), the probability of transferring an erroneous bit. One may find that the typical bit error rate target for high-speed I/Os is extremely low, ranging from 10^{-10} to 10^{-12}. It is because the high-speed I/Os have evolved from traditional digital I/Os which cannot tolerate any bit errors (e.g., no redundancy coding). In comparison, some other communication links such as wireless systems may aim at the higher rates of 10^{-6}. The stringent BER requirement makes the BER modeling, simulation, and measurement difficult and time-consuming because each error event is rare.

12.7.4 Synchronous Data Transmission

When transmitting a stream of bits from one chip to another, both sides need to agree on a convention that allows them to distinguish one bit from another. For example, suppose that the transmitter sends ten consecutive 1s. How can the receiver recognize that the string contained ten 1s rather than nine or eleven? Recall from Section 9.5.3 that, at slower speeds, the system can use *handshaking*: The transmitter notifies the receiver every time it is about to send a new bit and will only do so after the receiver acknowledges it and signals back to the transmitter that it is ready. One problem with handshaking, however, is that the data rate becomes limited by the channel delay. Thus, it cannot exploit the channel being a transmission line that can propagate the next pulse before the previous one reaches the far end.

Most high-speed I/Os instead use the *time* as the marker to tell the bits apart. In other words, the bits are transmitted at constant time intervals. For example, a 1 Gb/s link transmits a bit every 1 nanoseconds. Since no signals have to be exchanged between the transmitter and receiver for each bit, the signaling rate is no longer limited by the channel delay. However, this *synchronous transmission* poses a critical requirement on both the transmitter and receiver sides: the timing of each bit pulse being generated and detected must be precisely controlled. Since the uniform bit intervals are the only way to tell one bit from another, any deviation in the timings can cause data transmission errors.

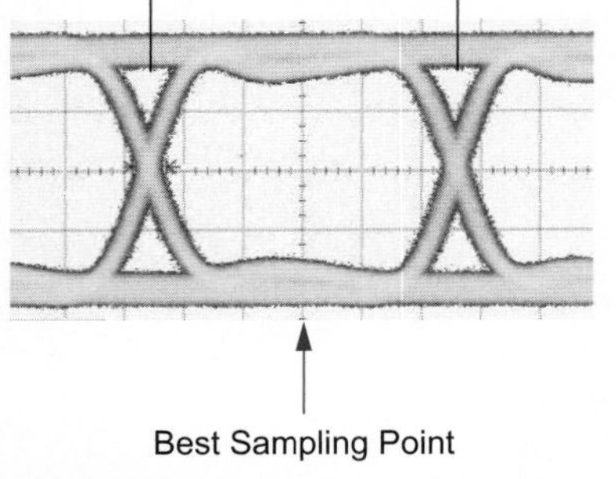

FIGURE 12.69 Eye diagram illustrating bit interval and best sampling point

For example, assume that logic 1 is represented by a high voltage level and logic 0 is by a low voltage level. (This is called *non-return-to-zero, NRZ*, signaling.) Figure 12.69 plots the signal as a function of its time offset within each bit period. This plot is called an *eye diagram* because if the bits are transmitted at constant bit intervals, the plot should have an opening in the middle where the signal never makes any transitions. Any nonuniformity in the bit intervals will reduce the opening in horizontal direction and narrow the time period in which the bits can be detected reliably. The receiver, on the other hand, must make the decision about each bit by sampling the signal at the position where the eye diagram has the largest opening. The central challenge of high-speed receiver design is to precisely identify the best time to sample the data stream.

The transmitter clock is typically generated by a PLL or DLL. As discussed in Section 12.5.1.5, the timing error (jitter) depends on the jitter of the input clock, the power supply noise, and the loop bandwidth. All the design considerations previously described to reduce the clock jitter apply here as well. However, design choices can differ depending on what type of jitter is being minimized. For example, in high-speed I/Os, the main interest is to minimize the deviation of each clock edge position from its nominal position in absolute time (i.e., *absolute jitter*). On the other hand, in many digital logic systems, the main interest is to reduce the change in the clock periods from one cycle to another (i.e., *cycle-to-cycle jitter*). For example, the jitter accumulation behavior of PLLs may make their cycle-to-cycle jitter low but the absolute jitter high.

The receiver must synchronize with the transmitter to sample the bit stream in the middle of the eye. The next three sections explore three different techniques for receiver clocking.

12.7.5 Clock Recovery in Source-Synchronous Systems

In source-synchronous systems, the transmitter sends a clock signal properly aligned with the data, as shown in Figure 12.70. Because such a *source-synchronous clock* consumes an additional I/O channel, it is often shared across multiple parallel data channels using the same timing. Any discrepancy in the transmission delays between channels shows up as timing error at the receiver. The parallel I/O interface for double-data rate (DDR) memory is a good example of source-synchronous clocking.

The data may be transmitted at the same rate or at twice the rate of the clock, as shown in Figure 12.71. In *single-data-rate* (SDR) systems, the receiver samples the data on the rising edge of the source-synchronous clock. The clock transitions at least twice as often as the data. If the transmitter or channel sets the maximum number of transitions per second and the clock operates at this rate, the data is carried at only half the system capacity. In *double-data-rate* systems, the receiver samples the data on both the rising and falling edges of the clock. DDR systems have a compelling advantage that the number of transitions per second is equal for the clock and data. Both can operate at the maximum bandwidth of the channel. However, the clock duty cycle must be maintained at 50%.

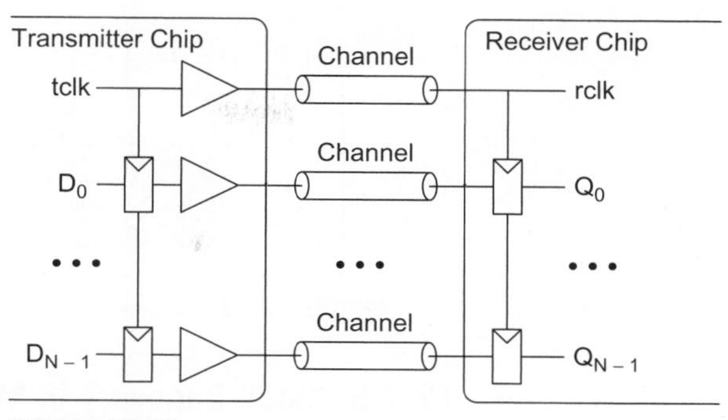

FIGURE 12.70 Source-synchronous system configuration

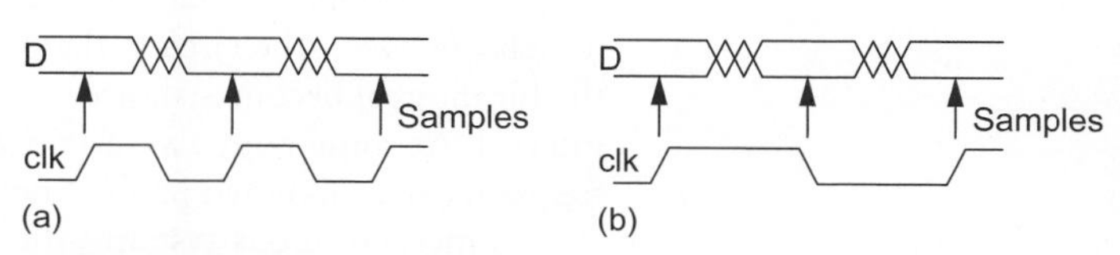

FIGURE 12.71 Clocking (a) single data rate, (b) double data rate

In principle, the receiver could simply sample the data using the source-synchronous clock. However, the clock often needs to be buffered, especially when controlling multiple parallel data channels. The buffer delay introduces skew, moving the clock away from the middle of the eye. Moreover, variations in the buffer delay caused by supply or substrate noise appear as further jitter. A common solution is to use a PLL or DLL to produce a receiver clock aligned with the source-synchronous clock, as shown in Figure 12.72.

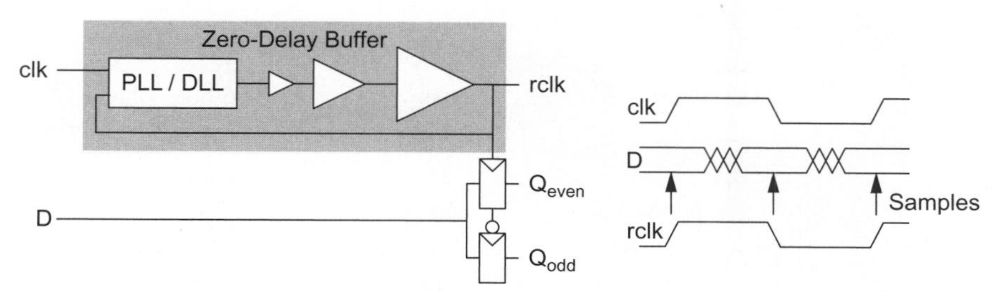

FIGURE 12.72 Clock recovery using zero-delay buffer

Source-synchronous clocks may not always be aligned with the center of the data eye. In fact, any phase relationship between the clock and data is possible as long as it is fixed and known. For example, in some applications, the transmitter may transmit a clock whose edges are aligned to those of the data. In this case, the PLL or DLL buffering the clock should also shift its phase by 90° to recenter the clock on the data eye. Figure 12.73 shows an example of a PLL that performs such 90° phase shift. The VCO generates four phases of the clock that are spaced by 90°. If one of the VCO clocks are aligned to the input clock, then the other clocks will have phases that are spaced by 90, 180, and 270°.

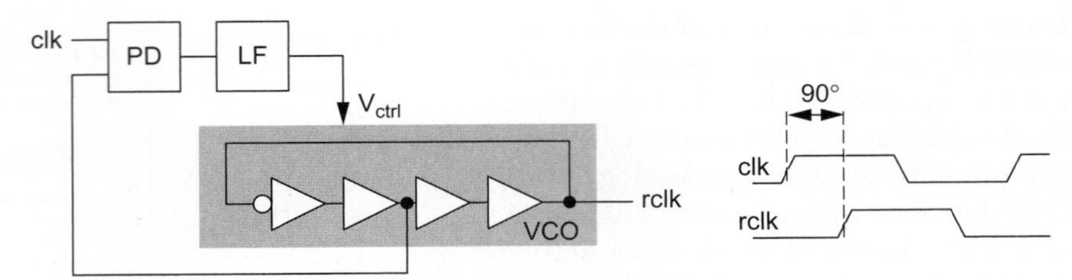

FIGURE 12.73 Clock recovery using PLL to shift phase by 90°

12.7.6 Clock Recovery in Mesochronous Systems

As one seeks higher speeds, it becomes more difficult to keep the clock and data aligned in source-synchronous systems. For example, a small difference in delay may exist between the clock and data channels due to random or systematic variations in their trace lengths, propagation velocities, characteristic impedances, etc. Moreover, there may be difference in delay between the circuits that drive the clock and data signals. As speeds increase and the bit interval becomes shorter, the difference in delay occupies a larger portion of the bit interval. At some point, we might as well consider that the clock and data have the same frequency but unknown phase. Such systems are called *mesochronous.*

In mesochronous systems, the receiver must realign the phase of the incoming clock before using it as a timing reference that triggers the data samplers. Since the clock still has the correct frequency, a circuit that can calibrate only its phase is sufficient. Figure 12.74 illustrate such a clock recovery loop. It is a feedback control loop which monitors the timing difference between the data and the recovered clock and adjusts the clock phase according to the difference. This is similar to the PLL and DLL architectures described in Section 12.5, except that the reference timing is embedded in a random data stream rather than indicated by a periodic clock. So, first we will examine the phase detectors that can compare timing between a clock and a data stream.

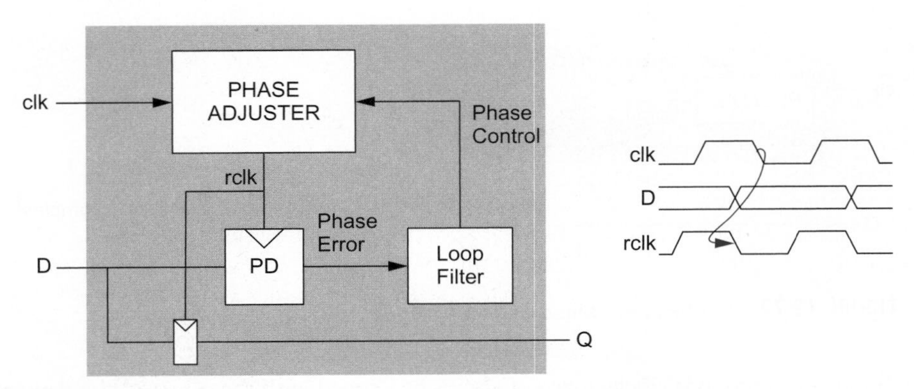

FIGURE 12.74 Phase calibration loop for mesochronous timing recovery

Phase detectors used in clock recovery loops are different from those discussed in Section 12.5.1.3 in that they operate on random data streams. One well-known implementation shown in Figure 12.75 is called *Hogge detector* [Hogge85]. The detector produces two outputs, UP and DN, whose net pulse width is proportional to the timing difference between the clock and data. Because this type of phase detector can detect the magnitude of the timing error as well as its polarity, it is called *linear* phase detector. It is contrasted to a different type of phase detectors which can detect the polarity only, called binary or bang-bang phase detectors.

In the Hogge detector, one of the outputs (DN) produces a pulse with a fixed width equal to a half of the bit interval while the other one (UP) produces a pulse whose width varies with the clock-to-data timing error. The UP pulse has the same width with DN when the clock and data phases are 90° apart, giving a net zero for the difference between the two pulse widths. These UP and DN pulses drive a charge pump, which adjusts the control voltage to a VCO. The Hogge detector will then align the clock to the center of the data eye.

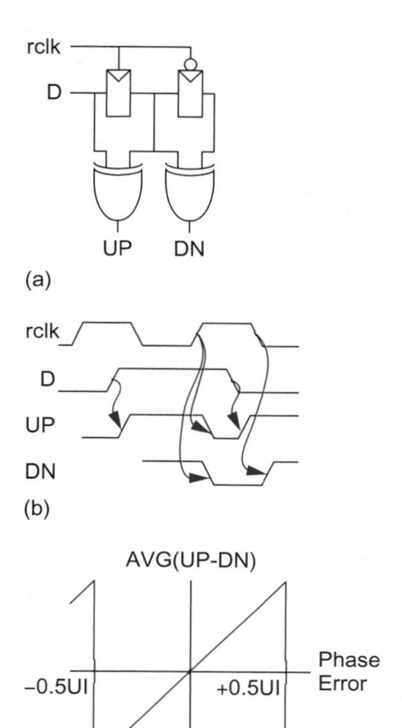

FIGURE 12.75 Hogge detector (a) circuit, (b) waveforms, (c) transfer curve

At high data rates, the Hogge detector face some limitations. First, the UP and DN pulses may become too narrow to be propagated to the charge pump. Remember that the typical bit intervals are already some fraction of the shortest period of a clock that can be fully propagated on the chip. Second, the Hogge detector may have skew between the nominal 90° locking point and the actual locking point. While such skews exist for any phase detectors, the problem lies in that the skew is likely to be different from those of the data samplers simply because they are different circuits.

Bang-bang phase detectors, on the other hand, use the exact same circuits for both timing and data detection, which makes them ideal for matching the skews. For these reasons, many clock recovery loops adopt bang-bang phase detectors. The *Alexander* or *bang-bang* phase detector shown in Figure 12.76 measures only the polarity of the timing error [Alexander75]. While it has the same UP and DN outputs with the Hogge detector, their pulse widths are fixed at the clock period and do not vary with the timing error. For each clock cycle, the UP signal is asserted when the clock is late compared to the data and the DN signal gets asserted when it is early. Neither output will be asserted if there is no transition in the data. The Alexander detector compares the clock and data timings by sampling the data stream twice within each bit interval. One sample is to read the data bit (data sample) and the other is to detect whether the transition between two adjacent bits has occurred or not (edge sample). By comparing the edge sample with the neighboring data samples, the phase detector can make a decision about the polarity of the timing error.

While binary phase detectors have advantages over the linear counterparts in that their output pulses are no narrower than the data signals themselves and that the systematic timing skews between the data and edge sampling circuits can be minimized, they lose all the magnitude information about the timing error. Because of this, the clock recovery loops with binary phase detectors cannot make timing adjustments that are proportional to the timing error. Instead, they can only make fixed-step adjustments based on the polarity. Choosing the right step size can be tricky. It should be large enough to reach lock quickly, yet small enough to limit jitter caused by dithering around the lock point.

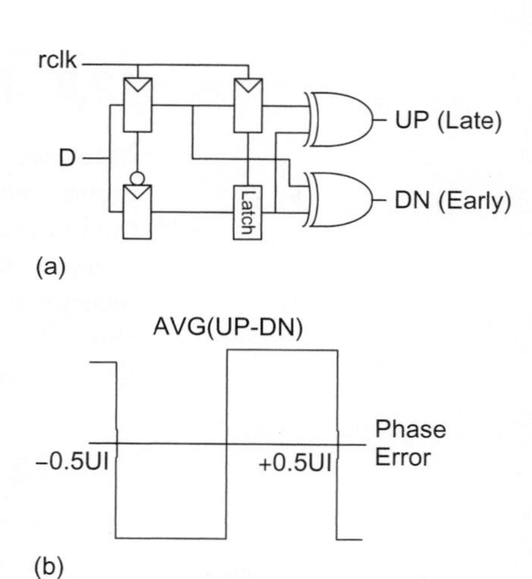

FIGURE 12.76 (a) Alexander detector, (b) transfer curves

Oversampling receivers sample the data more than twice during each bit interval [Yang96]. This provides more information to precisely adjust the receiver clock. The digital output can be processed by a digital loop filter. Fine sampling resolution, however, comes at the expense of area and power for all the samplers.

12.7.7 Clock Recovery in Pleisochronous Systems

In some applications, adding another channel for the source-synchronous clock may incur too much cost and it may be preferable to use a local clock reference for the receivers. While the local clock frequency can be accurately matched to the transmitted data rate, it may still have tiny errors (e.g., less than 200 ppm for quartz crystal oscillators). In comparison to the source-synchronous and mesochronous systems, the clock reference for the receiver not only have uncertain phase but also a small error in frequency. These types of systems are called *pleisiochronous*. In such systems, the clock recovery loop must be able to correct the frequency of the clock as well as its phase. The humble RS-232 serial port is a classic example of pleisochronous link in which the sender and receiver must agree on a baud rate.

A pleisochronous receiver commonly uses a PLL to generate a sampling clock centered on the eye of the data. The PLL may use a linear or binary phase detector as described in Section 12.7.6. One difference with the conventional PLLs, however, is that the phase detector can compare the timing only when the data has transitions and therefore large timing errors may result if the data stays at one value for a long period. To mitigate this, the data streams in pleisiochronous systems are often encoded with redundancy in order to maintain a minimum density in data transitions and constrain the timing error.

Another difference with a conventional PLL that operates on a periodic clock input is that the clock recovery PLL typically requires a frequency acquisition aid because its phase detector cannot detect a large error in frequency. For example, the phase detectors described in Section 12.7.6 cannot distinguish between the repeating patterns of 1010 at 1 Gb/s and 11001100 at 2 Gb/s. Therefore, it is necessary to use another means to ensure that the VCO is generating a correct frequency. One approach is to first lock the VCO clock to the local reference clock using a phase-frequency detector and then switch the loop to track the data timing. Once the VCO frequency is brought close enough to the desired frequency, the phase detector can keep the clock recovery loop in the locked state, as long as the data transitions often enough.

12.8 Random Circuits

Many security and authentication algorithms depend on randomness. For example, a Web browser encrypts your credit card number with a randomly generated key before sending the information over the Internet. Section 10.5.4 describes using linear feedback shift registers to generate pseudo-random bit sequences, but these are not good enough for strong security. Fortunately, nature provides various sources of random noise and variation on a chip. This section discusses true random number generators. It also examines chip identification using random variations.

12.8.1 True Random Number Generators

A *true random number generator* (TRNG) converts some source of physical randomness such as thermal noise into a random sequence of bits. Figure 12.77 shows a simple random number generator using thermal noise. The voltage across a resistor varies randomly with time due to the thermal excitation of electrons [Razavi03]. The amplified noise drives a voltage-controlled oscillator. The oscillator output is periodically sampled and stored in a shift register. The Sun Niagra2 processor includes a true random number generator with three independent thermal noise

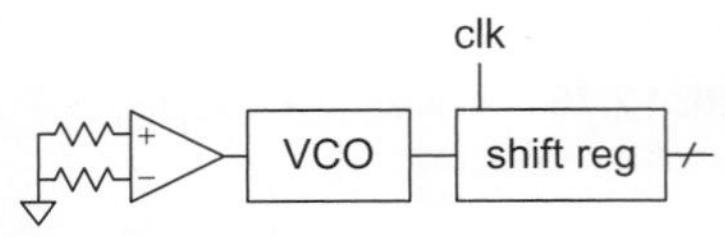

FIGURE 12.77 Thermal noise-based TRNG

modules XORed together [Nawathe08]. Other hardware implementations are described by [Kinniment02, Brederlow06, Tokunaga08].

Some hardware random number generators produce a biased pattern with an unequal probability of 0s and 1s. If the bits are uncorrelated, they can still be converted into an unbiased pattern at a lower data rate by applying *von Neumann's algorithm* [von Neumann51] to pairs of consecutive bits. The algorithm is summarized in Table 12.3.

TABLE 12.3 von Neumann's algorithm

Bit 1	Bit 2	Output
0	0	None
0	1	1
1	0	0
1	1	None

Evaluating the quality of a random sequence is subtle. The National Institute of Standards and Technologies publishes a standard statistical test method in the Federal Information Processing Standard (FIPS) 140.2 [NIST02].

12.8.2 Chip Identification

A *chip identification* (ID) number is a nonalterable bit sequence used to uniquely identify an integrated circuit or serve as a secret key. The simplest form of chip ID is a serial number encoded with fuses that are blown in the factory during manufacturing. Chip ID has many applications. A wireless sensor node or network interface card uses a unique address to differentiate itself from others. Manufacturers can use chip ID to detect rebranding or counterfeiting. Some cryptographic protocols use an ID for authentication. However, chip ID also raises serious privacy issues that tend to benefit governments and corporations at the expense of civil liberties, especially if the ID can be read by software without the consent of the user. For example, a textbook publisher might be able to use a chip ID to track the identity of a student using a pirated copy of an electronic book. A government might use the chip ID to identify an individual who visited censored Web sites.

Writing a chip ID at manufacturing time incurs some expense. Moreover, a counterfeiter could write the same ID for another chip. An alternative is to take advantage of process variation to provide a unique fingerprint for each chip. [Su08] identifies four characteristics of such a chip ID:

- The ID circuit must generate a binary ID code.
- The ID code must be repeatable and reliable over supply, temperature, aging, and thermal noise.
- The ID code length and stability must allow a high probability of correct identification of each die.
- The ID circuit must exhibit low power consumption and require no calibration.

Figure 12.78 shows an example of a bit cell in a chip ID array from [Su08]. When the cell is reset, nodes A and B are pulled to 0. When reset is released, the circuit behaves as a pair of cross-coupled inverters. Depending on the device mismatch and thermal noise, one node will be pulled high and the other low. The cell is tiled to form an array like an

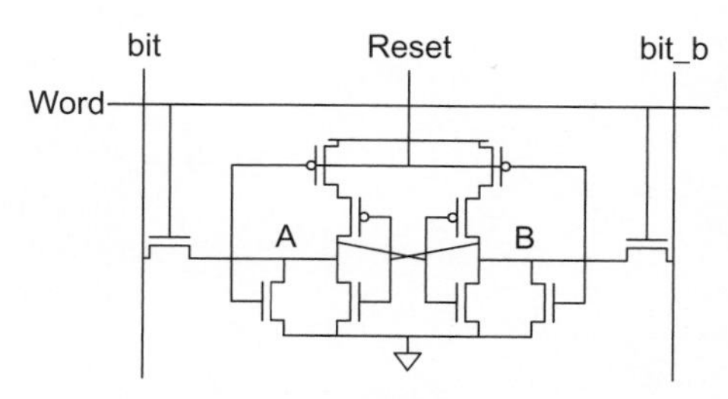

FIGURE 12.78 Chip identification bit cell

SRAM and the bits are read from the array in the same fashion. For example, an 8×16-bit array produces a 128-bit chip ID.

Each time the ID is read, noise may disturb some of the bits. If the noise is small compared to the typical mismatch, the number of differing bits (the *Hamming distance*) between the ID read and the true ID will be small. Therefore, two IDs are considered to correspond to the same chip if their IDs differ by up to d bits. d should be large enough that the probability of correctly identifying a chip is high, yet small enough that the probability of a different chip matching the same ID is low. Using a longer ID makes this easier. Other methods of chip identification involve measuring random differences in current [Lofstrom00] or delay [Lim05].

12.9 Pitfalls and Fallacies

Neglecting package parasitics

The resistance, capacitance, and inductance of the package have enormous impact on the power and I/O signal integrity of high-speed digital chips. They must be incorporated into modeling.

Using an inadequate power grid

A power grid should use generous amounts of the top two metal layers running in orthogonal directions. A mesh that mostly runs in only one direction is subject to excessive IR drops when many gates on a single wire switch simultaneously. It can also lead to serious inductive problems because of the huge current loops. The power grid should use many narrow wires interdigitated with the signals to provide a low S:R ratio rather than a few wide wires forming large current return loops. The grid should also avoid slots and other discontinuities that might lead to large current loops and high inductance.

Goofing your PLL/DLL

Phase-locked loops are notoriously difficult to design correctly. If poorly designed, they can oscillate at the wrong frequency, fail to acquire lock, or have excessive jitter. Careful circuit design is necessary to ensure they work across process variation and reject power supply noise. If the PLL does not work, testing the rest of the chip can be difficult or impossible. Most successful companies either have an in-house team that specializes in PLLs or they license their loops from a reputable supplier.

Top six ways to fool the masses about clock skew

1) **Calculate clock skew without using process variation data**

   Random skew depends entirely on the mismatch of transistors (especially L_e) and wires on a chip. This mismatch varies with distance and layout technique. The process corners model worst-case variation from chip to chip, which can be far greater than between two nearby transistors; this results in unacceptably conservative skew budgets. But reliable data for on-chip variation can be hard to obtain, especially for small ASIC design teams and universities. Unless this data is used, clock skew budgeting is largely a matter of guesswork.

2) **Claim "zero skew"**

   Many papers state that a system has *zero skew* when the writers really mean that it has zero systematic skew. These systems may have significant random skew as well as drift and jitter. The term *zero skew* is deceptive and is best avoided.

3) **Report only systematic skew**
   Many papers report only the systematic skew. In a well-balanced clock distribution network, systematic skew is often smaller than random skew and jitter.
4) **Ignore jitter**
   Jitter depends on time and space and is difficult to model or estimate. Unsophisticated clocking strategies sometimes ignore jitter. This results in unrealistic skew budgets. In particular, active deskew buffers increase clock distribution delay. Voltage noise on the buffers appears as jitter. Unless the supplies are unusually quiet, the buffers can increase jitter more than they decrease systematic or random skew.
5) **Report measured skew at only two elements over a brief period of time in a quiet environment**
   Measuring skew on a chip is difficult. Some papers measure clock interarrival times at only two or a few points on the chip for a brief period of time and report those as the skew. As a chip has many clocked elements, you are unlikely to find the worst-case skew by measuring just a few points. Moreover, measurements over a brief time interval are unlikely to capture worst-case jitter. The chip should be exercised through a variety of modes that cause large fluctuations in supply current to cause maximum power supply noise and clock jitter.
6) **Don't report the skew budget used during design**
   Designers often choose rather conservative clock skew budgets during design because they must ensure the design will operate correctly. Reporting a "measured" skew rather than a skew budget will give a smaller number.

Summary

This chapter has surveyed package, power distribution, clock, I/O, and random subsystem design. While each topic is a book in itself and a specialty design area, the short fat VLSI designer must understand enough about each area to optimize the system as a whole.

Packages connect the chip to the board or module, protect the chip, and are the first link in removing heat. They should offer plenty of connections, low thermal resistance, and low parasitics, while still being inexpensive to manufacture and test. Flip-chip packaging using solder bumps distributed across the die has become popular because of the large number of connections and low inductance.

The power distribution network consists of elements on the chip, package, and board. It must deliver a stable voltage across the chip under fluctuating current demands. Noise is caused by both average and peak current requirements. Multiple bypass capacitors offer low impedance to help filter high-frequency IR and L di/dt noise, but the DC supply resistance must be low enough to deliver the average current. V_{DD} and GND lines should be interdigitated in both directions with signal wires to provide small current return loops and low inductance. The supply wires must also have enough cross-sectional area to avoid electromigration problems. These requirements imply large amounts of metal and bypass capacitance, yet cost constraints dictate no more chip area than necessary.

A clocking subsystem includes clock generation, distribution, and gater elements. The clock generator can use a PLL to align the on-chip clock to an external reference for synchronous communication and to perform frequency multiplication. The clock distribution network should send the global clock to all clocked elements with low skew, yet not consume excessive power or area. The gaters perform local clock stopping or can produce multiple phases from the single global clock.

I/O signals include inputs, outputs, bidirectional signals, and analog signals. The I/O pads must deliver adequate bandwidth to large off-chip capacitances at voltage levels compatible with other chips. They must also protect the core circuitry against overvoltage and electrostatic discharge. High-speed parallel and serial links must account for the transmission line characteristics of the wires between chips. Their ultimate performance is limited by the ability to sample the received data at precisely the right time.

Chips are increasingly exploiting randomness for security applications. True random number generators can produce unguessable encryption keys. Random variations can also be used to uniquely identify an individual integrated circuit to serve as a serial number or to combat counterfeiting.

Exercises

12.1 Comment on the advantages and disadvantages of H-trees and clock grids. How does the hybrid tree/grid improve on a standard grid?

12.2 Explain how an electrostatic discharge event could cause latchup on a CMOS chip.

12.3 A ceramic PGA package with a good heat sink and fan has a thermal resistance to the ambient of 10 °C/W. The thermal resistance from the die to the package is 2 °C/W. If the package is in a chassis that will never exceed 50 °C and the maximum acceptable die temperature is 110 °C, how much power can the chip dissipate?

Methodology 13

13.1 Introduction

The manner in which you go about designing a particular system, chip, or circuit can have a profound impact on both the effort expended and the outcome of the design. IC designers have developed and adapted strategies from allied disciplines such as software engineering to form a cohesive set of principles to increase the likelihood of timely, successful designs. We will explore these principles in this chapter. While the broad principles of design have not changed in decades, the details of design styles and tools have evolved along with advances in technology and increasing levels of productivity. This chapter represents current CMOS design methods and provides an overview of a complex subject that could fill many books on its own. We encourage you to actively monitor the companies discussed and literature cited in the chapter to track the latest developments in this rapidly changing field.

As introduced in Section 1.6, an integrated circuit can be described in terms of three *domains*: (1) the *behavioral* domain, (2) the *structural* domain, and (3) the *physical* domain. The behavioral domain specifies what we wish to accomplish with a system. For instance, at the highest level, we might want to build an ultra-low-power radio for a distributed sensor network. The structural domain specifies the interconnection of components required to achieve the behavior we desire. Again, by way of example, our sensor radio might require a sensor, a radio transceiver, a processor and memory (with software), and a power source connected in a particular manner. Finally, the physical domain specifies how to arrange the components in order to connect them, which in turn allows the required behavior. Our example might start with the specification for an enclosure to hold the device, followed by a succession of physical drawings or specifications that may culminate in descriptions of geometry to be used to define a chip. Design flows from behavior to structure and ultimately to a physical implementation via a set of manual or automated transformations. At each transformation, the correctness of the transformation is tested by comparing the pre- and post-transformation design. For instance, if a power level is specified in the original behavioral description of the sensor radio, a test is run on the design in the structural domain with feedback from the physical domain to ensure this design goal is met.

In each of these domains there are a number of design options that can be selected to solve a particular problem. For instance, at the behavioral level, we can choose the wireless standard and the format in which data is transmitted by the sensor radio. In the structural domain, we can select which particular circuit style, logic family, or clocking strategy to use. At the physical level, we have many options about how the circuit is implemented in terms of chips, boards, and enclosures. These domains can further be hierarchically divided into different levels of *design abstraction*. Classically, these have included the following for digital chips:

- Architectural or functional level
- Logic or Register Transfer Level (RTL)
- Circuit level

For analog and RF circuits, the block diagram level replaces the logic level.

The relationship between description domains and levels of abstraction is elegantly shown by the *Gajski-Kuhn Y chart* in Figure 13.1 that was first introduced in Section 1.6.3. In this diagram, the three radial lines represent the behavioral, structural, and physical domains. Along each line are enumerated types of objects in that domain. In the behavioral domain, we have represented conventional software and hardware description language categories. As we move out along any of the radial axes, the increasing level of design abstraction is able to represent greater complexity. Thus, in the behavioral domain, the lowest level of abstraction is an instruction or a statement in software or HDL descriptions, respectively. Circles represent levels of similar design abstraction: the architectural, RTL, logic, and circuit levels. The particular abstraction levels and design objects may differ slightly depending on the design method.

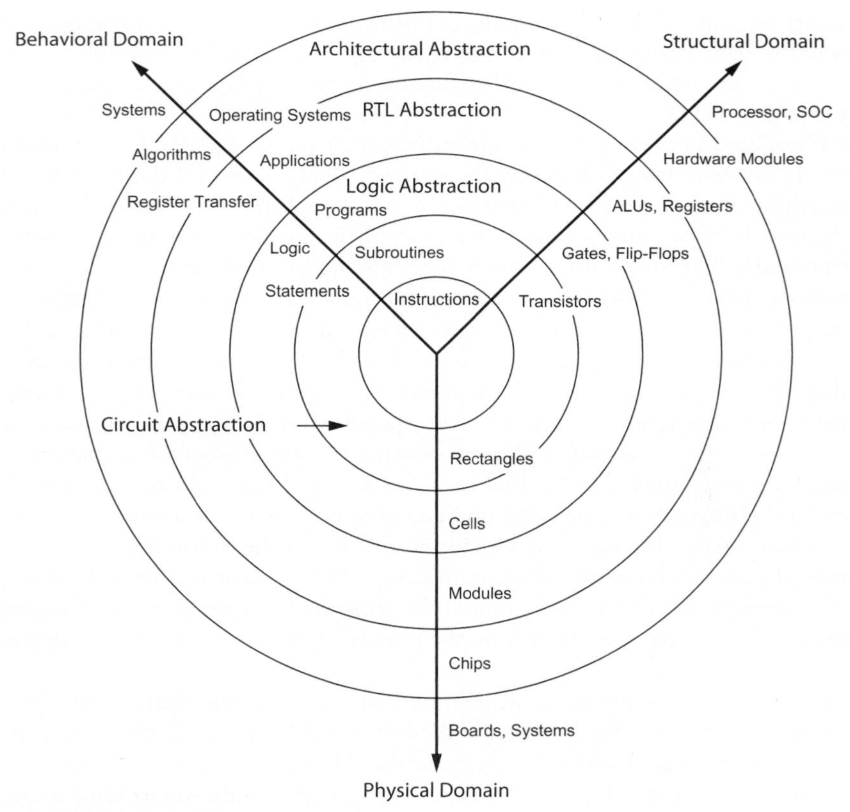

FIGURE 13.1 Gajski-Kuhn Y chart

In this chapter, we will examine how to transform a description from one domain into another while maintaining the integrity of the design. It is only in this way that we can start with a behavior and successfully build a product.

We begin by discussing some of the guiding principles that apply to most engineering projects. Then we survey the various design strategies available to the CMOS IC designer; these range from rapid prototyping or small-volume approaches to those suitable for high-volume digital, analog, or RF design. We then examine the economics of design, which can guide us to the right selection of an implementation strategy, and review documentation requirements.

13.2 Structured Design Strategies

The viability of an IC is in large part affected by the productivity that can be brought to bear on the design. This in turn depends on the efficiency with which the design can be converted from concept to architecture, to logic and memory, to circuit, and ultimately to physical layout. A good VLSI design system should provide for consistent descriptions in all three description domains (behavioral, structural, and physical) and at all relevant levels of abstraction (e.g., architecture, RTL/block, logic, and circuit). The means by which this is accomplished can be measured in various terms that differ in importance based on the application. These parameters can be summarized in terms of the following:

- Performance—speed, power, function, flexibility
- Size of die (hence, cost of die)
- Time to design (hence, cost of engineering and schedule)
- Ease of verification, test generation, and testability (hence, cost of engineering and schedule)

Design is a continuous trade-off to achieve adequate results for all of the above parameters. As such, the tools and methodologies used for a particular chip will be a function of these parameters. Certain end results have to be met (i.e., the chip must conform to certain performance specifications), but other constraints may depend on economics (i.e., size of die affecting yield) or even subjectivity (i.e., what one designer finds easy, another might find incomprehensible).

Given that the process of designing a system on silicon is complicated, the role of good VLSI-design aids is to reduce this complexity, increase productivity, and assure the designer of a working product. A good method of simplifying the approach to a design is by the use of constraints and abstractions. By using constraints, the tool designer has some hope of automating procedures and taking a lot of the "legwork" (effort) out of a design. By using abstractions, the designer can collapse details and arrive at a simpler object to handle.

In this chapter, we will examine design methodologies that allow a variation in the freedom available in the design strategy. The choice, assuming all styles are equally available, should be entirely economic. According to function, suitable design methods are selected. Following these steps, the required chip cost is estimated and the quickest means of achieving that chip should be chosen. We will focus on structured approaches to design since they offer the most appropriate method of dealing with design complexity.

The successful implementation of almost any integrated circuit requires attention to the details of the engineering design process. Over the years, a number of structured design techniques have been developed to deal with complex hardware and software projects. Not surprisingly, the techniques have a great deal of commonality. Rigorous application of these techniques can drastically alter the amount of effort that has to be expended on a given project and also, in all likelihood, the chances of successful conclusion.

13.2.1 A Software Radio—A System Example

To guide you through the process of structured design, we will use as an example a hypothetical "software radio," as illustrated in Figure 13.2. This device is used to transmit and receive radio frequency (RF) signals. Information is modulated onto an RF *carrier* to transmit data, voice, or video. The RF carrier is demodulated to receive information. An ideal software radio could receive any frequency and decode or encode any type of information at any data rate. Some day, this might be possible, but given the limitations of current processes, there are some bounds. To understand the impact of design methods on system solutions, we will examine the software radio in more detail. This system will then form the basis for discussion about structured approaches to design.

Figure 13.3 illustrates a typical transmit path for a generic radio transmitter, which is called an *IQ modulator*. An input data stream is encoded into *inphase* (*I*) and *quadrature* (*Q*)

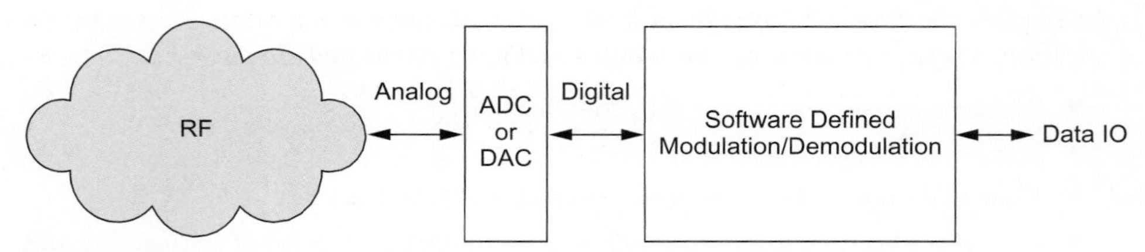

FIGURE 13.2 Software radio block diagram

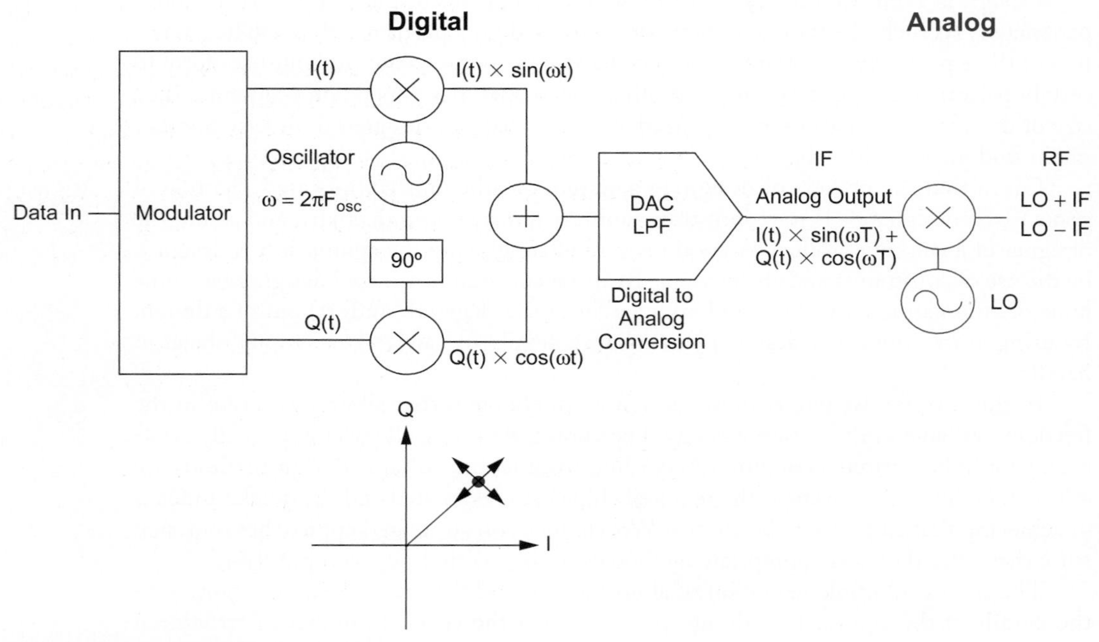

FIGURE 13.3 Software radio transmit path

signals. The I and Q represent signal amplitudes of a (voltage) vector that vary instantaneously in time as shown in the bottom of Figure 13.3. For appropriate I and Q values, any form of modulated carrier can be synthesized. I is multiplied by an oscillator (sine) operating at a frequency of F_{osc}. The quadrature (Q) signal is multiplied by the cosine of this frequency. The resultant signals are summed and passed to a digital-to-analog converter (DAC). In the design shown, this generates what we term an *Intermediate Frequency* or IF.

Typical IQ constellations are shown in Figure 13.4. *Amplitude Modulation* (AM), depicted in Figure 13.4(a), varies only in the magnitude of the carrier that varies in accordance with the amplitude of the modulation waveform. This is shown as a signal with an arbitrary phase angle (which we don't care about) and a vector that travels from the origin to a point on a circle that represents the maximum value of the carrier. In the case of an AM radio, the carrier frequency might be 800 KHz (in the AM band) and the modulation frequencies range from roughly 300 Hz to 6 KHz (voice and music frequencies). *Phase Modulation* is shown in Figure 13.4(b). Here, the vector travels around the maximum carrier amplitude circle varying the phase angle (δ) as the modulation changes. This is a constant amplitude modulation, which might be used with a carrier frequency of 100 MHz (in the FM broadcast band—we are loosely associating phase modulation with *frequency modulation* (FM) as they are closely related) and could have modulation frequencies of 200 Hz to 20 KHz (hi-fi audio). Finally, Figure 13.4(c) shows *Quadrature Phase Shift Keying* (QPSK) modulation, which is typical of data transmission systems. Two bits of data are encoded onto four phase points, as shown in the diagram. A typical carrier frequency might be 2.4 GHz in the Industrial Scientific and Medical (ISM) band and the modulation data rate might be 10 Mb/s.

Clearly, the ranges of carrier and modulation frequencies vary considerably. Generally, for high carrier frequencies, the modulation can be performed at a moderate frequency and then "mixed" up to a higher frequency by analog multiplication. This is completed in the analog domain and is illustrated by the blue components on the right side of Figure 13.3. An analog multiplier (called a *mixer* in RF terminology) takes an analog Local Oscillator (LO) and the Intermediate Frequency (IF) signal that we have generated and produces sum and difference frequencies. (It is also possible to generate the desired RF frequency directly, but in this design we will use an intermediate frequency approach.)

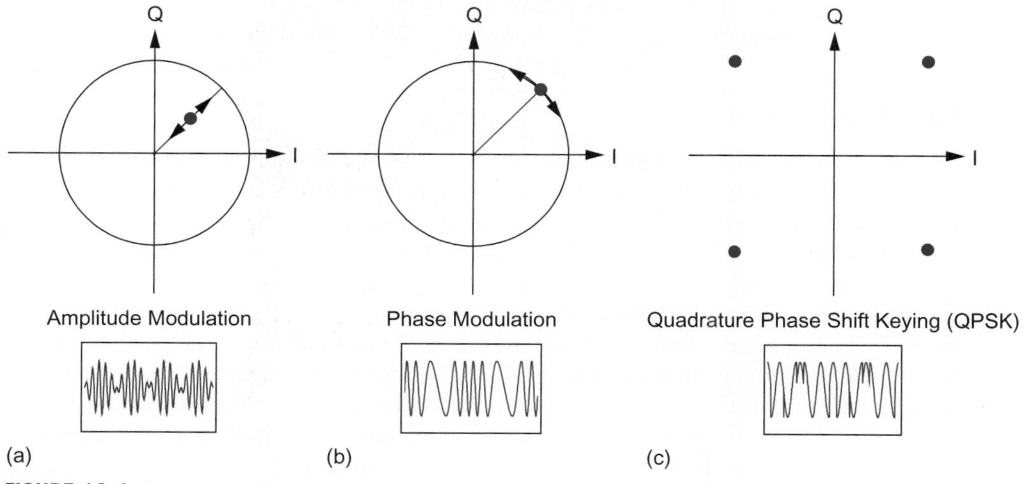

FIGURE 13.4 Examples of IQ modulation

Analog bandpass filtering or a slightly more sophisticated mixer can be used to select the mixing component (LO + IF or LO – IF) that we desire. For instance, if we generate a data signal on a 20 MHz IF and mix it with a 2.4 GHz LO, we can generate a 2.42 or 2.38 GHz data signal. This is called *upconversion*.

To complete the software radio, the receive path is shown in Figure 13.5. It is roughly the reverse of the transmit path. As in the transmit case, higher frequencies can be *down-converted* to lower IF frequencies that are suitable for processing by practical ADCs. The RF signal is mixed with the LO and low pass filtered to produce the difference frequency. For example, if a 2.4 GHz LO is mixed with the 2.42 GHz RF signal, the 20 MHz IF signal is restored. An analog-to-digital converter (ADC) converts the modulated IF carrier into a digital stream of data. This data is mixed (multiplied) in the digital domain by an oscillator operating at the IF frequency. After digital low pass filtering (LPF), the original *I* and *Q* signals can be reconstructed and passed to a demodulator. For further details on digital radio, consult a communications theory text such as [Haykin00].

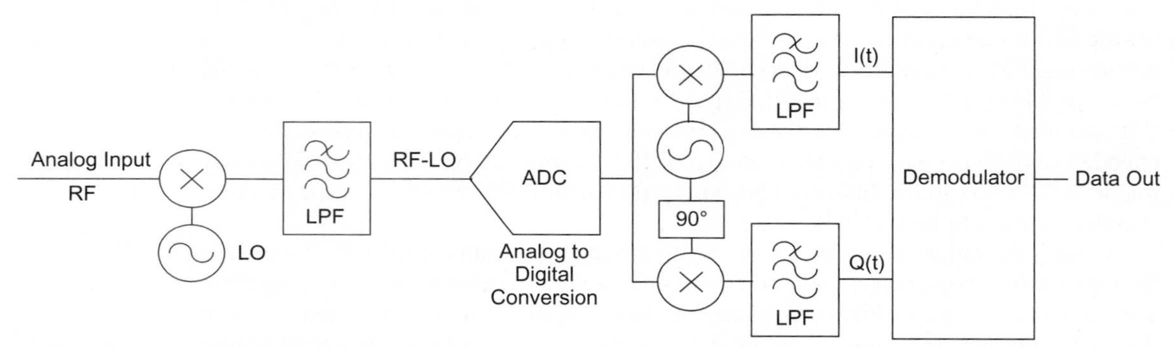

FIGURE 13.5 Software radio receive path

In summary, we see that multiplication, sine wave generation, and filtering are important for a software radio. While the modulation and demodulation have not been described in detail, operations can include equalization (multiplication), time to frequency conversion (fast Fourier transform), correlation, and other specialized coding operations. In the subsequent sections we will explore the design principles of hierarchy, regularity, modularity, and locality with concrete examples applied to the software radio.

13.2.2 Hierarchy

The use of hierarchy, or "divide and conquer," involves dividing a system into modules, then repeating this process on each module until the complexity of the submodules is at an appropriately comprehensible level of detail. This may entail stopping at a level where a prebuilt component is available for the particular function. The process parallels the software strategy in which large programs are split into smaller and smaller sections until simple subroutines with well-defined behavior and interfaces can be written. In the case of predefined modules, the design task involves using library code intended for the required function. The notion of "parallel hierarchy" can be used to aggregate descriptions in each of the behavioral, structural, and physical domains that represent a design (parallel hierarchy means a hierarchy—not necessarily identical—is used in each domain). Furthermore, equivalency tools can ensure the consistency of each domain. Because these tools can be applied hierarchically, you can progress in verification from the bottom to the top of a design, checking each level of hierarchy where domains are intended to correspond. For instance, a RISC processor core can have an HDL model that describes the behavior of the processor; a gate netlist that describes the type and interconnection of gates required to produce the processor; and a placement and routing description that describes how to physically build the processor in a given process. Later in the chapter, we will see how domain-to-domain comparisons are used to ensure consistency between domains.

Hierarchy allows the use of *virtual components*, soft versions of the more conventional packaged IC. Virtual components are placed into a chip design as pieces of code and come with support documentation such as verification scripts. They can be supplied by an independent *intellectual property* (IP) provider or can be reused from a previous product developed in your organization. Virtual components are discussed further in Section 13.5.7.

Example 13.1

The digital operations in the transmit path of the software radio (Figure 13.3) can be performed in software. Hence, a microprocessor can form the basis for the design. In this case, the design might have the hierarchy of a typical microprocessor, as shown in Figure 13.6. At the top level, the microprocessor contains an arithmetic logic unit (ALU), program counter (PC), register file, instruction decoder, and memory. The ALU can be further decomposed into an adder, a Boolean logic unit, and a shifter. The shifter and adder can together perform multiplication. The diagram illustrates how a relatively complex component can be rapidly decomposed into simple components within a few levels of hierarchy. Each level only has a few modules, which aids in the understanding of that level of the hierarchy.

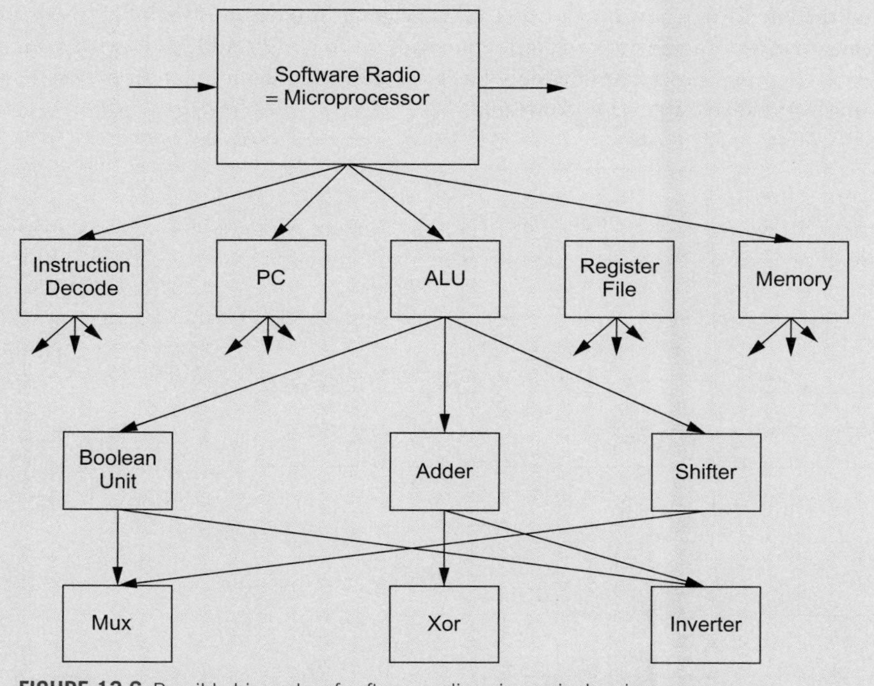

FIGURE 13.6 Possible hierarchy of software radio using a single microprocessor

Example 13.2

We can roughly estimate the performance required in the transmit path by noting that we require at least two multiplications, one addition, and two table lookups (sine and cosine). Another addition would be required to maintain a loop counter. An iterative multiply takes N cycles for an N-bit word, so for a 16-bit word width, the total number of cycles for the steps described would be approximately 16 + 16 + 1 + 2 + 2 + 1 (if table lookups take two clock cycles). This yields a total of roughly 40 clock cycles. For a 1 GHz processor, the fastest we could perform the IQ conversion would be approximately 40 ns, which, according to Nyquist's criteria ($F_{analog\_max} = F_{sample}/2$), would be capable of generating a 12.5 MHz IF signal. This is, of course, without any extra processing for modulating the carrier. While we could add another processor, this may be wasteful of area and power, given the operation that has to be performed.

A more power-efficient approach is to use dedicated hardware for the computationally intensive fixed-function blocks. The trick is to notice that the IQ modulator portion of the software radio transmit and receive path for a given DAC and ADC resolution has a relatively fixed architecture. For the transmit path, the hierarchy shown in Figure 13.7 can be used where the blue sections have been converted to fixed function blocks. This is a relatively safe bet because the IQ upconversion is a generic communications building block. In addition to the multipliers, a device called a *Numerically Controlled Oscillator* (NCO) has been introduced [Lu93, Lu93b, Hwang02]. The NCO, described in detail in the next section, generates sine or cosine waveforms at a speed determined by the delay through an N-bit adder where N is in the range of 16 to 32 for typical NCOs. The move to dedicated hardware for the IQ upconversion allows the circuit to produce a new value once every clock cycle. If we conservatively say that the arithmetic blocks operate at the same speed that the microprocessor ALU does, then the circuit will now operate at 1 GHz. Taking into account sampling theory, this means that we can generate analog frequencies up to almost 500 MHz with a suitable DAC. The microprocessor now only has to respond at the modulation data rate, providing IQ values to the IQ upconverter.

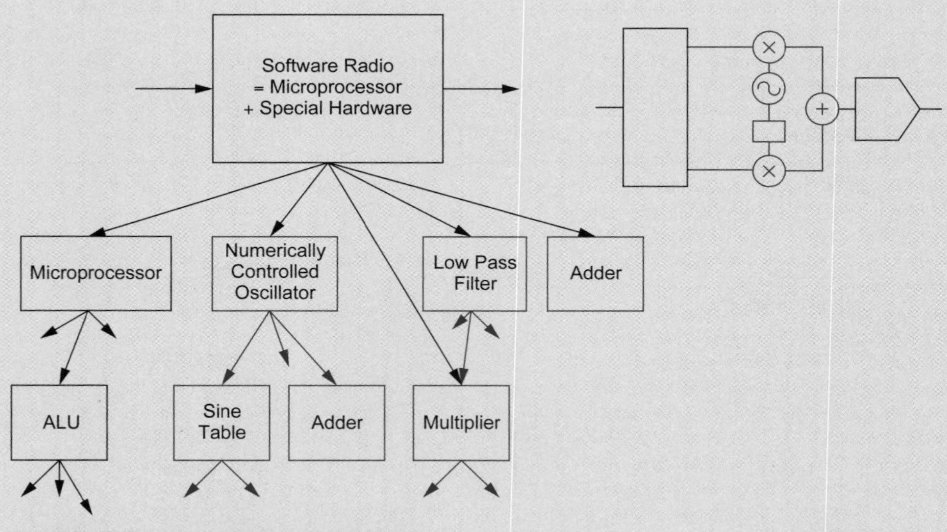

FIGURE 13.7 Transmit chain with dedicated IQ upconverter

13.2.3 Regularity

Hierarchy involves dividing a system into a set of submodules. However, hierarchy alone does not solve the complexity problem. For instance, we could repeatedly divide the hierarchy of a design into different submodules but still end up with a large number of different submodules. With regularity as a guide, the designer attempts to divide the hierarchy into a set of similar building blocks. Regularity can exist at all levels of the design hierarchy. At the circuit level, uniformly sized transistors can be used, while at the gate level, a finite library of fixed-height, variable-length logic gates can be used. At the logic level, parameterized RAMs and ROMs could be used in multiple places. At the architectural level, multiple identical processors can be used to boost performance.

Regularity aids in verification efforts by reducing the number of subcomponents to validate and by allowing formal verification programs (see Section 13.4.1.3) to operate more efficiently. Design reuse depends on the principle of regularity to use the same virtual component in multiple places or products.

Example 13.3

In an example of regularity applied to the software radio, we first look inside two of the blocks used in the designs shown in Figures 13.3 and 13.5 to assess what kinds of functions are required.

The NCO is shown in Figure 13.8(a). It is composed of a registered adder that is incremented every clock cycle by a phase increment register. This implements a phase counter, which is used to step through a ROM lookup table that provides phase-to-amplitude conversion. A phase offset can be added to the phase incrementer to perform phase modulation. With this structure, we are able to generate a digital sine wave.

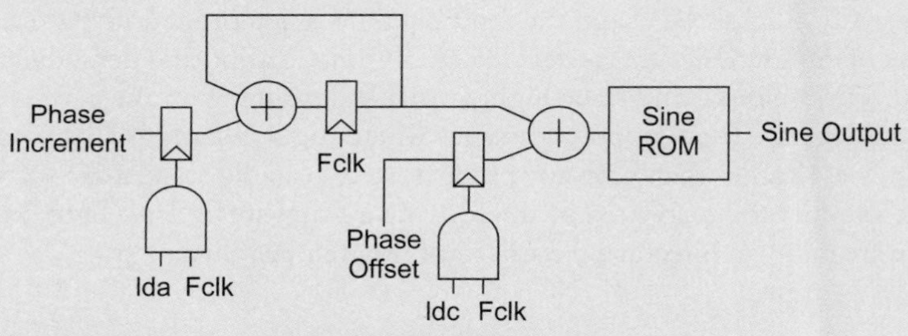

(a) Numerically Controlled Oscillator (NCO) Structure

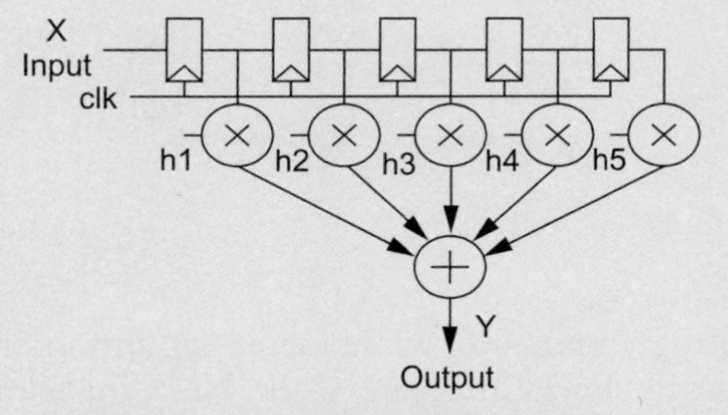

(b) Finite Impulse Response (FIR) Filter Structure

FIGURE 13.8 Structure of numerically controlled oscillator and low-pass filter (implemented as a finite impulse response (FIR) filter)

Turning to the low-pass filter shown in Figure 13.5, Figure 13.8(b) shows the structure for a commonly used low-pass filter implementation that is called a *Finite Impulse Response* (FIR) filter [Edwards93, Choi97]. The structure computes the function

$$Y[n] = \sum X[n-k]h[k] \quad \textbf{(13.1)}$$

where $X[n]$ is the sampled input, $h[k]$ are the filter coefficients that characterize the particular filter, and $Y[n]$ is the output. As the structure indicates, the filter is composed of registers, multipliers, and an adder. Filters are characterized by the number of *taps* (coefficients). More taps yield better filters approaching an ideal "brick wall" filter with steeper cutoff and low ripple. This, in turn, requires more registers and more multipliers.

Having examined the detail of these blocks, we notice that the common functions are registers, adders, and multipliers with precisions as yet undefined. Parallel N-bit adders can be composed of N single-bit full adders. Multipliers are also built from full adders. N-bit registers are built from 1-bit flip-flops. Thus, one form of regularity might be to use the same full adder for all parallel adders and multipliers. Similarly, the same flip-flop would be used in all locations.

Typically, the phase counter adder in the NCO would be of the order of 16–32 bits wide. The phase increment adder might be 8–16 bits wide. The sizes of the multipliers and adders in the FIR filter vary widely, but depend on the input data width. This typically varies from 1–12 bits.

Example 13.4

As illustrated in the previous section, IQ upconversion and downconversion can be converted to fixed hardware, as highlighted in blue in Figure 13.9. Whether the hardware is shared (i.e., the NCO and the multipliers) is a determination that can be made at the time of design. Once this is decided, the IQ modulation and demodulation is still undefined. These blocks tend to be highly variable depending on the particular system. Software radios have been proposed in areas where the standards are likely to evolve as time progresses. Rather than have any product fixed to an old standard, a software radio allows the product to be updated in the field via a firmware update. Thus, in our quest for a software radio architecture, we still want programmability.

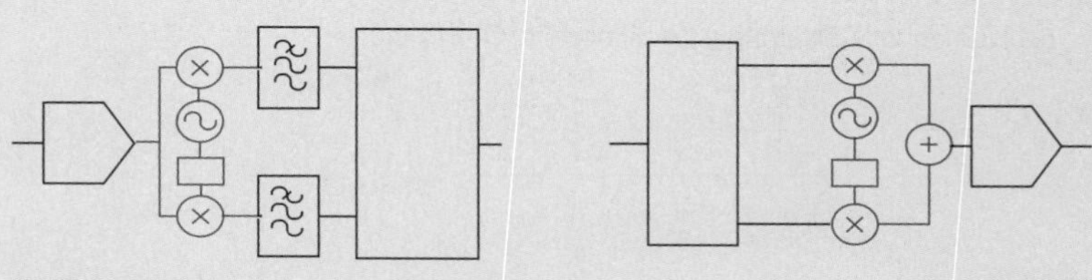

FIGURE 13.9 Common IQ blocks

A solution to maintaining programmability while increasing processing power might be to use a multiprocessor, as shown in Figure 13.10. Here, the hardware IQ up-and-down conversion has been retained and the IQ modulation/demodulation is performed by the four processors. The number of processors is arbitrary and would be ascertained by a detailed analysis of the required computational power.

Imagine that the computational power required slightly exceeds that provided by the four processors shown in Figure 13.10. Because multiplication is a frequently required operation in signal processing operations, it makes sense to build a multiplier into each microprocessor, as shown in Figure 13.11. Hence, we maintain regularity and improve processing power.

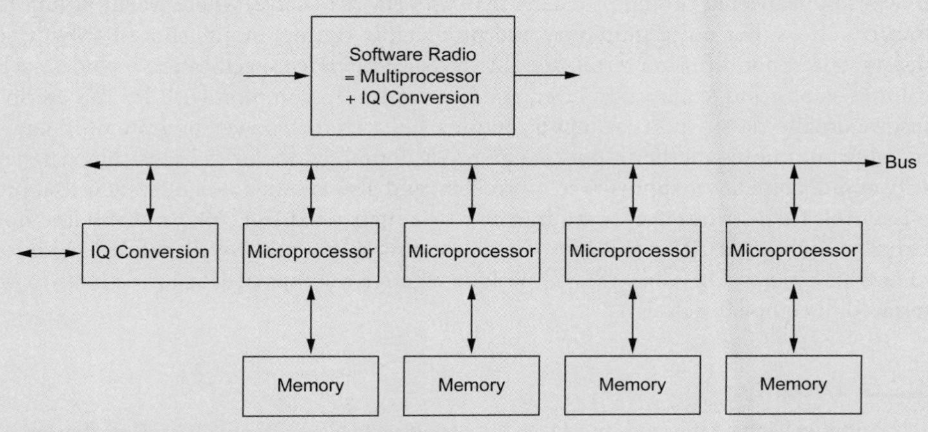

FIGURE 13.10 Software radio as a multiprocessor

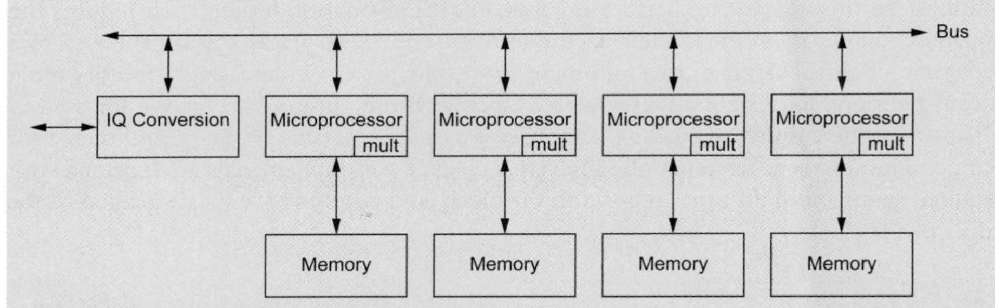

FIGURE 13.11 Enhanced multiprocessor for software radio

If the multiplication is a one-cycle operation, the throughput for multiplication-intensive operations can improve by a factor of up to M as compared to an M-bit processor with an iterative multiplication operation. This style of acceleration can be repeated for any operation that is computationally intensive. The application code is profiled, timing bottlenecks are identified, and custom hardware is added with appropriate instructions to access the hardware. In this manner, the overall solution remains programmable while the speed of processing increases markedly. Tensilica sells extensible processors using such an approach. However, adding functional units increases die size and power dissipation, so trade-offs are necessary.

13.2.4 Modularity

The tenet of modularity states that modules have well-defined functions and interfaces. If modules are "well-formed," the interaction with other modules can be well characterized. The notion of "well-formed" may differ from situation to situation, but a good starting point is the criteria placed on a "well-formed" software subroutine. First of all, a clearly defined interface is required. In the case of software, this is an argument list with typed variables. In the IC case, this corresponds to a clearly defined behavioral, structural, and physical interface that indicates the function as well as the name, signal type, and electrical and timing constraints of the ports on the design. Reasonable load capacitance and drive capability should be required for I/O ports. Too large a fanin or too small a drive capability can lead to unexpected timing problems that take effort to solve, where we are trying to minimize effort. For noise immunity and predictable timing, inputs should only drive transistor gates, not diffusion terminals. The physical interface specification includes such attributes as position, connection layer, and wire width. In common with HDL descriptions, we usually classify ports as inputs, outputs, bidirectional, power, or ground. In addition, we would note whether a port is analog or digital. Modularity helps the designer clarify and document an approach to a problem, and also allows a design system to more easily check the attributes of a module as it is constructed (i.e., that outputs are not shorted to each other). The ability to divide the task into a set of well-defined modules also aids in System-On-Chip (SOC) designs where a number of IP sources have to be interfaced to complete a design.

13.2.5 Locality

By defining well-characterized interfaces for a module, we are effectively stating that other than the specified external interfaces, the internals of the module are unimportant to other modules. In this way we are performing a form of "information hiding" that reduces the apparent complexity of the module. In the software and HDL world, this is paralleled by a reduction of global variables to a minimum (hopefully to zero). Increasingly, locality often means temporal locality or adherence to a clock or timing protocol. This is addressed in Chapter 9, where different clocking strategies are examined. One of the central themes of temporal locality is to reference all signals to a clock. Thus, input signals are specified with required setup and hold times relative to the clock, and outputs have delays related to the edges of the clock.

Example 13.5

In the example of the software radio, locality would probably be most evident in the floorplan of the chip. One example floorplan is shown in Figure 13.12. The analog blocks (ADC and DAC) are placed adjacent to the I/O pads. This is an example of physical locality because the analog blocks draw significant DC current and therefore the power busses have to be short and exhibit low resistance. Furthermore, the analog input and analog output signals can be routed to the pads without interference from digital signals. If necessary, the left edge of the chip can be guard-ringed and placed in a deep n-well if this process option is available. The digital IQ_upconversion module is placed near the DAC and ADC, and the four programmable processor/memory composites are arrayed across the chip.

An alternative floorplan is shown in Figure 13.13. Here, the analog blocks and IQ_ conversion module are placed at the top of the chip. The four processor/memory blocks are then arrayed around a centrally located bus. The area for both array possibilities is roughly the same, but the second floorplan is better because the bus connecting the processors is shorter and hence faster and potentially dissipates less power. This is an example of physical locality used to obtain good temporal performance.

FIGURE 13.12 A possible floorplan for software radio

13.2.6 Summary

There are strong parallels between the methods of design for software and hardware systems. Table 13.1 summarizes some of these parallels for the principles outlined above.

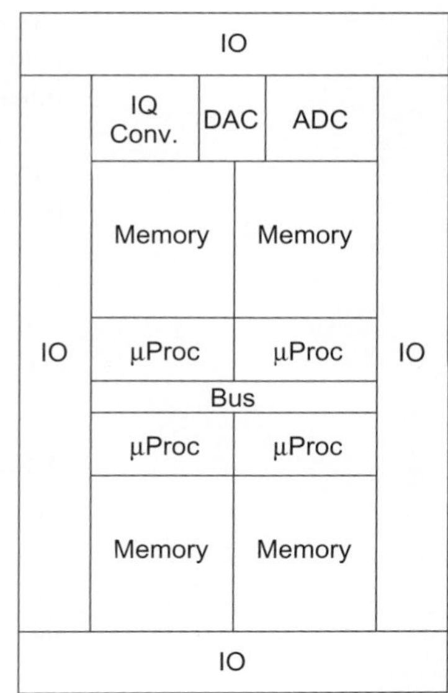

FIGURE 13.13 Alternate floorplan for software radio

TABLE 13.1 Structure software and VLSI hardware design

Design Principle	Software	Hardware
Hierarchy	Subroutines, libraries	Modules
Regularity	Iteration, code sharing, object-oriented procedures	Datapaths, module reuse, regular arrays, gate arrays, standard cells
Modularity	Well-defined subroutine interfaces	Well-defined module interfaces, timing and loading data for modules, registered inputs and outputs
Locality	Local scoping, no global variables	Local connections through floorplanning

13.3 Design Methods

In this section, we will examine a range of design methods that can be used to implement a CMOS system. This section will concentrate on the target of the design method, in contrast to the design flow used to build a chip. Design flows, which deal with how a design progresses through a set of tools, will be dealt with in the subsequent section. The base design methods are arranged roughly in order of "increased investment," which loosely relates to the time and cost it takes to design and implement the system. It is important to understand the costs, capabilities, and limitations of a given implementation technology to select the right solution. For instance, it is futile to design a custom chip when an off-the-shelf solution that meets the system criteria is available for the same or lower cost.

13.3.1 Microprocessor/DSP

Many times, the most practical method to solve a system design problem is to use a standard microprocessor or digital signal processor (DSP). There are many single-chip microprocessors with built-in RAM and EEROM/EPROM available in the market. For example, the PIC family of processors from Microchip offers a wide range of clock speeds, memory sizes, and analog I/O capability (ADCs) in a small package. For more signal-intensive problems, classical DSPs from vendors such as Analog Devices and Texas Instruments can be employed. Microprocessors provide great flexibility because systems can be upgraded in the field through software patches. Do not underestimate the cost of software development for microprocessor-based systems.

Even when you decide to build a system with an off-the-shelf microprocessor, you should consider the possibility of eventual integration. For example, if your product becomes very successful and you want to reduce costs by integrating it into a single system-on-chip rather than building it as a board with a microprocessor and various support chips, you will need a microprocessor that is available in embedded form so that you can keep your software. Examples of embedded commercial processor cores include ARM, MIPS, and IBM's PowerPC.

13.3.2 Programmable Logic

Often, the cost, speed, or power dissipation of a microprocessor may not meet system goals and an alternative solution is required. A variety of programmable chips are available that can be more efficient than general purpose microprocessors yet faster to develop than dedicated chips:

- Chips with programmable logic arrays
- Chips with programmable interconnect
- Chips with reprogrammable logic and interconnect

The system designer should be familiar with these options for two reasons:

1. It allows the designer to competently assess a particular system requirement for an IC and recommend a solution, given the system complexity, the speed of operation, cost goals, time-to-market goals, and any other top-level concerns.
2. It familiarizes the IC designer with methods of making any chip reprogrammable at the hardware level and hence both more useful and more widely applicable.

13.3.2.1 Programmable Logic Devices The devices covered in this section are descended from chips that implement two-level sum-of-product programmable logic arrays (PLAs) discussed in Section 11.7. They differ from the field-programmable gate arrays described in the next section in that they have limited routing capability. Historically, process densities did not allow the transistor count and routing resources found in modern field-programmable gate arrays. Programmable logic devices based on PLAs allowed a useful product to be fielded and well-established techniques allowed logic optimization to target PLA structures, so the associated CAD tools were relatively simple. They are still occasionally used because the regular array and interconnect make timing very predictable.

A PLA consists of an AND plane and an OR plane to compute any function expressed as a sum of products. Each transistor in the AND and OR plane must be capable of being programmed to be present or not. This can be achieved by fully populating the AND and OR plane with a NOR structure at each PLA location. Each node is programmed with a floating-gate transistor, a fusible link, or a RAM-controlled transistor, as illustrated in Figure 13.14. The first two versions were the way these types of devices were programmed when device densities were low. These devices, such as the Texas Instruments PAL16 family, are generally used for legacy applications.

13.3.2.2 Field-Programmable Gate Arrays (FPGAs) Field-Programmable Gate Arrays (FPGAs) use the high circuit densities in modern processes to construct ICs that, as their

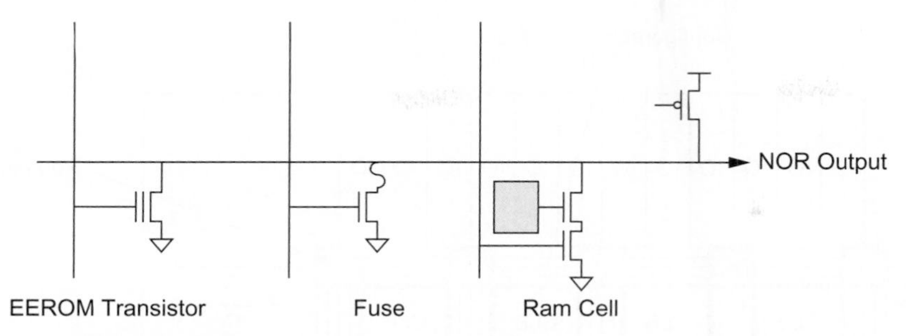

FIGURE 13.14 PLA NOR structure (one plane shown)

name suggests, are completely programmable even after a product is shipped or "in the field." Two basic versions exist. The first uses a special process option such as a fuse or antifuse to permanently program interconnect and personalize logic. These are one-time programmable. The second type uses static RAM or flash memory to configure routing and logic functions. In general, an FPGA chip consists of an array of logic cells surrounded by programmable routing resources.

As an example of the first type of FPGA, devices manufactured by Actel embed an array of logic modules within an interconnect matrix that is formed on the top metal layers. Successive routing channels run vertically or horizontally. A special one-time programmable contact, called an *antifuse*, is placed at the intersection of routing traces. These normally have high resistance (effectively an open circuit). Upon application of a special programming voltage across the contact, the resistance permanently drops to a few ohms. CMOS switches allow the programming voltage to be directed to any antifuse in the chip. The advantage of this type of routing is that the size of the programmable interconnect is tiny—the intersection area of two metal traces. Moreover, the on-resistance is low compared to a CMOS switch, so the circuit speed is not compromised. The disadvantage is that the interconnect is not reprogrammable, so once a chip is programmed, its function is fixed to the extent that the interconnect has been personalized.

Figure 13.15 shows the floorplan of a simplified FPGA. The chip is composed of an array of *configurable logic blocks* (CLBs). Metal routing tracks run vertically and horizontally between the array of CLBs. These terminate at the gray blocks, which are routing switches that can be implemented using antifuses, CMOS transmission gates, or tristate buffers. The routing resources can also be connected to the inputs and outputs of the adjacent CLBs. CLBs use programmable lookup tables to compute any function of several variables. Configurable I/O cells that can be used as input, output, or bidirectional pads surround the core array of CLBs.

A simple SRAM-based FPGA logic cell is shown in Figure 13.16. It is composed of a 16×1 static RAM as the logic element. This provides for any logic function of four variables merely by loading the RAM with the appropriate contents. Table 13.2 illustrates how the table should be loaded to perform various logic functions. A full adder can be implemented in two CLBs (one for carry and one for sum). The CLB shown also provides an optional output register. While it may seem inefficient or slow to use a RAM to perform logic, specially designed single-data line RAMs are small and fast in current processes, and resources such as the routing tend to dominate modern designs from a density and speed viewpoint.

FPGAs have matured to the stage where they boast millions of logic gate equivalents supported by megabits of RAM. I/Os can operate in excess of 10 GHz. FPGAs frequently

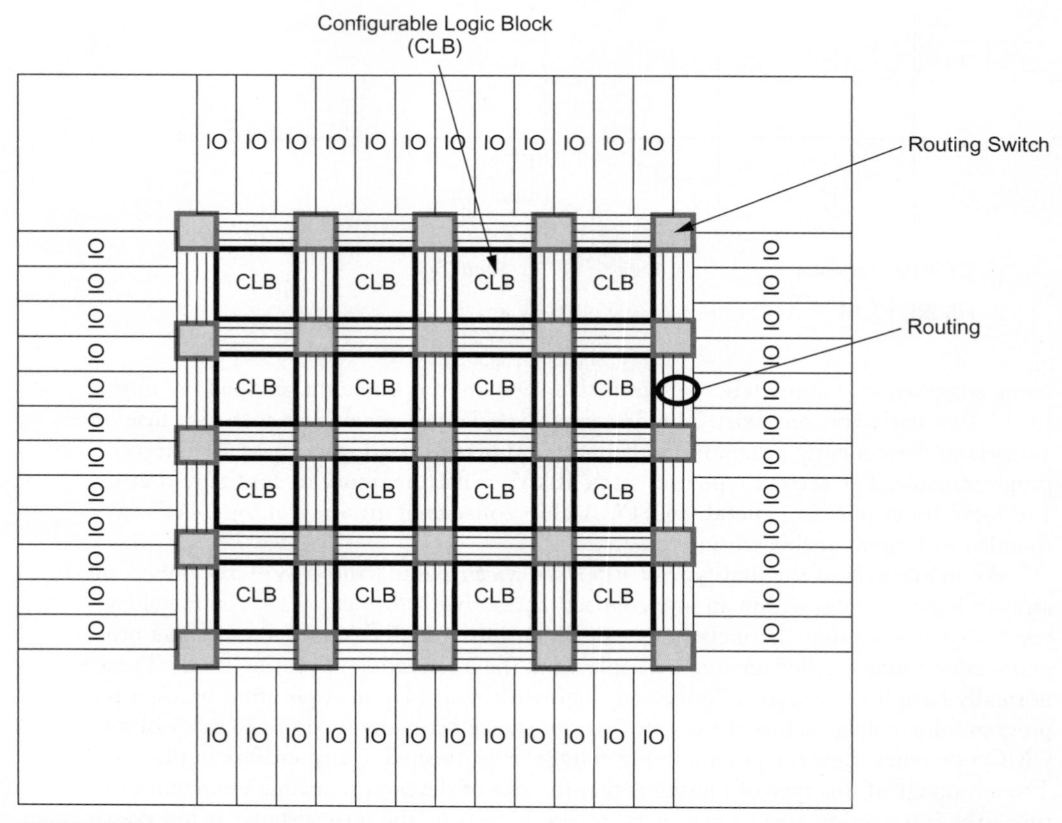

FIGURE 13.15 Simplified FPGA floorplan

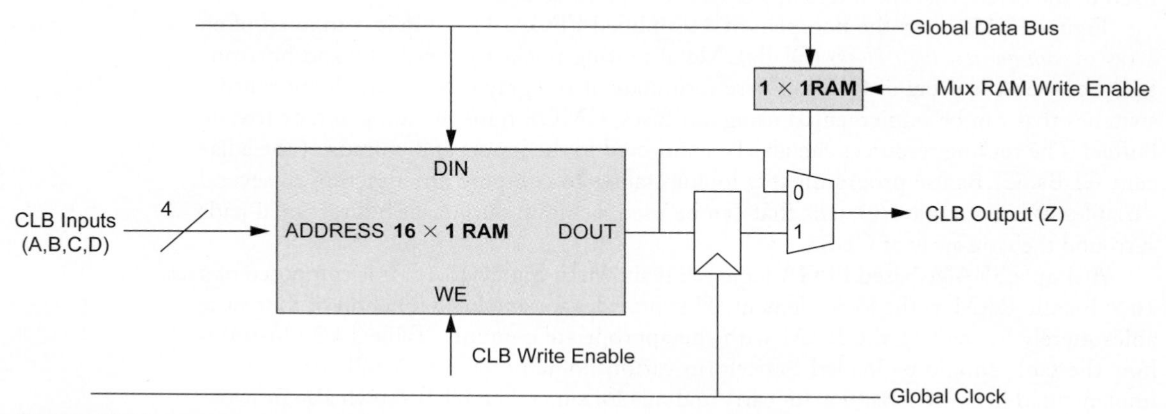

FIGURE 13.16 Simple FPGA logic cell

have embedded microprocessor cores and DSP accelerator hardware. Their low up-front cost and ease of correcting design errors makes them the best choice now for many low- to medium-volume custom logic applications.

TABLE 13.2 RAM CLB functions

Address	*ABCD*	$A \cdot B \cdot C \cdot D$	~*A*	SUM(*A*,*B*,*C*)
0	0000	0	1	0
1	1000	0	0	1
2	0100	0	1	1
3	1100	0	0	0
4	0010	0	1	1
5	1010	0	0	0
6	0110	0	1	0
7	1110	0	0	1
8	0001	0	1	0
9	1001	0	0	1
10	0101	0	1	1
11	1101	0	0	0
12	0011	0	1	1
13	1011	0	0	0
14	0111	0	1	0
15	1111	1	0	1

Note that (after sorting out the intellectual property rights with the appropriate patent holders) it is possible to implement FPGA blocks on any CMOS chip to provide some degree of programmability at the gate level.

13.3.3 Gate Array and Sea of Gates Design

The chips described in the previous section do not require a fabrication run. Designers typically strive to keep the non-recurring engineering cost (NRE, see Section 13.5.1) as low as possible. One method of doing this is to construct a common base array of transistors and personalize the chip by altering the metallization (metal and via masks) that is placed on top of the transistors. This style of chip is called a Gate Array (GA). A particular subclass of a gate array is known as a Sea-of-Gates (SOG) chip. Gate arrays used to be popular methods of designing semicustom ASICs.

It is still worthwhile understanding SOG techniques because they can also be used on custom chips to provide an area of reprogrammable logic on an otherwise fixed function chip. The system-on-chip can be comprised of a set of fixed functions (e.g., a processor, RAM, and dedicated accelerators), and an SOG area. Rows of nMOS and pMOS transistors are arrayed in the SOG portion of the chip. Each logic row consists of an n row and p row. Figure 13.17(a) shows an SOG structure, which features continuous rows of transistors. Grounding the gate of the nMOS transistor or connecting the gate of the pMOS transistor to the V_{DD} rail provides isolation between gates. Figure 13.17(b) shows a gate array structure that uses groups of three transistor pairs.

Figure 13.17(c) shows a portion of an SOG structure programmed to be a 3-input NAND gate. Note that the nMOS and pMOS transistors at each end isolate the gate, as described previously. Personalization of this SOG structure commences at contact and metal1 masks, and can continue up for all metal layers available in the process.

CAD tools have advanced to the point that reprogramming an SOG is barely easier than regenerating a cell-based layout. However, a small SOG area remains useful for

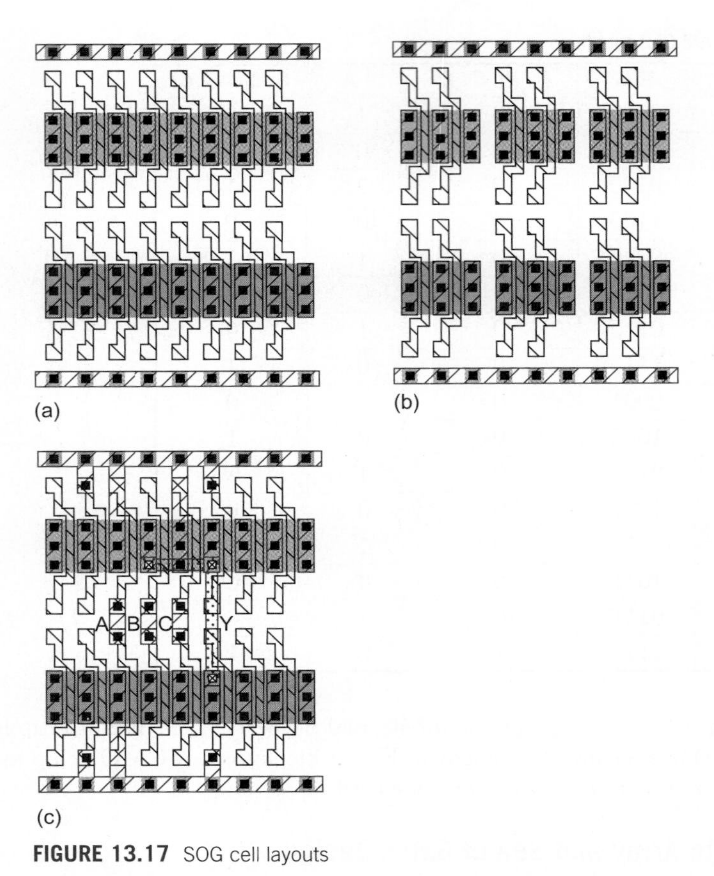

FIGURE 13.17 SOG cell layouts

correcting simple logic errors with metal-only fixes during debug and even during late design [Stolt08]. Moreover, process variability is driving designers toward restrictive design rules with regular structures that begin to resemble SOGs.

13.3.4 Cell-Based Design

Cell-based design uses a standard cell library as the basic building blocks of a chip. The cells are placed in appropriate positions, then their interconnections are routed. Cell-based design can deliver smaller, faster, and lower-power chips than FPGAs but has high NRE costs to produce the custom mask set. Therefore, it is only economical for high volume parts or when the performance commands a lucrative sales price. As compared to full-custom design, cell-based design offers much higher productivity because it uses predesigned cells with layouts. Foundries and library vendors supply cells with a wide range of functionality. These include the following:

- Small-scale integration (SSI) logic (NAND, NOR, XOR, AOI, OAI, inverters, buffers, registers)
- Memories (RAM, ROM, CAM, register files)
- System level modules such as processors, protocol processors, serial interfaces, and bus interfaces
- Possibility of mixed-signal and RF modules

Whereas Medium Scale Integration (MSI) functions such as adders, multipliers, and parity blocks used to be supplied as cells, synthesis engines commonly construct these from base-level Small Scale Integration (SSI) gates in current design systems.

A typical standard cell library is shown in Table 13.3. A 1x (normal power) cell commonly is defined to use the widest transistors that fit within the vertical pitch of the standard cell. 2x and larger (high power) cells use wider transistors to deliver more current. They must fold the transistors to fit within the cell; this comes at the expense of increased cell width. Gates are often available in low power versions as well. These cells use minimum-width transistors to reduce capacitance. Low-power cells tend to be slow because of the wire capacitance they must drive. Although they do not save area, they do reduce power consumption on noncritical paths.

Sophisticated libraries also generate memories of assorted sizes from a graphical user interface. The generators yield not only the physical layout but also a complete data sheet indicating access times, cycle times, and power dissipation.

In the event that a standard cell library may not be available for a process, it is worthwhile to review some of the approaches to standard cell design. Usually, standard cells are a fixed height with power and ground routed respectively at the top and bottom of the cells, as shown on the inside front cover. This allows the cells to be abutted end to end and to have the supply rails connect. A single row of nMOS transistors adjacent to GND (ground) and a single row of pMOS transistors adjacent to V_{DD} (power) are normally used. The polysilicon gate is connected from nMOS transistor to pMOS transistor and, in the case of multiplexers and registers, the

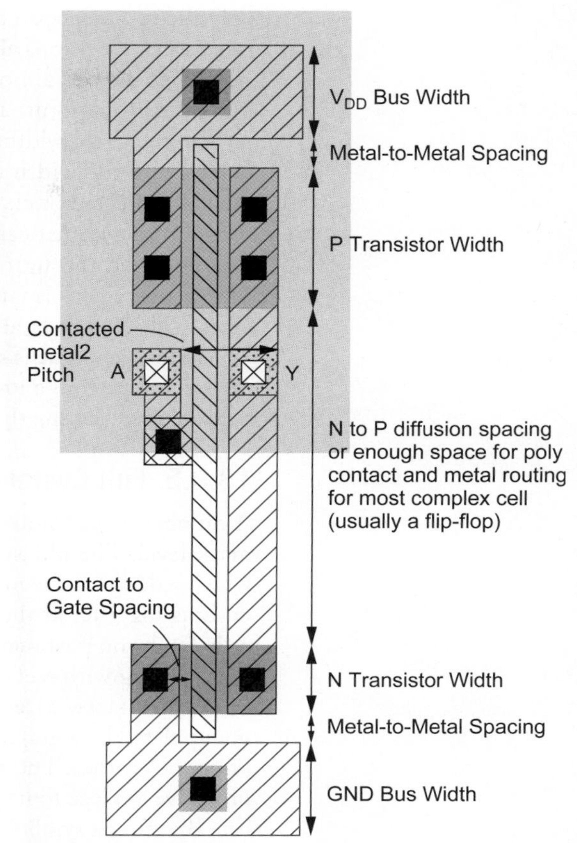

FIGURE 13.18 Typical standard cell layout with some of the constraints

TABLE 13.3 Typical standard cell library

Gate Type	Variations	Options
Inverter/Buffer/ Tristate Buffers		Wide range of power options, 1x, 2x, 4x, 8x, 16x, 32x, 64x minimum size inverter
NAND/AND	2–8 inputs	High, normal, low power
NOR/OR	2–8 inputs	High, normal, low power
XOR/XNOR		High, normal, low power
AOI/OAI	21, 22	High, normal, low power
Multiplexers	Inverting/noninverting	High, normal, low power
Adder/Half Adder		High, normal, low power
Latches		High, normal, low power
Flip-Flops	D, with and without synch/asynch set and reset, scan	High, normal, low power
I/O Pads	Input, output, tristate, bidirectional, boundary scan, slew rate limited, crystal oscillator	Various drive levels (1–16 mA) and logic levels

polysilicon connection has to be crossed between vertically coincident nMOS and pMOS transistors. Decisions about the sizes of transistors have to be made. Following this decision, the cells are almost completely defined by the process design rules. Figure 13.18 illustrates this point. The height of the cell is defined by the sum of the nMOS and pMOS transistor widths, the separation on n and p regions, the spacing to V_{DD} and GND busses, and the width of these busses. The horizontal pitch is defined by the poly-to-metal2 contacted pitch, as shown in the figure. It is relatively easy to construct a software program to automatically generate cells like the one shown in Figure 13.18. Cell delay is characterized through simulation to good agreement with silicon. Fabrication of such cells to prove performance is rarely required. Options to standard cells include routing the clock with the power and ground busses and routing multiple supply voltages to each cell. The latter technique is sometimes used to reduce power by connecting gates that are not in the critical path to a lower than normal supply voltage. Recall that the power drops with the square of the supply voltage.

13.3.5 Full Custom Design

A number of techniques can be used to design standard cells or larger circuit blocks at the mask level. The oldest and most traditional technique is termed *custom mask layout,* in which a designer sits in front of a graphics display running an interactive editor and pieces designs together at the geometry level one rectangle at a time. This work is sometimes called polygon pushing. A variation of custom mask design is called *symbolic layout.* Rather than dealing with rectangles and polygons on various mask levels, the primitives are transistors, contacts, wires, and ports (points of connection). These primitives can also be manipulated by a graphics editor. Some systems allow for a "design rule free" placement of symbolic entities. The actual placement occurs after a spacing process that compacts each primitive as close to its neighbor as possible according to the design rules of the process in use. By using a symbolic layout system, layout topologies can be transported from process to process without a huge amount of effort.

In these times of cell-based design, digital CMOS ICs use custom mask design only for the highest of volume parts such as microprocessor datapaths. However, analog and RF designs, cell libraries, memories, and I/O cells still frequently use custom design. There are a variety of custom MOS layout hints in Section 13.7. Custom design is also worthwhile pedagogically because it completes the link from transistors to systems.

From time to time, we have mentioned software generators as a method of generating physical layout. This kind of idea has been around for a long time and was often referred to as *silicon compilation.* Complete microprocessors were typical of layouts that were generated. A "correct by construction" method was used to build the layouts hierarchically. In other words, only the mask description was generated, with perhaps a high-level instruction level simulator being the behavioral model. Generators are the most common method used today for library generation.

With modern design flows, many different "views" of a design are required to integrate with the regular path through the design system. For instance, in addition to the behavioral model, a timing view would be needed for timing verification, a logic view might be required for simulation, and a circuit view for layout versus schematic or netlist comparisons would be needed. Software generators can be used to provide all of these views automatically.

Modern versions of the venerable "silicon compiler" can be built in a structured hierarchical manner to generate memories, register files, and other special-purpose structures that can benefit from a customized layout. One of the most straightforward approaches is to write custom placement routines that in essence "hand place" certain standard cells within the row structure of a standard cell design. For instance, you may prefer a certain adder design and have a datapath layout for the adder. An algorithm can be written to place the cells on the standard cell grid. In addition, a linked algorithm can be written to generate a gate netlist in an HDL. In this way, both the physical and structural design are captured. The behavior can be represented by an HDL function or module call. Such custom placement can shorten wire lengths and thus improve speed and power.

Custom-designed microprocessors routinely exceed 2 GHz in nanometer processes, while synthesized ASICs typically operate closer to 200–350 MHz. [Chinnery02] made a fascinating study of the differences between design methods that account for this gap. He identified microarchitecture, sequencing overhead, circuit families, logic design, cell design, layout, and design margining as the major differences. Since that study, CAD tools have improved, especially in the integration of synthesis and placement. Custom designs have become more conservative and now use static CMOS circuits and cell libraries similar to their ASIC cousins. Nevertheless, a wide gap still exists.

In a followup study, [Chinnery07] examines the gap between ASIC and custom design for power dissipation. Major factors for ASICs consuming more power than custom designs include microarchitecture, clock gating, logic style, logic design, technology mapping, cell and wire sizing, voltage scaling, floorplanning, process technology, and process variation. The study concludes that synthesizable designs typically consume 3–7× more power than custom designs but that better tools and cell libraries can close this gap to 2.6×.

13.3.6 Platform-Based Design—System on a Chip

As systems have become more complex, the use of predefined intellectual property (IP) blocks has become commonplace. Designs frequently use a number of common blocks such as RISC processors, memory, and I/O functions attached to common busses. A *platform* can be used to implement a design by using common structures such as busses and common high-level languages (such as C) to program the processors. To a large extent, the RISC processor and memories can be interchanged and the number and type of peripherals can be changed while maintaining good design and verification times because the modules have been predesigned and the test and verification scripts come with the IP blocks. The design task is to put the blocks together, design any application-specific blocks, and place and route a correctly operational chip. Note that the last step, while automated, still takes considerable engineering effort.

As many current chips feature one or more embedded microprocessors, the task of writing software is added to the task of designing logic. Moreover, platform-based design poses the problem of partitioning the complete solution between hardware (HDL, gates) and software (programmed on the processor/s). This tends to remain a somewhat manual task, but is increasingly automated by CAD tools.

Platform-based systems typically consist of a basic RISC processor, which can be extended with multipliers, floating point units, or specialized DSP units. In addition (e.g., in Tensilica's Xtensa system), by profiling the executable code, special hardware can be added that corresponds to hardware-assisted instructions, which are introduced into the instruction set. In theory, additional hardware or extra processors can deal with a wide range of computational loads.

Manual techniques for hardware-software codesign mirror this approach. That is, the design begins with a software simulation (ideally on the embedded processor). Timing estimates are gathered, and manual decisions about what to commit to hardware are made. Special simulators to deal with embedded processors and logic have been developed.

With platform-based design, we have in essence come full circle from the first design method suggested: programming a microprocessor. This is the reason processor selection is important when starting out on a product design that may eventually be integrated. As the software effort will often exceed the hardware effort, you don't want to repeat that effort.

13.3.7 Summary

In this section, we have summarized a range of CMOS design options ranging from a software-based microprocessor to full custom design. Table 13.4 summarizes these options in terms of a variety of criteria. Each category is ranked in relation to each design method from low to high.

TABLE 13.4 Comparison of CMOS design methods

Design Method	Non-Recurring Engineering	Unit Cost	Power Dissipation	Complexity of Implementation	Time to Market	Performance	Flexibility
Microprocessor/DSP	Low	Medium	High	Low	Low	Low	High
PLD	Low	Medium	Medium	Low	Low	Medium	Low
FPGA	Low	Medium	Medium	Medium	Low	High	High
Cell-Based	High	Low	Low	High	High	High	Low
Custom Design	High	Low	Low	High	High	Very High	Low
Platform-Based	High	Low	Low	High	High	High	Medium

The most cost-effective approach should be taken to hardware (or software) design given speed, power, and cost targets (occasionally, size will count as well). You should always use an off-the-shelf solution if system constraints are met, because the non-recurring engineering (NRE) costs are amortized over many units. The next most likely prospect is an FPGA design, especially for low-volume (100,000's) applications. Power and cost are the most likely attributes to be challenged in medium- to high-volume applications, and this is where standard cell designs will be used. Mixed-signal, RF, and high-speed digital designs require a cell-based or custom approach.

The NRE cost (predominantly mask cost) has reached a level where even industry prototypes must be done using multiproject chips, amortizing the mask cost over multiple designs. Designs must be as reprogrammable or adaptable as possible.

In 2006, there were an estimated 3000 to 5000 custom designers and 50,000 to 100,000 ASIC designers employed worldwide [Chinnery07]. The number of FPGA designers is even larger, and the number of designers using microcontrollers is greater still. CAD tool vendors cater to the most profitable markets, so most VLSI design tools are aimed at ASICs. Synopsys, Cadence, Mentor Graphics, and Magma are the largest suppliers, though many smaller companies offer specialty tools. The next section examines the design flows using these tools.

13.4 Design Flows

A design flow is a set of procedures that allows designers to progress from a specification for a chip to the final chip implementation in an error-free way. In the previous section, we discussed the basic CMOS design methods without mentioning how we actually design an FPGA, gate array, or cell-based system. In this section, we will summarize the main design flows in use today.

A general design flow is shown in Figure 13.19. Design starts at the behavioral level and then proceeds to the structural level (gates and registers). This step is called behavioral or *Register Transfer Level* (RTL) synthesis because the designs are captured at the RTL (memory elements and logic) level in an HDL. The description is then transformed to a physical description suitable for chip fabrication. This step is called *physical synthesis* (or *layout generation*). Normally, the synthesis steps are automated, albeit guided by human judgment. The verification steps are also shown.

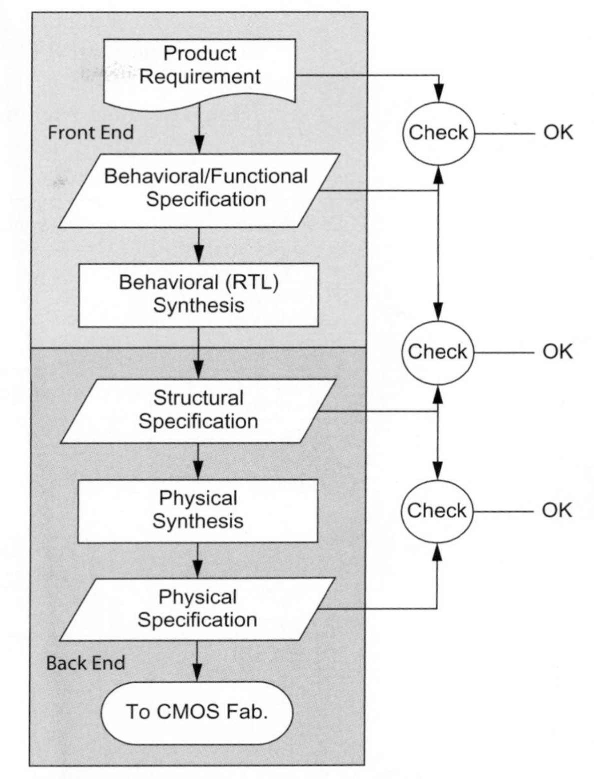

FIGURE 13.19 Generalized design flow

In Figure 13.19, the design has been partitioned into the *front end* stage at the behavioral level and the *back end* at the structural and physical levels. This is important because it illustrates a partitioning that is used to build *Application Specific Integrated Circuits* (ASICs). In an ASIC, the design can be developed at the HDL level and then passed to a company that completes the transition to an actual chip. In this way, the original design company does not have to invest the personnel or tools required to translate an HDL specification into a physical chip. Theoretically, in an ASIC flow, only a behavioral HDL needs to be designed and simulated (at the behavioral level). All subsequent operations can be completed by a third-party design service with only the final timing having to be verified by the back-end process. This is sometimes referred to as a "throw it over the wall" approach. While it works for moderately complex designs, the interaction between logic and layout is so important in more demanding circuits that such a flow becomes a schedule risk. Primarily, this occurs because the iteration time between logic design and physical placement takes too long when spread over two organizations. Multiple iterations are necessary because the prelayout timing estimates available to the HDL designer correlate poorly with the true postlayout timing because wire lengths are unpredictable before layout. Consider the case where the design cycle from logic to layout takes two hours when completed as an integrated task or one week if split into front-end and back-end tasks, as shown in the figure. If there are 100 iterations for the design, the integrated approach takes roughly 25 working days or five weeks, while the split approach takes two years (without vacations!). Having said this, companies are in business to make this approach work. If there are only 10 iterations, the times are much more reasonable.

The next two sections summarize each of the tools required to perform the automatic transformation. We also will examine the verification tools required to guarantee the correctness of the transformation and look at specific design flows. Then, we will describe a manual flow that is typical of a mixed-signal or RF design. Finally, we will outline a method of transforming directly from the behavioral to the physical level.

13.4.1 Behavioral Synthesis Design Flow (ASIC Design Flow)

At the behavioral level, the operation of the system is captured without having to specify the implementation. This level provides the most independence from implementation details and is the most dependent on the tool flow for a good design.

The most popular style of tools for behavioral synthesis are those that directly transform a behavioral RTL description to a structural gate-level netlist. A typical behavioral flow for an ASIC is shown in Figure 13.20. Tool suppliers include Synopsys, Cadence Design Systems, Mentor Graphics, and Synplicity.

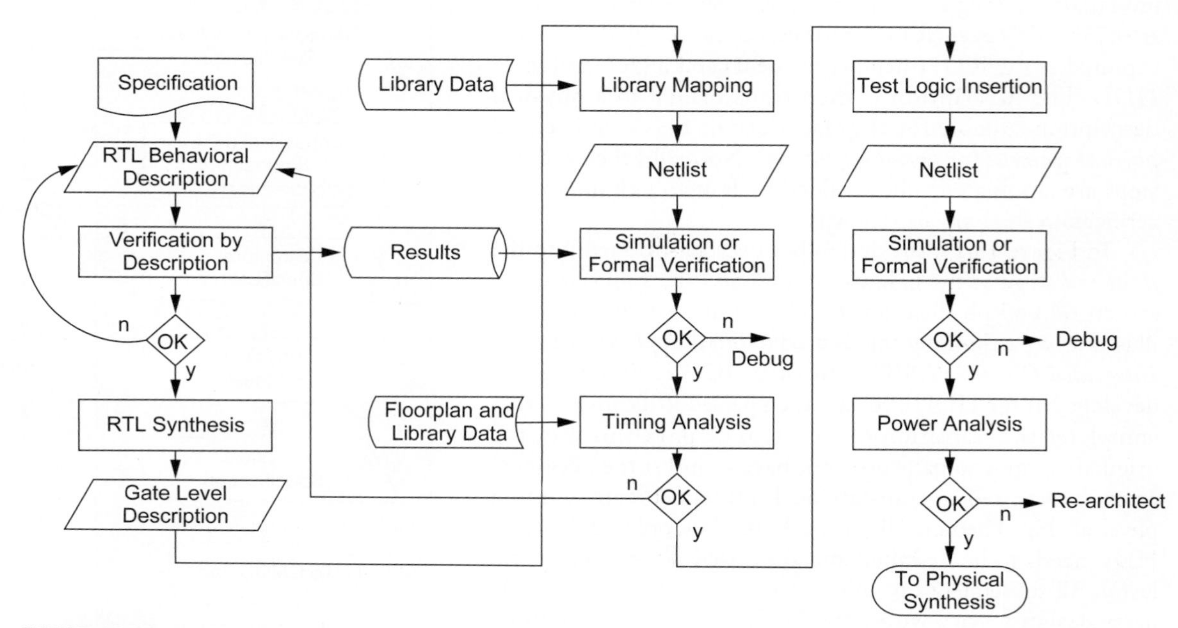

FIGURE 13.20 RTL synthesis flow

13.4.1.1 Logic Design and Verification The design starts with a specification, which might be a text description or a description in a system specification language. The designer(s) convert this to an RTL behavioral description in an HDL such as Verilog, or VHDL. A set of test benches are then constructed and the HDL is simulated to verify the correct behavior as defined by the specification and product requirements. Typical interactive design environments and simulators include NC-Verilog/SystemC/VHDL or Desktop Verilog/VHDL from Cadence Design Systems, VCS from Synopsys, ModelSim from Mentor Graphics and ActiveHDL from Aldec. Bear in mind that functional verification via simulation is usually carried out hierarchically. That is, after the overall architecture is defined, modules are successively built from the bottom up, verifying at each step. The design is iterated at this level until the correct behavior is evident. Test benches are covered further in Section 14.3.

Behavioral Verilog for an 8-bit implementation of the NCO previously introduced is presented below.

```
module nco #(parameter size = 8,
             counter_size = 16,
             table_size = 64)
            (input                          fclock, reset,
             input    [counter_size-1:0]   initial_phase, phase_increment,
             output   [size-1:0]           q);

   reg  [counter_size-1:0] phase;
```

```
   wire [size-3:0]        phase_part, inverted_adr, ROM_adr;
   wire [size-2:0]        ROM_data;
   wire [size-1:0]        wave_out;

   // numerically controlled oscillator
   // note that some constants are hardwired in the code below

   // phase counter
   always @(posedge fclock)
      if (reset) phase <= initial_phase;
      else phase <= phase + phase_increment;

   // add offset and determine ROM address
   assign phase_part = phase[counter_size-3:counter_size-8];
   assign inverted_adr = 7'3f - phase_part;
   assign ROM_adr = phase[counter_size-2] ? inverted_adr : phase_part;

   // look up data in ROM and negate if appropriate
   quarter_wave sine_table(ROM_adr, ROM_data);
   assign wave_out = phase[counter_size-1] ? ~ROM_data : ROM_data;
   assign q = wave_out + 8'h80 + phase[counter_size-1];
endmodule
```

13.4.1.2 RTL Synthesis The next step is to *synthesize* the behavioral description. This involves converting the RTL to generic gates and registers, optimizing the logic to improve speed and area, and mapping the generic gates to a standard cell library. Other steps involved at this stage are state machine decomposition, datapath optimization, and power optimization. Typical products include Design Compiler from Synopsys, RTL Compiler from Cadence, and Synplify Pro from Synplicity. The following description is a portion of the mapped generic Verilog for the NCO shown above.

```
module nco_struct_mapped(input          fclock, reset,
                         input  [15:0]  initial_phase, phase_increment,
                         output [7:0]   q);
   .
   .
   BUFX4 i_506(.A(n_355), .Y(q[7]));
   .
   MX2X1 i_00(.S0(reset), .B(initial_phase[15]), .A(nbus_1[15]),
      .Y(phase_0[15]));
   NAND2BX1 i_8(.AN(n_102), .B(n_101), .Y(n_104));
   XOR2X1 i_6(.A(phase[15]), .B(ROM_Table[6] ), .Y(n_103));
   .
   .
   DFFHQX1 phase_reg_0(.D(phase_0[15]), .CK(fclock), .Q(phase[15]));
   .
   .
   .
endmodule
```

13.4.1.3 Functional or Formal Verification We must now prove that the structural netlist performs the same function as the original behavioral HDL. Ideally, the netlist would be correct-by-construction, but ambiguities in HDLs sometimes cause the synthesizer to produce incorrect netlists from poorly written behavioral code. One verification strategy is to rerun the logic test benches and check that they produce exactly the same output for the behavioral and structural descriptions.

Another strategy is to use a *formal verification* program that compares the logical equivalence of the two descriptions. Formal verification tools are still maturing, but offer the advantage that they mathematically prove both descriptions have exactly the same Boolean functions [Anastasakis02, Perry05]. In contrast, simulation only is as good as the choice of test vectors. Formality from Synopsys and Incisive Conformal from Cadence are examples of formal verifiers.

Other types of verification that can be run are semantic and structural checks on the HDL. An example of a semantic check would be ensuring that all bus assignments match in bit width, while an example of a structural check would include making sure all outputs are connected.

13.4.1.4 Static Timing Analysis At this point, the functional equivalence of the gate-level description and the original behavioral description has been established. Now the temporal requirements of the design have to be checked. For example, the adder may add, but does it add fast enough? At the behavioral level, clock cycle time is an abstract notion, but at the structural level, an actual cycle time has to be met by a particular set of gates. A *timing analyzer* is used to verify the timing.

The timing analyzer is a critical analytical tool in the arsenal of the modern CMOS digital designer. Timing can be verified in a cursory manner using a timing simulator; i.e., a simulator in which the actual gate timings are used rather than a cycle-based or unit delay simulator. While useful, this approach is usually neither complete nor rigorous and can take an extraordinary amount of time to run.

Static timing analysis, on the other hand, runs quickly and exhaustively evaluates *all* timing paths. The inputs to the timing analyzer at this point are derived from the basic timing of the library gates due to intrinsic gate delays and routing loads that can be either estimated statistically or derived from floorplanning data. (See Section 13.4.2.2 for a description of floorplanning.) Timing analyzers check for both *max-delay* (will all flip-flops meet their setup time at the required cycle time?) and *min-delay* (will any flip-flop violate its hold time?).

Static timing analysis can suffer from *false path* problems. Typical of this problem might be a reset line in a circuit that has many clock cycles to operate. The timing analyzer might report that it cannot complete in one cycle. The designer must manually flag such multicycle paths for the timing analyzer.

Typical timing analyzers include ETS from Cadence and PrimeTime from Synopsys. Timing analysis reports will list a path from the output of a register to the input of another register. For each stage of logic, the delay of that stage and output arrival time are summarized. The paths are sorted by slack, with negative slack indicating critical paths that must be corrected.

13.4.1.5 Test Insertion Logic and registers are then inserted/modified to aid in manufacturing tests (see Section 14.6). Two basic techniques are used. One involves inserting scannable registers so that the state of a circuit can be set and monitored. Accompanying this option is a technique called *Automatic Test Pattern Generation* (ATPG), which is used to generate tests for a scannable design. The other technique, called *Built-In Self-Test* (BIST), modifies registers to allow *in situ* testing within the chip. Figure 13.21 shows the NCO after a test insertion program has run.

Typical commercially available test programs include DFT Max from Synopsys for scan insertion and Tetramax for ATPG. LogicVision markets ETLogic and ETMemory for built-in self-test.

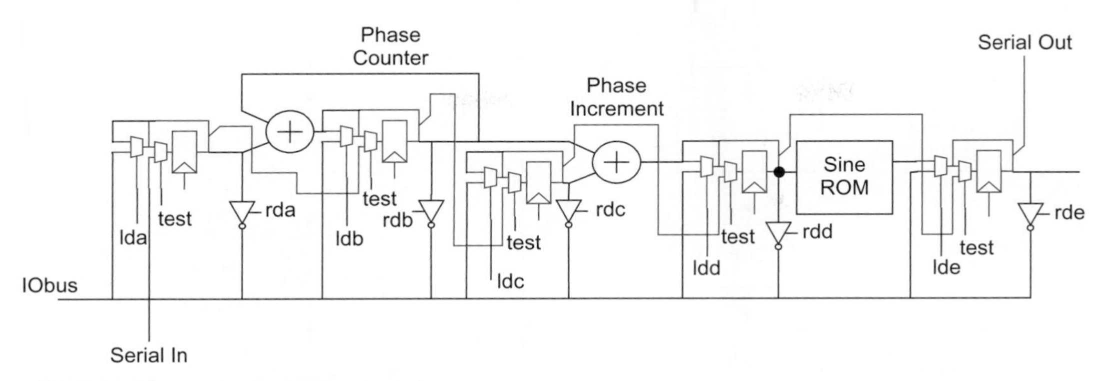

FIGURE 13.21 Scan register insertion for testing

13.4.1.6 Power Analysis The power consumption of the circuit is then estimated. Power consumption depends on the activity factors of the gates, which in turn depends on the inputs the chip receives. *Power analysis* can be performed for a particular set of test vectors by running a simulator and evaluating the total capacitance switched at each clock transition at each node. At this stage, if the power is too high, the design must return to the architectural level to rethink the solution. Commercial power analysis tools include PrimePower and Powermill from Synopsys.

13.4.1.7 Summary Apart from increasing design productivity, logic synthesis systems are useful for transforming between technologies. For instance, you might synthesize behavioral HDL onto multiple FPGAs and construct a prototype used to verify the operation of the circuit under real-world conditions. Then you can compile a single-chip version from the same HDL using a gate-array library.

13.4.2 Automated Layout Generation

Layout generation is the last step in the process of turning a design into a manufacturable database. It transforms a design from the structural to the physical domain. This step is sometimes called *physical synthesis* when the structural netlist is manipulated as the physical layout is generated.

Figure 13.22 shows a standard place & route layout generation design flow used in most ASICs. It begins with the structural netlist describing gates, flip-flops, and their interconnections. The netlist might be provided in the *Design Exchange Format* (DEF) as a Verilog netlist like the one in Section 13.4.1.2. The placement tool also takes a standard cell library definition describing cell dimensions and port locations, typically in the *Library Exchange Format* (LEF).

13.4.2.1 Placement The first step in Figure 13.22 is to place the standard cells. The key to automation of standard cell layouts is the use of constant-height, variable-width standard cells that are arrayed in rows across a chip, as shown in Figure 13.23. In contrast to SOG and gate array chips, standard cell chips can add application-specific custom blocks such as memories and analog blocks by allowing the standard cell rows to "flow" around the fixed-shape custom blocks. No separation has been shown between standard cell rows because routing takes place over the cells using multiple layers of metal. In older processes with two or three metal layers, a space between rows would be needed to allow routing. LEF summarizes the salient physical details of cells.

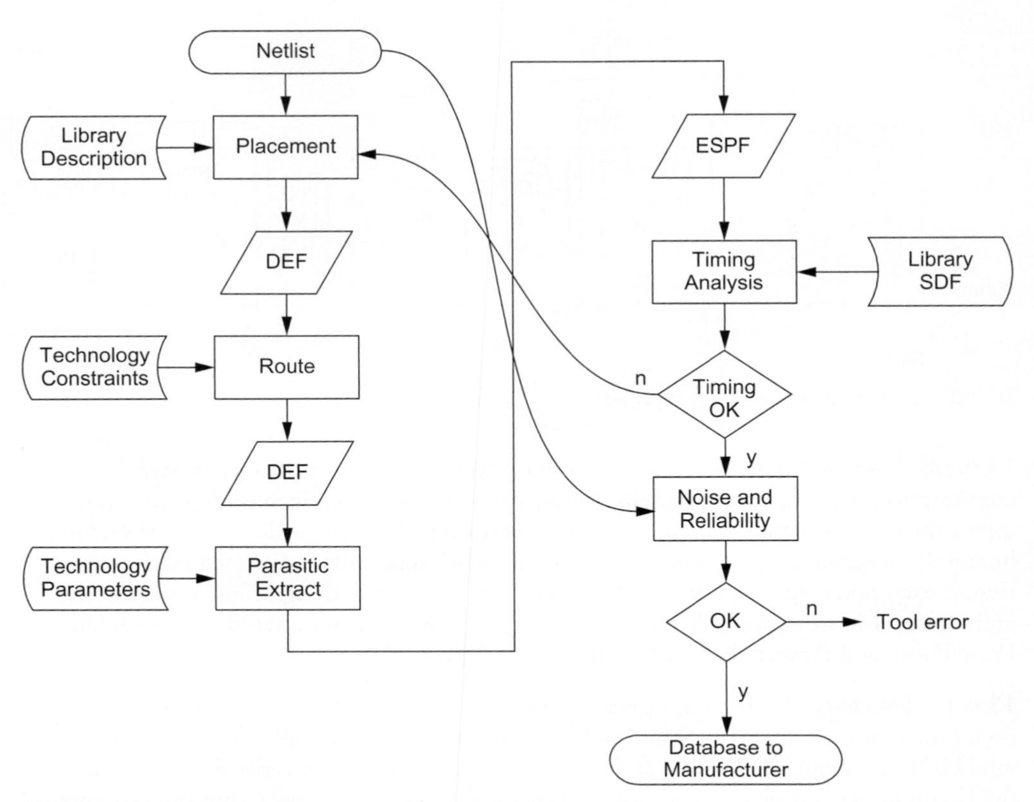

FIGURE 13.22 Standard cell place and route design flow

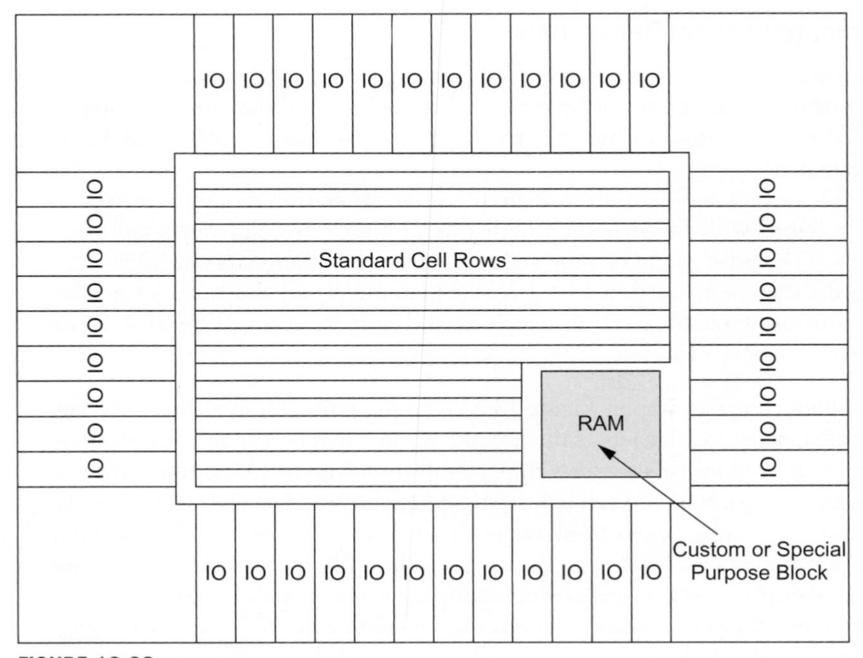

FIGURE 13.23 Standard cell chip layout

The objective of a simple placement algorithm is to minimize the length of wires. In *timing-driven placement*, the cost of wires is weighted to meet timing constraints. At the end of the placement phase, the cells have been fixed in position in the overall array. The placed design is saved in a standard format (e.g., DEF) for routing.

13.4.2.2 Floorplanning Increasingly a manual floorplanning step is required in the placement process. Rather than place a design "flat" (i.e., all cells at the same level of the hierarchy), modules are clustered in areas that are dictated by the need to communicate with other modules. Example 13.5 illustrated some floorplans for the software radio. This style of floorplanning might be completed prior to automatic placement.

13.4.2.3 Routing After placement of cells, the signal nets in the circuit need to be routed. Routing is normally divided into two steps: *global* routing and *detailed* routing.

A global router abstracts the routing problem to a notional set of abutting channels that cover the chip surface through which wires are routed. Routes are added to channels according to a cost function. Wires can be changed from channel to channel if the density of wires in a channel becomes too high. The detailed router places the actual geometry required to complete signal connections. Over time, a selection of detailed routers have been developed to automatically route signals. Older routers constrained signals to a grid of tracks, but newer *gridless* routers are more flexible for variable pitch wires. Moreover, they allow easy interface to foreign cells that may have I/O pin locations that are not on any specific routing grid. Routers also can route over the top of cells. LEF definitions are used to indicate obstructions on various layers in cell definitions. Advanced routers take into account manufacturability concerns such as redundant vias (more that one via inserted when space is available) and adjustable spacing (to separate wires and reduce coupling when there is room).

In the example of the flow shown in Figure 13.22, the router uses a technology file to specify routing layers and pitches for the process technology. It writes the results to another DEF file.

13.4.2.4 Parasitic Extraction The placed and routed design is then passed to the circuit *parasitic extractor*. In the example shown in Figure 13.22, the placed and routed design is provided to the extractor in DEF format and the output is an *Extended Standard Parasitic Format* (ESPF), *Reduced Standard Parasitic Format* (RSPF), or *Standard Parasitic Exchange Format* (SPEF) that describes the R's and C's associated with all nets in the layout. The extractor uses another technology file defining the interlayer capacitances and layer resistances.

The capacitance extractor can be a 2D, 2.5D, or 3D extractor. Two-dimensional (2D) extractors look at a cross-section assuming wires extend uniformly outside the section. A 2.5D extractor uses lookup tables to more accurately estimate capacitance near nonuniformities. A 3D extractor solves Maxwell's equations in three dimensions to precisely determine capacitance of complex geometries. 3D extraction used to be prohibitively time-consuming, but new statistical algorithms, such as those in QuickCap from Magma Design, deliver good accuracy with faster runtimes.

13.4.2.5 Timing Analysis Static timing analysis is now rerun with the actual routing loads placed on the gates. This is usually the bottleneck in the design process as the full reality of a physical realization is apparent. Multiple iterations of synthesis and placement & routing are usually necessary to converge on timing requirements.

Additionally, if possible (especially where dynamic circuits are used), a transistor-level timing simulation should be run. While this cannot usually be achieved using a SPICE-based simulator, a variety of transistor level simulators with "almost SPICE accuracy" have been in use since the late 1970s. These currently have the capacity to do whole-chip simulations at the transistor level, but at somewhat reduced transistor modeling accuracy. Nanosim from Synopsys and UltraSim from Cadence are examples of current simulators of this type.

13.4.2.6 Noise, V_{DD} Drop, and Electromigration Analysis Analyses are now run to check noise, IR drop in supply lines, and electromigration limits. Noise analysis is run to evaluate crosstalk due to interlayer routing capacitance. SignalStorm, ElectronStorm, and VoltageStorm from Cadence are examples of such tools.

13.4.2.7 Timing-Driven Placement The trouble with a place-then-route strategy is that after the layout is completed, the parasitic routing capacitance is extracted and the timing analysis is done to estimate timing. The timing is not known until the physical layout is complete. If timing problems are found, the cycle has to be repeated with some kind of constraint placed on the problematic paths. With complex designs this quickly gets out of control, to the point where changing something on one iteration could undo something fixed on a previous iteration. There are stories of designs that never were completed because of this problem.

The solution is to use a technique called *timing-driven placement*, which takes into account the timing (speed) of the circuit as cells are placed. Cells on critical paths are given priority to minimize wire delay. This approach, illustrated in Figure 13.24, has been successful and often results in a one-pass approach for many designs.

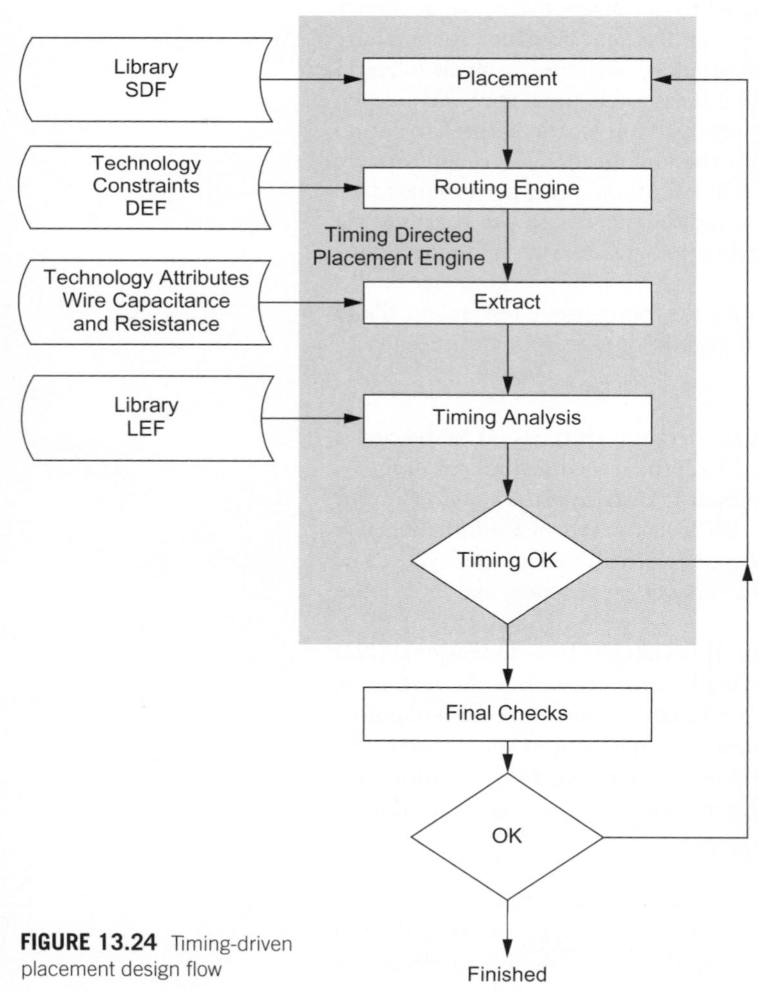

FIGURE 13.24 Timing-driven placement design flow

13.4.2.8 Clock-Tree Routing Central to modern high-speed designs is the clock distribution strategy. In Section 12.4.4, a number of these approaches are explained. To minimize skew, it is often best to preroute the clock and its buffers before the main logic placement and routing is completed. This task is performed with a *clock tree router*.

13.4.2.9 Power Analysis Power estimation can be repeated for the extracted design now that real wire capacitances are available. Similar techniques to those used during RTL synthesis are used.

13.4.3 Mixed-Signal or Custom-Design Flow

In the previous section, we described a flow that would be used for a purely digital chip in which the procedure for converting from HDL to layout is highly automated. This flow offers high productivity for most large digital chips with moderate performance requirements. But what of smaller analog, RF, and high-speed digital sections of a chip? For these sections, we use a custom-design flow, which is shown in Figure 13.25.

The designer begins by drawing a schematic (or possibly writing a netlist). An electrical rule check (ERC) verifies port connectivity and checks for unconnected inputs or outputs—the kind of simple connectivity errors that can occur easily in a manually drawn schematic. When the schematic is deemed correct, circuit simulation is then carried out using a SPICE-type simulator to verify DC, AC, transient, noise, and/or RF performance.

Once the circuit behavior of the module has been verified, the layout can commence, starting with the floorplan. Floorplanning can be an iterative process that is refined as actual module sizes and critical paths become known. Custom layout is a very time-consuming task; for example, a large microprocessor could keep a hundred mask design technicians busy for two years. Automating noncritical parts of the layout is essential for productivity. When the layout for the module is complete, a layout circuit extractor is invoked to determine the connectivity of primitives (MOS and bipolar transistors, diodes, resistors, capacitors, inductors) in the layout using rules like those illustrated in Section 15.5.2.

In the next step, the extracted netlist is compared to the schematic using a graph isomorphism program to determine whether the two netlists are identical in connectivity. This proceeds by assigning primitives to the nodes of a graph and the connections to the arcs in the graph. Graph coloring based on the connectivity and circuit parameters (i.e.,

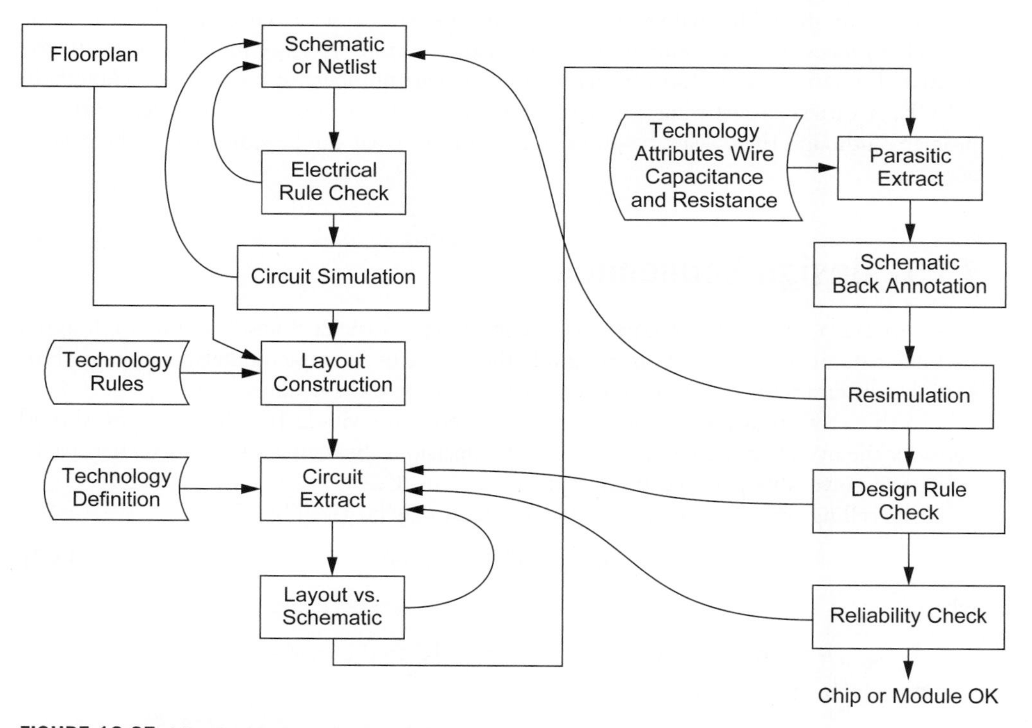

FIGURE 13.25 Mixed-signal or custom-design flow

transistor type, width, and length) determines the extent of the match. Once connectivity equivalence has been determined, each primitive attribute is checked for equivalence (i.e., capacitor or resistor value, transistor W/L). Discrepancies are reported to the user. Graphical feedback may be provided to help the designer find the source of any mismatch. This step is commonly called *layout versus schematic* (LVS).

Once the structural-to-physical equivalence has been established, the parasitic extract is completed. This adds the parasitic routing capacitance and resistance to the original primitive elements. In general, inductors are not extracted, but are dealt with by cookie cutting the inductor out of the layout and substituting a previously generated physical model. This is sometimes called *macro substitution*. The parasitic capacitance and resistance can be *back annotated* onto the schematic and the complete circuit resimulated. It must be pointed out that this step is extremely important. Matching simulated behavior to real device behavior is of critical importance in being able to accurately predict performance. It is too late when the circuit has been built!

The module layout can then be design-rule checked (DRC). Alternatively, this step can be completed just as the layout is completed. Normally, the AC performance is more important than tweaking the last design-rule error because running DRC on a circuit that does not meet performance goals is a waste of time.

Following this, a set of manufacturability verification steps needs to be completed. These can be manual or automated. In common with the standard cell design flow, power bus widths should be checked to ensure that they comply with metal migration and IR drop constraints. Power consumption can be found directly from circuit simulation. Adequate substrate and well contacts should be present in a bulk CMOS design, and all external I/O must be guard-ringed. At this stage, a check can also be made for substrate noise injection from digital to analog circuits. SubstrateStorm from Cadence performs this task.

This process can be completed hierarchically to build up large modules. Usually, the ultimate limitation comes from trying to simulate vast numbers of transistors accurately in SPICE. A variety of fast transistor-level simulators have been developed to deal with this problem, although there is always some upper limit to what can be simulated at the desired accuracy.

13.5 Design Economics

It is important for the IC designer to be able to predict the cost and the time to design a particular IC or sets of ICs. This can guide the choice of an implementation strategy. This section will summarize a simplified approach to estimate these values.

In this section, we will concentrate on the cost of a single IC, although you should consider the overall system when making such decisions. System-level issues such as packaging and power dissipation can affect the cost of an IC.

The selling price S_{total} of an integrated circuit may be given by

$$S_{total} = C_{total} / (1-m) \tag{13.2}$$

where

- C_{total} is the manufacturing cost of a single IC to the vendor.
- m is the desired profit margin.
- The margin has to be selected to ensure a profit after overhead (G&A) and the cost of sales (marketing and sales costs) have been considered.

The costs to produce an integrated circuit are generally divided into the following elements:

- Non-recurring engineering costs (NREs)
- Recurring costs
- Fixed costs

13.5.1 Non-Recurring Engineering Costs (NREs)

Non-recurring engineering costs are those that are spent once during the design of an integrated circuit. They include the following:

- Engineering design cost E_{total}
- Prototype manufacturing cost P_{total}

These costs are amortized over the total number of ICs sold. F_{total}, the total non-recurring cost, is given by

$$F_{total} = E_{total} + P_{total} \tag{13.3}$$

The NRE costs can be amortized over the lifetime volume of the chips. Alternatively, the non-recurring costs can be viewed as an investment for which there is a required rate of return. For instance, if $10M is invested in NRE for a chip, then $100M has to be generated for a rate of return of 10.

13.5.1.1 Engineering Costs The cost of designing the IC E_{total} hopefully will happen once during the chip design process. The costs include

- Personnel cost
- Support costs

The personnel costs might include the labor for

- Architectural design
- Logic capture
- Simulation for functionality
- Layout of modules and chip
- Timing verification
- DRC and tapeout procedures
- Test generation

The support costs amortized over the life of the equipment and the length of the design project include

- Computer costs
- CAD software costs
- Education or re-education costs

Costs can be drastically reduced by reusing modules or acquiring fully completed modules from an intellectual property vendor. As a guide the per annum costs might break down as follows (these figures are in U.S. dollars for engineers in the U.S. circa 2010):

Salary	\$50–\$100K
Overhead	\$10–\$30K
Computer	\$10K
CAD Tools (digital front end)	\$10K
CAD Tools (analog)	\$100K
CAD Tools (digital back end)	\$1M

The cost of the back-end tools clearly must be shared over the group designing the chips.

13.5.1.2 Prototype Manufacturing Costs These costs (P_{total}) are the fixed costs to get the first ICs from the vendor. They include

- The mask cost
- Test fixture costs
- Package tooling

The photo-mask cost depends on the number of steps used in the process and the precision required by each step. Masks on the metallization layers can be less expensive than on the lower layers because the pitch is not as tight. Figure 13.26 shows how mask costs have been exponentially increasing [Donovan02, LaPedus07]. The cost of a full set of masks in a 45 nm process is approximately \$5M.

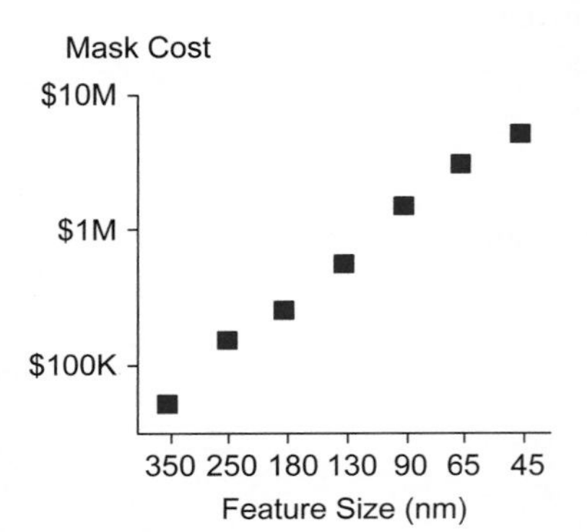

FIGURE 13.26
Approximate mask set cost

A test fixture consists of a printed wiring board probe assembly to probe individual die at the wafer level and interface to a tester. Costs range from \$1000 to \$50,000, depending on the complexity of the interface electronics.

If a custom package is required, it may have to be designed and manufactured (tooled). The time and expense of tooling a package depends on the sophistication of the package. Where possible, standard packages should be used.

An economical way of prototyping chips is to use a multiproject reticle that combines a number of different chip designs onto one mask set. Thus, if there were 200 sites available on a mask set and 20 projects were implemented, each project would get 10 die per wafer and the mask cost per project would be 1/20 of the cost of a complete mask set. This kind of service is provided by many of the silicon vendors and also MOSIS. For modest technology this can be quite cheap (~ \$1000 per mm^2 for 0.6 μm). Some commercial users worry about protection of intellectual property when they share a mask set.

Example 13.6

You are starting a company to commercialize your brilliant research idea. Estimate the cost to prototype a mixed-signal chip in a 45 nm process. Assume you have seven digital designers, three analog designers, and five support personnel and that the prototype takes two fabrication runs and two years.

SOLUTION: The seven digital designers will cost 7 × (\$70K + \$30K + \$10K + \$10K) = \$840K. The three analog designers will cost 3 × (\$100K + \$30K + \$10K + \$100K) = \$720K. The five support personnel cost 5 × (\$40K + \$20K + \$10K) = \$350K. One fabrication run with the back-end tools will cost \$6M. Thus, the cost is \$7.91M per year with one fab run. The total predicted cost here is nearly \$16M. The venture capitalists providing this money will want a good return for their risk so you'd better have a \$100M market for your idea. Typical chips at the 45 nm node require larger design teams and cost \$20–\$50M to design, so the markets must be even larger.

You may see ways to improve this. Clearly, you can reduce the number of people and the labor cost. You might reduce the CAD tool cost and the fabrication cost by doing multiproject chips. However, the latter approach will not get you to a pre-production version, because issues such as yield and behavior across process variations will not be proved. Your best bet may be to find a product niche that can be filled using a more mature and less expensive manufacturing process.

13.5.2 Recurring Costs

Once the development cost of an IC has been determined, the IC manufacturer will arrive at a price for the specific IC. A few large companies such as Intel, Toshiba, and IBM have in-house manufacturing divisions, but annual sales need to exceed about \$10B to justify the investments required to do your own manufacturing at the 45 nm node, and this figure continues to climb as processes advance. Many fabless semiconductor companies outsource their manufacturing to a silicon foundry such as TSMC, UMC, or IBM. In either case, manufacturing is a recurring cost; that is, it recurs every time an IC is sold. Another component of the recurring cost is the continuing cost to support the part from a technical viewpoint. Finally, there is what is called "the cost of sales," which is the marketing, sales force, and overhead costs associated with selling each IC. In a captive situation such as the IBM microelectronics division selling CPUs to the mainframe division, this might be zero.

The IC manufacturer will determine a part price for an IC based on the cost to produce that IC and a profit margin. The margin generally falls as the volume increases. An expression for the cost to fabricate an IC is as follows:

$$R_{\text{total}} = R_{\text{process}} + R_{\text{package}} + R_{\text{test}} \quad \textbf{(13.4)}$$

where

R_{package} = package cost

R_{test} = test cost—the cost to test an IC is usually proportional to the number of vectors and the time to test.

$$R_{\text{process}} = W/(N \times Y_w \times Y_{pa}) \quad \textbf{(13.5)}$$

where

W = wafer cost (\$500–\$5000 depending on process and wafer size)

N = gross die per wafer (the number of complete die on a wafer)

Y_w = die yield per wafer (should be ~70–90+% for moderate-sized dice in a mature process)

Y_{pa} = packaging yield (should be ~95–99%)

If a die has area A and is fabricated on a wafer with radius r, the gross number of dice per wafer is

$$N = \pi\left[\frac{r^2}{A} - \frac{2r}{\sqrt{2A}}\right] \tag{13.6}$$

where the second term accounts for wasted area around the edges of a circular wafer.

Example 13.7

Suppose your startup seeks a return on investment of 5. The wafers cost \$2000 and hold 400 gross die with a yield of 70%. If packaging, test, and fixed costs are negligible, how much do you need to charge per chip to have a 60% profit margin? How many chips do you need to sell to obtain a five-fold return on your \$16M investment?

SOLUTION: $R_{\text{total}} = R_{\text{process}} = \$2000/(400 \times 0.7) = \$7.14$. For a 60% margin, the chips are sold at $\$7.14/(1 - 0.6) = \$17.86$ with a profit of \$10.72 per unit. The desired ROI implies a profit of $\$16\text{M} \times 5 = \$80\text{M}$. Thus, $\$80\text{M}/\$10.72 = 7.4\text{M}$ chips must be sold.

The packaging yield is the number of chips that pass testing after the wafer has been diced and the parts packaged. The die yield is affected by defects randomly distributed around the wafer. The probability of a random defect causing a particular die to fail depends on the size of the die A and average number of defects per unit area D. If defects are distributed uniformly, then recall from EQ (6.23) that yield Y_w obeys a Poisson distribution given by [Seeds67]

$$Y_w = e^{-AD} \tag{13.7}$$

For small dice ($AD << 1$), Y_w is nearly 1 and R_{process} grows linearly with A. For large dice ($AD >> 1$), Y_w drops off rapidly because most chips will have defects and R_{process} grows exponentially with A.

Defect densities tend to be closely guarded trade secrets because they give competitors key information about the cost of manufacturing a chip. Figure 13.27 shows historical data indicating how manufacturing improvements have steadily improved the defect densities. Thus, chip makers now get better yields on larger chips than they did in the past, helping drive the incredible growth of the semiconductor market.

Example 13.8

If the defect density is 0.4 defects/cm^2, what is the yield on a 1 cm^2 die? How large can the die be if a 10% yield is required on a big new server chip?

SOLUTION: According to EQ (13.7), the yield on a 1 cm^2 die is 67%. A chip with an area of 5.75 cm^2 achieves a 10% yield.

13.5.3 Fixed Costs

Once a chip has been designed and put into manufacture, the cost to support that chip from an engineering viewpoint may have a few sources. *Data sheets* describing the characteristics of the IC have to be written, even for application-specific ICs that are not sold outside the company that developed them. From time to time, *application notes* describing how to use the IC may be needed. In addition, specific application support may have to be

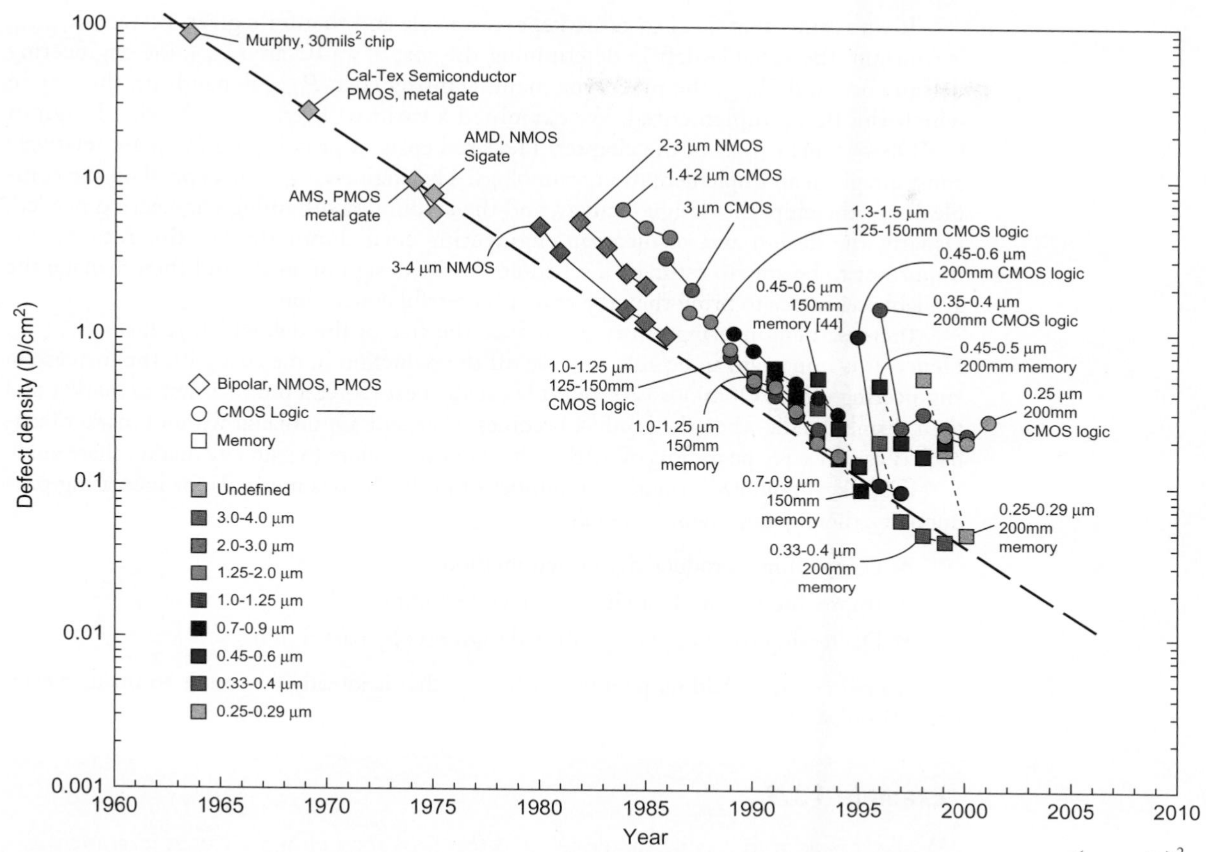

FIGURE 13.27 Defect density trends. Note that this data uses the Murphy Model rather than the Poisson Model: $Y = \left(\frac{1-e^{-AD}}{AD}\right)^2$ The Murphy Model predicts better yield at high defect density. (© 2002 IC Knowledge LLC, www.icknowledge.com reprinted with permission.)

provided to help particular users. This is especially true for ASICs, where the designer usually becomes the walking, talking, data sheet and application note. Another ongoing task may be failure or yield analysis if the part is in high volume and you want to increase the yield.

As a side comment, every chip or test chip designed should have accompanying documentation that explains what it is and how to use it. This even applies to chips designed in the academic environment because the time between design submission and fabricated chip can be quite large and can tax even the best memory.

13.5.4 Schedule

At the outset of a system design project involving newly designed ICs, it is important to estimate the design cost and design time for that system. Estimating the cost can help you determine the method by which the ICs will be designed. Estimating the schedule is essential to be able to select a strategy by which the ICs will be available in the right time and at the right price. This second task is usually the least well specified and requires some experience.

If we assume that fixed costs are kept reasonable and that for a given IC size, $R_{process}$ is constant, the variables left in determining the cost of an IC are E_{total}, the engineering design cost, and P_{total}, the prototype manufacturing cost. P_{total} depends on the way in which the IC is implemented. We examined a variety of strategies for the design of CMOS systems earlier in the chapter. The fixed costs of prototyping P_{total} are relatively constant, given an implementation technology. The engineering costs depend on the complexity of the chip, the design strategy, and the amount of sustaining engineering needed. Usually, the design and verification engineering costs dominate. For this reason, it is important to be able to estimate a schedule for the design of an IC and then manage the available resources to bring the project to a successful conclusion.

Increased engineering effort can reduce the size of the die, which reduces $R_{process}$. Hence, it is important to be able to trade off the reduction in die cost with the increase in engineering effort. Opinions vary, but it is usually best to get a product first to market and then shrink the die when the product becomes successful. Optimizing without market feedback is usually a recipe for loss of market share or even failure to gain any market share at all.

[Paraskevopoulos87] suggests a number of fairly obvious methods for increasing productivity, thereby improving schedules:

- Using a high-productivity design method
- Improving the productivity of a given technique
- Decreasing the complexity of the design task by partitioning

A final caution: Adding people to a project that is already late tends to make it even later [Brooks95].

Example 13.9

While it is hard to predict the design and test time for a chip, we can at least identify the main tasks and corresponding fixed periods in a chip design project. A representative Gantt chart is shown in Figure 13.28 for a project running over one year. The logic design time is shown as 12 weeks, which would be appropriate for an extremely simple chip. Double this time would be representative of moderately complex digital chips. The fixed times tend to be the fabrication time and packaging time, which are shown to be 10 weeks in the example. The design, debug, and test times will expand or contract to fit the complexity of the chip. And, if you are meticulous and lucky, you will not have to respin the chip.

ID	Task Name	Start	Finish	Duration	Q1 10 (Jan Feb Mar)	Q2 10 (Apr May Jun)	Q3 10 (Jul Aug Sep)	Q4 10 (Oct Nov Dec)	Jan
1	Specification	1/1/2010	1/28/2010	4w					
2	Digital Design	1/29/2010	4/21/2010	12w					
3	Place and Route	4/22/2010	6/16/2010	8w					
4	Fabrication	6/17/2010	8/11/2010	8w					
5	Packaging	8/12/2010	8/25/2010	2w					
6	Lab Test	8/26/2010	10/20/2010	8w					
7	Respin	10/21/2010	11/17/2010	4w					
8	Lab Test	11/18/2010	12/29/2010	6w					

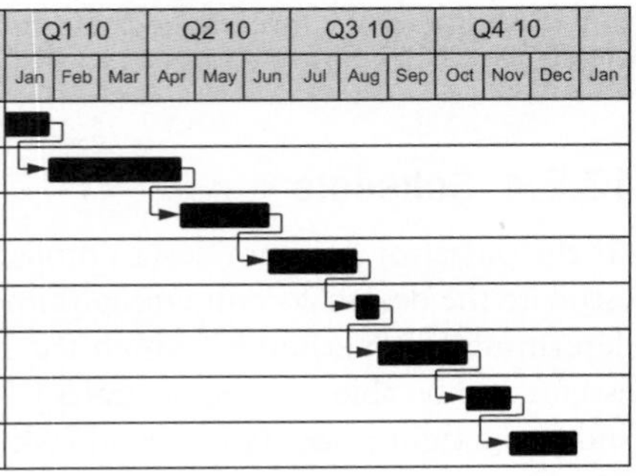

FIGURE 13.28 Gantt chart for simple chip

13.5.5 Personpower

To estimate the schedule, you must have some idea of the amount of effort required to complete the design. As we have seen, typical IC projects will involve the following tasks:

- Architectural design
- HDL capture
- Functional verification
- Place & route
- Timing verification, signal integrity, reliability verification
- DRC and tapeout procedures (ERC, LVS, mask generation)
- Test generation

While some researchers have attempted to derive analytical formulae for productivity, the best predictor of design schedule for a team is previous performance. Design time for a given team can be improved by design reuse or component-based design. It would seem that the time to design is proportional to the number of "modules" that are in the design raised to some power. That is, a four-module design is more than four times as complex as a single-module design. A module in this instance refers to a significant section of a chip such as a microprocessor, serial interface, or special functional unit.

Normally, projects are schedule-driven. In this case, it is important to make maximal use of design aids to meet the required schedule. Of importance is the cycle time of the so-called "edit-compile-debug" loop: i.e., the time it takes to make a change to the HDL; synthesize, place, and route it; and have a timing-verified final design. This can depend strongly on the efficiency of the design tools used, but if it is more than a day, design productivity can suffer. Ideally, the cycle is a few hours so that multiple bugs can be fixed each day.

Broadly speaking, schedules on the order of 18–24 months for a completely new chip seem to fit current average-complexity chips and state-of-the-art tools. For respins to slightly differentiate products, this can be reduced to six months or less, but there are certain fixed times such as IC fabrication and packaging that set hard limits on the complete design cycle time. Of course, for technologies such as FPGAs, design turnaround can be minutes (which is why FPGA verification is so important to ASIC or custom IC designs). New microprocessors seem to take three to five years, and most experience one or more schedule slips.

13.5.6 Project Management

Project management is the overall supervision of the project. Tasks include making certain sufficient resources are available at the appropriate time, ensuring communication between different groups assigned to the project, and summarizing progress and risks to management. The development of processes for the conception, design, and ultimate manufacture of products is also the purview of the project manager.

There are two main ways to manage a chip design. The first is what might be called the rapid prototyping approach that is typical of startup companies, where a full-time project manager may be a luxury (and probably is more aptly named "seat-of-the-pants project management"). In this approach, a time goal is set and the workload is set to fit the time available. It is vital to rapidly get to the point where a prototype of the design is working—in essence, the skeleton—and the meat (detail) is gradually added. This can be risky.

The more conventional approach, which is appropriate for large companies and the military, is to preplan everything, estimating task times and putting these into a project planning tool. This approach, while necessary for large groups, tends to be feature-driven and rarely delivers products in shorter time scales than the rapid prototyping approaches. It is suitable when the tasks are well-defined and have been done before (then you know what the task times should be). The approach is stable and, depending on the team, often delivers products within budget and on time.

13.5.7 Design Reuse

Rarely is an IC designed as a single event. Rather, companies wish to amortize the development effort of a particular IC over several generations of products. This normally means that the design has to be transferred between several different processes. When design was mainly manual and at the mask level, a great deal of effort was expended on techniques to allow porting of designs between processes with the minimum of human intervention. Techniques used here include the use of symbolic layout methods and mask resizing software.

With the emergence of cell-based design, design migration falls into two steps:

1. Acquiring or building a standard cell library in the new technology
2. Retargeting the HDL description to the new cell library

The design and test generation does not have to be redone, although timing analysis and regression test bench simulation should definitely be completed.

In design flows where these steps cannot be followed, strict use of structured design techniques and software generator technologies can markedly improve porting times. Maintaining accurate and clear documentation will alleviate many problems downstream.

With the maturation of cell-based design, especially standard cell libraries and the use of hardware description languages, the notion of virtual components has become important as a method of transferring and reusing designs. Virtual components, also called intellectual property (IP) blocks on an IC are notionally the same as discrete ICs used on a printed wiring board design. Each component has precisely defined behavior and a well-defined interface represented by a set of I/O pins and corresponding specifications for loading, setup and hold times, and delays. Components can be relatively simple or as complex as a RISC processor, MPEG decoder, or Wireless LAN modem. Virtual components can be classified as *hard*, *firm*, or *soft*. A hard module is normally defined at the mask level in a particular process. Thus, it will have a fixed floorplan, size, and a well-known set of timing parameters. A firm block will normally have a specific or generic netlist that describes each gate or register that must be used in the design (i.e., a 3-input NAND gate of normal power). This allows the design to be ported to multiple processes purely by netlist translation. The timing is dictated by the process and the final physical placement, however. A soft block is normally defined at the RTL level in the HDL. This captures the function of the block, but the detailed implementation is left to automated tools. Again, timing is dependent on the specific implementation. The Virtual Socket Interface Alliance monitors and encourages standards governing the implementation and use of virtual components.

Purchasing IP blocks is more like haggling for a used car than like buying breakfast cereal. It involves extensive negotiation with the vendors, and relationships are important. Assessing the quality of the IP block and its test bench is critical: a faulty IP block can sink your chip just as easily as a blown head gasket can leave you stranded in the Outback. Price sheets are not published, and licensing terms are generally kept confidential. As a very rough guideline, expect to pay on the order of $100K for a block such as a USB controller with its software stack and test fixture. Microprocessor cores may be offered on a

1% royalty basis. As a rule, if an IP block is available from a reputable source, purchasing the IP will normally be less expensive than redesigning it yourself.

13.6 Data Sheets and Documentation

A data sheet for an IC describes what it does and outlines the specifications for making the IC work in a system, such as power supply voltages, currents, input setup times, output delay times, and clock cycle times. The data sheet also includes package and pinout details.

A good habit to acquire is that of compiling a data sheet for any chip you might design. Not only is it the interface between the chip designer and the board-level designer, but also it is the interface to other members of the design team. In particular, it is good practice and is mandatory in industry to compile the data sheet for the chip and give it to the ultimate customer before the chip is fabricated. This prevents many undesirable scenarios that can arise when a perfectly designed chip meets a perfectly designed system. In this section, an outline of a typical data sheet will be reviewed by way of example.

13.6.1 The Summary

A summary of the chip includes the following details to orient the user:

- The designation and descriptive name of the chip
- A concise description of what the chip does
- A features list (optional for an internal product—but good for your ego!)
- A high-level block diagram of the chip function

13.6.2 Pinout

The pinout section should contain a description of the following pin attributes to document the external interface of the chip:

- Name of the pin
- Type of pin (i.e., whether input, output, tristate, digital, analog, etc.)
- A brief description of the pin function
- The package pin number

13.6.3 Description of Operation

This section should outline the operation of the chip as far as the user of the chip is interested. Programming options, data formats, and control options should be summarized.

13.6.4 DC Specifications

This section communicates the power dissipation and required voltages for the chip to correctly operate. The absolute maximum ratings should be stated for the following:

- Supply voltage
- Pin voltages
- Junction temperature

The style of each I/O (i.e., TTL, CMOS, LVDS, ECL) should be summarized and the following DC specifications should be given over the operating range (temperature and voltage—i.e., mins and maxes):

- V_{IL} and V_{IH} for each input
- V_{OL} and V_{IH} for each output (at a given maximum drive current level)
- The input loading for each input
- Quiescent current
- Leakage current
- Power-down current (if applicable)
- Any other relevant voltages and currents

13.6.5 AC Specifications

The following timing specifications should be presented:

- Setup and hold times on all inputs
- Clock (and all other relevant inputs) to output delay times
- Other critical timing such as minimum pulse widths

This data should be tabulated in table form and supported by a timing diagram where necessary. This is probably the most important section and an area where data provided ahead of the chip fabrication will aid the board designer. Designs are frequently snagged—for instance, when chip designers assume infinitely fast external memories and do not allow enough time between outputs changing and the next rising edge of the clock.

13.6.6 Package Diagram

A diagram of the package with the pin names attached should be supplied.

13.6.7 Principles of Operation Manual

Although the data sheet provides enough data to familiarize a user of a particular chip with the device, it is good practice to provide a Principles of Operation Manual for internal users that have to test the chip or build support systems.

13.6.8 User Manual

A User Manual should also be provided. This is designed for use outside the group that designed the chip and can be a "cut down" version of the Principles of Operation Manual.

13.7 Pitfalls and Fallacies

Inadequate design flow

In the past, universities and small companies could build interesting chips using open-source or inexpensive CAD tools. The MOSIS design rules provided a simple common denominator. This is no longer practical in nanometer processes where the design rules are so complex that industrial-strength DRC and extraction are necessary.

Insufficient verification

Synopsys found that 82% of design spins for chips with functional flaws were due to lack of verification [Schutten03]. Another 47% of respins had incorrect specifications. And 14% had errors in imported IP. This outlines the need for good specifications and a well-thought-out verification plan. Verification is further covered in Chapter 14.

Inaccurate parasitic extraction

Parasitic extractions programs output reams of data relating to C and R values in a design. Unless these are guaranteed by your vendor, it is prudent to do a small design and compare the values with hand-calculated values. You can never be too careful when it comes to designing a chip. When the chip comes back, compare a known path with what was predicted by the tool set.

Exercises

13.1 Estimate the die cost of a 4×4 mm die, with $Y_w = 80\%$ and $Y_{pa} = 98\%$ for an 8-inch wafer costing $2200 each. The die may be shrunk to 3.3×3.3 mm in a more advanced process that costs $3000 per wafer. Is it worth moving to the new process if the volume is large enough?

13.2 An FIR filter for a GSM receiver with sigma-delta converter as shown in Figure 13.8(b) has a single-bit input. To what structure do the multipliers degenerate? If the coefficients are a single bit and a 288-tap filter has to operate at 13 MHz, what architecture would you use for the overall design?

13.3 Sketch a stick diagram for a large inverter with an 80 λ pMOS transistor and 40 λ nMOS transistor. Fold the transistors so that no single transistor is wider than 20 λ.

13.4 Using the Sea of Gates structure from Figure 13.17(a), design the metallization for a 3-input NOR gate.

13.5 A fab house has a 180 nm process with a $500 cost per processed 8-inch wafer. If you do the design yourself using open-source tools and the mask cost is $250K, estimate the market size required to obtain 50% margin on a chip that is 3 mm on a side.

13.6 Explain the trade-offs between using a transmission gate or a tristate buffer to implement an FPGA routing block.

13.7 What kind of RAM cell would you use to control a configurable logic block in an FPGA? Design the cell and outline the reasons for your choice.

Test

14

14.1 Introduction

While in real estate the refrain is "Location! Location! Location!" the comparable advice in IC design should be "Testing! Testing! Testing!" For many chips, testing accounts for more effort than does design.

Tests fall into three main categories. The first set of tests verifies that the chip performs its intended function. These tests, called *functionality tests* or *logic verification*, are run before tapeout to verify the functionality of the circuit. The second set of tests are run on the first batch of chips that return from fabrication. These tests confirm that the chip operates as it was intended and help debug any discrepancies. They can be much more extensive than the logic verification tests because the chip can be tested at full speed in a system. For example, a new microprocessor can be placed in a prototype motherboard to try to boot the operating system. This *silicon debug* requires creative detective work to locate the cause of failures because the designer has much less visibility into the fabricated chip compared to during design verification. The third set of tests verify that every transistor, gate, and storage element in the chip functions correctly. These tests are conducted on each manufactured chip before shipping to the customer to verify that the silicon is completely intact. These are called *manufacturing tests*. In some cases, the same tests can be used for all three steps, but often it is better to use one set of tests to chase down logic bugs and another, separate set optimized to catch manufacturing defects.

In Section 13.5.2, we noted that the yield of a particular IC was the number of good die divided by the total number of die per wafer. Because of the complexity of the manufacturing process, not all die on a wafer function correctly. Dust particles and small imperfections in starting material or photomasking can result in bridged connections or missing features. These imperfections result in what is termed a *fault*. Later in the chapter, we will examine a number of fault mechanisms. The goal of a manufacturing test procedure is to determine which die are good and should be shipped to customers.

Testing a die (chip) can occur at the following levels:

- Wafer level
- Packaged chip level
- Board level
- System level
- Field level

By detecting a malfunctioning chip early, the manufacturing cost can be kept low. For instance, the approximate cost to a company of detecting a fault at the various levels [Williams86] is at least

- Wafer $0.01–$0.10
- Packaged chip $0.10–$1
- Board $1–$10
- System $10–$100
- Field $100–$1000

Obviously, if faults can be detected at the wafer level, the cost of manufacturing is lower. In an extreme example, Intel failed to correct a logic bug in the Pentium floating-point divider until more than 4 million units had shipped in 1994. IBM halted sales of Pentium-based computers and Intel was forced to recall the flawed chips. The mistake and lack of prompt response cost the company an estimated $450 million.

It is interesting to note that most failures of first-time silicon result from problems with the functionality of the design; i.e., the chip does exactly what the simulator said it would do, but for some reason (almost always human error) this functionality is not what the rest of the system expects.

The remainder of this section will provide an overview of the processes involved in logic verification, chip debug, and manufacturing test. Section 14.2 discusses the mechanics of testing and test programs. Sections 14.3 through 14.5 address the principles behind each phase of testing. If testing is not considered in advance, the manufacturing test can be extremely time consuming and hence expensive. Some chips have even proved impossible to debug because designers have so little visibility into the internal operation. Sections 14.6 and 14.7 focus on how to design chips to facilitate debug and manufacturing test at the chip and board level. [Wang08b] offers an entire book dedicated to test.

14.1.1 Logic Verification

Verification tests are usually the first ones a designer might construct as part of the design process. Does this adder add? Does this counter count? Does this state-machine yield the right outputs each cycle? Does this modem decode data correctly?

In Section 13.4.1.3, we noted that verification tests were required to prove that a synthesized gate description was functionally equivalent to the source RTL. Figure 14.1 shows that we may want to prove that the RTL is equivalent to the design specification at a higher behavioral or specification level of abstraction. The behavioral specification might be a verbal description, a plain language textual specification, a description in some high-level computer language such as C, a program in a system-modeling language such as SystemC, or a hardware description language such as VHDL or Verilog, or simply a table of inputs and required outputs. Often, designers produce a *golden model* in one of the previously mentioned formats and it becomes the reference against which all other representations are checked. Functional equivalence involves running a simulator on the two descriptions of the chip (e.g., one at the gate level and one at a functional level) and ensuring that the outputs are equivalent at some convenient check points in time for all inputs applied. This is most conveniently done in an HDL by employing a *test bench*; i.e., a wrapper that surrounds a module and provides for stimulus and automated checking. The most detailed check might be on a cycle-by-cycle basis. Increasingly, verification involves real-time or near real-time emulation in an FPGA-based system to confirm system-level performance *in situ*; i.e., in the actual system that will use the end chip. This is recommended because of the increasing level of complexity of chips and the systems they implement. As an example, in the area of wireless local area network chips, without a real-time emulation system, it is virtually impossible to simulate the unseen effects of an unreliable channel with out-of-band interferers.

You can check functional equivalence through simulation at various levels of the design hierarchy. If the description is at the RTL level, the behavior at a system level may be able to be fully verified. For instance, in the case of a microprocessor, you can boot the operating system and run key programs for the behavioral description. However, this might be impractical (due to long simulation times) for a gate-level model and even harder for a transistor-level model. The way out of this impasse is to use the hierarchy inherent within a system to verify chips and modules within chips. That, combined with well-defined modular interfaces, goes a long way in increasing the likelihood that a system composed of many VLSI chips will be first-time functional.

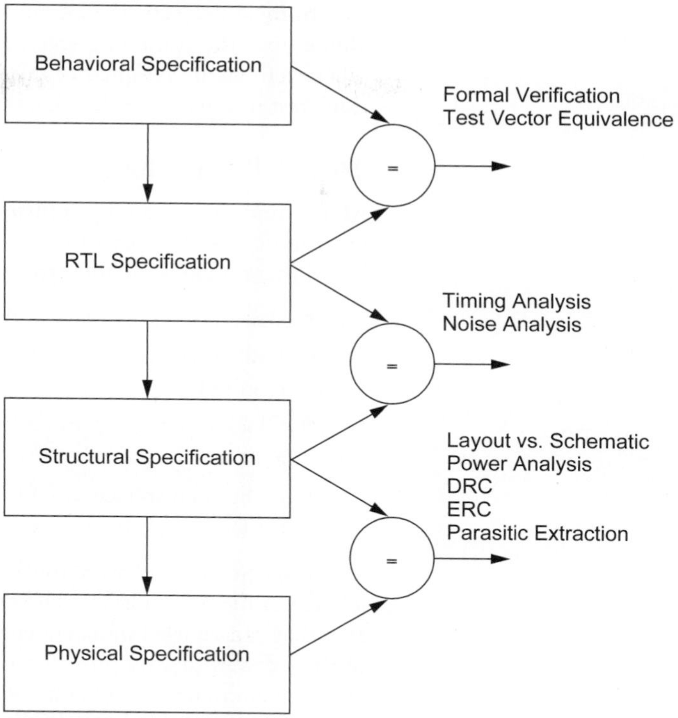

FIGURE 14.1 Functional equivalence at various levels of abstraction

The best advice with respect to writing functional tests is to simulate as closely as possible the way in which the chip or system will be used in the real world. Often, this is impractical due to slow simulation times and extremely long verification sequences. One approach is to move up the simulation hierarchy as modules become verified at lower levels. For instance, you could replace the gate-level adder and register modules in a video filter with functional models and then in turn replace the filter itself with a functional model. At each level, you can write small tests to verify the equivalence between the new higher-level functional model and the lower-level gate or functional level. At the top level, you can surround the filter functional model with a software environment that models the real-world use of the filter. For instance, you can feed a carefully selected subsample of a video frame to the filter and compare the output of the functional model with what the designer expected theoretically. You can also observe the video output on a video frame buffer to check that it looks correct (by no means an exhaustive test, but a confidence builder). Finally, if enough time is available, you can apply all or part of the functional test to the gate level and even the transistor level if transistor primitives have been used.

Verification at the top chip level using an FPGA emulator offers several advantages over simulation and, for that matter, the final chip implementation. Most noticeably, the emulation speed can be near if not real time. This means that the actual analog signals (if used) can be interfaced with the chip. Additionally, to assess system performance, you can introduce fine levels of observation and monitoring that might not be included in the final chip. For instance, you could include a bit-error rate circuit in a communication modem to aid performance optimization.

In most projects, the amount of verification effort greatly exceeds the design effort. Remember the following statement, culled from many years of IC design experience, whenever you are tempted to minimize verification effort to meet tight schedules: "If you don't test it, it won't work! (guaranteed)."

14.1.2 Debugging

Many times, when a chip returns from fabrication, the first set of tests are run in a lab environment, so you need to prepare for this event. You can begin by constructing a circuit board that provides the following attributes:

- Power for the IC with ability to vary V_{DD} and measure power dissipation
- Real-world signal connections (i.e., analog and digital inputs and outputs as required)
- Clock inputs as required (it is helpful to have a stable variable-frequency clock generator)
- A digital interface to a PC (either serial or parallel ports for slow data or PCI bus for fast data interchanges)

You can write software routines to interface with the chip through the serial or parallel port or the bus interface. The chip should have a serial UART port or some other interface that can be used independently of the normal operation of the chip. The lowest level of the software should provide for peeking (reading) and poking (writing) registers in the chip. An alternate or complementary approach is to provide interfaces for a logic analyzer. These are easily added to a PCB design in the form of multipinned headers. Figure 14.2 shows a typical test board, illustrating the *zero insertion force* (ZIF) socket for the chip (in the center of the board), an area for analog circuitry interface (on the left), a set of headers for logic analyzer connection (at the top and bottom) and a set of programmable power supplies (on the right). In addition, an interface is provided for control by a serial port of a PC (at the bottom left).

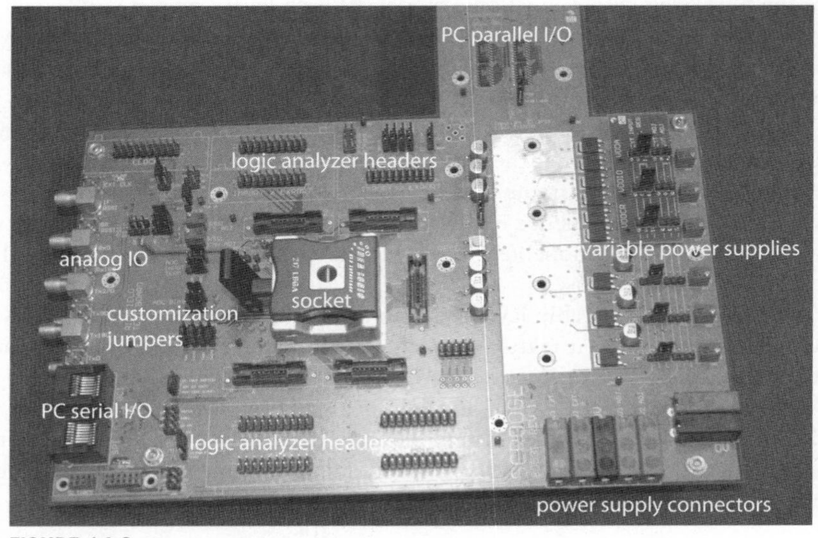

FIGURE 14.2 Typical test board

You should start with a "smoke test." This involves ramping the supply voltages from zero to V_{DD} while monitoring the current without any clocks running. For a fully static circuit, the current should remain at zero. Analog circuits will draw their quiescent current.

Following this, you can enable the clock(s); some dynamic current should be evident. Beware that many CMOS chips appear to operate when the clock is connected but the power supply is turned off because the clock may partially power the chip through the input protection diodes on the input pads. If possible, you should initially run the clock at reduced speed so that setup time failures are not the initial culprit in any debug operation.

In the case of a digital circuit, you should examine various registers for health using PC-based peek and poke software. This checks the integrity of the signal path from the PC to the chip. Often, designers place an ID in the register at address zero. Peeking at this register proves the read path from the chip. If the chip registers are reset to a known state, the registers can be read sequentially and compared with the design values. In the case of the logic analyzer, you can download the equivalent test pattern to exercise the chip. Frequently, these patterns can be automatically generated from the verification test bench. Up to this point, no functionality of the chip has been exercised apart from register reads and writes.

Where the chip has built-in self-test (see Sections 14.6 and 14.7), you can run the commercial software that provides for this functionality over a boundary scan interface. This type of system automatically runs a set of tests on the chip that completely verify the correct operation of all gates and registers as defined by the original RTL description. If this kind of a test interface was not used, you should pursue a manual effort in which the functionality of the chip is checked from the bottom-up. Of course, if you are a gambler, you can do a top-level test like running a piece of code or trying to boot the operating system right away. Experience shows that this often does not work, usually because of problems with the test fixture, and so you must revert to the bottom-up method to prove that one piece of the design works at a time.

If you detect anomalous behavior, you must go about debugging. The basic method is to postulate a method of failure, then test the hypothesis. Debug is an art in itself, but some pointers for sane debugging are as follows:

- Keep an annotated and dated logbook for all tests done.
- When postulating a cause for the bug and a test, do one change at a time and observe the result: Changing many things and then seeing if they work will not logically lead you to the bug and is commonly called the "shotgun approach."
- Check everything two or three times; never assume anything unless it is measured and logged in a notebook. Have someone independently check critical measurements.
- Check signals and supply voltages at the pins of the IC; frequently, new test boards have errors.
- Double-check the specified chip I/O and perform a continuity check from the IC pins to expected places (i.e., test pins, supplies) on the board.
- Never disregard a possible reason for a bug, however crazy, unless you can prove it isn't the cause.
- Use freeze spray or a heat gun to cool down or heat up a circuit to check for temperature problems.
- Check the state of any internal registers against that noted in the documentation.
- Evaluate the timing of any inputs and outputs with respect to the clock; often setup or hold times can be violated in a new test setup.

- When a bug is discovered and corrected, hunt for other portions of the design that might have a similar bug that hasn't been detected yet. Where there is one rat, there are many rats!
- Never assume anything—question everything—a slight touch of paranoia helps!!

[Agans06] cites nine "debugging rules" that bear repeating:

- *Understand the system.* If you are the designer, this should be self-evident. However, if you have been assigned to the task of debugging, follow this point keenly.
- *Make it fail.* Find a way to elicit the bug. A repeatable method is preferable.
- *Quit thinking and look.* Propose a test and investigate. You can start to eliminate possible sources of problems.
- *Divide and conquer.* Use hierarchy to eliminate known good parts of the system.
- *Change one thing at a time.* A very important rule.
- *Keep an audit trail.* No matter how good your memory is, a written account serves as a memory jog and something for someone else to look at to propose approaches.
- *Check the plug.* Check the complete test structure. More problems are found in new test harnesses than in the actual chip due to the level of verification used in each.
- *Get a fresh view.* Get a coworker involved. Take a break. Take a nap.
- *If you didn't fix it, it ain't fixed.* Problems do not mysteriously fix themselves. If you find a problem, verify it with simulation to prove your hypothesis of the failure mode.

After the chip is demonstrated to be operational, you can measure more subtle aspects of the design such as performance (power, speed, analog characteristics). This involves normal lab techniques of configure, measure, and record. Where possible, store all results as computer readable results (i.e., stored images from digital oscilloscope and screen dumps from logic analyzer) for communication with colleagues.

For the most part, if a digital chip simulates at the gate level and passes timing analysis checks during design, it will do exactly the same in silicon. Possible deviations from the simulated circuit occur in the following cases:

- Circuit is slower than predicted—fix—slow clock or raise V_{DD}
- Circuit has a race condition—fix—heat with heat gun if a logic gate caused race
- Circuit has dynamic logic problems—fix—don't do it again
- Gnarly crosstalk problems—fix—get better tools
- Wrong functionality—fix—do a better job of verification

With analog circuitry, a wide range of issues can affect performance over and above what was simulated. These include power and ground noise, substrate noise, and temperature and process effects. However, you can employ the same basic debug approaches.

14.1.3 Manufacturing Tests

Whereas verification or functionality tests seek to confirm the function of a chip as a whole, manufacturing tests are used to verify that every gate operates as expected. The need to do this arises from a number of manufacturing defects that might occur during either chip fabrication or accelerated life testing (where the chip is stressed by over-voltage and over-temperature operation). Typical defects include the following:

- Layer-to-layer shorts (e.g., metal-to-metal)
- Discontinuous wires (e.g., metal thins when crossing vertical topology jumps)
- Missing or damaged vias
- Shorts through the thin gate oxide to the substrate or well

These in turn lead to particular circuit maladies, including the following:

- Nodes shorted to power or ground
- Nodes shorted to each other
- Inputs floating/outputs disconnected

Tests are required to verify that each gate and register is operational and has not been compromised by a manufacturing defect. Tests can be carried out at the wafer level to cull out bad dies, or can be left until the parts are packaged. This decision would normally be determined by the yield and package cost. If the yield is high and the package cost low (i.e., a plastic package), then the part can be tested only once after packaging. However, if the wafer yield was lower and the package cost high (i.e., an expensive ceramic package), it is more economical to first screen bad dice at the wafer level. The length of the tests at the wafer level can be shortened to reduce test time based on experience with the test sequence.

Apart from the verification of internal gates, I/O integrity is also tested, with the following tests being completed:

- I/O levels (i.e., checking noise margin for TTL, ECL, or CMOS I/O pads)
- Speed test

With the use of on-chip test structures described in Section 14.6, full-speed wafer testing can be completed with a minimum of connected pins. This can be important in reducing the cost of the wafer test fixture.

In general, manufacturing test generation assumes the function of the circuit/chip is correct. It requires ways of exercising all gate inputs and monitoring all gate outputs.

Example 14.1

Consider testing the MIPS microprocessor from Chapter 1. Explain the difference between the tests you would use for logic verification or silicon debug and the tests you would use for manufacturing.

SOLUTION: Logic verification should test that each operation can be performed. For example, a test program might exercise all of the instructions to demonstrate that each one behaves as intended. Logic verification will not necessarily prove that the instruction works for all possible addresses and data values. In contrast, manufacturing tests must prove that every gate operates correctly. They ideally stimulate each gate to produce both a 0 and a 1 to ensure the gate is not damaged. The manufacturing tests may be the only tests applied to a microprocessor prior to it being placed in a system and used. Clearly, it is a challenge to devise a set of tests that is both complete enough that customers receive very few defective chips and short enough to keep testing economical.

14.2 Testers, Test Fixtures, and Test Programs

To test a chip after it is fabricated, you need a tester, a test fixture, and a test program.

14.2.1 Testers and Test Fixtures

A *tester* is a device that can apply a sequence of stimuli to a chip or system under test and monitor and/or record the results of those operations. Testers come in various shapes and sizes.

To test a chip, one or more of four general types of *test fixtures* may be required. These are as follows:

- A *probe card* to test at the wafer level or unpackaged die level with a chip tester
- A *load board* to test a packaged part with a chip tester
- A printed circuit board (PCB) for bench-level testing (with or without a tester)
- A PCB with the chip *in situ,* demonstrating the system application for which the chip is used

We will concentrate first on the cases where a general-purpose production tester is to be used. Production testers are usually expensive pieces of equipment with configurable I/O ports (drive current, output levels, input levels) and huge amounts of RAM behind each test pin. The tester drives input pins from this memory on a cycle-by-cycle basis and samples and stores the levels on output pins. Figure 14.3 shows a typical production tester. In the background, you can see the four-bay cabinet holding the drive electronics. To the right in the background is the controlling workstation. The test head is shown on the front center. This is where the chip is placed in the load board to be tested.

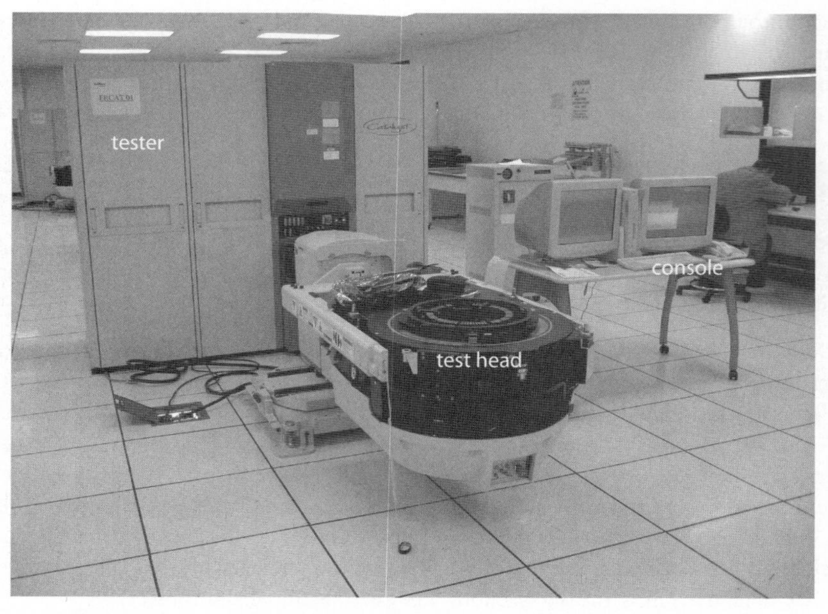

FIGURE 14.3 The Teradyne Catalyst: A typical production tester (Photo: John Haddy, Cisco Systems.)

The probe card or load board for the *device under test* (DUT) is connected to the tester, as shown in Figure 14.4. The test program is compiled and downloaded into the tester and the tests are applied to the bare die or packaged chip. The tester samples the chip outputs and compares the values with those provided by the test program. If there are any differences, the chip is marked as faulty (with an ink dot) and the failing tests may be displayed for reference and stored for later analysis. In the case of a probe card, the card is raised, moved to the next die on the wafer, lowered, and the test procedure repeated. In the case of a load board with automatic part handling, the tested part is removed from the board and sorted into a good or bad bin. A new part is fed to the load board and the test is repeated. In most cases, these procedures take a few seconds for each part tested.

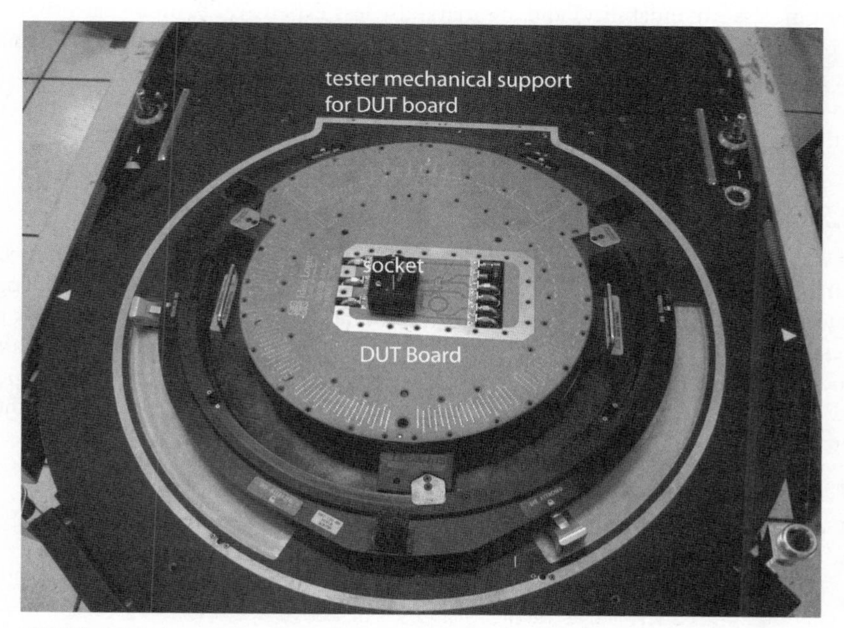

FIGURE 14.4 Tester load board in test head (Photo: John Haddy, Cisco Systems.)

The ability to vary the voltage and timing on a per-pin basis with a tester allows a process known as "shmooing" to be carried out. For instance, you could sweep V_{DD} from 3 V to 6 V on a 5 V part while varying the tester cycle time. This yields a graph called a *shmoo plot* that shows the speed sensitivity of the part with respect to voltage. Another shmoo that is frequently performed is to skew the timing on inputs with respect to the chip clock to look for setup and hold variations. Examples of shmoo plots and their interpretations are given in Section 14.4.

Testers can be very expensive, especially for high-frequency and/or analog/RF chips. Tester usage is charged by time, so the shorter a test runs, the cheaper a part is to test. Applying tests to check every node on the chip may be prohibitively costly, so some designs face a trade-off between test cost and the fraction of defective chips that slip through testing.

Example 14.2

Suppose a $5 million tester has an expected useful life of two years before it becomes inadequate to test faster next-generation parts. How much does the tester cost per second?

SOLUTION: Dividing the tester cost by the number of seconds in two years gives $0.08/second.

Testers are available that can be used to test an IC in a laboratory environment. They mirror large production testers, but generally have less functionality (e.g., slower, less memory per pin, less expandability) and are markedly less expensive. A probe card that allows wafer probing or a socketed load board is required for each design. A good logic analyzer with a pattern generator and a socketed test board can also be used to test a chip. Some groups effectively design their own logic analyzers by surrounding a chip with FPGAs and using the logic and RAM within the FPGA to apply and observe test patterns.

14.2.2 Test Programs

The tester requires a *test program* (in verification and test, this is an overloaded term). This program is normally written in a high-level language (for instance, the IMAGE language used by Teradyne is based on C) that supports a library of primitives for a particular tester. The test program specifies a set of input patterns and a set of output *assertions*. If an output does not match the asserted value at the corresponding time, the tester will report an error. Before the patterns and assertions are applied, the test program has to set up the various attributes of a tester such as the following:

- Set the supply voltages
- Assign mapping between stimulus file signal names and physical tester pins
- Set the pins on the tester to be inputs or outputs and their V_{OH}/V_{IH} levels
- Set the clock on the tester
- Set the input pattern and output assertion timing

And then on a per chip basis:

- Apply supply voltages
- Apply digital stimulus and record responses
- Check responses against assertions
- Report and log errors

A stimulus or pattern file can be derived from running a simulation on the design. Special *vector change descriptions* (VCDs) are used to compact simulation results. An example of a simple stimulus/pattern file for the case of a full adder follows:

```
        III     OO
                SC
                UA
                MR
                 R
        ABC      Y
0       000     00
1       001     10
2       010     10
3       011     01
4       100     10
5       101     01
6       110     01
7       111     11
```

The first line designates the signal directions and shows three inputs (I) and two outputs (O). Reading downward, the next five lines designate the signal names (A, B, C, SUM, CARRY). Thereafter, each line designates a new *test vector*. The first column is the test vector number. The next three columns are the binary value of the inputs and the following two columns are the expected output values. Each line represents a certain length clock cycle that is asserted by the tester. Signals change after a specified period in relation to an internal clock running at the required test period. Clock generation can be carried out in two different ways. First, the clock can be treated like any other signal, in which case, it takes two tester cycles to complete a single clock cycle: one for the clock low and one for the clock high. Alternatively, a timing generator can be used, which allows the clock rising edge (for instance) to be placed anywhere in the tester cycle. So for instance, if the inputs are changed at the start of the tester cycle, the clock might be programmed to rise at the middle of the cycle.

Each pin on the tester is connected to a function memory, which is used to either drive an input or check an output at a DUT pin. Multiple bits may be required per pin to control tristate input pins or mask outputs when they should be ignored.

The clock speed, T_c, is specified, as are supply voltage levels. The time at which pins are driven and sampled is also specified on a pin-by-pin basis (T_s). The format of the test data is usually chosen from Non Return to Zero (NRZ), Return To Zero (RTZ), or other formats such as Surround By Zero (SBZ).

14.2.3 Handlers

An IC handler is responsible for feeding ICs to a test fixture attached to a tester. Chutes or trays containing packaged chips can be used to gravity-feed the devices to the handler, which uses a variety of mechanical means to pick the chips up and place them in the test socket on the load board. The tester stimulus is then applied and chips are binned depending on whether or not they passed the test. It is possible to heat and cool a chuck to test the chip at temperature. However, package-level testing is not normally carried out at temperature because of the time it takes to temperature-cycle the chuck.

An example of a handler is shown in Figure 14.5. This is the NS-6040 from Seiko-Epson. The body of the machine holds the mechanical positioning equipment, while the upper central section

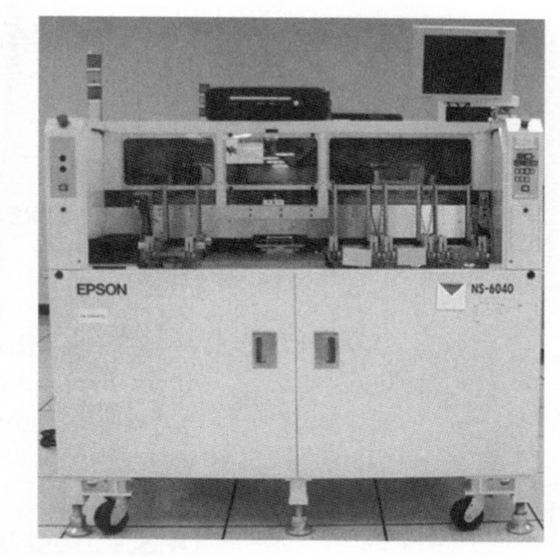

FIGURE 14.5 Photograph of an Epson NS-6040 IC handler (Photo: John Haddy, Cisco Systems.)

supports the test fixture. The light on top indicates a functioning or stopped machine and is designed to be visible across a production floor where many machines might be operating. A screen at the top right provides status information to the operator. The unit has wheels for easy movement, but also has firm footings, which are lowered when the machine is in use.

Handlers add a constant time to the test process, typically around 1 second. Thus, load boards and handlers are often constructed to deal with two or four chips at once to reduce the cost of testing. Because a load board must be designed to fit to a given handler, select the handler before starting design of the load board.

14.3 Logic Verification Principles

Figure 14.6(a) shows a combinational circuit with N inputs. To test this circuit exhaustively, a sequence of 2^N inputs (or test vectors) must be applied and observed to fully exercise the circuit. This combinational circuit is converted to a sequential circuit with addition of M registers, as shown in Figure 14.6(b). The state of the circuit is determined by the inputs and the previous state. A minimum of 2^{N+M} test vectors must be applied to exhaustively test the circuit. As observed by [Williams83] more than two decades ago,

> *With LSI, this may be a network with* $N = 25$ *and* $M = 50$*, or* 2^{75} *patterns, which is approximately* 3.8×10^{22}*. Assuming one had the patterns and applied them at an application rate of 1 μs per pattern, the test time would be over a billion years* (10^9)*.*

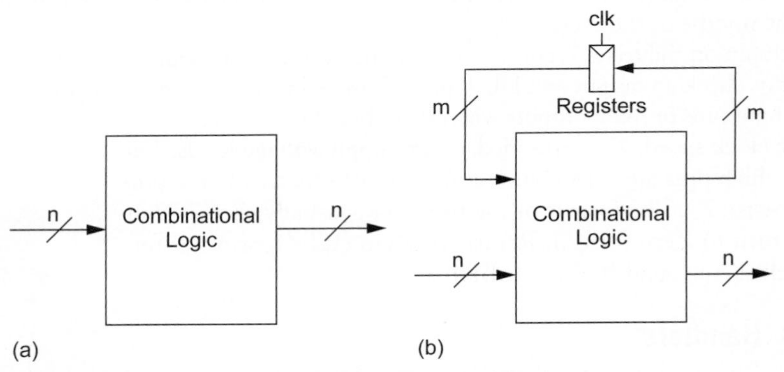

FIGURE 14.6 The combinational explosion in test vectors

Clearly, exhaustive testing is infeasible for most systems. Fortunately, the number of potentially nonfunctional nodes on a chip is much smaller than the number of states. The verification engineer must cleverly devise test vectors that detect any (or nearly any) defective node without requiring so many patterns.

14.3.1 Test Vectors

Test vectors are a set of patterns applied to inputs and a set of expected outputs. Both logic verification and manufacturing test require a good set of test vectors. The set should be large enough to catch all the logic errors and manufacturing defects, yet small enough to keep test time (and cost) reasonable.

Directed and *random* vectors are the most common types. Directed vectors are selected by an engineer who is knowledgeable about the system. Their purpose is to cover the corner cases where the system might be most likely to malfunction. For example, in a 32-bit datapath, likely corner cases include the following:

```
0x00000000  All zeros
0xFFFFFFFF  All ones
0x00000001  One in the lsb
0x80000000  One in the msb
0x55555555  Alternating 0's and 1's
0xAAAAAAAA  Alternating 1's and 0's
0x7A39D281  A random value
```

The circuit could be tested by applying all combinations of these directed vectors to the various inputs. Directed vectors are an efficient way to catch the most obvious design errors and a good logic designer will always run a set of directed tests on a new piece of RTL to ensure a minimum level of quality.

Applying a large number of random or semirandom vectors is a surprisingly good way to detect more subtle errors. The effectiveness of the set of vectors is measured by the fault coverage, which is discussed in Section 14.5.6. Automatic test pattern generation tools are good at producing high fault coverage for manufacturing test and are discussed in Section 14.5.7.

14.3.2 Testbenches and Harnesses

A verification *test bench* or *harness* is a piece of HDL code that is placed as a wrapper around a core piece of HDL to apply and check test vectors. In the simplest test bench, input vectors are applied to the module under test and at each cycle, the outputs are examined to determine whether they comply with a predefined expected data set. The expected outputs can be derived from the golden model and saved as a file or the value can be computed on the fly.

Simulators usually provide settable break points and single or multiple stepping abilities to allow the designer to step through a test sequence while debugging discrepancies.

14.3.3 Regression Testing

High-level language scripts are frequently used when running large testbenches, especially for *regression testing*. Regression testing involves performing a suite of simulations to automatically verify that no functionality has inadvertently changed in a module or set of modules. During a design, it is common practice to run a regression script every night after design activities have concluded to check that bug fixes or feature enhancements have not broken completed modules.

Example 14.3

Figure 13.11 showed a possible software radio architecture that used a combination of an IQ conversion block and a multiplier-based multiprocessor. The following regression testing might be done:

```
Test IQ Conversion
   Test Upconverter
      Test NCO
         Test Read and Write of All Registers
         Test Phase Incrementer
         Test Phase Adder
         Test Sine ROM (Read Contents)
         Test Overall NCO at a set of frequencies
      Test Multiplier
   Test Downconverter
      Test NCO
         ...
      Test Multiplier
         ...
      Test Low Pass Filter
         ...
Test Microprocessor Memory Core
   Test Microprocessor
      Test ALU
      Test Instruction Decode
      Test Program Counter
      Test Register File Read/Write
      Exhaustive Instruction Test
   Test Memory Read/Write
Test Interprocessor Bus IO
Test IQ Conversion to Processor pathways
Test Overall Software Radio Functionality
```

Note the way in which the correctness of modules is slowly built up by verifying lower-level models first. The low-level tests are gradually built up in complexity until the complete functionality can be verified. At low levels, it is easier to exhaustively verify that logic is correct. For instance, we can verify that the sine ROM is in fact generating a sine wave for one frequency. We then use this knowledge to postulate that it generates correct sine waves for all input frequencies when we verify at the levels above the NCO. At the chip level, we assume that IQ_conversion is correct for all combinations of signal frequency and local oscillator frequency even though we may only check a small subset. If we started at the top level and ran a simulation for a few frequencies, we could never have confidence that the lower levels were correct. In addition, if there is a problem, trying to locate the problem by debugging at the top level is futile. Running regression tests from the bottom up is designed to overcome this verification nightmare.

14.3.4 Version Control

Combined with regression testing is the use of versioning, that is, the orderly management of different design iterations. Unix/Linux tools such as CVS or Subversion are useful for this.

Example 14.4

In the software radio example, the regression testing halts at the ALU test in the example given above. Working late, the design leader, Vanessa Eagleeye, examines the CVS history and discovers that Fred Codechanger has made an edit to the ALU design to try a new adder during the day. She is able to revert the code to what was previously working and then rerun the regression test and have a peaceful night's sleep. Fred corrects his mistake the next day and is advised to remember to run the regression verification step before submitting such hurried edits.

14.3.5 Bug Tracking

Another important tool to use during verification (and in fact the whole design cycle) is a bug-tracking system. Bug-tracking systems such as the Unix/Linux based GNATS allow the management of a wide variety of bugs. In these systems, each bug is entered and the location, nature, and severity of the bug noted. The bug discoverer is noted, along with the perceived person responsible for fixing the bug.

Example 14.5

After Example 14.4, Vanessa enters a bug report describing the bug. She cites Fred as the person responsible and the level as severe. The next day, Fred fixes the problem and changes the bug status to fixed. The bug report is kept in the system, but does not appear in any listing of outstanding bugs. It is kept to track the re-introduction of bugs, as this might give managers an idea of a problem area in the design management.

Tracking the number of bugs can give you an idea of the rate at which a design is converging toward a finished state. If the trend is downward, the design is converging. On the other hand, an upward trend tends to indicate a design early in its verification cycle.

14.4 Silicon Debug Principles

The area of basic digital debugging was introduced in Section 14.1.2. A major challenge in silicon debugging is when the chip operates incorrectly, but you cannot ascertain the cause by making measurements at the chip pins or scan chain outputs (see Section 14.6.2).

There are a number of techniques for directly accessing the silicon. First, specific signals can be brought to the top of the chip as *probe points*. These are small squares (5–10 μm on a side) of top-level metal that connect to key points in the circuit that the designer has had the foresight to include before debug. The overglass cut mask should specify a hole in the passivation over the probe pads so the metal can be reliably contacted. Typical of these kinds of test points might be internal bias points in linear circuits or perhaps key points in a high-speed signal chain (be careful not to excessively load the circuit to be probed). The exposed squares can be probed with a picoprobe (fine-tipped probe) in a fixture under a microscope. During design, the load of the picoprobe has to be taken into account by providing buffers if necessary. The Model 35 probe from GGB Industries shown in Figure 14.7 has a capacitance of 50 fF, input resistance of 1.25 MΩ, and frequency response from DC to 26 GHz. It can probe down to a 10×10 μm window.

The die can also be probed electrically or optically if mechanical contact is not feasible. An *electron beam* (ebeam) probe uses a scanning electron microscope to produce a

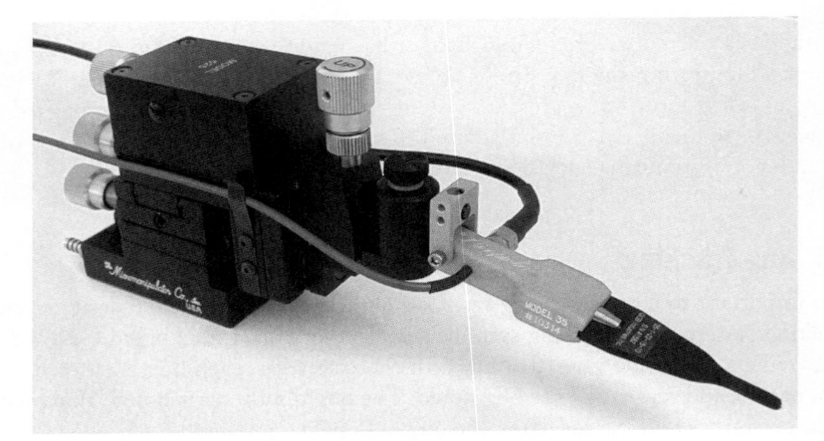

FIGURE 14.7 GGB Industries Model 15 picoprobe (© 2009 GGB Industries, reprinted with permission.)

tightly focused beam of electrons to measure on-chip voltages. Similarly, *Laser Voltage Probing* (LVP) [Lasserre99] involves shining a laser at a circuit and observing the reflected light. The reflections are modulated by the electric fields so switching waveforms can be deduced. However, the probing can be invasive; the stream of photons may disturb sensitive dynamic nodes. *Picosecond Imaging Circuit Analysis* (PICA) [Knebel98] captures faint light emission naturally produced by switching transistors and hence is noninvasive. Silicon is partially transparent to infrared light, so both LVP and PICA can be performed through the substrate from the backside of a chip in a flip-chip package.

On a more coarse scale, infrared (IR) imaging can be used to examine "hot spots" in a chip, which may be the source of problems (for instance, a resistive short between power rails). There are also liquid crystal materials, which can be "painted on" to a die to indicate temperature problems at a coarse resolution.

If the location of the fault is known, a *Focused Ion Beam* (FIB) can be used to cut wires or lay new conductors down. Even with plastic-packaged parts, the plastic can be carefully ground off and these repairs completed. The reason for this kind of tool is that normally in any chip project, time is of the essence and FIB runs are quicker (and cheaper for a few parts) than frequent mask changes. Laser cutting is also possible. Commercial providers such as MEFAS offer these services.

Example 14.6

A short between V_{DD} and GND has rendered a chip just back from tapeout nonfunctional. The position of the fault is known and it can be corrected by a cut to the top level metal. Several packaged parts are sent to the FIB house with a location from a given fiducial mark and an accompanying plot of the position of the metal to be cut. The FIB house exposes the die (i.e., by grinding a plastic package). The operator then locates the cut position manually using a microscope and runs the FIB machine. The modified packages are then returned to the designers, where hopefully they celebrate the successful test of an otherwise useless chip.

Debugging logic circuits will often involve extremely fast or novel circuits that are largely analog in nature. In this case, it is advisable to have a model of the circuit in question available in SPICE. Debugging analog circuits, as with purely digital circuits, involves making an assertion and then trying to prove the assertion is correct. This can begin with a SPICE simulation and then progress to silicon measurement.

Failures causes may be *manufacturing*, *functional*, or *electrical*. Manufacturing failures occur when a chip has a defect or is outside of the parametric specifications. Debug can reject chips with manufacturing problems, although circuits sensitive to weaknesses in the manufacturing process can be changed to improve yield, as will be discussed in Section 14.6.5. Functional failures are logic bugs or physical design errors that cause the chip to fail under all conditions. They arise from inadequate logic verification and are usually the easiest to fix. Electrical failures occur when the chip is logically correct, but malfunctions under certain conditions such as voltage, temperature, or frequency. Section 8.3 addressed many causes of electrical failures. Some electrical failures can be so severe that they appear as functional failures, while others occur rarely and are extremely difficult to reproduce and diagnose.

So-called shmoo plots can help to debug electrical failures in silicon [Baker97]. A shmoo plot is often made with voltage on one axis and speed on the other. The test vectors are applied at each combination of voltage and clock speed, and the success of the test is recorded. Often, only a set of vectors applicable to a particular module is applied to diagnose a problem in that module.

Figure 14.8 shows a shmoo from the Intel Atom microprocessor [Gerosa09]. Dots in the light gray area indicate correct operation, while different letters indicate different failure modes. The chip works at 1.25 GHz at 0.75 V and at 2.5 GHz at 1.15 V.

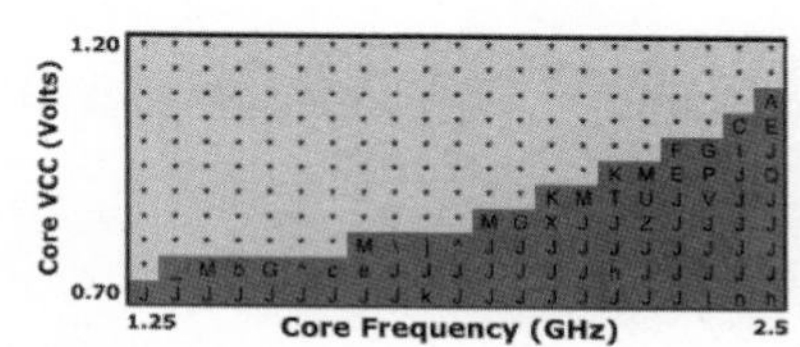

FIGURE 14.8 Shmoo for Intel Atom microprocessor (© IEEE 2009.)

The shmoo plots shown in Figure 14.9 illustrate a variety of conditions [Josephson02]. A healthy normal chip should operate at increasing frequency as the voltage increases. The brick wall pattern suggests that the chip may be randomly initialized in one of two states, only one of which is correct. For example, a register without a reset signal may randomly have an initial state of 0 or 1. The wall pattern in which the chip fails to operate at any frequency above or below a particular voltage can indicate charge sharing, coupling noise, or a race condition. The reverse speedpath behavior indicates a leakage problem in which a weakly held node leaks to an invalid level before the end of the cycle. At higher voltage, the leakage is exacerbated and appears at shorter clock periods. The floor is a variant on the leakage problem where the part fails at low frequency independent of the voltage. A finger indicates coupling problems dependent on the alignment of the aggressor and victim, where at certain frequencies the alignment always causes a failure.

A shmoo can also plot operating speed against temperature. At cold temperature, FETs are faster, have lower effective resistance, and have higher threshold voltages. A normal shmoo should show speed increasing as temperature decreases. Failures at low temperature could indicate coupling or charge sharing noise exacerbated by faster edge rates. Failures at high temperature could indicate excessive leakage or noise problems exacerbated by the lower threshold voltages. Walls at either temperature could indicate race conditions where the path that wins the race varies with temperature.

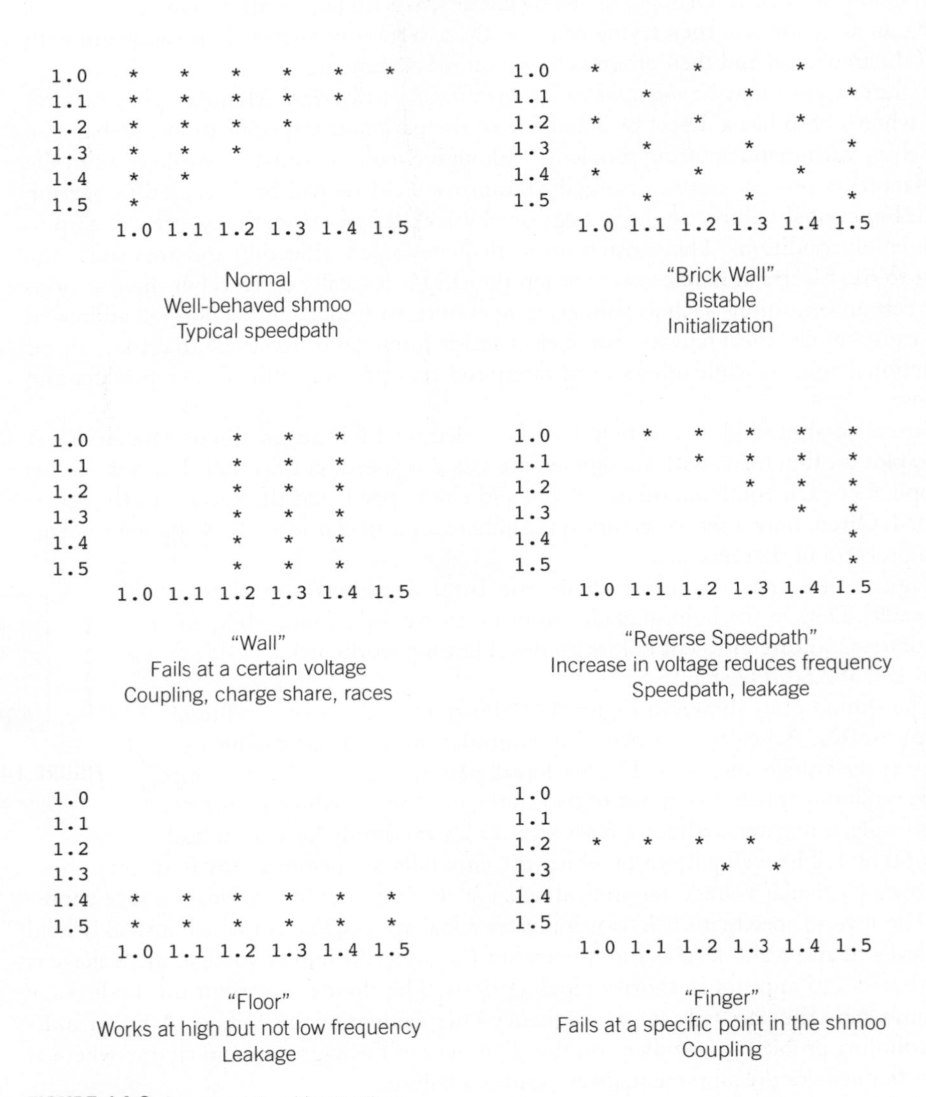

FIGURE 14.9 Shmoo plots with symptoms

14.5 Manufacturing Test Principles

As discussed in Section 13.5.2, integrated circuits have a yield of less than 100%. Figure 14.10 shows micrographs of some manufacturing defects.

The purpose of manufacturing test is to screen out most of the defective parts before they are shipped to customers. Typical commercial products target a defect rate of 350–1000 defects per million (DPM) chips shipped. The customer then assembles

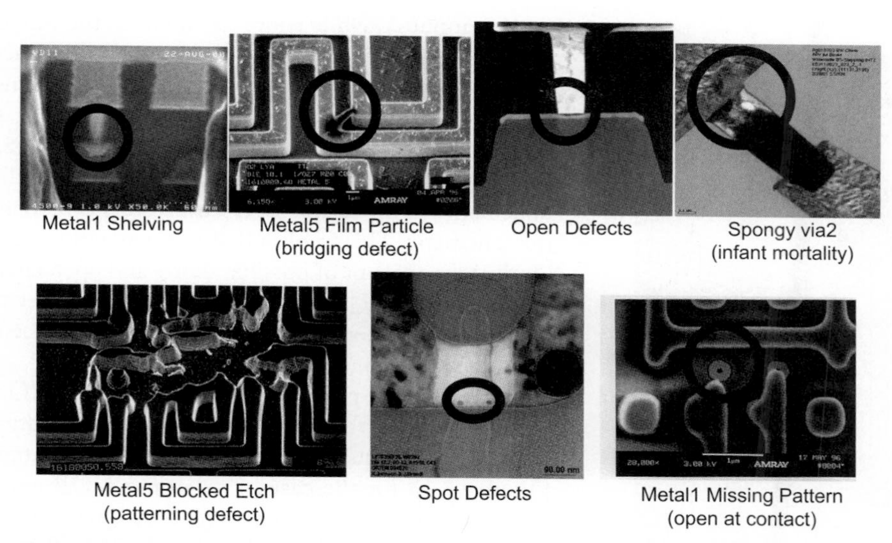

FIGURE 14.10 SEM images of manufacturing defects (Courtesy of Intel Corporation.)

systems from the chips, tests the systems, and discards or repairs defective systems. A high defect rate leads to unhappy customers.

A critical factor in all VLSI design is the need to incorporate methods of testing circuits. This task should proceed concurrently with architectural considerations and not be left until fabricated parts are available (as is a recurring temptation to designers).

14.5.1 Fault Models

To deal with the existence of good and bad parts, it is necessary to propose a *fault model*; i.e., a model for how faults occur and their impact on circuits. The most popular model is called the *Stuck-At* model. The *Short Circuit/ Open Circuit* model can be a closer fit to reality, but is harder to incorporate into logic simulation tools.

14.5.1.1 Stuck-At Faults In the Stuck-At model, a faulty gate input is modeled as a *stuck at zero* (Stuck-At-0, S-A-0) or *stuck at one* (Stuck-At-l, S-A-l). This model dates from board-level designs, where it was determined to be adequate for modeling faults. Figure 14.11 illustrates how an S-A-0 or S-A-1 fault might occur. These faults most frequently occur due to gate oxide shorts (the nMOS gate to GND or the pMOS gate to V_{DD}) or metal-to-metal shorts.

14.5.1.2 Short-Circuit and Open-Circuit Faults Other models include *stuck-open* or *shorted* models [Jayasumana91]. Two bridging or shorted faults are shown in Figure 14.12. The short $S1$ results in an S-A-0

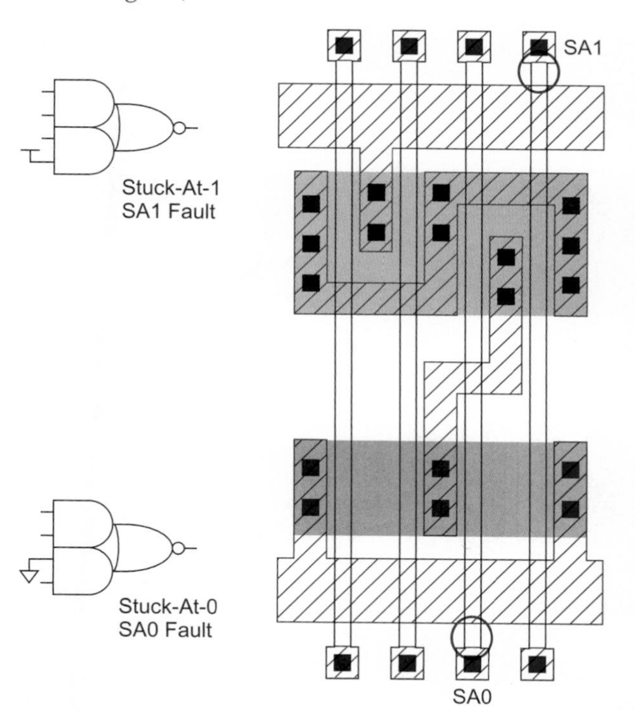

FIGURE 14.11 CMOS stuck-at faults

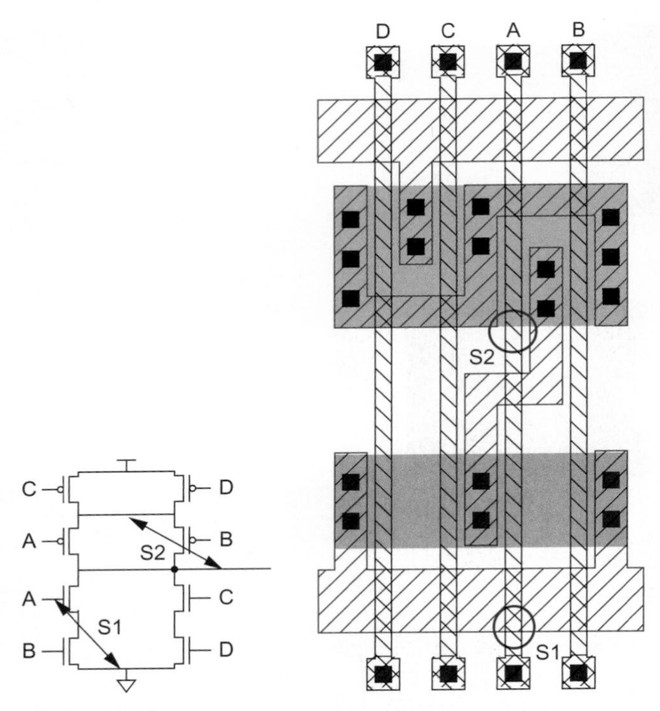

FIGURE 14.12 CMOS bridging faults

fault at input A, while short $S2$ modifies the function of the gate. It is evident that to ensure the most accurate modeling, faults should be modeled at the transistor level because it is only at this level that the complete circuit structure is known. For instance, in the case of a simple NAND gate, the intermediate node between the series nMOS transistors is hidden by the schematic. This implies that test generation should ideally take account of possible shorts and open circuits at the switch level [Galiay80]. Expediency dictates that most existing systems rely on Boolean logic representations of circuits and stuck-at fault modeling.

A particular problem that arises with CMOS is that it is possible for a fault to convert a combinational circuit into a sequential circuit. This is illustrated in Figure 14.13 for the case of a 2-input NOR gate in which one of the transistors is rendered ineffective. If nMOS transistor A is stuck open, then the function displayed by the gate will be

$$Z = \overline{A + B} + \bar{B}Z' \tag{14.1}$$

where Z' is the previous state of the gate. As another example, if either pMOS transistor is missing, the node would be arbitrarily charged (i.e., it might be high due to some weird charging sequence) until one of

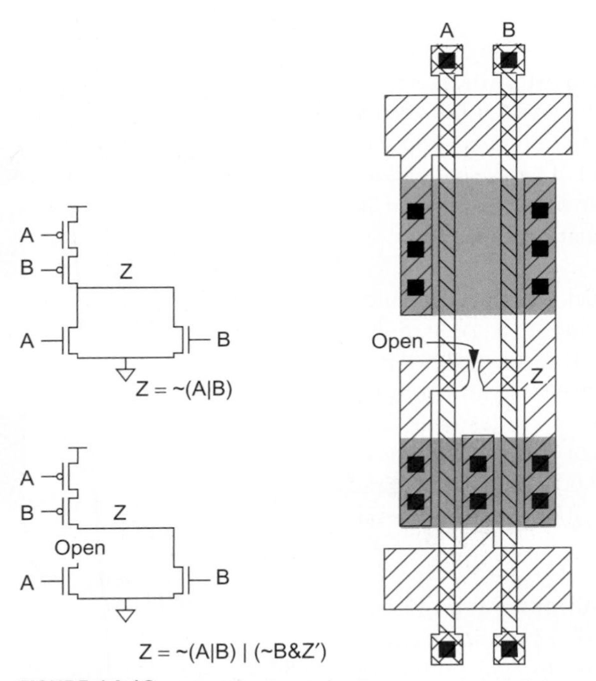

FIGURE 14.13 A CMOS open fault that causes sequential faults

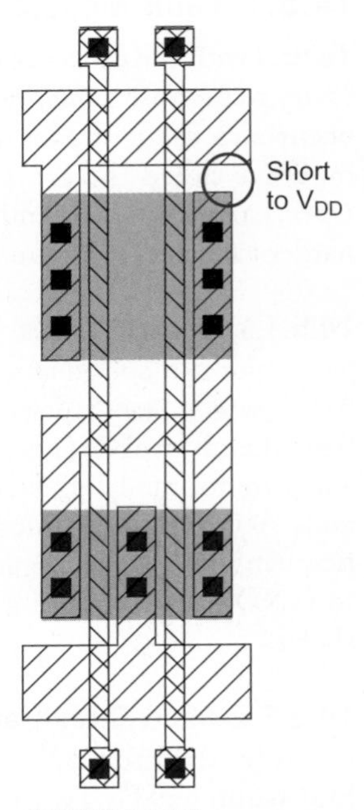

FIGURE 14.14 A defect that causes static I_{DD} current

the nMOS transistors discharged the node. Thereafter, it would remain at zero, barring charge leakage effects.

It is also possible for transistors to exhibit a stuck-open or stuck-closed state. Stuck-closed states can be detected by observing the static V_{DD} current (I_{DD}) while applying test vectors. Consider the fault shown in Figure 14.14, where the drain connection on a pMOS transistor in a 2-input NOR gate is shorted to V_{DD}. This could physically occur if stray metal (caused by a speck of dust at the photolithography stage) overlapped the V_{DD} line and drain connection as shown. If we apply the test vector 01 or 10 to the A and B inputs and measure the static I_{DD} current, we will notice that it rises to some value determined by size of the nMOS transistors.

14.5.2 Observability

The *observability* of a particular circuit node is the degree to which you can observe that node at the outputs of an integrated circuit (i.e., the pins). This metric is relevant when you want to measure the output of a gate within a larger circuit to check that it operates correctly. Given the limited number of nodes that can be directly observed, it is the aim of good chip designers to have easily observed gate outputs. Adoption of some basic design for test techniques can aid tremendously in this respect. Ideally, you should be able to observe directly or with moderate indirection (i.e., you may have to wait a few cycles) every gate output within an integrated circuit. While at one time this aim was hindered by the expense of extra test circuitry and a lack of design methodology, current processes and design practices allow you to approach this ideal. Section 14.6 examines a range of methods for increasing observability.

14.5.3 Controllability

The *controllability* of an internal circuit node within a chip is a measure of the ease of setting the node to a 1 or 0 state. This metric is of importance when assessing the degree of difficulty of testing a particular signal within a circuit. An easily controllable node would be directly settable via an input pad. A node with little controllability, such as the most significant bit of a counter, might require many hundreds or thousands of cycles to get it to the right state. Often, you will find it impossible to generate a test sequence to set a number of poorly controllable nodes into the right state. It should be the aim of good chip designers to make all nodes easily controllable. In common with observability, the adoption of some simple design for test techniques can aid in this respect tremendously. Making all flip-flops resettable via a global reset signal is one step toward good controllability.

14.5.4 Repeatability

The *repeatability* of system is the ability to produce the same outputs given the same inputs. Combinational logic and synchronous sequential logic is always repeatable when it is functioning correctly. However, certain asynchronous sequential circuits are nondeterministic. For example, an arbiter may select either input when both arrive at nearly the same time. Testing is much easier when the system is repeatable. Some systems with asynchronous interfaces have a lock-step mode to facilitate repeatable testing.

14.5.5 Survivability

The *survivability* of a system is the ability to continue function after a fault. For example, error-correcting codes provide survivability in the event of soft errors. Redundant rows and columns in memories and spare cores provide survivability in the event of manufacturing defects. Adaptive techniques provide survivability in the event of process variation. Some survivability features are invoked automatically by the hardware, while others are activated by blowing fuses after manufacturing test.

14.5.6 Fault Coverage

A measure of goodness of a set of test vectors is the amount of *fault coverage* it achieves. That is, for the vectors applied, what percentage of the chip's internal nodes were checked? Conceptually, the way in which the fault coverage is calculated is as follows. Each circuit node is taken in sequence and held to 0 (S-A-0), and the circuit is simulated with the test vectors comparing the chip outputs with a *known good machine*—a circuit with no nodes artificially set to 0 (or 1). When a discrepancy is detected between the *faulty machine* and the good machine, the fault is marked as detected and the simulation is stopped. This is repeated for setting the node to 1 (S-A-1). In turn, every node is stuck (artificially) at 1 and 0 sequentially. The fault coverage of a set of test vectors is the percentage of the total nodes that can be detected as faulty when the vectors are applied. To achieve world-class quality levels, circuits are required to have in excess of 98.5% fault coverage. The *Verification Methodology Manual* [Bergeron05] is the bible for fault coverage techniques.

14.5.7 Automatic Test Pattern Generation (ATPG)

Historically, in the IC industry, logic and circuit designers implemented the functions at the RTL or schematic level, mask designers completed the layout, and test engineers wrote the tests. In many ways, the test engineers were the Sherlock Holmes of the industry, reverse engineering circuits and devising tests that would test the circuits in an adequate manner. For the longest time, test engineers implored circuit designers to include extra circuitry to ease the burden of test generation. Happily, as processes have increased in density and chips have increased in complexity, the inclusion of test circuitry has become less of an overhead for both the designer and the manager worried about the cost of the die. In addition, as tools have improved, more of the burden for generating tests has fallen on the designer. To deal with this burden, *Automatic Test Pattern Generation* (ATPG) methods have been invented. The use of some form of ATPG is standard for most digital designs.

Commercial ATPG tools can achieve excellent fault coverage. However, they are computation-intensive and often must be run on servers or compute farms with many parallel processors. Some tools use statistical algorithms to predict the fault coverage of a set of vectors without performing as much simulation. Adding scan and built-in self-test, as described in Section 14.6, improves the observability of a system and can reduce the number of test vectors required to achieve a desired fault coverage.

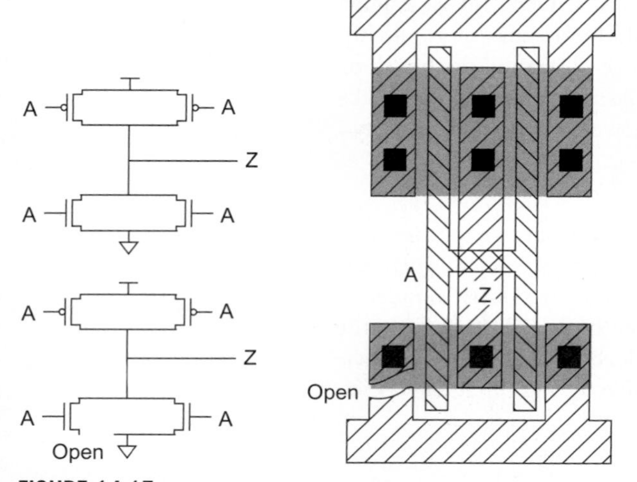

FIGURE 14.15 An example of a delay fault

14.5.8 Delay Fault Testing

The fault models dealt with until this point have neglected timing. Failures that occur in CMOS could leave the functionality of the circuit untouched, but affect the timing. For instance, consider the layout shown in Figure 14.15 for an inverter gate composed of paralleled nMOS and pMOS transistors. If an open circuit occurs in one of the nMOS transistor source connections to GND, then the gate would still function but with increased t_{pdf}. In addition, the fault now becomes sequential as the detection of the fault depends on the previous state of the gate.

Delay faults may be caused by crosstalk [Paul02]. Delay faults can also occur more often in SOI logic through the history effect. Software has been developed to model the effect of delay faults and is becoming more important as a failure mode as processes scale.

14.6 Design for Testability

The keys to designing circuits that are testable are controllability and observability. Restated, controllability is the ability to set (to 1) and reset (to 0) every node internal to the circuit. Observability is the ability to observe, either directly or indirectly, the state of any node in the circuit. Good observability and controllability reduce the cost of manufacturing testing because they allow high fault coverage with relatively few test vectors. Moreover, they can be essential to silicon debug because physically probing internal signals has become so difficult.

We will first cover three main approaches to what is commonly called *Design for Testability* (DFT). These may be categorized as follows:

- *Ad hoc* testing
- Scan-based approaches
- Built-in self-test (BIST)

14.6.1 *Ad Hoc* Testing

Ad hoc test techniques, as their name suggests, are collections of ideas aimed at reducing the combinational explosion of testing. They are summarized here for historical reasons. They are only useful for small designs where scan, ATPG, and BIST are not available. A complete scan-based testing methodology is recommended for all digital circuits. Having said that, the following are common techniques for ad hoc testing:

- Partitioning large sequential circuits
- Adding test points
- Adding multiplexers
- Providing for easy state reset

A technique classified in this category is the use of the bus in a bus-oriented system for test purposes. Each register has been made loadable from the bus and capable of being driven onto the bus. Here, the internal logic values that exist on a data bus are enabled onto the bus for testing purposes.

Frequently, multiplexers can be used to provide alternative signal paths during testing. In CMOS, transmission gate multiplexers provide low area and delay overhead.

Any design should always have a method of resetting the internal state of the chip within a single cycle or at most a few cycles. Apart from making testing easier, this also makes simulation faster as a few cycles are required to initialize the chip.

In general, *ad hoc* testing techniques represent a bag of tricks developed over the years by designers to avoid the overhead of a systematic approach to testing, as will be described in the next section. While these general approaches are still quite valid, process densities and chip complexities necessitate a structured approach to testing.

14.6.2 Scan Design

The *scan-design* strategy for testing has evolved to provide observability and controllability at each register. In designs with scan, the registers operate in one of two modes. In *normal mode*, they behave as expected. In *scan mode*, they are connected to form a giant shift register called a *scan chain* spanning the whole chip. By applying *N* clock pulses in scan mode, all *N* bits of state in the system can be shifted out and new *N* bits of state can be shifted in. Therefore, scan mode gives easy observability and controllability of every register in the system.

Modern scan is based on the use of scan registers, as shown in Figure 14.16. The scan register is a *D* flip-flop preceded by a multiplexer. When the *SCAN* signal is deasserted, the register behaves as a conventional register, storing data on the *D* input. When *SCAN* is asserted, the data is loaded from the *SI* pin, which is connected in shift register fashion to the previous register *Q* output in the scan chain.

For the circuit to load the scan chain, *SCAN* is asserted and *CLK* is pulsed eight times to load the first two ranks of 4-bit registers with data. *SCAN* is deasserted and *CLK* is asserted for one cycle to operate the circuit normally with predefined inputs. *SCAN* is then reasserted and *CLK* asserted eight times to read the stored data out. At the same time, the

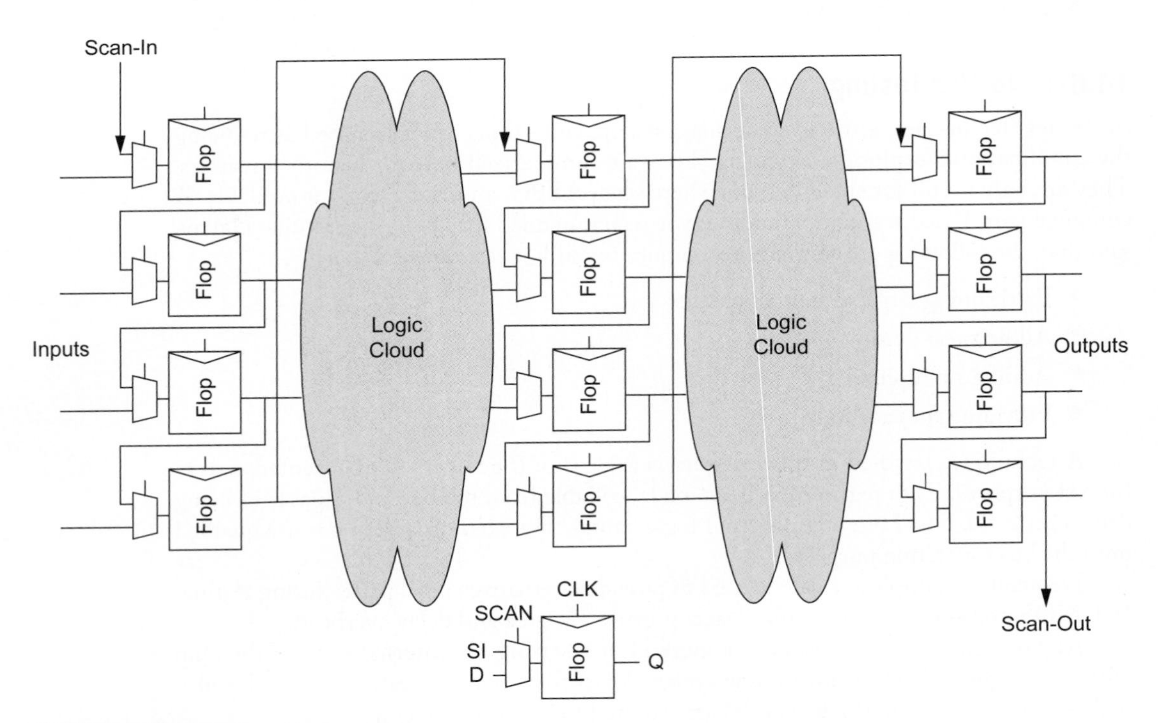

FIGURE 14.16 Scan-based testing

new register contents can be shifted in for the next test. Testing proceeds in this manner of serially clocking the data through the scan register to the right point in the circuit, running a single system clock cycle and serially clocking the data out for observation. In this scheme, every input to the combinational block can be controlled and every output can be observed. In addition, running a random pattern of 1s and 0s through the scan chain can test the chain itself.

Test generation for this type of test architecture can be highly automated. ATPG techniques can be used for the combinational blocks and, as mentioned, the scan chain is easily tested. The prime disadvantage is the area and delay impact of the extra multiplexer in the scan register. Designers (and managers alike) are in widespread agreement that this cost is more than offset by the savings in debug time and production test cost.

14.6.2.1 Parallel Scan You can imagine that serial scan chains can become quite long, and the loading and unloading can dominate testing time. A fairly simple idea is to split the chains into smaller segments. This can be done on a module-by-module basis or completed automatically to some specified scan length. Extending this to the limit yields an extension to serial scan called *random access scan* [Ando80]. To some extent, this is similar to that used inside FPGAs to load and read the control RAM.

The basic idea is shown in Figure 14.17. The figure shows a two-by-two register section. Each register receives a column (`column<m>`) and row (`row<n>`) access signal along with a row data line (`data<n>`). A global write signal (`write`) is connected to all registers. By asserting the row and column access signals in conjunction with the write signal, any register can be read or written in exactly the same method as a conventional RAM. The

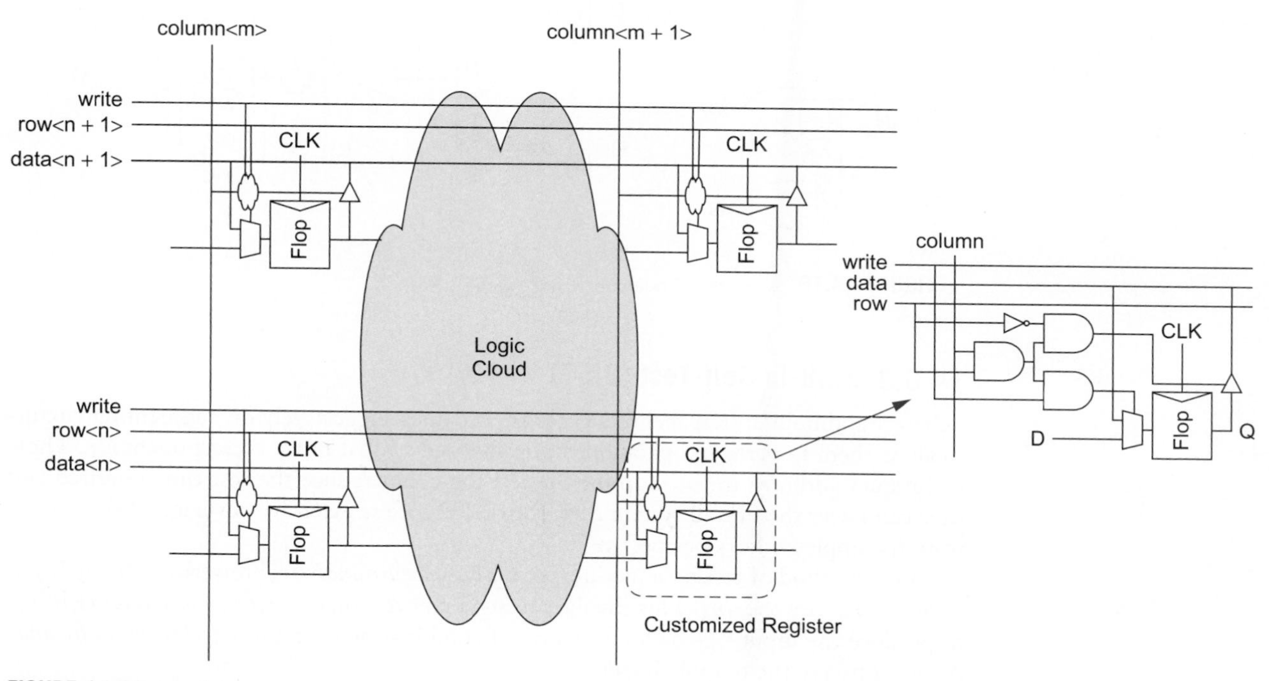

FIGURE 14.17 Parallel scan—basic structure

notional logic is shown to the right of the four registers. Implementing the logic required at the transistor level can reduce the overhead for each register.

14.6.2.2 Scannable Register Design As we have seen, an ordinary flip-flop can be made scannable by adding a multiplexer on the data input, as shown in Figure 14.18(a). Figure 14.18(b) shows a circuit design for such a scan register using a transmission-gate multiplexer. The setup time increases by the delay of the extra transmission gate in series with the D input as compared to the ordinary static flip-flop shown in Figure 9.19(b). Figure 14.18(c) shows a circuit using clock gating to obtain nearly the same setup time as the ordinary flip-flop. In either design, if a clock enable is used to stop the clock to unused portions of the chip, care must be taken that ϕ always toggles during scan mode.

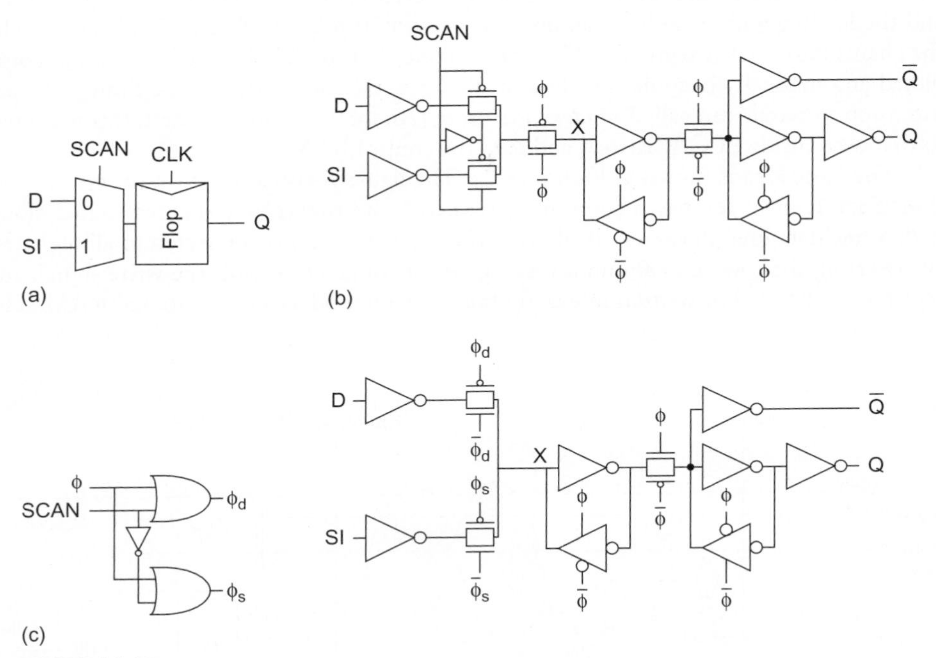

FIGURE 14.18 Scannable flip-flops

14.6.3 Built-In Self-Test (BIST)

Self-test and built-in test techniques, as their names suggest, rely on augmenting circuits to allow them to perform operations upon themselves that prove correct operation. These techniques add area to the chip for the test logic, but reduce the test time required and thus can lower the overall system cost. [Stroud02] offers extensive coverage of the subject from the implementer's perspective.

One method of testing a module is to use *signature analysis* [Frowerk77, Nadig77] or *cyclic redundancy checking*. This involves using a *pseudo-random sequence generator* (PRSG) to produce the input signals for a section of combinational circuitry and a *signature analyzer* to observe the output signals.

A PRSG of length n is constructed from a *linear feedback shift register* (LFSR), which in turn is made of n flip-flops connected in a serial fashion, as shown in Figure 14.19(a). The XOR of particular outputs are fed back to the input of the LFSR. An n-bit LFSR will cycle through 2^n-1 states before repeating the sequence. LFSRs are discussed further in Section 10.5.4. They are described by a *characteristic polynomial* indicating which bits are fed back. A *complete feedback shift register* (CFSR), shown in Figure 14.19(b), includes the zero state that may be required in some test situations [Wang86]. An n-bit LFSR is converted to an n-bit CFSR by adding an $n-1$ input NOR gate connected to all but the last bit. When in state 0…01, the next state is 0…00. When in state 0…00, the next state is 10…0. Otherwise, the sequence is the same. Alternatively, the bottom n bits of an $n+1$-bit LFSR can be used to cycle through the all zeros state without the delay of the NOR gate.

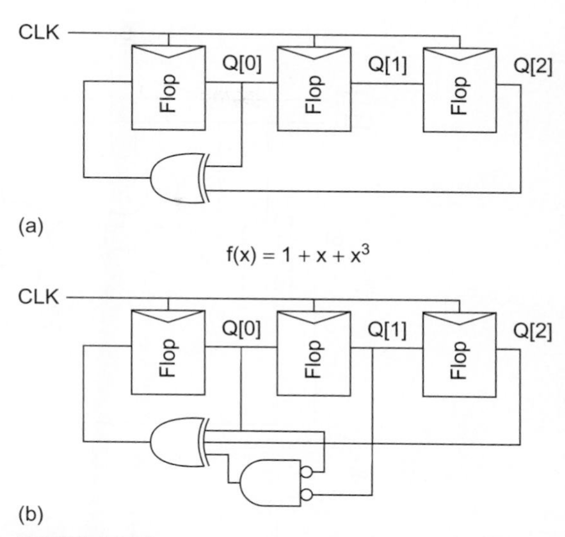

FIGURE 14.19 Pseudo-random sequence generator

A signature analyzer receives successive outputs of a combinational logic block and produces a *syndrome* that is a function of these outputs. The syndrome is reset to 0, and then XORed with the output on each cycle. The syndrome is swizzled each cycle so that a fault in one bit is unlikely to cancel itself out. At the end of a test sequence, the LFSR contains the syndrome that is a function of all previous outputs. This can be compared with the correct syndrome (derived by running a test program on the good logic) to determine whether the circuit is good or bad. If the syndrome contains enough bits, it is improbable that a defective circuit will produce the correct syndrome.

14.6.3.1 BIST The combination of signature analysis and the scan technique creates a structure known as *BIST*—for *Built-In Self-Test* or *BILBO*—for *Built-In Logic Block Observation* [Koenemann79]. The 3-bit BIST register shown in Figure 14.20 is a scannable, resettable register that also can serve as a pattern generator and signature analyzer. $C[1:0]$ specifies the mode of operation. In the reset mode (10), all the flip-flops are synchronously initialized to 0. In normal mode (11), the flip-flops behave normally with their D input and Q output. In scan mode (00), the flip-flops are configured as a 3-bit shift register between SI and SO. Note that there is an inversion between each stage. In test mode (01), the register behaves as a pseudo-random sequence generator or signature analyzer. If all the D inputs are held low, the Q outputs loop through a pseudo-random bit sequence, which can serve as the input to the combinational logic. If the D inputs are taken from the combinational logic output, they are swizzled with the existing state to produce the syndrome. In summary, BIST is performed by first resetting the syndrome in the output register. Then both registers are placed in the test mode to produce the pseudo-random inputs and calculate the syndrome. Finally, the syndrome is shifted out through the scan chain.

Various companies have commercial design aid packages that automatically replace ordinary registers with scannable BIST registers, check the fault coverage, and generate scripts for production testing. As an example, on a WLAN modem chip comprising roughly 1 million gates, a full at-speed test takes under a second with BIST. This comes with roughly a 7.3% overhead in the core area (but actually zero because the design was

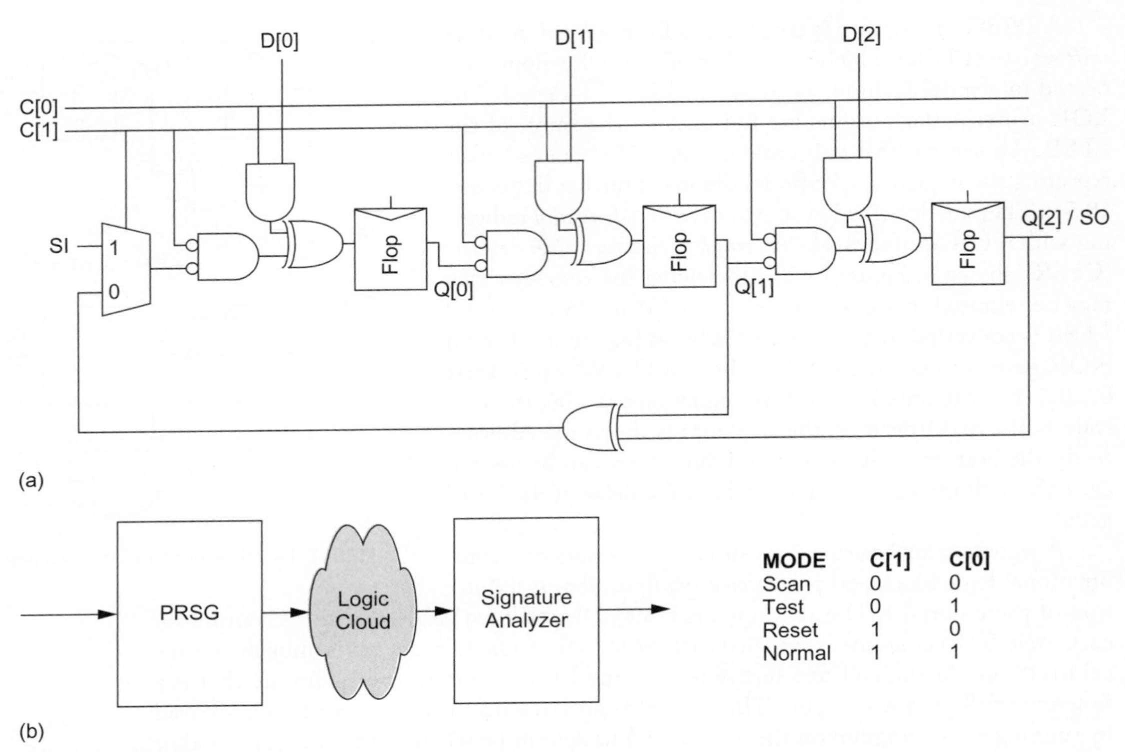

FIGURE 14.20 BIST (a) 3-bit register, (b) use in a system

pad limited) and a 99.7% fault coverage level. The WLAN modem parts designed in this way were fully tested in less than ten minutes on receipt of first silicon. This kind of test method is incredibly valuable for productivity in manufacturing test generation.

14.6.3.2 Memory BIST On many chips, memories account for the majority of the transistors. A robust testing methodology must be applied to provide reliable parts. In a typical MBIST scheme, multiplexers are placed on the address, data, and control inputs for the memory to allow direct access during test. During testing, a state machine uses these multiplexers to directly write a checkerboard pattern of alternating 1s and 0s. The data is read back, checked, then the inverse pattern is also applied and checked. ROM testing is even simpler: The contents are read out to a signature analyzer to produce a syndrome.

14.6.3.3 Other On-Chip Test Strategies On-chip speeds are usually so high that directly observing internal behavior for testing can be difficult or impossible. Designers have included on-chip logic analyzers and oscilloscopes to deal with this problem [Weinlader00, Lee06, Noguchi07]. Such systems typically require a trigger signal to initiate data collection, a high speed timing generator, analog or digital sampling, and a buffer to store the results until they can be off-loaded at lower speed. A drawback is that the nodes to be observed must be selected at design time, and these may not be the problem circuits. Nevertheless, probing major busses and critical analog/RF nodes can be helpful. Also, on-chip scopes have been used to characterize power supply noise [Alon05, Naffziger06] and clock jitter [Nose06].

Analog/digital converter testing requires real-time access to the digital output of the ADC. Providing parallel digital test ports by reassigning pins on the chip I/O can facilitate this testing. If this is impossible, a "capture RAM" on chip can be used to capture results in real-time and then the contents can be transferred off-chip at a slower rate for analysis.

If both ADCs and DACs are present, a loopback strategy can be employed, as shown in Figure 14.21. Both analog and digital signals can loop back. Communication and graphics systems frequently have I/O systems that can be configured as shown. It is often worthwhile to add a DAC and an ADC to a system to allow a level of analog self-test.

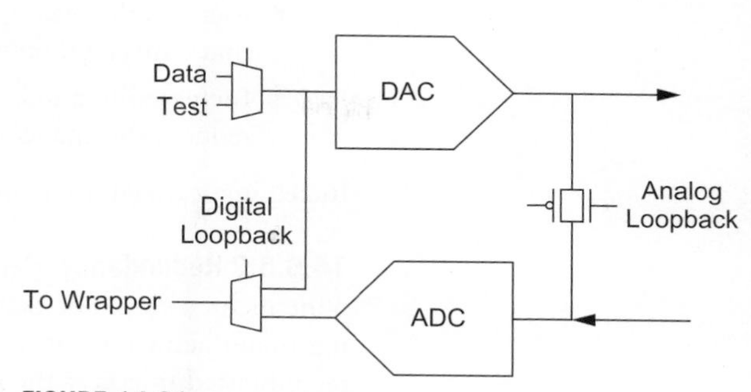

FIGURE 14.21 Analog and digital loopback

Providing on-chip debug circuitry involves quite a bit of imagination and forethought in terms of what might go wrong. It is often called "defensive design." Today, transistor counts and routing resources make it possible to include very sophisticated debug tools provided thought is given to the matter.

14.6.4 IDDQ Testing

Bridging faults were introduced in Section 14.5.1.2. A method of testing for bridging faults is called IDDQ test (V_{DD} supply current Quiescent) or supply current monitoring [Acken83, Lee92]. This relies on the fact that when a CMOS logic gate is not switching, it draws no DC current (except for leakage). When a bridging fault occurs, then for some combination of input conditions, a measurable DC I_{DD} will flow. Testing consists of applying the normal vectors, allowing the signals to settle, and then measuring I_{DD}. As potentially only one gate is affected, the IDDQ test has to be very sensitive. In addition, to be effective, any circuits that draw DC power such as pseudo-nMOS gates or analog circuits have to be disabled. Dynamic gates can also cause problems. As current measuring is slow, the tests must be run slower (of the order of 1 ms per vector) than normal, which increases the test time.

IDDQ testing can be completed externally to the chip by measuring the current drawn on the V_{DD} line or internally using specially constructed test circuits. This technique gives a form of indirect massive observability at little circuit overhead. However, as subthreshold leakage current increases, IDDQ testing ceases to be effective because variations in subthreshold leakage exceed currents caused by the faults.

14.6.5 Design for Manufacturability

Circuits can be optimized for manufacturability to increase their yield. This can be done in a number of different ways.

14.6.5.1 Physical At the physical level (i.e., mask level), the yield and hence manufacturability can be improved by reducing the effect of process defects. The design rules for particular processes will frequently have guidelines for improving yield. The following list is representative:

- Increase the spacing between wires where possible—this reduces the chance of a defect causing a short circuit.

- Increase the overlap of layers around contacts and vias—this reduces the chance that a misalignment will cause an aberration in the contact structure.
- Increase the number of vias at wire intersections beyond one if possible—this reduces the chance of a defect causing an open circuit.

Increasingly, design tools are dealing with these kinds of optimizations automatically.

14.6.5.2 Redundancy Redundant structures can be used to compensate for defective components on a chip. For example, memory arrays are commonly built with extra rows. During manufacturing test, if one of the words is found to be defective, the memory can be reconfigured to access the spare row instead. Laser-cut wires or electrically programmable fuses can be used for configuration. Similarly, if the memory has many banks and one or more are found to be defective, they can be disabled, possibly even under software control.

14.6.5.3 Power Elevated power can cause failure due to excess current in wires, which in turn can cause metal migration failures. In addition, high-power devices raise the die temperature, degrading device performance and, over time, causing device parameter shifts. The method of dealing with this component of manufacturability is to minimize power through design techniques described elsewhere in this text. In addition, a suitable package and heat sink should be chosen to remove excess heat.

14.6.5.4 Process Spread We have seen that process simulations can be carried out at different process corners. Monte Carlo analysis can provide better modeling for process spread and can help with centering a design within the process variations.

14.6.5.5 Yield Analysis When a chip has poor yield or will be manufactured in high volume, dice that fail manufacturing test can be taken to a laboratory for yield analysis to locate the root cause of the failure. If particular structures are determined to have caused many of the failures, the layout of the structures can be redesigned. For example, during volume production ramp-up for the Pentium microprocessor, the silicide over long thin polysilicon lines was found to crack and raise the wire resistance [Needham98]. This in turn led to slower-than-expected operation for the cracked chips. The layout was modified to widen polysilicon wires or strap them with metal wherever possible, boosting the yield at higher frequencies.

14.7 Boundary Scan

Up to this point we have concentrated on the methods of testing individual chips. Many system defects occur at the board level, including open or shorted printed circuit board traces and incomplete solder joints. At the board level, "bed-of-nails" testers historically were used to test boards. In this type of a tester, the board-under-test is lowered onto a set of test points (nails) that probe points of interest on the board. These can be sensed (the observable points) and driven (the controllable points) to test the complete board. At the chassis level, software programs are frequently used to test a complete board set. For instance, when a computer boots, it might run a memory test on the installed memory to detect possible faults.

The increasing complexity of boards and the movement to technologies such as surface mount technologies (with an absence of throughboard vias) resulted in system designers agreeing on a unified scan-based methodology called *boundary scan* for testing chips at the board (and system) level. Boundary scan was originally developed by the Joint Test Access Group and hence is commonly referred to as JTAG. Boundary scan has become a popular standard interface for controlling BIST features as well.

The IEEE 1149 boundary scan architecture [IEEE1149.1-01, Parker03] is shown in Figure 14.22. All of the I/O pins of each IC on the board are connected serially in a standardized scan chain accessed through the *Test Access Port* (TAP) so that every pin can be observed and controlled remotely through the scan chain. At the board level, ICs obeying the standard can be connected in series to form a scan chain spanning the entire board. Connections between ICs are tested by scanning values into the outputs of each chip and checking that those values are received at the inputs of the chips they drive. Moreover, chips with internal scan chains and BIST can access those features through boundary scan to provide a unified testing framework.

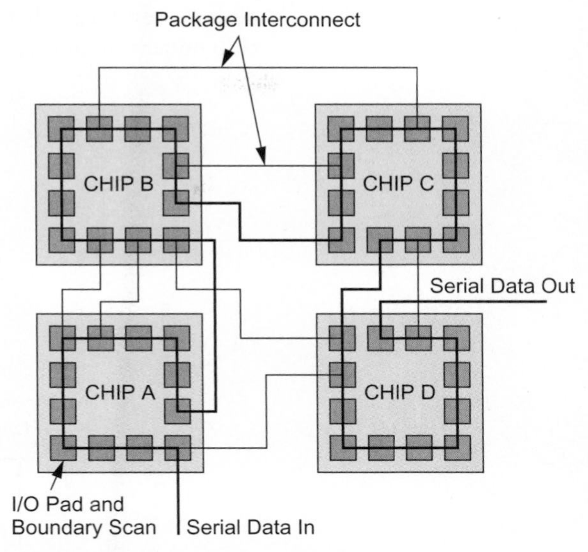

FIGURE 14.22 Boundary scan architecture

14.8 Testing in a University Environment

Industry environments are usually well-funded, and the appropriate testability tools are available to ensure a product-grade test effort. But what do you do in a university environment when the infrastructure might not be quite as affluent as in the industry setting? Not only may test tools be unavailable, but also the very act of building a test board can be a daunting extra amount of work on top of the chip design. The following are some tips that might help in this situation.

Taking the time to include circuitry to aid in testing on the chip is usually much easier than adding it at the board level. For a start, the integrated environment available for most IC design flows allows the designer to simulate the test circuitry. So, while it might seem superfluous to the task at hand, including test circuitry can save a huge amount of effort after the chip returns. Moreover, on-chip circuitry can often test at speeds that are impossible off-chip without extremely expensive production test machines. The main point is to think ahead.

Boundary scan and BIST greatly simplifies testing. If the chip has a standard boundary scan interface, it can be tested from a PC using a commercial boundary scan controller. For example, the Corelis NetUSB-1149.1/E can drive the scan chains at up to 80 MHz.

In the absence of BIST, there are several ways to test a chip. One is to breadboard or wirewrap a test board with switches for inputs and LEDs for outputs. This is tedious for all but the simplest chips. A custom-printed circuit board test fixture is even more labor-intensive, but often necessary for high-performance research chips. Another strategy is to use a logic analyzer with pattern generator. This approach requires a specialized test fixture to hold the chip and often has a steep learning curve for students, but it can perform tests at tens to hundreds of MHz. An increasingly popular method of testing digital chips is to design a test board that includes a large FPGA. The FPGA can drive test patterns to the chip under test and can store or analyze the responses. Figure 14.23 shows a typical setup.

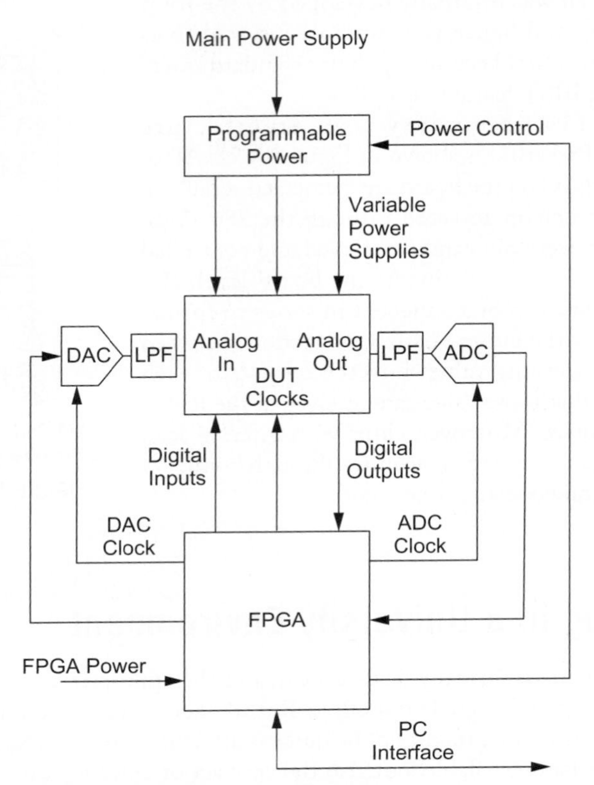

FIGURE 14.23 FPGA-assisted testing

14.9 Pitfalls and Fallacies

The following "war stories" are collected from real products at a wide variety of companies and published with permission, often under the condition of anonymity. They are presented to illustrate some of the pitfalls that can happen to smart people who are dealing with complex systems on a tight schedule. The skilled engineer learns from these mistakes; in most cases, the company extended their verification flow to ensure that similar problems would be caught before wreaking havoc on future products. Could one of these happen to you?

A product in the field hangs unpredictably

A microprocessor had been in the field for several years when reports began arriving from major customers that certain programs would cause the system to hang at unpredictable times with intervals of hours to days. The manufacturer appointed a tiger team to resolve the error. The hang rate proved to be insensitive to power supply voltage, operating temperature, and clock rate. It was observed on all versions of the chip regardless of foundry, manufacturing technology, or motherboard. The programs that failed all involved a mix of floating point and integer operations, not just integer codes.

After several months of work, the issue was isolated to a particular unit in the processor. By this point, 30 engineers were involved in chasing the problem. Picoprobing showed that when the hang occurred, an instruction was left stuck in the pipeline waiting to issue. A logic simulation of the RTL is much slower than running the actual code, but an engineer developed a simple test case that could trigger the hang on real hardware in a matter of seconds, and thus it could trigger the failure in simulation in a practical amount of time. Simulations showed that the RTL ran flawlessly, suggesting the error involved a circuit that did not match the RTL.

On this processor, the circuits had been verified against the RTL using a technique called "shadow-mode simulation." A "circuit understanding" tool parsed the transistor-level netlist into gates and identified the logic function of each gate. Circuits were verified to match the RTL by replacing a module of the RTL with the corresponding extracted circuit and simulating to check that the system produced identical results as the original RTL. The simulation is time-consuming, so each module is typically checked over tens of thousands of cycles, rather than the billions of cycles used in primary RTL verification.

A shadow-mode simulation using circuits from the failing unit still ran flawlessly. However, an engineer observed that a long wire crossing a large schematic was driven from both ends to reduce the RC delay. The signals X1 and X2 driving each end were intended to be identical (Figure 14.24). The engineer experimented with splitting the wire and checking that both drivers produced identical results, and on certain test cases they did not. This led to the wire experiencing contention and being driven to an indeterminate logic value. The invalid result propagated through other logic and hung the processor. Unfortunately, the circuit-understanding tool had incorrectly determined that the logic for the two ends was identical and had never detected the error. Even if the tool had been correct, the original test cases never would have exercised the patterns that caused the drivers to produce different results. A simple modification to the driver fixed the problem, but many units were already in the field. Fortunately, a software patch was developed to prevent the operations that caused the hang from ever being issued.

FIGURE 14.24 Long wire driven from both ends

Hanging is a serious problem, but not as severe as unknowingly calculating the wrong answer. After the problem was corrected, engineers spent several more weeks proving to customers that the failure mode would hang the machine but could never result in an incorrect calculation.

To avoid repeating this problem in the future, engineers have turned to formal verification tools that prove that RTL and schematics are equivalent in their Boolean function. Such tools are not susceptible to incomplete test patterns. However, the tools are often expensive, proprietary, and difficult to use.

A product fails after the manufacturing process matures

A team designing a data communications product was comfortable with a particular microprocessor that was at the end of its production run. The team negotiated to order several thousand units of the discontinued microprocessor before production was shut down. The data communications product became successful and was shipped in large quantity. After it had been in the field for some time, major customers reported that the product would crash in large networks. These customers included large financial, government, and Internet service provider organizations who were adversely affected by the crashes. It took the data communications company weeks to isolate the problem to hanging of the microprocessor, and then a team of engineers at the microprocessor company began investigating the issue.

The microprocessor team investigated potential signal and power supply integrity issues. Although no signal integrity problems were apparent, a shmoo plot showed unusual sensitivity of minimum clock period to supply voltage. An engineer had recently read the application note for the power regulator on the system board and had learned that it had a propensity for oscillation if not properly bypassed. The system board lacked the bypass capacitors recommended in the application note, so the engineer wrote a memo to the product manager suggesting a change to the board. The memo was misinterpreted as a solution to the problem and customers were informed that a fix was on its way. Unfortunately, further testing showed that bypassing the regulator did not fix the crashes.

When the system crashed, it wrote its state to a core file. An engineer began reading a hexadecimal dump of the file and noticed a pattern that led to solving the crash. The pattern was associated with simultaneous access to many banks in an eight-way associative instruction cache. The cache had fuses associated with each bank, so banks containing bad blocks could be disabled during manufacturing test. During original product debug, the manufacturing process was relatively immature and most processors only had five operational cache banks. However, the processors manufactured at the end of the production run were built on a more mature process and often had all eight banks functional. Simultaneous access to all the banks tickled a signal integrity problem, resulting in power supply droop from excessive *IR* drops caused by poor contacts to the V_{DD} plane. The solution was a software change to disable three of the banks at system startup.

Better power supply analysis is performed to avoid repeating this problem.

A wasted spin

A microprocessor was taped out and came back nearly fully operational. Minor changes were made to the layout and documentation was developed; then a second revision (colloquially called a second *spin* of the chip) was taped out. The second revision came back completely nonfunctional, with a short between power and ground. Optical inspection while manufacturing the polysilicon layer showed that there was no field oxide on the chip.

Inspection of the masks showed that the active area mask specified active area (i.e., diffusion) for the entire chip rather than just where transistors belonged. The layout tool assigned each layer—such as active area or metal1—a unique number. However, although the layout for active area layer was correct, the mask did not appear to match the active layer.

Layout documentation had been annotated on an unused layer by drawing rectangles and text to indicate functional blocks. A larger rectangle defined the entire chip area. Careful tracing of the mask-generation software found that the "unused" layer had been used for active area many years ago and that the documentation rectangles were merged with the true active area to form a blob of active covering the entire chip.

Another microprocessor from a different vendor also failed when it was first built. Visual inspection of the die showed that the entire cache was missing. The cache had been removed from the design database to speed up final verification because it had already been checked separately. An engineer neglected to put it back in before tapeout.

Both of these wasted fabrication runs could have been avoided by using more rigorous verification methods at both the design and mask fabrication facilities. Validation of dataset size by the designer would have caught the missing geometries. Use of the industry standard mask database inspection tools would have caught the error after mask build. Although in the past, fabrication of a modest number of parts for testing was a small part of the design cost, with the escalation of mask and wafer fabrication costs, these mistakes can be a multimillion-dollar error. The extra time to market has a large opportunity cost as well.

At high voltage, a chip only operates at low frequency

While booting the operating system during silicon debug, a microprocessor operated as expected at low voltage. At high voltage, the part only functioned at low frequency. The high-voltage roof is an indication of a potential coupling problem in which the coupling is exacerbated by the fast edge rates associated with high-voltage operation. Test cases revealed that the problem resulted from incorrect operation of the register file when certain instructions executed. When the designers inspected the scan latches, they found that the correct 0 value was sent to the register file to write, but that an incorrect 1 was read. This indicated that either read or write operation was failing at high voltage. Trying one operation at high voltage and the other at low voltage proved the problem was in the write path.

A schematic of the register file write circuitry is shown in Figure 14.25. The register file uses predischarged write bitlines that are conditionally pulled high, depending on the data. The appropriate cell is written by turning on the corresponding write access transistor. The register cell is intentionally unstable so that the value on the bitline can overpower the cell and write the appropriate value. A weak keeper holds the metal2 bitline low when writing a 0. However, the register file is large and the keeper is at the opposite end from the data transistor. The resistance of the long, thin wire further reduces the effectiveness of the keeper against noise on the bitline.

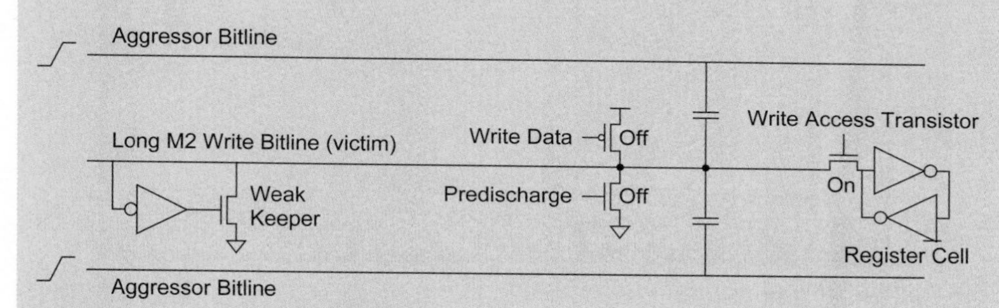

FIGURE 14.25 Register file write circuitry

When the neighboring bitlines switch high, they couple onto the victim line and tend to pull it high. The circuit fails if the aggressors introduce too much coupling noise. At high voltage, the aggressor drivers are stronger and cause a momentary glitch on the victim. At low frequency, the keeper is sufficient to restore the victim to a low level.

The coupling problem had been flagged during design by an automated noise-checking tool. However, the tool is conservative and the area of the register file would have increased significantly if the bitlines were spaced far enough apart to satisfy the tool. Therefore, the designer checked for excessive coupling with a SPICE simulation. The simulation apparently did not properly model the combination of circumstances that caused the failure. A second engineer cross-checked all circuits that waived the noise-checker warning, but also did not discover the excessive coupling. The problem was solved by placing a second keeper near the write data transistor to fight against the coupling.

Another funny shmoo

During silicon debug, a microprocessor cache only functioned correctly over the peculiar range of voltages and frequencies shown in the shmoo[1] in Figure 14.26. Test code exercising the

[1]A shmoo of this type is sometimes called a *flying saucer.*

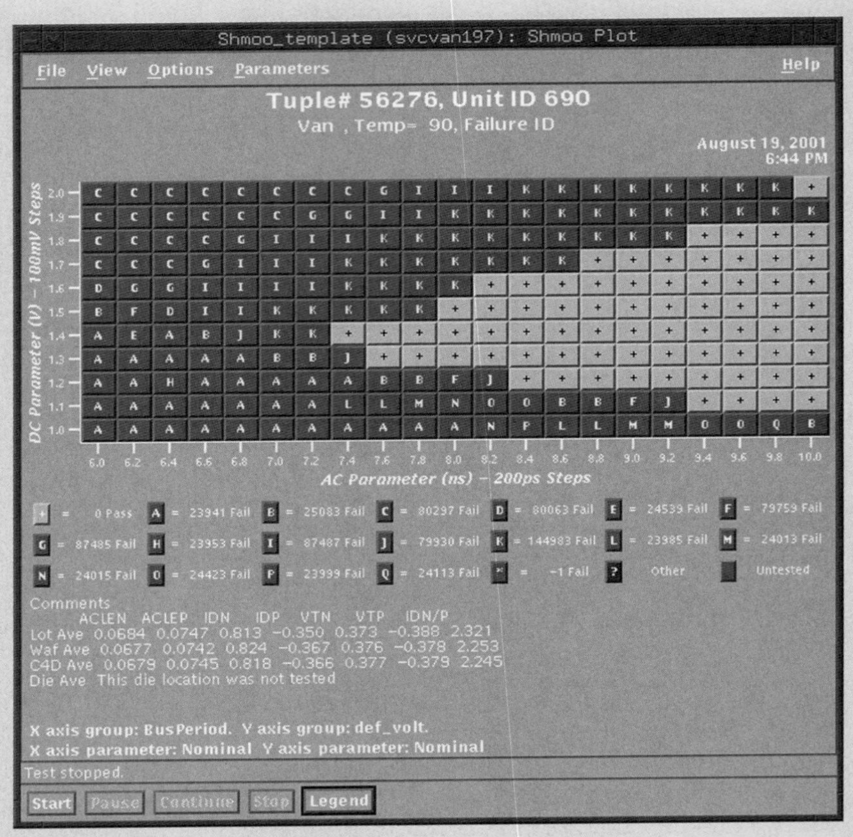

FIGURE 14.26 Flying saucer shmoo

cache revealed that failures were caused by bad data being read from the cache. Scan isolated the problem to a dynamic multiplexer choosing one of the global bitlines, as shown in Figure 14.27.

The multiplexer inputs were the NORs of dynamic metal3 global bitlines and corresponding select signals. The metal4 select lines were early and did not need to be dynamic, but were implemented as dynamic nodes anyway. All of the transistors in the dynamic multiplexer were supposed to remain OFF in this particular test case, leaving the multiplexer output high.

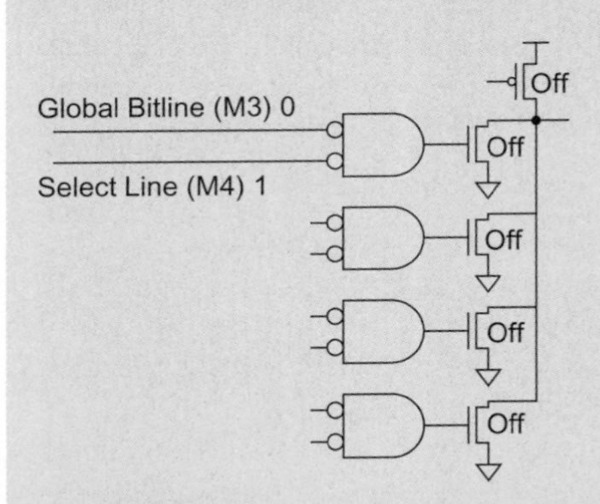

FIGURE 14.27 Dynamic bitline multiplexer

One input of the multiplexer had a low value on the global bitline, but was not selected, as shown. Therefore, the transistor should have been OFF. Nevertheless, the output of the multiplexer incorrectly discharged. One neighbor of the select line was ground; the other fell low. Coupling from a single neighbor is generally not enough to cause noise failure. However, many global bitlines ran over the top of the select line and also fell low. Laser voltage probing showed that the select line was incorrectly pulled low, apparently from coupling caused by these falling bitlines as well as the neighbor line. The odd shape of the shmoo happened because the failures only occurred when the neighbor and overhead lines both fell at about the same time; otherwise, the keeper on the select line was strong enough to recover from one noise event before the other arrived. Because the bitline and control paths were different, the noise events only happened simultaneously for certain voltages.

Noise analysis tools usually check only neighbors, and the single switching neighbor was not sufficient to trigger an error. In this circumstance, so many global bitlines ran over the top of the select wire that their coupling could not be neglected. The problem was fixed by converting the control line into a static signal more resistant to coupling noise. A better noise analyzer could have considered coupling from neighbors above and below, especially on dynamic nets. However, it is difficult to extract information about such orthogonal neighbors because they are often drawn at different levels of the layout hierarchy. Moreover, assuming all neighbors switch in the worst possible direction is usually pessimistic for long wires. Nevertheless, such a data-dependent failure mechanism is a source of nightmares for designers.

Incorrect operation at low temperature

A floating-point coprocessor was tested by running the LINPACK benchmark. The benchmark performs a series of floating-point operations and generates a checksum to verify the result. The chip would occasionally produce the wrong checksum. One of the engineers heated the coprocessor by removing the heat sink and found that the coprocessor became reliable at higher temperature.

This suggested that the problem might be caused by coupling, which is generally more serious at lower temperature where the edge rates are faster. The error was tracked to a long on-chip bus with many wires laid out on a tight pitch. Although the wires were subject to coupling noise, they were not on a critical path and should have had plenty of time to settle to the correct value. Unfortunately, they drove the diffusion input of a latch. When crosstalk drove an input below $-V_t$, it would turn on the pass transistor and incorrectly discharge the latch (see Section 8.3.9).

The floating-point unit bug was holding up lucrative product shipments. While a corrected coprocessor was being fabricated, the old unit was shipped in products with a bolt-on thermostat/heater unit used to guarantee a minimum operating temperature.

An obvious lesson of this experience is to avoid driving diffusion inputs with potentially noisy signals. More fundamentally, however, this bug demonstrated a marginal design of the cell library that should have been caught in the library review. Moreover, humans are inherently prone to errors. Electrical rules like no noisy diffusion inputs aren't worth the paper they are printed on unless computer code exists to enforce them.

Slower than expected performance

An application-specific integrated circuit (ASIC) was fabricated on a gate array by a third-party gate array manufacturer. Although static timing analysis predicted that the chip would function fast enough, the manufacturer found that most of the chips would not operate at the desired frequency and instead had to be derated by about 20%.

The designer examined a die plot, looking for the source of the unexpectedly slow performance. The plot showed that the horizontal power and ground lines were only strapped along the edges of the chip, as shown in Figure 14.28(a). Some rows of gates consumed large amounts of power, causing large *IR* drops along their power lines. Measurements showed that the power supply sometimes drooped below 2 V, despite the nominal 3.3 V power supply. When the wide vertical power supply straps were added, as shown in Figure 14.28(b), most chips met target speed.

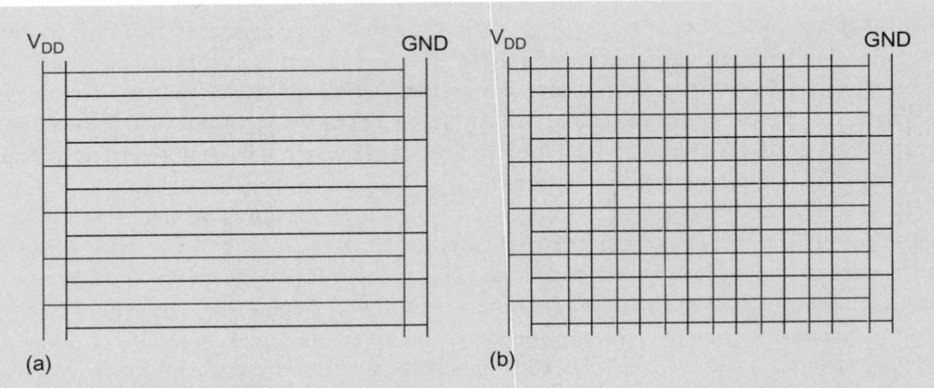

FIGURE 14.28 Power supply network

Modern chips require low-resistance on-chip power distribution networks and often use power and ground pads distributed across the die rather than just at the periphery to reduce the distance and resistance between the pads and the gates. Power integrity analysis should be performed to verify that the static or dynamic voltage droops remain within their budget everywhere on the chip.

Class chip failures

One of the authors has supervised a number of class project chips. The following are some of the reasons that chips have come back partially or completely nonfunctional:

- *Insufficient simulation*
  A ring oscillator was placed on the chip as a test structure to verify that the hardware was at least partially functional even if the rest of the chip might not work. It didn't oscillate. It had not been simulated because it was "too simple." Inspection during debug found that the oscillator had an even number of inverters!
  Another chip was designed with a new CAD tool that had a buggy simulator. Most of the chip operated correctly, but the chip as a whole would not simulate. The problem was attributed to a bug in the simulator and was taped out anyway. The chip came back nonfunctional.
- *Incomplete top-level verification*
  One year, a pad frame was used that was incompatible with the normal verification flow. The chip cores were verified, placed in the pad frame, and then routed to the pads. DRC and simulation were not performed on the connections to the pads, so students carefully scrutinized their routing by hand. Upon testing, three of the four different designs were found to have errors in the routing to the pads. No errors were found in the cores that had been verified. "If you don't test it, it won't work! (guaranteed)."
  — A neural network chip seemed to have a defective scan chain because the scan data out line never budged from 0 as configuration data was scanned into the chip. Testing found that the chip was correctly configured except in the last bit of the scan chain. Inspection of the layout revealed that the scan data out line (which came from the last bit of the scan chain) had been shorted to ground while being routed to the pads.

— A carry-lookahead adder produced incorrect results on certain input patterns. The least significant bits were always correct. Inspection of the layout revealed that the A[4] input was routed from the pad most of the way to the core but part of the wire was missing, probably because the designer accidentally hit UNDO after finishing the route.

— A GPS searcher chip had an inverter connected to a pair of pins to verify that the chip showed basic functionality. The output was stuck low. Inspection of the layout revealed that the input was attached to an output pad and the output to an input pad. The GPS searcher itself was fully operational.

While some of these may represent class situations, the same type of reasons for partial failure also plague industry chips. In particular, when time scales are stressed, the boundary conditions are often overlooked, which leads to problems when the chips are fabricated. Once a good verification methodology is put in place that includes a known-good pad frame, top-level DRC, and full-chip simulation, students have had a 100% success rate on class chips.

Summary

This chapter has summarized the important issues in CMOS chip testing and has provided some methods for incorporating test considerations into chips from the start of the design. Scan is now an indispensable technique to observe and control registers because probing signals directly has become extremely difficult. The importance of writing adequate tests for both the functional verification and manufacturing verification cannot be understated. It is probably the single most important activity in any CMOS chip design cycle and usually takes the longest time no matter what design methodology is used. If one message is left in your mind after reading this chapter, it should be that you are absolutely rigorous about the testing activity surrounding a chip project and it should rank first among any design trade-offs.

Exercises

14.1 You have to design an extremely fast divide by eight frequency divider that taxes the capabilities of the process you are using. What test strategy would you employ to test the divider? Explain the reasons for your choice.

14.2 Design a register that minimizes transistor count, but allows parallel scan to be implemented, as outlined in Figure 14.17.

14.3 Explain how a Pseudo-Random Sequence Generator (PRSG) can be used to test a 16-bit datapath. How would the outputs be collected and checked?

14.4 Design a block diagram of a test generator for a 4K × 32 static RAM.

14.5 Research the origin of the term "shmoo."

14.6 You have to test a large die (1 cm × 1 cm) that is housed in a package that costs $5. Would you do wafer testing? Why?

14.7 A verification script detects a single discrepancy between the golden model and your design out of 400,000 vectors. Would you proceed to fabrication? Explain your decision.

14.8 Explain what is meant by a Stuck-at-1 fault and a Stuck-at-0 fault.

14.9 How are sequential faults caused in CMOS? Give an example.

14.10 Explain the different kinds of physical faults that can occur on a CMOS chip and relate them to typical circuit failures.

14.11 Explain the terms controllability, observability, and fault coverage.

14.12 Why is it important to have a high fault coverage for a set of test vectors?

14.13 Explain how serial-scan testing is implemented.

14.14 Explain the principles of Built-In Self-Test (BIST). What are the advantages and disadvantages of BIST?

14.15 A circuit does not operate at the desired frequency. Cooling the circuit with freeze spray fixes the problem. A shmoo shows the circuit operates correctly at higher than nominal V_{DD}. What is the general nature of the likely problem and why?

Fabrication 15

15.1 Introduction

Chapter 1 summarized the steps in a basic CMOS process. These steps are expanded upon in this chapter. Where possible, the processing details are related to the way CMOS circuits and systems are designed. Modern CMOS processing is complex, and while coverage of every nuance is not within the scope of this book, we focus on the fundamental concepts that impact design.

A fair question from a designer would be "Why do I care how transistors are made?" In many cases, if designers understand the physical process, they will comprehend the reason for the underlying design rules and in turn use this knowledge to create a better design. Understanding the manufacturing steps is also important when debugging some difficult chip failures and improving yield.

Fabrication plants, or *fabs*, are enormously expensive to develop and operate. In the early days of the semiconductor industry, a few bright physicists and engineers could bring up a fabrication facility in an industrial building at a modest cost and most companies did their own manufacturing. Now, a fab processing 300 mm wafers in a 45 nm process costs about $3 billion. The research and development underlying the technology costs another $2.4 billion. Only a handful of companies in the world have the sales volumes to justify such a large investment. Even these companies are forming consortia to share the costs of technology development with their market rivals. Some companies, such as TSMC, UMC, Chartered, and IBM operate on a *foundry* model, selling space on their fab line to fabless semiconductor firms. Figure 15.1 shows workers and machinery in the cavernous clean room at IBM's East Fishkill 300 mm fab.

Recall that silicon in its pure or *intrinsic* state is a semiconductor, having bulk electrical resistance somewhere between that of a conductor and an insulator. The conductivity of silicon can be raised by several orders of magnitude by introducing *impurity* atoms into the silicon crystal lattice. These dopants can supply either free electrons or holes. Group III impurity elements such as boron that use up electrons are referred to as *acceptors* because they accept some of the electrons already in the silicon, leaving holes. Similarly, Group V *donor* elements such as arsenic and phosphorous provide electrons. Silicon that contains a majority of donors is known as *n-type*, while silicon that contains a majority of acceptors is known as *p-type*. When n-type and p-type materials are brought together, the region where the silicon changes from n-type to p-type is called a *junction*. By arranging junctions in certain physical structures and combining them with wires and insulators, various semiconductor devices can be constructed. Over the years, silicon semiconductor processing has evolved sophisticated techniques for building these junctions and other insulating and conducting structures.

FIGURE 15.1 IBM, East Fishkill, NY fab (Courtesy of International Business Machines Corporation. Unauthorized use not permitted.)

The chapter begins with the steps of a generic process characteristic of commercial 65 nm manufacturing. It also surveys a variety of process enhancements that benefit certain applications. The chapter examines layout design rules in more detail and discusses layout CAD issues such as design rule checking.

15.2 CMOS Technologies

CMOS processing steps can be broadly divided into two parts. Transistors are formed in the *Front-End-of-Line* (FEOL) phase, while wires are built in the *Back-End-of-Line* (BEOL) phase. This section examines the steps used through both phases of the manufacturing process.

15.2.1 Wafer Formation

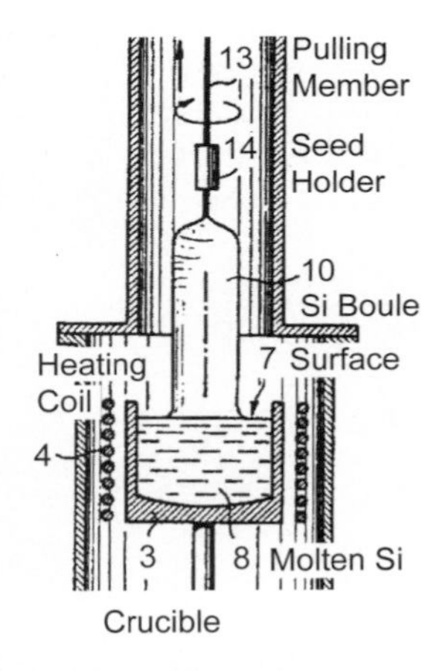

FIGURE 15.2 Czochralski system for growing Si boules (Adapted from [Schulmann98].)

The basic raw material used in CMOS fabs is a *wafer* or disk of silicon, roughly 75 mm to 300 mm (12″—a dinner plate!) in diameter and less than 1 mm thick. Wafers are cut from *boules*, cylindrical ingots of single-crystal silicon, that have been pulled from a crucible of pure molten silicon. This is known as the *Czochralski* method and is currently the most common method for producing single-crystal material. Controlled amounts of impurities are added to the melt to provide the crystal with the required electrical properties. A seed crystal is dipped into the melt to initiate crystal growth. The silicon ingot takes on the same crystal orientation as the seed. A graphite radiator heated by radio-frequency induction surrounds the quartz crucible and maintains the temperature a few degrees above the melting point of silicon (1425 °C). The atmosphere is typically helium or argon to prevent the silicon from oxidizing.

The seed is gradually withdrawn vertically from the melt while simultaneously being rotated, as shown in Figure 15.2. The molten silicon attaches itself to the seed and recrystallizes as it is withdrawn. The seed withdrawal and rotation rates determine the diameter of the ingot. Growth rates vary from 30 to 180 mm/hour.

15.2.2 Photolithography

Recall that regions of dopants, polysilicon, metal, and contacts are defined using masks. For instance, in places covered by the mask, ion implantation might not occur or the dielectric or metal layer might be left intact. In areas where the mask is absent, the implantation can occur, or dielectric or metal could be etched away. The patterning is achieved by a process called *photolithography*, from the Greek *photo* (light), *lithos* (stone), and *graphe* (picture), which literally means "carving pictures in stone using light." The primary method for defining areas of interest (i.e., where we want material to be present or absent) on a wafer is by the use of *photoresists*. The wafer is coated with the photoresist and subjected to selective illumination through the *photomask*. After the initial patterning of photoresist, other barrier layers such as polycrystalline silicon, silicon dioxide, or silicon nitride can be used as physical masks on the chip. This distinction will become more apparent as this chapter progresses.

A photomask is constructed with chromium (chrome) covered quartz glass. A UV light source is used to expose the photoresist. Figure 15.3 illustrates the lithography process. The photomask has chrome where light should be blocked. The UV light floods the mask from the backside and passes through the clear sections of the mask to expose the organic photoresist (PR) that has been coated on the wafer. A developer solvent is then used to dissolve the soluble unexposed photoresist, leaving islands of insoluble exposed photoresist. This is termed a *negative* photoresist. A *positive* resist is initially insoluble, and when exposed to UV becomes soluble. Positive resists provide for higher resolution than negative resists, but are less sensitive to light. As feature sizes become smaller, the photoresist layers have to be made thinner. In turn, this makes them less robust and more subject to failure which can impact the overall yield of a process and the cost to produce the chip.

The photomask is commonly called a *reticle* and is usually smaller than the wafer, e.g., 2 cm on a side. A *stepper* moves the reticle to successive locations to completely expose the wafer. *Projection printing* is normally used, in which lenses between the reticle and wafer focus the pattern on the wafer surface. Older techniques include *contact printing*, where the mask and wafer are in contact, and *proximity printing*, where the mask and wafer are close but not touching. The reticle can be the same size as the area to be patterned (1×) or larger. For instance, 2.5× and 5× steppers with optical reduction have been used in the industry.

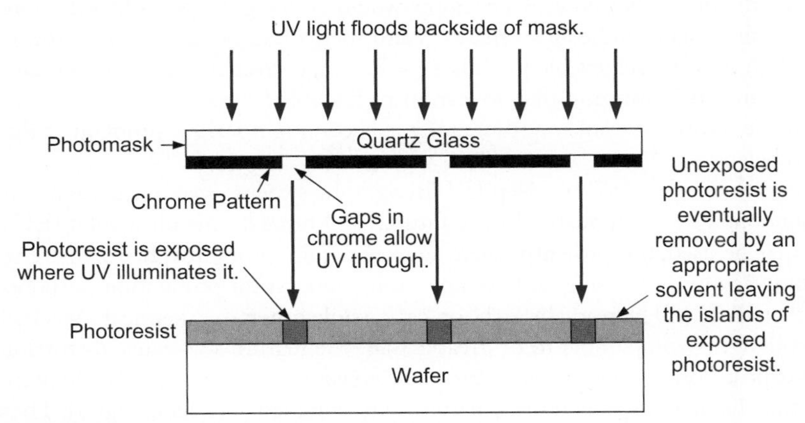

FIGURE 15.3 Photomasking with a negative resist (lens system between mask and wafer omitted to improve clarity and avoid diffracting the reader ☺)

The wavelength of the light source influences the minimum feature size that can be printed. Define the minimum *pitch* (width + spacing) of a process to be $2b$. The resolution of a lens depends on the wavelength λ of the light and the *numerical aperture* NA of the lens:

$$2b = k_1 \frac{\lambda}{\text{NA}} \tag{15.1}$$

The numerical aperture is

$$\text{NA} = n \sin \alpha \tag{15.2}$$

where n is the refractive index of the medium (1 for air, 1.33 for water, and up to 1.5 for oil), and α is the angle of acceptance of the lens. Increasing α requires larger optics. Lenses used in the 1970s had a numerical aperture of 0.2. Intel uses a numerical aperture of 0.92 for their 45 nm process [Mistry07]. Nikon and ASML broke the 1.0 barrier by introducing *immersion lithography* that takes advantage of water's higher refractive index [Geppert04], and in 2008, NA = 1.35 had been reached. All of these advances have come at the expense of multimillion dollar optics systems. k_1 depends on the coherence of the light, antireflective coatings, photoresist parameters, and resolution enhancement techniques. Presently, 0.8 is considered easy, while 0.5 is very hard.

The *depth of focus* is

$$\text{DOF} = \frac{k_2 \lambda}{\text{NA}^2} \tag{15.3}$$

where k_2 ranges from 0.5 to 1. Advanced lithography systems with short wavelengths and large numerical apertures have a very shallow depth of focus, requiring that the surface of the wafer be maintained extremely flat.

In the 1980s, mercury lamps with 436 nm or 365 nm wavelengths were used. At the 0.25 μm process generation, excimer lasers with 248 nm (deep ultraviolet) were adopted and have been used down to the 180 nm node. Currently, 193 nm argon-fluoride lasers are used for the *critical layers* down to the 45 nm node and beyond. The critical layers are those that define the device behavior. An example would be the gate (polysilicon), source/drain (diffusion), first metal, and contact masks. With such a laser, a numerical aperture of 1.35, and $k_1 = 0.5$, the best achievable pitch is $2b = 72$ nm, corresponding to a polysilicon half-pitch of 36 nm. It is amazing that we can print features so much smaller than the wavelength of the light, but even so, lithography is becoming a serious problem at the 45 nm node and below.

Efforts to develop 157 nm deep UV lithography systems were unsuccessful and have been abandoned by the industry. In the future, 13.5 nm extreme ultraviolet (EUV) light sources may be used, but presently, these sources require prohibitively expensive reflective optics and vacuum processing and are not strong enough for production purposes. Some predict that EUV will be ready by 2011 or 2012, while others are skeptical [Mack08].

Wavelengths comparable to or greater than the feature size cause distortion in the patterns exposed on the photoresist. *Resolution enhancement techniques* (RETs) precompensate for this distortion so the desired patterns are obtained [Schellenberg03]. These techniques involve modifying the amplitude, phase, or direction of the incoming light. The ends of a line in a layout receive less light than the center, causing nonuniform exposure. *Optical proximity correction* (OPC) makes small changes to the patterns on the masks to compensate for these local distortions. Figure 15.4 shows an example of printing with and without optical proximity correction. OPC predistorts the corners to reduce undesired rounding. *Phase shift masks* (PSM) takes advantage of the diffraction grating effect of parallel lines on a mask, varying the thickness of the mask to change the phase such that light from adjacent lines are out of phase and cancel where no light is desired. *Off-axis illumination* can also improve contrast for certain types of dense, repetitive patterns. *Double-patterning* is a sequence of two precisely aligned exposure steps with different masks for the same photoresist layer [Mack08]. OPC became necessary at the 180 nm node and all of these techniques are in heavy use by the 45 nm node.

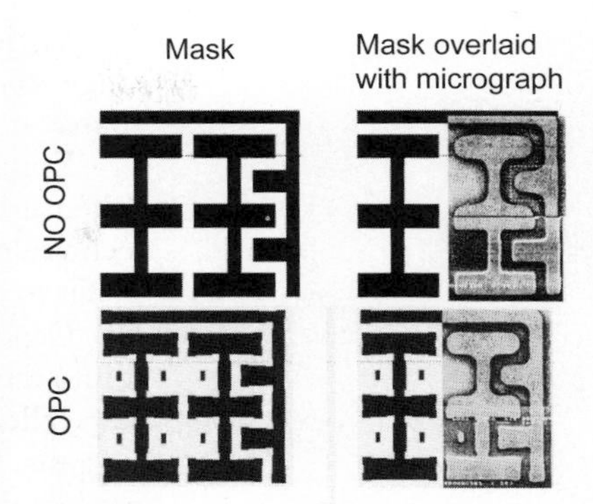

FIGURE 15.4 Subwavelength features printed with and without OPC. Predistortion of corners in OPC reduces undesired rounding. (Adapted from [Schellenberg98] with permission of SPIE.)

Each successive UV stepper is more expensive and the throughput of the stepper may decrease. This is just another contributory issue to the spiraling cost of chip manufacturing. The cost of masks is also skyrocketing, forcing chip designers to amortize design and mask expenses across the largest volume possible. This theme will be reinforced in Section 13.3.

15.2.3 Well and Channel Formation

The following are main CMOS technologies:

- n-well process
- p-well process
- twin-well process
- triple-well process

Silicon-on-insulator processes are also available through some manufacturers (see Section 15.4.1.2).

Chapter 1 outlined an n-well process. Historically, p-well processes preceded n-well processes. In a p-well process, the nMOS transistors are built in a p-well and the pMOS transistor is placed in the n-type substrate. p-well processes were used to optimize the pMOS transistor performance. Improved techniques allowed good pMOS transistors to be fabricated in an n-well and excellent nMOS transistors to be fabricated in the p-type substrate of an n-well process. In the n-well process, each group of pMOS transistors in an n-well shares the same body node but is isolated from the bodies of pMOS transistors in different wells. However, all the nMOS transistors on the chip share the same body, which is the substrate. Noise injected into the substrate by digital circuits can disturb sensitive analog or memory circuits. Twin-well processes accompanied the emergence of n-well processes. A twin-well process allows the optimization of each transistor type. A third well can be added to create a triple-well process. The triple-well process has emerged to provide good isolation between analog and digital blocks in mixed-signal chips; it is also used to isolate high-density dynamic memory from logic. Most fabrication lines provide a baseline twin-well process that can be upgraded to a triple-well process with the addition of a single mask level.

Wells and other features require regions of doped silicon. Varying proportions of donor and acceptor dopants can be achieved using *epitaxy*, *deposition*, or *implantation*.

Epitaxy involves growing a single-crystal film on the silicon surface (which is already a single crystal) by subjecting the silicon wafer surface to an elevated temperature and a source of dopant material.

Epitaxy can be used to produce a layer of silicon with fewer defects than the native wafer surface and also can help prevent latchup (see Section 6.3.6). Foundries may provide a choice of *epi* (with epitaxial layer) or non-epi wafers. Microprocessor designers usually prefer to use epi wafers for uniformity of device performance.

Deposition involves placing dopant material onto the silicon surface and then driving it into the bulk using a thermal diffusion step. This can be used to build deep junctions. A step called *chemical vapor deposition* (CVD) can be used for the deposition. As its name suggests, CVD occurs when heated gases react in the vicinity of the wafer and produce a product that is deposited on the silicon surface. CVD is also used to lay down thin films of material later in the CMOS process.

Ion implantation involves bombarding the silicon substrate with highly energized donor or acceptor atoms. When these atoms impinge on the silicon surface, they travel below the surface of the silicon, forming regions with varying doping concentrations. At elevated temperature (>800 °C) diffusion occurs between silicon regions having different densities of impurities, with impurities tending to diffuse from areas of high concentration to areas of low concentration. Therefore, it is important to keep the remaining process steps at as low a temperature as possible once the doped areas have been put into place. However, a high-temperature annealing step is often performed after ion implantation to redistribute dopants more uniformly. Ion implantation is the standard well and source/drain implant method used today. The placement of ions is a random process, so doping levels cannot be perfectly controlled, especially in tiny structures with relatively small numbers of dopant atoms. Statistical dopant fluctuations lead to variations in the threshold voltage that will be discussed in Section 6.5.2.2.

The first step in most CMOS processes is to define the well regions. In a triple-well process, a deep n-well is first driven into the p-type substrate, usually using high-energy Mega electron volt levels (MeV) ion implantation as opposed to a thermally diffused operation. This avoids the thermal cycling (i.e., the wafers do not have to be raised significantly in temperature), which improves throughput and reliability. A 2–3 MeV implantation can yield a 2.5–3.5 μm deep n-well. Such a well has a peak dopant concentration just under the surface and for this reason is called a *retrograde* well. This can enhance device performance by providing improved latchup characteristics and reduced susceptibility to vertical punch-through (see Section 6.3.5). A thick (3.5–5.5 μm) resist has to be used to block the high energy implantation where no well should be formed. Thick resists and deep implants necessarily lead to fairly coarse feature dimensions for wells, compared to the minimum feature size. Shallower n-well and p-well regions are then implanted. After the wells have been formed, the doping levels can be adjusted (using a *threshold implant*) to set the desired threshold voltages for both nMOS and pMOS transistors. With multiple threshold implant masks, multiple V_t options can be provided on the same chip. For a given gate and substrate material, the threshold voltage depends on the doping level in the substrate (N_A), the oxide thickness (t_{ox}), and the surface state charge (Q_{fc}). The implant can affect both N_A and Q_{fc} and hence V_t. Figure 15.5 shows a typical triple-well structure. As discussed, the nMOS transistor is situated in the p-well located in the deep n-well. Other nMOS transistors could be built in different p-wells so that they do not share the same body node. Transistors in a p-well in a triple-well process will have different characteristics than transistors in the substrate because of the different doping levels. The pMOS

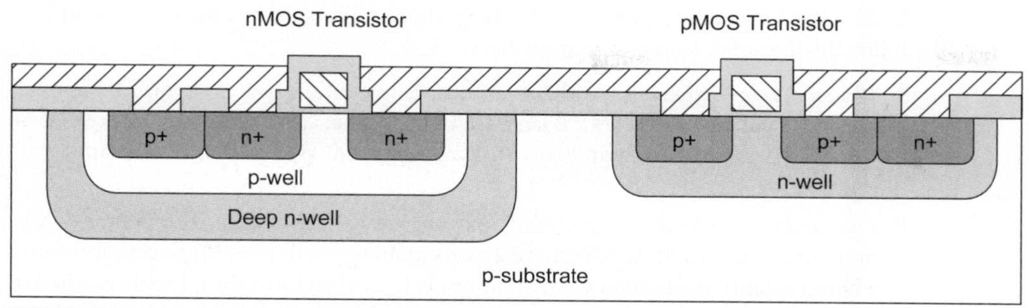

FIGURE 15.5 Well structure in triple-well process

transistors are located in the shallow (normal) n-well. The figure shows the cross-section of an inverter.

Wells are defined by separate masks. In the case of a twin-well process, only one mask need be defined because the other well by definition is its complement. Triple-well processes have to define at least two masks, one for the deep well and the other for either n-well or p-well.

Transistors near the edge of a retrograde well (e.g., within about 1 μm) may have different threshold voltages than those far from the edge because ions scatter off the photoresist mask into the edge of the well, as shown in Figure 15.6 [Hook03]. This is called the *well-edge proximity effect.*

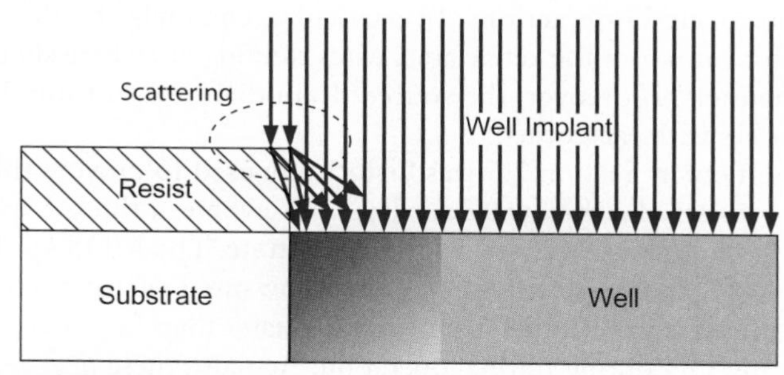

FIGURE 15.6 Well-edge proximity effect, in which dopants scattering off photoresist increase the doping level near the edge of a well (© IEEE 2003.)

15.2.4 Silicon Dioxide (SiO_2)

Many of the structures and manufacturing techniques used to make silicon integrated circuits rely on the properties of SiO_2. Therefore, reliable manufacture of SiO_2 is extremely important. In fact, unlike competing materials, silicon has dominated the industry because it has an easily processable oxide (i.e., it can be grown and etched). Various thicknesses of SiO_2 may be required, depending on the particular process. Thin oxides are required for transistor gates; thicker oxides might be required for higher voltage devices, while even thicker oxide layers might be required to ensure that transistors are not formed unintentionally in the silicon beneath polysilicon wires (see the next section).

Oxidation of silicon is achieved by heating silicon wafers in an oxidizing atmosphere. The following are some common approaches:

- *Wet oxidation*—when the oxidizing atmosphere contains water vapor. The temperature is usually between 900 °C and 1000 °C. This is also called *pyrogenic* oxidation when a 2:1 mixture of hydrogen and oxygen is used. Wet oxidation is a rapid process.
- *Dry oxidation*—when the oxidizing atmosphere is pure oxygen. Temperatures are in the region of 1200 °C to achieve an acceptable growth rate. Dry oxidation forms a better quality oxide than wet oxidation. It is used to form thin, highly controlled gate oxides, while wet oxidation may be used to form thick field oxides.
- *Atomic layer deposition (ALD)*—when a thin chemical layer (material A) is attached to a surface and then a chemical (material B) is introduced to produce a thin layer of the required layer (i.e., SiO_2—this can also be used for other various dielectrics and metals). The process is then repeated and the required layer is built up layer by layer. [George96, Klaus98].

The oxidation process normally consumes part of the silicon wafer (deposition and ALD do not). Since SiO_2 has approximately twice the volume of silicon, the SiO_2 layer grows almost equally in both vertical directions. Thus, after processing, the SiO_2 projects above and below the original unoxidized silicon surface.

15.2.5 Isolation

Individual devices in a CMOS process need to be isolated from one another so that they do not have unexpected interactions. In particular, channels should only be inverted beneath transistor gates over the active area; wires running elsewhere shouldn't create parasitic MOS channels. Moreover, the source/drain diffusions of unrelated transistors should not interfere with each other.

The process flow in Section 1.5 was historically used to provide this isolation. The transistor gate consists of a thin *gate oxide* layer. Elsewhere, a thicker layer of field oxide separates polysilicon and metal wires from the substrate. The MOS sandwich formed by the wire, thick oxide, and substrate behaves as an unwanted parasitic transistor. However, the thick oxide effectively sets a threshold voltage greater than V_{DD} that prevents the transistor from turning ON during normal operation. Actually, these *field devices* can be used for I/O protection and are discussed in Section 12.6.2. The source and drain of the transistors form reverse-biased p–n junctions with the substrate or well, isolating them from their neighbors.

The thick oxide used to be formed by a process called *Local Oxidation of Silicon* (LOCOS). A problem with LOCOS-based processes is the transition between thick and thin oxide, which extended some distance laterally to form a so-called *bird's beak*. The lateral distance is proportional to the oxide thickness, which limits the packing density of transistors.

Starting around the 0.35 μm node, *shallow trench isolation* (STI) was introduced to avoid the problems with LOCOS. STI forms insulating trenches of SiO_2 surrounding the transistors (everywhere except the active area). The trench width is independent of its depth, so transistors can be packed as closely as the lithography permits. The trenches isolate the wires from the substrate, preventing unwanted channel formation. They also reduce the sidewall capacitance and junction leakage current of the source and drain.

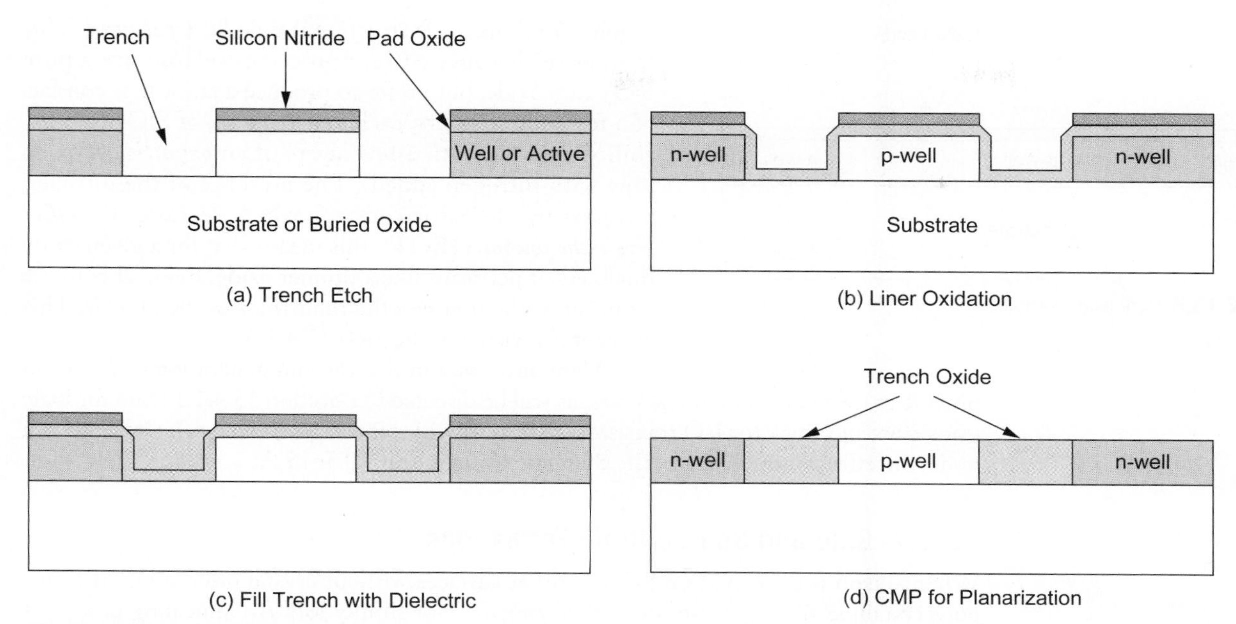

FIGURE 15.7 Shallow trench isolation

STI starts with a *pad oxide* and a silicon nitride layer, which act as the masking layers, as shown in Figure 15.7. Openings in the pad oxide are then used to etch into the well or substrate region (this process can also be used for source/drain diffusion). A liner oxide is then grown to cover the exposed silicon (Figure 15.7(b)). The trenches are filled with SiO_2 or other fillers using CVD that does not consume the underlying silicon (Figure 15.7(c)). The pad oxide and nitride are removed and a *Chemical Mechanical Polishing* (CMP) step is used to planarize the structure (Figure 15.7(d)). CMP, as its name suggests, combines a mechanical grinding action in which the rotating wafer is contacted by a stationary polishing head while an abrasive mixture is applied. The mixture also reacts chemically with the surface to aid in the polishing action. CMP is used to achieve flat surfaces, which are of central importance in modern processes with many layers.

From the designer's perspective, the presence of a deep n-well and/or trench isolation makes it easier to isolate noise-sensitive (analog or memory) portions of a chip from digital sections. Trench isolation also permits nMOS and pMOS transistors to be placed closer together because the isolation provides a higher source/drain *breakdown voltage*—the voltage at which a source or drain diode starts to conduct in the reverse-biased condition. The breakdown voltage must exceed the supply voltage (so junctions do not break down during normal operation) and is determined by the junction dimensions and doping levels of the junction formed. Deeper trenches increase the breakdown voltage.

15.2.6 Gate Oxide

The next step in the process is to form the gate oxide for the transistors. As mentioned, this is most commonly in the form of *silicon dioxide* (SiO_2).

In the case of STI-defined source/drain regions, the gate oxide is grown on top of the planarized structure that occurs at the stage shown in Figure 15.7(d). This is shown in

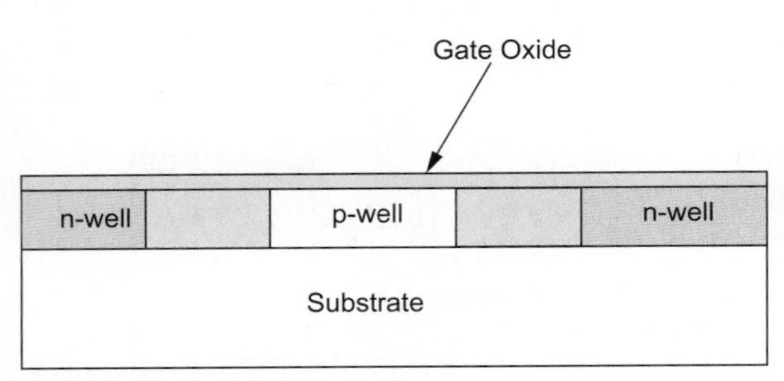

FIGURE 15.8 Gate oxide formation

Figure 15.8. The oxide structure is called the *gate stack*. This term arises because current processes seldom use a pure SiO_2 gate oxide, but prefer to produce a stack that consists of a few atomic layers, each 3–4 Å thick, of SiO_2 for reliability, overlaid with a few layers of an *oxynitrided* oxide (one with nitrogen added). The presence of the nitrogen increases the dielectric constant, which decreases the *effective oxide thickness* (EOT); this means that for a given oxide thickness, it performs like a thinner oxide. Being able to use a thicker oxide improves the robustness of the process. This concept is revisited in Section 15.4.1.3.

Many processes in the 180 nm generation and beyond provide at least two oxide thicknesses, as will be discussed in Section 15.4.1.1 (thin for logic transistors and thick for I/O transistors that must withstand higher voltages). At the 65 nm node, the effective thickness of the thin gate oxide is only 10.5–15 Å.

15.2.7 Gate and Source/Drain Formations

When silicon is deposited on SiO_2 or other surfaces without crystal orientation, it forms polycrystalline silicon, commonly called *polysilicon* or simply *poly*. An annealing process is used to control the size of the single crystal domains and to improve the quality of the polysilicon. Undoped polysilicon has high resistivity. The resistance can be reduced by implanting it with dopants and/or combining it with a refractory metal. The polysilicon gate serves as a mask to allow precise alignment of the source and drain on either side of the gate. This process is called a *self-aligned polysilicon gate* process. Aluminum could not be used because it would melt during formation of the source and drain.

As a historical note, early metal-gate processes first diffused source and drain regions, and then formed a metal gate. If the gate was misaligned, it could fail to cover the entire channel and lead to a transistor that never turned ON. To prevent this, the metal gate had to overhang the source and drain by more than the alignment tolerance of the process. This created large parasitic gate-to-source and gate-to-drain overlap capacitances that degraded switching speeds.

The steps to define the gate, source, and drain in a self-aligned polysilicon gate are as follows:

- Grow gate oxide wherever transistors are required (area = source + drain + gate)—elsewhere there will be thick oxide or trench isolation (Figure 15.9(a))
- Deposit polysilicon on chip (Figure 15.9(b))
- Pattern polysilicon (both gates and interconnect) (Figure 15.9(c))
- Etch exposed gate oxide—i.e., the area of gate oxide where transistors are required that was not covered by polysilicon; at this stage, the chip has windows down to the well or substrate wherever a source/drain diffusion is required (Figure 15.9(d))
- Implant pMOS and nMOS source/drain regions (Figure 15.9(e))

The source/drain implant density is relatively low, typically in the range 10^{18}–10^{20} cm^{-3} of impurity atoms. Such a *lightly doped drain* (LDD) structure reduces the electric field at the drain junction (the junction with the highest voltage), which improves the immunity of the device to hot electron damage (see Section 6.3.6) and suppresses short-channel effects. The LDD implants are shallow and lightly doped, so they exhibit low

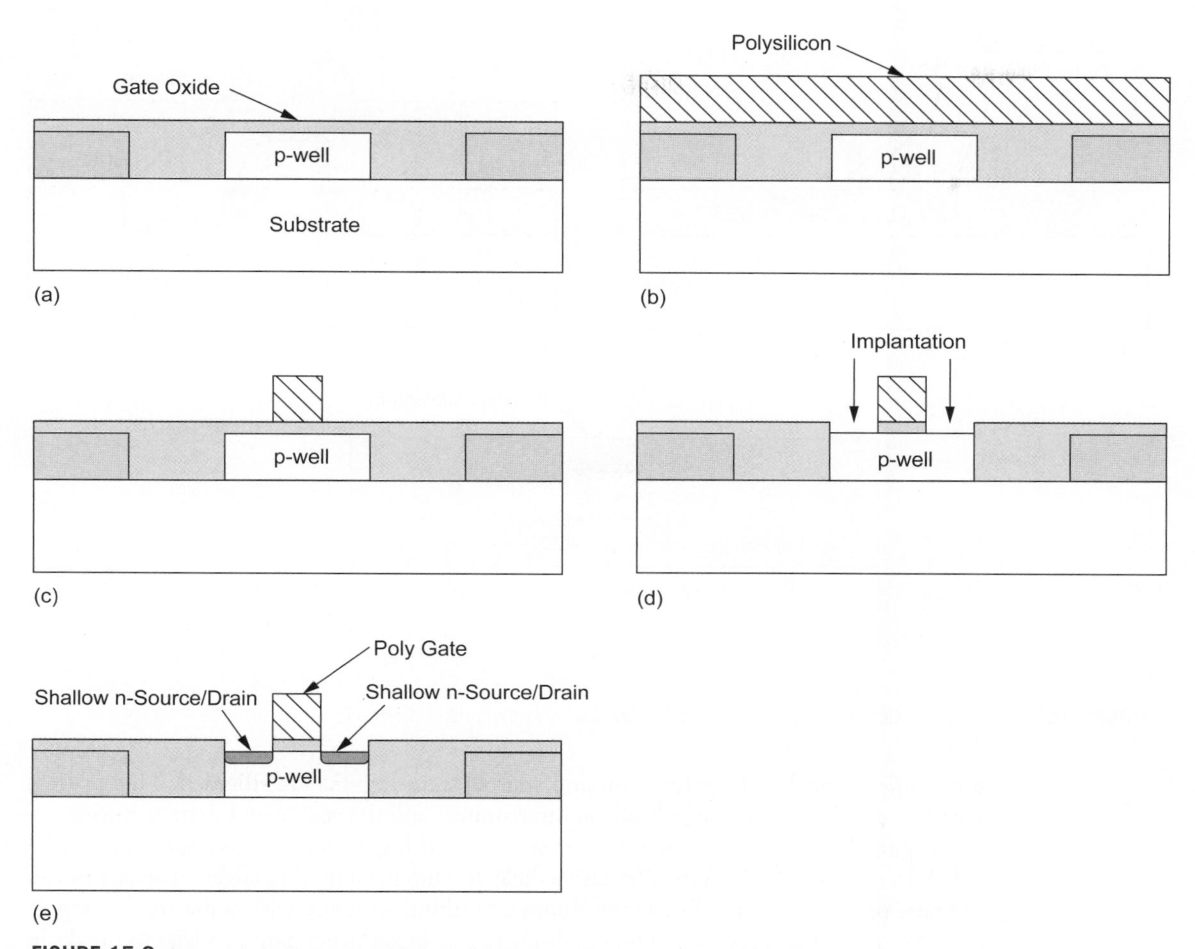

FIGURE 15.9 Gate and shallow source/drain definition

capacitance but high resistance. This reduces device performance somewhat because of the resistance in series with the transistor. Consequently, deeper, more heavily doped source/drain implants are needed in conjunction with the LDD implants to provide devices that combine hot electron suppression with low source/drain resistance. A silicon nitride (Si_3N_4) *spacer* along the edge of the gate serves as a mask to define the deeper diffusion regions, as shown in Figure 15.10(a). For in-depth coverage of various LDD structures, see [Ziegler02].

As mentioned, the polysilicon gate and source/drain diffusion have high resistance due to the resistivity of silicon and their extremely small dimensions. Modern processes form a surface layer of a refractory metal on the silicon to reduce the resistance. A *refractory metal* is one with a high melting point that will not be damaged during subsequent processing. Tantalum, nickel, molybdenum, titanium, or cobalt are commonly used. The metal is deposited on the silicon (specifically on the gate polysilicon and/or source/drain regions). A layer of *silicide* is formed when the two substances react at elevated temperatures. In a *polycide* process, only the gate polysilicon is silicided. In a silicide process (usually implemented as a self-aligned silicidization—from whence comes the synonymous

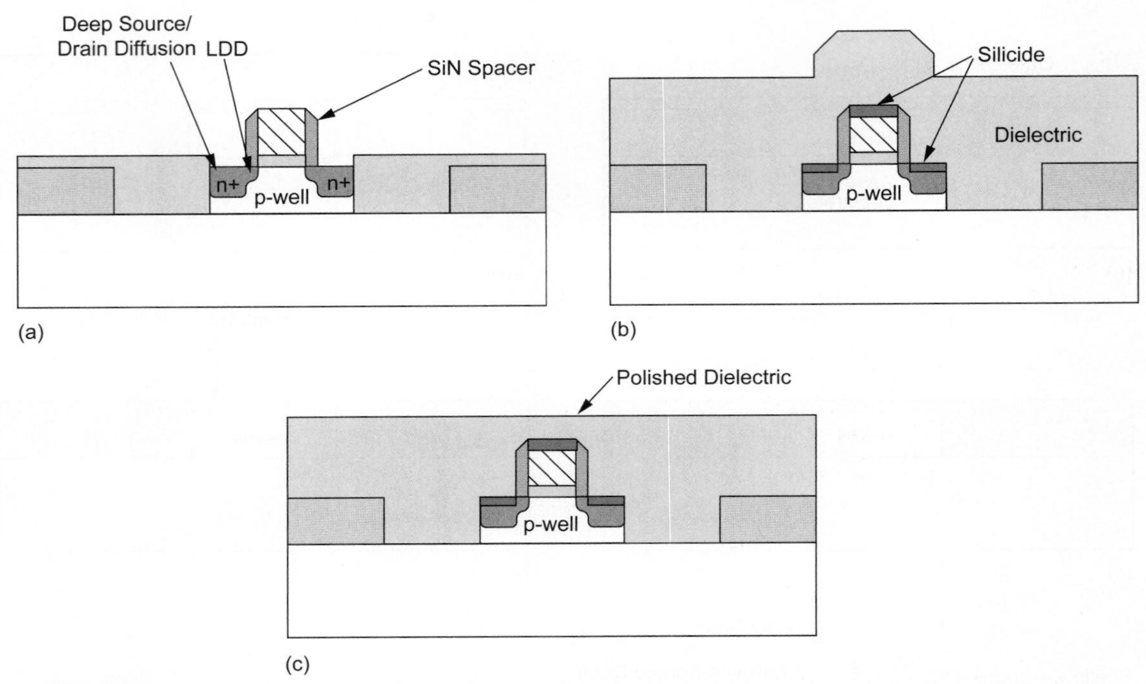

FIGURE 15.10 Transistor with LDD and deep diffusion, salicide, and planarized dielectric

term *salicide*) both gate polysilicon and source/drain regions are silicided. This process lowers the resistance of the polysilicon interconnect and the source and drain diffusion.

Figure 15.10(b) shows the resultant structure with gate and source/drain regions silicided. In addition, SiO_2 or an alternative dielectric has been used to cover all areas prior to the next processing steps. The figure shows a resulting structure with some vertical topology typical of older processes. The rapid transitions in surface height can lead to breaks in subsequent layers that fail to conform, or can entail a plethora of design rules that relate to metal edges. To avoid these problems, a CMP step is used to planarize the dielectric, leaving a flat surface for metallization as shown in Figure 15.10(c).

Nanometer processes involve another implantation step called *halo doping* that increases the doping of the substrate or well near the ends of the channels. The halo doping alleviates DIBL, short channel effects, and punchthrough but increases GIDL and BTBT leakage at the junction between the diffusion and channel [Roy03].

15.2.8 Contacts and Metallization

Contact cuts are made to source, drain, and gate according to the contact mask. These are holes etched in the dielectric after the source/drain step discussed in the previous section. Older processes commonly use aluminum (Al) for wires, although newer ones offer copper (Cu) for lower resistance. Tungsten (W) can be used as a *plug* to fill the contact holes (to alleviate problems of aluminum not conforming to small contacts). In some processes, the tungsten can also be used as a local interconnect layer.

Metallization is the process of building wires to connect the devices. As mentioned previously, conventional metallization uses aluminum. Aluminum can be deposited either by *evaporation* or *sputtering*. Evaporation is performed by passing a high electrical current through a thick aluminum wire in a vacuum chamber. Some of the aluminum atoms are vaporized and deposited on the wafer. An improved form of evaporation that suffers less from contamination focuses an electron beam at a container of aluminum to evaporate the metal. Sputtering is achieved by generating a gas plasma by ionizing an inert gas using an RF or DC electric field. The ions are focused on an aluminum target and the plasma dislodges metal atoms, which are then deposited on the wafer.

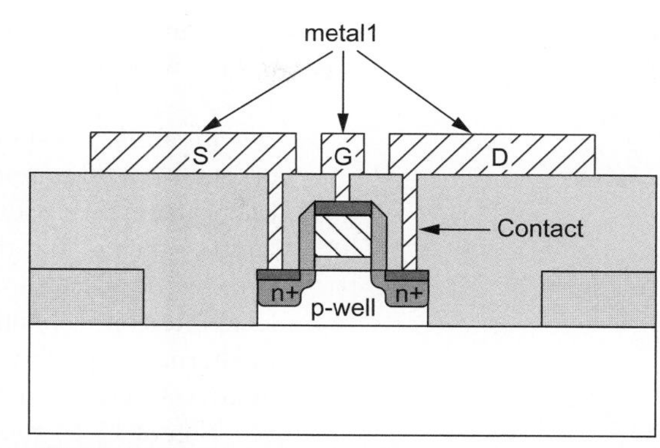

FIGURE 15.11 Aluminum metallization

Wet or *dry etching* can be used to remove unwanted metal. *Piranha solution* is a 3:1 to 5:1 mix of sulfuric acid and hydrogen peroxide that is used to clean wafers of organic and metal contaminants or photoresist after metal patterning. Plasma etching is a dry etch process with fluorine or chlorine gas used for metallization steps. The plasma charges the etch gas ions, which are attracted to the appropriately charged silicon surface. Very sharp etch profiles can be achieved using plasma etching. The result of the contact and metallization patterning steps is shown in Figure 15.11.

Subsequent intermetal vias and metallization are then applied. Some processes offer uniform metal dimensions for levels 2 to n–1, where n is the top level of metal. The top level is normally a thicker layer for use in power distribution and as such has relaxed width and spacing constraints. Other processes use successively thicker and wider metal for the upper layers, as will be explored in Section 5.1.2.

Polysilicon over diffusion normally forms a transistor gate, so a short metal1 wire is necessary to connect a diffusion output node to a polysilicon input. Some processes add tungsten (W) layer above polysilicon and below metal1; this layer is called *local interconnect* and can be drawn on a finer pitch than metal1. Local interconnect offers denser cell layouts, especially in static RAMs. Figure 15.12 shows a scanning electron micrograph of a partially completed SRAM array. The oxide has been removed to show the diffusion, polysilicon, local interconnect, and metal1. Local interconnect is used to connect the nMOS and pMOS transistors without rising up to metal1. SRAM cells are discussed further in Section 11.2.

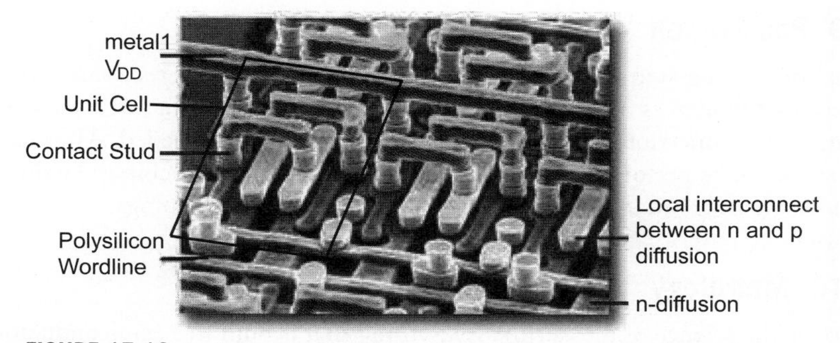

FIGURE 15.12 Partially completed 6-transistor SRAM array using local interconnect (Courtesy of International Business Machines Corporation. Unauthorized use not permitted.)

Contemporary logic processes use copper interconnects and low-k dielectrics to reduce wire resistance and capacitance. These enhancements are discussed in Section 15.4.2.

Figure 15.13 shows a cross-section of an IBM microprocessor showing the 11 layers of metal in a 90 nm process. The bottom level is tungsten local interconnect. The next five layers are on a 1× width and thickness (0.12 μm width and spacing). Metal 6–8 are on a 2× width, spacing, and thickness and metal 9–10 are 4×. These ten layers use copper wires with low-k dielectrics. The top level is aluminum and is primarily used for I/O pads. The local interconnect and metal1 are used in both directions, while the upper layers are used in alternating preferred directions. A pair of vias between metal 9 and 10 are visible. The interfaces between dielectric levels after each step of CMP are also visible.

Figure 15.14 shows a micrograph in which the oxide between metal layers has been stripped away to reveal the complex three-dimensional structure of chip wiring.

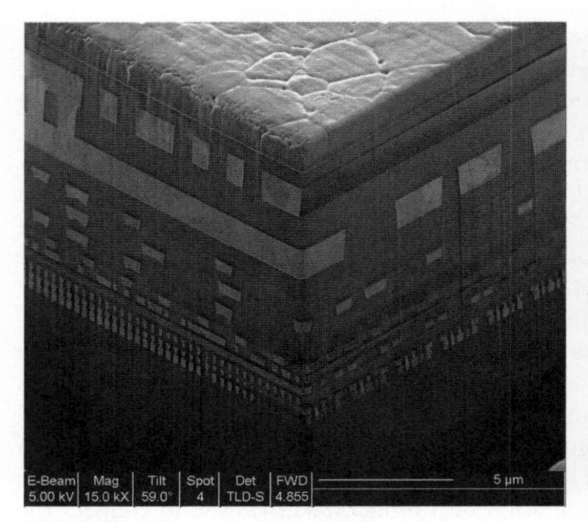

FIGURE 15.13 Cross-section showing 11 levels of metallization (Courtesy of International Business Machines Corporation. Unauthorized use not permitted.)

FIGURE 15.14 Micrograph of metallization in six-layer copper process (Courtesy of International Business Machines Corporation. Unauthorized use not permitted.)

15.2.9 Passivation

The final processing step is to add a protective glass layer called *passivation* or *overglass* that prevents the ingress of contaminants. Openings in the passivation layer, called *overglass cuts*, allow connection to I/O pads and test probe points if needed. After passivation, further steps can be performed such as *bumping*, which allows the chip to be directly connected to a circuit board using plated solder bumps in the pad openings.

15.2.10 Metrology

Metrology is the science of measuring. Everything that is built in a semiconductor process has to be measured to give feedback to the manufacturing process. This ranges from simple optical measurements of line widths to advanced techniques to measure thin films and defects such as voids in copper interconnect. A natural requirement exists for *in situ* real-time measurements so that the manufacturing process can be controlled in a direct feedback manner.

Optical microscopes are used to observe large structures and defects, but are no longer adequate for structures smaller than the wavelength of visible light (~0.5 μm). Scanning electron microscopy (SEM) is used to observe very small features. An SEM raster scans a structure under observation and observes secondary electron emission to produce an image of the surface of the structure. Energy Dispersive Spectroscopy (EDX) bombards a circuit with electrons causing x-ray emission. This can be used for imaging as well. A Transmission Electron Microscope (TEM), which observes the results of passing electrons through a sample (rather than bouncing them off the sample), is sometimes also used to measure structures.

15.3 Layout Design Rules

Layout rules, also referred to as *design rules*, were introduced in Chapter 1 and can be considered a prescription for preparing the photomasks that are used in the fabrication of integrated circuits. The rules are defined in terms of *feature sizes* (widths), *separations*, and *overlaps*. The main objective of the layout rules is to build reliably functional circuits in as small an area as possible. In general, design rules represent a compromise between performance and yield. The more conservative the rules are, the more likely it is that the circuit will function. However, the more aggressive the rules are, the greater the opportunity for improvements in circuit performance and size.

Design rules specify to the designer certain geometric constraints on the layout artwork so that the patterns on the processed wafer will preserve the topology and geometry of the designs. It is important to note that design rules do not represent some hard boundary between correct and incorrect fabrication. Rather, they represent a tolerance that ensures high probability of correct fabrication and subsequent operation. For example, you may find that a layout that violates design rules can still function correctly and vice versa. Nevertheless, any significant or frequent departure (*design rule waiver*) from design rules will seriously prejudice the success of a design.

Chapter 1 described a version of design rules based on the MOSIS CMOS scalable rules. The MOSIS rules are expressed in terms of λ. These rules allow some degree of scaling between processes, as in principle, you only need to reduce the value of λ and the designs will be valid in the next process down in size. Unfortunately, history has shown that processes rarely shrink uniformly. Thus, industry usually uses the actual micron design rules for layouts. At this time, custom layout is usually constrained to a number of often-used standard cells or memories, where the effort expended is amortized over many instances. Only for extremely high-volume chips is the cost savings of a smaller full-custom layout worth the labor cost of that layout.

15.3.1 Design Rule Background

We begin by examining the reasons for the most important design rules.

15.3.1.1 Well Rules The n-well is usually a deeper implant (especially a deep n-well) than the transistor source/drain implants, and therefore, it is necessary to provide sufficient clearance between the n-well edges and the adjacent *n*+ diffusions. The clearance between the well edge and an enclosed diffusion is determined by the transition of the field oxide across the well boundary. Processes that use STI may permit zero inside clearance. In older LOCOS processes, problems such as the bird's beak effect usually force substantial clearances. Being able to place nMOS and pMOS transistors closer together can significantly reduce the size of SRAM cells.

Because the n-well sheet resistance can be several kΩ per square, it is necessary to ground the well thoroughly by providing a sufficient number of well taps. This will prevent excessive voltage drops due to well currents. Guidelines on well and substrate taps are given in Section 6.3.6. Where wells are connected to different potentials (say in analog circuits), the spacing rules may differ from *equipotential* wells (all wells at the same voltage—the normal case in digital logic).

Mask Summary: The masks encountered for well specification may include n-well, p-well, and deep n-well. These are used to specify where the various wells are to be placed. Often only one well is specified in a twin-well process (i.e., n-well) and by default the p-well is in areas where the n-well isn't (i.e., p-well equals the logical NOT of the n-well).

15.3.1.2 Transistor Rules CMOS transistors are generally defined by at least four physical masks. These are active (also called diffusion, diff, thinox, OD, or RX), n-select (also called n-implant, nimp, or nplus), p-select (also called p-implant, pimp, or pplus) and polysilicon (also called poly, polyg, PO, or PC). The active mask defines all areas where either n- or p-type diffusion is to be placed *or* where the gates of transistors are to be placed. The gates of transistors are defined by the logical AND of the polysilicon mask and the active mask, i.e., where polysilicon crosses diffusion. The select layers define what type of diffusion is required. n-select surrounds active regions where n-type diffusion is required. p-select surrounds areas where p-type diffusion is required. n-diffusion areas inside p-well regions define nMOS transistors (or n-diffusion wires). n-diffusion areas inside n-well regions define n-well contacts. Likewise, p-diffusion areas inside n-wells define pMOS transistors (or p-diffusion wires). p-diffusion areas inside p-wells define substrate contacts (or p-well contacts). Frequently, design systems will define only n-diffusion (ndiff) and p-diffusion (pdiff) to reduce the complexity of the process. The appropriate selects are generated automatically. That is, ndiff will be converted automatically into active with an overlapping rectangle or polygon of n-select.

It is essential for the poly to cross active completely; otherwise the transistor that has been created will be shorted by a diffusion path between source and drain. Hence, poly is required to extend beyond the edges of the active area. This is often termed the *gate extension*. Active must extend beyond the poly gate so that diffused source and drain regions exist to carry charge into and out of the channel. Poly and active regions that should not form a transistor must be kept separated; this results in a spacing rule from active to polysilicon.

Figure 15.15(a) shows the mask construction for the final structures that appear in Figure 15.15(b).

Mask Summary: The basic masks (in addition to well masks) used to define transistors, diffusion interconnect (possibly resistors), and gate interconnect are active, n-select, p-select, and polysilicon. These may be called different names in some processes. Sometimes n-diffusion (ndiff) and p-diffusion (pdiff) masks are used in place of active to alleviate designer confusion.

15.3.1.3 Contact Rules There are several generally available contacts:

- Metal to p-active (p-diffusion)
- Metal to n-active (n-diffusion)

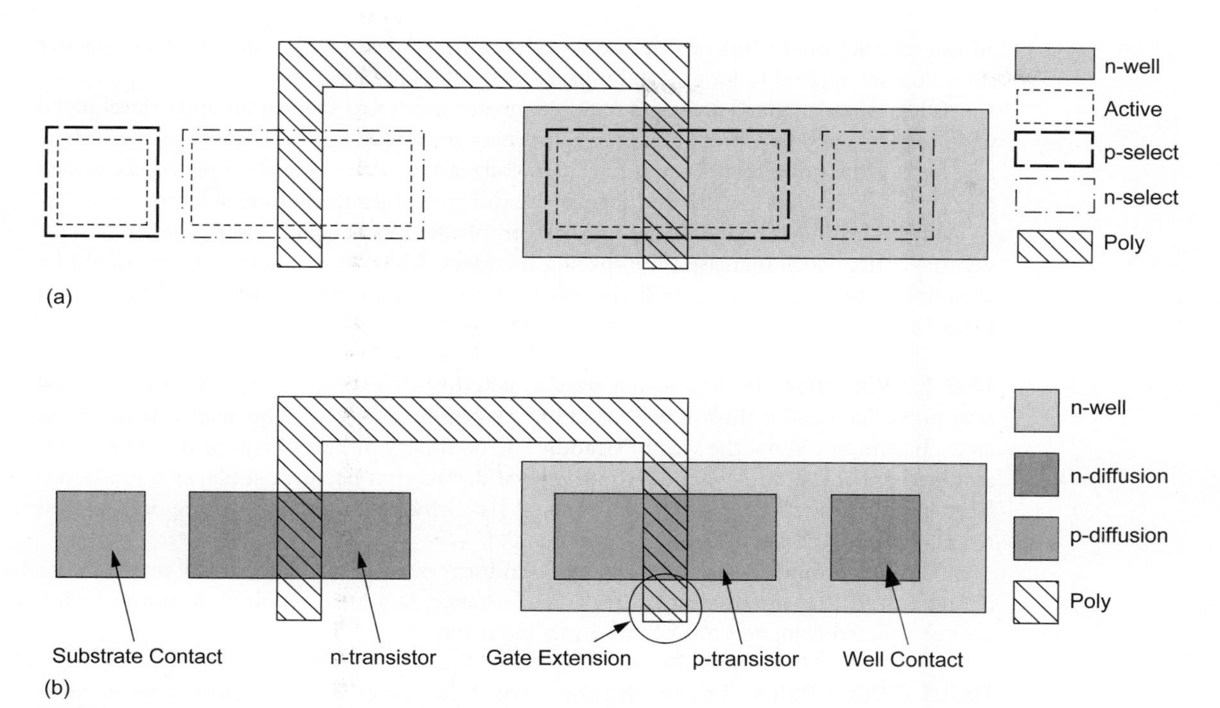

FIGURE 15.15 CMOS n-well process transistor and well/substrate contact construction

- Metal to polysilicon
- Metal to well or substrate

Depending on the process, other contacts such as *buried* polysilicon-active contacts may be allowed for local interconnect.

Because the substrate is divided into well regions, each isolated well must be tied to the appropriate supply voltage; i.e., the n-well must be tied to V_{DD} and the substrate or p-well must be tied to GND with well or substrate contacts. As mentioned in Section 1.5.1, metal makes a poor connection to the lightly doped substrate or well. Hence, a heavily doped active region is placed beneath the contact, as shown at the source of the nMOS transistor in Figure 15.16.

Whenever possible, use more than one contact at each connection. This significantly improves yield in many processes because the connection is still made even if one of the contacts is malformed.

Mask Summary: The only mask involved with contacts to active or poly is the contact mask, commonly called CONT or CA. Contacts are normally of uniform size to allow for consistent etching of very small features.

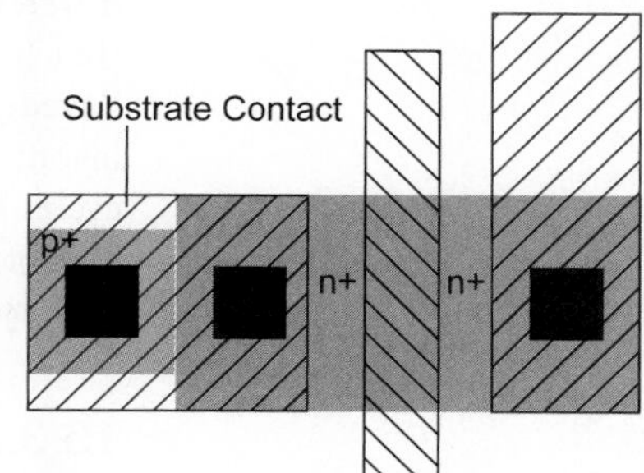

FIGURE 15.16 Substrate contact

15.3.1.4 Metal Rules Metal spacing may vary with the width of the metal line (so called fat-metal rules). That is, above some metal wire width, the minimum spacing may be increased. This is due to etch characteristics of small versus large metal wires. There may also be maximum metal width rules. That is, single metal wires cannot be greater than a certain width. If wider wires are desired, they are constructed by paralleling a number of smaller wires and adding checkerboard links to tie the wires together. Additionally, there may be spacing rules that are applied to long, closely spaced parallel metal lines.

Older nonplanarized processes required greater width and spacing on upper-level metal wires (e.g., metal3) to prevent breaks or shorts between adjoining wires caused by the vertical topology of the underlying layers. This is no longer a consideration for modern planarized processes. Nevertheless, width and spacing are still greater for thicker metal layers.

Mask Summary: Metal rules may be complicated by varying spacing dependent on width: As the width increases, the spacing increases. Metal overlap over contact might be zero or nonzero. Guidelines will also exist for electromigration, as discussed in Section 6.3.3.1.

15.3.1.5 Via Rules Processes may vary in whether they allow *stacked* vias to be placed over polysilicon and diffusion regions. Some processes allow vias to be placed within these areas, but do not allow the vias to straddle the boundary of polysilicon or diffusion. This results from the sudden vertical topology variations that occur at sublayer boundaries. Modern planarized processes permit stacked vias, which reduces the area required to pass from a lower-level metal to a high-level metal.

Mask Summary: Vias are normally of uniform size within a layer. They may increase in size toward the top of a metal stack. For instance, large vias required on power busses are constructed from an array of uniformly sized vias.

15.3.1.6 Other Rules The passivation or overglass layer is a protective layer of SiO_2 (glass) that covers the final chip. Appropriately sized openings are required at pads and any internal test points.

Some additional rules that might be present in some processes are as follows:

- Extension of polysilicon or metal beyond a contact or via
- Differing gate poly extensions depending on the device length
- Maximum width of a feature
- Minimum area of a feature (small pieces of photoresist can peel off and float away)
- Minimum notch sizes (small notches are rarely beneficial and can interfere with resolution enhancement techniques)

15.3.1.7 Summary Whereas earlier processes tended to be process driven and frequently had long and involved design rules, processes have become increasingly "designer friendly" or, more specifically, computer friendly (most of the mask geometries for designs are algorithmically produced). Companies sometimes create "generic" rules that span a number of different CMOS foundries that they might use. Some processes have design guidelines that feature structures to be avoided to ensure good yields. Traditionally, engineers followed yield-improvement cycles to determine the causes of defective chips and modify the layout to avoid the most common systematic failures. Time to market and product life cycles are now so short that yield improvement is only done for the highest volume parts. It is often better to reimplement a successful product in a new, smaller technology rather than to worry about improving the yield on the older, larger process.

15.3.2 Scribe Line and Other Structures

The *scribe line* surrounds the completed chip where it is cut with a diamond saw. The construction of the scribe line varies from manufacturer to manufacturer. It is designed to prevent the ingress of contaminants from the side of the chip (as opposed to the top of the chip, which is protected by the overglass).

Several other structures are included on a mask including the *alignment mark*, *critical dimension* structures, *vernier* structures, and *process check* structures [Hess94]. The mask alignment mark is usually placed by the foundry to align one mask to the next. Critical dimension test structures can be measured after processing to check proper etching of narrow polysilicon or metal lines. Vernier structures are used to judge the alignment between layers. A vernier is a set of closely spaced parallel lines on two layers. Misalignment between the two layers can be judged by the alignment of the two verniers. Test structures such as chains of contacts and vias, test transistors, and ring oscillators are used to evaluate contact resistance and transistor parameters. Often these structures can be placed along the scribe line so they do not consume useful wafer area.

15.3.3 MOSIS Scalable CMOS Design Rules

Class project designs often use the λ-based scalable CMOS design rules from MOSIS because they are simple and freely available. MOSIS once offered a wide variety of processes, from 2 μm to 180 nm, compatible with the scalable CMOS rules. Indeed, MOSIS also supports three variants of these rules: SCMOS, SUBM, and DEEP, which are progressively more conservative to support feature sizes down to 180 nm. Chips designed in the conservative DEEP rules could be fabricated on any of the MOSIS processes.

As time has passed, the older processes became obsolete and the newer processes have too many nuances to be compatible with scalable design rules. The MOSIS processes most commonly used today are the ON Semiconductor (formerly AMI) 0.5 μm process and the IBM 130, 90, 65, and 45 nm processes.

The 0.5 μm process is popular for university class projects because MOSIS Educational Program offers generous grants to cover fabrication costs for 1.5 mm × 1.5 mm "TinyChips." The best design rules for this process are the scalable SUBM rules[1] using $\lambda = 0.3\ \mu$m. Thus, a TinyChip is 5000 $\lambda \times$ 5000 λ. Polysilicon is drawn at 2 $\lambda = 0.6\ \mu$m, then biased by MOSIS by –0.1 μm prior to mask generation to give a true 0.5 μm gate length. When simulating circuits, be sure to use the biased channel lengths to model the transistor behavior accurately. In SPICE, the XL parameter is added to the specified transistor length to find the actual length. Thus, a SPICE deck could specify a drawn channel length of L = 0.6 μm for each transistor and include XL = -0.1μm in the model file to indicate a biased length of 0.5 μm. There is a tutorial at `www.cmosvlsi.com` on designing in this process with the Electric CAD tool suite. [Brunvand09] explains how to design in this process with the Cadence and Synopsys tool suites; this flow has a steeper learning curve but better mirrors industry practices.

Credible research chips need more advanced processes to reflect contemporary design challenges. The IBM processes are presently discounted for universities, and MOSIS offers certain research grants as well. The best way to design in these processes is with the Cadence and Synopsys tools using IBM's proprietary micron-based design rules. The design flow is presently poorly documented by MOSIS and ranges from difficult at the 130 nm node to worse at deeper nodes. Unfortunately, this presently limits access to these processes to highly sophisticated research groups.

[1]Technically, MOSIS has two sets of contact rules [MOSIS09]. The standard rules require polysilicon and active to overlap contacts by 1.5 λ. Half-lambda rules reduce productivity because they force the designer off a λ grid. The "alternate contact rules" are preferable because they require overlap by 1 λ, at the expense of more conservative spacing rules; these alternate rules are used in the examples in this text.

Section 1.5.3 introduced the SCMOS design rules. More extensive rules are illustrated and summarized on the inside back cover. Layouts consist of a set of rectangles on various layers such as polysilicon or metal. *Width* is the minimum width of a rectangle on a particular layer. *Spacing* is the minimum spacing between two rectangles on the same or different layers. *Overlap* specifies how much a rectangle must surround another on another layer. Dimensions are all specified in λ except for overglass cuts that do not scale well because they must contact large bond wires or probe tips. Select layers are often generated automatically and thus are not shown in the layout. If the active layer satisfies design rules, the select will too.

Contacts and vias must be exactly 2×2 λ. Larger connections are made from arrays of small vias to prevent current crowding at the periphery. The spacing rules of polysilicon or diffusion to arrays of multiple contacts is slightly larger than that to a single contact.

Section 1.5.5 estimated the pitch of lower-level metal to be 8 λ: 4 λ for the width and 4 λ for spacing. Technically, the minimum width and spacing are 3 λ, but the minimum metal contact size is 2×2 λ plus 1 λ surround on each side, for a width of 4 λ. Thus, the pitch for contacted metal lines can be reduced to 7 λ. Moreover, if the lines are drawn at 3 λ and the contacts are staggered so two adjacent lines never have adjacent contacts, the pitch reduces to 6.5 λ. Nevertheless, using a pitch of 8 λ for planning purposes is good practice and leaves a bit of "wiggle room" to solve difficult layout problems.

15.3.4 Micron Design Rules

Table 15.1 lists a set of micron design rules for a hypothetical 65 nm process representing an amalgamation of several real processes. Rule numbers reference the diagram on the inside back cover. Observe that the rules differ slightly but not immensely from lambda-based rules with $\lambda = 0.035$ μm. A complete set of micron design rules in this generation fills hundreds of pages. Note that upper level metal rules are highly variable depending on the metal thickness; thicker wires require greater widths and spacings and bigger vias.

TABLE 15.1 Micron design rules for 65 nm process

Layer	Rule	Description	65 nm Rule (μm)
Well	1.1	Width	0.5
	1.2	Spacing to well at different potential	0.7
	1.3	Spacing to well at same potential	0.7
Active (diffusion)	2.1	Width	0.10
	2.2	Spacing to active	0.12
	2.3	Source/drain surround by well	0.15
	2.4	Substrate/well contact surround by well	0.15
	2.5	Spacing to active of opposite type	0.25
Poly	3.1	Width	0.065
	3.2	Spacing to poly over field oxide	0.10
	3.2a	Spacing to poly over active	0.10
	3.3	Gate extension beyond active	0.10
	3.4	Active extension beyond poly	0.10
	3.5	Spacing of poly to active	0.07

TABLE 15.1 Micron design rules for 65 nm process (continued)

Layer	Rule	Description	65 nm Rule (μm)
Select	4.1	Spacing from substrate/well contact to gate	0.15
	4.2	Overlap of active	0.12
	4.3	Overlap of substrate/well contact	0.12
	4.4	Spacing to select	0.20
Contact (to poly or active)	5.1, 6.1	Width (exact)	0.08
	5.2b, 6.2b	Overlap by poly or active	0.01
	5.3, 6.3	Spacing to contact	0.10
	5.4	Spacing to gate	0.07
Metal1	7.1	Width	0.09
	7.2	Spacing to well metal1	0.09
	7.3, 8.3	Overlap of contact or via	0.01
	7.4	Spacing to metal for lines wider than 0.5 μm	0.30
Via1–Via6	8.1, 14.1, …	Width (exact)	0.10
	8.2, 14.2, …	Spacing to via on same layer	0.10
Metal2–Metal7	9.1, …	Width	0.10
	9.2, …	Spacing to same layer metal	0.10
	9.3, …	Overlap of via	0.01
	9.4, …	Spacing to metal for lines wider than 0.5 μm	0.30
Via7–8		Width	0.20
		Spacing	0.20
Metal8–9		Width	0.40
		Spacing to same layer metal	0.40
		Overlap of via	0.10
		Spacing to metal for lines wider than 1.0 μm	0.50

15.4 CMOS Process Enhancements

15.4.1 Transistors

15.4.1.1 Multiple Threshold Voltages and Oxide Thicknesses Some processes offer multiple threshold voltages and/or oxide thicknesses. Low-threshold transistors deliver more ON current, but also have greater subthreshold leakage. Providing two or more thresholds permits the designer to use low-V_t devices on critical paths and higher-V_t devices elsewhere to limit leakage power. Multiple masks and implantation steps are used to set the various thresholds. Alternatively, transistors with slightly longer channels can be used; these transistors naturally have higher thresholds because of the short channel effect (see Section 2.4.3.3) [Rohrer05].

Thin gate oxides also permit more ON current. However, they break down when exposed to the high voltages needed in I/O circuits. Oxides thinner than about 15 Å also

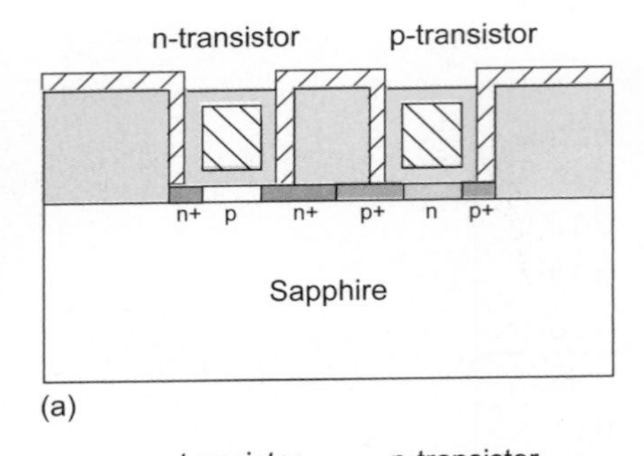

(a)

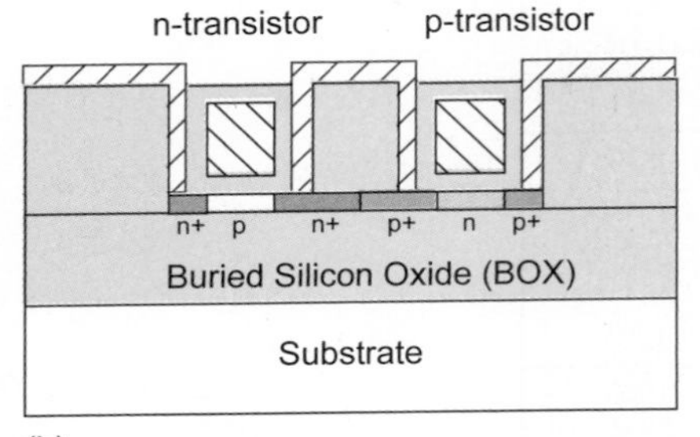

(b)

FIGURE 15.17 SOI types

contribute to large gate leakage currents. Many processes offer a second, thicker oxide for the I/O transistors (see Section 12.6). For example, 3.3 V I/O circuits commonly use 0.35 μm channel lengths and 7 nm gate oxides. When gate leakage is a problem and high-k dielectrics are unavailable, an intermediate oxide thickness may also be provided to reduce leakage. Again, multiple masks are used to define the different oxides.

15.4.1.2 Silicon on Insulator A variant of CMOS that has been available for many years is Silicon on Insulator (SOI). As the name suggests, this is a process where the transistors are fabricated on an insulator. SOI stands in contrast to conventional bulk processes in which the transistors are fabricated on a conductive substrate. Two main insulators are used: SiO_2 and sapphire. One major advantage of an insulating substrate is the elimination of the capacitance between the source/drain regions and body, leading to higher-speed devices. Another major advantage is lower subthreshold leakage due to steeper subthreshold slope resulting from a smaller n in EQ (2.44). The drawbacks are time-dependent threshold variations caused by the floating body.

Figure 15.17 shows two common types of SOI. Figure 15.17(a) illustrates a sapphire substrate. In this technology (for example, Peregrine Semiconductor's UltraCMOS), a thin layer of silicon is formed on the sapphire surface. The thin layer of silicon is selectively doped to define different threshold transistors. Gate oxide is grown on top of this and then polysilicon gates are defined. Following this, the nMOS and pMOS transistors are formed by implantation. Figure 15.17(b) shows a silicon-based SOI process. Here, a silicon substrate is used and a *buried oxide* (BOX) is grown on top of the silicon substrate. A thin silicon layer is then grown on top of the buried oxide and this is selectively implanted to form nMOS and pMOS transistor regions. Gate, source, and drain regions are then defined in a similar fashion to a bulk process. Sapphire is optically and RF transparent. As such, it can be of use in optoelectronic areas when merged with III-V based light emitters.

SOI devices and circuits are discussed further in Section 8.4.

15.4.1.3 High-k Gate Dielectrics MOS transistors need high gate capacitance to attract charge to the channel. This leads to very thin SiO_2 gate dielectrics (e.g., 10.5–12 Å, merely four atomic layers, in a 65 nm process). Gate leakage increases unacceptably below these thicknesses, which brings an end to classical scaling [Bai04]. Simple SiO_2 has a dielectric constant of $k = 3.9$. As shown in EQ (2.2), gates could use thicker dielectrics and hence leak less if a material with a higher dielectric constant were available.

A first step in this direction was the introduction of nitrogen to form *oxynitride* gate dielectrics, called SiON, around the 130 nm generation, providing k of about 4.1–4.2. High-k dielectrics entered commercial manufacturing in 2007, first with a hafnium-based material in Intel's 45 nm process [Auth08]. Hafnium oxide (HfO_2) has $k = 20$.

A depletion region forms at the interface of polysilicon and the gate dielectric. This effectively increases t_{ox}, which is undesirable for performance. Moreover, polysilicon gates can be incompatible with high-k dielectrics because of effects such as *threshold voltage pinning* and *phonon scattering*, which make it difficult to obtain low thresholds and reduce the mobility. The Intel 45 nm process returned to metal gates to solve these problems and also to reduce gate resistance, as shown in Figure 15.18 [Mistry07]. Thus, the term MOS is technically accurate again! nMOS and pMOS transistors use different types of metal with different work functions (energy required to free an electron from a solid) to set the threshold voltages. A second lower-resistance metal layer plays a role similar to a silicide.

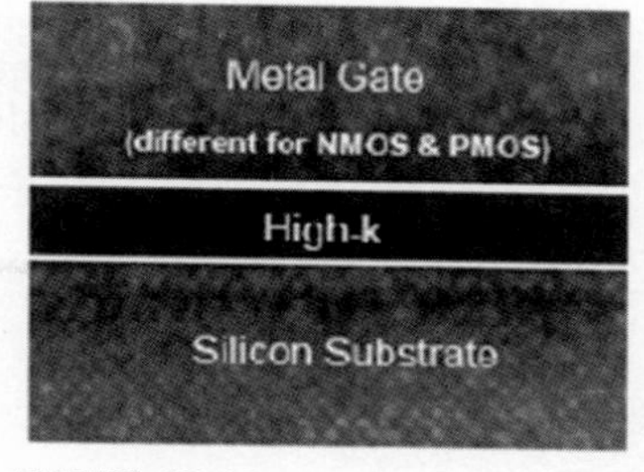

FIGURE 15.18 High-k gate stack TEM (© IEEE 2007.)

One of the challenges with metal gates is that they melt if exposed to the high temperature source/drain formation steps. But if the gate were formed after the source and drain, the self-alignment advantage would be lost. Intel sidesteps this conundrum by first building the transistor with a high-k dielectric and a standard polysilicon gate. After the transistor is complete and the interlayer dielectric is formed, the wafer is polished to expose the polysilicon gates and etched to remove the undesired poly. A thin metal gate is deposited in the trench. Different metals with different workfunctions are required for the nMOS and pMOS transistors. Finally, the trench is filled with a thicker layer of Al for low gate resistance, and the wafer is planarized again.

15.4.1.4 Higher Mobility Increasing the mobility (μ) of the semiconductor improves drive current and transistor speed. One way to improve the mobility is to introduce mechanical strain in the channel. This is called *strained silicon*.

Figure 15.19 shows strained nMOS and pMOS transistors in the Intel 65 nm process that achieve 40% and 100% higher mobility than unstrained transistors, respectively [Tyagi05, Thompson02, Thompson04]. The nMOS channel is under tensile stress created by an insulating film of silicon nitride (SiN) capping the gate. The pMOS channel is under compressive stress produced by etching a recess into the source and drain, then filling the slot with an epitaxial layer of silicon germanium (SiGe). Germanium is another group IV semiconductor with a larger atomic radius than silicon. When a small fraction of the silicon atoms are replaced by germanium, the lattice retains its shape but undergoes mechanical strain because of the larger atoms. Using separate strain mechanisms for the nMOS and pMOS transistors improves mobility of both electrons and holes. An alternative approach is to implant germanium atoms in the channel, introducing tensile stress that only improves electron mobility. STI also introduces stress that affects mobility, so the diffusion layout can impact performance [Topaloglu07].

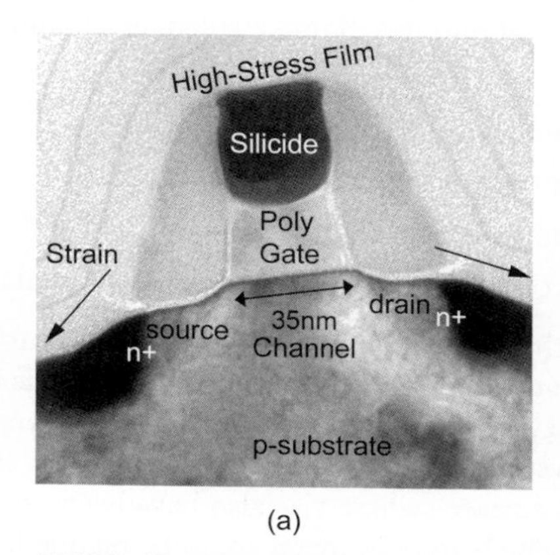

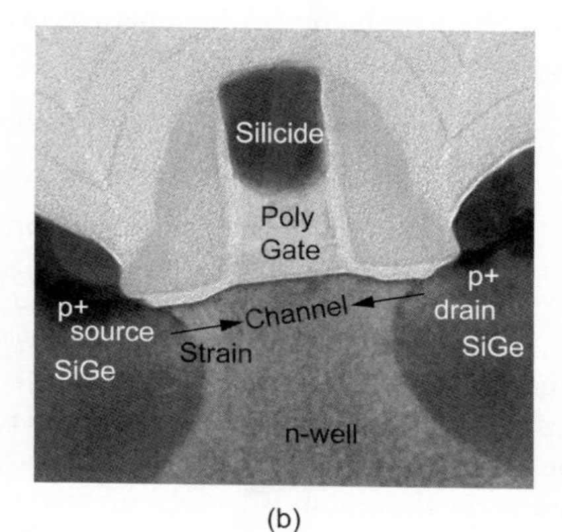

FIGURE 15.19 Strained silicon transistor micrographs: (a) nMOS, (b) pMOS (© IEEE 2005.)

SiGe is also used in high-performance bipolar transistors, especially for radio-frequency (RF) applications. SiGe bipolar devices can be combined with conventional CMOS on the same substrate, which is valuable for low-cost system-on-chip applications that require both digital and RF circuits [Hashimoto02, Harame01a, Harame01b].

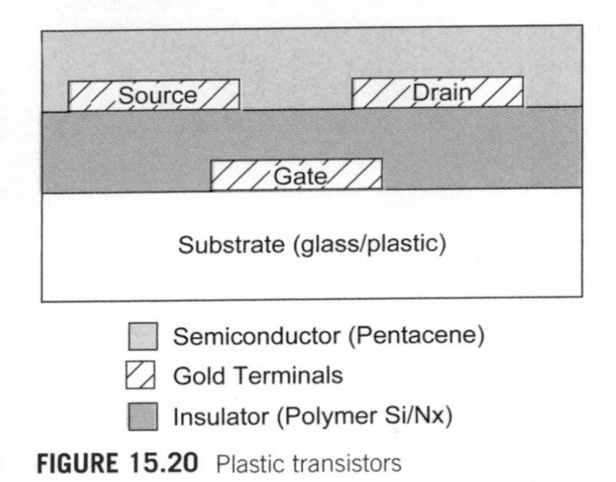

FIGURE 15.20 Plastic transistors

15.4.1.5 Plastic Transistors MOS transistors can be fabricated with organic chemicals. These transistors show promise in active matrix displays, flexible electronic paper, and radio-frequency ID tags because the devices can be manufactured from an inexpensive chemical solution [Huitema03, Myny09]. Figure 15.20 shows the structure of a plastic pMOS transistor. The transistor is built "upside down" with the gold gates and interconnect patterned first on the substrate. Then an organic insulator or silicon nitride is laid down, followed by the gold source and drain connections. Finally, the organic semiconductor (pentacene) is laid down. The mobility of the carriers in the plastic pMOS transistor is about 0.15 $cm^2/V \cdot s$. This is three orders of magnitude lower than that of a comparable silicon device, but is good enough for special applications. Typical lengths and widths are 5 μm and 400 μm, respectively.

15.4.1.6 High-Voltage Transistors High-voltage MOSFETs can also be integrated onto conventional CMOS processes for switching and high-power applications. Gate oxide thickness and channel length have to be larger than usual to prevent breakdown. Specialized process steps are necessary to achieve very high breakdown voltages.

15.4.2 Interconnect

Interconnect has advanced rapidly. While two or three metal layers were once the norm, CMP has enabled inexpensive processes to include seven or more layers. Copper metal and low-k dielectrics are almost universal to reduce the resistance and capacitance of these wires.

15.4.2.1 Copper Damascene Process While aluminum was long the interconnect metal of choice, copper has largely superseded it in nanometer processes. This is primarily due to the higher conductivity of copper compared to aluminum. Some challenges of adopting copper include the following [Merchant01]:

- Copper atoms diffuse into the silicon and dielectrics, destroying transistors.
- The processing required to etch copper wires is tricky.
- Copper oxide forms readily and interferes with good contacts.
- Care has to be taken not to introduce copper into the environment as a pollutant.

Barrier layers have to be used to prevent the copper from entering the silicon surface. A new metallization procedure called the *damascene process* was invented to form this barrier. The process gets its name from the medieval metallurgists of Damascus who crafted fine inlaid swords. In a conventional subtractive aluminum-based metallization step, as we have seen, aluminum is layered on the silicon surface (where vias also have been etched) and then a mask and resist are used to define which areas of metal are to be retained. The unneeded metal is etched away. A dielectric (SiO_2 or other) is then placed over the aluminum conductors and the process can be repeated.

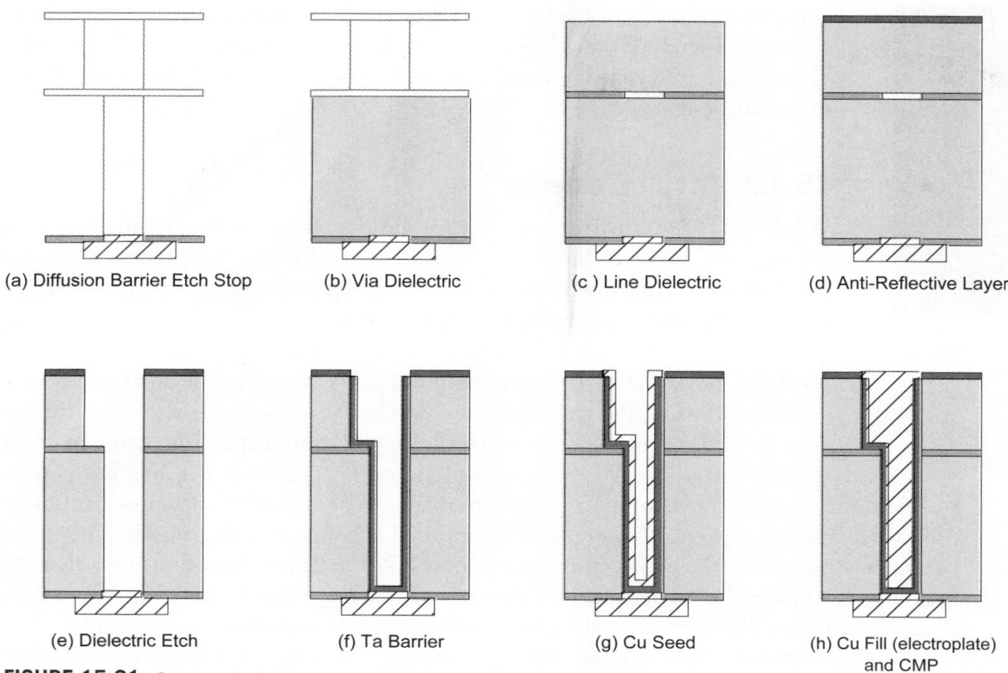

FIGURE 15.21 Copper dual damascene interconnect processing steps

A typical copper damascene process is shown in Figure 15.21, which is an adaptation of a dual damascene process flow from Novellus. Figure 15.21(a) shows a barrier layer over the prior metallization layer. This stops the copper from diffusing into the dielectric and silicon. The via dielectric is then laid down (Figure 15.21(b)). A further barrier layer can then be patterned, and the line dielectric is layered on top of the structure, as shown in Figure 15.21(c). An anti-reflective layer (which helps in the photolithographic process) is added to the top of the sandwich. The two dielectrics are then etched away where the lines and vias are required. A barrier layer such as 10 nm thick Ta or TaN film is then deposited to prevent the copper from diffusing into the dielectrics [Peng02]. As can be seen, a thin layer of the barrier remains at the bottom of the via so the barrier must be conductive. A copper seed layer is then coated over the barrier layer (Figure 15.21(g)). The resulting structure is electroplated full of copper, and finally the structure is ground flat with CMP, as shown in Figure 15.21(h).

15.4.2.2 Low-k Dielectrics SiO_2 has a dielectric constant of $k = 3.9–4.2$. Low-k dielectrics between wires are attractive because they decrease the wire capacitance [Brown03]. This reduces wire delay, noise, and power consumption. Adding fluorine to the silicon dioxide creates *fluorosilicate glass* (FSG or SiOF) with a dielectric constant of 3.6, widely used in 130 nm processes. Adding carbon to the oxide can reduce the dielectric constant to about 2.8–3; such SiCOH (also called *carbon-doped oxide*, CDO) is commonly used at the 90 and 65 nm generation. Alternatively, porous polymer-based dielectrics can deliver even lower dielectric constants. For example, SiLK, from Dow Chemical, has $k = 2.6$ and may scale to $k = 1.6–2.2$ by increasing the porosity. IBM has demonstrated air (or vacuum)

FIGURE 15.22 Micrograph showing air gap insulation between copper wires (Courtesy of International Business Machines Corporation. Unauthorized use not permitted.)

gaps, which have $k = 1$ where the dielectric has been eliminated entirely, as shown in Figure 15.22. Developing low-k dielectrics that can withstand the high temperatures during processing and the forces applied during CMP is a major challenge.

15.4.3 Circuit Elements

While CMOS transistors provide for almost complete digital functionality, the use of CMOS technology as the mixed signal and RF process of choice has driven the addition of special process options to enhance the performance of circuit elements required for these purposes.

15.4.3.1 Capacitors In a conventional CMOS process, a capacitor can be constructed using the gate and source/drain of an MOS transistor, a diffusion area (to ground or V_{DD}), or a parallel metal plate capacitor (using stacked metal layers). The MOS capacitor has good capacitance per area but is relatively nonlinear if operated over large voltage ranges. The diffusion capacitor cannot be used for a floating capacitor (where neither terminal is connected to ground). The metal parallel plate capacitor has low capacitance per area. Normally, the aim in using a floating capacitor is to have the highest ratio of desired capacitance value to stray capacitance (to ground normally). The bottom metal plate contributes stray capacitance to ground.

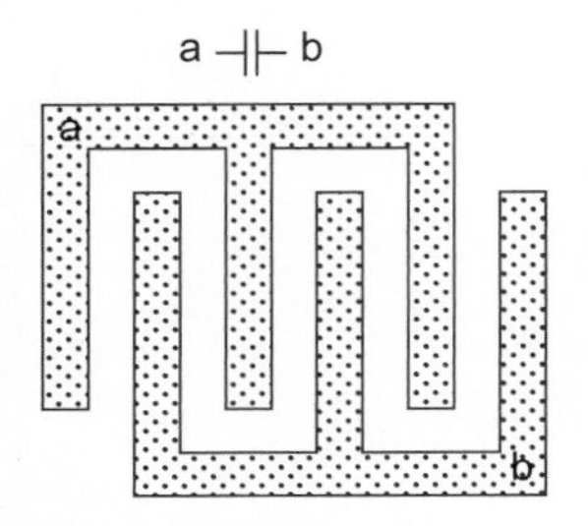

FIGURE 15.23 Fringe capacitor

Analog circuits frequently require capacitors in the range of 1 to 10 pF. The first method for doing this was to add a second polysilicon layer so that a *poly-insulator-poly* (PIP) capacitor could be constructed. A thin oxide was placed between the two polysilicon layers to achieve capacitance of approximately 1 fF/μm^2.

The most common capacitor used in CMOS processes today is a fringe capacitor, which consists of interdigitated fingers of metal, as shown in Figure 15.23. Multiple layers can be stacked to increase the capacitance per area.

15.4.3.2 Resistors In unaugmented processes, resistors can be built from any layer, with the final resistance depending on the resistivity of the layer. Building large resistances in a small area requires layers with high resistivity, particularly polysilicon, diffusion, and n-wells. Diffusion has a large parasitic capacitance to ground, making it unsuitable for high-frequency applications. Polysilicon gates are usually silicided to have low resistivity. The fix for this is to allow for *undoped* high-resistivity polysilicon. This is specified with a *silicide block* mask where high-value poly resistors are required. The resistivity can be tuned to around 300–1000 Ω/square, depending on doping levels. Another material used for precision resistors is nichrome, although this requires a special processing step.

A typical resistor layout is shown in Figure 15.24. This geometry is sometimes called a *meander* structure. A number of *unit* resistors have been used so that a variety of matched resistor values can be constructed. For instance, if 20 kΩ, and 15 kΩ resistors were required, a unit value of 5 kΩ could be used. Then three resistors (as shown) would construct a 15 kΩ resistor. The two resistors at the ends are called *dummy resistors* or fingers. They perform no circuit function, but replicate the proximity effects (such as etch and implant) that the interior resistors see during processing. This helps ensure that all resistors are matched.

The various resistor options have temperature and voltage coefficients. Foundry design manuals normally include these values.

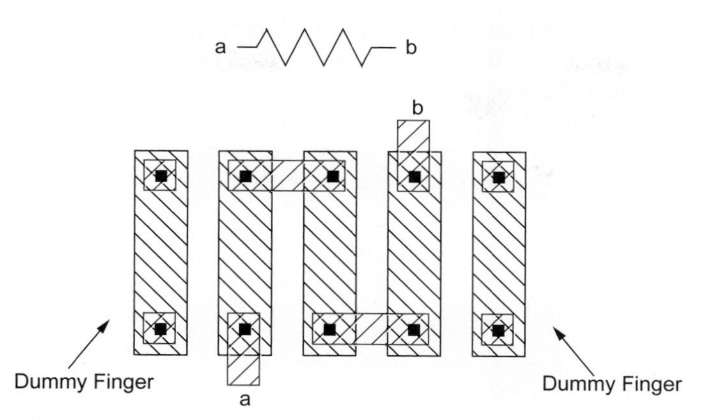

FIGURE 15.24 Resistor layout

15.4.3.3 Inductors The desire to integrate inductors on chips has increased radically with the upsurge in interest in RF circuits. The most common monolithic inductor is the *spiral* inductor, which is a spiral of upper-level metal. A typical inductor is shown in Figure 15.25(a). As the process is planar, an underpass connection has to be made to complete the inductor. A typical equivalent model is shown in Figure 15.25(b). In addition to the required inductance L, there are several parasitic components. R_s is the series resistance of the metal (and contacts) used to form the inductor. C_p is the parallel capacitance to ground due to the area of the metal wires forming the inductor. C_s is the shunt capacitance of the underpass. Finally, R_p is an element that models the loss incurred in the resistive substrate.

Usually, when considering an inductor, the parameters of interest to a designer are its inductance, the Q of the inductor, and the self-resonant frequency. High Qs are sought to create low phase-noise oscillators, narrow filters, and low-loss circuits in general. Q values for typical planar inductors on a bulk process range from 5 to 10.

The number of turns n required to achieve some inductance L if the wire pitch, in turns per meter, is $P = 1/2(W + S)$, is [Lee98]

$$n \approx \sqrt[3]{\frac{LP}{\mu_0}} \quad \textbf{(15.4)}$$

where $\mu_0 = 1.2 \times 10^{-6}$ H/m is the permeability of free space. Figure 15.25 has $n = 1.75$ turns. Higher-quality inductors can also be manufactured using bond wires between I/O pads. The inductance of a wire of length l and radius r is approximately

$$L \approx \frac{\mu_0 l}{2\pi}\left[\ln\frac{2l}{r} - 0.75\right] \quad \textbf{(15.5)}$$

or about 1 nH/mm for standard 1 mil (25 μm) bond wires.

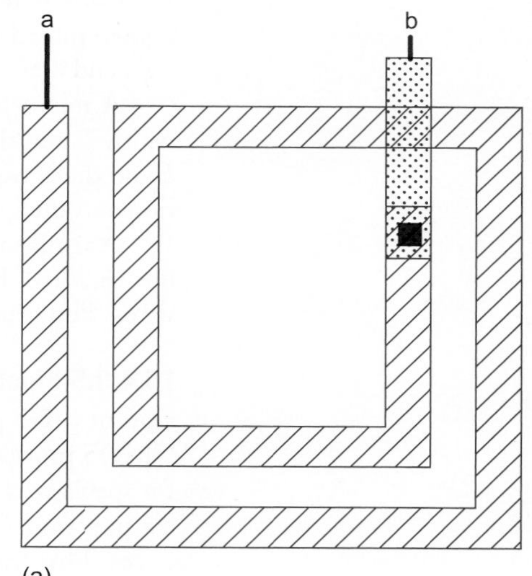

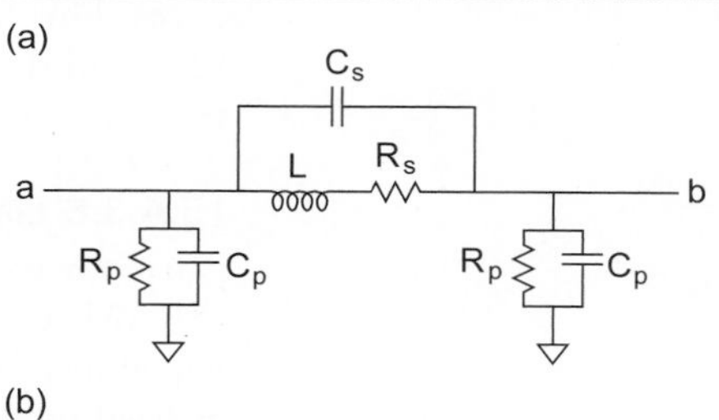

FIGURE 15.25 Typical spiral inductor and equivalent circuit [Rotella02]

Reduction in Q occurs because of the resistive loss in the conductors used to build the inductor (R_s), and the eddy current loss in the resistive silicon substrate (R_p). In an effort to increase Q, designers have resorted to removing the substrate below the inductor using MEMS techniques [Yoon02]. The easiest way to improve the Q of monolithic inductors is to increase the thickness of the top-level metal. The Q can also be improved by using a patterned ground shield in polysilicon under the inductor to decrease substrate losses.

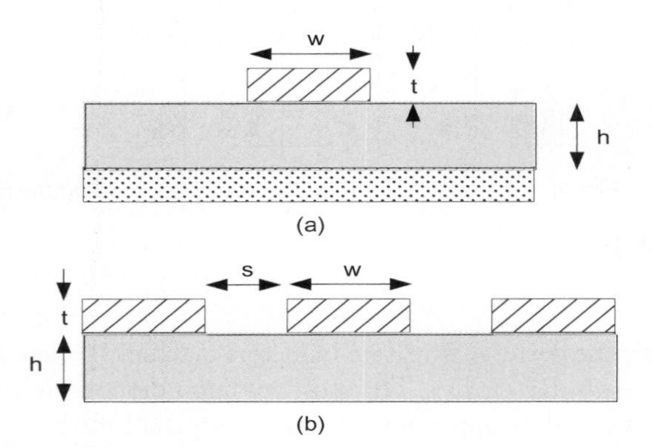

FIGURE 15.26 Microstrip and coplanar waveguide

15.4.3.4 Transmission Lines A *transmission line* can be used on a chip to provide a known impedance wire. Two basic kinds of transmission lines are commonly used: *microstrips* and *coplanar waveguides.*

A microstrip transmission line, as shown in Figure 15.26(a), is composed of a wire of width w and thickness t placed over a ground plane and separated by a dielectric of height h and dielectric constant k. In the chip case, the wire might be the top level of metallization and the ground plane the next metal layer down.

A coplanar waveguide does not require a sublayer ground plane and is shown in Figure 15.26(b). It consists of a wire of width w spaced s on each side from *coplanar* ground wires. The reader is referred to [Wadell91] for detailed design equations.

15.4.3.5 Bipolar Transistors Bipolar transistors were mentioned previously in our discussion of SiGe process options. Both *npn* and *pnp* bipolar transistors can be added to a CMOS process, which is then called a BiCMOS process. These processes tend to be used for specialized analog or high-voltage circuits. In a regular n-well process, a parasitic vertical pnp transistor is present that can be used for circuits such as bandgap voltage references. This transistor is shown in Figure 15.27 with the p-substrate collector, the n-well base, and the p-diffusion emitter. Both process cross-section and layout are shown. This transistor, in conjunction with a parasitic npn, is the cause of latchup (see Section 6.3.6).

15.4.3.6 Embedded DRAM Dynamic RAM (DRAM) uses a single transistor and a capacitor to store a bit of information. It is about five times denser than static RAM (SRAM) conventionally used on CMOS logic chips, so it can reduce the size of a chip containing large amounts of memory. DRAM was conventionally manufactured on specialized processes that produced low-performance logic transistors. DRAM requires specialized structures to build capacitors in a small area. One common structure is a *trench*,

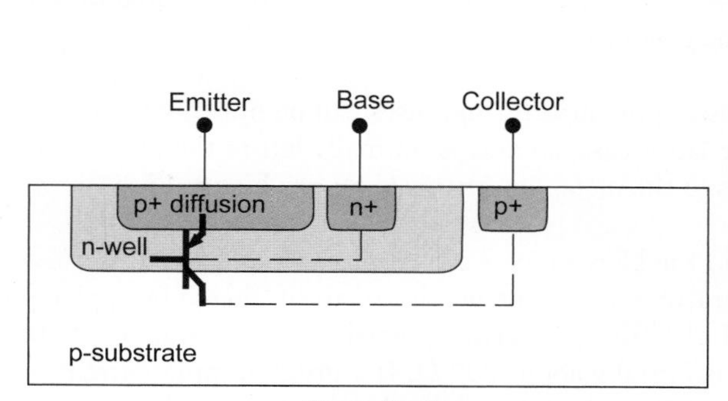

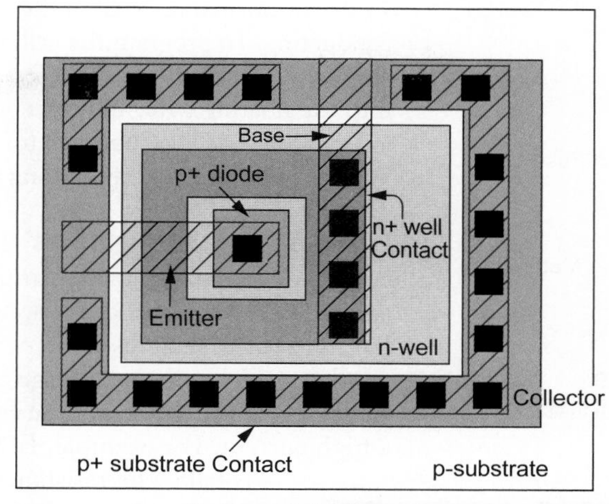

FIGURE 15.27 Vertical pnp bipolar transistor

which is etched down into the substrate. Some recent processes have introduced compact capacitor structures for building *embedded DRAM* alongside high-performance logic. Section 11.3 discusses DRAM in more depth.

15.4.3.7 Non-Volatile Memory *Non-volatile* memory (NVM) retains its state when the power is removed from the circuit. The simplest NVM is a *mask-programmed* ROM cell (see Section 11.4). This type of NVM is not reprogrammable or programmable after the device is manufactured. A *one-time programmable* (OTP) memory can be implemented using a fuse constructed of a thin piece of metal through which is passed a current that vaporizes the metal by exceeding the current density in the wire. The first reprogrammable memories used a stacked polysilicon gate structure and were programmed by applying a high voltage to the device in a manner that caused Fowler-Nordheim tunneling to store a charge on a floating gate. The whole memory could be erased by exposing it to UV light that knocked the charge off the gate. These memories evolved to *electrically erasable* memories, which are today represented by *Flash* memory.

A typical Flash memory transistor is shown in Figure 15.28 [She02]. The source and drain structures can vary considerably to allow for high-voltage operation, but the dual-gate structure is fairly common. The gate structure is a stacked configuration commencing with a thin tunnel oxide or nitride. A floating polysilicon gate sits on top of this oxide and a conventional gate oxide is placed on top of the floating gate. Finally, a polysilicon control gate is placed on top of the gate oxide. The operation of the cell is also shown in Figure

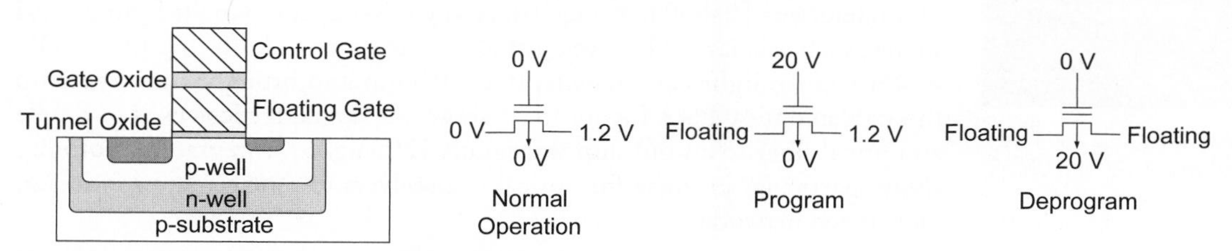

FIGURE 15.28 Flash memory construction and operation

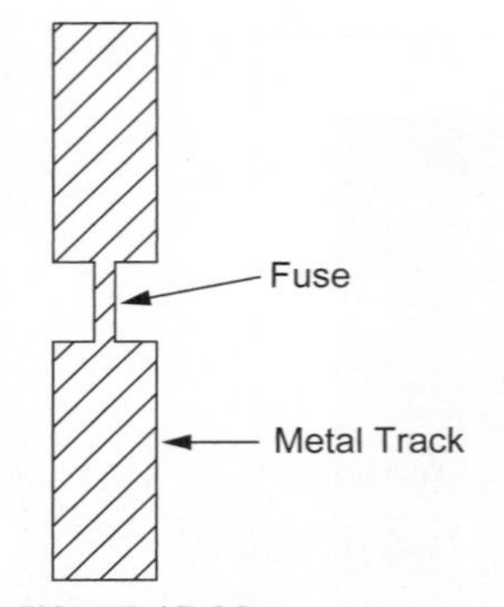

FIGURE 15.29
A typical metal fuse

15.28. In normal operation, the floating gate determines whether or not the transistor is conducting. To program the cell, the source is left floating and the control gate is raised to approximately 20 V (using an on-chip voltage multiplier). This causes electrons to tunnel into the floating gate, and thus program it. To deprogram a cell, the drain and source are left floating and the substrate (or well) is connected to 20 V. The electrons stored on the floating gate tunnel away, leaving the gate in an unprogrammed state.

15.4.3.8 Fuses and Antifuses During manufacturing, fuses can be blown with a high current or zapped by a laser. In the latter case, an area is normally left in the passivation oxide to allow the laser direct access to the metal link that is to be cut. Figure 15.29 shows the layout of a metal fuse.

Laser-blown fuses are large and the blow process can damage adjacent devices. *Electronic fuses* are structures whose characteristics can be nondestructively altered by applying a high current. For example, IBM eFUSEs are narrow polysilicon wires silicided with cobalt. The resistance is initially about 200 Ω. If a programming current of 10–15 mA is applied for 200 μs, the cobalt will migrate to the anode, as shown in Figure 15.30. This raises the resistance by an order of magnitude. Simple sense circuits are used to detect the state of the eFUSE. IBM uses fuses for chip serial numbers, thermal sensor calibration, and to reconfigure defective components [Rohrer05, Rizzolo07].

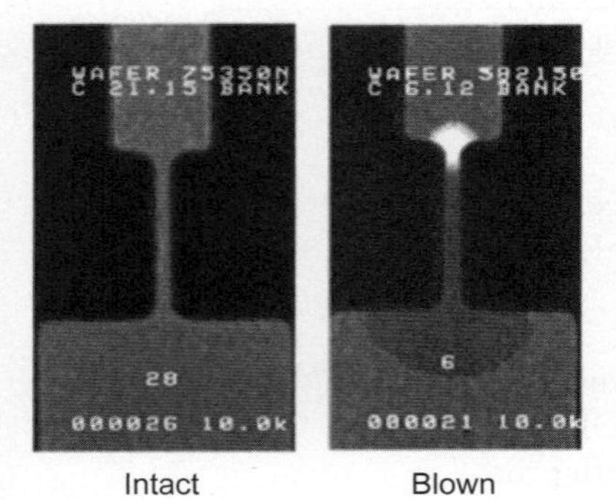

FIGURE 15.30 eFUSE (© IEEE 2005.)

An *antifuse* is a similar device that initially has a high resistivity but can become low resistance when a programming voltage is applied. This device requires special processing and is used in programmable logic devices (see Section 11.7).

15.4.3.9 Microelectromechanical Systems (MEMS) Semiconductor processes and especially CMOS processes have been used to construct tiny mechanical systems monolithically. A typical device is the well-known air-bag sensor, which is a small accelerometer consisting of an air bridge capacitor that can detect sudden changes in acceleration when co-integrated with some conditioning electronics. MEMS micromirrors on torsional hinges are used in inexpensive, high-resolution digital light projectors. Structures such as cantilevers, mechanical resonators, and even micromotors have been built. A full discussion of MEMS is beyond the scope of this book, but further material can be found in texts such as [Maluf04].

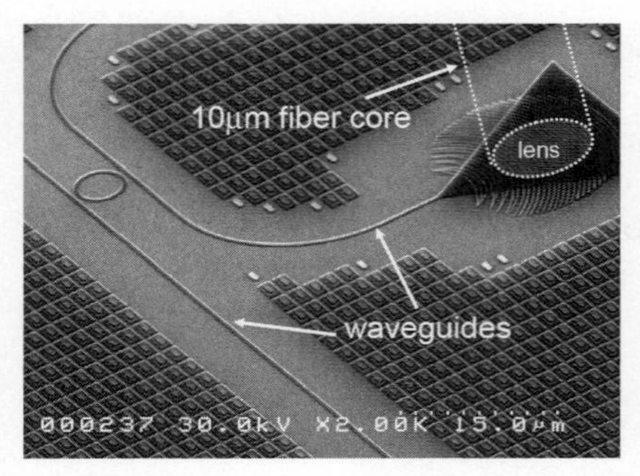

FIGURE 15.31 Optical waveguide and holographic lens integrated with a 130 nm CMOS process (© IEEE 2006.)

15.4.3.10 Integrated Photonics Although silicon is opaque at visible wavelengths, it is transparent in the infrared range used in optical fibers. Semiconductor photonic components are rapidly evolving. Components compatible with a conventional CMOS process include waveguides, modulators, and photodetectors [Salib04, Young10]. A key missing component is an optical source, such as a laser. However, just as V_{DD} is generated off-chip from a DC power supply, light can be generated off-chip and brought onto the chip through an optical fiber. Figure 15.31 shows a holographic lens used to couple an optical fiber to an on-chip waveguide [Huang06]. Integrated photonics shows particular promise for optical transceivers to replace copper wires in high-speed networks.

15.4.3.11 Three-Dimensional Integrated Circuits 3D ICs contain multiple layers of devices. Stacking transistors in layers can reduce wire lengths, improving speed and power. It also can permit heterogeneous technologies to be combined in one package; for example, logic, memory, and analog/RF chips can be stacked into one package.

IBM has described a process in which 200 mm wafers are ground down to a remarkable 20 μm thickness after fabrication [Topol06]. They are aligned to 1 μm tolerance, and then one is bonded on top of another using oxide-fusion or copper bonding. Tall skinny through-silicon vias (TSVs) between the wafers are etched and metallized; the aspect ratio of the vias and the thickness of the wafers sets the density of contacts between wafers. Densities of 10^4 TSVs/mm^2 or more can presently be achieved. Some of the challenges in 3D integration include wafer bowing, testing layers before they are bonded, and managing cooling and power delivery. Figure 15.32 shows two wafers bonded together. The bottom wafer has four levels of metal and the top wafer has two levels. The 8-μm wide landing pad on the top metal layer of the bottom wafer provides tolerances for misalignment with the 3D vias protruding from beneath the top wafer.

3D ICs are starting to move from research into production [Emma08]. An initial application is to stack multiple memory chips to provide a higher capacity in a standard form factor.

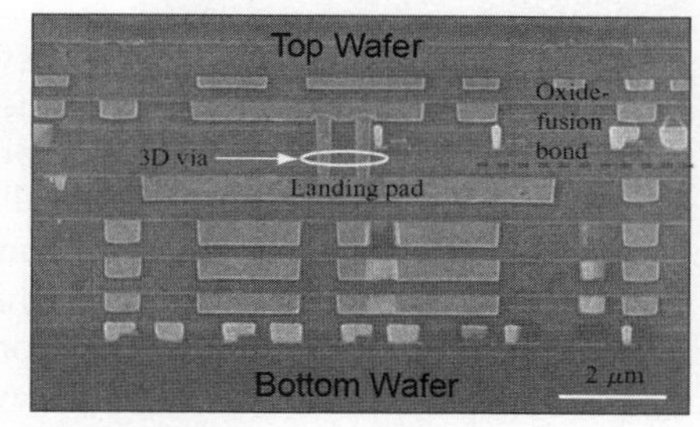

FIGURE 15.32 Scanning electron micrograph of 3-dimensional integration of two wafers (Reprinted from [Koester08]. Courtesy of International Business Machines Corporation. Unauthorized use not permitted.)

15.4.4 Beyond Conventional CMOS

A major problem with scaling bulk transistors is the subthreshold leakage from drain to source caused by the inability of the gate to turn off the channel completely. This can be improved by a gate structure where the gate is placed on two, three, or four sides of the channel to gain better control over the charge in the channel. A promising structure solves the problem by forming a vertical channel and constructing the gate in a pincer-like arrangement around three sides. These devices have been given the generic name *"finfets"* because the source/drain regions form fins on the silicon surface [Hisamoto98]. Figure 15.33(a) shows a 3D view of a finfet, while Figure 15.33(b) shows the cross-section and Figure 15.33(c) shows the top view. The gate wraps around three sides of the vertical

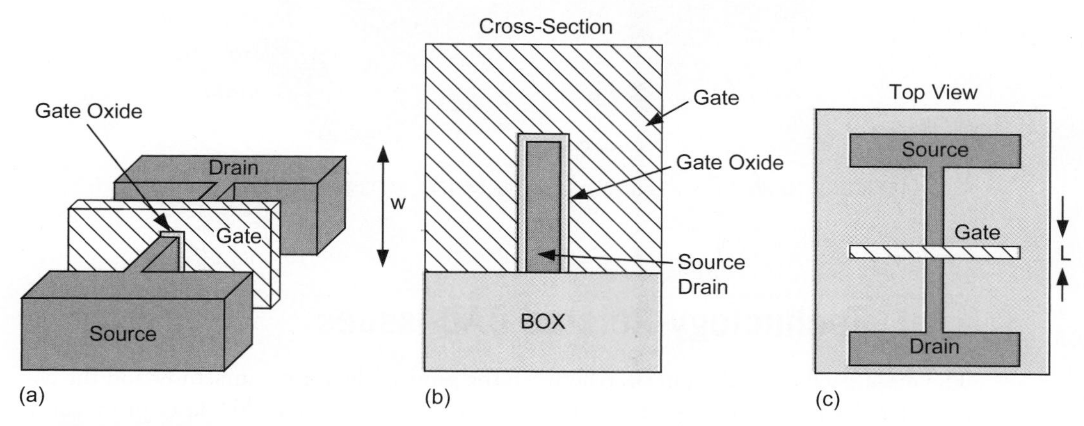

FIGURE 15.33 Finfet structure

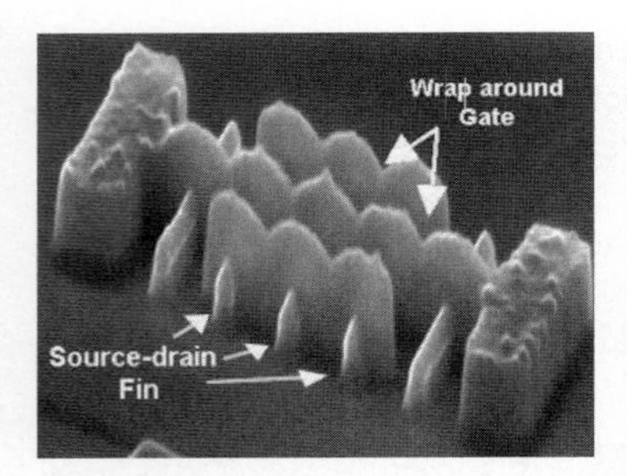

FIGURE 15.34 Trigate transistor (Reprinted with permission of Intel Corporation.)

source/drain fins. The width of the device is defined by the height of the fin, so wide devices are constructed by paralleling fins. Figure 15.34 shows a micrograph of a prototype finfet that Intel calls a *trigate transistor* [Kavalleros06].

Compounds from groups III and V of the periodic table, such as GaAs, offer electron mobilities up to 30 times higher than silicon. Such III-V materials have been research topics for decades. GaAs was once used for very high frequency applications, but has largely been replaced by advanced CMOS processes. However, III-V materials might be integrated into CMOS some day in the future.

Nanotechnology is presently a hot research area seeking alternative structures to replace CMOS when conventional scaling finally runs out of steam. Little obvious progress in radical new device structures has been made since the previous edition of the book, but conventional sub-100 nm CMOS transistors are now being called nanotechnology! Alternative technologies have a large hurdle to overcome competing with the hundreds of billions of dollars that have been invested in advancing CMOS over four decades.

Carbon nanotubes are one nanotechnology that have been used to demonstrate transistor behavior and build inverters [Liu01]. Nanotubes are cylinders with a diameter of a few nanometers. They are of interest because the nanotube is smaller than the predicted endpoint for CMOS gate lengths, and because the nanotubes offer high mobility. A theoretical nanotube transistor is shown in Figure 15.35 [Wong03]. Presently, the speeds are quite slow and the manufacturing techniques are limited, but they may be of interest in the future [Raychowdhury07, Patil09].

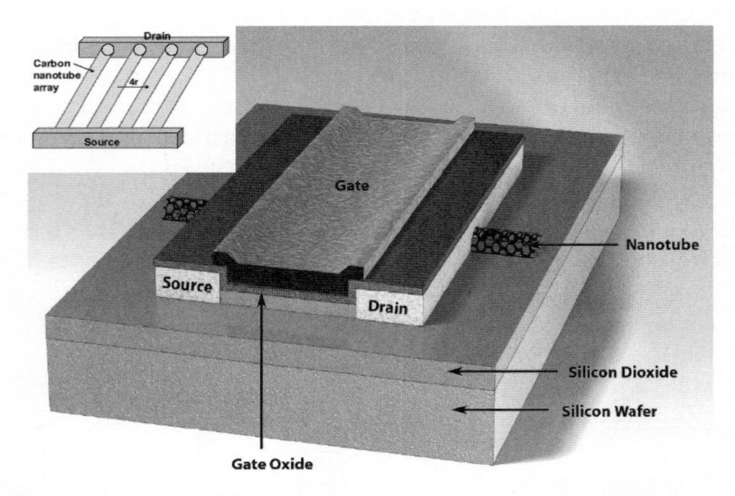

FIGURE 15.35 Carbon nanotube transistor (© IEEE 2003.)

15.5 Technology-Related CAD Issues

The mask database is the interface between the semiconductor manufacturer and the chip designer. Two basic checks have to be completed to ensure that this description can be turned into a working chip. First, the specified geometric design rules must be obeyed. Second, the interrelationship of the masks must, upon passing through the manufacturing process, produce the correct interconnected set of circuit elements. To check these two requirements, two basic CAD tools are required: a *Design Rule Check* (DRC) program and a mask circuit *extraction* program. The most common approach to implementing these tools is a set of subprograms that perform general geometry operations. A particular set of DRC rules or extraction rules for a given CMOS process (or any semiconductor process) defines the operations that must be performed on each mask and the inter-mask checks that must be completed. Accompanied by a written description, these *run sets* are usually the defining specification for a process.

In this section, we will examine a hypothetical DRC and extraction system to illustrate the nature of these run sets.

15.5.1 Design Rule Checking (DRC)

Although we can design the physical layout in a certain set of mask layers, the actual masks used in fabrication can be derived from the original specification. Similarly, when we want a program to determine what we have designed by examining the interrelationship of the various mask layers, it may be necessary to determine various logical combinations between masks.

To examine these concepts, let us posit the existence of the following functions (loosely based on the Cadence DRACULA DRC program), which we will apply to a geometric database (i.e., rectangles, polygons, and paths):

AND `layer1 layer2 -> layer3`
ANDs layer1 and layer2 together to produce layer3
(i.e., the intersection of the two input mask descriptions)

OR `layer1 layer2 -> layer3`
ORs layer1 and layer2 together to produce layer3
(i.e., the union of the two input mask descriptions)

NOT `layer1 layer2 -> layer3`
Subtracts layer2 from layer1 to produce layer3
(i.e., the difference of the two input mask descriptions)

WIDTH `layer > dimension -> layer3`
Checks that all geometry on layer is larger than dimension
Any geometry that is not is placed in layer3

SPACE `layer > dimension -> layer3`
Checks that all geometry on layer is spaced further than dimension
Any geometry that is not is placed in layer3

The following layers will be assumed as input:

```
nwell
active
p-select
n-select
poly
poly-contact
active-contact
metal
```

Typically, useful sublayers are generated initially. First, the four kinds of active area are isolated. The rule set to accomplish this is as follows:

```
NOT  all nwell -> substrate
AND  nwell active -> nwell-active
NOT  active nwell -> pwell-active
AND  nwell-active p-select -> pdiff
AND  nwell-active n-select -> vddn
AND  pwell-active n-select -> ndiff
AND  pwell-active p-select -> gndp
```

In the above specification, a number of new layers have been designated. For instance, the first rule states that wherever nwell is absent, a layer called substrate exists. The second rule states that all active areas within the nwell are nwell-active. A combination of nwell-active and p-select or n-select yields pdiff (p diffusion) or vddn (well tap).

To find the transistors, the following rule set is used:

```
AND poly ndiff -> ngates
AND poly pdiff -> pgates
```

The first rule states that the combination of poly and ndiff yields the ngates region—all of the n-transistor gates.

Typical design rule checks (DRC) might include the following:

```
WIDTH metal  < 0.13 -> metal-width-error
SPACE metal  < 0.13 -> metal-space-error
```

For instance, the first rule determines if any metal is narrower than 0.13 μm and places the errors in the metal-width-error layer. This layer might be interactively displayed to highlight the errors.

15.5.2 Circuit Extraction

Now imagine that we want to determine the electrical connectivity of a mask database. The following commands are required:

`CONNECT` `layer1 layer2`
Electrically connect layer1 and layer2.

`MOS` `name drain-layer gate-layer source-layer substrate-layer`
Define an MOS transistor in terms of the component terminal layers. (This is, admittedly, a little bit of magic.)

The connections between layers can be specified as follows:

```
CONNECT active-contact pdiff
CONNECT active-contact ndiff
CONNECT active-contact vddn
CONNECT active-contact gndp
CONNECT active-contact metal
CONNECT gndp substrate
CONNECT vddn nwell
CONNECT poly-contact poly
CONNECT poly-contact metal
```

The connections between the diffusions and metal are specified by the first seven statements. The last two statements specify how metal is connected to poly.

Finally, the active devices are specified in terms of the layers that we have derived:

```
MOS nmos ndiff ngates ndiff substrate
MOS pmos pdiff pgates pdiff nwell
```

An output statement might then be used to output the extracted transistors in some netlist format (i.e., SPICE format). The extracted netlist is often used to compare the layout against the intended schematic.

It is important to realize that the above run set is manually generated. The data you extract from such a program is only as good as the input. For instance, if parasitic routing capacitances are required, then each layer interaction must be coded. If parasitic resistance is important in determining circuit performance, it also must be specifically included in the extraction run set.

15.6 Manufacturing Issues

As processes have evolved, various rules and guidelines have emerged that reflect the complexity of the processing. These rules are often called *Design for Manufacturability* (DFM).

15.6.1 Antenna Rules

When a metal wire contacted to a transistor gate is plasma-etched, it can charge up to a voltage sufficient to zap the thin gate oxides. This is called *plasma-induced gate-oxide damage*, or simply the *antenna effect*. It can increase the gate leakage, change the threshold voltage, and reduce the life expectancy of a transistor. Longer wires accumulate more charge and are more likely to damage the gates.

During the high-temperature plasma etch process, the diodes formed by source and drain diffusions can conduct significant amounts of current. These diodes bleed off charge from wires before gate oxide is damaged.

Antenna rules specify the maximum area of metal that can be connected to a gate without a source or drain to act as a discharge element. Larger gates can withstand more charge buildup. The design rules normally define the maximum ratio of metal area to gate area such that charge on the metal will not damage the gate. The ratios can vary from 100:1 to 5000:1 depending on the thickness of the gate oxide (and hence breakdown voltage) of the transistor in question. Higher ratios apply to thicker gate oxide transistors (i.e., 3.3 V I/O transistors).

Figure 15.36 shows an antenna rule violation and two ways to fix it. In Figure 15.36(a), a long metal1 line is connected to a transistor gate. It has no connection to diffusion until metal2 is formed, so the gate may be damaged during the metal1 plasma etch. In Figure 15.36(b), the metal1 line is interrupted with a jumper to metal2. This reduces the amount of charge that could zap the gate during the metal1 etch and solves the problem. In Figure 15.36(c), an antenna diode is added, providing a discharge path during the etch. The diode is reverse-biased during normal operation and thus does not disturb circuit function (except for the area and capacitance that it contributes). Note that the problem could also have been solved by making the gate wider.

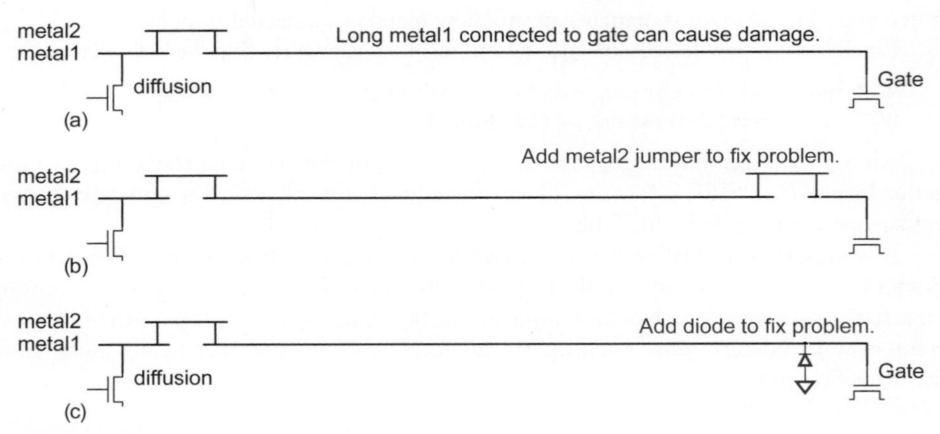

FIGURE 15.36 Antenna rule violation and fixes

For circuits requiring good matching, such as analog and memory cells, transistor gates should connect directly to diffusion with a short segment of metal1 to avoid gate damage that could introduce mismatches.

15.6.2 Layer Density Rules

Another set of rules that pertain to advanced processes are layer density rules, which specify a minimum and maximum density of a particular layer within a specified area. Etch rates have some sensitivity to the amount of material that must be removed. For example, if polysilicon density were too high or too low, transistor gates might end up over- or under-etched, resulting in channel-length variations. Similarly, the CMP process may cause dishing (excessive removal) of copper when the density is not uniform.

To prevent these issues, a metal layer might be required to have 30% minimum and 70% maximum density within a 100 μm by 100 μm area. For digital circuits, these density levels are normally reached with routine routing unless empty spaces exist. Analog and RF circuits, on the other hand, are almost by definition sparse. Thus, diffusion, polysilicon, and metal layers may have to be added manually or by a fill program after design has been completed. The fill can be grounded or left floating. Floating fill contributes lower total capacitance but more coupling capacitance to nearby wires. Grounded fill requires routing the ground net to the fill structures. Clever fill patterns such as staggered rectangles, plus-sign patterns, or diamonds result in lower and more predictable capacitance than do simple geometrical grids [Kahng08]. Designers must be aware of the fill so that it does not introduce unexpected parasitic capacitance to nearby wires.

15.6.3 Resolution Enhancement Rules

Some resolution enhancement techniques impose further design rules. For example, polysilicon typically uses the narrowest lines and thus needs the most enhancement. This can be simplest if polysilicon gates are only drawn in a single orientation (horizontal or vertical). Using a single orientation also reduces systematic process variability. Avoid small jogs and notches (those less than the minimum layer width), because such notches can interfere with proper OPC analysis.

The design community is presently debating a move toward *restrictive design rules* to facilitate RET and reduce manufacturing variability by limiting designers to a smaller set of uniform layout features. These rules might come at the expense of greater area. For example, Intel introduced restrictive design rules for polysilicon in the 45 nm process to control variation and facilitate 193 nm double-patterning lithography [Webb08]. Under these rules, polysilicon is limited to one pitch and direction in layout. This also simplified contact and metal1 rules: the contact pitch is the same as the gate pitch, and metal1 parallel to the gates also has the same pitch. Wide poly pads for contacts and orthogonal polysilicon routing were eliminated by introducing a *trench contact* suitable for local interconnect. Intel found that the restrictive rules did not impact standard cell density and that excellent yield is achieved.

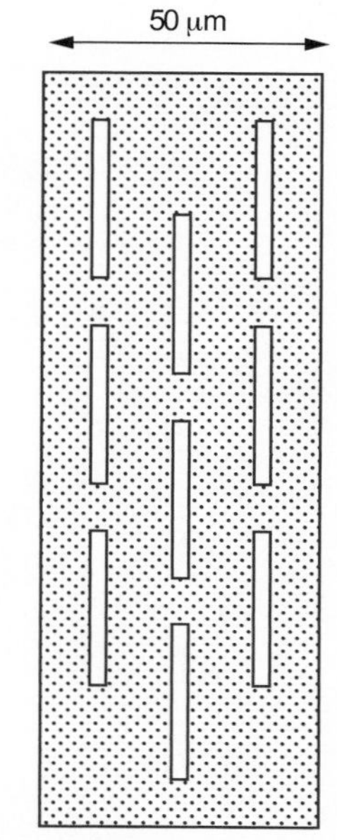

FIGURE 15.37 Slots in wide metal power bus

15.6.4 Metal Slotting Rules

Some processes have special rules requiring that wide (e.g. > 10–40 μm) metal wires have slots. Slots are long slits, on the order of 3 μm wide, in the wire running parallel to the direction of current flow, as shown in Figure 15.37. They provide stress relief, help keep the wire in place, and reduce the risk of electromigration failure (see Section 6.3.3.1). Design rules vary widely between manufacturers.

15.6.5 Yield Enhancement Guidelines

To improve yield, some processes recommend increasing certain widths and spacings where they do not impact area or performance. For example, increasing the polysilicon gate extension slightly reduces the risk of transistor failures from poly/diffusion mask misalignment. Increasing space between metal lines where possible reduces the risk of shorts and also reduces wire capacitance. Other good practices to improve yield include the following:

- Space out wires to reduce risk of short circuits and reduce capacitance.
- Use non-minimum-width wires to reduce risk of open circuits and to reduce resistance.
- Use at least two vias for every connection to avoid open circuits if one via is malformed, and to reduce electromigration wearout.
- Surround contacts and vias by landing pads with more than the minimum overlap to reduce resistance variation and open circuits caused by misaligned contacts.
- Use wider-than-minimum transistors; minimum-width transistors are subject to greater variability and tend not to perform as well.
- Avoid non-rectangular shapes such as 45-degree angles and circles. For specialized circuits such as RAMs that strongly benefit from 45-degree angles, verify masks after optical proximity correction analysis.
- Place dummy transistors or cells at the edge of arrays and sensitive circuits to improve uniformity and matching.
- If it looks nice, it will work better.

15.7 Pitfalls and Fallacies

Targeting a bleeding-edge process

There is a fine balance when you are deciding whether or not to move to a new process for a new design. On the one hand, you are tempted by increased density and speed. On the other hand, support for the new process can initially be expensive (becoming familiar with process rules, CAD tool scripts, porting analog and RF designs, locating logic libraries, etc.). In addition, CMOS foundries frequently tune their processes in the first few months of production, and often yield improvement steps can reflect back to design rule changes that impact designs late in their tapeout schedule. For this reason, it is frequently prudent not to jump immediately into a new process when it becomes available. On the other hand, if you are limited in speed or some other attribute that is solved by the new process, then you don't have much choice but to bite the bullet.

Using lambda design rules on commercial designs

Lambda rules have been used in this text for ease of explanation and consistency. They are usable for class designs. However, they are not very useful for production designs for deep submicron processes. Of particular concern are the metal width and spacing rules, which are too conservative for most production processes.

Failing to account for the parasitic effects of metal fill

With area density rules, particularly in metal, most design flows include an automatic fill step to achieve the correct metal density. Particularly in analog and RF circuits, it is important to either exclude the automatic fill operation from that area or check circuit performance after the fill by completing a full parasitic extract and rerunning the verification simulation scripts.

Failing to include process calibration test structures

In the discussion on scribe line structures, it was mentioned that test structures are frequently inserted here by the silicon manufacturer. Documentation is often unavailable, so it is prudent for designers (particularly in academic designs, which receive less support from a foundry) to include their own test structures such as transistors or ring oscillators. This allows designers to calibrate the silicon against simulation models.

Waiving design rules

Sometimes it is tempting to ignore a design rule when you are certain it does not apply. For example, consider two wires separated by only 2 λ. This violates a design rule because the wires might short together during manufacturing. If the wires are actually connected elsewhere, one might ignore the rule because further shorting is harmless. However, it is possible that the "antifeature" between the wires would produce a narrow strip of photoresist that could break off and float around during manufacturing, damaging some other structure. Moreover, even if the rule violation is safe, keeping track of all the legitimate exceptions is too much work, especially on a large design. It is better to simply fix the design rule error.

Logos on the image sensor for the Spirit Mars rover (Reprinted from Molecular Expressions Silicon Zoo, `micro.magnet.fsu.edu/creatures`, with permission of Michael Davidson.)

Placing cute logos on a chip

Designers have a tradition of hiding their initials on the chip or embedding cute logos in an unused corner of the die. Some automatic wafer inspection tools find that the logos look more like a spec of dust than a legitimate chip structure and mark all of the chips as defective! Some companies now ban the inclusion of layout that is not essential to the operation of the device. Others require placing the logo in the corner of the chip and covering it with a special pseudolayer called LOGO to tell RET and wafer inspection tools to ignore the logo.

15.8 Historical Perspective

In the first days of integrated circuits, layout editors and design rule checkers were humans with knives and magnifying lenses. [Volk01] tells a captivating story of design at Intel in the early 1970s. Mask designers drew layout with sharp colored pencils on very large sheets of Mylar graph paper, as shown in Figure 15.38(a). Engineers and technicians then scrutinized the drawings to see if all of the design rules were satisfied and if the connections matched the schematic. Most chips at the time were probably manufactured with minor design rule errors, but correct wiring was essential. For example, two engineers each checked all 20,000 transistors on the 8086 in 1977 by hand. Both found 19 of the same 20 errors, giving confidence that the design was correct.

Technicians working at a light table then cut each level of layout onto sheets of *rubylith* to make the masks, as shown in Figure 15.38(b). Rubylith is a two-layered material with a base of heavy transparent Mylar and a thin film of red cellophane-like material. The red film was then peeled away where transistors or wires should be formed. The designer and technician spent days inspecting the rubylith for peeling errors and unintended cuts. The sheets had to be handled with great care to avoid rubbing off pieces. Corrections were performed with a surgical scalpel and metal ruler to add new wires, or with red tape to remove objects. The final result was checked with a 7 times magnifying glass. Finally, the rubylith sheets were sent to a mask vendor to be optically reduced to form the masks. Despite all this care, the initial version of Intel's first product, the 3101 64-bit RAM, was actually a 63-bit RAM because of an error peeling the rubylith. Designers today still gripe at their tools, but the industry has come a long way.

Advances in semiconductor devices are usually presented at the International Electron Devices Meeting (IEDM). Table 15.2 summarizes key characteristics from Intel and

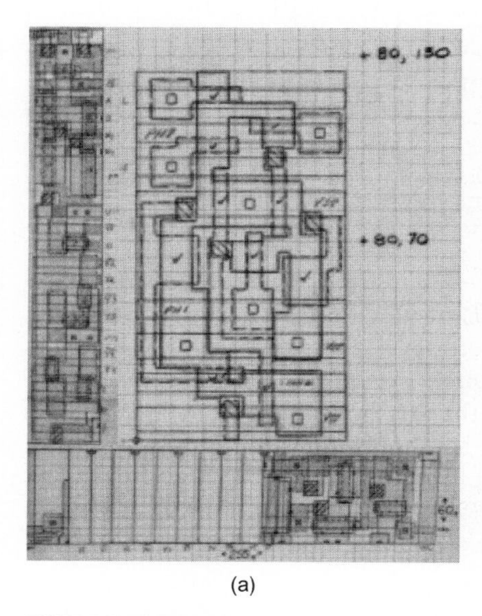

(a)

(b)

FIGURE 15.38 Hand-drawn layout: (a) standard cell, (b) cutting patterns onto rubylith (Reprinted from [Volk01] with permission of Intel Corporation.)

IBM. Process development has become so expensive that IBM has formed the Common Platform alliance with partners including Chartered Semiconductor, Samsung, Infineon, and STMicro, to share the R&D costs. IBM offers both SOI and bulk processes; Table 15.2 focuses on their SOI devices that have better I_{dsat} / I_{off} ratios. All of the processes in this table are considered *high-performance processes* that focus on a high I_{dsat}. Many manufacturers also offer *low-power processes* with higher threshold voltages and thicker oxides to reduce leakage, especially in battery-powered communications devices.

TABLE 15.2 CMOS process characteristics

Manufacturer		Intel							IBM			
Feature Size f	nm	250	180	130	90	65	45	32	130	90	65	45
Reference		[Bohr96]	[Yang98]	[Tyagi00]	[Thompson02]	[Bai04]	[Mistry07]	[Natarajan08]	[Sleight01]	[Khare02]	[Lee05]	[Narasimha06]
V_{DD}	V	1.8	1.5	1.3	1.2	1.2	1	1	1.2	1	1	1
L_{gate}	nm	180	140	70	50	35	35	30	60	45	40	35
t_{ox}	nm	4.1	3	1.5	1.2	1.2	1	0.9	2.3	1.85	1.05	1.15
I_{dsat-n}	μA/μm	700	940	1170	1449	1460	1360	1550	915	1000	1137	1140
I_{dsat-p}	μA/μm	320	420	600	725	880	1070	1210	520	480	700	800
I_{off}	nA/μm	1	3	100	400	100	100	100	100	200	200	200
Strain		no	no	no	yes	yes	yes	yes	no	no	yes	yes
High-k Gates		no	no	no	no	no	yes	yes	no	no	no	no
Gate Pitch	nm	640	480	336	260	220	160	112.5	325	245		190
Metal1 Pitch	nm	640	500	350	220	210	150	112.5	350	245		140
Metal Layers		5	6	6	7	8	9	9	8	10	10	10
Material		Al	Al	Cu	Cu	Cu	Cu	Cu	Cu	Cu	Cu	Cu
Low-k Dielectric		none	FSG	FSG	CDO	CDO	CDO	CDO	SiLK	SiLK	CDO	Porous
k		3.9	3.55	3.6	2.9	2.9					2.75	2.4
SRAM Cell Size	μm^2	10.26	5.59	2.09	1	0.57	0.346	0.171	1.8	0.99	0.65	0.37

The transistor characteristics are listed for low-V_t transistors. Since the 130 nm generation, nearly all processes have offered a regular-V_t transistor offering an order of magnitude lower I_{off} at the expense of a 15% reduction in I_{dsat}. Some low-power processes provide a high-V_t transistor to reduce leakage by another order of magnitude. Most manufacturers use a separate implant mask to specify the threshold voltage, but Intel reduces manufacturing cost by using a slightly (~10%) longer channel length instead, which increases V_t on account of the short-channel effect [Rusu07].

Reported subthreshold slopes range from 85–100 mV/decade. DIBL coefficients range from 100–130 mV/V and tend to get larger with technology scaling.

Summary

CMOS process options and directions can greatly influence design decisions. Frequently, the combination of performance and cost possibilities in a new process can provide new product opportunities that were not available previously. Similarly, venerable processes can offer good opportunities with the right product.

One issue that has to be kept in mind is the ever-increasing cost of having a CMOS design fabricated in a leading-edge process. Mask cost for critical layers is in the vicinity of $100K per mask. A full mask set for a 65 nm process can exceed $1M in cost, and the price has been roughly doubling at each technology node. This in turn is reflected in the types of design and approaches to design that are employed for CMOS chips of the future. For instance, making a design programmable so that it can have a longer product life is a good first start. Chapter 13 covers these approaches in depth.

For more advanced reading on silicon processing, consult textbooks such as [Wolf00].

Exercises

15.1 Explain the difference between a polycide and a salicide CMOS process. Which would be likely to have higher performance and why?

15.2 Calculate the minimum contacted pitch as shown in Figure 15.39 for metal1 in terms of λ using the SUBM rules. Is there a wiring strategy that can reduce this pitch?

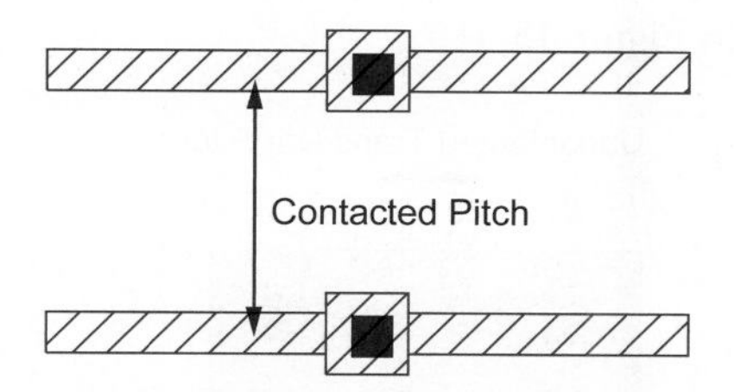

FIGURE 15.39 Contacted metal pitch

15.3 A 248 nm UV step and scan machine costs $10M and can produce 80 300 mm diameter, 90 nm node wafers per hour. A 193 nm UV step and scan machine costs $40M and can process 20 300 mm diameter, 50 nm node wafers per hour. If the machines have a depreciation period of four years, what is the difference in the cost per chip for a chip that occupies 50 square mm at 90 nm resolution if the stepper is used 10 times per process run for the critical layers?

15.4 If the gate oxide thickness in a SiO_2-based structure is 2 nm, what would be the thickness of an HfO_2-based dielectric providing the same capacitance?

15.5 Design a metal6 fuse ROM cell in a process where the minimum metal width is 0.5 μm and the maximum current density is 2 mA/μm. A fuse current of less than 10 mA is desired.

15.6 Draw the layout for a pMOS transistor in an n-well process that has active, p-select, n-select, polysilicon, contact, and metal1 masks. Include the well contact to V_{DD}.

15.7 What is the lowest resistance metal for interconnect? Why isn't it used?

15.8 Using Figure 15.40 and the SUBM design rules, calculate the minimum n to p pitch and the minimum inverter height with and without the poly contact to the gate (in). If an SOI process has 2 λ spacing between n and p diffusion, to what are the two pitches reduced?

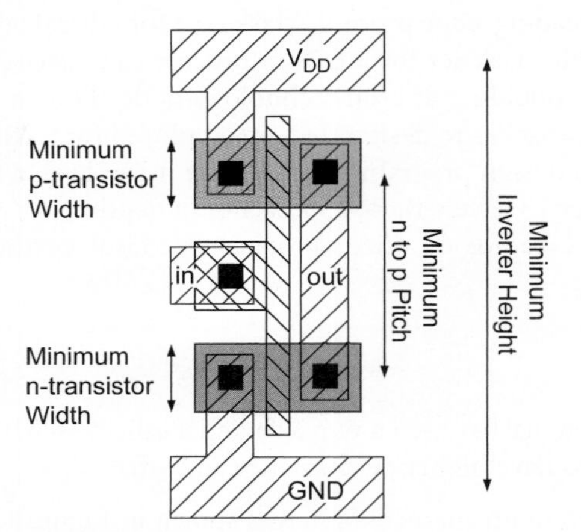

FIGURE 15.40 Minimum inverter height

15.9 Using the SUBM rules, calculate the minimum uncontacted and contacted transistor pitch, as shown in Figure 15.41.

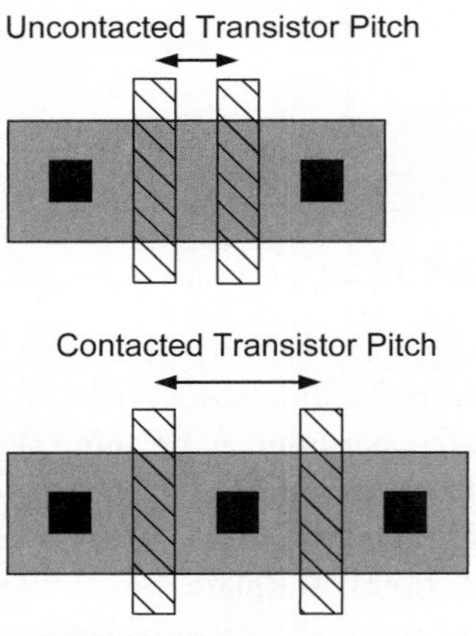

FIGURE 15.41 Uncontacted and contacted transistor pitch

References

A majority of references are from IEEE publications and can be obtained from `ieeexplore.ieee.org`. An electronic version of this bibliography with hyperlinks is available at `www.cmosvlsi.com`. The *IEEE Journal of Solid-State Circuits* is cited heavily and is abbreviated as *JSSC*.

[Abdollahi04] A. Abdollahi, F. Fallah, and M. Pedram, "Leakage current reduction in CMOS VLSI circuits by input vector control," *IEEE Trans. VLSI*, vol. 12, no. 2, Feb. 2004, pp. 140–154.

[Acken83] J. Acken, "Testing for bridging faults (shorts) in CMOS circuits," *Proc. Design Automation Conf.*, 1983, pp. 717–718.

[Afghahi90] M. Afghahi and C. Svensson, "A unified single-phase clocking scheme for VLSI systems," *JSSC*, vol. 25, no. 1, Feb. 1990, pp. 225–233.

[Agans06] D. Agans, *Debugging*, New York: Amacon, 2006, `www.debuggingrules.com`.

[Agarwal01] V. Agarwal, S. Keckler, and D. Burger, "The effect of technology scaling on microarchitectural structures," *Computer Architecture and Technology Laboratory Technical Report TR2000–02*, University of Texas at Austin, 2001.

[Agarwal04] A. Agarwal, V. Zolotov, and D. Blaauw, "Statistical clock skew analysis considering intra-die process variations," *IEEE Trans. CAD*, vol. 23, no. 8, Aug. 2004, pp. 1231–1242.

[Agarwal07] A. Agarwal, K. Kang, S. Bhunia, J. Gallagher, and K. Roy, "Device-aware yield-centric dual-V_t design under parameter variations in nanoscale technologies," *IEEE Trans. VLSI*, vol. 15, no. 6, Jun. 2007, pp. 660–671.

[Agarwal07b] K. Agarwal, R. Rao, D. Sylvester, and R. Brown, "Parametric yield analysis and optimization in leakage dominated technologies," *IEEE Trans. VLSI*, vol. 15, no. 6, Jun. 2007, pp. 613–623.

[Agrawal08] B. Agrawal and T. Sherwood, "Ternary CAM power and delay model: extensions and uses," *IEEE Trans. VLSI*, vol. 16, no. 5, May 2008, pp. 554–564.

[Aisaka02] K. Aisaka et al., "Design rule for frequency-voltage cooperative power control and its application to an MPEG-4 decoder," *Proc. VLSI Circuits Symp.*, 2002, pp. 216–217.

[Alexander75] J. Alexander, "Clock recovery from random binary signals," *Electronics Letters*, vol. 11, no. 22, Oct. 30, 1975, pp. 541–542.

[Allam00] M. Allam, M. Anis, and M. Elmasry, "High-speed dynamic logic styles for scaled-down CMOS and MTCMOS technologies," *Proc. Intl. Symp. Low Power Electronics and Design*, 2000, pp. 155–160.

[Alon05] E. Alon, V. Stojanovic, and M. Horowitz, "Circuits and techniques for high-resolution measurement of on-chip power supply noise," *JSSC*, vol. 40, no. 4, Apr. 2005, pp. 820–828.

[Alvandpour02] A. Alvandpour, R. Krishnamurthy, K. Soumyanath, and S. Borkar, "A sub-130-nm conditional keeper technique," *JSSC*, vol. 37, no. 5, May 2002, pp. 633–638.

[Amrutur98] B. Amrutur and M. Horowitz, "A replica technique for wordline and sense control in low-power SRAM's," *JSSC*, vol. 33, no. 8, Aug. 1998, pp. 1208–1219.

[Amrutur00] B. Amrutur and M. Horowitz, "Speed and power scaling of SRAM's," *JSSC*, vol. 35, no. 2, Feb. 2000, pp. 175–185.

[Amrutur01] B. Amrutur and M. Horowitz, "Fast low-power decoders for RAMs," *JSSC*, vol. 36, no. 10, Oct. 2001, pp. 1506–1515.

[Anastasakis02] D. Anastasakis, R. Damiano, H. Ma, and T. Stanion, "A practical and efficient method for compare-point matching," *Proc. Design Automation Conf.*, Jun. 2002, pp. 305–310.

[Anderson02] F. Anderson, J. Wells, and E. Berta, "The core clock system on the next generation Itanium microprocessor," *Proc. IEEE Intl. Solid-State Circuits Conf.*, Feb. 2002, pp. 146–147, 453.

[Ando80] H. Ando, "Testing VLSI with random access scan," *Digest of Papers COMPCON 80*, Feb. 1980, pp. 50–52.

[Anis03] M. Anis and M. Elmasry, *Multi-Threshold CMOS Digital Circuits*, Norwell, MA: Kluwer, 2003.

[Arnaud08] F. Arnaud et al., "32 nm general purpose bulk CMOS technology for high performance applications at low voltage," *Proc. Intl. Electron Devices Meeting*, Dec. 2008, pp. 1–4.

[Artisan02] Artisan Components, *TSMC 0.18 μm Process 1.8-Volt SAGE-X Standard Cell Library Databook*, Release 4.0, Feb. 2002.

[Asenov07] A. Asenov, "Simulation of statistical variability in nano MOSFETs," *Proc. VLSI Technology Symp.*, Jun. 2007, pp. 86–87.

[Auth08] C. Auth et al., "45 nm high-k+ metal gate strain-enhanced transistors," *Intel Technology Journal*, vol. 12, no. 2, Jun. 2008, pp. 77–85.

[Baghini02] M. Baghini and M. Desai, "Impact of technology scaling on metastability performance of CMOS synchronizing latches," *Proc. Intl. Conf. VLSI Design*, 2002, pp. 317–322.

[Bai04] P. Bai et al., "A 65 nm logic technology featuring 35 nm gate lengths, enhanced channel strain, 8 Cu interconnect layers, low-k ILD and 0.57 μm^2 SRAM cell," *Proc. Intl. Electron Devices Meeting*, Dec. 2004, pp. 657–660.

[Bailey98] D. Bailey and B. Benschneider, "Clocking design and analysis for a 600-MHz Alpha microprocessor," *JSSC*, vol. 33, no. 11, Nov. 1998, pp. 1627–1633.

[Baker97] K. Baker and J. van Beers, "Shmoo plotting: the black art of IC testing," *IEEE Design and Test of Computers*, vol. 14, no. 3, Jul.–Sep. 1997, pp. 90–97.

[Bakoglu90] H. Bakoglu, *Circuits, Interconnections, and Packaging for VLSI*, Reading, MA: Addison-Wesley, 1990.

[Barke88] E. Barke, "Line-to-ground capacitance calculation for VLSI: a comparison," *IEEE Trans. Computer-Aided Design*, vol. 7, no. 2, Feb. 1988, pp. 295–298.

[Barry03] J. Barry, E. Lee, and D. Messerschmitt, *Digital Communication*, 3rd ed., New York: Springer, 2003.

[Barth08] J. Barth et al., "A 500 MHz random cycle, 1.5 ns latency, SOI embedded DRAM macro featuring a three-transistor micro sense amplifier," *JSSC*, vol. 43, no. 1, Jan. 2008, pp. 86–95.

[Baugh73] C. Baugh and B. Wooley, "A two's complement parallel array multiplication algorithm," *IEEE Trans. Computers*, vol. C-22, no. 12, Dec. 1973, pp. 1045–1047.

[Baumann01] R. Baumann, "Soft errors in advanced semiconductor devices—part I: the three radiation sources," *IEEE Trans. Device & Materials Reliability*, vol. 1, no. 1, Mar. 2001, pp. 17–22.

[Baumann05] R. Baumann, "Soft errors in advanced computer systems," *IEEE Design & Test of Computers*, vol. 22, no. 3, May–Jun. 2005, pp. 258–266.

[Beaumont-Smith99] A. Beaumont-Smith, N. Burgess, S. Lefrere, and C. Lim, "Reduced latency IEEE floating-point adder architectures," *Proc. IEEE Symp. Computer Arithmetic*, Apr. 1999, pp. 35–42.

[Beaumont-Smith01] A. Beaumont-Smith and C. Lim, "Parallel prefix adder design," *Proc. IEEE Symp. Computer Arithmetic*, 2001, pp. 218–225.

[Bedrij62] O. Bedrij, "Carry-select adder," *IRE Trans. Electronic Computers*, vol. 11, Jun. 1962, pp. 340–346.

[Belluomini05] W. Belluomini et al., "An 8GHz floating-point multiply," *Proc. Intl. Solid-State Circuits Conf.*, Feb. 2005, pp. 374–375, 604.

[Bergeron05] J. Bergeron, E. Cerny, A. Hunter, and A. Nightingale, *Verification Methodology Manual for SystemVerilog*, New York: Springer, 2005.

[Bernstein99] K. Bernstein, K. Carrig, C. Durham, P. Hansen, D. Hogenmiller, E. Nowak, and N. Roher, *High Speed CMOS Design Styles*, Boston: Kluwer Academic Publishers, 1999.

[Bernstein00] K. Bernstein and N. Rohrer, *SOI Circuit Design Concepts*, Boston: Kluwer Academic Publishers, 2000.

[Bernstein06] K. Bernstein et al., "High-performance CMOS variability in the 65-nm regime and beyond," *IBM J. Research and Dev.*, vol. 50, no. 4/5, Jul./Sep. 2006, pp. 433–449.

[Bewick94] G. Bewick, "Fast Multiplication: Algorithms and Implementation," Ph.D. thesis, Stanford University, CSL-TR-94-617, 1994.

[Bhavnagarwala01] A. Bhavnagarwala, Xinghai Tang, and J. Meindl, "The impact of intrinsic device fluctuations on CMOS SRAM cell stability," *JSSC*, vol. 36, no. 4, Apr. 2001, pp. 658–665.

[Bhavnagarwala04] A. Bhavnagarwala et al., "A transregional CMOS SRAM with single, logic V_{DD} and dynamic power rails," *Proc. VLSI Circuits Symp.*, Jun. 2004, pp. 292–293.

[Bhavnagarwala05] A. Bhavnagarwala et al., "Fluctuation limits & scaling opportunities for CMOS SRAM cells," *Proc. Intl. Electron Devices Meeting*, Dec. 2005, pp. 659–662.

[Black69] J. Black, "Electromigration—A brief survey and some recent results," *IEEE Trans. Electron Devices*, vol. ED-16, no. 4, Apr. 1969, pp. 338–347.

[Blackburn96] J. Blackburn, L. Arndt, and E. Swartzlander, "Optimization of spanning tree carry lookahead adders," *Proc. 30th Asilomar Conf. Signals, Systems, and Computers*, vol. 1, 1996, pp. 177–181.

[Bohr95] M. Bohr, "Interconnect scaling-the real limiter to high performance ULSI," *Proc. Intl. Electron Devices Meeting*, Dec. 1995, pp. 241–244.

[Bohr96] M. Bohr et al., "A high performance 0.25 μm logic technology optimized for 1.8 V operation," *Proc. Intl. Electron Devices Meeting*, Dec. 1996, pp. 847–850.

[Booth51] A. Booth, "A signed binary multiplication technique," *Quarterly J. Mechanics and Applied Mathematics*, vol. IV, pt. 2, Jun. 1951, pp. 236–240.

[Bowhill95] W. Bowhill et al., "Circuit implementation of a 300-MHz 64-bit second-generation CMOS Alpha CPU," *Digital Technical Journal*, vol. 7, no. 1, 1995, pp. 100–115.

[Bowman02] K. Bowman, S. Duvall, and J. Meindl, "Impact of die-to-die and within-die parameter fluctuations on the maximum clock frequency distribution for gigascale integration," *JSSC*, vol. 37, no. 2, Feb. 2002, pp. 183–190.

[Bowman09] K. Bowman et al., "Energy-efficient and metastability-immune resilient circuits for dynamic variation tolerance," *JSSC*, vol. 44, no. 1, Jan. 2009, pp. 49–63.

[Brederlow06] R. Brederlow, R. Prakash, C. Paulus, and R. Thewes, "A low-power true random number generator using random telegraph noise of single oxide-traps," *Proc. Intl. Solid-State Circuits Conf.*, Feb. 2006, pp. 1666–1675.

[Brent82] R. Brent and H. Kung, "A regular layout for parallel adders," *IEEE Trans. Computers*, vol. C-31, no. 3, Mar. 1982, pp. 260–264.

[Brewer08] J. Brewer and M. Gill, eds., *Nonvolatile Memory Technology with Emphasis on Flash*, Piscataway, NJ: IEEE Press, 2008.

[Brooks95] F. Brooks, *The Mythical Man-Month*, Boston: Addison-Wesley, 1995.

[Brown03] A. Brown, "Fast films," *IEEE Spectrum*, vol. 40, no. 2, Feb. 2003, pp. 36–40.

[Brunvand09] E. Brunvand, *Digital VLSI Chip Design with Cadence and Synopsys CAD Tools*, Boston: Addison Wesley, 2009.

[Brusamarello08] L. Brusamarello, R. da Silva, G. Wirth, and R. Reis, "Probabilistic approach for yield analysis of dynamic logic circuits," *IEEE Trans. Circuits & Systems*, vol. 55, no. 8, Sep. 2008, pp. 2238–2248.

[Budnik06] M. Budnik and K. Roy, "A power delivery and decoupling network minimizing ohmic loss and supply voltage variation in silicon nanoscale technologies," *IEEE Trans. VLSI*, vol. 14, no. 12, Dec. 2006, pp. 1336–1346.

[Burd00] T. Burd, T. Pering, A. Stratakos, and R. Brodersen, "A dynamic voltage scaled microprocessor system," *JSSC*, vol. 35, no. 11, Nov. 2000, pp. 1571–1580.

[Burgess02] N. Burgess, "The flagged prefix adder and its applications in integer arithmetic," *J. VLSI Signal Processing*, 2002, pp. 263–271.

[Burgess09] N. Burgess, "Implementation of recursive Ling adders in CMOS VLSI," *Proc. Asilomar Conf. Signals, Systems, and Computers*, 2009.

[Burks46] A. Burks, H. Goldstine, and J. von Neumann, *Preliminary discussion of the logical design of an electronic computing instrument, part 1*, vol. 1, Inst. Advanced Study, Princeton, NJ, 1946.

[Burleson98] W. Burleson, M. Ciesielski, F. Klass, and W. Liu, "Wave-pipelining: a tutorial and research survey," *IEEE Trans. VLSI*, vol. 6, no. 3, Sep. 1998, pp. 464–474.

[Calhoun04] B. Calhoun, F. Honore, and A. Chandrakasan, "A leakage reduction methodology for distributed MTCMOS," *JSSC*, vol. 39, no. 5, May 2004, pp. 818–826.

[Calhoun05] B. Calhoun, A. Wang, and A. Chandrakasan, "Modeling and sizing for minimum energy operation in subthreshold circuits," *JSSC*, vol. 40, no. 9, Sep. 2005, pp. 1778–1786.

[Calhoun06a] B. Calhoun and A. Chandrakasan, "Ultra-dynamic voltage scaling (UDVS) using sub-threshold operation and local voltage dithering," *JSSC*, vol. 41, no. 1, Jan. 2006, pp. 238–245.

[Calhoun06b] B. Calhoun and A. Chandrakasan, "Static noise margin variation for sub-threshold SRAM in 65-nm CMOS," *JSSC*, vol. 41, no. 7, Jul. 2006, pp. 1673–1679.

[Calhoun07] B. Calhoun and A. Chandrakasan, "A 256-kb 65-nm sub-threshold SRAM design for ultra-low-voltage operation," *JSSC*, vol. 42, no. 3, Mar. 2007, pp. 680–688.

[Calin96] T. Calin, M. Nicolaidis, and R. Velazco, "Upset hardened memory design for submicron CMOS technology," *IEEE Trans. Nuclear Science*, vol. 43, no. 6, Dec. 1996, pp. 2874–2878.

[Calma84] Calma Corporation, GDS II Stream Format, Jul. 1984.

[Carr72] W. Carr and J. Mize, *MOS/LSI Design and Application*, New York: McGraw-Hill, 1972.

[Celik02] M. Celik, L. Pileggi, and A. Odabasioglu, *IC Interconnect Analysis*, Boston: Kluwer Academic Publishers, 2002.

[Cenker79] R. Cenker, D. Clemons, W. Huber, J. Petrizzi, F. Procyk, and G. Trout, "A fault-tolerant 64K dynamic random-access memory," *IEEE Trans. Electron Devices*, vol. 26, no. 6, Jun. 1979, pp. 853–860.

[Chan90] P. Chan and M. Schlag, "Analysis and design of CMOS Manchester adders with variable carry-skip," *IEEE Trans. Computers*, vol. 39, no. 8, Aug. 1990, pp. 983–992.

[Chan05] S. Chan, K. Shepard, and P. Restle, "Uniform-phase uniform-amplitude resonant-load global clock distributions," *JSSC*, vol. 40, no. 1, Jan. 2005, pp. 102–109.

[Chan09] S. Chan et al., "A resonant global clock distribution for the cell broadband engine processor," *JSSC*, vol. 44, no. 1, Jan. 2009, pp. 64–72.

[Chandrakasan92] A. Chandrakasan, S. Sheng, and R. Brodersen, "Low-power CMOS digital design," *JSSC*, vol. 27, no. 4, Apr. 1992, pp. 473–484.

[Chandrakasan01] A. Chandrakasan, W. Bowhill, and F. Fox, ed., *Design of High-Performance Microprocessor Circuits*, Piscataway, NJ: IEEE Press, 2001.

[Chaney73] T. Chaney and C. Molnar, "Anomalous behavior of synchronizer and arbiter circuits," *IEEE Trans. Computers*, vol. C-22, Apr. 1973, pp. 421–422.

[Chaney83] T. Chaney, "Measured flip-flop responses to marginal triggering," *IEEE Trans. Computers*, vol. C-32, no. 12, Dec. 1983, pp. 1207–1209.

[Chang05] J. Chang et al., "A 130-nm triple-V_t 9-MB third-level on-die cache for the 1.7-GHz Itanium 2 processor," *JSSC*, vol. 40, no. 1, Jan. 2005, pp. 195–203.

[Chang07] J. Chang et al., "The 65-nm 16-MB shared on-die L3 cache for the dual-core Intel Xeon processor 7100 Series," *JSSC*, vol. 42, no. 4, Apr. 2007, pp. 846–852.

[Chang08] L. Chang et al., "An 8T-SRAM for variability tolerance and low-voltage operation in high-performance caches," *JSSC*, vol. 43, no. 4, Apr. 2008, pp. 956–963.

[Chao89] H. Chao and C. Johnston, "Behavior analysis of CMOS D flip-flops," *JSSC*, vol. 24, no. 5, Oct. 1989, pp. 1454–1458.

[Chappell91] T. Chappell, B. Chappell, S. Schuster, J. Allan, S. Klepner, R. Joshi, and R. Franch, "A 2-ns cycle, 3.8-ns access 512-kb CMOS ECL SRAM with a fully pipelined architecture," *JSSC*, vol. 26, no. 11, Nov. 1991, pp. 1577–1585.

[Chen96] K. Chen, H. Wann, J. Dunster, P. Ko, C. Hu, and M. Yoshida, "MOSFET carrier mobility model based on gate oxide thickness, threshold and gate voltages," *Solid-State Electronics*, vol. 39, no. 10, 1996, pp. 1515–1518.

[Chen97] K. Chen, C. Hu, P. Fang, M. Lin, and D. Wollesen, "Predicting CMOS speed with gate oxide and voltage scaling and interconnect loading effects," *IEEE Trans. Electron Devices*, vol. 44, no. 11, Nov. 1997, pp. 1951–1957.

[Chen03] T. Chen and S. Naffziger, "Comparison of adaptive body bias (ABB) and adaptive supply voltage (ASV) for improving delay and leakage under the presence of process variation," *IEEE Trans. VLSI*, vol. 11, no. 5, Oct. 2003, pp. 888–899.

[Chen06] J. Chen, L. Clark, and T. Chen, "An ultra-low-power memory with a subthreshold power supply voltage," *JSSC*, vol. 41, no. 10, Oct. 2006, pp. 2344–2353.

[Chen07] G. Chen, D. Blaauw, T. Mudge, D. Sylvester, and Nam Sung Kim, "Yield-driven near-threshold SRAM design," *Proc. Intl. Conf. Computer-Aided Design*, Nov. 2007, pp. 660–666.

[Cheng99] Y. Cheng and C. Hu, *MOSFET Modeling & BSIM3 User's Guide*, Boston: Kluwer Academic Publishers, 1999.

[Cheng00] Y. Cheng, C. Tsai, C. Teng, and S. Kang, *Electrothermal Analysis of VLSI Systems*, Boston: Kluwer Academic Publishers, 2000.

[Chern92] J. Chern, J. Huang, L. Arledge, P. Li, and P. Yang, "Multilevel metal capacitance models for CAD design synthesis systems," *IEEE Electron Device Letters*, vol. 13, no. 1, Jan. 1992, pp. 32–34.

[Childs84] R. Childs, J. Crawford, D. House, and R. Noyce, "A processor family for personal computers," *Proc. IEEE*, vol. 72, no. 3, Mar. 1984, pp. 363–376.

[Chinnery02] D. Chinnery and K. Keutzer, *Closing the Gap Between ASIC and Custom: Tools and techniques for high-performance ASIC design*, Boston: Kluwer Academic Publishers, 2002.

[Chinnery07] D. Chinnery & K. Keutzer, *Closing the Power Gap Between ASIC and Custom*, New York: Springer, 2007.

[Choi97] J. Choi, L. Jang, S. Jung, and J. Choi, "Structured design of a 288-tap FIR filter by optimized partial product tree compression," *JSSC*, vol. 32, no. 3, Mar. 1997, pp. 468–476.

[Chong06] K. Chong, L. McMurchie, and C. Sechen, "A 64b adder using self-calibrating differential output prediction logic," *Proc. Intl. Solid-State Circuits Conf.*, Feb. 2006, pp. 1745–1754.

[Chopra06] K. Chopra, C. Kashyap, H. Su, and D. Blaauw, "Current source driver model synthesis and worst-case alignment for accurate timing and noise analysis," *Intl. Workshop on Timing in Synthesis and Specification* (TAU), Feb. 2006.

[Choudhury97] M. Choudhury and J. Miller, "A 300 MHz CMOS microprocessor with multi-media technology," *Proc. IEEE Intl. Solid-State Circuits Conf.*, 1997, pp. 170–171.

[Christiansen06] C. Christiansen, J. Gambino, J. Therrien, D. Hunt, and J. Gill, "Effect of wire thickness on electromigration and stress migration lifetime of Cu," *Proc. Failure Analysis of Integrated Circuits*, Jul. 2006, pp. 349–354.

[Chu86] K. Chu and D. Pulfrey, "Design procedures for differential cascode voltage switch circuits," *JSSC*, vol. SC-21, no. 6, Dec. 1986, pp. 1082–1087.

[Chu87] K. Chu and D. Pulfrey, "A comparison of CMOS circuit techniques: differential cascode voltage switch logic versus conventional logic," *JSSC*, vol. SC-22, no. 4, Aug. 1987, pp. 528–532.

[Clark61] C. Clark, "The greatest of a finite set of random variables," *Operations Research*, vol. 9, no. 2, Mar.–Apr. 1961, pp. 145–162.

[Clark01] L. Clark et al., "An embedded 32-b microprocessor core for low-power and high-performance applications," *JSSC*, vol. 36, no. 11, Nov. 2001, pp. 1599–1608.

[Clark02] L. Clark, S. Demmons, N. Deutscher, and F. Ricci, "Standby power management for a 0.18 μm microprocessor," *Proc. Intl. Symp. Low Power Electronics and Design*, Aug. 2002, pp. 7–12.

[Cobbold66] R. Cobbold, "Temperature effects on M.O.S. transistors," *Electronics Letters*, vol. 2, no. 6, Jun. 1966, pp. 190–192.

[Cobbold70] R. Cobbold, *Theory and Application of Field Transistors*, New York: Wiley Interscience, 1970.

[Colwell95] R. Colwell and R. Steck, "A 0.6 μm BiCMOS processor with dynamic execution," *Proc. IEEE Solid-State Circuits Conf.*, 1995, pp. 176–177.

[Colwell06] R. Colwell, *The Pentium Chronicles: The People, Passion, and Politics Behind Intel's Landmark Chips*, New York: Wiley, 2006.

[Cooley07] J. Cooley, "Verilog vs. VHDL," `www.deepchip.com/items/dvcon07-02.html`, Apr. 2007.

[Cortadella92] J. Cortadella and J. Llabería, "Evaluation of A + B = K conditions without carry propagation," *IEEE Trans. Computers*, vol. 41, no. 11, Nov. 1992, pp. 1484–1487.

[Crews03] M. Crews and Y. Yuenyongsgool, "Practical design for transferring signals between clock domains," *EDN Magazine*, Feb. 20, 2003, pp. 65–71.

[Curran02] B. Curran et al., "IBM eServer z900 high-frequency microprocessor technology, circuits, and design methodology," *IBM J. Research and Development*, vol. 46, no. 4/5, Jul./Sep. 2002, pp. 631–644.

[Dabral98] S. Dabral and T. Maloney, *Basic ESD and I/O Design*, New York: John Wiley & Sons, 1998.

[Dadda65] L. Dadda, "Some schemes for parallel multipliers," *Alta Frequenza*, vol. 34, no. 5, May 1965, pp. 349–356.

[Dally98] W. Dally and J. Poulton, *Digital Systems Engineering*, Cambridge, UK: Cambridge University Press, 1998.

[Das06] S. Das et al., "A self-tuning DVS processor using delay-error detection and correction," *JSSC*, vol. 41, no. 4, Apr. 2006, pp. 792–804.

[Das09] S. Das et al., "RazorII: in situ error detection and correction for PVT and SER tolerance," *JSSC*, vol. 44, no. 1, Jan. 2009, pp. 32–48.

[Davari99] B. Davari, "CMOS technology: present and future," *Symp. VLSI Circuits Digest Tech. Papers*, 1999, pp. 5–10.

[Davis69] R. Davis, "The ILLIAC IV processing element," *IEEE Trans. Computers*, vol. C-18, no. 9, Sep. 1969, pp. 800–816.

[Degalahal05] V. Degalahal, L. Li, V. Narayanan, M. Kandemir, and M. Irwin, "Soft errors issues in low-power caches," *IEEE Trans. VLSI*, vol. 13, no. 10, Oct. 2005, pp. 1157–1166.

[DeHon93] A. DeHon, T. Knight Jr., and T. Simon, "Automatic impedance control," *Proc. Intl. Solid-State Circuits Conf.*, Feb. 1993, pp. 164–165, 283.

[Deleganes02] D. Deleganes, J. Douglas, B. Kommandur, and M. Patyra, "Designing a 3GHz, 130nm, Intel Pentium 4 processor," *Symp. VLSI Circuits Digest Tech. Papers*, 2002, pp. 130–133.

[Deleganes04] D. Deleganes et al., "LVS technology for the Intel Pentium 4 processor on 90nm technology," *Intel Techology Journal*, vol. 8, no. 1, Feb. 2004, pp. 43–53.

[Deleganes05] D. Deleganes et al., "Low-voltage swing logic circuits for a Pentium 4 processor integer core," *JSSC*, vol. 40, no. 1, Jan. 2005, pp. 36–43.

[Delgado-Frias00] J. Delgado-Frias and J. Nyathi, "A high-performance encoder with priority lookahead," *IEEE Trans. Circuits and Systems I*, vol. 47, no. 9, Sep. 2000, pp. 1390–1393.

[Dennard68] R. Dennard, "Field-effect transistor memory," US Patent 3,387,286, 1968.

[Dennard74] R. Dennard et al., "Design of ion-implanted MOSFET's with very small physical dimensions," *JSSC*, vol. SC-9, no. 5, Oct. 1974, pp. 256–268.

[Dhanesha95] H. Dhanesha, K. Falakshahi, and M. Horowitz, "Array-of-arrays architecture for parallel floating point multiplication," *Proc. Conf. Advanced Research in VLSI*, 1995, pp. 150–157.

[Dickson76] J. Dickson, "On-chip high-voltage generation in MNOS integrated circuits using an improved voltage multiplier technique," *JSSC*, vol. 11, no. 3, Jun. 1976, pp. 374–378.

[Dike99] C. Dike and E. Burton, "Miller and noise effects in a synchronizing flip-flop," *JSSC*, vol. 34, no. 6, Jun. 1999, pp. 849–855.

[Dobbalaere95] I. Dobbalaere, M. Horowitz, and A. El Gamal, "Regenerative feedback repeaters for programmable interconnect," *JSSC*, vol. 30, no. 11, Nov. 1995, pp. 1246–1253.

[Dobberpuhl92] D. Dobberpuhl et al., "A 200-MHz 64-b dual-issue CMOS microprocessor," *JSSC*, vol. 27, no. 11, Nov. 1992, pp. 1555–1567.

[Dobson95] J. Dobson and G. Blair, "Fast two's complement VLSI adder design," *Electronics Letters*, vol. 31, no. 20, Sep. 1995, pp. 1721–1722.

[Donnay03] S. Donnay and G. Gielen, eds., *Substrate Noise Coupling in Mixed-Signal ASICs*, Boston: Kluwer Academic Publishers, 2003.

[Donovan02] C. Donovan and M. Flynn, "A 'digital' 6-bit ADC in 0.25 μm CMOS," *JSSC*, vol. 37, no. 3, Mar. 2002, pp. 432–437.

[Doran88] R. Doran, "Variants of an improved carry look-ahead adder," *IEEE Trans. Computers*, vol. 37, no. 9, Sep. 1988, pp. 1110–1113.

[Doyle91] B. Doyle, B. Fishbein, and K. Mistry, "NBTI-enhanced hot carrier damage in p-channel MOSFETs," *Proc. Intl. Electron Devices Meeting*, 1991, pp. 529–532A.

[Drake07] A. Drake et al., "A distributed critical-path timing monitor for a 65nm high-performance microprocessor," *Proc. Intl. Solid-State Circuits Conf.*, Feb. 2007, pp. 398–399.

[Draper97] D. Draper et al., "Circuit techniques in a 266-MHz MMX-enabled processor," *JSSC*, vol. 32, no. 11, Nov. 1997, pp. 1650–1664.

[Dunga07] M. Dunga et al., *BSIM 4.6.1 MOSFET Model User's Manual*, Department of Electrical Engineering and Computer Sciences, UC Berkeley, 2007.

[Earle65] J. Earle, "Latched carry-save adder," *IBM Tech. Disclosure Bulletin*, vol. 7, no. 10, Mar. 1965, pp. 909–910.

[Edwards93] B. Edwards, A. Corry, N. Weste, and C. Greenberg, "A single-chip video ghost canceller," *JSSC*, vol. 28, no. 3, Mar. 1993, pp. 379–383.

[Eichelberger78] E. Eichelberger and T. Williams, "A logic design structure for LSI testability," *J. Design Automation and Fault Tolerant Computing*, vol. 2, no. 2, May 1978, pp. 165–178.

[Elmore48] W. Elmore, "The transient response of damped linear networks with particular regard to wideband amplifiers," *J. Applied Physics*, vol. 19, no. 1, Jan. 1948, pp. 55–63.

[Emma08] P. Emma and E. Kursan, "Is 3D chip technology the next growth engine for performance improvement?" *IBM J. Research & Dev.*, vol. 52, no. 6, Nov. 2008, pp. 541–552.

[EPA07] US Environmental Protection Agency, *Report to Congress on Server and Data Center Energy Efficiency*, Public Law 109-431, Aug. 2, 2007.

[Ercegovac89] M. Ercegovac and T. Lang, "Binary counter with counting period of one half adder independent of counter size," *IEEE Trans. Circuits & Systems*, vol. 36, no. 6, Jun. 1989, pp. 924–926.

[Ernst03] D. Ernst et al., "Razor: a low-power pipeline based on circuit-level timing speculation," *Proc. Intl. Symp. Microarchitecture*, Dec. 2003, pp. 7–18.

[Estreich82] D. Estreich and R. Dutton, "Modeling latch-up in CMOS integrated circuits," *IEEE Trans. Computer-Aided Design*, vol. CAD-1, no. 4, Oct. 1982, pp. 157–162.

[Evans95] R. Evans and P. Franzon, "Energy consumption modeling and optimization for SRAM's," *JSSC*, vol. 30, no. 5, May 1995, pp. 571–579.

[Faggin96] F. Faggin, M. Hoff, S. Mazor, and M. Shima, "The history of the 4004," *IEEE Micro*, vol. 16, no. 6, Dec. 1996, pp. 10–20.

[Fetzer02] E. Fetzer, M. Gibson, A. Klein, N. Calick, C. Zhu, E. Busta, and B. Mohammad, "A fully bypassed six-issue integer datapath and register file on the Itanium-2 microprocessor," *JSSC*, vol. 37, no. 11, Nov. 2002, pp. 1433–1440.

[Fetzer06] E. Fetzer, D. Dahle, C. Little, and K. Safford, "The parity protected, multithreaded register files on the 90-nm Itanium microprocessor," *JSSC*, vol. 41, no. 1, Jan. 2006, pp. 246–255.

[Fischer06] T. Fischer, J. Desai, B. Doyle, S. Naffziger, and B. Patella, "A 90-nm variable frequency clock system for a power-managed Itanium architecture processor," *JSSC*, vol. 41, no. 1, Jan. 2006, pp. 218–228.

[Fishburn85] J. Fishburn and A. Dunlop, "TILOS: A posynomial programming approach to transistor sizing," *Proc. Intl. Conf. Computer-Aided Design*, Nov. 1985, pp. 326–328.

[Flannagan85] S. Flannagan, "Synchronization reliability in CMOS technology," *JSSC*, vol. SC-20, no. 4, Aug. 1985, pp. 880–882.

[Foty96] D. Foty, *MOSFET Modeling with SPICE: Principles and Practices*, Upper Saddle River, NJ: Prentice Hall, 1996.

[Friedman84] V. Friedman and S. Liu, "Dynamic logic CMOS circuits," *JSSC*, vol. SC-19, no. 2, Apr. 1984, pp. 263–266.

[Frohman69] D. Frohman-Bentchkowsky and A. Grove, "Conductance of MOS transistors in saturation," *IEEE Trans. Electron Devices*, vol. ED-16, no. 1, Jan. 1969, pp. 108–113.

[Frowerk77] R. Frowerk, "Signature Analysis: A New Digital Field Service Method," *Hewlett Packard Journal*, May 1977, pp. 2–8.

[Fujiwara06] E. Fujiwara, *Code Design for Dependable Systems: Theory and Practical Applications*, New York: Wiley, 2006.

[Gabara92] T. Gabara and S. Knauer, "Digitally adjustable resistors in CMOS for high-performance applications," *JSSC*, vol. 27, no. 8, Aug. 1992, pp. 1176–1185.

[Gago93] A. Gago, R. Escano, and J. Hidalgo, "Reduced implementation of D-type DET flip-flops," *JSSC*, vol. 28, no. 3, Mar. 1993, pp. 400–402.

[Gajski83] D. Gajski and R. Kuhn, "New VLSI tools," *Computer*, vol. 16, no. 12, Dec. 1983, pp. 11–14.

[Galiay80] J. Galiay, Y. Crouzet, and M. Verginiault, "Physical versus logical fault models MOS LSI circuits: impact on their testability," *IEEE Trans. Computers*, vol. C-29, no. 6, Jun. 1980, pp. 527–531.

[Gauthier02] C. Gauthier and B. Amick, "Inductance: Implications and solutions for high-speed digital circuits: the chip electrical interface," *Proc. IEEE Intl. Solid-State Circuits Conf.*, vol. 2, 2002, pp. 563–565.

[Geannopoulos98] G. Geannopoulos and X. Dai, "An adaptive digital deskewing circuit for clock distribution networks," *Proc. IEEE Intl. Solid-State Circuits Conf.*, 1998, pp. 400–401.

[Gelsinger01] P. Gelsinger, "Microprocessors for the new millennium: challenges, opportunities, and new frontiers," *Proc. IEEE Intl. Solid-State Circuits Conf.*, 2001, pp. 22–25.

[George96] S. George, A. Ott, and J. Klaus, "Surface chemistry for atomic layer growth," *J. Phys. Chem.*, vol. 100, 1996, pp. 13121–13131.

[George07] V. George et al., "Penryn: 45-nm next generation Intel Core 2 processor," *Proc. Intl. Solid-State Circuits Conf.*, Nov. 2007, pp. 14–17.

[Geppert04] L. Geppert, "Chip making's wet new world," *IEEE Spectrum*, vol. 41, no. 5, May 2004, pp. 29–33.

[Gerosa94] G. Gerosa et al., "A 2.2 W, 80 MHz superscalar RISC microprocessor," *JSSC*, vol. 29, no. 12, Dec. 1994, pp. 1440–1452.

[Gerosa09] G. Gerosa et al., "A sub-2 W low power IA processor for mobile internet devices in 45 nm high-k metal gate CMOS," *JSSC*, vol. 44, no. 1, Jan. 2009, pp. 73–82.

[Ginosar03] R. Ginosar, "Fourteen ways to fool your synchronizer," *Proc. Intl. Symp. Asynchronous Circuits and Systems*, May 2003, pp. 89–96.

[Glasser85] L. Glasser and D. Dobberpuhl, *The Design and Analysis of VLSI Circuits*, Reading, MA: Addison Wesley, 1985.

[Gochman03] S. Gochman et al., "The Intel Pentium M processor: microarchitecture and performance," *Intel Technology Journal*, vol. 7, no. 2, May 2003, pp. 21–36.

[Golden99] M. Golden et al., "A seventh-generation x86 microprocessor," *JSSC*, vol. 34, no. 11, Nov. 1999, pp. 1466–1477.

[Golomb81] S. Golomb, *Shift Register Sequences*, Revised Edition, Laguna Hills, CA: Aegean Park Press, 1981.

[Gonclaves83] N. Gonclaves and H. DeMan, "NORA: a racefree dynamic CMOS technique for pipelined logic structures," *JSSC*, vol. SC-18, no. 3, Jun. 1983, pp. 261–266.

[Gonzalez96] R. Gonzalez and M. Horowitz, "Energy dissipation in general purpose microprocessors," *JSSC*, vol. 31, no. 9, Sep. 1996, pp. 1277–1284.

[Gonzalez97] R. Gonzalez, B. Gordon, and M. Horowitz, "Supply and threshold voltage scaling for low power CMOS," *JSSC*, vol. 32, no. 8, Aug. 1997, pp. 1210–1216.

[Goto92] G. Goto, T. Sato, M. Nakajima, and T. Sukemura, "A 54×54-b regularly structured tree multiplier," *JSSC*, vol. 27, no. 9, Sep. 1992, pp. 1229–1236.

[Goto97] G. Goto et al., "A 4.1-ns compact 54×54-b multiplier utilizing sign-select Booth encoders," *JSSC*, vol. 32, no. 11, Nov. 1997, pp. 1676–1682.

[Grad04] J. Grad and J. Stine, "A hybrid Ling carry-select adder," *Proc. Asilomar Conf. Signals, Systems, and Computers*, Nov. 2004, pp. 1363–1367.

[Gray53] F. Gray, "Pulse code communications," US Patent 2,632,058, 1953.

[Gray01] P. Gray, P. Hurst, S. Lewis, and R. Meyer, *Analysis and Design of Analog Integrated Circuits*, 4th ed., New York: John Wiley & Sons, 2001.

[Greenhill97] D. Greenhill et al., "A 330 MHz 4-way superscalar microprocessor," *Proc. Intl. Solid-State Circuits Conf.*, Feb. 1997, pp. 166–167, 449.

[Gregg07] J. Gregg and T. Chen, "Post silicon power/performance optimization in the presence of process variations using individual well-adaptive body biasing," *IEEE Trans. VLSI*, vol. 15, no. 3, Mar. 2007, pp. 366–376.

[Griffin83] W. Griffin and J. Hiltebeitel, "CMOS 4-way XOR circuit," *IBM Technical Disclosure Bulletin*, vol. 25, no. 11B, Apr. 1983, pp. 6066–6067.

[Gronowski96] P. Gronowski et al., "A 433-MHz 64-b quad-issue RISC microprocessor," *JSSC*, vol. 31, no. 11, Nov. 1996, pp. 1687–1696.

[Gronowski98] P. Gronowski, W. Bowhill, R. Preston, M. Gowan, and R. Allmon, "High-performance microprocessor design," *JSSC*, vol. 33, no. 5, May 1998, pp. 676–686.

[Grotjohn86] T. Grotjohn and B. Hoefflinger, "Sample-set differential logic (SSDL) for complex high-speed VLSI," *JSSC*, vol. SC-21, no. 2, Apr. 1986, pp. 367–369.

[Guilar09] N. Guilar, T. Kleeburg, A. Chen, D. Yankelevich, and R. Amirtharajah, "Integrated solar energy harvesting and storage," *IEEE Trans. VLSI*, vol. 17, no. 5, May 2009, pp. 627–637.

[Gunning92] B. Gunning, L. Yuan, T. Nguyen, and T. Wong, "A CMOS low-voltage-swing transmission-line transceiver," *Proc. Intl. Solid-State Circuits Conf.*, Feb. 1992, pp. 58–59.

[Guo05] X. Guo and C. Sechen, "High speed redundant adder and divider in output prediction logic," *Proc. Computer Society Symp. VLSI*, May 2005, pp. 34–41.

[Gutierrez01] E. Gutierrez, J. Deen, and C. Claeys (eds.), *Low Temperature Electronics: Physics, Devices, Circuits, and Applications*, New York: Academic Press, 2001.

[Gutnik97] V. Gutnik and A. Chandrakasan, "Embedded power supply for low-power DSP," *IEEE Trans. VLSI*, vol. 5, no. 4, Dec. 1997, pp. 425–435.

[Guyot87] A. Guyot, B. Hochet, and J. Muller, "A way to build efficient carry-skip adders," *IEEE Trans. Computers*, vol. 36, no. 10, Oct. 1987, pp. 1144–1152.

[Guyot97] A. Guyot and S. Abou-Samra, "Modeling power consumption in arithmetic operators," *Microelectronic Engineering*, vol. 39, 1997, pp. 245–253.

[Hall00] S. Hall, G. Hall, and J. McCall, *High-Speed Digital System Design*, New York: Wiley, 2000.

[Hamada98] M. Hamada et al., "A top-down low power design technique using clustered voltage scaling with variable supply-voltage scheme," *Proceedings of the IEEE*, May 11–14, 1998, pp. 495–498.

[Hamming50] R. Hamming, "Error Detecting and Error Correcting Codes," *Bell Systems Technical Journal*, vol. 29, pp. 147–160.

[Hamzaoglu02] F. Hamzaoglu and M. Stan, "Circuit-level techniques to control gate leakage for sub-100 nm CMOS," *Proc. Intl. Symp. Low Power Electronics and Design*, 2002, pp. 60–63.

[Hamzaoglu09] F. Hamzaoglu et al., "A 3.8 GHz 153 Mb SRAM design with dynamic stability enhancement and leakage reduction in 45 nm high-k metal gate CMOS technology," *JSSC*, vol. 44, no. 1, Jan. 2009, pp. 148–154.

[Han87] T. Han and D. Carlson, "Fast area-efficient VLSI adders," *Proc. IEEE Symp. Computer Arithmetic*, 1987, pp. 49–56.

[Hanson06] S. Hanson et al., "Ultralow-voltage minimum-energy CMOS," *IBM J. Research & Dev.*, vol. 50, no. 4/5, Jul./Sep. 2006, pp. 469–490.

[Hanson09] S. Hanson et al., "A low-voltage processor for sensing applications with picowatt standby mode," *JSSC*, vol. 44, no. 4, Apr. 2009, pp. 1145–1155.

[Harame01a] D. Harame and B. Meyerson, "The early history of IBM's SiGe mixed signal technology," *IEEE Transactions on Electron Devices*, vol. 48, no. 11, Nov. 2001, pp. 2555–2567.

[Harame01b] D. Harame et al., "Current status and future trends of SiGe BiCMOS technology," *IEEE Transactions on Electron Devices*, vol. 48, no. 11, Nov. 2001, pp. 2575–2594.

[Haring96] R. Haring et al., "Self-resetting logic register and incrementer," *Symp. VLSI Circuits Digest Tech. Papers*, 1996, pp. 18–19.

[Harris97] D. Harris and M. Horowitz, "Skew-tolerant domino circuits," *JSSC*, vol. 32, no. 11, Nov. 1997, pp. 1702–1711.

[Harris99] D. Harris, M. Horowitz, and D. Liu, "Timing analysis including clock skew," *IEEE Trans. Computer-Aided Design*, vol. 18, no. 11, Nov. 1999, pp. 1608–1618.

[Harris01a] D. Harris, *Skew-Tolerant Circuit Design*, San Francisco, CA: Morgan Kaufmann, 2001.

[Harris01b] D. Harris and S. Naffziger, "Statistical clock skew modeling with data delay variations," *IEEE Trans. VLSI*, vol. 9, no. 6, Dec. 2001, pp. 888–898.

[Harris03] D. Harris, "A taxonomy of prefix networks," *Proc. 37th Asilomar Conf. Signals, Systems, and Computers*, 2003, pp. 2213–2217.

[Harris04] D. Harris, "Logical effort of higher valency adders," *Proc. Asilomar Conf. Signals, Systems, and Computers*, Nov. 2004, pp. 1358–1362.

[Harris07] D. Harris and S. Harris, *Digital Design and Computer Architecture*, San Francisco: Morgan Kaufmann Publishers, 2007.

[Hart06] J. Hart et al., "Implementation of a fourth-generation 1.8-GHz dual-core SPARC V9 microprocessor," *JSSC*, vol. 41, no. 1, Jan. 2006, pp. 210–217.

[Hashemian92] R. Hashemian and C. Chen, "A new parallel technique for design of decrement/increment and two's complement circuits," *Proc. IEEE Midwest Symp. Circuits and Systems*, vol. 2, 1992, pp. 887–890.

[Hashimoto02] T. Hashimoto et al., "Integration of a 0.13-μm CMOS and a high performance self-aligned SiGe HBT featuring low base resistance," *Proc. Intl. Electron Devices Meeting*, Dec. 2002, pp. 779–782.

[Hatamian86] M. Hatamian and G. Cash, "A 70-MHz 8-bit × 8-bit parallel pipelined multiplier in 2.5-μm CMOS," *JSSC*, vol. 21, no. 4, Aug. 1986, pp. 505–513.

[Haykin00] S. Haykin, *Digital Communications*, New York: John Wiley & Sons, 2000.

[Hazucha00] P. Hazucha, C. Svensson, and S. Wender, "Cosmic-ray soft error rate characterization of a standard 0.6-μm CMOS process," *JSSC*, vol. 35, no. 10, Oct. 2000, pp. 1422–1429.

[Hazucha04] P. Hazucha et al., "Measurements and analysis of SER-tolerant latch in a 90-nm dual-V_t CMOS process," *JSSC*, vol. 39, no. 9, Sep. 2004, pp. 1536–1543.

[Heald93] R. Heald and J. Holst, "A 6-ns cycle 256 kb cache memory and memory management unit," *JSSC*, vol. 28, no. 11, Nov. 1993, pp. 1078–1083.

[Heald98] R. Heald et al., "64-Kbyte sum-addressed-memory cache with 1.6-ns cycle and 2.6-ns latency," *JSSC*, vol. 33, no. 11, Nov. 1998, pp. 1682–1689.

[Heald00] R. Heald et al., "A third-generation SPARC v9 64-b microprocessor," *JSSC*, vol. 35, no. 11, Nov. 2000, pp. 1526–1538.

[Hedenstierna87] N. Hedenstierna and K. Jeppson, "CMOS circuit speed and buffer optimization," *IEEE Trans. Computer-Aided Design*, vol. CAD-6, no. 2, Mar. 1987, pp. 270–281.

[Heikes94] C. Heikes, "A 4.5mm^2 multiplier array for a 200MFLOP pipelined coprocessor," *Proc. IEEE Intl. Solid-State Circuits Conf.*, 1994, pp. 290–291.

[Heller84] L. Heller, W. Griffin, J. Davis, and N. Thoma, "Cascode voltage switch logic: a differential CMOS logic family," *Proc. IEEE Intl. Solid-State Circuits Conf.*, 1984, pp. 16–17.

[Hennessy90] J. Hennessy and D. Patterson, *Computer Architecture: A Quantitative Approach*, San Mateo, CA: Morgan Kaufmann Publishers, Inc., 1990.

[Heo02] S. Heo, K. Barr, M. Hampton, and K. Asanovic, "Dynamic fine-grain leakage reduction using leakage-biased bitlines," *Proc. Intl. Symp. Computer Architecture*, 2002, pp. 137–147.

[Hess94] C. Hess and L. Weiland, "Drop-in process control checkerboard test structure for efficient online process characterization and defect problem debugging," *Proc. IEEE Int. Conf. Microelectronic Test Structures*, vol. 7, Mar., 1994, pp. 152–159.

[Hicks08] J. Hicks, "45nm transistor reliability," *Intel Technology Journal*, vol. 12, no. 2, Jun. 2008, pp. 131–144.

[Hidaka89] H. Hidaka, K. Fujishima, Y. Matsuda, M. Asakura, and T. Yoshihara, "Twisted bit-line architectures for multi-megabit DRAM's," *JSSC*, vol. 24, no. 1, Feb. 1989, pp. 21–27.

[Hilewitz04] Y. Hilewitz, Z. Shi, and R. Lee, "Comparing fast implementations of bit permutation instructions," *Proc. Asilomar Conf. Signals, Systems, and Computers*, Nov. 2004, pp. 1856–1863.

[Hilewitz07] Y. Hilewitz and R. Lee, "Performing advanced bit manipulations efficiently in general-purpose processors," *Proc. Computer Arithmetic Symp.*, June 2007, pp. 251–260.

[Hill68] C. Hill, "Noise margin and noise immunity in logic circuits," *Microelectronics*, vol. 1, Apr. 1968, pp. 16–21.

[Hinton01] G. Hinton et al., "A 0.18-μm CMOS IA-32 processor with a 4-GHz integer execution unit," *JSSC*, vol. 36, no. 11, Nov. 2001, pp. 1617–1627.

[Hisamoto98] D. Hisamoto et al., "A folded-channel MOSFET for deep-sub-tenth micron era," *Tech. Digest Intl. Electron Devices Meeting*, San Francisco, Dec. 1998, pp. 1032–1034.

[Ho01] R. Ho, K. Mai, and M. Horowitz, "The future of wires," *Proc. IEEE*, vol. 89, no. 4, Apr. 2001, pp. 490–504.

[Ho03a] R. Ho, K. Mai, and M. Horowitz, "Efficient on-chip global interconnects," *Symp. VLSI Circuits Digest Tech. Papers*, 2003, pp. 271–274.

[Ho03b] R. Ho, K. Mai, and M. Horowitz, "Managing wire scaling: a circuit perspective," *Proc. IEEE Interconnect Technology Conf.*, 2003, pp. 177–179.

[Ho07] R. Ho, "Dealing with issues in VLSI interconnect scaling," *Intl. Solid-State Circuits Conf. Tutorial*, Feb. 2007.

[Hoeneisen72] B. Hoeneisen and C. Mead, "Fundamental limitations in Microelectronics-I. MOS technology," *Solid-State Electronics*, vol. 15, 1972, pp. 819–829.

[Hogge85] C. Hogge Jr., "A self correcting clock recovery circuit," *IEEE Trans. Electron Devices*, vol. 32, no. 12, Dec. 1985, pp. 2704–2706.

[Hook03] T. Hook et al., "Lateral ion implant straggle and mask proximity effect," *IEEE Trans. Electron Devices*, vol. 50, no. 9, Sep. 2003, pp. 1946–1951.

[Horowitz83] M. Horowitz and R. Dutton, "Resistance extraction from mask layout data," *IEEE Trans. Computer-Aided Design*, vol. CAD-2, no. 3, Jul. 1983, pp. 145–150.

[Horowitz87] M. Horowitz et al., "MIPS-X: a 20-MIPS peak, 32-bit microprocessor with on-chip cache," *JSSC*, vol. SC-22, no. 5, Oct. 1987, pp. 790–799.

[Horowitz04] M. Horowitz and W. Dally, "How scaling will change processor architecture," *Proc. Intl. Solid-State Circuits Conf.*, Feb. 2004, pp. 132–133.

[Horstmann89] J. Horstmann, H. Eichel, and R. Coates, "Metastability behavior of CMOS ASIC flip-flops in theory and test," *JSSC*, vol. 24, no. 1, Feb. 1989, pp. 146–157.

[Hrishikesh02] M. Hrishikesh et al., "The optimal logic depth per pipeline stage is 6 to 8 FO4 inverter delays," *Proc. Intl. Symp. Computer Architecture*, 2002, pp. 14–24.

[Hsiao70] M. Hsiao, "A class of optimal minimum odd-weight-column SEC-DED codes," *IBM J. Research & Dev.*, vol. 14, no. 4, Jul. 1970, pp. 395–401.

[Hsu91] W. Hsu, B. Sheu, and S. Gowda, "Design of reliable VLSI circuits using simulation techniques," *JSSC*, vol. 26, no. 3, Mar. 1991, pp. 452–457.

[Hsu92] W. Hsu, B. Sheu, S. Gowda, and C. Hwang, "Advanced integrated-circuit reliability simulation including dynamic stress effects," *JSSC*, vol. 27, no. 3, Mar. 1992, pp. 247–257.

[Hsu06a] S. Hsu et al., "A 110 GOPS/W 16-bit multiplier and reconfigurable PLA loop in 90-nm CMOS," *JSSC*, vol. 41, no. 1, Jan. 2006, pp. 256–264.

[Hsu06b] S. Hsu, A. Agarwal, M. Anders, S. Mathew, R. Krishnamurthy, and S. Borkar, "An 8.8GHz 198mW 16 × 64b 1R/1W variation tolerant register file in 65nm CMOS," *Proc. Intl. Solid-State Circuits Conf.*, Feb. 2006, pp. 1785–1797.

[Hu90] Y. Hu and S. Chen, "GM_Plan: A Gate Matrix Layout Algorithm Based on Artificial Intelligence Planning Techniques," *IEEE Trans. Computer-Aided Design*, vol. 9, no. 8, Aug. 1990, pp. 836–845.

[Hu92] C. Hu, "IC Reliability Simulation," *JSSC*, vol. 27, no. 3, Mar. 1992, pp. 241–246.

[Hu95] C. Hu, K. Rodbell, T. Sullivan, K. Lee, and D. Bouldin, "Electromigration and stress-induced voiding in fine Al and Al-alloy thin-film lines," *IBM J. Research and Development*, vol. 39, no. 4, Jul. 1995, pp. 465–497.

[Huang00] Z. Huang and M. Ercegovac, "Effect of wire delay on the design of prefix adders in deep-submicron technology," *Proc. 34th Asilomar Conf. Signals, Systems, and Computers*, vol. 2, 2000, pp. 1713–1717.

[Huang02] C. Huang, J. Wang, and Y. Huang, "Design of high-performance CMOS priority encoders and incrementer/decrementers using multilevel lookahead and multilevel folding techniques," *JSSC*, vol. 37, no. 1, Jan. 2002, pp. 63–76.

[Huang03] X. Huang et al., "Loop-based interconnect modeling and optimization approach for multigigahertz clock network design," *JSSC*, vol. 38, no. 3, Mar. 2003, pp. 457–463.

[Huang05] Z. Huang and M. Ercegovac, "High-performance low-power left-to-right array multiplier design," *IEEE Trans. Computers*, vol. 54, no. 3, Mar. 2005, pp. 272–283.

[Huang06] A. Huang et al., "A 10Gb/s photonic modulator and WDM MUX/DEMUX integrated with electronics in 0.13/spl mu/m SOI CMOS," *Proc. Intl. Solid-State Circuits Conf.*, Feb. 2006, pp. 922–929.

[Huh98] Y. Huh, Y. Sung, and S. Kang, "A study of hot-carrier-induced mismatch drift: a reliability issue for VLSI circuits," *JSSC*, vol. 33, no. 6, Jun. 1998, pp. 921–927.

[Huitema03] E. Huitema et al., "Plastic transistors in active-matrix displays," *Proc. IEEE Intl. Solid-State Circuits Conf.*, Feb. 2003, pp. 380–381.

[Huntzicker08] S. Huntzicker, M. Dayringer, J. Soprano, A. Weerasinghe, D. Harris, and D. Patil, "Energy-delay tradeoffs in 32-bit static shifter designs," *Proc. Intl. Conf. Computer Design*, Oct. 2008, pp. 626–632.

[Hwang89] I. Hwang and A. Fisher, "Ultrafast compact 32-bit CMOS adders in multiple-output domino logic," *JSSC*, vol. 24, no. 2, Apr. 1989, pp. 358–369.

[Hwang99a] W. Hwang, R. Joshi, and W. Henkels, "A 500-MHz, 32-Word × 64-bit, eight-port self-resetting CMOS register file," *JSSC*, vol. 34, no. 1, Jan. 1999, pp. 56–67.

[Hwang99b] W. Hwang, G. Gristede, P. Sanda, S. Wang, and D. Heidel, "Implementation of a self-resetting CMOS 64-bit parallel adder with enhanced testability," *JSSC*, vol. 34, no. 8, Aug. 1999, pp. 1108–1117.

[Hwang02] D. Hwang, F. Dengwei, and A. Willson Jr, "A 400-MHz processor for the efficient conversion of rectangular to polar coordinates for digital communications applications," *Symp. VLSI Circuits Digest Tech. Papers*, Jun. 2002, pp. 248–251.

[ICKnowledge02] IC Knowledge, "Defect density trends," 2002, `www.icknowledge.com/trends/defects.pdf`.

[IEEE1076-08] IEEE Standard 1076-2008 (Revision of IEEE Standard 1076-2002), *VHDL Language Reference Manual*, 2009.

[IEEE1149.1-01] IEEE Standard 1149.1-2001, *Test Access Port and Boundary-Scan Architecture*, 2001.

[IEEE1364-01] IEEE Standard 1364-2001, *Verilog Hardware Description Language*, 2001.

[IEEE 1800-2009] IEEE Standard 1800-2009, *System Verilog-Unified Hardware Design, Specification, and Verification Language*, 2009.

[Intel10] Intel Corporation, *Microprocessor Quick Reference Guide*, `www.intel.com/pressroom/kits/quickreffam.htm`, 2010.

[Isaac08] R. Isaac, "The remarkable story of the DRAM industry," *IEEE SSCS News*, Winter 2008, pp. 45–49.

[Ishihara04] F. Ishihara, F. Sheikh, and B. Nikolic, "Level conversion for dual-supply systems," *IEEE Trans. VLSI*, vol. 12, no. 2, Feb. 2004, pp. 185–195.

[Ismail99] Y. Ismail, E. Friedman, and J. Neves, "Figures of merit to characterize the importance of on-chip interconnect," *IEEE Trans. VLSI*, vol. 7, no. 4, Dec. 1999, pp. 442–449.

[Itoh96] K. Itoh, A. Fridi, A. Bellaouar, and M. Elmasry, "A deep sub-V, single power-supply SRAM cell with multi-V_t, boosted storage node and dynamic load," *Proc. VLSI Circuits Symp.*, Jun. 1996, pp. 132–133.

[Itoh97] K. Itoh, Y. Nakagome, S. Kimura, and T. Watanabe, "Limitations and challenges of multigigabit DRAM chip design," *JSSC*, vol. 32, no. 5, May 1997, pp. 624–634.

[Itoh01] N. Itoh, Y. Naemura, H. Makino, Y. Nakase, T. Yoshihara, and Y. Horiba, "A 600-MHz 54×54-bit multiplier with rectangular-styled Wallace tree," *JSSC*, vol. 36, no. 2, Feb. 2001, pp. 249–257.

[Itoh01k] K. Itoh, *VLSI Memory Chip Design*, Berlin: Springer-Verlag, 2001.

[Itoh01n] N. Itoh et al., "A 600-MHz 54 × 54-bit multiplier with rectangular-styled Wallace tree," *JSSC*, vol. 36, no. 2, Feb. 2001, pp. 249–257.

[Itoh09] K. Itoh, "Adaptive circuits for the 0.5-V nanoscale CMOS era," *Proc. Intl. Solid-State Circuits Conf.*, Feb. 2009, pp. 14–20.

[Jackson04] R. Jackson and S. Talwar, "High speed binary addition," *Proc. Asilomar Conf. Signals, Systems, and Computers*, Nov. 2004, pp. 1350–1353.

[Jacoboni77] C. Jacoboni, C. Canali, G. Ottaviani, and A. Alberigi Quaranta, "A review of some charge transport properties of silicon," *Solid-State Electronics*, vol. 20, 1977, pp. 77–89.

[Jayasumana91] A. Jayasumana, Y. Malaiya, and R. Rajsuman, "Design of CMOS circuits for stuck-open fault testability," *JSSC*, vol. 26, no. 1, Jan. 1991, pp. 58–61.

[Ji-ren87] Y. Ji-ren, I. Karlsson, and C. Svensson, "A true single-phase-clock dynamic CMOS circuit technique," *JSSC*, vol. SC-22, no. 5, Oct. 1987, pp. 899–901.

[Johnson88] M. Johnson, "A symmetric CMOS NOR gate for high-speed applications," *JSSC*, vol. SC-23, no. 5, Oct. 1988, pp. 1233–1236.

[Johnson91] B. Johnson, T. Quarles, A. Newton, D. Pederson, and A. Sangiovanni-Vincentelli, *SPICE3 Version 3e User's Manual*, UC Berkeley, Apr. 1991.

[Johnston96] A. Johnston, "The influence of VLSI technology evolution on radiation-induced latchup in space systems," *IEEE Trans. Nuclear Science*, vol. 43, no. 2, Apr. 1996, pp. 505–521.

[Josephson02] D. Josephson, "The manic depression of microprocessor debug," *Proc. Intl. Test Conf.*, 2002, pp. 657–663.

[Jung01] S. Jung, S. Yoo, K. Kim, and S. Kang, "Skew-tolerant high-speed (STHS) domino logic," *Proc. IEEE Intl. Symp. Circuits and Systems*, 2001, pp. 154–157.

[Kahng08] A. Kahng and K. Samadi, "CMP fill synthesis: A survey of recent studies," *IEEE Trans. CAD*, vol. 27, no. 1, Jan. 2008, pp. 3–19.

[Kamon94] M. Kamon, J. Tsuk, and J. White, "FASTHENRY: a multipole-accelerated 3-D inductance extraction program," *IEEE Trans. Microwave Theory and Techniques*, vol. 42, no. 9, Sep. 1994, pp. 1750–1758.

[Kanamoto07] T. Kanamoto et al., "Impact of well edge proximity effect on timing," *Proc. European Solid State Device Research Conf.*, Sep. 2007, pp. 115–118.

[Kanda02] K. Kanda, T. Miyazaki, M. Sik, H. Kawaguchi, and T. Sakurai, "Two orders of magnitude leakage power reduction of low voltage SRAMs by row-by-row dynamic V_{DD} control (RRDV) scheme," *Proc. Intl. ASIC/SOC Conf.*, Sep. 2002, pp. 381–385.

[Kang03] S. Kang and Y. Leblebici, *CMOS Digital Integrated Circuits*, 3rd ed., Boston: McGraw Hill, 2003.

[Kanj06] R. Kanj, R. Joshi, and S. Nassif, "Mixture importance sampling and its application to the analysis of SRAM designs in the presence of rare failure events," *Proc. Design Automation Conf.*, 2006, pp. 69–72.

[Kanno07] Y. Kanno et al., "Hierarchical power distribution with power tree in dozens of power domains for 90-nm low-power multi-CPU SoCs," *JSSC*, vol. 42, no. 1, Jan. 2007, pp. 74–83.

[Kantabutra91] V. Kantabutra, "Designing optimum carry-skip adders," *Proc. IEEE Symp. Computer Arithmetic*, 1991, pp. 146–153.

[Kantabutra93] V. Kantabutra, "A recursive carry-lookahead/carry-select hybrid adder," *IEEE Trans. Computers*, vol. 42, no. 12, Dec. 1993, pp. 1495–1499.

[Kapur02] P. Kapur, J. McVittie, and K. Saraswat, "Technology and reliability constrained future copper interconnects. I. Resistance modeling," *IEEE Trans. Electron Devices*, vol. 49, no. 4, Apr. 2002, pp. 590–597.

[Karnik01] T. Karnik, B. Bloechel, K. Soumyanath, V. De, and S. Borkar, "Scaling trends of cosmic rays induced soft errors in static latches beyond 0.18 m," *Symp. VLSI Circuits Digest Tech. Papers*, 2001, pp. 61–62.

[Kavaleros06] J. Kavalieros et al., "Tri-gate transistor architecture with high-k gate dielectrics, metal gates and strain engineering," *Proc. VLSI Technology Symp.*, 2006, pp. 50–51.

[Keating07] M. Keating, D. Flynn, R. Aitken, A. Gibbons, and K. Shi, *Low Power Methodology Manual*, New York: Springer, 2007.

[Keeth07] B. Keeth, J. Baker, B. Johnson, F. Lin, *DRAM Circuit Design: Fundamental and High-Speed Topics*, IEEE Press, 2007.

[Keshavarzi01] A. Keshavarzi et al., "Effectiveness of reverse body bias for leakage control in scaled dual V_t CMOS ICs," *Proc. Intl. Symp. Low Power Electronics and Design*, 2001, pp. 207–212.

[Keyes70] R. Keyes, E. Harris, and K. Konnerth, "The role of low temperatures in the operation of logic circuitry," *Proc. IEEE*, vol. 58, no. 12, Dec. 1970, pp. 1914–1932.

[Keys75] R. Keyes, "The effect of randomness in the distribution of impurity atoms on FET thresholds," *Applied Physics*, 8, 1975, pp. 251–259.

[Khalil08] D. Khalil, M. Khellah, Nam-Sung Kim, Y. Ismail, T. Karnik, and V. De, "Accurate estimation of SRAM dynamic stability," *IEEE Trans. VLSI*, vol. 16, no. 12, Dec. 2008, pp. 1639–1647.

[Khare02] M. Khare et al., "A high performance 90nm SOI technology with 0.992 μm^2 6T-SRAM cell," *Proc. Intl. Electron Devices Meeting*, 2002, pp. 407–410.

[Khellah06] M. Khellah et al., "Wordline & bitline pulsing schemes for improving SRAM cell stability in low-V_{cc} 65nm CMOS designs," *Proc. VLSI Circuits Symp.*, 2006, pp. 9–10.

[Khellah07] M. Khellah et al., "A 256-Kb dual-VCC SRAM building block in 65-nm CMOS process with actively clamped sleep transistor," *JSSC*, vol. 42, no. 1, Jan. 2007, pp. 233–242.

[Khellah09] M. Khellah et al., "Process, temperature, and supply-noise tolerant 45 nm dense cache arrays with diffusion-notch-free (DNF) 6T SRAM cells and dynamic multi-Vcc circuits," *JSSC*, vol. 44, no. 4, Apr. 2009, pp. 1199–1208.

[Kielkowski95] R. Kielkowski, *SPICE: Practical Device Modeling*, Boston: McGraw-Hill, 1995.

[Kilburn59] T. Kilburn, D. Edwards, and D. Aspinall, "Parallel addition in digital computers—a new fast 'carry' circuit," *Proc. IEE*, vol. 106B, 1959, pp. 460–464.

[Kim00] J. Kim, Y. Jang, and H. Park, "CMOS sense amplifier-based flip-flop with two N-C2MOS output latches," *Electronics Letters*, vol. 36, no. 6, Mar. 16, 2000, pp. 498–500.

[Kim03] C. Kim, K. Roy, S. Hsu, A. Alvandpour, R. Krishnamurthy, and S. Borkar, "A process variation compensating technique for sub-90 nm dynamic circuits," *Proc. VLSI Circuits Symp.*, Jun. 2003, pp. 205–206.

[Kim03b] J. Kim, M. Horowitz, and Gu-Yeon Wei, "Design of CMOS adaptive-bandwidth PLL/DLLs: a general approach," *IEEE Trans. Circuits & Systems*, vol. 50, no. 11, Nov. 2003, pp. 860–869.

[Kim05] C. Kim, J. Kim, S. Mukhopadhyay, and K. Roy, "A forward body-biased low-leakage SRAM cache: device, circuit and architecture considerations," *IEEE Trans. VLSI*, vol. 13, no. 3, Mar. 2005, pp. 349–357.

[Kim07] J. Kim, K. Jones, and M. Horowitz, "Variable domain transformation for linear PAC analysis of mixed-signal systems," *Proc. Intl. Conf. Computer-Aided Design*, Nov. 2007, pp. 887–894.

[Kim09] T. Kim, J. Liu, and C. Kim, "A voltage scalable 0.26 V, 64 kb 8T SRAM with V_{min} lowering techniques and deep sleep mode," *JSSC*, vol. 44, no. 6, Jun. 2009, pp. 1785–1795.

[Kinniment02] D. Kinniment and E. Chester, "Design of an on-chip random number generator using metastability," *Proc. Intl. Solid-State Circuits Conf.*, Sep. 2002, pp. 595–598.

[Kio01] S. Kio, L. McMurchie, and C. Sechen, "Application of output prediction logic to differential CMOS," *Proc. IEEE Computer Society Workshop on VLSI*, 2001, pp. 57–65.

[Kitsukawa93] G. Kitsukawa et al., "256-Mb DRAM circuit technologies for file applications," *JSSC*, vol. 28, no. 11, Nov. 1993, pp. 1105–1113.

[Klass99] F. Klass et al., "A new family of semidynamic and dynamic flip-flops with embedded logic for high-performance processors," *JSSC*, vol. 34, no. 5, May 1999, pp. 712–716.

[Klaus98] J. Klaus, A. Ott, A. Dillon, and S. George, "Atomic layer controlled growth of Si_3N_4 films using sequential surface reactions," *Surf. Sci.*, vol. 418, 1998, pp. L14–L19.

[Knebel98] D. Knebel et al., "Diagnosis and characterization of timing-related defects by time-dependent light emission," *IEEE Intl. Test Conf.*, 1998, pp. 733–739.

[Knowles01] S. Knowles, "A family of adders," *Proc. IEEE Symp. Computer Arithmetic*, 2001, pp. 277–284.

[Koenemann79] B. Koenemann, J. Mucha, and G. Zwiehoff, "Built-in logic block observation techniques," *Proc. Intl. Test Conf.*, Oct. 1979, pp. 37–41.

[Koester08] S. Koester et al., "Wafer-level 3D integration technology," *IBM J. Research and Dev.*, vol. 52, no. 6, Nov. 2008, pp. 583–597.

[Kogge73] P. Kogge and H. Stone, "A parallel algorithm for the efficient solution of a general class of recurrence equations," *IEEE Trans. Computers*, vol. C-22, no. 8, Aug. 1973, pp. 786–793.

[Koh01] M. Koh et al., "Limit of gate oxide thickness scaling in MOSFETs due to apparent threshold voltage fluctuation induced by tunnel leakage current," *IEEE Trans. Electron Devices*, vol. 48, no. 2, Feb. 2001, pp. 259–264.

[Konstadinidis09] G. Konstadinidis et al., "Architecture and physical implementation of a third generation 65 nm, 16 core, 32 thread chip-multithreading SPARC processor," *JSSC*, vol. 44, no. 1, Jan. 2009, pp. 7–17.

[Kozu96] S. Kozu et al., "A 100 MHz 0.4W RISC processor with 200 MHz multiply-adder, using pulse-register technique," *Proc. IEEE Intl. Solid-State Circuits Conf.*, 1996, pp. 140–141.

[Krambeck82] R. Krambeck, C. Lee, and H. Law, "High speed compact circuits with CMOS," *JSSC*, vol. SC-17, no. 3, Jun. 1982, pp. 614–619.

[Kuang05] J. Kuang et al., "A double-precision multiplier with fine-grained clock-gating support for a first-generation CELL processor," *Proc. Intl. Solid-State Circuits Conf.*, Feb. 2005, pp. 378–605.

[Kulkarni04] S. Kulkarni and D. Sylvester, "High performance level conversion for dual V_{DD} design," *IEEE Trans. VLSI*, vol. 12, no. 9, Sep. 2004, pp. 926–936.

[Kumar94] R. Kumar, "ACMOS: an adaptive CMOS high performance logic," *Electronics Letters*, vol. 30, no. 6, Mar. 1994, pp. 483–484.

[Kumar01] R. Kumar, "Interconnect and noise immunity design for the Pentium 4 processor," *Intel Technology Journal*, vol. 5, no. 1, Q1 2001, pp. 1–12.

[Kumar06] R. Kumar and V. Kursun, "Reversed temperature-dependent propagation delay characteristics in nanometer CMOS circuits," *IEEE Trans. Circuits & Systems*, vol. 53, no. 10, Oct. 2006, pp. 1078–1082.

[Kumar09] R. Kumar and G. Hinton, "A family of 45nm IA processors," *Proc. Intl. Solid-State Circuits Conf.*, Feb. 2009, pp. 58–59.

[Kuo01] J. Kuo and S. Lin, *Low-Voltage SOI CMOS VLSI Devices and Circuits*, New York: Wiley Interscience, 2001.

[Kurd01] N. Kurd, J. Barkarullah, R. Dizon, T. Fletcher, and P. Madland, "A multigigahertz clocking scheme for the Pentium 4 microprocessor," *JSSC*, vol. 36, no. 11, Nov. 2001, pp. 1647–1653.

[Kuroda96] T. Kuroda et al., "A 0.9-V, 150-MHz, 10-mW, 4 mm^2, 2-D discrete cosine transform core processor with variable threshold-voltage (VT) scheme," *JSSC*, vol. 31, no. 11, Nov. 1996, pp. 1770–1779.

[Kwong06] J. Kwong and A. Chandrakasan, "Variation-driven device sizing for minimum energy sub-threshold circuits," *Proc. Intl. Symp. Low Power Electronics & Design*, Oct. 2006, pp. 8–13.

[Kwong09] J. Kwong, Y. Ramadass, N. Verma, and A. Chandrakasan, "A 65 nm sub-V_t microcontroller with integrated SRAM and switched capacitor DC-DC converter," *JSSC*, vol. 44, no. 1, Jan. 2009, pp. 115–126.

[Kynett88] V. Kynett et al., "An in-system reprogrammable 256k CMOS flash memory," *Proc. Intl. Solid-State Circuits Conf.*, Feb. 1988, pp. 132–133, 330.

[Ladner80] R. Ladner and M. Fischer, "Parallel prefix computation," *J. ACM*, vol. 27, no. 4, Oct. 1980, pp. 831–838.

[Lai97] F. Lai and W. Hwang, "Design and implementation of differential cascode voltage switch with pass-gate (DCVSPG) logic for high-performance digital systems," *JSSC*, vol. 32, no. 4, Apr. 1997, pp. 563–573.

[LaPedus07] M. LaPedus, "Costs cast ICs into Darwinian struggle," *EE Times*, Mar. 30, 2007.

[Larsson94] P. Larsson and C. Svensson, "Impact of clock slope on true single phase clocked (TSPC) CMOS circuits," *JSSC*, vol. 29, no. 6, Jun. 1994, pp. 723–726.

[Lasserre99] F. Lasserre et al., "Laser beam backside probing of CMOS integrated circuits," *Microelectronics and Reliability*, Jun. 1999, vol. 39, no. 6, pp. 957–961.

[Le06] T. Le, J. Han, A. von Jouanne, K. Mayaram, and T. Fiez, "Piezoelectric micro-power generation interface circuits," *JSSC*, vol. 41, no. 6, Jun. 2006, pp. 1411–1420.

[Leblebici96] Y. Leblebici, "Design considerations for CMOS digital circuits with improved hot-carrier reliability," *JSSC*, vol. 31, no. 7, Jul. 1996, pp. 1014–1024.

[Lee86] C. Lee and E. Szeto, "Zipper CMOS," *IEEE Circuits and Systems Magazine*, May 1986, pp. 10–16.

[Lee92] K. Lee and M. Breuer, "Design and test rules for CMOS circuits to facilitate IDDQ testing of bridging faults," *IEEE Trans. On CAD of Integrated circuits*, vol. 11, no. 5, May 1992, pp. 659–670.

[Lee98] M. Lee, "A multilevel parasitic interconnect capacitance modeling and extraction for reliable VLSI on-chip clock delay evaluation," *JSSC*, vol. 33, no. 4, Apr. 1998, pp. 657–661.

[Lee03] D. Lee, W. Kwong, D. Blaauw, and D. Sylvester, "Analysis and minimization techniques for total leakage considering gate oxide leakage," *Proc. Design Automation Conf.*, Jun. 2003, pp. 175–180.

[Lee05] W. Lee et al., "High performance 65 nm SOI technology with enhanced transistor strain and advanced-low-K BEOL," *Proc. Intl. Electron Devices Meeting*, Dec. 2005.

[Lee06] L. Lee, D. Weinlader, and C.K. Yang, "A sub-10-ps multiphase sampling system using redundancy," *JSSC*, vol. 41, no. 1, Jan. 2006, pp. 265–273.

[Lehman61] M. Lehman and N. Burla, "Skip technique for high-speed carry-propagation in binary arithmetic units," *IRE Trans. Electronic Computers*, vol. 10, Dec. 1961, pp. 691–698.

[Leighton92] F. Leighton, *Introduction to Parallel Algorithms and Architectures: Arrays; Trees; Hypercubes*, San Francisco: Morgan Kaufmann, 1992.

[Leon07] A. Leon, K. Tam, J. Shin, D. Weisner, and F. Schumacher, "A power-efficient high-throughput 32-thread SPARC processor," *JSSC*, vol. 42, no. 1, Jan. 2007, pp. 7–16.

[Lhermet08] H. Lhermet, C. Condemine, M. Plissonnier, R. Salot, P. Audebert, and M. Rosset, "Efficient power management circuit: from thermal energy harvesting to above-IC microbattery energy storage," *JSSC*, vol. 43, no. 1, Jan. 2008, pp. 246–255.

[Liew90] B. Liew, N. Cheung, and C. Hu, "Projecting interconnect electromigration lifetime for arbitrary current waveforms," *IEEE Trans. Electron Devices*, vol. 37, no. 5, May 1990, pp. 1343–1351.

[Lih07] Y. Lih, N. Tzartzanis, and W. Walker, "A leakage current replica keeper for dynamic circuits," *JSSC*, vol. 42, no. 1, Jan. 2007, pp. 48–55.

[Lim72] R. Lim, "A barrel switch design," *Computer Design*, Aug. 1972, pp. 76–78.

[Lim05] D. Lim, J. Lee, B. Gassend, G. Suh, M. van Dijk, and S. Devadas, "Extracting secret keys from integrated circuits," *IEEE Trans. VLSI*, vol. 13, no. 10, Oct. 2005, pp. 1200–1205.

[Lin83] S. Lin and D. Costello, *Error Control Coding: Fundamentals and Applications*, Upper Saddle River, NJ: Prentice Hall, 1983.

[Linderman04] M. Linderman, D. Harris, and D. Diaz, "Bounding bus delay and noise effects of on-chip inductance," *Proc. Workshop on Signal Propagation on Interconnects*, May 2004, pp. 167–170.

[Ling81] H. Ling, "High-speed binary adder," *IBM J. Research and Development*, vol. 25, no. 3, May 1981, pp. 156–166.

[Liu01] X. Liu, C. Lee, C. Zhou, and J. Han, "Carbon nanotube field-effect inverters," *Appl. Phys. Letters*, vol. 79, no. 20, Nov. 2001, pp. 3329–3331.

[Lofstrom00] K. Lofstrom, W. Daasch, and D. Taylor, "IC identification circuit using device mismatch," *Proc. Intl. Solid-State Circuits Conf.*, 2000, pp. 372–373.

[Lohstroh79] J. Lohstroh, "Static and dynamic noise margins of logic circuits," *JSSC*, vol. SC-14, no. 3, Jun. 1979, pp. 591–598.

[Lohstroh83] J. Lohstroh, E. Seevinck, and J. de Groot, "Worst-case static noise margin criteria for logic circuits and their mathematical equivalence," *JSSC*, vol. SC-18, no. 6, Dec. 1983, pp. 803–807.

[Lu88] S. Lu, "Implementation of iterative networks with CMOS differential logic," *JSSC*, vol. 23, no. 4, Aug. 1988, pp. 1013–1017.

[Lu91] S. Lu and M. Ercegovac, "Evaluation of two-summand adders implemented in ECDL CMOS differential logic," *JSSC*, vol. 26, no. 8, Aug. 1991, pp. 1152–1160.

[Lu93] F. Lu, H. Samueli, J. Yuan, and C. Svensson, "A 700 MHz 24-b pipelined accumulator in 1.2-μm CMOS for application as a numerically controlled oscillator," *JSSC*, vol. 28, no. 8, Aug. 1993, pp. 878–886.

[Lu93b] F. Lu and H. Samueli, "A 200-MHz CMOS pipelined multiplier-accumulator using a quasi-domino dynamic full-adder cell design," *JSSC*, vol. 28, no. 2, Feb. 1993, pp. 123–132.

[Lu08] S. Lu, S. Hsu, and D. Somasekhar, "Memory arrays circuits for computer architects," *IEEE Micro Tutorial*, 2008.

[Lynch92] T. Lynch and E. Swartzlander, "A spanning tree carry lookahead adder," *IEEE Trans. Computers*, vol. 41, no. 8, Aug. 1992, pp. 931–939.

[Lyon87] R. Lyon and R. Schediwy, "CMOS static memory with a new four-transistor memory cell," *Proc. Advanced Research in VLSI*, Mar. 1987, pp. 111–132.

[Lyons62] R. Lyons and W. Vanderkulk, "The use of triple-modular redundancy to improve computer reliability," *IBM Journal*, Apr. 1962, pp. 200–209.

[Ma94] S. Ma and P. Franzon, "Energy control and accurate delay estimation in the design of CMOS buffers," *JSSC*, vol. 29, no. 9, Sep. 1994, pp. 1150–1153.

[Mack08] C. Mack, "Seeing double," *IEEE Spectrum*, vol. 45, no. 11, Nov. 2008, pp. 46–51.

[MacSorley61] O. MacSorley, "High-Speed arithmetic in binary computers," *Proc. IRE*, vol. 49, pt. 1, Jan. 1961, pp. 67–91.

[Mahalingam85] M. Mahalingam, "Thermal management in semiconductor device packages," *Proc. IEEE Custom Integrated Circuits Conf.*, 1985, pp. 46–49.

[Mai05] K. Mai et al., "Architecture and circuit techniques for a 1.1-GHz 16-kb reconfigurable memory in 0.18-μm CMOS," *JSSC*, vol. 40, no. 1, Jan. 2005, pp. 261–275.

[Maier97] C. Maier et al., "A 533-MHz BiCMOS superscalar RISC microprocessor," *JSSC*, vol. 32, no. 11, Nov. 1997, pp. 1625–1634.

[Majerski67] S. Majerski, "On determination of optimal distributions of carry skips in adders," *IEEE Trans. Electronic Computers*, vol. EC-16, no. 1, 1967, pp. 45–58.

[Maksimović00] D. Maksimović, V. Oklobdzija, B. Nikolic, and K. Current, "Clocked CMOS adiabatic logic with integrated single-phase power-clock supply," *IEEE Trans. VLSI*, vol. 8, no. 4, Aug. 2000, pp. 460–463.

[Maluf04] N. Maluf and K. Williams, *An Introduction to Microelectromechanical Systems*, 2nd ed., Norwood, MA: Artech House, 2004.

[Maneatis03] J. Maneatis, I. McClatchie, J. Maxey, and M. Shankaradas, "Self-biased high-bandwidth low-jitter 1-to-4096 multiplier clock generator PLL," *JSSC,* vol. 38, no. 11, Nov. 2003, pp. 1795–1803.

[Markovic04] D. Markovic, V. Stojanovic, B. Nikolic, M. Horowitz, and R. Brodersen, "Methods for true energy-performance optimization," *JSSC*, vol. 39, no. 8, Aug. 2004, pp. 1282–1293.

[Masuoka84] F. Masuoka, M. Asano, H. Iwahashi, T. Komuro, and S. Tanaka, "A new flash E2PROM cell using triple polysilicon technology," *Proc. Intl. Electron Devices Meeting*, 1984, pp. 464–467.

[Masuoka87] F. Masuoka, M. Momodomi, Y. Iwata, and R. Shirota, "New ultra high density EPROM and flash EEPROM with NAND structure cell," *Proc. Intl. Electron Devices Meeting*, 1987, pp. 552–555.

[Mathew03] S. Mathew, M. Anders, R. Krishnamurthy, and S. Borkar, "A 4-GHz 130-nm address generation unit with 32-bit sparse-tree adder core," *JSSC*, vol. 38, no. 5, May 2003, pp. 689–695.

[Mathew05] S. Mathew, M. Anders, B. Bloechel, Trang Nguyen, R. Krishnamurthy, and S. Borkar, "A 4-GHz 300-mW 64-bit integer execution ALU with dual supply voltages in 90-nm CMOS," *JSSC*, vol. 40, no. 1, Jan. 2005, pp. 44–51.

[Matsui94] M. Matsui et al., "A 200 MHz 13 mm^2 2-D DCT macrocell using sense-amplifier pipeline flip-flop scheme," *JSSC*, vol. 29, no. 12, Dec. 1994, pp. 1482–1490.

[May79] T. May and M. Woods, "Alpha-particle-induced soft errors in dynamic memories," *IEEE Trans. Electron Devices*, vol. ED-26, no. 1, Jan. 1979, pp. 2–9.

[McGowen06] R. McGowen et al., "Power and temperature control on a 90-nm Itanium family processor," *JSSC*, vol. 41, no. 1, Jan. 2006, pp. 229–237.

[McMurchie00] L. McMurchie, S. Kio, G. Yee, T. Thorp, and C. Sechen, "Output prediction logic: a high-performance CMOS design technique," *Proc. Intl. Conf. Computer Design*, 2000, pp. 247–254.

[Mead80] C. Mead and L. Conway, *Introduction to VLSI Systems*, Reading, MA: Addison-Wesley, 1980.

[Mears96] J. Mears, "Transmission line RAPIDDESIGNER operation and applications guide," National Semiconductor Application Note 905, May 1996, `www.national.com/an/AN/AN-905.pdf`.

[Mehta99] G. Mehta, D. Harris, and D. Singh, "Pulsed Domino Latches," US Patent 5,880,608, 1999.

[Meier99] N. Meier, T. Marieb, P. Flinn, R. Gleixner, and J. Bravman, "In-situ studies of electromigration voiding in passivated copper interconnects," *AIP Conf. Proc. 491*, Fifth Intl. Workshop on Stress-Induced Phenomena in Metallization, Jun. 1999, p. 180.

[Meijs84] N. van der Meijs, and J. Fokkema, "VLSI circuit reconstruction from mask topology," *Integration, The VLSI Journal,* vol. 2, no. 2, Jun. 1984, pp. 85–119.

[Meindl00] J. Meindl and J. Davis, "The fundamental limit on binary switching energy for terascale integration (TSI)," *JSSC*, vol. 35, no. 10, Oct. 2000, pp. 1515–1516.

[Meng08] X. Meng, R. Saleh, and K. Arabi, "Layout of decoupling capacitors in IP blocks for 90-nm CMOS," *IEEE Trans. VLSI*, vol. 16, no. 11, Nov. 2008, pp. 1581–1588.

[Merchant01] S. Merchant, S. Kang, M. Sanganeria, B. van Schravendijk, and T. Mountsier, "Copper interconnects for semiconductor devices," *JOM: Journal of the Minerals, Metals, and Materials Society*, vol. 53, no. 6, Jun. 2001, pp. 43–48

[Messerschmitt90] D. Messerschmitt, "Synchronization in digital system design," *IEEE J. Selected Areas Communications*, vol. 8, no. 8, Oct. 1990, pp. 1404–1419.

[Min06] K. Min, H. Choi, H. Choi, H. Kawaguchi, and T. Sakurai, "Leakage-suppressed clock-gating circuit with Zigzag Super Cut-off CMOS (ZSCCMOS) for leakage-dominant sub-70-nm and sub-1-V-V_{DD} LSIs," *IEEE Trans. VLSI*, vol. 14, no. 4, Apr. 2006, pp. 430–435.

[Misaka96] A. Misaka, A. Goda, K. Matsuoka, H. Umimoto, and S. Odanaka, "A statistical critical dimension control at CMOS cell level," *Proc. Intl. Electron Devices Meeting*, Dec. 1996, pp. 631–634.

[Mistry07] K. Mistry et al., "A 45nm logic technology with high-k+ metal gate transistors, strained silicon, 9 Cu interconnect layers, 193nm dry patterning, and 100% Pb-free packaging," *Proc. Intl. Electron Devices Meeting*, Dec. 2007, pp. 247–250.

[Mitra05] S. Mitra, T. Karnik, N. Seifert, and Ming Zhang, "Logic soft errors in sub-65nm technologies design and CAD challenges," *Proc. Design Automation Conf.*, Jun. 2005, pp. 2–4.

[Mizuno94] T. Mizuno, J. Okumtura, and A. Toriumi, "Experimental study of threshold voltage fluctuation due to statistical variation of chanel dopant number in MOSFET's," *IEEE Trans. Electron Devices*, vol. 41, no. 11, Nov. 1994, pp. 2216–2221.

[Moazzami90] R. Moazzami and C. Hu, "Projecting gate oxide reliability and optimizing reliability screens," *IEEE Trans. Electron Devices*, vol. 37, no. 7, Jul. 1990, pp. 1643–1650.

[Monsieur01] F. Monsieur, E. Vincent, D. Roy, S. Bruyre, G. Pananakakis, and G. Ghibaudo, "Time to breakdown and voltage to breakdown modeling for ultra-thin oxides (Tox<32Å)," *Proc. Intl. Integrated Reliability Workshop*, 2001, pp. 20–25.

[Montanaro96] J. Montanaro et al., "A 160-MHz, 32-b, 0.5-W CMOS RISC microprocessor," *JSSC*, vol. 31, no. 11, Nov. 1996, pp. 1703–1714.

[Montoye90] R. Montoye, P. Cook, E. Hokenek, and R. Havreluk, "An 18 ns 56-bit multiply-adder circuit," *Proc. Intl. Solid-State Circuits Conf.*, Feb. 1990, pp. 46–47.

[Moon08] P. Moon et al., "Process and electrical results for the on-die interconnect stack for Intel's 45nm process generation," *Intel Technology Journal*, vol. 12, no. 2, Jun. 2008, pp. 87–92.

[Moore65] G. Moore, "Cramming more components onto integrated circuits," *Electronics*, vol. 38, no. 8, Apr. 1965.

[Moore03] G. Moore, "No exponential is forever: but 'forever' can be delayed!" *Proc. IEEE Intl. Solid-State Circuits Conf.*, 2003, pp. 1–19.

[Morgan59] C. Morgan and D. Jarvis, "Transistor logic using current switching routing techniques and its application to a fast carry-propagation adder," *Proc. IEE*, vol. 106B, 1959, pp. 467–468.

[Morgenshtein09] A. Morgenshtein, E. Friedman, R. Ginosar, and A. Kolodny, "Unified logical effort—a method for delay evaluation and minimization in logic paths with RC interconnect," to appear in *IEEE Trans. VLSI*, 2010.

[Mori91] J. Mori et al., "A 10 ns 54×54 b parallel structured full array multiplier with 0.5 μm CMOS technology," *JSSC*, vol. 26, no. 4, Apr. 1991, pp. 600–606.

[Morita06] Y. Morita et al., "A V_{th}-variation-tolerant SRAM with 0.3-V minimum operation voltage for memory-rich SoC under DVS environment," *Proc. VLSI Circuits Symp.*, 2006, pp. 13–14.

[Morrison61] P. Morrison and E. Morrison, eds., *Charles Babbage: On the Principles and Development of the Calculator*, New York: Dover, 1961.

[Morton99] S. Morton, "On-chip inductance issues in multiconductor systems," *Proc. Design Automation Conf.*, 1999, pp. 921–926.

[Mou90] Z. Mou and F. Jutand, "A class of close-to-optimum adder trees allowing regular and compact layout," *Proc. IEEE Intl. Conf. on Computer Design*, 1990, pp. 251–254.

[Mukhopadhyay03] S. Mukhopadhyay, C. Neau, R. Cakici, A. Agarwal, C. Kim, and K. Roy, "Gate leakage reduction for scaled devices using transistor stacking," *IEEE Trans. VLSI*, vol. 11, no. 4, Aug. 2003, pp. 716–730.

[Mukhopadhyay05] S. Mukhopadhyay, A. Raychowdhury, and K. Roy, "Accurate estimation of total leakage in nanometer-scale bulk CMOS circuits based on device geometry and doping profile," *IEEE Trans. CAD*, vol. 24, no. 3, Mar. 2005, pp. 363–381.

[Mule02] A. Mule, E. Glytsis, T. Gaylord, and J. Meindl, "Electrical and optical clock distribution networks for gigascale microprocessors," *IEEE Trans. VLSI*, vol. 10, no. 5, Oct. 2002, pp. 582–594.

[Muller03] R. Muller, T. Kamins, and M. Chan, *Device Electronics for Integrated Circuits*, 3rd ed., New York: John Wiley & Sons, 2003.

[Muller08] M. Muller, "Embedded processing at the heart of life and style," *Proc. Intl. Solid-State Circuits Conf.*, Feb. 2008, pp. 32–37.

[Murabayashi96] F. Murabayashi et al., "2.5 V CMOS circuit techniques for a 200 MHz superscalar RISC processor," *JSSC*, vol. 31, no. 7, Jul. 1996, pp. 972–980.

[Mutoh95] S. Mutoh et al., "1-V power supply high-speed digital circuit technology with multithreshold-voltage CMOS," *JSSC*, vol. 30, no. 8, Aug. 1995, pp. 847–854.

[Mutoh99] S. Mutoh, S. Shigematsu, Y. Gotoh, and S. Konaka, "Design method of MTCMOS power switch for low-voltage high-speed LSIs," *Proc. Design Automation Conf.*, Jan. 1999, pp. 113–116.

[Myny09] K. Myny et al., "A 128b organic RFID transponder chip, including Manchester encoding and ALOHA anti-collision protocol, operating with a data rate of 1529b/s," *Proc. Intl. Solid-State Circuits Conf.*, Feb. 2009, pp. 206–207.

[Na02] M. Na, E. Nowak, W. Haensch, and J. Cai, "The effective drive current in CMOS inverters," *Proc. Intl. Electron Devices Meeting*, 2002, pp. 121–124.

[Nabors92] K. Nabors, S. Kim, and J. White, "Fast capacitance extraction of general three-dimensional structures," *IEEE Trans. Microwave Theory and Techniques*, vol. 40, no. 7, Jul. 1992, pp. 1496–1506.

[Nadig77] H. Nadig, "Signature analysis—concepts, examples and guidelines," *Hewlett Packard Journal*, vol. 28, no. 9, May 1977, pp. 15–21.

[Naffziger96] S. Naffziger, "A subnanosecond 0.5μm 64b adder design," *Proc. IEEE Intl. Solid-State Circuits Conf.*, 1996, pp. 362–363.

[Naffziger98] S. Naffziger, "High speed addition using Ling's equations and dynamic CMOS logic," US Patent 5,719,803, 1998.

[Naffziger02] S. Naffziger, G. Colon-Bonet, T. Fischer, R. Riedlinger, T. Sullivan, and T. Grutkowski, "The implementation of the Itanium 2 microprocessor," *JSSC*, vol. 37, no. 11, Nov. 2002, pp. 1448–1460.

[Naffziger06] S. Naffziger et al., "The implementation of a 2-core, multi-threaded Itanium family processor," *JSSC*, vol. 41, no. 1, Jan. 2006, pp. 197–209.

[Naffziger06b] S. Naffziger, "High-performance processors in a power-limited world," *Proc. VLSI Circuits Symp.*, 2006, pp. 93–97.

[Nagel75] L. Nagel, *SPICE2: a computer program to simulate semiconductor circuits*, Memo ERL-M520, Dept. of Electrical Engineering and Computer Science, University of California at Berkeley, May 9, 1975.

[Najm07] F. Najm, N. Menezes, and I. Ferzli, "A yield model for integrated circuits and its application to statistical timing analysis," *IEEE Trans. CAD*, vol. 26, no. 3, Mar. 2007, pp. 574–591.

[Nakagome03] Y. Nakagome, M. Horiguchi, T. Kawahara, and K. Itoh, "Review and future prospects of low-voltage RAM circuits," *IBM J. Research and Dev.*, vol. 47, no. 5/6, Sep./Nov. 2003, pp. 525–552.

[Nalamalpu02] A. Nalamalpu, S. Srinivasan, and W. Burleson, "Boosters for driving long on-chip interconnects—design issues, interconnect synthesis, and comparison with repeaters," *IEEE Trans. Computer-Aided Design*, vol. 21, no. 1, Jan. 2002, pp. 50–62.

[Nambu98] H. Nambu et al., "A 1.8-ns access, 550-MHz, 4.5-Mb CMOS SRAM," *JSSC*, vol. 33, no. 11, Nov. 1998, pp. 1650–1658.

[Narasimha06] S. Narasimha et al., "High performance 45-nm SOI technology with enhanced strain, porous low-k BEOL, and immersion lithography," *Proc. Intl. Electron Devices Meeting*, Dec. 2006, pp. 1–4.

[Narayanan96] V. Narayanan, B. Chappell, and B. Fleischer, "Static timing analysis for self-resetting circuits," *Proc. Intl. Conf. Computer-Aided Design*, 1996, pp. 119–126.

[Narendra99] S. Narendra, D. Antoniadis, and V. De, "Impact of using adaptive body bias to compensate die-to-die V_t variation on within-die V_t variation," *Proc. Intl. Symp. Low Power Electronics and Design*, 1999, pp. 229–232.

[Narendra01] S. Narendra, S. Borkar, V. De, D. Antoniadis, and A. Chandrakasan, "Scaling of stack effect and its application for leakage reduction," *Proc. Intl. Symp. Low Power Electronics and Design*, 2001, pp. 195–200.

[Narendra03] S. Narendra, A. Keshavarzi, B. Bloechel, S. Borkar, and V. De, "Forward body bias for microprocessors in 130-nm technology generation and beyond," *JSSC*, vol. 38, no. 5, May 2003, pp. 696–701.

[Narendra06] S. Narendra and A. Chandrakasan, *Leakage in Nanometer CMOS Technologies*, New York: Springer, 2006.

[Natarajan08] S. Natarajan et al., "A 32 nm logic technology featuring 2nd-generation high-k + metal-gate transistors, enhanced channel strain and 0.171 μm^2 SRAM cell size in a 291 Mb array," *Proc. Intl. Electron Devices Meeting*, Dec. 2008, pp. 1–3.

[National08] National Semiconductor, *LVDS Owner's Manual*, 4th ed., 2008, `www.national.com/LVDS`.

[Nawathe08] U. Nawathe, M. Hassan, K. Yen, A. Kumar, A. Ramachandran, and D. Greenhill, "Implementation of an 8-core, 64-thread, power-efficient SPARC server on a chip," *JSSC*, vol. 43, no. 1, Jan. 2008, pp. 6–20.

[Needham98] W. Needham, C. Prunty, and E. Yeoh, "High volume microprocessor test escapes, an analysis of defects our tests are missing," *Proc. Intl. Test Conf.*, 1998, pp. 25–34.

[Ng96] P. Ng, P. Balsara, and D. Steiss, "Performance of CMOS differential circuits," *JSSC*, vol. 31, no. 6, Jun. 1996, pp. 841–846.

[Nii04] K. Nii et al., "A 90-nm low-power 32-kB embedded SRAM with gate leakage suppression circuit for mobile applications," *JSSC*, vol. 39, no. 4, Apr. 2004, pp. 684–693.

[Nikolić00] B. Nikolić, V. Oklobdzija, V. Stojanović, W. Jia, J. Chiu, and M. Leung, "Improved sense-amplifier-based flip-flop: design and measurements," *JSSC*, vol. 35, no. 6, Jun. 2000, pp. 876–884.

[NIST02] National Institute of Standards and Technology, "Security requirements for cryptographic modules," FIPS 140-2, 2001.

[Noguchi07] K. Noguchi and M. Nagata, "An on-chip multichannel waveform monitor for diagnosis of systems-on-a-chip integration," *IEEE Trans. VLSI*, vol. 15, no. 10, Oct. 2007, pp. 1101–1110.

[Noice83] D. Noice, *A clocking discipline for two-phase digital integrated circuits*, Stanford University Technical Report, Jan. 1983.

[Northrop99] G. Northrop et al., "609 MHz G5 S/399 microprocessor," *Proc. Intl. Solid-State Circuits Conf.*, 1999, pp. 88–89.

[Nose00a] K. Nose and T. Sakurai, "Analysis and future trend of short-circuit power," *IEEE Trans. CAD*, vol. 19, no. 9, Sep. 2000, pp. 1023–1030.

[Nose00b] K. Nose, Soo-Ik Chae, and T. Sakurai, "Voltage dependent gate capacitance and its impact in estimating power and delay of CMOS digital circuits with low supply voltage," *Proc. Intl. Symp. Low Power Electronics & Design*, 2000, pp. 228–230.

[Nose00c] K. Nose and T. Sakurai, "Optimization of V_{DD} and V_{TH} for low-power and high-speed applications," *Proc. Design Automation Conf.*, 2000, pp. 469–474.

[Nose06] K. Nose, M. Kajita, and M. Mizuno, "A 1ps-resolution jitter-measurement macro using interpolated jitter oversampling," *Proc. Intl. Solid-State Circuits Conf.*, Feb. 2006, pp. 2112–2121.

[Nowka98] K. Nowka and T. Galambos, "Circuit design techniques for a gigahertz integer microprocessor," *Proc. Intl. Conf. Computer Design*, 1998, pp. 11–16.

[Oh06] H. Oh et al., "A fully pipelined single-precision floating-point unit in the synergistic processor element of a CELL processor," *JSSC*, vol. 41, no. 4, Apr. 2006, pp. 759–771.

[Ohbayashi07] S. Ohbayashi et al., "A 65-nm SoC embedded 6T-SRAM designed for manufacturability with read and write operation stabilizing circuits," *JSSC*, vol. 42, no. 4, Apr. 2007, pp. 820–829.

[Ohkubo95] N. Ohkubo et al., "A 4.4 ns CMOS 54 × 54-b multiplier using pass-transistor multiplexer," *JSSC*, vol. 30, no. 3, Mar. 1995, pp. 251–257.

[Oklobdzija96] V. Oklobdzija, D. Villeger, and S. Liu, "A method for speed optimized partial product reduction and generation of fast parallel multipliers using an algorithmic approach ," *IEEE Trans. Computers*, vol. 45, no. 3, Mar. 1996, pp. 294–306.

[Oklobdzija85] V. Oklobdzija and E. Barnes, "Some optimal schemes for ALU implementation in VLSI technology," *Proc. Computer Arithmetic Symp.*, 1985, pp. 137–143.

[Oklobdzija86] V. Oklobdzija and R. Montoye, "Design-performancce trade-offs in CMOS-domino logic," *JSSC*, vol. SC-21, no. 2, April 1986, pp. 304–309.

[Oklobdzija05] V. Oklobdzija, B. Zeydel, H. Dao, S. Mathew, and R. Krishnamurthy, "Comparison of high-performance VLSI adders in the energy-delay space," *IEEE Trans. VLSI*, vol. 13, no. 6, Jun. 2005, pp. 754–758.

[Ortiz-Conde02] A. Ortiz-Conde, F. Sánchez, J. Liou, A. Cerdeira, M. Estrada, and Y. Yue, "A review of recent MOSFET threshold voltage extraction methods," *Microelectronics Reliability*, vol. 42, 2002, pp. 583–596.

[Osada01] K. Osada et al., "Universal-V_{DD} 0.65-2.0-V 32-kB cache using a voltage-adapted timing-generation scheme and a lithographically symmetrical cell," *JSSC*, vol. 36, no. 11, Nov. 2001, pp. 1738–1744.

[Osada04] K. Osada, K. Yamaguchi, Y. Saitoh, and T. Kawahara, "SRAM immunity to cosmic-ray-induced multierrors based on analysis of an induced parasitic bipolar effect," *JSSC*, vol. 39, no. 5, May 2004, pp. 827–833.

[Osada06] K. Osada, "Reviews and prospects of nanoscale SRAMs," *Proc. Intl Conf. Integrated Circuit Design & Tech.*, 2006, pp. 1–8.

[Pagiamtzis06] K. Pagiamtzis and A. Sheikholeslami, "Content-addressable memory (CAM) circuits and architectures: a tutorial and survey," *JSSC*, vol. 41, no. 3, Mar. 2006, pp. 712–727.

[Paik96] W. Paik, H. Ki, and S. Kim, "Push-pull pass-transistor logic family for low voltage and low power," *Proc. 22nd European Solid-State Circuits Conf.*, 1996, pp. 116–119.

[Paik08] P. Paik, V. Pamula, and K. Chakrabarty, "Adaptive cooling of integrated circuits using digital microfluidics," *IEEE Trans. VLSI*, vol. 16, no. 4, Apr. 2008, pp. 432–443.

[Parameswar96] A. Parameswar, H. Hara, and T. Sakurai, "A swing restored pass-transistor logic-based multiply and accumulate circuit for multimedia applications," *JSSC*, vol. 31, no. 6, Jun. 1996, pp. 804–809.

[Paraskevopoulos87] D. Paraskevopoulos and C. Fey, "Studies in LSI technology economics III: design schedules for application-specific integrated circuits," *JSSC*, vol. SC-22, no. 2, Apr. 1987, pp. 223–229.

[Park00] J. Park, H. Ngo, J. Silberman, and S. Dhong, "470 ps 64-bit parallel binary adder," *Proc. VLSI Circuits Symp.*, 2000, pp. 192–193.

[Parker03] K. Parker, *The Boundary-Scan Handbook*, Boston: Kluwer Academic Publishers, 2003.

[Partovi96] H. Partovi et al., "Flow-through latch and edge-triggered flip-flop hybrid elements," *Proc. IEEE Intl. Solid-State Circuits Conf.*, 1996, pp. 138–139.

[Pasternak87] J. Pasternak, A. Shubat, and C. Salama, "CMOS differential pass-transistor logic design," *JSSC*, vol. SC-22, no. 2, Apr. 1987, pp. 216–222.

[Pasternak91] J. Pasternak and C. Salama, "Design of submicrometer CMOS differential pass-transistor logic circuits," *JSSC*, vol. 26, no. 9, Sep. 1991, pp. 1249–1258.

[Patil07] D. Patil, O. Azizi, M. Horowitz, R. Ho, and R. Ananthraman, "Robust energy-efficient adder topologies," *Proc. Computer Arithmetic Symp.*, Jun. 2007, pp. 16–28.

[Patil09] N. Patil, Jie Deng, S. Mitra, and H. Wong, "Circuit-level performance benchmarking and scalability analysis of carbon nanotube transistor circuits," *IEEE Trans. Nanotechnology*, vol. 8, no. 1, Jan. 2009, pp. 37–45.

[Patterson04] D. Patterson and J. Hennessy, *Computer Organization and Design*, 3rd ed., San Francisco, CA: Morgan Kaufmann, 2004.

[Paul02] B. Paul and K. Roy. Testing cross-talk induced delay faults in static CMOS circuit through dynamic timing analysis. *Proc. Intl. Test Conf.*, Oct. 2002, pp. 384–390.

[Paul07] B. Paul, Kunhyuk Kang, H. Kufluoglu, M. Alam, and K. Roy, "Negative bias temperature instability: estimation and design for improved reliability of nanoscale circuits," *IEEE Trans. CAD*, vol. 26, no. 4, Apr. 2007, pp. 743–751.

[Pedroni10] V. Pedroni, *Circuit Design and Simulation with VHDL*, 2nd ed., Cambridge, MA: MIT Press, 2010.

[Pelgrom89] M. Pelgrom, A. Duinmaijer, and A. Welbers, "Matching properties of MOS transistors," *JSSC*, vol. 24, no. 5, Oct. 1989, pp. 1433–1440.

[Peng02] C. Peng et al., "A 90 nm generation copper dual damascene technology with ALD TaN barrier," *Tech. Digest Intl. Electron Devices Meeting*, Dec. 2002, pp. 603–606.

[Penney72] W. Penney and L. Lau, *MOS Integrated Circuits*, New York: Van Nostrand Reinhold, 1972.

[Perry05] D. Perry and H. Foster, *Applied Formal Verification*, New York: McGraw-Hill, 2005.

[Pertijs06] M. Pertijs and J. Huijsing, *Precision Temperature Sensors in CMOS Technology*, New York: Springer, 2006.

[Petegem94] W. van Petegem, B. Geeraerts, W. Sansen, and B. Graindourze, "Electrothermal simulation and design of integrated circuits," *JSSC*, vol. 29, no. 2, Feb. 1994, pp. 143–146.

[Pfennings85] L. Pfennings, W. Mol, J. Bastiens, and J. van Dijk, "Differential split-level CMOS logic for subnanosecond speeds," *JSSC*, vol. SC-20, no. 5, Oct. 1985, pp. 1050–1055.

[Pham06] D. Pham et al., "Overview of the architecture, circuit design, and physical implementation of a first-generation cell processor," *JSSC*, vol. 41, no. 1, Jan. 2006, pp. 179–196.

[Pihl98] J. Pihl, "Single-ended swing restoring pass transistor cells for logic synthesis and optimization," *Proc. IEEE Intl. Symp. Circuits and Systems*, vol. 2, 1998, pp. 41–44.

[Piña02] C. Piña, "Evolution of the MOSIS VLSI educational program," *Proc. Electronic Design, Test, and Applications Workshop*, 2002, pp. 187–191.

[Plass07] D. Plass and Y. Chan, "IBM POWER6 SRAM arrays," *IBM J. Research and Dev.*, vol. 51, no. 6, Nov. 2007, pp. 747–756.

[Pollack99] F. Pollack, "New microarchitectural challenges in the coming generations of CMOS process technologies," *Intl. Symp. Microarchitecture*, Keynote address, 1999.

[Pretorius86] J. Pretorius, A. Shubat, and A. Salama, "Latched domino CMOS logic," *JSSC*, vol. SC-21, no. 4, Aug. 1986, pp. 514–522.

[Price95] D. Price, "Pentium FDIV flaw—lessons learned," *IEEE Micro*, vol. 15, no. 2, Apr. 1995, pp. 86–88.

[Proakis08] J. Proakis and M. Salehi, *Digital Communications*, New York: McGraw Hill, 2008.

[Proebsting91] R. Proebsting, "Speed enhancement technique for CMOS circuits," US Patent 4,985,643, 1991.

[Quach92] N. Quach and M. Flynn, "High-speed addition in CMOS," *IEEE Trans. Computers*, vol. 41, no. 12, Dec. 1992, pp. 1612–1615.

[Quader94] K. Quader, E. Minami, W. Huang, P. Ko, and C. Hu, "Hot-carrier-reliability design guidelines for CMOS logic circuits," *JSSC*, vol. 29, no. 3, Mar. 1994, pp. 253–262.

[Ramadass10] Y. Ramadass and A. Chandrakasan, "An efficient piezoelectric energy harvesting interface circuit using a bias-flip rectifier and shared inductor," *JSSC*, vol. 45, no. 1, Jan. 2010, pp. 189–204.

[Rao03] R. Rao, A. Srivastava, D. Blaauw, and D. Sylvester, "Statistical estimation of leakage current considering inter- and intra-die process variation," *Proc. Intl. Symp. Low Power Electronics & Design*, Aug. 2003, pp. 84–89.

[Rao07] R. R. Rao, K. Chopra, D. Blaauw, and D. Sylvester, "Computing the soft error rate of a combinational logic circuit using parameterized descriptors," *IEEE Trans. CAD*, vol. 26, no. 3, Mar. 2007, pp. 468–479.

[Raychowdhury07] A. Raychowdhury and K. Roy, "Carbon nanotube electronics: design of high-performance and low-power digital circuits," *IEEE Trans. Circuits & Systems*, vol. 54, no. 11, Nov. 2007, pp. 2391–2401.

[Razavi03] B. Razavi, *Design of Analog CMOS Integrated Circuits*, McGraw Hill, 2003.

[Reddy02] V. Reddy et al., "Impact of negative bias temperature instability on digital circuit reliability," *Proc. 40th IEEE Intl. Reliability Physics Symp.*, 2002, pp. 248–254.

[Restle01] P. Restle et al., "A clock distribution network for microprocessors," *JSSC*, vol. 36, no. 5, May 2001, pp. 792–799.

[Restle98] P. Restle and A. Deutsch, "Designing the best clock distribution network," *Symp. VLSI Circuits Digest Tech. Papers*, 1998, pp. 2–5.

[Riordan97] M. Riordan and L. Hoddeson, *Crystal Fire: The Invention of the Transistor and the Birth of the Information Age*, New York: W. W. Norton & Co, 1998.

[Rizzolo07] R. Rizzolo, "IBM System z9 eFUSE applications and methodology," *IBM J. Research and Dev.*, vol. 51, no. 1/2, Jan./Mar. 2007, pp. 65–75.

[Rohrer05] N. Rohrer et al., "A 64-bit microprocessor in 130-nm and 90-nm technologies with power management features," *JSSC*, vol. 40, no. 1, Jan. 2005, pp. 19–27.

[Rotella02] F. Rotella, V. Blaschke, and D. Howard, "A broad-band scalable lumped-element inductor model using analytic expressions to incorporate skin effect, substrate loss, and proximity effect," *Tech. Digest Intl. Electron Devices Meeting*, Dec. 2002, pp. 471–474.

[Roy03] K. Roy, S. Mukhopadhyay, and H. Mahmoodi-Meimand, "Leakage current mechanisms and leakage reduction techniques in deep-submicrometer CMOS circuits," *Proceedings of the IEEE*, vol. 91, no. 2, Feb. 2003, pp. 305–327.

[Ruehli73] A. Ruehli and P. Brennan, "Efficient capacitance calculations for three-dimensional multiconductor systems," *IEEE Trans. Microwave Theory and Techniques*, vol. MTT-21, no. 2, Feb. 1973, pp. 76–82.

[Rusu00] S. Rusu and G. Singer, "The first IA-64 microprocessor," *JSSC*, vol. 35, no. 11, Nov. 2000, pp. 1539–1544.

[Rusu03] S. Rusu, J. Stinson, S. Tam, J. Leung, H. Muljono, and B. Cherkauer, "A 1.5-GHz 130-nm Itanium 2 processor with 6-MB on-die L3 cache," *JSSC*, vol. 38, no. 11, Nov. 2003, pp. 1887–1895.

[Rusu07] S. Rusu et al., "A 65-nm dual-core multithreaded Xeon processor with 16-MB L3 cache," *JSSC*, vol. 42, no. 1, Jan. 2007, pp. 17–25.

[Rusu10] S. Rusu et al., "A 45 nm 8-core Enterprise Xeon Processor," *JSSC*, vol. 45, no. 1, Jan. 2010, pp. 7–14.

[Rzepka98] S. Rzepka, K. Banerjee, E. Meusel, and C. Hu, "Characterization of self-heating in advanced VLSI interconnect lines based on thermal finite element simulation," *IEEE Trans. Components, Packaging, and Manufacturing Technology—Part A*, vol. 21, no. 3, Sep. 1998, pp. 406–411.

[Sah64] C. Sah, "Characteristics of the Metal-Oxide-Semiconductor Transistors," *IEEE Trans. Electron Devices*, ED-11, Jul. 1964, pp. 324–345.

[Saint02] C. Saint and J. Saint, *IC Mask Design: Essential Layout Techniques*, New York: McGraw-Hill, 2002.

[Sakran07] N. Sakran, M. Yuffe, M. Mehalel, J. Doweck, E. Knoll, and A. Kovacs, "The implementation of the 65nm dual-core 64b Merom processor," *Proc. Intl. Solid-State Circuits Conf.*, Feb. 2007, pp. 106–107, 590.

[Sakurai83] T. Sakurai, "Approximation of wiring delay in MOSFET LSI," *JSSC*, vol. SC-18, no. 4, Aug. 1983, pp. 418–426.

[Sakurai86] T. Sakurai, K. Nogami, M. Kakumu, and T. Iizuka, "Hot-carrier generation in submicrometer VLSI environment," *JSSC*, vol. SC-21, no. 1, Feb. 1986, pp. 187–192.

[Sakurai90] T. Sakurai and R. Newton, "Alpha-Power Law MOSFET Model and its Applications to CMOS Inverter Delay and Other Formulas," *JSSC*, vol. 25, no. 2, April 1990, pp. 584–594.

[Sakurai91] T. Sakurai and A. Newton, "Delay analysis of series-connected MOSFET circuits," *JSSC*, vol. 26, no. 2, Feb. 1991, pp. 122–131.

[Salib04] M. Salib et al., "Silicon photonics," *Intel Technology Journal*, vol. 08, no. 2, May 2004, pp. 143–160.

[Samson08] G. Samson, N. Ananthapadmanabhan, S. Badrudduza, and L. Clark, "Low-power dynamic memory word line decoding for static random access memories," *JSSC*, vol. 43, no. 11, Nov. 2008, pp. 2524–2532.

[Samson09] G. Samson and L. Clark, "Low-power race-free programmable logic arrays," *JSSC*, vol. 44, no. 3, Mar. 2009, pp. 935–946.

[Santoro89] M. Santoro, "Design and Clocking of VLSI Multipliers," Ph.D. thesis, Stanford University, CSL-TR-89-397, 1989.

[Sathe07] V. Sathe, J. Chueh, and M. Papaefthymiou, "Energy-efficient GHz-class charge-recovery logic," *JSSC*, vol. 42, no. 1, Jan. 2007, pp. 38–47.

[Schellenberg98] F. Schellenberg, H. Zhang, and J. Morrow, "SEMATECH J111 Project: OPC validation," *Proc. Optical Microlithography XI*, SPIE vol. 3334, 1998, pp. 892–911.

[Schellenberg03] F. Schellenberg, "A little light magic," *IEEE Spectrum*, vol. 40, no. 9, Sep. 2003, pp. 34–39.

[Schmitt38] O. Schmitt, "A thermionic trigger," *J. Scientific Instruments*, vol. 15, Jan. 1938, pp. 24–26.

[Schulmann98] W. Schulmann, F. Thimm, and H. Kaiser, "Rotating head for crystal pulling systems for carrying out the Czochralski process," US Patent 5766348, 1998.

[Schultz90] K. Schultz, R. Francis, and K. Smith, "Ganged CMOS: trading standby power for speed," *JSSC*, vol. SC-25, no. 3, Jun. 1990, pp. 870–873.

[Schutten03] R. Schutten, T. Fitzpatrick, "Design for verification—blueprint for productivity and product quality," Synopsys white paper, 2003.

[Schutz98] J. Schutz and R. Wallace, "A 450 MHz IA32 P6 family microprocessor," *Proc. Intl. Solid-State Circuits Conf.*, Feb. 1998, pp. 236–237.

[Seeds67] R. Seeds, "Yield and cost analysis of bipolar LSI," *Intl. Electron Device Meeting*, Oct. 1967.

[Seevinck87] E. Seevinck, F. List, and J. Lohstroh, "Static-noise margin analysis of MOS SRAM cells," *JSSC*, vol. 22, no. 5, Oct. 1987, pp. 748–754.

[Segura04] J. Segura and C. Hawkins, *CMOS Electronics: How it Works, How it Fails*, Hoboken, NJ: John Wiley & Sons (IEEE Press), 2004.

[Shahidi02] G. Shahidi, "SOI technology for the GHz era," *IBM J. Research and Development*, vol. 46, no. 2/3, Mar./May 2002, pp. 121–131.

[Sharifkhani09] M. Sharifkhani and M. Sachdev, "SRAM cell stability: a dynamic perspective," *JSSC*, vol. 44, no. 2, Feb. 2009, pp. 609–619.

[She02] M. She et al., "JVD silicon nitride as tunnel dielectric in p-channel flash memory," *IEEE Electron Device Letters*, vol. 23, no. 2, Feb. 2002, pp. 91–93.

[Shepard99] K. Shepard, V. Narayanan, and R. Rose, "Harmony: static noise analysis of deep submicron digital integrated circuits," *IEEE Trans. Computer-Aided Design*, vol. 18, no. 8, Aug. 1999, pp. 1132–1150.

[Sheu87] B. Sheu, D. Scharfetter, P. Ko, and M. Jeng, "BSIM: Berkeley short-channel IGFET model for MOS transistors," *JSSC*, vol. SC-22, no. 4, Aug. 1987, pp. 558–566.

[Shichman68] H. Shichman and D. Hodges, "Modeling and simulation of insulated-gate field-effect transistor switching circuits," *JSSC*, vol. SC-3, no. 3, Sep. 1968, pp. 285–289.

[Shigematsu97] S. Shigematsu, S. Mutoh, Y. Matsuya, Y. Tanabe, and J. Yamada, "A 1-V high-speed MTCMOS circuit scheme for power-down application circuits," *JSSC*, vol. 32, no. 6, Jun. 1997, pp. 861–869.

[Shin05] J. Shin, B. Petrick, M. Singh, and A. Leon, "Design and implementation of an embedded 512-KB level-2 cache subsystem," *JSSC*, vol. 40, no. 9, Sep. 2005, pp. 1815–1820.

[Shockley52] W. Shockley, "A unipolar 'field-effect' transistor," *Proc. IRE*, vol. 40, 1952, pp. 1365–1376.

[Shoji82] M. Shoji, "Electrical design of BELLMAC-32a microprocessor," *Proc. IEEE Intl. Conf. Circuits and Computers,* Sep. 1982, pp. 112–115.

[Shoji86] M. Shoji, "Elimination of process-dependent clock skew in CMOS VLSI," *JSSC*, vol. SC-21, no. 5, Oct. 1986, pp. 875–880.

[SIA97] Semiconductor Industry Association, *International Technology Roadmap for Semiconductors*, 1997.

[SIA07] Semiconductor Industry Association, *International Technology Roadmap for Semiconductors*, 2007.

[Silberman98] J. Silberman et al., "A 1.0-GHz single-issue 64-bit PowerPC integer microprocessor," *JSSC*, vol. 33, no. 11, Nov. 1998, pp. 1600–1608.

[Singh08] P. Singh, J. Seo, D. Blaauw, and D. Sylvester, "Self-timed regenerators for high-speed and low-power on-chip global interconnect," *IEEE Trans. VLSI*, vol. 16, no. 6, Jun. 2008, pp. 673–677.

[Sklansky60] J. Sklansky "Conditional-sum addition logic," *IRE Trans. Electronic Computers*, vol. EC-9, Jun. 1960, pp. 226–231.

[Sklar01] B. Sklar, *Digital Communications: Fundamentals and Applications*, 2nd ed., Upper Saddle River, NJ: Prentice Hall, 2001.

[Sleight01] J. Sleight et al., "A high performance 0.13 μm SOI CMOS technology with a 70 nm silicon film and with a second generation low-k Cu BEOL," *Proc. Intl. Electron Devices Meeting*, 2001, pp. 11.3.1–11.3.4.

[Smith99] L. Smith, R. Anderson, D. Forehand, T. Pelc, and T. Roy, "Power distribution system design methodology and capacitor selection for modern CMOS technology," *IEEE Trans. Advanced Packaging*, vol. 22, no. 3, Aug. 1999, pp. 284–291.

[Sodini84] C. Sodini, Ping-Keung Ko, and J. Moll, "The effect of high fields on MOS device and circuit performance," *IEEE Trans. Electron Devices*, vol. 31, no. 10, Oct. 1984, pp. 1386–1393.

[Solomatnikov00] A. Solomatnikov, D. Somasekhar, K. Roy, and C. Koh, "Skewed CMOS: noise-immune high-performance low-power static circuit family," *Proc. IEEE Intl. Conf. Computer Design*, 2000, pp. 241–246.

[Somasekhar96] D. Somasekhar and K. Roy, "Differential current switch logic: a low power DCVS logic family," *JSSC*, vol. 31, no. 7, Jul. 1996, pp. 981–991.

[Somasekhar98] D. Somasekhar and K. Roy, "LVDCSL: a high fan-in, high-performance, low-voltage differential current switch logic family," *IEEE Trans. VLSI*, vol. 6, no. 4, Dec. 1998, pp. 573–577.

[Somasekhar00] D. Somasekhar, S. Choi, K. Roy, Y. Ye, and V. De, "Dynamic noise analysis in precharge-evaluate circuits," *Proc. Design Automation Conf.*, 2000, pp. 243–246.

[Song96] M. Song, G. Kang, S. Kim, and B. Kang, "Design methodology for high speed and low power digital circuits with energy economized pass-transistor logic (EEPL)," *Proc. 22nd European Solid-State Circuits Conf.*, 1996, pp. 120–123.

[Song01] S. Song et al., "On the gate oxide scaling of high performance CMOS transistors," *Proc. Intl. Electron Devices Meeting*, 2001, pp. 3.2.1–3.2.4.

[Sparsø01] J. Sparsø and S. Furber, eds., *Principles of Asynchronous Circuit Design: A Systems Perspective*, Boston: Kluwer Academic Publishers, 2001.

[Srinivas92] H. Srinivas and K. Parhi, "A fast VLSI adder architecture," *JSSC*, vol. 27, no. 5, May 1992, pp. 761–767.

[Srinivasan02] V. Srinivasan et al., "Optimizing pipelines for power and performance," *Proc. Intl. Symp. Microarchitecture*, 2002, pp. 333–344.

[Stackhouse09] B. Stackhouse et al., "A 65 nm 2-billion transistor quad-core Itanium processor," *JSSC*, vol. 44, no. 1, Jan. 2009, pp. 18–31.

[Stan98] M. Stan, A. Tenca, and M. Ercegovac, "Long and fast up/down counters," *IEEE Trans. Computers*, vol. 47, no. 7, Jul. 1998, pp. 722–735.

[Stan99] M. Stan, "Optimal voltages and sizing for low power [CMOS VLSI]," *Proc. Intl. Conf. VLSI Design*, Jan. 1999, pp. 428–433.

[Stelling98] P. Stelling, C. Martel, V. Oklobdzija, and R. Ravi, "Optimal circuits for parallel multipliers," *IEEE Trans. Computers*, vol. 47, no. 3, Mar. 1998, pp. 273–285.

[Stinson03] J. Stinson and S. Rusu, "A 1.5 GHz third generation Itanium processor," *Proc. Design Automation Conf.*, 2003, pp. 706–709.

[Stojanovic99] V. Stojanovic and V. Oklobdûija, "Comparative analysis of master-slave latches and flip-flops for high-performance and low-power systems," *JSSC*, vol. 34, no. 4, Apr. 1999, pp. 536–548.

[Stolk98] P. Stolk, F. Widdershoven, and D. Klaassen, "Modeling statistical dopant fluctuations in MOS transistors," *IEEE Trans. Electron Devices*, vol. 45, no. 9, Sep. 1998, pp. 1960–1971.

[Stolt08] B. Stolt et al., "Design and Implementation of the POWER6 microprocessor," *JSSC*, vol. 43, no. 1, Jan. 2008, pp. 21–28.

[Strollo05] A. Strollo, D. De Caro, E. Napoli, and N. Petra, "A novel high-speed sense-amplifier-based flip-flop," *IEEE Trans. VLSI*, vol. 13, no. 11, Nov. 2005, pp. 1266–1274.

[Stroud02] C. Stroud, *A Designer's Guide to Built-in Self-Test*, Boston: Kluwer Academic Publishers, 2002.

[Su03] H. Su, F. Liu, A. Devgan, E. Acar, and S. Nassif, "Full chip leakage-estimation considering power supply and temperature variations," *Proc. Intl. Symp. Low Power Electronics & Design*, Aug. 2003, pp. 78–83.

[Su08] Y. Su, J. Holleman, and B. Otis, "A digital 1.6 pJ/bit chip identification circuit using process variations," *JSSC*, vol. 43, no. 1, Jan. 2008, pp. 69–77.

[Sun87] J. Sun, Y. Taur, R. Dennard, and S. Klepner, "Submicrometer-channel CMOS for low-temperature operation," *IEEE Trans. Electron Devices*, vol. ED-34, no. 1, Jan. 1987, pp. 19–26.

[Sutherland99] I. Sutherland, B. Sproull, and D. Harris, *Logical Effort: Designing Fast CMOS Circuits*, San Francisco, CA: Morgan Kaufmann, 1999.

[Suzuki73] Y. Suzuki, K. Odagawa and T. Abe, "Clocked CMOS calculator circuitry," *JSSC*, vol. SC-8, no. 6, Dec. 1973, pp. 462–469.

[Suzuki93] M. Suzuki, N. Ohkubo, T. Shinbo, T. Yamanaka, A. Shimizu, K. Sasaki, and Y. Nakagome, "A 1.5-ns 32-b CMOS ALU in double pass-transistor logic," *JSSC*, vol. 28, no. 11, Nov. 1993, pp. 1145–1151.

[Svensson03] C. Svensson, "Forty years of feature-size predictions (1962–2002)," *Proc. Intl. Solid-State Circuits Conf.*, 2003, pp. 35–36.

[Swanson72] R. Swanson and J. Meindl, "Ion-implanted complementary MOS transistors in low-voltage circuits," *JSSC*, vol. 7, no. 2, Apr. 1972, pp. 146–153.

[Sweeney02] P. Sweeney, *Error Control Coding: From Theory to Practice*, New York: John Wiley & Sons, 2002.

[Sylvester98] D. Sylvester and K. Keutzer, "Getting to the bottom of deep submicron," *Proc. IEEE/ACM Intl. Conf. Computer-Aided Design*, 1998, pp. 203–211.

[Takahashi98] M. Takahashi et al., "A 60-mW MPEG4 video codec using clustered voltage scaling with variable supply-voltage scheme," *JSSC*, vol. 33, no. 11, Nov. 1998, pp. 1772–1780.

[Takeuchi94] K. Takeuchi and M. Fukuma, "Effects of the velocity saturated region on MOSFET characteristics," *IEEE Trans. Electron Devices*, vol. 41, no. 9, Sep. 1994, pp. 1623–1627.

[Tam00] S. Tam, S. Rusu, U. Desai, R. Kim, J. Zhang, and I. Young, "Clock generation and distribution for the first IA-64 microprocessor," *JSSC*, vol. 35, no. 11, Nov. 2000, pp. 1545–1552.

[Tam04] S. Tam, R. Limaye, and U. Desai, "Clock generation and distribution for the 130-nm Itanium 2 processor with 6-MB on-die L3 cache," *JSSC*, vol. 39, no. 4, Apr. 2004.

[Tang97] X. Tang, V. De, and J. Meindl, "Intrinsic MOSFET parameter fluctuations due to random dopant placement," *IEEE Trans. VLSI*, vol. 5, no. 4, Dec 1997, pp. 369–376.

[Tawfik09] S. Tawfik and V. Kursun, "Low power and high speed multi threshold voltage interface circuits," *IEEE Trans. VLSI*, vol. 17, no. 5, May 2009, pp. 638–645.

[Tharakan92] G. Tharakan and S. Kang, "A new design of a fast barrel switch network," *JSSC*, vol. 27, no. 2, Feb. 1992, pp. 217–221.

[Thompson02] S. Thompson et al., "A 90 nm logic technology featuring 50 nm strained silicon channel transistors, 7 layers of Cu interconnects, low k ILD, and 1 μm^2 SRAM cell," *Proc. Intl. Electron Devices Meeting*, 2002, pp. 61–64.

[Thompson04] S. Thompson et al., "A logic nanotechnology featuring strained-silicon," *IEEE Electron Device Letters*, vol. 25, no. 4, Apr. 2004, pp. 191–193.

[Thorp99] T. Thorp, G. Yee, and C. Sechen, "Design and synthesis of monotonic circuits," *Proc. IEEE Intl. Conf. Computer Design*, 1999, pp. 569–572.

[Tierno08] J. Tierno, A. Rylyakov, and D. Friedman, "A wide power supply range, wide tuning range, all static CMOS all digital PLL in 65 nm SOI," *JSSC*, vol. 43, no. 1, Jan. 2008, pp. 42–51.

[Tobias95] P. Tobias and D. Trindade, *Applied Reliability*, 2nd ed., New York: Van Nostrand Reinhold, 1995.

[Toh88] K. Toh, P. Ko, and R. Meyer, "An engineering model for short-channel MOS devices," *JSSC*, vol. 23, no. 4, Aug. 1988, pp. 950–958.

[Tokunaga08] C. Tokunaga, D. Blaauw, and T. Mudge, "True random number generator with a metastability-based quality control," *JSSC*, vol. 43, no. 1, Jan. 2008, pp. 78–85.

[Topaloglu07] R. Topaloglu, "Standard cell and custom circuit optimization using dummy diffusions through STI width stress effect utilization," *Proc. Custom Integrated Circuits Conf.*, Sep. 2007, pp. 619–622.

[Topol06] A. Topol et al., "Three-dimensional integrated circuits," *IBM J. Research and Dev.*, vol. 50, no. 4/5, Jul./Sep. 2006, pp. 491–506.

[Trinh09] C. Trinh et al., "A 5.6MB/s 64Gb 4b/cell NAND flash memory in 43nm CMOS," *Proc. Intl. Solid-State Circuits Conf.*, Feb. 2009, pp. 246–247, 247a.

[Troutman86] R. Troutman, *Latchup in CMOS Technology: The Problem and its Cure*, Boston: Kluwer Academic Publishers, 1986.

[Tschanz01] J. Tschanz, S. Narendra, Zhanping Chen, S. Borkar, M. Sachdev, and Vivek De, "Comparative delay and energy of single edge-triggered and dual edge-triggered pulsed flip-flops for high-performance microprocessors," *Proc. Intl. Symp. Low Power Electronics & Design*, 2001, pp. 147–152.

[Tschanz02] J. Tschanz et al., "Adaptive body bias for reducing impacts of die-to-die and within-die parameter variations on microprocessor frequency and leakage," *JSSC*, vol. 37, no. 11, Nov. 2002, pp. 1396–1402.

[Tschanz03] J. Tschanz, S. Narendra, Y. Ye, B. Bloechel, S. Borkar, and V. De, "Dynamic sleep transistor and body bias for active leakage power control of microprocessors," *JSSC*, vol. 38, no. 11, Nov. 2003, pp. 1838–1845.

[Tschanz03b] J. Tschanz, S. Narendra, R. Nair, and V. De, "Effectiveness of adaptive supply voltage and body bias for reducing impact of parameter variations in low power and high performance microprocessors," *JSSC*, vol. 38, no. 5, May 2003, pp. 826–829.

[Tschanz07] J. Tschanz et al., "Adaptive frequency and biasing techniques for tolerance to dynamic temperature-voltage variations and aging," *Proc. Intl. Solid-State Circuits Conf.*, Feb. 2007, pp. 292–604.

[Tsividis99] Y. Tsividis, *Operation and Modeling of the MOS Transistor*, 2nd ed., Boston: McGraw-Hill, 1999.

[Tyagi93] A. Tyagi, "A reduced-area scheme for carry-select adders," *JSSC*, vol. 42, no. 10, Oct. 1993, pp. 1163–1170.

[Tyagi00] S. Tyagi et al., "A 130 nm generation logic technology featuring 70 nm transistors, dual V_t transistors and 6 layers of Cu interconnects," *Proc. Intl. Electron Devices Meeting*, 2000, pp. 567–570.

[Tyagi05] S. Tyagi et al., "An advanced low power, high performance, strained channel 65nm technology," *Proc. Intl. Electron Devices Meeting*, Dec. 2005, pp. 245–247.

[Uehara81] T. Uehara and W. van Cleemput, "Optimal layout of CMOS functional arrays," *IEEE Trans. Computers*, vol. C-30, no. 5, May 1981, pp. 305–312.

[Unger86] S. Unger and C. Tan, "Clocking schemes for high-speed digital systems," *IEEE Trans. Computers*, vol. 35, no. 10, Oct. 1986, pp. 880–895.

[Usami95] K. Usami and M. Horowitz, "Clustered voltage scaling for low-power design," *Proc. Intl. Symp. Low Power Electronics and Design*, 1995, pp. 3–8.

[Vadasz66] L. Vadasz and A. Grove, "Temperature dependence of MOS transistor characteristics below saturation," *IEEE. Trans. Electron Devices*, vol. ED-13, no. 13, 1966, pp. 863–866.

[Vadasz69] L. Vadasz, A. Grove, T. Rowe, and G. Moore, "Silicon-gate technology," *IEEE Spectrum*, vol. 6, no. 10, Oct. 1969, pp. 28–35.

[van Berkel99] C. van Berkel and C. Molnar, "Beware the three-way arbiter," *JSSC*, vol. 34, no. 6, Jun. 1999, pp. 840–848.

[Vangal02] S. Vangal et al., "5-GHz 32-bit integer execution core in 130-nm dual-V_T CMOS," *JSSC*, vol. 37, no. 11, Nov. 2002, pp. 1421–1432.

[Veendrick80] H. Veendrick, "The behavior of flip-flops used as synchronizers and prediction of their failure rate," *JSSC*, vol. SC-15, no. 2, Apr. 1980, pp. 169–176.

[Veendrick84] H. Veendrick, "Short-circuit dissipation of static CMOS circuitry and its impact on the design of buffer circuits," *JSSC*, vol. SC-19, no. 4, Aug. 1984, pp. 468–473.

[Vittal99] A. Vittal et al., "Crosstalk in VLSI interconnections," *IEEE Trans. Computer-Aided Design*, vol. 18, no. 12, Dec. 1999, pp. 1817–1824.

[Vittoz72] E. Vittoz, B. Gerber, and F. Leuenberger, "Silicon-gate CMOS frequency divider for electronic wrist watch," *JSSC*, vol. 7, no. 2, Apr. 1972, pp. 100–104.

[Volk01] A. Volk, P. Stoll, and P. Metrovich, "Recollections of early chip development at Intel," *Intel Technology Journal*, Q1 2001, pp. 1–12.

[Vollertsen99] R. Vollertsen, "Burn-in," *IEEE Integrated Reliability Workshop Final Report*, 1999, pp. 167–173.

[von Arnim05] K. von Armin, et al., "Efficiency of body biasing in 90-nm CMOS for low-power digital circuits," *JSSC*, vol. 40, no. 7, July 2005, pp. 1549–1556.

[von Neumann51] J. von Neumann, "Various techniques used in connection with random digits," notes by Forsythe, G., *National Bureau of Standards Applied Math Series*, 1951, vol. 12, pp. 36–38. Reprinted in *von Neumann's Collected Works*, vol. 5, Pergamon Press, 1963, pp. 768–770.

[von Neumann66] J. von Neumann, *Theory of Self-Reproducing Automata*, Urbana, IL: Univ. Illinois Press, 1966.

[Wadell01] B. Wadell, *Transmission Line Design Handbook*, Norwood, MA: Artech House, 1991.

[Wakerly00] J. Wakerly, *Digital Design Principles and Practices*, 3rd ed., Upper Saddle River, NJ: Prentice Hall, 2000.

[Wallace64] C. Wallace, "A suggestion for a fast multiplier," *IEEE Trans. Electronic Computers*, Feb. 1964, pp. 14–17.

[Wang86] L. Wang and E. McCluskey, "Complete feedback shift register design for built-in self test," *Proc. Design Automation Conf.*, Nov. 1986, pp. 56–59.

[Wang89] J. Wang, C. Wu, and M. Tsai, "CMOS nonthreshold logic (NTL) and cascode nonthreshold logic (CNTL) for high-speed applications," *JSSC*, vol. 24, no. 3, Jun. 1989, pp. 779–786.

[Wang93] Z. Wang, G. Jullien, W. Miller, and J. Wang, "New concepts for the design of carry-lookahead adders," *Proc. IEEE Intl. Symp. Circuits and Systems*, vol. 3, 1993, pp. 1837–1840.

[Wang94] J. Wang, S. Fang, and W. Feng, "New efficient designs for XOR and XNOR functions on the transistor level," *JSSC*, vol. 29, no. 7, Jul. 1994, pp. 780–786.

[Wang97] Z. Wang, G. Jullien, W. Miller, J. Wang, and S. Bizzan, "Fast adders using enhanced multiple-output domino logic," *JSSC*, vol. 32, no. 2, Feb. 1997, pp. 206–214.

[Wang00] J. Wang and C. Huang, "High-speed and low-power CMOS priority encoders," *JSSC*, vol. 35, no. 10, Oct. 2000, pp. 1511–1514.

[Wang01] J. Wang, C. Chang, and C. Yeh, "Analysis and design of high-speed and low-power CMOS PLAs," *JSSC*, vol. 36, no. 8, Aug. 2001, pp. 1250–1262.

[Wang02] A. Wang, A. Chandrakasan, and S. Kosonocky, "Optimal supply and threshold scaling for subthreshold CMOS circuits," *Proc. Computer Society Symp. VLSI*, 2002, pp. 5–9.

[Wang06] G. Wang et al., "A 0.127 μm^2 high performance 65 nm SOI based embedded DRAM for on-processor applications," *Proc. Intl. Electron Devices Meeting*, Dec. 2006, pp. 1–4.

[Wang07] C. Wang, C. Lee, and W. Lin, "A 4-kb low-power SRAM design with negative word-line scheme," *IEEE Trans. Circuits & Systems*, vol. 54, no. 5, May 2007, pp. 1069–1076.

[Wang08a] A. Wang and S. Naffziger, eds., *Adaptive Techniques for Dynamic Processor Optimization: Theory and Practice*, New York: Springer, 2008.

[Wang08b] L. Wang, C. Stroud, and N. Touba, eds., *System-on-Chip Test Architectures: Nanometer Design for Testability*, Elsevier, 2008.

[Wanlass63] F. Wanlass and C. Sah, "Nanowatt logic using field effect metal-oxide semiconductor triodes," *Proc. IEEE Intl. Solid-State Circuits Conf.*, 1963, pp. 32–33.

[Warnock06] J. Warnock et al., "Circuit design techniques for a first-generation cell broadband engine processor," *JSSC*, vol. 41, no. 8, Aug. 2006, pp. 1692–1706.

[Webb97] C. Webb et al., "A 400-MHz S/390 microprocessor," *JSSC*, vol. 32, no. 11, Nov. 1997, pp. 1665–1675.

[Webb08] C. Webb, "45 nm design for manufacturing," *Intel Technology Journal*, vol. 12, no. 2, Jun. 2008, pp. 121–130.

[Wei98] L. Wei, Z. Chen, M. Johnson, K. Roy, and V. De, "Design and optimization of low voltage high performance dual threshold CMOS circuits," *Proc. Design Automation Conf.*, 1998, pp. 489–494.

[Wei00] G. Wei, J. Kim, D. Liu, S. Sidiropoulos, and M. Horowitz, "A variable-frequency parallel I/O interface with adaptive power-supply regulation," *JSSC*, vol. 35, no. 11, Nov. 2000, pp. 1600–1610.

[Weinberger58] A. Weinberger and J. Smith, "A logic for high-speed addition," *System Design of Digital Computer at the National Bureau of Standards: Methods for High-Speed Addition and Multiplication*, National Bureau of Standards, circular 591, section 1, Feb. 1958, pp. 3–12.

[Weinberger81] A. Weinberger, "4-2 carry-save adder module," *IBM Technical Disclosure Bulletin*, vol. 23, no. 8, Jan. 1981, pp. 3811–3814.

[Weinlader00] D. Weinlader, Ron Ho, Chih-Kong Keng Yang, and M. Horowitz, "An eight channel 35 GSample/s CMOS timing analyzer," *Proc. Intl. Solid-State Circuits Conf.*, 2000, pp. 170–171.

[Weiss02] D. Weiss, J. Wuu, and V. Chin, "The on-chip 3-MB subarray-based third-level cache on an Itanium microprocessor," *JSSC*, vol. 37, no. 11, Nov. 2002, pp. 1523–1529.

[Widmer83] A. Widmer and P. Franaszek, "A DC-balanced partitioned-block 8B/10B transmission code," *IBM J. Research and Dev.*, vol. 27, no. 5, Sep. 1983, pp. 440–451.

[Wijeratne07] S. Wijeratne et al., "A 9-GHz 65-nm Intel Pentium 4 processor integer execution unit," *JSSC*, vol. 42, no. 1, Jan. 2007, pp. 26–37.

[Williams83] T. Williams and K. Parker, "Design for Testability—A Survey," *Proc. IEEE*, vol. 71, no. 1, Jan. 1983, pp. 98–112.

[Williams86] T. Williams, "Design for testability," *Proc. NATO Advanced Study Inst. Computer Design Aids for VLSI Circuits*, (P. Antognetti et al., eds.), NATO ASI Series, Martinus Nijhoff Publishers, 1986, pp. 359–416.

[Williams91] T. Williams and M. Horowitz, "A zero-overhead self-timed 160-ns 54-b CMOS divider," *JSSC*, vol. 26, no. 11, Nov. 1991, pp. 1651–1661.

[Wilton96] S. Wilton and N. Jouppi, "CACTI: an enhanced cache access and cycle time model," *JSSC*, vol. 31, no. 5, May 1996, pp. 677–688.

[Wing82] O. Wing, "Automated gate matrix layout," *Proc. IEEE Intl. Symp. Circuits and Systems*, vol. 2, 1982, pp. 681–685.

[Wolf00] S. Wolf and R. Tauber, *Silicon Processing for the VLSI Era*, 2nd ed., Sunset Beach, CA: Lattice Press, 2000.

[Wong03] H. Wong, J. Appenzeller, V. Derycke, R. Martel, S. Wind, and P. Avouris, "Carbon nanotube field effect transistors—fabrication, device physics, and circuit implications," *Proc. Intl. Solid-State Circuits Conf.*, 2003, pp. 370–500.

[Wong06] K. Wong, T. Rahal-Arabi, M. Ma, and G. Taylor, "Enhancing microprocessor immunity to power supply noise with clock-data compensation," *JSSC*, vol. 41, no. 4, Apr. 2006, pp. 749–758.

[Wu91] C. Wu and K. Cheng, "Latched CMOS differential logic (LCDL) for complex high-speed VLSI," *JSSC*, vol. 26, no. 9, Sep. 1991, pp. 1324–1328.

[Xu08] J. Xu et al., "A band-limited active damping circuit with 13 dB power supply resonance reduction," *JSSC*, vol. 43, no. 1, Jan. 2008, pp. 61–68.

[Yabuuchi07] M. Yabuuchi et al., "A 45nm low-standby-power embedded SRAM with improved immunity against process and temperature variations," *Proc. Intl. Solid-State Circuits Conf.*, Feb. 2007, pp. 326–606.

[Yamada95] H. Yamada, T. Hotta, T. Nishiyama, F. Murabayashi, T. Yamauchi, and H. Sawamoto, "A 13.3 ns double-precision floating-point ALU and multiplier," *Proc. Intl. Conf. Computer Design*, 1995, pp. 466–470.

[Yamaoka04] M. Yamaoka, K. Osada, and K. Ishibashi, "0.4-V logic-library-friendly SRAM array using rectangular-diffusion cell and delta-boosted-array voltage scheme," *JSSC*, vol. 39, no. 6, Jun. 2004, pp. 934–940.

[Yamaoka04b] M. Yamaoka et al., "A 300MHz 25 μA/Mb leakage on-chip SRAM module featuring process-variation immunity and low-leakage-active mode for mobile-phone application processor," *Proc. Intl. Solid-State Circuits Conf.*, Feb. 2004, pp. 494–542.

[Yamaoka06] M. Yamaoka et al., "90-nm process-variation adaptive embedded SRAM modules with power-line-floating write technique," *JSSC*, vol. 41, no. 3, Mar. 2006, pp. 705–711.

[Yang96] C. K. Yang and M. Horowitz, "A 0.8-μm CMOS 2.5 Gb/s oversampling receiver and transmitter for serial links," *JSSC*, vol. 31, no. 12, Dec. 1996, pp. 2015–2023.

[Yang98] S. Yang et al., "A high performance 180 nm generation logic technology," *Tech. Digest Intl. Electron Device Meeting*, Dec. 1998, pp. 197–200.

[Yano90] K. Yano, T. Yamanaka, T. Nishida, M. Saito, K. Shimohigashi, and A. Shimizu, "A 3.8-ns 16 × 16-b multiplier using complementary pass-transistor logic," *JSSC*, vol. 25, no. 2, Apr. 1990, pp. 388–395.

[Yano96] K. Yano, Y. Sasaki, K. Rikino, and K. Seki, "Top-down pass-transistor logic design," *JSSC*, vol. 31, no. 6, Jun. 1996, pp. 792–803.

[Ye98] Y. Ye, S. Borkar, and V. De, "A new technique for standby leakage reduction in high-performance circuits," *Symp. VLSI Circuits Digest Tech. Papers*, 1998, pp. 40–41.

[Ye00] Y. Ye, J. Tschanz, S. Narendra, S. Borkar, M. Stan, and V. De, "Comparative delay, noise and energy of high-performance domino adders with stack node preconditioning (SNP)," *Symp. VLSI Circuits Digest Tech. Papers*, 2000, pp. 188–191.

[Yee00] G. Yee and C. Sechen, "Clock-delayed domino for dynamic circuit design," *IEEE Trans. VLSI*, vol. 8, no. 4, Aug. 2000, pp. 425–430.

[Yoon02] J. Yoon et al., "CMOS-compatible surface-micromachined suspended-spiral inductors for multi-GHz silicon RF ICs," *IEEE Electron Device Letters*, vol. 23, no. 10, Oct. 2002, pp. 591–593.

[Yoshimoto83] M. Yoshimoto et al., "A divided word-line structure in the static RAM and its application to a 64K full CMOS RAM," *JSSC*, vol. SC-18, no. 5, Oct. 1983, pp. 479–485.

[Young00] K. Young et al., "A 0.13 μm CMOS technology with 193 nm lithography and Cu/Low-k for high performance applications," *Proc. Intl. Electron Devices Meeting*, 2000, pp. 563–566.

[Young10] I. Young et al., "Optical I/O technology for tera-scale computing," *JSSC*, vol. 45, no. 1, Jan. 2010, pp. 235–248.

[Yuan82] C. Yuan and T. Trick, "A simple formula for the estimation of the capacitance of two-dimensional interconnects in VLSI circuits," *IEEE Electron Device Letters*, vol. EDL-3, Dec. 1982, pp. 391–393.

[Yuan89] J. Yuan and C. Svensson, "High-speed CMOS circuit technique," *JSSC*, vol. 24, no. 1, Feb. 1989, pp. 62–70.

[Zerbe01] J. Zerbe et al., "A 2 Gb/s/pin 4-PAM parallel bus interface with transmit crosstalk cancellation, equalization, and integrating receivers," *Proc. Intl. Solid-State Circuits Conf.*, 2001, pp. 66–67, 432.

[Zhai05a] B. Zhai, D. Blaauw, D. Sylvester, and K. Flautner, "The limit of dynamic voltage scaling and insomniac dynamic voltage scaling," *IEEE Trans. VLSI*, vol. 13, no. 11, Nov. 2005, pp. 1239–1252.

[Zhai05b] B. Zhai, S. Hanson, D. Blaauw, and D. Sylvester, "Analysis and mitigation of variability in subthreshold design," *Proc. Intl. Symp. Low Power Electronics & Design*, Aug. 2005, pp. 20–25.

[Zhai08] B. Zhai, S. Hanson, D. Blaauw, and D. Sylvester, "A variation-tolerant sub-200 mV 6-T subthreshold SRAM," *JSSC*, vol. 43, no. 10, Oct. 2008, pp. 2338–2348.

[Zhang05] K. Zhang et al., "SRAM design on 65-nm CMOS technology with dynamic sleep transistor for leakage reduction," *JSSC*, vol. 40, no. 4, Apr. 2005, pp. 895–901.

[Zhang06] K. Zhang et al., "A 3-GHz 70-mb SRAM in 65-nm CMOS technology with integrated column-based dynamic power supply," *JSSC*, vol. 41, no. 1, Jan. 2006, pp. 146–151.

[Zhao07] P. Zhao, J. McNeely, P. Golconda, M. Bayoumi, R. Barcenas, and Weidong Kuang, "Low-power clock branch sharing double-edge triggered flip-flop," *IEEE Trans. VLSI*, vol. 15, no. 3, Mar. 2007, pp. 338–345.

[Zhao09] P. Zhao et al., "Low-power clocked-pseudo-NMOS flip-flop for level conversion in dual supply systems," *IEEE Trans. VLSI*, vol. 17, no. 9, Sep. 2009, pp. 1196–1202.

[Zhou99] X. Zhou, K. Lim, and D. Lim, "A simple and unambiguous definition of threshold voltage and its implications in deep-submicron MOS device modeling," *IEEE Trans. Electron Devices*, vol. 46, no. 4, Apr. 1999, pp. 807–809.

[Ziegler96] J. Ziegler, "Terrestrial cosmic rays," *IBM J. Research and Development*, vol. 40, no. 1, Jan. 1996, pp. 19–39.

[Ziegler02] J. Ziegler, *Ion-Implantation—Science and Technology,* 2002 ed., IIT Press, 2002.

[Zimmermann96] R. Zimmermann, "Non-heuristic optimization and synthesis of parallel-prefix adders," *Proc. Intl. Workshop on Logic and Architecture Synthesis,* Dec. 1996, pp. 123–132.

[Zimmermann97] R. Zimmermann and W. Fichtner, "Low-power logic styles: CMOS versus pass-transistor logic," *JSSC,* vol. 32, no. 7, Jul. 1997, pp. 1079–1090.

[Zlatanovici09] R. Zlatanovici, S. Kao, and B. Nikolic, "Energy-delay optimization of 64-bit carry-lookahead adders with a 240 ps 90 nm CMOS design example," *JSSC,* vol. 44, no. 2, Feb. 2009, pp. 569–583.

[Zuras86] D. Zuras and W. McAllister, "Balanced delay trees and combinatorial division in VLSI," *JSSC,* vol. SC-21, no. 5, Oct. 1986, pp. 814–819.

Index

Credits

Figure 1.1 © 1997 Semiconductor Industry Association. All Rights Reserved.
Figure 1.2a Courtesy AT&T.
Figure 1.2b Courtesy Texas Instruments
Figure 1.3a © Copyright 1967 IEEE. All Rights Reserved.
Figure 1.3b Courtesy of Intel Corporation.
Figure 1.48 Copyright © 2003 The McGraw-Hill Companies, Inc. All Rights Reserved.
Figure 1.71 Courtesy of Intel Corporation.
Figure 3.25 Copyright © ARM. All Rights Reserved.
Figure 3.41 Copyright © ARM. All Rights Reserved
Figure 4.11 © Copyright 2004 IEEE—All Rights Reserved.
Figure 4.19 © Copyright 2007 IEEE—All Rights Reserved.
Figure 4.21 © Copyright 2007 IEEE—All Rights Reserved.
Figure 4.26 © Copyright 2005 IEEE—All Rights Reserved.
Figure 4.27 © Copyright 2002 IEEE—All Rights Reserved.
Figure 4.28 © Copyright 1997 IEEE—All Rights Reserved.
Figure 4.30 © Copyright 2009 IEEE—All Rights Reserved.
Figure 4.33 Courtesy of Intel Corporation.
Figure 5.2 (a) © Copyright 2010 IEEE—All Rights Reserved.
Figure 5.2 (b) Courtesy of Intel Corporation.
Figure 5.3 Courtesy of Intel Corporation.
Figure 6.2 © Copyright IBM Corporation. All Rights Reserved.
Figure 6.3 © Copyright 2001 IEEE—All Rights Reserved.
Figure 6.4 From M. Pelgrom, "Nanometer CMOS: An Analog Design Challenge!" IEEE Distinguished Lecture, Denver 2006. Reprinted by permission of the author. Figure courtesy of Boris Ljevar (NXP). All Rights Reserved.
Figure 6.7 © Copyright 2010 IEEE—All Rights Reserved.
Figure 6.8 © Copyright 2010 IEEE—All Rights Reserved.
Figure 6.11 © Copyright 2010 IEEE—All Rights Reserved.
Figure 6.15 © Copyright 2010 IEEE—All Rights Reserved.
Figure 6.16 © 1997 Semiconductor Industry Association. All Rights Reserved.
Figure 6.21 Courtesy Larry Pileggi.
Figure 6.22 Courtesy of Texas Instruments.
Figure 6.23 © Copyright 2006 IBM Corporation. All rights reserved.
Figure 6.27 © Copyright 2010 IEEE—All Rights Reserved.
Figures 6.30–6.38 Courtesy of Intel Corporation.
Figure 8.62 Courtesy International Business Machines Corporation. Unauthorized use not permitted.
Figure 9.6 © Copyright 2002 IEEE—All Rights Reserved.
Figure 10.38 © Copyright 2007 IEEE—All Rights Reserved
Figure 10.39 © Copyright 2009 IEEE—All Rights Reserved.
Figure 11.16 (a) © Copyright 2000 IEEE—All Rights Reserved.
Figure 11.16 (b) © Copyright 2002 IEEE—All Rights Reserved.
Figure 11.16 (c) © Copyright 2004 IEEE—All Rights Reserved.
Figure 11.16 (d) © Copyright 2007 IEEE—All Rights Reserved.
Figure 11.16 (e) © Copyright 2008 IEEE—All Rights Reserved.
Figure 11.36 © Copyright 2005 IEEE—All Rights Reserved.
Figure 11.50 © Copyright 2008 IEEE—All Rights Reserved.
Figure 11.62 © Copyright 2009 IEEE—All Rights Reserved.
Figure 12.3 Courtesy of Intel Corporation.
Figure 12.6 © Copyright Sun Microsystems, Inc.
Figure 12.12 © Copyright 2008 IEEE—All Rights Reserved.
Figure 12.13 © Copyright 2006 IEEE—All Rights Reserved.
Figure 13.27 Copyright © 2000–2010 IC Knowledge. All Rights Reserved.
Figure 14.7 © 2009 GGB Industries, reprinted with permision.
Figure 14.9 © Copyright 2009 IEEE—All Rights Reserved.
Figure 14.10 Courtesy of Intel Corporation.
Figure 15.1 Courtesy of International Business Machines Corporation. Unauthorized use not permitted.
Figure 15.4 Copyright © 1998 SPIE. All Rights Reserved.
Figure 15.6 © Copyright 2008 IEEE—All Rights Reserved.
Figure 15.12 Courtesy of International Business Machines Corporation. Unauthorized use not permitted.
Figure 15.13 Courtesy of International Business Machines Corporation. Unauthorized use not permitted.
Figure 15.14 Courtesy of International Business Machines Corporation. Unauthorized use not permitted.
Figure 15.18 © Copyright 2007 IEEE—All Rights Reserved.
Figure 15.19 (a)(b) © Copyright 2005 IEEE—All Rights Reserved.
Figure 15.22 Courtesy of International Business Machines Corporation. Unauthorized use not permitted.
Figure 15.30 © Copyright 2005 IEEE—All Rights Reserved.
Figure 15.31 © Copyright 2006 IEEE—All Rights Reserved.
Figure 15.32 © Copyright 2008 IBM Corporation. All rights reserved.
Figure 15.34 Courtesy of Intel Corporation.
Figure 15.35 © Copyright 2003 IEEE—All Rights Reserved.
Figure 15.38 Courtesy of Intel Corporation.
Artwork p. 694 © 1995–2010 by Michael W. Davidson and The Florida State University. All Rights Reserved. Reproduced with permission.

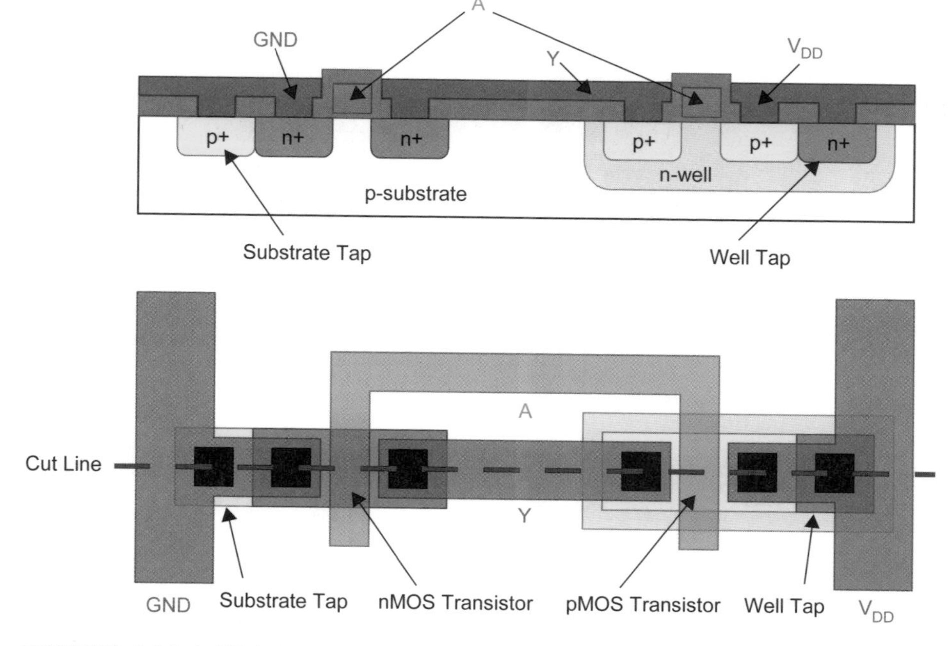

FIGURES 1.34–1.35(a) Inverter Cross-Section and Top View

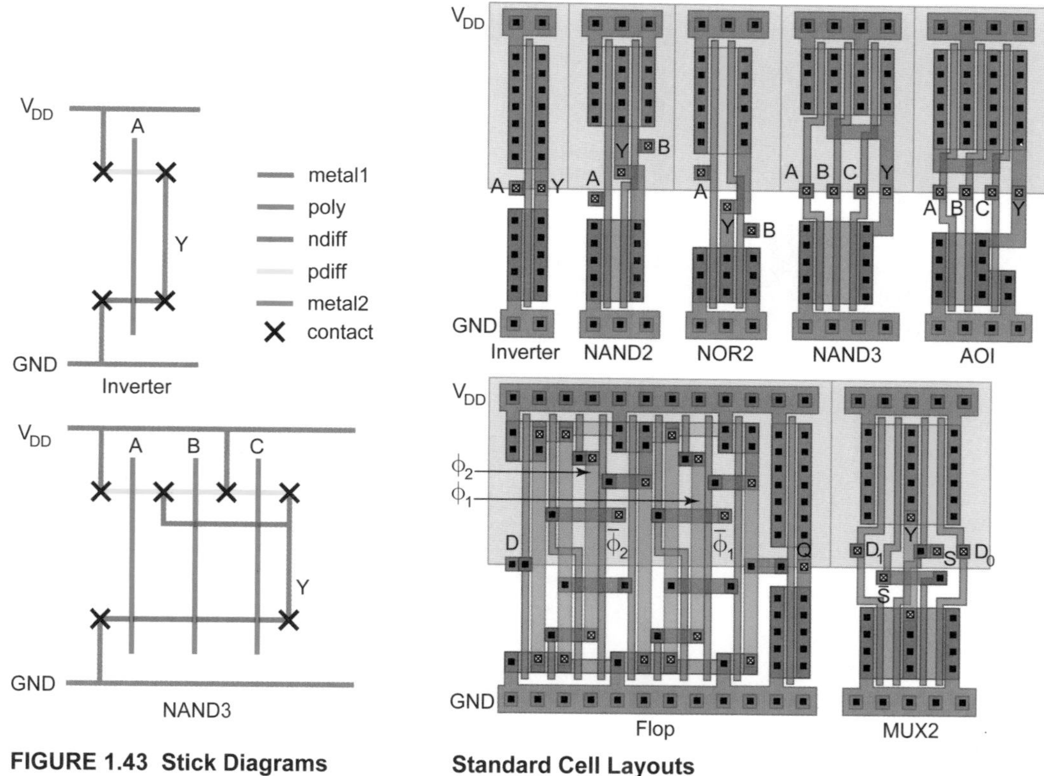

FIGURE 1.43 Stick Diagrams

Standard Cell Layouts

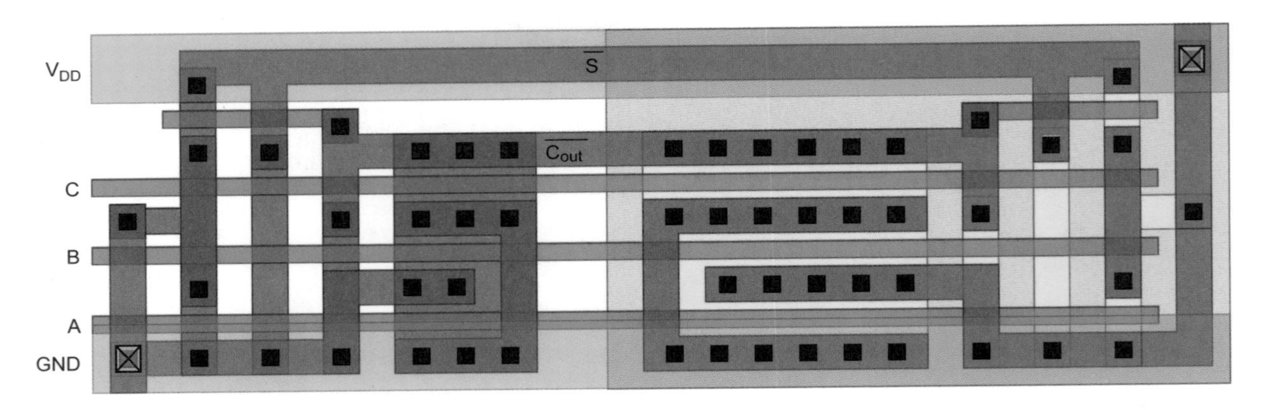

Full Adder Layout

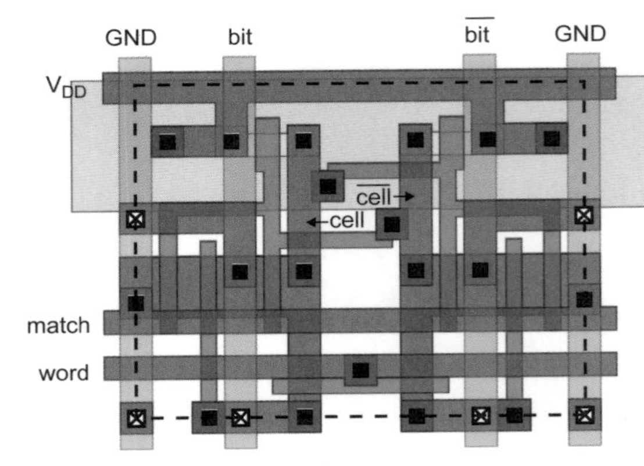

CAM Layout

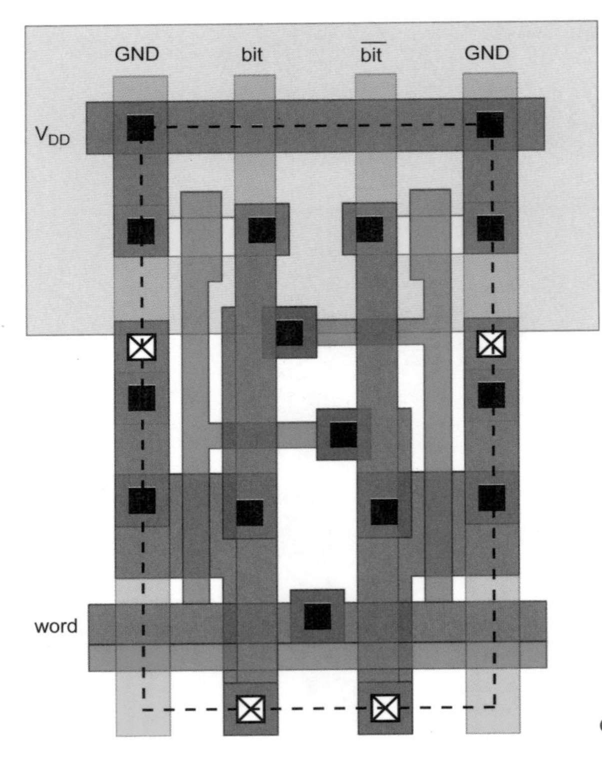

6T SRAM Layout

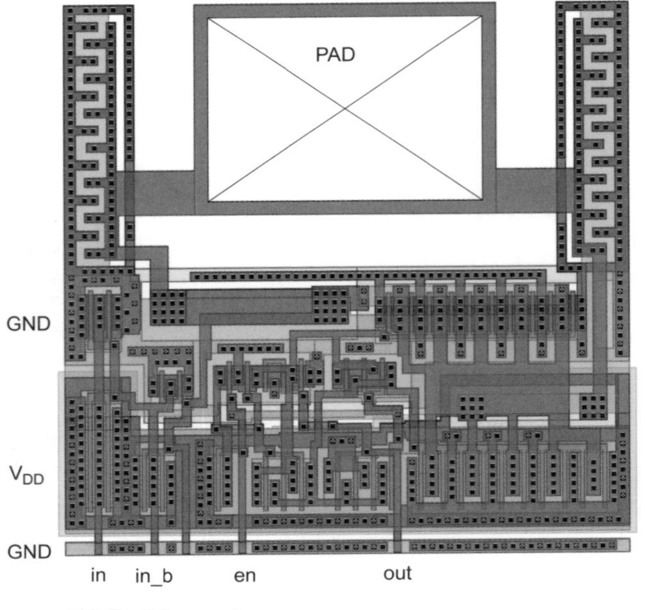

I/O Pad Layout